# Conversion Factors

## Length

1 in. = 2.54 cm

1 ft = 0.3048 m

1 mi = 5280 ft = 1.609 km

1 m = 3.281 ft

1 km = 0.6214 mi

1 angstrom (Å) = $10^{-10}$ m

## Mass

1 slug = 14.59 kg

1 kg = 1000 grams = $6.852 \times 10^{-2}$ slug

1 atomic mass unit (u) = $1.6605 \times 10^{-27}$ kg

(1 kg has a weight of 2.205 lb where the
   acceleration due to gravity is 32.174 ft/s$^2$)

## Time

1 day = 24 h = $1.44 \times 10^3$ min = $8.64 \times 10^4$ s

1 yr = 365.24 days = $3.156 \times 10^7$ s

## Speed

1 mi/h = 1.609 km/h = 1.467 ft/s = 0.4470 m/s

1 km/h = 0.6214 mi/h = 0.2778 m/s = 0.9113 ft/s

## Force

1 lb = 4.448 N

1 N = $10^5$ dynes = 0.2248 lb

## Work and Energy

1 J = 0.7376 ft·lb = $10^7$ ergs

1 kcal = 4186 J

1 Btu = 1055 J

1 kWh = $3.600 \times 10^6$ J

1 eV = $1.602 \times 10^{-19}$ J

## Power

1 hp = 550 ft·lb/s = 745.7 W

1 W = 0.7376 ft·lb/s

## Pressure

1 Pa = 1 N/m$^2$ = $1.450 \times 10^{-4}$ lb/in.$^2$

1 lb/in.$^2$ = $6.895 \times 10^3$ Pa

1 atm = $1.013 \times 10^5$ Pa = 1.013 bar =
   14.70 lb/in.$^2$ = 760 torr

## Volume

1 liter = $10^{-3}$ m$^3$ = 1000 cm$^3$ = 0.03531 ft$^3$

1 ft$^3$ = 0.02832 m$^3$ = 7.481 U.S. gallons

1 U.S. gallon = $3.785 \times 10^{-3}$ m$^3$ = 0.1337 ft$^3$

## Angle

1 radian = 57.30°

1° = 0.01745 radian

# Standard Prefixes Used to Denote Multiples of Ten

| Prefix | Symbol | Factor |
|---|---|---|
| Tera | T | $10^{12}$ |
| Giga | G | $10^{9}$ |
| Mega | M | $10^{6}$ |
| Kilo | k | $10^{3}$ |
| Hecto | h | $10^{2}$ |
| Deka | da | $10^{1}$ |
| Deci | d | $10^{-1}$ |
| Centi | c | $10^{-2}$ |
| Milli | m | $10^{-3}$ |
| Micro | $\mu$ | $10^{-6}$ |
| Nano | n | $10^{-9}$ |
| Pico | p | $10^{-12}$ |
| Femto | f | $10^{-15}$ |

# Basic Mathematical Formulae

Area of a circle = $\pi r^2$

Circumference of a circle = $2\pi r$

Surface area of a sphere = $4\pi r^2$

Volume of a sphere = $\frac{4}{3}\pi r^3$

Pythagorean theorem: $h^2 = h_o^2 + h_a^2$

Sine of an angle: $\sin\theta = h_o/h$

Cosine of an angle: $\cos\theta = h_a/h$

Tangent of an angle: $\tan\theta = h_o/h_a$

Law of cosines: $c^2 = a^2 + b^2 - 2ab\cos\gamma$

Law of sines: $a/\sin\alpha = b/\sin\beta = c/\sin\gamma$

Quadratic formula:

If $ax^2 + bx + c = 0$, then, $x = (-b \pm \sqrt{b^2 - 4ac})/(2a)$

# eGrade Plus

## www.wiley.com/college/cutnell

## Based on the Activities You Do Every Day

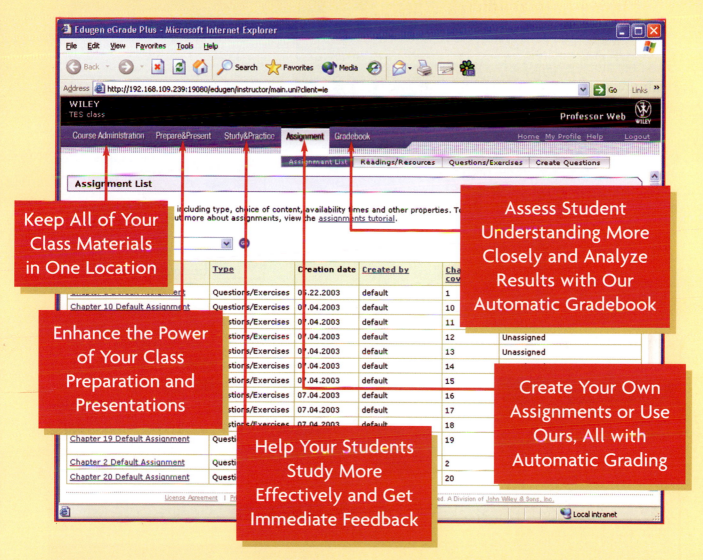

**Keep All of Your Class Materials in One Location**

**Enhance the Power of Your Class Preparation and Presentations**

**Assess Student Understanding More Closely and Analyze Results with Our Automatic Gradebook**

**Create Your Own Assignments or Use Ours, All with Automatic Grading**

**Help Your Students Study More Effectively and Get Immediate Feedback**

**All the content and tools you need, all in one location, in an easy-to-use browser format.**

**Choose the resources you need, or rely on the arrangement supplied by us.**

Now, many of Wiley's textbooks are available with eGrade Plus, a powerful online tool that provides a completely integrated suite of teaching and learning resources in one easy-to-use website. eGrade Plus integrates Wiley's world-renowned content with media, including a multimedia version of the text, PowerPoint slides, and more. Upon adoption of eGrade Plus, you can begin to customize your course with the resources shown here.

**See for yourself!**

**Go to www.wiley.com/college/egradeplus for an online demonstration of this powerful new software.**

# Students,
# eGrade Plus Allows You to:

## Study More Effectively

## Get Immediate Feedback When You Practice on Your Own

Our website links directly to **electronic book content,** so that you can review the text while you study and complete homework online. Additional resources include **Concept Simulations, MCAT problems,** and **Interactive Learningware** with step by step problem solving tutorials to help you review key topics.

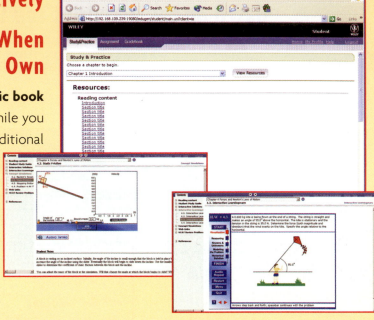

## Complete Assignments / Get Help with Problem Solving

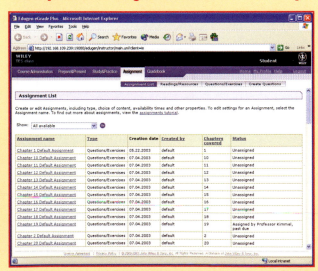

An **"Assignment"** area keeps all your assigned work in one location, making it easy for you to stay on task. In addition, many homework problems contain a **link** to the relevant section of the **electronic book,** providing you with a text explanation to help you conquer problem-solving obstacles as they arise.

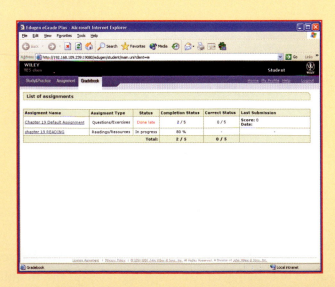

## Keep Track of How You're Doing

A **Personal Gradebook** allows you to view your results from past assignments at any time.

# John D. Cutnell
# Kenneth W. Johnson

Southern Illinois University at Carbondale

# Essentials
## of Physics

WILEY

## John Wiley & Sons, Inc.

| | |
|---|---|
| ACQUISITION EDITOR | *Stuart Johnson* |
| DEVELOPMENT EDITOR | *Georgia Kamvosoulis Mederer* |
| MARKETING MANAGER | *Amanda Wygal* |
| PRODUCTION EDITOR | *Elizabeth Swain* |
| COVER AND TEXT DESIGNER | *Madelyn Lesure* |
| SENIOR ILLUSTRATION EDITOR | *Sigmund Malinowski* |
| ELECTRONIC ILLUSTRATIONS | *Precision Graphics* |
| PHOTO EDITOR | *Hilary Newman* |
| | |
| COVER PHOTO | *© John Kelly/The Image Bank/Getty Images.* |

This book was set in 10/12 Times Roman by Progressive Information Technologies and printed and bound by Von Hoffmann Press, Inc. The cover was printed by Von Hoffmann Press.

This book is printed on acid free paper.

ISBN 0-471-71398-8

Printed in the United States of America

10 9 8 7 6 5 4 3 2

# Contents

v

# Preface

We frequently hear from both instructors and students that physics textbooks today are too long, too cluttered, and too expensive. We have written *Essentials of Physics* to offer an alternative to the standard texts. We have tried to present the core topics of introductory physics in as straightforward a manner as possible. We hope that this approach will help students more easily see that physics is a discipline built on a set of unifying principles and not just an overwhelming collection of disconnected equations to be memorized.

*Essentials* is an adaptation of our well-known text, *Physics*.

## Goals

Helping students develop a conceptual understanding of physics principles is a primary goal of this text, as it is of *Physics*. This is a challenging task, because physics is often regarded as a collection of equations that can be used blindly to solve problems. However, a good problem-solving technique does not begin with equations. It starts with a firm grasp of the concepts and how they fit together to provide a coherent description of the physical world.

The ability to reason in an organized manner is essential to solving problems, and helping students to improve their reasoning ability is also one of our primary goals. To this end, we have included *Explicit Reasoning Steps* in all examples. A strong reasoning ability, combined with firm conceptual understanding, is the legacy that we wish to leave to our students.

Finally, we want to show students that physics principles come into play over and over again in their lives. It is always easier to learn something new if it is directly related to day-to-day living. Direct applications of physics principles are identified in the margin by the label *The Physics of....* Many of these applications are biomedical in nature and deal with human physiology. Others, such as CDs and DVDs and digital photography, deal with modern technology. Still others deal with things that we take for granted in our lives, such as household plumbing. We have also incorporated real-world situations into many of the worked-out examples and the homework material at the end of each chapter.

## Features of *Essentials of Physics*

**Explicit Reasoning Steps**  Because reasoning is the cornerstone of problem solving, we believe that students will benefit from seeing the reasoning stated explicitly. Therefore, the format in which examples are worked out includes an explicit reasoning step. In this step we explain what motivates our procedure for solving the problem before any algebraic or numerical work is done.

**Conceptual Examples**  A good problem-solving technique begins with a foundation of conceptual understanding. Therefore, every chapter includes examples that are entirely conceptual in nature. These examples are worked out in a rigorous, but qualitative, fashion, with no (or very few) equations. We have added art to many of these as a way to help students visualize the concepts. Our intent is to provide students with explicit models of how to "think through" a problem before attempting to solve it numerically. The *Conceptual Examples* deal with a wide range of topics and a large number of issues that often confuse students. Wherever possible, we have focused on real-world situations and have structured the examples so that they lead naturally to homework material found at the end of each chapter. The related homework material is explicitly identified, and that material contains a cross reference that encourages students to review the pertinent *Conceptual Example*.

**Concept Summaries**  All chapter-ending summaries are written so that they present an abridged, but comprehensive, version of the material on a section-by-section basis, including important equations. To help students review the material, the discussion of each section is flanked by two lists. On one side, a list of topics provides a quick-reference option for locating a given topic. On the other side, a color-coded list of the text examples and the online learning aids (Concept Simulations, Interactive LearningWare, and Interactive Solutions) shows students the related learning aids. The summary also identifies the online Self-Assessment Tests that are keyed to the sections, so that students can evaluate their understanding of the chapter topics.

*The Physics of . . .*  This edition contains 230 real-world applications that reflect our commitment to show students how prevalent physics is in their lives. Each application is identified in the margin with the label *The Physics of. . .*, and those that deal with biomedical material are further marked with an icon in the shape of a caduceus.

*Problem Solving Insights*  To reinforce the problem-solving techniques illustrated in the worked-out examples, we have included short statements in the margins identified by the label *Problem Solving Insight*. These statements help students to develop good problem-solving skills by providing the kind of advice that a teacher gives when explaining a calculation in detail.

*Homework Problems* *Essentials of Physics* provides a large selection of homework material. In writing the problems, we have used a wide variety of real-world situations with realistic data. The problems are ranked according to difficulty, with the most difficult problems marked with a double asterisk (**) and those of intermediate difficulty marked with a single asterisk (*). The easiest problems are unmarked. Those whose solutions appear in the *Student Solutions Manual* are identified with the label **ssm**. Those whose solutions are available on the Web (**www.wiley.com/college/cutnell**) are marked with the label **www**. Problems and questions that are biomedical in nature are identified with an icon in the shape of a caduceus.

## Student Web Site for *Essentials of Physics:* www.wiley.com/college/cutnell

*Self-Assessment Tests*  The self-assessment tests are online tests that are linked directly to specific sections of the book. There are about two self-assessment tests per chapter, each containing 8–10 questions and problems. The tests are designed so that students, after studying the designated sections, can immediately assess their understanding of the material. Each contains thought-provoking conceptual questions as well as calculational problems. The Self-Assessment tests are user friendly, for they can be taken as often as the student wishes, and they provide extensive feedback in the form of helpful hints and suggestions when incorrect answers are selected.

*Concept Simulations*  Often, many concepts in physics—such as relative velocities, collisions, and ray tracing—can be better understood when they are simulated. With each simulation, the student can experiment, because one or more parameters are under user control. Many of the simulations provide references to related homework material, which, in turn, is cross-referenced to the simulation.

*Interactive Solutions*  Often, students experience difficulty in getting started with the more difficult homework problems. The Interactive Solutions feature consists of 111 online problems. Each Interactive Solution is worked out by the student in an interactive manner, and the solution is designed to serve as a model for the associated homework problem. Those homework problems that are associated with Interactive Solutions are identified in the text.

*Interactive LearningWare*  This feature consists of interactive calculational examples. Each example is designed in an interactive format that consists of five steps. We believe that these steps are essential to the development of problem-solving skills and that students will benefit from seeing them explicitly identified in each example.

1. The *Visualization* step uses animations and drawings to help students understand the general nature of the situation being analyzed.

2. The *Reasoning* step establishes the conceptual basis for the solution with the aid of multiple-choice questions and answers.

3. The *Knowns-and-Unknowns* step deals with the important process of identifying all the variables and the symbols that represent them. This information is not simply given to the students but, rather, is interactively developed in the form of tables that evolve as the example proceeds.

4. In the *Modeling* step, the student, using the results of the previous steps, models the problem with equations developed in the text.

5. Finally, in the *Solution* step, the solution to the problem is obtained.

In summary, the hallmark of these Interactive LearningWare examples is a user-friendly clarity that allows the student to see how conceptual understanding and quantitative analysis work together in a problem-solving environment. Related homework material is identified and contains a cross-reference to the LearningWare.

• • •

In spite of our best efforts to produce an error-free book, errors no doubt remain. They are solely our responsibility, and we would appreciate hearing of any that you find. We hope that this text makes learning and teaching physics easier and more enjoyable and look forward to hearing about your experiences with it. Please feel free to write us care of Physics Editor, Higher Education Division, John Wiley & Sons, Inc., 111 River Street, Hoboken NJ 07030-5774, or contact us at **www.wiley.com/college/cutnell.**

# Supplements

An extensive package of supplements to accompany *Essentials of Physics* is available to assist both the teacher and the student.

## Instructor's Supplements

*Helping Teachers Teach*

**Instructor's Resource Guide** by David T. Marx, Southern Illinois University of Carbondale. This guide contains an extensive listing of Web-based physics education resources. It also includes teaching ideas, lecture notes, demonstration suggestions, alternative syllabi for courses of different lengths and emphasis, as well as conversion notes that allow instructors to use their class notes from other texts.

**Instructor's Solutions Manual** by John D. Cutnell and Kenneth W. Johnson. This manual provides worked-out solutions for all end-of-chapter Problems. Volume 1 contains the solutions for Chapters 1–17. Volume 2 contains the solutions for Chapters 18–32.

**Test Bank** by David T. Marx, Southern Illinois University at Carbondale. This manual includes more than 2200 multiple-choice questions. These items are also available in the *Computerized Test Bank* (see below).

**Instructor's Resource CD-ROM** This CD contains:

- The entire *Instructor's Solutions Manual* in both Microsoft Word © (Windows and Macintosh) and PDF files.

- A *Computerized Test Bank,* for use with both Windows and Macintosh PC's with full editing features to help you customize tests.

- Multiple-choice format for 1200 of the end-of-chapter problems in the text. This material allows instructors to assign homework in a machine-gradable format.

- All text illustrations, suitable for classroom projection, printing, and web posting.

***Wiley Physics Simulations CD-ROM*** This CD contains 50 interactive simulations (Java applets) that can be used for classroom demonstrations.

***On-line Homework and Quizzing*** *Essentials of Physics* supports WebAssign, CAPA, and eGrade, which are programs that give instructors the ability to deliver and grade homework and quizzes over the Internet.

***WebCT and Blackboard*** Many of the instructor and student supplements have been coded for easy integration into WebCT and Blackboard, which are course management programs that allow instructors to set up a complete on-line course with chat rooms, bulletin boards, quizzing, built-in student tracking tools, etc.

## Student's Supplements
*Helping Students Learn*

***Student Web Site*** This Web site (**www.wiley.com/college/cutnell**) is designed to assist students further in the study of physics. At this site, students can access the following resources; the first few pages of the text show some examples of these resources.

- Self-Assessment Tests

- Solutions to selected end-of-chapter problems that are identified with a **www** icon in the text.

- Interactive Solutions

- Concept Simulations

- Interactive LearningWare examples

- Review quizzes for the MCAT exam

- Links to other Web sites that offer physics tutorial help

***Student Study Guide*** by John D. Cutnell and Kenneth W. Johnson. This student study guide consists of traditional print materials; with the Student Web site, it provides a rich, interactive environment for review and study.

***Student Solutions Manual*** by John D. Cutnell and Kenneth W. Johnson. This manual provides students with complete worked-out solutions for approximately 600 of the odd-numbered end-of-chapter problems. These problems are indicated in the text with an **ssm** icon.

# Chapter 1 Introduction and Mathematical Concepts

## 1.1 The Nature of Physics

The science of physics has developed out of the efforts of men and women to explain our physical environment. These efforts have been so successful that the laws of physics now encompass a remarkable variety of phenomena, including planetary orbits, radio and TV waves, magnetism, and lasers, to name just a few.

The exciting feature of physics is its capacity for predicting how nature will behave in one situation on the basis of experimental data obtained in another situation. Such predictions place physics at the heart of modern technology and, therefore, can have a tremendous impact on our lives. Rocketry and the development of space travel have their roots firmly planted in the physical laws of Galileo Galilei (1564–1642) and Isaac Newton (1642–1727). The transportation industry relies heavily on physics in the development of engines and the design of aerodynamic vehicles. Entire electronics and computer industries owe their existence to the invention of the transistor, which grew directly out of the laws of physics that describe the electrical behavior of solids. The telecommunications industry depends extensively on electromagnetic waves, whose existence was predicted by James Clerk Maxwell (1831–1879) in his theory of electricity and magnetism. The medical profession uses X-ray, ultrasonic, and magnetic resonance methods for obtaining images of the interior of the human body, and physics lies at the core of all these. Perhaps the most widespread impact in modern technology is that due to the laser. Fields ranging from space exploration to medicine benefit from this incredible device, which is a direct application of the principles of atomic physics.

Because physics is so fundamental, it is a required course for students in a wide range of major areas. We welcome you to the study of this fascinating topic. You will learn how to see the world through the "eyes" of physics and to reason as a physicist does. In the process, you will learn how to apply physics principles to a wide range of problems. We hope that you will come to recognize that physics has important things to say about your environment.

## 1.2 Units

Physics experiments involve the measurement of a variety of quantities, and a great deal of effort goes into making these measurements as accurate and reproducible as possible. The first step toward ensuring accuracy and reproducibility is defining the units in which the measurements are made.

In this text, we emphasize the system of units known as *SI units,* which stands for the French phrase "Le Système International d'Unités." By international agreement, this system employs the *meter* (m) as the unit of length, the *kilogram* (kg) as the unit of mass, and the *second* (s) as the unit of time. Two other systems of units are also in use, however. The CGS system utilizes the centimeter (cm), the gram (g), and the second for length, mass, and time, respectively, and the BE or British Engineering system (the gravitational version) uses the foot (ft), the slug (sl), and the second. Table 1.1 summarizes the units used for length, mass, and time in the three systems.

**Table 1.1**  *Units of Measurement*

|  | System | | |
|---|---|---|---|
|  | SI | CGS | BE |
| Length | meter (m) | centimeter (cm) | foot (ft) |
| Mass | kilogram (kg) | gram (g) | slug (sl) |
| Time | second (s) | second (s) | second (s) |

**Figure 1.1** The standard platinum–iridium meter bar. (Courtesy Bureau International des Poids et Mesures, France)

Originally, the meter was defined in terms of the distance measured along the earth's surface between the north pole and the equator. Eventually, a more accurate measurement standard was needed, and by international agreement the meter became the distance between two marks on a bar of platinum–iridium alloy (see Figure 1.1) kept at a temperature of 0 °C. Today, to meet further demands for increased accuracy, the meter is defined as the distance that light travels in a vacuum in a time of 1/299 792 458 second. This definition arises because the speed of light is a universal constant that is defined to be 299 792 458 m/s.

The definition of a kilogram as a unit of mass has also undergone changes over the years. As Chapter 4 discusses, the mass of an object indicates the tendency of the object to continue in motion with a constant velocity. Originally, the kilogram was expressed in terms of a specific amount of water. Today, one kilogram is defined to be the mass of a standard cylinder of platinum–iridium alloy, like the one in Figure 1.2.

As with the units for length and mass, the present definition of the second as a unit of time is different from the original definition. Originally, the second was defined according to the average time for the earth to rotate once about its axis, one day being set equal to 86 400 seconds. The earth's rotational motion was chosen because it is naturally repetitive, occurring over and over again. Today, we still use a naturally occurring repetitive phenomenon to define the second, but of a very different kind. We use the electromagnetic waves emitted by cesium-133 atoms in an atomic clock like that in Figure 1.3. One second is defined as the time needed for 9 192 631 770 wave cycles to occur.*

The units for length, mass, and time, along with a few other units that will arise later, are regarded as *base* SI units. The word "base" refers to the fact that these units are used along with various laws to define additional units for other important physical quantities, such as force and energy. The units for such other physical quantities are referred to as *derived* units, since they are combinations of the base units. Derived units will be introduced from time to time, as they arise naturally along with the related physical laws.

The value of a quantity in terms of base or derived units is sometimes a very large or very small number. In such cases, it is convenient to introduce larger or smaller units that are related to the normal units by multiples of ten. Table 1.2 summarizes the prefixes that are used to denote multiples of ten. For example, 1000 or $10^3$ meters are referred to as 1 kilometer (km), and 0.001 or $10^{-3}$ meter is called 1 millimeter (mm). Similarly, 1000 grams and 0.001 gram are referred to as 1 kilogram (kg) and 1 milligram (mg), respectively. Appendix A contains a discussion of scientific notation and powers of ten, such as $10^3$ and $10^{-3}$.

**Figure 1.2** The standard platinum–iridium kilogram is kept at the International Bureau of Weights and Measures in Sèvres, France. (Courtesy Bureau International des Poids et Mesures, France)

## 1.3  *The Role of Units in Problem Solving*
### THE CONVERSION OF UNITS

Since any quantity, such as length, can be measured in several different units, it is important to know how to convert from one unit to another. For instance, the foot can be used to express the distance between the two marks on the standard platinum–iridium meter bar. There are 3.281 feet in one meter, and this number can be used to convert from meters to feet, as the following example demonstrates.

### Example 1  The World's Highest Waterfall

The highest waterfall in the world is Angel Falls in Venezuela, with a total drop of 979.0 m (see Figure 1.4). Express this drop in feet.

**Reasoning** When converting between units, we write down the units explicitly in the calculations and treat them like any algebraic quantity. In particular, we will take advantage of the following algebraic fact: Multiplying or dividing an equation by a factor of 1 does not alter an equation.

---

* See Chapter 16 for a discussion of waves in general and Chapter 24 for a discussion of electromagnetic waves in particular.

**Figure 1.3** This atomic clock, the NIST-F1, is considered one of the world's most accurate clocks. It keeps time with an uncertainty of about one second in twenty million years. (© Geoffrey Wheeler)

**Figure 1.4** Angel Falls in Venezuela is the highest waterfall in the world. (© Kevin Schafer/The Image Bank/Getty Images)

**Solution** Since 3.281 feet = 1 meter, it follows that (3.281 feet)/(1 meter) = 1. Using this factor of 1 to multiply the equation "Length = 979.0 meters," we find that

$$\text{Length} = (979.0 \text{ m})(1) = (979.0 \text{ meters}) \left( \frac{3.281 \text{ feet}}{1 \text{ meter}} \right) = \boxed{3212 \text{ feet}}$$

The colored lines emphasize that the units of meters behave like any algebraic quantity and cancel when the multiplication is performed, leaving only the desired unit of feet to describe the answer. In this regard, note that 3.281 feet = 1 meter also implies that (1 meter)/(3.281 feet) = 1. However, we chose not to multiply by a factor of 1 in this form, because the units of meters would not have canceled.

   A calculator gives the answer as 3212.099 feet. Standard procedures for significant figures, however, indicate that the answer should be rounded off to four significant figures, since the value of 979.0 meters is accurate to only four significant figures. In this regard, the "1 meter" in the denominator does not limit the significant figures of the answer, because this number is precisely one meter by definition of the conversion factor. Appendix B contains a review of significant figures.

*In any conversion, if the units do not combine algebraically to give the desired result, the conversion has not been carried out properly.* With this in mind, the next example stresses the importance of writing down the units and illustrates a typical situation in which several conversions are required.

**Table 1.2** *Standard Prefixes Used to Denote Multiples of Ten*

| Prefix | Symbol | Factor[a] |
|---|---|---|
| Tera | T | $10^{12}$ |
| Giga[b] | G | $10^{9}$ |
| Mega | M | $10^{6}$ |
| Kilo | k | $10^{3}$ |
| Hecto | h | $10^{2}$ |
| Deka | da | $10^{1}$ |
| Deci | d | $10^{-1}$ |
| Centi | c | $10^{-2}$ |
| Milli | m | $10^{-3}$ |
| Micro | $\mu$ | $10^{-6}$ |
| Nano | n | $10^{-9}$ |
| Pico | p | $10^{-12}$ |
| Femto | f | $10^{-15}$ |

[a] Appendix A contains a discussion of powers of ten and scientific notation.

[b] Pronounced jig′a.

## Example 2   Interstate Speed Limit

Express the speed limit of 65 miles/hour in terms of meters/second.

**Reasoning** As in Example 1, it is important to write down the units explicitly in the calculations and treat them like any algebraic quantity. Here, two well-known relationships come into play—namely, 5280 feet = 1 mile and 3600 seconds = 1 hour. As a result, (5280 feet)/(1 mile) = 1 and (3600 seconds)/(1 hour) = 1. In our solution we will use the fact that multiplying and dividing by these factors of unity does not alter an equation.

**Solution** Multiplying and dividing by factors of unity, we find the speed limit in feet per second as shown below:

$$\text{Speed} = \left(65\ \frac{\text{miles}}{\text{hour}}\right)(1)(1) = \left(65\ \frac{\text{miles}}{\text{hour}}\right)\left(\frac{5280\ \text{feet}}{1\ \text{mile}}\right)\left(\frac{1\ \text{hour}}{3600\ \text{s}}\right) = 95\ \frac{\text{feet}}{\text{second}}$$

To convert feet into meters, we use the fact that (1 meter)/(3.281 feet) = 1:

$$\text{Speed} = \left(95\ \frac{\text{feet}}{\text{second}}\right)(1) = \left(95\ \frac{\text{feet}}{\text{second}}\right)\left(\frac{1\ \text{meter}}{3.281\ \text{feet}}\right) = \boxed{29\ \frac{\text{meters}}{\text{second}}}$$

In addition to their role in guiding the use of conversion factors, units serve a useful purpose in solving problems. They can provide an internal check to eliminate errors, if they are carried along during each step of a calculation and treated like any algebraic factor. In particular, remember that ***only quantities with the same units can be added or subtracted.*** Thus, at one point in a calculation, if you find yourself adding 12 miles to 32 kilometers, stop and reconsider. Either miles must be converted into kilometers or kilometers must be converted into miles before the addition can be carried out.

## DIMENSIONAL ANALYSIS

We have seen that many quantities are denoted by specifying both a number and a unit. For example, the distance to the nearest telephone may be 8 meters, or the speed of a car might be 25 meters/second. Each quantity, according to its physical nature, requires a certain *type* of unit. Distance must be measured in a length unit such as meters, feet, or miles, and a time unit will not do. Likewise, the speed of an object must be specified as a length unit divided by a time unit. In physics, the term ***dimension*** is used to refer to the physical nature of a quantity and the type of unit used to specify it. Distance has the dimension of length, which is symbolized as [L], while speed has the dimensions of length [L] divided by time [T], or [L/T]. Many physical quantities can be expressed in terms of a combination of fundamental dimensions such as length [L], time [T], and mass [M]. Later on, we will encounter certain other quantities, such as temperature, which are also fundamental. A fundamental quantity like temperature cannot be expressed as a combination of the dimensions of length, time, mass, or any other fundamental dimension.

Dimensional analysis is used to check mathematical relations for the consistency of their dimensions. As an illustration, consider a car that starts from rest and accelerates to a speed $v$ in a time $t$. Suppose we wish to calculate the distance $x$ traveled by the car but are not sure whether the correct relation is $x = \frac{1}{2}vt^2$ or $x = \frac{1}{2}vt$. We can decide by checking the quantities on both sides of the equals sign to see whether they have the same dimensions. If the dimensions are not the same, the relation is incorrect. For $x = \frac{1}{2}vt^2$, we use the dimensions for distance [L], time [T], and speed [L/T] in the following way:

$$x = \tfrac{1}{2}vt^2$$

*Dimensions*

$$[L] \overset{?}{=} \left[\frac{L}{\cancel{T}}\right][T]^2 = [L][T]$$

Dimensions cancel just like algebraic quantities, and pure numerical factors like $\frac{1}{2}$ have no dimensions, so they can be ignored. The dimension on the left of the equals sign does not match those on the right, so the relation $x = \frac{1}{2}vt^2$ cannot be correct. On the other hand, applying dimensional analysis to $x = \frac{1}{2}vt$, we find that

$$x = \tfrac{1}{2}vt$$

*Dimensions*

$$[L] \overset{?}{=} \left[\frac{L}{\cancel{T}}\right][\cancel{T}] = [L]$$

The dimension on the left of the equals sign matches that on the right, so this relation is dimensionally correct. If we know that one of our two choices is the right one, then $x = \frac{1}{2}vt$ is it. In the absence of such knowledge, however, dimensional analysis cannot identify the correct relation. It can only identify which choices *may be* correct, since it does not account for numerical factors like $\frac{1}{2}$ or for the manner in which an equation was derived from physics principles.

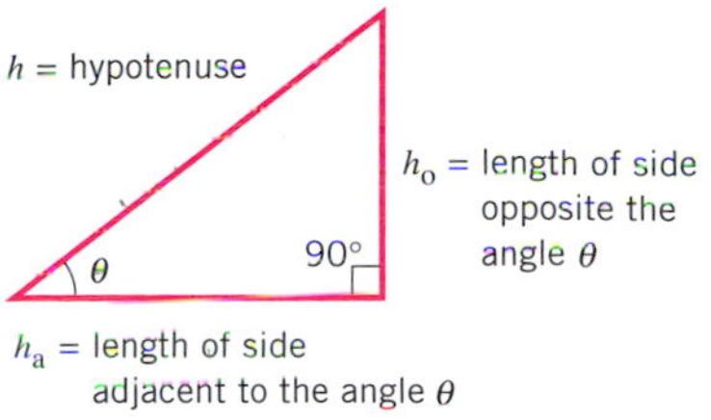

*h* = hypotenuse

$h_o$ = length of side opposite the angle $\theta$

$h_a$ = length of side adjacent to the angle $\theta$

**Figure 1.5** A right triangle.

# 1.4 Trigonometry

Scientists use mathematics to help them describe how the physical universe works, and trigonometry is an important branch of mathematics. Three trigonometric functions are utilized throughout this text. They are the sine, the cosine, and the tangent of the angle $\theta$ (Greek theta), abbreviated as sin $\theta$, cos $\theta$, and tan $\theta$, respectively. These functions are defined below in terms of the symbols given along with the right triangle in Figure 1.5.

---

■ **DEFINITION OF SIN $\theta$, COS $\theta$, AND TAN $\theta$**

$$\sin \theta = \frac{h_o}{h} \tag{1.1}$$

$$\cos \theta = \frac{h_a}{h} \tag{1.2}$$

$$\tan \theta = \frac{h_o}{h_a} \tag{1.3}$$

$h$ = length of the **hypotenuse** of a right triangle
$h_o$ = length of the side **opposite** the angle $\theta$
$h_a$ = length of the side **adjacent** to the angle $\theta$

---

The sine, cosine, and tangent of an angle are numbers without units, because each is the ratio of the lengths of two sides of a right triangle. Example 3 illustrates a typical application of Equation 1.3.

## Example 3   Using Trigonometric Functions

On a sunny day, a tall building casts a shadow that is 67.2 m long. The angle between the sun's rays and the ground is $\theta = 50.0°$, as Figure 1.6 shows. Determine the height of the building.

**Reasoning** We want to find the height of the building. Therefore, we begin with the colored right triangle in Figure 1.6 and identify the height as the length $h_o$ of the side opposite the angle $\theta$. The length of the shadow is the length $h_a$ of the side that is adjacent to the angle $\theta$. The ratio of the length of the opposite side to the length of the adjacent side is the tangent of the angle $\theta$, which can be used to find the height of the building.

**Solution** We use the tangent function in the following way, with $\theta = 50.0°$ and $h_a = 67.2$ m:

$$\tan \theta = \frac{h_o}{h_a} \tag{1.3}$$

$$h_o = h_a \tan \theta = (67.2 \text{ m})(\tan 50.0°) = (67.2 \text{ m})(1.19) = \boxed{80.0 \text{ m}}$$

The value of tan 50.0° is found by using a calculator.

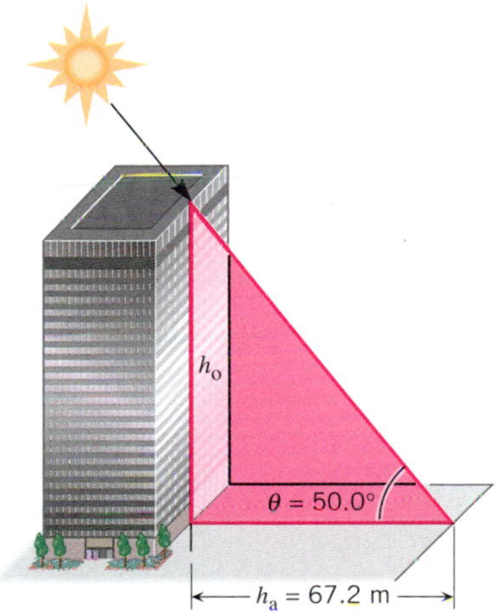

**Figure 1.6** From a value for the angle $\theta$ and the length $h_a$ of the shadow, the height $h_o$ of the building can be found using trigonometry.

The sine, cosine, or tangent may be used in calculations such as that in Example 3, depending on which side of the triangle has a known value and which side is asked for. However, *the choice of which side of the triangle to label $h_o$ (opposite) and which to label $h_a$ (adjacent) can be made only after the angle $\theta$ is identified.*

Often the values for two sides of the right triangle in Figure 1.5 are available, and the value of the angle $\theta$ is unknown. The concept of *inverse trigonometric functions* plays an

**Problem solving insight**

important role in such situations. Equations 1.4–1.6 give the inverse sine, inverse cosine, and inverse tangent in terms of the symbols used in the drawing. For instance, Equation 1.4 is read as "$\theta$ equals the angle whose sine is $h_o/h$."

$$\theta = \sin^{-1}\left(\frac{h_o}{h}\right) \tag{1.4}$$

$$\theta = \cos^{-1}\left(\frac{h_a}{h}\right) \tag{1.5}$$

$$\theta = \tan^{-1}\left(\frac{h_o}{h_a}\right) \tag{1.6}$$

The use of "$-1$" as an exponent in Equations 1.4–1.6 *does not mean* "take the reciprocal." For instance, $\tan^{-1}(h_o/h_a)$ does not equal $1/\tan(h_o/h_a)$. Another way to express the inverse trigonometric functions is to use arc sin, arc cos, and arc tan instead of $\sin^{-1}$, $\cos^{-1}$, and $\tan^{-1}$. Example 4 illustrates the use of an inverse trigonometric function.

### Example 4   Using Inverse Trigonometric Functions

A lakefront drops off gradually at an angle $\theta$, as Figure 1.7 indicates. For safety reasons, it is necessary to know how deep the lake is at various distances from the shore. To provide some information about the depth, a lifeguard rows straight out from the shore a distance of 14.0 m and drops a weighted fishing line. By measuring the length of the line, the lifeguard determines the depth to be 2.25 m. (a) What is the value of $\theta$? (b) What would be the depth $d$ of the lake at a distance of 22.0 m from the shore?

**Reasoning** Near the shore, the lengths of the opposite and adjacent sides of the right triangle in Figure 1.7 are $h_o = 2.25$ m and $h_a = 14.0$ m, relative to the angle $\theta$. Having made this identification, we can use the inverse tangent to find the angle in part (a). For part (b) the opposite and adjacent sides farther from the shore become $h_o = d$ and $h_a = 22.0$ m. With the value for $\theta$ obtained in part (a), the tangent function can be used to find the unknown depth. Considering the way in which the lake bottom drops off in Figure 1.7, we expect the unknown depth to be greater than 2.25 m.

**Solution**

(a) Using the inverse tangent given in Equation 1.6, we find that

$$\theta = \tan^{-1}\left(\frac{h_o}{h_a}\right) = \tan^{-1}\left(\frac{2.25 \text{ m}}{14.0 \text{ m}}\right) = \boxed{9.13°}$$

(b) With $\theta = 9.13°$, the tangent function given in Equation 1.3 can be used to find the unknown depth farther from the shore, where $h_o = d$ and $h_a = 22.0$ m. Since $\tan \theta = h_o/h_a$, it follows that

$$h_o = h_a \tan \theta$$
$$d = (22.0 \text{ m})(\tan 9.13°) = \boxed{3.54 \text{ m}}$$

which is greater than 2.25 m, as expected.

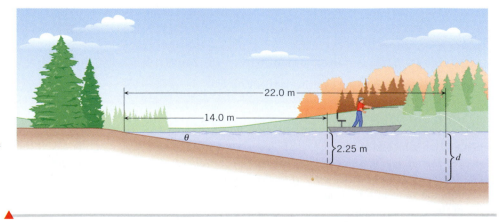

**Figure 1.7** If the distance from the shore and the depth of the water at any one point are known, the angle $\theta$ can be found with the aid of trigonometry. Knowing the value of $\theta$ is useful, because then the depth $d$ at another point can be determined.

The right triangle in Figure 1.5 provides the basis for defining the various trigonometric functions according to Equations 1.1–1.3. These functions always involve an angle and two sides of the triangle. There is also a relationship among the lengths of the three sides of a right triangle. This relationship is known as the **Pythagorean theorem** and is used often in this text.

---

■ **PYTHAGOREAN THEOREM**

The square of the length of the hypotenuse of a right triangle is equal to the sum of the squares of the lengths of the other two sides:

$$h^2 = h_o^2 + h_a^2 \qquad (1.7)$$

---

## 1.5 *Scalars and Vectors*

The volume of water in a swimming pool might be 50 cubic meters, or the winning time of a race could be 11.3 seconds. In cases like these, only the size of the numbers matters. In other words, *how much* volume or time is there? The "50" specifies the amount of water in units of cubic meters, while the "11.3" specifies the amount of time in seconds. Volume and time are examples of scalar quantities. A **scalar quantity** is one that can be described with a single number (including any units) giving its size or magnitude. Some other common scalars are temperature (e.g., 20 °C) and mass (e.g., 85 kg).

While many quantities in physics are scalars, there are also many that are not, and for these quantities the magnitude tells only part of the story. Consider Figure 1.8, which depicts a car that has moved 2 km along a straight line from start to finish. When describing the motion, it is incomplete to say that "the car moved a distance of 2 km." This statement would indicate only that the car ends up somewhere on a circle whose center is at the starting point and whose radius is 2 km. A complete description must include the direction along with the distance, as in the statement "the car moved a distance of 2 km in a direction 30° north of east." A quantity that deals inherently with both magnitude and direction is called a **vector quantity.** Because direction is an important characteristic of vectors, arrows are used to represent them; *the direction of the arrow gives the direction of the vector.* The colored arrow in Figure 1.8, for example, is called the displacement vector, because it shows how the car is displaced from its starting point. Chapter 2 discusses this particular vector.

The length of the arrow in Figure 1.8 represents the magnitude of the displacement vector. If the car had moved 4 km instead of 2 km from the starting point, the arrow would have been drawn twice as long. **By convention, the length of a vector arrow is proportional to the magnitude of the vector.**

In physics there are many important kinds of vectors, and the practice of using the length of an arrow to represent the magnitude of a vector applies to each of them. All forces, for instance, are vectors. In common usage a force is a push or a pull, and the direction in which a force acts is just as important as the strength or magnitude of the force. The magnitude of a force is measured in SI units called newtons (N). An arrow representing a force of 20 newtons is drawn twice as long as one representing a force of 10 newtons.

The fundamental distinction between scalars and vectors is the characteristic of direction. Vectors have it, and scalars do not. Conceptual Example 5 helps to clarify this distinction and explains what is meant by the "direction" of a vector.

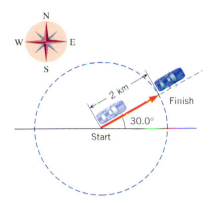

**Figure 1.8** A vector quantity has a magnitude and a direction. The arrow in this drawing represents a displacement vector.

### Conceptual Example 5
### Vectors, Scalars, and the Role of Plus and Minus Signs

▼

There are places where the temperature is +20 °C at one time of the year and −20 °C at another time. Do the plus and minus signs that signify positive and negative temperatures imply that temperature is a vector quantity?

**Reasoning and Solution** A vector has a physical direction associated with it, due east or due west, for example. The question, then, is whether such a direction is associated with temperature. In particular, do the plus and minus signs that go along with temperature imply this kind of direction? On a thermometer, the algebraic signs simply mean that the temperature is a number less than or greater than zero on the scale and have nothing to do with east, west, or any other physical direction. Temperature, then, is not a vector. It is a scalar, and scalars can sometimes be negative. *The fact that a quantity is positive or negative does not necessarily mean that the quantity is a scalar or a vector.*

Often, for the sake of convenience, quantities such as volume, time, displacement, and force are represented by symbols. This text follows the usual practice of writing vectors in boldface symbols* **(this is boldface)** and writing scalars in italic symbols (*this is italic*). Thus, a displacement vector is written as "**A** = 750 m, due east," where the **A** is a boldface symbol. By itself, however, separated from the direction, the magnitude of this vector is a scalar quantity. Therefore, the magnitude is written as "*A* = 750 m," where the *A* is an italic symbol.

## 1.6 *Vector Addition and Subtraction*

### ADDITION

Often it is necessary to add one vector to another, and the process of addition must take into account both the magnitude and the direction of the vectors. The simplest situation occurs when the vectors point along the same direction—that is, when they are colinear, as in Figure 1.9. Here, a car first moves along a straight line, with a displacement vector **A** of 275 m, due east. Then, the car moves again in the same direction, with a displacement vector **B** of 125 m, due east. These two vectors add to give the total displacement vector **R**, which would apply if the car had moved from start to finish in one step. The symbol **R** is used because the total vector is often called the ***resultant vector***. With the tail of the second arrow located at the head of the first arrow, the two lengths simply add to give the length of the total displacement. This kind of vector addition is identical to the familiar addition of two scalar numbers (2 + 3 = 5) *and can be carried out here only because the vectors point along the same direction*. In such cases we add the individual magnitudes to get the magnitude of the total, knowing in advance what the direction must be. Formally, the addition is written as follows:

$$\mathbf{R} = \mathbf{A} + \mathbf{B}$$
$$\mathbf{R} = 275 \text{ m, due east} + 125 \text{ m, due east} = 400 \text{ m, due east}$$

Perpendicular vectors are frequently encountered, and Figure 1.10 indicates how they can be added. This figure applies to a car that first travels with a displacement vector **A** of 275 m, due east, and then with a displacement vector **B** of 125 m, due north. The two vectors add to give a resultant displacement vector **R**. Once again, the vectors to be added are arranged in a tail-to-head fashion, and the resultant vector points from the tail of the first to the head of the last vector added. The resultant displacement is given by the vector equation

$$\mathbf{R} = \mathbf{A} + \mathbf{B}$$

The addition in this equation cannot be carried out by writing *R* = 275 m + 125 m, because the vectors have different directions. Instead, we take advantage of the fact that the triangle in Figure 1.10 is a right triangle and use the Pythagorean theorem (Equation 1.7). According to this theorem, the magnitude of **R** is

$$R = \sqrt{(275 \text{ m})^2 + (125 \text{ m})^2} = 302 \text{ m}$$

The angle $\theta$ in Figure 1.10 gives the direction of the resultant vector. Since the lengths of all three sides of the right triangle are now known, either sin $\theta$, cos $\theta$, or tan $\theta$ can be used to determine $\theta$. Noting that tan $\theta = B/A$ and using the inverse trigonometric function, we find that:

$$\theta = \tan^{-1}\left(\frac{B}{A}\right) = \tan^{-1}\left(\frac{125 \text{ m}}{275 \text{ m}}\right) = 24.4°$$

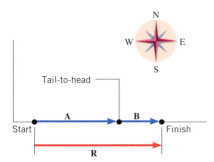

**Figure 1.9** Two colinear displacement vectors **A** and **B** add to give the resultant displacement vector **R**.

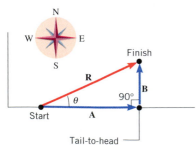

**Figure 1.10** The addition of two perpendicular displacement vectors **A** and **B** gives the resultant vector **R**.

---

\* A vector quantity can also be represented without boldface symbols, by including an arrow above the symbol—e.g., $\vec{A}$.

Thus, the resultant displacement of the car has a magnitude of 302 m and points north of east at an angle of 24.4°. This displacement would bring the car from the start to the finish in Figure 1.10 in a single straight-line step.

When two vectors to be added are not perpendicular, the tail-to-head arrangement does not lead to a right triangle, and the Pythagorean theorem cannot be used. Figure 1.11*a* illustrates such a case for a car that moves with a displacement **A** of 275 m, due east, and then with a displacement **B** of 125 m in a direction 55.0° north of west. As usual, the resultant displacement vector **R** is directed from the tail of the first to the head of the last vector added. The vector addition is still given according to

$$\mathbf{R} = \mathbf{A} + \mathbf{B}$$

However, the magnitude of **R** is not $R = \sqrt{A^2 + B^2}$, because the vectors **A** and **B** are not perpendicular and the Pythagorean theorem does not apply. Some other means must be used to find the magnitude and direction of the resultant vector.

One approach uses a graphical technique. In this method, a diagram is constructed in which the arrows are drawn tail to head. The lengths of the vector arrows are drawn to scale, and the angles are drawn accurately (with a protractor, perhaps). Then, the length of the arrow representing the resultant vector is measured with a ruler. This length is converted to the magnitude of the resultant vector by using the scale factor with which the drawing is constructed. In Figure 1.11*b*, for example, a scale of one centimeter of arrow length for each 10.0 m of displacement is used, and it can be seen that the length of the arrow representing **R** is 22.8 cm. Since each centimeter corresponds to 10.0 m of displacement, the magnitude of **R** is 228 m. The angle $\theta$, which gives the direction of **R**, can be measured with a protractor to be $\theta = 26.7°$ north of east.

## SUBTRACTION

The subtraction of one vector from another is carried out in a way that depends on the following fact. *When a vector is multiplied by −1, the magnitude of the vector remains the same, but the direction of the vector is reversed.* Conceptual Example 6 illustrates the meaning of this statement.

### *Conceptual Example 6* Multiplying a Vector by −1

Consider the two vectors described as follows:

1. A woman climbs 1.2 m up a ladder, so that her displacement vector **D** is 1.2 m, upward along the ladder, as in Figure 1.12*a*.

2. A man is pushing with 450 N of force on his stalled car, trying to move it eastward. The force vector **F** that he applies to the car is 450 N, due east, as in Figure 1.13*a*.

What are the physical meanings of the vectors −**D** and −**F**?

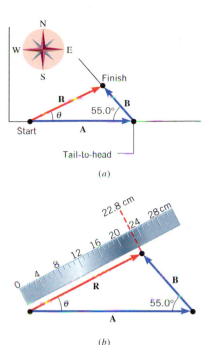

(a)

(b)

**Figure 1.11** (*a*) The two displacement vectors **A** and **B** are neither colinear nor perpendicular but even so they add to give the resultant vector **R**. (*b*) In one method for adding them together, a graphical technique is used.

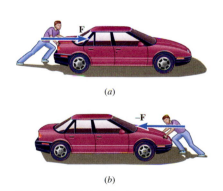

(a)

(b)

**Figure 1.13** (*a*) The force vector for a man pushing on a car with 450 N of force in a direction due east is **F**. (*b*) The force vector for a man pushing on a car with 450 N of force in a direction due west is −**F**.

(a)                                      (b)

**Figure 1.12** (*a*) The displacement vector for a woman climbing 1.2 m up a ladder is **D**. (*b*) The displacement vector for a woman climbing 1.2 m down a ladder is −**D**.

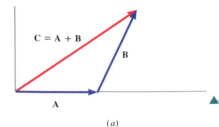

*(a)*

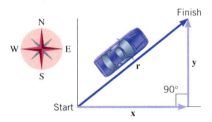

*(b)*

**Figure 1.14** (*a*) Vector addition according to **C** = **A** + **B**. (*b*) Vector subtraction according to **A** = **C** − **B** = **C** + (−**B**).

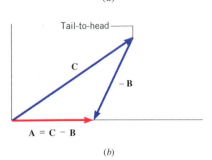

**Figure 1.15** The displacement vector **r** and its vector components **x** and **y**.

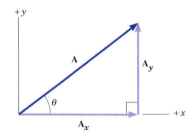

**Figure 1.16** An arbitrary vector **A** and its vector components **A**$_x$ and **A**$_y$.

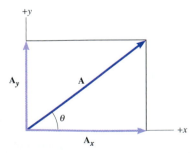

**Figure 1.17** This alternative way of drawing the vector **A** and its vector components is completely equivalent to that shown in Figure 1.16.

**Reasoning and Solution** A displacement vector of −**D** is (−1)**D** and has the same magnitude as the vector **D**, but is opposite in direction. Thus, −**D** would represent the displacement of a woman climbing 1.2 m down the ladder, as in Figure 1.12*b*. Similarly, a force vector of −**F** has the same magnitude as the vector **F** but has the opposite direction. As a result, −**F** would represent a force of 450 N applied to the car in a direction of due west, as in Figure 1.13*b*.

In practice, vector subtraction is carried out exactly like vector addition, except that one of the vectors added is multiplied by a scalar factor of −1. To see why, look at the two vectors **A** and **B** in Figure 1.14*a*. These vectors add together to give a third vector **C**, according to **C** = **A** + **B**. Therefore, we can calculate vector **A** as **A** = **C** − **B**, which is an example of vector subtraction. However, we can also write this result as **A** = **C** + (−**B**) and treat it as vector addition. Figure 1.14*b* shows how to calculate vector **A** by adding the vectors **C** and −**B**. Notice that vectors **C** and −**B** are arranged tail to head and that any suitable method of vector addition can be employed to determine **A**.

## 1.7 *The Components of a Vector*

### VECTOR COMPONENTS

Suppose a car moves along a straight line from start to finish in Figure 1.15, the corresponding displacement vector being **r**. The magnitude and direction of the vector **r** give the distance and direction traveled along the straight line. However, the car could also arrive at the finish point by first moving due east, turning through 90°, and then moving due north. This alternative path is shown in the drawing and is associated with the two displacement vectors **x** and **y**. The vectors **x** and **y** are called the *x* vector component and the *y* vector component of **r**.

Vector components are very important in physics and have two basic features that are apparent in Figure 1.15. One is that the components add together to equal the original vector:

$$\mathbf{r} = \mathbf{x} + \mathbf{y}$$

The components **x** and **y**, when added vectorially, convey exactly the same meaning as does the original vector **r**: they indicate how the finish point is displaced relative to the starting point. In general, *the components of any vector can be used in place of the vector itself in any calculation where it is convenient to do so.* The other feature of vector components that is apparent in Figure 1.15 is that **x** and **y** are not just any two vectors that add together to give the original vector **r**: they are perpendicular vectors. This perpendicularity is a valuable characteristic, as we will soon see.

Any type of vector may be expressed in terms of its components, in a way similar to that illustrated for the displacement vector in Figure 1.15. Figure 1.16 shows an arbitrary vector **A** and its vector components **A**$_x$ and **A**$_y$. The components are drawn parallel to convenient *x* and *y* axes and are perpendicular. They add vectorially to equal the original vector **A**:

$$\mathbf{A} = \mathbf{A}_x + \mathbf{A}_y$$

There are times when a drawing such as Figure 1.16 is not the most convenient way to represent vector components, and Figure 1.17 presents an alternative method. The disadvantage of this alternative is that the tail-to-head arrangement of **A**$_x$ and **A**$_y$ is missing, an arrangement that is a nice reminder that **A**$_x$ and **A**$_y$ add together to equal **A**.

The definition that follows summarizes the meaning of vector components:

■ **DEFINITION OF VECTOR COMPONENTS**

In two dimensions, the vector components of a vector **A** are two perpendicular vectors **A**$_x$ and **A**$_y$ that are parallel to the *x* and *y* axes, respectively, and add together vectorially so that **A** = **A**$_x$ + **A**$_y$.

The values calculated for vector components depend on the orientation of the vector relative to the axes used as a reference. Figure 1.18 illustrates this fact for a vector **A** by showing two sets of axes, one set being rotated clockwise relative to the other. With respect to the black axes, vector **A** has perpendicular vector components $\mathbf{A}_x$ and $\mathbf{A}_y$; with respect to the colored rotated axes, vector **A** has different vector components $\mathbf{A}_x{}'$ and $\mathbf{A}_y{}'$. The choice of which set of components to use is purely a matter of convenience.

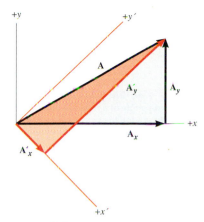

**Figure 1.18** The vector components of the vector depend on the orientation of the axes used as a reference.

## SCALAR COMPONENTS

It is often easier to work with the **scalar components,** $A_x$ and $A_y$ (note the italic symbols), rather than the vector components $\mathbf{A}_x$ and $\mathbf{A}_y$. Scalar components are positive or negative numbers (with units) that are defined as follows. The component $A_x$ has a magnitude that is equal to that of $\mathbf{A}_x$ and is given a positive sign if $\mathbf{A}_x$ points along the $+x$ axis and a negative sign if $\mathbf{A}_x$ points along the $-x$ axis. The component $A_y$ is defined in a similar manner. The following table shows an example of vector and scalar components:

| Vector Components | Scalar Components | Unit Vectors |
|---|---|---|
| $\mathbf{A}_x = 8$ meters, directed along the $+x$ axis | $A_x = +8$ meters | $\mathbf{A}_x = (+8 \text{ meters})\,\hat{\mathbf{x}}$ |
| $\mathbf{A}_y = 10$ meters, directed along the $-y$ axis | $A_y = -10$ meters | $\mathbf{A}_y = (-10 \text{ meters})\,\hat{\mathbf{y}}$ |

In this text, when we use the term "component," we will be referring to a scalar component, unless otherwise indicated.

Another method of expressing vector components is to use unit vectors. A **unit vector** is a vector that has a magnitude of 1, but no dimensions. We will use a caret (^) to distinguish it from other vectors. Thus,

$\hat{\mathbf{x}}$ is a dimensionless unit vector of length 1 that points in the positive $x$ direction, and

$\hat{\mathbf{y}}$ is a dimensionless unit vector of length 1 that points in the positive $y$ direction.

These unit vectors are illustrated in Figure 1.19. With the aid of unit vectors, the vector components of an arbitrary vector **A** can be written as $\mathbf{A}_x = A_x\,\hat{\mathbf{x}}$ and $\mathbf{A}_y = A_y\,\hat{\mathbf{y}}$, where $A_x$ and $A_y$ are its scalar components (see the drawing and third column of the table above). The vector **A** is then written as $\mathbf{A} = A_x\,\hat{\mathbf{x}} + A_y\,\hat{\mathbf{y}}$.

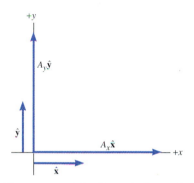

**Figure 1.19** The dimensionless unit vectors $\hat{\mathbf{x}}$ and $\hat{\mathbf{y}}$ have magnitudes equal to 1, and they point in the $+x$ and $+y$ directions, respectively. Expressed in terms of unit vectors, the vector components of the vector **A** are $A_x\,\hat{\mathbf{x}}$ and $A_y\,\hat{\mathbf{y}}$.

## RESOLVING A VECTOR INTO ITS COMPONENTS

If the magnitude and direction of a vector are known, it is possible to find the components of the vector. The process of finding the components is called "resolving the vector into its components." As Example 7 illustrates, this process can be carried out with the aid of trigonometry, because the two perpendicular vector components and the original vector form a right triangle.

### Example 7  Finding the Components of a Vector

A displacement vector **r** has a magnitude of $r = 175$ m and points at an angle of $50.0°$ relative to the $x$ axis in Figure 1.20. Find the $x$ and $y$ components of this vector.

**Reasoning** We will base our solution on the fact that the triangle formed in Figure 1.20 by the vector **r** and its components **x** and **y** is a right triangle. This fact enables us to use the trigonometric sine and cosine functions, as defined in Equations 1.1 and 1.2.

**Solution 1** The $y$ component can be obtained using the $50.0°$ angle and Equation 1.1, $\sin \theta = y/r$:

$$y = r \sin \theta = (175 \text{ m})(\sin 50.0°) = \boxed{134 \text{ m}}$$

In a similar fashion, the $x$ component can be obtained using the $50.0°$ angle and Equation 1.2, $\cos \theta = x/r$:

$$x = r \cos \theta = (175 \text{ m})(\cos 50.0°) = \boxed{112 \text{ m}}$$

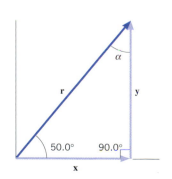

**Figure 1.20** The $x$ and $y$ components of the displacement vector **r** can be found using trigonometry.

**Solution 2** The angle $\alpha$ in Figure 1.20 can also be used to find the components. Since $\alpha + 50.0° = 90.0°$, it follows that $\alpha = 40.0°$. The solution using $\alpha$ yields the same answers as in Solution 1:

$$\cos \alpha = \frac{y}{r}$$

$$y = r \cos \alpha = (175 \text{ m})(\cos 40.0°) = \boxed{134 \text{ m}}$$

$$\sin \alpha = \frac{x}{r}$$

$$x = r \sin \alpha = (175 \text{ m})(\sin 40.0°) = \boxed{112 \text{ m}}$$

**Problem solving insight**
Either acute angle of a right triangle can be used to determine the components of a vector. The choice of angle is a matter of convenience.

**Problem solving insight**
You can check to see whether the components of a vector are correct by substituting them into the Pythagorean theorem and verifying that the result is the magnitude of the original vector.

Since the vector components and the original vector form a right triangle, the Pythagorean theorem can be applied to check the validity of calculations such as those in Example 7. Thus, with the components obtained in Example 7, the theorem can be used to verify that the magnitude of the original vector is indeed 175 m, as given initially:

$$r = \sqrt{(112 \text{ m})^2 + (134 \text{ m})^2} = 175 \text{ m}$$

**Problem solving insight**

It is possible for one of the components of a vector to be zero. This does not mean that the vector itself is zero, however. *For a vector to be zero, every vector component must individually be zero.* Thus, in two dimensions, saying that $\mathbf{A} = 0$ is equivalent to saying that $\mathbf{A}_x = 0$ and $\mathbf{A}_y = 0$. Or, stated in terms of scalar components, if $\mathbf{A} = 0$, then $A_x = 0$ and $A_y = 0$.

**Problem solving insight**

*Two vectors are equal if, and only if, they have the same magnitude and direction.* Thus, if one displacement vector points east and another points north, they are *not* equal, even if each has the same magnitude of 480 m. In terms of vector components, two vectors, $\mathbf{A}$ and $\mathbf{B}$, are equal if, and only if, each vector component of one is equal to the corresponding vector component of the other. In two dimensions, if $\mathbf{A} = \mathbf{B}$, then $\mathbf{A}_x = \mathbf{B}_x$ and $\mathbf{A}_y = \mathbf{B}_y$. Alternatively, using scalar components, we write that $A_x = B_x$ and $A_y = B_y$.

## 1.8  Addition of Vectors by Means of Components

The components of a vector provide the most convenient and accurate way of adding (or subtracting) any number of vectors. For example, suppose that vector $\mathbf{A}$ is added to vector $\mathbf{B}$. The resultant vector is $\mathbf{C}$, where $\mathbf{C} = \mathbf{A} + \mathbf{B}$. Figure 1.21$a$ illustrates this vector addition, along with the $x$ and $y$ vector components of $\mathbf{A}$ and $\mathbf{B}$. In part $b$ of the drawing, the vectors $\mathbf{A}$ and $\mathbf{B}$ have been removed, because we can use the vector components of these

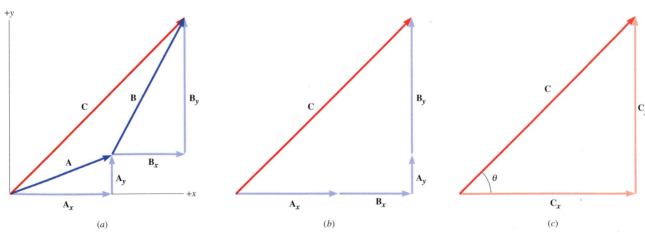

**Figure 1.21** ($a$) The vectors $\mathbf{A}$ and $\mathbf{B}$ add together to give the resultant vector $\mathbf{C}$. The $x$ and $y$ vector components of $\mathbf{A}$ and $\mathbf{B}$ are also shown. ($b$) The drawing illustrates that $\mathbf{C}_x = \mathbf{A}_x + \mathbf{B}_x$ and $\mathbf{C}_y = \mathbf{A}_y + \mathbf{B}_y$. ($c$) Vector $\mathbf{C}$ and its components form a right triangle.

vectors in place of them. The vector component $\mathbf{B}_x$ has been shifted downward and arranged tail to head with the vector component $\mathbf{A}_x$. Similarly, the vector component $\mathbf{A}_y$ has been shifted to the right and arranged tail to head with the vector component $\mathbf{B}_y$. The $x$ components are colinear and add together to give the $x$ component of the resultant vector $\mathbf{C}$. In like fashion, the $y$ components are colinear and add together to give the $y$ component of $\mathbf{C}$. In terms of scalar components, we can write

$$C_x = A_x + B_x \quad \text{and} \quad C_y = A_y + B_y$$

The vector components $\mathbf{C}_x$ and $\mathbf{C}_y$ of the resultant vector form the sides of the right triangle shown in Figure 1.21c. Thus, we can find the magnitude of $\mathbf{C}$ by using the Pythagorean theorem:

$$C = \sqrt{C_x^2 + C_y^2}$$

The angle $\theta$ that $\mathbf{C}$ makes with the $x$ axis is given by $\theta = \tan^{-1}(C_y/C_x)$. Example 8 illustrates how to add several vectors using the component method.

## Example 8 The Component Method of Vector Addition

A jogger runs 145 m in a direction 20.0° east of north (displacement vector $\mathbf{A}$) and then 105 m in a direction 35.0° south of east (displacement vector $\mathbf{B}$). Determine the magnitude and direction of the resultant vector $\mathbf{C}$ for these two displacements.

**Reasoning** Figure 1.22a shows the vectors $\mathbf{A}$ and $\mathbf{B}$, assuming that the $y$ axis corresponds to the direction due north. Since the vectors are not given in component form, we will begin by using the given magnitudes and directions to find the components. Then, the components of $\mathbf{A}$ and $\mathbf{B}$ can be used to find the components of the resultant $\mathbf{C}$. Finally, with the aid of the Pythagorean theorem and trigonometry, the components of $\mathbf{C}$ can be used to find its magnitude and direction.

**Solution** The first two rows of the following table give the $x$ and $y$ components of the vectors $\mathbf{A}$ and $\mathbf{B}$. Note that the component $B_y$ is negative, because $\mathbf{B}_y$ points downward, in the negative $y$ direction in the drawing.

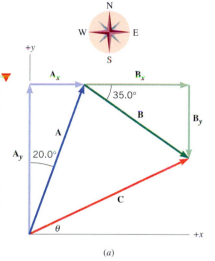

| Vector | $x$ Component | $y$ Component |
|--------|--------------|--------------|
| **A** | $A_x = (145 \text{ m}) \sin 20.0° = 49.6 \text{ m}$ | $A_y = (145 \text{ m}) \cos 20.0° = 136 \text{ m}$ |
| **B** | $B_x = (105 \text{ m}) \cos 35.0° = 86.0 \text{ m}$ | $B_y = -(105 \text{ m}) \sin 35.0° = -60.2 \text{ m}$ |
| **C** | $C_x = A_x + B_x = 135.6 \text{ m}$ | $C_y = A_y + B_y = 76 \text{ m}$ |

The third row in the table gives the $x$ and $y$ components of the resultant vector $\mathbf{C}$: $C_x = A_x + B_x$ and $C_y = A_y + B_y$. Part $b$ of the drawing shows $\mathbf{C}$ and its vector components. The magnitude of $\mathbf{C}$ is given by the Pythagorean theorem as

$$C = \sqrt{C_x^2 + C_y^2} = \sqrt{(135.6 \text{ m})^2 + (76 \text{ m})^2} = \boxed{155 \text{ m}}$$

The angle $\theta$ that $\mathbf{C}$ makes with the $x$ axis is

$$\theta = \tan^{-1}\left(\frac{C_y}{C_x}\right) = \tan^{-1}\left(\frac{76 \text{ m}}{135.6 \text{ m}}\right) = \boxed{29°}$$

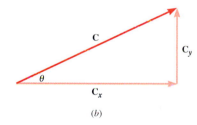

**Figure 1.22** (*a*) The vectors $\mathbf{A}$ and $\mathbf{B}$ add together to give the resultant vector $\mathbf{C}$. The vector components of $\mathbf{A}$ and $\mathbf{B}$ are also shown. (*b*) The resultant vector $\mathbf{C}$ can be obtained once its components have been found.

# Concept Summary

This summary presents an abridged version of the chapter, including the important equations and all available learning aids. For convenient reference, the learning aids (including the text's examples) are placed next to or immediately after the relevant equation or discussion. The following learning aids may be found on-line at **www.wiley.com/college/cutnell**:

**Interactive LearningWare** examples are solved according to a five-step interactive format that is designed to help you develop problem-solving skills.

**Concept Simulations** are animated versions of text figures or animations that illustrate important concepts. You can control parameters that affect the display, and we encourage you to experiment.

**Interactive Solutions** offer specific models for certain types of problems in the chapter homework. The calculations are carried out interactively.

**Self-Assessment Tests** include both qualitative and quantitative questions. Extensive feedback is provided for both incorrect and correct answers, to help you evaluate your understanding of the material.

| Topic | Discussion | Learning Aids |
|---|---|---|

### 1.2 Units

The SI system of units includes the meter (m), the kilogram (kg), and the second (s) as the base units for length, mass, and time, respectively.

**Meter**

One meter is the distance that light travels in a vacuum in a time of 1/299 792 458 second.

**Kilogram**

One kilogram is the mass of a standard cylinder of platinum–iridium alloy kept at the International Bureau of Weights and Measures.

**Second**

One second is the time for a certain type of electromagnetic wave emitted by cesium-133 atoms to undergo 9 192 631 770 wave cycles.

### 1.3 The Role of Units in Problem Solving

**Conversion of units**

To convert a number from one unit to another, multiply the number by the ratio of the two units. For instance, to convert 979 meters to feet, multiply 979 meters by the factor (3.281 foot/1 meter).    **Examples 1, 2**

**Dimension**

The dimension of a quantity represents its physical nature and the type of unit used to specify it. Three such dimensions are length [L], mass [M], time [T].

**Dimensional analysis**

A method for checking mathematical relations for the consistency of their dimensions.

### 1.4 Trigonometry

**Sine, cosine, and tangent of an angle $\theta$**

The sine, cosine, and tangent functions of an angle $\theta$ are defined in terms of a right triangle that contains $\theta$:    **Example 3**
**Interactive Solution 1.17**

$$\sin\theta = \frac{h_o}{h} \ (1.1) \qquad \cos\theta = \frac{h_a}{h} \ (1.2) \qquad \tan\theta = \frac{h_o}{h_a} \ (1.3)$$

where $h_o$ and $h_a$ are, respectively, the lengths of the sides opposite and adjacent to the angle $\theta$, and $h$ is the length of the hypotenuse.

**Inverse trigonometric functions**

The inverse sine, inverse cosine, and inverse tangent functions are    **Example 4**

$$\theta = \sin^{-1}\left(\frac{h_o}{h}\right) \ (1.4) \qquad \theta = \cos^{-1}\left(\frac{h_a}{h}\right) \ (1.5)$$

$$\theta = \tan^{-1}\left(\frac{h_o}{h_a}\right) \ (1.6)$$

**Pythagorean theorem**

The Pythagorean theorem states that the square of the length of the hypotenuse of a right triangle is equal to the sum of the squares of the lengths of the other two sides:

$$h^2 = h_o^2 + h_a^2 \qquad\qquad (1.7)$$

### 1.5 Scalars and Vectors

**Scalars and vectors**

A scalar quantity is described completely by its size, which is also called its magnitude. A vector quantity has both a magnitude and a direction. Vectors are often represented by arrows, the length of the arrow being proportional to the magnitude of the vector and the direction of the arrow indicating the direction of the vector.    **Example 5**

### 1.6 Vector Addition and Subtraction

**Graphical method of vector addition and subtraction**

One procedure for adding vectors utilizes a graphical technique, in which the vectors to be added are arranged in a tail-to-head fashion. The resultant vector is drawn from the tail of the first vector to the head of the last vector.

The subtraction of a vector is treated as the addition of a vector that has been multiplied by a scalar factor of $-1$. Multiplying a vector by $-1$ reverses the direction of the vector.    **Example 6**
**Interactive Solution 1.29**

    Use **Self-Assessment Test 1.1** to evaluate your understanding of Sections 1.5 and 1.6.

| Topic | Discussion | Learning Aids |
|---|---|---|

### 1.7 The Components of a Vector

**Vector components**

In two dimensions, the vector components of a vector $\mathbf{A}$ are two perpendicular vectors $\mathbf{A}_x$ and $\mathbf{A}_y$ that are parallel to the $x$ and $y$ axes, respectively, and that add together vectorially so that $\mathbf{A} = \mathbf{A}_x + \mathbf{A}_y$.

**Scalar components**

The scalar component $A_x$ has a magnitude that is equal to that of $\mathbf{A}_x$ and is given a positive sign if $\mathbf{A}_x$ points along the $+x$ axis and a negative sign if $\mathbf{A}_x$ points along the $-x$ axis. The scalar component $A_y$ is defined in a similar manner.

**Interactive Solution 1.37**
**Example 7**

**Condition for a vector to be zero**

A vector is zero if, and only if, each of its vector components is zero.

**Condition for two vectors to be equal**

Two vectors are equal if, and only if, they have the same magnitude and direction. Alternatively, two vectors are equal in two dimensions if the $x$ vector components of each are equal and the $y$ vector components of each are equal.

### 1.8 Addition of Vectors by Means of Components

If two vectors $\mathbf{A}$ and $\mathbf{B}$ are added to give a resultant vector $\mathbf{C}$ such that $\mathbf{C} = \mathbf{A} + \mathbf{B}$, then

**Example 8**
**Concept Simulation 1.1**

$$C_x = A_x + B_x \quad \text{and} \quad C_y = A_y + B_y$$

**Interactive Solution 1.49**

where $C_x$, $A_x$, and $B_x$ are the scalar components of the vectors along the $x$ direction, and $C_y$, $A_y$, and $B_y$ are the scalar components of the vectors along the $y$ direction.

 *Use Self-Assessment Test 1.2 to evaluate your understanding of Sections 1.7 and 1.8.*

# Problems

*Problems that are not marked with a star are considered the easiest to solve. Problems that are marked with a single star (\*) are more difficult, while those marked with a double star (\*\*) are the most difficult.*

**ssm** Solution is in the Student Solutions Manual.      **www** Solution is available on the World Wide Web at www.wiley.com/college/cutnell

☤ This icon represents a biomedical application.

### Section 1.2 Units, Section 1.3 The Role of Units in Problem Solving

**1. ssm** The mass of the parasitic wasp *Caraphractus cintus* can be as small as $5 \times 10^{-6}$ kg. What is this mass in (a) grams (g), (b) milligrams (mg), and (c) micrograms ($\mu$g)?

**2.** Vesna Vulovic survived the longest fall on record without a parachute when her plane exploded and she fell 6 miles, 551 yards. What is this distance in meters?

**3.** How many seconds are there in (a) one hour and thirty-five minutes and (b) one day?

**4.** Bicyclists in the Tour de France reach speeds of 34.0 miles per hour (mi/h) on flat sections of the road. What is this speed in (a) kilometers per hour (km/h) and (b) meters per second (m/s)?

**5. ssm** The largest diamond ever found had a size of 3106 carats. One carat is equivalent to a mass of 0.200 g. Use the fact that 1 kg (1000 g) has a weight of 2.205 lb under certain conditions, and determine the weight of this diamond in pounds.

**6.** A bottle of wine known as a magnum contains a volume of 1.5 liters. A bottle known as a jeroboam contains 0.792 U.S. gallons. How many magnums are there in one jeroboam?

**7.** The following are dimensions of various physical parameters that will be discussed later on in the text. Here [L], [T], and [M] denote, respectively, dimensions of length, time, and mass.

| | Dimension | | Dimension |
|---|---|---|---|
| Distance $(x)$ | [L] | Acceleration $(a)$ | $[L]/[T]^2$ |
| Time $(t)$ | [T] | Force $(F)$ | $[M][L]/[T]^2$ |
| Mass $(m)$ | [M] | Energy $(E)$ | $[M][L]^2/[T]^2$ |
| Speed $(v)$ | $[L]/[T]$ | | |

Which of the following equations are dimensionally correct?

(a) $F = ma$      (d) $E = max$

(b) $x = \frac{1}{2}at^3$      (e) $v = \sqrt{Fx/m}$

(c) $E = \frac{1}{2}mv$

**8.** The variables $x$, $v$, and $a$ have the dimensions of [L], [L]/[T], and $[L]/[T]^2$, respectively. These variables are related by an equation that has the form $v^n = 2ax$, where $n$ is an integer constant (1, 2, 3, etc.) without dimensions. What must be the value of $n$, so that both sides of the equation have the same dimensions? Explain your reasoning.

**\*9. ssm** The depth of the ocean is sometimes measured in fathoms (1 fathom = 6 feet). Distance on the surface of the ocean is sometimes measured in nautical miles (1 nautical mile = 6076 feet). The water beneath a surface rectangle 1.20 nautical miles by 2.60 nautical miles has a depth of 16.0 fathoms. Find the volume of water (in cubic meters) beneath this rectangle.

\* **10.** A spring is hanging down from the ceiling, and an object of mass $m$ is attached to the free end. The object is pulled down, thereby stretching the spring, and then released. The object oscillates up and down, and the time $T$ required for one complete up-and-down oscillation is given by the equation $T = 2\pi\sqrt{m/k}$, where $k$ is known as the spring constant. What must be the dimension of $k$ for this equation to be dimensionally correct?

**Section 1.4 Trigonometry**

**11.** You are driving into St. Louis, Missouri, and in the distance you see the famous Gateway-to-the-West arch. This monument rises to a height of 192 m. You estimate your line of sight with the top of the arch to be 2.0° above the horizontal. Approximately how far (in kilometers) are you from the base of the arch?

**12.** An observer, whose eyes are 1.83 m above the ground, is standing 32.0 m away from a tree. The ground is level, and the tree is growing perpendicular to it. The observer's line of sight with the treetop makes an angle of 20.0° above the horizontal. How tall is the tree?

**13. ssm www** A highway is to be built between two towns, one of which lies 35.0 km south and 72.0 km west of the other. What is the shortest length of highway that can be built between the two towns, and at what angle would this highway be directed with respect to due west?

**14.** The two hot-air balloons in the drawing are 48.2 and 61.0 m above the ground. A person in the left balloon observes that the right balloon is 13.3° above the horizontal. What is the horizontal distance $x$ between the two balloons?

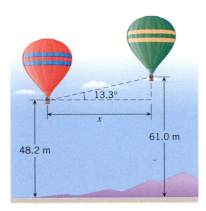

*Problem 14*

**15.** The silhouette of a Christmas tree is an isosceles triangle. The angle at the top of the triangle is 30.0°, and the base measures 2.00 m across. How tall is the tree?

\* **16.** The drawing shows sodium and chlorine ions positioned at the corners of a cube that is part of the crystal structure of sodium chloride (common table salt). The edge of the cube is 0.281 nm    (1 nm = 1    nanometer = $10^{-9}$ m) in length. Find the distance (in nanometers) between the sodium ion located at one corner of the cube and the chlorine ion located on the diagonal at the opposite corner.

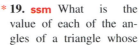

\* **17. Interactive Solution 1.17** at **www.wiley.com/college/cutnell** presents a method for modeling this problem. What is the value of the angle $\theta$ in the drawing that accompanies problem 16?

\* **18.** A person is standing at the edge of the water and looking out at the ocean (see the drawing). The height of the person's eyes above the water is $h = 1.6$ m, and the radius of the earth is $R = 6.38 \times 10^6$ m. (a) How far is it to the horizon? In other words, what is the distance $d$ from the person's eyes to the horizon? (*Note: At the horizon the angle between the line of sight and the radius of the earth is 90°.*) (b) Express this distance in miles.

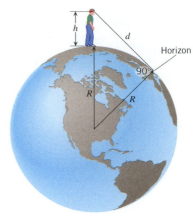

*Problem 18*

\* **19. ssm** What is the value of each of the angles of a triangle whose sides are 95, 150, and 190 cm in length? (*Hint: Consider using the law of cosines given in Appendix E.*)

\*\* **20.** A regular tetrahedron is a three-dimensional object that has four faces, each of which is an equilateral triangle. Each of the edges of such an object has a length $L$. The height $H$ of a regular tetrahedron is the perpendicular distance from one corner to the center of the opposite triangular face. Show that the ratio between $H$ and $L$ is $H/L = \sqrt{2/3}$.

**Section 1.6 Vector Addition and Subtraction**

**21.** A force vector $\mathbf{F}_1$ points due east and has a magnitude of 200 newtons. A second force $\mathbf{F}_2$ is added to $\mathbf{F}_1$. The resultant of the two vectors has a magnitude of 400 newtons and points along the east/west line. Find the magnitude and direction of $\mathbf{F}_2$. Note that there are two answers.

**22.** (a) Two workers are trying to move a heavy crate. One pushes on the crate with a force $\mathbf{A}$, which has a magnitude of 445 newtons and is directed due west. The other pushes with a force $\mathbf{B}$, which has a magnitude of 325 newtons and is directed due north. What are the magnitude and direction of the resultant force $\mathbf{A} + \mathbf{B}$ applied to the crate? (b) Suppose that the second worker applies a force $-\mathbf{B}$ instead of $\mathbf{B}$. What then are the magnitude and direction of the resultant force $\mathbf{A} - \mathbf{B}$ applied to the crate? In both cases express the direction relative to due west.

**23. ssm** Displacement vector $\mathbf{A}$ points due east and has a magnitude of 2.00 km. Displacement vector $\mathbf{B}$ points due north and has a magnitude of 3.75 km. Displacement vector $\mathbf{C}$ points due west and has a magnitude of 2.50 km. Displacement vector $\mathbf{D}$ points due south and has a magnitude of 3.00 km. Find the magnitude and direction (relative to due west) of the resultant vector $\mathbf{A} + \mathbf{B} + \mathbf{C} + \mathbf{D}$.

**24.** The drawing shows a triple jump on a checkerboard, starting at the center of square A and ending on the center of square B. Each side of a square measures 4.0 cm. What is the magnitude of the displacement of the colored checker during the triple jump?

**25. ssm www** One displacement vector $\mathbf{A}$ has a magnitude of 2.43 km and points due north. A second displacement vector $\mathbf{B}$ has a magnitude of 7.74 km and also points due north. (a) Find the magnitude

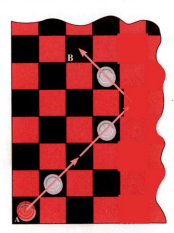

*Problem 24*

and direction of **A − B**. (b) Find the magnitude and direction of **B − A**.

**26.** Two bicyclists, starting at the same place, are riding toward the same campground by two different routes. One cyclist rides 1080 m due east and then turns due north and travels another 1430 m before reaching the campground. The second cyclist starts out by heading due north for 1950 m and then turns and heads directly toward the campground. (a) At the turning point, how far is the second cyclist from the campground? (b) What direction (measured relative to due east) must the second cyclist head during the last part of the trip?

*27. **ssm www** A car is being pulled out of the mud by two forces that are applied by the two ropes shown in the draw-

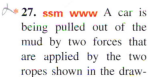

ing. The dashed line in the drawing bisects the 30.0° angle. The magnitude of the force applied by each rope is 2900 newtons. Arrange the force vectors tail to head and use the graphical technique to answer the following questions. (a) How much force would a single rope need to apply to accomplish the same effect as the two forces added together? (b) How would the single rope be directed relative to the dashed line?

*28. In wandering, a grizzly bear makes a displacement of 1563 m due west, followed by a displacement of 3348 m in a direction 32.0° north of west. What are (a) the magnitude and (b) the direction of the displacement needed for the bear to *return to its starting point?* Specify the direction relative to due east.

*29. Before starting this problem, review **Interactive Solution 1.29** at **www.wiley.com/college/cutnell**. Vector **A** has a magnitude of 12.3 units and points due west. Vector **B** points due north. (a) What is the magnitude of **B** if **A + B** has a magnitude of 15.0 units? (b) What is the direction of **A + B** relative to due west? (c) What is the magnitude of **B** if **A − B** has a magnitude of 15.0 units? (d) What is the direction of **A − B** relative to due west?

*30. At a picnic, there is a contest in which hoses are used to shoot water at a beach ball from three directions. As a result, three forces act on the ball, **F₁**, **F₂**, and **F₃** (see the drawing). The magnitudes of **F₁** and **F₂** are $F_1 = 50.0$ newtons and $F_2 = 90.0$ newtons. Using a scale drawing and the graphical technique, determine

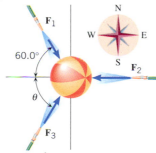

(a) the magnitude of **F₃** and (b) the angle $\theta$ such that the resultant force acting on the ball is zero.

### Section 1.7 The Components of a Vector

**31. ssm** The speed of an object and the direction in which it moves constitute a vector quantity known as the velocity. An ostrich is running at a speed of 17.0 m/s in a direction of 68.0° north of west. What is the magnitude of the ostrich's velocity component that is directed (a) due north and (b) due west?

**32.** Your friend has slipped and fallen. To help her up, you pull with a force **F**, as the drawing shows. The vertical component of this force is 130 newtons, and the horizontal component is 150 newtons. Find (a) the magnitude of **F** and (b) the angle $\theta$.

**33.** An ocean liner leaves New York City and travels 18.0° north of east for 155 km. How far east and how far north has it gone? In other words, what are the magnitudes of the components of the ship's displacement vector in the directions (a) due east and (b) due north?

**34.** Soccer player #1 is 8.6 m from the goal, as the drawing shows. If she kicks the ball directly into the net, the ball has a displacement labeled **A**. If, on the other hand, she first kicks it to player #2, who then kicks it into the net, the ball undergoes two successive displacements, **Aᵧ** and **Aₓ**. What are the magnitude and direction of **Aₓ** and **Aᵧ**?

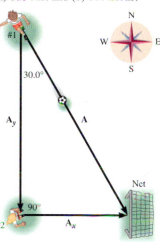

*Problem 34*

**35. ssm** Two ropes are attached to a heavy box to pull it along the floor. One rope applies a force of 475 newtons in a direction due west; the other applies a force of 315 newtons in a direction due south. As we will see later in the text, force is a vector quantity. (a) How much force should be applied by a single rope, and (b) in what direction (relative to due west), if it is to accomplish the same effect as the two forces added together?

**36.** On takeoff, an airplane climbs with a speed of 180 m/s at an angle of 34° above the horizontal. The speed and direction of the airplane constitute a vector quantity known as the velocity. The sun is shining directly overhead. How fast is the shadow of the plane moving along the ground? (That is, what is the magnitude of the horizontal component of the plane's velocity?)

*37. To review the solution to a similar problem, consult **Interactive Solution 1.37** at **www.wiley.com/college/cutnell**. The magnitude of the force vector **F** is 82.3 newtons. The *x* component of this vector is directed along the +*x* axis and has a magnitude of 74.6 newtons. The *y* component points along the +*y* axis. (a) Find the direction of **F** relative to the +*x* axis. (b) Find the component of **F** along the +*y* axis.

*38. A force vector points at an angle of 52° above the +*x* axis. It has a *y* component of +290 newtons. Find (a) the magnitude and (b) the *x* component of the force vector.

**39. ssm www** The drawing shows a force vector that has a magnitude of 475 newtons. Find the (a) *x*, (b) *y*, and (c) *z* components of the vector.

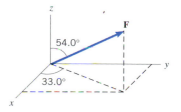

### Section 1.8 Addition of Vectors by Means of Components

**40.** As an aid in working this problem, consult **Concept Simulation 1.1** at **www.wiley.com/college/cutnell**. Two forces are applied to a tree stump to pull it out of the ground. Force **Fₐ** has a magnitude of 2240 newtons and points 34.0° south of east, while force **F_B** has a magnitude of 3160 newtons and points due south. Using the component method, find the magnitude and direction of the resultant force **Fₐ + F_B** that is applied to the stump. Specify the direction with respect to due east.

**41. ssm** A golfer, putting on a green, requires three strokes to "hole the ball." During the first putt, the ball rolls 5.0 m due east. For the second putt, the ball travels 2.1 m at an angle of 20.0° north of east. The third putt is 0.50 m due north. What displacement (magnitude and direction relative to due east) would have been needed to "hole the ball" on the very first putt?

**42.** You are on a treasure hunt and your map says "Walk due west for 52 paces, then walk 30.0° north of west for 42 paces, and finally walk due north for 25 paces." What is the magnitude of the component of your displacement in the direction (a) due north and (b) due west?

**43.** Find the resultant of the three displacement vectors in the drawing by means of the component method. The magnitudes of the vectors are $A = 5.00$ m, $B = 5.00$ m, and $C = 4.00$ m.

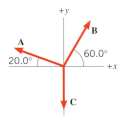

*Problem 43*

**44.** A baby elephant is stuck in a mud hole. To help pull it out, game keepers use a rope to apply force $\mathbf{F_A}$, as part *a* of the drawing shows. By itself, however, force $\mathbf{F_A}$ is insufficient. Therefore, two additional forces $\mathbf{F_B}$ and $\mathbf{F_C}$ are applied, as in part *b* of the drawing. Each of these additional forces has the same magnitude $F$. The magnitude of the resultant force acting on the elephant in part *b* of the drawing is twice that in part *a*. Find the ratio $F/F_A$.

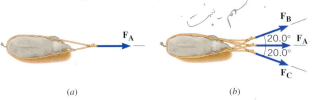

(a)                    (b)

**45.** As preparation for this problem, consult **Concept Simulation 1.1** at **www.wiley.com/college/cutnell.** On a safari, a team of naturalists sets out toward a research station located 4.8 km away in a direction 42° north of east. After traveling in a straight line for 2.4 km, they stop and discover that they have been traveling 22° north of east, because their guide misread his compass. What are (a) the magnitude and (b) the direction (relative to due east) of the displacement vector now required to bring the team to the research station?

**46.** Three forces are applied to an object, as indicated in the drawing. Force $\mathbf{F_1}$ has a magnitude of 21.0 newtons (21.0 N) and is directed 30.0° to the left of the +y axis. Force $\mathbf{F_2}$ has a magnitude of 15.0 N and points along the +x axis. What must be the magnitude and direction (specified by the angle $\theta$ in the drawing) of the third force $\mathbf{F_3}$ such that the vector sum of the three forces is 0 N?

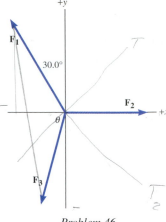

*Problem 46*

**47. ssm** Vector **A** has a magnitude of 6.00 units and points due east. Vector **B** points due north. (a) What is the magnitude of **B**, if the vector **A + B** points 60.0° north of east? (b) Find the magnitude of **A + B**.

**48.** The route followed by a hiker consists of three displacement vectors **A**, **B**, and **C**. Vector **A** is along a measured trail and is 1550 m in a direction 25.0° north of east. Vector **B** is not along a measured trail, but the hiker uses a compass and knows that the direction is 41.0° east of south. Similarly, the direction of vector **C** is 35.0° north of west. The hiker ends up back where she started, so the resultant displacement is zero, or **A + B + C = 0**. Find the magnitudes of (a) vector **B** and (b) vector **C**.

**49. Interactive Solution 1.49** at **www.wiley.com/college/cutnell** presents the solution to a problem that is similar to this one. Vector **A** has a magnitude of 145 units and points 35.0° north of west. Vector **B** points 65.0° east of north. Vector **C** points 15.0° west of south. These three vectors add to give a resultant vector that is zero. Using components, find the magnitudes of (a) vector **B** and (b) vector **C**.

**50.** A grasshopper makes four jumps. The displacement vectors are (1) 27.0 cm, due west; (2) 23.0 cm, 35.0° south of west; (3) 28.0 cm, 55.0° south of east; and (4) 35.0 cm, 63.0° north of east. Find the magnitude and direction of the resultant displacement. Express the direction with respect to due west.

# Chapter 2 Kinematics in One Dimension

حبیش شناسی

علم احب ام مورک

تعریف حان

## 2.1 Displacement

There are two aspects to any motion. In a purely descriptive sense, there is the movement itself. Is it rapid or slow, for instance? Then, there is the issue of what causes the motion or what changes it, which requires that forces be considered. **Kinematics** deals with the concepts that are needed to describe motion, without any reference to forces. The present chapter discusses these concepts as they apply to motion in one dimension, and the next chapter treats two-dimensional motion. **Dynamics** deals with the effect that forces have on motion, a topic that is considered in Chapter 4. Together, kinematics and dynamics form the branch of physics known as **mechanics.** We turn now to the first of the kinematics concepts to be discussed, which is displacement.

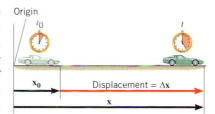

To describe the motion of an object, we must be able to specify the location of the object at all times, and Figure 2.1 shows how to do this for one-dimensional motion. In this drawing, the initial position of a car is indicated by the vector labeled $x_0$. The length of $x_0$ is the distance of the car from an arbitrarily chosen origin. At a later time the car has moved to a new position, which is indicated by the vector $x$. The **displacement** of the car $\Delta x$ (read as "delta x" or "the change in x") is a vector drawn from the initial position to the final position. Displacement is a vector quantity in the sense discussed in Section 1.5, for it conveys both a magnitude (the distance between the initial and final positions) and a direction. The displacement can be related to $x_0$ and $x$ by noting from the drawing that

**Figure 2.1** The displacement $\Delta x$ is a vector that points from the initial position $x_0$ to the final position $x$.

$$x_0 + \Delta x = x \quad \text{or} \quad \Delta x = x - x_0$$

Thus, the displacement $\Delta x$ is the difference between $x$ and $x_0$, and the Greek letter delta ($\Delta$) is used to signify this difference. It is important to note that the change in any variable is always the final value minus the initial value.

---

■ **DEFINITION OF DISPLACEMENT**

The displacement is a vector that points from an object's initial position to its final position and has a magnitude that equals the shortest distance between the two positions.

**SI Unit of Displacement:** meter (m)

---

The SI unit for displacement is the meter (m), but there are other units as well, such as the centimeter and the inch. When converting between centimeters (cm) and inches (in.), remember that 2.54 cm = 1 in.

Often, we will deal with motion along a straight line. In such a case, a displacement in one direction along the line is assigned a positive value, and a displacement in the opposite direction is assigned a negative value. For instance, assume that a car is moving along an east/west direction and that a positive (+) sign is used to denote a direction due east. Then, $\Delta x = +500$ m represents a displacement that points to the east and has a magnitude of 500 meters. Conversely, $\Delta x = -500$ m is a displacement that has the same magnitude but points in the opposite direction, due west.

## 2.2 Speed and Velocity

### AVERAGE SPEED

One of the most obvious features of an object in motion is how fast it is moving. If a car travels 200 meters in 10 seconds, we say its average speed is 20 meters per second, the *average speed* being the distance traveled divided by the time required to cover the distance:

$$\text{Average speed} = \frac{\text{Distance}}{\text{Elapsed time}} \tag{2.1}$$

Equation 2.1 indicates that the unit for average speed is the unit for distance divided by the unit for time, or meters per second (m/s) in SI units. Example 1 illustrates how the idea of average speed is used.

### Example 1   Distance Run by a Jogger

How far does a jogger run in 1.5 hours (5400 s) if his average speed is 2.22 m/s?

**Reasoning**  The average speed of the jogger is the average distance per second that he travels. Thus, the distance covered by the jogger is equal to the average distance per second (his average speed) multiplied by the number of seconds (the elapsed time) that he runs.

**Solution**  To find the distance run, we rewrite Equation 2.1 as

$$\text{Distance} = (\text{Average speed})(\text{Elapsed time}) = (2.22 \text{ m/s})(5400 \text{ s}) = \boxed{12\ 000 \text{ m}}$$

Speed is a useful idea, because it indicates how fast an object is moving. However, speed does not reveal anything about the direction of the motion. To describe both how fast an object moves and the direction of its motion, we need the vector concept of velocity.

## AVERAGE VELOCITY

Suppose that the initial position of the car in Figure 2.1 is $x_0$ when the time is $t_0$. A little later the car arrives at the final position $x$ at the time $t$. The difference between these times is the time required for the car to travel between the two positions. We denote this difference by the shorthand notation $\Delta t$ (read as "delta $t$"), where $\Delta t$ represents the final time $t$ minus the initial time $t_0$:

$$\Delta t = \underbrace{t - t_0}_{\text{Elapsed time}}$$

Note that $\Delta t$ is defined in a manner analogous to $\Delta x$, which is the final position minus the initial position ($\Delta x = x - x_0$). Dividing the displacement $\Delta x$ of the car by the elapsed time $\Delta t$ gives the *average velocity* of the car. It is customary to denote the average value of a quantity by placing a horizontal bar above the symbol representing the quantity. The average velocity, then, is written as $\overline{v}$, as specified in Equation 2.2:

> ■ **DEFINITION OF AVERAGE VELOCITY**
>
> $$\text{Average velocity} = \frac{\text{Displacement}}{\text{Elapsed time}}$$
>
> $$\overline{v} = \frac{x - x_0}{t - t_0} = \frac{\Delta x}{\Delta t} \tag{2.2}$$
>
> *SI Unit of Average Velocity:* meter per second (m/s)

Equation 2.2 indicates that the unit for average velocity is the unit for length divided by the unit for time, or meters per second (m/s) in SI units. Velocity can also be expressed in other units, such as kilometers per hour (km/h) or miles per hour (mi/h).

Average velocity is a vector that points in the same direction as the displacement in Equation 2.2. Figure 2.2 illustrates that the velocity of a car confined to move along a line can point either in one direction or in the opposite direction. As with displacement, we will use plus and minus signs to indicate the two possible directions. If the displacement points in the positive direction, the average velocity is positive. Conversely, if the dis-

**Figure 2.2** In this time-lapse photo of traffic on the Los Angeles Freeway in California, the velocity of a car in the left lane (white headlights) is opposite to that of an adjacent car in the right lane (red taillights). (© Peter Essick/Aurora & Quanta Productions)

placement points in the negative direction, the average velocity is negative. Example 2 illustrates these features of average velocity.

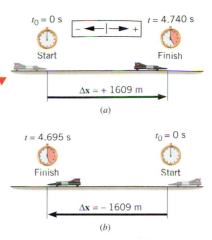

### Example 2   The World's Fastest Jet-Engine Car

Andy Green in the car *ThrustSSC* set a world record of 341.1 m/s (763 mi/h) in 1997. The car was powered by two jet engines, and it was the first one officially to exceed the speed of sound. To establish such a record, the driver makes two runs through the course, one in each direction, to nullify wind effects. Figure 2.3a shows that the car first travels from left to right and covers a distance of 1609 m (1 mile) in a time of 4.740 s. Figure 2.3b shows that in the reverse direction, the car covers the same distance in 4.695 s. From these data, determine the average velocity for each run.

**Reasoning** Average velocity is defined as the displacement divided by the elapsed time. In using this definition we recognize that the displacement is not the same as the distance traveled. Displacement takes the direction of the motion into account, and distance does not. During both runs, the car covers the same distance of 1609 m. However, for the first run the displacement is $\Delta x = +1609$ m, while for the second it is $\Delta x = -1609$ m. The plus and minus signs are essential, because the first run is to the right, which is the positive direction, and the second run is in the opposite or negative direction.

**Figure 2.3** The arrows in the box at the top of the drawing indicate the positive and negative directions for the displacements of the car, as explained in Example 2.

**Solution** According to Equation 2.2, the average velocities are

*Run 1*
$$\overline{v} = \frac{\Delta x}{\Delta t} = \frac{+1609 \text{ m}}{4.740 \text{ s}} = \boxed{+339.5 \text{ m/s}}$$

*Run 2*
$$\overline{v} = \frac{\Delta x}{\Delta t} = \frac{-1609 \text{ m}}{4.695 \text{ s}} = \boxed{-342.7 \text{ m/s}}$$

In these answers the algebraic signs convey the directions of the velocity vectors. In particular, for Run 2 the minus sign indicates that the average velocity, like the displacement, points to the left in Figure 2.3b. The magnitudes of the velocities are 339.5 and 342.7 m/s. The average of these numbers is 341.1 m/s and is recorded in the record book.

## INSTANTANEOUS VELOCITY

Suppose the magnitude of your average velocity for a long trip was 20 m/s. This value, being an average, does not convey any information about how fast you were moving at any instant during the trip. Surely there were times when your car traveled faster than 20 m/s and times when it traveled more slowly. The ***instantaneous velocity*** v of the car indicates how fast the car moves and the direction of the motion at each instant of time. The magnitude of the instantaneous velocity is called the ***instantaneous speed,*** and it is the number (with units) indicated by the speedometer.

The instantaneous velocity at any point during a trip can be obtained by measuring the time interval $\Delta t$ for the car to travel a *very small* displacement $\Delta x$. We can then compute the average velocity over this interval. If the time $\Delta t$ is small enough, the instantaneous velocity does not change much during the measurement. Then, the instantaneous velocity v at the point of interest is approximately equal to ($\approx$) the average velocity $\overline{v}$ computed over the interval, or $v \approx \overline{v} = \Delta x/\Delta t$ (for sufficiently small $\Delta t$). In fact, in the limit that $\Delta t$ becomes infinitesimally small, the instantaneous velocity and the average velocity become equal, so that

$$v = \lim_{\Delta t \to 0} \frac{\Delta x}{\Delta t} \tag{2.3}$$

The notation $\lim_{\Delta t \to 0} (\Delta x/\Delta t)$ means that the ratio $\Delta x/\Delta t$ is defined by a limiting process in which smaller and smaller values of $\Delta t$ are used, so small that they approach zero. As smaller values of $\Delta t$ are used, $\Delta x$ also becomes smaller. However, the ratio $\Delta x/\Delta t$ does *not* become zero but, rather, approaches the value of the instantaneous velocity. For brevity, we will use the word *velocity* to mean "instantaneous velocity" and *speed* to mean "instantaneous speed."

In a wide range of motions, the velocity changes from moment to moment. To describe the manner in which it changes, the concept of acceleration is needed.

## 2.3  *Acceleration*

The velocity of a moving object may change in a number of ways. For example, it may increase, as it does when the driver of a car steps on the gas pedal to pass the car ahead. Or it may decrease, as it does when the driver applies the brakes to stop at a red light. In either case, the change in velocity may occur over a short or a long time interval.

To describe how the velocity of an object changes during a given time interval, we now introduce the new idea of acceleration; this idea depends on two concepts that we have previously encountered, velocity and time. Specifically, the notion of acceleration emerges when the *change* in the velocity is combined with the time during which the change occurs.

The meaning of **average acceleration** can be illustrated by considering a plane during takeoff. Figure 2.4 focuses attention on how the plane's velocity changes along the runway. During an elapsed time interval $\Delta t = t - t_0$, the velocity changes from an initial value of $\mathbf{v_0}$ to a final value of $\mathbf{v}$. The change $\Delta \mathbf{v}$ in the plane's velocity is its final velocity minus its initial velocity, so that $\Delta \mathbf{v} = \mathbf{v} - \mathbf{v_0}$. The average acceleration $\bar{\mathbf{a}}$ is defined in the following manner, to provide a measure of how much the velocity changes per unit of elapsed time.

---

■ **DEFINITION OF AVERAGE ACCELERATION**

$$\text{Average acceleration} = \frac{\text{Change in velocity}}{\text{Elapsed time}}$$

$$\bar{\mathbf{a}} = \frac{\mathbf{v} - \mathbf{v_0}}{t - t_0} = \frac{\Delta \mathbf{v}}{\Delta t} \tag{2.4}$$

*SI Unit of Average Acceleration:* meter per second squared $(\text{m/s}^2)$

---

The average acceleration $\bar{\mathbf{a}}$ is a vector that points in the same direction as $\Delta \mathbf{v}$, the change in the velocity. Following the usual custom, plus and minus signs indicate the two possible directions for the acceleration vector when the motion is along a straight line.

We are often interested in an object's acceleration at a particular instant of time. The **instantaneous acceleration a** can be defined by analogy with the procedure used in Section 2.2 for instantaneous velocity:

$$\mathbf{a} = \lim_{\Delta t \to 0} \frac{\Delta \mathbf{v}}{\Delta t} \tag{2.5}$$

Equation 2.5 indicates that the instantaneous acceleration is a limiting case of the average acceleration. When the time interval $\Delta t$ for measuring the acceleration becomes extremely small (approaching zero in the limit), the average acceleration and the instantaneous acceleration become equal. Moreover, in many situations the acceleration is constant, so the acceleration has the same value at any instant of time. In the future, we will use the word *acceleration* to mean "instantaneous acceleration." Example 3 deals with the acceleration of a plane during takeoff.

**Figure 2.4**  During takeoff, the plane accelerates from an initial velocity $\mathbf{v_0}$ to a final velocity $\mathbf{v}$ during the time interval $\Delta t = t - t_0$.

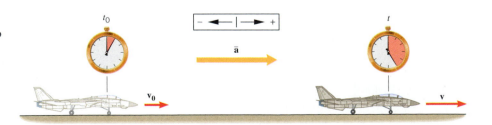

## *Example 3* Acceleration and Increasing Velocity

Suppose the plane in Figure 2.4 starts from rest ($\mathbf{v_0}$ = 0 m/s) when $t_0$ = 0 s. The plane accelerates down the runway and at $t$ = 29 s attains a velocity of $\mathbf{v}$ = +260 km/h, where the plus sign indicates that the velocity points to the right. Determine the average acceleration of the plane.

**Reasoning** The average acceleration of the plane is defined as the change in its velocity divided by the elapsed time. The change in the plane's velocity is its final velocity $\mathbf{v}$ minus its initial velocity $\mathbf{v_0}$, or $\mathbf{v} - \mathbf{v_0}$. The elapsed time is the final time $t$ minus the initial time $t_0$, or $t - t_0$.

*Problem solving insight*
The change in any variable is the final value minus the initial value: for example, the change in velocity is $\Delta\mathbf{v} = \mathbf{v} - \mathbf{v_0}$, and the change in time is $\Delta t = t - t_0$.

**Solution** The average acceleration is expressed by Equation 2.4 as

$$\overline{\mathbf{a}} = \frac{\mathbf{v} - \mathbf{v_0}}{t - t_0} = \frac{260 \text{ km/h} - 0 \text{ km/h}}{29 \text{ s} - 0 \text{ s}} = \boxed{+9.0 \frac{\text{km/h}}{\text{s}}}$$

The average acceleration calculated in Example 3 is read as "nine kilometers per hour per second." Assuming the acceleration of the plane is constant, a value of $+9.0 \frac{\text{km/h}}{\text{s}}$ means the velocity changes by +9.0 km/h during each second of the motion. During the first second, the velocity increases from 0 to 9.0 km/h; during the next second, the velocity increases by another 9.0 km/h to 18 km/h, and so on. Figure 2.5 illustrates how the velocity changes during the first two seconds. By the end of the 29th second, the velocity is 260 km/h.

It is customary to express the units for acceleration solely in terms of SI units. One way to obtain SI units for the acceleration in Example 3 is to convert the velocity units from km/h to m/s:

$$\left(260 \frac{\text{km}}{\text{h}}\right)\left(\frac{1000 \text{ m}}{1 \text{ km}}\right)\left(\frac{1 \text{ h}}{3600 \text{ s}}\right) = 72 \frac{\text{m}}{\text{s}}$$

The average acceleration then becomes

$$\overline{\mathbf{a}} = \frac{72 \text{ m/s} - 0 \text{ m/s}}{29 \text{ s} - 0 \text{ s}} = +2.5 \text{ m/s}^2$$

where we have used $2.5 \frac{\text{m/s}}{\text{s}} = 2.5 \frac{\text{m}}{\text{s} \cdot \text{s}} = 2.5 \frac{\text{m}}{\text{s}^2}$. An acceleration of $2.5 \frac{\text{m}}{\text{s}^2}$ is read as "2.5 meters per second per second" (or "2.5 meters per second squared") and means that the velocity changes by 2.5 m/s during each second of the motion.

Example 4 deals with a case where the motion becomes slower as time passes.

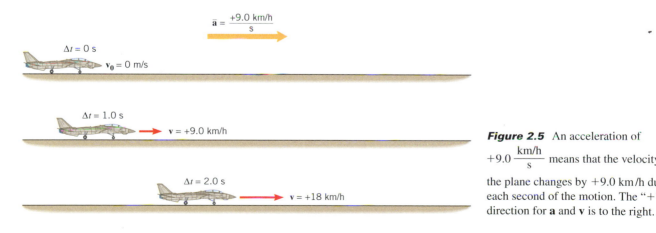

*Figure 2.5* An acceleration of $+9.0 \frac{\text{km/h}}{\text{s}}$ means that the velocity of the plane changes by +9.0 km/h during each second of the motion. The "+" direction for **a** and **v** is to the right.

## *Example 4* Acceleration and Decreasing Velocity

A drag racer crosses the finish line, and the driver deploys a parachute and applies the brakes to slow down, as Figure 2.6 illustrates. The driver begins slowing down when $t_0$ = 9.0 s and

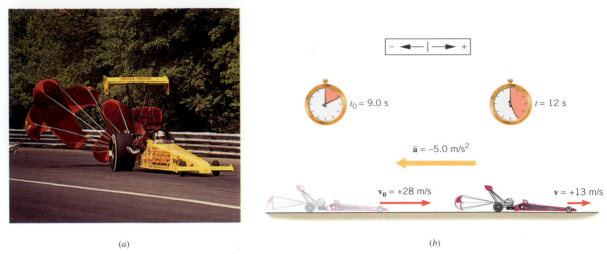

(a)                (b)

**Figure 2.6** (a) To slow down, a drag racer deploys a parachute and applies the brakes. (b) The velocity of the car is decreasing, giving rise to an average acceleration $\bar{\mathbf{a}}$ that points opposite to the velocity. (© Geoff Stunkard/ quartermilestones.com)

the car's velocity is $\mathbf{v_0} = +28$ m/s. When $t = 12.0$ s, the velocity has been reduced to $\mathbf{v} = +13$ m/s. What is the average acceleration of the dragster?

**Reasoning** The average acceleration of an object is always specified as its change in velocity, $\mathbf{v} - \mathbf{v_0}$, divided by the elapsed time, $t - t_0$. This is true whether the final velocity is less than the initial velocity or greater than the initial velocity.

**Solution** The average acceleration is, according to Equation 2.4,

$$\bar{\mathbf{a}} = \frac{\mathbf{v} - \mathbf{v_0}}{t - t_0} = \frac{13 \text{ m/s} - 28 \text{ m/s}}{12.0 \text{ s} - 9.0 \text{ s}} = \boxed{-5.0 \text{ m/s}^2}$$

Figure 2.7 shows how the velocity of the dragster changes during the braking, assuming that the acceleration is constant throughout the motion. The acceleration calculated in Example 4 is negative, indicating that the acceleration points to the left in the drawing. As a result, the acceleration and the velocity point in *opposite* directions. ***Whenever the acceleration and velocity vectors have opposite directions, the object slows down and is said to be "decelerating."*** In contrast, the acceleration and velocity vectors in Figure 2.5 point in the *same* direction, and the object speeds up.

## 2.4 *Equations of Kinematics for Constant Acceleration*

In discussing the equations of kinematics, it will be convenient to assume that the object is located at the origin $\mathbf{x_0} = 0$ m when $t_0 = 0$ s. With this assumption, the displacement

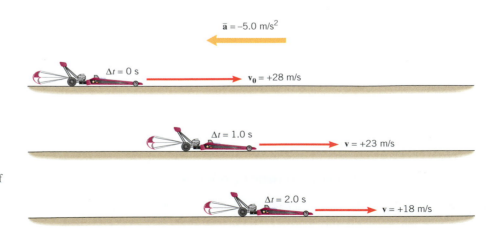

**Figure 2.7** Here, an acceleration of $-5.0$ m/s$^2$ means the velocity decreases by 5.0 m/s during each second of elapsed time.

$\Delta\mathbf{x} = \mathbf{x} - \mathbf{x_0}$ becomes $\Delta\mathbf{x} = \mathbf{x}$. Furthermore, it is customary to dispense with the use of boldface symbols for the displacement, velocity, and acceleration vectors in the equations that follow. We will, however, continue to convey the directions of these vectors with plus or minus signs.

Consider an object that has an initial velocity of $v_0$ at time $t_0 = 0$ s and moves for a time $t$ with a constant acceleration $a$. For a complete description of the motion, it is also necessary to know the final velocity and displacement at time $t$. The final velocity $v$ can be obtained directly from Equation 2.4:

$$\bar{a} = a = \frac{v - v_0}{t} \quad \text{or} \quad v = v_0 + at \quad \text{(constant acceleration)} \quad (2.4)$$

The displacement $x$ at time $t$ can be obtained from Equation 2.2, if a value for the average velocity $\bar{v}$ can be obtained. Considering the assumption that $\mathbf{x_0} = 0$ m at $t_0 = 0$ s, we have

$$\bar{v} = \frac{x - x_0}{t - t_0} = \frac{x}{t} \quad \text{or} \quad x = \bar{v}t \quad (2.2)$$

Because the acceleration is constant, the velocity increases at a constant rate. Thus, the average velocity $\bar{v}$ is midway between the initial and final velocities:

$$\bar{v} = \tfrac{1}{2}(v_0 + v) \quad \text{(constant acceleration)} \quad (2.6)$$

Equation 2.6, like Equation 2.4, applies only if the acceleration is constant and cannot be used when the acceleration is changing. The displacement at time $t$ can now be determined as

$$x = \bar{v}t = \tfrac{1}{2}(v_0 + v)t \quad \text{(constant acceleration)} \quad (2.7)$$

Notice in Equations 2.4 ($v = v_0 + at$) and 2.7 [$x = \tfrac{1}{2}(v_0 + v)t$] that there are five kinematic variables:

1. $x$ = displacement
2. $a = \bar{a}$ = acceleration (constant)
3. $v$ = final velocity at time $t$
4. $v_0$ = initial velocity at time $t_0 = 0$ s
5. $t$ = time elapsed since $t_0 = 0$ s

Each of the two equations contains four of these variables, so if three of them are known, the fourth variable can always be found. Example 5 illustrates how Equations 2.4 and 2.7 are used to describe the motion of an object.

## Example 5   The Displacement of a Speedboat

The speedboat in Figure 2.8 has a constant acceleration of $+2.0$ m/s$^2$. If the initial velocity of the boat is $+6.0$ m/s, find its displacement after 8.0 seconds.

**Figure 2.8** (a) An accelerating speedboat. (b) The boat's displacement $x$ can be determined if the boat's acceleration, initial velocity, and time of travel are known. (© Onne van der Wal/Corbis Images)

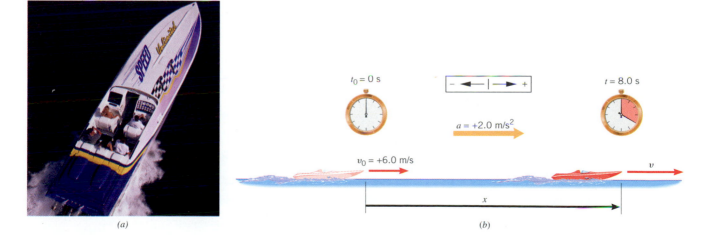

(a)

(b)

**Reasoning** Numerical values for the three known variables are listed in the data table below. We wish to determine the displacement $x$ of the speedboat, so it is an unknown variable. Therefore, we have placed a question mark in the displacement column of the data table.

| | | Speedboat Data | | |
|---|---|---|---|---|
| $x$ | $a$ | $v$ | $v_0$ | $t$ |
| ? | $+2.0$ m/s$^2$ | | $+6.0$ m/s | $8.0$ s |

We can use $x = \frac{1}{2}(v_0 + v)t$ to find the displacement of the boat if a value for the final velocity $v$ can be found. To find the final velocity, it is necessary to use the value given for the acceleration, because it tells us how the velocity changes, according to $v = v_0 + at$.

**Solution** The final velocity is

$$v = v_0 + at = 6.0 \text{ m/s} + (2.0 \text{ m/s}^2)(8.0 \text{ s}) = +22 \text{ m/s} \tag{2.4}$$

The displacement of the boat can now be obtained:

$$x = \frac{1}{2}(v_0 + v)t = \frac{1}{2}(6.0 \text{ m/s} + 22 \text{ m/s})(8.0 \text{ s}) = \boxed{+110 \text{ m}} \tag{2.7}$$

A calculator would give the answer as 112 m, but this number must be rounded to 110 m, since the data are accurate to only two significant figures.

The solution to Example 5 involved two steps: finding the final velocity $v$ and then calculating the displacement $x$. It would be helpful if we could find an equation that allows us to determine the displacement in a single step. Using Example 5 as a guide, we can obtain such an equation by substituting the final velocity $v$ from Equation 2.4 ($v = v_0 + at$) into Equation 2.7 [$x = \frac{1}{2}(v_0 + v)t$]:

$$x = \frac{1}{2}(v_0 + v)t = \frac{1}{2}(v_0 + \boxed{v_0 + at})t = \frac{1}{2}(2v_0 t + at^2)$$

$$x = v_0 t + \frac{1}{2}at^2 \qquad \text{(constant acceleration)} \tag{2.8}$$

You can verify that Equation 2.8 gives the displacement of the speedboat directly without the intermediate step of determining the final velocity. The first term ($v_0 t$) on the right side of this equation represents the displacement that would result if the acceleration were zero and the velocity remained constant at its initial value of $v_0$. The second term ($\frac{1}{2}at^2$) gives the additional displacement that arises because the velocity changes ($a$ is not zero) to values that are different from its initial value. We now turn to another example of accelerated motion.

**The physics of catapulting a jet from an aircraft carrier.**

### *Example 6*   Catapulting a Jet

A jet is taking off from the deck of an aircraft carrier, as Figure 2.9 shows. Starting from rest, the jet is catapulted with a constant acceleration of $+31$ m/s$^2$ along a straight line and reaches a velocity of $+62$ m/s. Find the displacement of the jet.

**Figure 2.9**  (*a*) A plane is being launched from an aircraft carrier. (*b*) During the launch, a catapult accelerates the jet down the flight deck. (© George Hall/Corbis Images)

(a)

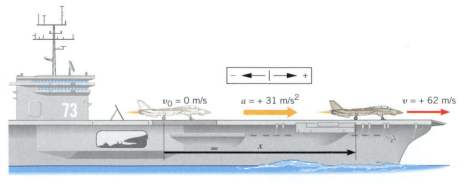

(b)

**Reasoning** The data are as follows:

| Jet Data | | | | |
|---|---|---|---|---|
| $x$ | $a$ | $v$ | $v_0$ | $t$ |
| ? | $+31$ m/s$^2$ | $+62$ m/s | $0$ m/s | |

The initial velocity $v_0$ is zero, since the jet starts from rest. The displacement $x$ of the aircraft can be obtained from $x = \frac{1}{2}(v_0 + v)t$, if we can determine the time $t$ during which the plane is being accelerated. But $t$ is controlled by the value of the acceleration. With larger accelerations, the jet reaches its final velocity in shorter times, as can be seen by solving Equation 2.4 ($v = v_0 + at$) for $t$.

**Solution** Solving Equation 2.4 for $t$, we find

$$t = \frac{v - v_0}{a} = \frac{62 \text{ m/s} - 0 \text{ m/s}}{31 \text{ m/s}^2} = 2.0 \text{ s}$$

Since the time is now known, the displacement can be found by using Equation 2.7:

$$x = \tfrac{1}{2}(v_0 + v)t = \tfrac{1}{2}(0 \text{ m/s} + 62 \text{ m/s})(2.0 \text{ s}) = \boxed{+62 \text{ m}} \qquad (2.7)$$

When $a$, $v$, and $v_0$ are known, but the time $t$ is not known, as in Example 6, it is possible to calculate the displacement $x$ in a single step. Solving Equation 2.4 for the time $[t = (v - v_0)/a]$ and then substituting into Equation 2.7 $[x = \frac{1}{2}(v_0 + v)t]$ reveals that

$$x = \tfrac{1}{2}(v_0 + v)t = \tfrac{1}{2}(v_0 + v)\boxed{\frac{v - v_0}{a}} = \frac{v^2 - v_0^2}{2a}$$

Solving for $v^2$ shows that

$$v^2 = v_0^2 + 2ax \qquad \text{(constant acceleration)} \qquad (2.9)$$

It is a straightforward exercise to verify that Equation 2.9 can be used to find the displacement of the jet in Example 6 without having to solve first for the time.

Table 2.1 presents a summary of the equations that we have been considering. These equations are called the *equations of kinematics.* Each equation contains four variables, as indicated by the check marks (✓) in the table. The next section shows how to apply the equations of kinematics.

# 2.5 Applications of the Equations of Kinematics

The equations of kinematics can be used for any moving object, as long as the acceleration of the object is constant. However, to avoid errors when using these equations, it helps to follow a few sensible guidelines and to be alert for a few situations that can arise during your calculations.

**Table 2.1** *Equations of Kinematics for Constant Acceleration*

| Equation Number | Equation | Variables | | | | |
|---|---|---|---|---|---|---|
| | | $x$ | $a$ | $v$ | $v_0$ | $t$ |
| (2.4) | $v = v_0 + at$ | — | ✓ | ✓ | ✓ | ✓ |
| (2.7) | $x = \frac{1}{2}(v_0 + v)t$ | ✓ | — | ✓ | ✓ | ✓ |
| (2.8) | $x = v_0 t + \frac{1}{2}at^2$ | ✓ | ✓ | — | ✓ | ✓ |
| (2.9) | $v^2 = v_0^2 + 2ax$ | ✓ | ✓ | ✓ | ✓ | — |

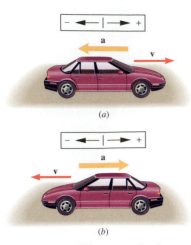

*(a)*

*(b)*

**Figure 2.10** When a car decelerates along a straight road, the acceleration vector points opposite to the velocity vector, as Conceptual Example 7 discusses.

*Decide at the start which directions are to be called positive (+) and negative (−) relative to a conveniently chosen coordinate origin.* This decision is arbitrary, but important because displacement, velocity, and acceleration are vectors, and their directions must always be taken into account. In the examples that follow, the positive and negative directions will be shown in the drawings that accompany the problems. It does not matter which direction is chosen to be positive. However, once the choice is made, it should not be changed during the course of the calculation.

*As you reason through a problem before attempting to solve it, be sure to interpret the terms "decelerating" or "deceleration" correctly, should they occur in the problem statement.* These terms are the source of frequent confusion, and Conceptual Example 7 offers help in understanding them.

### Conceptual Example 7    Deceleration Versus Negative Acceleration

A car is traveling along a straight road and is decelerating. Does the car's acceleration $a$ necessarily have a negative value?

**Reasoning and Solution** We begin with the meaning of the term "decelerating," which has nothing to do with whether the acceleration $a$ is positive or negative. The term means only that the acceleration vector points opposite to the velocity vector and indicates that the moving object is slowing down. When a moving object slows down, its instantaneous speed (the magnitude of the instantaneous velocity) decreases. One possibility is that the velocity vector of the car points to the right, in the positive direction, as Figure 2.10a shows. The term "decelerating" implies that the acceleration vector points opposite, or to the left, which is the negative direction. Here, the value of the acceleration $a$ would indeed be negative. However, there is another possibility. The car could be traveling to the left, as in Figure 2.10b. Now, since the velocity vector points to the left, the acceleration vector would point opposite or to the right, according to the meaning of the term "decelerating." But right is the positive direction, so the acceleration $a$ would have a positive value in Figure 2.10b. We see, then, that *a decelerating object does not necessarily have a negative acceleration.*

**Related Homework:** *Problems 20, 38*

*Sometimes there are two possible answers to a kinematics problem, each answer corresponding to a different situation.* Example 8 discusses one such case.

### Example 8    An Accelerating Spacecraft

**The physics of the acceleration caused by a retrorocket.**

The spacecraft shown in Figure 2.11a is traveling with a velocity of +3250 m/s. Suddenly the retrorockets are fired, and the spacecraft begins to slow down with an acceleration whose magnitude is 10.0 m/s². What is the velocity of the spacecraft when the displacement of the craft is +215 km, relative to the point where the retrorockets began firing?

**Reasoning** Since the spacecraft is slowing down, the acceleration must be opposite to the velocity. The velocity points to the right in the drawing, so the acceleration points to the left, in the negative direction; thus, $a = -10.0$ m/s². The three known variables are listed as follows:

| Spacecraft Data | | | | |
|---|---|---|---|---|
| $x$ | $a$ | $v$ | $v_0$ | $t$ |
| +215 000 m | −10.0 m/s² | ? | +3250 m/s | |

The final velocity $v$ of the spacecraft can be calculated using Equation 2.9, since it contains the four pertinent variables.

**Solution** From Equation 2.9 ($v^2 = v_0^2 + 2ax$), we find that

$$v = \pm\sqrt{v_0^2 + 2ax} = \pm\sqrt{(3250 \text{ m/s})^2 + 2(-10.0 \text{ m/s}^2)(215\,000 \text{ m})}$$
$$= \boxed{+2500 \text{ m/s}} \quad \text{and} \quad \boxed{-2500 \text{ m/s}}$$

Both of these answers correspond to the *same* displacement ($x = +215$ km), but each arises in a different part of the motion. The answer $v = +2500$ m/s corresponds to the situation in Fig-

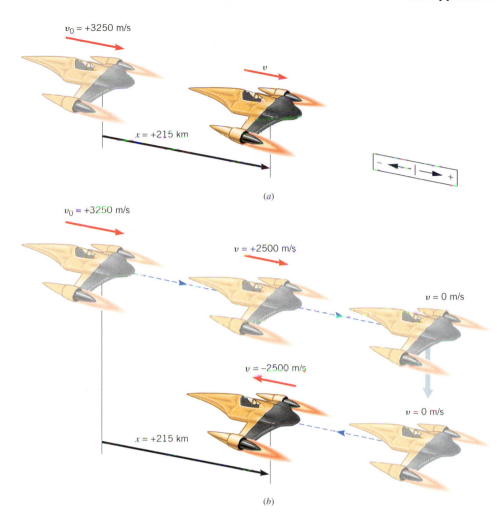

$v_0 = +3250$ m/s

$v$

$x = +215$ km

$(a)$

$v_0 = +3250$ m/s

$v = +2500$ m/s

$v = 0$ m/s

$v = -2500$ m/s

$v = 0$ m/s

$x = +215$ km

$(b)$

**Figure 2.11** (*a*) Because of an acceleration of $-10.0$ m/s$^2$, the spacecraft changes its velocity from $v_0$ to $v$. (*b*) Continued firing of the retrorockets changes the direction of the craft's motion.

ure 2.11*a*, where the spacecraft has slowed to a speed of 2500 m/s, but is still traveling to the right. The answer $v = -2500$ m/s arises because the retrorockets eventually bring the spacecraft to a momentary halt and cause it to reverse its direction. Then it moves to the left, and its speed increases due to the continually firing rockets. After a time, the velocity of the craft becomes $v = -2500$ m/s, giving rise to the situation in Figure 2.11*b*. In both parts of the drawing the spacecraft has the same displacement, but a greater travel time is required in part *b* compared to part *a*.

*The motion of two objects may be interrelated, so they share a common variable. The fact that the motions are interrelated is an important piece of information. In such cases, data for only two variables need be specified for each object.*

*Often the motion of an object is divided into segments, each with a different acceleration. When solving such problems, it is important to realize that the final velocity for one segment is the initial velocity for the next segment,* as Example 9 illustrates.

### Example 9  A Motorcycle Ride

A motorcycle, starting from rest, has an acceleration of $+2.6$ m/s$^2$. After the motorcycle has traveled a distance of 120 m, it slows down with an acceleration of $-1.5$ m/s$^2$ until its velocity is $+12$ m/s (see Figure 2.12). What is the total displacement of the motorcycle?

**Reasoning**  The total displacement is the sum of the displacements for the first ("speeding up") and second ("slowing down") segments. The displacement for the first segment is $+120$ m. The displacement for the second segment can be found if the initial velocity for this segment can be determined, since values for two other variables are already known ($a = -1.5$ m/s$^2$ and $v = +12$ m/s). The initial velocity for the second segment can be determined, since it is the final velocity of the first segment.

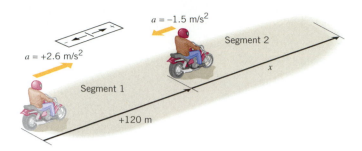

**Figure 2.12** This motorcycle ride consists of two segments, each with a different acceleration.

**Solution** Recognizing that the motorcycle starts from rest ($v_0 = 0$ m/s), we can determine the final velocity $v$ of the first segment from the given data:

| Segment 1 Data | | | | |
|---|---|---|---|---|
| $x$ | $a$ | $v$ | $v_0$ | $t$ |
| $+120$ m | $+2.6$ m/s$^2$ | ? | 0 m/s | |

From Equation 2.9 ($v^2 = v_0^2 + 2ax$), it follows that

$$v = +\sqrt{v_0^2 + 2ax} = +\sqrt{(0 \text{ m/s})^2 + 2(2.6 \text{ m/s}^2)(120 \text{ m})} = +25 \text{ m/s}$$

Now we can use $+25$ m/s as the initial velocity for the second segment, along with the remaining data listed below:

| Segment 2 Data | | | | |
|---|---|---|---|---|
| $x$ | $a$ | $v$ | $v_0$ | $t$ |
| ? | $-1.5$ m/s$^2$ | $+12$ m/s | $+25$ m/s | |

The displacement for segment 2 can be obtained by solving $v^2 = v_0^2 + 2ax$ for $x$:

$$x = \frac{v^2 - v_0^2}{2a} = \frac{(12 \text{ m/s})^2 - (25 \text{ m/s})^2}{2(-1.5 \text{ m/s}^2)} = +160 \text{ m}$$

The total displacement of the motorcycle is 120 m + 160 m = $\boxed{280 \text{ m}}$.

▲

# 2.6 *Freely Falling Bodies*

Everyone has observed the effect of gravity as it causes objects to fall downward. In the absence of air resistance, it is found that all bodies at the same location above the earth fall vertically with the same acceleration. Furthermore, if the distance of the fall is small compared to the radius of the earth, the acceleration remains essentially constant throughout the descent. This idealized motion, in which air resistance is neglected and the acceleration is nearly constant, is known as *free-fall.* Since the acceleration is constant in free-fall, the equations of kinematics can be used.

The acceleration of a freely falling body is called the ***acceleration due to gravity,*** and its magnitude (without any algebraic sign) is denoted by the symbol $g$. The acceleration due to gravity is directed downward, toward the center of the earth. Near the earth's surface, $g$ is approximately

$$g = 9.80 \text{ m/s}^2 \quad \text{or} \quad 32.2 \text{ ft/s}^2$$

Unless circumstances warrant otherwise, we will use either of these values for $g$ in subsequent calculations. In reality, however, $g$ decreases with increasing altitude and varies slightly with latitude.

Figure 2.13*a* shows the well-known phenomenon of a rock falling faster than a sheet of paper. The effect of air resistance is responsible for the slower fall of the paper, for when air is removed from the tube, as in Figure 2.13*b*, the rock and the paper have exactly

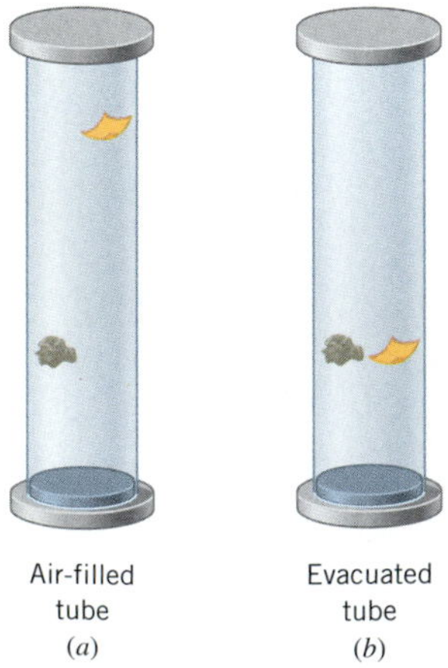

| Air-filled tube | Evacuated tube |
|---|---|
| (a) | (b) |

**Figure 2.13** (*a*) In the presence of air resistance, the acceleration of the rock is greater than that of the paper. (*b*) In the absence of air resistance, both the rock and the paper have the same acceleration.

the same acceleration due to gravity. In the absence of air, the rock and the paper both exhibit free-fall motion. Free-fall is closely approximated for objects falling near the surface of the moon, where there is no air to retard the motion. A nice demonstration of lunar free-fall was performed by astronaut David Scott, who dropped a hammer and a feather simultaneously from the same height. Both experienced the same acceleration due to lunar gravity and consequently hit the ground at the same time. The accceleration due to gravity near the surface of the moon is approximately one-sixth as large as that on the earth.

When the equations of kinematics are applied to free-fall motion, it is natural to use the symbol $y$ for the displacement, since the motion occurs in the vertical or $y$ direction. Thus, when using the equations in Table 2.1 for free-fall motion, we will simply replace $x$ with $y$. There is no significance to this change. The equations have the same algebraic form for either the horizontal or vertical direction, provided that the acceleration remains constant during the motion. We now turn our attention to several examples that illustrate how the equations of kinematics are applied to freely falling bodies.

## Example 10   A Falling Stone

A stone is dropped from rest from the top of a tall building, as Figure 2.14 indicates. After 3.00 s of free-fall, what is the displacement $y$ of the stone?

**Reasoning** The upward direction is chosen as the positive direction. The three known variables are shown in the box below. The initial velocity $v_0$ of the stone is zero, because the stone is dropped from rest. The acceleration due to gravity is negative, since it points downward in the negative direction.

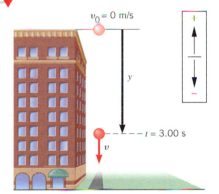

**Figure 2.14** The stone, starting with zero velocity at the top of the building, is accelerated downward by gravity.

**Stone Data**

| $y$ | $a$ | $v$ | $v_0$ | $t$ |
|-----|-----|-----|-------|-----|
| ? | $-9.80 \text{ m/s}^2$ | | $0 \text{ m/s}$ | $3.00 \text{ s}$ |

Equation 2.8 contains the appropriate variables and offers a direct solution to the problem. Since the stone moves downward, and upward is the positive direction, we expect the displacement $y$ to have a negative value.

**Solution** Using Equation 2.8, we find that

$$y = v_0 t + \tfrac{1}{2}at^2 = (0 \text{ m/s})(3.00 \text{ s}) + \tfrac{1}{2}(-9.80 \text{ m/s}^2)(3.00 \text{ s})^2 = \boxed{-44.1 \text{ m}}$$

The answer for $y$ is negative, as expected.

## Example 11   The Velocity of a Falling Stone

After 3.00 s of free-fall, what is the velocity $v$ of the stone in Figure 2.14?

**Reasoning** Because of the acceleration due to gravity, the magnitude of the stone's downward velocity increases by 9.80 m/s during each second of free-fall. The data for the stone are the same as in Example 10, and Equation 2.4 offers a direct solution for the final velocity. Since the stone is moving downward in the negative direction, the value determined for $v$ should be negative.

**Solution** Using Equation 2.4, we obtain

$$v = v_0 + at = 0 \text{ m/s} + (-9.80 \text{ m/s}^2)(3.00 \text{ s}) = \boxed{-29.4 \text{ m/s}}$$

The velocity is negative, as expected.

The acceleration due to gravity is always a downward-pointing vector. It describes how the speed increases for an object that is falling freely downward. This same acceleration also describes how the speed decreases for an object moving upward under the influence of gravity alone, in which case the object eventually comes to a momentary halt and then falls back to earth. Examples 12 and 13 show how the equations of kinematics are applied to an object that is moving upward under the influence of gravity.

## Example 12 How High Does It Go?

**Problem solving insight**
"Implied data" are important. In Example 12, for instance, the phrase "how high does the coin go" refers to the maximum height, which occurs when the final velocity $v$ in the vertical direction is $v = 0$ m/s.

A football game customarily begins with a coin toss to determine who kicks off. The referee tosses the coin up with an initial speed of 5.00 m/s. In the absence of air resistance, how high does the coin go above its point of release?

**Reasoning** The coin is given an upward initial velocity, as in Figure 2.15. But the acceleration due to gravity points downward. Since the velocity and acceleration point in opposite directions, the coin slows down as it moves upward. Eventually, the velocity of the coin becomes $v = 0$ m/s at the highest point. Assuming that the upward direction is positive, the data can be summarized as shown below:

| Coin Data | | | | |
|---|---|---|---|---|
| $y$ | $a$ | $v$ | $v_0$ | $t$ |
| ? | $-9.80$ m/s$^2$ | 0 m/s | $+5.00$ m/s | |

With these data, we can use Equation 2.9 ($v^2 = v_0^2 + 2ay$) to find the maximum height y.

**Solution** Rearranging Equation 2.9, we find that the maximum height of the coin above its release point is

$$y = \frac{v^2 - v_0^2}{2a} = \frac{(0 \text{ m/s})^2 - (5.00 \text{ m/s})^2}{2(-9.80 \text{ m/s}^2)} = \boxed{1.28 \text{ m}}$$

## Example 13 How Long Is It in the Air?

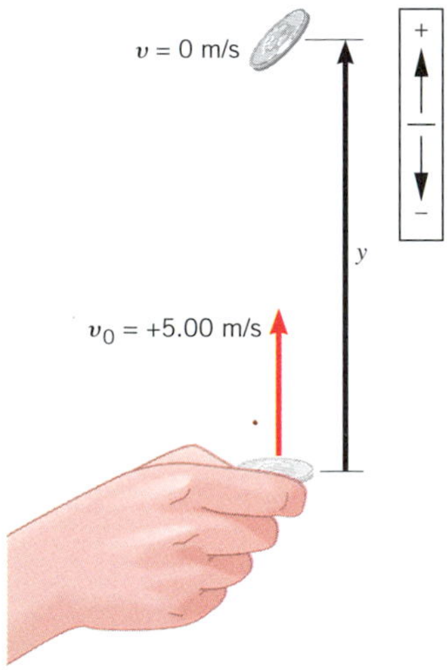

In Figure 2.15, what is the total time the coin is in the air before returning to its release point?

**Reasoning** During the time the coin travels upward, gravity causes its speed to decrease to zero. On the way down, however, gravity causes the coin to regain the lost speed. Thus, the time for the coin to go up is equal to the time for it to come down. In other words, the total travel time is twice the time for the upward motion. The data for the coin during the upward trip are the same as in Example 12. With these data, we can use Equation 2.4 ($v = v_0 + at$) to find the upward travel time.

**Solution** Rearranging Equation 2.4, we find that

$$t = \frac{v - v_0}{a} = \frac{0 \text{ m/s} - 5.00 \text{ m/s}}{-9.80 \text{ m/s}^2} = 0.510 \text{ s}$$

The total up-and-down time is twice this value, or $\boxed{1.02 \text{ s}}$.

It is possible to determine the total time by another method. When the coin is tossed upward and returns to its release point, the displacement for the *entire trip* is y = 0 m. With this value for the displacement, Equation 2.8 ($y = v_0 t + \frac{1}{2}at^2$) can be used to find the time for the entire trip directly.

**Figure 2.15** At the start of a football game, a referee tosses a coin upward with an initial velocity of $v_0 = +5.00$ m/s. The velocity of the coin is momentarily zero when the coin reaches its maximum height.

Examples 12 and 13 illustrate that the expression "freely falling" does not necessarily mean an object is falling down. A freely falling object is any object moving either upward or downward under the influence of gravity alone. In either case, the object always experiences the same *downward acceleration* due to gravity, a fact that is the focus of the next example.

## Conceptual Example 14 Acceleration Versus Velocity

There are three parts to the motion of the coin in Figure 2.15. On the way up, the coin has a velocity vector that is directed upward and has a decreasing magnitude. At the top of its path, the coin momentarily has a zero velocity. On the way down, the coin has a downward-pointing velocity vector with an increasing magnitude. In the absence of air resistance, does the acceleration of the coin, like the velocity, change from one part of the motion to another?

**Reasoning and Solution** Since air resistance is absent, the coin is in free-fall motion. Therefore, the acceleration vector is that due to gravity and has the same magnitude and the same direction at all times. It has a magnitude of 9.80 m/s$^2$ and points downward during both the upward and downward portions of the motion. Furthermore, just because the coin's instantaneous velocity is zero at the top of the motional path, don't think that the acceleration vector is also zero there. Acceleration is the rate at which velocity changes, and the velocity at the top is changing, even though at one instant it is zero. In fact, the acceleration at the top has the same magnitude of 9.80 m/s$^2$ and the same downward direction as during the rest of the motion. Thus, *the coin's velocity vector changes from moment to moment, but its acceleration vector does not change.*

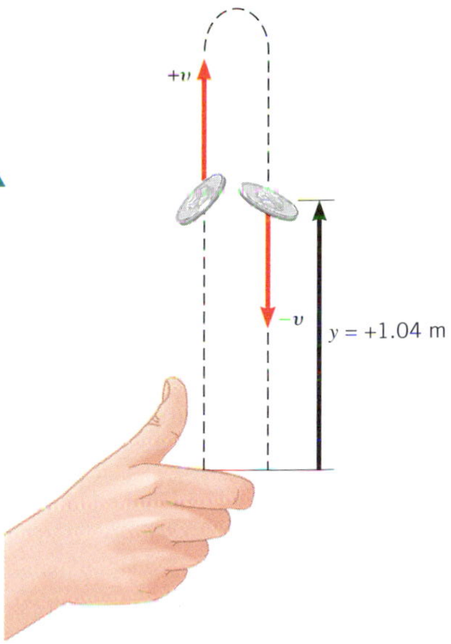

The motion of an object that is thrown upward and eventually returns to earth contains a symmetry that is useful to keep in mind from the point of view of problem solving. The calculations just completed indicate that a time symmetry exists in free-fall motion, in the sense that the time required for the object to reach maximum height equals the time for it to return to its starting point.

A type of symmetry involving the speed also exists. Figure 2.16 shows the coin considered in Examples 12 and 13. At any displacement $y$ above the point of release, the coin's speed during the upward trip equals the speed at the same point during the downward trip. For instance, when $y = +1.04$ m, Equation 2.9 gives two possible values for the final velocity $v$, assuming that the initial velocity is $v_0 = +5.00$ m/s:

$$v^2 = v_0^2 + 2ay = (5.00 \text{ m/s})^2 + 2(-9.80 \text{ m/s}^2)(1.04 \text{ m}) = 4.62 \text{ m}^2/\text{s}^2$$

$$v = \pm 2.15 \text{ m/s}$$

The value $v = +2.15$ m/s is the velocity of the coin on the upward trip, and $v = -2.15$ m/s is the velocity on the downward trip. The speed in both cases is identical and equals 2.15 m/s. Likewise, the speed just as the coin returns to its point of release is 5.00 m/s, which equals the initial speed. This symmetry involving the speed arises because the coin loses 9.80 m/s in speed each second on the way up and gains back the same amount each second on the way down. In Conceptual Example 15, we use just this kind of symmetry to guide our reasoning as we analyze the motion of a pellet shot from a gun.

**Figure 2.16** For a given displacement along the motional path, the upward speed of the coin is equal to its downward speed, but the two velocities point in opposite directions.

## Conceptual Example 15 Taking Advantage of Symmetry

Figure 2.17$a$ shows a pellet, having been fired from a gun, moving straight upward from the edge of a cliff. The initial speed of the pellet is 30 m/s. It goes up and then falls back down, eventually hitting the ground beneath the cliff. In Figure 2.17$b$ the pellet has been fired straight downward at the same initial speed. In the absence of air resistance, does the pellet in part $b$ strike the ground beneath the cliff with a smaller, a greater, or the same speed as the pellet in part $a$?

**Reasoning and Solution** Because air resistance is absent, the motion is that of free-fall, and the symmetry inherent in free-fall motion offers an immediate answer to the question. Figure 2.17$c$ shows why. This part of the drawing shows the pellet after it has been fired upward and then fallen back down to its starting point. Symmetry indicates that the speed in part $c$ is the same as in part $a$—namely, 30 m/s. Thus, part $c$ is just like part $b$, where the pellet is actually fired downward with a speed of 30 m/s. Consequently, whether the pellet is fired as in part $a$ or part $b$, it starts to move downward from the cliff edge at a speed of 30 m/s. In either case, there is the same acceleration due to gravity and the same displacement from the cliff edge to the ground below. Under these conditions, *the pellet reaches the ground with the same speed no matter in which vertical direction it is fired initially.*

**Related Homework:** *Problems 43, 46*

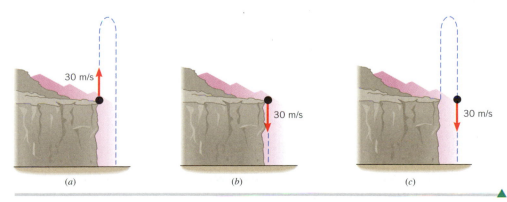

(a)  (b)  (c)

**Figure 2.17** (*a*) From the edge of a cliff, a pellet is fired straight upward from a gun. The pellet's initial speed is 30 m/s. (*b*) The pellet is fired straight downward with an initial speed of 30 m/s. (*c*) In Conceptual Example 15 this drawing plays the central role in reasoning that is based on symmetry.

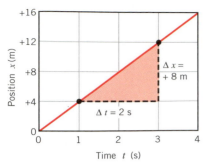

**Figure 2.18** A graph of position vs. time for an object moving with a constant velocity of $v = \Delta x/\Delta t = +4$ m/s.

## 2.7  *Graphical Analysis of Velocity and Acceleration*

Graphical techniques are helpful in understanding the concepts of velocity and acceleration. Suppose a bicyclist is riding with a constant velocity of $v = +4$ m/s. The position $x$ of the bicycle can be plotted along the vertical axis of a graph, while the time $t$ is plotted along the horizontal axis. Since the position of the bike increases by 4 m every second, the graph of $x$ versus $t$ is a straight line. Furthermore, if the bike is assumed to be at $x = 0$ m when $t = 0$ s, the straight line passes through the origin, as Figure 2.18 shows. Each point on this line gives the position of the bike at a particular time. For instance, at $t = 1$ s the position is 4 m, while at $t = 3$ s the position is 12 m.

In constructing the graph in Figure 2.18, we used the fact that the velocity was $+4$ m/s. Suppose, however, that we were given this graph, but did not have prior knowledge of the velocity. The velocity could be determined by considering what happens to the bike between the times of 1 and 3 s, for instance. The change in time is $\Delta t = 2$ s. During this time interval, the position of the bike changes from $+4$ to $+12$ m, and the change in position is $\Delta x = +8$ m. The ratio $\Delta x/\Delta t$ is called the **slope** of the straight line.

$$\text{Slope} = \frac{\Delta x}{\Delta t} = \frac{+8 \text{ m}}{2 \text{ s}} = +4 \text{ m/s}$$

Notice that the slope is equal to the velocity of the bike. This result is no accident, because $\Delta x/\Delta t$ is the definition of average velocity (see Equation 2.2). Thus, for an object moving with a constant velocity, the slope of the straight line in a position–time graph gives the velocity. Since the position–time graph is a straight line, any time interval $\Delta t$ can be chosen to calculate the velocity. Choosing a different $\Delta t$ will yield a different $\Delta x$, but the velocity $\Delta x/\Delta t$ will not change. In the real world, objects rarely move with a constant velocity at all times, as the next example illustrates.

### Example 16   A Bicycle Trip

A bicyclist maintains a constant velocity on the outgoing leg of a trip, zero velocity while stopped, and another constant velocity on the way back. Figure 2.19 shows the corresponding position–time graph. Using the time and position intervals indicated in the drawing, obtain the velocities for each segment of the trip.

**Reasoning** The average velocity $\overline{v}$ is equal to the displacement $\Delta x$ divided by the elapsed time $\Delta t$, $\overline{v} = \Delta x/\Delta t$. The displacement is the final position minus the initial position, which is a positive number for segment 1 and a negative number for segment 3. Note for segment 2 that $\Delta x = 0$ m, since the bicycle is at rest. The drawing shows values for $\Delta x$ and $\Delta t$ for each of the three segments.

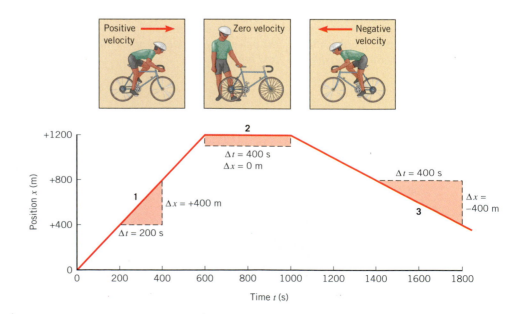

**Figure 2.19** This position-vs.-time graph consists of three straight-line segments, each corresponding to a different constant velocity.

**Solution** The average velocities for the three segments are

**Segment 1**   $\overline{v} = \dfrac{\Delta x}{\Delta t} = \dfrac{800 \text{ m} - 400 \text{ m}}{400 \text{ s} - 200 \text{ s}} = \dfrac{+400 \text{ m}}{200 \text{ s}} = \boxed{+2 \text{ m/s}}$

**Segment 2**   $\overline{v} = \dfrac{\Delta x}{\Delta t} = \dfrac{1200 \text{ m} - 1200 \text{ m}}{1000 \text{ s} - 600 \text{ s}} = \dfrac{0 \text{ m}}{400 \text{ s}} = \boxed{0 \text{ m/s}}$

**Segment 3**   $\overline{v} = \dfrac{\Delta x}{\Delta t} = \dfrac{400 \text{ m} - 800 \text{ m}}{1800 \text{ s} - 1400 \text{ s}} = \dfrac{-400 \text{ m}}{400 \text{ s}} = \boxed{-1 \text{ m/s}}$

In the second segment of the journey the velocity is zero, reflecting the fact that the bike is stationary. Since the position of the bike does not change, segment 2 is a horizontal line that has a zero slope. In the third part of the motion the velocity is negative, because the position of the bike decreases from $x = +800$ m to $x = +400$ m during the 400-s interval shown in the graph. As a result, segment 3 has a negative slope, and the velocity is negative.

▲

If the object is accelerating, its velocity is changing. When the velocity is changing, the $x$-versus-$t$ graph is not a straight line, but is a curve, perhaps like that in Figure 2.20. This curve was drawn using Equation 2.8 ($x = v_0 t + \frac{1}{2}at^2$), assuming an acceleration of $a = 0.26$ m/s² and an initial velocity of $v_0 = 0$ m/s. The velocity at any instant of time can be determined by measuring the slope of the curve at that instant. The slope at any point along the curve is defined to be the slope of the tangent line drawn to the curve at that point. For instance, in Figure 2.20 a tangent line is drawn at $t = 20.0$ s. To determine the slope of the tangent line, a triangle is constructed using an arbitrarily chosen time interval of $\Delta t = 5.0$ s. The change in $x$ associated with this time interval can be read from the tangent line as $\Delta x = +26$ m. Therefore,

$$\text{Slope of tangent line} = \frac{\Delta x}{\Delta t} = \frac{+26 \text{ m}}{5.0 \text{ s}} = +5.2 \text{ m/s}$$

The slope of the tangent line is the instantaneous velocity, which in this case is $v = +5.2$ m/s. This graphical result can be verified by using Equation 2.4 with $v_0 = 0$ m/s: $v = at = (+0.26 \text{ m/s}^2)(20.0 \text{ s}) = +5.2$ m/s.

Insight into the meaning of acceleration can also be gained with the aid of a graphical representation. Consider an object moving with a constant acceleration of $a = +6$ m/s². If the object has an initial velocity of $v_0 = +5$ m/s, its velocity at any time is represented by Equation 2.4 as

$$v = v_0 + at = 5 \text{ m/s} + (6 \text{ m/s}^2)t$$

This relation is plotted as the velocity-versus-time graph in Figure 2.21. The graph of $v$ versus $t$ is a straight line that intercepts the vertical axis at $v_0 = 5$ m/s. The slope of this straight line can be calculated from the data shown in the drawing:

$$\text{Slope} = \frac{\Delta v}{\Delta t} = \frac{+12 \text{ m/s}}{2 \text{ s}} = +6 \text{ m/s}^2$$

The ratio $\Delta v/\Delta t$ is, by definition, equal to the average acceleration (Equation 2.4), so the slope of the straight line in a velocity–time graph is the average acceleration.

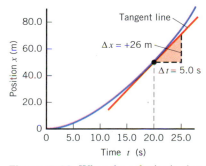

**Figure 2.20** When the velocity is changing, the position-vs.-time graph is a curved line. The slope $\Delta x/\Delta t$ of the tangent line drawn to the curve at a given time is the instantaneous velocity at that time.

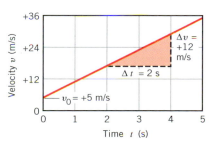

**Figure 2.21** A velocity-vs.-time graph that applies to an object with an acceleration of $\Delta v/\Delta t = +6$ m/s². The initial velocity is $v_0 = +5$ m/s when $t = 0$ s.

# Concept Summary

This summary presents an abridged version of the chapter, including the important equations and all available learning aids. For convenient reference, the learning aids (including the text's examples) are placed next to or immediately after the relevant equation or discussion. The following learning aids may be found on-line at **www.wiley.com/college/cutnell**:

| | |
|---|---|
| **Interactive LearningWare** examples are solved according to a five-step interactive format that is designed to help you develop problem-solving skills. | **Concept Simulations** are animated versions of text figures or animations that illustrate important concepts. You can control parameters that affect the display, and we encourage you to experiment. |
| **Interactive Solutions** offer specific models for certain types of problems in the chapter homework. The calculations are carried out interactively. | **Self-Assessment Tests** include both qualitative and quantitative questions. Extensive feedback is provided for both incorrect and correct answers, to help you evaluate your understanding of the material. |

| Topic | Discussion | Learning Aids |
|---|---|---|
| | **2.1 Displacement** | |
| **Displacement** | Displacement is a vector that points from an object's initial position to its final position. The magnitude of the displacement is the shortest distance between the two positions. | |
| | **2.2 Speed and Velocity** | |
| | The average speed of an object is the distance traveled by the object divided by the time required to cover the distance: | |
| **Average speed** | $$\text{Average speed} = \frac{\text{Distance}}{\text{Elapsed time}} \qquad (2.1)$$ | **Example 1** |
| | The average velocity $\bar{\mathbf{v}}$ of an object is the object's displacement $\Delta\mathbf{x}$ divided by the elapsed time $\Delta t$: | |
| **Average velocity** | $$\bar{\mathbf{v}} = \frac{\Delta\mathbf{x}}{\Delta t} \qquad (2.2)$$ | **Example 2** |
| | Average velocity is a vector that has the same direction as the displacement. When the elapsed time becomes infinitesimally small, the average velocity becomes equal to the instantaneous velocity $\mathbf{v}$, the velocity at an instant of time: | |
| **Instantaneous velocity** | $$\mathbf{v} = \lim_{\Delta t \to 0} \frac{\Delta\mathbf{x}}{\Delta t} \qquad (2.3)$$ | |
| | **2.3 Acceleration** | |
| | The average acceleration $\bar{\mathbf{a}}$ is a vector. It equals the change $\Delta\mathbf{v}$ in the velocity divided by the elapsed time $\Delta t$, the change in the velocity being the final minus the initial velocity: | |
| **Average acceleration** | $$\bar{\mathbf{a}} = \frac{\Delta\mathbf{v}}{\Delta t} \qquad (2.4)$$ | **Examples 3, 4** |
| | When $\Delta t$ becomes infinitesimally small, the average acceleration becomes equal to the instantaneous acceleration $\mathbf{a}$: | **Interactive Solution 2.17** |
| **Instantaneous acceleration** | $$\mathbf{a} = \lim_{\Delta t \to 0} \frac{\Delta\mathbf{v}}{\Delta t} \qquad (2.5)$$ | |
| | Acceleration is the rate at which the velocity is changing. | |

 **Use Self-Assessment Test 2.1 to evaluate your understanding of Sections 2.1–2.3.**

| Topic | Discussion | Learning Aids |
|---|---|---|
| | **2.4 Equations of Kinematics for Constant Acceleration** | |
| | **2.5 Applications of the Equations of Kinematics** | |
| | The equations of kinematics apply when an object moves with a constant acceleration along a straight line. These equations relate the displacement $x - x_0$, the acceleration $a$, the final velocity $v$, the initial velocity $v_0$, and the elapsed time $t - t_0$. Assuming that $x_0 = 0$ m at $t_0 = 0$ s, the equations of kinematics are | **Examples 5–9** |
| | | **Concept Simulations 2.1, 2.2** |
| **Equations of kinematics** | $$v = v_0 + at \qquad (2.4)$$ | **Interactive LearningWare 2.1, 2.2** |
| | $$x = \tfrac{1}{2}(v_0 + v)t \qquad (2.7)$$ | |
| | $$x = v_0 t + \tfrac{1}{2}at^2 \qquad (2.8)$$ | **Interactive Solutions 2.29, 2.31** |
| | $$v^2 = v_0^2 + 2ax \qquad (2.9)$$ | |

| Topic | Discussion | Learning Aids |
|---|---|---|

Use *Self-Assessment Test 2.2* to evaluate your understanding of Sections 2.4 and 2.5.

### 2.6 Freely Falling Bodies

**Acceleration due to gravity**

In free-fall motion, an object experiences negligible air resistance and a constant acceleration due to gravity. All objects at the same location above the earth have the same acceleration due to gravity. The acceleration due to gravity is directed toward the center of the earth and has a magnitude of approximately $9.80 \text{ m/s}^2$ near the earth's surface.

**Examples 10–15**

**Concept Stimulation 2.3**

**Interactive Solutions 2.47, 2.49**

### 2.7 Graphical Analysis of Velocity and Acceleration

The slope of a plot of position versus time for a moving object gives the object's velocity. The slope of a plot of velocity versus time gives the object's acceleration.

**Example 16**

**Concept Simulation 2.4**

Use *Self-Assessment Test 2.3* to evaluate your understanding of Sections 2.6 and 2.7.

# Problems

**ssm** Solution is in the Student Solutions Manual.     **www** Solution is available on the World Wide Web at www.wiley.com/college/cutnell
🍵 This icon represents a biomedical application.

### Section 2.1 Displacement, Section 2.2 Speed and Velocity

**1. ssm** A plane is sitting on a runway, awaiting takeoff. On an adjacent parallel runway, another plane lands and passes the stationary plane at a speed of 45 m/s. The arriving plane has a length of 36 m. By looking out of a window (very narrow), a passenger on the stationary plane can see the moving plane. For how long a time is the moving plane visible?

**2.** One afternoon, a couple walks three-fourths of the way around a circular lake, the radius of which is 1.50 km. They start at the west side of the lake and head due south to begin with. (a) What is the distance they travel? (b) What are the magnitude and direction (relative to due east) of the couple's displacement?

**3. ssm** A whale swims due east for a distance of 6.9 km, turns around and goes due west for 1.8 km, and finally turns around again and heads 3.7 km due east. (a) What is the total distance traveled by the whale? (b) What are the magnitude and direction of the displacement of the whale?

**4.** The Space Shuttle travels at a speed of about $7.6 \times 10^3$ m/s. The blink of an astronaut's eye lasts about 110 ms. How many football fields (length = 91.4 m) does the Shuttle cover in the blink of an eye?

**5.** As the earth rotates through one revolution, a person standing on the equator traces out a circular path whose radius is equal to the radius of the earth ($6.38 \times 10^6$ m). What is the average speed of this person in (a) meters per second and (b) miles per hour?

**6.** In 1954 the English runner Roger Bannister broke the four-minute barrier for the mile with a time of 3:59.4 s (3 min and 59.4 s). In 1999 the Moroccan runner Hicham el-Guerrouj set a record of 3:43.13 s for the mile. If these two runners had run in the same race, each running the entire race at the average speed that earned him a place in the record books, el-Guerrouj would have won. By how many meters?

**7.** A tourist being chased by an angry bear is running in a straight line toward his car at a speed of 4.0 m/s. The car is a distance $d$ away. The bear is 26 m behind the tourist and running at 6.0 m/s. The tourist reaches the car safely. What is the maximum possible value for $d$?

**\* 8.** In reaching her destination, a backpacker walks with an average velocity of 1.34 m/s, due west. This average velocity results because she hikes for 6.44 km with an average velocity of 2.68 m/s, due west, turns around, and hikes with an average velocity of 0.447 m/s, due east. How far east did she walk?

**\* 9. ssm www** A woman and her dog are out for a morning run to the river, which is located 4.0 km away. The woman runs at 2.5 m/s in a straight line. The dog is unleashed and runs back and forth at 4.5 m/s between his owner and the river, until she reaches the river. What is the total distance run by the dog?

**\* 10.** A car makes a trip due north for three-fourths of the time and due south one-fourth of the time. The average northward velocity has a magnitude of 27 m/s, and the average southward velocity has a magnitude of 17 m/s. What is the average velocity, magnitude and direction, for the entire trip?

**\*\* 11.** You are on a train that is traveling at 3.0 m/s along a level straight track. Very near and parallel to the track is a wall that slopes upward at a 12° angle with the horizontal. As you face the window (0.90 m high, 2.0 m wide) in your compartment, the train is moving to the left, as the drawing indicates. The top edge of the wall first appears at window corner $A$ and eventually disappears at window corner $B$. How much time passes between appearance and disappearance of the upper edge of the wall?

## Section 2.3 Acceleration

**12.** For a standard production car, the highest road-tested acceleration ever reported occurred in 1993, when a Ford RS200 Evolution went from zero to 26.8 m/s (60 mi/h) in 3.275 s. Find the magnitude of the car's acceleration.

**13. ssm** A motorcycle has a constant acceleration of 2.5 m/s². Both the velocity and acceleration of the motorcycle point in the same direction. How much time is required for the motorcycle to change its speed from (a) 21 to 31 m/s, and (b) 51 to 61 m/s?

**14.** NASA has developed *Deep-Space 1* (DS-1), a spacecraft that is scheduled to rendezvous with the asteroid named 1992 KD (which orbits the sun millions of miles from the earth). The propulsion system of DS-1 works by ejecting high-speed argon ions out the rear of the engine. The engine slowly increases the velocity of DS-1 by about +9.0 m/s per day. (a) How much time (in days) will it take to increase the velocity of DS-1 by +2700 m/s? (b) What is the acceleration of DS-1 (in m/s²)?

**15. ssm** A runner accelerates to a velocity of 5.36 m/s due west in 3.00 s. His average acceleration is 0.640 m/s², also directed due west. What was his velocity when he began accelerating?

**16.** The land speed record of 13.9 m/s (31 mi/h) for birds is held by the Australian emu. An emu running due south in a straight line at this speed slows down to a speed of 11.0 m/s in 3.0 s. (a) What is the direction of the bird's acceleration? (b) Assuming that the acceleration remains the same, what is the bird's velocity after an additional 4.0 s has elapsed?

***17.** Consult **Interactive Solution 2.17** at **www.wiley.com/college/cutnell** before beginning this problem. A car is traveling along a straight road at a velocity of +36.0 m/s when its engine cuts out. For the next twelve seconds the car slows down, and its average acceleration is $\bar{a}_1$. For the next six seconds the car slows down further, and its average acceleration is $\bar{a}_2$. The velocity of the car at the end of the eighteen-second period is +28.0 m/s. The ratio of the average acceleration values is $\bar{a}_1/\bar{a}_2 = 1.50$. Find the velocity of the car at the end of the initial twelve-second interval.

****18.** Two motorcycles are traveling due east with different velocities. However, four seconds later, they have the same velocity. During this four-second interval, motorcycle A has an average acceleration of 2.0 m/s² due east, while motorcycle B has an average acceleration of 4.0 m/s² due east. By how much did the speeds *differ* at the beginning of the four-second interval, and which motorcycle was moving faster?

## Section 2.4 Equations of Kinematics for Constant Acceleration, Section 2.5 Applications of the Equations of Kinematics

**19.** In getting ready to slam-dunk the ball, a basketball player starts from rest and sprints to a speed of 6.0 m/s in 1.5 s. Assuming that the player accelerates uniformly, determine the distance he runs.

**20.** Review Conceptual Example 7 as background for this problem. A car is traveling to the left, which is the negative direction. The direction of travel remains the same throughout this problem. The car's initial speed is 27.0 m/s, and during a 5.0-s interval, it changes to a final speed of (a) 29.0 m/s and (b) 23.0 m/s. In each case, find the acceleration (magnitude and algebraic sign) and state whether or not the car is decelerating.

**21. ssm** A VW Beetle goes from 0 to 60.0 mi/h with an acceleration of +2.35 m/s². (a) How much time does it take for the Beetle to reach this speed? (b) A top-fuel dragster can go from 0 to 60.0 mi/h in 0.600 s. Find the acceleration (in m/s²) of the dragster.

**22.** (a) What is the magnitude of the average acceleration of a skier who, starting from rest, reaches a speed of 8.0 m/s when going down a slope for 5.0 s? (b) How far does the skier travel in this time?

**23.** The left ventricle of the heart accelerates blood from rest to a velocity of +26 cm/s. (a) If the displacement of the blood during the acceleration is +2.0 cm, determine its acceleration (in cm/s²). (b) How much time does blood take to reach its final velocity?

**24.** Consult **Concept Simulation 2.1** at **www.wiley.com/college/cutnell** for help in preparing for this problem. A cheetah is hunting. Its prey runs for 3.0 s at a constant velocity of +9.0 m/s. Starting from rest, what constant acceleration must the cheetah maintain in order to run the same distance as its prey runs in the same time?

**25. ssm** A jetliner, traveling northward, is landing with a speed of 69 m/s. Once the jet touches down, it has 750 m of runway in which to reduce its speed to 6.1 m/s. Compute the average acceleration (magnitude and direction) of the plane during landing.

**26.** Consult **Concept Simulation 2.1** at **www.wiley.com/college/cutnell** before starting this problem. The Kentucky Derby is held at the Churchill Downs track in Louisville, Kentucky. The track is one and one-quarter miles in length. One of the most famous horses to win this event was Secretariat. In 1973 he set a Derby record that has never been broken. His average acceleration during the last four quarter-miles of the race was +0.0105 m/s². His velocity at the start of the final mile (x = +1609 m) was about +16.58 m/s. The acceleration, although small, was very important to his victory. To assess its effect, determine the difference between the time he would have taken to run the final mile at a constant velocity of +16.58 m/s and the time he actually took. Although the track is oval in shape, assume it is straight for the purpose of this problem.

**27. ssm www** A speed ramp at an airport is basically a large conveyor belt on which you can stand and be moved along. The belt of one ramp moves at a constant speed such that a person who stands still on it leaves the ramp 64 s after getting on. Clifford is in a real hurry, however, and skips the speed ramp. Starting from rest with an acceleration of 0.37 m/s², he covers the same distance as the ramp does, but in one-fourth the time. What is the speed at which the belt of the ramp is moving?

***28.** A drag racer, starting from rest, speeds up for 402 m with an acceleration of +17.0 m/s². A parachute then opens, slowing the car down with an acceleration of −6.10 m/s². How fast is the racer moving 3.50 × 10² m after the parachute opens?

***29.** Review **Interactive Solution 2.29** at **www.wiley.com/college/cutnell** in preparation for this problem. Suppose a car is traveling at 20.0 m/s, and the driver sees a traffic light turn red. After 0.530 s has elapsed (the reaction time), the driver applies the brakes, and the car decelerates at 7.00 m/s². What is the stopping distance of the car, as measured from the point where the driver first notices the red light?

***30.** A speedboat starts from rest and accelerates at +2.01 m/s² for 7.00 s. At the end of this time, the boat continues for an additional 6.00 s with an acceleration of +0.518 m/s². Following this, the boat accelerates at −1.49 m/s² for 8.00 s. (a) What is the velocity of the boat at t = 21.0 s? (b) Find the total displacement of the boat.

***31. Interactive Solution 2.31** at **www.wiley.com/college/cutnell** offers help in modeling this problem. A car is traveling at a constant speed of 33 m/s on a highway. At the instant this car passes an entrance ramp, a second car enters the highway from the ramp. The second car starts from rest and has a constant acceleration. What acceleration must it maintain, so that the two cars meet for the first time at the next exit, which is 2.5 km away?

***32.** A cab driver picks up a customer and delivers her 2.00 km away, on a straight route. The driver accelerates to the speed limit and, on reaching it, begins to decelerate at once. The magnitude of the deceleration is three times the magnitude of the acceleration. Find the lengths of the acceleration and deceleration phases.

**\* 33.** Along a straight road through town, there are three speed-limit signs. They occur in the following order: 55, 35, and 25 mi/h, with the 35-mi/h sign being midway between the other two. Obeying these speed limits, the smallest possible time $t_A$ that a driver can spend on this part of the road is to travel between the first and second signs at 55 mi/h and between the second and third signs at 35 mi/h. More realistically, a driver could slow down from 55 to 35 mi/h with a constant deceleration and then do a similar thing from 35 to 25 mi/h. This alternative requires a time $t_B$. Find the ratio $t_B/t_A$.

**\*\* 34.** A Boeing 747 "Jumbo Jet" has a length of 59.7 m. The runway on which the plane lands intersects another runway. The width of the intersection is 25.0 m. The plane decelerates through the intersection at a rate of 5.70 m/s$^2$ and clears it with a final speed of 45.0 m/s. How much time is needed for the plane to clear the intersection?

**\*\* 35. ssm** A train has a length of 92 m and starts from rest with a constant acceleration at time $t = 0$ s. At this instant, a car just reaches the end of the train. The car is moving with a constant velocity. At a time $t = 14$ s, the car just reaches the front of the train. Ultimately, however, the train pulls ahead of the car, and at time $t = 28$ s, the car is again at the rear of the train. Find the magnitudes of (a) the car's velocity and (b) the train's acceleration.

**\*\* 36.** In the one-hundred-meter dash a sprinter accelerates from rest to a top speed with an acceleration whose magnitude is 2.68 m/s$^2$. After achieving top speed, he runs the remainder of the race without speeding up or slowing down. If the total race is run in 12.0 s, how far does he run during the acceleration phase?

**Section 2.6 Freely Falling Bodies**

**37. ssm** A penny is dropped from rest from the top of the Sears Tower in Chicago. Considering that the height of the building is 427 m and ignoring air resistance, find the speed with which the penny strikes the ground.

**38.** In preparation for this problem, review Conceptual Example 7. From the top of a cliff, a person uses a slingshot to fire a pebble straight downward, which is the negative direction. The initial speed of the pebble is 9.0 m/s. (a) What is the acceleration (magnitude and direction) of the pebble during the downward motion? Is the pebble decelerating? Explain. (b) After 0.50 s, how far beneath the cliff top is the pebble?

**39. Concept Simulation 2.3** at **www.wiley.com/college/cutnell** offers a useful review of the concepts central to this problem. An astronaut on a distant planet wants to determine its acceleration due to gravity. The astronaut throws a rock straight up with a velocity of +15 m/s and measures a time of 20.0 s before the rock returns to his hand. What is the acceleration (magnitude and direction) due to gravity on this planet?

**40.** The drawing shows a device that you can make with a piece of cardboard, which can be used to measure a person's reaction time. Hold the card at the top and suddenly drop it. Ask a friend to try to catch the card between his or her thumb and index finger. Initially, your friend's fingers must be level with the asterisks at the bottom. By noting where your friend catches the card, you can determine his or her reaction time in milliseconds (ms). Calculate the distances $d_1$, $d_2$, and $d_3$.

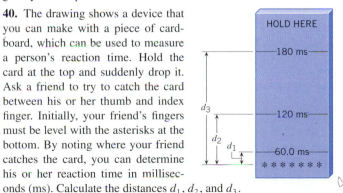

**41. ssm** From her bedroom window a girl drops a water-filled balloon to the ground, 6.0 m below. If the balloon is released from rest, how long is it in the air?

**42.** Review **Concept Simulation 2.3** at **www.wiley.com/college/cutnell** before attempting this problem. At the beginning of a basketball game, a referee tosses the ball straight up with a speed of 4.6 m/s. A player cannot touch the ball until after it reaches its maximum height and begins to fall down. What is the minimum time that a player must wait before touching the ball?

**43.** Review Conceptual Example 15 before attempting this problem. Two identical pellet guns are fired simultaneously from the edge of a cliff. These guns impart an initial speed of 30.0 m/s to each pellet. Gun A is fired straight upward, with the pellet going up and then falling back down, eventually hitting the ground beneath the cliff. Gun B is fired straight downward. In the absence of air resistance, how long after pellet B hits the ground does pellet A hit the ground?

**44.** A diver springs upward with an initial speed of 1.8 m/s from a 3.0-m board. (a) Find the velocity with which he strikes the water. [*Hint: When the diver reaches the water, his displacement is $y = -3.0$ m (measured from the board), assuming that the downward direction is chosen as the negative direction.*] (b) What is the highest point he reaches above the water?

**45. ssm** A wrecking ball is hanging at rest from a crane when suddenly the cable breaks. The time it takes for the ball to fall halfway to the ground is 1.2 s. Find the time it takes for the ball to fall from rest all the way to the ground.

**46.** Before working this problem, review Conceptual Example 15. A pellet gun is fired straight downward from the edge of a cliff that is 15 m above the ground. The pellet strikes the ground with a speed of 27 m/s. How far above the cliff edge would the pellet have gone had the gun been fired straight upward?

**47.** Consult **Interactive Solution 2.47** at **www.wiley.com/college/cutnell** before beginning this problem. A ball is thrown straight upward and rises to a maximum height of 12.0 m above its launch point. At what height above its launch point has the speed of the ball decreased to one-half of its initial value?

**\* 48.** Two arrows are shot vertically upward. The second arrow is shot after the first one, but while the first is still on its way up. The initial speeds are such that both arrows reach their maximum heights at the same instant, although these heights are different. Suppose that the initial speed of the first arrow is 25.0 m/s and that the second arrow is fired 1.20 s after the first. Determine the initial speed of the second arrow.

**\* 49.** Review **Interactive Solution 2.49** at **www.wiley.com/college/cutnell** before beginning this problem. A woman on a bridge 75.0 m high sees a raft floating at a constant speed on the river below. She drops a stone from rest in an attempt to hit the raft. The stone is released when the raft has 7.00 m more to travel before passing under the bridge. The stone hits the water 4.00 m in front of the raft. Find the speed of the raft.

**\* 50.** Consult **Concept Simulation 2.3** at **www.wiley.com/college/cutnell** to review the concepts on which this problem is based. Two students, Anne and Joan, are bouncing straight up and down on a trampoline. Anne bounces twice as high as Joan does. Assuming both are in free-fall, find the ratio of the time Anne spends between bounces to the time Joan spends.

**\* 51. ssm** A log is floating on swiftly moving water. A stone is dropped from rest from a 75-m-high bridge and lands on the log as it passes under the bridge. If the log moves with a constant speed of 5.0 m/s, what is the horizontal distance between the log and the bridge when the stone is released?

*52. (a) Just for fun, a person jumps from rest from the top of a tall cliff overlooking a lake. In falling through a distance $H$, she acquires a certain speed $v$. Assuming free-fall conditions, how much farther must she fall in order to acquire a speed of $2v$? Express your answer in terms of $H$. (b) Would the answer to part (a) be different if this event were to occur on another planet where the acceleration due to gravity had a value other than 9.80 m/s²? Explain.

*53. **ssm www** A spelunker (cave explorer) drops a stone from rest into a hole. The speed of sound is 343 m/s in air, and the sound of the stone striking the bottom is heard 1.50 s after the stone is dropped. How deep is the hole?

*54. A ball is thrown upward from the top of a 25.0-m-tall building. The ball's initial speed is 12.0 m/s. At the same instant, a person is running on the ground at a distance of 31.0 m from the building. What must be the average speed of the person if he is to catch the ball at the bottom of the building?

**55. A ball is dropped from rest from the top of a cliff that is 24 m high. From ground level, a second ball is thrown straight upward at the same instant that the first ball is dropped. The initial speed of the second ball is exactly the same as that with which the first ball eventually hits the ground. In the absence of air resistance, the motions of the balls are just the reverse of each other. Determine how far below the top of the cliff the balls cross paths.

**56. Review **Interactive LearningWare 2.2** at **www.wiley.com/college/cutnell** as an aid in solving this problem. A hot air balloon is ascending straight up at a constant speed of 7.0 m/s. When the balloon is 12.0 m above the ground, a gun fires a pellet straight up from ground level with an initial speed of 30.0 m/s. Along the paths of the balloon and the pellet, there are two places where each of them has the same altitude at the same time. How far above ground level are these places?

**Section 2.7 Graphical Analysis of Velocity and Acceleration**

57. **ssm** For the first 10.0 km of a marathon, a runner averages a velocity that has a magnitude of 15.0 km/h. For the next 15.0 km, he averages 10.0 km/h, and for the last 15.0 km, he averages 5.0 km/h. Construct, to scale, the position–time graph for the runner.

58. A bus makes a trip according to the position–time graph shown in the drawing. What is the average velocity (magnitude and direction) of the bus during each of the segments labeled $A$, $B$, and $C$? Express your answers in km/h.

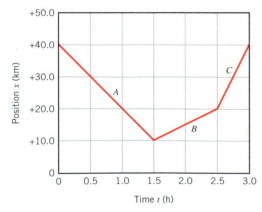

59. **Concept Simulation 2.5** at **www.wiley.com/college/cutnell** provides a review of the concepts that play a role in this problem. A snowmobile moves according to the velocity–time graph shown in the drawing (see top of right column). What is the snowmobile's average acceleration during each of the segments $A$, $B$, and $C$?

60. A person who walks for exercise produces the position–time graph given with this problem. (a) Without doing any calculations, decide which segments of the graph ($A$, $B$, $C$, or $D$) indicate positive,

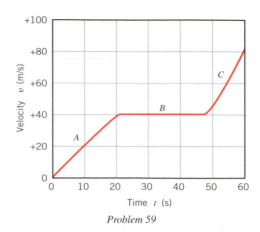

Problem 59

negative, and zero average velocities. (b) Calculate the average velocity for each segment to verify your answers to part (a).

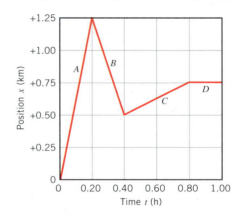

*61. **ssm** A bus makes a trip according to the position–time graph shown in the illustration. What is the average acceleration (in km/h²) of the bus for the entire 3.5-h period shown in the graph?

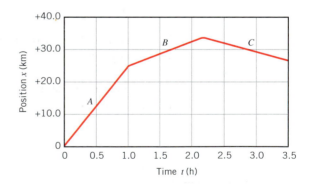

*62. A runner is at the position $x = 0$ m when time $t = 0$ s. One hundred meters away is the finish line. Every ten seconds, this runner runs half the remaining distance to the finish line. During each ten-second segment, the runner has a constant velocity. For the first forty seconds of the motion, construct (a) the position–time graph and (b) the velocity–time graph.

**63. Two runners start one hundred meters apart and run toward each other. Each runs ten meters during the first second. During each second thereafter, each runner runs ninety percent of the distance he ran in the previous second. Thus, the velocity of each person changes from second to second. However, during any one second, the velocity remains constant. Make a position–time graph for one of the runners. From this graph, determine (a) how much time passes before the runners collide and (b) the speed with which each is running at the moment of collision.

# Chapter 3 Kinematics in Two Dimensions

## 3.1 Displacement, Velocity, and Acceleration

In Chapter 2 the concepts of displacement, velocity, and acceleration are used to describe an object moving in one dimension. There are also situations in which the motion is along a curved path that lies in a plane. Such two-dimensional motion can be described using the same concepts. In Grand Prix racing, for example, the course follows a curved road, and Figure 3.1 shows a race car at two different positions along it. These positions are identified by the vectors $\mathbf{r}$ and $\mathbf{r_0}$, which are drawn from an arbitrary coordinate origin. The *displacement* $\Delta\mathbf{r}$ of the car is the vector drawn from the initial position $\mathbf{r_0}$ at time $t_0$ to the final position $\mathbf{r}$ at time $t$. The magnitude of $\Delta\mathbf{r}$ is the shortest distance between the two positions. In the drawing, the vectors $\mathbf{r_0}$ and $\Delta\mathbf{r}$ are drawn tail to head, so it is evident that $\mathbf{r}$ is the vector sum of $\mathbf{r_0}$ and $\Delta\mathbf{r}$. (See Sections 1.5 and 1.6 for a review of vectors and vector addition.) This means that $\mathbf{r} = \mathbf{r_0} + \Delta\mathbf{r}$, or

$$\text{Displacement} = \Delta\mathbf{r} = \mathbf{r} - \mathbf{r_0}$$

The displacement here is defined as it is in Chapter 2. Now, however, the displacement vector can lie anywhere in a plane, rather than just along a straight line.

The average velocity $\overline{\mathbf{v}}$ of the car between two positions is defined in a manner similar to that in Equation 2.2, as the displacement $\Delta\mathbf{r} = \mathbf{r} - \mathbf{r_0}$ divided by the elapsed time $\Delta t = t - t_0$:

$$\overline{\mathbf{v}} = \frac{\mathbf{r} - \mathbf{r_0}}{t - t_0} = \frac{\Delta\mathbf{r}}{\Delta t} \tag{3.1}$$

Since both sides of Equation 3.1 must agree in direction, the average velocity vector has the same direction as the displacement. The velocity of the car at an instant of time is its *instantaneous velocity* $\mathbf{v}$. The average velocity becomes equal to the instantaneous velocity $\mathbf{v}$ in the limit that $\Delta t$ becomes infinitesimally small:

$$\mathbf{v} = \lim_{\Delta t \to 0} \frac{\Delta\mathbf{r}}{\Delta t}$$

Figure 3.2 illustrates that the instantaneous velocity $\mathbf{v}$ is tangent to the path of the car. The drawing also shows the vector components $\mathbf{v}_x$ and $\mathbf{v}_y$ of the velocity, which are parallel to the $x$ and $y$ axes, respectively.

The *average acceleration* $\overline{\mathbf{a}}$ is defined just as it is for one-dimensional motion — namely, as the change in velocity, $\Delta\mathbf{v} = \mathbf{v} - \mathbf{v_0}$, divided by the elapsed time $\Delta t$:

$$\overline{\mathbf{a}} = \frac{\mathbf{v} - \mathbf{v_0}}{t - t_0} = \frac{\Delta\mathbf{v}}{\Delta t} \tag{3.2}$$

The average acceleration vector has the same direction as the change in velocity. In the limit that the elapsed time becomes infinitesimally small, the average acceleration becomes equal to the *instantaneous acceleration* $\mathbf{a}$:

$$\mathbf{a} = \lim_{\Delta t \to 0} \frac{\Delta\mathbf{v}}{\Delta t}$$

The acceleration has a vector component $\mathbf{a}_x$ along the $x$ direction and a vector component $\mathbf{a}_y$ along the $y$ direction.

## 3.2 Equations of Kinematics in Two Dimensions

To understand how displacement, velocity, and acceleration are applied to two-dimensional motion, consider a spacecraft equipped with two engines that are mounted perpendicular to each other. These engines produce the only forces that the craft experiences,

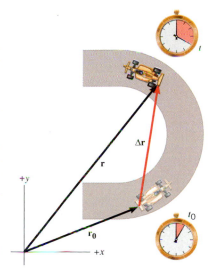

**Figure 3.1** The displacement $\Delta\mathbf{r}$ of the car is a vector that points from the initial position of the car at time $t_0$ to the final position at time $t$. The magnitude of $\Delta\mathbf{r}$ is the shortest distance between the two positions.

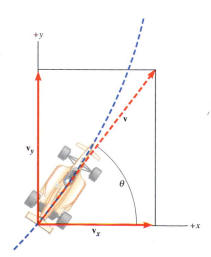

**Figure 3.2** The instantaneous velocity $\mathbf{v}$ and its two vector components $\mathbf{v}_x$ and $\mathbf{v}_y$.

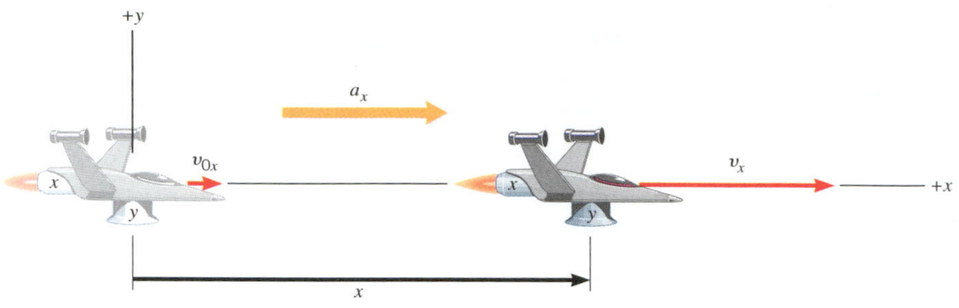

**Figure 3.3** The spacecraft is moving with a constant acceleration $a_x$ parallel to the $x$ axis. There is no motion in the $y$ direction, and the $y$ engine is turned off.

and the spacecraft is assumed to be at the coordinate origin when $t_0 = 0$ s, so that $\mathbf{r_0} = 0$ m. At a later time $t$, the spacecraft's displacement is $\Delta\mathbf{r} = \mathbf{r} - \mathbf{r_0} = \mathbf{r}$. Relative to the $x$ and $y$ axes, the displacement $\mathbf{r}$ has vector components of $\mathbf{x}$ and $\mathbf{y}$, respectively.

In Figure 3.3 only the engine oriented along the $x$ direction is firing, and the vehicle accelerates along this direction. It is assumed that the velocity in the $y$ direction is zero, and it remains zero, since the $y$ engine is turned off. The motion of the spacecraft along the $x$ direction is described by the five kinematic variables $x$, $a_x$, $v_x$, $v_{0x}$, and $t$. Here the symbol "$x$" reminds us that we are dealing with the $x$ components of the displacement, velocity, and acceleration vectors. (See Sections 1.7 and 1.8 for a review of vector components.) The variables $x$, $a_x$, $v_x$, and $v_{0x}$ are scalar components (or "components," for short). As Section 1.7 discusses, these components are positive or negative numbers (with units), depending on whether the associated vector components point along the $+x$ or the $-x$ axis. If the spacecraft has a constant acceleration along the $x$ direction, the motion is exactly like that described in Chapter 2, and the equations of kinematics can be used. For convenience, these equations are written in the left column of Table 3.1.

Figure 3.4 is analogous to Figure 3.3, except that now only the $y$ engine is firing, and the spacecraft accelerates along the $y$ direction. Such a motion can be described in terms of the kinematic variables $y$, $a_y$, $v_y$, $v_{0y}$, and $t$. And if the acceleration along the $y$ direction is constant, these variables are related by the equations of kinematics, as written in the right column of Table 3.1. Like their counterparts in the $x$ direction, the scalar components, $y$, $a_y$, $v_y$, and $v_{0y}$, may be positive $(+)$ or negative $(-)$ numbers (with units).

If both engines of the spacecraft are firing *at the same time,* the resulting motion takes place in part along the $x$ axis and in part along the $y$ axis, as Figure 3.5 illustrates. The thrust of each engine gives the vehicle a corresponding acceleration component. The $x$ engine accelerates the ship in the $x$ direction and causes a change in the $x$ component of the velocity. Likewise, the $y$ engine causes a change in the $y$ component of the velocity. *It is important to realize that the x part of the motion occurs exactly as it would if the y part did not occur at all. Similarly, the y part of the motion occurs exactly as it would if the x part of the motion did not exist.* In other words, the $x$ and $y$ motions are independent of each other.

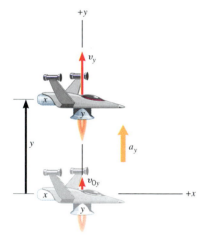

**Figure 3.4** The spacecraft is moving with a constant acceleration $a_y$ parallel to the $y$ axis. There is no motion in the $x$ direction, and the $x$ engine is turned off.

**Table 3.1   *Equations of Kinematics for Constant Acceleration in Two-Dimensional Motion***

| $x$ Component | | Variable | $y$ Component | |
|---|---|---|---|---|
| $x$ | | Displacement | $y$ | |
| $a_x$ | | Acceleration | $a_y$ | |
| $v_x$ | | Final velocity | $v_y$ | |
| $v_{0x}$ | | Initial velocity | $v_{0y}$ | |
| $t$ | | Elapsed time | $t$ | |
| $v_x = v_{0x} + a_x t$ | (3.3a) | | $v_y = v_{0y} + a_y t$ | (3.3b) |
| $x = \frac{1}{2}(v_{0x} + v_x)t$ | (3.4a) | | $y = \frac{1}{2}(v_{0y} + v_y)t$ | (3.4b) |
| $x = v_{0x}t + \frac{1}{2}a_x t^2$ | (3.5a) | | $y = v_{0y}t + \frac{1}{2}a_y t^2$ | (3.5b) |
| $v_x^2 = v_{0x}^2 + 2a_x x$ | (3.6a) | | $v_y^2 = v_{0y}^2 + 2a_y y$ | (3.6b) |

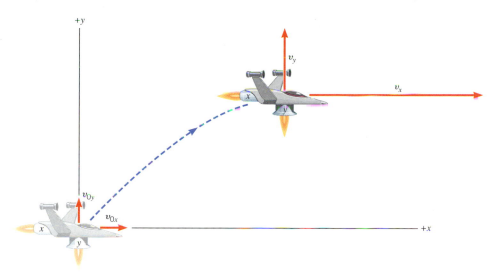

**Figure 3.5** The two-dimensional motion of the spacecraft can be viewed as the combination of the separate $x$ and $y$ motions.

## Example 1  A Moving Spacecraft

In the $x$ direction, the spacecraft in Figure 3.5 has an initial velocity component of $v_{0x} = +22$ m/s and an acceleration component of $a_x = +24$ m/s$^2$. In the $y$ direction, the analogous quantities are $v_{0y} = +14$ m/s and $a_y = +12$ m/s$^2$. The directions to the right and upward have been chosen as the positive directions. Find (a) $x$ and $v_x$, (b) $y$ and $v_y$, and (c) the final velocity (magnitude and direction) of the spacecraft at time $t = 7.0$ s.

**Reasoning** The motion in the $x$ direction and the motion in the $y$ direction can be treated separately, each as a one-dimensional motion. We will follow this approach in parts (a) and (b) to obtain the location and velocity components of the spacecraft. Then, in part (c) the velocity components will be combined to give the final velocity.

**Solution** (a) The data for the motion in the $x$ direction are listed below.

| | *x*-Direction Data | | | |
|---|---|---|---|---|
| $x$ | $a_x$ | $v_x$ | $v_{0x}$ | $t$ |
| ? | +24 m/s$^2$ | ? | +22 m/s | 7.0 s |

The $x$ component of the craft's displacement can be found by using Equation 3.5a.

$$x = v_{0x}t + \tfrac{1}{2}a_x t^2 = (22 \text{ m/s})(7.0 \text{ s}) + \tfrac{1}{2}(24 \text{ m/s}^2)(7.0 \text{ s})^2 = \boxed{+740 \text{ m}}$$

The velocity component $v_x$ can be calculated with the aid of Equation 3.3a:

$$v_x = v_{0x} + a_x t = (22 \text{ m/s}) + (24 \text{ m/s}^2)(7.0 \text{ s}) = \boxed{+190 \text{ m/s}}$$

**(b)** The data for the motion in the $y$ direction are listed below.

| | *y*-Direction Data | | | |
|---|---|---|---|---|
| $y$ | $a_y$ | $v_y$ | $v_{0y}$ | $t$ |
| ? | +12 m/s$^2$ | ? | +14 m/s | 7.0 s |

Proceeding in the same manner as in part (a), we find that

$$\boxed{y = +390 \text{ m}} \quad \text{and} \quad \boxed{v_y = +98 \text{ m/s}}$$

**(c)** Figure 3.6 shows the velocity of the vehicle and its components $v_x$ and $v_y$. The magnitude $v$ of the velocity can be found by using the Pythagorean theorem:

$$v = \sqrt{v_x{}^2 + v_y{}^2} = \sqrt{(190 \text{ m/s})^2 + (98 \text{ m/s})^2} = \boxed{210 \text{ m/s}}$$

*Problem solving insight*
When the motion is two-dimensional, the time variable $t$ has the same value for both the $x$ and $y$ directions.

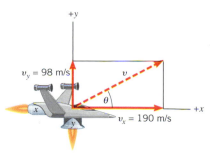

**Figure 3.6** The magnitude of the velocity vector gives the speed of the spacecraft, and the angle $\theta$ gives the direction of travel relative to the positive $x$ direction.

The direction of the velocity vector is given by the angle $\theta$ in the drawing:

$$\tan \theta = \frac{v_y}{v_x} \quad \text{or} \quad \theta = \tan^{-1}\left(\frac{v_y}{v_x}\right) = \tan^{-1}\left(\frac{98 \text{ m/s}}{190 \text{ m/s}}\right) = \boxed{27°}$$

After 7.0 s, the spacecraft has a velocity of 210 m/s in a direction of 27° above the positive $x$ axis. The craft is 740 m to the right and 390 m above the origin, as in Figure 3.5.

## 3.3 *Projectile Motion*

The biggest thrill in baseball is a home run. The motion of the ball on its curving path into the stands is a common type of two-dimensional motion called "projectile motion." A good description of such motion can often be obtained with the assumption that air resistance is absent.

### Example 2    A Falling Care Package

Figure 3.7 shows an airplane moving horizontally with a constant velocity of +115 m/s at an altitude of 1050 m. The directions to the right and upward have been chosen as the positive directions. The plane releases a "care package" that falls to the ground along a curved trajectory. Ignoring air resistance, determine the time required for the package to hit the ground.

**Reasoning** The time required for the package to hit the ground is the time it takes for the package to fall through a vertical distance of 1050 m. In falling, it moves to the right, as well as downward, but these two parts of the motion occur independently. Therefore, we can focus solely on the vertical part. We note that the package is moving initially in the horizontal or $x$ direction, not in the $y$ direction, so that $v_{0y} = 0$ m/s. Furthermore, when the package hits the ground, the $y$ component of its displacement is $y = -1050$ m, as the drawing shows. The acceleration is that due to gravity, so $a_y = -9.80$ m/s$^2$. These data are summarized as follows:

| y-Direction Data |||||
| --- | --- | --- | --- | --- |
| $y$ | $a_y$ | $v_y$ | $v_{0y}$ | $t$ |
| $-1050$ m | $-9.80$ m/s$^2$ | | 0 m/s | ? |

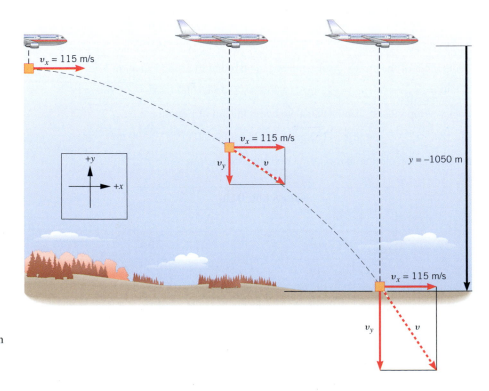

**Figure 3.7** The package falling from the plane is an example of projectile motion, as Examples 2 and 3 discuss.

With these data, Equation 3.5b ($y = v_{0y}t + \frac{1}{2}a_yt^2$) can be used to find the fall time.

**Solution** Since $v_{0y} = 0$ m/s, it follows from Equation 3.5b that $y = \frac{1}{2}a_yt^2$ and

$$t = \sqrt{\frac{2y}{a_y}} = \sqrt{\frac{2(-1050 \text{ m})}{-9.80 \text{ m/s}^2}} = \boxed{14.6 \text{ s}}$$

**Problem solving insight**
The variables $y$, $a_y$, $v_y$, and $v_{0y}$ are scalar components. Therefore, an algebraic sign (+ or −) must be included with each one to denote direction.

The freely falling package in Example 2 picks up vertical speed on the way down. The horizontal component of the velocity, however, retains its initial value of $v_{0x} = +115$ m/s throughout the entire descent. Since the plane also travels at a constant horizontal velocity of $+115$ m/s, it remains directly above the falling package. The pilot always sees the package directly beneath the plane, as the dashed vertical lines in Figure 3.7 show. This result is a direct consequence of the fact that the package has no acceleration in the horizontal direction. In reality, air resistance would slow down the package, and it would not remain directly beneath the plane during the descent. Figure 3.8 further clarifies this point by illustrating what happens to two packages that are released simultaneously from the same height. Package B is given an initial velocity component of $v_{0x} = +115$ m/s in the horizontal direction, as in Example 2, and the package follows the path shown in the figure. Package A, on the other hand, is dropped from a stationary balloon and falls straight down toward the ground, since $v_{0x} = 0$ m/s. Both packages hit the ground at the same time.

Not only do the packages in Figure 3.8 reach the ground at the same time, but the $y$ components of their velocities are also equal at all points on the way down. However, package B does hit the ground with a greater speed than does package A. Remember, speed is the magnitude of the velocity vector, and the velocity of B has an $x$ component, whereas the velocity of A does not. The magnitude and direction of the velocity vector for package B at the instant just before the package hits the ground is computed in Example 3.

### Example 3   The Velocity of the Care Package

For the situation shown in Figure 3.8, find the speed of package B and the direction of the velocity vector just before package B hits the ground.

**Reasoning** Since the speed $v$ of the package is given by $v = \sqrt{v_x^2 + v_y^2}$, it is necessary to know values for $v_x$ and $v_y$ at the instant before impact. The component $v_x$ is constant and has a value of 115 m/s. The component $v_y$ can be determined by using Equation 3.3b and the data from Example 2 ($a_y = -9.80$ m/s$^2$, $v_{0y} = 0$ m/s, $t = 14.6$ s). Thus, we expect that the final speed $v$ of package B will be greater than 115 m/s.

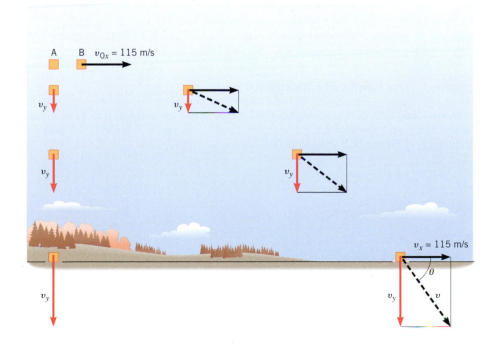

**Figure 3.8** Package A and package B are released simultaneously at the same height and strike the ground at the same time because their $y$ variables ($y$, $a_y$, and $v_{0y}$) are the same.

**Solution** From Equation 3.3b, it follows that

$$v_y = v_{0y} + a_y t = 0 \text{ m/s} + (-9.80 \text{ m/s}^2)(14.6 \text{ s}) = -143 \text{ m/s}$$

The speed of package B at the instant before impact is

$$v = \sqrt{(115 \text{ m/s})^2 + (-143 \text{ m/s})^2} = \boxed{184 \text{ m/s}}$$

which is greater than 115 m/s, as expected. The velocity vector makes an angle $\theta$ with the horizontal, as Figure 3.8 indicates:

$$\cos \theta = \frac{v_x}{v} \quad \text{or} \quad \theta = \cos^{-1}\left(\frac{v_x}{v}\right) = \cos^{-1}\left(\frac{115 \text{ m/s}}{184 \text{ m/s}}\right) = \boxed{51.3°}$$

An important feature of projectile motion is that there is no acceleration in the horizontal or $x$ direction. Conceptual Example 4 discusses an interesting implication of this feature.

### Conceptual Example 4   I Shot a Bullet into the Air. . . .

Suppose you are driving in a convertible with the top down. The car is moving to the right at a constant velocity. As Figure 3.9 illustrates, you point a rifle straight upward and fire it. In the absence of air resistance, where would the bullet land—behind you, ahead of you, or in the barrel of the rifle?

**Reasoning and Solution** If air resistance were present, it would slow down the bullet and cause it to land behind you, toward the rear of the car. However, air resistance is absent, so we must consider the bullet's motion more carefully. Before the rifle is fired, the bullet, rifle, and car are moving together, so the bullet and rifle have the same horizontal velocity as the car. When the rifle is fired, the bullet is given an additional velocity component in the vertical direction; the bullet retains the velocity of the car as its initial horizontal velocity component, since the rifle is pointed straight up. Because there is no air resistance to slow it down, the bullet experiences no horizontal acceleration. Thus, the bullet's horizontal velocity component does not change. It retains its initial value, and remains matched to that of the rifle and the car. As a result, *the bullet remains directly above the rifle at all times and would fall directly back into the barrel of the rifle,* as the drawing indicates. This situation is analogous to that in Figure 3.7, where the care package, as it falls, remains directly below the plane.

**Related Homework:** *Problem 34*

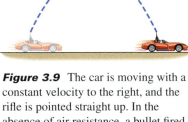

**Figure 3.9** The car is moving with a constant velocity to the right, and the rifle is pointed straight up. In the absence of air resistance, a bullet fired from the rifle has no acceleration in the horizontal direction. As a result, the bullet would land back in the barrel of the rifle.

Often projectiles, like footballs and baseballs, are sent into the air at an angle with respect to the ground. From a knowledge of the projectile's initial velocity, a wealth of information can be obtained about the motion. For instance, Example 5 demonstrates how to calculate the maximum height reached by the projectile.

### Example 5   The Height of a Kickoff

A placekicker kicks a football at an angle of $\theta = 40.0°$ above the horizontal axis, as Figure 3.10 shows. The initial speed of the ball is $v_0 = 22$ m/s. Ignore air resistance and find the maximum height $H$ that the ball attains.

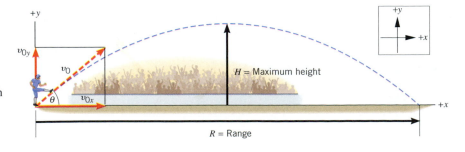

**Figure 3.10** A football is kicked with an initial speed of $v_0$ at an angle of $\theta$ above the ground. The ball attains a maximum height $H$ and a range $R$.

**Reasoning** The maximum height is a characteristic of the vertical part of the motion, which can be treated separately from the horizontal part. In preparation for making use of this fact, we calculate the vertical component of the initial velocity:

$$v_{0y} = v_0 \sin\theta = +(22 \text{ m/s}) \sin 40.0° = +14 \text{ m/s}$$

The vertical component of the velocity, $v_y$, decreases as the ball moves upward, Eventually, $v_y = 0$ m/s at the maximum height $H$. The data below can be used in Equation 3.6b ($v_y{}^2 = v_{0y}{}^2 + 2a_y y$) to find the maximum height:

<table>
<tr><th colspan="5">y-Direction Data</th></tr>
<tr><th>$y$</th><th>$a_y$</th><th>$v_y$</th><th>$v_{0y}$</th><th>$t$</th></tr>
<tr><td>$H = ?$</td><td>$-9.80$ m/s$^2$</td><td>$0$ m/s</td><td>$+14$ m/s</td><td></td></tr>
</table>

**Problem solving insight**
When a projectile reaches maximum height, the vertical component of its velocity is momentarily zero ($v_y = 0$ m/s). However, the horizontal component of its velocity is not zero.

**Solution** From Equation 3.6b, we find that

$$y = H = \frac{v_y{}^2 - v_{0y}{}^2}{2a_y} = \frac{(0 \text{ m/s})^2 - (14 \text{ m/s})^2}{2(-9.80 \text{ m/s}^2)} = \boxed{+10 \text{ m}}$$

The height $H$ depends only on the $y$ variables; the same height would have been reached had the ball been thrown *straight up* with an initial velocity of $v_{0y} = +14$ m/s.

It is also possible to find the total time or "hang time" during which the football in Figure 3.10 is in the air. Example 6 shows how to determine this time.

### Example 6   The Time of Flight of a Kickoff

For the motion illustrated in Figure 3.10, ignore air resistance and use the data from Example 5 to determine the time of flight between kickoff and landing.

*The physics of the "hang time" of a football.*

**Reasoning** Given the initial velocity, it is the acceleration due to gravity that determines how long the ball stays in the air. Thus, to find the time of flight we deal with the vertical part of the motion. Since the ball starts at and returns to ground level, the displacement in the $y$ direction is zero. The initial velocity component in the $y$ direction is the same as that in Example 5; that is, $v_{0y} = +14$ m/s. Therefore, we have

<table>
<tr><th colspan="5">y-Direction Data</th></tr>
<tr><th>$y$</th><th>$a_y$</th><th>$v_y$</th><th>$v_{0y}$</th><th>$t$</th></tr>
<tr><td>$0$ m</td><td>$-9.80$ m/s$^2$</td><td></td><td>$+14$ m/s</td><td>?</td></tr>
</table>

The time of flight can be determined from Equation 3.5b ($y = v_{0y}t + \frac{1}{2}a_y t^2$).

**Solution** Using Equation 3.5b, we find

$$0 \text{ m} = (14 \text{ m/s})t + \tfrac{1}{2}(-9.80 \text{ m/s}^2)t^2 = [(14 \text{ m/s}) + \tfrac{1}{2}(-9.80 \text{ m/s}^2)t]t$$

There are two solutions to this equation. One is given by

$$(14 \text{ m/s}) + \tfrac{1}{2}(-9.80 \text{ m/s}^2)t = 0 \quad \text{or} \quad t = 2.9 \text{ s}$$

The other is given by $t = 0$ s. The solution we seek is $\boxed{t = 2.9 \text{ s}}$, because $t = 0$ s corresponds to the initial kickoff.

Another important feature of projectile motion is called the "range." The range, as Figure 3.10 shows, is the horizontal distance traveled between launching and landing, assuming the projectile returns to the *same vertical level* at which it was fired. Example 7 shows how to obtain the range.

### Example 7   The Range of a Kickoff

For the motion shown in Figure 3.10 and discussed in Examples 5 and 6, ignore air resistance and calculate the range $R$ of the projectile.

**Reasoning** The range is a characteristic of the horizontal part of the motion. Thus, our starting point is to determine the horizontal component of the initial velocity:

$$v_{0x} = v_0 \cos \theta = +(22 \text{ m/s}) \cos 40.0° = +17 \text{ m/s}$$

Recall from Example 6 that the time of flight is $t = 2.9$ s. Since there is no acceleration in the $x$ direction, $v_x$ remains constant, and the range is simply the product of $v_x = v_{0x}$ and the time.

**Solution** The range is

$$x = R = v_{0x}t = +(17 \text{ m/s})(2.9 \text{ s}) = \boxed{+49 \text{ m}}$$

The range in the previous example depends on the angle $\theta$ at which the projectile is fired above the horizontal. When air resistance is absent, the maximum range results when $\theta = 45°$.

The examples considered thus far have used information about the initial location and velocity of a projectile to determine the final location and velocity. Example 8 deals with the opposite situation and illustrates how the final parameters can be used with the equations of kinematics to determine the initial parameters.

### Example 8   A Home Run

A baseball player hits a home run, and the ball lands in the left-field seats, 7.5 m above the point at which it was hit. It lands with a velocity of 36 m/s at an angle of 28° below the horizontal (see Figure 3.11). Ignoring air resistance, find the initial velocity with which the ball leaves the bat.

**Reasoning** To find the initial velocity, we must determine its magnitude (the initial speed $v_0$) and its direction (the angle $\theta$ in the drawing). These quantities are related to the horizontal and vertical components of the initial velocity ($v_{0x}$ and $v_{0y}$) by the relations

$$v_0 = \sqrt{v_{0x}^2 + v_{0y}^2} \quad \text{and} \quad \theta = \tan^{-1}\left(\frac{v_{0y}}{v_{0x}}\right)$$

Therefore, it is necessary to find $v_{0x}$ and $v_{0y}$, which we will do with the equations of kinematics.

**Solution** Since air resistance is being ignored, the horizontal component of the velocity $v_x$ remains constant throughout the motion. Thus,

$$v_{0x} = v_x = +(36 \text{ m/s}) \cos 28° = +32 \text{ m/s}$$

The value for $v_{0y}$ can be obtained from Equation 3.6b and the data displayed below (see Figure 3.11 for the positive and negative directions):

| | | y-Direction Data | | |
|---|---|---|---|---|
| $y$ | $a_y$ | $v_y$ | $v_{0y}$ | $t$ |
| +7.5 m | −9.80 m/s² | (−36 sin 28°) m/s | ? | |

$$v_y^2 = v_{0y}^2 + 2a_y y \quad \text{or} \quad v_{0y} = +\sqrt{v_y^2 - 2a_y y}$$

$$v_{0y} = +\sqrt{[(-36 \sin 28°) \text{ m/s}]^2 - 2(-9.80 \text{ m/s}^2)(7.5 \text{ m})} = +21 \text{ m/s}$$

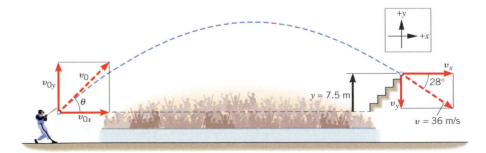

**Figure 3.11** The velocity and location of the baseball upon landing can be used to determine its initial velocity, as Example 8 illustrates.

In determining $v_{0y}$ we choose the plus sign for the square root, because the vertical component of the initial velocity points upward in Figure 3.11, which is the positive direction. The initial speed $v_0$ and angle $\theta$ of the baseball are

$$v_0 = \sqrt{v_{0x}^2 + v_{0y}^2} = \sqrt{(32 \text{ m/s})^2 + (21 \text{ m/s})^2} = \boxed{38 \text{ m/s}}$$

$$\theta = \tan^{-1}\left(\frac{v_{0y}}{v_{0x}}\right) = \tan^{-1}\left(\frac{21 \text{ m/s}}{32 \text{ m/s}}\right) = \boxed{33°}$$

In projectile motion, the magnitude of the acceleration due to gravity affects the trajectory in a significant way. For example, a baseball or a golf ball would travel much farther and higher on the moon than on the earth, when launched with the same initial velocity. The reason is that the moon's gravity is only about one-sixth as strong as the earth's.

Section 2.6 points out that certain types of symmetry with respect to time and speed are present for freely falling bodies. These symmetries are also found in projectile motion, since projectiles are falling freely in the vertical direction. In particular, the time required for a projectile to reach its maximum height $H$ is equal to the time spent returning to the ground. In addition, Figure 3.12 shows that the speed $v$ of the object at any height above the ground on the upward part of the trajectory is equal to the speed $v$ at the same height on the downward part. Although the two speeds are the same, the velocities are different, because they point in different directions. Conceptual Example 9 shows how to use this type of symmetry in your reasoning.

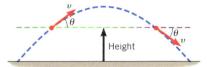

**Figure 3.12** The speed $v$ of a projectile at a given height above the ground is the same on the upward and downward parts of the trajectory. The velocities are different, however, since they point in different directions.

## Conceptual Example 9   Two Ways to Throw a Stone

From the top of a cliff overlooking a lake, a person throws two stones. The stones have identical initial speeds $v_0$, but stone 1 is thrown downward at an angle $\theta$ below the horizontal, while stone 2 is thrown upward at the same angle above the horizontal, as Figure 3.13 shows. Neglect air resistance and decide which stone, if either, strikes the water with the greater velocity.

**Reasoning and Solution** We might guess that stone 1, being hurled downward, would strike the water with the greater velocity. To show that this is not true, let's follow the path of stone 2 as it rises to its maximum height and falls back to earth. Notice point $P$ in the drawing, where stone 2 returns to its initial height; here the speed of stone 2 is $v_0$, but its velocity is directed at an angle $\theta$ below the horizontal. This is exactly the type of projectile symmetry illustrated in Figure 3.12. At this point, then, stone 2 has a velocity that is identical to the velocity with which stone 1 is thrown downward from the top of the cliff. From this point on, the velocity of stone 2 changes in exactly the same way as that for stone 1, so *both stones strike the water with the same velocity.*

**Related Homework:** *Problem 37*

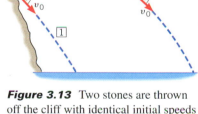

**Figure 3.13** Two stones are thrown off the cliff with identical initial speeds $v_0$, but at equal angles $\theta$ that are below and above the horizontal. Conceptual Example 9 compares the velocities with which the stones hit the water below.

In all the examples in this section, the projectiles follow a curved trajectory. In general, if the only acceleration is that due to gravity, the shape of the path can be shown to be a *parabola*.

# Concept Summary

This summary presents an abridged version of the chapter, including the important equations and all available learning aids. For convenient reference, the learning aids (including the text's examples) are placed next to or immediately after the relevant equation or discussion. The following learning aids may be found on-line at **www.wiley.com/college/cutnell**:

| | |
|---|---|
| **Interactive LearningWare** examples are solved according to a five-step interactive format that is designed to help you develop problem-solving skills. | **Concept Simulations** are animated versions of text figures or animations that illustrate important concepts. You can control parameters that affect the display, and we encourage you to experiment. |
| **Interactive Solutions** offer specific models for certain types of problems in the chapter homework. The calculations are carried out interactively. | **Self-Assessment Tests** include both qualitative and quantitative questions. Extensive feedback is provided for both incorrect and correct answers, to help you evaluate your understanding of the material. |

| Topic | Discussion | Learning Aids |
|---|---|---|

### 3.1 Displacement, Velocity, and Acceleration

**Displacement vector**

The position of an object is located with a vector $\mathbf{r}$ drawn from the coordinate origin to the object. The displacement $\Delta\mathbf{r}$ of the object is defined as $\Delta\mathbf{r} = \mathbf{r} - \mathbf{r_0}$, where $\mathbf{r}$ and $\mathbf{r_0}$ specify its final and initial positions, respectively.

**Average velocity**

The average velocity $\bar{\mathbf{v}}$ of an object moving between two positions is defined as its displacement $\Delta\mathbf{r} = \mathbf{r} - \mathbf{r_0}$ divided by the elapsed time $\Delta t = t - t_0$:   **Interactive Solution 3.11**

$$\bar{\mathbf{v}} = \frac{\mathbf{r} - \mathbf{r_0}}{t - t_0} = \frac{\Delta\mathbf{r}}{\Delta t} \qquad (3.1)$$

**Instantaneous velocity**

The instantaneous velocity $\mathbf{v}$ is the velocity at an instant of time. The average velocity becomes equal to the instantaneous velocity in the limit that the elapsed time $\Delta t$ becomes infinitesimally small ($\Delta t \rightarrow 0$ s):

$$\mathbf{v} = \lim_{\Delta t \to 0} \frac{\Delta\mathbf{r}}{\Delta t}$$

**Average acceleration**

The average acceleration $\bar{\mathbf{a}}$ of an object is the change in its velocity $\Delta\mathbf{v} = \mathbf{v} - \mathbf{v_0}$ divided by the elapsed time $\Delta t = t - t_0$:   **Interactive LearningWare 3.1**

$$\bar{\mathbf{a}} = \frac{\mathbf{v} - \mathbf{v_0}}{t - t_0} = \frac{\Delta\mathbf{v}}{\Delta t} \qquad (3.2)$$

**Instantaneous acceleration**

The instantaneous acceleration $\mathbf{a}$ is the acceleration at an instant of time. The average acceleration becomes equal to the instantaneous acceleration in the limit that the elapsed time $\Delta t$ becomes infinitesimally small:

$$\mathbf{a} = \lim_{\Delta t \to 0} \frac{\Delta\mathbf{v}}{\Delta t}$$

### 3.2 Equations of Kinematics in Two Dimensions

Motion in two dimensions can be described in terms of the time $t$ and the $x$ and $y$ components of four vectors: the displacement, the acceleration, and the initial and final velocities.   **Example 1**

**Independence of the x and y parts of the motion**

The $x$ part of the motion occurs exactly as it would if the $y$ part did not occur at all. Similarly, the $y$ part of the motion occurs exactly as it would if the $x$ part of the motion did not exist. The motion can be analyzed by treating the $x$ and $y$ components of the four vectors separately and realizing that the time $t$ is the same for each component.

**Equations of kinematics for constant acceleration**

When the acceleration is constant, the $x$ components of the displacement, the acceleration, and the initial and final velocities are related by the equations of kinematics, and so are the $y$ components:

| $x$ Component | | $y$ Component | |
|---|---|---|---|
| $v_x = v_{0x} + a_x t$ | (3.3a) | $v_y = v_{0y} + a_y t$ | (3.3b) |
| $x = \frac{1}{2}(v_{0x} + v_x)t$ | (3.4a) | $y = \frac{1}{2}(v_{0y} + v_y)t$ | (3.4b) |
| $x = v_{0x}t + \frac{1}{2}a_x t^2$ | (3.5a) | $y = v_{0y}t + \frac{1}{2}a_y t^2$ | (3.5b) |
| $v_x^2 = v_{0x}^2 + 2a_x x$ | (3.6a) | $v_y^2 = v_{0y}^2 + 2a_y y$ | (3.6b) |

The directions of these components are conveyed by assigning a plus ($+$) or minus ($-$) sign to each one.

### 3.3 Projectile Motion

**Acceleration in projectile motion**

Projectile motion is an idealized kind of motion that occurs when a moving object (the projectile) experiences only the acceleration due to gravity, which acts vertically downward. If the trajectory of the projectile is near the earth's surface, $a_y$ has a magnitude of 9.80 m/s$^2$. The acceleration has no horizontal component ($a_x = 0$ m/s$^2$), the effects of air resistance being negligible.   **Examples 2–8**  **Concept Simulations 3.1, 3.2**  **Interactive LearningWare 3.2**  **Interactive Solutions 3.35, 3.39, 3.67**

**Symmetries in projectile motion**

There are several symmetries in projectile motion: (1) The time to reach maximum height from any point is equal to the time spent returning from the maxi-   **Example 9**

| Topic | Discussion | Learning Aids |
|-------|-----------|---------------|

mum height to that point. (2) The speed of a projectile depends only on its height above its launch point, and not on whether it is moving upward or downward.

 Use *Self-Assessment Test 3.1* to evaluate your understanding of Sections 3.1–3.3.

# Problems

ssm Solution is in the Student Solutions Manual.   www Solution is available on the World Wide Web at www.wiley.com/college/cutnell
⚕ This icon represents a biomedical application.

## Section 3.1 Displacement, Velocity, and Acceleration

**1.** ssm In diving to a depth of 750 m, an elephant seal also moves 460 m due east of his starting point. What is the magnitude of the seal's displacement?

**2.** A baseball player hits a triple and ends up on third base. A baseball "diamond" is a square, each side of length 27.4 m, with home plate and the three bases on the four corners. What is the magnitude of his displacement?

**3.** A mountain-climbing expedition establishes two intermediate camps, labeled A and B in the drawing, above the base camp. What is the magnitude $\Delta r$ of the displacement between camp A and camp B?

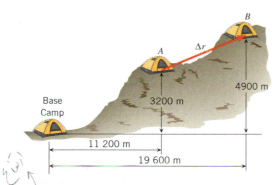

**4.** The altitude of a hang glider is increasing at a rate of 6.80 m/s. At the same time, the shadow of the glider moves along the ground at a speed of 15.5 m/s when the sun is directly overhead. Find the magnitude of the glider's velocity.

**5.** ssm A jetliner is moving at a speed of 245 m/s. The vertical component of the plane's velocity is 40.6 m/s. Determine the magnitude of the horizontal component of the plane's velocity.

**6.** In a football game a kicker attempts a field goal. The ball remains in contact with the kicker's foot for 0.050 s, during which time it experiences an acceleration of 340 m/s². The ball is launched at an angle of 51° above the ground. Determine the horizontal and vertical components of the launch velocity.

**7.** A dolphin leaps out of the water at an angle of 35° above the horizontal. The horizontal component of the dolphin's velocity is 7.7 m/s. Find the magnitude of the vertical component of the velocity.

**8.** A skateboarder, starting from rest, rolls down a 12.0-m ramp. When she arrives at the bottom of the ramp her speed is 7.70 m/s. (a) Determine the magnitude of her acceleration, assumed to be constant. (b) If the ramp is inclined at 25.0° with respect to the ground, what is the component of her acceleration that is parallel to the ground?

**9.** ssm www In a mall, a shopper rides up an escalator between floors. At the top of the escalator, the shopper turns right and walks 9.00 m to a store. The magnitude of the shopper's displacement from the bottom of the escalator is 16.0 m. The vertical distance between the floors is 6.00 m. At what angle is the escalator inclined above the horizontal?

**\* 10.** Interactive LearningWare 3.1 at www.wiley.com/college/cutnell reviews the approach taken in problems such as this one. A bird watcher meanders through the woods, walking 0.50 km due east, 0.75 km due south, and 2.15 km in a direction 35.0° north of west. The time required for this trip is 2.50 h. Determine the magnitude and direction (relative to due west) of the bird watcher's (a) displacement and (b) average velocity. Use kilometers and hours for distance and time, respectively.

**\* 11.** Interactive Solution 3.11 at www.wiley.com/college/cutnell presents a model for solving this problem. The earth moves around the sun in a nearly circular orbit of radius $1.50 \times 10^{11}$ m. During the three summer months (an elapsed time of $7.89 \times 10^6$ s), the earth moves one-fourth of the distance around the sun. (a) What is the average speed of the earth? (b) What is the magnitude of the average velocity of the earth during this period?

## Section 3.2 Equations of Kinematics in Two Dimensions,
## Section 3.3 Projectile Motion

**12.** A spacecraft is traveling with a velocity of $v_{0x} = 5480$ m/s along the +x direction. Two engines are turned on for a time of 842 s. One engine gives the spacecraft an acceleration in the +x direction of $a_x = 1.20$ m/s², while the other gives it an acceleration in the +y direction of $a_y = 8.40$ m/s². At the end of the firing, find (a) $v_x$ and (b) $v_y$.

**13.** A tennis ball is struck such that it leaves the racket horizontally with a speed of 28.0 m/s. The ball hits the court at a horizontal distance of 19.6 m from the racket. What is the height of the tennis ball when it leaves the racket?

**14.** A volleyball is spiked so that it has an initial velocity of 15 m/s directed downward at an angle of 55° below the horizontal. What is the horizontal component of the ball's velocity when the opposing player fields the ball?

**15.** ssm A golf ball rolls off a horizontal cliff with an initial speed of 11.4 m/s. The ball falls a vertical distance of 15.5 m into a lake below. (a) How much time does the ball spend in the air? (b) What is the speed $v$ of the ball just before it strikes the water?

**16.** A rock climber throws a small first aid kit to another climber who is higher up the mountain. The initial velocity of the kit is 11 m/s at an angle of 65° above the horizontal. At the instant when the kit is caught, it is traveling horizontally, so its vertical speed is zero. What is the vertical height between the two climbers?

$V = \sqrt{(14)^2 + (10.2)^2} = 14.1$

$Vx = 1.20 \quad Y = 10.0m \Rightarrow V_y = V_{0y} + 2a x = 2(-9.8)(-10) = 14$

**17.** **ssm** **www** A diver runs horizontally with a speed of 1.20 m/s off a platform that is 10.0 m above the water. What is his speed just before striking the water?   $14.1 \, m/s$

**18.** **Concept Simulation 3.2** at **www.wiley.com/college/cutnell** reviews the concepts that are important in this problem. A golfer imparts a speed of 30.3 m/s to a ball, and it travels the maximum possible distance before landing on the green. The tee and the green are at the same elevation. (a) How much time does the ball spend in the air? (b) What is the longest "hole in one" that the golfer can make, if the ball does not roll when it hits the green?

**19.** Review **Interactive LearningWare 3.2** at **www.wiley.com/college/cutnell** in preparation for this problem. The acceleration due to gravity on the moon has a magnitude of 1.62 m/s². Examples 5–7 deal with a placekicker kicking a football. Assume that the ball is kicked on the moon instead of on the earth. Find (a) the maximum height *H* and (b) the range that the ball would attain on the moon.

**20.** Michael Jordan, formerly of the Chicago Bulls basketball team, has some fanatic fans. They claim that he is able to jump and remain in the air for two full seconds from launch to landing. Evaluate this claim by calculating the maximum height that such a jump would attain. For comparison, Jordan's maximum jump height has been estimated at about one meter.

$0 = V_0 \sin 35 + (-9.8)t \Rightarrow \frac{-V_0 \sin 35}{-9.8} \Rightarrow x = V_{0x} t \Rightarrow (V_0 \cos 35 s) t$

$V_{0y} = V_0$
$V_{0z} = 25$
$V_y = 0$
$t = ?$

$t = \sqrt{\frac{2(54)}{9.8}}$

**21.** **ssm** A fire hose ejects a stream of water at an angle of 35.0° above the horizontal. The water leaves the nozzle with a speed of 25.0 m/s. Assuming that the water behaves like a projectile, how far from a building should the fire hose be located to hit the highest possible fire?

**22.** A car drives straight off the edge of a cliff that is 54 m high. The police at the scene of the accident note that the point of impact is 130 m from the base of the cliff. How fast was the car traveling when it went over the cliff?   $t = 3.3 \quad , \quad V = 39.4$

$t = 3.3$
$V = \frac{130}{3.3} = 39.4$

**23.** **Interactive LearningWare 3.2** at **www.wiley.com/college/cutnell** provides a review of the concepts that are important in this problem. On a distant planet, golf is just as popular as it is on earth. A golfer tees off and drives the ball 3.5 times as far as he would have on earth, given the same initial velocities on both planets. The ball is launched at a speed of 45 m/s at an angle of 29° above the horizontal. When the ball lands, it is at the same level as the tee. On the distant planet, what is (a) the maximum height and (b) the range of the ball?

**24.** The 1994 Winter Olympics included the aerials competition in skiing. In this event skiers speed down a ramp that slopes sharply upward at the end. The sharp upward slope launches them into the air, where they perform acrobatic maneuvers. In the women's competition, the end of a typical launch ramp is directed 63° above the horizontal. With this launch angle, a skier attains a height of 13 m above the end of the ramp. What is the skier's launch speed?

**25.** **ssm** An eagle is flying horizontally at 6.0 m/s with a fish in its claws. It accidentally drops the fish. (a) How much time passes before the fish's speed doubles? (b) How much additional time would be required for the fish's speed to double again?

**26.** A major-league pitcher can throw a baseball in excess of 41.0 m/s. If a ball is thrown horizontally at this speed, how much will it drop by the time it reaches a catcher who is 17.0 m away from the point of release?

$a_x = 0$
$a_y = 9.8$
$V_{0x} = 41$
$V_{0y} = 0$
$t = ?$

$t = \frac{x}{V_x} = .415 \quad Y = \frac{1}{2}(-9.8)(.415)^2 = -.84$

**27.** A motorcycle daredevil is attempting to jump across as many buses as possible (see the drawing at the top of the right column). The takeoff ramp makes an angle of 18.0° above the horizontal, and the landing ramp is identical to the takeoff ramp. The buses are parked side by side, and each bus is 2.74 m wide. The cyclist leaves the ramp with a speed of 33.5 m/s. What is the maximum number of buses over which the cyclist can jump?   $V_0 = 33.5$

$Y = \sin 18 (33.5) = 10.4$

$R = \frac{V_0^2 \sin 2\theta}{g} = \frac{(33.5)^2 (.588)}{9.8} = 67.335 / 2.74 = 24$

33.5 m/s

18.0°        2.74 m        18.0°

$V_{0x} = 2.7$
$75 \sqrt{V_y} \quad \tan \frac{y}{x} \quad$ Same 21

**28.** Suppose the water at the top of Niagara Falls has a horizontal speed of 2.7 m/s just before it cascades over the edge of the falls. At what vertical distance below the edge does the velocity vector of the water point downward at a 75° angle below the horizontal?

**29.** **ssm** A hot-air balloon is rising straight up with a speed of 3.0 m/s. A ballast bag is released from rest relative to the balloon when it is 9.5 m above the ground. How much time elapses before the ballast bag hits the ground?   $V_{0b} = 0 \quad -9.5 = 3t + \frac{1}{2}(-9.8)t^2$
$-4.9t^2 + 3t + 9.5 \Rightarrow t = 1.73s$

**30.** A rocket is fired at a speed of 75.0 m/s from ground level, at an angle of 60.0° above the horizontal. The rocket is fired toward an 11.0-m-high wall, which is located 27.0 m away. By how much does the rocket clear the top of the wall?

**31.** A horizontal rifle is fired at a bull's-eye. The muzzle speed of the bullet is 670 m/s. The barrel is pointed directly at the center of the bull's-eye, but the bullet strikes the target 0.025 m below the center. What is the horizontal distance between the end of the rifle and the bull's-eye?   $t = .071 \quad d = Vt \Rightarrow 670(.071) = 47.57$

$V_{0y} = 0$
$V_{0x}$
$y = -.025$
$a = 9.8$

**32.** An airplane with a speed of 97.5 m/s is climbing upward at an angle of 50.0° with respect to the horizontal. When the plane's altitude is 732 m, the pilot releases a package. (a) Calculate the distance along the ground, measured from a point directly beneath the point of release, to where the package hits the earth. (b) Relative to the ground, determine the angle of the velocity vector of the package just before impact.   $V_0^2 = \frac{Rg}{\sin 2\theta} = \frac{(8.7)(-9.8)}{\sin(46)}$

**\* 33.** **ssm** An Olympic long jumper leaves the ground at an angle of 23° and travels through the air for a horizontal distance of 8.7 m before landing. What is the takeoff speed of the jumper?   $11 \, m/s$

**\* 34.** Review **Conceptual Example 4** and **Concept Simulation 3.1** at **www.wiley.com/college/cutnell** before beginning this problem. You are traveling in a convertible with the top down. The car is moving at a constant velocity of 25 m/s, due east along flat ground. You throw a tomato straight upward at a speed of 11 m/s. How far has the car moved when you get a chance to catch the tomato?

**\* 35.** As an aid in working this problem, consult **Interactive Solution 3.35** at **www.wiley.com/college/cutnell**. A soccer player kicks the ball toward a goal that is 16.8 m in front of him. The ball leaves his foot at a speed of 16.0 m/s and an angle of 28.0° above the ground. Find the speed of the ball when the goalie catches it in front of the net.

**\* 36.** In the javelin throw at a track-and-field event, the javelin is launched at a speed of 29 m/s at an angle of 36° above the horizontal. As the javelin travels upward, its velocity points above the horizontal at an angle that decreases as time passes. How much time is required for the angle to be reduced from 36° at launch to 18°?

**\* 37.** **ssm** **www** As preparation for this problem, review Conceptual Example 9. The two stones described there have identical initial speeds of $v_0 = 13.0$ m/s and are thrown at an angle $\theta = 30.0°$, one below the horizontal and one above the horizontal. What is the distance *between* the points where the stones strike the ground?

**\* 38.** Stones are thrown horizontally with the same velocity from the tops of two different buildings. One stone lands twice as far from the base of the building from which it was thrown as does the other stone. Find the ratio of the height of the taller building to the height of the shorter building.

* **39.** Before beginning this problem, review **Interactive Solution 3.39** at **www.wiley.com/college/cutnell.** After leaving the end of a ski ramp, a ski jumper lands downhill at a point that is displaced 51.0 m horizontally from the end of the ramp. His velocity, just before landing, is 23.0 m/s and points in a direction 43.0° below the horizontal. Neglecting air resistance and any lift he experiences while airborne, find his initial velocity (magnitude and direction) when he left the end of the ramp. Express the direction as an angle relative to the horizontal.

* **40.** The lob in tennis is an effective tactic when your opponent is near the net. It consists of lofting the ball over his head, forcing him to move quickly away from the net (see the drawing). Suppose that you loft the ball with an initial speed of 15.0 m/s, at an angle of 50.0° above the horizontal. At this instant your opponent is 10.0 m away from the ball. He begins moving away from you 0.30 s later, hoping to reach the ball and hit it back at the moment that it is 2.10 m above its launch point. With what minimum average speed must he move? (Ignore the fact that he can stretch, so that his racket can reach the ball before he does.)

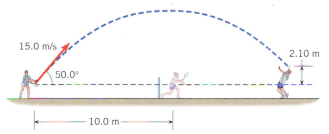

** **41. ssm** The drawing shows an exaggerated view of a rifle that has been "sighted in" for a 91.4-meter target. If the muzzle speed of the bullet is $v_0 = 427$ m/s, what are the two possible angles $\theta_1$ and $\theta_2$ between the rifle barrel and the horizontal such that the bullet will hit the target? One of these angles is so large that it is never used in target shooting. (*Hint: The following trigonometric identity may be useful:* $2 \sin \theta \cos \theta = \sin 2\theta$.)

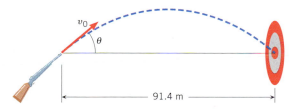

** **42.** A placekicker is about to kick a field goal. The ball is 26.9 m from the goalpost. The ball is kicked with an initial velocity of 19.8 m/s at an angle $\theta$ above the ground. Between what two angles, $\theta_1$ and $\theta_2$, will the ball clear the 2.74-m-high crossbar? (*Hint: The following trigonometric identities may be useful:* sec $\theta = 1/(\cos \theta)$ *and* sec$^2 \theta = 1 + \tan^2 \theta$.)

*** **43. ssm** From the top of a tall building, a gun is fired. The bullet leaves the gun at a speed of 340 m/s, parallel to the ground. As the

drawing shows, the bullet puts a hole in a window of another building and hits the wall that faces the window. Using the data in the drawing, determine the distances $D$ and $H$, which locate the point where the gun was fired. Assume that the bullet does not slow down as it passes through the window.

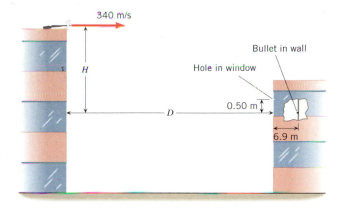

** **44.** A baseball is hit into the air at an initial speed of 36.6 m/s and an angle of 50.0° above the horizontal. At the same time, the center fielder starts running away from the batter and catches the ball 0.914 m above the level at which it was hit. If the center fielder is initially $1.10 \times 10^2$ m from home plate, what must be his average speed?

** **45.** Two cannons are mounted as shown in the drawing and rigged to fire simultaneously. They are used in a circus act in which two clowns serve as human cannonballs. The clowns are fired toward each other and collide at a height of 1.00 m above the muzzles of the cannons. Clown A is launched at a 75.0° angle, with a speed of 9.00 m/s. The horizontal separation between the clowns as they leave the cannons is 6.00 m. Find the launch speed $v_{0B}$ and the launch angle $\theta_B (>45.0°)$ for clown B.

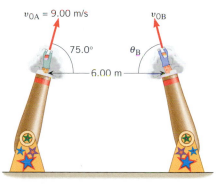

** **46.** A small can is hanging from the ceiling. A rifle is aimed directly at the can, as the figure illustrates. At the instant the gun is fired, the can is released. Ignore air resistance and show that the bullet will always strike the can, regardless of the initial speed of the bullet. Assume that the bullet strikes the can before the can reaches the ground.

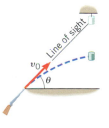

# Chapter 4 Forces and Newton's Laws of Motion

## 4.1 The Concepts of Force and Mass

In common usage, a *force* is a push or a pull, as the examples in Figure 4.1 illustrate. In basketball, a player launches a shot by pushing on the ball. The tow bar attached to a speeding boat pulls a water skier. Forces such as those that launch the basketball or pull the skier are called *contact forces,* because they arise from the physical contact between two objects. There are circumstances, however, in which two objects exert forces on one another even though they are not touching. Such forces are referred to as *noncontact forces* or *action-at-a-distance forces.* One example of such a noncontact force occurs when a skydiver is pulled toward the earth because of the force of gravity. The earth exerts this force even when it is not in direct contact with the skydiver. In Figure 4.1, arrows are used to represent the forces. It is appropriate to use arrows, because a force is a vector quantity and has both a magnitude and a direction. The direction of the arrow gives the direction of the force, and the length is proportional to its strength or magnitude.

The word *mass* is just as familiar as the word force. A massive supertanker, for instance, is one that contains an enormous amount of mass. As we will see in the next section, it is difficult to set such a massive object into motion and difficult to bring it to a halt once it is moving. In comparison, a penny does not contain much mass. The emphasis here is on the amount of mass, and the idea of direction is of no concern. Therefore, mass is a scalar quantity.

During the seventeenth century, Isaac Newton, starting with the work of Galileo, developed three important laws that deal with force and mass. Collectively they are called "Newton's laws of motion" and provide the basis for understanding the effect that forces have on an object. Because of the importance of these laws, a separate section will be devoted to each one.

## 4.2 Newton's First Law of Motion

### THE FIRST LAW

To gain some insight into Newton's first law, think about the game of ice hockey (Figure 4.2). If a player does not hit a stationary puck, it will remain at rest on the ice. After the puck is struck, however, it coasts on its own across the ice, slowing down only slightly because of friction. Since ice is very slippery, there is only a relatively small amount of friction to slow down the puck. In fact, if it were possible to remove all friction and wind resistance, and if the rink were infinitely large, the puck would coast forever in a straight line at a constant speed. Left on its own, the puck would lose none of the velocity imparted to it at the time it was struck. This is the essence of Newton's first law of motion:

> ■ **NEWTON'S FIRST LAW OF MOTION**
> An object continues in a state of rest or in a state of motion at a constant speed along a straight line, unless compelled to change that state by a net force.

In the first law the phrase "net force" is crucial. Often, several forces act simultaneously on a body, and *the net force is the vector sum of all of them.* Individual forces matter only to the extent that they contribute to the total. For instance, if friction and other opposing forces were absent, a car could travel forever at 30 m/s in a straight line, without using any gas after it has come up to speed. In reality gas is needed, but only so that

**Figure 4.1** The arrow labeled **F** represents the force that acts on the basketball, the water skier, and the skydiver. (*a.* © AFP/Corbis Images, *b.* © Rick Doyle/Corbis Images, *c.* © Taxi/ Getty Images)

the engine can produce the necessary force to cancel opposing forces such as friction. This cancellation ensures that there is no net force to change the state of motion of the car.

When an object moves at a constant speed along a straight line, its velocity is constant. Newton's first law indicates that a state of rest (zero velocity) and a state of constant velocity are completely equivalent, in the sense that neither one requires the application of a net force to sustain it. *The purpose served when a net force acts on an object is not to sustain the object's velocity, but, rather, to change it.*

## INERTIA AND MASS

A greater net force is required to change the velocity of some objects than of others. For instance, a net force that is just enough to cause a bicycle to pick up speed will cause only an imperceptible change in the motion of a freight train. In comparison to the bicycle, the train has a much greater tendency to remain at rest. Accordingly, we say that the train has more *inertia* than the bicycle. Quantitatively, the inertia of an object is measured by its *mass*. The following definition of inertia and mass indicates why Newton's first law is sometimes called the law of inertia:

> ■ **DEFINITION OF INERTIA AND MASS**
>
> Inertia is the natural tendency of an object to remain at rest or in motion at a constant speed along a straight line. The mass of an object is a quantitative measure of inertia.
>
> *SI Unit of Inertia and Mass:* kilogram (kg)

The SI unit for mass is the kilogram (kg), whereas the units in the CGS system and the BE system are the gram (g) and the slug (sl), respectively. Conversion factors between these units are given on the page facing the inside of the front cover. Figure 4.3 gives the masses of various objects, ranging from a penny to a supertanker. The larger the mass, the greater is the inertia. Often the words "mass" and "weight" are used interchangeably, but this is incorrect. Mass and weight are different concepts, and Section 4.7 will discuss the distinction between them.

Figure 4.4 on page 56 shows a useful application of inertia. Automobile seat belts unwind freely when pulled gently, so they can be buckled. But in an accident, they hold you safely in place. One seat-belt mechanism consists of a ratchet wheel, a locking bar, and a pendulum. The belt is wound around a spool mounted on the ratchet wheel. While the car is at rest or moving at a constant velocity, the pendulum hangs straight down, and the locking bar rests horizontally, as the gray part of the drawing shows. Consequently, nothing prevents the ratchet wheel from turning, and the seat belt can be pulled out easily. When the car suddenly slows down in an accident, however, the relatively massive lower part of the pendulum keeps moving forward because of its inertia. The pendulum swings on its pivot into the position shown in color and causes the locking bar to block the rotation of the ratchet wheel, thus preventing the seat belt from unwinding.

## AN INERTIAL REFERENCE FRAME

Newton's first law (and also the second law) can appear to be invalid to certain observers. Suppose, for instance, that you are a passenger riding in a friend's car. While the car moves at a constant speed along a straight line, you do not feel the seat pushing against your back to any unusual extent. This experience is consistent with the first law, which indicates that in the absence of a net force you should move with a constant velocity. Suddenly the driver floors the gas pedal. Immediately you feel the seat pressing against your back as the car accelerates. Therefore, you sense that a force is being applied to you. The first law leads you to believe that your motion should change, and, relative to the ground outside, your motion does change. But *relative to the car,* you can see that your motion does *not* change, because you remain stationary with respect to the car. Clearly, Newton's

**Figure 4.2** The game of ice hockey can give some insight into Newton's laws of motion. (© Getty Images News and Sport Services)

**The physics of** seat belts.

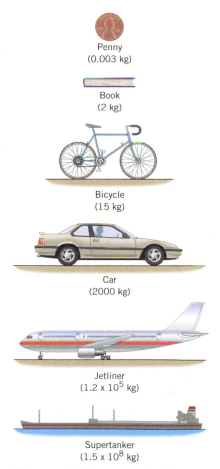

Penny
(0.003 kg)

Book
(2 kg)

Bicycle
(15 kg)

Car
(2000 kg)

Jetliner
($1.2 \times 10^5$ kg)

Supertanker
($1.5 \times 10^8$ kg)

**Figure 4.3** The masses of various objects.

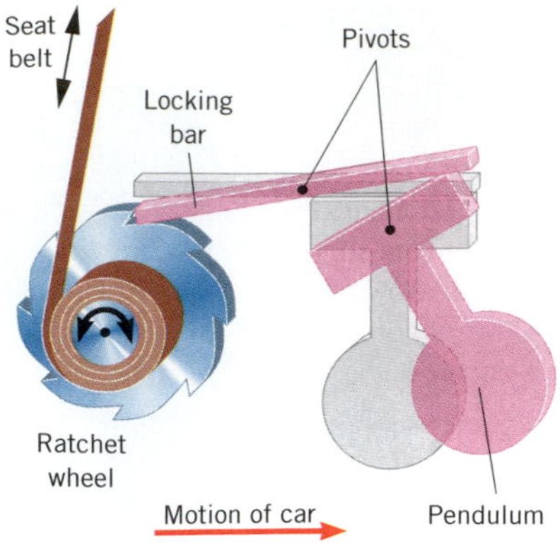

**Figure 4.4** Inertia plays a central role in one seat-belt mechanism. The gray part of the drawing applies when the car is at rest or moving at a constant velocity. The colored parts show what happens when the car suddenly slows down, as in an accident.

first law does not hold for observers who use the accelerating car as a frame of reference. As a result, such a reference frame is said to be noninertial. All accelerating reference frames are noninertial. In contrast, observers for whom the law of inertia is valid are said to be using *inertial reference frames* for their observations, as defined below:

■ **DEFINITION OF AN INERTIAL REFERENCE FRAME**

An inertial reference frame is one in which Newton's law of inertia is valid.

The acceleration of an inertial reference frame is zero, so it moves with a constant velocity. All of Newton's laws of motion are valid in inertial reference frames, and when we apply these laws, we will be assuming such a reference frame. In particular, the earth itself is a good approximation of an inertial reference frame.

## 4.3 Newton's Second Law of Motion

Newton's first law indicates that if no net force acts on an object, then the velocity of the object remains unchanged. The second law deals with what happens when a net force does act. Consider a hockey puck once again. When a player strikes a stationary puck, he causes the velocity of the puck to change. In other words, he makes the puck accelerate. The cause of the acceleration is the force that the hockey stick applies. As long as this force acts, the velocity increases, and the puck accelerates. Now, suppose another player strikes the puck and applies twice as much force as the first player does. The greater force produces a greater acceleration. In fact, if the friction between the puck and the ice is negligible, and if there is no wind resistance, the acceleration of the puck is directly proportional to the force. Twice the force produces twice the acceleration. Moreover, the acceleration is a vector quantity, just as the force is, and points in the same direction as the force.

Often, several forces act on an object simultaneously. Friction and wind resistance, for instance, do have some effect on a hockey puck. In such cases, it is the net force, or the vector sum of all the forces acting, that is important. Mathematically, the net force is written as $\Sigma\mathbf{F}$, where the Greek capital letter $\Sigma$ (sigma) denotes the vector sum. Newton's second law states that the acceleration is proportional to the net force acting on the object.

In Newton's second law, the net force is only one of two factors that determine the acceleration. The other is the inertia or mass of the object. After all, the same net force that imparts an appreciable acceleration to a hockey puck (small mass) will impart very little acceleration to a semitrailer truck (large mass). Newton's second law states that for a given net force, the magnitude of the acceleration is inversely proportional to the mass. Twice the mass means one-half the acceleration, if the same net force acts on both objects. Thus, the second law shows how the acceleration depends on both the net force and the mass, as given in Equation 4.1.

■ **NEWTON'S SECOND LAW OF MOTION**

When a net external force $\Sigma\mathbf{F}$ acts on an object of mass $m$, the acceleration $\mathbf{a}$ that results is directly proportional to the net force and has a magnitude that is inversely proportional to the mass. The direction of the acceleration is the same as the direction of the net force.

$$\mathbf{a} = \frac{\Sigma\mathbf{F}}{m} \quad \text{or} \quad \Sigma\mathbf{F} = m\mathbf{a} \tag{4.1}$$

*SI Unit of Force:* $\text{kg}\cdot\text{m/s}^2$ = newton (N)

Note that the net force in Equation 4.1 includes only the forces that the environment exerts on the object of interest. Such forces are called *external forces.* In contrast, *internal forces* are forces that one part of an object exerts on another part of the object and are not included in Equation 4.1.

**Table 4.1**  *Units for Mass, Acceleration, and Force*

| System | Mass | Acceleration | Force |
|--------|------|--------------|-------|
| SI | kilogram (kg) | meter/second$^2$ (m/s$^2$) | newton (N) |
| CGS | gram (g) | centimeter/second$^2$ (cm/s$^2$) | dyne (dyn) |
| BE | slug (sl) | foot/second$^2$ (ft/s$^2$) | pound (lb) |

According to Equation 4.1, the SI unit for force is the unit for mass (kg) times the unit for acceleration (m/s$^2$), or

$$\text{SI unit for force} = (\text{kg})\left(\frac{\text{m}}{\text{s}^2}\right) = \frac{\text{kg}\cdot\text{m}}{\text{s}^2}$$

The combination of kg · m/s$^2$ is called a *newton* (N) and is a derived SI unit, not a base unit; 1 newton = 1 N = 1 kg · m/s$^2$.

In the CGS system, the procedure for establishing the unit of force is the same as with SI units, except that mass is expressed in grams (g) and acceleration in cm/s$^2$. The resulting unit for force is the *dyne*; 1 dyne = 1 g · cm/s$^2$.

In the BE system, the unit for force is defined to be the pound (lb),* and the unit for acceleration is ft/s$^2$. With this procedure, Newton's second law can then be used to obtain the unit for mass:

$$\text{BE unit for force} = \text{lb} = (\text{unit for mass})\left(\frac{\text{ft}}{\text{s}^2}\right)$$

$$\text{Unit for mass} = \frac{\text{lb}\cdot\text{s}^2}{\text{ft}}$$

The combination of lb · s$^2$/ft is the unit for mass in the BE system and is called the *slug* (sl); 1 slug = 1 sl = 1 lb · s$^2$/ft.

Table 4.1 summarizes the various units for mass, acceleration, and force. Conversion factors between force units from different systems are provided on the page facing the inside of the front cover.

When using the second law to calculate the acceleration, it is necessary to determine the net force that acts on the object. In this determination a *free-body diagram* helps enormously. A free-body diagram is a diagram that represents the object and the forces that act on it. Only the forces that *act on the object* appear in a free-body diagram. Forces that the object exerts on its environment are not included. Example 1 illustrates the use of a free-body diagram.

*Problem solving insight*
A free-body diagram is very helpful when applying Newton's second law. Always start a problem by drawing the free-body diagram.

## *Example 1*  Pushing a Stalled Car

Two people are pushing a stalled car, as Figure 4.5*a* indicates. The mass of the car is 1850 kg. One person applies a force of 275 N to the car, while the other applies a force of 395 N. Both forces act in the same direction. A third force of 560 N also acts on the car, but in a direction opposite to that in which the people are pushing. This force arises because of friction and the extent to which the pavement opposes the motion of the tires. Find the acceleration of the car.

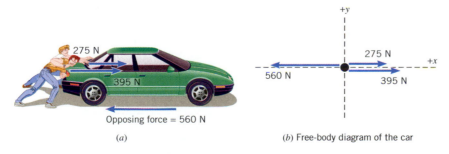

(a)

(b) Free-body diagram of the car

**Figure 4.5** (*a*) Two people push a stalled car, in opposition to a force created by friction and the pavement. (*b*) A free-body diagram that shows the horizontal forces acting on the car.

* We refer here to the gravitational version of the BE system, in which a force of one pound is defined to be the pull of the earth on a certain standard body at a location where the acceleration due to gravity is 32.174 ft/s$^2$.

**Reasoning** According to Newton's second law, the acceleration is the net force divided by the mass of the car. To determine the net force, we use the free-body diagram in Figure 4.5b. In this diagram, the car is represented as a dot, and its motion is along the +x axis. The diagram makes it clear that the forces all act along one direction. Therefore, they can be added as colinear vectors to obtain the net force.

**Solution** The net force is

$$\Sigma F = +275 \text{ N} + 395 \text{ N} - 560 \text{ N} = +110 \text{ N}$$

The acceleration can now be obtained:

$$a = \frac{\Sigma F}{m} = \frac{+110 \text{ N}}{1850 \text{ kg}} = \boxed{+0.059 \text{ m/s}^2} \qquad (4.1)$$

The plus sign indicates that the acceleration points along the +x axis, in the same direction as the net force.

## 4.4 The Vector Nature of Newton's Second Law of Motion

When a football player throws a pass, the direction of the force he applies to the ball is important. Both the force and the resulting acceleration of the ball are vector quantities, as are all forces and accelerations. The directions of these vectors can be taken into account in two dimensions by using x and y components. The net force $\Sigma \mathbf{F}$ in Newton's second law has components $\Sigma F_x$ and $\Sigma F_y$, while the acceleration $\mathbf{a}$ has components $a_x$ and $a_y$. Consequently, Newton's second law, as expressed in Equation 4.1, can be written in an equivalent form as two equations, one for the x components and one for the y components:

$$\Sigma F_x = ma_x \qquad (4.2a)$$
$$\Sigma F_y = ma_y \qquad (4.2b)$$

This procedure is similar to that employed in Chapter 3 for the equations of two-dimensional kinematics (see Table 3.1). The components in Equations 4.2a and 4.2b are scalar components and will be either positive or negative numbers, depending on whether they point along the positive or negative x or y axis. The remainder of this section deals with examples that show how these equations are used.

### Example 2   Applying Newton's Second Law Using Components

A man is stranded on a raft (mass of man and raft = 1300 kg), as shown in Figure 4.6a. By paddling, he causes an average force **P** of 17 N to be applied to the raft in a direction due east (the +x direction). The wind also exerts a force **A** on the raft. This force has a magnitude of 15 N and points 67° north of east. Ignoring any resistance from the water, find the x and y components of the raft's acceleration.

**Reasoning** Since the mass of the man and the raft is known, Newton's second law can be used to determine the acceleration components from the given forces. According to the form of the second law in Equations 4.2a and 4.2b, the acceleration component in a given direction is the component of the net force in that direction divided by the mass. As an aid in determining the components $\Sigma F_x$ and $\Sigma F_y$ of the net force, we use the free-body diagram in Figure 4.6b. In this diagram, the direction due east is the +x direction.

**Solution** Figure 4.6b shows the force components:

| Force | x Component | y Component |
|-------|-------------|-------------|
| **P** | +17 N | 0 N |
| **A** | $+(15 \text{ N}) \cos 67° = +6 \text{ N}$ | $+(15 \text{ N}) \sin 67° = +14 \text{ N}$ |
| | $\Sigma F_x = +17 \text{ N} + 6 \text{ N} = +23 \text{ N}$ | $\Sigma F_y = +14 \text{ N}$ |

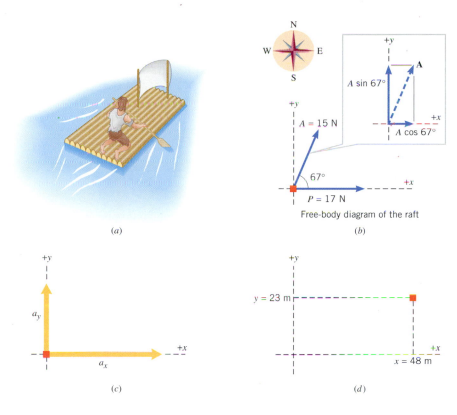

(a)

(b)

Free-body diagram of the raft

(c)

(d)

**Figure 4.6** (a) A man is paddling a raft, as in Examples 2 and 3. (b) The free-body diagram shows the forces **P** and **A** that act on the raft. Forces acting on the raft in a direction perpendicular to the surface of the water play no role in the examples and are omitted for clarity. (c) The raft's acceleration components $a_x$ and $a_y$. (d) In 65 s, the components of the raft's displacement are $x = 48$ m and $y = 23$ m.

The plus signs indicate that $\Sigma F_x$ points in the direction of the $+x$ axis and $\Sigma F_y$ points in the direction of the $+y$ axis. The $x$ and $y$ components of the acceleration point in the directions of $\Sigma F_x$ and $\Sigma F_y$, respectively, and can now be calculated:

$$a_x = \frac{\Sigma F_x}{m} = \frac{+23 \text{ N}}{1300 \text{ kg}} = \boxed{+0.018 \text{ m/s}^2} \tag{4.2a}$$

$$a_y = \frac{\Sigma F_y}{m} = \frac{+14 \text{ N}}{1300 \text{ kg}} = \boxed{+0.011 \text{ m/s}^2} \tag{4.2b}$$

These acceleration components are shown in Figure 4.6c.

## Example 3 The Displacement of a Raft

At the moment the forces **P** and **A** begin acting on the raft in Example 2, the velocity of the raft is 0.15 m/s, in a direction due east (the $+x$ direction). Assuming that the forces are maintained for 65 s, find the $x$ and $y$ components of the raft's displacement during this time interval.

**Reasoning** Once the net force acting on an object and the object's mass have been used in Newton's second law to determine the acceleration, it becomes possible to use the equations of kinematics to describe the resulting motion. We know from Example 2 that the acceleration components are $a_x = +0.018$ m/s$^2$ and $a_y = +0.011$ m/s$^2$, and it is given here that the initial velocity components are $v_{0x} = +0.15$ m/s and $v_{0y} = 0$ m/s. Thus, Equation 3.5a ($x = v_{0x}t + \frac{1}{2}a_x t^2$) and Equation 3.5b ($y = v_{0y}t + \frac{1}{2}a_y t^2$) can be used with $t = 65$ s to determine the $x$ and $y$ components of the raft's displacement.

**Solution** According to Equations 3.5a and 3.5b, the $x$ and $y$ components of the displacement are

$$x = v_{0x}t + \tfrac{1}{2}a_x t^2 = (0.15 \text{ m/s})(65 \text{ s}) + \tfrac{1}{2}(0.018 \text{ m/s}^2)(65 \text{ s})^2 = \boxed{48 \text{ m}}$$

$$y = v_{0y}t + \tfrac{1}{2}a_y t^2 = (0 \text{ m/s})(65 \text{ s}) + \tfrac{1}{2}(0.011 \text{ m/s}^2)(65 \text{ s})^2 = \boxed{23 \text{ m}}$$

Figure 4.6d shows the final location of the raft.

**Figure 4.7** The astronaut pushes on the spacecraft with a force $+\mathbf{P}$. According to Newton's third law, the spacecraft simultaneously pushes back on the astronaut with a force $-\mathbf{P}$.

## 4.5  *Newton's Third Law of Motion*

Imagine you are in a football game. You line up facing your opponent, the ball is snapped, and the two of you crash together. No doubt, you feel a force. But think about your opponent. He too feels something, for while he is applying a force to you, you are applying a force to him. In other words, there isn't just one force on the line of scrimmage; there is a pair of forces. Newton was the first to realize that all forces occur in pairs and there is no such thing as an isolated force, existing all by itself. His third law of motion deals with this fundamental characteristic of forces.

> ■ **NEWTON'S THIRD LAW OF MOTION**
>
> Whenever one body exerts a force on a second body, the second body exerts an oppositely directed force of equal magnitude on the first body.

The third law is often called the "action–reaction" law, because it is sometimes quoted as follows: "For every action (force) there is an equal, but opposite, reaction."

Figure 4.7 illustrates how the third law applies to an astronaut who is drifting just outside a spacecraft and who pushes on the spacecraft with a force **P**. According to the third law, the spacecraft pushes back on the astronaut with a force $-\mathbf{P}$ that is equal in magnitude but opposite in direction. In Example 4, we examine the accelerations produced by each of these forces.

### Example 4  The Accelerations Produced by Action and Reaction Forces

Suppose that the mass of the spacecraft in Figure 4.7 is $m_S = 11\,000$ kg and that the mass of the astronaut is $m_A = 92$ kg. In addition, assume that the astronaut exerts a force of $\mathbf{P} = +36$ N on the spacecraft. Find the accelerations of the spacecraft and the astronaut.

**Reasoning** According to Newton's third law, when the astronaut applies the force $\mathbf{P} = +36$ N to the spacecraft, the spacecraft applies a reaction force $-\mathbf{P} = -36$ N to the astronaut. As a result, the spacecraft and the astronaut accelerate in opposite directions. Although the action and reaction forces have the same magnitude, they do not create accelerations of the same magnitude, because the spacecraft and the astronaut have different masses. According to Newton's second law, the astronaut, having a much smaller mass, will experience a much larger acceleration. In applying the second law, we note that the net force acting on the spacecraft is $\Sigma\mathbf{F} = \mathbf{P}$, while the net force acting on the astronaut is $\Sigma\mathbf{F} = -\mathbf{P}$.

**Solution** Using the second law, we find that the acceleration of the spacecraft is

$$\mathbf{a}_S = \frac{\mathbf{P}}{m_S} = \frac{+36 \text{ N}}{11\,000 \text{ kg}} = \boxed{+0.0033 \text{ m/s}^2}$$

The acceleration of the astronaut is

$$\mathbf{a}_A = \frac{-\mathbf{P}}{m_A} = \frac{-36 \text{ N}}{92 \text{ kg}} = \boxed{-0.39 \text{ m/s}^2}$$

**Problem solving insight**
Even though the magnitudes of the action and reaction forces are always equal, these forces do not necessarily produce accelerations that have equal magnitudes, since each force acts on a different object that may have a different mass.

**The physics of**
automatic trailer brakes.

There is a clever application of Newton's third law in some rental trailers. As Figure 4.8 illustrates, the tow bar connecting the trailer to the rear bumper of a car contains a

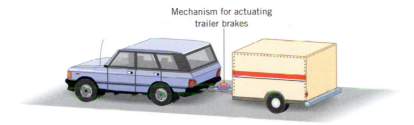

Mechanism for actuating trailer brakes

**Figure 4.8** Some rental trailers include an automatic brake-actuating mechanism.

mechanism that can automatically actuate brakes on the trailer wheels. This mechanism works without the need for electrical connections between the car and the trailer. When the driver applies the car brakes, the car slows down. Because of inertia, however, the trailer continues to roll forward and begins pushing against the bumper. In reaction, the bumper pushes back on the tow bar. The reaction force is used by the mechanism in the tow bar to "push the brake pedal" for the trailer.

## 4.6 *Types of Forces: An Overview*

In nature there are two general types of forces, fundamental and nonfundamental. Fundamental forces are the ones that are truly unique, in the sense that all other forces can be explained in terms of them. Only three fundamental forces have been discovered:

1. Gravitational force
2. Strong nuclear force
3. Electroweak force

The gravitational force is discussed in the next section. The strong nuclear force plays a primary role in the stability of the nucleus of the atom (see Section 31.2). The electroweak force is a single force that manifests itself in two ways (see Section 32.6). One manifestation is the electromagnetic force that electrically charged particles exert on one another (see Sections 18.5, 21.2, and 21.8). The other manifestation is the so-called weak nuclear force that plays a role in the radioactive disintegration of certain nuclei (see Section 31.5).

Except for the gravitational force, all of the forces discussed in this chapter are nonfundamental, because they are related to the electromagnetic force. They arise from the interactions between the electrically charged particles that comprise atoms and molecules. Our understanding of which forces are fundamental, however, is continually evolving. For instance, in the 1860s and 1870s James Clerk Maxwell showed that the electric force and the magnetic force could be explained as manifestations of a single electromagnetic force. Then, in the 1970s, Sheldon Glashow (1932– ), Abdus Salam (1926–1996), and Steven Weinberg (1933– ) presented the theory that explains how the electromagnetic force and the weak nuclear force are related to the electroweak force. They received a Nobel prize in 1979 for their achievement. Today, efforts continue that have the goal of reducing further the number of fundamental forces.

## 4.7 *The Gravitational Force*

### NEWTON'S LAW OF UNIVERSAL GRAVITATION

Objects fall downward because of gravity, and Chapters 2 and 3 discuss how to describe the effects of gravity by using a value of $g = 9.80$ m/s$^2$ for the downward acceleration it causes. However, nothing has been said about why $g$ is 9.80 m/s$^2$. The reason is fascinating, as we will now see.

The acceleration due to gravity is like any other acceleration, and Newton's second law indicates that it must be caused by a net force. In addition to his famous three laws of motion, Newton also provided a coherent understanding of the ***gravitational force***. His "law of universal gravitation" is stated as follows:

**Figure 4.9** The two particles, whose masses are $m_1$ and $m_2$, are attracted by gravitational forces $+\mathbf{F}$ and $-\mathbf{F}$.

■ **NEWTON'S LAW OF UNIVERSAL GRAVITATION**

Every particle in the universe exerts an attractive force on every other particle. A particle is a piece of matter, small enough in size to be regarded as a mathematical point. For two particles that have masses $m_1$ and $m_2$ and are separated by a distance $r$, the force that each exerts on the other is directed along the line joining the particles (see Figure 4.9) and has a magnitude given by

$$F = G\frac{m_1 m_2}{r^2} \tag{4.3}$$

The symbol $G$ denotes the universal gravitational constant, whose value is found experimentally to be

$$G = 6.673 \times 10^{-11}\ \text{N·m}^2/\text{kg}^2$$

The constant $G$ that appears in Equation 4.3 is called the ***universal gravitational constant,*** because it has the same value for all pairs of particles anywhere in the universe, no matter what their separation. The value for $G$ was first measured in an experiment by the English scientist Henry Cavendish (1731–1810), more than a century after Newton proposed his law of universal gravitation.

To see the main features of Newton's law of universal gravitation, look at the two particles in Figure 4.9. They have masses $m_1$ and $m_2$ and are separated by a distance $r$. In the picture, it is assumed that a force pointing to the right is positive. The gravitational forces point along the line joining the particles and are

$+\mathbf{F}$, the gravitational force exerted on particle 1 by particle 2

$-\mathbf{F}$, the gravitational force exerted on particle 2 by particle 1

These two forces have equal magnitudes and opposite directions. They act on different bodies, causing them to be mutually attracted. In fact, these forces are an action–reaction pair, as required by Newton's third law. Example 5 shows that the magnitude of the gravitational force is extremely small for ordinary values of the masses and the distance between them.

## *Example 5*   Gravitational Attraction

What is the magnitude of the gravitational force that acts on each particle in Figure 4.9, assuming $m_1 = 12$ kg (approximately the mass of a bicycle), $m_2 = 25$ kg, and $r = 1.2$ m?

**Reasoning and Solution**   The magnitude of the gravitational force can be found using Equation 4.3:

$$F = G\frac{m_1 m_2}{r^2} = (6.67 \times 10^{-11}\ \text{N·m}^2/\text{kg}^2)\frac{(12\ \text{kg})(25\ \text{kg})}{(1.2\ \text{m})^2} = \boxed{1.4 \times 10^{-8}\ \text{N}}$$

For comparison, you exert a force of about 1 N when pushing a doorbell, so that the gravitational force is exceedingly small in circumstances such as those here. This result is due to the fact that $G$ itself is very small. However, if one of the bodies has a large mass, like that of the earth ($5.98 \times 10^{24}$ kg), the gravitational force can be large.

*Problem solving insight*
**When applying Newton's gravitation law to uniform spheres of matter, remember that the distance $r$ is between the centers of the spheres, not between the surfaces.**

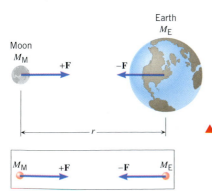

**Figure 4.10** The gravitational force that each uniform sphere of matter exerts on the other is the same as if each sphere were a particle with its mass concentrated at its center. The earth (mass $M_E$) and the moon (mass $M_M$) approximate such uniform spheres.

As expressed by Equation 4.3, Newton's law of gravitation applies only to particles. However, most familiar objects are too large to be considered particles. Nevertheless, the law of universal gravitation can be applied to such objects with the aid of calculus. Newton was able to prove that an object of finite size can be considered to be a particle for purposes of using the gravitation law, provided the mass of the object is distributed with spherical symmetry about its center. Thus, Equation 4.3 can be applied when each object is a sphere whose mass is spread uniformly over its entire volume. Figure 4.10 shows this kind of application, assuming that the earth and the moon are such uniform spheres of matter. In this case, $r$ is the distance *between the centers of the spheres* and not the distance between the outer surfaces. The gravitational forces that the spheres exert on each other are the same as if the entire mass of each were concentrated at its center. Even if the objects are not uniform spheres, Equation 4.3 can be used to a good degree of approximation if the sizes of the objects are small relative to the distance of separation $r$.

## WEIGHT

The weight of an object arises because of the gravitational pull of the earth.

> ■ **DEFINITION OF WEIGHT**
>
> The weight of an object on or above the earth is the gravitational force that the earth exerts on the object. The weight always acts downward, toward the center of the earth. On or above another astronomical body, the weight is the gravitational force exerted on the object by that body.
>
> **SI Unit of Weight:** newton (N)

Using $W$ for the magnitude of the weight,* $m$ for the mass of the object, and $M_E$ for the mass of the earth, it follows from Equation 4.3 that

$$W = G\frac{M_E m}{r^2} \tag{4.4}$$

Equation 4.4 and Figure 4.11 both emphasize that an object has weight whether or not it is resting on the earth's surface, because the gravitational force is acting even when the distance $r$ is not equal to the radius $R_E$ of the earth. However, the gravitational force becomes weaker as $r$ increases, since $r$ is in the denominator of Equation 4.4. Figure 4.12, for example, shows how the weight of the Hubble Space Telescope becomes smaller as the distance $r$ from the center of the earth increases. In Example 6 the telescope's weight is determined when it is on earth and in orbit.

## Example 6   The Hubble Space Telescope

The mass of the Hubble Space Telescope is 11 600 kg. Determine the weight of the telescope (a) when it was resting on the earth and (b) as it is in its orbit 598 km above the earth's surface.

**Reasoning** The weight of the Hubble Space Telescope is the gravitational force exerted on it by the earth. According to Equation 4.4, the weight varies inversely as the square of the radial distance $r$. Thus, we expect the telescope's weight on the earth's surface ($r$ smaller) to be greater than its weight in orbit ($r$ larger).

**Solution**

(a) On the earth's surface, the weight is given by Equation 4.4 with $r = 6.38 \times 10^6$ m (the earth's radius):

$$W = G\frac{M_E m}{r^2} = \frac{(6.67 \times 10^{-11}\text{ N}\cdot\text{m}^2/\text{kg}^2)(5.98 \times 10^{24}\text{ kg})(11\,600\text{ kg})}{(6.38 \times 10^6\text{ m})^2}$$

$$\boxed{W = 1.14 \times 10^5\text{ N}}$$

(b) When the telescope is 598 km above the surface, its distance from the center of the earth is

$$r = 6.38 \times 10^6\text{ m} + 598 \times 10^3\text{ m} = 6.98 \times 10^6\text{ m}$$

The weight now can be calculated as in part (a), except the new value of $r$ must be used: $\boxed{W = 0.950 \times 10^5\text{ N}}$. As expected, the weight is less in orbit.

---

* Often, the word "weight" and the phrase "magnitude of the weight" are used interchangeably, even though weight is a vector. Generally, the context makes it clear when the direction of the weight vector must be taken into account.

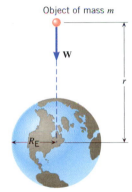

Object of mass $m$

**W**

$R_E$

Mass of earth = $M_E$

**Figure 4.11** On or above the earth, the weight **W** of an object is the gravitational force exerted on the object by the earth.

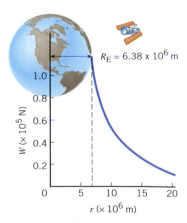

$R_E = 6.38 \times 10^6$ m

**Figure 4.12** The weight of the Hubble Space Telescope decreases as it gets farther from the earth. The distance from the center of the earth to the telescope is $r$.

The space age has forced us to broaden our ideas about weight. For instance, an astronaut weighs only about one-sixth as much on the moon as on the earth. To obtain his weight on the moon from Equation 4.4, it is only necessary to replace $M_E$ by $M_M$ (the mass of the moon) and let $r = R_M$ (the radius of the moon).

## RELATION BETWEEN MASS AND WEIGHT

Although massive objects weigh a lot on the earth, mass and weight are not the same quantity. As Section 4.2 discusses, mass is a quantitative measure of inertia. As such, mass is an intrinsic property of matter and does not change as an object is moved from one location to another. Weight, in contrast, is the gravitational force acting on the object and can vary, depending on how far the object is above the earth's surface or whether it is located near another body such as the moon.

The relation between weight $W$ and mass $m$ can be written in one of two ways:

**Problem solving insight**
Mass and weight are different quantities. They cannot be interchanged when solving problems.

$$W = \boxed{G\frac{M_E}{r^2}}\, m \tag{4.4}$$

$$W = m\,\boxed{g} \tag{4.5}$$

Equation 4.4 is Newton's law of universal gravitation, and Equation 4.5 is Newton's second law (net force equals mass times acceleration) incorporating the acceleration $g$ due to gravity. These expressions make the distinction between mass and weight stand out. The weight of an object whose mass is $m$ depends on the values for the universal gravitational constant $G$, the mass $M_E$ of the earth, and the distance $r$. These three parameters together determine the acceleration $g$ due to gravity. The specific value of $g = 9.80 \text{ m/s}^2$ applies only when $r$ equals the radius $R_E$ of the earth. For larger values of $r$, as would be the case on top of a mountain, the effective value of $g$ is less than $9.80 \text{ m/s}^2$. The fact that $g$ decreases as the distance $r$ increases means that the weight likewise decreases. The mass of the object, however, does not depend on these effects and does not change. Conceptual Example 7 further explores the difference between mass and weight.

### *Conceptual Example 7* Mass Versus Weight

A vehicle is being designed for use in exploring the moon's surface and is being tested on earth, where it weighs roughly six times more than it will on the moon. In one test, the acceleration of the vehicle along the ground is measured. To achieve the same acceleration on the moon, will the net force acting on the vehicle be greater than, less than, or the same as that required on earth?

**Reasoning and Solution** The net force $\Sigma\mathbf{F}$ required to accelerate the vehicle is specified by Newton's second law as $\Sigma\mathbf{F} = m\mathbf{a}$, where $m$ is the vehicle's mass and $\mathbf{a}$ is the acceleration along the ground. For a given acceleration, the net force depends only on the mass. But the mass is an intrinsic property of the vehicle and is the same on the moon as it is on the earth. Therefore, the same net force would be required for a given acceleration on the moon as on the earth. Do not be misled by the fact that the vehicle weighs more on earth. The greater weight occurs only because the earth's mass and radius are different than the moon's. In any event, *in Newton's second law, the net force is proportional to the vehicle's mass, not its weight.*

**Related Homework:** *Problems 21, 22*

## 4.8 *The Normal Force*

### THE DEFINITION AND INTERPRETATION OF THE NORMAL FORCE

In many situations, an object is in contact with a surface, such as a tabletop. Because of the contact, there is a force acting on the object. The present section discusses only one component of this force, the component that acts perpendicular to the surface. The next

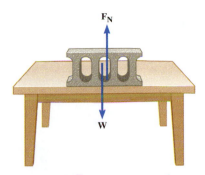

section discusses the component that acts parallel to the surface. The perpendicular component is called the *normal force.*

■ **DEFINITION OF THE NORMAL FORCE**

The normal force $\mathbf{F_N}$ is one component of the force that a surface exerts on an object with which it is in contact—namely, the component that is perpendicular to the surface.

Figure 4.13 shows a block resting on a horizontal table and identifies the two forces that act on the block, the weight $\mathbf{W}$ and the normal force $\mathbf{F_N}$. To understand how an inanimate object, such as a tabletop, can exert a normal force, think about what happens when you sit on a mattress. Your weight causes the springs in the mattress to compress. As a result, the compressed springs exert an upward force (the normal force) on you. In a similar manner, the weight of the block causes invisible "atomic springs" in the surface of the table to compress, thus producing a normal force on the block.

Newton's third law plays an important role in connection with the normal force. In Figure 4.13, for instance, the block exerts a force on the table by pressing down on it. Consistent with the third law, the table exerts an oppositely directed force of equal magnitude on the block. This reaction force is the normal force. The magnitude of the normal force indicates how hard the two objects press against each other.

If an object is resting on a horizontal surface and there are no vertically acting forces except the object's weight and the normal force, the magnitudes of these two forces are equal; that is, $F_N = W$. This is the situation in Figure 4.13. The weight must be balanced by the normal force for the object to remain at rest on the table. If the magnitudes of these forces were not equal, there would be a net force acting on the block, and the block would accelerate either upward or downward, in accord with Newton's second law.

If other forces in addition to $\mathbf{W}$ and $\mathbf{F_N}$ act in the vertical direction, the magnitudes of the normal force and the weight are no longer equal. In Figure 4.14a, for instance, a box whose weight is 15 N is being pushed downward against a table. The pushing force has a magnitude of 11 N. Thus, the total downward force exerted on the box is 26 N, and this must be balanced by the upwardacting normal force if the box is to remain at rest. In this situation, then, the normal force is 26 N, which is considerably larger than the weight of the box.

Figure 4.14b illustrates a different situation. Here, the box is being pulled upward by a rope that applies a force of 11 N. The net force acting on the box due to its weight and the rope is only 4 N, downward. To balance this force, the normal force needs to be only 4 N. It is not hard to imagine what would happen if the force applied by the rope were increased to 15 N—exactly equal to the weight of the box. In this situation, the normal force would become zero. In fact, the table could be removed, since the block would be supported entirely by the rope. The situations in Figure 4.14 are consistent with the idea that the magnitude of the normal force indicates how hard two objects press against each other. Clearly, the box and the table press against each other harder in part *a* of the picture than in part *b*.

Like the box and the table in Figure 4.14, various parts of the human body press against one another and exert normal forces. Example 8 illustrates the remarkable ability of the human skeleton to withstand a wide range of normal forces.

## Example 8   A Balancing Act

In a circus balancing act, a woman performs a headstand on top of a man's head, as Figure 4.15a on page 66 illustrates. The woman weighs 490 N, and the man's head and neck weigh 50 N. It is primarily the seventh cervical vertebra in the spine that supports all the weight above the shoulders. What is the normal force that this vertebra exerts on the neck and head of the man (a) before the act and (b) during the act?

**Reasoning** To begin, we draw a free-body diagram for the neck and head of the man. Before the act, there are only two forces, the weight of the man's head and neck, and the normal

**Figure 4.13** Two forces act on the block, its weight $\mathbf{W}$ and the normal force $\mathbf{F_N}$ exerted by the surface of the table.

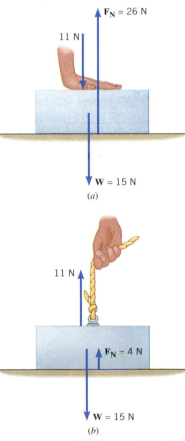

**Figure 4.14** (*a*) The normal force $\mathbf{F_N}$ is greater than the weight of the box, because the box is being pressed downward with an 11-N force. (*b*) The normal force is smaller than the weight, because the rope supplies an upward force of 11 N that partially supports the box.

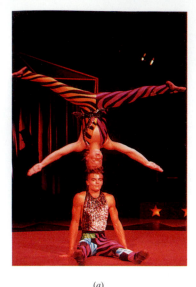

(a)

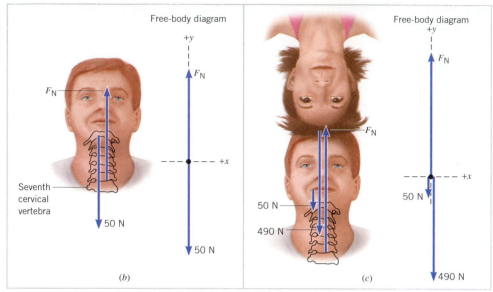

Free-body diagram

Seventh cervical vertebra

$F_N$

50 N

50 N

50 N

(b)

Free-body diagram

50 N

490 N

$F_N$

50 N

490 N

(c)

**Figure 4.15** (a) A balancing act and free-body diagrams for the man's body above the shoulders (b) before the act and (c) during the act. For convenience, the scales used for the vectors in parts b and c are different. (© Lester Sloan/ Woodfin Camp & Associates)

**The physics of** the human skeleton.

force. During the act, an additional force is present due to the woman's weight. In both cases, the upward and downward forces must balance for the head and neck to remain at rest. This condition of balance will lead us to values for the normal force.

**Solution**

**(a)** Figure 4.15b shows the free-body diagram for the man's head and neck before the act. The only forces acting are the normal force $\mathbf{F_N}$ and the 50-N weight. These two forces must balance for the man's head and neck to remain at rest. Therefore, the seventh cervical vertebra exerts a normal force of $\boxed{F_N = 50 \text{ N}}$.

**(b)** Figure 4.15c shows the free-body diagram that applies during the act. Now, the total downward force exerted on the man's head and neck is 50 N + 490 N = 540 N, which must be balanced by the upward normal force, so that $\boxed{F_N = 540 \text{ N}}$.

In summary, the normal force does not necessarily have the same magnitude as the weight of the object. The value of the normal force depends on what other forces are present. It also depends on whether the objects in contact are accelerating. In one situation that involves accelerating objects, the magnitude of the normal force can be regarded as a kind of "apparent weight," as we will now see.

## APPARENT WEIGHT

Usually, the weight of an object can be determined with the aid of a scale. However, even though a scale is working properly, there are situations in which it does not give the correct weight. In such situations, the reading on the scale gives only the "apparent" weight, rather than the gravitational force or "true" weight. The apparent weight is the force that the object exerts on the scale with which it is in contact.

To see the discrepancies that can arise between true weight and apparent weight, consider the scale in the elevator in Figure 4.16. The reasons for the discrepancies will be explained shortly. A person whose true weight is 700 N steps on the scale. If the elevator is at rest or moving with a constant velocity (either upward or downward), the scale registers the true weight, as Figure 4.16a illustrates.

If the elevator is accelerating, the apparent weight and the true weight are not equal. When the elevator accelerates upward, the apparent weight is greater than the true weight, as Figure 4.16b shows. Conversely, if the elevator accelerates downward, as in part c, the apparent weight is less than the true weight. In fact, if the elevator falls freely, so its acceleration is equal to the acceleration due to gravity, the apparent weight becomes zero, as part d indicates. In a situation such as this, where the apparent weight is zero, the per-

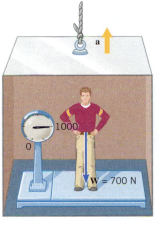

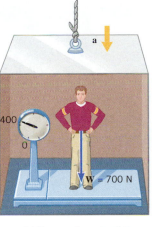

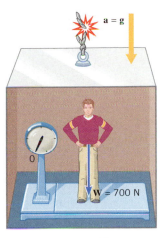

(*a*) No acceleration (**v** = constant)   (*b*) Upward acceleration   (*c*) Downward acceleration   (*d*) Free-fall

son is said to be "weightless." The apparent weight, then, does not equal the true weight if the scale and the person on it are accelerating.

The discrepancies between true weight and apparent weight can be understood with the aid of Newton's second law. Figure 4.17 shows a free-body diagram of the person in the elevator. The two forces that act on him are the true weight $\mathbf{W} = m\mathbf{g}$ and the normal force $\mathbf{F_N}$ exerted by the platform of the scale. Applying Newton's second law in the vertical direction gives

$$\Sigma F_y = +F_N - mg = ma$$

where $a$ is the acceleration of the elevator and person. In this result, the symbol $g$ stands for the magnitude of the acceleration due to gravity and can never be a negative quantity. However, the acceleration $a$ may be either positive or negative, depending on whether the elevator is accelerating upward (+) or downward (−). Solving for the normal force $F_N$ shows that

$$F_N = \underbrace{mg}_{\substack{\text{True} \\ \text{weight}}} + ma \qquad (4.6)$$

$$\underbrace{F_N}_{\substack{\text{Apparent} \\ \text{weight}}}$$

In Equation 4.6, $F_N$ is the magnitude of the normal force exerted on the person by the scale. But in accord with Newton's third law, $F_N$ is also the magnitude of the downward force that the person exerts on the scale—namely, the apparent weight.

Equation 4.6 contains all the features shown in Figure 4.16. If the elevator is not accelerating, $a = 0$ m/s², and the apparent weight equals the true weight. If the elevator accelerates upward, $a$ is positive, and the equation shows that the apparent weight is greater than the true weight. If the elevator accelerates downward, $a$ is negative, and the apparent weight is less than the true weight. If the elevator falls freely, $a = -g$, and the apparent weight is zero. The apparent weight is zero because when both the person and the scale fall freely, they cannot push against one another. In this text, when the weight is given, it is assumed to be the true weight, unless stated otherwise.

# 4.9 *Static and Kinetic Frictional Forces*

When an object is in contact with a surface, there is a force acting on the object. The previous section discusses the component of this force that is perpendicular to the surface, which is called the normal force. When the object moves or attempts to move along the surface, there is also a component of the force that is parallel to the surface. This parallel force component is called the ***frictional force,*** or simply ***friction.***

In many situations considerable engineering effort is expended trying to reduce friction. For example, oil is used to reduce the friction that causes wear and tear in the pistons and cylinder walls of an automobile engine. Sometimes, however, friction is

**Figure 4.16** (*a*) When the elevator is not accelerating, the scale registers the true weight ($W = 700$ N) of the person. (*b*) When the elevator accelerates upward, the apparent weight (1000 N) exceeds the true weight. (*c*) When the elevator accelerates downward, the apparent weight (400 N) is less than the true weight. (*d*) The apparent weight is zero if the elevator falls freely—that is, if it falls with the acceleration due to gravity.

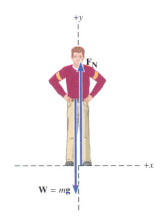

**Figure 4.17** A free-body diagram showing the forces acting on the person riding in the elevator of Figure 4.16. **W** is the true weight, and $\mathbf{F_N}$ is the normal force exerted on the person by the platform of the scale.

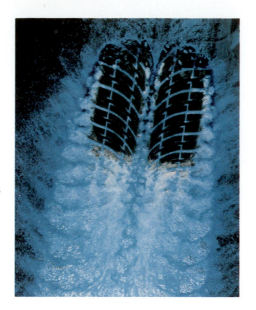

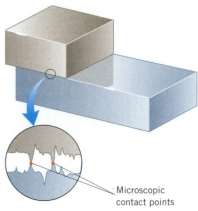

**Figure 4.18** This photo, shot from underneath a transparent surface, shows a tire rolling under wet conditions. The channels in the tire collect and divert the water away from the regions where the tire contacts the surface, thus providing better traction. (Courtesy The Goodyear Tire & Rubber Company. Used with permission.)

Microscopic
contact points

**Figure 4.19** Even when two highly polished surfaces are in contact, they touch only at a relatively few points.

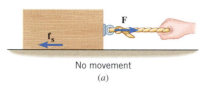

No movement
(a)

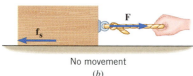

No movement
(b)

Just when movement begins
(c)

**Figure 4.20** Applying a small force **F** to the block, as in parts *a* and *b*, produces no movement, because the static frictional force **f**$_s$ exactly balances the applied force. (*c*) The block just begins to move when the applied force is slightly greater than the maximum static frictional force **f**$_s^{MAX}$.

**Figure 4.21** The maximum static frictional force **f**$_s^{MAX}$ would be the same, no matter which side of the block is in contact with the table.

absolutely necessary. Without friction, car tires could not provide the traction needed to move the car. In fact, the raised tread on a tire is designed to maintain friction. On a wet road, the spaces in the tread pattern (see Figure 4.18) provide channels for the water to collect and be diverted away. Thus, these channels largely prevent the water from coming between the tire surface and the road surface, where it would reduce friction and allow the tire to skid.

Surfaces that appear to be highly polished can actually look quite rough when examined under a microscope. Such an examination reveals that two surfaces in contact touch only at relatively few spots, as Figure 4.19 illustrates. The microscopic area of contact for these spots is substantially less than the apparent macroscopic area of contact between the surfaces—perhaps thousands of times less. At these contact points the molecules of the different bodies are close enough together to exert strong attractive intermolecular forces on one another, leading to what are known as "cold welds." Frictional forces are associated with these welded spots, but the exact details of how frictional forces arise are not well understood. However, some empirical relations have been developed that make it possible to account for the effects of friction.

Figure 4.20 helps to explain the main features of the type of friction known as *static friction.* The block in this drawing is initially at rest on a table, and as long as there is no attempt to move the block, there is no static frictional force. Then, a horizontal force **F** is applied to the block by means of a rope. If **F** is small, as in part *a*, experience tells us that the block still does not move. Why? It does not move because the static frictional force **f**$_s$ exactly cancels the effect of the applied force. The direction of **f**$_s$ is opposite to that of **F**, and the magnitude of **f**$_s$ equals the magnitude of the applied force, $f_s = F$. Increasing the applied force in Figure 4.20 by a small amount still does not cause the block to move. There is no movement because the static frictional force also increases by an amount that cancels out the increase in the applied force (see part *b* of the drawing). If the applied force continues to increase, however, there comes a point when the block finally "breaks away" and begins to slide. The force just before breakaway represents the *maximum static frictional force* **f**$_s^{MAX}$ that the table can exert on the block (see part *c* of the drawing). Any applied force that is greater than **f**$_s^{MAX}$ cannot be balanced by static friction, and the resulting net force accelerates the block to the right.

Experimental evidence shows that, to a good degree of approximation, the maximum static frictional force between a pair of dry, unlubricated surfaces has two main characteristics. It is independent of the apparent macroscopic area of contact between the objects, provided that the surfaces are hard or nondeformable. For instance, in Figure 4.21 the maximum static frictional force that the surface of the table can exert on a block is the same, whether the block is resting on its largest or its smallest side. The other main characteristic of **f**$_s^{MAX}$ is that its magnitude is proportional to the magnitude of the normal force $\mathbf{F_N}$. As Section 4.8 points out, the magnitude of the normal force indicates how hard two surfaces are being pressed together. The harder they are pressed, the larger is $f_s^{MAX}$, presumably because the number of "cold-welded," microscopic contact points is increased. Equation 4.7 expresses the proportionality between $f_s^{MAX}$ and $F_N$ with the aid of a proportionality constant $\mu_s$, which is called the *coefficient of static friction.*

---

■ **STATIC FRICTIONAL FORCE**

The magnitude $f_s$ of the static frictional force can have any value from zero up to a maximum value of $f_s^{MAX}$, depending on the applied force. In other words, $f_s \leq f_s^{MAX}$, where the symbol "$\leq$" is read as "less than or equal to." The equality holds only when $f_s$ attains its maximum value, which is

$$f_s^{MAX} = \mu_s F_N \tag{4.7}$$

In Equation 4.7, $\mu_s$ is the coefficient of static friction, and $F_N$ is the magnitude of the normal force.

---

It should be emphasized that Equation 4.7 relates only the magnitudes of $\mathbf{f_s^{MAX}}$ and $\mathbf{F_N}$, *not the vectors themselves.* This equation does not imply that the directions of the vectors are the same. In fact, $\mathbf{f_s^{MAX}}$ is parallel to the surface, while $\mathbf{F_N}$ is perpendicular to it.

The coefficient of static friction, being the ratio of the magnitudes of two forces ($\mu_s = f_s^{MAX}/F_N$), has no units. Also, it depends on the type of material from which each surface is made (steel on wood, rubber on concrete, etc.), the condition of the surfaces (polished, rough, etc.), and other variables such as temperature. Typical values for $\mu_s$ range from about 0.01 for smooth surfaces to about 1.5 for rough surfaces. Example 9 illustrates the use of Equation 4.7 for determining the maximum static frictional force.

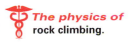 *The physics of* **rock climbing.**

### Example 9  The Force Needed to Start a Sled Moving

A sled is resting on a horizontal patch of snow, and the coefficient of static friction is $\mu_s = 0.350$. The sled and its rider have a total mass of 38.0 kg. What is the magnitude of the maximum horizontal force that can be applied to the sled before it just begins to move?

**Reasoning** The maximum horizontal force occurs when its magnitude equals that of the maximum force of static friction, as indicated in Figure 4.20c. Equation 4.7 ($f_s^{MAX} = \mu_s F_N$) specifies how the magnitude $f_s^{MAX}$ of the maximum static frictional force is related to the magnitude $F_N$ of the normal force. We can determine $F_N$ by noting that the sled does not accelerate in the vertical direction. Thus, the net force acting vertically on the sled must be zero. Consequently, the normal force and the weight of the sled and its rider must balance, so that $F_N = mg$.

**Solution** The magnitude of the maximum horizontal force is

$$f_s^{MAX} = \mu_s F_N = \mu_s mg = (0.350)(38.0 \text{ kg})(9.80 \text{ m/s}^2) = \boxed{130 \text{ N}}$$

---

Static friction is often very useful, as it is to the rock climber in Figure 4.22, for instance. He presses outward against the walls of the rock formation with his hands and feet to create sufficiently large normal forces, so that the static frictional forces can support his weight.

Once two surfaces begin sliding over one another, the static frictional force is no longer of any concern. Instead, a type of friction known as *kinetic\* friction* comes into play. The kinetic frictional force opposes the relative sliding motion. If you have ever pushed an object across a floor, you may have noticed that it takes less force to keep the object sliding than it takes to get it going in the first place. In other words, the kinetic frictional force is usually less than the static frictional force.

\* The word "kinetic" is derived from the Greek word *kinetikos,* meaning "of motion."

**Figure 4.22** In maneuvering his way up Devil's Tower in Wyoming, this rock climber uses the static frictional forces between his hands and feet and the vertical rock walls to support his weight. (© Brian Bailey/Image State)

Experimental evidence indicates that the kinetic frictional force $\mathbf{f_k}$ has three main characteristics, to a good degree of approximation. It is independent of the apparent area of contact between the surfaces (see Figure 4.21). It is independent of the speed of the sliding motion, if the speed is small. And lastly, the magnitude of the kinetic frictional force is proportional to the magnitude of the normal force. Equation 4.8 expresses this proportionality with the aid of a proportionality constant $\mu_k$, which is called the ***coefficient of kinetic friction.***

■ **KINETIC FRICTIONAL FORCE**

The magnitude $f_k$ of the kinetic frictional force is given by

$$f_k = \mu_k F_N \tag{4.8}$$

In Equation 4.8, $\mu_k$ is the coefficient of kinetic friction, and $F_N$ is the magnitude of the normal force.

Equation 4.8, like Equation 4.7, is a relationship between only the magnitudes of the frictional and normal forces. The directions of these forces are perpendicular to each other. Moreover, like the coefficient of static friction, the coefficient of kinetic friction is a number without units and depends on the type and condition of the two surfaces that are in contact. Values for $\mu_k$ are typically less than those for $\mu_s$, reflecting the fact that kinetic friction is generally less than static friction. The next example illustrates the effect of kinetic friction.

### Example 10  Sled Riding

A sled is traveling at 4.00 m/s along a horizontal stretch of snow, as Figure 4.23*a* illustrates. The coefficient of kinetic friction is $\mu_k = 0.0500$. How far does the sled go before stopping?

**Reasoning** The sled comes to a halt because the kinetic frictional force opposes the motion and causes the sled to slow down. Therefore, we will determine the kinetic frictional force and use it in Newton's second law to find the acceleration of the sled. Knowing the acceleration, we can determine the stopping distance by employing the appropriate equation of kinematics, as discussed in Chapter 3.

**Solution** To determine the magnitude $f_k$ of the kinetic frictional force, it is necessary to know the magnitude $F_N$ of the normal force, since $f_k = \mu_k F_N$. Part *b* of Figure 4.23 shows the free-body diagram. Since the sled does not accelerate in the vertical direction, there can be no net force acting vertically on the sled. As a result, the normal force and the weight $\mathbf{W}$ must balance, so the magnitude of the normal force is $F_N = mg$. The magnitude of the kinetic frictional force is

$$f_k = \mu_k F_N = \mu_k mg \tag{4.8}$$

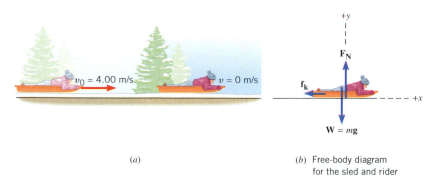

(a)

(b) Free-body diagram for the sled and rider

**Figure 4.23**  (a) The moving sled decelerates because of the kinetic frictional force. (b) Three forces act on the moving sled, the weight $\mathbf{W}$ of the sled and its rider, the normal force $\mathbf{F_N}$, and the kinetic frictional force $\mathbf{f_k}$. The free-body diagram for the sled shows these forces.

The kinetic frictional force is the only force acting on the sled in the $x$ direction, so it is the net force. Newton's second law then gives the acceleration as

$$a_x = \frac{-f_k}{m} = \frac{-\mu_k mg}{m} = -\mu_k g \qquad (4.2a)$$

The minus sign before $f_k$ arises because the kinetic frictional force opposes the sliding motion of the sled and is directed to the left, or along the $-x$ axis, in Figure 4.23b. Therefore, the acceleration also points along the $-x$ axis. Notice that the acceleration does not depend on the mass of the sled and rider, since the mass is eliminated algebraically. The stopping distance $x$ can be obtained with the aid of Equation 3.6a from the equations of kinematics ($v_x^2 = v_{0x}^2 + 2a_x x$) as

$$x = \frac{v_x^2 - v_{0x}^2}{2a_x}$$

where $v_x$ and $v_{0x}$ are the final and initial velocities, respectively. Using the fact that $a_x = -\mu_k g$, we find

$$x = \frac{v_x^2 - v_{0x}^2}{-2\,\mu_k g} = \frac{(0 \text{ m/s})^2 - (4.00 \text{ m/s})^2}{-2(0.0500)(9.80 \text{ m/s}^2)} = \boxed{16.3 \text{ m}}$$

Static friction opposes the impending relative motion between two objects, while kinetic friction opposes the relative sliding motion that actually does occur. In either case, *relative motion* is opposed. However, this opposition to relative motion does not mean that friction prevents or works against the motion of *all* objects. For instance, the foot of a person walking exerts a force on the earth, and the earth exerts a reaction force on the foot. This reaction force is a static frictional force, and it opposes the impending backward motion of the foot, propelling the person forward in the process. Kinetic friction can also cause an object to move, all the while opposing relative motion, as it does in Example 10. In this example the kinetic frictional force acts on the sled and opposes the relative motion of the sled and the earth. Newton's third law indicates, however, that since the earth exerts the kinetic frictional force on the sled, the sled must exert a reaction force on the earth. In response, the earth accelerates, but because of the earth's huge mass, the motion is too slight to be noticed.

**The physics of** walking.

## 4.10  The Tension Force

Forces are often applied by means of cables or ropes that are used to pull an object. For instance, Figure 4.24a shows a force **T** being applied to the right end of a rope attached to a box. Each particle in the rope in turn applies a force to its neighbor. As a result, the force is applied to the box, as part b of the drawing shows.

In situations such as that in Figure 4.24, we say that "the force **T** is applied to the box because of the tension in the rope," meaning that the tension and the force applied to the box have the same magnitude. However, the word "tension" is commonly used to mean the tendency of the rope to be pulled apart. To see the relationship between these two uses of the word "tension," consider the left end of the rope, which applies the force **T** to the box. In accordance with Newton's third law, the box applies a reaction force to the rope. The reaction force has the same magnitude as **T** but is oppositely directed. In other words, a force $-\mathbf{T}$ acts on the left end of the rope. Thus, forces of equal magnitude act on opposite ends of the rope, as in Figure 4.24c, and tend to pull it apart.

In the previous discussion, we have used the concept of a "massless" rope ($m = 0$ kg) without saying so. In reality, a massless rope does not exist, but it is useful as an idealization when applying Newton's second law. According to the second law, a net force is required to accelerate an object that has mass. In contrast, no net force is needed to accelerate a massless rope, since $\Sigma \mathbf{F} = m\mathbf{a}$ and $m = 0$ kg. Thus, when a force **T** is applied to one end of a massless rope, none of the force is needed to accelerate the rope. As a result, the force **T** is also applied undiminished to the object attached at the other end,

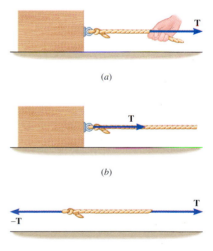

**Figure 4.24** (a) A force **T** is being applied to the right end of a rope. (b) The force is transmitted to the box. (c) Forces are applied to both ends of the rope. These forces have equal magnitudes and opposite directions.

**Figure 4.25** The force **T** applied at one end of a massless rope is transmitted undiminished to the other end, even when the rope bends around a pulley, provided the pulley is also massless and friction is absent.

as we assumed in Figure 4.24.* If the rope had mass, however, some of the force **T** would have to be used to accelerate the rope. The force applied to the box would then be less than **T**, and the tension would be different at different locations along the rope. In this text we will assume that a rope connecting one object to another is massless, unless stated otherwise. The ability of a massless rope to transmit tension undiminished from one end to the other is not affected when the rope passes around objects such as the pulley in Figure 4.25 (provided the pulley itself is massless and frictionless).

# 4.11 *Equilibrium Applications of Newton's Laws of Motion*

Have you ever been so upset that it took days to recover your "equilibrium?" In this context, the word "equilibrium" refers to a balanced state of mind, one that is not changing wildly. In physics, the word "equilibrium" also refers to a lack of change, but in the sense that the velocity of an object isn't changing. If its velocity doesn't change, an object is not accelerating. Our definition of equilibrium, then, is as follows:

■ **DEFINITION OF EQUILIBRIUM†**

An object is in equilibrium when it has zero acceleration.

Since the acceleration is zero for an object in equilibrium, all of the acceleration components are also zero. In two dimensions, this means that $a_x = 0$ m/s² and $a_y = 0$ m/s². Substituting these values into the second law ($\Sigma F_x = ma_x$ and $\Sigma F_y = ma_y$) shows that the $x$ component and the $y$ component of the net force must each be zero. Thus, in two dimensions, the equilibrium condition is expressed by two equations:

$$\Sigma F_x = 0 \tag{4.9a}$$

$$\Sigma F_y = 0 \tag{4.9b}$$

In other words, the forces acting on an object in equilibrium must balance.

Example 11 deals with a traction device in which three forces act together to bring about the equilibrium.

**The physics of**
**traction for a foot injury.**

### Example 11  Traction for the Foot

Figure 4.26*a* shows a traction device used with a foot injury. The weight of the 2.2-kg object creates a tension in the rope that passes around the pulleys. Therefore, tension forces **T₁** and **T₂** are applied to the pulley on the foot. It may seem surprising that the rope applies a force to either side of the foot pulley. A similar effect occurs when you place a finger inside a rubber band and push downward. You can feel each side of the rubber band pulling upward on the finger. The foot pulley is kept in equilibrium because the foot also applies a force **F** to it. This force arises in reaction (Newton's third law) to the pulling effect of the forces **T₁** and **T₂**. Ignoring the weight of the foot, find the magnitude of **F**.

**Reasoning** The forces **T₁**, **T₂**, and **F** keep the pulley on the foot at rest. The pulley, therefore, has no acceleration and is in equilibrium. As a result, the sum of the $x$ components and the sum of the $y$ components of the three forces must each be zero. Figure 4.26*b* shows the free-body diagram of the pulley on the foot. The $x$ axis is chosen to be along the direction of force **F**, and the components of the forces are indicated in the drawing. (See Section 1.7 for a review of vector components.)

---

* If a rope is not accelerating, **a** is zero in the second law, and $\Sigma \mathbf{F} = m\mathbf{a} = 0$, regardless of the mass of the rope. Then, the rope can be ignored, no matter what mass it has.
† In this discussion of equilibrium we ignore rotational motion, which is discussed in Chapters 8 and 9. In Section 9.2 a more complete treatment of the equilibrium of a rigid object is presented and takes into account the concept of torque and the fact that objects can rotate.

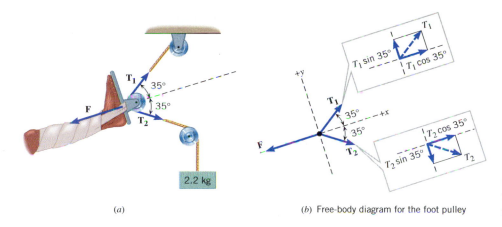

(a)

(b) Free-body diagram for the foot pulley

**Figure 4.26** (a) A traction device for the foot. (b) The free-body diagram for the pulley on the foot.

**Solution** Since the sum of the y components of the forces is zero, it follows that

$$\Sigma F_y = +T_1 \sin 35° - T_2 \sin 35° = 0 \qquad (4.9b)$$

or $T_1 = T_2$. In other words, the magnitudes of the tension forces are equal. In addition, the sum of the x components of the forces is zero, so we have that

$$\Sigma F_x = +T_1 \cos 35° + T_2 \cos 35° - F = 0 \qquad (4.9a)$$

Solving for $F$ and letting $T_1 = T_2 = T$, we find that $F = 2T \cos 35°$. However, the tension $T$ in the rope is determined by the weight of the 2.2-kg object: $T = mg$, where $m$ is its mass and $g$ is the acceleration due to gravity. Therefore, the magnitude of $\mathbf{F}$ is

$$F = 2T \cos 35° = 2mg \cos 35° = 2(2.2 \text{ kg})(9.80 \text{ m/s}^2) \cos 35° = \boxed{35 \text{ N}}$$

**Problem solving insight**
Choose the orientation of the x, y axes for convenience. In Example 11, the axes have been rotated so the force F points along the x axis. Since F does not have a component along the y axis, the analysis is simplified.

Example 12 presents another situation in which three forces are responsible for the equilibrium of an object. However, in this example all the forces have different magnitudes.

## Example 12   Replacing an Engine

An automobile engine has a weight $\mathbf{W}$, whose magnitude is $W = 3150$ N. This engine is being positioned above an engine compartment, as Figure 4.27a illustrates. To position the engine, a

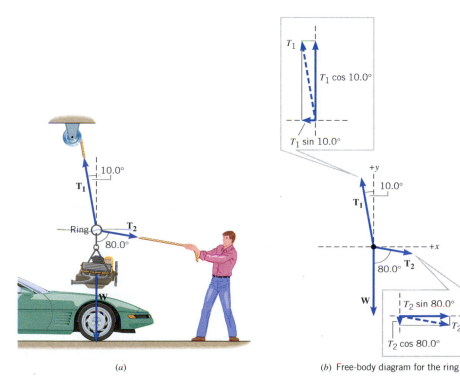

(a)

(b) Free-body diagram for the ring

**Figure 4.27** (a) The ring is in equilibrium because of the three forces $\mathbf{T}_1$ (the tension force in the supporting cable), $\mathbf{T}_2$ (the tension force in the positioning rope), and $\mathbf{W}$ (the weight of the engine). (b) The free-body diagram for the ring.

worker is using a rope. Find the tension $\mathbf{T}_1$ in the supporting cable and the tension $\mathbf{T}_2$ in the positioning rope.

**Reasoning** Under the influence of the forces $\mathbf{W}$, $\mathbf{T}_1$, and $\mathbf{T}_2$ the ring in Figure 4.27*a* is at rest and, therefore, in equilibrium. Consequently, the sum of the *x* components and the sum of the *y* components of these forces must each be zero; $\Sigma F_x = 0$ and $\Sigma F_y = 0$. By using these relations, we can find $T_1$ and $T_2$. Figure 4.27*b* shows the free-body diagram of the ring and the force components for a suitable *x*, *y* axis system.

**Solution** The free-body diagram shows the components for each of the three forces, and the components are listed in the following table:

| Force | x Component | y Component |
|-------|-------------|-------------|
| $\mathbf{T_1}$ | $-T_1 \sin 10.0°$ | $+T_1 \cos 10.0°$ |
| $\mathbf{T_2}$ | $+T_2 \sin 80.0°$ | $-T_2 \cos 80.0°$ |
| $\mathbf{W}$ | $0$ | $-W$ |

**Problem solving insight**
When an object is in equilibrium, as here in Example 12, the net force is zero, $\Sigma F = 0$. This does not mean that each individual force is zero. It means that the vector sum of all the forces is zero.

The plus signs in the table denote components that point along the positive axes, and the minus signs denote components that point along the negative axes. Setting the sum of the *x* components and the sum of the *y* components equal to zero leads to the following two equations:

$$\Sigma F_x = -T_1 \sin 10.0° + T_2 \sin 80.0° = 0 \qquad (4.9a)$$

$$\Sigma F_y = +T_1 \cos 10.0° - T_2 \cos 80.0° - W = 0 \qquad (4.9b)$$

Solving the first of these equations for $T_1$ shows that

$$T_1 = \left(\frac{\sin 80.0°}{\sin 10.0°}\right) T_2$$

Substituting this expression for $T_1$ into the second equation gives

$$\left(\frac{\sin 80.0°}{\sin 10.0°}\right) T_2 \cos 10.0° - T_2 \cos 80.0° - W = 0$$

$$T_2 = \frac{W}{\left(\dfrac{\sin 80.0°}{\sin 10.0°}\right) \cos 10.0° - \cos 80.0°}$$

Setting $W = 3150$ N in this result yields $\boxed{T_2 = 582 \text{ N}}$.

Since $T_1 = \left(\dfrac{\sin 80.0°}{\sin 10.0°}\right) T_2$ and $T_2 = 582$ N, it follows that $\boxed{T_1 = 3.30 \times 10^3 \text{ N}}$.

An object can be moving and still be in equilibrium, provided there is no acceleration. Example 13 illustrates such a case.

### Example 13 Equilibrium at Constant Velocity

A jet plane is flying with a constant speed along a straight line, at an angle of 30.0° above the horizontal, as Figure 4.28*a* indicates. The plane has a weight $\mathbf{W}$ whose magnitude is $W = 86\,500$ N, and its engines provide a forward thrust $\mathbf{T}$ of magnitude $T = 103\,000$ N. In addition, the lift force $\mathbf{L}$ (directed perpendicular to the wings) and the force $\mathbf{R}$ of air resistance (directed opposite to the motion) act on the plane. Find $\mathbf{L}$ and $\mathbf{R}$.

**Problem solving insight**
A moving object is in equilibrium if it moves with a constant velocity; then its acceleration is zero. A zero acceleration is the fundamental characteristic of an object in equilibrium.

**Reasoning** Figure 4.28*b* shows the free-body diagram of the plane, including the forces $\mathbf{W}$, $\mathbf{L}$, $\mathbf{T}$, and $\mathbf{R}$. Since the plane is not accelerating, it is in equilibrium, and the sum of the *x* components and the sum of the *y* components of these forces must be zero. The lift force $\mathbf{L}$ and the force $\mathbf{R}$ of air resistance can be obtained from these equilibrium conditions. To calculate the components, we have chosen axes in the free-body diagram that are rotated by 30.0° from their usual horizontal–vertical positions. This has been done purely for convenience, since the weight $\mathbf{W}$ is then the only force that does not lie along either axis.

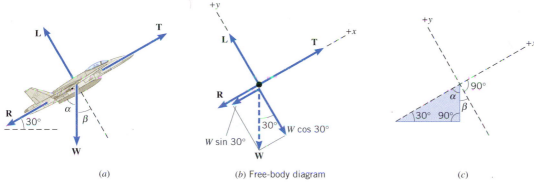

**Figure 4.28** (*a*) A plane moves with a constant velocity at an angle of 30.0° above the horizontal due to the action of four forces, the weight **W**, the lift **L**, the engine thrust **T**, and the air resistance **R**. (*b*) The free-body diagram for the plane. (*c*) This geometry occurs often in physics.

**Solution** When determining the components of the weight, it is necessary to realize that the angle $\beta$ in Figure 4.28*a* is 30.0°. Part *c* of the drawing focuses attention on the geometry that is responsible for this fact. There it can be seen that $\alpha + \beta = 90°$ and $\alpha + 30° = 90°$, with the result that $\beta = 30°$. The table below lists the components of the forces that act on the jet.

| Force | *x* Component | *y* Component |
|---|---|---|
| **W** | $-W \sin 30.0°$ | $-W \cos 30.0°$ |
| **L** | 0 | $+L$ |
| **T** | $+T$ | 0 |
| **R** | $-R$ | 0 |

Setting the sum of the *x* component of the forces to zero gives

$$\Sigma F_x = -W \sin 30.0° + T - R = 0 \qquad (4.9a)$$

$$R = T - W \sin 30.0° = 103\ 000\ \text{N} - (86\ 500\ \text{N}) \sin 30.0° = \boxed{59\ 800\ \text{N}}$$

Setting the sum of the *y* component of the forces to zero gives

$$\Sigma F_y = -W \cos 30.0° + L = 0 \qquad (4.9b)$$

$$L = W \cos 30.0° = (86\ 500\ \text{N}) \cos 30.0° = \boxed{74\ 900\ \text{N}}$$

# 4.12 *Nonequilibrium Applications of Newton's Laws of Motion*

When an object is accelerating, it is not in equilibrium. The forces acting on it are not balanced, so the net force is not zero in Newton's second law. Since the object is now accelerating, the representation of Newton's second law in Equations 4.2a and 4.2b applies instead of Equations 4.9a and 4.9b:

$$\Sigma F_x = ma_x \qquad (4.2a) \qquad \text{and} \qquad \Sigma F_y = ma_y \qquad (4.2b)$$

Example 14 uses these equations in a situation where the forces are applied in directions similar to those in Example 11, except that now an acceleration is present.

## Example 14  Towing a Supertanker

A supertanker of mass $m = 1.50 \times 10^8$ kg is being towed by two tugboats, as in Figure 4.29*a* (on page 76). The tensions in the towing cables apply the forces $\mathbf{T}_1$ and $\mathbf{T}_2$ at equal angles of 30.0° with respect to the tanker's axis. In addition, the tanker's engines produce a forward drive force **D**, whose magnitude is $D = 75.0 \times 10^3$ N. Moreover, the water applies an opposing force **R**, whose magnitude is $R = 40.0 \times 10^3$ N. The tanker moves forward with an

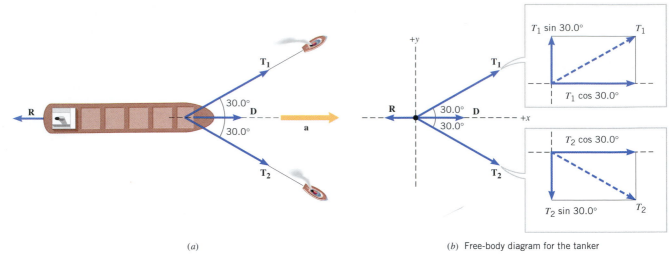

(a)                                                              (b) Free-body diagram for the tanker

**Figure 4.29** (a) Four forces act on a supertanker: $\mathbf{T_1}$ and $\mathbf{T_2}$ are the tension forces due to the towing cables, $\mathbf{D}$ is the forward drive force produced by the tanker's engines, and $\mathbf{R}$ is the force with which the water opposes the tanker's motion. (b) The free-body diagram for the tanker.

acceleration that points along the tanker's axis and has a magnitude of $2.00 \times 10^{-3}$ m/s$^2$. Find the magnitudes of the tensions $\mathbf{T_1}$ and $\mathbf{T_2}$.

**Reasoning** The unknown forces $\mathbf{T_1}$ and $\mathbf{T_2}$ contribute to the net force that accelerates the tanker. To determine $\mathbf{T_1}$ and $\mathbf{T_2}$, therefore, we analyze the net force, which we will do using components. The various force components can be found by referring to the free-body diagram for the tanker in Figure 4.29b, where the ship's axis is chosen as the $x$ axis. We will then use Newton's second law in its component form, $\Sigma F_x = ma_x$ and $\Sigma F_y = ma_y$, to obtain the magnitudes of $\mathbf{T_1}$ and $\mathbf{T_2}$.

**Solution** The individual force components are summarized as follows:

| Force | $x$ Component | $y$ Component |
|:-----:|:-------------:|:-------------:|
| $\mathbf{T_1}$ | $+T_1 \cos 30.0°$ | $+T_1 \sin 30.0°$ |
| $\mathbf{T_2}$ | $+T_2 \cos 30.0°$ | $-T_2 \sin 30.0°$ |
| $\mathbf{D}$ | $+D$ | $0$ |
| $\mathbf{R}$ | $-R$ | $0$ |

Since the acceleration points along the $x$ axis, there is no $y$ component of the acceleration ($a_y = 0$ m/s$^2$). Consequently, the sum of the $y$ components of the forces must be zero:

$$\Sigma F_y = +T_1 \sin 30.0° - T_2 \sin 30.0° = 0$$

This result shows that the magnitudes of the tensions in the cables are equal, $T_1 = T_2$. Since the ship accelerates along the $x$ direction, the sum of the $x$ components of the forces is not zero. The second law indicates that

$$\Sigma F_x = T_1 \cos 30.0° + T_2 \cos 30.0° + D - R = ma_x$$

Since $T_1 = T_2$, we can replace the two separate tension symbols by a single symbol $T$, the magnitude of the tension. Solving for $T$ gives

$$T = \frac{ma_x + R - D}{2 \cos 30.0°}$$

$$= \frac{(1.50 \times 10^8 \text{ kg})(2.00 \times 10^{-3} \text{ m/s}^2) + 40.0 \times 10^3 \text{ N} - 75.0 \times 10^3 \text{ N}}{2 \cos 30.0°}$$

$$= \boxed{1.53 \times 10^5 \text{ N}}$$

It often happens that two objects are connected somehow, perhaps by a drawbar like that used when a truck pulls a trailer. If the tension in the connecting device is of no interest, the objects can be treated as a single composite object when applying Newton's second law. However, if it is necessary to find the tension, as in the next example, then the second law must be applied separately to at least one of the objects.

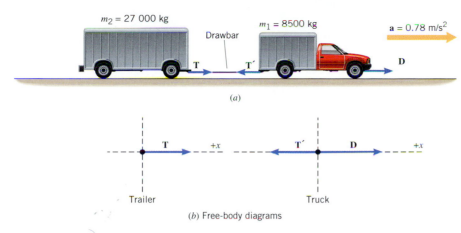

(a)

(b) Free-body diagrams

**Figure 4.30** (*a*) The force **D** acts on the truck and propels it forward. The drawbar exerts the tension force **T'** on the truck and the tension force **T** on the trailer. (*b*) The free-body diagrams for the truck and the trailer, ignoring the vertical forces.

## Example 15 Hauling a Trailer

A truck is hauling a trailer along a level road, as Figure 4.30*a* illustrates. The mass of the truck is $m_1 = 8500$ kg and that of the trailer is $m_2 = 27\,000$ kg. The two move along the *x* axis with an acceleration of $a_x = 0.78$ m/s$^2$. Ignoring the retarding forces of friction and air resistance, determine (a) the tension **T** in the horizontal drawbar between the trailer and the truck and (b) the force **D** that propels the truck forward.

**Reasoning** Since the truck and the trailer accelerate along the horizontal direction and friction is being ignored, only forces that have components in the horizontal direction are of interest. Therefore, Figure 4.30 omits the weight and the normal force, which act vertically. To determine the tension force **T** in the drawbar, we draw the free-body diagram for the trailer and apply Newton's second law, $\Sigma F_x = ma_x$. Similarly, we can determine the propulsion force **D** by drawing the free-body diagram for the truck and applying Newton's second law.

**Solution**

(a) The free-body diagram for the trailer is shown in Figure 4.30*b*. There is only one horizontal force acting on the trailer, the tension force **T** due to the drawbar. Therefore, it is straightforward to obtain the tension from $\Sigma F_x = m_2 a_x$, since the mass of the trailer and the acceleration are known:

$$\Sigma F_x = T = m_2 a_x = (27\,000 \text{ kg})(0.78 \text{ m/s}^2) = \boxed{21\,000 \text{ N}}$$

(b) Two horizontal forces act on the truck, as the free-body diagram in Figure 4.30*b* shows. One is the desired force **D**. The other is the force **T'**. According to Newton's third law, **T'** is the force with which the trailer pulls back on the truck, in reaction to the truck pulling forward. If the drawbar has negligible mass, the magnitude of **T'** is equal to the magnitude of **T**—namely, 21 000 N. Since the magnitude of **T'**, the mass of the truck, and the acceleration are known, $\Sigma F_x = m_1 a_x$ can be used to determine the drive force:

$$\Sigma F_x = +D - T' = m_1 a_x$$

$$D = m_1 a_x + T' = (8500 \text{ kg})(0.78 \text{ m/s}^2) + 21\,000 \text{ N} = \boxed{28\,000 \text{ N}}$$

In Section 4.11 we examined situations where the net force acting on an object is zero, and in this section we have considered two examples where the net force is not zero. Conceptual Example 16 illustrates a common situation where the net force is zero at certain times but is not zero at other times.

## Conceptual Example 16 The Motion of a Water Skier

Figure 4.31 (on page 78) shows a water skier at four different moments:

(a) The skier is floating motionless in the water.

(b) The skier is being pulled out of the water and up onto the skis.

(c) The skier is moving at a constant speed along a straight line.

(d) The skier has let go of the tow rope and is slowing down.

For each moment, explain whether the net force acting on the skier is zero.

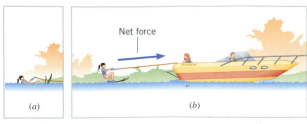
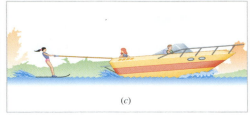

(a)           (b)           (c)           (d)

*Figure 4.31* A water skier (*a*) floating in water, (*b*) being pulled up by the boat, (*c*) moving at a constant velocity, and (*d*) slowing down.

**Reasoning and Solution** According to Newton's second law, if an object has zero acceleration, the net force acting on it is zero. In such a case, the object is in equilibrium. In contrast, if the object has an acceleration, the net force acting on it is not zero. Such an object is not in equilibrium. We will apply this criterion to each of the four phases of the motion to decide whether the net force is zero.

(**a**) The skier is floating motionless in the water, so her velocity and acceleration are both zero. Therefore, the net force acting on her is zero, and she is in equilibrium.

(**b**) As the skier is being pulled up and out of the water, her velocity is increasing. Thus, she is accelerating, and the net force acting on her is not zero. The skier is not in equilibrium. The direction of the net force is shown in Figure 4.31*b*.

(**c**) The skier is now moving at a constant speed along a straight line, so her velocity is constant. Since her velocity is constant, her acceleration is zero. Thus, the net force acting on her is zero, and she is again in equilibrium, even though she is moving.

(**d**) After the skier lets go of the tow rope, her speed decreases, so she is decelerating. Thus, the net force acting on her is not zero, and she is not in equilibrium. The direction of the net force is shown in Figure 4.31*d*, and it is opposite to that in (*b*).

The force of gravity is often present among the forces that affect the acceleration of an object. Examples 17–19 deal with typical situations.

### *Example 17* Hauling a Crate

A flatbed truck is carrying a crate up a 10.0° hill, as Figure 4.32*a* illustrates. The coefficient of static friction between the truck bed and the crate is $\mu_s = 0.350$. Find the maximum acceleration that the truck can attain before the crate begins to slip backward relative to the truck.

**Reasoning** The crate will not slip as long as it has the same acceleration as the truck. Therefore, a net force must act on the crate to accelerate it, and the static frictional force $\mathbf{f}_s$ contributes in a major way to this net force. Since the crate tends to slip backward, the static frictional force must be directed forward, up the hill. As the acceleration of the truck increases, $\mathbf{f}_s$ must also increase to produce a corresponding increase in the acceleration of the crate. However, the static frictional force can increase only until its maximum magnitude $f_s^{\text{MAX}} = \mu_s F_N$ is reached, at which point the crate and the truck have the maximum acceleration $\mathbf{a}^{\text{MAX}}$. If the acceleration increases even more, the crate will slip. To find $\mathbf{a}^{\text{MAX}}$, we focus our attention on the crate, and part *b* of the drawing shows its free-body diagram. The three forces acting on the crate at the instant slipping begins are its weight $\mathbf{W} = m\mathbf{g}$, the normal force $\mathbf{F}_N$ exerted by the truck bed, and the maximum static frictional force $\mathbf{f}_s^{\text{MAX}}$.

*Figure 4.32* (*a*) A crate on a truck is kept from slipping by the static frictional force $\mathbf{f}_s^{\text{MAX}}$. The other forces that act on the crate are its weight $\mathbf{W}$ and the normal force $\mathbf{F}_N$. (*b*) The free-body diagram of the crate.

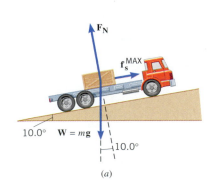

(*a*)

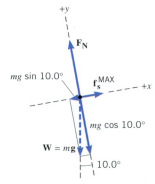

(*b*) Free-body diagram of the crate

**Solution** Using the $x$ components of the forces (see the free-body diagram for the magnitudes of the components) and Newton's second law ($\Sigma F_x = ma_x$), we have

$$\Sigma F_x = -mg \sin 10.0° + \mu_s F_N = ma^{MAX}$$

$$a^{MAX} = \frac{-mg \sin 10.0° + \mu_s F_N}{m}$$

Before this equation can be used to calculate $a^{MAX}$, however, a value is needed for $F_N$. This can be obtained by considering the force components along the $y$ axis. Since the crate does not accelerate along this axis, the sum of the $y$ components must be zero according to Newton's second law ($\Sigma F_y = ma_y = 0$):

$$\Sigma F_y = -mg \cos 10.0° + F_N = 0 \quad \text{or} \quad F_N = mg \cos 10.0°$$

Substituting this expression for $F_N$ into the equation for $a^{MAX}$ and algebraically eliminating the mass $m$ reveals that

$$a^{MAX} = -g \sin 10.0° + \mu_s g \cos 10.0°$$

$$= -(9.80 \text{ m/s}^2) \sin 10.0° + (0.350)(9.80 \text{ m/s}^2) \cos 10.0° = \boxed{1.68 \text{ m/s}^2}$$

*Problem solving insight*
The magnitude $F_N$ of the normal force is not necessarily equal to the weight of the object.

## *Example 18* Accelerating Blocks

Block 1 (mass $m_1 = 8.00$ kg) is moving on a frictionless 30.0° incline. This block is connected to block 2 (mass $m_2 = 22.0$ kg) by a massless cord that passes over a massless and frictionless pulley (see Figure 4.33*a*). Find the acceleration of each block and the tension in the cord.

**Reasoning** Since both blocks accelerate, there must be a net force acting on each one. The key to this problem is to realize that Newton's second law can be used separately for each block to relate the net force and the acceleration. Note also that both blocks have accelerations of the same magnitude $a$, since they move as a unit. We assume that block 1 accelerates up the incline and choose this direction to be the $+x$ axis. If block 1 in reality accelerates down the incline, then the value obtained for the acceleration will be a negative number.

**Solution** Three forces act on block 1: (1) $W_1$ is its weight [$W_1 = m_1g = (8.00$ kg) $\times$ (9.80 m/s$^2$) = 78.4 N], (2) $T$ is the force applied because of the tension in the cord, and (3) $F_N$ is the normal force that the incline exerts. Figure 4.33*b* shows the free-body diagram for block 1. The weight is the only force that does not point along the $x$, $y$ axes, and its $x$ and $y$ components are given in the diagram. Applying Newton's second law ($\Sigma F_x = m_1 a_x$) to block 1 shows that

$$\Sigma F_x = -W_1 \sin 30.0° + T = m_1 a$$

where we have set $a_x = a$. This equation cannot be solved as it stands, since both $T$ and $a$ are unknown quantities. To complete the solution, we next consider block 2.

Two forces act on block 2, as the free-body diagram in Figure 4.33*b* indicates: (1) $W_2$ is its weight [$W_2 = m_2g = (22.0$ kg)(9.80 m/s$^2$) = 216 N] and (2) $T'$ is exerted as a result of

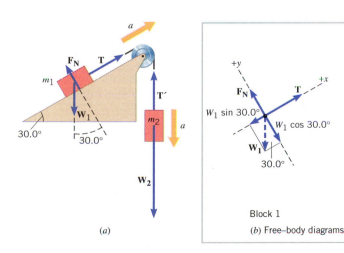

(a)

(b) Free–body diagrams

*Figure 4.33* (*a*) Three forces act on block 1: its weight $W_1$, the normal force $F_N$, and the force $T$ due to the tension in the cord. Two forces act on block 2: its weight $W_2$ and the force $T'$ due to the tension. The acceleration is labeled $a$. (*b*) The free–body diagrams for the two blocks.

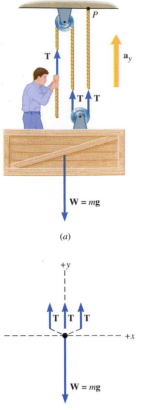

(a)

+y

+x

W = mg

(b) Free-body diagram of the unit

**Figure 4.34** (*a*) A window washer pulls down on the rope to hoist the scaffold up the side of a building. The force **T** results from the effort of the window washer and acts on him and the scaffold in three places, as discussed in Example 19. (*b*) The free-body diagram of the unit comprising the man and the scaffold.

block 1 pulling back on the connecting cord. Since the cord and the frictionless pulley are massless, the magnitudes of **T'** and **T** are the same: $T' = T$. Applying Newton's second law ($\Sigma F_y = m_2 a_y$) to block 2 reveals that

$$\Sigma F_y = T - W_2 = m_2(-a)$$

The acceleration $a_y$ has been set equal to $-a$ since block 2 moves downward along the $-y$ axis in the free-body diagram, consistent with the assumption that block 1 moves up the incline. Now there are two equations in two unknowns, and they may be solved simultaneously (see Appendix C) to give $T$ and $a$:

$$\boxed{T = 86.3 \text{ N}} \quad \text{and} \quad \boxed{a = 5.89 \text{ m/s}^2}$$

### Example 19  Hoisting a Scaffold

A window washer on a scaffold is hoisting the scaffold up the side of a building by pulling downward on a rope, as in Figure 4.34*a*. The magnitude of the pulling force is 540 N, and the combined mass of the worker and the scaffold is 155 kg. Find the upward acceleration of the unit.

**Reasoning** The worker and the scaffold form a single unit, on which the rope exerts a force in three places. The left end of the rope exerts an upward force **T** on the worker's hands. This force arises because he pulls downward with a 540-N force, and the rope exerts an oppositely directed force of equal magnitude on him, in accord with Newton's third law. Thus, the magnitude $T$ of the upward force is $T = 540$ N and is the magnitude of the tension in the rope. If the masses of the rope and each pulley are negligible and if the pulleys are friction-free, the tension is transmitted undiminished along the rope. Then, a 540-N tension force **T** acts upward on the left side of the scaffold pulley (see part *a* of the drawing). A tension force is also applied to the point P, where the rope attaches to the roof. The roof pulls back on the rope in accord with the third law, and this pull leads to the 540-N tension force **T** that acts on the right side of the scaffold pulley. In addition to the three upward forces, the weight of the unit must be taken into account [$W = mg = (155 \text{ kg})(9.80 \text{ m/s}^2) = 1520$ N]. Part *b* of the drawing shows the free-body diagram.

**Solution** Newton's second law ($\Sigma F_y = ma_y$) can be applied to calculate the acceleration $a_y$:

$$\Sigma F_y = +T + T + T - W = ma_y$$

$$a_y = \frac{3T - W}{m} = \frac{3(540 \text{ N}) - 1520 \text{ N}}{155 \text{ kg}} = \boxed{0.65 \text{ m/s}^2}$$

## Concept Summary

This summary presents an abridged version of the chapter, including the important equations and all available learning aids. For convenient reference, the learning aids (including the text's examples) are placed next to or immediately after the relevant equation or discussion. The following learning aids may be found on-line at **www.wiley.com/college/cutnell**:

| | |
|---|---|
| **Interactive LearningWare** examples are solved according to a five-step interactive format that is designed to help you develop problem-solving skills. | **Concept Simulations** are animated versions of text figures or animations that illustrate important concepts. You can control parameters that affect the display, and we encourage you to experiment. |
| **Interactive Solutions** offer specific models for certain types of problems in the chapter homework. The calculations are carried out interactively. | **Self-Assessment Tests** include both qualitative and quantitative questions. Extensive feedback is provided for both incorrect and correct answers, to help you evaluate your understanding of the material. |

| Topic | Discussion | Learning Aids |
|---|---|---|

### 4.1 The Concepts of Force and Mass

| | | |
|---|---|---|
| Contact and noncontact forces | A force is a push or a pull and is a vector quantity. Contact forces arise from the physical contact between two objects. Noncontact forces are also called action-at-a-distance forces, because they arise without physical contact between two objects. | |
| Mass | Mass is a property of matter that determines how difficult it is to accelerate or decelerate an object. Mass is a scalar quantity. | |

| Topic | Discussion | Learning Aids |
|-------|-----------|---------------|

### 4.2 Newton's First Law of Motion

**Newton's first law**

Newton's first law of motion, sometimes called the law of inertia, states that an object continues in a state of rest or in a state of motion at a constant speed along a straight line unless compelled to change that state by a net force.

**Inertia**
**Mass**

**Inertial reference frame**

Inertia is the natural tendency of an object to remain at rest or in motion at a constant speed along a straight line. The mass of a body is a quantitative measure of inertia and is measured in an SI unit called the kilogram (kg). An inertial reference frame is one in which Newton's law of inertia is valid.

### 4.3 Newton's Second Law of Motion

### 4.4 The Vector Nature of Newton's Second Law of Motion

Newton's second law of motion states that when a net force $\Sigma\mathbf{F}$ acts on an object of mass $m$, the acceleration $\mathbf{a}$ of the object can be obtained from the following equation:

**Newton's second law (vector form)**

$$\Sigma\mathbf{F} = m\mathbf{a} \qquad (4.1)$$

This is a vector equation and, for motion in two dimensions, is equivalent to the following two equations:

**Newton's second law (component form)**

$$\Sigma F_x = ma_x \qquad (4.2a)$$
$$\Sigma F_y = ma_y \qquad (4.2b)$$

In these equations the $x$ and $y$ subscripts refer to the scalar components of the force and acceleration vectors. The SI unit of force is the Newton (N).

**Free-body diagram**

When determining the net force, a free-body diagram is helpful. A free-body diagram is a diagram that represents the object and the forces acting on it.

**Concept Simulation 4.1**

**Examples 1–3**

**Interactive LearningWare 4.1**

**Interactive Solution 4.11**

### 4.5 Newton's Third Law of Motion

**Newton's third law of motion**

Newton's third law of motion, often called the action–reaction law, states that whenever one object exerts a force on a second object, the second object exerts an oppositely directed force of equal magnitude on the first object.

**Example 4**

 **Use Self-Assessment Test 4.1 to evaluate your understanding of Sections 4.1–4.5.**

### 4.6 Types of Forces: An Overview

**Fundamental forces**

Only three fundamental forces have been discovered: the gravitational force, the strong nuclear force, and the electroweak force. The electroweak force manifests itself as either the electromagnetic force or the weak nuclear force.

### 4.7 The Gravitational Force

Newton's law of universal gravitation states that every particle in the universe exerts an attractive force on every other particle. For two particles that are separated by a distance $r$ and have masses $m_1$ and $m_2$, the law states that the magnitude of this attractive force is

**Newton's law of universal gravitation**

$$F = G\frac{m_1 m_2}{r^2} \qquad (4.3)$$

The direction of this force lies along the line between the particles. The constant $G$ has a value of $G = 6.673 \times 10^{-11} \ \text{N} \cdot \text{m}^2/\text{kg}^2$ and is called the universal gravitational constant.

The weight $W$ of an object on or above the earth is the gravitational force that the earth exerts on the object and can be calculated from the mass $m$ of the object and the acceleration $g$ due to the earth's gravity according to

**Weight and mass**

$$W = mg \qquad (4.5)$$

**Examples 5, 6**

**Example 7**

### 4.8 The Normal Force

**Normal force**

The normal force $\mathbf{F}_N$ is one component of the force that a surface exerts on an object with which it is in contact—namely, the component that is perpendicular to the surface.

**Example 8**

| Topic | Discussion | Learning Aids |
|---|---|---|

The apparent weight is the force that an object exerts on the platform of a scale and may be larger or smaller than the true weight $mg$ if the object and the scale have an acceleration $a$ (+ if upward, − if downward). The apparent weight is

**Apparent weight**

$$\text{Apparent weight} = mg + ma \qquad (4.6)$$

### 4.9 *Static and Kinetic Frictional Forces*

A surface exerts a force on an object with which it is in contact. The component of the force perpendicular to the surface is called the normal force. The component parallel to the surface is called friction.

**Friction**

The force of static friction between two surfaces opposes any impending relative motion of the surfaces. The magnitude of the static frictional force depends on the magnitude of the applied force and can assume any value up to a maximum of **Example 9**

$$f_s^{\text{MAX}} = \mu_s F_N \qquad (4.7)$$

**Maximum static frictional force**

where $\mu_s$ is the coefficient of static friction and $F_N$ is the magnitude of the normal force.

**Concept Simulations 4.2, 4.4**

**Interactive LearningWare 4.2**

**Example 10**

The force of kinetic friction between two surfaces sliding against one another opposes the relative motion of the surfaces. This force has a magnitude given by **Concept Simulation 4.3**

**Kinetic frictional force**

$$f_k = \mu_k F_N \qquad (4.8)$$

**Interactive Solutions 4.41, 4.101**

where $\mu_k$ is the coefficient of kinetic friction.

### 4.10 *The Tension Force*

The word "tension" is commonly used to mean the tendency of a rope to be pulled apart due to forces that are applied at each end. Because of tension, a rope transmits a force from one end to the other. When a rope is accelerating, the force is transmitted undiminished only if the rope is massless.

 **Use *Self-Assessment Test 4.2* to evaluate your understanding of Sections 4.6–4.10.**

### 4.11 *Equilibrium Applications of Newton's Laws of Motion*

**Definition of equilibrium**

An object is in equilibrium when the object has zero acceleration, or, in other words, when it moves at a constant velocity (which may be zero). The sum of the forces that act on an object in equilibrium is zero. Under equilibrium conditions in two dimensions, the separate sums of the force components in the $x$ direction and in the $y$ direction must each be zero:

**The equilibrium condition**

$$\Sigma F_x = 0 \qquad (4.9a)$$
$$\Sigma F_y = 0 \qquad (4.9b)$$

**Examples 11, 12, 13**

**Interactive LearningWare 4.3**

### 4.12 *Nonequilibrium Applications of Newton's Laws of Motion*

If an object is not in equilibrium, then Newton's second law must be used to account for the acceleration:

$$\Sigma F_x = ma_x \qquad (4.2a)$$
$$\Sigma F_y = ma_y \qquad (4.2b)$$

**Examples 14–19**

**Interactive LearningWare 4.4**

**Interactive Solution 4.77**

 **Use *Self-Assessment Test 4.3* to evaluate your understanding of Sections 4.11 and 4.12.**

# Problems

## Section 4.3 Newton's Second Law of Motion

**1.** An airplane has a mass of $3.1 \times 10^4$ kg and takes off under the influence of a constant net force of $3.7 \times 10^4$ N. What is the net force that acts on the plane's 78-kg pilot?   *a₂ 1.193 , Fₚ₂93.ₐ*

**2. Concept Simulation 4.1** at **www.wiley.com/college/cutnell** reviews the central idea in this problem. A boat has a mass of 6800 kg. Its engines generate a drive force of 4100 N, due west, while the wind exerts a force of 800 N, due east, and the water exerts a resistive force of 1200 N due east. What is the magnitude and direction of the boat's acceleration?

**3.** In the amusement park ride known as Magic Mountain Superman, powerful magnets accelerate a car and its riders from rest to 45 m/s (about 100 mi/h) in a time of 7.0 s. The mass of the car and riders is $5.5 \times 10^3$ kg. Find the average net force exerted on the car and riders by the magnets.   *3·5×10⁴*

**4.** Scientists are experimenting with a kind of gun that may eventually be used to fire payloads directly into orbit. In one test, this gun accelerates a 5.0-kg projectile from rest to a speed of $4.0 \times 10^3$ m/s. The net force accelerating the projectile is $4.9 \times 10^5$ N. How much time is required for the projectile to come up to speed?   *a₂ ꜰ⁄ₘ ₂ 9.8×10⁴ ⇒ t₂ ᵛ⁻ᵛᵒ⁄ₐ ₂.₁₁₁ₛ*

**5. ssm** When a 58-g tennis ball is served, it accelerates from rest to a speed of 45 m/s. The impact with the racket gives the ball a constant acceleration over a distance of 44 cm. What is the magnitude of the net force acting on the ball?

**6. Interactive LearningWare 4.1** at **www.wiley.com/college/cutnell** reviews the approach taken in problems such as this one. A 1580-kg car is traveling with a speed of 15.0 m/s. What is the magnitude of the horizontal net force that is required to bring the car to a halt in a distance of 50.0 m?   *V²ᵤ V₀² +2ax   ax 1580₂⁻³³₅ₛ*

**7. ssm** A person with a black belt in karate has a fist that has a mass of 0.70 kg. Starting from rest, this fist attains a velocity of 8.0 m/s in 0.15 s. What is the magnitude of the average net force applied to the fist to achieve this level of performance?

**\* 8.** An arrow, starting from rest, leaves the bow with a speed of 25.0 m/s. If the average force exerted on the arrow by the bow were doubled, all else remaining the same, with what speed would the arrow leave the bow?

**\* 9. ssm www** Two forces $\mathbf{F_A}$ and $\mathbf{F_B}$ are applied to an object whose mass is 8.0 kg. The larger force is $\mathbf{F_A}$. When both forces point due east, the object's acceleration has a magnitude of 0.50 m/s². However, when $\mathbf{F_A}$ points due east and $\mathbf{F_B}$ points due west, the acceleration is 0.40 m/s², due east. Find (a) the magnitude of $\mathbf{F_A}$ and (b) the magnitude of $\mathbf{F_B}$.

## Section 4.4 The Vector Nature of Newton's Second Law of Motion, Section 4.5 Newton's Third Law of Motion

**10.** A force vector has a magnitude of 720 N and a direction of 38° north of east. Determine the magnitude and direction of the components of the force that point along the north–south line and along the east–west line.

**11.** Review **Interactive Solution 4.11** at **www.wiley.com/college/cutnell** before starting this problem. Two forces, $\mathbf{F_1}$ and $\mathbf{F_2}$, act on

the 7.00-kg block shown in the drawing. The magnitudes of the forces are $F_1 = 59.0$ N and $F_2 = 33.0$ N. What is the horizontal acceleration (magnitude and direction) of the block?

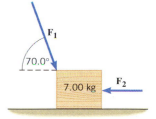

*Problem 11*

**12.** A 350-kg sailboat has an acceleration of 0.62 m/s² at an angle of 64° north of east. Find the magnitude and direction of the net force that acts on the sailboat.

**13. ssm** Only two forces act on an object (mass = 3.00 kg), as in the drawing. Find the magnitude and direction (relative to the $x$ axis) of the acceleration of the object.

*Problem 13*

**14.** Airplane flight recorders must be able to survive catastrophic crashes. Therefore, they are typically encased in crash-resistant steel or titanium boxes that are subjected to rigorous testing. One of the tests is an impact shock test, in which the box must survive being thrown at high speeds against a barrier. A 41-kg box is thrown at a speed of 220 m/s and is brought to a halt in a collision that lasts for a time of 6.5 ms. What is the magnitude of the average net force that acts on the box during the collision?

**\* 15.** A duck has a mass of 2.5 kg. As the duck paddles, a force of 0.10 N acts on it in a direction due east. In addition, the current of the water exerts a force of 0.20 N in a direction of 52° south of east. When these forces begin to act, the velocity of the duck is 0.11 m/s in a direction due east. Find the magnitude and direction (relative to due east) of the displacement that the duck undergoes in 3.0 s while the forces are acting.

**\*\* 16.** At a time when mining asteroids has become feasible, astronauts have connected a line between their 3500-kg space tug and a 6200-kg asteroid. Using their ship's engine, they pull on the asteroid with a force of 490 N. Initially the tug and the asteroid are at rest, 450 m apart. How much time does it take for the ship and the asteroid to meet?

**\*\* 17. ssm www** A 325-kg boat is sailing 15.0° north of east at a speed of 2.00 m/s. Thirty seconds later, it is sailing 35.0° north of east at a speed of 4.00 m/s. During this time, three forces act on the boat: a 31.0-N force directed 15.0° north of east (due to an auxiliary engine), a 23.0-N force directed 15.0° south of west (resistance due to the water), and $\mathbf{F_W}$ (due to the wind). Find the magnitude and direction of the force $\mathbf{F_W}$. Express the direction as an angle with respect to due east.

## Section 4.7 The Gravitational Force

**18.** A bowling ball (mass = 7.2 kg, radius = 0.11 m) and a billiard ball (mass = 0.38 kg, radius = 0.028 m) may each be treated as uniform spheres. What is the magnitude of the maximum gravitational force that each can exert on the other?

**19.** On earth, two parts of a space probe weigh 11 000 N and 3400 N. These parts are separated by a center-to-center distance of

12 m and may be treated as uniform spherical objects. Find the magnitude of the gravitational force that each part exerts on the other out in space, far from any other objects.

**20.** A rock of mass 45 kg accidentally breaks loose from the edge of a cliff and falls straight down. The magnitude of the air resistance that opposes its downward motion is 250 N. What is the magnitude of the acceleration of the rock?

**21. ssm** In preparation for this problem, review Conceptual Example 7. A space traveler whose mass is 115 kg leaves earth. What are his weight and mass (a) on earth and (b) in interplanetary space where there are no nearby planetary objects?

**22.** Review Conceptual Example 7 in preparation for this problem. In tests on earth a lunar surface exploration vehicle (mass = $5.90 \times 10^3$ kg) achieves a forward acceleration of 0.220 m/s². To achieve this same acceleration on the moon, the vehicle's engines must produce a drive force of $1.43 \times 10^3$ N. What is the magnitude of the frictional force that acts on the vehicle on the moon?

**23. ssm** Synchronous communications satellites are placed in a circular orbit that is $3.59 \times 10^7$ m above the surface of the earth. What is the magnitude of the acceleration due to gravity at this distance?

**24.** The drawing (not to scale) shows one alignment of the sun, earth, and moon. The gravitational force $\mathbf{F}_{SM}$ that the sun exerts on the moon is perpendicular to the force $\mathbf{F}_{EM}$ that the earth exerts on the moon. The masses are: mass of sun = $1.99 \times 10^{30}$ kg, mass of earth = $5.98 \times 10^{24}$ kg, mass of moon = $7.35 \times 10^{22}$ kg. The distances shown in the drawing are $r_{SM} = 1.50 \times 10^{11}$ m and $r_{EM} = 3.85 \times 10^8$ m. Determine the magnitude of the net gravitational force on the moon.

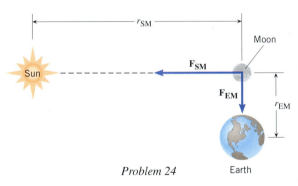

*Problem 24*

**25.** The mass of a robot is 5450 kg. This robot weighs 3620 N more on planet A than it does on planet B. Both planets have the same radius of $1.33 \times 10^7$ m. What is the difference $M_A - M_B$ in the masses of these planets?

**26.** The weight of an object is the same on two different planets. The mass of planet A is only sixty percent that of planet B. Find $r_A/r_B$, which is the ratio of the radii of the planets.

**27. ssm** Mars has a mass of $6.46 \times 10^{23}$ kg and a radius of $3.39 \times 10^6$ m. (a) What is the acceleration due to gravity on Mars? (b) How much would a 65-kg person weigh on this planet?

**\*28.** Three uniform spheres are located at the corners of an equilateral triangle. Each side of the triangle has a length of 1.20 m. Two of the spheres have a mass of 2.80 kg each. The third sphere (mass unknown) is released from rest. Considering only the gravitational forces that the spheres exert on each other, what is the magnitude of the initial acceleration of the third sphere?

**\*29. ssm www** Several people are riding in a hot-air balloon. The combined mass of the people and balloon is 310 kg. The balloon is motionless in the air, because the downward-acting weight of the people and balloon is balanced by an upward-acting "buoyant"

force. If the buoyant force remains constant, how much mass should be dropped overboard so the balloon acquires an upward acceleration of 0.15 m/s²?

**\*30.** The sun is more massive than the moon, but the sun is farther from the earth. Which one exerts a greater gravitational force on a person standing on the earth? Give your answer by determining the ratio $F_{sun}/F_{moon}$ of the magnitudes of the gravitational forces. Use the data on the inside of the front cover.

**\*31.** At a distance $H$ above the surface of a planet, the true weight of a remote probe is one percent less than its true weight on the surface. The radius of the planet is $R$. Find the ratio $H/R$.

**\*32.** Jupiter is the largest planet in our solar system, having a mass and radius that are, respectively, 318 and 11.2 times that of earth. Suppose that an object falls from rest near the surface of each planet and that the acceleration due to gravity remains constant during the fall. Each object falls the same distance before striking the ground. Determine the ratio of the time of fall on Jupiter to that on earth.

**\*\*33.** Two particles are located on the x axis. Particle 1 has a mass $m$ and is at the origin. Particle 2 has a mass $2m$ and is at $x = +L$. A third particle is placed between particles 1 and 2. Where on the x axis should the third particle be located so that the magnitude of the gravitational force on *both* particle 1 and particle 2 doubles? Express your answer in terms of $L$.

**Section 4.8 The Normal Force, Section 4.9 Static and Kinetic Frictional Forces**

**34.** A 35-kg crate rests on a horizontal floor, and a 65-kg person is standing on the crate. Determine the magnitude of the normal force that (a) the floor exerts on the crate and (b) the crate exerts on the person.

**35. ssm** A rocket blasts off from rest and attains a speed of 45 m/s in 15 s. An astronaut has a mass of 57 kg. What is the astronaut's apparent weight during takeoff?

**36.** A woman stands on a scale in a moving elevator. Her mass is 60.0 kg, and the combined mass of the elevator and scale is an additional 815 kg. Starting from rest, the elevator accelerates upward. During the acceleration, the hoisting cable applies a force of 9410 N. What does the scale read during the acceleration?

**37.** A block whose weight is 45.0 N rests on a horizontal table. A horizontal force of 36.0 N is applied to the block. The coefficients of static and kinetic friction are 0.650 and 0.420, respectively. Will the block move under the influence of the force, and, if so, what will be the block's acceleration? Explain your reasoning.

**38.** A cup of coffee is sitting on a table in an airplane that is flying at a constant altitude and a constant velocity. The coefficient of static friction between the cup and the table is 0.30. Suddenly, the plane accelerates, its altitude remaining constant. What is the maximum acceleration that the plane can have without the cup sliding backward on the table?

**39. ssm** A 20.0-kg sled is being pulled across a horizontal surface at a constant velocity. The pulling force has a magnitude of 80.0 N and is directed at an angle of 30.0° above the horizontal. Determine the coefficient of kinetic friction.

**40.** A 6.00-kg box is sliding across the horizontal floor of an elevator. The coefficient of kinetic friction between the box and the floor is 0.360. Determine the kinetic frictional force that acts on the box when the elevator is (a) stationary, (b) accelerating upward with an acceleration whose magnitude is 1.20 m/s², and (c) accelerating downward with an acceleration whose magnitude is 1.20 m/s².

**41.** Review **Interactive Solution 4.41** at **www.wiley.com/college/cutnell** in preparation for this problem. An 81-kg baseball player slides into second base. The coefficient of kinetic friction between the player and the ground is 0.49. (a) What is the magnitude of the frictional force? (b) If the player comes to rest after 1.6 s, what was his initial velocity?

\* **42.** One block rests upon a horizontal surface. A second identical block rests upon the first one. The coefficient of static friction between the blocks is the same as the coefficient of static friction between the lower block and the horizontal surface. A horizontal force is applied to the upper block, and its magnitude is slowly increased. When the force reaches 47.0 N, the upper block just begins to slide. The force is then removed from the upper block, and the blocks are returned to their original configuration. What is the magnitude of the horizontal force that should be applied to the lower block, so that it just begins to slide out from under the upper block?

\* **43.** **ssm** A skater with an initial speed of 7.60 m/s is gliding across the ice. Air resistance is negligible. (a) The coefficient of kinetic friction between the ice and the skate blades is 0.100. Find the deceleration caused by kinetic friction. (b) How far will the skater travel before coming to rest?

\* **44.** Refer to **Concept Simulation 4.4** at **www.wiley.com/college/cutnell** for background relating to this problem. The drawing shows a large cube (mass = 25 kg) being accelerated across a horizontal frictionless surface by a horizontal force **P**. A small cube (mass = 4.0 kg) is in contact with the front surface of the large cube and will slide downward unless **P** is sufficiently large. The coefficient of static friction between the cubes is 0.71. What is the smallest magnitude that **P** can have in order to keep the small cube from sliding downward?

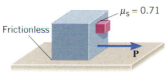

Frictionless $\mu_s = 0.71$ **P**

\*\* **45.** While moving in, a new homeowner is pushing a box across the floor at a constant velocity. The coefficient of kinetic friction between the box and the floor is 0.41. The pushing force is directed downward at an angle $\theta$ below the horizontal. When $\theta$ is greater than a certain value, it is not possible to move the box, no matter how large the pushing force is. Find that value of $\theta$.

**Section 4.10 The Tension Force, Section 4.11 Equilibrium Applications of Newton's Laws of Motion**

**46.** Part *a* of the drawing shows a bucket of water suspended from the pulley of a well; the tension in the rope is 92.0 N. Part *b* shows the same bucket of water being pulled up from the well at a constant velocity. What is the tension in the rope in part *b*?

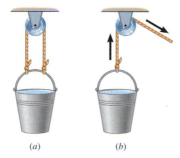

(*a*)        (*b*)

*Problem 46*

**47.** A supertanker (mass = $1.70 \times 10^8$ kg) is moving with a constant velocity. Its engines generate a forward thrust of $7.40 \times 10^5$ N. Determine (a) the magnitude of the resistive force exerted on the tanker by the water and (b) the magnitude of the upward buoyant force exerted on the tanker by the water.

**48.** Review **Interactive LearningWare 4.3** at **www.wiley.com/college/cutnell** in preparation for this problem. The helicopter in the drawing is moving horizontally to the right at a constant velocity. The weight of the helicopter is $W = 53\,800$ N. The lift force **L** generated by the rotating blade makes an angle of 21.0° with respect to the vertical. (a) What is the magnitude of the lift force? (b) Determine the magnitude of the air resistance **R** that opposes the motion.

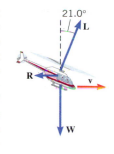

21.0°
**L**
**R**        **v**
**W**

**49.** **ssm** A 1.40-kg bottle of vintage wine is lying horizontally in the rack shown in the drawing. The two surfaces on which the bottle rests are 90.0° apart, and the right surface makes an angle of 45.0° with respect to the ground. Each surface exerts a force on the bottle that is perpendicular to the surface. What is the magnitude of each of these forces?

90.0°

**50.** The steel I-beam in the drawing has a weight of 8.00 kN and is being lifted at a constant velocity. What is the tension in each cable attached to its ends?

**51.** A stuntman is being pulled along a rough road at a constant velocity, by a cable attached to a moving truck. The cable is parallel to the ground. The mass of the stuntman is 109 kg, and the coefficient of kinetic friction between the road and him is 0.870. Find the tension in the cable.

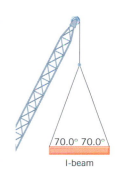

70.0° 70.0°

I-beam

*Problem 50*

**52.** As preparation for this problem, review Example 13. Suppose that the pilot suddenly jettisons 2800 N of fuel. If the plane is to continue moving with the same velocity under the influence of the same air resistance **R**, by how much does the pilot have to reduce (a) the thrust and (b) the lift?

**53.** **ssm** Three forces act on a moving object. One force has a magnitude of 80.0 N and is directed due north. Another has a magnitude of 60.0 N and is directed due west. What must be the magnitude and direction of the third force, such that the object continues to move with a constant velocity?

**54.** The drawing shows a circus clown who weighs 890 N. The coefficient of static friction between the clown's feet and the ground is 0.53. He pulls vertically downward on a rope that passes around three pulleys and is tied around his feet. What is the minimum pulling force that the clown must exert to yank his feet out from under himself?

\* **55.** The drawing shows box 1 resting on a table, with box 2 resting on top of box 1. A massless rope passes over a massless, frictionless pulley. One end of the rope is connected to box 2 and the other end is connected to box 3. The weights of the three boxes are $W_1 = 55$ N, $W_2 = 35$ N, and $W_3 = 28$ N. Determine the magnitude of the normal force that the table exerts on box 1.

* **56.** **Interactive LearningWare 4.3** at www.wiley.com/college/cut-nell reviews the principles that play a role in this problem. During a storm, a tree limb breaks off and comes to rest across a barbed wire fence at a point that is not in the middle between two fence posts. The limb exerts a downward force of 151 N on the wire. The left section of the wire makes an angle of 14.0° relative to the horizontal and sustains a tension of 447 N. Find the magnitude and direction of the tension that the right section of the wire sustains.

* **57.** **ssm** A person is trying to judge whether a picture (mass = 1.10 kg) is properly positioned by temporarily pressing it against a wall. The pressing force is perpendicular to the wall. The coefficient of static friction between the picture and the wall is 0.660. What is the minimum amount of pressing force that must be used?

* **58.** A mountain climber, in the process of crossing between two cliffs by a rope, pauses to rest. She weighs 535 N. As the drawing shows, she is closer to the left cliff than to the right cliff, with the result that the tensions in the left and right sides of the rope are not the same. Find the tensions in the rope to the left and to the right of the mountain climber.

* **59.** A skier is pulled up a slope at a constant velocity by a tow bar. The slope is inclined at 25.0° with respect to the horizontal. The force applied to the skier by the tow bar is parallel to the slope. The skier's mass is 55.0 kg, and the coefficient of kinetic friction between the skis and the snow is 0.120. Find the magnitude of the force that the tow bar exerts on the skier.

** **60.** The weight of the block in the drawing is 88.9 N. The coefficient of static friction between the block and the vertical wall is 0.560. (a) What minimum force **F** is required to prevent the block from sliding down the wall? (*Hint: The static frictional force exerted on the block is directed upward, parallel to the wall.*) (b) What minimum force is required to start the block moving up the wall? (*Hint: The static frictional force is now directed down the wall.*)

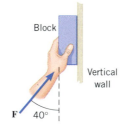

** **61.** **ssm** A bicyclist is coasting straight down a hill at a constant speed. The mass of the rider and bicycle is 80.0 kg, and the hill is inclined at 15.0° with respect to the horizontal. Air resistance opposes the motion of the cyclist. Later, the bicyclist climbs the same hill at the same constant speed. How much force (directed parallel to the hill) must be applied to the bicycle in order for the bicyclist to climb the hill?

** **62.** A damp washcloth is hung over the edge of a table to dry. Thus, part (mass = $m_{on}$) of the washcloth rests on the table and part (mass = $m_{off}$) does not. The coefficient of static friction between the table and the washcloth is 0.40. Determine the maximum fraction $[m_{off}/(m_{on} + m_{off})]$ that can hang over the edge without causing the whole washcloth to slide off the table.

## Section 4.12 Nonequilibrium Applications of Newton's Laws of Motion

**63.** Only two forces act on an object (mass = 4.00 kg), as in the drawing. Find the magnitude and direction (relative to the x axis) of the acceleration of the object.

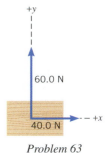

*Problem 63*

**64.** **Concept Simulation 4.1** at www.wiley.com/college/cutnell reviews the concepts that are important in this problem. The speed of a bobsled is increasing because it has an acceleration of 2.4 m/s². At a given instant in time, the forces resisting the motion, including kinetic friction and air resistance, total 450 N. The mass of the bobsled and its riders is 270 kg. (a) What is the magnitude of the force propelling the bobsled forward? (b) What is the magnitude of the net force that acts on the bobsled?

**65.** **ssm** A 1380-kg car is moving due east with an initial speed of 27.0 m/s. After 8.00 s the car has slowed down to 17.0 m/s. Find the magnitude and direction of the net force that produces the deceleration.

**66.** A falling skydiver has a mass of 110 kg. (a) What is the magnitude of the skydiver's acceleration when the upward force of air resistance has a magnitude that is equal to one-third of his weight? (b) After the parachute opens, the skydiver descends at a constant velocity. What is the force of air resistance (magnitude and direction) that acts on the skydiver?

**67.** In the drawing, the weight of the block on the table is 422 N and that of the hanging block is 185 N. Ignoring all frictional effects and assuming the pulley to be massless, find (a) the acceleration of the two blocks and (b) the tension in the cord.

**68.** A 292-kg motorcycle is accelerating up along a ramp that is inclined 30.0° above the horizontal. The propulsion force pushing the motorcycle up the ramp is 3150 N, and air resistance produces a force of 250 N that opposes the motion. Find the magnitude of the motorcycle's acceleration.

**69.** **ssm www** A student is skateboarding down a ramp that is 6.0 m long and inclined at 18° with respect to the horizontal. The initial speed of the skateboarder at the top of the ramp is 2.6 m/s. Neglect friction and find the speed at the bottom of the ramp.

**70.** A rescue helicopter is lifting a man (weight = 822 N) from a capsized boat by means of a cable and harness. (a) What is the tension in the cable when the man is given an initial upward acceleration of 1.10 m/s²? (b) What is the tension during the remainder of the rescue when he is pulled upward at a constant velocity?

**71.** **ssm** In a supermarket parking lot, an employee is pushing ten empty shopping carts, lined up in a straight line. The acceleration of the carts is 0.050 m/s². The ground is level, and each cart has a mass of 26 kg. (a) What is the net force acting on any one of the carts? (b) Assuming friction is negligible, what is the force exerted by the fifth cart on the sixth cart?

**72.** **Interactive LearningWare 4.4** at www.wiley.com/college/cutnell provides a review of the concepts that are important in this problem. A rocket of mass 4.50 × 10⁵ kg is in flight. Its thrust is directed at an angle of 55.0° above the horizontal and has a magnitude of 7.50 × 10⁶ N. Find the magnitude and direction of the rocket's acceleration. Give the direction as an angle above the horizontal.

**73.** A $1.14 \times 10^4$-kg lunar landing craft is about to touch down on the surface of the moon, where the acceleration due to gravity is 1.60 m/s². At an altitude of 165 m the craft's downward velocity is 18.0 m/s. To slow down the craft, a retrorocket is firing to provide an upward thrust. Assuming the descent is vertical, find the magnitude of the thrust needed to reduce the velocity to zero at the instant when the craft touches the lunar surface.

* **74.** Consult **Concept Simulation 4.2** at **www.wiley.com/college/cutnell** in preparation for this problem. A crate is resting on a ramp that is inclined at an angle $\theta$ above the horizontal. As $\theta$ is increased, the crate remains in place until $\theta$ reaches a value of 38.0°. Then the crate begins to slide down the slope. (a) Determine the coefficient of static friction between the crate and the ramp surface. (b) The coefficient of kinetic friction between the crate and the ramp surface is 0.600. Find the acceleration of the moving crate.

* **75. ssm** To hoist himself into a tree, a 72.0-kg man ties one end of a nylon rope around his waist and throws the other end over a branch of the tree. He then pulls downward on the free end of the rope with a force of 358 N. Neglect any friction between the rope and the branch, and determine the man's upward acceleration.

* **76.** A 205-kg log is pulled up a ramp by means of a rope that is parallel to the surface of the ramp. The ramp is inclined at 30.0° with respect to the horizontal. The coefficient of kinetic friction between the log and the ramp is 0.900, and the log has an acceleration of 0.800 m/s². Find the tension in the rope.

* **77.** Review **Interactive Solution 4.77** at **www.wiley.com/college/cutnell** before starting this problem. The drawing shows Robin Hood (mass = 77.0 kg) about to escape from a dangerous situation. With one hand, he is gripping the rope that holds up a chandelier (mass = 195 kg). When he cuts the rope where it is tied to the floor, the chandelier will fall, and he will be pulled up toward a balcony above. Ignore the friction between the rope and the beams over which it slides, and find (a) the acceleration with which Robin is pulled upward and (b) the tension in the rope while Robin escapes.

* **78.** A train consists of 50 cars, each of which has a mass of $6.8 \times 10^3$ kg. The train has an acceleration of $+8.0 \times 10^{-2}$ m/s². Ignore friction and determine the tension in the coupling (a) between the 30th and 31st cars and (b) between the 49th and 50th cars.

* **79. ssm** A box is sliding up an incline that makes an angle of 15.0° with respect to the horizontal. The coefficient of kinetic friction between the box and the surface of the incline is 0.180. The initial speed of the box at the bottom of the incline is 1.50 m/s. How far does the box travel along the incline before coming to rest?

* **80.** A girl is sledding down a slope that is inclined at 30.0° with respect to the horizontal. A moderate wind is aiding the motion by providing a steady force of 105 N that is parallel to the motion of the sled. The combined mass of the girl and sled is 65.0 kg, and the coefficient of kinetic friction between the runners of the sled and the snow is 0.150. How much time is required for the sled to travel down a 175-m slope, starting from rest?

* **81.** At an airport, luggage is unloaded from a plane into the three cars of a luggage carrier, as the drawing shows. The acceleration of the carrier is 0.12 m/s², and friction is negligible. The coupling bars

have negligible mass. By how much would the tension in *each* of the coupling bars A, B, and C change if 39 kg of luggage were removed from car 2 and placed in (a) car 1 and (b) car 3? If the tension changes, specify whether it increases or decreases.

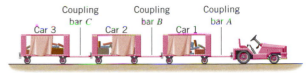

* **82.** Consult **Interactive LearningWare 4.2** at **www.wiley.com/college/cutnell** before beginning this problem. A truck is traveling at a speed of 25.0 m/s along a level road. A crate is resting on the bed of the truck, and the coefficient of static friction between the crate and the truck bed is 0.650. Determine the shortest distance in which the truck can come to a halt without causing the crate to slip forward relative to the truck.

** **83. ssm** A penguin slides at a constant velocity of 1.4 m/s down an icy incline. The incline slopes above the horizontal at an angle of 6.9°. At the bottom of the incline, the penguin slides onto a horizontal patch of ice. The coefficient of kinetic friction between the penguin and the ice is the same for the incline as for the horizontal patch. How much time is required for the penguin to slide to a halt after entering the horizontal patch of ice?

** **84.** As part *a* of the drawing shows, two blocks are connected by a rope that passes over a set of pulleys. One block has a weight of 412 N, and the other has a weight of 908 N. The rope and the pulleys are massless and there is no friction. (a) What is the acceleration of the lighter block? (b) Suppose that the heavier block is removed, and a downward force of 908 N is provided by someone pulling on the rope, as part *b* of the drawing shows. Find the acceleration of the remaining block. (c) Explain why the answers in (a) and (b) are different.

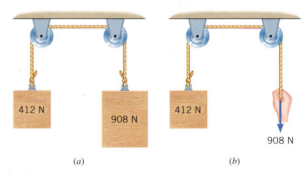

(a)　　　　　　　　　　　　　(b)

** **85.** In the drawing, the rope and the pulleys are massless, and there is no friction. Find (a) the tension in the rope and (b) the acceleration of the 10.0-kg block. (*Hint: The larger mass moves twice as far as the smaller mass.*)

** **86.** A 5.00-kg block is placed on top of a 12.0-kg block that rests on a frictionless table. The coefficient of static friction between the two blocks is 0.600. What is the maximum horizontal force that can be applied before the 5.00-kg block begins to slip relative to the 12.0-kg block, if the force is applied to (a) the more massive block and (b) the less massive block?

*Problem 85*

# Chapter 5 Dynamics of Uniform Circular Motion

## 5.1 Uniform Circular Motion

There are many examples of motion on a circular path. Of the many possibilities, we single out those that satisfy the following definition:

> ■ **DEFINITION OF UNIFORM CIRCULAR MOTION**
>
> Uniform circular motion is the motion of an object traveling at a constant (uniform) speed on a circular path.

As an example of uniform circular motion, Figure 5.1 shows a model airplane on a guideline. The speed of the plane is the magnitude of the velocity vector **v**, and since the speed is constant, the vectors in the drawing have the same magnitude at all points on the circle.

Sometimes it is more convenient to describe uniform circular motion by specifying the period of the motion, rather than the speed. The **period T** is the time required to travel once around the circle—that is, to make one complete revolution. There is a relationship between period and speed, since speed $v$ is the distance traveled (circumference of the circle $= 2\pi r$) divided by the time $T$:

$$v = \frac{2\pi r}{T} \tag{5.1}$$

If the radius is known, as in Example 1, the speed can be calculated from the period or vice versa.

### Example 1   A Tire-Balancing Machine

The wheel of a car has a radius of $r = 0.29$ m and is being rotated at 830 revolutions per minute (rpm) on a tire-balancing machine. Determine the speed (in m/s) at which the outer edge of the wheel is moving.

**Reasoning** The speed $v$ can be obtained directly from $v = 2\pi r/T$, but first the period $T$ is needed. The period is the time for one revolution, and it must be expressed in seconds, because the problem asks for the speed in meters per second.

**Solution** Since the tire makes 830 revolutions in one minute, the number of minutes required for a single revolution is

$$\frac{1}{830 \text{ revolutions/min}} = 1.2 \times 10^{-3} \text{ min/revolution}$$

Therefore, the period is $T = 1.2 \times 10^{-3}$ min, which corresponds to 0.072 s. Equation 5.1 can now be used to find the speed:

$$v = \frac{2\pi r}{T} = \frac{2\pi(0.29 \text{ m})}{0.072 \text{ s}} = \boxed{25 \text{ m/s}}$$

The definition of uniform circular motion emphasizes that the speed, or the magnitude of the velocity vector, is constant. It is equally significant that the direction of the vector is *not constant*. In Figure 5.1, for instance, the velocity vector changes direction as the plane moves around the circle. Any change in the velocity vector, even if it is only a change in direction, means that an acceleration is occurring. This particular acceleration is called "centripetal acceleration," because it points toward the center of the circle, as the next section explains.

**Figure 5.1** The motion of an airplane flying at a constant speed on a horizontal circular path is an example of uniform circular motion.

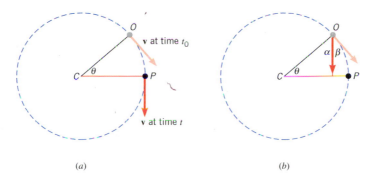

(a)                                          (b)

**Figure 5.2** (*a*) For an object (●) in uniform circular motion, the velocity **v** has different directions at different places on the circle. (*b*) The velocity vector has been removed from point *P*, shifted parallel to itself, and redrawn with its tail at point *O*.

# 5.2 *Centripetal Acceleration*

In this section we determine how the magnitude $a_c$ of the centripetal acceleration depends on the speed $v$ of the object and the radius $r$ of the circular path. We will see that $a_c = v^2/r$.

In Figure 5.2*a* an object (symbolized by a dot ●) is in uniform circular motion. At time $t_0$ the velocity is tangent to the circle at point $O$, and at a later time $t$ the velocity is tangent at point $P$. As the object moves from $O$ to $P$, the radius traces out the angle $\theta$, and the velocity vector changes direction. To emphasize the change, part *b* of the picture shows the velocity vector removed from point $P$, shifted parallel to itself, and redrawn with its tail at point $O$. The angle $\beta$ between the two vectors indicates the change in direction. Since the radii $CO$ and $CP$ are perpendicular to the tangents at points $O$ and $P$, respectively, it follows that $\alpha + \beta = 90°$ and $\alpha + \theta = 90°$. Therefore, angle $\beta$ and angle $\theta$ are equal.

As always, acceleration is the change $\Delta\mathbf{v}$ in velocity divided by the elapsed time $\Delta t$, or $\mathbf{a} = \Delta\mathbf{v}/\Delta t$. Figure 5.3*a* shows the two velocity vectors oriented at the angle $\theta$ with respect to one another, together with the vector $\Delta\mathbf{v}$ that represents the change in velocity. The change $\Delta\mathbf{v}$ is the increment that must be added to the velocity at time $t_0$, so that the resultant velocity has the new direction after an elapsed time $\Delta t = t - t_0$. Figure 5.3*b* shows the sector of the circle $COP$. In the limit that $\Delta t$ is very small, the arc length $OP$ is approximately a straight line whose length is the distance $v\Delta t$ traveled by the object. In this limit, $COP$ is an isosceles triangle, as is the triangle in part *a* of the drawing. Since both triangles have equal apex angles $\theta$, they are similar, so that

$$\frac{\Delta v}{v} = \frac{v\,\Delta t}{r}$$

This equation can be solved for $\Delta v/\Delta t$, to show that the magnitude $a_c$ of the centripetal acceleration is given by $a_c = v^2/r$.

Centripetal acceleration is a vector quantity and, therefore, has a direction as well as a magnitude. The direction is toward the center of the circle, and Conceptual Example 2 helps us to set the stage for explaining this important fact.

## Conceptual Example 2   Which Way Will the Object Go?

In Figure 5.4 an object, such as a model airplane on a guideline, is in uniform circular motion. The object is symbolized by a dot (●), and at point $O$ it is released suddenly from its circular path. For instance, the guideline for a model plane is cut suddenly. Does the object move along the straight tangent line between points $O$ and $A$ or along the circular arc between points $O$ and $P$?

**Reasoning and Solution** Newton's first law of motion guides our reasoning. An object continues in a state of rest or in a state of motion at a constant speed along a straight line unless compelled to change that state by a net force. When the object is suddenly released from its circular path, there is no longer a net force being applied to the object. In the case of a model airplane, the guideline cannot apply a force, since it is cut. Gravity certainly acts on the plane, but the wings provide a lift force that balances the weight of the plane. In the absence of a net force, then, the plane or any object would continue to move at a constant speed along a

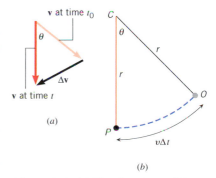

(a)

(b)

**Figure 5.3** (*a*) The directions of the velocity vector at times $t$ and $t_0$ differ by the angle $\theta$. (*b*) When the object moves along the circle from $O$ to $P$, the radius $r$ traces out the same angle $\theta$. Here, the sector $COP$ has been rotated clockwise by 90° relative to its orientation in Figure 5.2.

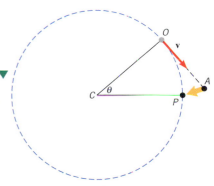

**Figure 5.4** If an object (●) moving on a circular path were released from its path at point $O$, it would move along the straight tangent line $OA$ in the absence of a net force.

straight line in the direction it had at the time of release. This speed and direction are given in Figure 5.4 by the velocity vector **v**. As a result, *the object would move along the straight line between points O and A, not on the circular arc between points O and P.*

**Related Homework:**  *Problem 4*

As Example 2 discusses, the object in Figure 5.4 would travel on a tangent line if it were released from its circular path suddenly at point *O*. It would move in a straight line to point *A* in the time it would have taken to travel on the circle to point *P*. It is as if the object drops through the distance *AP* in the process of remaining on the circle, and *AP* is directed toward the center of the circle in the limit that the angle $\theta$ is small. Thus, the object accelerates toward the center of the circle at every moment. The acceleration is called *centripetal acceleration,* because the word "centripetal" means "center-seeking."

> ■ **CENTRIPETAL ACCELERATION**
>
> *Magnitude:*  The centripetal acceleration of an object moving with a speed $v$ on a circular path of radius $r$ has a magnitude $a_c$ given by
>
> $$a_c = \frac{v^2}{r} \tag{5.2}$$
>
> *Direction:*  The centripetal acceleration vector always points toward the center of the circle and continually changes direction as the object moves.

The following example illustrates the effect of the radius $r$ on the centripetal acceleration.

### Example 3  The Effect of Radius on Centripetal Acceleration

**The physics of a bobsled track.**

The bobsled track at the 1994 Olympics in Lillehammer, Norway, contained turns with radii of 33 m and 24 m, as Figure 5.5 illustrates. Find the centripetal acceleration at each turn for a speed of 34 m/s, a speed that was achieved in the two-man event. Express the answers as multiples of $g = 9.8$ m/s$^2$.

**Reasoning**  In each case, the magnitude of the centripetal acceleration can be obtained from $a_c = v^2/r$. Since the radius $r$ is in the denominator on the right side of this expression, we expect the acceleration to be smaller when $r$ is larger.

**Solution**  From $a_c = v^2/r$ it follows that

*Radius = 33 m*          $a_c = \dfrac{(34 \text{ m/s})^2}{33 \text{ m}} = 35 \text{ m/s}^2 = \boxed{3.6 \, g}$

*Radius = 24 m*          $a_c = \dfrac{(34 \text{ m/s})^2}{24 \text{ m}} = 48 \text{ m/s}^2 = \boxed{4.9 \, g}$

The centripetal acceleration is indeed smaller when the radius is larger. In fact, with $r$ in the denominator on the right of $a_c = v^2/r$, the acceleration approaches zero when the radius becomes very large. Uniform circular motion along the arc of an infinitely large circle entails no acceleration, because it is just like motion at a constant speed along a straight line.

In Section 4.11 we learned that an object is in equilibrium when it has zero acceleration. Conceptual Example 4 discusses whether an object undergoing uniform circular motion can ever be at equilibrium.

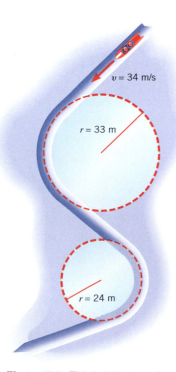

**Figure 5.5**  This bobsled travels at the same speed around two curves with different radii. For the turn with the larger radius, the sled has a smaller centripetal acceleration.

*v* = 34 m/s

*r* = 33 m

*r* = 24 m

### Conceptual Example 4  Uniform Circular Motion and Equilibrium

A car moves at a constant speed, and there are three parts to the motion. It moves along a straight line toward a circular turn, goes around the turn, and then moves away along a straight line. In each of these parts, is the car in equilibrium?

**Reasoning and Solution** An object in equilibrium has no acceleration, according to the definition given in Section 4.11. As the car approaches the turn, both the speed and direction of the motion are constant. Thus, the velocity vector does not change, and there is no acceleration. The same is true as the car moves away from the turn. For these parts of the motion, then, the car is in equilibrium. As the car goes around the turn, however, the direction of travel changes, so the car has a centripetal acceleration that is characteristic of uniform circular motion. Because of this acceleration, the car is not in equilibrium during the turn. In general, ***an object that is in uniform circular motion can never be in equilibrium.***

We have seen that going around tight turns (smaller *r*) and gentle turns (larger *r*) at the same speed entails different centripetal accelerations. And most drivers know that such turns "feel" different. This feeling is associated with the force that is present in uniform circular motion, and we now turn to this topic.

## 5.3 *Centripetal Force*

Newton's second law indicates that whenever an object accelerates, there must be a net force to create the acceleration. Thus, in uniform circular motion there must be a net force to produce the centripetal acceleration. The second law gives this net force as the product of the object's mass *m* and its acceleration $v^2/r$. The net force causing the centripetal acceleration is called the ***centripetal force*** $F_c$ and points in the same direction as the acceleration—that is, toward the center of the circle.

> ■ **CENTRIPETAL FORCE**
>
> *Magnitude:* The centripetal force is the name given to the net force required to keep an object of mass *m*, moving at a speed *v*, on a circular path of radius *r*, and it has a magnitude of
>
> $$F_c = \frac{mv^2}{r} \tag{5.3}$$
>
> *Direction:* The centripetal force always points toward the center of the circle and continually changes direction as the object moves.

The phrase "centripetal force" does not denote a new and separate force created by nature. The phrase merely labels the net force pointing toward the center of the circular path, and this net force is the vector sum of all the force components that point along the radial direction.

In some cases, it is easy to identify the source of the centripetal force, as when a model airplane on a guideline flies in a horizontal circle. The only force pulling the plane inward is the tension in the line, so this force alone (or a component of it) is the centripetal force. Example 5 illustrates the fact that higher speeds require greater tensions.

### Example 5 The Effect of Speed on Centripetal Force

The model airplane in Figure 5.6 has a mass of 0.90 kg and moves at a constant speed on a circle that is parallel to the ground. The path of the airplane and its guideline lie in the same horizontal plane, because the weight of the plane is balanced by the lift generated by its wings. Find the tension *T* in the guideline (length = 17 m) for speeds of 19 and 38 m/s.

**Reasoning** Since the plane flies on a circular path, it experiences a centripetal acceleration that is directed toward the center of the circle. According to Newton's second law of motion, this acceleration is produced by a net force that acts on the plane and this net force is called the centripetal force. The centripetal force is also directed toward the center of the circle. Since the tension *T* in the guideline is the only force pulling the plane inward, it must be the centripetal force.

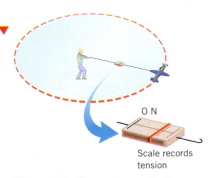

**Figure 5.6** The scale records the tension in the guideline. See Example 5.

**Solution**  Equation 5.3 gives the tension directly: $F_c = T = mv^2/r$.

*Speed = 19 m/s*     $T = \dfrac{(0.90 \text{ kg})(19 \text{ m/s})^2}{17 \text{ m}} = \boxed{19 \text{ N}}$

*Speed = 38 m/s*     $T = \dfrac{(0.90 \text{ kg})(38 \text{ m/s})^2}{17 \text{ m}} = \boxed{76 \text{ N}}$

Conceptual Example 6 deals with another case where it is easy to identify the source of the centripetal force.

### Conceptual Example 6  A Trapeze Act

In a circus, a man hangs upside down from a trapeze, legs bent over the bar and arms downward, holding his partner (see Figure 5.7). Is it harder for the man to hold his partner when the partner hangs straight down and is stationary or when the partner is swinging through the straight-down position?

**Reasoning and Solution**  When the man and his partner are stationary, the man's arms must support his partner's weight. When the two are swinging, however, the man's arms must do an additional job. Then, the partner is moving on a circular arc and has a centripetal acceleration. The man's arms must exert an additional pull so that there will be sufficient centripetal force to produce this acceleration. Because of the additional pull, *it is harder for the man to hold his partner while swinging than while stationary.*

**Figure 5.7**  Does the man hanging upside down from the trapeze have a harder job holding his partner when the team is swinging through the straight-down position than when they are hanging straight down and stationary? (Courtesy Ringling Brothers and Barnum & Bailey®, The Greatest Show on Earth®)

When a car moves at a steady speed around an unbanked curve, the centripetal force keeping the car on the curve comes from the static friction between the road and the tires, as Figure 5.8 indicates. It is static, rather than kinetic friction, because the tires are not slipping with respect to the radial direction. If the static frictional force is insufficient, given the speed and the radius of the turn, the car will skid off the road. Example 7 shows how an icy road can limit safe driving.

### Example 7  Centripetal Force and Safe Driving

Compare the maximum speeds at which a car can safely negotiate an unbanked turn (radius = 50.0 m) in dry weather (coefficient of static friction = 0.900) and icy weather (coefficient of static friction = 0.100).

**Reasoning**  At the maximum speed, the maximum centripetal force acts on the tires, and static friction must provide it. The magnitude of the maximum force of static friction is specified by Equation 4.7 as $f_s^{\text{MAX}} = \mu_s F_N$, where $\mu_s$ is the coefficient of static friction and $F_N$ is the magnitude of the normal force. Our strategy, then, is to find the normal force, substitute it into the expression for the maximum force of static friction, and then equate the result to $mv^2/r$. Experience indicates that the maximum speed should be greater for the dry road than for the icy road.

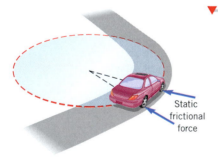

**Figure 5.8**  When the car moves without skidding around a curve, static friction between the road and the tires provides the centripetal force to keep the car on the road.

Static frictional force

**Solution**  Since the car does not accelerate in the vertical direction, the weight $mg$ of the car is balanced by the normal force, so $F_N = mg$. From Equations 4.7 and 5.3 it follows that

$$F_c = \mu_s F_N = \mu_s mg = \frac{mv^2}{r}$$

Consequently, $\mu_s g = v^2/r$, and

$$v = \sqrt{\mu_s gr}$$

The mass $m$ of the car has been eliminated algebraically from this result. All cars, heavy or light, have the same maximum speed:

*Dry road ($\mu_s = 0.900$)*     $v = \sqrt{(0.900)(9.80 \text{ m/s}^2)(50.0 \text{ m})} = \boxed{21.0 \text{ m/s}}$

*Icy road ($\mu_s = 0.100$)*     $v = \sqrt{(0.100)(9.80 \text{ m/s}^2)(50.0 \text{ m})} = \boxed{7.00 \text{ m/s}}$

As expected, the dry road allows the greater maximum speed.

**Problem solving insight**
When using an equation to obtain a numerical answer, algebraically solve for the unknown variable in terms of the known variables. Then substitute in the numbers for the known variables, as this example shows.

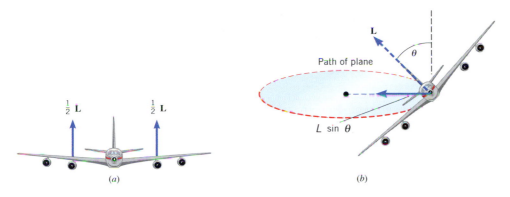

**Figure 5.9** (*a*) The air exerts an upward lifting force $\frac{1}{2}$**L** on each wing. (*b*) When a plane executes a circular turn, the plane banks at an angle $\theta$. The lift component $L \sin \theta$ is directed toward the center of the circle and provides the centripetal force.

The passenger in Figure 5.8 must also experience a centripetal force to remain on the circular path. However, if the upholstery is very slippery, there may not be enough static friction to keep him in place as the car makes a tight turn at high speed. Then, when viewed from inside the car, he appears to be thrown toward the outside of the curve. What really happens is that the passenger slides off on a tangent to the circle, until he encounters a source of centripetal force to keep him in place while the car turns. This occurs when the passenger bumps into the side of the car, which pushes on him with the necessary force.

Sometimes the source of the centripetal force is not obvious. A pilot making a turn, for instance, banks or tilts the plane at an angle to create the centripetal force. As a plane flies, the air pushes upward on the wing surfaces with a net lifting force **L** that is perpendicular to the wing surfaces, as Figure 5.9*a* shows. Part *b* of the drawing illustrates that when the plane is banked at an angle $\theta$, a component $L \sin \theta$ of the lifting force is directed toward the center of the turn. It is this component that provides the centripetal force. Greater speeds and/or tighter turns require greater centripetal forces. In such situations, the pilot must bank the plane at a larger angle, so that a larger component of the lift points toward the center of the turn. The technique of banking into a turn also has an application in the construction of high-speed roadways, where the road itself is banked to achieve a similar effect, as the next section discusses.

*The physics of*
**flying an airplane in a banked turn.**

## 5.4 *Banked Curves*

When a car travels without skidding around an unbanked curve, the static frictional force between the tires and the road provides the centripetal force. The reliance on friction can be eliminated completely for a given speed, however, if the curve is banked at an angle relative to the horizontal, much in the same way that a plane is banked while making a turn.

Figure 5.10*a* shows a car going around a friction-free banked curve. The radius of the curve is $r$, where $r$ is measured parallel to the horizontal and not to the slanted surface. Part *b* shows the normal force $\mathbf{F_N}$ that the road applies to the car, the normal force being

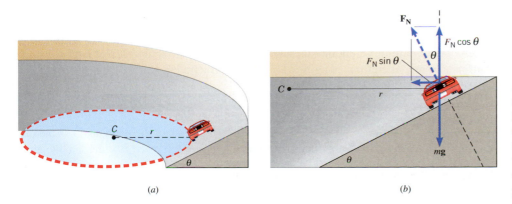

**Figure 5.10** (*a*) A car travels on a circle of radius $r$ on a frictionless banked road. The banking angle is $\theta$, and the center of the circle is at $C$. (*b*) The forces acting on the car are its weight $m\mathbf{g}$ and the normal force $\mathbf{F_N}$. A component $F_N \sin \theta$ of the normal force provides the centripetal force.

perpendicular to the road. Because the roadbed makes an angle $\theta$ with respect to the horizontal, the normal force has a component $F_N \sin \theta$ that points toward the center $C$ of the circle and provides the centripetal force:

$$F_c = F_N \sin \theta = \frac{mv^2}{r}$$

The vertical component of the normal force is $F_N \cos \theta$ and, since the car does not accelerate in the vertical direction, this component must balance the weight $mg$ of the car. Therefore, $F_N \cos \theta = mg$. Dividing this equation into the previous one shows that

$$\frac{F_N \sin \theta}{F_N \cos \theta} = \frac{mv^2/r}{mg}$$

$$\tan \theta = \frac{v^2}{rg} \tag{5.4}$$

Equation 5.4 indicates that, for a given speed $v$, the centripetal force needed for a turn of radius $r$ can be obtained from the normal force by banking the turn at an angle $\theta$, independent of the mass of the vehicle. Greater speeds and smaller radii require more steeply banked curves—that is, larger values of $\theta$. At a speed that is too small for a given $\theta$, a car would slide down a frictionless banked curve; at a speed that is too large, a car would slide off the top. The next example deals with a famous banked curve.

### Example 8  The Daytona 500

**The physics of**
**the Daytona International Speedway.**

The Daytona 500 is the major event of the NASCAR (National Association for Stock Car Auto Racing) season. It is held at the Daytona International Speedway in Daytona, Florida. The turns in this oval track have a maximum radius (at the top) of $r = 316$ m and are banked steeply, with $\theta = 31°$ (see Figure 5.10). Suppose these maximum-radius turns were frictionless. At what speed would the cars have to travel around them?

**Reasoning** In the absence of friction, the horizontal component of the normal force that the track exerts on the car must provide the centripetal force. Therefore, the speed of the car is given by Equation 5.4.

**Solution** From Equation 5.4, it follows that

$$v = \sqrt{rg \tan \theta} = \sqrt{(316 \text{ m})(9.80 \text{ m/s}^2) \tan 31°} = \boxed{43 \text{ m/s (96 mph)}}$$

Drivers actually negotiate the turns at speeds up to 195 mph, however, which requires a greater centripetal force than that implied by Equation 5.4 for frictionless turns. Static friction provides the additional force.

## 5.5  *Satellites in Circular Orbits*

Today there are many satellites in orbit about the earth. The ones in circular orbits are examples of uniform circular motion. Like a model airplane on a guideline, each satellite is kept on its circular path by a centripetal force. The gravitational pull of the earth provides the centripetal force and acts like an invisible guideline for the satellite.

*There is only one speed that a satellite can have if the satellite is to remain in an orbit with a fixed radius.* To see how this fundamental characteristic arises, consider the gravitational force acting on the satellite of mass $m$ in Figure 5.11. Since the gravitational force is the only force acting on the satellite in the radial direction, it alone provides the centripetal force. Therefore, using Newton's law of gravitation (Equation 4.3), we have

$$F_c = G \frac{mM_E}{r^2} = \frac{mv^2}{r}$$

where $G$ is the universal gravitational constant, $M_E$ is the mass of the earth, and $r$ is the distance from the center of the earth to the satellite. Solving for the speed $v$ of the satellite gives

$$v = \sqrt{\frac{GM_E}{r}} \tag{5.5}$$

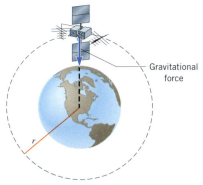

*Gravitational force*

**Figure 5.11** For a satellite in circular orbit around the earth, the gravitational force provides the centripetal force.

If the satellite is to remain in an orbit of radius $r$, the speed must have precisely this value. Note that the radius $r$ of the orbit is in the denominator in Equation 5.5. This means that the closer the satellite is to the earth, the smaller is the value for $r$ and the greater the orbital speed must be.

The mass $m$ of the satellite does not appear in Equation 5.5, having been eliminated algebraically. **_Consequently, for a given orbit, a satellite with a large mass has exactly the same orbital speed as a satellite with a small mass._** However, more effort is certainly required to lift the larger-mass satellite into orbit. The orbital speed of one famous artificial satellite is determined in the following example.

### Example 9 Orbital Speed of the Hubble Space Telescope

Determine the speed of the Hubble Space Telescope (see Figure 5.12) orbiting at a height of 598 km above the earth's surface.

**Reasoning** Before Equation 5.5 can be applied, the orbital radius $r$ must be determined _relative to the center of the earth_. Since the radius of the earth is approximately $6.38 \times 10^6$ m, and the height of the telescope above the earth's surface is $0.598 \times 10^6$ m, the orbital radius is $r = 6.98 \times 10^6$ m.

**Solution** The orbital speed is

$$v = \sqrt{\frac{GM_E}{r}} = \sqrt{\frac{(6.67 \times 10^{-11}\,\text{N}\cdot\text{m}^2/\text{kg}^2)(5.98 \times 10^{24}\,\text{kg})}{6.98 \times 10^6\,\text{m}}}$$

$$v = \boxed{7.56 \times 10^3\,\text{m/s (16 900 mi/h)}}$$

**Figure 5.12** The Hubble Space Telescope orbits the earth, after being released from the Space Shuttle _Discovery._ (Courtesy NASA)

**_The physics of_**
**the Hubble Space Telescope.**

Many applications of satellite technology affect our lives. One is a network of 24 satellites called the Global Positioning System (GPS), which can be used to determine the position of an object to within 15 m or less. Figure 5.13 illustrates how the system works. Each GPS satellite carries a highly accurate atomic clock, whose time is transmitted to the ground continually by means of radio waves. In the drawing, a car carries a computerized GPS receiver that can detect the waves and is synchronized to the satellite clock. The receiver can, therefore, determine the distance between the car and a satellite from a knowledge of the travel time of the waves and the speed at which they move. This speed, as we will see in Chapter 24, is the speed of light and is known with great precision. A measurement using a single satellite locates the car somewhere on a circle, as Figure 5.13a shows, while a measurement using a second satellite locates the car on another

**_Problem solving insight_**
The orbital radius _r_ that appears in the relation $v = \sqrt{GM_E/r}$ is the distance from the satellite to the center of the earth (not to the surface of the earth).

**_The physics of_**
**the Global Positioning System.**

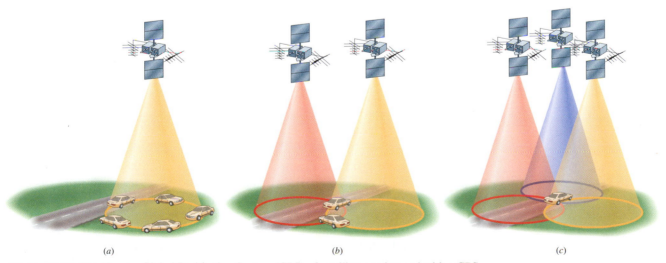

**Figure 5.13** The Navstar Global Positioning System (GPS) of satellites can be used with a GPS receiver to locate an object, such as a car, on the earth. (_a_) One satellite identifies the car as being somewhere on a circle. (_b_) A second places it on another circle, which identifies two possibilities for the exact spot. (_c_) A third provides the means for deciding where the car is.

circle. The intersection of the circles reveals two possible positions for the car, as in Figure 5.13*b*. With the aid of a third satellite, a third circle can be established, which intersects the other two and identifies the car's exact position, as in Figure 5.13*c*. The use of ground-based radio beacons to provide additional reference points leads to a system called Differential GPS, which can locate objects even more accurately than the satellite-based system alone. Navigational systems for automobiles and portable systems that tell hikers and people with visual impairments where they are located are two of the many uses for the GPS technique. GPS applications are so numerous that they have developed into a multibillion dollar industry.

Equation 5.5 applies to man-made earth satellites or to natural satellites like the moon. It also applies to circular orbits about any astronomical object, provided $M_E$ is replaced by the mass of the object on which the orbit is centered. Example 10, for instance, shows how scientists have applied this equation to conclude that a supermassive black hole is probably located at the center of the galaxy known as M87. This galaxy is located at a distance of about 50 million light-years away from the earth. (One light-year is the distance that light travels in a year, or $9.5 \times 10^{15}$ m.)

### Example 10    A Supermassive Black Hole

The Hubble telescope has detected the light being emitted from different regions of galaxy M87, which is shown in Figure 5.14. The black circle identifies the center of the galaxy. From the characteristics of this light, astronomers have determined an orbiting speed of $7.5 \times 10^5$ m/s for matter located at a distance of $5.7 \times 10^{17}$ m from the center. Find the mass $M$ of the object located at the galactic center.

**Reasoning and Solution** Replacing $M_E$ in Equation 5.5 with $M$ gives $v = \sqrt{GM/r}$, which can be solved to show that

$$M = \frac{v^2 r}{G} = \frac{(7.5 \times 10^5 \text{ m/s})^2 (5.7 \times 10^{17} \text{ m})}{6.67 \times 10^{-11} \text{ N} \cdot \text{m}^2 / \text{kg}^2} = \boxed{4.8 \times 10^{39} \text{ kg}}$$

The ratio of this incredibly large mass to the mass of our sun is $(4.8 \times 10^{39} \text{ kg})/(2.0 \times 10^{30} \text{ kg}) = 2.4 \times 10^9$. Thus, matter equivalent to 2.4 billion suns is located at the center of galaxy M87. Considering that the volume of space in which this matter is located contains relatively few visible stars, researchers have concluded that the data provide strong evidence for the existence of a supermassive black hole. The term "black hole" is used because the tremendous mass prevents even light from escaping. The light that forms the image in Figure 5.14 comes not from the black hole itself, but from matter that surrounds it.

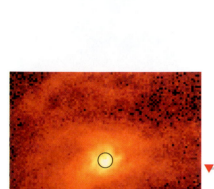

**Figure 5.14** This image of the ionized gas (yellow) at the heart of galaxy M87 was obtained by the Hubble Space Telescope. The circle identifies the center of the galaxy, at which a black hole is thought to exist. (Courtesy NASA and Space Telescope Science Institute)

**The physics of**
**locating a black hole.**

The period $T$ of a satellite is the time required for one orbital revolution. As in any uniform circular motion, the period is related to the speed of the motion by $v = 2\pi r/T$. Substituting $v$ from Equation 5.5 shows that

$$\sqrt{\frac{GM_E}{r}} = \frac{2\pi r}{T}$$

Solving this expression for the period $T$ gives

$$T = \frac{2\pi r^{3/2}}{\sqrt{GM_E}} \tag{5.6}$$

Although derived for earth orbits, Equation 5.6 can also be used for calculating the periods of those planets in nearly circular orbits about the sun, if $M_E$ is replaced by the mass $M_S$ of the sun and $r$ is interpreted as the distance between the center of the planet and the center of the sun. The fact that the period is proportional to the three-halves power of the orbital radius is known as Kepler's third law, and it is one of the laws discovered by Johannes Kepler (1571–1630) during his studies of planetary motion. Kepler's third law also holds for elliptical orbits, which will be discussed in Chapter 9.

An important application of Equation 5.6 occurs in the field of communications, where "synchronous satellites" are put into a circular orbit that is in the plane of the equator, as Figure 5.15 shows. The orbital period is chosen to be one day, which is also the time it takes for the earth to turn once about its axis. Therefore, these satellites move around their orbits in a way that is synchronized with the rotation of the earth. For earth-based observers, synchronous satellites have the useful characteristic of appearing in fixed positions in the sky and can serve as "stationary" relay stations for communication signals sent up from the earth's surface. This is exactly what is done in the digital satellite systems that are a popular alternative to cable TV. As the blowup in Figure 5.15 indicates, a small "dish" antenna on your house picks up the digital TV signals relayed to earth by the satellite. After being decoded, these signals are delivered to your TV set. All synchronous satellites are in orbit at the same height above the earth's surface, as Example 11 shows.

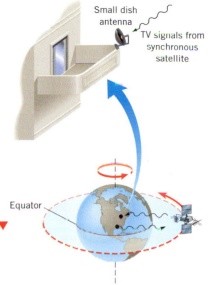

### Example 11   The Orbital Radius for Synchronous Satellites

What is the height $H$ above the earth's surface at which all synchronous satellites (regardless of mass) must be placed in orbit?

**Reasoning**  The period $T$ of a synchronous satellite is one day, so we can use Equation 5.6 to find the distance $r$ from the center of the earth. To find the height $H$ of the satellite above the earth's surface we will have to take into account the fact that the earth itself has a radius of $6.38 \times 10^6$ m.

**Solution**  A period of one day* corresponds to $T = 8.64 \times 10^4$ s. In using this value it is convenient to rearrange the equation $T = 2\pi r^{3/2}/\sqrt{GM_E}$ as follows:

$$r^{3/2} = \frac{T\sqrt{GM_E}}{2\pi} = \frac{(8.64 \times 10^4 \text{ s})\sqrt{(6.67 \times 10^{-11} \text{ N} \cdot \text{m}^2/\text{kg}^2)(5.98 \times 10^{24} \text{ kg})}}{2\pi}$$

By squaring and then taking the cube root, we find that $r = 4.23 \times 10^7$ m. Since the radius of the earth is approximately $6.38 \times 10^6$ m, the height of the satellite above the earth's surface is

$$H = 4.23 \times 10^7 \text{ m} - 0.64 \times 10^7 \text{ m} = \boxed{3.59 \times 10^7 \text{ m (22 300 mi)}}$$

**Figure 5.15**  A synchronous satellite orbits the earth once per day on a circular path that lies in the plane of the equator. Digital satellite system television uses such satellites as relay stations for TV signals that are sent up from the earth's surface and then rebroadcast down toward your own small dish antenna.

*The physics of*
**digital satellite system TV.**

# 5.6  *Apparent Weightlessness and Artificial Gravity*

The idea of life on board an orbiting satellite conjures up visions of astronauts floating around in a state of "weightlessness," as in Figure 5.16. Actually, this state should be called "apparent weightlessness," because it is similar to the condition of zero apparent weight that occurs in an elevator during free-fall. Conceptual Example 12 explores this similarity.

*The physics of*
**apparent weightlessness.**

### Conceptual Example 12   Apparent Weightlessness and Free-Fall

Figure 5.17 shows a person on a scale in a freely falling elevator and in a satellite in a circular orbit. In each case, what apparent weight is recorded by the scale?

**Reasoning and Solution**  As Section 4.8 discusses, apparent weight is the force that an object exerts on the platform of a scale. In the freely falling elevator in Figure 5.17a, the apparent weight is zero. The reason is that both the scale and the person on it fall together and, therefore, cannot push against one another. In the orbiting satellite in Figure 5.17b, both the person and the scale are in uniform circular motion. Objects in uniform circular motion continually accelerate or "fall" toward the center of the circle, in order to remain on the circular path. Consequently, the scale in Figure 5.17b and the person on it both "fall" with the same acceleration toward the center of the orbit and cannot push against one another. Thus, *the*

**Figure 5.16**  As she orbits the earth, astronaut Janet Kavandi floats around in a state of apparent weightlessness. (Courtesy NASA)

---

* Successive appearances of the sun define the solar day of 24 h or $8.64 \times 10^4$ s. The sun moves against the background of the stars, however, and the time required for the earth to turn once on its axis relative to the fixed stars is 23 h 56 min, which is called the sidereal day. The sidereal day should be used in Example 11, but the neglect of this effect introduces an error of less than 0.4% in the answer.

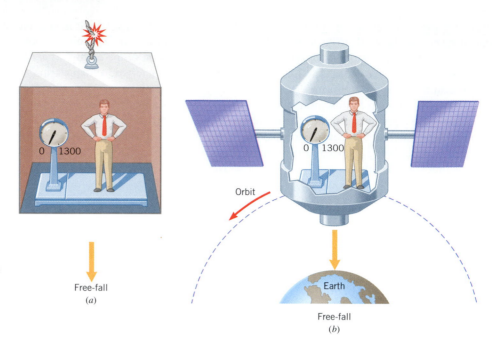

**Figure 5.17** (a) During free-fall, the elevator accelerates downward with the acceleration due to gravity, and the apparent weight of the person is zero. (b) The orbiting space station is also in free-fall toward the center of the earth.

Free-fall
(a)

Orbit

Earth

Free-fall
(b)

*apparent weight in the satellite is zero, just as it is in the freely falling elevator.* The only difference between the satellite and the elevator is that the satellite moves on a circle, so that its "falling" does not bring it closer to the earth. In contrast to the apparent weight, the true weight is the gravitational force ($F = GmM_E/r^2$) that the earth exerts on an object and is not zero in a freely falling elevator or aboard an orbiting satellite.

**The physics of artificial gravity.**

The physiological effects of prolonged apparent weightlessness are only partially known. To minimize such effects, it is likely that artificial gravity will be provided in large space stations of the future. To help explain artificial gravity, Figure 5.18 shows a space station rotating about an axis. Because of the rotational motion, any object located at a point $P$ on the interior surface of the station experiences a centripetal force directed toward the axis. The surface of the station provides this force by pushing on the feet of an astronaut, for instance. The centripetal force can be adjusted to match the astronaut's earth weight by properly selecting the rotational speed of the space station, as Examples 13 and 14 illustrate.

### Example 13    Artificial Gravity

At what speed must the surface of the space station ($r = 1700$ m) move in Figure 5.18, so that the astronaut at point $P$ experiences a push on his feet that equals his earth weight?

**Reasoning** The floor of the rotating space station exerts a normal force on the feet of the astronaut. This is the centripetal force ($F_c = mv^2/r$) that keeps the astronaut moving in a circular path. Since the magnitude of the normal force equals the astronaut's earth weight, we can determine the speed $v$ of the space station.

**Solution** The earth weight of the astronaut (mass = $m$) is $mg$, and with this substitution, Equation 5.3 can be used to determine the required speed: $F_c = mg = mv^2/r$. Solving this equation for the speed, we find that

$$v = \sqrt{rg} = \sqrt{(1700 \text{ m})(9.80 \text{ m/s}^2)} = \boxed{130 \text{ m/s}}$$

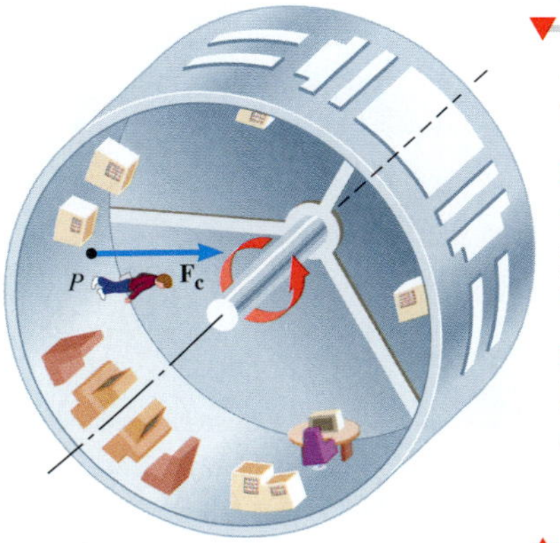

**Figure 5.18** The surface of the rotating space station pushes on an object with which it is in contact and thereby provides the centripetal force that keeps the object moving on a circular path.

### Example 14    A Rotating Space Laboratory

A space laboratory is rotating to create artificial gravity, as Figure 5.19 indicates. Its period of rotation is chosen so the outer ring ($r_O = 2150$ m) simulates the acceleration due to gravity on earth ($9.80 \text{ m/s}^2$). What should be the radius $r_1$ of the inner ring, so it simulates the acceleration due to gravity on the surface of Mars ($3.72 \text{ m/s}^2$)?

**Reasoning** The value given for either acceleration corresponds to the centripetal acceleration $a_c = v^2/r$ in the corresponding ring. Moreover, the speed $v$ and radius $r$ are related according to $v = 2\pi r/T$ (Equation 5.1), where $T$ is the period of the motion. Although no value is given for $T$, we do know that the laboratory is rigid. This is an important observation, because all points on a rigid object make one revolution in the same time. Therefore, both rings have the same period, a fact that will make a solution possible.

**Solution** Substituting Equation 5.1 into the expression for the centripetal acceleration gives

$$a_c = \frac{v^2}{r} = \frac{\left(\dfrac{2\pi r}{T}\right)^2}{r} = \frac{4\pi^2 r}{T^2}$$

Applying this result to both rings shows that

$$\underbrace{9.80 \text{ m/s}^2 = \frac{4\pi^2 (2150 \text{ m})}{T^2}}_{\text{Outer ring}} \quad \text{and} \quad \underbrace{3.72 \text{ m/s}^2 = \frac{4\pi^2 r_1}{T^2}}_{\text{Inner ring}}$$

Dividing the inner-ring expression by the outer-ring expression and eliminating the common algebraic factors of $4\pi^2$ and $T^2$ gives

$$\frac{3.72 \text{ m/s}^2}{9.80 \text{ m/s}^2} = \frac{r_1}{2150 \text{ m}} \quad \text{or} \quad r_1 = \boxed{816 \text{ m}}$$

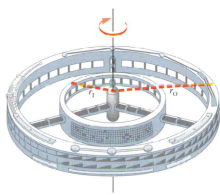

**Figure 5.19** The outer ring (radius = $r_O$) of this rotating space laboratory simulates gravity on earth, while the inner ring (radius = $r_1$) simulates gravity on Mars.

**Problem solving insight**
All points on a rotating rigid body have the same value for the period $T$ of the motion.

# Concept Summary

This summary presents an abridged version of the chapter, including the important equations and all available learning aids. For convenient reference, the learning aids (including the text's examples) are placed next to or immediately after the relevant equation or discussion. The following Learning Aids may be found on-line at **www.wiley.com/college/cutnell**:

| | |
|---|---|
| **Interactive LearningWare** examples are solved according to a five-step interactive format that is designed to help you develop problem-solving skills. | **Concept Simulations** are animated versions of text figures or animations that illustrate important concepts. You can control parameters that affect the display, and we encourage you to experiment. |
| **Interactive Solutions** offer specific models for certain types of problems in the chapter homework. The calculations are carried out interactively. | **Self-Assessment Tests** include both qualitative and quantitative questions. Extensive feedback is provided for both incorrect and correct answers, to help you evaluate your understanding of the material. |

| Topic | Discussion | Learning Aids |
|---|---|---|
| | **5.1 Uniform Circular Motion** | |
| | Uniform circular motion is the motion of an object traveling at a constant (uniform) speed on a circular path. | |
| Period of the motion | The period $T$ is the time required for the object to travel once around the circle. The speed $v$ of the object is related to the period and the radius $r$ of the circle by | **Example 1** |

$$v = \frac{2\pi r}{T} \qquad (5.1)$$

| Topic | Discussion | Learning Aids |
|---|---|---|
| | **5.2 Centripetal Acceleration** | |
| | An object in uniform circular motion experiences an acceleration, known as centripetal acceleration. The magnitude $a_c$ of the centripetal acceleration is | **Examples 2, 3, 4**<br><br>**Interactive LearningWare 5.1** |
| Centripetal acceleration (magnitude) | | **Interactive Solution 5.3** |

$$a_c = \frac{v^2}{r} \qquad (5.2)$$

| Topic | Discussion |
|---|---|
| | where $v$ is the speed of the object and $r$ is the radius of the circle. The direction |
| Centripetal acceleration (direction) | of the centripetal acceleration vector always points toward the center of the circle and continually changes as the object moves. |

| Topic | Discussion | Learning Aids |
|-------|------------|---------------|

### 5.3 Centripetal Force

To produce a centripetal acceleration, a net force pointing toward the center of the circle is required. This net force is called the centripetal force, and its magnitude $F_c$ is

**Centripetal force (magnitude)**

$$F_c = \frac{mv^2}{r}$$    (5.3)

where $m$ and $v$ are the mass and speed of the object, and $r$ is the radius of the

**Centripetal force (direction)**

circle. The direction of the centripetal force vector, like that of the centripetal acceleration vector, always points toward the center of the circle.

**Examples 5, 6, 7**

**Interactive LearningWare 5.2**

**Concept Simulations 5.1, 5.2**

**Interactive Solution 5.19**

 **Use *Self-Assessment Test 5.1* to evaluate your understanding of Sections 5.1–5.3.**

### 5.4 Banked Curves

A vehicle can negotiate a circular turn without relying on static friction to provide the centripetal force, provided the turn is banked at an angle relative to the horizontal. The angle $\theta$ at which a friction-free curve must be banked is related to the speed $v$ of the vehicle, the radius $r$ of the curve, and the magnitude $g$ of the acceleration due to gravity by

**Example 8**

**Interactive Solution 5.23**

**Angle of a banked turn**

$$\tan \theta = \frac{v^2}{rg}$$    (5.4)

### 5.5 Satellites in Circular Orbits

When a satellite orbits the earth, the gravitational force provides the centripetal force that keeps the satellite moving in a circular orbit. The speed $v$ and period $T$ of a satellite depend on the mass $M_E$ of the earth and the radius $r$ of the orbit according to

**Orbital speed**

$$v = \sqrt{\frac{GM_E}{r}}$$    (5.5)

**Examples 9, 10, 11**

**Orbital period**

$$T = \frac{2\pi r^{3/2}}{\sqrt{GM_E}}$$    (5.6)

**Interactive Solution 5.31**

where $G$ is the universal gravitational constant.

### 5.6 Apparent Weightlessness and Artificial Gravity

The apparent weight of an object is the force that it exerts on a scale with which it is in contact. All objects, including people, on board an orbiting satellite are in free-fall, since they experience negligible air resistance and they have an acceleration that is equal to the acceleration due to gravity. When a person is in free-fall, his or her apparent weight is zero, because both the person and the scale fall freely and cannot push against one another.

**Examples 12, 13, 14**

 **Use *Self-Assessment Test 5.2* to evaluate your understanding of Sections 5.4–5.6.**

## Problems

ssm   Solution is in the Student Solutions Manual.        www   Solution is available on the World Wide Web at www.wiley.com/college/cutnell
This icon represents a biomedical application.

**Section 5.1 Uniform Circular Motion, Section 5.2 Centripetal Acceleration**

**1.** ssm How long does it take a plane, traveling at a constant speed of 110 m/s, to fly once around a circle whose radius is 2850 m?

**2.** A car travels at a constant speed around a circular track whose radius is 2.6 km. The car goes once around the track in 360 s. What is the magnitude of the centripetal acceleration of the car?

**3.** Interactive Solution 5.3 at www.wiley.com/college/cutnell presents a method for modeling this problem. The blade of a windshield wiper moves through an angle of 90.0° in 0.40 s. The tip of the blade moves on the arc of a circle that has a radius of 0.45 m. What is the magnitude of the centripetal acceleration of the tip of the blade?

**4.** Review Conceptual Example 2 in preparation for this problem. In Figure 5.4, an object, after being released from its circular path, travels the distance *OA* in the same time it would have moved from *O* to *P* on the circle. The speed of the object on and off the circle remains constant at the same value. Suppose that the radius of the circle in Figure 5.4 is 3.6 m and the angle θ is 25°. What is the distance *OA*?

**5. ssm** Computer-controlled display screens provide drivers in the Indianapolis 500 with a variety of information about how their cars are performing. For instance, as a car is going through a turn, a speed of 221 mi/h (98.8 m/s) and a centripetal acceleration of 3.00*g* (three times the acceleration due to gravity) are displayed. Determine the radius of the turn (in meters).

**6.** There is a clever kitchen gadget for drying lettuce leaves after you wash them. It consists of a cylindrical container mounted so that it can be rotated about its axis by turning a hand crank. The outer wall of the cylinder is perforated with small holes. You put the wet leaves in the container and turn the crank to spin off the water. The radius of the container is 12 cm. When the cylinder is rotating at 2.0 revolutions per second, what is the magnitude of the centripetal acceleration at the outer wall?

**7.** A bicycle chain is wrapped around a rear sprocket (*r* = 0.039 m) and a front sprocket (*r* = 0.10 m). The chain moves with a speed of 1.4 m/s around the sprockets, while the bike moves at a constant velocity. Find the magnitude of the acceleration of a chain link that is in contact with (a) the rear sprocket, (b) neither sprocket, and (c) the front sprocket.

* **8.** Each of the space shuttle's main engines is fed liquid hydrogen by a high-pressure pump. Turbine blades inside the pump rotate at 617 rev/s. A point on one of the blades traces out a circle with a radius of 0.020 m as the blade rotates. (a) What is the magnitude of the centripetal acceleration that the blade must sustain at this point? (b) Express this acceleration as a multiple of *g* = 9.80 m/s².

* **9. ssm** The large blade of a helicopter is rotating in a horizontal circle. The length of the blade is 6.7 m, measured from its tip to the center of the circle. Find the ratio of the centripetal acceleration at the end of the blade to that which exists at a point located 3.0 m from the center of the circle.

* **10.** A centrifuge is a device in which a small container of material is rotated at a high speed on a circular path. Such a device is used in medical laboratories, for instance, to cause the more dense red blood cells to settle through the less dense blood serum and collect at the bottom of the container. Suppose the centripetal acceleration of the sample is 6.25 × 10³ times as large as the acceleration due to gravity. How many revolutions per minute is the sample making, if it is located at a radius of 5.00 cm from the axis of rotation?

**Section 5.3 Centripetal Force**

**11. ssm** A 0.015-kg ball is shot from the plunger of a pinball machine. Because of a centripetal force of 0.028 N, the ball follows a circular arc whose radius is 0.25 m. What is the speed of the ball?

**12.** In a skating stunt known as "crack-the-whip," a number of skaters hold hands and form a straight line. They try to skate so that the line rotates about the skater at one end, who acts as the pivot. The skater farthest out has a mass of 80.0 kg and is 6.10 m from the pivot. He is skating at a speed of 6.80 m/s. Determine the magnitude of the centripetal force that acts on him.

**13. Concept Simulation 5.1** at **www.wiley.com/college/cutnell** reviews the concepts that are involved in this problem. A child is twirling a 0.0120-kg ball on a string in a horizontal circle whose radius is 0.100 m. The ball travels once around the circle in 0.500 s. (a) Determine the centripetal force acting on the ball. (b) If the

speed is doubled, does the centripetal force double? If not, by what factor does the centripetal force increase?

**14.** At an amusement park there is a ride in which cylindrically shaped chambers spin around a central axis. People sit in seats facing the axis, their backs against the outer wall. At one instant the outer wall moves at a speed of 3.2 m/s, and an 83-kg person feels a 560-N force pressing against his back. What is the radius of a chamber?

**15. Concept Simulation 5.2** at **www.wiley.com/college/cutnell** reviews the concepts that play a role in this problem. A car is safely negotiating an unbanked circular turn at a speed of 21 m/s. The maximum static frictional force acts on the tires. Suddenly a wet patch in the road reduces the maximum static frictional force by a factor of three. If the car is to continue safely around the curve, to what speed must the driver slow the car?

**16.** A stone has a mass of 6.0 × 10⁻³ kg and is wedged into the tread of an automobile tire, as the drawing shows. The coefficient of static friction between the stone and each side of the tread channel is 0.90. When the tire surface is rotating at 13 m/s, the stone flies out of the tread. The magnitude $F_N$ of the normal force that each side of the tread channel exerts on the stone is 1.8 N. Assume that only static friction supplies the centripetal force, and determine the radius *r* of the tire.

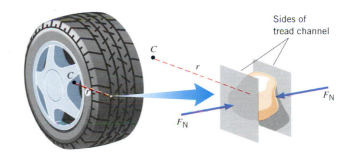

* **17. ssm www** The hammer throw is a track-and-field event in which a 7.3-kg ball (the "hammer") is whirled around in a circle several times and released. It then moves upward on the familiar curving path of projectile motion and eventually returns to earth some distance away. The world record for this distance is 86.75 m, achieved in 1986 by Yuriy Sedykh. Ignore air resistance and the fact that the ball is released above the ground rather than at ground level. Furthermore, assume that the ball is whirled on a circle that has a radius of 1.8 m and that its velocity at the instant of release is directed 41° above the horizontal. Find the magnitude of the centripetal force acting on the ball just prior to the moment of release.

* **18.** A block is hung by a string from the inside roof of a van. When the van goes straight ahead at a speed of 28 m/s, the block hangs vertically down. But when the van maintains this same speed around an unbanked curve (radius = 150 m), the block swings toward the outside of the curve. Then the string makes an angle θ with the vertical. Find θ.

* **19. Interactive Solution 5.19** at **www.wiley.com/college/cutnell** illustrates a method for modeling this problem. A "swing" ride at a carnival consists of chairs that are swung in a circle by 15.0-m cables attached to a vertical rotating pole, as the drawing shows. Suppose the total mass of a chair and its occupant is 179 kg. (a) Deter-

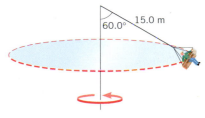

mine the tension in the cable attached to the chair. (b) Find the speed of the chair.

## Section 5.4 Banked Curves

**20.** At what angle should a curve of radius 150 m be banked, so cars can travel safely at 25 m/s without relying on friction?

**21.** **ssm** A curve of radius 120 m is banked at an angle of 18°. At what speed can it be negotiated under icy conditions where friction is negligible?

**22.** On a banked race track, the smallest circular path on which cars can move has a radius of 112 m, while the largest has a radius of 165 m, as the drawing illustrates. The height of the outer wall is 18 m. Find (a) the smallest and (b) the largest speed at which cars can move on this track without relying on friction.

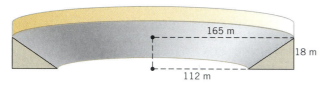

**23.** As an aid in working this problem, consult **Interactive Solution 5.23** at **www.wiley.com/college/cutnell.** A racetrack has the shape of an inverted cone, as the drawing shows. On this surface the cars race in circles that are parallel to the ground. For a speed of 34.0 m/s, at what value of the distance *d* should a driver locate his car if he wishes to stay on a circular path without depending on friction?

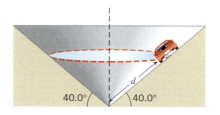

**24.** Before attempting this problem, review Examples 7 and 8. Two curves on a highway have the same radii. However, one is unbanked and the other is banked at an angle $\theta$. A car can safely travel along the unbanked curve at a maximum speed $v_0$ under conditions when the coefficient of static friction between the tires and the road is $\mu_s = 0.81$. The banked curve is frictionless, and the car can negotiate it at the same maximum speed $v_0$. Find the angle $\theta$ of the banked curve.

**25.** **ssm www** A jet ($m = 2.00 \times 10^5$ kg), flying at 123 m/s, banks to make a horizontal circular turn. The radius of the turn is 3810 m. Calculate the necessary lifting force.

**26.** The drawing shows a baggage carousel at an airport. Your suitcase has not slid all the way down the slope and is going around at a constant speed on a circle ($r = 11.0$ m) as the carousel turns. The coefficient of static fric-

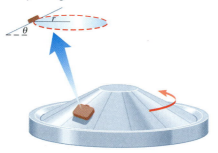

tion between the suitcase and the carousel is 0.760, and the angle $\theta$ in the drawing is 36.0°. How much time is required for your suitcase to go around once?

## Section 5.5 Satellites in Circular Orbits, Section 5.6 Apparent Weightlessness and Artificial Gravity

**27.** **ssm www** A satellite is in a circular orbit around an unknown planet. The satellite has a speed of $1.70 \times 10^4$ m/s, and the radius of the orbit is $5.25 \times 10^6$ m. A second satellite also has a circular orbit around this same planet. The orbit of this second satellite has a radius of $8.60 \times 10^6$ m. What is the orbital speed of the second satellite?

**28.** A rocket is used to place a synchronous satellite in orbit about the earth. What is the speed of the satellite in orbit?

**29.** A satellite is placed in orbit $6.00 \times 10^5$ m above the surface of Jupiter. Jupiter has a mass of $1.90 \times 10^{27}$ kg and a radius of $7.14 \times 10^7$ m. Find the orbital speed of the satellite.

**30.** The moon orbits the earth at a distance of $3.85 \times 10^8$ m. Assume that this distance is between the centers of the earth and the moon and that the mass of the earth is $5.98 \times 10^{24}$ kg. Find the period for the moon's motion around the earth. Express the answer in days and compare it to the length of a month.

**31.** Review **Interactive Solution 5.31** at **www.wiley.com/college/cutnell** before beginning this problem. Two satellites, A and B, are in different circular orbits about the earth. The orbital speed of satellite A is three times that of satellite B. Find the ratio $(T_A/T_B)$ of the periods of the satellites.

**\*32.** A satellite is in a circular earth orbit that has a radius of $6.7 \times 10^6$ m. A model airplane is flying on a 15-m guideline in a horizontal circle. The guideline is parallel to the ground. Find the speed of the plane such that the plane and the satellite have the same centripetal acceleration.

**\*33.** **ssm** A satellite has a mass of 5850 kg and is in a circular orbit $4.1 \times 10^5$ m above the surface of a planet. The period of the orbit is two hours. The radius of the planet is $4.15 \times 10^6$ m. What is the true weight of the satellite when it is at rest on the planet's surface?

**\*34.** The earth orbits the sun once a year at a distance of $1.50 \times 10^{11}$ m. Venus orbits the sun at a distance of $1.08 \times 10^{11}$ m. These distances are between the centers of the planets and the sun. How long (in earth days) does it take for Venus to make one orbit around the sun?

**\*\*35.** To create artificial gravity, the space station shown in the drawing is rotating at a rate of 1.00 rpm. The radii of the cylindrically shaped chambers have the ratio $r_A/r_B = 4.00$. Each chamber A simulates an acceleration due to gravity of 10.0 m/s². Find values for (a) $r_A$, (b) $r_B$, and (c) the acceleration due to gravity that is simulated in chamber B.

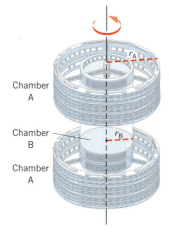

# Chapter 6 Work and Energy

## 6.1 Work Done by a Constant Force

Work is a familiar concept. For example, it takes work to push a stalled car. In fact, more work is done when the pushing force is greater or when the displacement of the car is greater. Force and displacement are, in fact, the two essential elements of work, as Figure 6.1 illustrates. The drawing shows a constant pushing force **F** that points in the same direction as the resulting displacement **s**.* In such a case, the work $W$ is defined as the magnitude $F$ of the force times the magnitude $s$ of the displacement: $W = Fs$. The work done to push a car is the same whether the car is moved north to south or east to west, provided that the amount of force used and the distance moved are the same. Since work does not convey directional information, it is a scalar quantity.

The equation $W = Fs$ indicates that the unit of work is the unit of force times the unit of distance, or the newton·meter in SI units. One newton·meter is referred to as a *joule* (J) (rhymes with "cool"), in honor of James Joule (1818–1889) and his research into the nature of work, energy, and heat. Table 6.1 on p. 104 summarizes the units for work in several systems of measurement.

The definition of work as $W = Fs$ does have one surprising feature: If the distance $s$ is zero, the work is zero, even if a force is applied. Pushing on an immovable object, such as a brick wall, may tire your muscles, but there is no work done of the type we are discussing. In physics, the idea of work is intimately tied up with the idea of motion. If the object does not move, the force acting on the object does no work.

Often, the force and displacement do not point in the same direction. For instance, Figure 6.2a on p. 104 shows a suitcase-on-wheels being pulled to the right by a force that is applied along the handle. The force is directed at an angle $\theta$ relative to the displacement. In such a case, only the component of the force along the displacement is used in defining work. As part *b* of the drawing illustrates, this component is $F \cos \theta$, and it appears in the general definition of work as follows:

---

■ **DEFINITION OF WORK DONE BY A CONSTANT[†] FORCE**

The work done on an object by a constant force **F** is

$$W = (F \cos \theta)s \qquad (6.1)$$

where $F$ is the magnitude of the force, $s$ is the magnitude of the displacement, and $\theta$ is the angle between the force and the displacement.

**SI Unit of Work:** newton·meter = joule (J)

---

When the force points in the same direction as the displacement, then $\theta = 0°$, and Equation 6.1 reduces to $W = Fs$. The following example illustrates how Equation 6.1 is used to calculate work.

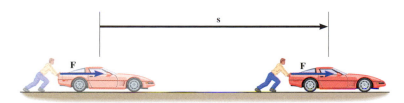

**Figure 6.1** Work is done when a force **F** pushes a car through a displacement **s**.

---

* When discussing work, it is customary to use the symbol **s** for the displacement, rather than **x** or **y**.
† Section 6.9 considers the work done by a variable force.

**Table 6.1**   *Units of Measurement for Work*

| System | Force | × | Distance | = | Work |
|--------|-------|---|----------|---|------|
| SI | newton (N) | | meter (m) | | joule (J) |
| CGS | dyne (dyn) | | centimeter (cm) | | erg |
| BE | pound (lb) | | foot (ft) | | foot · pound (ft · lb) |

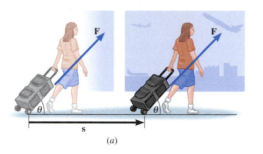

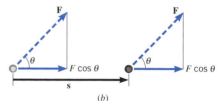

**Figure 6.2** (*a*) Work can be done by a force **F** that points at an angle *θ* relative to the displacement **s**. (*b*) The force component that points along the displacement is *F* cos *θ*.

(*a*)

(*b*)

## Example 1   Pulling a Suitcase-on-Wheels

Find the work done by a 45.0-N force in pulling the suitcase in Figure 6.2*a* at an angle $\theta = 50.0°$ for a distance $s = 75.0$ m.

**Reasoning**   The pulling force causes the suitcase to move a distance of 75.0 m and does work. However, the force makes an angle of 50.0° with the displacement, and we must take this angle into account by using the definition of work given by Equation 6.1.

**Solution**   The work done by the 45.0-N force is

$$W = (F \cos \theta)s = [(45.0 \text{ N}) \cos 50.0°](75.0 \text{ m}) = \boxed{2170 \text{ J}}$$

The answer is expressed in newton · meters or joules (J).

The definition of work in Equation 6.1 takes into account only the component of the force in the direction of the displacement. The force component perpendicular to the displacement does no work. To do work, there must be a force *and* a displacement, and since there is no displacement in the perpendicular direction, there is no work done by the perpendicular component of the force. If the entire force is perpendicular to the displacement, the angle *θ* in Equation 6.1 is 90°, and the force does no work at all.

Work can be either positive or negative, depending on whether a component of the force points in the same direction as the displacement or in the opposite direction. Example 2 illustrates how positive and negative work arise.

**Figure 6.3** (*a*) In the bench press, work is done during both the lifting and lowering phases. (©Patrick Bennet/Corbis Images) (*b*) During the lifting phase, the force **F** does positive work. (*c*) During the lowering phase, the force does negative work.

## Example 2   Bench-Pressing

The weight lifter in Figure 6.3*a* is bench-pressing a barbell whose weight is 710 N. In part *b* of the figure, he raises the barbell a distance of 0.65 m above his chest, and in part *c* he lowers it

(*a*)

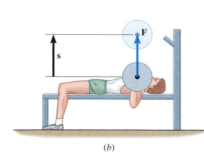

(*b*)

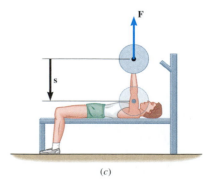

(*c*)

the same distance. The weight is raised and lowered at a constant velocity. Determine the work done on the barbell by the weight lifter during (a) the lifting phase and (b) the lowering phase.

**Reasoning** To calculate the work, it is necessary to know the force exerted by the weight lifter. The barbell is raised and lowered at a constant velocity and, therefore, is in equilibrium. Consequently, the force **F** exerted by the weight lifter must balance the weight of the barbell, so $F = 710$ N. During the lifting phase, the force **F** and displacement **s** are in the same direction, as Figure 6.3b shows. The angle between them is $\theta = 0°$. When the barbell is lowered, however, the force and displacement are in opposite directions, as in Figure 6.3c. The angle between the force and the displacement is now $\theta = 180°$. With these observations, we can find the work.

**Solution**

(a) During the lifting phase, the work done by the force **F** is given by Equation 6.1 as

$$W = (F \cos \theta)s = [(710 \text{ N}) \cos 0°](0.65 \text{ m}) = 460 \text{ J}$$

(b) The work done during the lowering phase is

$$W = (F \cos \theta)s$$
$$= [(710 \text{ N}) \cos 180°](0.65 \text{ m}) = \boxed{-460 \text{ J}}$$

since $\cos 180° = -1$. The work is negative, because the force is opposite to the displacement. Weight lifters call each complete up-and-down movement of the barbell a repetition, or "rep." The lifting of the weight is referred to as the positive part of the rep, and the lowering is known as the negative part.

*The physics of positive and negative "reps" in weight lifting.*

Example 3 deals with the work done by a static frictional force when it acts on a crate that is resting on the bed of an accelerating truck.

## Example 3   Accelerating a Crate

Figure 6.4a shows a 120-kg crate on the flatbed of a truck that is moving with an acceleration of $a = +1.5$ m/s$^2$ along the positive $x$ axis. The crate does not slip with respect to the truck, as the truck undergoes a displacement whose magnitude is $s = 65$ m. What is the total work done on the crate by all of the forces acting on it?

**Reasoning** The free-body diagram in Figure 6.4b shows the forces that act on the crate: (1) the weight **W** of the crate, (2) the normal force **F**$_N$ exerted by the flatbed, and (3) the static frictional force **f**$_s$, which is exerted by the flatbed in the forward direction and keeps the crate from slipping backward. The weight and the normal force are perpendicular to the displacement, so they do no work. Only the static frictional force does work, since it acts in the $x$ direction. To determine the frictional force, we note that the crate does not slip and, therefore, must have the same acceleration of $a = +1.5$ m/s$^2$ as does the truck. The force creating this acceleration is the static frictional force, and knowing the mass of the crate and its acceleration, we can use Newton's second law to obtain its magnitude. Then, knowing the frictional force and the displacement, we can determine the total work done on the crate.

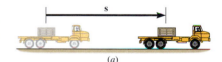

(a)

**Solution** From Newton's second law, we find that the magnitude $f_s$ of the static frictional force is

$$f_s = ma = (120 \text{ kg})(1.5 \text{ m/s}^2) = 180 \text{ N}$$

The total work is that done by the static frictional force and is

$$W = (f_s \cos \theta)s = (180 \text{ N})(\cos 0°)(65 \text{ m}) = \boxed{1.2 \times 10^4 \text{ J}} \tag{6.1}$$

The work is positive, because the frictional force is in the same direction as the displacement ($\theta = 0°$).

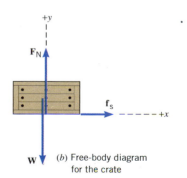

(b) Free-body diagram for the crate

**Figure 6.4** (a) The truck and crate are accelerating to the right for a distance of $s = 65$ m. (b) The free-body diagram for the crate.

# 6.2  The Work–Energy Theorem and Kinetic Energy

Most people expect that if you do work, you get something as a result. In physics, when a net force performs work on an object, there is always a result from the effort. The result is a change in the **kinetic energy** of the object. As we will now see, the relationship that re-

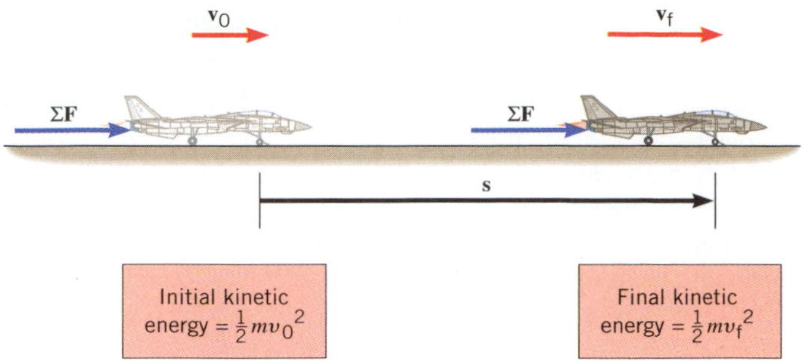

**Figure 6.5** A constant net external force $\Sigma\mathbf{F}$ acts over a displacement $\mathbf{s}$ and does work on the plane. As a result of the work done, the plane's kinetic energy changes.

lates work to the change in kinetic energy is known as the **work–energy theorem.** This theorem is obtained by bringing together three basic concepts that we've already learned about. First we'll apply Newton's second law of motion, $\Sigma F = ma$, which relates the net force $\Sigma F$ to the acceleration $a$ of an object. Then, we'll determine the work done by the net force when the object moves through a certain distance. Finally, we'll use Equation 2.9, one of the equations of kinematics, to relate the distance and acceleration to the initial and final speeds of the object. The result of this approach will be the work–energy theorem.

To gain some insight into the idea of kinetic energy and the work–energy theorem, look at Figure 6.5, where a constant net external force $\Sigma\mathbf{F}$ acts on an airplane of mass $m$. This net force is the vector sum of all the external forces acting on the plane, and, for simplicity, it is assumed to have the same direction as the displacement $\mathbf{s}$. According to Newton's second law, the net force produces an acceleration $a$, given by $a = \Sigma F/m$. Consequently, the speed of the plane changes from an initial value of $v_0$ to a final value of $v_f$.* Multiplying both sides of $\Sigma F = ma$ by the distance $s$ gives

$$\underbrace{(\Sigma F)s}_{\substack{\text{Work done by} \\ \text{net ext. force}}} = mas$$

The left side of this equation is the work done by the net external force. The term $as$ on the right side can be related to $v_0$ and $v_f$ by using Equation 2.9 ($v_f^2 = v_0^2 + 2as$) from the equations of kinematics. Solving this equation to give $as = \frac{1}{2}(v_f^2 - v_0^2)$ and substituting into $(\Sigma F)s = mas$ shows that

$$\underbrace{(\Sigma F)s}_{\substack{\text{Work done by} \\ \text{net ext. force}}} = \underbrace{\tfrac{1}{2}mv_f^2}_{\substack{\text{Final} \\ \text{KE}}} - \underbrace{\tfrac{1}{2}mv_0^2}_{\substack{\text{Initial} \\ \text{KE}}}$$

This expression is the work–energy theorem. Its left side is the work $W$ done by the net external force, while its right side involves the difference between two terms, each of which has the form $\frac{1}{2}(\text{mass})(\text{speed})^2$. The quantity $\frac{1}{2}(\text{mass})(\text{speed})^2$ is called kinetic energy (KE) and plays a significant role in physics, as we will soon see.

■ **DEFINITION OF KINETIC ENERGY**

The kinetic energy KE of an object with mass $m$ and speed $v$ is given by

$$\text{KE} = \tfrac{1}{2}mv^2 \tag{6.2}$$

*SI Unit of Kinetic Energy:* joule (J)

The SI unit of kinetic energy is the same as the unit for work, the joule. Kinetic energy, like work, is a scalar quantity. These are not surprising observations, for work and kinetic energy are closely related, as is clear from the following statement of the work–energy theorem.

*For extra emphasis, the final speed is now represented by the symbol $v_f$, rather than $v$.

■ **THE WORK–ENERGY THEOREM**

When a net external force does work $W$ on an object, the kinetic energy of the object changes from its initial value of $KE_0$ to a final value of $KE_f$, the difference between the two values being equal to the work:

$$W = KE_f - KE_0 = \tfrac{1}{2}mv_f^2 - \tfrac{1}{2}mv_0^2 \tag{6.3}$$

The work–energy theorem may be derived for any direction of the force relative to the displacement, not just the situation in Figure 6.5. In fact, the force may even vary from point to point along a path that is curved rather than straight, and the theorem remains valid. According to the work–energy theorem, a moving object has kinetic energy, because work was done to accelerate the object from rest to a speed $v_f$.* Conversely, an object with kinetic energy can perform work, if it is allowed to push or pull on another object. Example 4 illustrates the work–energy theorem and considers a single force that does work to change the kinetic energy of a space probe.

## Example 4 *Deep Space 1*

The space probe *Deep Space 1* was launched October 24, 1998. Its mass was 474 kg. The goal of the mission was to test a new kind of engine called an ion propulsion drive, which generates only a weak thrust, but can do so for long periods of time using only small amounts of fuel. The mission has been spectacularly successful. Consider the probe traveling at an initial speed of $v_0 = 275$ m/s. No forces act on it except the 56.0-mN thrust of its engine. This external force **F** is directed parallel to the displacement **s** of magnitude $2.42 \times 10^9$ m (Figure 6.6). Determine the final speed of the probe, assuming that the mass remains nearly constant.

*The physics of* **an ion propulsion drive.**

**Reasoning** Since the force **F** is the only force acting on the probe, it is the net external force, and the work it does causes the kinetic energy of the spacecraft to change. The work can be calculated directly from Equation 6.1 and is positive, because **F** points in the same direction as the displacement **s**. According to the work–energy theorem ($W = KE_f - KE_0$), a positive value for $W$ means that the kinetic energy increases. We can use the theorem to find the final kinetic energy and, hence, the final speed. Since the kinetic energy increases, we expect the final speed to be greater than the initial speed.

**Solution** From the definition of work (Equation 6.1), we have

$$W = (F \cos \theta)s = [(56.0 \times 10^{-3} \text{ N}) \cos 0°](2.42 \times 10^9 \text{ m}) = 1.36 \times 10^8 \text{ J}$$

where $\theta = 0°$, because the force and displacement are in the same direction (see the drawing). Since $W = KE_f - KE_0$ according to the work–energy theorem, the final kinetic energy of the probe is

$$KE_f = W + KE_0 = (1.36 \times 10^8 \text{ J}) + \tfrac{1}{2}(474 \text{ kg})(275 \text{ m/s})^2 = 1.54 \times 10^8 \text{ J}$$

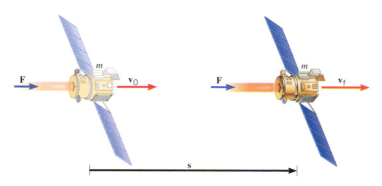

**Figure 6.6** The engine of *Deep Space 1* generates a single force **F** that points in the same direction as the displacement **s**. The force performs positive work, causing the probe to gain kinetic energy.

* Strictly speaking, the work–energy theorem, as given by Equation 6.3, applies only to a single particle, which occupies a mathematical point in space. A macroscopic object, however, is a collection or system of particles and is spread out over a region of space. Therefore, when a force is applied to a macroscopic object, the point of application of the force may be anywhere on the object. To take into account this and other factors, a discussion of work and energy is required that is beyond the scope of this text. The interested reader may refer to A. B. Arons, *The Physics Teacher,* October 1989, p. 506.

The final kinetic energy is $KE_f = \frac{1}{2}mv_f^2$, so the final speed is

$$v_f = \sqrt{\frac{2(KE_f)}{m}} = \sqrt{\frac{2(1.54 \times 10^8 \text{ J})}{474 \text{ kg}}} = \boxed{806 \text{ m/s}}$$

As expected, the final speed is greater than the initial speed.

In Example 4 only the force of the engine does work. If several external forces act on an object, they must be added together vectorially to give the net force. The work done by the net force can then be related to the change in the object's kinetic energy by using the work–energy theorem, as in the next example.

### *Example 5*   Downhill Skiing

A 58-kg skier is coasting down a 25° slope, as Figure 6.7*a* shows. A kinetic frictional force of magnitude $f_k = 70$ N opposes her motion. Near the top of the slope, the skier's speed is $v_0 = 3.6$ m/s. Ignoring air resistance, determine the speed $v_f$ at a point that is displaced 57 m downhill.

**Reasoning** As in Example 4, we will use the work–energy theorem to find the final speed. Note that the work done by the net external force is needed, not just the work done by any single force. The net force is the vector sum of the forces shown in the free-body diagram in Figure 6.7*b*. In this diagram, the normal force $\mathbf{F_N}$ is balanced by the component of the skier's weight ($mg \cos 25°$) perpendicular to the slope, because no acceleration occurs along that direction. Therefore, the net force points along the *x* axis.

**Solution** The net external force in Figure 6.7*b* points along the *x* axis and is

$$\Sigma F = mg \sin 25° - f_k = (58 \text{ kg})(9.80 \text{ m/s}^2) \sin 25° - 70 \text{ N} = +170 \text{ N}$$

The work done by the net force is

$$W = (\Sigma F \cos \theta)s = [(170 \text{ N}) \cos 0°](57 \text{ m}) = 9700 \text{ J} \qquad (6.1)$$

where $\theta = 0°$ because the net force and the displacement point in the same direction down the slope. From the work–energy theorem ($W = KE_f - KE_0$), it follows that the final kinetic energy of the skier is

$$KE_f = W + KE_0$$
$$= 9700 \text{ J} + \frac{1}{2}(58 \text{ kg})(3.6 \text{ m/s})^2 = 10\ 100 \text{ J}$$

Since the final kinetic energy is $KE_f = \frac{1}{2}mv_f^2$, the final speed of the skier is

$$v_f = \sqrt{\frac{2(KE_f)}{m}} = \sqrt{\frac{2(10\ 100 \text{ J})}{58 \text{ kg}}} = \boxed{19 \text{ m/s}}$$

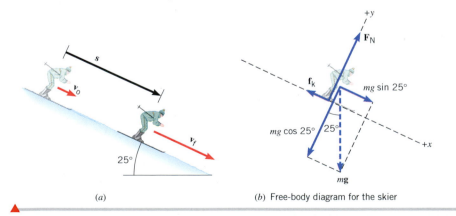

**Figure 6.7** (*a*) A skier coasting downhill. (*b*) The free-body diagram for the skier.

(*a*)

(*b*) Free-body diagram for the skier

**Problem solving insight**

Example 5 emphasizes that *the work–energy theorem deals with the work done by the net external force. The work–energy theorem does not apply to the work done by an individual force,* unless that force happens to be the only one present, in which case it is the net force. If the work done by the net force is *positive,* as in Example 5, the kinetic energy of the object *increases.* If the work done is *negative,* the kinetic energy *decreases.* If the work is zero, the kinetic energy remains the same. Conceptual Example 6 explores these ideas further.

### Conceptual Example 6  Work and Kinetic Energy

Figure 6.8 illustrates a satellite moving about the earth in a circular orbit and in an elliptical orbit. The only external force that acts on the satellite is the gravitational force. For these two orbits, determine whether the kinetic energy of the satellite changes during the motion.

**Reasoning and Solution** For the circular orbit in Figure 6.8*a* the gravitational force **F** does no work on the satellite, since the force is perpendicular to the instantaneous displacement **s** at all times. The gravitational force is the only external force acting on the satellite, so it is the net force. Thus, the work done by the net force is zero, and according to the work–energy theorem (Equation 6.3), the kinetic energy of the satellite (and, hence, its speed) remains the same everywhere on the orbit.

In contrast, the gravitational force does do work on a satellite in an elliptical orbit. For example, as the satellite moves toward the earth in the top part of Figure 6.8*b*, there is a component of the gravitational force that points in the same direction as the displacement. Consequently, the gravitational force does positive work during this part of the orbit, and the kinetic energy of the satellite increases. When the satellite moves away from the earth, as in the lower part of Figure 6.8*b*, the gravitational force has a component that points opposite to the displacement. Now, the gravitational force does negative work, and the kinetic energy of the satellite decreases.

Can you identify the two places on an elliptical orbit where the gravitational force does no work?

**Related Homework:** *Problem 16*

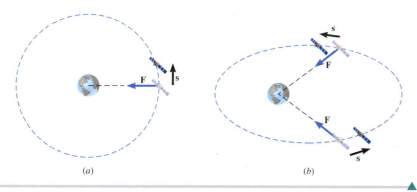

(a)                                    (b)

**Figure 6.8** (*a*) In a circular orbit, the gravitational force **F** is always perpendicular to the displacement **s** of the satellite and does no work. (*b*) In an elliptical orbit, there can be a component of the force along the displacement, and, consequently, work can be done.

## 6.3  *Gravitational Potential Energy*
### WORK DONE BY THE FORCE OF GRAVITY

The gravitational force is a well-known force that can do positive or negative work, and Figure 6.9 helps to show how the work can be determined. This drawing depicts a basketball of mass $m$ moving vertically downward, the force of gravity $mg$ being the only force acting on the ball. The initial height of the ball is $h_0$, and the final height is $h_f$, both distances measured from the earth's surface. The displacement **s** is downward and has a magnitude of $s = h_0 - h_f$. To calculate the work $W_{gravity}$ done on the ball by the force of gravity, we use $W = (F \cos \theta)s$ with $F = mg$ and $\theta = 0°$, since the force and displacement are in the same direction:

$$W_{gravity} = (mg \cos 0°)(h_0 - h_f) = mg(h_0 - h_f) \qquad (6.4)$$

Equation 6.4 is valid for *any path* taken between the initial and final heights, and not just for the straight-down path shown in Figure 6.9. For example, the same expression can be derived for both paths shown in Figure 6.10. Thus, only the *difference in vertical dis-*

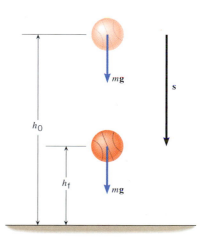

**Figure 6.9** Gravity exerts a force $mg$ on the basketball. Work is done by the gravitational force as the basketball falls from a height of $h_0$ to a height of $h_f$.

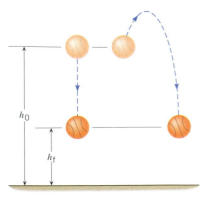

**Figure 6.10** An object can move along different paths in going from an initial height of $h_0$ to a final height of $h_f$. In each case, the work done by the gravitational force is the same $[W_{gravity} = mg(h_0 - h_f)]$, since the change in vertical distance $(h_0 - h_f)$ is the same.

tances $(h_0 - h_f)$ need be considered when calculating the work done by gravity. Since the difference in the vertical distances is the same for each path in the drawing, the work done by gravity is the same in each case. We are assuming here that the difference in heights is small compared to the radius of the earth, so that the magnitude $g$ of the acceleration due to gravity is the same at every height. Moreover, for positions close to the earth's surface, we can use the value of $g = 9.80 \text{ m/s}^2$.

Since only the difference between $h_0$ and $h_f$ appears in Equation 6.4, the vertical distances themselves need not be measured from the earth. For instance, they could be measured relative to a zero level that is one meter above the ground, and $h_0 - h_f$ would still have the same value. Example 7 illustrates how the work done by gravity is used in conjunction with the work–energy theorem.

### Example 7   A Gymnast on a Trampoline

A gymnast springs vertically upward from a trampoline as in Figure 6.11a. The gymnast leaves the trampoline at a height of 1.20 m and reaches a maximum height of 4.80 m before falling back down. All heights are measured with respect to the ground. Ignoring air resistance, determine the initial speed $v_0$ with which the gymnast leaves the trampoline.

**Reasoning** We can find the initial speed of the gymnast (mass = $m$) by using the work–energy theorem, provided the work done by the net external force can be determined. Since only the gravitational force acts on the gymnast in the air, it is the net force, and we can evaluate the work by using the relation $W_{gravity} = mg(h_0 - h_f)$.

**Solution** Figure 6.11b shows the gymnast moving upward. The initial and final heights are $h_0 = 1.20 \text{ m}$ and $h_f = 4.80 \text{ m}$, respectively. The initial speed is $v_0$ and the final speed is $v_f = 0 \text{ m/s}$, since the gymnast comes to a momentary halt at the highest point. Since $v_f = 0 \text{ m/s}$, the final kinetic energy is $KE_f = 0 \text{ J}$, and the work–energy theorem becomes $W = KE_f - KE_0 = -KE_0$. The work $W$ is that due to gravity, so this theorem reduces to $W_{gravity} = mg(h_0 - h_f) = -\frac{1}{2}mv_0^2$. Solving for $v_0$ gives

$$v_0 = \sqrt{-2g(h_0 - h_f)} = \sqrt{-2(9.80 \text{ m/s}^2)(1.20 \text{ m} - 4.80 \text{ m})} = \boxed{8.40 \text{ m/s}}$$

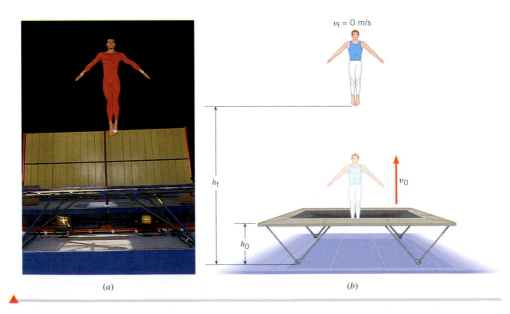

**Figure 6.11** (a) A gymnast bounces on a trampoline. (©James D. Wilson Liaison/Getty Images) (b) The gymnast moves upward with an initial speed $v_0$ and reaches maximum height with a final speed of zero.

## GRAVITATIONAL POTENTIAL ENERGY

We have seen that an object in motion has kinetic energy. There are also other types of energy. For example, an object may possess energy by virtue of its position relative to the earth; such an object is said to have gravitational potential energy. A pile driver, for instance, is used by construction workers to pound "piles," or structural support beams, into the ground. The pile driver contains a massive hammer that is raised to a height $h$ and then dropped (see Figure 6.12). As a result, the hammer has the potential to do the work of driving the pile into the ground. The greater the height of the hammer, the greater is the potential for doing work, and the greater is the gravitational potential energy.

Now, let's obtain an expression for the gravitational potential energy. Our starting point is Equation 6.4 for the work done by the gravitational force as an object moves from an initial height $h_0$ to a final height $h_f$:

$$W_{\text{gravity}} = \underbrace{mgh_0}_{\substack{\text{Initial} \\ \text{gravitational} \\ \text{potential energy} \\ \text{PE}_0}} - \underbrace{mgh_f}_{\substack{\text{Final} \\ \text{gravitational} \\ \text{potential energy} \\ \text{PE}_f}} \qquad (6.4)$$

This equation indicates that the work done by the gravitational force is equal to the difference between the initial and final values of the quantity $mgh$. The value of $mgh$ is larger when the height is larger and smaller when the height is smaller. We are led, then, to identify the quantity $mgh$ as the **gravitational potential energy.** The concept of potential energy is associated only with a type of force known as a "conservative" force, as we will discuss in Section 6.4.

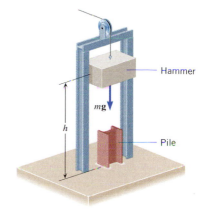

**Figure 6.12** In a pile driver, the gravitational potential energy of the hammer relative to the ground is PE = $mgh$.

---

■ **DEFINITION OF GRAVITATIONAL POTENTIAL ENERGY**

The gravitational potential energy PE is the energy that an object of mass $m$ has by virtue of its position relative to the surface of the earth. That position is measured by the height $h$ of the object relative to an arbitrary zero level:

$$\text{PE} = mgh \qquad (6.5)$$

*SI Unit of Gravitational Potential Energy:* joule (J)

---

Gravitational potential energy, like work and kinetic energy, is a scalar quantity and has the same SI unit as they do, the joule. It is the *difference* between two potential energies that is related by Equation 6.4 to the work done by the force of gravity. Therefore, the zero level for the heights can be taken anywhere, as long as both $h_0$ and $h_f$ are measured relative to the same zero level. The gravitational potential energy depends on both the object and the earth ($m$ and $g$, respectively), as well as the height $h$. Therefore, the gravitational potential energy belongs to the object and the earth as a system, although one often speaks of the object alone as possessing the gravitational potential energy.

# 6.4 *Conservative Versus Nonconservative Forces*

The gravitational force has an interesting property that when an object is moved from one place to another, the work done by the gravitational force does not depend on the choice of path. In Figure 6.10, for instance, an object moves from an initial height $h_0$ to a final height $h_f$ along two different paths. As Section 6.3 discusses, the work done by gravity depends only on the initial and final heights, and not on the path between these heights. For this reason, the gravitational force is called a **conservative force,** according to version 1 of the following definition:

---

■ **DEFINITION OF A CONSERVATIVE FORCE**

*Version 1* A force is conservative when the work it does on a moving object is independent of the path between the object's initial and final positions.
*Version 2* A force is conservative when it does no net work on an object moving around a closed path, starting and finishing at the same point.

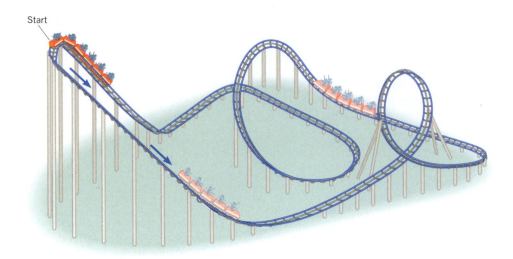

Start

**Figure 6.13** A roller coaster track is an example of a closed path.

The gravitational force is our first example of a conservative force. Later, we will encounter others, such as the elastic force of a spring and the electrical force of electrically charged particles. With each conservative force we will associate a potential energy, as we have already done in the gravitational case (see Equation 6.5). For other conservative forces, however, the algebraic form of the potential energy will differ from that in Equation 6.5.

Figure 6.13 helps us to illustrate version 2 of the definition of a conservative force. The picture shows a roller coaster car racing through dips and double dips, ultimately returning to its starting point. This kind of path, which begins and ends at the same place, is called a *closed* path. Gravity provides the only force that does work on the car, assuming that there is no friction or air resistance. Of course, the track exerts a normal force, but this force is always directed perpendicular to the motion and does no work. On the downward parts of the trip, the gravitational force does positive work, increasing the car's kinetic energy. Conversely, on the upward parts of the motion, the gravitational force does negative work, decreasing the car's kinetic energy. Over the entire trip, the gravitational force does as much positive work as negative work, so the net work is zero, and the car returns to its starting point with the same kinetic energy it had at the start. Therefore, consistent with version 2 of the definition of a conservative force, $W_{\text{gravity}} = 0 \, \text{J}$ for a closed path.

Not all forces are conservative forces. A force is nonconservative if the work it does on an object moving between two points depends on the path of the motion between the points. The kinetic frictional force is one example of a nonconservative force. When an object slides over a surface, the kinetic frictional force points opposite to the sliding motion and does negative work. Between any two points, greater amounts of work are done over longer paths between the points, so that the work depends on the choice of path. Thus, the kinetic frictional force is nonconservative. Air resistance is another nonconservative force. The concept of potential energy is not defined for a nonconservative force.

For a closed path, the total work done by a nonconservative force is not zero as it is for a conservative force. In Figure 6.13, for instance, a frictional force would oppose the motion and slow down the car. Unlike gravity, friction would do negative work on the car throughout the entire trip, on *both* the up and down parts of the motion. Assuming that the car makes it back to the starting point, the car would have *less* kinetic energy than it had originally. Table 6.2 gives some examples of conservative and nonconservative forces.

In normal situations, conservative forces (such as gravity) and nonconservative forces (such as friction and air resistance) act simultaneously on an object. Therefore, we write the work $W$ done by the net external force as $W = W_{\text{c}} + W_{\text{nc}}$, where $W_{\text{c}}$ is the work done by the conservative forces and $W_{\text{nc}}$ is the work done by the nonconservative forces. According to the work–energy theorem, the work done by the net external force is equal to the change in the object's kinetic energy, or $W_{\text{c}} + W_{\text{nc}} = \frac{1}{2}mv_{\text{f}}^2 - \frac{1}{2}mv_0^2$. If the only con-

**Table 6.2   Some Conservative and Nonconservative Forces**

**Conservative Forces**
  Gravitational force (Ch. 4)
  Elastic spring force (Ch. 10)
  Electric force (Ch. 18, 19)

**Nonconservative Forces**
  Static and kinetic frictional forces
  Air resistance
  Tension
  Normal force
  Propulsion force of a rocket

servative force acting is the gravitational force, then $W_c = W_{gravity} = mg(h_0 - h_f)$, and the work–energy theorem becomes

$$mg(h_0 - h_f) + W_{nc} = \tfrac{1}{2}mv_f^2 - \tfrac{1}{2}mv_0^2$$

The work done by the gravitational force can be moved to the right side of this equation, with the result that

$$W_{nc} = (\tfrac{1}{2}mv_f^2 - \tfrac{1}{2}mv_0^2) + (mgh_f - mgh_0) \tag{6.6}$$

In terms of kinetic and potential energies, we find that

$$\underbrace{W_{nc}}_{\substack{\text{Net work done by} \\ \text{nonconservative} \\ \text{forces}}} = \underbrace{(KE_f - KE_0)}_{\substack{\text{Change in} \\ \text{kinetic energy}}} + \underbrace{(PE_f - PE_0)}_{\substack{\text{Change in} \\ \text{gravitational} \\ \text{potential energy}}} \tag{6.7a}$$

Equation 6.7a states that the net work $W_{nc}$ done by all the external nonconservative forces equals the change in the object's kinetic energy plus the change in its gravitational potential energy. It is customary to use the delta symbol ($\Delta$) to denote such changes; thus, $\Delta KE = (KE_f - KE_0)$ and $\Delta PE = (PE_f - PE_0)$. With the delta notation, the work–energy theorem takes the form

$$W_{nc} = \Delta KE + \Delta PE \tag{6.7b}$$

In the next two sections, we will show why the form of the work–energy theorem expressed by Equations 6.7a and 6.7b is useful.

**Problem solving insight**
The change in both the kinetic and the potential energy is always the final value minus the initial value: $\Delta KE = KE_f - KE_0$ and $\Delta PE = PE_f - PE_0$.

## 6.5 The Conservation of Mechanical Energy

The concept of work and the work–energy theorem have led us to the conclusion that an object can possess two kinds of energy: kinetic energy, KE, and gravitational potential energy, PE. The sum of these two energies is called the **total mechanical energy** $E$, so that $E = KE + PE$. The concept of total mechanical energy will be extremely useful in describing the motion of objects in this and other chapters.

By rearranging the terms on the right side of Equation 6.7a, the work–energy theorem can be expressed in terms of the total mechanical energy:

$$W_{nc} = (KE_f - KE_0) + (PE_f - PE_0) \tag{6.7a}$$

$$= \underbrace{(KE_f + PE_f)}_{E_f} - \underbrace{(KE_0 + PE_0)}_{E_0}$$

or

$$W_{nc} = E_f - E_0 \tag{6.8}$$

Remember: Equation 6.8 is just another form of the work–energy theorem. It states that $W_{nc}$, the net work done by external nonconservative forces, changes the total mechanical energy from an initial value of $E_0$ to a final value of $E_f$.

The conciseness of the work–energy theorem in the form $W_{nc} = E_f - E_0$ allows an important basic principle of physics to stand out. This principle is known as the conservation of mechanical energy. Suppose that the net work $W_{nc}$ done by external nonconservative forces is zero, so $W_{nc} = 0$ J. Then, Equation 6.8 reduces to

$$E_f = E_0 \tag{6.9a}$$

$$\underbrace{\tfrac{1}{2}mv_f^2 + mgh_f}_{E_f} = \underbrace{\tfrac{1}{2}mv_0^2 + mgh_0}_{E_0} \tag{6.9b}$$

Equation 6.9a indicates that the final mechanical energy is equal to the initial mechanical energy. Consequently, the total mechanical energy *remains constant all along the path between the initial and final points,* never varying from the initial value of $E_0$. A quantity that remains constant throughout the motion is said to be "conserved." The fact that the total mechanical energy is conserved when $W_{nc} = 0$ J is called the ***principle of conservation of mechanical energy.***

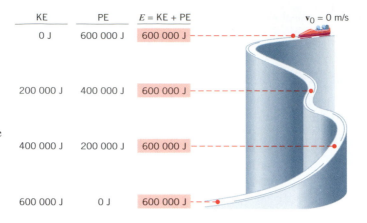

| KE | PE | $E = $ KE + PE | | $v_0 = 0$ m/s |
|---|---|---|---|---|
| 0 J | 600 000 J | 600 000 J | | |
| 200 000 J | 400 000 J | 600 000 J | | |
| 400 000 J | 200 000 J | 600 000 J | | |
| 600 000 J | 0 J | 600 000 J | | |

**Figure 6.14** If friction and wind resistance are ignored, a bobsled run illustrates how kinetic and potential energy can be interconverted, while the total mechanical energy remains constant. The total mechanical energy is 600 000 J, being all potential energy at the top and all kinetic energy at the bottom.

■ **THE PRINCIPLE OF CONSERVATION OF MECHANICAL ENERGY**

The total mechanical energy ($E = $ KE + PE) of an object remains constant as the object moves, provided that the net work done by external nonconservative forces is zero, $W_{nc} = 0$ J.

The principle of conservation of mechanical energy offers keen insight into the way in which the physical universe operates. While the sum of the kinetic and potential energies at any point is conserved, the two forms may be interconverted or transformed into one another. Kinetic energy of motion is converted into potential energy of position, for instance, when a moving object coasts up a hill. Conversely, potential energy is converted into kinetic energy when an object is allowed to fall. Figure 6.14 illustrates such transformations of energy for a bobsled run, assuming that nonconservative forces, such as friction and wind resistance, can be ignored. The normal force, being directed perpendicular to the path, does no work. Only the force of gravity does work, so the total mechanical energy $E$ remains constant at all points along the run. The conservation principle is well known for the ease with which it can be applied, as in the following examples.

### Example 8   A Daredevil Motorcyclist

A motorcyclist is trying to leap across the canyon shown in Figure 6.15 by driving horizontally off the cliff at a speed of 38.0 m/s. Ignoring air resistance, find the speed with which the cycle strikes the ground on the other side.

**Reasoning** Once the cycle leaves the cliff, no forces other than gravity act on the cycle, since air resistance is being ignored. Thus, the work done by external nonconservative forces is zero, $W_{nc} = 0$ J. Accordingly, the principle of conservation of mechanical energy holds, so the

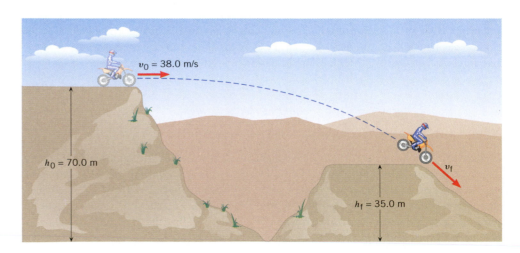

$v_0 = 38.0$ m/s

$h_0 = 70.0$ m

$h_f = 35.0$ m

$v_f$

**Figure 6.15** A daredevil jumping a canyon.

total mechanical energy is the same at the final and initial positions of the motorcycle. We will use this important observation to determine the final speed of the cyclist.

**Solution**  The principle of conservation of mechanical energy is written as

$$\underbrace{\tfrac{1}{2}mv_{\mathrm{f}}^2 + mgh_{\mathrm{f}}}_{E_{\mathrm{f}}} = \underbrace{\tfrac{1}{2}mv_0^2 + mgh_0}_{E_0} \qquad (6.9b)$$

The mass $m$ of the rider and cycle can be eliminated algebraically from this equation, since $m$ appears as a factor in every term. Solving for $v_{\mathrm{f}}$ gives

$$v_{\mathrm{f}} = \sqrt{v_0^2 + 2g(h_0 - h_{\mathrm{f}})}$$

$$v_{\mathrm{f}} = \sqrt{(38.0 \text{ m/s})^2 + 2(9.80 \text{ m/s}^2)(70.0 \text{ m} - 35.0 \text{ m})} = \boxed{46.2 \text{ m/s}}$$

**Problem solving insight**
Be on the alert for factors, such as the mass $m$ here in Example 8, that sometimes can be eliminated algebraically when using the conservation of mechanical energy.

Examples 9 and 10 emphasize that the principle of conservation of mechanical energy can be applied even when forces act perpendicular to the path of a moving object.

## Conceptual Example 9   The Favorite Swimming Hole

A rope is tied to a tree limb and used by a swimmer to swing into the water below. The person starts from rest with the rope held in the horizontal position, as in Figure 6.16, swings downward, and then lets go of the rope. Three forces act on him: his weight, the tension in the rope, and the force due to air resistance. His initial height $h_0$ and final height $h_{\mathrm{f}}$ are known. Considering the nature of these forces, conservative versus nonconservative, can we use the principle of conservation of mechanical energy to find his speed $v_{\mathrm{f}}$ at the point where he lets go of the rope?

**Reasoning and Solution**  The principle of conservation of mechanical energy can be used only if the net work $W_{\mathrm{nc}}$ done by nonconservative forces is zero, $W_{\mathrm{nc}} = 0$ J. The tension and the force due to air resistance are nonconservative forces (see Table 6.2), so we need to inquire whether the net work done by these forces is zero. The tension **T** is always perpendicular to the circular path of the motion, as shown in the drawing. Thus, the angle $\theta$ between the tension and the displacement is 90°. According to Equation 6.1, the work depends on the cosine of this angle, or cos 90°, which is zero. Thus, the work done by the tension is zero. On the other hand, the force due to air resistance is directed opposite to the motion of the swinging person. The angle $\theta$ between this force and the displacement is 180°. The work done by the air resistance is not zero, because the cosine of 180° is not zero. Since the net work done by the two nonconservative forces is not zero, $W_{\mathrm{net}} \neq 0$ J, we cannot use the principle of conservation of mechanical energy to determine the final speed. On the other hand, if the force due to air resistance is very small, then the work done by this force is negligible. In this case, the net work done by the nonconservative forces is effectively zero, and the principle of conservation of mechanical energy can be used to determine the final speed of the swimmer.

**Problem solving insight**
When nonconservative forces are perpendicular to the motion, we can still use the principle of conservation of mechanical energy, because such "perpendicular" forces do no work.

**Related Homework:** *Problem 43*

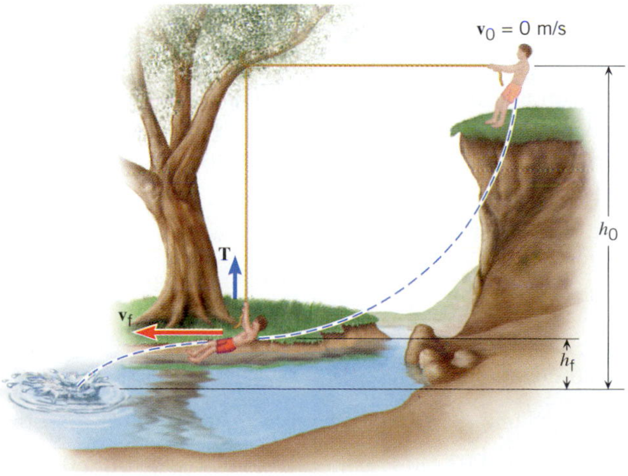

**Figure 6.16**  During the downward swing, the tension **T** in the rope acts perpendicular to the circular arc and, hence, does no work on the person.

**Figure 6.17** The Steel Dragon roller coaster in Mie, Japan, is a giant. It includes a vertical drop of 93.5 m. (©AFP/Corbis Images)

**The physics of**
a giant roller coaster.

The next example illustrates how the conservation of mechanical energy is applied to the breathtaking drop of a roller coaster.

### Example 10 The Steel Dragon

The tallest and fastest roller coaster in the world is now the Steel Dragon in Mie, Japan (Figure 6.17). The ride includes a vertical drop of 93.5 m. The coaster has a speed of 3.0 m/s at the top of the drop. Neglect friction and find the speed of the riders at the bottom.

**Reasoning** Since we are neglecting friction, we may set the work done by the frictional force equal to zero. A normal force from the seat acts on each rider, but this force is perpendicular to the motion, so it does not do any work. Thus, the work done by external nonconservative forces is zero, $W_{nc} = 0$ J, and we may use the principle of conservation of mechanical energy to find the speed of the riders at the bottom.

**Solution** The principle of conservation of mechanical energy states that

$$\underbrace{\tfrac{1}{2}mv_f^2 + mgh_f}_{E_f} = \underbrace{\tfrac{1}{2}mv_0^2 + mgh_0}_{E_0} \tag{6.9b}$$

The mass $m$ of the rider appears as a factor in every term in this equation and can be eliminated algebraically. Solving for the final speed gives

$$v_f = \sqrt{v_0^2 + 2g(h_0 - h_f)}$$

$$v_f = \sqrt{(3.0 \text{ m/s})^2 + 2(9.80 \text{ m/s})(93.5 \text{ m})} = \boxed{42.9 \text{ m/s (about 96 mi/h)}}$$

where the vertical drop is $h_0 - h_f = 93.5$ m.

## 6.6 Nonconservative Forces and the Work–Energy Theorem

Most moving objects experience nonconservative forces, such as friction, air resistance, and propulsive forces, and the work $W_{nc}$ done by the net external nonconservative force is not zero. In these situations, the difference between the final and initial total mechanical energies is equal to $W_{nc}$, according to $W_{nc} = E_f - E_0$ (Equation 6.8). Consequently, the total mechanical energy is not conserved. The next two examples illustrate how Equation 6.8 is used when nonconservative forces are present and do work.

### Example 11 The Steel Dragon, Revisited

In Example 10, we ignored nonconservative forces, such as friction. In reality, however, such forces are present when the roller coaster descends. The actual speed of the riders at the bottom is 41.0 m/s, which is less than that determined in Example 10. Assuming again that the coaster has a speed of 3.0 m/s at the top, find the work done by nonconservative forces on a 55.0-kg rider during the descent from a height $h_0$ to a height $h_f$, where $h_0 - h_f = 93.5$ m.

**Reasoning** Since the speed at the top, the final speed, and the vertical drop are given, we can determine the initial and final total mechanical energies of the rider. The work–energy theorem, $W_{nc} = E_f - E_0$, can then be used to determine the work $W_{nc}$ done by the nonconservative forces.

**Problem solving insight**
As illustrated here and in Example 3, a nonconservative force such as friction can do negative or positive work. It does negative work when it has a component opposite to the displacement and slows down the object. It does positive work when it has a component in the direction of the displacement and speeds up the object.

**Solution** The work–energy theorem is

$$W_{nc} = \underbrace{(\tfrac{1}{2}mv_f^2 + mgh_f)}_{E_f} - \underbrace{(\tfrac{1}{2}mv_0^2 + mgh_0)}_{E_0} \tag{6.8}$$

Rearranging the terms on the right side of this equation gives

$$W_{nc} = \tfrac{1}{2}m(v_f^2 - v_0^2) - mg(h_0 - h_f)$$

$$W_{nc} = \tfrac{1}{2}(55.0 \text{ kg})[(41.0 \text{ m/s})^2 - (3.0 \text{ m/s})^2] - (55.0 \text{ kg})(9.80 \text{ m/s}^2)(93.5 \text{ m}) = \boxed{-4400 \text{ J}}$$

## Example 12  Fireworks

A 0.20-kg rocket in a fireworks display is launched from rest and follows an erratic flight path to reach the point $P$, as Figure 6.18 shows. Point $P$ is 29 m above the starting point. In the process, 425 J of work is done on the rocket by the nonconservative force generated by the burning propellant. Ignoring air resistance and the mass lost due to the burning propellant, find the speed $v_f$ of the rocket at the point $P$.

**Reasoning**  The only nonconservative force acting on the rocket is the force generated by the burning propellant, and the work done by this force is $W_{nc} = 425$ J. Because work is done by a nonconservative force, we use the work–energy theorem in the form $W_{nc} = E_f - E_0$ to find the final speed $v_f$ of the rocket.

**Solution**  From the work–energy theorem we have

$$W_{nc} = (\tfrac{1}{2}mv_f^2 + mgh_f) - (\tfrac{1}{2}mv_0^2 + mgh_0) \tag{6.8}$$

Solving this expression for the final speed of the rocket, we get

$$v_f = \sqrt{\frac{2[W_{nc} + \tfrac{1}{2}mv_0^2 - mg(h_f - h_0)]}{m}}$$

$$v_f = \sqrt{\frac{2[425 \text{ J} + \tfrac{1}{2}(0.20 \text{ kg})(0 \text{ m/s})^2 - (0.20 \text{ kg})(9.80 \text{ m/s}^2)(29 \text{ m})]}{0.20 \text{ kg}}} = \boxed{61 \text{ m/s}}$$

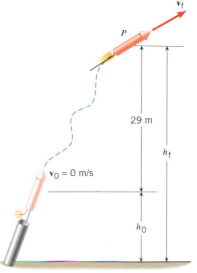

**Figure 6.18**  A fireworks rocket, moving along an erratic flight path, reaches a point $P$ that is 29 m above the launch point.

## 6.7  Power

In many situations, the time it takes to do work is just as important as the amount of work that is done. Consider two automobiles that are identical in all respects (e.g., same mass), except that one has a "souped-up" engine. The car with the "souped-up" engine can go from 0 to 27 m/s (60 mph) in 4 seconds, while the other car requires 8 seconds to achieve the same speed. Each engine does work in accelerating its car, but one does it more quickly. Where cars are concerned, we associate the quicker performance with an engine that has a larger horsepower rating. A large horsepower rating means that the engine can do a large amount of work in a short time. In physics, the horsepower rating is just one way to measure an engine's ability to generate power. The idea of **power** incorporates both the concepts of work and time, for power is work done per unit time.

■ **DEFINITION OF AVERAGE POWER**

Average power $\overline{P}$ is the average rate at which work $W$ is done, and it is obtained by dividing $W$ by the time $t$ required to perform the work:

$$\overline{P} = \frac{\text{Work}}{\text{Time}} = \frac{W}{t} \tag{6.10a}$$

**SI Unit of Power:** joule/s = watt (W)

The definition of average power presented in Equation 6.10a involves the work that is done by the net force. However, the work–energy theorem relates this work to the change in the energy of the object (see, for example, Equations 6.3 and 6.8). Therefore, we can also define average power as the rate at which the energy is changing, or as the change in energy divided by the time during which the change occurs:

$$\overline{P} = \frac{\text{Change in energy}}{\text{Time}} \tag{6.10b}$$

Since work, energy, and time are scalar quantities, power is also a scalar quantity. The unit in which power is expressed is that of work divided by time, or a joule per second in SI units. One joule per second is called a watt (W), in honor of James Watt (1736–1819), the developer of the steam engine. The unit of power in the BE system is the foot-pound per second (ft·lb/s), although the familiar horsepower (hp) unit is used

**Table 6.3**   *Units of Measurement for Power*

| System | Work | ÷ | Time | = | Power |
|---|---|---|---|---|---|
| SI | joule (J) | | second (s) | | watt (W) |
| CGS | erg | | second (s) | | erg per second (erg/s) |
| BE | foot · pound (ft · lb) | | second (s) | | foot · pound per second (ft · lb/s) |

frequently for specifying the power generated by electric motors and internal combustion engines:

$$1 \text{ horsepower} = 550 \text{ foot} \cdot \text{pounds/second} = 745.7 \text{ watts}$$

Table 6.3 summarizes the units for power in the various systems of measurement.

Equation 6.10b provides the basis for understanding the production of power in the human body. In this context the "Change in energy" on the right-hand side of the equation refers to the energy produced by metabolic processes, which, in turn, is derived from the food we eat. Table 6.4 gives typical metabolic rates of energy production needed to sustain various activities. Running at 15 km/h (9.3 mi/h), for example, requires metabolic power sufficient to operate eighteen 75-watt light bulbs, and the metabolic power used in sleeping would operate a single 75-watt bulb.

An alternative expression for power can be obtained from Equation 6.1, which indicates that the work $W$ done when a constant net force of magnitude $F$ points in the same direction as the displacement is $W = (F \cos 0°)s = Fs$. Dividing both sides of this equation by the time $t$ it takes for the force to move the object through the distance $s$, we obtain

$$\frac{W}{t} = \frac{Fs}{t}$$

But $W/t$ is the average power $\overline{P}$, and $s/t$ is the average speed $\overline{v}$, so that

$$\overline{P} = F\overline{v} \tag{6.11}$$

The next example illustrates the use of Equation 6.11.

**The physics of human metabolism.**

**Table 6.4**   *Human Metabolic Rates[a]*

| Activity | Rate (watts) |
|---|---|
| Running (15 km/h) | 1340 W |
| Skiing | 1050 W |
| Biking | 530 W |
| Walking (5 km/h) | 280 W |
| Sleeping | 77 W |

[a]For a young 70-kg male.

### *Example 13*   The Power to Accelerate a Car

A $1.10 \times 10^3$-kg car, starting from rest, accelerates for 5.00 s. The magnitude of the acceleration is $a = 4.60$ m/s². Determine the average power generated by the net force that accelerates the vehicle.

**Reasoning**  We can find the average power by using the relation $\overline{P} = F\overline{v}$, provided the magnitude $F$ of the net force acting on the car and the average speed $\overline{v}$ of the car can be determined. The net force can be obtained from Newton's second law, and the average speed can be calculated from the equations of kinematics.

**Solution**  According to Newton's second law, the magnitude of the net force is

$$F = ma = (1.10 \times 10^3 \text{ kg})(4.60 \text{ m/s}^2) = 5060 \text{ N}$$

Since the car starts from rest ($v_0 = 0$ m/s) and has a constant acceleration, the average speed $\overline{v}$ of the car is one-half of its final speed $\overline{v}_f$:

$$\overline{v} = \tfrac{1}{2}(v_0 + v_f) = \tfrac{1}{2}v_f \tag{2.6}$$

Because the initial speed of the car is zero, the final speed of the car after 5.00 s is the product of its acceleration and time:

$$v_f = v_0 + at = (0 \text{ m/s}) + (4.60 \text{ m/s}^2)(5.00 \text{ s}) = 23.0 \text{ m/s} \tag{2.4}$$

Thus, the average speed is $\overline{v} = 11.5$ m/s, and the average power is

$$\overline{P} = F\overline{v} = (5060 \text{ N})(11.5 \text{ m/s}) = \boxed{5.82 \times 10^4 \text{ W } (78.0 \text{ hp})}$$

# 6.8 Other Forms of Energy and the Conservation of Energy

Up to now, we have considered only two types of energy, kinetic energy and gravitational potential energy. There are many other types, however. Electrical energy is used to run electrical appliances. Energy in the form of heat is utilized in cooking food. Moreover, the work done by the kinetic frictional force often appears as heat, as you can experience by rubbing your hands back and forth. When gasoline is burned, some of the stored chemical energy is released and does the work of moving cars, airplanes, and boats. The chemical energy stored in food provides the energy needed for metabolic processes.

One of the most controversial forms of energy is nuclear energy. The research of many scientists, most notably Albert Einstein, led to the discovery that mass itself is one manifestation of energy. Einstein's famous equation, $E_0 = mc^2$, describes how mass $m$ and energy $E_0$ are related, where $c$ is the speed of light in a vacuum and has a value of $3.00 \times 10^8$ m/s. Because the speed of light is so large, this equation implies that very small masses are equivalent to large amounts of energy. The relationship between mass and energy will be discussed further in Chapter 28.

We have seen that kinetic energy can be converted into gravitational potential energy and vice versa. In general, energy of all types can be converted from one form to another. Part of the chemical energy stored in food, for example, is transformed into gravitational potential energy when a hiker climbs a mountain. Suppose a 65-kg hiker eats a 250-Calorie* snack, which contains $1.0 \times 10^6$ J of chemical energy. If this were 100% converted into potential energy $mg(h_f - h_0)$, the change in height would be

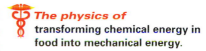

*The physics of* transforming chemical energy in food into mechanical energy.

$$h_f - h_0 = \frac{1.0 \times 10^6 \text{ J}}{(65 \text{ kg})(9.8 \text{ m/s}^2)} = 1600 \text{ m}$$

At a more realistic conversion efficiency of 50%, the change in height would be 800 m. Similarly, in a moving car the chemical energy of gasoline is converted into kinetic energy, as well as into electrical energy and heat.

Whenever energy is transformed from one form to another, it is found that no energy is gained or lost in the process; the total of all the energies before the process is equal to the total of the energies after the process. This observation leads to the following important principle:

*The physics of* the compound bow.

■ **THE PRINCIPLE OF CONSERVATION OF ENERGY**

Energy can neither be created nor destroyed, but can only be converted from one form to another.

(a)

Learning how to convert energy from one form to another more efficiently is one of the main goals of modern science and technology.

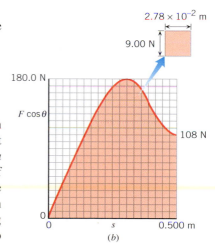

# 6.9 Work Done by a Variable Force

The work $W$ done by a constant force (constant in both magnitude and direction) is given by Equation 6.1 as $W = (F \cos \theta)s$. Quite often, situations arise in which the force is not constant but changes with the displacement of the object. For instance, Figure 6.19a shows an archer using a high-tech compound bow. This type of bow consists of a series of pulleys and strings that produce a force-versus-displacement graph like that in Figure 6.19b. One of the key features of the compound bow is that the force rises to a maximum as the string is drawn back, and then falls to 60% of this maximum value when the string is fully drawn. The reduced force at $s = 0.500$ m makes it much easier for the archer to hold the fully drawn bow while aiming the arrow.

**Figure 6.19** (a) A compound bow. (©Ron Chappel) (b) A plot of $F \cos \theta$ versus $s$ as the bowstring is drawn back.

*Energy content in food is typically given in units called Calories, which we will discuss in Section 12.7.

**Figure 6.20** (*a*) The work done by the average-force component $(F \cos \theta)_1$ during the small displacement $\Delta s_1$ is $(F \cos \theta)_1 \Delta s_1$, which is the area of the colored rectangle. (*b*) The work done by a variable force is equal to the colored area under the $F \cos \theta$-versus-$s$ curve.

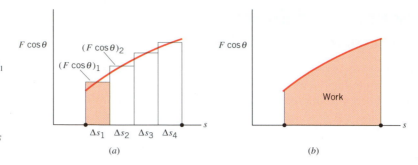

When the force varies with the displacement, as in Figure 6.19*b*, we cannot use the relation $W = (F \cos \theta)s$ to find the work, because this equation is valid only when the force is constant. However, we can use a graphical method. In this method we divide the total displacement into very small segments, $\Delta s_1$, $\Delta s_2$, and so on (see Figure 6.20*a*). For each segment, the *average value* of the force component is indicated by a short horizontal line. For example, the short horizontal line for segment $\Delta s_1$ is labeled $(F \cos \theta)_1$ in Figure 6.20*a*. We can then use this average value as the constant-force component in Equation 6.1 and determine an approximate value for the work $\Delta W_1$ done during the first segment: $\Delta W_1 = (F \cos \theta)_1 \Delta s_1$. But this work is just the area of the colored rectangle in the drawing. The word "area" here refers to the area of a rectangle that has a width of $\Delta s_1$ and a height of $(F \cos \theta)_1$; it does not mean an area in square meters, such as the area of a parcel of land. In a like manner, we can calculate an approximate value for the work for each segment. Then we add the results for the segments to get, approximately, the work $W$ done by the variable force:

$$W \approx (F \cos \theta)_1 \Delta s_1 + (F \cos \theta)_2 \Delta s_2 + \cdots$$

The symbol "$\approx$" means "approximately equal to." The right side of this equation is the sum of all the rectangular areas in Figure 6.20*a* and is an approximate value for the area shaded in color under the graph in Figure 6.20*b*. If the rectangles are made narrower and narrower by decreasing each $\Delta s$, the right side of this equation eventually becomes equal to the area under the graph. Thus, we define the work done by a variable force as follows: ***The work done by a variable force in moving an object is equal to the area under the graph of $F \cos \theta$ versus $s$.*** Example 14 illustrates how to use this graphical method to determine the approximate work done when a high-tech compound bow is drawn.

### Example 14    Work and the Compound Bow

Find the work that the archer must do in drawing back the string of the compound bow in Figure 6.19 from 0 to 0.500 m.

**Reasoning** The work is equal to the colored area under the curved line in Figure 6.19*b*. For convenience, this area is divided into a number of small squares, each having an area of $(9.00 \text{ N})(2.78 \times 10^{-2} \text{ m}) = 0.250 \text{ J}$. The area can be found by counting the number of squares under the curve and multiplying by the area per square.

**Solution** We estimate that there are 242 colored squares in the drawing. Since each square represents 0.250 J of work, the total work done is

$$W = (242 \text{ squares})\left(0.250 \frac{\text{J}}{\text{square}}\right) = \boxed{60.5 \text{ J}}$$

When the arrow is fired, part of this work is imparted to it as kinetic energy.

# Concept Summary

This summary presents an abridged version of the chapter, including the important equations and all available learning aids. For convenient reference, the learning aids (including the text's examples) are placed next to or immediately after the relevant equation or discussion. The following learning aids may be found on-line at **www.wiley.com/college/cutnell**:

| | |
|---|---|
| **Interactive LearningWare** examples are solved according to a five-step interactive format that is designed to help you develop problem-solving skills. | **Concept Simulations** are animated versions of text figures or animations that illustrate important concepts. You can control parameters that affect the display, and we encourage you to experiment. |
| **Interactive Solutions** offer specific models for certain types of problems in the chapter homework. The calculations are carried out interactively. | **Self-Assessment Tests** include both qualitative and quantitative questions. Extensive feedback is provided for both incorrect and correct answers, to help you evaluate your understanding of the material. |

| Topic | Discussion | Learning Aids |
|---|---|---|

### 6.1 Work Done by a Constant Force

The work $W$ done by a constant force acting on an object is

**Work done by constant force**
$$W = (F \cos \theta)s \qquad (6.1)$$ **Example 1**

where $F$ is the magnitude of the force, $s$ is the magnitude of the displacement, and $\theta$ is the angle between the force and the displacement vectors. Work is a scalar quantity and can be positive or negative, depending on whether the force **Examples 2, 3** has a component that points, respectively, in the same direction as the displacement or in the opposite direction. The work is zero if the force is perpendicular **Interactive LearningWare 6.1** ($\theta = 90°$) to the displacement.

### 6.2 The Work–Energy Theorem and Kinetic Energy

The kinetic energy KE of an object of mass $m$ and speed $v$ is

**Kinetic energy**
$$\text{KE} = \tfrac{1}{2}mv^2 \qquad (6.2)$$

The work–energy theorem states that the work $W$ done by the net external **Examples 4, 5, 6** force acting on an object equals the difference between the object's final kinetic energy $\text{KE}_f$ and initial kinetic energy $\text{KE}_0$: **Concept Simulation 6.1**

**Work–energy theorem**
$$W = \text{KE}_f - \text{KE}_0 \qquad (6.3)$$ **Interactive LearningWare 6.2**

The kinetic energy increases when the net force does positive work and decreases when the net force does negative work.

 **Use Self-Assessment Test 6.1 to evaluate your understanding of Sections 6.1 and 6.2.**

### 6.3 Gravitational Potential Energy

The work done by the force of gravity on an object of mass $m$ is

**Work done by force of gravity**
$$W_{\text{gravity}} = mg(h_0 - h_f) \qquad (6.4)$$ **Example 7**

where $h_0$ and $h_f$ are the initial and final heights of the object, respectively.

Gravitational potential energy PE is the energy that an object has by virtue of its position. For an object near the surface of the earth, the gravitational potential energy is given by

**Gravitational potential energy**
$$\text{PE} = mgh \qquad (6.5)$$

where $h$ is the height of the object relative to an arbitrary zero level.

### 6.4 Conservative versus Nonconservative Forces

**Conservative force**
A conservative force is one that does the same work in moving an object between two points, independent of the path taken between the points. Alternatively, a force is conservative if the work it does in moving an object around any closed path is zero. A force is nonconservative if the work it does on an **Nonconservative force** object moving between two points depends on the path of the motion between the points.

### 6.5 The Conservation of Mechanical Energy

The total mechanical energy $E$ is the sum of the kinetic energy and potential energy:

**Total mechanical energy**
$$E = \text{KE} + \text{PE}$$

| Topic | Discussion | Learning Aids |
|---|---|---|
| **Alternate form for work–energy theorem** | The work–energy theorem can be expressed in an alternate form as $$W_{nc} = E_f - E_0 \qquad (6.8)$$ where $W_{nc}$ is the net work done by the external nonconservative forces, and $E_f$ and $E_0$ are the final and initial total mechanical energies, respectively. | |
| **Principle of conservation of mechanical energy** | The principle of conservation of mechanical energy states that the total mechanical energy $E$ remains constant along the path of an object, provided that the net work done by external nonconservative forces is zero. Whereas $E$ is constant, KE and PE may be transformed into one another. | **Examples 8, 9, 10** Interactive LearningWare 6.3 Interactive Solutions 6.33, 6.39 |

 **Use Self-Assessment Test 6.2 to evaluate your understanding of Sections 6.3–6.5.**

### 6.6 Nonconservative Forces and the Work–Energy Theorem

**Examples 11, 12**
Interactive Solutions 6.51, 6.75

### 6.7 Power

Average power $\overline{P}$ is the work done per unit time or the rate at which work is done:

| **Average power** | $$\overline{P} = \frac{\text{Work}}{\text{Time}} \qquad (6.10a)$$ |
|---|---|

It is also the rate at which energy changes:

$$\overline{P} = \frac{\text{Change in energy}}{\text{Time}} \qquad (6.10b)$$

**Example 13**
Interactive Solution 6.55

When a force of magnitude $F$ acts on an object moving with an average speed $\overline{v}$, the average power is given by

$$\overline{P} = F\overline{v} \qquad (6.11)$$

### 6.8 Other Forms of Energy and the Conservation of Energy

**Principle of conservation of energy** — The principle of conservation of energy states that energy cannot be created or destroyed but can only be transformed from one form to another.

### 6.9 Work Done by a Variable Force

**Work done by a variable force** — The work done by a variable force of magnitude $F$ in moving an object through a displacement of magnitude $s$ is equal to the area under the graph of $F \cos \theta$ versus $s$. The angle $\theta$ is the angle between the force and displacement vectors.

**Example 14**

 **Use Self-Assessment Test 6.3 to evaluate your understanding of Sections 6.6–6.9.**

# Problems

*Problems that are not marked with a star are considered the easiest to solve. Problems that are marked with a single star (*) are more difficult, while those marked with a double star (**) are the most difficult.*

**ssm** Solution is in the Student Solutions Manual.    **www** Solution is available on the World Wide Web at www.wiley.com/college/cutnell
⚕ This icon represents a biomedical application.

### Section 6.1 Work Done by a Constant Force

**1. ssm** The brakes of a truck cause it to slow down by applying a retarding force of $3.0 \times 10^3$ N to the truck over a distance of 850 m. What is the work done by this force on the truck? Is the work positive or negative? Why?

**2.** The drawing (at right) shows a boat being pulled by two locomotives through a canal of length 2.00 km. The tension in each cable is $5.00 \times 10^3$ N, and $\theta = 20.0°$. What is the net work done on the boat by the two locomotives?

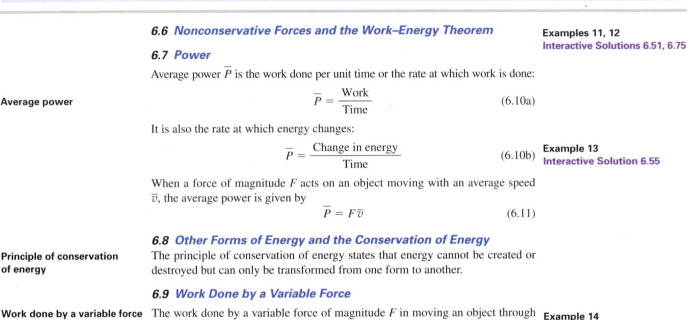

**3.** A person pulls a toboggan for a distance of 35.0 m along the snow with a rope directed 25.0° above the snow. The tension in the rope is 94.0 N. (a) How much work is done on the toboggan by the tension force? (b) How much work is done if the same tension is directed parallel to the snow?   a) 1834 J   b) 1654 J

**4.** A 75.0-kg man is riding an escalator in a shopping mall. The escalator moves the man at a constant velocity from ground level to the floor above, a vertical height of 4.60 m. What is the work done on the man by (a) the gravitational force and (b) the escalator?   ~ -3007   3007

**5. ssm** Suppose in Figure 6.2 that $+1.10 \times 10^3$ J of work are done by the force **F** (magnitude = 30.0 N) in moving the suitcase a distance of 50.0 m. At what angle $\theta$ is the force oriented with respect to the ground?

**6.** Consult **Interactive LearningWare 6.1** at **www.wiley.com/college/cutnell** for background pertinent to this problem. The drawing shows a plane diving toward the ground and then climbing back upward. During each of these motions, the lift force **L** acts perpendicular to the displacement **s**, which has the same magnitude of $1.7 \times 10^3$ m in each case. The engines of the plane exert a thrust **T**, which points in the direction of the displacement and has the same magnitude during the dive and the climb. The weight **W** of the plane has a magnitude of $5.9 \times 10^4$ N. In both motions, net work is performed due to the combined action of the forces **L**, **T**, and **W**. (a) Is more net work done during the dive or the climb? Explain. (b) Find the difference between the net work done during the dive and the climb.

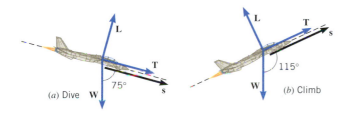

(a) Dive    (b) Climb

**7.** A person pushes a 16.0-kg shopping cart at a constant velocity for a distance of 22.0 m. She pushes in a direction 29.0° below the horizontal. A 48.0-N frictional force opposes the motion of the cart. (a) What is the magnitude of the force that the shopper exerts? Determine the work done by (b) the pushing force, (c) the frictional force, and (d) the gravitational force.

**\* 8.** A 55-kg box is being pushed a distance of 7.0 m across the floor by a force **P** whose magnitude is 150 N. The force **P** is parallel to the displacement of the box. The coefficient of kinetic friction is 0.25. Determine the work done on the box by each of the *four* forces that act on the box. Be sure to include the proper plus or minus sign for the work done by each force.

**\* 9. ssm** A husband and wife take turns pulling their child in a wagon along a horizontal sidewalk. Each exerts a constant force and pulls the wagon through the same displacement. They do the same amount of work, but the husband's pulling force is directed 58° above the horizontal, and the wife's pulling force is directed 38° above the horizontal. The husband pulls with a force whose magnitude is 67 N. What is the magnitude of the pulling force exerted by his wife?

**\* 10.** A $1.00 \times 10^2$-kg crate is being pushed across a horizontal floor by a force **P** that makes an angle of 30.0° below the horizontal. The coefficient of kinetic friction is 0.200. What should be the magnitude of **P**, so that the net work done by it and the kinetic frictional force is zero?

**\*\* 11.** A 1200-kg car is being driven up a 5.0° hill. The frictional force is directed opposite to the motion of the car and has a magnitude of

$f = 524$ N. A force **F** is applied to the car by the road and propels the car forward. In addition to these two forces, two other forces act on the car: its weight **W** and the normal force $\mathbf{F}_N$ directed perpendicular to the road surface. The length of the road up the hill is 290 m. What should be the magnitude of **F**, so that the net work done by all the forces acting on the car is +150 kJ?

### Section 6.2 The Work–Energy Theorem and Kinetic Energy

**12.** A fighter jet is launched from an aircraft carrier with the aid of its own engines and a steam-powered catapult. The thrust of its engines is $2.3 \times 10^5$ N. In being launched from rest it moves through a distance of 87 m and has a kinetic energy of $4.5 \times 10^7$ J at lift-off. What is the work done on the jet by the catapult?

**13. ssm** The hammer throw is a track-and-field event in which a 7.3-kg ball (the "hammer"), starting from rest, is whirled around in a circle several times and released. It then moves upward on the familiar curving path of projectile motion. In one throw, the hammer is given a speed of 29 m/s. For comparison, a .22 caliber bullet has a mass of 2.6 g and, starting from rest, exits the barrel of a gun with a speed of 410 m/s. Determine the work done to launch the motion of (a) the hammer and (b) the bullet.

**14.** Refer to **Concept Simulation 6.1** at **www.wiley.com/college/cutnell** for a review of the concepts with which this problem deals. A 0.075-kg arrow is fired horizontally. The bowstring exerts an average force of 65 N on the arrow over a distance of 0.90 m. With what speed does the arrow leave the bow?

**15.** When a 0.045-kg golf ball takes off after being hit, its speed is 41 m/s. (a) How much work is done on the ball by the club? (b) Assume that the force of the golf club acts parallel to the motion of the ball and that the club is in contact with the ball for a distance of 0.010 m. Ignore the weight of the ball and determine the average force applied to the ball by the club.

**16.** As background for this problem, review Conceptual Example 6. A 7420-kg satellite has an elliptical orbit, as in Figure 6.8b. The point on the orbit that is farthest from the earth is called the *apogee* and is at the far right side of the drawing. The point on the orbit that is closest to the earth is called the *perigee* and is at the left side of the drawing. Suppose that the speed of the satellite is 2820 m/s at the apogee and 8450 m/s at the perigee. Find the work done by the gravitational force when the satellite moves from (a) the apogee to the perigee and (b) the perigee to the apogee.

**17. ssm www** Two cars, A and B, are traveling with the same speed of 40.0 m/s, each having started from rest. Car A has a mass of $1.20 \times 10^3$ kg, and car B has a mass of $2.00 \times 10^3$ kg. Compared to the work required to bring car A up to speed, how much *additional* work is required to bring car B up to speed?

**18. Interactive LearningWare 6.2** at **www.wiley.com/college/cutnell** provides a useful review of the concepts that play a role in this problem. A $5.0 \times 10^4$-kg space probe is traveling at a speed of 11 000 m/s through deep space. Retrorockets are fired along the line of motion to reduce the probe's speed. The retrorockets generate a force of $4.0 \times 10^5$ N over a distance of 2500 km. What is the final speed of the probe?

**\* 19. ssm** A sled is being pulled across a horizontal patch of snow. Friction is negligible. The pulling force points in the same direction as the sled's displacement, which is along the +x axis. As a result, the kinetic energy of the sled increases by 38%. By what percentage would the sled's kinetic energy have increased if this force had pointed 62° above the +x axis?

**\* 20.** A 16-kg sled is being pulled along the horizontal snow-covered ground by a horizontal force of 24 N. Starting from rest, the sled attains a speed of 2.0 m/s in 8.0 m. Find the coefficient of kinetic friction between the runners of the sled and the snow.

*　**21. ssm www** A 6200-kg satellite is in a circular earth orbit that has a radius of $3.3 \times 10^7$ m. A net external force must act on the satellite to make it change to a circular orbit that has a radius of $7.0 \times 10^6$ m. What work must the net external force do?

*　**22.** A rescue helicopter lifts a 79-kg person straight up by means of a cable. The person has an upward acceleration of 0.70 m/s$^2$ and is lifted from rest through a distance of 11 m. (a) What is the tension in the cable? How much work is done by (b) the tension in the cable and (c) the person's weight? (d) Use the work–energy theorem and find the final speed of the person.

*　**23.** An extreme skier, starting from rest, coasts down a mountain that makes an angle of 25.0° with the horizontal. The coefficient of kinetic friction between her skis and the snow is 0.200. She coasts for a distance of 10.4 m before coming to the edge of a cliff. Without slowing down, she skis off the cliff and lands downhill at a point whose vertical distance is 3.50 m below the edge. How fast is she going just before she lands?

**　**24.** The model airplane in Figure 5.6 is flying at a speed of 22 m/s on a horizontal circle of radius 16 m. The mass of the plane is 0.90 kg. The person holding the guideline pulls it in until the radius of the circle becomes 14 m. The plane speeds up, and the tension in the guideline becomes four times greater. What is the net work done on the plane?

## Section 6.3 Gravitational Potential Energy
### Section 6.4 Conservative versus Nonconservative Forces

**25. ssm** A bicyclist rides 5.0 km due east, while the resistive force from the air has a magnitude of 3.0 N and points due west. The rider then turns around and rides 5.0 km due west, back to her starting point. The resistive force from the air on the return trip has a magnitude of 3.0 N and points due east. (a) Find the work done by the resistive force during the round trip. (b) Based on your answer to part (a), is the resistive force a conservative force? Explain.

**26.** A 0.60-kg basketball is dropped out of a window that is 6.1 m above the ground. The ball is caught by a person whose hands are 1.5 m above the ground. (a) How much work is done on the ball by its weight? What is the gravitational potential energy of the basketball, relative to the ground, when it is (b) released and (c) caught? (d) How is the change ($PE_f - PE_0$) in the ball's gravitational potential energy related to the work done by its weight?

**27.** Relative to the ground, what is the gravitational potential energy of a 55.0-kg person who is at the top of the Sears Tower, a height of 443 m above the ground?

**28.** A shot-putter puts a shot (weight = 71.1 N) that leaves his hand at a distance of 1.52 m above the ground. (a) Find the work done by the gravitational force when the shot has risen to a height of 2.13 m above the ground. Include the correct sign for the work. (b) Determine the change ($\Delta PE = PE_f - PE_0$) in the gravitational potential energy of the shot.

**29.** A 75.0-kg skier rides a 2830-m-long lift to the top of a mountain. The lift makes an angle of 14.6° with the horizontal. What is the change in the skier's gravitational potential energy?

**30.** When an 81.0-kg adult uses a spiral staircase to climb to the second floor of his house, his gravitational potential energy increases by $2.00 \times 10^3$ J. By how much does the potential energy of an 18.0-kg child increase when the child climbs a normal staircase to the second floor?

**31. ssm** "Rocket man" has a propulsion unit strapped to his back. He starts from rest on the ground, fires the unit, and is propelled straight upward. At a height of 16 m, his speed is 5.0 m/s. His mass, including the propulsion unit, has the approximately constant value of 136 kg. Find the work done by the force generated by the propulsion unit.

## Section 6.5 The Conservation of Mechanical Energy

**32.** Consult **Interactive LearningWare 6.3** at **www.wiley.com/college/cutnell** for a review of the concepts on which this problem is based. A gymnast is swinging on a high bar. The distance between his waist and the bar is 1.1 m, as the drawing shows. At the top of the swing his speed is momentarily zero. Ignoring friction and treating the gymnast as if all of his mass is located at his waist, find his speed at the bottom of the swing.

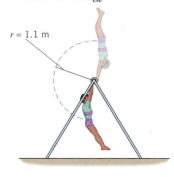

$r = 1.1$ m

**33.** **Interactive Solution 6.33** at **www.wiley.com/college/cutnell** presents a model for solving this problem. A slingshot fires a pebble from the top of a building at a speed of 14.0 m/s. The building is 31.0 m tall. Ignoring air resistance, find the speed with which the pebble strikes the ground when the pebble is fired (a) horizontally, (b) vertically straight up, and (c) vertically straight down.

**34.** A water-skier lets go of the tow rope upon leaving the end of a jump ramp at a speed of 14.0 m/s. As the drawing indicates, the skier has a speed of 13.0 m/s at the highest point of the jump. Ignoring air resistance, determine the skier's height $H$ above the *top of the ramp* at the highest point.

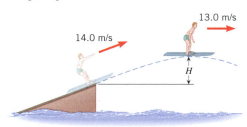

14.0 m/s　13.0 m/s　$H$

**35. ssm** A pole-vaulter approaches the takeoff point at a speed of 9.00 m/s. Assuming that only this speed determines the height to which he can rise, find the maximum height at which the vaulter can clear the bar.

**36.** A 47.0-g golf ball is driven from the tee with an initial speed of 52.0 m/s and rises to a height of 24.6 m. (a) Neglect air resistance and determine the kinetic energy of the ball at its highest point. (b) What is its speed when it is 8.0 m below its highest point?

**37.** A cyclist approaches the bottom of a gradual hill at a speed of 11 m/s. The hill is 5.0 m high, and the cyclist estimates that she is going fast enough to coast up and over it without pedaling. Ignoring air resistance and friction, find the speed at which the cyclist crests the hill.

*　**38.** A particle, starting from point A in the drawing, is projected down the curved runway. Upon leaving the runway at point B, the particle is traveling straight upward and reaches a height of 4.00 m above the floor before falling back down. Ignoring friction and air resistance, find the speed of the particle at point A.

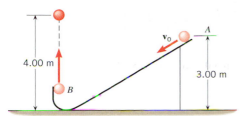

**39.** Review **Interactive Solution 6.39** at **www.wiley.com/college/cutnell** for background on this problem. A wrecking ball swings at the end of a 12.0-m cable on a vertical circular arc. The crane operator manages to give the ball a speed of 5.00 m/s as the ball passes through the lowest point of its swing and then gives the ball no further assistance. Friction and air resistance are negligible. What speed $v_f$ does the ball have when the cable makes an angle of 20.0° with respect to the vertical?

**40.** The drawing shows a skateboarder moving at 5.4 m/s along a horizontal section of a track that is slanted upward by 48° above the horizontal at its end, which is 0.40 m above the ground. When she leaves the track, she follows the characteristic path of projectile motion. Ignoring friction and air resistance, find the maximum height $H$ to which she rises above the end of the track.

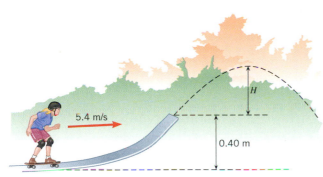

**41. ssm** A water slide is constructed so that swimmers, starting from rest at the top of the slide, leave the end of the slide traveling horizontally. As the drawing shows, one person hits the water 5.00 m from the end of the slide in a time of 0.500 s after leaving the slide. Ignoring friction and air resistance, find the height $H$ in the drawing.

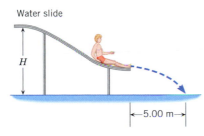

Water slide

**42.** A skier starts from rest at the top of a hill. The skier coasts down the hill and up a second hill, as the drawing illustrates. The crest of the second hill is circular, with a radius of $r = 36$ m. Neglect friction and air resistance. What must be the height $h$ of the first hill so that the skier just loses contact with the snow at the crest of the second hill?

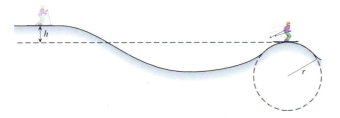

**43.** Conceptual Example 9 provides background for this problem. A swing is made from a rope that will tolerate a maximum tension of $8.00 \times 10^2$ N without breaking. Initially, the swing hangs vertically. The swing is then pulled back at an angle of 60.0° with respect to the vertical and released from rest. What is the mass of the heaviest person who can ride the swing?

**44.** The drawing shows a version of the loop-the-loop trick for a small car. If the car is given an initial speed of 4.0 m/s, what is the largest value that the radius $r$ can have if the car is to remain in contact with the circular track at all times?

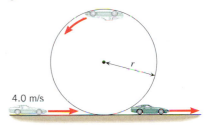

### Section 6.6 Nonconservative Forces and the Work–Energy Theorem

**45. ssm** A roller coaster (375 kg) moves from $A$ (5.00 m above the ground) to $B$ (20.0 m above the ground). Two nonconservative forces are present: friction does $-2.00 \times 10^4$ J of work on the car, and a chain mechanism does $+3.00 \times 10^4$ J of work to help the car up a long climb. What is the change in the car's kinetic energy, $\Delta KE = KE_f - KE_0$, from $A$ to $B$?

**46.** A basketball player makes a jump shot. The 0.600-kg ball is released at a height of 2.00 m above the floor with a speed of 7.20 m/s. The ball goes through the net 3.10 m above the floor at a speed of 4.20 m/s. What is the work done on the ball by air resistance, a nonconservative force?

**47.** A $5.00 \times 10^2$-kg hot-air balloon takes off from rest at the surface of the earth. The nonconservative wind and lift forces take the balloon up, doing $+9.70 \times 10^4$ J of work on the balloon in the process. At what height above the surface of the earth does the balloon have a speed of 8.00 m/s?

**48.** A projectile of mass 0.750 kg is shot straight up with an initial speed of 18.0 m/s. (a) How high would it go if there were no air friction? (b) If the projectile rises to a maximum height of only 11.8 m, determine the magnitude of the average force due to air resistance.

**49. ssm** The (nonconservative) force propelling a $1.50 \times 10^3$-kg car up a mountain road does $4.70 \times 10^6$ J of work on the car. The car starts from rest at sea level and has a speed of 27.0 m/s at an altitude of $2.00 \times 10^2$ m above sea level. Obtain the work done on the car by the combined forces of friction and air resistance, both of which are nonconservative forces.

**50.** A pitcher throws a 0.140-kg baseball, and it approaches the bat at a speed of 40.0 m/s. The bat does $W_{nc} = 70.0$ J of work on the ball in hitting it. Ignoring air resistance, determine the speed of the ball after the ball leaves the bat and is 25.0 m above the point of impact.

**51. Interactive Solution 6.51** at **www.wiley.com/college/cutnell** offers help in modeling this problem. A basketball of mass 0.60 kg is dropped from rest from a height of 1.05 m. It rebounds to a height of 0.57 m. (a) How much mechanical energy was lost during the collision with the floor? (b) A basketball player dribbles the ball from a height of 1.05 m by exerting a constant downward force on it for a distance of 0.080 m. In dribbling, the player compensates for the mechanical energy lost during each bounce. If the ball now returns to a height of 1.05 m, what is the magnitude of the force?

**52.** In attempting to pass the puck to a teammate, a hockey player gives it an initial speed of 1.7 m/s. However, this speed is inadequate to compensate for the kinetic friction between the puck and the ice. As a result, the puck travels only one-half the distance between the players before sliding to a halt. What minimum initial speed should the puck have been given so that it reached the teammate, assuming that the same force of kinetic friction acted on the puck everywhere between the two players?

**\* 53. ssm** At a carnival, you can try to ring a bell by striking a target with a 9.00-kg hammer. In response, a 0.400-kg metal piece is sent upward toward the bell, which is 5.00 m above. Suppose that 25.0% of the hammer's kinetic energy is used to do the work of sending the metal piece upward. How fast must the hammer be moving when it strikes the target so that the bell just barely rings?

**\*\* 54.** A 3.00-kg model rocket is launched straight up. It reaches a maximum height of $1.00 \times 10^2$ m above where its engine cuts out, even though air resistance performs $-8.00 \times 10^2$ J of work on the rocket. What would have been this height if there were no air resistance?

### Section 6.7 Power

**55.** Interactive Solution 6.55 at **www.wiley.com/college/cutnell** offers a model for solving this problem. A car accelerates uniformly from rest to 20.0 m/s in 5.6 s along a level stretch of road. Ignoring friction, determine the average power required to accelerate the car if (a) the weight of the car is $9.0 \times 10^3$ N, and (b) the weight of the car is $1.4 \times 10^4$ N.

**56.** A person is making homemade ice cream. She exerts a force of magnitude 22 N on the free end of the crank handle, and this end moves in a circular path of radius 0.28 m. The force is always applied parallel to the motion of the handle. If the handle is turned once every 1.3 s, what is the average power being expended?

**57. ssm** One kilowatt·hour (kWh) is the amount of work or energy generated when one kilowatt of power is supplied for a time of one hour. A kilowatt·hour is the unit of energy used by power companies when figuring your electric bill. Determine the number of joules of energy in one kilowatt·hour.

**58.** A $3.00 \times 10^2$-kg piano is being lifted at a steady speed from ground level straight up to an apartment 10.0 m above the ground. The crane that is doing the lifting produces a steady power of $4.00 \times 10^2$ W. How much time does it take to lift the piano?

**\* 59.** In 2.0 minutes, a ski lift raises four skiers at constant speed to a height of 140 m. The average mass of each skier is 65 kg. What is the average power provided by the tension in the cable pulling the lift?

**\* 60.** A motorcycle (mass of cycle plus rider = $2.50 \times 10^2$ kg) is traveling at a steady speed of 20.0 m/s. The force of air resistance acting on the cycle and rider is $2.00 \times 10^2$ N. Find the power necessary to sustain this speed if (a) the road is level and (b) the road is sloped upward at 37.0° with respect to the horizontal.

**\*\* 61. ssm** The motor of a ski boat generates an average power of $7.50 \times 10^4$ W when the boat is moving at a constant speed of 12 m/s. When the boat is pulling a skier at the same speed, the engine must generate an average power of $8.30 \times 10^4$ W. What is the tension in the tow rope that is pulling the skier?

**\*\* 62.** A 1900-kg car experiences a combined force of air resistance and friction that has the same magnitude whether the car goes up or down a hill at 27 m/s. Going up a hill, the car's engine needs to produce 47 hp more power to sustain the constant velocity than it does going down the same hill. At what angle is the hill inclined above the horizontal?

### Section 6.9 Work Done by a Variable Force

**63.** The graph shows the net external force component $F \cos \theta$ along the displacement as a function of the magnitude $s$ of the displacement. The graph applies to a 65-kg ice skater. How much work does the net force component do on the skater from (a) 0 to 3.0 m and (b) 3.0 m to 6.0 m? (c) If the initial speed of the skater is 1.5 m/s when $s = 0$ m, what is the speed when $s = 6.0$ m?

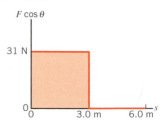

**64.** The graph shows how the force component $F \cos \theta$ along the displacement varies with the magnitude $s$ of the displacement. Find the work done by the force. (*Hint: Recall how the area of a triangle is related to the triangle's base and height.*)

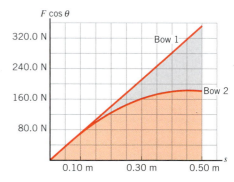

**65. ssm** The drawing shows the force-versus-displacement graph for two different bows. These graphs give the force that an archer must apply to draw the bowstring. (a) For which bow is more work required to draw the bow fully from $s = 0$ to $s = 0.50$ m? Give your reasoning. (b) Estimate the additional work required for the bow identified in part (a) compared to the other bow.

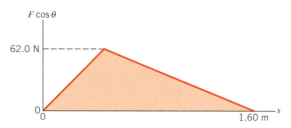

**66.** Review Example 14, in which the work done in drawing the bowstring in Figure 6.19 from $s = 0$ to $s = 0.500$ m is determined. In part *b* of the figure, the force component $F \cos \theta$ reaches a maximum at $s = 0.306$ m. Find the percentage of the total work that is done when the bowstring is moved (a) from $s = 0$ to 0.306 m and (b) from $s = 0.306$ to 0.500 m.

**67.** A net external force is applied to a 6.00-kg object that is initially at rest. The net force component along the displacement of the object varies with the magnitude of the displacement as shown in the drawing. (a) How much work is done by the net force? (b) What is the speed of the object at $s = 20.0$ m?

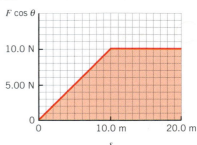

# Chapter 7 Impulse and Momentum

كان زدن

مقدار حركت - تيزى حركت آتي

## 7.1 The Impulse–Momentum Theorem

There are many situations in which the force acting on an object is not constant, but varies with time. For instance, Figure 7.1a shows a baseball being hit, and part b of the figure illustrates approximately how the force applied to the ball by the bat changes during the time of contact. The magnitude of the force is zero at the instant $t_0$ just before the bat touches the ball. During contact, the force rises to a maximum and then returns to zero at the time $t_f$ when the ball leaves the bat. The time interval $\Delta t = t_f - t_0$ during which the bat and ball are in contact is quite short, being only a few-thousandths of a second, although the maximum force can be very large, often exceeding thousands of newtons. For comparison, the graph also shows the magnitude $\overline{F}$ of the average force exerted on the ball during the time of contact. Figure 7.2 depicts other situations in which a time-varying force is applied to a ball.

If a baseball is to be hit well, both the magnitude of the force and the time of contact are important. When a large average force acts on the ball for a long enough time, the ball is hit solidly. To describe such situations, we bring together the average force and the time of contact, calling the product of the two the *impulse* of the force.

> ■ **DEFINITION OF IMPULSE**
>
> The impulse **J** of a force is the product of the average force $\overline{F}$ and the time interval $\Delta t$ during which the force acts:
>
> $$\mathbf{J} = \overline{\mathbf{F}}\,\Delta t \qquad (7.1)$$
>
> Impulse is a vector quantity and has the same direction as the average force.
>
> *SI Unit of Impulse:* newton · second (N · s)

When a ball is hit, it responds to the value of the impulse. A large impulse produces a large response; that is, the ball departs from the bat with a large velocity. However, we know from experience that the more massive the ball, the less velocity it has after leaving the bat. Both mass and velocity play a role in how an object responds to a given impulse, and the effect of each of them is included in the concept of *linear momentum,* which is defined as follows:

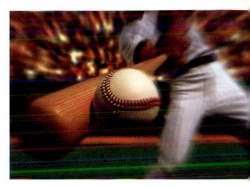

(a)

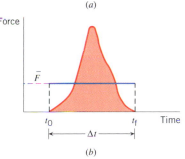

(b)

**Figure 7.1** (a) The collision time between a bat and a ball is very short, often less than a millisecond, but the force can be quite large. (© Chuck Savage/Corbis Images) (b) When the bat strikes the ball, the magnitude of the force exerted on the ball rises to a maximum value and then returns to zero when the ball leaves the bat. The time interval during which the force acts is $\Delta t$, and the magnitude of the average force is $\overline{F}$.

**Figure 7.2** In each of these situations, the force applied to the ball varies with time. The time of contact is small, but the maximum force can be large. (*left,* © Tommy Hindley/Aurora Photos; *right,* © Allsport/Getty Images)

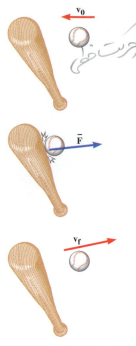

**Figure 7.3** When a bat hits a ball, an average force $\overline{\mathbf{F}}$ is applied to the ball by the bat. As a result, the ball's velocity changes from an initial value of $\mathbf{v_0}$ (top drawing) to a final value of $\mathbf{v_f}$ (bottom drawing).

■ **DEFINITION OF LINEAR MOMENTUM**

The linear momentum **p** of an object is the product of the object's mass $m$ and velocity **v**:

$$\mathbf{p} = m\mathbf{v} \tag{7.2}$$

Linear momentum is a vector quantity that points in the same direction as the velocity.

***SI Unit of Linear Momentum:*** kilogram·meter/second (kg·m/s)

Newton's second law of motion can now be used to reveal a relationship between impulse and momentum. Figure 7.3 shows a ball with an initial velocity of $\mathbf{v_0}$ approaching a bat, being struck by the bat, and then departing with a final velocity of $\mathbf{v_f}$. When the velocity of an object changes from $\mathbf{v_0}$ to $\mathbf{v_f}$ during a time interval $\Delta t$, the average acceleration $\overline{\mathbf{a}}$ is given by Equation 2.4 as

$$\overline{\mathbf{a}} = \frac{\mathbf{v_f} - \mathbf{v_0}}{\Delta t}$$

According to Newton's second law, $\Sigma\overline{\mathbf{F}} = m\overline{\mathbf{a}}$, the average acceleration is produced by the net average force $\Sigma\overline{\mathbf{F}}$. Here $\Sigma\overline{\mathbf{F}}$ represents the vector sum of all the average forces that act on the object. Thus,

$$\Sigma\overline{\mathbf{F}} = m\left(\frac{\mathbf{v_f} - \mathbf{v_0}}{\Delta t}\right) = \frac{m\mathbf{v_f} - m\mathbf{v_0}}{\Delta t} \tag{7.3}$$

In this result, the numerator on the far right is the final momentum minus the initial momentum, which is the change in momentum. Thus, the net average force is given by the change in momentum per unit of time.* Multiplying both sides of Equation 7.3 by $\Delta t$ yields Equation 7.4, which is known as the ***impulse–momentum theorem.***

■ **IMPULSE–MOMENTUM THEOREM**

When a net force acts on an object, the impulse of this force is equal to the change in momentum of the object:

$$\underbrace{(\Sigma\overline{\mathbf{F}})\,\Delta t}_{\text{Impulse}} = \underbrace{m\mathbf{v_f}}_{\substack{\text{Final} \\ \text{momentum}}} - \underbrace{m\mathbf{v_0}}_{\substack{\text{Initial} \\ \text{momentum}}} \tag{7.4}$$

**Impulse = Change in momentum**

During a collision, it is often difficult to measure the net average force $\Sigma\overline{\mathbf{F}}$, so it is not easy to determine the impulse, $(\Sigma\overline{\mathbf{F}})\,\Delta t$, directly. On the other hand, it is usually straightforward to measure the mass and velocity of an object, so that its momentum just after the collision $m\mathbf{v_f}$ and just before it $m\mathbf{v_0}$ can be found. Thus, the impulse–momentum theorem allows us to gain information about the impulse indirectly by measuring the change in momentum that the impulse causes. Then, armed with a knowledge of the contact time $\Delta t$, we can evaluate the net average force. Examples 1 and 2 illustrate how the theorem is used in this way.

### *Example 1*   A Well-Hit Ball

A baseball ($m = 0.14$ kg) has an initial velocity of $\mathbf{v_0} = -38$ m/s as it approaches a bat. We have chosen the direction of approach as the negative direction. The bat applies an average force $\overline{\mathbf{F}}$ that is much larger than the weight of the ball, and the ball departs from the bat with a final velocity of $\mathbf{v_f} = +58$ m/s. (a) Determine the impulse applied to the ball by the bat. (b) Assuming that the time of contact is $\Delta t = 1.6 \times 10^{-3}$ s, find the average force exerted on the ball by the bat.

*The equality between the net force and the change in momentum per unit time is the version of the second law of motion presented originally by Newton.

**Reasoning** Two forces act on the ball during impact, and together they constitute the net average force: the average force $\overline{\mathbf{F}}$ exerted by the bat, and the weight of the ball. Since $\overline{\mathbf{F}}$ is much greater than the weight of the ball, we neglect the weight. Thus, the net average force is equal to $\overline{\mathbf{F}}$, or $\Sigma \overline{\mathbf{F}} = \overline{\mathbf{F}}$. In hitting the ball, the bat imparts an impulse to it. We cannot use Equation 7.1 ($\mathbf{J} = \overline{\mathbf{F}} \Delta t$) to determine the impulse $\mathbf{J}$ directly, since $\overline{\mathbf{F}}$ is not known. We can find the impulse indirectly, however, by turning to the impulse–momentum theorem, which states that the impulse is equal to the ball's final momentum minus its initial momentum. With values for the impulse and the time of contact, Equation 7.1 can be used to determine the average force applied to the ball by the bat.

**Solution**

**(a)** According to the impulse–momentum theorem, the impulse $\mathbf{J}$ applied to the ball is

$$\mathbf{J} = m\mathbf{v_f} - m\mathbf{v_0}$$

$$= \underbrace{(0.14 \text{ kg})(+58 \text{ m/s})}_{\text{Final momentum}} - \underbrace{(0.14 \text{ kg})(-38 \text{ m/s})}_{\text{Initial momentum}} = \boxed{+13.4 \text{ kg} \cdot \text{m/s}}$$

**(b)** Now that the impulse is known, the contact time can be used in Equation 7.1 to find the average force $\overline{\mathbf{F}}$ exerted by the bat on the ball:

$$\overline{\mathbf{F}} = \frac{\mathbf{J}}{\Delta t} = \frac{+13.4 \text{ kg} \cdot \text{m/s}}{1.6 \times 10^{-3} \text{ s}} = \boxed{+8400 \text{ N}}$$

The force is positive, indicating that it points opposite to the velocity of the approaching ball. A force of 8400 N corresponds to 1900 lb, such a large value being necessary to change the ball's momentum during the brief contact time.

> **Problem solving insight**
> Momentum is a vector and, as such, has a magnitude and a direction. For motion in one dimension, be sure to indicate the direction by assigning a plus or minus sign to it, as in this example.

## Example 2   A Rain Storm

During a storm, rain comes straight down with a velocity of $\mathbf{v_0} = -15$ m/s and hits the roof of a car perpendicularly (see Figure 7.4). The mass of rain per second that strikes the car roof is 0.060 kg/s. Assuming that the rain comes to rest upon striking the car ($\mathbf{v_f} = 0$ m/s), find the average force exerted by the rain on the roof.

**Reasoning** This example differs from Example 1 in an important way. Example 1 gives information about the ball and asks for the force applied to the ball. In contrast, the present example gives information about the rain but asks for the force acting on the roof. However, the force exerted on the roof by the rain and the force exerted on the rain by the roof have equal magnitudes and opposite directions, according to Newton's law of action and reaction (see Section 4.5). Thus, we will find the force exerted on the rain and then apply the law of action and reaction to obtain the force on the roof. Two forces act on the rain while it impacts with the roof: the average force $\overline{\mathbf{F}}$ exerted by the roof, and the weight of the rain. These two forces constitute the net average force. By comparison, however, the force $\overline{\mathbf{F}}$ is much greater than the weight, so we may neglect the weight. Thus, the net average force becomes equal to $\overline{\mathbf{F}}$, or $\Sigma \overline{\mathbf{F}} = \overline{\mathbf{F}}$. The value of $\overline{\mathbf{F}}$ can be obtained by applying the impulse–momentum theorem to the rain.

**Solution** The average force $\overline{\mathbf{F}}$ needed to reduce the rain's velocity from $\mathbf{v_0} = -15$ m/s to $\mathbf{v_f} = 0$ m/s is given by Equation 7.4 as

$$\overline{\mathbf{F}} = \frac{m\mathbf{v_f} - m\mathbf{v_0}}{\Delta t} = -\left(\frac{m}{\Delta t}\right)\mathbf{v_0}$$

The term $m/\Delta t$ is the mass of rain per second that strikes the roof, so that $m/\Delta t = 0.060$ kg/s. Thus, the average force exerted on the rain by the roof is

$$\overline{\mathbf{F}} = -(0.060 \text{ kg/s})(-15 \text{ m/s}) = +0.90 \text{ N}$$

This force is in the positive or upward direction, which is reasonable since the roof exerts an upward force on each falling drop in order to bring it to rest. According to the action–reaction law, the force exerted on the roof by the rain also has a magnitude of 0.90 N but points downward: Force on roof = $\boxed{-0.90 \text{ N}}$.

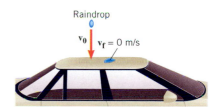

**Figure 7.4** A raindrop falling on a car roof has an initial velocity of $\mathbf{v_0}$ just before striking the roof. The final velocity of the raindrop is $\mathbf{v_f} = 0$ m/s, because it comes to rest on the roof.

As you reason through problems such as those in Examples 1 and 2, take advantage of the impulse–momentum theorem. It is a powerful statement that can lead to significant insights. The following Conceptual Example further illustrates its use.

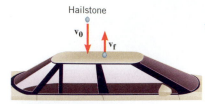

Hailstone

$v_0$  $v_f$

**Figure 7.5** Hailstones have a downward velocity of $v_0$ just before striking this car roof. They rebound with an upward velocity of $v_f$.

## Conceptual Example 3  Hailstones Versus Raindrops

In Example 2 rain is falling on the roof of a car and exerts a force on it. Instead of rain, suppose hail is falling. The hail comes straight down at a mass rate of $m/\Delta t = 0.060$ kg/s and an initial velocity of $v_0 = -15$ m/s and strikes the roof perpendicularly, just as the rain does in Example 2. However, unlike rain, hail usually does not come to rest after striking a surface. Instead, the hailstones bounce off the roof of the car. If hail fell instead of rain, would the force on the roof be smaller than, equal to, or greater than that calculated in Example 2?

**Reasoning and Solution** The raindrops and the hailstones fall in exactly the same way. That is, they both fall with the same initial velocity and mass rate, and they both strike the roof perpendicularly. However, there is an important difference: the raindrops come to rest (see Figure 7.4), while the hailstones bounce upward after striking the roof (see Figure 7.5). According to the impulse–momentum theorem, the impulse that acts on an object is given by the change in the momentum of the object. This change is $mv_f - mv_0 = m\,\Delta v$ and is proportional to the change in velocity $\Delta v$. For a hailstone, the change in velocity is from $v_0$ (downward) to $v_f$ (upward). This is a larger change than for the raindrop, for which the change is only from $v_0$ (downward) to zero. Therefore, the change in momentum is greater for the hailstone, and, correspondingly, a greater impulse must act on it. But an impulse is the product of the average force and the time interval $\Delta t$. Since the same amount of mass falls in the same time interval in either case, $\Delta t$ is the same for hailstones as for raindrops. The greater impulse acting on the hailstones, then, means that the car roof must exert a greater force on the hailstones than on the raindrops. Conversely, according to Newton's action–reaction law, *the car roof experiences a greater force from the hailstones than from the raindrops.*

**Related Homework:** *Problem 6*

# 7.2 The Principle of Conservation of Linear Momentum

It is worthwhile to compare the impulse–momentum theorem to the work–energy theorem discussed in Chapter 6. The impulse–momentum theorem states that the impulse produced by a net force is equal to the change in the object's momentum, while the work–energy theorem states that the work done by a net force is equal to the change in the object's kinetic energy. The work–energy theorem leads directly to the principle of conservation of mechanical energy and, as we will see, the impulse–momentum theorem also leads to a conservation principle, known as the conservation of linear momentum.

We begin by applying the impulse–momentum theorem to a midair collision between two objects. The two objects (masses $m_1$ and $m_2$) are approaching each other with initial velocities $v_{01}$ and $v_{02}$, as Figure 7.6a shows. The collection of objects being studied is referred to as the "system." In this case, the system contains only the two objects. They interact during the collision in part *b* and then depart with the final velocities $v_{f1}$ and $v_{f2}$ shown in part *c*. Because of the collision, the initial and final velocities are not the same.

Two types of forces act on the system:

1. *Internal forces* — Forces that the objects within the system exert on each other.

2. *External forces* — Forces exerted on the objects by agents external to the system.

During the collision in Figure 7.6b, $F_{12}$ is the force exerted on object 1 by object 2, while $F_{21}$ is the force exerted on object 2 by object 1. These forces are action–reaction forces that are equal in magnitude but opposite in direction, so $F_{12} = -F_{21}$. They are also internal forces, since they are forces that the two objects within the system exert on each other. The force of gravity also acts on the objects, their weights being $W_1$ and $W_2$. These weights, however, are external forces, because they are applied by the earth, which is outside the system. Friction and air resistance would also be considered external forces, although these forces are ignored here for the sake of simplicity. The impulse–momentum theorem, as applied to each object, gives the following results:

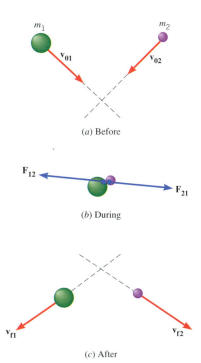

$m_1$   $m_2$

$v_{01}$   $v_{02}$

(a) Before

$F_{12}$   $F_{21}$

(b) During

$v_{f1}$   $v_{f2}$

(c) After

**Figure 7.6** (a) The velocities of the two objects before the collision are $v_{01}$ and $v_{02}$. (b) During the collision, each object exerts a force on the other. These forces are $F_{12}$ and $F_{21}$. (c) The velocities after the collision are $v_{f1}$ and $v_{f2}$.

*Object 1*
$$( \underbrace{\mathbf{W}_1}_{\substack{\text{External} \\ \text{force}}} + \underbrace{\overline{\mathbf{F}}_{12}}_{\substack{\text{Internal} \\ \text{force}}} ) \Delta t = m_1\mathbf{v}_{f1} - m_1\mathbf{v}_{01}$$

*Object 2*
$$( \underbrace{\mathbf{W}_2}_{\substack{\text{External} \\ \text{force}}} + \underbrace{\overline{\mathbf{F}}_{21}}_{\substack{\text{Internal} \\ \text{force}}} ) \Delta t = m_2\mathbf{v}_{f2} - m_2\mathbf{v}_{02}$$

Adding these equations produces a single result for the system as a whole:

$$(\underbrace{\mathbf{W}_1 + \mathbf{W}_2}_{\substack{\text{External} \\ \text{forces}}} + \underbrace{\overline{\mathbf{F}}_{12} + \overline{\mathbf{F}}_{21}}_{\substack{\text{Internal} \\ \text{forces}}}) \Delta t = \underbrace{(m_1\mathbf{v}_{f1} + m_2\mathbf{v}_{f2})}_{\substack{\text{Total final} \\ \text{momentum } \mathbf{P}_f}} - \underbrace{(m_1\mathbf{v}_{01} + m_2\mathbf{v}_{02})}_{\substack{\text{Total initial} \\ \text{momentum } \mathbf{P}_0}}$$

On the right side of this equation, the quantity $m_1\mathbf{v}_{f1} + m_2\mathbf{v}_{f2}$ is the vector sum of the final momenta for each object, or the total final momentum $\mathbf{P}_f$ of the system. Likewise, $m_1\mathbf{v}_{01} + m_2\mathbf{v}_{02}$ is the total initial momentum $\mathbf{P}_0$. Therefore, the result above can be rewritten as

$$\left( \begin{array}{c} \textbf{Sum of average} \\ \textbf{external forces} \end{array} + \begin{array}{c} \textbf{Sum of average} \\ \textbf{internal forces} \end{array} \right) \Delta t = \mathbf{P}_f - \mathbf{P}_0 \qquad (7.5)$$

The advantage of the internal/external force classification is that the internal forces always add together to give zero, as a consequence of Newton's law of action–reaction; $\overline{\mathbf{F}}_{12} = -\overline{\mathbf{F}}_{21}$ so that $\overline{\mathbf{F}}_{12} + \overline{\mathbf{F}}_{21} = 0$. Cancellation of the internal forces occurs no matter how many parts there are to the system and allows us to ignore the internal forces, as Equation 7.6 indicates:

$$\textbf{(Sum of average external forces) } \Delta t = \mathbf{P}_f - \mathbf{P}_0 \qquad (7.6)$$

We developed this result with gravity as the only external force. But, in general, the sum of the external forces on the left includes *all* external forces.

With the aid of Equation 7.6, it is possible to see how the conservation of linear momentum arises. Suppose that the sum of the external forces is zero. A system for which this is true is called an ***isolated system.*** Then Equation 7.6 indicates that

$$0 = \mathbf{P}_f - \mathbf{P}_0 \quad \text{or} \quad \mathbf{P}_f = \mathbf{P}_0 \qquad (7.7a)$$

In other words, the final total momentum of the isolated system after the objects in Figure 7.6 collide is the same as the initial total momentum.* Explicitly writing out the final and initial momenta for the two-body collision, we obtain for Equation 7.7a that

$$\underbrace{m_1\mathbf{v}_{f1} + m_2\mathbf{v}_{f2}}_{\mathbf{P}_f} = \underbrace{m_1\mathbf{v}_{01} + m_2\mathbf{v}_{02}}_{\mathbf{P}_0} \qquad (7.7b)$$

This result is an example of a general principle known as the **principle of conservation of linear momentum.**

---

■ **PRINCIPLE OF CONSERVATION OF LINEAR MOMENTUM**

The total linear momentum of an isolated system remains constant (is conserved). An isolated system is one for which the vector sum of the average external forces acting on the system is zero.

---

This principle applies to a system containing any number of objects, regardless of the internal forces, provided the system is isolated. Whether the system is isolated depends on whether the vector sum of the external forces is zero. Judging whether a force is internal or external depends on which objects are included in the system, as Conceptual Example 4 illustrates.

---

*Technically, the initial and final momenta are equal when the impulse of the sum of the external forces is zero—that is, when the left-hand side of Equation 7.6 is zero. Sometimes, however, the initial and final momenta are very nearly equal even when the sum of the external forces is not zero. This occurs when the time $\Delta t$ during which the forces act is so short that it is effectively zero. Then, the left-hand side of Equation 7.6 is approximately zero.

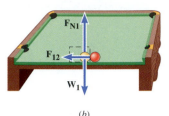

(a)

(b)

**Figure 7.7** (a) Two billiard balls collide. They have weights $W_1$ and $W_2$. The dashed box emphasizes that both balls are included in the system. $F_{N1}$ and $F_{N2}$ are the normal forces that the table applies to the balls. (b) Now only ball 1 is included in the system.

## *Conceptual Example 4* Is the Total Momentum Conserved?

Imagine two balls colliding on a billiard table that is friction-free. Use the momentum conservation principle in answering the following questions. (a) Is the total momentum of the two-ball system the same before and after the collision? (b) Answer part (a) for a system that contains only one of the two colliding balls.

### Reasoning and Solution

**(a)** Figure 7.7a shows the balls colliding, and the dashed outline emphasizes that both balls are part of the system. The collision forces are not shown, because they are internal forces and cannot cause the total momentum of the system to change. The external forces include the weights $W_1$ and $W_2$ of the balls. There are two additional external forces, however: the upward-pointing normal force that the table exerts on each ball. These normal forces are $F_{N1}$ and $F_{N2}$. Since the balls do not accelerate in the vertical direction, the normal forces must balance the weights, so that the vector sum of the four external forces in Figure 7.7a is zero. Furthermore, the table is friction-free. Thus, there is no net external force to change the total momentum of the two-ball system, and it is the same before and after the collision. *The total momentum of this two-ball system is conserved.*

**(b)** In Figure 7.7b only one ball is included in the system, as the dashed outline indicates. The forces acting on this system are all external and include the weight $W_1$ of the ball and the normal force $F_{N1}$. As in part a of the drawing, these two forces balance. However, there is a third external force to consider. Ball 2 is outside the system, so the force $F_{12}$ that it applies to the system during the collision is now an external force. As a result, the vector sum of the three external forces is not zero, and the net external force causes the total momentum of the one-ball system to be different after the collision than it was before the collision. *The total momentum of this one-ball system is not conserved.*

Next, we apply the principle of conservation of linear momentum to the problem of assembling a freight train.

## *Example 5* Assembling a Freight Train

A freight train is being assembled in a switching yard, and Figure 7.8 shows two boxcars. Car 1 has a mass of $m_1 = 65 \times 10^3$ kg and moves at a velocity of $v_{01} = +0.80$ m/s. Car 2, with a mass of $m_2 = 92 \times 10^3$ kg and a velocity of $v_{02} = +1.3$ m/s, overtakes car 1 and couples to it. Neglecting friction, find the common velocity $v_f$ of the cars after they become coupled.

**Reasoning** The two boxcars constitute the system. The sum of the external forces acting on the system is zero, because the weight of each car is balanced by a corresponding normal force, and friction is being neglected. Thus, the system is isolated, and the principle of conservation of linear momentum applies. The coupling forces that each car exerts on the other are internal forces and do not affect the applicability of this principle.

**Solution** Momentum conservation indicates that

$$\underbrace{(m_1 + m_2)v_f}_{\substack{\text{Total momentum} \\ \text{after collision}}} = \underbrace{m_1v_{01} + m_2v_{02}}_{\substack{\text{Total momentum} \\ \text{before collision}}}$$

This equation can be solved for $v_f$, the common velocity of the two cars after the collision:

$$
\begin{aligned}
v_f &= \frac{m_1v_{01} + m_2v_{02}}{m_1 + m_2} \\
&= \frac{(65 \times 10^3 \text{ kg})(0.80 \text{ m/s}) + (92 \times 10^3 \text{ kg})(1.3 \text{ m/s})}{(65 \times 10^3 \text{ kg} + 92 \times 10^3 \text{ kg})} = \boxed{+1.1 \text{ m/s}}
\end{aligned}
$$

*Problem solving insight*
The conservation of linear momentum is applicable only when the net external force acting on the system is zero. Therefore, the first step in applying momentum conservation is to be sure that the net external force is zero.

**Figure 7.8** (a) The boxcar on the left eventually catches up with the other boxcar and (b) couples to it. The coupled cars move together with a common velocity after the collision.

(a) Before          (b) After

In the previous example it can be seen that the velocity of car 1 increases, while the velocity of car 2 decreases as a result of the collision. The acceleration and deceleration arise at the moment the cars become coupled, because the cars exert internal forces on each other. The powerful feature of the momentum conservation principle is that it allows us to determine the changes in velocity without knowing what the internal forces are. Example 6 further illustrates this feature.

## *Example 6* Ice Skaters

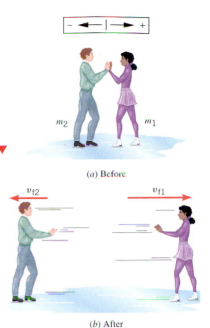

(a) Before

(b) After

**Figure 7.9** (a) In the absence of friction, two skaters pushing on each other constitute an isolated system. (b) As the skaters move away, the total linear momentum of the system remains zero, which is what it was initially.

Starting from rest, two skaters "push off" against each other on smooth level ice, where friction is negligible. As Figure 7.9a shows, one is a woman ($m_1 = 54$ kg), and one is a man ($m_2 = 88$ kg). Part b of the drawing shows that the woman moves away with a velocity of $v_{f1} = +2.5$ m/s. Find the "recoil" velocity $v_{f2}$ of the man.

**Reasoning** For a system consisting of the two skaters on level ice, the sum of the external forces is zero. This is because the weight of each skater is balanced by a corresponding normal force and friction is negligible. The skaters, then, constitute an isolated system, and the principle of conservation of linear momentum applies. We expect the man to have a smaller recoil speed for the following reason. The internal forces that the man and woman exert on each other during pushoff have equal magnitudes but opposite directions, according to Newton's action–reaction law. The man, having the larger mass, experiences a smaller acceleration according to Newton's second law. Hence, he acquires a smaller recoil speed.

**Solution** The total momentum of the skaters before they push on each other is zero, since they are at rest. Momentum conservation requires that the total momentum remains zero after the skaters have separated, as in part b of the drawing:

$$\underbrace{m_1 v_{f1} + m_2 v_{f2}}_{\substack{\text{Total momentum} \\ \text{after pushing}}} = \underbrace{0}_{\substack{\text{Total momentum} \\ \text{before pushing}}}$$

Solving for the recoil velocity of the man gives

$$v_{f2} = \frac{-m_1 v_{f1}}{m_2} = \frac{-(54 \text{ kg})(+2.5 \text{ m/s})}{88 \text{ kg}} = \boxed{-1.5 \text{ m/s}}$$

The minus sign indicates that the man moves to the left in the drawing. After the skaters separate, the total momentum of the system remains zero, because momentum is a vector quantity, and the momenta of the man and the woman have equal magnitudes but opposite directions.

*It is important to realize that the total linear momentum may be conserved even when the kinetic energies of the individual parts of a system change.* In Example 6, for instance, the initial kinetic energy is zero, since the skaters are stationary. But after they push off, the skaters are moving, so each has kinetic energy. The kinetic energy changes because work is done by the internal force that each skater exerts on the other. However, internal forces cannot change the total linear momentum of a system, since the total linear momentum of an isolated system is conserved in the presence of such forces.

*Problem solving insight*

## 7.3 *Collisions in One Dimension*

As discussed in the last section, the total linear momentum is conserved when two objects collide, provided they constitute an isolated system. When the objects are atoms or subatomic particles, the total kinetic energy of the system is often conserved also. In other words, the total kinetic energy of the particles before the collision equals the total kinetic energy of the particles after the collision, so that kinetic energy gained by one particle is lost by another.

In contrast, when two macroscopic objects collide, such as two cars, the total kinetic energy after the collision is generally less than that before the collision. Kinetic energy is lost mainly in two ways: (1) it can be converted into heat because of friction, and (2) it is spent in creating permanent distortion or damage, as in an automobile collision. With very hard objects, such as a solid steel ball and a marble floor, the permanent distortion suffered upon collision is much smaller than with softer objects and, consequently, less kinetic energy is lost.

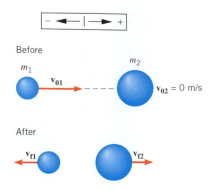

**Figure 7.10** A 0.250-kg ball, traveling with an initial velocity of $v_{01} = +5.00$ m/s, undergoes an elastic collision with a 0.800-kg ball that is initially at rest.

**Problem solving insight**
As long as the net external force is zero, the conservation of linear momentum applies to any type of collision. This is true whether the collision is elastic or inelastic.

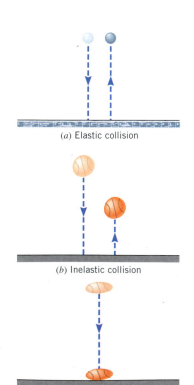

(a) Elastic collision

(b) Inelastic collision

(c) Completely inelastic collision

**Figure 7.11** (a) A hard steel ball would rebound to its original height after striking a hard marble surface if the collision were elastic. (b) A partially deflated basketball has little bounce on a soft asphalt surface. (c) A deflated basketball has no bounce at all.

Collisions are often classified according to whether the total kinetic energy changes during the collision:

1. *Elastic collision*—One in which the total kinetic energy of the system after the collision is equal to the total kinetic energy before the collision.

2. *Inelastic collision*—One in which the total kinetic energy of the system is *not* the same before and after the collision; if the objects stick together after colliding, the collision is said to be completely inelastic.

The boxcars coupling together in Figure 7.8 are an example of a completely inelastic collision. When a collision is completely inelastic, the greatest amount of kinetic energy is lost. Example 7 shows how one particular elastic collision is described using the conservation of linear momentum and the fact that no kinetic energy is lost.

## Example 7   A Collision in One Dimension  (26)

As in Figure 7.10, a ball of mass $m_1 = 0.250$ kg and velocity $v_{01} = +5.00$ m/s collides head-on with a ball of mass $m_2 = 0.800$ kg that is initially at rest ($v_{02} = 0$ m/s). No external forces act on the balls. If the collision is elastic, what are the velocities of the balls after the collision?

**Reasoning** The total linear momentum of the two-ball system is conserved, because no external forces act on the system. Momentum conservation applies whether or not the collision is elastic:

$$\underbrace{m_1 v_{f1} + m_2 v_{f2}}_{\substack{\text{Total momentum} \\ \text{after collision}}} = \underbrace{m_1 v_{01} + 0}_{\substack{\text{Total momentum} \\ \text{before collision}}}$$

Before and after an elastic collision, the total kinetic energy is the same.

$$\underbrace{\tfrac{1}{2}m_1 v_{f1}^2 + \tfrac{1}{2}m_2 v_{f2}^2}_{\substack{\text{Total kinetic energy} \\ \text{after collision}}} = \underbrace{\tfrac{1}{2}m_1 v_{01}^2 + 0}_{\substack{\text{Total kinetic energy} \\ \text{before collision}}}$$

We expect that ball 1, having the smaller mass, will rebound to the left after striking ball 2, which is more massive. Ball 2 will be driven to the right in the process. One solution to the equations above, then, should give a value for $v_{f1}$ that is negative (the left direction in Figure 7.10) and a value for $v_{f2}$ that is positive.

**Solution** The equations above are simultaneous equations containing the two unknown quantities $v_{f1}$ and $v_{f2}$. To solve them, we begin by rearranging the equation expressing momentum conservation to show that $v_{f2} = m_1(v_{01} - v_{f1})/m_2$. Substituting this result into the equation expressing the conservation of kinetic energy leads to the following equation for $v_{f1}$:

$$v_{f1} = \left(\frac{m_1 - m_2}{m_1 + m_2}\right)v_{01} \qquad (7.8a)$$

The expression for $v_{f1}$ in Equation 7.8a can be substituted into either of the equations obtained in the Reasoning to show that

$$v_{f2} = \left(\frac{2m_1}{m_1 + m_2}\right)v_{01} \qquad (7.8b)$$

With the given values for $m_1$, $m_2$, and $v_{01}$, Equations 7.8 yield the following values for $v_{f1}$ and $v_{f2}$:

$$\boxed{v_{f1} = -2.62 \text{ m/s}} \quad \text{and} \quad \boxed{v_{f2} = +2.38 \text{ m/s}}$$

The negative value for $v_{f1}$ indicates that ball 1 rebounds to the left after the collision in Figure 7.10, while the positive value for $v_{f2}$ indicates that ball 2 moves to the right, as expected.

We can get a feel for an elastic collision by dropping a steel ball onto a hard surface, such as a marble floor. If the collision is elastic, the ball will rebound to its original height, as Figure 7.11a illustrates. In contrast, a partially deflated basketball exhibits little rebound from a relatively soft asphalt surface, as in part b, indicating that a fraction of the ball's kinetic energy is dissipated during the inelastic collision. The completely deflated basketball in part *c* has no bounce at all, and a maximum amount of kinetic energy is lost during the completely inelastic collision.

The next example illustrates a completely inelastic collision in a device called a "ballistic pendulum." This device can be used to measure the speed of a bullet.

### Example 8    A Ballistic Pendulum

A ballistic pendulum can be used to measure the speed of a projectile, such as a bullet. The ballistic pendulum shown in Figure 7.12*a* consists of a block of wood (mass $m_2$ = 2.50 kg) suspended by a wire of negligible mass. A bullet (mass $m_1$ = 0.0100 kg) is fired with a speed $v_{01}$. Just after the bullet collides with it, the block (with the bullet in it) has a speed $v_f$ and then swings to a maximum height of 0.650 m above the initial position (see part *b* of the drawing). Find the speed $v_{01}$ of the bullet, assuming that air resistance is negligible.

**Reasoning** The physics of the ballistic pendulum can be divided into two parts. The first is the completely inelastic collision between the bullet and the block. The second is the resulting motion of the block and bullet as they swing upward. The total momentum of the system (block plus bullet) is conserved during the collision, because the suspension wire supports the system's weight, which means that the sum of the external forces acting on the system is nearly zero. Furthermore, as the system swings upward, the principle of conservation of mechanical energy applies, because nonconservative forces do no work. The tension force in the wire does no work because it acts perpendicular to the motion. Since air resistance is negligible, we can ignore the work it does.

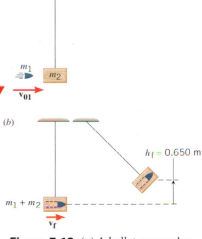

**Figure 7.12**    (*a*) A bullet approaches a ballistic pendulum. (*b*) The block and bullet swing upward after the collision.

**The physics of**
**measuring the speed of a bullet.**

**Solution** Applying the momentum conservation principle gives

$$\underbrace{(m_1 + m_2)v_f}_{\substack{\text{Total momentum} \\ \text{after collision}}} = \underbrace{m_1v_{01}}_{\substack{\text{Total momentum} \\ \text{before collision}}}$$

This equation can be solved for the initial speed $v_{01}$ of the bullet:

$$v_{01} = \frac{m_1 + m_2}{m_1}v_f$$

A value is now needed for the speed $v_f$ immediately after the collision. It can be obtained from the maximum height to which the system swings, by using the principle of conservation of mechanical energy:

$$\underbrace{(m_1 + m_2)gh_f}_{\substack{\text{Total mechanical energy} \\ \text{at the top of the swing,} \\ \text{all potential}}} = \underbrace{\tfrac{1}{2}(m_1 + m_2)v_f^2}_{\substack{\text{Total mechanical energy} \\ \text{at the bottom of the} \\ \text{swing, all kinetic}}}$$

Solving this expression for $v_f$ gives $v_f = \sqrt{2gh_f}$. Substituting this result into the equation for $v_{01}$, we find that

$$v_{01} = \left(\frac{m_1 + m_2}{m_1}\right)\sqrt{2gh_f} = \left(\frac{0.0100 \text{ kg} + 2.50 \text{ kg}}{0.0100 \text{ kg}}\right)\sqrt{2(9.80 \text{ m/s}^2)(0.650 \text{ m})}$$

$$= \boxed{+896 \text{ m/s}}$$

## 7.4    *Collisions in Two Dimensions*

The collisions discussed so far have been "head-on," or one-dimensional, because the velocities of the objects all point along a single line before and after contact. Collisions often occur, however, in two or three dimensions. Figure 7.13 shows a two-dimensional case in which two balls collide on a horizontal frictionless table.

For the system consisting of the two balls, the external forces include the weights of the balls and the corresponding normal forces produced by the table. Since each weight is balanced by a normal force, the sum of the external forces is zero, and the total momentum of the system is conserved, as Equation 7.7b indicates. Momentum is a vector quantity, however, and in two dimensions the *x* and *y* components of the total momentum are

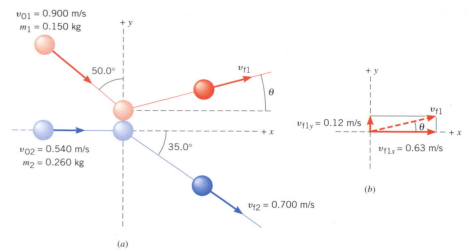

**Figure 7.13** (*a*) Top view of two balls colliding on a horizontal frictionless table. (*b*) This part of the drawing shows the *x* and *y* components of the velocity of ball 1 after the collision.

conserved separately. In other words, Equation 7.7b is equivalent to the following two equations:

*x Component*
$$\underbrace{m_1 v_{f1x} + m_2 v_{f2x}}_{\mathbf{P}_{fx}} = \underbrace{m_1 v_{01x} + m_2 v_{02x}}_{\mathbf{P}_{0x}} \qquad (7.9a)$$

*y Component*
$$\underbrace{m_1 v_{f1y} + m_2 v_{f2y}}_{\mathbf{P}_{fy}} = \underbrace{m_1 v_{01y} + m_2 v_{02y}}_{\mathbf{P}_{0y}} \qquad (7.9b)$$

These equations are written for a system that contains two objects. If a system contains more than two objects, a mass-times-velocity term must be included for each additional object on either side of Equations 7.9a and 7.9b. Example 9 shows how to deal with a two-dimensional collision when the total linear momentum of the system is conserved.

### Example 9 A Collision in Two Dimensions

For the data given in Figure 7.13, use momentum conservation to determine the magnitude and direction of the final velocity of ball 1 after the collision.

**Reasoning** The magnitude and direction of the final velocity of ball 1 can be obtained once the components $v_{f1x}$ and $v_{f1y}$ of this velocity are known. We will begin by determining these components with the aid of the momentum conservation principle.

**Solution** Applying momentum conservation in the *x*-direction (Equation 7.9a), we find that

*x Component*
$$\underbrace{(0.150 \text{ kg})(v_{f1x})}_{\text{Ball 1, after}} + \underbrace{(0.260 \text{ kg})(0.700 \text{ m/s})(\cos 35.0°)}_{\text{Ball 2, after}}$$
$$= \underbrace{(0.150 \text{ kg})(0.900 \text{ m/s})(\sin 50.0°)}_{\text{Ball 1, before}} + \underbrace{(0.260 \text{ kg})(0.540 \text{ m/s})}_{\text{Ball 2, before}}$$

This equation can be solved to show that $v_{f1x} = +0.63$ m/s. Applying momentum conservation in the *y*-direction (Equation 7.9b), we find that

**Problem solving insight**
Momentum, being a vector quantity, has a magnitude and a direction. In two dimensions, take into account the direction by using vector components and assigning a plus or minus sign to each component, as illustrated in this example.

*y Component*
$$\underbrace{(0.150 \text{ kg})(v_{f1y})}_{\text{Ball 1, after}} + \underbrace{(0.260 \text{ kg})[-(0.700 \text{ m/s})(\sin 35.0°)]}_{\text{Ball 2, after}}$$
$$= \underbrace{(0.150 \text{ kg})[-(0.900 \text{ m/s})(\cos 50.0°)]}_{\text{Ball 1, before}} + \underbrace{0}_{\text{Ball 2, before}}$$

The solution to this equation reveals that $v_{f1y} = +0.12$ m/s.

Figure 7.13*b* shows the *x* and *y* components of the final velocity of ball 1. The magnitude of the velocity is

$$v_{f1} = \sqrt{(0.63 \text{ m/s})^2 + (0.12 \text{ m/s})^2} = \boxed{0.64 \text{ m/s}}$$

The direction of the velocity is given by the angle $\theta$:

$$\theta = \tan^{-1}\left(\frac{0.12 \text{ m/s}}{0.63 \text{ m/s}}\right) = \boxed{11°}$$

## 7.5 *Center of Mass*

In previous sections, we have encountered situations in which objects interact with one another, such as the two skaters pushing off in Example 6. In these situations, the mass of the system is located in several places, and the various objects move relative to each other before, after, and even during the interaction. It is possible, however, to speak of a kind of average location for the total mass by introducing a concept known as the *center of mass* (abbreviated as "cm"). With the aid of this concept, we will be able to gain additional insight into the principle of conservation of linear momentum.

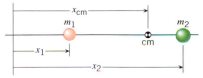

**Figure 7.14** The center of mass cm of the two particles is located on a line between them and lies closer to the more massive one.

The center of mass is a point that represents the average location for the total mass of a system. Figure 7.14, for example, shows two particles of mass $m_1$ and $m_2$ that are located on the *x* axis at the positions $x_1$ and $x_2$, respectively. The position $x_{cm}$ of the center-of-mass point from the origin is defined to be

*Center of mass*
$$x_{cm} = \frac{m_1 x_1 + m_2 x_2}{m_1 + m_2} \qquad (7.10)$$

Each term in the numerator of this equation is the product of a particle's mass and position, while the denominator is the total mass of the system. If the two masses are equal, we expect the average location of the total mass to be midway between the particles. With $m_1 = m_2 = m$, Equation 7.10 becomes $x_{cm} = (mx_1 + mx_2)/(m + m) = \frac{1}{2}(x_1 + x_2)$, which indeed corresponds to the point midway between the particles. Alternatively, suppose that $m_1 = 5.0$ kg and $x_1 = 2.0$ m, while $m_2 = 12$ kg and $x_2 = 6.0$ m. Then we expect the average location of the total mass to be located closer to particle 2, since it is more massive. Equation 7.10 is also consistent with this expectation, for it gives

$$x_{cm} = \frac{(5.0 \text{ kg})(2.0 \text{ m}) + (12 \text{ kg})(6.0 \text{ m})}{5.0 \text{ kg} + 12 \text{ kg}} = 4.8 \text{ m}$$

If a system contains more than two particles, the center-of-mass point can be determined by generalizing Equation 7.10. For three particles, for instance, the numerator would contain a third term $m_3 x_3$, and the total mass in the denominator would be $m_1 + m_2 + m_3$. For a macroscopic object, which contains many, many particles, the center-of-mass point is located at the geometric center of the object, provided that the mass is distributed symmetrically about the center. Such would be the case for a billiard ball. For objects such as a golf club, the mass is not distributed symmetrically and the center-of-mass point is not located at the geometric center of the club. The driver used to launch a golf ball from the tee, for instance, has more mass in the club head than in the handle, so the center-of-mass point is closer to the head than to the handle.

To see how the center-of-mass concept is related to momentum conservation, suppose that the two particles in a system are moving, as they would be during a collision. With the aid of Equation 7.10, we can determine the velocity $v_{cm}$ of the center-of-mass point. During a time interval $\Delta t$, the particles experience displacements of $\Delta x_1$ and $\Delta x_2$, as Figure 7.15 shows. They have different displacements during this time because they have different velocities. Equation 7.10 can be used to find the displacement $\Delta x_{cm}$ of the center of mass by replacing $x_{cm}$ by $\Delta x_{cm}$, $x_1$ by $\Delta x_1$, and $x_2$ by $\Delta x_2$:

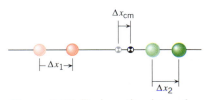

**Figure 7.15** During a time interval $\Delta t$, the displacements of the particles are $\Delta x_1$ and $\Delta x_2$, while the displacement of the center of mass is $\Delta x_{cm}$.

$$\Delta x_{cm} = \frac{m_1 \Delta x_1 + m_2 \Delta x_2}{m_1 + m_2}$$

Now we divide both sides of this equation by the time interval $\Delta t$. In the limit as $\Delta t$ becomes infinitesimally small, the ratio $\Delta x_{cm}/\Delta t$ becomes equal to the instantaneous velocity $v_{cm}$ of the center of mass. (See Section 2.2 for a review of instantaneous velocity.) Likewise, the ratios $\Delta x_1/\Delta t$ and $\Delta x_2/\Delta t$ become equal to the instantaneous velocities $v_1$ and $v_2$, respectively. Thus, we have

*Velocity of center of mass*
$$v_{cm} = \frac{m_1 v_1 + m_2 v_2}{m_1 + m_2} \tag{7.11}$$

The numerator $(m_1 v_1 + m_2 v_2)$ on the right-hand side in Equation 7.11 is the momentum of particle 1 $(m v_1)$ plus the momentum of particle 2 $(m_2 v_2)$, which is the total linear momentum of the system. In an isolated system, the total linear momentum does not change because of an interaction such as a collision. Therefore, Equation 7.11 indicates that the velocity $v_{cm}$ of the center of mass does not change either. To emphasize this important point, consider the collision discussed in Example 7. With the data from that example, we can apply Equation 7.11 to determine the velocity of the center of mass before and after the collision:

*Before collision*
$$v_{cm} = \frac{(0.250\text{ kg})(+5.00\text{ m/s}) + (0.800\text{ kg})(0\text{ m/s})}{0.250\text{ kg} + 0.800\text{ kg}} = +1.19\text{ m/s}$$

*After collision*
$$v_{cm} = \frac{(0.250\text{ kg})(-2.62\text{ m/s}) + (0.800\text{ kg})(+2.38\text{ m/s})}{0.250\text{ kg} + 0.800\text{ kg}} = +1.19\text{ m/s}$$

Thus, the velocity of the center of mass is the same before and after the objects interact during a collision in which the total linear momentum is conserved.

# Concept Summary

This summary presents an abridged version of the chapter, including the important equations and all available learning aids. For convenient reference, the learning aids (including the text's examples) are placed next to or immediately after the relevant equation or discussion. The following learning aids may be found on-line at **www.wiley.com/college/cutnell**:

| | |
|---|---|
| **Interactive LearningWare** examples are solved according to a five-step interactive format that is designed to help you develop problem-solving skills. | **Concept Simulations** are animated versions of text figures or animations that illustrate important concepts. You can control parameters that affect the display, and we encourage you to experiment. |
| **Interactive Solutions** offer specific models for certain types of problems in the chapter homework. The calculations are carried out interactively. | **Self-Assessment Tests** include both qualitative and quantitative questions. Extensive feedback is provided for both incorrect and correct answers, to help you evaluate your understanding of the material. |

| Topic | Discussion | Learning Aids |
|---|---|---|

### 7.1 The Impulse–Momentum Theorem

The impulse $\mathbf{J}$ of a force is the product of the average force $\overline{\mathbf{F}}$ and the time interval $\Delta t$ during which the force acts:

**Impulse**
$$\mathbf{J} = \overline{\mathbf{F}}\,\Delta t \tag{7.1}$$

Impulse is a vector that points in the same direction as the average force.

The linear momentum $\mathbf{p}$ of an object is the product of the object's mass $m$ and velocity $\mathbf{v}$:

**Linear momentum**
$$\mathbf{p} = m\mathbf{v} \tag{7.2}$$

Linear momentum is a vector that points in the same direction as the velocity. The total linear momentum of a system of objects is the vector sum of the momenta of the individual objects.

**Examples 1, 2, 3**

The impulse–momentum theorem states that when a net force $\Sigma\overline{\mathbf{F}}$ acts on an object, the impulse of the net force is equal to the change in momentum of the object:

**Impulse–momentum theorem**
$$(\Sigma\,\overline{\mathbf{F}})\,\Delta t = m\mathbf{v}_f - m\mathbf{v}_0 \tag{7.4}$$

**Interactive LearningWare 7.1**

**Interactive Solution 7.13**

| Topic | Discussion | Learning Aids |
|-------|-----------|---------------|

### 7.2 The Principle of Conservation of Linear Momentum

External forces are those that agents external to the system exert on objects within the system. An isolated system is one for which the vector sum of the average external forces acting on the system is zero.

The principle of conservation of linear momentum states that the total linear momentum of an isolated system remains constant. For a two-body collision, the conservation of linear momentum can be written as

**Conservation of linear momentum**

$$\underbrace{m_1\mathbf{v_{f1}} + m_2\mathbf{v_{f2}}}_{\substack{\text{Final total linear} \\ \text{momentum}}} = \underbrace{m_1\mathbf{v_{01}} + m_2\mathbf{v_{02}}}_{\substack{\text{Initial total linear} \\ \text{momentum}}} \tag{7.7b}$$

**Examples 4, 5, 6**

**Concept Simulation 7.1**

**Interactive LearningWare 7.2**

where $m_1$ and $m_2$ are the masses, $\mathbf{v_{f1}}$ and $\mathbf{v_{f2}}$ are the final velocities, and $\mathbf{v_{01}}$ and $\mathbf{v_{02}}$ are the initial velocities of the objects.

**Interactive Solution 7.19**

 *Use Self-Assessment Test 7.1 to evaluate your understanding of Sections 7.1 and 7.2.*

### 7.3 Collisions in One Dimension

**Elastic collision**

An elastic collision is one in which the total kinetic energy of the system after the collision is equal to the total kinetic energy of the system before the collision.

**Examples 7, 8**

**Concept Simulation 7.2**

**Inelastic collision**

An inelastic collision is one in which the total kinetic energy of the system is not the same before and after the collision. If the objects stick together after the collision, the collision is said to be completely inelastic.

**Interactive Solutions 7.33, 7.55**

### 7.4 Collisions in Two Dimensions

When the total linear momentum is conserved in a two-dimensional collision, the $x$ and $y$ components of the total linear momentum are conserved separately. For a collision between two objects, the conservation of total linear momentum can be written as

**Conservation of linear momentum**

$$\underbrace{m_1v_{f1x} + m_2v_{f2x}}_{\substack{x \text{ component of final} \\ \text{total linear momentum}}} = \underbrace{m_1v_{01x} + m_2v_{02x}}_{\substack{x \text{ component of initial} \\ \text{total linear momentum}}} \tag{7.9a}$$

**Example 9**

**Interactive LearningWare 7.3**

$$\underbrace{m_1v_{f1y} + m_2v_{f2y}}_{\substack{y \text{ component of final} \\ \text{total linear momentum}}} = \underbrace{m_1v_{01y} + m_2v_{02y}}_{\substack{y \text{ component of initial} \\ \text{total linear momentum}}} \tag{7.9b}$$

### 7.5 Center of Mass

The location of the center of mass of two particles lying on the $x$ axis is given by

**Location of center of mass**

$$x_{cm} = \frac{m_1x_1 + m_2x_2}{m_1 + m_2} \tag{7.10}$$

**Interactive Solution 7.43**

where $m_1$ and $m_2$ are the masses of the particles and $x_1$ and $x_2$ are their positions relative to the coordinate origin. If the particles move with velocities $v_1$ and $v_2$, the velocity $v_{cm}$ of the center of mass is

**Velocity of center of mass**

$$v_{cm} = \frac{m_1v_1 + m_2v_2}{m_1 + m_2} \tag{7.11}$$

If the total linear momentum of a system of particles remains constant during an interaction such as a collision, the velocity of the center of mass also remains constant.

 *Use Self-Assessment Test 7.2 to evaluate your understanding of Sections 7.3–7.5.*

# Problems

**ssm**  Solution is in the Student Solutions Manual.     **www**  Solution is available on the World Wide Web at www.wiley.com/college/cutnell

☤  This icon represents a biomedical application.

### Section 7.1 The Impulse–Momentum Theorem

**1. ssm** A volleyball is spiked so that its incoming velocity of $+4.0$ m/s is changed to an outgoing velocity of $-21$ m/s. The mass of the volleyball is 0.35 kg. What impulse does the player apply to the ball?

**2. Interactive LearningWare 7.1** at **www.wiley.com/college/cutnell** provides a review of the concepts that are involved in this problem. A 62.0-kg person, standing on a diving board, dives straight down into the water. Just before striking the water, her speed is 5.50 m/s. At a time of 1.65 s after she enters the water, her speed is reduced to 1.10 m/s. What is the net average force (magnitude and direction) that acts on her when she is in the water?

**3.** A golfer, driving a golf ball off the tee, gives the ball a velocity of $+38$ m/s. The mass of the ball is 0.045 kg, and the duration of the impact with the golf club is $3.0 \times 10^{-3}$ s. (a) What is the change in momentum of the ball? (b) Determine the average force applied to the ball by the club.

**4.** A baseball ($m = 149$ g) approaches a bat horizontally at a speed of 40.2 m/s (90 mi/h) and is hit straight back at a speed of 45.6 m/s (102 mi/h). If the ball is in contact with the bat for a time of 1.10 ms, what is the average force exerted on the ball by the bat? Neglect the weight of the ball, since it is so much less than the force of the bat. Choose the direction of the incoming ball as the positive direction.

**5. ssm** A soccer player kicks a ball with an average force of $+1400$ N, and her foot remains in contact with the ball for a time of $7.9 \times 10^{-3}$ s. What is the impulse (magnitude and direction) of this force?

**6.** Before starting this problem, review Conceptual Example 3. Suppose that the hail described there bounces off the roof of the car with a velocity of $+15$ m/s. Ignoring the weight of the hailstones, calculate the force exerted by the hail on the roof. Compare your answer to that obtained in Example 2 for the rain, and verify that your answer is consistent with the conclusion reached in Conceptual Example 3.

**7.** A 46-kg skater is standing still in front of a wall. By pushing against the wall she propels herself backward with a velocity of $-1.2$ m/s. Her hands are in contact with the wall for 0.80 s. Ignore friction and wind resistance. Find the magnitude and direction of the average force she exerts on the wall (which has the same magnitude, but opposite direction, as the force that the wall applies to her).

**8.** ☤ When jumping straight down, you can be seriously injured if you land stiff-legged. One way to avoid injury is to bend your knees upon landing to reduce the force of the impact. A 75-kg man just before contact with the ground has a speed of 6.4 m/s. (a) In a stiff-legged landing he comes to a halt in 2.0 ms. Find the average net force that acts on him during this time. (b) When he bends his knees, he comes to a halt in 0.10 s. Find the average net force now. (c) During the landing, the force of the ground on the man points upward, while the force due to gravity points downward. The average net force acting on the man includes both of these forces. Taking into account the directions of the forces, find the force of the ground on the man in parts (a) and (b).

**\*9. ssm www** A golf ball strikes a hard, smooth floor at an angle of 30.0° and, as the drawing shows, rebounds at the same angle. The

mass of the ball is 0.047 kg, and its speed is 45 m/s just before and after striking the floor. What is the magnitude of the impulse applied to the golf ball by the floor? (*Hint: Note that only the vertical component of the ball's momentum changes during impact with the floor, and ignore the weight of the ball.*)

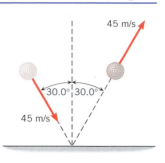

**\*10.** A student ($m = 63$ kg) falls freely from rest and strikes the ground. During the collision with the ground, he comes to rest in a time of 0.010 s. The average force exerted on him by the ground is $+18\,000$ N, where the upward direction is taken to be the positive direction. From what height did the student fall? Assume that the only force acting on him during the collision is that due to the ground.

**\*11.** A 0.500-kg ball is dropped from rest at a point 1.20 m above the floor. The ball rebounds straight upward to a height of 0.700 m. What are the magnitude and direction of the impulse of the net force applied to the ball during the collision with the floor?

**\*12.** An 85-kg jogger is heading due east at a speed of 2.0 m/s. A 55-kg jogger is heading 32° north of east at a speed of 3.0 m/s. Find the magnitude and direction of the sum of the momenta of the two joggers.

**\*13.** Consult **Interactive Solution 7.13** at **www.wiley.com/college/cutnell** for a review of problem-solving skills that are involved in this problem. A stream of water strikes a stationary turbine blade horizontally, as the drawing illustrates. The incident water stream has a velocity of $+16.0$ m/s, while the exiting water stream has a velocity of $-16.0$ m/s. The mass of water per second that strikes the blade is 30.0 kg/s. Find the magnitude of the average force exerted on the water by the blade.

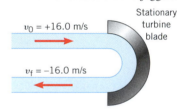

**\*\*14.** A dump truck is being filled with sand. The sand falls straight downward from rest from a height of 2.00 m above the truck bed, and the mass of sand that hits the truck per second is 55.0 kg/s. The truck is parked on the platform of a weight scale. By how much does the scale reading exceed the weight of the truck and sand?

### Section 7.2 The Principle of Conservation of Linear Momentum

**15.** ☤ **Interactive LearningWare 7.2** at **www.wiley.com/college/cutnell** provides a review of the concepts that are important in this problem. For tests using a *ballistocardiograph*, a patient lies on a horizontal platform that is supported on jets of air. Because of the air jets, the friction impeding the horizontal motion of the platform is negligible. Each time the heart beats, blood is pushed out of the heart in a direction that is nearly parallel to the platform. Since momentum must be conserved, the body and the platform recoil, and this recoil can be detected to provide information about the heart. For each beat, suppose that 0.050 kg of blood is pushed out of the heart with a velocity of $+0.25$ m/s and that the mass of the patient and platform is 85 kg. Assuming that the patient does not slip with respect to the platform, and that the patient and platform start from rest, determine the recoil velocity.

**16.** A 55-kg swimmer is standing on a stationary 210-kg floating raft. The swimmer then runs off the raft horizontally with a velocity of +4.6 m/s relative to the shore. Find the recoil velocity that the raft would have if there were no friction and resistance due to the water.

**17. ssm Concept Simulation 7.1** at **www.wiley.com/college/cutnell** illustrates the physics principles in this problem. In a science fiction novel two enemies, Bonzo and Ender, are fighting in outer space. From stationary positions they push against each other. Bonzo flies off with a velocity of +1.5 m/s, while Ender recoils with a velocity of −2.5 m/s. (a) Without doing any calculations, decide which person has the greater mass. Give your reasoning. (b) Determine the ratio of the masses ($m_{Bonzo}/m_{Ender}$) of these two people.

**18.** Consult **Concept Simulation 7.1** at **www.wiley.com/college/cutnell** in preparation for this problem. Two friends, Al and Jo, have a combined mass of 168 kg. At an ice skating rink they stand close together on skates, at rest and facing each other, with a compressed spring between them. The spring is kept from pushing them apart because they are holding each other. When they release their arms, Al moves off in one direction at a speed of 0.90 m/s, while Jo moves off in the opposite direction at a speed of 1.2 m/s. Assuming that friction is negligible, find Al's mass.

**\* 19.** To view an interactive solution to a problem that is very similar to this one, go to **www.wiley.com/college/cutnell** and select **Interactive Solution 7.19**. A fireworks rocket is moving at a speed of 45.0 m/s. The rocket suddenly breaks into two pieces of equal mass, which fly off with velocities $v_1$ and $v_2$, as shown in the drawing. What is the magnitude of (a) $v_1$ and (b) $v_2$?

**\* 20.** Two ice skaters have masses $m_1$ and $m_2$ and are initially stationary. Their skates are identical. They push against one another, as in Figure 7.9, and move in opposite directions with different speeds. While they are pushing against each other, any kinetic frictional forces acting on their skates can be ignored. However, once the skaters separate, kinetic frictional forces eventually bring them to a halt. As they glide to a halt, the magnitudes of their accelerations are equal, and skater 1 glides twice as far as skater 2. What is the ratio $m_1/m_2$ of their masses?

**\* 21. ssm** By accident, a large plate is dropped and breaks into three pieces. The pieces fly apart parallel to the floor. As the plate falls, its momentum has only a vertical component, and no component parallel to the floor. After the collision, the component of the total momentum parallel to the floor must remain zero, since the net external force acting on the plate has no component parallel to the floor. Using the data shown in the drawing, find the masses of pieces 1 and 2.

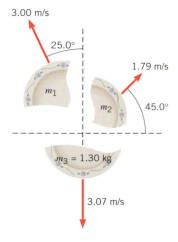

**\* 22.** The lead female character in the movie *Diamonds Are Forever* is standing at the edge of an offshore oil rig. As she fires a gun, she is driven back over the edge and into the sea. Suppose the mass of a bullet is 0.010 kg and its velocity is +720 m/s. Her mass (including the gun) is 51 kg. (a) What recoil velocity does she acquire in response to a single shot from a stationary position, assuming that no external force keeps her in place? (b) Under the same assumption, what would be her recoil velocity if, instead, she shoots a blank cartridge that ejects a mass of $5.0 \times 10^{-4}$ kg at a velocity of +720 m/s?

**\*\* 23. ssm www** A cannon of mass $5.80 \times 10^3$ kg is rigidly bolted to the earth so it can recoil only by a negligible amount. The cannon fires an 85.0-kg shell horizontally with an initial velocity of +551 m/s. Suppose the cannon is then unbolted from the earth, and no external force hinders its recoil. What would be the velocity of a shell fired by this loose cannon? (*Hint: In both cases assume that the burning gunpowder imparts the same kinetic energy to the system.*)

**\*\* 24.** A wagon is coasting at a speed $v_A$ along a straight and level road. When ten percent of the wagon's mass is thrown off the wagon, parallel to the ground and in the forward direction, the wagon is brought to a halt. If the direction in which this mass is thrown is exactly reversed, but the speed of this mass relative to the wagon remains the same, the wagon accelerates to a new speed $v_B$. Calculate the ratio $v_B/v_A$.

**Section 7.3 Collisions in One Dimension, Section 7.4 Collisions in Two Dimensions**

**25.** In a football game, a receiver is standing still, having just caught a pass. Before he can move, a tackler, running at a velocity of +4.5 m/s, grabs him. The tackler holds onto the receiver, and the two move off together with a velocity of +2.6 m/s. The mass of the tackler is 115 kg. Assuming that momentum is conserved, find the mass of the receiver.

**26.** A 1055-kg van, stopped at a traffic light, is hit directly in the rear by a 715-kg car traveling with a velocity of +2.25 m/s. Assume that the transmission of the van is in neutral, the brakes are not being applied, and the collision is elastic. What is the final velocity of (a) the car and (b) the van?

**27. ssm** A golf ball bounces down a flight of steel stairs, striking each stair once on the way down. The ball starts at the top step with a vertical velocity component of zero. If all the collisions with the stairs are elastic, and if the vertical height of the staircase is 3.00 m, determine the bounce height when the ball reaches the bottom of the stairs. Neglect air resistance.

**28.** A 2.50-g bullet, traveling at a speed of 425 m/s, strikes the wooden block of a ballistic pendulum, such as that in Figure 7.12. The block has a mass of 215 g. (a) Find the speed of the bullet/block combination immediately after the collision. (b) How high does the combination rise above its initial position?

**29.** A cue ball (mass = 0.165 kg) is at rest on a frictionless pool table. The ball is hit dead center by a pool stick, which applies an impulse of +1.50 N·s to the ball. The ball then slides along the table and makes an elastic head-on collision with a second ball of equal mass that is initially at rest. Find the velocity of the second ball just after it is struck.

**30.** The drawing shows a collision between two pucks on an air-hockey table. Puck A has a mass of 0.025 kg and is moving along the x axis with a velocity of +5.5 m/s. It makes a collision with puck B, which has a mass of 0.050 kg and is initially at rest. The collision is not head-on. After the collision, the two pucks fly apart with the angles shown in the drawing. Find the final speed of (a) puck A and (b) puck B.

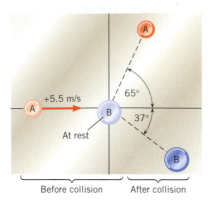

Before collision       After collision

**31. ssm www** A 5.00-kg ball, moving to the right at a velocity of +2.00 m/s on a frictionless table, collides head-on with a stationary 7.50-kg ball. Find the final velocities of the balls if the collision is (a) elastic and (b) completely inelastic.

**32.** A 0.150-kg projectile is fired with a velocity of +715 m/s at a 2.00-kg wooden block that rests on a frictionless table. The velocity of the block, immediately after the projectile passes through it, is +40.0 m/s. Find the velocity with which the projectile exits from the block.

**\*33. Interactive Solution 7.33** at **www.wiley.com/college/cutnell** illustrates how to model a problem similar to this one. An automobile has a mass of 2100 kg and a velocity of +17 m/s. It makes a rear-end collision with a stationary car whose mass is 1900 kg. The cars lock bumpers and skid off together with the wheels locked. (a) What is the velocity of the two cars just after the collision? (b) Find the impulse (magnitude and direction) that acts on the skidding cars from just after the collision until they come to a halt. (c) If the coefficient of kinetic friction between the wheels of the cars and the pavement is $\mu_k = 0.68$, determine how far the cars skid before coming to rest.

**\*34.** A mine car, whose mass is 440 kg, rolls at a speed of 0.50 m/s on a horizontal track, as the drawing shows. A 150-kg chunk of coal has a speed of 0.80 m/s when it leaves the chute. Determine the velocity of the car/coal system after the coal has come to rest in the car.

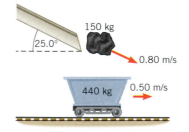

**\*35. ssm** A 50.0-kg skater is traveling due east at a speed of 3.00 m/s. A 70.0-kg skater is moving due south at a speed of 7.00 m/s. They collide and hold on to each other after the collision, managing to move off at an angle $\theta$ south of east, with a speed of $v_f$. Find (a) the angle $\theta$ and (b) the speed $v_f$, assuming that friction can be ignored.

**\*36.** An electron collides elastically with a stationary hydrogen atom. The mass of the hydrogen atom is 1837 times that of the electron. Assume that all motion, before and after the collision, occurs along the same straight line. What is the ratio of the kinetic energy of the hydrogen atom after the collision to that of the electron before the collision?

**\*37.** A 60.0-kg person, running horizontally with a velocity of +3.80 m/s, jumps onto a 12.0-kg sled that is initially at rest. (a) Ignoring the effects of friction during the collision, find the velocity of the sled and person as they move away. (b) The sled and person coast

30.0 m on level snow before coming to rest. What is the coefficient of kinetic friction between the sled and the snow?

**\*\*38. Concept Simulation 7.2** at **www.wiley.com/college/cutnell** provides a view of this elastic collision. Two identical balls are traveling toward each other with velocities of −4.0 and +7.0 m/s, and they experience an elastic head-on collision. Obtain the velocities (magnitude and direction) of each ball after the collision.

**\*\*39. ssm** Starting with an initial speed of 5.00 m/s at a height of 0.300 m, a 1.50-kg ball swings downward and strikes a 4.60-kg ball that is at rest, as the drawing shows. (a) Using the principle of conservation of mechanical energy, find the speed of the 1.50-kg ball just before impact. (b) Assuming that the collision is elastic, find the velocities (magnitude and direction) of both balls just after the collision. (c) How high does each ball swing after the collision, ignoring air resistance?

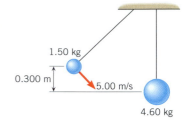

**\*\*40.** A ball is dropped from rest at the top of a 6.10-m-tall building, falls straight downward, collides inelastically with the ground, and bounces back. The ball loses 10.0% of its kinetic energy every time it collides with the ground. How many bounces can the ball make and still reach a windowsill that is 2.44 m above the ground?

### Section 7.5 Center of Mass

**41. ssm** The earth and moon are separated by a center-to-center distance of $3.85 \times 10^8$ m. The mass of the earth is $5.98 \times 10^{24}$ kg and that of the moon is $7.35 \times 10^{22}$ kg. How far does the center of mass lie from the center of the earth?

**42.** Consider the two moving boxcars in Example 5. Determine the velocity of their center of mass (a) before and (b) after the collision. (c) Should your answer in part (b) be less than, greater than, or equal to the common velocity $v_f$ of the two coupled cars after the collision? Justify your answer.

**43. Interactive Solution 7.43** at **www.wiley.com/college/cutnell** presents a method for modeling this problem. The carbon monoxide molecule (CO) consists of a carbon atom and an oxygen atom separated by a distance of $1.13 \times 10^{-10}$ m. The mass $m_C$ of the carbon atom is 0.750 times the mass $m_O$ of the oxygen atom: $m_C = 0.750\, m_O$. Determine the location of the center of mass of this molecule relative to the carbon atom.

**\*44.** The drawing shows the bond lengths and angles in the nitric acid ($HNO_3$) molecule, which is planar. The masses of the atoms are $m_H = 1.67 \times 10^{-27}$ kg, $m_N = 23.3 \times 10^{-27}$ kg, and $m_O = 26.6 \times 10^{-27}$ kg. Locate the center of mass of this molecule relative to the hydrogen atom. (*Hint: The oxygen atoms are located symmetrically on either side of the H−O−N line.*)

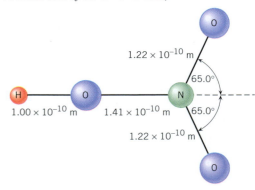

# Chapter 8  Rotational Kinematics

## 8.1  Rotational Motion and Angular Displacement

In the simplest kind of rotation, points on a rigid object move on circular paths. In Figure 8.1, for example, we see the circular paths for points $A$, $B$, and $C$ on a spinning skater. The centers of all such circular paths define a line, called the **axis of rotation.**

The angle through which a rigid object rotates about a fixed axis is called the **angular displacement.** Figure 8.2 shows how the angular displacement is measured for a rotating compact disc (CD). Here, the axis of rotation passes through the center of the disc and is perpendicular to its surface. On the surface of the CD we draw a radial line, which is a line that intersects the axis of rotation perpendicularly. As the CD turns, we observe the angle through which this line moves relative to a convenient reference line that does not rotate. The radial line moves from its initial orientation at angle $\theta_0$ to a final orientation at angle $\theta$ (Greek letter theta). In the process, the line sweeps out the angle $\theta - \theta_0$. As with other differences that we have encountered ($\Delta x = x - x_0$, $\Delta v = v - v_0$, and $\Delta t = t - t_0$), it is customary to denote the difference between the final and initial angles by the notation $\Delta\theta$ (read as "delta theta"): $\Delta\theta = \theta - \theta_0$. The angle $\Delta\theta$ is the angular displacement. A rotating object may rotate either counterclockwise or clockwise, and standard convention calls a counterclockwise displacement positive and a clockwise displacement negative.

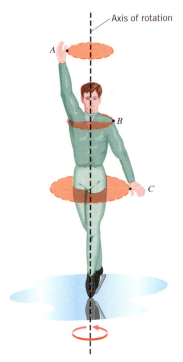

**Figure 8.1** When an object rotates, points on the object, such as $A$, $B$, or $C$, move on circular paths. The centers of the circles form a line that is the axis of rotation.

---

■ **DEFINITION OF ANGULAR DISPLACEMENT**

When a rigid body rotates about a fixed axis, the angular displacement is the angle $\Delta\theta$ swept out by a line passing through any point on the body and intersecting the axis of rotation perpendicularly. By convention, the angular displacement is positive if it is counterclockwise and negative if it is clockwise.

**SI Unit of Angular Displacement:** radian (rad)*

---

Angular displacement is often expressed in one of three units. The first is the familiar **degree,** and it is well known that there are 360 degrees in a circle. The second unit is the **revolution (rev),** one revolution representing one complete turn of 360°. The most useful unit from a scientific viewpoint is the SI unit called the **radian (rad).** Figure 8.3 shows how the radian is defined, again using a CD as an example. The picture focuses attention on a point $P$ on the disc. This point starts out on the stationary reference line, so that $\theta_0 = 0$ rad, and the angular displacement is $\Delta\theta = \theta - \theta_0 = \theta$. As the disc rotates, the point traces out an arc of length $s$, which is measured along a circle of radius $r$. Equation 8.1 defines the angle $\theta$ in radians:

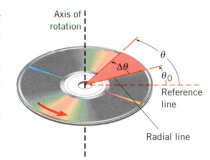

**Figure 8.2** The angular displacement of a CD is the angle $\Delta\theta$ swept out by a radial line as the disc turns about its axis of rotation.

$$\theta \text{ (in radians)} = \frac{\text{Arc length}}{\text{Radius}} = \frac{s}{r} \tag{8.1}$$

According to this definition, an angle in radians is the ratio of two lengths, for example, meters/meters. In calculations, therefore, the radian is treated as a number without units and has no effect on other units that it multiplies or divides.

To convert between degrees and radians, it is only necessary to remember that the arc length of an entire circle of radius $r$ is the circumference $2\pi r$. Therefore, according to Equation 8.1, *the number of radians that corresponds to 360°, or one revolution, is*

$$\theta = \frac{2\pi r}{r} = 2\pi \text{ rad}$$

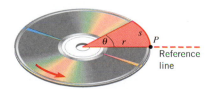

**Figure 8.3** In radian measure, the angle $\theta$ is defined to be the arc length $s$ divided by the radius $r$.

*The radian is neither a base SI unit nor a derived one. It is regarded as a supplementary SI unit.

Since $2\pi$ rad corresponds to $360°$, the number of degrees in one radian is

$$1 \text{ rad} = \frac{360°}{2\pi} = 57.3°$$

It is useful to express an angle $\theta$ in radians, because then the arc length $s$ subtended at any radius $r$ can be calculated by multiplying $\theta$ by $r$. Example 1 illustrates this point and also shows how to convert between degrees and radians.

### Example 1    Adjacent Synchronous Satellites

**The physics of communications satellites.**

Synchronous or "stationary" communications satellites are put into an orbit whose radius is $r = 4.23 \times 10^7$ m. The orbit is in the plane of the equator, and two adjacent satellites have an angular separation of $\theta = 2.00°$, as Figure 8.4 illustrates. Find the arc length $s$ (see the drawing) that separates the satellites.

**Reasoning** Since the radius $r$ and the angle $\theta$ are known, we may find the arc length $s$ by using the relation $\theta$ (in radians) $= s/r$. But first, the angle must be converted to radians from degrees.

**Solution** To convert $2.00°$ into radians, we use the fact that $2\pi$ radians is equivalent to $360°$:

$$2.00° = (2.00 \text{ degrees}) \left( \frac{2\pi \text{ radians}}{360 \text{ degrees}} \right) = 0.0349 \text{ radians}$$

From Equation 8.1, it follows that the arc length between the satellites is

$$s = r\theta = (4.23 \times 10^7 \text{ m})(0.0349 \text{ rad}) = \boxed{1.48 \times 10^6 \text{ m (920 miles)}}$$

The radian, being a unitless quantity, is dropped from the final result, leaving the answer expressed in meters.

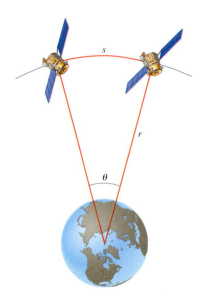

**Figure 8.4** Two adjacent synchronous satellites have an angular separation of $\theta = 2.00°$. The distances and angles have been exaggerated for clarity.

Conceptual Example 2 takes advantage of the radian as a unit for measuring angles and explains the spectacular phenomenon of a total solar eclipse.

### Conceptual Example 2    A Total Eclipse of the Sun

The diameter of the sun is about 400 times greater than that of the moon. By coincidence, the sun is also about 400 times farther from the earth than is the moon. For an observer on earth, compare the angle subtended by the moon to the angle subtended by the sun, and explain why this result leads to a total solar eclipse.

**Reasoning and Solution** Figure 8.5a shows a person on earth viewing the moon and the sun. The angle $\theta_{\text{moon}}$ subtended by the moon is given by Equation 8.1 as the arc length $s_{\text{moon}}$ divided by the distance $r_{\text{moon}}$ from the earth to the moon: $\theta_{\text{moon}} = s_{\text{moon}}/r_{\text{moon}}$. The earth and moon are sufficiently far apart that the arc length $s_{\text{moon}}$ is very nearly equal to the moon's diameter. Similar reasoning applies for the sun, so that the angle subtended by the sun is $\theta_{\text{sun}} = s_{\text{sun}}/r_{\text{sun}}$. But since $s_{\text{sun}} \approx 400 \, s_{\text{moon}}$ (where the symbol "$\approx$" means "approximately equal to") and $r_{\text{sun}} \approx 400 \, r_{\text{moon}}$, the two angles are approximately equal: $\theta_{\text{moon}} \approx \theta_{\text{sun}}$. Figure 8.5b shows what happens when the moon comes between the sun and the earth. *Since the angle subtended by the moon is nearly equal to the angle subtended by the sun, the moon blocks most of the sun's light from reaching the observer's eyes,* and a total solar eclipse like that in Figure 8.5c occurs.

**The physics of a total solar eclipse.**

**Related Homework:** *Problem 14*

## 8.2  *Angular Velocity and Angular Acceleration*

### ANGULAR VELOCITY

In Section 2.2 we introduced the idea of linear velocity to describe how fast an object moves and the direction of its motion. We now introduce the analogous idea of angular velocity to describe the motion of a rigid object rotating about an axis.

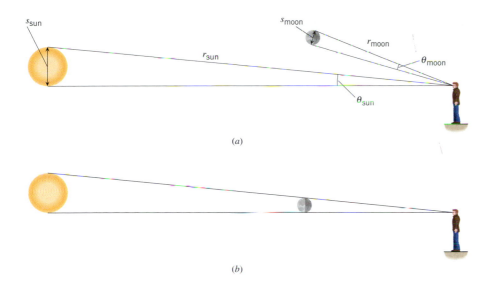

(a)

(b)

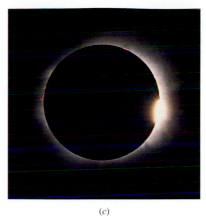

(c)

**Figure 8.5** (a) The angles subtended by the moon and sun at the eyes of the observer are $\theta_{\text{moon}}$ and $\theta_{\text{sun}}$. (The distances and angles are exaggerated for the sake of clarity.) (b) Since the moon and sun subtend approximately the same angle, the moon blocks nearly all the sun's light from reaching the observer's eyes. (c) The result is a total solar eclipse. (© Roger Ressmeyer/Corbis Images)

According to Equation 2.2 ($\overline{\mathbf{v}} = \Delta\mathbf{x}/\Delta t$), the average linear velocity is the linear displacement of the object divided by the time required for the displacement to occur. For rotational motion about a fixed axis, the *average angular velocity* $\overline{\omega}$ (Greek letter omega) is obtained in an analogous way, as the angular displacement divided by the elapsed time during which the displacement occurs.

■ **DEFINITION OF AVERAGE ANGULAR VELOCITY**

$$\frac{\text{Average angular}}{\text{velocity}} = \frac{\text{Angular displacement}}{\text{Elapsed time}}$$

$$\overline{\omega} = \frac{\theta - \theta_0}{t - t_0} = \frac{\Delta\theta}{\Delta t} \tag{8.2}$$

*SI Unit of Angular Velocity:* radian per second (rad/s)

The SI unit for angular velocity is the radian per second (rad/s), although other units such as revolutions per minute (rev/min or rpm) are also used. In agreement with the sign convention adopted for angular displacement, angular velocity is positive when the rotation is counterclockwise and negative when it is clockwise. Example 3 shows how the concept of average angular velocity is applied to a gymnast.

## *Example 3* Gymnast on a High Bar

A gymnast on a high bar swings through two revolutions in a time of 1.90s, as Figure 8.6 suggests. Find the average angular velocity (in rad/s) of the gymnast.

**Reasoning** The average angular velocity of the gymnast in rad/s is the angular displacement in radians divided by the elapsed time. However, the angular displacement is given as two revolutions, so we begin by converting this value into radian measure.

**Solution** The angular displacement (in radians) of the gymnast is

$$\Delta\theta = -2.00 \text{ revolutions} \left( \frac{2\pi \text{ radians}}{1 \text{ revolution}} \right) = -12.6 \text{ radians}$$

where the minus sign denotes that the gymnast rotates clockwise (see the drawing). The average angular velocity is

$$\overline{\omega} = \frac{\Delta\theta}{\Delta t} = \frac{-12.6 \text{ rad}}{1.90 \text{ s}} = \boxed{-6.63 \text{ rad/s}} \tag{8.2}$$

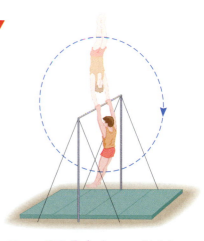

**Figure 8.6** Swinging on a high bar.

The *instantaneous angular velocity* $\omega$ is the angular velocity that exists at any given instant. To measure it, we follow the same procedure used in Chapter 2 for the instantaneous linear velocity. In this procedure, a small angular displacement $\Delta\theta$ occurs during a small time interval $\Delta t$. The time interval is so small that it approaches zero ($\Delta t \to 0$), and in this limit, the measured average angular velocity, $\overline{\omega} = \Delta\theta/\Delta t$, becomes the instantaneous angular velocity $\omega$:

$$\omega = \lim_{\Delta t \to 0} \overline{\omega} = \lim_{\Delta t \to 0} \frac{\Delta\theta}{\Delta t} \tag{8.3}$$

The magnitude of the instantaneous angular velocity, without reference to whether it is a positive or negative quantity, is called the *instantaneous angular speed.* If a rotating object has a constant angular velocity, the instantaneous value and the average value are the same.

## ANGULAR ACCELERATION

In linear motion, a changing velocity means that an acceleration is occurring. Such is also the case in rotational motion; a changing angular velocity means that an *angular acceleration* is occurring. There are many examples of angular acceleration. For instance, as a compact disc recording is played, the disc turns with an angular velocity that is continually decreasing. And when the push buttons of an electric blender are changed from a lower setting to a higher setting, the angular velocity of the blades increases.

When the linear velocity of an object changes, Equation 2.4 ($\overline{\mathbf{a}} = \Delta\mathbf{v}/\Delta t$) defines the average linear acceleration as the change in velocity per unit time. When the angular velocity changes from an initial value of $\omega_0$ at time $t_0$ to a final value of $\omega$ at time $t$, the average angular acceleration $\overline{\alpha}$ (Greek letter alpha) is defined similarly:

■ **DEFINITION OF AVERAGE ANGULAR ACCELERATION**

$$\text{Average angular acceleration} = \frac{\text{Change in angular velocity}}{\text{Elapsed time}}$$

$$\overline{\alpha} = \frac{\omega - \omega_0}{t - t_0} = \frac{\Delta\omega}{\Delta t} \tag{8.4}$$

*SI Unit of Average Angular Acceleration:* radian per second squared ($\text{rad/s}^2$)

The SI unit for average angular acceleration is the unit for angular velocity divided by the unit for time, or $(\text{rad/s})/\text{s} = \text{rad/s}^2$. An angular acceleration of $+5$ $\text{rad/s}^2$, for example, means that the angular velocity of the rotating object increases by $+5$ radians per second during each second of acceleration.

The *instantaneous angular acceleration* $\alpha$ is the angular acceleration at a given instant. In discussing linear motion, we assumed a condition of constant acceleration, so that the average and instantaneous accelerations were identical ($\overline{\mathbf{a}} = \mathbf{a}$). Similarly, we assume that the angular acceleration is constant, so that the instantaneous angular acceleration $\alpha$ and the average angular acceleration $\overline{\alpha}$ are the same ($\overline{\alpha} = \alpha$). The next example illustrates the concept of angular acceleration.

## Example 4   A Jet Revving Its Engines

A jet awaiting clearance for takeoff is momentarily stopped on the runway. As seen from the front of one engine, the fan blades are rotating with an angular velocity of $-110$ rad/s, where the negative sign indicates a clockwise rotation (see Figure 8.7). As the plane takes off, the angular velocity of the blades reaches $-330$ rad/s in a time of 14 s. Find the angular acceleration, assuming it to be constant.

**Reasoning** Since the angular acceleration is constant, it is equal to the average angular acceleration. The average acceleration is the change in the angular velocity, $\omega - \omega_0$, divided by the elapsed time, $t - t_0$.

**Figure 8.7** The fan blades of a jet engine have an angular velocity in a clockwise direction.

**Solution** Applying the definition of average angular acceleration given in Equation 8.4, we find that

$$\bar{\alpha} = \frac{\omega - \omega_0}{t - t_0} = \frac{(-330 \text{ rad/s}) - (-110 \text{ rad/s})}{14 \text{ s}} = \boxed{-16 \text{ rad/s}^2}$$

Thus, the magnitude of the angular velocity increases by 16 rad/s during each second that the blades are accelerating. The negative sign in the answer indicates that the direction of the angular acceleration is also in the clockwise direction.

▲

# 8.3 *The Equations of Rotational Kinematics*

In Chapters 2 and 3 the concepts of displacement, velocity, and acceleration were introduced. We combined these concepts and developed a set of equations called the equations of kinematics for constant acceleration (see Tables 2.1 and 3.1). These equations are a great aid in solving problems involving linear motion in one and two dimensions.

We now take a similar approach for rotational motion. We will bring together the ideas of angular displacement, angular velocity, and angular acceleration to produce a set of equations called the equations of kinematics for constant angular acceleration. These equations, like those developed in Chapters 2 and 3, will prove very useful in solving problems that involve rotational motion.

A complete description of rotational motion requires values for the angular displacement $\Delta\theta$, the angular acceleration $\alpha$, the final angular velocity $\omega$, the initial angular velocity $\omega_0$, and the elapsed time $\Delta t$. In Example 4, for instance, only the angular displacement of the fan blades during the 14-s interval is missing. Such missing information can be calculated, however. For convenience in the calculations, we assume that the orientation of the rotating object is given by $\theta_0 = 0$ rad at time $t_0 = 0$ s. Then, the angular displacement becomes $\Delta\theta = \theta - \theta_0 = \theta$, and the time interval becomes $\Delta t = t - t_0 = t$.

In Example 4, the angular velocity of the fan blades changes at a constant rate from an initial value of $\omega_0 = -110$ rad/s to a final value of $\omega = -330$ rad/s. Therefore, the average angular velocity is midway between the initial and final values:

$$\bar{\omega} = \tfrac{1}{2}[(-110 \text{ rad/s}) + (-330 \text{ rad/s})] = -220 \text{ rad/s}$$

In other words, when the angular acceleration is constant, the average angular velocity is given by

$$\bar{\omega} = \tfrac{1}{2}(\omega_0 + \omega) \tag{8.5}$$

With a value for the average angular velocity, Equation 8.2 can be used to obtain the angular displacement of the fan blades:

$$\theta = \bar{\omega}t = (-220 \text{ rad/s})(14 \text{ s}) - -3100 \text{ rad}$$

In general, when the angular acceleration is constant, the angular displacement can be obtained from

$$\theta = \bar{\omega}t = \tfrac{1}{2}(\omega_0 + \omega)t \tag{8.6}$$

This equation and Equation 8.4 provide a complete description of rotational motion under the condition of constant angular acceleration. Equation 8.4 (with $t_0 = 0$ s) and Equation 8.6 are compared with the analogous results for linear motion in the first two rows of Table 8.1. The purpose of this comparison is to emphasize that the mathematical forms of

**Table 8.1** *The Equations of Kinematics for Rotational and Linear Motion*

| Rotational Motion ($\alpha$ = constant) | | Linear Motion ($a$ = constant) | |
|---|---|---|---|
| $\omega = \omega_0 + \alpha t$ | (8.4) | $v = v_0 + at$ | (2.4) |
| $\theta = \tfrac{1}{2}(\omega_0 + \omega)t$ | (8.6) | $x = \tfrac{1}{2}(v_0 + v)t$ | (2.7) |
| $\theta = \omega_0 t + \tfrac{1}{2}\alpha t^2$ | (8.7) | $x = v_0 t + \tfrac{1}{2}at^2$ | (2.8) |
| $\omega^2 = \omega_0^2 + 2\alpha\theta$ | (8.8) | $v^2 = v_0^2 + 2ax$ | (2.9) |

| Rotational Motion | Quantity | Linear Motion |
|:---:|:---|:---:|
| $\theta$ | Displacement | $x$ |
| $\omega_0$ | Initial velocity | $v_0$ |
| $\omega$ | Final velocity | $v$ |
| $\alpha$ | Acceleration | $a$ |
| $t$ | Time | $t$ |

Axis of rotation

**Figure 8.8** The angular velocity of the blades in an electric blender changes each time a different push button is chosen.

**Problem solving insight**
Each equation of rotational kinematics contains four of the five kinematic variables, $\theta$, $\alpha$, $\omega$, $\omega_0$, and $t$. Therefore, it is necessary to have values for three of these variables if one of the equations is to be used to determine a value for the unknown variable.

**Problem solving insight**

**The physics of** "crack-the-whip."

Equations 8.4 and 2.4 are identical, as are the forms of Equations 8.6 and 2.7. Of course, the symbols used for the rotational variables are different from those used for the linear variables, as Table 8.2 indicates.

In Chapter 2, Equations 2.4 and 2.7 are used to derive the remaining two equations of kinematics (Equations 2.8 and 2.9). These additional equations convey no new information but are convenient to have when solving problems. Similar derivations can be carried out here. The results are listed as Equations 8.7 and 8.8 below and in Table 8.1; they can be inferred directly from their counterparts in linear motion by making the substitution of symbols indicated in Table 8.2:

$$\theta = \omega_0 t + \tfrac{1}{2}\alpha t^2 \tag{8.7}$$

$$\omega^2 = \omega_0^2 + 2\alpha\theta \tag{8.8}$$

The four equations in the left column of Table 8.1 are called the **equations of rotational kinematics for constant angular acceleration.** The following example illustrates that they are used in the same fashion as the equations of linear kinematics.

### Example 5 Blending with a Blender

The blades of an electric blender are whirling with an angular velocity of $+375$ rad/s while the "puree" button is pushed in, as Figure 8.8 shows. When the "blend" button is pressed, the blades accelerate and reach a greater angular velocity after the blades have rotated through an angular displacement of $+44.0$ rad (seven revolutions). The angular acceleration has a constant value of $+1740$ rad/s$^2$. Find the final angular velocity of the blades.

**Reasoning** The three known variables are listed in the table below, along with a question mark indicating that a value for the final angular velocity $\omega$ is being sought.

| $\theta$ | $\alpha$ | $\omega$ | $\omega_0$ | $t$ |
|:---:|:---:|:---:|:---:|:---:|
| $+44.0$ rad | $+1740$ rad/s$^2$ | ? | $+375$ rad/s | |

We can use Equation 8.8, because it relates the angular variables $\theta$, $\alpha$, $\omega$, and $\omega_0$.

**Solution** From Equation 8.8 ($\omega^2 = \omega_0^2 + 2\alpha\theta$) it follows that

$$\omega = +\sqrt{\omega_0^2 + 2\alpha\theta} = \sqrt{(375 \text{ rad/s})^2 + 2(1740 \text{ rad/s}^2)(44.0 \text{ rad})} = \boxed{+542 \text{ rad/s}}$$

The negative root is disregarded, since the blades do not reverse their direction of rotation.

*The equations of rotational kinematics can be used with any self-consistent set of units for $\theta$, $\alpha$, $\omega$, $\omega_0$, and $t$.* Radians are used in Example 5 only because data are given in terms of radians. Had the data for $\theta$, $\alpha$, and $\omega_0$ been provided in rev, rev/s$^2$, and rev/s, respectively, then Equation 8.8 could have been used to determine the answer for $\omega$ directly in rev/s.

## 8.4 Angular Variables and Tangential Variables

In the familiar ice-skating stunt known as "crack-the-whip," a number of skaters attempt to maintain a straight line as they skate around the one person (the pivot) who remains in place. Figure 8.9 shows each skater moving on a circular arc and includes the corresponding velocity vector at the instant portrayed in the picture. For every individual skater, the vector is drawn tangent to the appropriate circle and, therefore, is called the **tangential velocity** $\mathbf{v}_T$. The magnitude of the tangential velocity is referred to as the **tangential speed.**

Of all the skaters involved in the stunt, the one farthest from the pivot has the hardest job. Why? Because, in keeping the line straight, this skater covers more distance than anyone else. To accomplish this, he must skate faster than anyone else and, thus, must have the largest tangential speed. In fact, the line remains straight only if each person skates with the correct tangential speed. The skaters closer to the pivot must move with smaller tangential speeds than those farther out, as indicated by the magnitudes of the tangential velocity vectors in Figure 8.9.

With the aid of Figure 8.10, it is possible to show that the tangential speed of any skater is directly proportional to his distance $r$ from the pivot, assuming a given angular speed for the rotating line. When the line rotates as a rigid unit for a time $t$, it sweeps out the angle $\theta$ shown in the drawing. The distance $s$ through which a skater moves along a circular arc can be calculated from Equation 8.1, $s = r\theta$, provided $\theta$ is measured in radians. Dividing both sides of this equation by $t$ gives $s/t = r(\theta/t)$. The term $s/t$ is the tangential speed $v_T$ (e.g., in meters/second) of the skater, while $\theta/t$ is the angular speed $\omega$ (in radians/second) of the line:

$$v_T = r\omega \qquad (\omega \text{ in rad/s}) \tag{8.9}$$

In this expression, the terms $v_T$ and $\omega$ refer to the magnitudes of the tangential and angular velocities, respectively, and are numbers without algebraic signs.

It is important to emphasize that the angular speed $\omega$ in Equation 8.9 must be expressed in radian measure (e.g., in rad/s); no other units, such as revolutions per second, are acceptable. This restriction arises because the equation was derived by using the definition of radian measure, $s = r\theta$.

The real challenge for the "crack-the-whip" skaters is to keep the line straight while making it pick up angular speed—that is, while giving it an angular acceleration. To make the angular speed of the line increase, each skater must increase his tangential speed, since the two speeds are related according to $v_T = r\omega$. Of course, the fact that a skater must skate faster and faster means that he must accelerate, and his tangential acceleration $a_T$ can be related to the angular acceleration $\alpha$ of the line. If time is measured relative to $t_0 = 0$ s, the definition of linear acceleration is given by Equation 2.4 as $a_T = (v_T - v_{T0})/t$, where $v_T$ and $v_{T0}$ are the final and initial tangential speeds, respectively. Substituting $v_T = r\omega$ for the tangential speed shows that

$$a_T = \frac{v_T - v_{T0}}{t} = \frac{(r\omega) - (r\omega_0)}{t} = r\left(\frac{\omega - \omega_0}{t}\right)$$

Since $\alpha = (\omega - \omega_0)/t$ according to Equation 8.4, it follows that

$$a_T = r\alpha \qquad (\alpha \text{ in rad/s}^2) \tag{8.10}$$

This result shows that, for a given value of $\alpha$, the tangential acceleration $a_T$ is proportional to the radius $r$, so the skater farthest from the pivot must have the largest tangential acceleration. In this expression, the terms $a_T$ and $\alpha$ refer to the magnitudes of the numbers involved, without reference to any algebraic sign. Moreover, as is the case for $\omega$ in $v_T = r\omega$, only radian measure can be used for $\alpha$ in Equation 8.10.

There is an advantage to using the angular velocity $\omega$ and the angular acceleration $\alpha$ to describe the rotational motion of a rigid object. The advantage is that these angular quantities describe the motion of the *entire object*. In contrast, the tangential quantities $v_T$ and $a_T$ describe only the motion of a single point on the object, and Equations 8.9 and 8.10 indicate that different points located at different distances $r$ have different tangential velocities and accelerations. Example 6 stresses this advantage.

### Example 6 A Helicopter Blade

A helicopter blade has an angular speed of $\omega = 6.50$ rev/s and an angular acceleration of $\alpha = 1.30$ rev/s$^2$. For points 1 and 2 on the blade in Figure 8.11, find the magnitudes of (a) the tangential speeds and (b) the tangential accelerations.

**Reasoning** Since the radius $r$ for each point and the angular speed $\omega$ of the helicopter blade are known, we can find the tangential speed $v_T$ for each point by using the relation $v_T = r\omega$. However, since this equation can be used only with radian measure, the angular speed $\omega$ must be converted to rad/s from rev/s. In a similar manner, the tangential acceleration $a_T$ for points 1 and 2 can be found using $a_T = r\alpha$, provided the angular acceleration $\alpha$ is expressed in rad/s$^2$ rather than in rev/s$^2$.

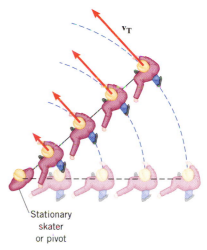

Stationary skater or pivot

**Figure 8.9** When doing a stunt known as "crack-the-whip," each skater along the radial line moves on a circular arc. The tangential velocity $\mathbf{v}_T$ of each skater is represented by an arrow that is tangent to each arc.

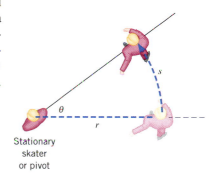

Stationary skater or pivot

**Figure 8.10** During a time $t$, the line of skaters sweeps through an angle $\theta$. An individual skater, located at a distance $r$ from the stationary skater, moves through a distance $s$ on a circular arc.

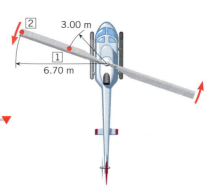

3.00 m

6.70 m

**Figure 8.11** Points 1 and 2 on the rotating blade of the helicopter have the same angular speed and acceleration, but they have *different* tangential speeds and accelerations.

**Solution**

**(a)** Converting the angular speed $\omega$ to rad/s from rev/s, we obtain

$$\omega = \left(6.50\ \frac{\text{rev}}{\text{s}}\right)\left(\frac{2\pi\ \text{rad}}{1\ \text{rev}}\right) = 40.8\ \frac{\text{rad}}{\text{s}}$$

The tangential speed for each point is

| | | |
|---|---|---|
| *Point 1* | $v_T = r\omega = (3.00\ \text{m})(40.8\ \text{rad/s}) = $ $\boxed{122\ \text{m/s (273 mph)}}$ | (8.9) |
| *Point 2* | $v_T = r\omega = (6.70\ \text{m})(40.8\ \text{rad/s}) = $ $\boxed{273\ \text{m/s (611 mph)}}$ | (8.9) |

The rad unit, being dimensionless, does not appear in the final answers.

**(b)** Converting the angular acceleration $\alpha$ to rad/s$^2$ from rev/s$^2$, we find

$$\alpha = \left(1.30\ \frac{\text{rev}}{\text{s}^2}\right)\left(\frac{2\pi\ \text{rad}}{1\ \text{rev}}\right) = 8.17\ \frac{\text{rad}}{\text{s}^2}$$

The tangential accelerations can now be determined:

| | | |
|---|---|---|
| *Point 1* | $a_T = r\alpha = (3.00\ \text{m})(8.17\ \text{rad/s}^2) = $ $\boxed{24.5\ \text{m/s}^2}$ | (8.10) |
| *Point 2* | $a_T = r\alpha = (6.70\ \text{m})(8.17\ \text{rad/s}^2) = $ $\boxed{54.7\ \text{m/s}^2}$ | (8.10) |

## 8.5 Centripetal Acceleration and Tangential Acceleration

When an object picks up speed as it moves around a circle, it has a tangential acceleration, as discussed in the last section. In addition, the object also has a centripetal acceleration, as emphasized in Chapter 5. That chapter deals with *uniform circular motion,* in which a particle moves at a constant tangential speed on a circular path. The tangential speed $v_T$ is the magnitude of the tangential velocity vector. Even when the magnitude of the tangential velocity is constant, an acceleration is present, since the direction of the velocity changes continually. Because the resulting acceleration points toward the center of the circle, it is called the centripetal acceleration. Figure 8.12a shows the centripetal acceleration $\mathbf{a_c}$ for a model airplane flying in uniform circular motion on a guide wire. The magnitude of $\mathbf{a_c}$ is

$$a_c = \frac{v_T^2}{r} \qquad (5.2)$$

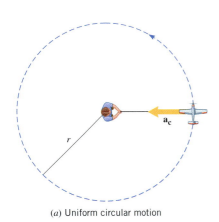

*(a)* Uniform circular motion

The subscript "T" has now been included in this equation as a reminder that it is the tangential speed that appears in the numerator.

The centripetal acceleration can be expressed in terms of the angular speed $\omega$ by using $v_T = r\omega$ (Equation 8.9):

$$a_c = \frac{v_T^2}{r} = \frac{(r\omega)^2}{r} = r\omega^2 \qquad (\omega \text{ in rad/s}) \qquad (8.11)$$

Only radian measure, such as rad/s, can be used for $\omega$ in this result, since the relation $v_T = r\omega$ presumes radian measure.

While considering uniform circular motion in Chapter 5, we ignored the details of how the motion is established in the first place. In Figure 8.12b, for instance, the engine of the plane produces a thrust in the tangential direction, and this force leads to a tangential acceleration. In response, the tangential speed of the plane increases from moment to moment, until the situation shown in the drawing results. While the tangential speed is changing, the motion is called *nonuniform circular motion.*

Figure 8.12b illustrates an important feature of nonuniform circular motion. Since the direction and the magnitude of the tangential velocity are both changing, the airplane experiences two acceleration components simultaneously. The changing direction means that there is a centripetal acceleration $\mathbf{a_c}$. The magnitude of $\mathbf{a_c}$ at any moment can be calculated using the value of the instantaneous angular speed and the radius: $a_c = r\omega^2$. The

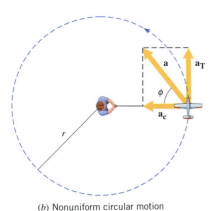

*(b)* Nonuniform circular motion

**Figure 8.12** *(a)* If a model airplane flying on a guide wire has a constant tangential speed, the motion is uniform circular motion, and the plane experiences only a centripetal acceleration $\mathbf{a_c}$. *(b)* Nonuniform circular motion occurs when the tangential speed changes. Then there is a tangential acceleration $\mathbf{a_T}$ in addition to the centripetal acceleration.

fact that the magnitude of the tangential velocity is changing means that there is also a tangential acceleration $\mathbf{a}_T$. The magnitude of $\mathbf{a}_T$ can be determined from the angular acceleration $\alpha$ according to $a_T = r\alpha$, as the previous section explains. If the magnitude $F_T$ of the net tangential force and the mass $m$ are known, $a_T$ also can be calculated using Newton's second law, $F_T = ma_T$. Figure 8.12b shows the two acceleration components. The total acceleration is given by the vector sum of $\mathbf{a}_c$ and $\mathbf{a}_T$. Since $\mathbf{a}_c$ and $\mathbf{a}_T$ are perpendicular, the magnitude of the total acceleration $\mathbf{a}$ can be obtained from the Pythagorean theorem as $a = \sqrt{a_c^2 + a_T^2}$, while the angle $\phi$ in the drawing can be determined from $\tan \phi = a_T/a_c$. The next example applies these concepts to a discus thrower.

### Example 7 A Discus Thrower

Discus throwers often warm up by standing with both feet flat on the ground and throwing the discus with a twisting motion of their bodies. Figure 8.13a illustrates a top view of such a warm-up throw. Starting from rest, the thrower accelerates the discus to a final angular velocity of $+15.0$ rad/s in a time of $0.270$ s before releasing it. During the acceleration, the discus moves on a circular arc of radius $0.810$ m. Find (a) the magnitude $a$ of the total acceleration of the discus just before it is released and (b) the angle $\phi$ that the total acceleration makes with the radius at this moment.

**Reasoning** Since the tangential speed of the discus is increasing, the discus simultaneously experiences a tangential acceleration $\mathbf{a}_T$ and a centripetal acceleration $\mathbf{a}_c$ that are oriented at right angles to each other. The magnitude of the total acceleration is $a = \sqrt{a_c^2 + a_T^2}$, where $a_c$ and $a_T$ are the magnitudes of the centripetal and tangential accelerations. The angle $\phi$ in Figure 8.13b is given by $\phi = \tan^{-1}(a_T/a_c)$. The magnitude of the centripetal acceleration can be evaluated from $a_c = r\omega^2$. The magnitude of the tangential acceleration follows from $a_T = r\alpha$, where $\alpha$ can be found from the definition of angular acceleration in Equation 8.4.

**Solution**

(a) Just before the moment of release, the magnitude of the total acceleration of the discus is $a = \sqrt{a_c^2 + a_T^2}$. According to Equation 8.11, the magnitude of the centripetal acceleration is $a_c = r\omega^2$, where the radius is $r = 0.810$ m and the final angular velocity is $\omega = +15.0$ rad/s. According to Equation 8.10, the magnitude of the tangential acceleration is $a_T = r\alpha$, where the angular acceleration $\alpha$ is not given. However, it can be obtained from the definition of angular acceleration in Equation 8.4 as $\alpha = (\omega - \omega_0)/t$, where the time is $t = 0.270$ s and the initial angular velocity is $\omega_0 = 0$ rad/s, since the discus starts from rest. With these substitutions we find that

$$a = \sqrt{a_c^2 + a_T^2} = \sqrt{(r\omega^2)^2 + (r\alpha)^2} = r\sqrt{\omega^4 + \alpha^2}$$

$$= r\sqrt{\omega^4 + \frac{(\omega - \omega_0)^2}{t^2}} = (0.810 \text{ m})\sqrt{(15.0 \text{ rad/s})^4 + \left(\frac{15.0 \text{ rad/s} - 0 \text{ rad/s}}{0.270 \text{ s}}\right)^2}$$

$$= \boxed{188 \text{ m/s}^2}$$

(b) The angle $\phi$ in Figure 8.13b is given by

$$\phi = \tan^{-1}\left(\frac{a_T}{a_c}\right) = \tan^{-1}\left(\frac{r\alpha}{r\omega^2}\right) = \tan^{-1}\left(\frac{\alpha}{\omega^2}\right)$$

$$= \tan^{-1}\left[\frac{(\omega - \omega_0)/t}{\omega^2}\right] = \tan^{-1}\left[\frac{(15.0 \text{ rad/s} - 0 \text{ rad/s})/(0.270 \text{ s})}{(15.0 \text{ rad/s})^2}\right] = \boxed{13.9°}$$

## 8.6 Rolling Motion

Rolling motion is a familiar situation that involves rotation, as Figure 8.14 illustrates for the case of an automobile tire. The essence of rolling motion is that there is *no slipping* at the point of contact where the tire touches the ground. To a good approximation, the tires on a normally moving automobile roll and do not slip. On the other hand, the squealing tires that accompany the start of a drag race are rotating, but they are not rolling while they rapidly spin and slip against the ground.

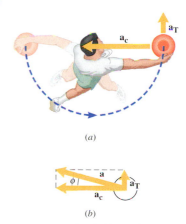

(a)

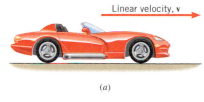

(b)

**Figure 8.13** (a) A discus thrower and the centripetal acceleration $\mathbf{a}_c$ and tangential acceleration $\mathbf{a}_T$ that act on the discus. (b) The total acceleration $\mathbf{a}$ of the discus just before the discus is released is the vector sum of $\mathbf{a}_c$ and $\mathbf{a}_T$.

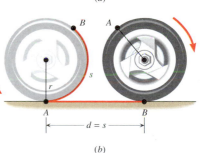

(a)

(b)

**Figure 8.14** (a) An automobile moves with a linear speed $v$. (b) If the tires roll and do not slip, the distance $d$, through which an axle moves, equals the circular arc length $s$ along the outer edge of a tire.

When the tires in Figure 8.14 roll, there is a relationship between the angular speed at which the tires rotate and the linear speed (assumed constant) at which the car moves forward. In part *b* of the drawing, consider the points labeled *A* and *B* on the left tire. Between these points we apply a coat of red paint to the tread of the tire; the length of this circular arc of paint is *s*. The tire then rolls to the right until point *B* comes in contact with the ground. As the tire rolls, all the paint comes off the tire and sticks to the ground, leaving behind the horizontal red line shown in the drawing. The axle of the wheel moves through a linear distance *d*, which is equal to the length of the horizontal strip of paint. Since the tire does not slip, the distance *d* must be equal to the circular arc length *s*, measured along the outer edge of the tire: $d = s$. Dividing both sides of this equation by the elapsed time *t* shows that $d/t = s/t$. The term $d/t$ is the speed at which the axle moves parallel to the ground—namely, the linear speed *v* of the car. The term $s/t$ is the tangential speed $v_T$ at which a point on the outer edge of the tire moves relative to the axle. In addition, $v_T$ is related to the angular speed $\omega$ about the axle according to $v_T = r\omega$ (Equation 8.9). Therefore, it follows that

$$\underbrace{v}_{\substack{\text{Linear} \\ \text{speed}}} = \underbrace{r\omega}_{\substack{\text{Tangential} \\ \text{speed, } v_T}} \qquad (\omega \text{ in rad/s}) \qquad (8.12)$$

If the car in Figure 8.14 has a linear acceleration **a** parallel to the ground, a point on the tire's outer edge experiences a tangential acceleration $\mathbf{a_T}$ relative to the axle. The same kind of reasoning used in the last paragraph reveals that the magnitudes of these accelerations are the same and that they are related to the angular acceleration $\alpha$ of the wheel relative to the axle:

$$\underbrace{a}_{\substack{\text{Linear} \\ \text{acceleration}}} = \underbrace{r\alpha}_{\substack{\text{Tangential} \\ \text{acceleration, } a_T}} \qquad (\alpha \text{ in rad/s}^2) \qquad (8.13)$$

Equations 8.12 and 8.13 may be applied to any rolling motion, because the object does not slip against the surface on which it is rolling. Example 8 illustrates the basic features of rolling motion.

### *Example 8*  An Accelerating Car

An automobile starts from rest and for 20.0 s has a constant linear acceleration of 0.800 m/s² to the right, as in Figure 8.14. During this period, the tires do not slip. The radius of the tires is 0.330 m. At the end of the 20.0-s interval, what is the angle through which each wheel has rotated?

**Reasoning**  As the car accelerates, the tires rotate faster and faster. Thus, each tire has an angular acceleration, which must be taken into account when we determine the angular displacement of each wheel. Since the tires roll and do not slip, the magnitude $\alpha$ of the angular acceleration is related to the magnitude *a* of the linear acceleration of the car by $a = r\alpha$ (Equation 8.13). Therefore, we find that

$$\alpha = \frac{a}{r} = \frac{0.800 \text{ m/s}^2}{0.330 \text{ m}} = 2.42 \text{ rad/s}^2$$

Since the tires rotate faster and faster in the clockwise, or negative, direction as the car moves to the right (see Figure 8.14), the angular acceleration is negative. Taking into account the negative value of the acceleration, the angular data are as follows:

| $\theta$ | $\alpha$ | $\omega$ | $\omega_0$ | $t$ |
|---|---|---|---|---|
| ? | $-2.42$ rad/s² | | 0 rad/s | 20.0 s |

The angular displacement $\theta$ is given in terms of $\alpha$, $\omega_0$, and *t* by Equation 8.7.

**Solution**  From Equation 8.7, we find that

$$\theta = \omega_0 t + \tfrac{1}{2}\alpha t^2 = (0 \text{ rad/s})(20.0 \text{ s}) + \tfrac{1}{2}(-2.42 \text{ rad/s}^2)(20.0 \text{ s})^2 = \boxed{-484 \text{ rad}}$$

The angular displacement $\theta$ is negative, because the wheels rotate in the clockwise direction.

# Concept Summary

This summary presents an abridged version of the chapter, including the important equations and all available learning aids. For convenient reference, the learning aids (including the text's examples) are placed next to or immediately after the relevant equation or discussion. The following learning aids may be found on-line at **www.wiley.com/college/cutnell**:

| | |
|---|---|
| **Interactive LearningWare** examples are solved according to a five-step interactive format that is designed to help you develop problem-solving skills. | **Concept Simulations** are animated versions of text figures or animations that illustrate important concepts. You can control parameters that affect the display, and we encourage you to experiment. |
| **Interactive Solutions** offer specific models for certain types of problems in the chapter homework. The calculations are carried out interactively. | **Self-Assessment Tests** include both qualitative and quantitative questions. Extensive feedback is provided for both incorrect and correct answers, to help you evaluate your understanding of the material. |

| Topic | Discussion | Learning Aids |
|---|---|---|
| | **8.1 Rotational Motion and Angular Displacement** | |
| Angular displacement | When a rigid body rotates about a fixed axis, the angular displacement is the angle swept out by a line passing through any point on the body and intersecting the axis of rotation perpendicularly. By convention, the angular displacement is positive if it is counterclockwise and negative if it is clockwise. | |
| Radian | The radian (rad) is the SI unit of angular displacement. In radians, the angle $\theta$ is defined as the circular arc length $s$ traveled by a point on the rotating body divided by the radial distance $r$ of the point from the axis: | |
| | $$\theta \text{ (in radians)} = \frac{s}{r} \qquad (8.1)$$ | **Examples 1, 2** |
| | **8.2 Angular Velocity and Angular Acceleration** | |
| | The average angular velocity $\overline{\omega}$ is the angular displacement $\Delta\theta$ divided by the elapsed time $\Delta t$: | |
| Average angular velocity | $$\overline{\omega} = \frac{\Delta\theta}{\Delta t} \qquad (8.2)$$ | **Example 3** |
| Instantaneous angular velocity | As $\Delta t$ approaches zero, the average angular velocity becomes equal to the instantaneous angular velocity $\omega$. The magnitude of the instantaneous angular velocity is called the instantaneous angular speed. | |
| | The average angular acceleration $\overline{\alpha}$ is the change $\Delta\omega$ in the angular velocity divided by the elapsed time $\Delta t$: | |
| Average angular acceleration | $$\overline{\alpha} = \frac{\Delta\omega}{\Delta t} \qquad (8.4)$$ | **Example 4** |
| Instantaneous angular acceleration | As $\Delta t$ approaches zero, the average angular acceleration becomes equal to the instantaneous angular acceleration $\alpha$. | |
| | **8.3 The Equations of Rotational Kinematics** | |
| | The equations of rotational kinematics apply when a rigid body rotates with a constant angular acceleration about a fixed axis. These equations relate the angular displacement $\theta - \theta_0$, the angular acceleration $\alpha$, the final angular velocity $\omega$, the initial angular velocity $\omega_0$, and the elapsed time $t - t_0$. Assuming that $\theta_0 = 0$ rad at $t_0 = 0$ s, the equations of rotational kinematics are | |
| | $$\omega = \omega_0 + \alpha t \qquad (8.4)$$ | **Example 5** |
| Equations of rotational kinematics | $$\theta = \tfrac{1}{2}(\omega + \omega_0)t \qquad (8.6)$$ | **Interactive LearningWare 8.1** |
| | $$\theta = \omega_0 t + \tfrac{1}{2}\alpha t^2 \qquad (8.7)$$ | |
| | $$\omega^2 = \omega_0^2 + 2\alpha\theta \qquad (8.8)$$ | **Interactive Solution 8.25** |
| | These equations may be used with any self-consistent set of units and are not restricted to radian measure. | |

 **Use Self-Assessment Test 8.1 to evaluate your understanding of Sections 8.1–8.3.**

| Topic | Discussion | Learning Aids |
|-------|-----------|---------------|

### 8.4 *Angular Variables and Tangential Variables*

When a rigid body rotates through an angle $\theta$ about a fixed axis, any point on the body moves on a circular arc of length $s$ and radius $r$. Such a point has a tangential velocity (magnitude $= v_T$) and, possibly, a tangential acceleration (magnitude $= a_T$). The angular and tangential variables are related by the following equations:

**Relations between angular and tangential variables**

$$s = r\theta \qquad (\theta \text{ in rad}) \qquad (8.1) \quad \textbf{Example 6}$$

$$v_T = r\omega \qquad (\omega \text{ in rad/s}) \qquad (8.9)$$

$$a_T = r\alpha \qquad (\alpha \text{ in rad/s}^2) \qquad (8.10) \quad \text{Interactive Solutions 8.31, 8.61}$$

These equations refer to the magnitudes of the variables involved, without reference to positive or negative signs, and only radian measure can be used when applying them.

### 8.5 *Centripetal Acceleration and Tangential Acceleration*

The magnitude $a_c$ of the centripetal acceleration of a point on an object rotating with uniform or nonuniform circular motion can be expressed in terms of the radial distance $r$ of the point from the axis and the angular speed $\omega$:

**Centripetal acceleration**

$$a_c = r\omega^2 \qquad (\omega \text{ in rad/s}) \qquad (8.11) \quad \text{Interactive LearningWare 8.2}$$

**Example 7**

**Total acceleration**

This point experiences a total acceleration $\mathbf{a}$ that is the vector sum of two perpendicular acceleration components, the centripetal acceleration $\mathbf{a_c}$ and the tangential acceleration $\mathbf{a_T}$; $\mathbf{a} = \mathbf{a_c} + \mathbf{a_T}$.

### 8.6 *Rolling Motion*

The essence of rolling motion is that there is no slipping at the point where the object touches the surface upon which it is rolling. As a result, the tangential speed $v_T$ of a point on the outer edge of a rolling object, measured relative to the axis through the center of the object, is equal to the linear speed $v$ with which the object moves parallel to the surface. In other words, we have

**Concept Simulation 8.1**

$$v = v_T = r\omega \qquad (\omega \text{ in rad/s}) \qquad (8.12)$$

The magnitudes of the tangential acceleration $a_T$ and the linear acceleration $a$ of a rolling object are similarly related:

**Interactive Solution 8.49**

$$a = a_T = r\alpha \qquad (\alpha \text{ in rad/s}^2) \qquad (8.13) \quad \textbf{Example 8}$$

 **Use *Self-Assessment Test 8.2* to evaluate your understanding of Sections 8.4–8.6.**

# Problems

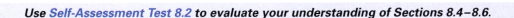

**ssm** Solution is in the Student Solutions Manual.      **www** Solution is available on the World Wide Web at www.wiley.com/college/cutnell

☤ This icon represents a biomedical application.

**Section 8.1 Rotational Motion and Angular Displacement,
Section 8.2 Angular Velocity and Angular Acceleration**

**1. ssm** A diver completes $3\frac{1}{2}$ somersaults in 1.7 s. What is the average angular speed (in rad/s) of the diver?

**2.** A pitcher throws a curveball that reaches the catcher in 0.60 s. The ball curves because it is spinning at an average angular velocity of 330 rev/min (assumed constant) on its way to the catcher's mitt. What is the angular displacement of the baseball (in radians) as it travels from the pitcher to the catcher?

**3. ssm** In Europe, surveyors often measure angles in *grads*. There are 100 grads in one-quarter of a circle. How many grads are there in one radian?

**4.** A pulsar is a rapidly rotating neutron star that continuously emits a beam of radio waves in a searchlight manner. Each time the pulsar makes one revolution, the rotating beam sweeps across the earth, and the earth receives a pulse of radio waves. For one particular pulsar, the time between two successive pulses is 0.033 s. Determine the average angular speed (in rad/s) of this pulsar.

**5.** A CD has a playing time of 74 minutes. When the music starts, the CD is rotating at an angular speed of 480 revolutions per minute (rpm). At the end of the music, the CD is rotating at 210 rpm. Find the magnitude of the average angular acceleration of the CD. Express your answer in rad/s$^2$.

**6.** A Ferris wheel rotates at an angular velocity of 0.24 rad/s. Starting from rest, it reaches its operating speed with an average angular acceleration of 0.030 rad/s². How long does it take the wheel to come up to operating speed?

**7. ssm** An electric circular saw is designed to reach its final angular speed, starting from rest, in 1.50 s. Its average angular acceleration is 328 rad/s². Obtain its final angular speed.

**\* 8.** A floor polisher has a rotating disk of radius 15 cm. The disk rotates at a constant angular velocity of 1.4 rev/s and is covered with a soft material that does the polishing. An operator holds the polisher in one place for 45 s, in order to buff an especially scuffed area of the floor. How far (in meters) does a spot on the outer edge of the disk move during this time?

**\* 9.** A space station consists of two donut-shaped living chambers, A and B, that have the radii shown in the drawing. As the station rotates, an astronaut in chamber A is moved $2.40 \times 10^2$ m along a circular arc. How far along a circular arc is an astronaut in chamber B moved during the same time?

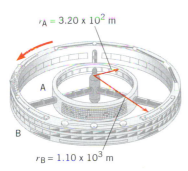

$r_A = 3.20 \times 10^2$ m

$r_B = 1.10 \times 10^3$ m

**\* 10.** The drawing shows a graph of the angular velocity of a rotating wheel as a function of time. Although not shown in the graph, the angular velocity continues to increase at the same rate until $t = 8.0$ s. What is the angular displacement of the wheel from 0 to 8.0 s?

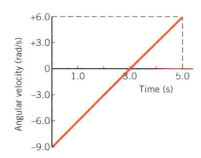

**\* 11. ssm** The drawing shows a device that can be used to measure the speed of a bullet. The device consists of two rotating disks, separated by a distance of $d = 0.850$ m, and rotating with an angular speed of 95.0 rad/s. The bullet first passes through the left disk and then through the right disk. It is found that the angular displacement between the two bullet holes is $\theta = 0.240$ rad. From these data, determine the speed of the bullet.

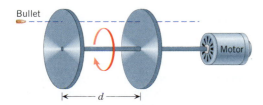

Bullet

Motor

$d$

**\* 12.** A stroboscope is a light that flashes on and off at a constant rate. It can be used to illuminate a rotating object, and if the flashing rate

is adjusted properly, the object can be made to appear stationary. (a) What is the shortest time between flashes of light that will make a three-bladed propeller appear stationary when it is rotating with an angular speed of 16.7 rev/s? (b) What is the next shortest time?

**\* 13.** A baton twirler throws a spinning baton directly upward. As it goes up and returns to the twirler's hand, the baton turns through four revolutions. Ignoring air resistance and assuming that the average angular speed of the baton is 1.80 rev/s, determine the height to which the center of the baton travels above the point of release.

**\*\* 14.** Review Conceptual Example 2 before attempting this problem. The moon has a diameter of $3.48 \times 10^6$ m and is a distance of $3.85 \times 10^8$ m from the earth. The sun has a diameter of $1.39 \times 10^9$ m and is $1.50 \times 10^{11}$ m from the earth. (a) Determine (in radians) the angles subtended by the moon and the sun, as measured by a person standing on the earth. (b) Based on your answers to part (a), decide whether a total eclipse of the sun is really "total." Give your reasoning. (c) Determine the ratio, expressed as a percentage, of the apparent circular area of the moon to the apparent circular area of the sun.

**\*\* 15.** A quarterback throws a pass that is a perfect spiral. In other words, the football does not wobble, but spins smoothly about an axis passing through each end of the ball. Suppose the ball spins at 7.7 rev/s. In addition, the ball is thrown with a linear speed of 19 m/s at an angle of 55° with respect to the ground. If the ball is caught at the same height at which it left the quarterback's hand, how many revolutions has the ball made while in the air?

**Section 8.3 The Equations of Rotational Kinematics**

**16.** A gymnast is performing a floor routine. In a tumbling run she spins through the air, increasing her angular velocity from 3.00 to 5.00 rev/s while rotating through one-half of a revolution. How much time does this maneuver take?

**17. ssm** An electric fan is running on HIGH. After the LOW button is pressed, the angular speed of the fan decreases to 83.8 rad/s in 1.75 s. The deceleration is 42.0 rad/s². Determine the initial angular speed of the fan.

**18.** A basketball player is balancing a spinning basketball on the tip of his finger. The angular velocity of the ball slows down from 18.5 to 14.1 rad/s. During the slow-down, the angular displacement is 85.1 rad. Determine the time it takes for the ball to slow down.

**19. ssm www** A flywheel has a constant angular deceleration of 2.0 rad/s². (a) Find the angle through which the flywheel turns as it comes to rest from an angular speed of 220 rad/s. (b) Find the time required for the flywheel to come to rest.

**20.** The angular speed of the rotor in a centrifuge increases from 420 to 1420 rad/s in a time of 5.00 s. (a) Obtain the angle through which the rotor turns. (b) What is the magnitude of the angular acceleration?

**\* 21.** A top is a toy that is made to spin on its pointed end by pulling on a string wrapped around the body of the top. The string has a length of 64 cm and is wrapped around the top at a place where its radius is 2.0 cm. The thickness of the string is negligible. The top is initially at rest. Someone pulls the free end of the string, thereby unwinding it and giving the top an angular acceleration of +12 rad/s². What is the final angular velocity of the top when the string is completely unwound?

**\* 22. Interactive LearningWare 8.1** at **www.wiley.com/college/cutnell** reviews the approach that is necessary for solving problems such as this one. A motorcyclist is traveling along a road and accelerates for 4.50 s to pass another cyclist. The angular acceleration of each wheel is +6.70 rad/s², and, just after passing, the angular velocity of each is +74.5 rad/s, where the plus signs indicate counterclockwise directions. What is the angular displacement of each wheel during this time?

\* **23. ssm** A spinning wheel on a fireworks display is initially rotating in a counterclockwise direction. The wheel has an angular acceleration of $-4.00$ rad/s$^2$. Because of this acceleration, the angular velocity of the wheel changes from its initial value to a final value of $-25.0$ rad/s. While this change occurs, the angular displacement of the wheel is zero. (Note the similarity to that of a ball being thrown vertically upward, coming to a momentary halt, and then falling downward to its initial position.) Find the time required for the change in the angular velocity to occur.

\* **24.** A dentist causes the bit of a high-speed drill to accelerate from an angular speed of $1.05 \times 10^4$ rad/s to an angular speed of $3.14 \times 10^4$ rad/s. In the process, the bit turns through $1.88 \times 10^4$ rad. Assuming a constant angular acceleration, how long would it take the bit to reach its maximum speed of $7.85 \times 10^4$ rad/s, starting from rest?

\* **25. Interactive Solution 8.25** at **www.wiley.com/college/cutnell** offers a model for solving this problem. The drive propeller of a ship starts from rest and accelerates at $2.90 \times 10^{-3}$ rad/s$^2$ for $2.10 \times 10^3$ s. For the next $1.40 \times 10^3$ s the propeller rotates at a constant angular speed. Then it decelerates at $2.30 \times 10^{-3}$ rad/s$^2$ until it slows (without reversing direction) to an angular speed of 4.00 rad/s. Find the total angular displacement of the propeller.

\* **26.** At the local swimming hole, a favorite trick is to run horizontally off a cliff that is 8.3 m above the water. One diver runs off the edge of the cliff, tucks into a "ball," and rotates on the way down with an average angular speed of 1.6 rev/s. Ignore air resistance and determine the number of revolutions she makes while on the way down.

\*\* **27. ssm www** A child, hunting for his favorite wooden horse, is running on the ground around the edge of a stationary merry-go-round. The angular speed of the child has a constant value of 0.250 rad/s. At the instant the child spots the horse, one-quarter of a turn away, the merry-go-round begins to move (in the direction the child is running) with a constant angular acceleration of 0.0100 rad/s$^2$. What is the shortest time it takes for the child to catch up with the horse?

### Section 8.4 Angular Variables and Tangential Variables

**28.** Our sun rotates in a circular orbit about the center of the Milky Way galaxy. The radius of the orbit is $2.2 \times 10^{20}$ m, and the angular speed of the sun is $1.2 \times 10^{-15}$ rad/s. (a) What is the tangential speed of the sun? (b) How long (in years) does it take for the sun to make one revolution around the center?

**29. ssm** A disk (radius = 2.00 mm) is attached to a high-speed drill at a dentist's office and is turning at $7.85 \times 10^4$ rad/s. Determine the tangential speed of a point on the outer edge of this disk.

**30.** A string trimmer is a tool for cutting grass and weeds; it utilizes a length of nylon "string" that rotates about an axis perpendicular to one end of the string. The string rotates at an angular speed of 47 rev/s, and its tip has a tangential speed of 54 m/s. What is the length of the rotating string?

**31. Interactive Solution 8.31** at **www.wiley.com/ college/cutnell** offers one approach to solving this problem. The drawing shows the blade of a chain saw. The rotating sprocket tip at the end of the guide bar has a radius of $4.0 \times 10^{-2}$ m. The linear speed of a chain link at point A is 5.6 m/s. Find the angular speed of the sprocket tip in rev/s.

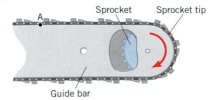

**32.** An auto race is held on a circular track. A car completes one lap in a time of 18.9 s, with an average tangential speed of 42.6 m/s. Find (a) the average angular speed and (b) the radius of the track.

**33.** The earth has a radius of $6.38 \times 10^6$ m and turns on its axis once every 23.9 h. (a) What is the tangential speed (in m/s) of a person living in Ecuador, a country that lies on the equator? (b) At what latitude (i.e., the angle $\theta$ in the drawing) is the tangential speed one-third that of a person living in Ecuador?

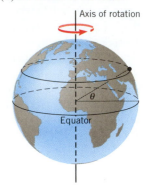

*Problem 33*

\* **34.** A baseball pitcher throws a baseball horizontally at a linear speed of 42.5 m/s (about 95 mi/h). Before being caught, the baseball travels a horizontal distance of 16.5 m and rotates through an angle of 49.0 rad. The baseball has a radius of 3.67 cm and is rotating about an axis as it travels, much like the earth does. What is the tangential speed of a point on the "equator" of the baseball?

\* **35. ssm** A compact disc (CD) contains music on a spiral track. Music is put onto a CD with the assumption that, during playback, the music will be detected at a *constant tangential speed* at any point. Since $v_T = r\omega$, a CD rotates at a smaller angular speed for music near the outer edge and a larger angular speed for music near the inner part of the disc. For music at the outer edge ($r = 0.0568$ m), the angular speed is 3.50 rev/s. Find (a) the constant tangential speed at which music is detected and (b) the angular speed (in rev/s) for music at a distance of 0.0249 m from the center of a CD.

\* **36.** A thin rod (length = 1.50 m) is oriented vertically, with its bottom end attached to the floor by means of a frictionless hinge. The mass of the rod may be ignored, compared to the mass of the object fixed to the top of the rod. The rod, starting from rest, tips over and rotates downward. (a) What is the angular speed of the rod just before it strikes the floor? (*Hint: Consider using the principle of conservation of mechanical energy.*) (b) What is the magnitude of the angular acceleration of the rod just before it strikes the floor?

\*\* **37.** One type of slingshot can be made from a length of rope and a leather pocket for holding the stone. The stone can be thrown by whirling it rapidly in a horizontal circle and releasing it at the right moment. Such a slingshot is used to throw a stone from the edge of a cliff, the point of release being 20.0 m above the base of the cliff. The stone lands on the ground below the cliff at a point X. The horizontal distance of point X from the base of the cliff (directly beneath the point of release) is thirty times the radius of the circle on which the stone is whirled. Determine the angular speed of the stone at the moment of release.

### Section 8.5 Centripetal Acceleration and Tangential Acceleration

**38.** A ceiling fan has two different angular speed settings: $\omega_1 = 440$ rev/min and $\omega_2 = 110$ rev/min. What is the ratio $a_1/a_2$ of the centripetal accelerations of a given point on a fan blade?

**39. ssm** A race car travels with a constant tangential speed of 75.0 m/s around a circular track of radius 625 m. Find (a) the magnitude of the car's total acceleration and (b) the direction of its total acceleration relative to the radial direction.

**40.** The earth orbits the sun once a year ($3.16 \times 10^7$ s) in a nearly circular orbit of radius $1.50 \times 10^{11}$ m. With respect to the sun, determine (a) the angular speed of the earth, (b) the tangential speed of the earth, and (c) the magnitude and direction of the earth's centripetal acceleration.

**41. ssm** A 220-kg speedboat is negotiating a circular turn (radius = 32 m) around a buoy. During the turn, the engine causes a net tangential force of magnitude 550 N to be applied to the boat. The initial tangential speed of the boat going into the turn is 5.0 m/s. (a) Find the tangential acceleration. (b) After the boat is 2.0 s into the turn, find the centripetal acceleration.

* **42. Interactive LearningWare 8.2** at **www.wiley.com/college/ cutnell** provides a review of the concepts that are important in this problem. A race car, starting from rest, travels around a circular turn of radius 23.5 m. At a certain instant, the car is still accelerating, and its angular speed is 0.571 rad/s. At this time, the total acceleration (centripetal plus tangential) makes an angle of 35.0° with respect to the radius. (The situation is similar to that in Figure 8.12b.) What is the magnitude of the total acceleration?

* **43.** A rectangular plate is rotating with a constant angular acceleration about an axis that passes perpendicularly through one corner, as the drawing shows. The tangential acceleration measured at corner A has twice the magnitude of that measured at corner B. What is the ratio $L_1/L_2$ of the lengths of the sides of the rectangle?

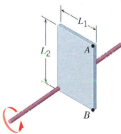

* **44.** A centrifuge is used for training astronauts to withstand large accelerations. It consists of a chamber (in which the astronaut sits) that is fixed to the end of a long horizontal and rigid pole. The arrangement is rotated about an axis perpendicular to the pole's free end. Such a centrifuge starts from rest and has an angular acceleration of 0.25 rad/s². The chamber is 3.0 m from the axis of rotation. Through what angle has the device rotated when the centripetal acceleration experienced by an astronaut in the chamber is four times the acceleration due to the earth's gravity?

** **45. ssm** An electric drill starts from rest and rotates with a constant angular acceleration. After the drill has rotated through a certain angle, the magnitude of the centripetal acceleration of a point on the drill is twice the magnitude of the tangential acceleration. What is the angle?

### Section 8.6 Rolling Motion

*Note: All problems in this section assume that there is no slipping of the surfaces in contact during the rolling motion.*

**46.** An automobile tire has a radius of 0.330 m, and its center moves forward with a linear speed of $v$ = 15.0 m/s. (a) Determine the angular speed of the wheel. (b) Relative to the axle, what is the tangential speed of a point located 0.175 m from the axle?

**47. ssm www** A motorcycle accelerates uniformly from rest and reaches a linear speed of 22.0 m/s in a time of 9.00 s. The radius of each tire is 0.280 m. What is the magnitude of the angular acceleration of each tire?

**48.** Suppose you are riding a stationary exercise bicycle, and the electronic meter indicates that the wheel is rotating at 9.1 rad/s. The wheel has a radius of 0.45 m. If you ride the bike for 35 min, how far would you have gone if the bike could move?

**49.** Refer to **Interactive Solution 8.49** at **www.wiley.com/college/ cutnell** in preparation for this problem. A car is traveling with a speed of 20.0 m/s along a straight horizontal road. The wheels have a radius of 0.300 m. If the car speeds up with a linear acceleration of 1.50 m/s² for 8.00 s, find the angular displacement of each wheel during this period.

**50. Concept Simulation 8.1** at **www.wiley.com/college/cutnell** reviews the concept that plays the central role in this problem. The warranty on a new tire says that an automobile can travel for a distance of 96 000 km before the tire wears out. The radius of the tire is 0.31 m. How many revolutions does the tire make before wearing out?

* **51. ssm www** The drawing shows a view (from beneath) of the platter on a belt-drive turntable. The platter has an angular speed of 3.49 rad/s. The pulley on the motor shaft has a radius of 1.27 cm. Assuming that the belt does not slip, determine the angular speed of the motor shaft.

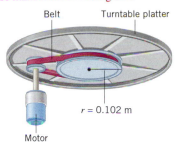

* **52.** The penny-farthing is a bicycle that was popular between 1870 and 1890. As the drawing shows, this type of bicycle has a large front wheel and a small rear wheel. On a Sunday ride in the park the front wheel (radius = 1.20 m) makes 276 revolutions. How many revolutions does the rear wheel (radius = 0.340 m) make?

* **53.** A ball of radius 0.200 m rolls along a horizontal table top with a constant linear speed of 3.60 m/s. The ball rolls off the edge and falls a vertical distance of 2.10 m before hitting the floor. What is the angular displacement of the ball while the ball is in the air?

* **54.** The two-gear combination shown in the drawing is being used to hoist the load $L$ with a constant upward speed of 2.50 m/s. The rope attached to the load is being wound onto a cylinder behind the big gear. The depth of the teeth of the gears is negligible compared to the radii. Determine the angular velocity (magnitude and direction) of (a) the larger gear and (b) the smaller gear.

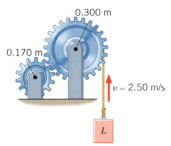

** **55.** A bicycle is rolling down a circular hill that has a radius of 9.00 m. As the drawing illustrates, the angular displacement of the bicycle is 0.960 rad. The radius of each wheel is 0.400 m. What is the angle (in radians) through which each tire rotates?

# Chapter 9  Rotational Dynamics

## 9.1  The Action of Forces and Torques on Rigid Objects

The mass of most rigid objects, such as a propeller or a wheel, is spread out and not concentrated at a single point. These objects can move in a number of ways. Figure 9.1a illustrates one possibility called translational motion, in which all points on the body travel on parallel paths (not necessarily straight lines). In pure translation there is no rotation of any line in the body. Because translational motion can occur along a curved line, it is often called curvilinear motion or linear motion. Another possibility is rotational motion, which may occur in combination with translational motion, as is the case for the somersaulting gymnast in Figure 9.1b.

We have seen many examples of how a net force affects linear motion by causing an object to accelerate. We now need to take into account the possibility that a rigid object can also have an angular acceleration. A net external force causes linear motion to change, but what causes rotational motion to change? For example, something causes the rotational velocity of a speedboat's propeller to change when the boat accelerates. Is it simply the net force? As it turns out, it is not the net external force, but rather the net external torque that causes the rotational velocity to change. Just as greater net forces cause greater linear accelerations, greater net torques cause greater rotational or angular accelerations.

Figure 9.2 helps to explain the idea of torque. When you push on a door with a force **F**, as in part a, the door opens more quickly when the force is larger. Other things being equal, a larger force generates a larger torque. However, the door does not open as quickly if you apply the same force at a point closer to the hinge, as in part b, because the force now produces less torque. Furthermore, if your push is directed nearly at the hinge, as in part c, you will have a hard time opening the door at all, because the torque is nearly zero. In summary, the torque depends on the magnitude of the force, on the point where the force is applied relative to the axis of rotation (the hinge in Figure 9.2), and on the direction of the force.

For simplicity, we deal with situations in which the force lies in a plane that is perpendicular to the axis of rotation. In Figure 9.3, for instance, the axis is perpendicular to the page and the force lies in the plane of the paper. The drawing shows the line of action and the lever arm of the force, two concepts that are important in the definition of torque. The *line of action* is an extended line drawn colinear with the force. The *lever arm* is the distance $\ell$ between the line of action and the axis of rotation, measured on a line that is perpendicular to both. The torque is represented by the symbol $\tau$ (Greek letter *tau*), and its magnitude is defined as the magnitude of the force times the lever arm:

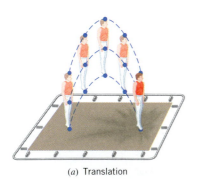

(a) Translation

(b) Combined translation and rotation

**Figure 9.1** An example of (a) translational motion and (b) combined translational and rotational motions.

---

■ **DEFINITION OF TORQUE**

Torque = (Magnitude of the force) × (Lever arm)

$$\tau = F\ell \tag{9.1}$$

**Direction:** The torque is positive when the force tends to produce a counterclockwise rotation about the axis, and negative when the force tends to produce a clockwise rotation.

**SI Unit of Torque:** newton · meter (N · m)

---

Equation 9.1 indicates that forces of the same magnitude can produce *different* torques, depending on the value of the lever arm, and Example 1 illustrates this important feature.

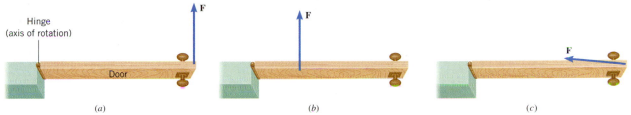

**Figure 9.2** It is easier to open a door with a force of a given magnitude by (a) pushing at the door's outer edge than by (b) pushing closer to the axis of rotation (the hinge). (c) Pushing nearly into the hinge makes it very difficult to open the door.

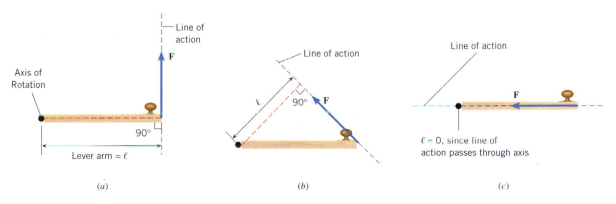

**Figure 9.3** In this top view, the hinges of a door appear as a black dot (●) and define the axis of rotation. The line of action and lever arm ℓ are illustrated for a force applied to the door (a) perpendicularly and (b) at an angle. (c) The lever arm is zero because the line of action passes through the axis of rotation.

## Example 1 Different Lever Arms, Different Torques

In Figure 9.3 a force whose magnitude is 55 N is applied to a door. However, the lever arms are different in the three parts of the drawing: (a) $\ell = 0.80$ m, (b) $\ell = 0.60$ m, and (c) $\ell = 0$ m. Find the magnitude of the torque in each case.

**Reasoning** In each case the lever arm is the perpendicular distance between the axis of rotation and the line of action of the force. In part a this perpendicular distance is equal to the width of the door. In parts b and c, however, the lever arm is less than the width. Because the lever arm is different in each case, the torque is different, even though the magnitude of the applied force is the same.

**Solution** Equation 9.1 gives the following values for the torques:

**(a)** $\tau = F\ell = (55 \text{ N})(0.80 \text{ m}) = \boxed{44 \text{ N} \cdot \text{m}}$

**(b)** $\tau = F\ell = (55 \text{ N})(0.60 \text{ m}) = \boxed{33 \text{ N} \cdot \text{m}}$

**(c)** $\tau = F\ell = (55 \text{ N})(0 \text{ m}) = \boxed{0 \text{ N} \cdot \text{m}}$

In part c the line of action of **F** passes through the axis of rotation (the hinge). Hence, the lever arm is zero, and the torque is zero.

In our bodies, muscles and tendons produce torques about various joints. Example 2 illustrates how the Achilles tendon produces a torque about the ankle joint.

## Example 2 The Achilles Tendon

Figure 9.4a shows the ankle joint and the Achilles tendon attached to the heel at point P. The tendon exerts a force of magnitude $F = 720$ N, as Figure 9.4b indicates. Determine the torque (magnitude and direction) of this force about the ankle joint, which is located $3.6 \times 10^{-2}$ m away from point P.

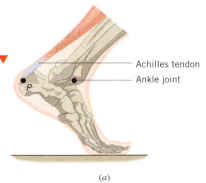

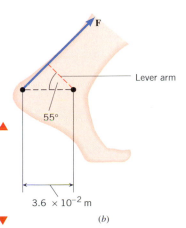

**Figure 9.4** The force **F** generated by the Achilles tendon produces a clockwise (negative) torque about the ankle joint.

**Reasoning** To calculate the magnitude of the torque, it is necessary to have a value for the lever arm $\ell$. However, the lever arm is not the given distance of $3.6 \times 10^{-2}$ m. Instead, the lever arm is the perpendicular distance between the axis of rotation at the ankle joint and the line of action of the force **F**. In Figure 9.4b this distance is indicated by the dashed red line.

**Solution** From the drawing, it can be seen that the lever arm is $\ell = (3.6 \times 10^{-2}$ m) cos 55°. The magnitude of the torque is

$$\tau = F\ell = (720 \text{ N})(3.6 \times 10^{-2} \text{ m}) \cos 55° = 15 \text{ N} \cdot \text{m} \tag{9.1}$$

The force **F** tends to produce a clockwise rotation about the ankle joint, so the torque is negative: $\boxed{\tau = -15 \text{ N} \cdot \text{m}}$.

*The physics of* the Achilles tendon.

## 9.2 *Rigid Objects in Equilibrium*

If a rigid body is in equilibrium, neither its linear motion nor its rotational motion changes. This lack of change leads to certain equations that apply for rigid-body equilibrium. For instance, an object whose linear motion is not changing has no acceleration **a**. Therefore, the net force $\Sigma\mathbf{F}$ applied to the object must be zero, since $\Sigma\mathbf{F} = m\mathbf{a}$ and $\mathbf{a} = 0$. For two-dimensional motion the $x$ and $y$ components of the net force are separately zero: $\Sigma F_x = 0$ and $\Sigma F_y = 0$ (Equations 4.9a and 4.9b). In calculating the net force, we include only forces from external agents, or *external forces*.* In addition to linear motion, we must consider rotational motion, which also does not change under equilibrium conditions. This means that the net external torque acting on the object must be zero, because it is what causes rotational motion to change. Using the symbol $\Sigma\tau$ to represent the net external torque (the sum of all positive and negative torques), we have

$$\Sigma\tau = 0 \tag{9.2}$$

We define rigid-body equilibrium, then, in the following way.

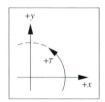

■ **EQUILIBRIUM OF A RIGID BODY**

A rigid body is in equilibrium if it has zero translational acceleration and zero angular acceleration. In equilibrium, the sum of the externally applied forces is zero, and the sum of the externally applied torques is zero:

$$\Sigma F_x = 0 \quad \text{and} \quad \Sigma F_y = 0 \tag{4.9a and 4.9b}$$
$$\Sigma\tau = 0 \tag{9.2}$$

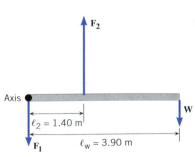

(a)

### Example 3   A Diving Board

A woman whose weight is 530 N is poised at the right end of a diving board with a length of 3.90 m. The board has negligible weight and is bolted down at the left end, while being supported 1.40 m away by a fulcrum, as Figure 9.5a shows. Find the forces $\mathbf{F}_1$ and $\mathbf{F}_2$ that the bolt and the fulcrum, respectively, exert on the board.

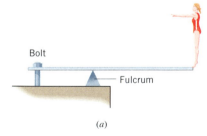

(b) Free-body diagram of the diving board

**Figure 9.5** (a) A diver stands at the end of a diving board. (b) The free-body diagram for the diving board. The box at the upper left shows the positive $x$ and $y$ directions for the forces, as well as the positive (counterclockwise) direction for the torques.

**Reasoning** Part *b* of the figure shows the free-body diagram of the diving board. Three forces act on the board: $\mathbf{F}_1$, $\mathbf{F}_2$, and the force due to the diver's weight **W**. In choosing the directions of $\mathbf{F}_1$ and $\mathbf{F}_2$ we have used our intuition: $\mathbf{F}_1$ points downward, because the bolt must pull in that direction to counteract the tendency of the board to rotate clockwise about the fulcrum; $\mathbf{F}_2$ points upward, because the board pushes downward against the fulcrum, which, in reaction, pushes upward on the board. Since the board is stationary, it is in equilibrium.

**Solution** Since the board is in equilibrium, the sum of the vertical forces must be zero:

$$\Sigma F_y = -F_1 + F_2 - W = 0 \tag{4.9b}$$

* We ignore internal forces that one part of an object exerts on another part, because they occur in action–reaction pairs, each of which consists of oppositely directed forces of equal magnitude. The effect of one force cancels the effect of the other, as far as the acceleration of the entire object is concerned.

Similarly, the sum of the torques must be zero, $\Sigma \tau = 0$. For calculating torques, we select an axis that passes through the left end of the board and is perpendicular to the page. (We will see shortly that this choice is arbitrary.) The force $\mathbf{F_1}$ produces no torque since it passes through the axis and has a zero lever arm, while $\mathbf{F_2}$ creates a counterclockwise (positive) torque, and $\mathbf{W}$ produces a clockwise (negative) torque. The free-body diagram shows the lever arms for the torques:

$$\Sigma \tau = +F_2 \ell_2 - W \ell_W = 0 \qquad (9.2)$$

Solving this equation for $F_2$ yields

$$F_2 = \frac{W \ell_W}{\ell_2} = \frac{(530 \text{ N})(3.90 \text{ m})}{1.40 \text{ m}} = \boxed{1480 \text{ N}}$$

This value for $F_2$, along with $W = 530$ N, can be substituted into Equation 4.9b above to show that $\boxed{F_1 = 950 \text{ N}}$.

In Example 3 the sum of the external torques is calculated using an axis that passes through the left end of the diving board. *However, the choice of the axis is completely arbitrary, because if an object is in equilibrium, it is in equilibrium with respect to any axis whatsoever.* Thus, the sum of the external torques is zero, no matter where the axis is placed. One usually chooses the location so that the lines of action of one or more of the unknown forces pass through the axis. Such a choice simplifies the torque equation, because the torques produced by these forces are zero. For instance, in Example 3 the torque due to the force $F_1$ does not appear in Equation 9.2, because the lever arm of this force is zero.

In a calculation of torque, the lever arm of the force must be determined relative to the axis of rotation. In Example 3 the lever arms are obvious, but sometimes a little care is needed in determining them, as in the next example.

### Example 4 Fighting a Fire

In Figure 9.6a an 8.00-m ladder of weight $W_L = 355$ N leans against a smooth vertical wall. The term "smooth" means that the wall can exert only a normal force directed perpendicular to the wall and cannot exert a frictional force parallel to it. A firefighter, whose weight is $W_F = 875$ N, stands 6.30 m from the bottom of the ladder. Assume that the ladder's weight acts at the ladder's center and neglect the hose's weight. Find the forces that the wall and the ground exert on the ladder.

**Reasoning** Part *b* of the figure shows the free-body diagram of the ladder. The following forces act on the ladder:

1. Its weight $\mathbf{W_L}$

2. A force due to the weight $\mathbf{W_F}$ of the firefighter

3. The force $\mathbf{P}$ applied to the top of the ladder by the wall and directed perpendicular to the wall

4. The forces $\mathbf{G_x}$ and $\mathbf{G_y}$, which are the horizontal and vertical components of the force exerted by the ground on the bottom of the ladder

The ground, unlike the wall, is not smooth, so that the force $\mathbf{G_x}$ is produced by static friction and prevents the ladder from slipping. The force $\mathbf{G_y}$ is the normal force applied to the ladder by the ground. The ladder is in equilibrium, so the sum of these forces and the sum of the torques produced by them must be zero.

**Solution** Since the net force acting on the ladder is zero, we have

$$\Sigma F_x = G_x - P = 0 \qquad (4.9a)$$

$$\Sigma F_y = G_y - W_L - W_F = 0 \qquad (4.9b)$$

$$\text{or} \quad G_y = W_L + W_F$$

$$= 355 \text{ N} + 875 \text{ N} = \boxed{1230 \text{ N}}$$

Equation 4.9a cannot be solved as it stands, because it contains two unknown variables. However, another equation can be obtained from the fact that the net torque acting on an

*Problem solving insight*

(a)

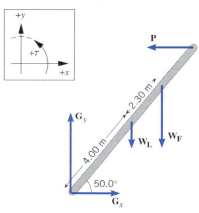

(b) Free-body diagram of the ladder

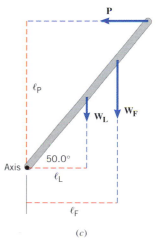

(c)

**Figure 9.6** (a) A ladder leaning against a smooth wall. (b) The free-body diagram for the ladder. (c) Three of the forces that act on the ladder and their lever arms. The axis of rotation is at the lower end of the ladder and is perpendicular to the page.

object in equilibrium is zero. In calculating torques, it is convenient to use an axis at the left end of the ladder, directed perpendicular to the page, as Figure 9.6c indicates. This axis is convenient, because $G_x$ and $G_y$ produce no torques about it, their lever arms being zero. Consequently, these forces will not appear in the equation representing the balance of torques. The lever arms for the remaining forces are shown in part c as red dashed lines. The following list summarizes these forces, their lever arms, and the torques:

| Force | Lever Arm | Torque |
|---|---|---|
| $W_L = 355$ N | $\ell_L = (4.00 \text{ m}) \cos 50.0°$ | $-W_L\ell_L$ |
| $W_F = 875$ N | $\ell_F = (6.30 \text{ m}) \cos 50.0°$ | $-W_F\ell_F$ |
| $P$ | $\ell_P = (8.00 \text{ m}) \sin 50.0°$ | $+P\ell_P$ |

Setting the sum of the torques equal to zero gives

$$\Sigma\tau = -W_L\ell_L - W_F\ell_F + P\ell_P = 0 \qquad (9.2)$$

Solving this equation for $P$ gives

$$P = \frac{W_L\ell_L + W_F\ell_F}{\ell_P}$$

$$= \frac{(355 \text{ N})(4.00 \text{ m}) \cos 50.0° + (875 \text{ N})(6.30 \text{ m}) \cos 50.0°}{(8.00 \text{ m}) \sin 50.0°} = \boxed{727 \text{ N}}$$

Substituting $P = 727$ N into Equation 4.9a indicates that $G_x = P = \boxed{727 \text{ N}}$.

---

To a large extent the directions of the forces acting on an object in equilibrium can be deduced using intuition. Sometimes, however, the direction of an unknown force is not obvious, and it is inadvertently drawn reversed in the free-body diagram. This kind of mistake causes no difficulty. ***Choosing the direction of an unknown force backward in the free-body diagram simply means that the value determined for the force will be a negative number,*** as the next example illustrates.

*Problem solving insight*

 ***The physics of bodybuilding.***

### *Example 5*   Bodybuilding

**Figure 9.7** (a) The fully extended, horizontal arm of a bodybuilder supports a dumbbell. (b) The free-body diagram for the arm. (c) Three of the forces that act on the arm and their lever arms. The axis of rotation at the left end of the arm is perpendicular to the page. Force vectors are not to scale.

A bodybuilder holds a dumbbell of weight $\mathbf{W_d}$ as in Figure 9.7a. His arm is horizontal and weighs $W_a = 31.0$ N. The deltoid muscle is assumed to be the only muscle acting and is attached to the arm as shown. The maximum force $\mathbf{M}$ that the deltoid muscle can supply has a magnitude of 1840 N. Figure 9.7b shows the distances that locate where the various forces act on the arm. What is the weight of the heaviest dumbbell that can be held, and what are the horizontal and vertical force components, $\mathbf{S_x}$ and $\mathbf{S_y}$, that the shoulder joint applies to the left end of the arm?

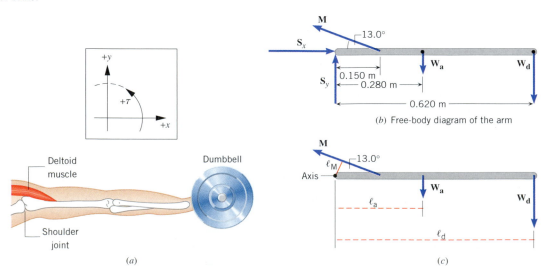

(b) Free-body diagram of the arm

(a)

(c)

**Reasoning** Figure 9.7*b* is the free-body diagram for the arm. Note that $\mathbf{S}_x$ is directed to the right, because the deltoid muscle pulls the arm in toward the shoulder joint, and the joint pushes back in accordance with Newton's third law. The direction of the force $\mathbf{S}_y$, however, is less obvious, and we are alert for the possibility that the direction chosen in the free-body diagram is backward. If so, the value obtained for $\mathbf{S}_y$ will be negative.

**Solution** The arm is in equilibrium, so the net force acting on it is zero:

$$\Sigma F_x = S_x - M \cos 13.0° = 0 \qquad (4.9a)$$

$$\text{or} \quad S_x = M \cos 13.0° = (1840 \text{ N}) \cos 13.0° = \boxed{1790 \text{ N}}$$

$$\Sigma F_y = S_y + M \sin 13.0° - W_a - W_d = 0 \qquad (4.9b)$$

Equation 4.9b cannot be solved at this point, because it contains two unknowns, $S_y$ and $W_d$. However, since the arm is in equilibrium, the torques acting on the arm must balance, and this fact provides another equation. To calculate torques, we choose an axis through the left end of the arm and perpendicular to the page. With this axis, the torques due to $\mathbf{S}_x$ and $\mathbf{S}_y$ are zero, because the line of action of each force passes through the axis and the lever arm of each force is zero. The list below summarizes the remaining forces, their lever arms (see Figure 9.7*c*), and the torques.

| Force | Lever Arm | Torque |
|---|---|---|
| $W_a = 31.0 \text{ N}$ | $\ell_a = 0.280 \text{ m}$ | $-W_a \ell_a$ |
| $W_d$ | $\ell_d = 0.620 \text{ m}$ | $-W_d \ell_d$ |
| $M = 1840 \text{ N}$ | $\ell_M = (0.150 \text{ m}) \sin 13.0°$ | $+M \ell_M$ |

The condition specifying a zero net torque is

$$\Sigma \tau = -W_a \ell_a - W_d \ell_d + M \ell_M = 0 \qquad (9.2)$$

Solving this equation for $W_d$ yields

$$W_d = \frac{-W_a \ell_a + M \ell_M}{\ell_d}$$

$$= \frac{-(31.0 \text{ N})(0.280 \text{ m}) + (1840 \text{ N})(0.150 \text{ m}) \sin 13.0°}{0.620 \text{ m}} = \boxed{86.1 \text{ N}}$$

Substituting this value for $W_d$ into Equation 4.9b above and solving for $S_y$ gives us $\boxed{S_y = -297 \text{ N}}$. The minus sign indicates that the choice of direction for $S_y$ in the free-body diagram is wrong. In reality, $S_y$ has a magnitude of 297 N but is directed downward, not upward.

*Problem solving insight*
When a force is negative, such as $S_y = -297$ N in this example, it means that the direction of the force is opposite to that chosen originally.

# 9.3 *Center of Gravity*

Often, it is important to know the torque produced by the weight of an *extended* body. In Examples 4 and 5, for instance, it is necessary to determine the torques caused by the weight of the ladder and the arm, respectively. In both cases the weight is considered to act at a definite point for the purpose of calculating the torque. This point is called the *center of gravity* (abbreviated "cg").

---

■ **DEFINITION OF CENTER OF GRAVITY**

The center of gravity of a rigid body is the point at which its weight can be considered to act when the torque due to the weight is being calculated.

---

When an object has a symmetrical shape and its weight is distributed uniformly, the center of gravity lies at its geometrical center. For instance, Figure 9.8 shows a thin, uniform, horizontal rod of length $L$ attached to a vertical wall by a hinge. The center of gravity of the rod is located at the geometrical center. The lever arm for the weight $\mathbf{W}$ is $L/2$, and the magnitude of the torque is $\tau = W(L/2)$. In a similar fashion, the center of gravity of any symmetrically shaped and uniform object, such as a sphere, disk, cube, or cylinder, is located at its geometrical center. However, this does not mean that the center of gravity

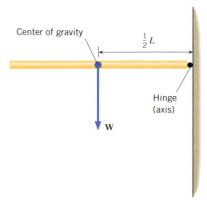

**Figure 9.8** A thin, uniform, horizontal rod of length $L$ is attached to a vertical wall by a hinge. The center of gravity of the rod is at its geometrical center.

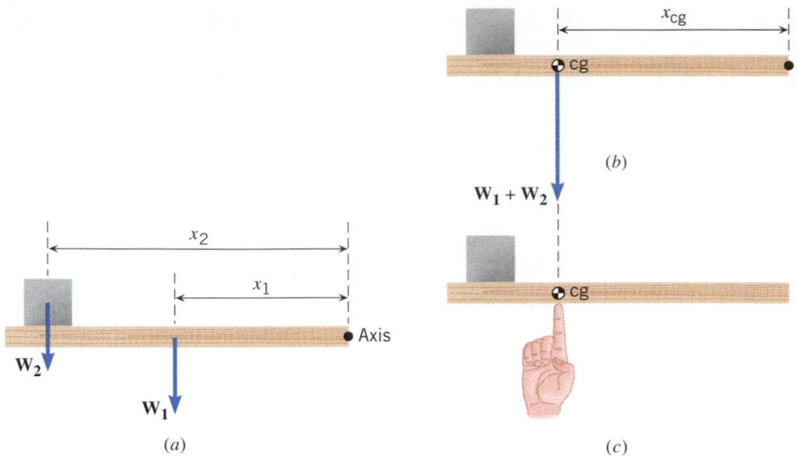

**Figure 9.9** (a) A box rests near the left end of a horizontal board. (b) The total weight ($\mathbf{W_1} + \mathbf{W_2}$) acts at the center of gravity of the group. (c) The group can be balanced by applying an external force (due to the index finger) at the center of gravity.

must lie within the object itself. The center of gravity of a compact disc recording, for instance, lies at the center of the hole in the disc and is, therefore, "outside" the object.

Suppose we have a group of objects, with known weights and centers of gravity, and it is necessary to know the center of gravity for the group as a whole. As an example, Figure 9.9a shows a group composed of two parts: a horizontal uniform board (weight $\mathbf{W_1}$) and a uniform box (weight $\mathbf{W_2}$) near the left end of the board. The center of gravity can be determined by calculating the net torque created by the board and box about an axis that is picked arbitrarily to be at the right end of the board. Part *a* of the figure shows the weights $\mathbf{W_1}$ and $\mathbf{W_2}$ and their corresponding lever arms $x_1$ and $x_2$. The net torque is $\Sigma\tau = W_1 x_1 + W_2 x_2$. It is also possible to calculate the net torque by treating the total weight $\mathbf{W_1} + \mathbf{W_2}$ as if it were located at the center of gravity and had the lever arm $x_{cg}$, as part *b* of the drawing indicates: $\Sigma\tau = (W_1 + W_2)x_{cg}$. The two values for the net torque must be the same, so that

$$W_1 x_1 + W_2 x_2 = (W_1 + W_2)x_{cg}$$

This expression can be solved for $x_{cg}$, which locates the center of gravity relative to the axis:

**Center of gravity**
$$x_{cg} = \frac{W_1 x_1 + W_2 x_2 + \cdots}{W_1 + W_2 + \cdots} \qquad (9.3)$$

The notation "$+ \cdots$" indicates that Equation 9.3 can be extended to account for any number of weights distributed along a horizontal line. Figure 9.9c illustrates that the group can be balanced by a single external force (due to the index finger), if the line of action of the force passes through the center of gravity, and if the force is equal in magnitude, but opposite in direction, to the weight of the group. Example 6 demonstrates how to calculate the center of gravity for the human arm.

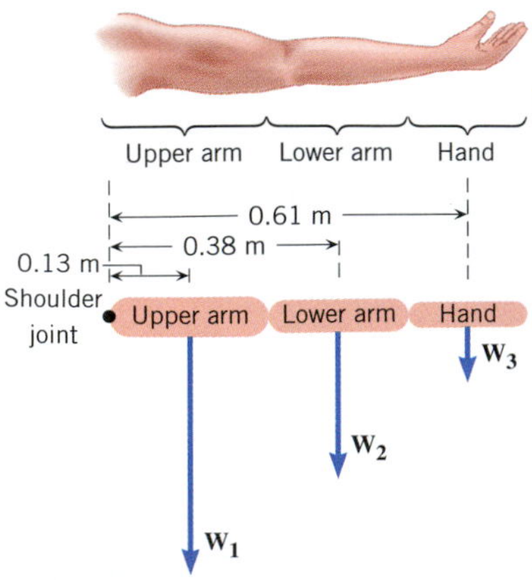

**Figure 9.10** The three parts of a human arm, and the weight and center of gravity for each.

### Example 6 The Center of Gravity of an Arm

The horizontal arm in Figure 9.10 is composed of three parts: the upper arm (weight $W_1 = 17$ N), the lower arm ($W_2 = 11$ N), and the hand ($W_3 = 4.2$ N). The drawing shows the center of gravity of each part, measured with respect to the shoulder joint. Find the center of gravity of the entire arm, relative to the shoulder joint.

**Reasoning and Solution** The coordinate $x_{cg}$ of the center of gravity is given by

$$x_{cg} = \frac{W_1 x_1 + W_2 x_2 + W_3 x_3}{W_1 + W_2 + W_3} \qquad (9.3)$$

$$= \frac{(17\text{ N})(0.13\text{ m}) + (11\text{ N})(0.38\text{ m}) + (4.2\text{ N})(0.61\text{ m})}{17\text{ N} + 11\text{ N} + 4.2\text{ N}} = \boxed{0.28\text{ m}}$$

The center of gravity plays an important role in determining whether a group of objects remains in equilibrium as the weight distribution within the group changes. A change in the weight distribution causes a change in the position of the center of gravity, and if the change is too great, the group will not remain in equilibrium. Conceptual Example 7 discusses a shift in the center of gravity that led to an embarrassing result.

## Conceptual Example 7    Overloading a Cargo Plane

Figure 9.11*a* shows a stationary cargo plane with its front landing gear 9 meters off the ground. This accident occurred because the plane was overloaded toward the rear. How did a shift in the center of gravity of the loaded plane cause the accident?

**Reasoning and Solution**   Figure 9.11*b* shows a drawing of a correctly loaded plane, with the center of gravity located between the front and the rear landing gears. The weight **W** of the plane and cargo acts downward at the center of gravity, and the normal forces **F**$_{N1}$ and **F**$_{N2}$ act upward at the front and at the rear landing gear, respectively. With respect to an axis at the rear landing gear, the counterclockwise torque due to **F**$_{N1}$ balances the clockwise torque due to **W**, and the plane remains in equilibrium. Figure 9.11*c* shows the plane with too much cargo loaded toward the rear, just after the plane has begun to rotate counterclockwise. Because of the overloading, the center of gravity has shifted behind the rear landing gear. The torque due to **W** is now counterclockwise and is not balanced by any clockwise torque. Due to the unbalanced counterclockwise torque, the plane rotates until its tail hits the ground, which applies an upward force to the tail. The clockwise torque due to this upward force balances the counterclockwise torque due to **W**, and the plane comes again into an equilibrium state, this time with the front landing gear 9 meters off the ground.

**Related Homework:**  *Problems 12, 19*

**Figure 9.11**  (*a*) This stationary cargo plane is sitting on its tail at Los Angeles International Airport, after being overloaded toward the rear. (© AP/Wide World Photos) (*b*) In a correctly loaded plane, the center of gravity is between the front and the rear landing gear. (*c*) When the plane is overloaded toward the rear, the center of gravity shifts behind the rear landing gear, and the accident in part *a* occurs.

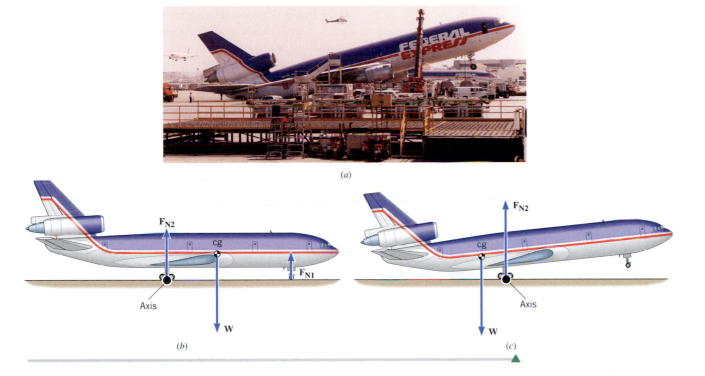

(*a*)

(*b*)                                                                      (*c*)

The center of gravity of an object with an irregular shape and a nonuniform weight distribution can be found by suspending the object from two different points $P_1$ and $P_2$, one at a time. Figure 9.12*a* shows the object at the moment of release, when its weight **W**, acting at the center of gravity, has a nonzero lever arm $\ell$ relative to the axis shown in the drawing. At this instant the weight produces a torque about the axis. The tension force **T** applied to the object by the suspension cord produces no torque because its line of action passes through the axis. Hence, in part *a* there is a net torque applied to the object, and the object begins to rotate. Friction eventually brings the object to rest as in part *b*, where the center of gravity lies directly below the point of suspension. In such an orientation, the line of action of the weight passes through the axis, so there is no

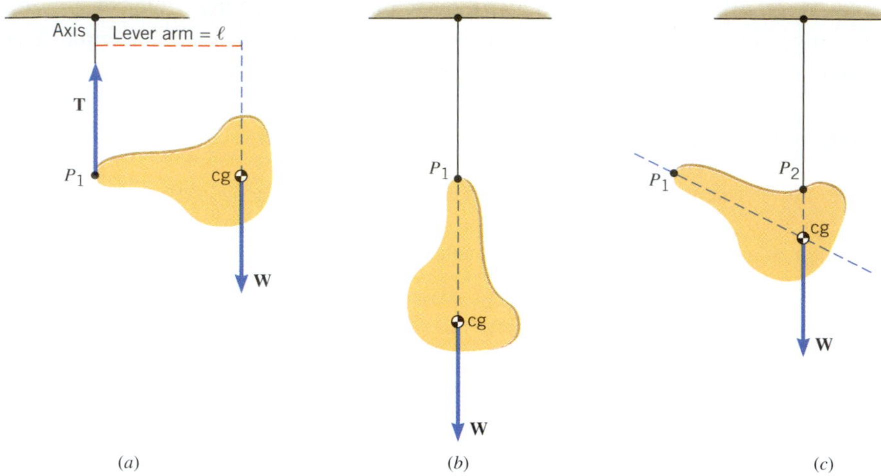

**Figure 9.12** The center of gravity (cg) of an object can be located by suspending the object from two different points, $P_1$ and $P_2$, one at a time.

(a)

(b)

(c)

longer any net torque. In the absence of a net torque the object remains at rest. By suspending the object from a second point $P_2$ (see Figure 9.12c), a second line through the object can be established, along which the center of gravity must also lie. The center of gravity, then, must be at the intersection of the two lines.

The center of gravity is closely related to the center-of-mass concept discussed in Section 7.5. To see why they are related, let's replace each occurrence of the weight in Equation 9.3 by $W = mg$, where $m$ is the mass of a given object and $g$ is the acceleration due to gravity at the location of the object. Suppose that $g$ has the same value everywhere that the objects are located. Then it can be algebraically canceled from each term on the right side of Equation 9.3. The resulting equation, which contains only masses and distances, is the same as Equation 7.10, which defines the location of the center of mass. Thus, the two points are identical. For ordinary-sized objects, like cars and boats, the center of gravity coincides with the center of mass.

## 9.4 *Newton's Second Law for Rotational Motion About a Fixed Axis*

The goal of this section is to put Newton's second law into a form suitable for describing the rotational motion of a rigid object about a fixed axis. We begin by considering a particle moving on a circular path. Figure 9.13 presents a good approximation of this situation by using a small model plane on a guideline of negligible mass. The plane's engine produces a net external tangential force $\mathbf{F}_T$ that gives the plane a tangential acceleration $a_T$. In accord with Newton's second law, it follows that $F_T = ma_T$. The torque $\tau$ produced by this force is $\tau = F_T r$, where the radius $r$ of the circular path is also the lever arm. As a result, the torque is $\tau = ma_T r$. But the tangential acceleration is related to the angular acceleration $\alpha$ according to $a_T = r\alpha$ (Equation 8.10), where $\alpha$ must be expressed in rad/s$^2$. With this substitution for $a_T$, the torque becomes

$$\tau = \underbrace{(mr^2)}_{\substack{\text{Moment} \\ \text{of inertia } I}}\alpha \qquad (9.4)$$

Equation 9.4 is the form of Newton's second law we have been seeking. It indicates that the net external torque $\tau$ is directly proportional to the angular acceleration $\alpha$. The constant of proportionality is $I = mr^2$, which is called the ***moment of inertia of the particle.*** The SI unit for moment of inertia is kg · m$^2$.

If all objects were single particles, it would be just as convenient to use the second law in the form $F_T = ma_T$ as in the form $\tau = I\alpha$. The advantage in using $\tau = I\alpha$ is that it can be applied to any rigid body rotating about a fixed axis, and not just to a particle. To

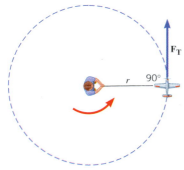

**Figure 9.13** A model airplane on a guideline has a mass $m$ and is flying on a circle of radius $r$ (top view). A net tangential force $\mathbf{F}_T$ acts on the plane.

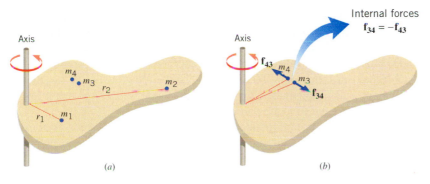

**Figure 9.14** (a) A rigid body consists of a large number of particles, four of which are shown. (b) The internal forces that particles 3 and 4 exert on each other obey Newton's law of action and reaction.

illustrate how this advantage arises, Figure 9.14a shows a flat sheet of material that rotates about an axis perpendicular to the sheet. The sheet is composed of a number of mass particles, $m_1$, $m_2$, . . . , $m_N$, where $N$ is very large. Only four particles are shown for the sake of clarity. Each particle behaves in the same way as the model airplane in Figure 9.13 and obeys the relation $\tau = (mr^2)\alpha$:

$$\tau_1 = (m_1 r_1^2)\alpha$$
$$\tau_2 = (m_2 r_2^2)\alpha$$
$$\vdots$$
$$\tau_N = (m_N r_N^2)\alpha$$

In these equations each particle has the same angular acceleration $\alpha$, since the rotating object is assumed to be rigid. Adding together the $N$ equations and factoring out the common value of $\alpha$, we find that

$$\underbrace{\Sigma\tau}_{\substack{\text{Net} \\ \text{external torque}}} = \underbrace{(\Sigma mr^2)}_{\substack{\text{Moment} \\ \text{of inertia}}}\alpha \qquad (9.5)$$

where $\Sigma\tau = \tau_1 + \tau_2 + \cdots + \tau_N$ is the sum of the external torques, and $\Sigma mr^2 = m_1 r_1^2 + m_2 r_2^2 + \cdots + m_N r_N^2$ represents the sum of the individual moments of inertia. The latter quantity is the ***moment of inertia I of the body:***

**Moment of inertia of a body**
$$I = \Sigma mr^2 \qquad (9.6)$$

In this equation, $r$ is the perpendicular radial distance of each particle from the axis of rotation. Combining Equation 9.6 with Equation 9.5 gives the following result:

■ **ROTATIONAL ANALOG OF NEWTON'S SECOND LAW FOR A RIGID BODY ROTATING ABOUT A FIXED AXIS**

$$\text{Net external torque} = \left(\begin{array}{c}\text{Moment of} \\ \text{inertia}\end{array}\right) \times \left(\begin{array}{c}\text{Angular} \\ \text{acceleration}\end{array}\right)$$

$$\Sigma\tau = I\alpha \qquad (9.7)$$

***Requirement:*** $\alpha$ must be expressed in rad/s².

The form of the second law for rotational motion, $\Sigma\tau = I\alpha$, is similar to that for translational (linear) motion, $\Sigma F = ma$, and is valid only in an inertial frame. The moment of inertia $I$ plays the same role for rotational motion that the mass $m$ does for translational motion. Thus, $I$ is a measure of the rotational inertia of a body. When using Equation 9.7, $\alpha$ must be expressed in rad/s², because the relation $a_T = r\alpha$ (which requires radian measure) was used in the derivation.

When calculating the sum of torques in Equation 9.7, it is necessary to include only the *external torques,* those applied by agents outside the body. The torques produced by internal forces need not be considered, because they always combine to produce a net torque of zero. Internal forces are those that one particle within the body exerts on another particle. They always occur in pairs of oppositely directed forces of equal magnitude, in accord with Newton's third law (see $m_3$ and $m_4$ in Figure 9.14*b*). The forces in such a pair have the same line of action, so they have identical lever arms and produce torques of equal magnitudes. One torque is counterclockwise, while the other is clockwise, the net torque from the pair being zero.

It can be seen from Equation 9.6 that the moment of inertia depends on both the mass of each particle and its distance from the axis of rotation. The farther a particle is from the axis, the greater is its contribution to the moment of inertia. Therefore, although a rigid object possesses a unique total mass, it does not have a unique moment of inertia, for **the moment of inertia depends on the location and orientation of the axis relative to the particles that make up the object.** Example 8 shows how the moment of inertia can change when the axis of rotation changes.

**Problem solving insight**

### Example 8  The Moment of Inertia Depends on Where the Axis Is

Two particles each have a mass $m$ and are fixed to the ends of a thin rigid rod, whose mass can be ignored. The length of the rod is $L$. Find the moment of inertia when this object rotates relative to an axis that is perpendicular to the rod at (a) one end and (b) the center. (See Figure 9.15.)

**Reasoning** When the axis of rotation changes, the distance $r$ between the axis and each particle changes. In determining the moment of inertia using $I = \Sigma mr^2$, we must be careful to use the distances that apply for each axis.

**Solution**

**(a)** Particle 1 lies on the axis, as part *a* of the drawing shows, and has a zero radial distance: $r_1 = 0$. In contrast, particle 2 moves on a circle whose radius is $r_2 = L$. Noting that $m_1 = m_2 = m$, we find that the moment of inertia is

$$I = \Sigma mr^2 = m_1 r_1^2 + m_2 r_2^2 = \boxed{mL^2} \qquad (9.6)$$

**(b)** Part *b* of the drawing shows that particle 1 no longer lies on the axis but now moves on a circle of radius $r_1 = L/2$. Particle 2 moves on a circle with the same radius, $r_2 = L/2$. Therefore,

$$I = \Sigma mr^2 = m_1 r_1^2 + m_2 r_2^2 = m(L/2)^2 + m(L/2)^2 = \boxed{\tfrac{1}{2}mL^2}$$

This value differs from that in part (a) because the axis of rotation is different.

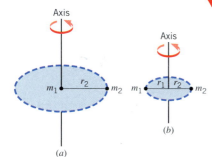

**Figure 9.15** Two particles, masses $m_1$ and $m_2$, are attached to the ends of a massless rigid rod. The moment of inertia of this object is different, depending on whether the rod rotates about an axis through (*a*) the end or (*b*) the center of the rod.

The procedure illustrated in Example 8 can be extended using integral calculus to evaluate the moment of inertia of a rigid object with a continuous mass distribution, and Table 9.1 gives some typical results. These results depend on the total mass of the object, its shape, and the location and orientation of the axis.

When forces act on a rigid object, they can affect its motion in two ways. They can produce a translational acceleration $a$ (components $a_x$ and $a_y$). The forces can also produce torques, which can produce an angular acceleration $\alpha$. In general, we can deal with the resulting combined motion by using Newton's second law. For the translational motion, we use the law in the form $\Sigma F = ma$. For the rotational motion, we use the law in the form $\Sigma \tau = I\alpha$. When $a$ (both components) and $\alpha$ are zero, there is no acceleration of any kind, and the object is in equilibrium. This is the situation already discussed in Section 9.2. If any component of $a$ or $\alpha$ is nonzero, we have accelerated motion, and the object is not in equilibrium. Examples 9, 10, and 11 deal with this type of situation.

### Example 9  The Torque of an Electric Saw Motor

The motor in an electric saw brings the circular blade from rest up to the rated angular velocity of 80.0 rev/s in 240.0 rev. One type of blade has a moment of inertia of $1.41 \times 10^{-3}$ kg·m². What net torque (assumed constant) must the motor apply to the blade?

**Reasoning** Newton's second law for rotational motion (Equation 9.7, $\Sigma\tau = I\alpha$) can be used to find the net torque $\Sigma\tau$, once the angular acceleration $\alpha$ is determined. The angular acceleration can be calculated from the data in the following table and the appropriate equation of rotational kinematics:

| $\theta$ | $\alpha$ | $\omega$ | $\omega_0$ | $t$ |
|---|---|---|---|---|
| 1508 rad (240.0 rev) | ? | 503 rad/s (80.0 rev/s) | 0 rad/s | |

The data for the angular displacement $\theta$ and the angular velocity $\omega$ have been converted to radian measure because $\Sigma\tau = I\alpha$ requires that $\alpha$ be expressed in radian measure.

**Solution** From Equation 8.8 ($\omega^2 = \omega_0^2 + 2\alpha\theta$) it follows that

$$\alpha = \frac{\omega^2 - \omega_0^2}{2\theta}$$

Newton's second law for rotational motion (Equation 9.7) can now be used to obtain the net torque:

$$\Sigma\tau = I\alpha = I\left(\frac{\omega^2 - \omega_0^2}{2\theta}\right)$$

$$= (1.41 \times 10^{-3}\ \text{kg} \cdot \text{m}^2)\left[\frac{(503\ \text{rad/s})^2 - (0\ \text{rad/s})^2}{2(1508\ \text{rad})}\right] = \boxed{0.118\ \text{N} \cdot \text{m}}$$

To accelerate a wheelchair, the rider applies a force to a handrail attached to each wheel. The torque generated by this force is the product of the magnitude of the force and the lever arm. As Figure 9.16 illustrates, the lever arm is just the radius of the circular rail, which is designed to be as large as possible. Thus, a relatively large torque can be generated for a given force, allowing the rider to accelerate quickly.

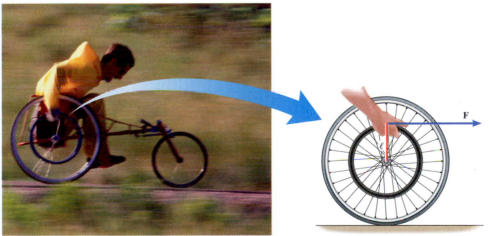

**Figure 9.16** A rider applies a force **F** to the circular handrail. The torque produced by this force is the product of its magnitude and the lever arm $\ell$ about the axis of rotation. (© Dan Coffee/The Image Bank/Getty Images)

Example 9 shows how Newton's second law for rotational motion is used when design considerations demand an adequately large angular acceleration. There are also situations when it is desirable to have as little angular acceleration as possible, and Conceptual Example 10 deals with one of them.

## Conceptual Example 10   Archery and Bow Stabilizers

Archers can shoot with amazing accuracy, especially using modern bows such as the one in Figure 9.17. Notice the bow stabilizer, a long, thin rod that extends from the front of the bow

**Table 9.1   Moments of Inertia for Various Rigid Objects of Mass M**

Thin-walled hollow cylinder or hoop

$I = MR^2$

Solid cylinder or disk

$I = \frac{1}{2}MR^2$

Thin rod, axis perpendicular to rod and passing through center

$I = \frac{1}{12}ML^2$

Thin rod, axis perpendicular to rod and passing through one end

$I = \frac{1}{3}ML^2$

Solid sphere, axis through center

$I = \frac{2}{5}MR^2$

Solid sphere, axis tangent to surface

$I = \frac{7}{5}MR^2$

Thin-walled spherical shell, axis through center

$I = \frac{2}{3}MR^2$

Thin rectangular sheet, axis parallel to one edge and passing through center of other edge

$I = \frac{1}{12}ML^2$

Thin rectangular sheet, axis along one edge

$I = \frac{1}{3}ML^2$

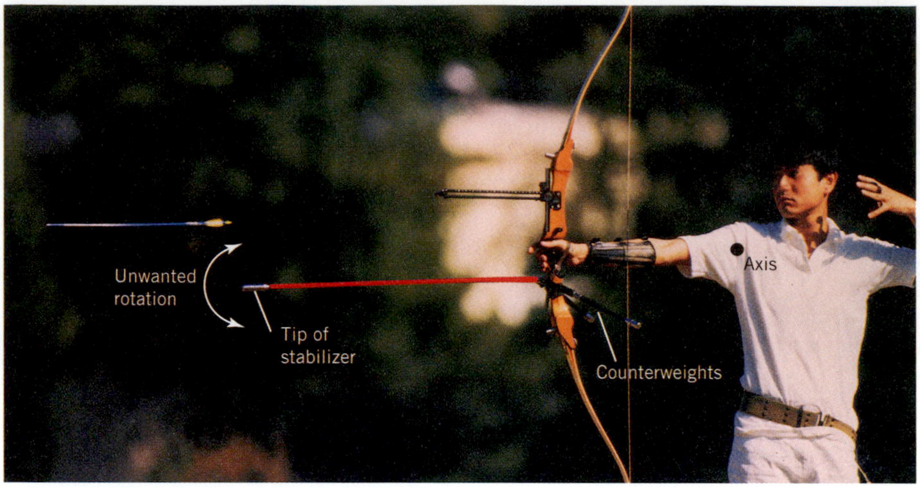

**Figure 9.17** The stabilizer helps to steady the archer's aim, as Conceptual Example 10 discusses. Relative to an axis through the archer's shoulder, the moment of inertia of the bow is larger with the stabilizer than without it. The counterweights below the archer's hand help to keep the center of gravity of the bow near the hand. (© Amwell/Stone/Getty Images)

**The physics of archery and bow stabilizers.**

and has a relatively massive cylinder at the tip. Advertisements claim that the stabilizer helps to steady the archer's aim. Could there be any truth to this claim? Explain.

**Reasoning and Solution** To help explain why the stabilizer works, we have added to the photograph an axis for rotation that passes through the archer's shoulder and is perpendicular to the plane of the paper. Any angular acceleration $\alpha$ about this axis will lead to a rotation of the bow that will degrade the archer's aim. The acceleration will be created by any unbalanced torques that occur while the archer's tensed muscles try to hold the drawn bow. Newton's second law for rotation indicates, however, that the angular acceleration is $\alpha = (\Sigma\tau)/I$. The moment of inertia $I$ is in the denominator on the right side of this equation. Therefore, to the extent that $I$ is larger, a given net torque $\Sigma\tau$ will create a smaller angular acceleration and less disturbance of the aim. It is to increase the moment of inertia of the bow that the stabilizer has been added. The relatively massive cylinder is particularly effective in increasing the moment of inertia, because it is placed at the tip of the stabilizer, far from the axis of rotation (a large value of $r$ in the equation $I = \Sigma mr^2$).

▲

Rotational motion and translational motion sometimes occur together. The next example deals with an interesting situation in which both angular acceleration and translational acceleration must be considered.

## Example 11 | Hoisting a Crate

A crate that weighs 4420 N is being lifted by the mechanism shown in Figure 9.18a. The two cables are wrapped around their respective pulleys, which have radii of 0.600 and 0.200 m. The pulleys are fastened together to form a dual pulley and turn as a single unit about the center axle, relative to which the combined moment of inertia is $I = 50.0$ kg·m². A tension of magnitude $T_1 = 2150$ N is maintained in the cable attached to the motor. Find the angular acceleration of the dual pulley and the tension in the cable connected to the crate.

**Reasoning** To determine the angular acceleration of the dual pulley and the tension in the cable attached to the crate, we will apply Newton's second law to the pulley and the crate separately. Three external forces act on the dual pulley, as its free-body diagram shows (Figure 9.18b). These are (1) the tension $\mathbf{T_1}$ in the cable connected to the motor, (2) the tension $\mathbf{T_2}$ in the cable attached to the crate, and (3) the reaction force $\mathbf{P}$ exerted on the dual pulley by the axle. The force $\mathbf{P}$ arises because the two cables pull the pulley down and to the left into the axle, and the axle pushes back, thus keeping the pulley in place. The net torque that results from these forces obeys Newton's second law for rotational motion (Equation 9.7). Two external forces act on the crate, as its free-body diagram indicates (Figure 9.18c). These are (1) the cable tension $\mathbf{T_2}'$ and (2) the weight $\mathbf{W}$ of the crate. The net force that results from these forces obeys Newton's second law for translational motion (Equation 4.2b).

**Solution** Using the lever arms $\ell_1$ and $\ell_2$ shown in part $b$ of the figure, we can apply the second law to the rotational motion of the pulley. We note that the force $\mathbf{P}$ has a zero lever arm,

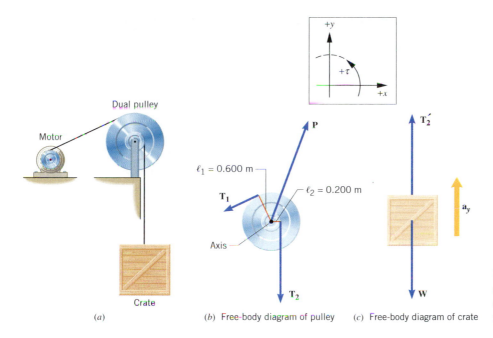

Figure 9.18 (a) The crate is lifted upward by the motor and pulley arrangement. The free-body diagram for (b) the dual pulley and (c) the crate.

(a)

(b) Free-body diagram of pulley

(c) Free-body diagram of crate

Crate

Dual pulley

Motor

$\ell_1 = 0.600$ m

$\ell_2 = 0.200$ m

Axis

since the line of action of **P** passes directly through the axle:

$$\Sigma\tau = T_1\ell_1 - T_2\ell_2 = I\alpha \tag{9.7}$$

$$(2150\ \text{N})(0.600\ \text{m}) - T_2(0.200\ \text{m}) = (50.0\ \text{kg}\cdot\text{m}^2)\alpha$$

This equation contains two unknown quantities, so a second equation is needed and may be obtained by applying Newton's second law to the upward translational motion of the crate. In doing so, we note that the magnitude of the tension in the cable attached to the crate is $T_2' = T_2$ and that the mass of the crate is $m = (4420\ \text{N})/(9.80\ \text{m/s}^2) = 451$ kg:

$$\Sigma F_y = ma_y \tag{4.2b}$$

$$T_2 - (4420\ \text{N}) = (451\ \text{kg})\,a_y$$

Because the cable attached to the crate rolls on the pulley without slipping, the linear acceleration $a_y$ of the crate is related to the angular acceleration $\alpha$ of the pulley via Equation 8.13: $a_y = r\alpha = (0.200\ \text{m})\alpha$. With this substitution for $a_y$, Equation 4.2b becomes

$$T_2 - (4420\ \text{N}) = (451\ \text{kg})(0.200\ \text{m})\alpha$$

This result and Equation 9.7 can be solved simultaneously to yield

$$\boxed{T_2 = 4960\ \text{N}} \quad \text{and} \quad \boxed{\alpha = 6.0\ \text{rad/s}^2}$$

We have seen that Newton's second law for rotational motion, $\Sigma\tau = I\alpha$, has the same form as that for translational motion, $\Sigma F = ma$, so each rotational variable has a translational analog: torque $\tau$ and force $F$ are analogous quantities, as are moment of inertia $I$ and mass $m$, and angular acceleration $\alpha$ and linear acceleration $a$. The other physical concepts developed for studying translational motion, such as kinetic energy and momentum, also have rotational analogs. For future reference, Table 9.2 on p. 172 itemizes these concepts and their rotational analogs.

# 9.5 *Rotational Work and Energy*

Work and energy are among the most fundamental and useful concepts in physics. Chapter 6 discusses their application to translational motion. These concepts are equally useful for rotational motion, provided they are expressed in terms of angular variables.

The work $W$ done by a constant force that points in the same direction as the displacement is $W = Fs$ (Equation 6.1), where $F$ and $s$ are the magnitudes of the force and

**Table 9.2**    *Analogies Between Rotational and Translational Concepts*

| Physical Concept | Rotational | Translational |
|---|---|---|
| Displacement | $\theta$ | $s$ |
| Velocity | $\omega$ | $v$ |
| Acceleration | $\alpha$ | $a$ |
| The cause of acceleration | Torque $\tau$ | Force $F$ |
| Inertia | Moment of inertia $I$ | Mass $m$ |
| Newton's second law | $\Sigma\tau = I\alpha$ | $\Sigma F = ma$ |
| Work | $\tau\theta$ | $Fs$ |
| Kinetic energy | $\frac{1}{2}I\omega^2$ | $\frac{1}{2}mv^2$ |
| Momentum | $L = I\omega$ | $p = mv$ |

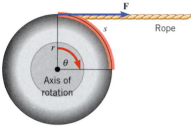

**Figure 9.19** The force **F** does work in rotating the wheel through the angle $\theta$.

displacement, respectively. To see how this expression can be rewritten using angular variables, consider Figure 9.19. Here a rope is wrapped around a wheel and is under a constant tension $F$. If the rope is pulled out a distance $s$, the wheel rotates through an angle $\theta = s/r$ (Equation 8.1), where $r$ is the radius of the wheel and $\theta$ is in radians. Thus, $s = r\theta$, and the work done by the tension force in turning the wheel is $W = Fs = Fr\theta$. But $Fr$ is the torque $\tau$ applied to the wheel by the tension, so the rotational work can be written as follows:

---

■ **DEFINITION OF ROTATIONAL WORK**

The rotational work $W_R$ done by a constant torque $\tau$ in turning an object through an angle $\theta$ is

$$W_R = \tau\theta \qquad (9.8)$$

*Requirement:* $\theta$ must be expressed in radians.

**SI Unit of Rotational Work:** joule (J)

---

Section 6.2 discusses the work–energy theorem and kinetic energy. There we saw that the work done on an object by a net external force causes the translational kinetic energy ($\frac{1}{2}mv^2$) of the object to change. In an analogous manner, the rotational work done by a net external torque causes the rotational kinetic energy to change. A rotating body possesses kinetic energy, because its constituent particles are moving. If the body is rotating with an angular speed $\omega$, the tangential speed $v_T$ of a particle at a distance $r$ from the axis is $v_T = r\omega$ (Equation 8.9). Figure 9.20 shows two such particles. If a particle's mass is $m$, its kinetic energy is $\frac{1}{2}mv_T^2 = \frac{1}{2}mr^2\omega^2$. The kinetic energy of the entire rotating body, then, is the sum of the kinetic energies of the particles:

$$\text{Rotational KE} = \Sigma(\tfrac{1}{2}mr^2\omega^2) = \tfrac{1}{2}\underbrace{(\Sigma mr^2)}_{\substack{\text{Moment of}\\\text{inertia, } I}}\omega^2$$

In this result, the angular speed $\omega$ is the same for all particles in a rigid body and, therefore, has been factored outside the summation. According to Equation 9.6, the term in parentheses is the moment of inertia, $I = \Sigma mr^2$, so the rotational kinetic energy takes the following form:

---

■ **DEFINITION OF ROTATIONAL KINETIC ENERGY**

The rotational kinetic energy $KE_R$ of a rigid object rotating with an angular speed $\omega$ about a fixed axis and having a moment of inertia $I$ is

$$KE_R = \tfrac{1}{2}I\omega^2 \qquad (9.9)$$

*Requirement:* $\omega$ must be expressed in rad/s.

**SI Unit of Rotational Kinetic Energy:** joule (J)

---

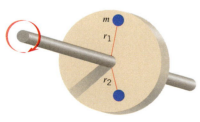

**Figure 9.20** The rotating wheel is composed of many particles, two of which are shown.

Kinetic energy is one part of an object's total mechanical energy. The total mechanical energy is the sum of the kinetic and potential energies and obeys the principle of conservation of mechanical energy (see Section 6.5). We need to remember that translational and rotational motion can occur simultaneously. When a bicycle coasts down a hill, for instance, its tires are both translating and rotating. An object such as a rolling bicycle tire has both translational and rotational kinetic energies, so that the total mechanical energy is

$$ \underbrace{E}_{\substack{\text{Total} \\ \text{mechanical} \\ \text{energy}}} = \underbrace{\tfrac{1}{2}mv^2}_{\substack{\text{Translational} \\ \text{kinetic energy}}} + \underbrace{\tfrac{1}{2}I\omega^2}_{\substack{\text{Rotational} \\ \text{kinetic energy}}} + \underbrace{mgh}_{\substack{\text{Gravitational} \\ \text{potential energy}}} $$

Here $m$ is the mass of the object, $v$ is the translational speed of its center of mass, $I$ is its moment of inertia about an axis through the center of mass, $\omega$ is its angular speed, and $h$ is the height of the object's center of mass relative to an arbitrary zero level. Mechanical energy is conserved if $W_{nc}$, the net work done by external nonconservative forces and torques, is zero. If the total mechanical energy is conserved as an object moves, its final total mechanical energy $E_f$ equals its initial total mechanical energy $E_0$: $E_f = E_0$.

Example 12 illustrates the effect of combined translational and rotational motion, but in the context of how the total mechanical energy of a cylinder is conserved as it rolls down an incline.

### Example 12  Rolling Cylinders

A thin-walled hollow cylinder (mass = $m_h$, radius = $r_h$) and a solid cylinder (mass = $m_s$, radius = $r_s$) start from rest at the top of an incline (Figure 9.21). Both cylinders start at the same vertical height $h_0$. All heights are measured relative to an arbitrarily chosen zero level that passes through the center of mass of a cylinder when it is at the bottom of the incline (see the drawing). Ignoring energy losses due to retarding forces, determine which cylinder has the greatest translational speed upon reaching the bottom.

**Reasoning** Only the conservative force of gravity does work on the cylinders, so the total mechanical energy is conserved as they roll down. The total mechanical energy $E$ at any height $h$ above the zero level is the sum of the translational kinetic energy ($\tfrac{1}{2}mv^2$), the rotational kinetic energy ($\tfrac{1}{2}I\omega^2$), and the gravitational potential energy ($mgh$):

$$ E = \tfrac{1}{2}mv^2 + \tfrac{1}{2}I\omega^2 + mgh $$

As the cylinders roll down, potential energy is converted into kinetic energy, but the kinetic energy is shared between the translational form ($\tfrac{1}{2}mv^2$) and the rotational form ($\tfrac{1}{2}I\omega^2$). The object with more of its kinetic energy in the translational form will have the greater translational speed at the bottom of the incline. We expect the solid cylinder to have the greater translational speed, because more of its mass is located near the rotational axis and, thus, possesses less rotational kinetic energy.

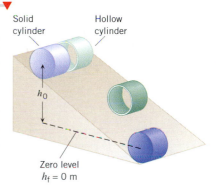

**Figure 9.21** A hollow cylinder and a solid cylinder start from rest and roll down the incline plane. The conservation of mechanical energy can be used to show that the solid cylinder, having the greater translational speed, reaches the bottom first.

**Solution** The total mechanical energy $E_f$ at the bottom ($h_f = 0$ m) is the same as the total mechanical energy $E_0$ at the top ($h = h_0$, $v_0 = 0$ m/s, $\omega_0 = 0$ rad/s):

$$ \tfrac{1}{2}mv_f^2 + \tfrac{1}{2}I\omega_f^2 + mgh_f = \tfrac{1}{2}mv_0^2 + \tfrac{1}{2}I\omega_0^2 + mgh_0 $$

$$ \tfrac{1}{2}mv_f^2 + \tfrac{1}{2}I\omega_f^2 = mgh_0 $$

Since each cylinder rolls without slipping, the final rotational speed $\omega_f$ and the final translational speed $v_f$ of its center of mass are related according to Equation 8.12, $\omega_f = v_f/r$, where $r$ is the radius of the cylinder. Substituting this expression for $\omega_f$ into the energy-conservation equation and solving for $v_f$ yields

$$ v_f = \sqrt{\frac{2mgh_0}{m + I/r^2}} $$

Setting $m = m_h$, $r = r_h$ and $I = m r_h^2$ for the hollow cylinder and then setting $m = m_s$, $r = r_s$ and $I = \tfrac{1}{2}m r_s^2$ for the solid cylinder (see Table 9.1), we find that the two cylinders have the following translational speeds at the bottom of the incline:

**Hollow cylinder**
$$ v_f = \sqrt{gh_0} $$

**Solid cylinder**
$$ v_f = \sqrt{\frac{4gh_0}{3}} = 1.15\sqrt{gh_0} $$

The solid cylinder, having the greater translational speed, arrives at the bottom first.

## 9.6  *Angular Momentum*

In Chapter 7 the linear momentum $p$ of an object is defined as the product of its mass $m$ and linear velocity $v$; that is, $p = mv$. For rotational motion the analogous concept is called the ***angular momentum*** $L$. The mathematical form of angular momentum is analogous to that of linear momentum, with the mass $m$ and the linear velocity $v$ being replaced with their rotational counterparts, the moment of inertia $I$ and the angular velocity $\omega$.

---

■ **DEFINITION OF ANGULAR MOMENTUM**

The angular momentum $L$ of a body rotating about a fixed axis is the product of the body's moment of inertia $I$ and its angular velocity $\omega$ with respect to that axis:

$$L = I\omega \tag{9.10}$$

*Requirement:* $\omega$ must be expressed in rad/s.

*SI Unit of Angular Momentum:* $\text{kg} \cdot \text{m}^2/\text{s}$

---

Linear momentum is an important concept in physics because the total linear momentum of a system is conserved when the sum of the average external forces acting on the system is zero. Then, the final total linear momentum $P_\text{f}$ and the initial total linear momentum $P_0$ are the same: $P_\text{f} = P_0$. Recall that each translational variable is analogous to a corresponding rotational variable. Therefore, if we replace "force" by "torque" and "linear momentum" by "angular momentum," we arrive at the following: When the sum of the average external torques is zero, then the final and initial angular momenta are the same: $L_\text{f} = L_0$, which is the ***principle of conservation of angular momentum.***

---

■ **PRINCIPLE OF CONSERVATION OF ANGULAR MOMENTUM**

The total angular momentum of a system remains constant (is conserved) if the net average external torque acting on the system is zero.

---

Example 13 illustrates an interesting consequence of the conservation of angular momentum.

**The physics of** a spinning ice skater.

### Conceptual Example 13   A Spinning Skater

In Figure 9.22a an ice skater is spinning with both arms and a leg outstretched. In Figure 9.22b she pulls her arms and leg inward. As a result of this maneuver, her spinning motion changes dramatically. Using the principle of conservation of angular momentum, explain how and why it changes.

**Reasoning and Solution**  Choosing the skater as the system, we can apply the conservation principle provided that the net external torque produced by air resistance and by friction between the skates and the ice is negligibly small. We assume that it is. Then the skater in Figure 9.22a would spin forever at the same angular velocity, since her angular momentum is conserved in the absence of a net external torque. In Figure 9.22b the inward movement of her arms and leg involves internal, not external, torques and, therefore, does not change her angular momentum. But angular momentum is the product of the moment of inertia $I$ and angular velocity $\omega$. By moving the mass of her arms and leg inward, the skater decreases the distance $r$ of the mass from the axis of rotation and, consequently, decreases her moment of inertia $I$ ($I = \Sigma mr^2$). If the product of $I$ and $\omega$ is to remain constant, then $\omega$ must increase. Thus, the consequence of pulling her arms and leg inward is that ***she spins with a larger angular velocity.***

**Related Homework:** *Problem 52*

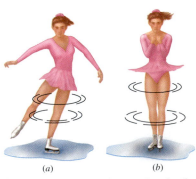

*(a)*              *(b)*

**Figure 9.22**  (a) A skater spins slowly on one skate, with both arms and one leg outstretched. (b) As she pulls her arms and leg in toward the rotational axis, her moment of inertia $I$ decreases, and her angular velocity $\omega$ increases.

The next example involves a satellite and illustrates another application of the principle of conservation of angular momentum.

## Example 14   A Satellite in an Elliptical Orbit

An artificial satellite is placed into an elliptical orbit about the earth, as in Figure 9.23. Telemetry data indicate that its point of closest approach (called the *perigee*) is $r_P = 8.37 \times 10^6$ m from the center of the earth, and its point of greatest distance (called the *apogee*) is $r_A = 25.1 \times 10^6$ m from the center of the earth. The speed of the satellite at the perigee is $v_P = 8450$ m/s. Find its speed $v_A$ at the apogee.

**The physics of a satellite in orbit about the earth.**

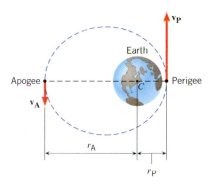

**Reasoning** The only force of any significance that acts on the satellite is the gravitational force of the earth. However, at any instant, this force is directed toward the center of the earth and passes through the axis about which the satellite instantaneously rotates. Therefore, the gravitational force exerts *no torque* on the satellite (the lever arm is zero). Consequently, the angular momentum of the satellite remains constant at all times.

**Solution** Since the angular momentum is the same at the apogee (A) and the perigee (P), it follows that $I_A\omega_A = I_P\omega_P$. Furthermore, the orbiting satellite can be considered a point mass, so its moment of inertia is $I = mr^2$ (see Equation 9.4). In addition, the angular speed $\omega$ of the satellite is related to its tangential speed $v_T$ by $\omega = v_T/r$ (Equation 8.9). If these relations are used at the apogee and perigee, the conservation of angular momentum gives the following result:

$$I_A\omega_A = I_P\omega_P \quad \text{or} \quad (mr_A^2)\left(\frac{v_A}{r_A}\right) = (mr_P^2)\left(\frac{v_P}{r_P}\right)$$

$$v_A = \frac{r_P v_P}{r_A} = \frac{(8.37 \times 10^6 \text{ m})(8450 \text{ m/s})}{25.1 \times 10^6 \text{ m}} = \boxed{2820 \text{ m/s}}$$

**Figure 9.23** A satellite is moving in an elliptical orbit about the earth. The gravitational force exerts no torque on the satellite, so the angular momentum of the satellite is conserved.

The answer is independent of the mass of the satellite. The satellite behaves just like the skater in Figure 9.22, because its speed is greater at the perigee, where the moment of inertia is smaller.

The result in Example 14 indicates that a satellite does not have a constant speed in an elliptical orbit. The speed changes from a maximum at the perigee to a minimum at the apogee; the closer the satellite comes to the earth, the faster it travels. Planets moving around the sun in elliptical orbits exhibit the same kind of behavior, and Johannes Kepler (1571–1630) formulated his famous second law based on observations of such characteristics of planetary motion. Kepler's second law states that, in a given amount of time, a line joining any planet to the sun sweeps out the same amount of area no matter where the planet is on its elliptical orbit, as Figure 9.24 illustrates. The conservation of angular momentum can be used to show why the law is valid, by means of a calculation similar to that in Example 14.

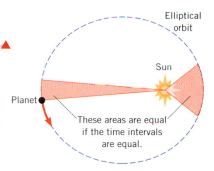

**Figure 9.24** Kepler's second law of planetary motion states that a line joining a planet to the sun sweeps out equal areas in equal time intervals.

# Concept Summary

This summary presents an abridged version of the chapter, including the important equations and all available learning aids. For convenient reference, the learning aids (including the text's examples) are placed next to or immediately after the relevant equation or discussion. The following learning aids may be found on-line at **www.wiley.com/college/cutnell**:

| | |
|---|---|
| **Interactive LearningWare** examples are solved according to a five-step interactive format that is designed to help you develop problem-solving skills. | **Concept Simulations** are animated versions of text figures or animations that illustrate important concepts. You can control parameters that affect the display, and we encourage you to experiment. |
| **Interactive Solutions** offer specific models for certain types of problems in the chapter homework. The calculations are carried out interactively. | **Self-Assessment Tests** include both qualitative and quantitative questions. Extensive feedback is provided for both incorrect and correct answers, to help you evaluate your understanding of the material. |

| Topic | Discussion | Learning Aids |
|---|---|---|

### 9.1   The Action of Forces and Torques on Rigid Objects

**Line of action**
**Lever arm**
The line of action of a force is an extended line that is drawn colinear with the force. The lever arm $\ell$ is the distance between the line of action and the axis of rotation, measured on a line that is perpendicular to both.

| Topic | Discussion | Learning Aids |
|---|---|---|

**Torque**

The torque of a force has a magnitude $\tau$ that is given by the magnitude $F$ of the force times the lever arm $\ell$:

$$\tau = F\ell \qquad (9.1)$$

The torque is positive when the force tends to produce a counterclockwise rotation about the axis, and negative when the force tends to produce a clockwise rotation.

**Examples 1, 2**
**Interactive LearningWare 9.1**
**Interactive Solution 9.3**

### 9.2 Rigid Objects in Equilibrium

**Equilibrium of a rigid body**

A rigid body is in equilibrium if it has zero translational acceleration and zero angular acceleration. In equilibrium, the net external force and the net external torque acting on the body are zero:

$$\Sigma F_x = 0 \quad \text{and} \quad \Sigma F_y = 0 \qquad (4.9a \text{ and } 4.9b)$$
$$\Sigma \tau = 0 \qquad (9.2)$$

**Examples 3, 4, 5**
**Interactive LearningWare 9.2**
**Concept Simulation 9.1**
**Interactive Solution 9.17**

### 9.3 Center of Gravity

**Definition of center of gravity**

The center of gravity of a rigid object is the point where its entire weight can be considered to act when calculating the torque due to the weight. For a symmetrical body with uniformly distributed weight, the center of gravity is at the geometrical center of the body. When a number of objects whose weights are $W_1$, $W_2$, ... are distributed along the $x$ axis at locations $x_1$, $x_2$, ... , the center of gravity $x_{cg}$ is located at

$$x_{cg} = \frac{W_1 x_1 + W_2 x_2 + \cdots}{W_1 + W_2 + \cdots} \qquad (9.3)$$

The center of gravity is identical to the center of mass, provided the acceleration due to gravity does not vary over the physical extent of the objects.

**Examples 6, 7**
**Concept Simulation 9.2**

Use **Self-Assessment Test 9.1** to evaluate your understanding of Sections 9.1–9.3.

### 9.4 Newton's Second Law for Rotational Motion About a Fixed Axis

**Moment of inertia**

The moment of inertia $I$ of a body composed of $N$ particles is

$$I = m_1 r_1^2 + m_2 r_2^2 + \cdots + m_N r_N^2 = \Sigma m r^2 \qquad (9.6)$$

where $m$ is the mass of a particle and $r$ is the perpendicular distance of the particle from the axis of rotation.

**Example 8**

**Newton's second law for rotational motion**

For a rigid body rotating about a fixed axis, Newton's second law for rotational motion is

$$\Sigma \tau = I\alpha \qquad (\alpha \text{ in rad/s}^2) \qquad (9.7)$$

where $\Sigma \tau$ is the net external torque applied to the body, $I$ is the moment of inertia of the body, and $\alpha$ is its angular acceleration.

**Examples 9, 10, 11**
**Concept Simulations 9.3, 9.4**
**Interactive Solution 9.35**

### 9.5 Rotational Work and Energy

**Rotational work**

The rotational work $W_R$ done by a constant torque $\tau$ in turning a rigid body through an angle $\theta$ is

$$W_R = \tau\theta \qquad (\theta \text{ in radians}) \qquad (9.8)$$

**Rotational kinetic energy**

The rotational kinetic energy $KE_R$ of a rigid object rotating with an angular speed $\omega$ about a fixed axis and having a moment of inertia $I$ is

$$KE_R = \tfrac{1}{2}I\omega^2 \qquad (9.9)$$

**Total mechanical energy**

The total mechanical energy $E$ of a rigid body is the sum of its translational kinetic energy ($\tfrac{1}{2}mv^2$), its rotational kinetic energy ($\tfrac{1}{2}I\omega^2$), and its gravitational potential energy ($mgh$):

$$E = \tfrac{1}{2}mv^2 + \tfrac{1}{2}I\omega^2 + mgh$$

where $m$ is the mass of the object, $v$ is the translational speed of its center of mass, $I$ is its moment of inertia about an axis through the center of mass, $\omega$ is its angular speed, and $h$ is the height of the object's center of mass relative to an arbitrary zero level.

**Interactive LearningWare 9.3**
**Interactive Solution 9.47**

| Topic | Discussion | Learning Aids |
|---|---|---|
| Conservation of total mechanical energy | The total mechanical energy is conserved if the net work done by external non-conservative forces and torques is zero. When the total mechanical energy is conserved, the final total mechanical energy $E_f$ equals the initial total mechanical energy $E_0$: $E_f = E_0$. | Example 12 |

### 9.6 Angular Momentum

The angular momentum of a rigid body rotating with an angular velocity $\omega$ about a fixed axis and having a moment of inertia $I$ with respect to that axis is

| Angular momentum | | |
|---|---|---|

$$L = I\omega \quad (\omega \text{ in rad/s}) \quad (9.10)$$

| Conservation of angular momentum | The principle of conservation of angular momentum states that the total angular momentum of a system remains constant (is conserved) if the net average external torque acting on the system is zero. When the total angular momentum is conserved, the final angular momentum $L_f$ equals the initial angular momentum $L_0$: $L_f = L_0$. | Examples 13, 14 <br> Interactive Solution 9.55 |

*Use Self-Assessment Test 9.2 to evaluate your understanding of Sections 9.4–9.6.*

# Problems

## Section 9.1 The Action of Forces and Torques on Rigid Objects

**1. ssm www** A square, 0.40 m on a side, is mounted so that it can rotate about an axis that passes through the center of the square. The axis is perpendicular to the plane of the square. A force of 15 N lies in this plane and is applied to the square. What is the magnitude of the maximum torque that such a force could produce?

**2.** You are installing a new spark plug in your car, and the manual specifies that it be tightened to a torque that has a magnitude of 45 N·m. Using the data in the drawing, determine the magnitude $F$ of the force that you must exert on the wrench.

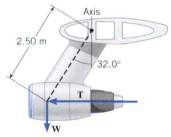

*Problem 2*

**3. Interactive Solution 9.3** at **www.wiley.com/college/cutnell** presents a model for solving this problem. The wheel of a car has a radius of 0.350 m. The engine of the car applies a torque of 295 N·m to this wheel, which does not slip against the road surface. Since the wheel does not slip, the road must be applying a force of static friction to the wheel that produces a countertorque. Moreover, the car has a constant velocity, so this countertorque balances the applied torque. What is the magnitude of the static frictional force?

**4.** In San Francisco a very simple technique is used to turn around a cable car when it reaches the end of its route. The car rolls onto a turntable, which can rotate about a vertical axis through its center. Then, two people push perpendicularly on the car, one at each end, as in the drawing.

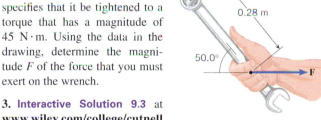

The turntable is rotated one-half of a revolution to turn the car around. If the length of the car is 9.20 m and each person pushes with a 185-N force, what is the magnitude of the net torque applied to the car?

**5. ssm** The steering wheel of a car has a radius of 0.19 m, while the steering wheel of a truck has a radius of 0.25 m. The same force is applied in the same direction to each. What is the ratio of the torque produced by this force in the truck to the torque produced in the car?

**6.** The drawing shows a jet engine suspended beneath the wing of an airplane. The weight **W** of the engine is 10 200 N and acts as shown in the drawing. In flight the engine produces a thrust **T** of 62 300 N that is parallel to the ground. The rotational axis in the drawing is perpendicular to the plane of the paper. With respect to this axis, find the magnitude of the torque due to (a) the weight and (b) the thrust.

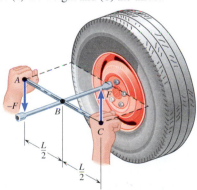

***7.** A pair of forces with equal magnitudes, opposite directions, and different lines of action is called a "couple." When a couple acts on a rigid object, the couple produces a torque that does *not* depend on the location of the axis. The drawing shows a couple acting on a tire wrench, each force being perpendicular to the wrench. Determine an expression for the torque produced by the couple when the axis is perpendicular to the tire and passes through (a) point $A$, (b) point $B$, and (c) point $C$. Express your answers in terms of the magnitude $F$ of the force and the length $L$ of the wrench.

* **8. Interactive LearningWare 9.1** at **www.wiley.com/college/cutnell** reviews the concepts that are important in this problem. One end of a meter stick is pinned to a table, so the stick can rotate freely in a plane parallel to the tabletop (see the drawing). Two forces, both parallel to the tabletop, are applied to the stick in such a way that the net torque is zero. One force has a magnitude of 4.00 N and is applied perpendicular to the stick at the free end. The other force has a magnitude of 6.00 N and acts at a 60.0° angle with respect to the stick. Where along the stick is the 6.00-N force applied? Express this distance with respect to the axis of rotation.

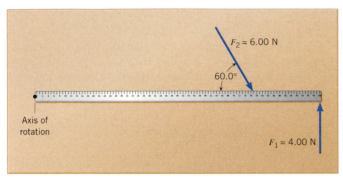

Table (top view)

* **9. ssm** One end of a meter stick is pinned to a table, so the stick can rotate freely in a plane parallel to the tabletop. Two forces, both parallel to the tabletop, are applied to the stick in such a way that the net torque is zero. One force has a magnitude of 2.00 N and is applied perpendicular to the length of the stick at the free end. The other force has a magnitude of 6.00 N and acts at a 30.0° angle with respect to the length of the stick. Where along the stick is the 6.00-N force applied? Express this distance with respect to the end that is pinned.

** **10.** A rotational axis is directed perpendicular to the plane of a square and is located as shown in the drawing. Two forces, $F_1$ and $F_2$, are applied to diagonally opposite corners, and act along the sides of the square, first as shown in part *a* and then as shown in part *b* of the drawing. In each case the net torque produced by the forces is zero. The square is one meter on a side, and the magnitude of $F_2$ is three times that of $F_1$. Find the distances *a* and *b* that locate the axis.

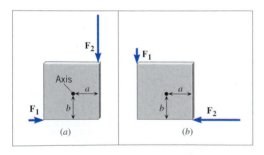

(a)      (b)

### Section 9.2 Rigid Objects in Equilibrium, Section 9.3 Center of Gravity

**11.** The drawing shows a person whose weight is $W = 584$ N doing push-ups. Find the normal force exerted by the floor on *each* hand and *each* foot, assuming that the person holds this position.

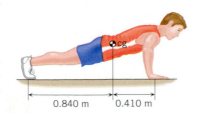

0.840 m     0.410 m

**12.** Review Conceptual Example 7 before starting this problem. A uniform plank of length 5.0 m and weight 225 N rests horizontally on two supports, with 1.1 m of the plank hanging over the right support (see the drawing). To what distance *x* can a person who weighs 450 N walk on the overhanging part of the plank before it just begins to tip?

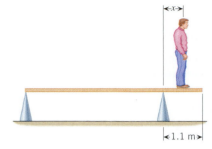

**13. ssm** Review Example 4 before attempting this problem. What is the minimum value for the coefficient of static friction between the ladder and the ground, so that the ladder does not slip?

**14. Concept Simulation 9.1** at **www.wiley.com/college/cutnell** illustrates how the forces can vary in problems of this type. A hiker, who weighs 985 N, is strolling through the woods and crosses a small horizontal bridge. The bridge is uniform, weighs 3610 N, and rests on two concrete supports, one at each end. He stops one-fifth of the way along the bridge. What is the magnitude of the force that a concrete support exerts on the bridge (a) at the near end and (b) at the far end?

**15. ssm** A uniform door (0.81 m wide and 2.1 m high) weighs 140 N and is hung on two hinges that fasten the long left side of the door to a vertical wall. The hinges are 2.1 m apart. Assume that the lower hinge bears all the weight of the door. Find the magnitude and direction of the horizontal component of the force applied *to the door* by (a) the upper hinge and (b) the lower hinge. Determine the magnitude and direction of the force applied *by the door* to (c) the upper hinge and (d) the lower hinge.

**16.** A lunch tray is being held in one hand, as the drawing illustrates. The mass of the tray itself is 0.200 kg, and its center of gravity is located at its geometrical center. On the tray is a 1.00-kg plate of food and a 0.250-kg cup of coffee. Obtain the force **T** exerted by the thumb and the force **F** exerted by the four fingers. Both forces act perpendicular to the tray, which is being held parallel to the ground.

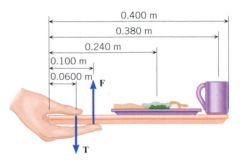

**17. Interactive Solution 9.17** at **www.wiley.com/college/cutnell** illustrates how to model this type of problem. A person exerts a horizontal force of 190 N in the test apparatus shown in the drawing. Find the horizontal force **M** (magnitude and direction) that his flexor muscle exerts on his forearm.

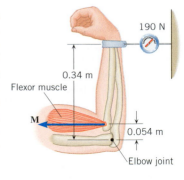

**18.** ⚕ In an isometric exercise a person places a hand on a scale and pushes vertically downward, keeping the forearm horizontal. This is possible because the triceps muscle applies an upward force **M** perpendicular to the arm, as the drawing indicates. The forearm weighs 22.0 N and has a center of gravity as indicated. The scale registers 111 N. Determine the magnitude of **M**.

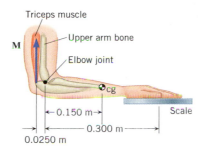

Triceps muscle
Upper arm bone
Elbow joint
cg
Scale
0.150 m
0.300 m
0.0250 m

**19.** Review **Concept Stimulation 9.2** at **www.wiley.com/college/cutnell** and Conceptual Example 7 as background material for this problem. A jet transport has a weight of $1.00 \times 10^6$ N and is at rest on the runway. The two rear wheels are 15.0 m behind the front wheel, and the plane's center of gravity is 12.6 m behind the front wheel. Determine the normal force exerted by the ground on (a) the front wheel and on (b) *each* of the two rear wheels.

**20.** The drawing shows a bicycle wheel resting against a small step whose height is $h = 0.120$ m. The weight and radius of the wheel are $W = 25.0$ N and $r = 0.340$ m. A horizontal force **F** is applied to the axle of the wheel. As the magnitude of **F** increases, there comes a time when the wheel just begins to rise up and loses contact with the ground. What is the magnitude of the force when this happens?

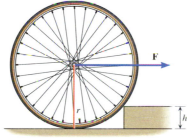

**21.** **ssm** A uniform board is leaning against a smooth vertical wall. The board is at an angle $\theta$ above the horizontal ground. The coefficient of static friction between the ground and the lower end of the board is 0.650. Find the smallest value for the angle $\theta$, such that the lower end of the board does not slide along the ground.

**22.** ⚕ A person is sitting with one leg outstretched so that it makes an angle of 30.0° with the horizontal, as the drawing indicates. The weight of the leg below the knee is 44.5 N with the center of gravity located below the knee joint. The leg is being held in this position because of the force **M** applied by the quadriceps muscle, which is attached 0.100 m below the knee joint (see the drawing). Obtain the magnitude of **M**.

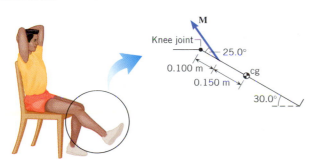

M
Knee joint
25.0°
0.100 m
0.150 m
cg
30.0°

**23.** **ssm** ⚕ A man holds a 178-N ball in his hand, with the forearm horizontal (see the drawing). He can support the

ball in this position because of the flexor muscle force **M**, which is applied perpendicular to the forearm. The forearm weighs 22.0 N and has a center of gravity as indicated. Find (a) the magnitude of **M** and (b) the magnitude and direction of the force applied by the upper arm bone to the forearm at the elbow joint.

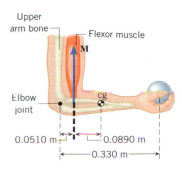

Upper arm bone
Flexor muscle
M
Elbow joint
cg
0.0510 m
0.0890 m
0.330 m

**24.** A woman who weighs $5.00 \times 10^2$ N is leaning against a smooth vertical wall, as the drawing shows. Find (a) the force $F_N$ (directed perpendicular to the wall) exerted on her shoulders by the wall and the (b) horizontal and (c) vertical components of the force exerted on her shoes by the ground.

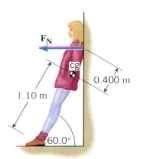

$F_N$
cg
0.400 m
1.10 m
60.0°

**25.** **ssm www** An inverted "V" is made of uniform boards and weighs 356 N. Each side has the same length and makes a 30.0° angle with the vertical, as the drawing shows. Find the magnitude of the static frictional force that acts on the lower end of each leg of the "V."

30.0° 30.0°

**26.** The drawing shows an Λ-shaped ladder. Both sides of the ladder are equal in length. This ladder is standing on a frictionless horizontal surface, and only the crossbar (which has a negligible mass) of the "A" keeps the ladder from collapsing. The ladder is uniform and has a mass of 20.0 kg. Determine the tension in the crossbar of the ladder.

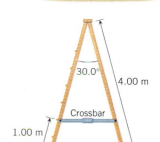

30.0°
4.00 m
Crossbar
1.00 m

**27.** Two vertical walls are separated by a distance of 1.5 m, as the drawing shows. Wall 1 is smooth, while wall 2 is not smooth. A uniform board is propped between them. The coefficient of static friction between the board and wall 2 is 0.98. What is the length of the longest board that can be propped between the walls?

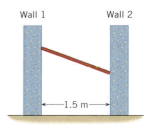

Wall 1
Wall 2
1.5 m

**Section 9.4 Newton's Second Law for Rotational Motion About a Fixed Axis**

**28.** A solid circular disk has a mass of 1.2 kg and a radius of 0.16 m. Each of three identical thin rods has a mass of 0.15 kg. The rods are attached perpendicularly to the plane of the disk at its outer edge

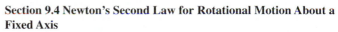

to form a three-legged stool (see the drawing). Find the moment of inertia of the stool with respect to an axis that is perpendicular to the plane of the disk at its center. *(Hint: When considering the moment of inertia of each rod, note that all of the mass of each rod is located at the same perpendicular distance from the axis.)*

**29.** **ssm** A clay vase on a potter's wheel experiences an angular acceleration of 8.00 rad/s² due to the application of a 10.0-N·m net torque. Find the total moment of inertia of the vase and potter's wheel.

**30.** A ceiling fan is turned on and a net torque of 1.8 N·m is applied to the blades. The blades have a total moment of inertia of 0.22 kg·m². What is the angular acceleration of the blades?

**31.** A uniform solid disk with a mass of 24.3 kg and a radius of 0.314 m is free to rotate about a frictionless axle. Forces of 90.0 and 125 N are applied to the disk, as the drawing illustrates. What is (a) the net torque produced by the two forces and (b) the angular acceleration of the disk?

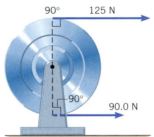

**32.** A rotating door is made from four rectangular glass panes, as shown in the drawing. The mass of each pane is 85 kg. A person pushes on the outer edge of one pane with a force of F = 68 N that is directed perpendicular to the pane. Determine the magnitude of the door's angular acceleration.

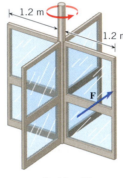

*Problem 32*

**33.** **ssm** A bicycle wheel has a radius of 0.330 m and a rim whose mass is 1.20 kg. The wheel has 50 spokes, each with a mass of 0.010 kg. (a) Calculate the moment of inertia of the rim about the axle. (b) Determine the moment of inertia of *any one spoke*, assuming it to be a long, thin rod that can rotate about one end. (c) Find the *total* moment of inertia of the wheel, including the rim and all 50 spokes.

**34.** A CD has a mass of 17 g and a radius of 6.0 cm. When inserted into a player, the CD starts from rest and accelerates to an angular velocity of 21 rad/s in 0.80 s. Assuming the CD is a uniform solid disk, determine the net torque acting on it.

**35.** **Interactive Solution 9.35** at **www.wiley.com/college/cutnell** presents a method for modeling this problem. A cylinder is rotating about an axis that passes through the center of each circular end piece. The cylinder has a radius of 0.0830 m, an angular speed of 76.0 rad/s, and a moment of inertia of 0.615 kg·m². A brake shoe presses against the surface of the cylinder and applies a tangential frictional force to it. The frictional force reduces the angular speed of the cylinder by a factor of two during a time of 6.40 s. (a) Find the magnitude of the angular deceleration of the cylinder. (b) Find the magnitude of the force of friction applied by the brake shoe.

*  **36.** Two thin rectangular sheets (0.20 m × 0.40 m) are identical. In the first sheet the axis of rotation lies along the 0.20-m side, and in the second it lies along the 0.40-m side. The same torque is applied to each sheet. The first sheet, starting from rest, reaches its final angular velocity in 8.0 s. How long does it take for the second sheet, starting from rest, to reach the same angular velocity?

* **37.** **ssm** A stationary bicycle is raised off the ground, and its front wheel (m = 1.3 kg) is rotating at an angular velocity of 13.1 rad/s (see the drawing). The front brake is then applied for 3.0 s, and the wheel slows down to 3.7 rad/s. Assume that all the mass of the wheel is concentrated in the rim, the radius of which is 0.33 m. The coefficient of kinetic friction between each brake pad and the rim is $\mu_k = 0.85$. What is the magnitude of the normal force that *each* brake pad applies to the rim?

* **38.** The drawing shows a model for the motion of the human forearm in throwing a dart. Because of the force **M** applied by the triceps muscle, the forearm can rotate about an axis at the elbow joint. Assume that the forearm has the dimensions shown in the drawing and a moment of inertia of 0.065 kg·m² (including the effect of the dart) relative to the axis at the elbow. Assume also that the force **M** acts perpendicular to the forearm. Ignoring the effect of gravity and any frictional forces, determine the magnitude of the force **M** needed to give the dart a tangential speed of 5.0 m/s in 0.10 s, starting from rest.

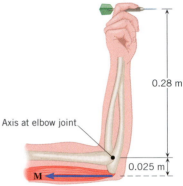

* **39.** **ssm** A thin, rigid, uniform rod has a mass of 2.00 kg and a length of 2.00 m. (a) Find the moment of inertia of the rod relative to an axis that is perpendicular to the rod at one end. (b) Suppose all the mass of the rod were located at a single point. Determine the perpendicular distance of this point from the axis in part (a), such that this point particle has the same moment of inertia as the rod. This distance is called the **radius of gyration** of the rod.

* **40.** The **parallel axis theorem** provides a useful way to calculate the moment of inertia I about an arbitrary axis. The theorem states that $I = I_{cm} + Mh^2$, where $I_{cm}$ is the moment of inertia of the object relative to an axis that passes through the center of mass and is parallel to the axis of interest, M is the total mass of the object, and h is the perpendicular distance between the two axes. Use this theorem and information to determine an expression for the moment of inertia of a solid cylinder of radius R relative to an axis that lies on the surface of the cylinder and is perpendicular to the circular ends.

* **41.** The drawing shows the top view of two doors. The doors are uniform and identical. Door A rotates about an axis through its left edge, and door B rotates about an axis through the center. The same force **F** is applied perpendicular to each door at its right edge, and the force remains perpendicular as the door turns. Starting from rest, door A rotates through a certain angle in 3.00 s. How long does it take door B to rotate through the same angle?

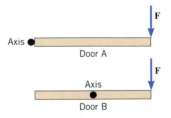

** **42.** See **Concept Simulation 9.4** at www.wiley.com/college/cutnell to review the principles involved in this problem. By means of a rope whose mass is negligible, two blocks are suspended over a pulley, as the drawing shows. The pulley can be treated as a uniform solid cylindrical disk. The downward acceleration of the 44.0-kg block is observed to be exactly one-half the acceleration due to gravity. Noting that the tension in the rope is not the same on each side of the pulley, find the mass of the pulley.

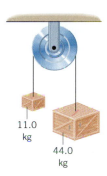

11.0 kg

44.0 kg

## Section 9.5 Rotational Work and Energy

**43. ssm** Three objects lie in the $x$, $y$ plane. Each rotates about the $z$ axis with an angular speed of 6.00 rad/s. The mass $m$ of each object and its perpendicular distance $r$ from the $z$ axis are: (1) $m_1 = 6.00$ kg and $r_1 = 2.00$ m, (2) $m_2 = 4.00$ kg and $r_2 = 1.50$ m, (3) $m_3 = 3.00$ kg and $r_3 = 3.00$ m. (a) Find the tangential speed of each object. (b) Determine the total kinetic energy of this system using the expression $\text{KE} = \frac{1}{2}m_1v_1^2 + \frac{1}{2}m_2v_2^2 + \frac{1}{2}m_3v_3^2$. (c) Obtain the moment of inertia of the system. (d) Find the rotational kinetic energy of the system using the relation $\frac{1}{2}I\omega^2$ to verify that the answer is the same as that in (b).

**44.** Calculate the kinetic energy that the earth has because of (a) its rotation about its own axis and (b) its motion around the sun. Assume that the earth is a uniform sphere and that its path around the sun is circular. For comparison, the total energy used in the United States in one year is about $9.3 \times 10^{19}$ J.

**45. ssm** A flywheel is a solid disk that rotates about an axis that is perpendicular to the disk at its center. Rotating flywheels provide a means for storing energy in the form of rotational kinetic energy and are being considered as a possible alternative to batteries in electric cars. The gasoline burned in a 300-mile trip in a typical midsize car produces about $1.2 \times 10^9$ J of energy. How fast would a 13-kg flywheel with a radius of 0.30 m have to rotate to store this much energy? Give your answer in rev/min.

**46.** A helicopter has two blades (see Figure 8.11), each of which has a mass of 240 kg and can be approximated as a thin rod of length 6.7 m. The blades are rotating at an angular speed of 44 rad/s. (a) What is the total moment of inertia of the two blades about the axis of rotation? (b) Determine the rotational kinetic energy of the spinning blades.

* **47. Interactive Solution 9.47** at www.wiley.com/college/cutnell offers a model for solving problems of this type. A solid sphere is rolling on a surface. What fraction of its total kinetic energy is in the form of rotational kinetic energy about the center of mass?

* **48.** A thin uniform rod is initially positioned in the vertical direction, with its lower end attached to a frictionless axis that is mounted on the floor. The rod has a length of 2.00 m and is allowed to fall, starting from rest. Find the tangential speed of the free end of the rod, just before the rod hits the floor after rotating through 90°.

* **49. ssm** Review Example 12 before attempting this problem. A solid cylinder and a thin-walled hollow cylinder (see Table 9.1) have the same mass and radius. They are rolling horizontally toward the bottom of an incline. The center of mass of each has the same translational speed. The cylinders roll up the incline and reach their highest points. Calculate the ratio of the distances ($s_{\text{solid}}/s_{\text{hollow}}$) along the incline through which each center of mass moves.

* **50.** Review Example 12 before attempting this problem. A bowling ball encounters a 0.760-m vertical rise on the way back to the ball rack, as the drawing illustrates. Ignore frictional losses and assume that the mass of the ball is distributed uniformly. The translational speed of the ball is 3.50 m/s at the bottom of the rise. Find the translational speed at the top.

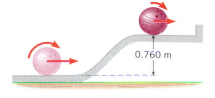

0.760 m

** **51.** A tennis ball, starting from rest, rolls down the hill in the drawing. At the end of the hill the ball becomes airborne, leaving at an angle of 35° with respect to the ground. Treat the ball as a thin-walled spherical shell, and determine the range $x$.

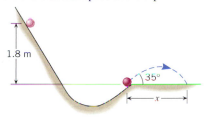

1.8 m

35°

$x$

## Section 9.6 Angular Momentum

**52.** Before starting this problem, review Conceptual Example 13. A woman stands at the center of a platform. The woman and the platform rotate with an angular speed of 5.00 rad/s. Friction is negligible. Her arms are outstretched, and she is holding a dumbbell in each hand. In this position the total moment of inertia of the rotating system (platform, woman, and dumbbells) is 5.40 kg·m². By pulling in her arms, she reduces the moment of inertia to 3.80 kg·m². Find her new angular speed.

**53. ssm www** Two disks are rotating about the same axis. Disk A has a moment of inertia of 3.4 kg·m² and an angular velocity of +7.2 rad/s. Disk B is rotating with an angular velocity of −9.8 rad/s. The two disks are then linked together without the aid of any external torques, so that they rotate as a single unit with an angular velocity of −2.4 rad/s. The axis of rotation for this unit is the same as that for the separate disks. What is the moment of inertia of disk B?

**54.** A solid disk rotates at an angular velocity of 0.067 rad/s with respect to an axis perpendicular to the disk at its center. The moment of inertia of the disk is 0.10 kg·m². From above, sand is dropped straight down onto this rotating disk, so that a thin uniform ring of sand is formed at a distance of 0.40 m from the axis. The sand in the ring has a mass of 0.50 kg. After all the sand is in place, what is the angular velocity of the disk?

**55. Interactive Solution 9.55** at www.wiley.com/college/cutnell illustrates one way of solving a problem similar to this one. A thin rod has a length of 0.25 m and rotates in a circle on a frictionless tabletop. The axis is perpendicular to the length of the rod at one of its ends. The rod has an angular velocity of 0.32 rad/s and a moment of inertia of $1.1 \times 10^{-3}$ kg·m². A bug standing on the axis decides to crawl out to the other end of the rod. When the bug (mass = $4.2 \times 10^{-3}$ kg) gets where it's going, what is the angular velocity of the rod?

**56.** A flat uniform circular disk (radius = 2.00 m, mass = $1.00 \times 10^2$ kg) is initially stationary. The disk is free to rotate in the horizontal plane about a frictionless axis perpendicular to the center of the disk. A 40.0-kg person, standing 1.25 m from the axis, begins to run on the disk in a circular path and has a tangential speed of 2.00 m/s relative to the ground. Find the resulting angular speed of the disk (in rad/s) and describe the direction of the rotation.

* **57. ssm** A cylindrically shaped space station is rotating about the axis of the cylinder to create artificial gravity. The radius of the cylinder is 82.5 m. The moment of inertia of the station without people is

$3.00 \times 10^9$ kg·m². Suppose 500 people, with an average mass of 70.0 kg each, live on this station. As they move radially from the outer surface of the cylinder toward the axis, the angular speed of the station changes. What is the maximum possible percentage change in the station's angular speed due to the radial movement of the people?

* **58.** A thin uniform rod is rotating at an angular velocity of 7.0 rad/s about an axis that is perpendicular to the rod at its center. As the drawing indicates, the rod is hinged at two places, one-quarter of the length

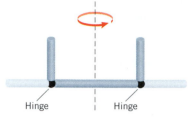

Hinge    Hinge

from each end. Without the aid of external torques, the rod suddenly assumes a "u" shape, with the arms of the "u" parallel to the rotation axis. What is the angular velocity of the rotating "u"?

** **59.** A platform is rotating at an angular speed of 2.2 rad/s. A block is resting on this platform at a distance of 0.30 m from the axis. The coefficient of static friction between the block and the platform is 0.75. Without any external torque acting on the system, the block is moved toward the axis. Ignore the moment of inertia of the platform and determine the smallest distance from the axis at which the block can be relocated and still remain in place as the platform rotates.

** **60.** A small 0.500-kg object moves on a frictionless horizontal table in a circular path of radius 1.00 m. The angular speed is 6.28 rad/s. The object is attached to a string of negligible mass that passes through a small hole in the table at the center of the circle. Someone under the table begins to pull the string downward to make the circle smaller. If the string will tolerate a tension of no more than 105 N, what is the radius of the smallest possible circle on which the object can move?

# Chapter 10 Simple Harmonic Motion and Elasticity

## 10.1 The Ideal Spring and Simple Harmonic Motion

Springs are familiar objects that have many applications, ranging from push-button switches on electronic components, to automobile suspension systems, to mattresses. They are useful because they can be stretched or compressed. For example, the top drawing in Figure 10.1 shows a spring being stretched. Here a hand applies a pulling force $F_{\text{Applied}}$ to the spring. In response, the spring stretches and undergoes a displacement of $x$ from its original, or "unstrained," length. The bottom drawing in Figure 10.1 illustrates the spring being compressed. Now the hand applies a pushing force to the spring, and it again undergoes a displacement from its unstrained length.

Experiment reveals that for relatively small displacements, the force $F_{\text{Applied}}$ required to stretch or compress a spring is directly proportional to the displacement $x$, or $F_{\text{Applied}} \propto x$. As is customary, this proportionality may be converted into an equation by introducing a proportionality constant $k$:

$$F_{\text{Applied}} = kx \qquad (10.1)$$

The constant $k$ is called the **spring constant,** and Equation 10.1 shows that it has the dimensions of force per unit length (N/m). A spring that behaves according to $F_{\text{Applied}} = kx$ is said to be an **ideal spring.** Example 1 illustrates one application of such a spring.

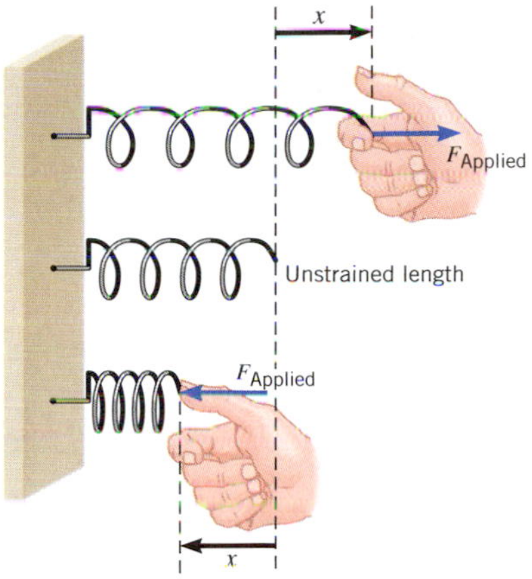

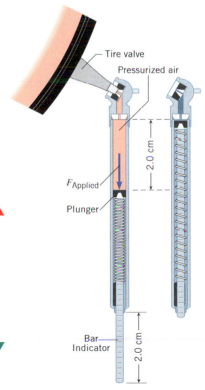

**Figure 10.1** An ideal spring obeys the equation $F_{\text{Applied}} = kx$, where $F_{\text{Applied}}$ is the force applied to the spring, $x$ is the displacement of the spring from its unstrained length, and $k$ is the spring constant.

### Example 1 A Tire Pressure Gauge

In a tire pressure gauge, the air in the tire pushes against a plunger attached to a spring when the gauge is pressed against the tire valve, as in Figure 10.2. Suppose the spring constant of the spring is $k = 320$ N/m and the bar indicator of the gauge extends 2.0 cm when the gauge is pressed against the tire valve. What force does the air in the tire apply to the spring?

**Reasoning** We assume that the spring is an ideal spring, so that the relation $F_{\text{Applied}} = kx$ is obeyed. The spring constant $k$ is known, as is the displacement $x$. Therefore, we can determine the force applied to the spring.

**Solution** The force needed to compress the spring is given by Equation 10.1 as

$$F_{\text{Applied}} = kx = (320 \text{ N/m})(0.020 \text{ m}) = \boxed{6.4 \text{ N}}$$

Thus, the exposed length of the bar indicator indicates the force that the air pressure in the tire exerts on the spring. We will see later that pressure is force per unit area, so force is pressure times area. Since the area of the plunger surface is fixed, the bar indicator can be marked in units of pressure.

**The physics of a tire pressure gauge.**

**Figure 10.2** In a tire pressure gauge, the pressurized air from the tire exerts a force $F_{\text{Applied}}$ that compresses a spring.

Sometimes the spring constant $k$ is referred to as the **stiffness** of the spring, because a large value for $k$ means the spring is "stiff," in the sense that a large force is required to stretch or compress it. Conceptual Example 2 examines what happens to the stiffness of a spring when the spring is cut into two shorter pieces.

### Conceptual Example 2 Are Shorter Springs Stiffer Springs?

Figure 10.3$a$ shows a 10-coil spring that has a spring constant $k$. If this spring is cut in half, so there are two 5-coil springs, what is the spring constant of each of the smaller springs?

**Reasoning and Solution** It is tempting to say that each 5-coil spring has a spring constant of $\frac{1}{2}k$, or one-half the value for the 10-coil spring. However, the spring constant of each 5-coil spring is, in fact, $2k$. Here's the logic. When a force $F_{\text{Applied}}$ is applied to the

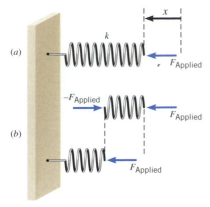

**Figure 10.3** (*a*) The 10-coil spring has a spring constant $k$. The applied force is $F_{\text{Applied}}$, and the displacement of the spring from its unstrained length is $x$. (*b*) The spring in part *a* is divided in half, so that the forces acting on the two 5-coil springs can be analyzed.

spring, the displacement of the spring from its unstrained length is $x$, as Figure 10.3*a* illustrates. Now consider part *b* of the drawing, which shows the spring divided in half between the fifth and sixth coil (counting from the right). The spring is in equilibrium, so that the net force acting on the right half (coils 1–5) must be zero. Thus, as part *b* shows, a force of $-F_{\text{Applied}}$ must act on coil 5 in order to balance the force $F_{\text{Applied}}$ that acts on coil 1. It is the adjacent coil 6 that exerts the force $-F_{\text{Applied}}$ on coil 5, and Newton's action–reaction law now comes into play. It tells us that coil 5, in response, exerts an oppositely directed force of equal magnitude on coil 6. In other words, the force $F_{\text{Applied}}$ is also exerted on the left half of the spring, as part *b* also indicates. As a result, the left half compresses by an amount that is one-half the displacement $x$ experienced by the 10-coil spring. In summary, we have a 10-coil spring for which the force $F_{\text{Applied}}$ causes a displacement $x$ and a 5-coil spring for which the same force causes a displacement $\frac{1}{2}x$. We conclude, then, that the 5-coil spring must be twice as stiff as the 10-coil spring. When the 10-coil spring is cut in half, the spring constant of each resulting 5-coil spring is $2k$. In general, the spring constant is inversely proportional to the number of coils in the spring, so ***shorter springs are stiffer springs,*** all other things being equal.

**Related Homework:** *Problems 8, 16*

To stretch or compress a spring, a force must be applied to it. In accord with Newton's third law, the spring exerts an oppositely directed force of equal magnitude. This reaction force is applied by the spring to the agent that does the pulling or pushing. In other words, the reaction force is applied to the object attached to the spring. The reaction force is also called a "restoring force," for a reason that will be clarified shortly. The restoring force of an ideal spring is obtained from the relation $F_{\text{Applied}} = kx$ by including the minus sign required by Newton's action–reaction law, as indicated in Equation 10.2.

■ **HOOKE'S LAW\* RESTORING FORCE OF AN IDEAL SPRING**

The restoring force of an ideal spring is

$$F = -kx \tag{10.2}$$

where $k$ is the spring constant and $x$ is the displacement of the spring from its unstrained length. The minus sign indicates that the restoring force always points in a direction opposite to the displacement of the spring.

Figure 10.4 helps to explain why the phrase "restoring force" is used. In the picture, an object of mass $m$ is attached to a spring on a frictionless table. In part $A$, the spring has been stretched to the right, so it exerts the leftward-pointing force $F$. When

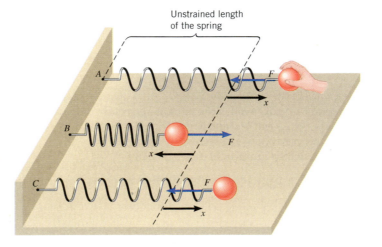

**Figure 10.4** The restoring force (see blue arrows) produced by an ideal spring always points opposite to the displacement (see black arrows) of the spring and leads to a back-and-forth motion of the object.

\* As we will see in Section 10.8, Equation 10.2 is similar to a relationship first discovered by Robert Hooke (1635–1703).

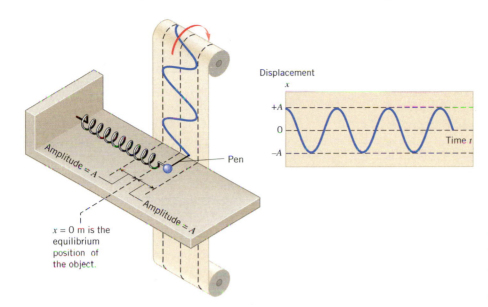

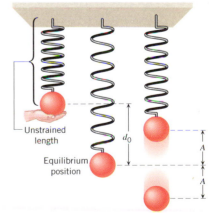

**Figure 10.5** When an object moves in simple harmonic motion, a graph of its position as a function of time has a sinusoidal shape with an amplitude *A*. A pen attached to the object records the graph.

the object is released, this force pulls it to the left, restoring it toward its equilibrium position. However, consistent with Newton's first law, the moving object has inertia and coasts beyond the equilibrium position, compressing the spring as in part *B*. The force exerted by the spring now points to the right and, after bringing the object to a momentary halt, acts to restore the object to its equilibrium position. But the object's inertia again carries it beyond the equilibrium position, this time stretching the spring and leading to the restoring force *F* shown in part *C*. The back-and-forth motion illustrated in the drawing then repeats itself, continuing forever, since no friction acts on the object or the spring.

*When the restoring force has the mathematical form given by F = −kx, the type of friction-free motion illustrated in Figure 10.4 is designated as "simple harmonic motion."* By attaching a pen to the object and moving a strip of paper past it at a steady rate, we can record the position of the vibrating object as time passes. Figure 10.5 illustrates the resulting graphical record of simple harmonic motion. The maximum excursion from equilibrium is the **amplitude** *A* of the motion. The shape of this graph is characteristic of simple harmonic motion and is called "sinusoidal," because it has the shape of a trigonometric sine or cosine function.

The restoring force also leads to simple harmonic motion when the object is attached to a vertical spring, just as it does when the spring is horizontal. When the spring is vertical, however, the weight of the object causes the spring to stretch, and the motion occurs with respect to the equilibrium position of the object on the stretched spring, as Figure 10.6 indicates. The amount of initial stretching $d_0$ due to the weight can be calculated by equating the weight to the magnitude of the restoring force that supports it; thus, $mg = kd_0$, which gives $d_0 = mg/k$.

# 10.2 *Simple Harmonic Motion and the Reference Circle*

Simple harmonic motion, like any motion, can be described in terms of displacement, velocity, and acceleration, and the model in Figure 10.7 is helpful in explaining these characteristics. The model consists of a small ball attached to the top of a rotating turntable. The ball is moving in uniform circular motion (see Section 5.1) on a path known as the *reference circle.* As the ball moves, its shadow falls on a strip of film, which is moving upward at a steady rate and records where the shadow is. A comparison of the film with the paper in Figure 10.5 reveals the same kind of patterns, suggesting that the shadow of the ball is a good model for simple harmonic motion.

**Figure 10.6** The weight of an object on a vertical spring stretches the spring by an amount $d_0$. Simple harmonic motion of amplitude *A* occurs with respect to the equilibrium position of the object on the stretched spring.

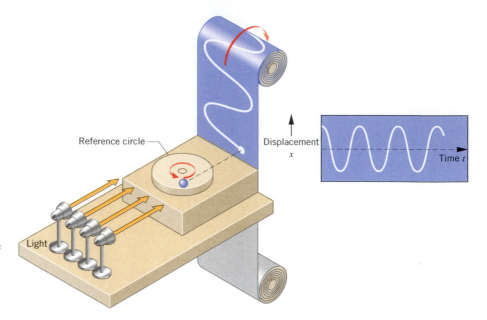

**Figure 10.7** The ball mounted on the turntable moves in uniform circular motion, and its shadow, projected on a moving strip of film, executes simple harmonic motion.

## DISPLACEMENT

Figure 10.8 takes a closer look at the reference circle (radius = $A$) and indicates how to determine the displacement of the shadow on the film. The ball starts on the $x$ axis at $x = +A$ and moves through the angle $\theta$ in a time $t$. Since the circular motion is uniform, the ball moves with a constant angular speed $\omega$ (in rad/s). Therefore, the angle has a value (in rad) of $\theta = \omega t$. The displacement $x$ of the shadow is just the projection of the radius $A$ onto the $x$ axis:

$$x = A \cos \theta = A \cos \omega t \qquad (10.3)$$

Figure 10.9 shows a graph of this equation. As time passes, the shadow of the ball oscillates between the values of $x = +A$ and $x = -A$, corresponding to the limiting values of $+1$ and $-1$ for the cosine of an angle. The radius $A$ of the reference circle, then, is the amplitude of the simple harmonic motion.

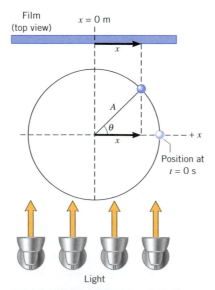

**Figure 10.8** A top view of a ball on a turntable. The ball's shadow on the film has a displacement $x$ that depends on the angle $\theta$ through which the ball has moved on the reference circle.

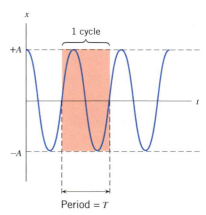

**Figure 10.9** For simple harmonic motion, the graph of displacement $x$ versus time $t$ is a sinusoidal curve. The period $T$ is the time required for one complete motional cycle.

As the ball moves one revolution or cycle around the reference circle, its shadow executes one cycle of back-and-forth motion. For any object in simple harmonic motion, the time required to complete one cycle is the *period T*, as Figure 10.9 indicates. The value of $T$ depends on the angular speed $\omega$ of the ball because the greater the angular speed, the shorter the time it takes to complete one revolution. We can obtain the relationship between $\omega$ and $T$ by recalling that $\omega = \Delta\theta/\Delta t$ (Equation 8.2), where $\Delta\theta$ is the angular displacement of the ball and $\Delta t$ is the time. For one cycle, $\Delta\theta = 2\pi$ rad and $\Delta t = T$, so that

$$\omega = \frac{2\pi}{T} \qquad (\omega \text{ in rad/s}) \qquad (10.4)$$

Often, instead of the period, it is more convenient to speak of the *frequency f* of the motion, the frequency being just the number of cycles of the motion per second. For example, if an object on a spring completes 10 cycles in one second, the frequency is $f = 10$ cycles/s. The period $T$, or the time for one cycle, would be $\frac{1}{10}$ s. Thus, frequency and period are related according to

$$f = \frac{1}{T} \qquad (10.5)$$

Usually one cycle per second is referred to as one hertz (Hz), the unit being named after Heinrich Hertz (1857–1894). One thousand cycles per second is called one kilohertz (kHz). Thus, five thousand cycles per second, for instance, can be written as 5 kHz.

Using the relationships $\omega = 2\pi/T$ and $f = 1/T$, we can relate the angular speed $\omega$ (in rad/s) to the frequency $f$ (in cycles/s or Hz):

$$\omega = \frac{2\pi}{T} = 2\pi f \qquad (\omega \text{ in rad/s}) \qquad (10.6)$$

Because $\omega$ is directly proportional to the frequency $f$, $\omega$ is often called the *angular frequency*.

## VELOCITY

The reference circle model can also be used to determine the velocity of an object in simple harmonic motion. Figure 10.10 shows the tangential velocity $\mathbf{v_T}$ of the ball on the reference circle. The drawing indicates that the velocity $\mathbf{v}$ of the shadow is just the $x$ component of the vector $\mathbf{v_T}$; that is, $v = -v_T \sin\theta$, where $\theta = \omega t$. The minus sign is necessary, since $\mathbf{v}$ points to the left, in the direction of the negative $x$ axis. Since the tangential speed $v_T$ is related to the angular speed $\omega$ by $v_T = r\omega$ (Equation 8.9) and since $r = A$, it follows that $v_T = A\omega$. Therefore, the velocity in simple harmonic motion is given by

$$v = -A\omega \sin\theta = -A\omega \sin\omega t \qquad (\omega \text{ in rad/s}) \qquad (10.7)$$

This velocity is *not* constant, but varies between maximum and minimum values as time passes. When the shadow changes direction at either end of the oscillatory motion, the velocity is momentarily zero. When the shadow passes through the $x = 0$ m position, the velocity has a maximum magnitude of $A\omega$, since the sine of an angle is between $+1$ and $-1$:

$$v_{max} = A\omega \qquad (\omega \text{ in rad/s}) \qquad (10.8)$$

Both the amplitude $A$ and the angular frequency $\omega$ determine the maximum velocity, as Example 3 emphasizes.

### Example 3 The Maximum Speed of a Loudspeaker Diaphragm

The diaphragm of a loudspeaker moves back and forth in simple harmonic motion to create sound, as in Figure 10.11. The frequency of the motion is $f = 1.0$ kHz and the amplitude is $A = 0.20$ mm. (a) What is the maximum speed of the diaphragm? (b) Where in the motion does this maximum speed occur?

**Reasoning** The maximum speed $v_{max}$ of an object vibrating in simple harmonic motion is $v_{max} = A\omega$ ($\omega$ in rad/s), according to Equation 10.8. The angular frequency $\omega$ is related to the frequency $f$ by $\omega = 2\pi f$, according to Equation 10.6.

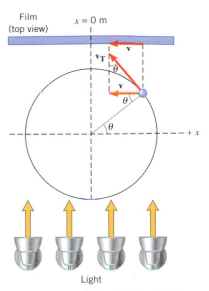

Film (top view)

**Figure 10.10** The velocity $\mathbf{v}$ of the ball's shadow is the $x$ component of the tangential velocity $\mathbf{v_T}$ of the ball on the reference circle.

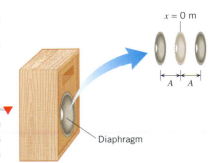

**Figure 10.11** The diaphragm of a loudspeaker generates a sound by moving back and forth in simple harmonic motion.

**Solution**

(a) Using Equations 10.8 and 10.6, we find that the maximum speed of the vibrating diaphragm is

$$v_{\text{max}} = A\omega = A(2\pi f) = (0.20 \times 10^{-3}\ \text{m})(2\pi)(1.0 \times 10^3\ \text{Hz}) = \boxed{1.3\ \text{m/s}}$$

(b) The speed of the diaphragm is zero when the diaphragm momentarily comes to rest at either end of its motion: $x = +A$ and $x = -A$. Its maximum speed occurs midway between these two positions, or at $x = 0$ m.

Simple harmonic motion is not just any kind of vibratory motion. It is a very specific kind and, among other things, must have the velocity given by Equation 10.7. For instance, advertising signs often use a "moving light" display to grab your attention. Conceptual Example 4 examines the back-and-forth motion in one such display, to see whether it is simple harmonic motion.

### Conceptual Example 4    Moving Lights

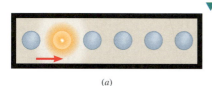

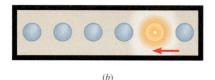

*(a)*

*(b)*

**Figure 10.12** The motion of a lighted bulb is from (*a*) left to right and then from (*b*) right to left.

Over the entrance to a restaurant is mounted a strip of equally spaced light bulbs, as Figure 10.12*a* illustrates. Starting at the left end, each bulb turns on in sequence for one-half second. Thus, a lighted bulb appears to move from left to right. Once the apparent motion of a lighted bulb reaches the right side of the sign, the motion reverses. The lighted bulb then appears to move to the left, as part *b* of the drawing indicates. Thus, the lighted bulb appears to oscillate back and forth. Is the apparent motion simple harmonic motion?

**Reasoning and Solution** Since the bulbs are equally spaced and each bulb remains lit for the same amount of time, the apparent motion of a lighted bulb across the sign occurs at a constant speed. If the motion were simple harmonic motion, however, it would not have a constant speed (see Equation 10.7), because it would have zero speed at each end of the sign and increase to a maximum speed at the center of the sign. Therefore, although the apparent motion of a lighted bulb is oscillatory, *it is not simple harmonic motion.*

## ACCELERATION

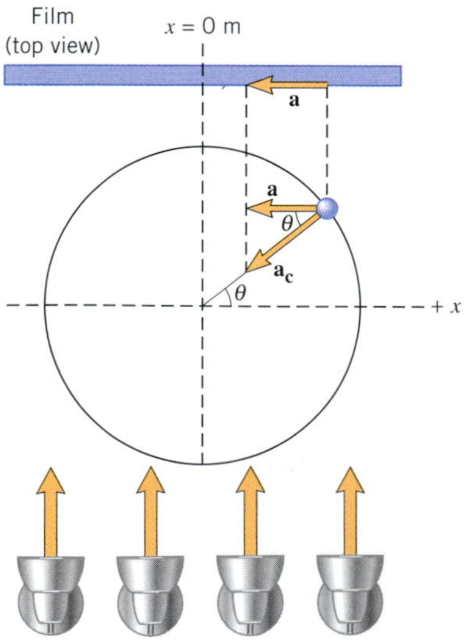

**Figure 10.13** The acceleration **a** of the ball's shadow is the *x* component of the centripetal acceleration **a**$_c$ of the ball on the reference circle.

In simple harmonic motion, the velocity is not constant; consequently, there must be an acceleration. This acceleration can also be determined with the aid of the reference-circle model. As Figure 10.13 shows, the ball on the reference circle moves in uniform circular motion, and, therefore, has a centripetal acceleration $\mathbf{a}_c$ that points toward the center of the circle. The acceleration **a** of the shadow is the *x* component of the centripetal acceleration; $a = -a_c \cos\theta$. The minus sign is needed because the acceleration of the shadow points to the left. Recalling that the centripetal acceleration is related to the angular speed $\omega$ by $a_c = r\omega^2$ (Equation 8.11) and using $r = A$, we find that $a_c = A\omega^2$. With this substitution, the acceleration in simple harmonic motion becomes

$$a = -A\omega^2 \cos\theta = -A\omega^2 \cos\omega t \qquad (\omega \text{ in rad/s}) \qquad (10.9)$$

The acceleration, like the velocity, does *not* have a constant value as time passes. The maximum magnitude of the acceleration is

$$a_{\text{max}} = A\omega^2 \qquad (\omega \text{ in rad/s}) \qquad (10.10)$$

Although both the amplitude $A$ and the angular frequency $\omega$ determine the maximum value, the frequency has a particularly strong effect, because it is squared. Example 5 shows that the acceleration can be remarkably large in a practical situation.

**The physics of a loudspeaker diaphragm.**

### Example 5    The Loudspeaker Revisited—The Maximum Acceleration

The loudspeaker diaphragm in Figure 10.11 is vibrating at a frequency of $f = 1.0$ kHz, and the amplitude of the motion is $A = 0.20$ mm. (a) What is the maximum acceleration of the diaphragm, and (b) where does this maximum acceleration occur?

**Reasoning** The maximum acceleration $a_{max}$ of an object vibrating in simple harmonic motion is $a_{max} = A\omega^2$ ($\omega$ in rad/s), according to Equation 10.10. Equation 10.6 shows that the angular frequency $\omega$ is related to the frequency $f$ by $\omega = 2\pi f$.

**Solution**

(a) Using Equations 10.10 and 10.6, we find that the maximum acceleration of the vibrating diaphragm is

$$a_{max} = A\omega^2 = A(2\pi f)^2 = (0.20 \times 10^{-3}\ \text{m})[2\pi(1.0 \times 10^3\ \text{Hz})]^2$$
$$= \boxed{7.9 \times 10^3\ \text{m/s}^2}$$

**Problem solving insight**
Do not confuse the vibrational frequency $f$ with the angular frequency $\omega$. The value for $f$ is in hertz (cycles per second). The value for $\omega$ is in radians per second. The two are related by $\omega = 2\pi f$.

This is an incredible acceleration, being more than 800 times the acceleration due to gravity, and the diaphragm must be built to withstand it.

(b) The maximum acceleration occurs when the force acting on the diaphragm is a maximum. The maximum force arises when the diaphragm is at the ends of its path, where the displacement is greatest. Thus, the maximum acceleration occurs at $x = +A$ and $x = -A$ in Figure 10.11. ▲

## FREQUENCY OF VIBRATION

With the aid of Newton's second law ($\Sigma F = ma$), it is possible to determine the frequency at which an object of mass $m$ vibrates on a spring. We assume that the mass of the spring itself is negligible and that the only force acting on the object in the horizontal direction is due to the spring—that is, the Hooke's law restoring force. Thus, the net force is $\Sigma F = -kx$, and Newton's second law becomes $-kx = ma$, where $a$ is the acceleration of the object. The displacement and acceleration of an oscillating spring are, respectively, $x = A \cos \omega t$ (Equation 10.3) and $a = -A\omega^2 \cos \omega t$ (Equation 10.9). Substituting these expressions for $x$ and $a$ into the relation $-kx = ma$, we find that

$$-k(A \cos \omega t) = m(-A\omega^2 \cos \omega t)$$

which yields

$$\omega = \sqrt{\frac{k}{m}} \qquad (\omega \text{ in rad/s}) \qquad (10.11)$$

In this expression, the angular frequency $\omega$ must be in radians per second. Larger spring constants $k$ and smaller masses $m$ result in larger frequencies. Example 6 illustrates an application of Equation 10.11.

### Example 6   A Body Mass Measurement Device

Astronauts who spend long periods of time in orbit periodically measure their body masses as part of their health-maintenance programs. On earth, it is simple to measure body weight $W$ with a scale and convert it to mass $m$ using the acceleration due to gravity, since $W = mg$. However, this procedure does not work in orbit, because both the scale and the astronaut are in free-fall and cannot press against each other (see Conceptual Example 12 in Chapter 5). Instead, astronauts use a body mass measurement device, as Figure 10.14 illustrates. This device consists of a spring-mounted chair in which the astronaut sits. The chair is then started oscillating in simple harmonic motion. The period of the motion is measured electronically and is automatically converted into a value of the astronaut's mass, after the mass of the chair is taken into account. The spring used in one such device has a spring constant of 606 N/m, and the mass of the chair is 12.0 kg. The measured oscillation period is 2.41 s. Find the mass of the astronaut.

**The physics of** a body mass measurement device.

**Reasoning** The relation $\omega = \sqrt{k/m}$ (Equation 10.11) can be solved for the mass $m$ in terms of the spring constant $k$ and the angular frequency $\omega$. The spring constant is known. Although the angular frequency is not known, it can be related to the given oscillation period of $T = 2.41$ s by using $\omega = 2\pi/T$ (Equation 10.4). The mass calculated using Equation 10.11 is the total mass of the astronaut and the chair, so that it will be necessary to subtract the mass of the chair to obtain the mass of the astronaut.

**Solution** Using Equations 10.11 and 10.4 gives

$$\omega = \frac{2\pi}{T} = \sqrt{\frac{k}{m}}$$

**Figure 10.14** Astronaut Tamara Jernigan uses a body mass measurement device to measure her mass while in orbit. (Courtesy NASA.)

Solving for the mass $m$, we find that

$$m = \frac{kT^2}{4\pi^2} = \frac{(606 \text{ N/m})(2.41 \text{ s})^2}{4\pi^2} = 89.2 \text{ kg}$$

Accounting for the 12.0-kg mass of the chair reveals that the mass of the astronaut is

$$m_{\text{Astronaut}} = 89.2 \text{ kg} - 12.0 \text{ kg} = \boxed{77.2 \text{ kg}}$$

**The physics of**
**detecting and measuring small**
**amounts of chemicals.**

Example 6 indicates that the mass of the vibrating object influences the frequency of simple harmonic motion. Electronic sensors are being developed that take advantage of this effect in detecting and measuring small amounts of chemicals. These sensors utilize tiny quartz crystals that vibrate when an electric current passes through them. If the crystal is coated with a substance that absorbs a particular chemical, then its mass increases as the chemical is absorbed and, according to the relation $f = \frac{1}{2\pi}\sqrt{k/m}$ (Equations 10.6 and 10.11), the frequency of the simple harmonic motion decreases. The change in frequency is detected electronically, and the sensor is calibrated to give the mass of the absorbed chemical.

## 10.3   Energy and Simple Harmonic Motion

**The physics of**
**a door-closing unit.**

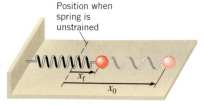

We saw in Chapter 6 that an object above the surface of the earth has gravitational potential energy. Therefore, when the object is allowed to fall, like the hammer of the pile driver in Figure 6.12, it can do work. A spring also has potential energy when the spring is stretched or compressed, which we refer to as *elastic potential energy.* Because of elastic potential energy, a stretched or compressed spring can do work on an object that is attached to the spring. For instance, Figure 10.15 shows a door-closing unit that is often found on screen doors. When the door is opened, a spring inside the unit is compressed and has elastic potential energy. When the door is released, the compressed spring expands and does the work of closing the door.

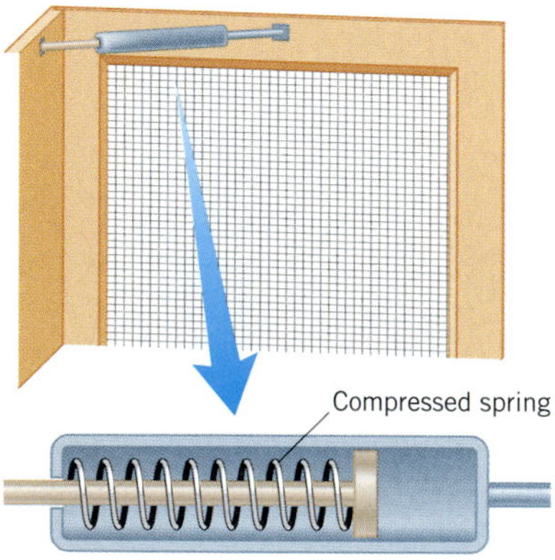

Compressed spring

To find an expression for the elastic potential energy, we will determine the work done by the spring force on an object. Figure 10.16 shows an object attached to one end of a stretched spring. When the object is released, the spring contracts and pulls the object from its initial position $x_0$ to its final position $x_f$. The work $W$ done by a constant force is given by Equation 6.1 as $W = (F \cos \theta)s$, where $F$ is the magnitude of the force, $s$ is the magnitude of the displacement ($s = x_0 - x_f$), and $\theta$ is the angle between the force and the displacement. The magnitude of the spring force is not constant, however. Equation 10.2 gives the spring force as $F = -kx$, and as the spring contracts, the magnitude of this force changes from $kx_0$ to $kx_f$. In using Equation 6.1 to determine the work, we can account for the changing magnitude by using an average magnitude $\overline{F}$ in place of the constant magnitude $F$. Because the dependence of the spring force on $x$ is linear, the magnitude of the average force is just one-half the sum of the initial and final values, or $\overline{F} = \frac{1}{2}(kx_0 + kx_f)$. The work $W_{\text{elastic}}$ done by the average spring force is, then,

**Figure 10.15**  A door-closing unit. The elastic potential energy stored in the compressed spring is used to close the door.

$$W_{\text{elastic}} = (\overline{F} \cos \theta)s = \tfrac{1}{2}(kx_0 + kx_f)\cos 0° (x_0 - x_f)$$

$$W_{\text{elastic}} = \underbrace{\tfrac{1}{2}kx_0^2}_{\substack{\text{Initial elastic} \\ \text{potential energy}}} - \underbrace{\tfrac{1}{2}kx_f^2}_{\substack{\text{Final elastic} \\ \text{potential energy}}} \tag{10.12}$$

Position when
spring is
unstrained

$x_f$
$x_0$

**Figure 10.16**  When the object is released, its displacement changes from an initial value of $x_0$ to a final value of $x_f$.

In the calculation above, $\theta$ is 0°, since the spring force points to the left in Figure 10.16, which is the same direction as the displacement. Equation 10.12 indicates that the work done by the spring force is equal to the difference between the initial and final values of the quantity $\frac{1}{2}kx^2$. The quantity $\frac{1}{2}kx^2$ is analogous to the quantity $mgh$, which we identified in Section 6.3 as the gravitational potential energy. Here, we identify the quantity $\frac{1}{2}kx^2$ as the elastic potential energy. Equation 10.13 indicates that the elastic potential energy is a maximum for a fully stretched or compressed spring and zero for a spring that is neither stretched nor compressed ($x = 0$ m).

■ **DEFINITION OF ELASTIC POTENTIAL ENERGY**

The elastic potential energy $PE_{elastic}$ is the energy that a spring has by virtue of being stretched or compressed. For an ideal spring that has a spring constant $k$ and is stretched or compressed by an amount $x$ relative to its unstrained length, the elastic potential energy is

$$PE_{elastic} = \tfrac{1}{2}kx^2 \qquad (10.13)$$

*SI Unit of Elastic Potential Energy:* joule (J)

The total mechanical energy $E$ is a familiar idea that we originally defined to be the sum of the translational kinetic energy and the gravitational potential energy. Then, we included the rotational kinetic energy.

In Equation 10.14 the elastic potential energy is now included as part of the total mechanical energy:

$$E \quad = \quad \tfrac{1}{2}mv^2 \quad + \quad \tfrac{1}{2}I\omega^2 \quad + \quad mgh \quad + \quad \tfrac{1}{2}kx^2 \qquad (10.14)$$

| Total mechanical energy | Translational kinetic energy | Rotational kinetic energy | Gravitational potential energy | Elastic potential energy |

As Section 6.5 discusses, the total mechanical energy is conserved when external nonconservative forces (such as friction) do no net work; that is, when $W_{nc} = 0$ J. Then, the final and initial values of $E$ are the same: $E_f = E_0$. The principle of conservation of total mechanical energy is the subject of the next example.

## Example 7 An Object on a Horizontal Spring

Figure 10.17 shows an object of mass $m = 0.200$ kg that is vibrating on a horizontal frictionless table. The spring has a spring constant $k = 545$ N/m. It is stretched initially to $x_0 = 4.50$ cm and then released from rest (see part A of the drawing). Determine the final translational speed $v_f$ of the object when the final displacement of the spring is (a) $x_f = 2.25$ cm and (b) $x_f = 0$ cm.

**Reasoning** The conservation of mechanical energy indicates that, in the absence of friction (a nonconservative force), the final and initial total mechanical energies are the same:

$$E_f = E_0$$

$$\tfrac{1}{2}mv_f^2 + \tfrac{1}{2}I\omega_f^2 + mgh_f + \tfrac{1}{2}kx_f^2 = \tfrac{1}{2}mv_0^2 + \tfrac{1}{2}I\omega_0^2 + mgh_0 + \tfrac{1}{2}kx_0^2$$

Since the object is moving on a horizontal table, the final and initial heights are the same: $h_f = h_0$. The object is not rotating, so its angular speed is zero: $\omega_f = \omega_0 = 0$ rad/s. And, as the problem states, the initial translational speed of the object is zero, $v_0 = 0$ m/s. With these substitutions, the conservation-of-energy equation becomes

$$\tfrac{1}{2}mv_f^2 + \tfrac{1}{2}kx_f^2 = \tfrac{1}{2}kx_0^2$$

from which we can obtain $v_f$:

$$v_f = \sqrt{\frac{k}{m}(x_0^2 - x_f^2)}$$

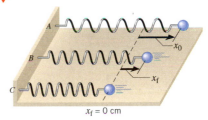

**Figure 10.17** The total mechanical energy of this system is entirely elastic potential energy (A), partly elastic potential energy and partly kinetic energy (B), and entirely kinetic energy (C).

## Solution

**(a)** Since $x_0 = 0.0450$ m and $x_f = 0.0225$ m, the final translational speed is

$$v_f = \sqrt{\frac{545 \text{ N/m}}{0.200 \text{ kg}}[(0.0450 \text{ m})^2 - (0.0225 \text{ m})^2]} = \boxed{2.03 \text{ m/s}}$$

The total mechanical energy at this point is partly translational kinetic energy ($\tfrac{1}{2}mv_f^2 = 0.414$ J) and partly elastic potential energy ($\tfrac{1}{2}kx_f^2 = 0.138$ J). The total mechanical energy $E$ is the sum of these two energies: $E = 0.414$ J $+ 0.138$ J $= 0.552$ J. Because the total mechanical energy remains constant during the motion, this value equals the initial total mechanical energy when the object is stationary and the energy is entirely elastic potential energy ($E_0 = \frac{1}{2}kx_0^2 = 0.552$ J).

**(b)** When $x_0 = 0.0450$ m and $x_f = 0$ m, we have

$$v_f = \sqrt{\frac{k}{m}(x_0^2 - x_f^2)} = \sqrt{\frac{545 \text{ N/m}}{0.200 \text{ kg}}[(0.0450 \text{ m})^2 - (0 \text{ m})^2]} = \boxed{2.35 \text{ m/s}}$$

Now the total mechanical energy is due entirely to the translational kinetic energy ($\frac{1}{2}mv_f^2 = 0.552$ J), since the elastic potential energy is zero. Note that the total mechanical energy is the same as it is in part (a). In the absence of friction, the simple harmonic motion of a spring converts the different types of energy between one form and another, the total always remaining the same.

Conceptual Example 8 takes advantage of energy conservation to illustrate what happens to the maximum speed, amplitude, and angular frequency of a simple harmonic oscillator when its mass is changed suddenly at a certain point in the motion.

### *Conceptual Example 8*
### Changing the Mass of a Simple Harmonic Oscillator

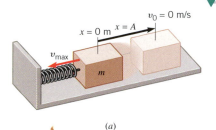

(a)

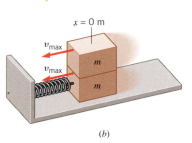

(b)

**Figure 10.18** (*a*) A box of mass *m*, starting from rest at *x* = *A*, undergoes simple harmonic motion about *x* = 0 m. (*b*) When *x* = 0 m, a second box, with the same mass and speed, is attached.

Figure 10.18*a* shows a box of mass *m* attached to a spring that has a force constant *k*. The box rests on a horizontal, frictionless surface. The spring is initially stretched to *x* = *A* and then released from rest. The box then executes simple harmonic motion that is characterized by a maximum speed $v_{max}$, an amplitude *A*, and an angular frequency $\omega$. When the box is passing through the point where the spring is unstrained (*x* = 0 m), a second box of the same mass *m* and speed $v_{max}$ is attached to it, as in part *b* of the drawing. Discuss what happens to (a) the maximum speed, (b) the amplitude, and (c) the angular frequency of the subsequent simple harmonic motion.

**Reasoning and Solution**

**(a)** The maximum speed of an object in simple harmonic motion occurs when the object is passing through the point where the spring is unstrained (*x* = 0 m), as in Figure 10.18*b*. Since the second box is attached at this point with the same speed, *the maximum speed of the two-box system remains the same as that of the one-box system.*

**(b)** At the same speed, the maximum kinetic energy of the two boxes is twice that of a single box, since the mass is twice as much. Subsequently, when the two boxes move to the left and compress the spring, their kinetic energy is converted into elastic potential energy. Since the two boxes have twice as much kinetic energy as one box alone, the two will have twice as much elastic potential energy when they come to a halt at the extreme left. Here, we are using the principle of conservation of mechanical energy, which applies since friction is absent. But the elastic potential energy is proportional to the amplitude squared ($A^2$) of the motion, *so the amplitude of the two-box system is greater than that of the one-box system by a factor of $\sqrt{2}$.*

**(c)** The angular frequency $\omega$ of a simple harmonic oscillator is given by Equation 10.11 as $\omega = \sqrt{k/m}$. Since the mass of the two-box system is twice that of the one-box system, *the angular frequency of the two-box system is smaller than that of the one-box system by a factor of $\sqrt{2}$.*

**Related Homework:** *Problem 32*

In the previous two examples, gravitational potential energy plays no role because the spring is horizontal. The next example illustrates that gravitational potential energy must be taken into account when a spring is oriented vertically.

### *Example 9*   A Falling Ball on a Vertical Spring

A 0.20-kg ball is attached to a vertical spring, as in Figure 10.19. The spring constant of the spring is 28 N/m. The ball, supported initially so that the spring is neither stretched nor compressed, is released from rest. In the absence of air resistance, how far does the ball fall before being brought to a momentary stop by the spring?

**Reasoning** Since air resistance is absent, only the conservative forces of gravity and the spring act on the ball. Therefore, the principle of conservation of mechanical energy applies:

$$E_f = E_0$$
$$\tfrac{1}{2}mv_f^2 + \tfrac{1}{2}I\omega_f^2 + mgh_f + \tfrac{1}{2}kx_f^2 = \tfrac{1}{2}mv_0^2 + \tfrac{1}{2}I\omega_0^2 + mgh_0 + \tfrac{1}{2}kx_0^2$$

The problem states that the final and initial translational speeds of the ball are zero: $v_f = v_0 = 0$ m/s. The ball and spring do not rotate, so the final and initial angular speeds are also zero: $\omega_f = \omega_0 = 0$ rad/s. As Figure 10.19 indicates, the initial height of the ball is $h_0$, and the final height is $h_f = 0$ m. In addition, the spring is unstrained ($x_0 = 0$ m) to begin with, and so it has no elastic potential energy initially. With these substitutions, the conservation of mechanical energy equation reduces to

$$\tfrac{1}{2}kx_f^2 = mgh_0$$

This result shows that the initial gravitational potential energy ($mgh_0$) is converted into elastic potential energy ($\tfrac{1}{2}kx_f^2$). When the ball falls to its lowest point, its displacement is $x_f = -h_0$, where the minus sign indicates that the displacement is downward. Substituting this result into the equation above and solving for $h_0$ yields $h_0 = 2mg/k$.

**Solution** The distance that the ball falls before coming to a momentary halt is

$$h_0 = \frac{2mg}{k} = \frac{2(0.20 \text{ kg})(9.8 \text{ m/s}^2)}{28 \text{ N/m}} = \boxed{0.14 \text{ m}}$$

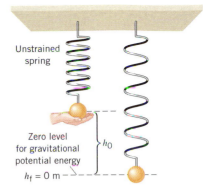

**Figure 10.19** The ball is supported initially so that the spring is unstrained. After being released from rest, the ball falls through the distance $h_0$ before being momentarily stopped by the spring.

**Problem solving insight**
When evaluating the total mechanical energy $E$, always include a potential energy term for every conservative force acting on the system. In Example 9 there are two such terms, gravitational and elastic.

## 10.4 The Pendulum

As Figure 10.20 shows, a *simple pendulum* consists of a particle of mass $m$, attached to a frictionless pivot $P$ by a cable of length $L$ and negligible mass. When the particle is pulled away from its equilibrium position by an angle $\theta$ and released, it swings back and forth. By attaching a pen to the bottom of the swinging particle and moving a strip of paper beneath it at a steady rate, we can record the position of the particle as time passes. The graphical record reveals a pattern that is similar (but not identical) to the sinusoidal pattern for simple harmonic motion.

The force of gravity is responsible for the back-and-forth rotation about the axis at $P$. The rotation speeds up as the particle approaches the lowest point on the arc and slows down on the upward part of the swing. Eventually the angular speed is reduced to zero, and the particle swings back. As Section 9.4 discusses, a net torque is required to change the angular speed. The gravitational force $mg$ produces this torque. (The tension $\mathbf{T}$ in the cable creates no torque, because it points directly at the pivot $P$ and, therefore, has a zero lever arm.) According to Equation 9.1, the magnitude of the torque $\tau$ is the product of the magnitude $mg$ of the gravitational force and the lever arm $\ell$, so that $\tau = -(mg)\ell$. The minus sign is included since the torque is a restoring torque: that is, it acts to reduce the angle $\theta$ [the angle $\theta$ is positive (counterclockwise), while the torque is negative (clockwise)]. The lever arm $\ell$ is the perpendicular distance between the line of action of $mg$ and the pivot $P$. From Figure 10.20 it can be seen that $\ell$ is very nearly equal to the arc length $s$ of the circular path when the angle $\theta$ is small (about 10° or less). Furthermore, if $\theta$ is expressed in radians, the arc length and the radius $L$ of the circular path are related, according to $s = L\theta$ (Equation 8.1). Under these conditions, it follows that $\ell \approx s = L\theta$, and the torque created by gravity is

$$\tau \approx \underbrace{-mgL}_{k'}\, \theta$$

In the equation above, the term $mgL$ has a constant value $k'$, independent of $\theta$. *For small angles*, then, the torque that restores the pendulum to its vertical equilibrium position is proportional to the angular displacement $\theta$. The expression $\tau = -k'\theta$ has the same form as the Hooke's law restoring force for an ideal spring, $F = -kx$. Therefore, we expect the frequency of the back-and-forth movement of the pendulum to be given by an equation analogous to Equation 10.11 ($\omega = 2\pi f = \sqrt{k/m}$). In place of the spring constant

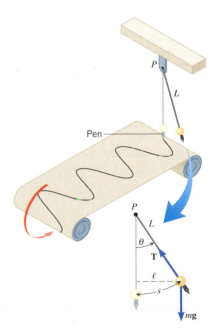

**Figure 10.20** A simple pendulum swinging back and forth about the pivot $P$. If the angle $\theta$ is small, the swinging is approximately simple harmonic motion.

$k$, the constant $k' = mgL$ will appear, and, as usual in rotational motion, in place of the mass $m$, the moment of inertia $I$ will appear:

$$\omega = 2\pi f = \sqrt{\frac{mgL}{I}} \qquad \text{(small angles only)} \qquad (10.15)$$

The moment of inertia of a particle of mass $m$, rotating at a radius $r = L$ about an axis, is given by $I = mL^2$ (Equation 9.6). Substituting this expression for $I$ into Equation 10.15 reveals for a simple pendulum that

**Simple pendulum** $\qquad \omega = 2\pi f = \sqrt{\dfrac{g}{L}} \qquad \text{(small angles only)} \qquad (10.16)$

The mass of the particle has been eliminated algebraically from this expression, so only the length $L$ and the acceleration $g$ due to gravity determine the frequency of a simple pendulum. If the angle of oscillation is large, the pendulum does not exhibit simple harmonic motion, and Equation 10.16 does not apply. Equation 10.16 provides the basis for using a pendulum to keep time, as Example 10 demonstrates.

### Example 10   Keeping Time

Determine the length of a simple pendulum that will swing back and forth in simple harmonic motion with a period of 1.00 s.

**Reasoning** When a simple pendulum is swinging back and forth in simple harmonic motion, its frequency $f$ is given by Equation 10.16 as $f = \frac{1}{2\pi}\sqrt{g/L}$, where $g$ is the acceleration due to gravity and $L$ is the length of the pendulum. We also know from Equation 10.5 that the frequency is the reciprocal of the period $T$, so $f = 1/T$. Thus, the equation above becomes $1/T = \frac{1}{2\pi}\sqrt{g/L}$. We can solve this equation for the length $L$ of the pendulum.

**Solution** The length of the pendulum is

$$L = \frac{T^2 g}{4\pi^2} = \frac{(1.00\ \text{s})^2 (9.80\ \text{m/s}^2)}{4\pi^2} = \boxed{0.248\ \text{m}}$$

Figure 10.21 shows a clock that uses a pendulum to keep time.

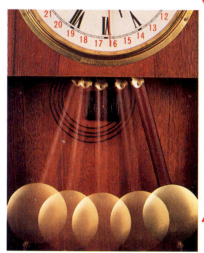

**Figure 10.21** This pendulum clock keeps time as the pendulum swings back and forth. (© Robert Mathena/ Fundamental Photographs)

It is not necessary that the object in Figure 10.20 be a particle. It may be an extended object, in which case the pendulum is called a ***physical pendulum.*** For small oscillations, Equation 10.15 still applies, but the moment of inertia $I$ is no longer $mL^2$. The proper value for the rigid object must be used. (See Section 9.4 for a discussion of moment of inertia.) In addition, the length $L$ for a physical pendulum is the distance between the axis at $P$ and the center of gravity of the object. The next example deals with an important type of physical pendulum.

### Example 11   Pendulum Motion and Walking

When we walk, our legs alternately swing forward about the hip joint as a pivot. In this motion the leg is acting approximately as a physical pendulum. Treating the leg as a uniform rod of length $D = 0.80$ m, find the time it takes for the leg to swing forward.

**The physics of pendulum motion and walking.**

**Reasoning** The time it takes for the leg to swing forward is one-half of the period $T$, which is related to the frequency $f$ by $f = 1/T$ (Equation 10.5). For a physical pendulum the frequency is given by $f = \frac{1}{2\pi}\sqrt{mgL/I}$ (Equation 10.15), where the moment of inertia for a thin rod of length $D$ rotating about an axis perpendicular to one end is given in Table 9.1 as $I = \frac{1}{3}mD^2$. In Equation 10.15 the length $L$ is the distance between the pivot at the hip and the center of gravity of the leg. Since we are treating the leg as a thin uniform rod, the center of gravity is at the center and $L = 0.40$ m.

**Solution** Using Equations 10.5 and 10.15, we find that the period is

$$f = \frac{1}{T} = \frac{1}{2\pi}\sqrt{\frac{mgL}{I}} \qquad \text{or} \qquad T = 2\pi\sqrt{\frac{I}{mgL}}$$

Substituting the moment of inertia $I = \frac{1}{3}mD^2$ of the thin rod into this result gives

$$T = 2\pi\sqrt{\frac{\frac{1}{3}mD^2}{mgL}} = \frac{2\pi D}{\sqrt{3gL}} = \frac{2\pi(0.80\text{ m})}{\sqrt{3(9.8\text{ m/s}^2)(0.40\text{ m})}} = 1.5\text{ s}$$

The desired time is one-half of the period or $\boxed{0.75\text{ s}}$.

Shock absorber

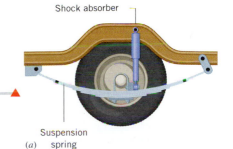

Suspension
(a)    spring

# 10.5  Damped Harmonic Motion

In simple harmonic motion, an object oscillates with a constant amplitude, because there is no mechanism for dissipating energy. In reality, however, friction or some other energy-dissipating mechanism is always present. In the presence of energy dissipation, the amplitude of oscillation decreases as time passes, and the motion is no longer simple harmonic motion. Instead, it is referred to as *damped harmonic motion,* the decrease in amplitude being called "damping."

   One widely used application of damped harmonic motion is in the suspension system of an automobile. Figure 10.22a shows a shock absorber attached to a main suspension spring of a car. A shock absorber is designed to introduce damping forces, which reduce the vibrations associated with a bumpy ride. As part b of the drawing shows, a shock absorber consists of a piston in a reservoir of oil. When the piston moves in response to a bump in the road, holes in the piston head permit the piston to pass through the oil. Viscous forces that arise during this movement cause the damping.

   Figure 10.23 illustrates the different degrees of damping that can exist. As applied to the example of a car's suspension system, these graphs show the vertical position of the chassis after it has been pulled upward by an amount $A_0$ at time $t_0 = 0$ s and then released. Part a of the figure compares undamped or simple harmonic motion in curve 1 (red) to slightly damped motion in curve 2 (green). In damped harmonic motion, the chassis oscillates with decreasing amplitude and it eventually comes to rest. As the degree of damping is increased from curve 2 to curve 3 (gold), the car makes fewer oscillations before coming to a halt. Part b of the drawing shows that as the degree of damping is increased further, there comes a point when the car does not oscillate at all after it is released but, rather, settles directly back to its equilibrium position, as in curve 4 (blue). The smallest degree of damping that completely eliminates the oscillations is termed "critical damping," and the motion is said to be *critically damped.*

   Figure 10.23b also shows that the car takes the longest time to return to its equilibrium position in curve 5 (purple), where the degree of damping is above the value for critical damping. When the damping exceeds the critical value, the motion is said to be *overdamped.* In contrast, when the damping is less than the critical level, the motion is said to be *underdamped* (curves 2 and 3). Typical automobile shock absorbers are designed to produce underdamped motion somewhat like that in curve 3.

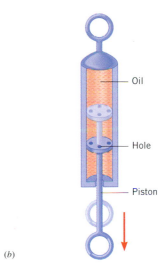

Oil

Hole

Piston

(b)

**Figure 10.22**  (a) A shock absorber mounted in the suspension system of an automobile and (b) a simplified, cutaway view of the shock absorber.

**The physics of**
**a shock absorber.**

**Figure 10.23**  Damped harmonic motion. The degree of damping increases from curve 1 to curve 5. Curve 1 represents undamped or simple harmonic motion. Curves 2 and 3 show underdamped motion. Curve 4 represents critically damped harmonic motion. Curve 5 shows overdamped motion.

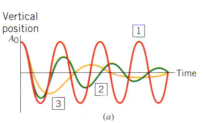
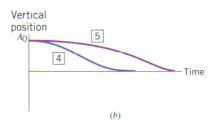

# 10.6  Driven Harmonic Motion and Resonance

In damped harmonic motion, a mechanism such as friction dissipates or reduces the energy of an oscillating system, with the result that the amplitude of the motion decreases in time. This section discusses the opposite effect—namely, the increase in amplitude that results when energy is continually added to an oscillating system.

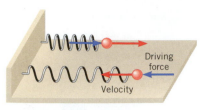

**Figure 10.24** Resonance occurs when the frequency of the driving force (blue arrows) matches a frequency at which the object naturally vibrates. The red arrows represent the velocity of the object.

To set an object on an ideal spring into simple harmonic motion, some agent must apply a force that stretches or compresses the spring initially. Suppose that this force is applied at all times, not just for a brief initial moment. The force could be provided, for example, by a person who simply pushes and pulls the object back and forth. The resulting motion is known as **driven harmonic motion,** because the additional force drives or controls the behavior of the object to a large extent. The additional force is identified as the **driving force.**

Figure 10.24 illustrates one particularly important example of driven harmonic motion. Here, the driving force has the same frequency as the spring system and always points in the direction of the object's velocity. The frequency of the spring system is $f = (1/2\pi)\sqrt{k/m}$ and is called a natural frequency, because it is the frequency at which the spring system naturally oscillates. Since the driving force and the velocity always have the same direction, positive work is done on the object at all times, and the total mechanical energy of the system increases. As a result, the amplitude of the vibration becomes larger and will increase without limit, if there is no damping force to dissipate the energy being added by the driving force. The situation depicted in Figure 10.24 is known as **resonance.**

■ **RESONANCE**

Resonance is the condition in which a time-dependent force can transmit large amounts of energy to an oscillating object, leading to a large amplitude motion. In the absence of damping, resonance occurs when the frequency of the force matches a natural frequency at which the object will oscillate.

The role played by the frequency of a driving force is a critical one. The matching of this frequency with a natural frequency of vibration allows even a relatively weak force to produce a large amplitude vibration, because the effect of each push–pull cycle is cumulative.

**The physics of**
**high tides at the Bay of Fundy.**

Resonance can occur with any object that can oscillate, and springs need not be involved. The greatest tides in the world occur in the Bay of Fundy, which lies between the Canadian provinces of New Brunswick and Nova Scotia. Figure 10.25 shows the enormous difference between the water level at high and low tides, a difference that in some locations averages about 15 m. This phenomenon is partly due to resonance. The time, or period, that it takes for the tide to flow into and ebb out of a bay depends on the size of the bay, the topology of the bottom, and the configuration of the shoreline. The ebb and flow of the water in the Bay of Fundy has a period of 12.5 hours, which is very close to the lunar tidal period of 12.42 hours. The tide then "drives" water into and out of the Bay of Fundy at a frequency (once per 12.42 hours) that nearly matches the natural frequency of the bay (once per 12.5 hours). The result is the extraordinary high tide in the bay. (You can create a similar effect in a bathtub full of water by moving back and forth in synchronism with the waves you're causing.)

**Figure 10.25** The Bay of Fundy at (*a*) high tide and (*b*) low tide. In some places the water level changes by almost 15 m. (© Everett Johnson/Leo de Wys, Inc.)

(*a*)

(*b*)

# 10.7 Elastic Deformation
## STRETCHING, COMPRESSION, AND YOUNG'S MODULUS

We have seen that a spring returns to its original shape when the force compressing or stretching it is removed. In fact, all materials become distorted in some way when they are squeezed or stretched, and many of them, such as rubber, return to their original shape when the squeezing or stretching is removed. Such materials are said to be "elastic." From an atomic viewpoint, elastic behavior has its origin in the forces that atoms exert on each other, and Figure 10.26 symbolizes these forces with the aid of springs. It is because of these atomic-level "springs" that a material tends to return to its initial shape once the forces that cause the deformation are removed.

The interatomic forces that hold the atoms of a solid together are particularly strong, so considerable force must be applied to stretch a solid object. Experiments have shown that the magnitude of the force can be expressed by the following relation, provided that the amount of stretching is small compared to the original length of the object:

$$F = Y \left( \frac{\Delta L}{L_0} \right) A \qquad (10.17)$$

As Figure 10.27 shows, $F$ denotes the magnitude of the stretching force applied perpendicularly to the surface at the end, $A$ is the cross-sectional area of the rod, $\Delta L$ is the increase in length, and $L_0$ is the original length. The term $Y$ is a proportionality constant called **Young's modulus**, after Thomas Young (1773–1829). Solving Equation 10.17 for $Y$ shows that Young's modulus has units of force per unit area (N/m²). *It should be noted that the magnitude of the force in Equation 10.17 is proportional to the fractional increase in length $\Delta L/L_0$, rather than the absolute increase $\Delta L$.* The magnitude of the force is also proportional to the cross-sectional area $A$, which need not be circular, but can have any shape (e.g., rectangular).

Table 10.1 reveals that the value of Young's modulus depends on the nature of the material. The values for metals are much larger than those for bone, for example. Equation 10.17 indicates that, for a given force, the material with the greater value of $Y$ undergoes the smaller change in length. This difference between the changes in length is the reason why surgical implants (e.g., artificial hip joints), which are often made from stainless steel or titanium alloys, can lead to chronic deterioration of the bone that is in contact with the implanted prosthesis.

Forces that are applied as in Figure 10.27 and cause stretching are called "tensile" forces, because they create a tension in the material, much like the tension in a rope. Equation 10.17 also applies when the force compresses the material along its length. In this situation, the force is applied in a direction opposite to that shown in Figure 10.27, and $\Delta L$ stands for the amount by which the original length $L_0$ decreases. Table 10.1 indicates, for example, that bone has different values of Young's modulus for compression and tension, the value for tension being greater. Such differences are related to the structure of the material. The solid part of bone consists of collagen fibers (a protein material) distributed throughout hydroxyapatite (a mineral). The collagen acts like the steel rods in reinforced concrete and increases the value of $Y$ for tension relative to that for compression.

Most solids have Young's moduli that are rather large, reflecting the fact that a large force is needed to change the length of a solid object by even a small amount, as Example 12 illustrates.

## Example 12  Bone Compression

In a circus act, a performer supports the combined weight (1640 N) of a number of colleagues (see Figure 10.28). Each thighbone (femur) of this performer has a length of 0.55 m and an effective cross-sectional area of $7.7 \times 10^{-4}$ m². Determine the amount by which each thighbone compresses under the extra weight.

**Reasoning** The additional weight supported by each thighbone is $F = \frac{1}{2}(1640 \text{ N}) = 820 \text{ N}$, and Table 10.1 indicates that Young's modulus for bone compression is $9.4 \times 10^9$ N/m².

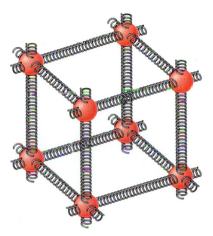

**Figure 10.26** The forces between atoms act like springs. The atoms are represented by red spheres, and the springs between some atoms have been omitted for clarity.

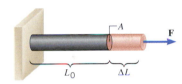

**Figure 10.27** In this diagram, **F** denotes the stretching force, $A$ the cross-sectional area, $L_0$ the original length of the rod, and $\Delta L$ the amount of stretch.

**The physics of surgical implants.**

**The physics of bone structure.**

**Table 10.1  Values for the Young's Modulus of Solid Materials**

| Material | Young's Modulus $Y$ (N/m²) |
|---|---|
| Aluminum | $6.9 \times 10^{10}$ |
| Bone | |
|    Compression | $9.4 \times 10^{9}$ |
|    Tension | $1.6 \times 10^{10}$ |
| Brass | $9.0 \times 10^{10}$ |
| Brick | $1.4 \times 10^{10}$ |
| Copper | $1.1 \times 10^{11}$ |
| Mohair | $2.9 \times 10^{9}$ |
| Nylon | $3.7 \times 10^{9}$ |
| Pyrex glass | $6.2 \times 10^{10}$ |
| Steel | $2.0 \times 10^{11}$ |
| Teflon | $3.7 \times 10^{8}$ |
| Titanium | $1.2 \times 10^{11}$ |
| Tungsten | $3.6 \times 10^{11}$ |

**Figure 10.28** The entire weight of the balanced group is supported by the legs of the performer who is lying on his back. (Image courtesy of Ringling Brothers and Barnum & Bailey®, THE GREATEST SHOW ON EARTH®.)

**Table 10.2** *Values for the Shear Modulus of Solid Materials*

| Material | Shear Modulus $S$ (N/m$^2$) |
|---|---|
| Aluminum | $2.4 \times 10^{10}$ |
| Bone | $1.2 \times 10^{10}$ |
| Brass | $3.5 \times 10^{10}$ |
| Copper | $4.2 \times 10^{10}$ |
| Lead | $5.4 \times 10^{9}$ |
| Nickel | $7.3 \times 10^{10}$ |
| Steel | $8.1 \times 10^{10}$ |
| Tungsten | $1.5 \times 10^{11}$ |

Since the length and cross-sectional area of the thighbone are also known, we may use Equation 10.17 to find the amount by which the additional weight compresses the thighbone.

**Solution** The amount of compression $\Delta L$ of each thighbone is

$$\Delta L = \frac{FL_0}{YA} = \frac{(820 \text{ N})(0.55 \text{ m})}{(9.4 \times 10^9 \text{ N/m}^2)(7.7 \times 10^{-4} \text{ m}^2)} = \boxed{6.2 \times 10^{-5} \text{ m}}$$

This is a very small change, the fractional decrease being $\Delta L/L_0 = 0.000\ 11$.

## SHEAR DEFORMATION AND THE SHEAR MODULUS

It is possible to deform a solid object in a way other than stretching or compressing it. For instance, place a book on a rough table and push on the top cover, as in Figure 10.29a. Notice that the top cover, and the pages below it, become shifted relative to the stationary bottom cover. The resulting deformation is called a *shear deformation* and occurs because of the combined effect of the force **F** applied (by the hand) to the top of the book and the force $-$**F** applied (by the table) to the bottom of the book. The directions of the forces are parallel to the covers of the book, each of which has an area $A$, as illustrated in part $b$ of the drawing. These two forces have equal magnitudes, but opposite directions, so the book remains in equilibrium. Equation 10.18 gives the magnitude $F$ of the force needed to produce an amount of shear $\Delta X$ for an object with thickness $L_0$:

$$F = S\left(\frac{\Delta X}{L_0}\right)A \qquad (10.18)$$

This equation is very similar to Equation 10.17. The constant of proportionality $S$ is called the *shear modulus* and, like Young's modulus, has units of force per unit area (N/m$^2$). The value of $S$ depends on the nature of the material, and Table 10.2 gives some representative values. Example 13 illustrates how to determine the shear modulus of a favorite dessert.

### *Example 13* J-E-L-L-O

A block of Jell-O is resting on a plate. Figure 10.30a gives the dimensions of the block. You are bored, impatiently waiting for dinner, and push tangentially across the top surface with a force of $F = 0.45$ N, as in part $b$ of the drawing. The top surface moves a distance $\Delta X = 6.0 \times 10^{-3}$ m relative to the bottom surface. Use this idle gesture to measure the shear modulus of Jell-O.

**Reasoning** The finger applies a force that is parallel to the top surface of the Jell-O block. The shape of the block changes, because the top surface moves a distance $\Delta X$ relative to the bottom surface. The magnitude of the force required to produce this change in shape is given by Equation 10.18 as $F = S(\Delta X/L_0)A$. We know the values for all the variables in this relation except $S$, which can be determined.

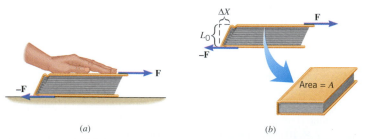

**Figure 10.29** (a) An example of a shear deformation. The shearing forces **F** and $-$**F** are applied parallel to the top and bottom covers of the book. In general, shearing forces cause a solid object to change its shape. (b) The shear deformation is $\Delta X$. The area of each cover is $A$, and the thickness of the book is $L_0$.

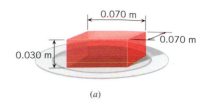

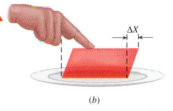

**Solution** Solving Equation 10.18 for the shear modulus $S$, we find that $S = FL_0/(A\,\Delta X)$, where $A = (0.070\text{ m})(0.070\text{ m})$ is the area of the top surface, and $L_0 = 0.030$ m is the thickness of the block:

$$S = \frac{FL_0}{A\,\Delta X} = \frac{(0.45\text{ N})(0.030\text{ m})}{(0.070\text{ m})(0.070\text{ m})(6.0 \times 10^{-3}\text{ m})} = \boxed{460\text{ N/m}^2}$$

Jell-O can be deformed easily, so its shear modulus is significantly less than that of a more rigid material like steel (see Table 10.2).

**Figure 10.30** (a) A block of Jell-O and (b) a shearing force applied to it.

Although Equations 10.17 and 10.18 are algebraically similar, they refer to different kinds of deformations. The tensile force in Figure 10.27 is perpendicular to the surface whose area is $A$, whereas the shearing force in Figure 10.29 is parallel to that surface. Furthermore, the ratio $\Delta L/L_0$ in Equation 10.17 is different from the ratio $\Delta X/L_0$ in Equation 10.18. The distances $\Delta L$ and $L_0$ are parallel, whereas $\Delta X$ and $L_0$ are perpendicular. Young's modulus refers to a *change in length* of one dimension of a solid object as a result of tensile or compressive forces. The shear modulus refers to a *change in shape* of a solid object as a result of shearing forces.

## VOLUME DEFORMATION AND THE BULK MODULUS

When a compressive force is applied along one dimension of a solid, the length of that dimension decreases. It is also possible to apply compressive forces so that the size of every dimension (length, width, and depth) decreases, leading to a decrease in volume, as Figure 10.31 illustrates. This kind of overall compression occurs, for example, when an object is submerged in a liquid, and the liquid presses inward everywhere on the object. The forces acting in such situations are applied perpendicular to every surface, and it is more convenient to speak of the perpendicular force per unit area, rather than the amount of any one force in particular. The magnitude of the perpendicular force per unit area is called the *pressure P.*

■ **DEFINITION OF PRESSURE**

The pressure $P$ is the magnitude $F$ of the force acting perpendicular to a surface divided by the area $A$ over which the force acts:

$$P = \frac{F}{A} \qquad (10.19)$$

*SI Unit of Pressure:* N/m$^2$ = pascal (Pa)

Equation 10.19 indicates that the SI unit for pressure is the unit of force divided by the unit of area, or newton/meter$^2$ (N/m$^2$). This unit of pressure is often referred to as a *pascal* (Pa), named after the French scientist Blaise Pascal (1623–1662).

Suppose we change the pressure on an object by an amount $\Delta P$, where, as usual, the "delta" notation $\Delta P$ represents the final pressure $P$ minus the initial pressure $P_0$: $\Delta P = P - P_0$. Because of this change in pressure, the volume of the object changes by an amount $\Delta V = V - V_0$, where $V$ and $V_0$ are the final and initial volumes, respectively. Such a pressure change occurs, for example, when a swimmer dives deeper into the water. Experiment reveals that the change $\Delta P$ in pressure needed to change the volume by an amount $\Delta V$ is directly proportional to the fractional change $\Delta V/V_0$ in the volume:

$$\Delta P = -B\left(\frac{\Delta V}{V_0}\right) \qquad (10.20)$$

This relation is analogous to Equations 10.17 and 10.18, except that the area $A$ in those equations does not appear here explicitly; the area is already taken into account by the concept of pressure (force per unit area). The proportionality constant $B$ is known as the *bulk modulus.* The minus sign occurs because an increase in pressure ($\Delta P$ positive) al-

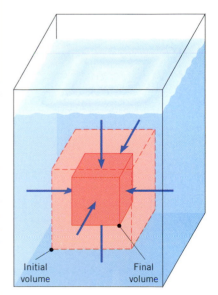

**Figure 10.31** The arrows denote the forces that push perpendicularly on every surface of an object immersed in a liquid. The force per unit area is the pressure. When the pressure increases, the volume of the object decreases.

**Table 10.3**  *Values for the Bulk Modulus of Solid and Liquid Materials*

| Material | Bulk Modulus $B$ (N/m²) |
|---|---|
| **Solids** | |
| Aluminum | $7.1 \times 10^{10}$ |
| Brass | $6.7 \times 10^{10}$ |
| Copper | $1.3 \times 10^{11}$ |
| Lead | $4.2 \times 10^{10}$ |
| Nylon | $6.1 \times 10^{9}$ |
| Pyrex glass | $2.6 \times 10^{10}$ |
| Steel | $1.4 \times 10^{11}$ |
| **Liquids** | |
| Ethanol | $8.9 \times 10^{8}$ |
| Oil | $1.7 \times 10^{9}$ |
| Water | $2.2 \times 10^{9}$ |

ways creates a decrease in volume ($\Delta V$ negative), and $B$ is given as a positive quantity. Like Young's modulus and the shear modulus, the bulk modulus has units of force per unit area (N/m²), and its value depends on the nature of the material. Table 10.3 gives representative values of the bulk modulus.

## 10.8  Stress, Strain, and Hooke's Law

Equations 10.17, 10.18, and 10.20 specify the amount of force needed for a given amount of elastic deformation, and they are repeated in Table 10.4 to emphasize their common features. The left side of each equation is the magnitude of the force per unit area required to cause an elastic deformation. In general, the ratio of the magnitude of the force to the area is called the **stress.** The right side of each equation involves the change in a quantity ($\Delta L$, $\Delta X$, or $\Delta V$) divided by a quantity ($L_0$ or $V_0$) relative to which the change is compared. The terms $\Delta L/L_0$, $\Delta X/L_0$, and $\Delta V/V_0$ are unitless ratios, and each is referred to as the **strain** that results from the applied stress. In the case of stretch and compression, the strain is the fractional change in length, whereas in volume deformation it is the fractional change in volume. In shear deformation the strain refers to a change in shape of the object. Experiments show that these three equations, with constant values for Young's modulus, the shear modulus, and the bulk modulus, apply to a wide range of materials. Therefore, stress and strain are directly proportional to one another, a relationship first discovered by Robert Hooke (1635–1703) and now referred to as **Hooke's law.**

■ **HOOKE'S LAW FOR STRESS AND STRAIN**

Stress is directly proportional to strain.

**SI Unit of Stress:** newton per square meter = pascal (Pa)

**SI Unit of Strain:** Strain is a unitless quantity.

In reality, materials obey Hooke's law only up to a certain limit, as Figure 10.32 shows. As long as stress remains proportional to strain, a plot of stress versus strain is a straight line. The point on the graph where the material begins to deviate from straight-line behavior is called the "proportionality limit." Beyond the proportionality limit stress and strain are no longer directly proportional. However, if the stress does not exceed the "elastic limit" of the material, the object will return to its original size and shape once the stress is removed. The "elastic limit" is the point beyond which the object no longer returns to its original size and shape when the stress is removed; the object remains permanently deformed.

**Table 10.4**  *Stress and Strain Relations for Elastic Behavior*

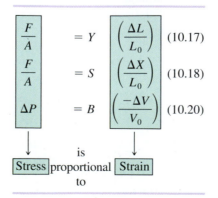

$$\frac{F}{A} = Y\left(\frac{\Delta L}{L_0}\right) \quad (10.17)$$

$$\frac{F}{A} = S\left(\frac{\Delta X}{L_0}\right) \quad (10.18)$$

$$\Delta P = B\left(\frac{-\Delta V}{V_0}\right) \quad (10.20)$$

Stress  is proportional to  Strain

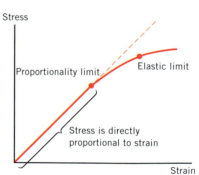

**Figure 10.32** Hooke's law (stress is directly proportional to strain) is valid only up to the proportionality limit of a material. Beyond this limit, Hooke's law no longer applies. Beyond the elastic limit, the material remains deformed even when the stress is removed.

# Concept Summary

This summary presents an abridged version of the chapter, including the important equations and all available learning aids. For convenient reference, the learning aids (including the text's examples) are placed next to or immediately after the relevant equation or discussion. The following learning aids may be found on-line at **www.wiley.com/college/cutnell**:

| | |
|---|---|
| **Interactive LearningWare** examples are solved according to a five-step interactive format that is designed to help you develop problem-solving skills. | **Concept Simulations** are animated versions of text figures or animations that illustrate important concepts. You can control parameters that affect the display, and we encourage you to experiment. |
| **Interactive Solutions** offer specific models for certain types of problems in the chapter homework. The calculations are carried out interactively. | **Self-Assessment Tests** include both qualitative and quantitative questions. Extensive feedback is provided for both incorrect and correct answers, to help you evaluate your understanding of the material. |

| Topic | Discussion | Learning Aids |
|---|---|---|
| | ### 10.1 The Ideal Spring and Simple Harmonic Motion | |
| | The force that must be applied to stretch or compress an ideal spring is | **Examples 1, 2** |
| **Force applied to an ideal spring** | $$F_{\text{Applied}} = kx \qquad (10.1)$$ where $k$ is the spring constant and $x$ is the displacement of the spring from its unstrained length. | **Interactive Solution 10.9** |
| | A spring exerts a restoring force on an object attached to the spring. The restoring force $F$ produced by an ideal spring is | |
| **Restoring force of an ideal spring** | $$F = -kx \qquad (10.2)$$ where the minus sign indicates that the restoring force points opposite to the displacement of the spring. | |
| **Simple harmonic motion** | Simple harmonic motion is the oscillatory motion that occurs when a restoring force of the form $F = -kx$ acts on an object. A graphical record of position versus time for an object in simple harmonic motion is sinusoidal. The ampli- | |
| **Amplitude** | tude $A$ of the motion is the maximum distance that the object moves away from its equilibrium position. | |
| | ### 10.2 Simple Harmonic Motion and the Reference Circle | |
| | The period $T$ of simple harmonic motion is the time required to complete one cycle of the motion, and the frequency $f$ is the number of cycles per second that occurs. Frequency and period are related according to | |
| **Period and frequency** | $$f = \frac{1}{T} \qquad (10.5)$$ | |
| **Angular frequency** | The frequency $f$ (in Hz) is related to the angular frequency $\omega$ (in rad/s) according to $$\omega = 2\pi f \qquad (\omega \text{ in rad/s}) \qquad (10.6)$$ | |
| | The maximum speed of an object in simple harmonic motion is | |
| **Maximum speed** | $$v_{\text{max}} = A\omega \qquad (\omega \text{ in rad/s}) \qquad (10.8)$$ where $A$ is the amplitude of the motion. | **Examples 3, 4** |
| | The maximum acceleration of an object in simple harmonic motion is | **Example 5** |
| **Maximum acceleration** | $$a_{\text{max}} = A\omega^2 \qquad (\omega \text{ in rad/s}) \qquad (10.10)$$ | |
| | The angular frequency of simple harmonic motion is | **Example 6** |
| **Angular frequency of simple harmonic motion** | $$\omega = \sqrt{\frac{k}{m}} \qquad (\omega \text{ in rad/s}) \qquad (10.11)$$ | **Interactive LearningWare 10.1** |

 **Use Self-Assessment Test 10.1 to evaluate your understanding of Sections 10.1 and 10.2.**

| Topic | Discussion | Learning Aids |
|---|---|---|
| | ### 10.3 Energy and Simple Harmonic Motion | |
| | The elastic potential energy of an object attached to an ideal spring is | |
| **Elastic potential energy** | $$\text{PE}_{\text{elastic}} = \tfrac{1}{2}kx^2 \qquad (10.13)$$ | |

| Topic | Discussion | Learning Aids |
|---|---|---|

| **Total mechanical energy** | The total mechanical energy $E$ of such a system is the sum of its translational and rotational kinetic energies, gravitational potential energy, and elastic potential energy: $$E = \tfrac{1}{2}mv^2 + \tfrac{1}{2}I\omega^2 + mgh + \tfrac{1}{2}kx^2 \quad (10.14)$$ | Examples 7, 8, 9 <br> **Interactive LearningWare 10.2** <br> **Concept Simulation 10.1** |
| **Conservation of mechanical energy** | If external nonconservative forces like friction do no net work, the total mechanical energy of the system is conserved: $$E_f = E_0$$ | **Interactive Solutions 10.31, 10.35** |

### 10.4 The Pendulum

A simple pendulum is a particle of mass $m$ attached to a frictionless pivot by a cable whose length is $L$ and whose mass is negligible. The small-angle ($\leq 10°$) back-and-forth swinging of a simple pendulum is simple harmonic motion, but large-angle movement is not. The frequency $f$ of the motion is given by

**Concept Simulation 10.2**

| **Frequency of a simple pendulum** | $$2\pi f = \sqrt{\frac{g}{L}} \quad \text{(small angles only)} \quad (10.16)$$ | **Example 10** |

A physical pendulum consists of a rigid object, with moment of inertia $I$ and mass $m$, suspended from a frictionless pivot. For small-angle displacements, the frequency $f$ of simple harmonic motion for a physical pendulum is given by

| **Frequency of a physical pendulum** | $$2\pi f = \sqrt{\frac{mgL}{I}} \quad \text{(small angles only)} \quad (10.15)$$ | **Example 11** |

where $L$ is the distance between the axis of rotation and the center of gravity of the rigid object.

### 10.5 Damped Harmonic Motion

| **Damped harmonic motion** <br> **Critical damping** | Damped harmonic motion is motion in which the amplitude of oscillation decreases as time passes. Critical damping is the minimum degree of damping that eliminates any oscillations in the motion as the object returns to its equilibrium position. | **Concept Simulation 10.3** |

### 10.6 Driven Harmonic Motion and Resonance

| **Driven harmonic motion** <br> **Resonance** | Driven harmonic motion occurs when a driving force acts on an object along with the restoring force. Resonance is the condition under which the driving force can transmit large amounts of energy to an oscillating object, leading to large-amplitude motion. In the absence of damping, resonance occurs when the frequency of the driving force matches a natural frequency at which the object oscillates. | |

### 10.7 Elastic Deformation

One type of elastic deformation is stretch and compression. The magnitude $F$ of the force required to stretch or compress an object of length $L_0$ and cross-sectional area $A$ by an amount $\Delta L$ is (see Figure 10.27)

| **Young's modulus** | $$F = Y\left(\frac{\Delta L}{L_0}\right)A \quad (10.17)$$ | **Example 12** |

where $Y$ is a constant called Young's modulus.

Another type of elastic deformation is shear. The magnitude $F$ of the shearing force required to create an amount of shear $\Delta X$ for an object of thickness $L_0$ and cross-sectional area $A$ is (see Figure 10.29)

| **Shear modulus** | $$F = S\left(\frac{\Delta X}{L_0}\right)A \quad (10.18)$$ | **Example 13** |

where $S$ is a constant called the shear modulus.

A third type of elastic deformation is volume deformation, which has to do with pressure. The pressure $P$ is the magnitude $F$ of the force acting perpendicular to a surface divided by the area $A$ over which the force acts:

| **Pressure** | $$P = \frac{F}{A} \quad (10.19)$$ | |

The SI unit for pressure is N/m$^2$, a unit known as a pascal (Pa): 1 Pa = 1 N/m$^2$.

| Topic | Discussion | Learning Aids |
|---|---|---|

The change $\Delta P$ in pressure needed to change the volume $V_0$ of an object by an amount $\Delta V$ is (see Figure 10.31)

**Bulk modulus**

$$\Delta P = -B\left(\frac{\Delta V}{V_0}\right) \qquad (10.20)$$

where $B$ is a constant known as the bulk modulus.

### 10.8 Stress, Strain, and Hooke's Law

**Stress and strain**   Stress is the magnitude of the force per unit area applied to an object and causes strain. For stretch/compression, the strain is the fractional change $\Delta L/L_0$ in length. For shear, the strain reflects the change in shape of the object and is given by $\Delta X/L_0$ (see Figure 10.29). For volume deformation, the strain is the fractional change in volume $\Delta V/V_0$. Hooke's law states that stress is directly proportional to strain.

**Hooke's law**

 **Use Self-Assessment Test 10.2 to evaluate your understanding of Sections 10.3–10.8.**

# Problems

*Note: Unless otherwise indicated, the values for Young's modulus Y, the shear modulus S, and the bulk modulus B are given, respectively, in Table 10.1, Table 10.2, and Table 10.3.*

### Section 10.1 The Ideal Spring and Simple Harmonic Motion

**1. ssm** A hand exerciser utilizes a coiled spring. A force of 89.0 N is required to compress the spring by 0.0191 m. Determine the force needed to compress the spring by 0.0508 m.

**2.** When the rubber band in a slingshot is stretched, it obeys Hooke's law. Suppose that the "spring constant" for the rubber band is $k = 44$ N/m. When the rubber band is pulled back with a force of 9.6 N, how far does it stretch?

**3.** A spring has a spring constant of 248 N/m. Find the magnitude of the force needed (a) to stretch the spring by $3.00 \times 10^{-2}$ m from its unstrained length and (b) to compress the spring by the same amount.

**4.** The graph shows the force $F$ that an archer applies to the string of a long bow versus the string's displacement $x$. Drawing back this bow is analogous to stretching a spring. From the data in the graph determine the effective spring constant of the bow.

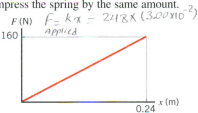

$F = kx = 248 \times (3.00 \times 10^{-2})$

Applied

F (N)

160

0.24   $x$ (m)

**5. ssm** A car is hauling a 92-kg trailer, to which it is connected by a spring. The spring constant is 2300 N/m. The car accelerates with an acceleration of 0.30 m/s$^2$. By how much does the spring stretch?

**6.** A 0.70-kg block is hung from and stretches a spring that is attached to the ceiling. A second block is attached to the first one, and the amount that the spring stretches from its unstrained length triples. What is the mass of the second block?

**\* 7. ssm** A small ball is attached to one end of a spring that has an unstrained length of 0.200 m. The spring is held by the other end, and the ball is whirled around in a horizontal circle at a speed of 3.00 m/s. The spring remains nearly parallel to the ground during the motion and is observed to stretch by 0.010 m. By how much would the spring stretch if it were attached to the ceiling and the ball allowed to hang straight down, motionless?

**\* 8.** Review Conceptual Example 2 as an aid in solving this problem. An object is attached to the lower end of a 100-coil spring that is hanging from the ceiling. The spring stretches by 0.160 m. The spring is then cut into two identical springs of 50 coils each. As the drawing shows, each spring is attached between the ceiling and the object. By how much does each spring stretch?

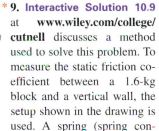

50-coil spring

**\* 9. Interactive Solution 10.9** at **www.wiley.com/college/cutnell** discusses a method used to solve this problem. To measure the static friction coefficient between a 1.6-kg block and a vertical wall, the setup shown in the drawing is used. A spring (spring constant = 510 N/m) is attached to the block. Someone pushes on the end of the spring in a direction perpendicular to the wall until the block does not slip downward. If the spring in such a setup is compressed by 0.039 m, what is the coefficient of static friction?

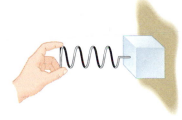

**\* 10.** A 10.1-kg uniform board is wedged into a corner and held by a spring at a 50.0° angle, as the drawing shows. The spring has a spring constant of 176 N/m and is parallel to the floor. Find the amount by which the spring is stretched from its unstrained length.

50.0°

*Problem 10*

**\* 11. ssm** In 0.750 s, a 7.00-kg block is pulled through a distance of 4.00 m on a frictionless horizontal surface, starting from rest. The block has a constant acceleration and is pulled by means of a horizontal spring that is attached to the block. The spring constant of the spring is 415 N/m. By how much does the spring stretch?

** **12.** A 30.0-kg block is resting on a flat horizontal table. On top of this block is resting a 15.0-kg block, to which a horizontal spring is attached, as the drawing illustrates. The spring constant of the spring is 325 N/m. The

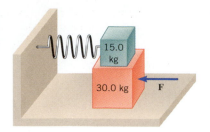

coefficient of kinetic friction between the lower block and the table is 0.600, and the coefficient of static friction between the two blocks is 0.900. A horizontal force **F** is applied to the lower block as shown. This force is increasing in such a way as to keep the blocks moving at a *constant speed*. At the point where the upper block begins to slip on the lower block, determine (a) the amount by which the spring is compressed and (b) the magnitude of the force **F**.

** **13.** A 15.0-kg block rests on a horizontal table and is attached to one end of a massless, horizontal spring. By pulling horizontally on the other end of the spring, someone causes the block to accelerate uniformly and reach a speed of 5.00 m/s in 0.500 s. In the process, the spring is stretched by 0.200 m. The block is then pulled at a *constant speed* of 5.00 m/s, during which time the spring is stretched by only 0.0500 m. Find (a) the spring constant of the spring and (b) the coefficient of kinetic friction between the block and the table.

### Section 10.2 Simple Harmonic Motion and the Reference Circle

**14.** A loudspeaker diaphragm is producing a sound for 2.5 s by moving back and forth in simple harmonic motion. The angular frequency of the motion is $7.54 \times 10^4$ rad/s. How many times does the diaphragm move back and forth?

**15.** **ssm** The shock absorbers in the suspension system of a car are in such bad shape that they have no effect on the behavior of the springs attached to the axles. Each of the identical springs attached to the front axle supports 320 kg. A person pushes down on the middle of the front end of the car and notices that it vibrates through five cycles in 3.0 s. Find the spring constant of either spring.

**16.** Refer to Conceptual Example 2 as an aid in solving this problem. A 100-coil spring has a spring constant of 420 N/m. It is cut into four shorter springs, each of which has 25 coils. One end of a 25-coil spring is attached to a wall. An object of mass 46 kg is attached to the other end of the spring, and the system is set into horizontal oscillation. What is the angular frequency of the motion?

**17.** **Concept Simulation 10.3** at **www.wiley.com/college/cutnell** illustrates the concepts pertinent to this problem. An 0.80-kg object is attached to one end of a spring, as in

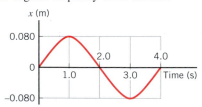

Figure 10.5, and the system is set into simple harmonic motion. The displacement $x$ of the object as a function of time is shown in the drawing. With the aid of these data, determine (a) the amplitude $A$ of the motion, (b) the angular frequency $\omega$, (c) the spring constant $k$, (d) the speed of the object at $t = 1.0$ s, and (e) the magnitude of the object's acceleration at $t = 1.0$ s.

**18.** A computer to be used in a satellite must be able to withstand accelerations of up to 25 times the acceleration due to gravity. In a test to see whether it meets this specification, the computer is bolted to a frame that is vibrated back and forth in simple harmonic motion at a frequency of 9.5 Hz. What is the minimum amplitude of vibration that must be used in this test?

* **19.** Objects of equal mass are oscillating up and down in simple harmonic motion on two different vertical springs. The spring constant of spring 1 is 174 N/m. The motion of the object on spring 1 has twice the amplitude as the motion of the object on spring 2. The magnitude of the maximum velocity is the same in each case. Find the spring constant of spring 2.

* **20.** When an object of mass $m_1$ is hung on a vertical spring and set into vertical simple harmonic motion, its frequency is 12.0 Hz. When another object of mass $m_2$ is hung on the spring along with $m_1$, the frequency of the motion is 4.00 Hz. Find the ratio $m_2/m_1$ of the masses.

* **21.** **ssm Interactive LearningWare 10.1** at **www.wiley.com/college/cutnell** reviews the concepts involved in this problem. A spring stretches by 0.018 m when a 2.8-kg object is suspended from its end. How much mass should be attached to this spring so that its frequency of vibration is $f = 3.0$ Hz?

* **22.** A 3.0-kg block is placed between two horizontal springs. Neither spring is strained when the block is located at the position labeled $x = 0$ m in the drawing. The block is then displaced a distance of 0.070 m from the position where $x = 0$ m and released from rest. (a) What is the speed of the block when it passes back through the $x = 0$ m position? (b) Determine the angular frequency $\omega$ of this system.

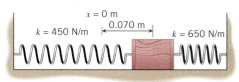

** **23.** A tray is moved horizontally back and forth in simple harmonic motion at a frequency of $f = 2.00$ Hz. On this tray is an empty cup. Obtain the coefficient of static friction between the tray and the cup, given that the cup begins slipping when the amplitude of the motion is $5.00 \times 10^{-2}$ m.

### Section 10.3 Energy and Simple Harmonic Motion

**24.** An archer pulls the bowstring back for a distance of 0.470 m before releasing the arrow. The bow and string act like a spring whose spring constant is 425 N/m. (a) What is the elastic potential energy of the drawn bow? (b) The arrow has a mass of 0.0300 kg. How fast is it traveling when it leaves the bow?

**25.** **ssm Concept Simulation 10.1** at **www.wiley.com/college/cutnell** allows you to explore the concepts to which this problem relates. A 2.00-kg object is hanging from the end of a vertical spring. The spring constant is 50.0 N/m. The object is pulled 0.200 m downward and released from rest. Complete the table below by calculating the translational kinetic energy, the gravitational potential energy, the elastic potential energy, and the total mechanical energy $E$ for each of the vertical positions indicated. The vertical positions $h$ indicate distances above the point of release, where $h = 0$ m.

| $h$ (meters) | KE | PE (gravity) | PE (elastic) | $E$ |
|---|---|---|---|---|
| 0 | | | | |
| 0.200 | | | | |
| 0.400 | | | | |

**26.** In preparation for shooting a ball in a pinball machine, a spring ($k = 675$ N/m) is compressed by 0.0650 m relative to its unstrained length. The ball ($m = 0.0585$ kg) is at rest against the spring at point A. When the spring is released, the ball slides (without rolling) to point B, which is 0.300 m higher than point A. How fast is the ball moving at B?

**27.** A spring is hung from the ceiling. A 0.450-kg block is then attached to the free end of the spring. When released from rest, the block drops 0.150 m before momentarily coming to rest. (a) What is the spring constant of the spring? (b) Find the angular frequency of the block's vibrations.

**28.** A vertical spring with a spring constant of 450 N/m is mounted on the floor. From directly above the spring, which is unstrained, a 0.30-kg block is dropped from rest. It collides with and sticks to the spring, which is compressed by 2.5 cm in bringing the block to a momentary halt. Assuming air resistance is negligible, from what height (in cm) above the compressed spring was the block dropped?

**29. ssm** A $1.00 \times 10^{-2}$-kg block is resting on a horizontal frictionless surface and is attached to a horizontal spring whose spring constant is 124 N/m. The block is shoved parallel to the spring axis and is given an initial speed of 8.00 m/s, while the spring is initially unstrained. What is the amplitude of the resulting simple harmonic motion?

**30.** A 3.2-kg block is hanging stationary from the end of a vertical spring that is attached to the ceiling. The elastic potential energy of this spring/mass system is 1.8 J. What is the elastic potential energy of the system when the 3.2-kg block is replaced by a 5.0-kg block?

**31.** Refer to **Interactive Solution 10.31** at **www.wiley.com/college/cutnell** for help in solving this problem. A heavy-duty stapling gun uses a 0.140-kg metal rod that rams against the staple to eject it. The rod is pushed by a stiff spring called a "ram spring" ($k = 32\ 000$ N/m). The mass of this spring may be ignored. Squeezing the handle of the gun first compresses the ram spring by $3.0 \times 10^{-2}$ m from its unstrained length and then releases it. Assuming that the ram spring is oriented vertically and is still compressed by $0.8 \times 10^{-2}$ m when the downward-moving ram hits the staple, find the speed of the ram at the instant of contact.

**\* 32.** Review Conceptual Example 8 before starting this problem. A block is attached to a horizontal spring and oscillates back and forth on a frictionless horizontal surface at a frequency of 3.00 Hz. The amplitude of the motion is $5.08 \times 10^{-2}$ m. At the point where the block has its maximum speed, it suddenly splits into two identical parts, only one part remaining attached to the spring. (a) What is the amplitude and the frequency of the simple harmonic motion that exists after the block splits? (b) Repeat part (a), assuming that the block splits when it is at one of its extreme positions.

**\* 33. ssm** A 1.1-kg object is suspended from a vertical spring whose spring constant is 120 N/m. (a) Find the amount by which the spring is stretched from its unstrained length. (b) The object is pulled straight down by an additional distance of 0.20 m and released from rest. Find the speed with which the object passes through its original position on the way up.

**\* 34.** An 86.0-kg climber is scaling the vertical wall of a mountain. His safety rope is made of nylon that, when stretched, behaves like a spring with a spring constant of $1.20 \times 10^3$ N/m. He accidentally slips and falls freely for 0.750 m before the rope runs out of slack. How much is the rope stretched when it breaks his fall and momentarily brings him to rest?

**\* 35.** Refer to **Interactive Solution 10.35** at **www.wiley.com/college/cutnell** to review a method by which this problem can be solved. An 11.2-kg block and a 21.7-kg block are resting on a horizontal frictionless surface. Between the two is squeezed a spring (spring constant = 1330 N/m). The spring is compressed by 0.141 m from its unstrained length and is not attached permanently to either block.

With what speed does each block move away after the mechanism keeping the spring squeezed is released and the spring falls away?

**\*\* 36.** A $1.00 \times 10^{-2}$-kg bullet is fired horizontally into a 2.50-kg wooden block attached to one end of a massless, horizontal spring ($k = 845$ N/m). The other end of the spring is fixed in place, and the spring is unstrained initially. The block rests on a horizontal, frictionless surface. The bullet strikes the block perpendicularly and quickly comes to a halt within it. As a result of this completely inelastic collision, the spring is compressed along its axis and causes the block/bullet to oscillate with an amplitude of 0.200 m. What is the speed of the bullet?

**\*\* 37. ssm** A 70.0-kg circus performer is fired from a cannon that is elevated at an angle of 40.0° above the horizontal. The cannon uses strong elastic bands to propel the performer, much in the same way that a slingshot fires a stone. Setting up for this stunt involves stretching the bands by 3.00 m from their unstrained length. At the point where the performer flies free of the bands, his height above the floor is the same as that of the net into which he is shot. He takes 2.14 s to travel the horizontal distance of 26.8 m between this point and the net. Ignore friction and air resistance and determine the effective spring constant of the firing mechanism.

**\*\* 38. Interactive LearningWare 10.2** at **www.wiley.com/college/cutnell** explores the approach taken in problems such as this one. A spring is mounted vertically on the floor. The mass of the spring is negligible. A certain object is placed on the spring to compress it. When the object is pushed further down by just a bit and then released, one up/down oscillation cycle occurs in 0.250 s. However, when the object is pushed down by $5.00 \times 10^{-2}$ m to point P and then released, the object flies entirely off the spring. To what height above point P does the object rise in the absence of air resistance?

**Section 10.4 The Pendulum**

**39.** If the period of a simple pendulum is to be 2.0 s, what should be its length?

**40.** A simple pendulum is made from a 0.65-m-long string and a small ball attached to its free end. The ball is pulled to one side through a small angle and then released from rest. After the ball is released, how much time elapses before it attains its greatest speed?

**41. ssm Concept Simulation 10.2** at **www.wiley.com/college/cutnell** allows you to explore the effect of the acceleration due to gravity on pendulum motion, which is the focus of this problem. Astronauts on a distant planet set up a simple pendulum of length 1.2 m. The pendulum executes simple harmonic motion and makes 100 complete vibrations in 280 s. What is the acceleration due to gravity?

**42.** A pendulum clock can be approximated as a simple pendulum of length 1.00 m and keeps accurate time at a location where $g = 9.83$ m/s². In a location where $g = 9.78$ m/s², what must be the new length of the pendulum, such that the clock continues to keep accurate time (that is, its period remains the same)?

**\* 43. ssm** Pendulum A is a physical pendulum made from a thin, rigid, and uniform rod whose length is $d$. One end of this rod is attached to the ceiling by a frictionless hinge, so the rod is free to swing back and forth. Pendulum B is a simple pendulum whose length is also $d$. Obtain the ratio $T_A/T_B$ of their periods for small-angle oscillations.

**\* 44.** A pendulum is constructed from a thin, rigid, and uniform rod with a small sphere attached to the end opposite the pivot. This arrangement is a good approximation to a simple pendulum (period = 0.66 s), because the mass of the sphere (lead) is much greater than the mass of the rod (aluminum). When the sphere is removed, the pendulum no longer is a simple pendulum, but is then a physical pendulum. What is the period of the physical pendulum?

**\*\* 45.** A point on the surface of a solid sphere (radius = $R$) is attached directly to a pivot on the ceiling. The sphere swings back and forth as a physical pendulum with a small amplitude. What is the length of a simple pendulum that has the same period as this physical pendulum? Give your answer in terms of $R$.

### Section 10.7 Elastic Deformation, Section 10.8 Stress, Strain, and Hooke's Law

**46.** A tow truck is pulling a car out of a ditch by means of a steel cable that is 9.1 m long and has a radius of 0.50 cm. When the car just begins to move, the tension in the cable is 890 N. How much has the cable stretched?

**47. ssm www** A 3500-kg statue is placed on top of a cylindrical concrete ($Y = 2.3 \times 10^{10}$ N/m$^2$) stand. The stand has a cross-sectional area of $7.3 \times 10^{-2}$ m$^2$ and a height of 1.8 m. By how much does the statue compress the stand?

**48.** A copper cube, 0.30 m on a side, is subjected to two shearing forces, each of magnitude $F = 6.0 \times 10^6$ N (see the drawing). Find the angle $\theta$ (in degrees), which is one measure of how the shape of the block has been altered by shear deformation.

**49. ssm** Two metal beams are joined together by four rivets, as the drawing indicates. Each rivet has a radius of $5.0 \times 10^{-3}$ m and is to be exposed to a shearing stress of no more than $5.0 \times 10^8$ Pa. What is the maximum tension **T** that can be applied to each beam, assuming that each rivet carries one-fourth of the total load?

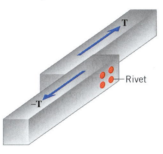

**50.** An 1800-kg car, being lifted at a steady speed by a crane, hangs at the end of a cable whose radius is $6.0 \times 10^{-3}$ m. The cable is 15 m in length and stretches by $8.0 \times 10^{-3}$ m because of the weight of the car. Determine (a) the stress, (b) the strain, and (c) Young's modulus for the cable.

**51.** The pressure increases by $1.0 \times 10^4$ N/m$^2$ for every meter of depth beneath the surface of the ocean. At what depth does the volume of a Pyrex glass cube, $1.0 \times 10^{-2}$ m on an edge at the ocean's surface, decrease by $1.0 \times 10^{-10}$ m$^3$?

**52.** The femur is a bone in the leg whose minimum cross-sectional area is about $4.0 \times 10^{-4}$ m$^2$. A compressional force in excess of $6.8 \times 10^4$ N will fracture this bone. (a) Find the maximum stress that this bone can withstand. (b) What is the strain that exists under a maximum-stress condition?

**53. ssm** A piece of aluminum is surrounded by air at a pressure of $1.01 \times 10^5$ Pa. The aluminum is placed in a vacuum chamber where the pressure is reduced to zero. Determine the fractional change $\Delta V/V_0$ in the volume of the aluminum.

**54.** When subjected to a force of compression, the length of a bone decreases by $2.7 \times 10^{-5}$ m. When this same bone is subjected to a tensile force of the same magnitude, by how much does it stretch?

**55.** The shovel of a backhoe is controlled by hydraulic cylinders that are moved by oil under pressure. Determine the volume strain $\Delta V/V_0$ (including the algebraic sign) experienced by the oil, when the pressure increases from $1.8 \times 10^5$ Pa to $6.5 \times 10^5$ Pa while the shovel is digging a trench.

**56.** A copper cylinder and a brass cylinder are stacked end to end, as in the drawing. Each cylinder has a radius of 0.25 cm. A compressive force of $F = 6500$ N is applied to the right end of the brass cylinder. Find the amount by which the length of the stack decreases.

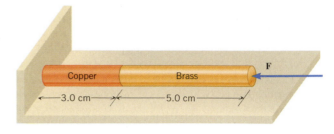

**\* 57.** A die is designed to punch holes of radii $1.00 \times 10^{-2}$ m in a metal sheet that is $3.0 \times 10^{-3}$ m thick, as the drawing illustrates. To punch through the sheet, the die must exert a shearing stress of $3.5 \times 10^8$ Pa. What force **F** must be applied to the die?

*Problem 57*

**\* 58.** A gymnast does a one-arm handstand. The humerus, which is the upper arm bone between the elbow and the shoulder joint, may be approximated as a 0.30-m-long cylinder with an outer radius of $1.00 \times 10^{-2}$ m and a hollow inner core with a radius of $4.0 \times 10^{-3}$ m. Excluding the arm, the mass of the gymnast is 63 kg. (a) What is the compressional strain of the humerus? (b) By how much is the humerus compressed?

**\* 59. ssm** A helicopter is lifting a 2100-kg jeep. The steel suspension cable is 48 m long and has a radius of $5.0 \times 10^{-3}$ m. (a) Find the amount that the cable is stretched when the jeep is suspended motionless in the air. (b) What is the amount of cable stretch when the jeep is hoisted upward with an acceleration of 1.5 m/s$^2$?

**\* 60.** A block of copper is securely fastened to the floor. A force of 1800 N is applied to the top surface of the block, as the drawing shows. Find (a) the amount by which the height of the block is changed and (b) the shear deformation of the block.

**\* 61. ssm www** An 8.0-kg stone at the end of a steel wire is being whirled in a circle at a constant tangential speed of 12 m/s. The stone is moving on the surface of a frictionless horizontal table. The wire is 4.0 m long and has a radius of $1.0 \times 10^{-3}$ m. Find the strain in the wire.

*Problem 60*

**\* 62.** Two rods are identical in all respects except one: one rod is made from aluminum, and the other from tungsten. The rods are joined end to end, in order to make a single rod that is twice as long as either the aluminum or tungsten rod. What is the effective value of Young's modulus for this composite rod? That is, what value $Y_{Composite}$ of Young's modulus should be used in Equation 10.17 when applied to the composite rod? Note that the change $\Delta L_{Composite}$ in the length of the composite rod is the sum of the changes in length of the aluminum and tungsten rods.

* **63.** A $1.0 \times 10^{-3}$-kg spider is hanging vertically by a thread that has a Young's modulus of $4.5 \times 10^9$ N/m$^2$ and a radius of $13 \times 10^{-6}$ m. Suppose that a 95-kg person is hanging vertically on an aluminum wire. What is the radius of the wire that would exhibit the same strain as the spider's thread, when the thread is stressed by the full weight of the spider?

* **64.** A square plate is $1.0 \times 10^{-2}$ m thick, measures $3.0 \times 10^{-2}$ m on a side, and has a mass of $7.2 \times 10^{-2}$ kg. The shear modulus of the material is $2.0 \times 10^{10}$ N/m$^2$. One of the square faces rests on a flat horizontal surface, and the coefficient of static friction between the plate and the surface is 0.90. A force is applied to the top of the plate, as in Figure 10.29a. Determine (a) the maximum possible amount of shear stress, (b) the maximum possible amount of shear strain, and (c) the maximum possible amount of shear deformation $\Delta X$ (see Figure 10.29b) that can be created by the applied force just before the plate begins to move.

** **65. ssm www** A solid brass sphere is subjected to a pressure of $1.0 \times 10^5$ Pa due to the earth's atmosphere. On Venus the pressure due to the atmosphere is $9.0 \times 10^6$ Pa. By what fraction $\Delta r/r_0$ (including the algebraic sign) does the radius of the sphere change when it is exposed to the Venusian atmosphere? Assume that the change in radius is very small relative to the initial radius.

** **66.** A cylindrically shaped piece of collagen (a substance found in the body in connective tissue) is being stretched by a force that increases from 0 to $3.0 \times 10^{-2}$ N. The length and radius of the collagen are, respectively, 2.5 and 0.091 cm, and Young's modulus is $3.1 \times 10^6$ N/m$^2$. (a) If the stretching obeys Hooke's law, what is the spring constant $k$ for collagen? (b) How much work is done by the variable force that stretches the collagen? (See Section 6.9 for a discussion of the work done by a variable force.)

# Chapter 11 Fluids

## 11.1 Mass Density

Fluids are materials that can flow, and they include both gases and liquids. Air is the most common gas, and moves from place to place as wind. Water is the most familiar liquid and has many uses, from generating hydroelectric power to white-water rafting. The **mass density** of a liquid or gas is an important factor that determines its behavior as a fluid. As indicated below, the mass density is the mass per unit volume and is denoted by the Greek letter rho ($\rho$).

> ■ **DEFINITION OF MASS DENSITY**
>
> The mass density $\rho$ is the mass $m$ of a substance divided by its volume $V$:
>
> $$\rho = \frac{m}{V} \tag{11.1}$$
>
> **SI Unit of Mass Density:** $kg/m^3$

Equal volumes of different substances generally have different masses, so the density depends on the nature of the material, as Table 11.1 indicates. Gases have the smallest densities because gas molecules are relatively far apart and a gas contains a large fraction of empty space. In contrast, the molecules are much more tightly packed in liquids and solids, and the tighter packing leads to larger densities. The densities of gases are very sensitive to changes in temperature and pressure. However, for the range of temperatures and pressures encountered in this text, the densities of liquids and solids do not differ much from the values in Table 11.1.

It is the mass of a substance, not its weight, that enters into the definition of density. In situations where weight is needed, it can be calculated from the mass density, the volume, and the acceleration due to gravity, as Example 1 illustrates.

### Example 1  Blood as a Fraction of Body Weight

The body of a man whose weight is 690 N contains about $5.2 \times 10^{-3}$ m$^3$ (5.5 qt) of blood. (a) Find the blood's weight and (b) express it as a percentage of the body weight.

**Reasoning**  To find the weight $W$ of the blood, we need the mass $m$, since $W = mg$, where $g$ is the magnitude of the acceleration due to gravity. According to Table 11.1, the density of blood is 1060 kg/m$^3$, so the mass of the blood can be found by using the given volume of $5.2 \times 10^{-3}$ m$^3$ in Equation 11.1.

**Solution**

(a) The mass and the weight of the blood are

$$m = \rho V = (1060 \text{ kg/m}^3)(5.2 \times 10^{-3} \text{ m}^3) = 5.5 \text{ kg} \tag{11.1}$$

$$W = mg = (5.5 \text{ kg})(9.80 \text{ m/s}^2) = \boxed{54 \text{ N}} \tag{4.5}$$

(b) The percentage of body weight contributed by the blood is

$$\text{Percentage} = \frac{54 \text{ N}}{690 \text{ N}} \times 100 = \boxed{7.8\%}$$

A convenient way to compare densities is to use the concept of **specific gravity.** The specific gravity of a substance is its density divided by the density of a standard reference material, usually chosen to be water at 4 °C.

$$\text{Specific gravity} = \frac{\text{Density of substance}}{\text{Density of water at 4 °C}} = \frac{\text{Density of substance}}{1.000 \times 10^3 \text{ kg/m}^3} \tag{11.2}$$

**Table 11.1  Mass Densities[a] of Common Substances**

| Substance | Mass Density $\rho$ (kg/m$^3$) |
|---|---|
| **Solids** | |
| Aluminum | 2 700 |
| Brass | 8 470 |
| Concrete | 2 200 |
| Copper | 8 890 |
| Diamond | 3 520 |
| Gold | 19 300 |
| Ice | 917 |
| Iron (steel) | 7 860 |
| Lead | 11 300 |
| Quartz | 2 660 |
| Silver | 10 500 |
| Wood (yellow pine) | 550 |
| **Liquids** | |
| Blood (whole, 37 °C) | 1 060 |
| Ethyl alcohol | 806 |
| Mercury | 13 600 |
| Oil (hydraulic) | 800 |
| Water (4 °C) | $1.000 \times 10^3$ |
| **Gases** | |
| Air | 1.29 |
| Carbon dioxide | 1.98 |
| Helium | 0.179 |
| Hydrogen | 0.0899 |
| Nitrogen | 1.25 |
| Oxygen | 1.43 |

[a] Unless otherwise noted, densities are given at 0 °C and 1 atm pressure.

Being the ratio of two densities, specific gravity has no units. For example, Table 11.1 reveals that diamond has a specific gravity of 3.52, since the density of diamond is 3.52 times greater than the density of water at 4 °C.

The next two sections deal with the important concept of pressure. We will see that the density of a fluid is one factor determining the pressure that a fluid exerts.

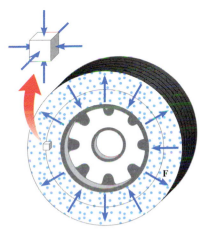

**Figure 11.1** In colliding with the inner walls of the tire, the air molecules (blue dots) exert a force on every part of the wall surface. If a small cube were inserted inside the tire, the cube would experience forces (blue arrows) acting perpendicular to each of its six faces.

## 11.2 Pressure

People who have fixed a flat tire know something about pressure. The final step in the job is to reinflate the tire to the proper pressure. The underinflated tire is soft because it contains an insufficient number of air molecules to push outward against the rubber and give the tire that solid feel. When air is added from a pump, the number of molecules and the collective force they exert are increased. The air molecules within a tire are free to wander throughout its entire volume, and in the course of their wandering they collide with one another and with the inner walls of the tire. The collisions with the walls allow the air to exert a force against every part of the wall surface, as Figure 11.1 shows. The pressure $P$ exerted by a fluid is defined in Section 10.7 (Equation 10.19) as the magnitude $F$ of the force acting perpendicular to a surface divided by the area $A$ over which the force acts:

$$P = \frac{F}{A} \tag{11.3}$$

The SI unit for pressure is a newton/meter$^2$ (N/m$^2$), a combination that is referred to as a pascal (Pa). A pressure of 1 Pa is a very small amount. Many common situations involve pressures of approximately $10^5$ Pa, an amount referred to as one *bar* of pressure. Alternatively, force can be measured in pounds and area in square inches, so another unit for pressure is pounds per square inch (lb/in.$^2$), often abbreviated as "psi."

Because of its pressure, the air in a tire applies a force to any surface with which the air is in contact. Suppose, for instance, that a small cube is inserted inside the tire. As Figure 11.1 shows, the air pressure causes a force to act perpendicularly on each face of the cube. In a similar fashion, a liquid such as water also exerts pressure. A swimmer, for example, feels the water pushing perpendicularly inward everywhere on her body, as Figure 11.2 illustrates. In general, a static fluid cannot produce a force parallel to a surface, for if it did, the surface would apply a reaction force to the fluid, consistent with Newton's action–reaction law. In response, the fluid would flow and would not then be static.

While fluid pressure can generate a force, pressure itself is not a vector quantity, as is the force. In the definition of pressure, $P = F/A$, the symbol $F$ refers only to the magnitude of the force, so that pressure has no directional characteristic. The force generated by the pressure of a static fluid is always perpendicular to the surface that the fluid contacts, as Example 2 illustrates.

### Example 2 The Force on a Swimmer

Suppose the pressure acting on the back of a swimmer's hand is $1.2 \times 10^5$ Pa, a realistic value near the bottom of the diving end of a pool. The surface area of the back of the hand is $8.4 \times 10^{-3}$ m$^2$. (a) Determine the magnitude of the force that acts on it. (b) Discuss the direction of the force.

**Reasoning** From the definition of pressure in Equation 11.3, we can see that the magnitude of the force is the pressure times the area. The direction of the force is always perpendicular to the surface that the water contacts.

**Solution**

(a) A pressure of $1.2 \times 10^5$ Pa is $1.2 \times 10^5$ N/m$^2$. From Equation 11.3, we find

$$F = PA = (1.2 \times 10^5 \text{ N/m}^2)(8.4 \times 10^{-3} \text{ m}^2) = \boxed{1.0 \times 10^3 \text{ N}}$$

This is a rather large force, about 230 lb.

(b) In Figure 11.2, the hand (palm downward) is oriented parallel to the bottom of the pool. Since the water pushes perpendicularly against the back of the hand, the force **F** is directed

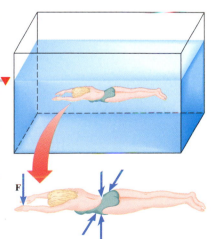

**Figure 11.2** Water applies a force perpendicular to each surface within the water, including the walls and bottom of the swimming pool, and all parts of the swimmer's body.

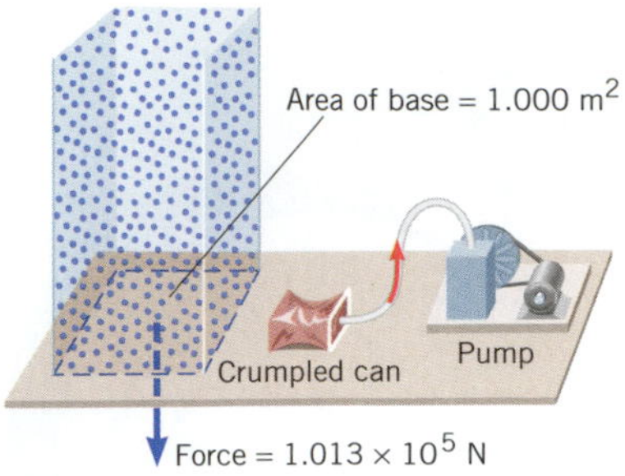

Area of base = 1.000 m$^2$

Crumpled can    Pump

Force = $1.013 \times 10^5$ N

**Figure 11.3** Atmospheric pressure at sea level is $1.013 \times 10^5$ Pa, which is sufficient to crumple a can if the inside air is pumped out.

**The physics of lynx paws.**

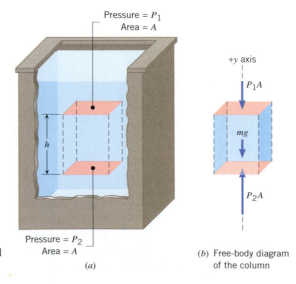

**Figure 11.4** Lynx have large paws that act as natural snowshoes. (© Daniel J. Cox/Stone/Getty Images)

downward in the drawing. This downward-acting force is balanced by an upward-acting force on the palm, so that the hand is in equilibrium. If the hand were rotated by 90°, the directions of these forces would also be rotated by 90°, always being perpendicular to the hand.

A person need not be under water to experience the effects of pressure. Walking about on land, we are at the bottom of the earth's atmosphere, which is a fluid and pushes inward on our bodies just like the water in a swimming pool. As Figure 11.3 indicates, there is enough air above the surface of the earth to create the following pressure at sea level:

**Atmospheric pressure at sea level**      $1.013 \times 10^5$ Pa = 1 atmosphere

This amount of pressure corresponds to 14.70 lb/in.$^2$ and is referred to as one *atmosphere (atm)*. One atmosphere of pressure is a significant amount. Look, for instance, in Figure 11.3 at the results of pumping out the air from within a gasoline can. With no internal air to push outward, the inward push of the external air is unbalanced and is strong enough to crumple the can.

In contrast to the situation in Figure 11.3, reducing the pressure is sometimes beneficial. Lynx, for example, are especially well suited for hunting on snow because of their oversize paws (see Figure 11.4). The large paws function as snowshoes that distribute the weight over a large area. Thus, they reduce the weight per unit area, or the pressure that the cat applies to the surface, which helps to keep it from sinking into the snow.

## 11.3 Pressure and Depth in a Static Fluid

The deeper an underwater swimmer goes, the more strongly the water pushes on his body and the greater is the pressure that he experiences. To determine the relation between pressure and depth, we turn to Newton's second law. In using the second law, we will focus on two external forces that act on the fluid. One is the gravitational force— that is, the weight of the fluid. The other is the collisional force that is responsible for fluid pressure, as the previous section discusses. Since the fluid is at rest, its acceleration is zero ($\mathbf{a} = 0$ m/s$^2$), and it is in equilibrium. By applying the second law in the form $\Sigma \mathbf{F} = 0$, we will derive a relation between pressure and depth. This relation is especially important because it leads to Pascal's principle (Section 11.5) and Archimedes' principle (Section 11.6), both of which are essential in describing the properties of static fluids.

Figure 11.5 shows a container of fluid and focuses attention on one column of the fluid. The free-body diagram in the figure shows all the vertical forces acting on the column. On the top face (area = $A$), the fluid pressure $P_1$ generates a downward force whose magnitude is $P_1 A$. Similarly, on the bottom face, the pressure $P_2$ generates an upward

Pressure = $P_1$
Area = $A$

+y axis

$P_1 A$

$mg$

$h$

$P_2 A$

**Figure 11.5** (*a*) A container of fluid in which one column of the fluid is outlined. The fluid is at rest. (*b*) The free-body diagram, showing the vertical forces acting on the column.

Pressure = $P_2$
Area = $A$
(*a*)

(*b*) Free-body diagram of the column

force of magnitude $P_2A$. The pressure $P_2$ is greater than the pressure $P_1$ because the bottom face supports the weight of more fluid than the upper one does. In fact, the excess weight supported by the bottom face is exactly the weight of the fluid within the column. As the free-body diagram indicates, this weight is $mg$, where $m$ is the mass of the fluid and $g$ is the magnitude of the acceleration due to gravity. Since the column is in equilibrium, we can set the sum of the vertical forces equal to zero and find that

$$\Sigma F_y = P_2A - P_1A - mg = 0 \quad \text{or} \quad P_2A = P_1A + mg$$

The mass $m$ is related to the density $\rho$ and the volume $V$ of the column by $m = \rho V$. Since the volume is the cross-sectional area $A$ times the vertical dimension $h$, we have $m = \rho Ah$. With this substitution, the condition for equilibrium becomes $P_2A = P_1A + \rho Ahg$. The area $A$ can be eliminated algebraically from this expression, with the result that

$$P_2 = P_1 + \rho gh \tag{11.4}$$

Equation 11.4 indicates that if the pressure $P_1$ is known at a higher level, the larger pressure $P_2$ at a deeper level can be calculated by adding the increment $\rho gh$. In determining the pressure increment $\rho gh$, we assumed that the density $\rho$ is the same at any vertical distance $h$ or, in other words, the fluid is incompressible. The assumption is reasonable for liquids, since the bottom layers can support the upper layers with little compression. In a gas, however, the lower layers are compressed markedly by the weight of the upper layers, with the result that the density varies with vertical distance. For example, the density of our atmosphere is larger near the earth's surface than it is at higher altitudes. When applied to gases, the relation $P_2 = P_1 + \rho gh$ can be used only when $h$ is small enough that any variation in $\rho$ is negligible.

A significant feature of Equation 11.4 is that the pressure increment $\rho gh$ is affected by the vertical distance $h$, but not by any horizontal distance within the fluid. Conceptual Example 3 helps to clarify this feature.

## Conceptual Example 3   The Hoover Dam

Lake Mead is the largest wholly artificial reservoir in the United States and was formed after the completion of the Hoover Dam in 1936. As Figure 11.6a suggests, the water in the reservoir backs up behind the dam for a considerable distance (about 200 km or 120 miles). Suppose that all the water were removed, except for a relatively narrow vertical column in contact with the dam. Figure 11.6b shows a side view of this hypothetical situation, in which the water against the dam has the same depth as in Figure 11.6a. Would the Hoover Dam still be needed to contain the water in this hypothetical reservoir, or could a much less massive structure do the job?

**Reasoning and Solution** Since our hypothetical reservoir contains much less water than Lake Mead, it is tempting to say that a much less massive structure than the Hoover Dam would be needed. Not so, however. Hoover Dam would still be needed to contain our hypothetical reservoir. To see why, imagine a small square on the inner face of the dam, located beneath the water. The magnitude of the force on this square is the product of its area and the pressure of the water. But the pressure depends on the depth—that is, on the vertical distance $h$ in Equation 11.4. The horizontal distance of the water behind the dam does not appear in this equation and, therefore, has no effect on the pressure. Consequently, at a given depth the small square experiences a force that is the same in both parts of Figure 11.6. Certainly the dam experiences greater forces generated by greater water pressures at greater depths. But no matter where on the inner face of the dam the square is, the force that the water applies to it depends only on the depth, not on the amount of water backed up behind the dam. *Thus, the dam for our imaginary reservoir would sustain the same forces that the Hoover Dam sustains and would need to be equally large.*

The next example deals further with the relationship between pressure and depth given by Equation 11.4.

## Example 4   The Swimming Hole

Figure 11.7 shows the cross section of a swimming hole. Points $A$ and $B$ are both located at a distance of $h = 5.50$ m below the surface of the water. Find the pressure at each of these two points.

(a)

(b)

**Figure 11.6** (a) The Hoover Dam in Nevada and Lake Mead behind it. (b) This drawing shows a hypothetical reservoir formed by removing most of the water from Lake Mead. Conceptual Example 3 compares the dam needed for this hypothetical reservoir with the Hoover Dam. (© Jim Richardson/Corbis Images)

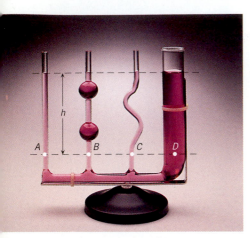

**Figure 11.8** Since points *A, B, C,* and *D* are at the same distance *h* beneath the liquid surface, the pressure at each of them is the same. (© Richard Megna/ Fundamental Photographs)

**Problem solving insight**
The pressure at any point in a fluid depends on the vertical distance *h* of the point beneath the surface. However, for a given vertical distance, the pressure is the same, no matter where the point is located horizontally in the fluid.

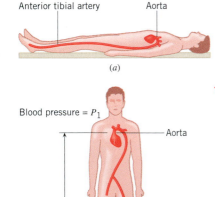

(a)

Blood pressure = $P_1$

Aorta

1.35 m

Anterior tibial artery

Blood pressure = $P_2$

(b)

**Figure 11.9** The blood pressure in the feet can exceed the blood pressure in the heart, depending on whether a person is (a) reclining horizontally or (b) standing.

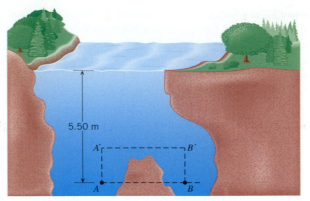

5.50 m

**Figure 11.7** The pressures at points *A* and *B* are the same, since both points are located at the same vertical distance of 5.50 m beneath the surface of the water.

**Reasoning** The pressure at point *B* is the same as that at point *A*, since both are located at the *same vertical distance* beneath the surface and only the vertical distance *h* affects the pressure increment $\rho g h$ in Equation 11.4. To understand this important feature more clearly, consider the path $AA'B'B$ in Figure 11.7. The pressure decreases on the way up along the vertical segment $AA'$ and increases by the same amount on the way back down along segment $B'B$. Since no change in pressure occurs along the horizontal segment $A'B'$, the pressure is the same at *A* and *B*.

**Solution** The pressure acting on the surface of the water is the atmospheric pressure of $1.01 \times 10^5$ Pa. Using this value as $P_1$ in Equation 11.4, we can determine a value for the pressure $P_2$ at either point *A* or *B*, both of which are located 5.50 m under the water. Table 11.1 gives the density of water as $1.000 \times 10^3$ kg/m³.

$$P_2 = P_1 + \rho g h$$
$$P_2 = 1.01 \times 10^5 \text{ Pa} + (1.000 \times 10^3 \text{ kg/m}^3)(9.80 \text{ m/s}^2)(5.50 \text{ m}) = \boxed{1.55 \times 10^5 \text{ Pa}}$$

Figure 11.8 shows an irregularly shaped container of liquid. Reasoning similar to that used in Example 4 leads to the conclusion that the pressure is the same at points *A, B, C,* and *D*, since each is at the same vertical distance *h* beneath the surface. In effect, the arteries in our bodies constitute an irregularly shaped "container" for the blood. The next example examines the blood pressure at different places in this "container."

### *Example 5*   Blood Pressure

Blood in the arteries is flowing, but as a first approximation, the effects of this flow can be ignored and the blood can be treated as a static fluid. Estimate the amount by which the blood pressure $P_2$ in the anterior tibial artery at the foot exceeds the blood pressure $P_1$ in the aorta at the heart when a person is (a) reclining horizontally as in Figure 11.9a and (b) standing as in Figure 11.9b.

**Reasoning and Solution**

**(a)** When the body is horizontal, there is little or no vertical separation between the feet and the heart. Since $h = 0$ m,

$$P_2 - P_1 = \rho g h = \boxed{0 \text{ Pa}} \qquad (11.4)$$

**(b)** When an adult is standing up, the vertical separation between the feet and the heart is about 1.35 m, as Figure 11.9b indicates. Table 11.1 gives the density of blood as 1060 kg/m³, so that

$$P_2 - P_1 = \rho g h = (1060 \text{ kg/m}^3)(9.80 \text{ m/s}^2)(1.35 \text{ m}) = \boxed{1.40 \times 10^4 \text{ Pa}}$$

Sometimes fluid pressure places limits on how a job can be done. Conceptual Example 6 illustrates how fluid pressure restricts the height to which water can be pumped.

## Conceptual Example 6  Pumping Water

Figure 11.10 shows two methods for pumping water from a well. In one method, the pump is submerged in the water at the bottom of the well, while in the other, it is located at ground level. If the well is shallow, either technique can be used. However, if the well is very deep, only one of the methods works. Which is it?

**Reasoning and Solution** To answer this question, we need to examine the nature of the job done by each pump. The pump at the bottom of the well literally pushes water up the pipe. For a very deep well, the column of water becomes very tall, and the pressure at the bottom of the pipe becomes large, due to the pressure increment $\rho gh$ in Equation 11.4. However, as long as the pump can push with sufficient strength to overcome the large pressure, it can shove the next increment of water into the pipe, so the method can be used for very deep wells. In contrast, the pump at ground level does not push water at all. It removes air from the pipe. As the pump reduces the air pressure within the pipe (see point $A$), the greater air pressure outside (see point $B$) pushes water up the pipe. But even the strongest pump can only remove *all* of the air. Once the air is completely removed, an increase in pump strength does not increase the height to which the water is pushed by the external air pressure. ***Thus, a ground-level pump can only cause water to rise to a certain maximum height, so it cannot be used for very deep wells.*** (Incidentally, the pump at ground level works just the way you do when you drink through a straw. You draw some of the air out of the straw, and the external air pressure pushes the liquid up into it.)

**Related Homework:** *Problem 25*

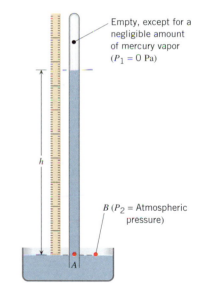

**Figure 11.10** A water pump can be placed at the bottom of a well or at ground level. Conceptual Example 6 discusses how the depth of the well influences the choice of which method to use.

# 11.4 *Pressure Gauges*

One of the simplest pressure gauges is the mercury barometer used for measuring atmospheric pressure. This device is a tube sealed at one end, filled completely with mercury, and then inverted, so that the open end is under the surface of a pool of mercury (see Figure 11.11). Except for a negligible amount of mercury vapor, the space above the mercury in the tube is empty, and the pressure $P_1$ is nearly zero there. The pressure $P_2$ at point $A$ at the bottom of the mercury column is the same as that at point $B$—namely, atmospheric pressure—for these two points are at the same level. With $P_1 = 0$ Pa and $P_2 = P_{atm}$, it follows from Equation 11.4 that $P_{atm} = 0$ Pa $+ \rho gh$. Thus, the atmospheric pressure can be determined from the height $h$ of the mercury in the tube, the density $\rho$ of mercury, and the acceleration due to gravity. Usually weather forecasters report the pressure in terms of the height $h$, expressing it in millimeters or inches of mercury. For instance, using $P_{atm} = 1.013 \times 10^5$ Pa and $\rho = 13.6 \times 10^3$ kg/m$^3$ for the density of mercury, we find that $h = P_{atm}/(\rho g) = 760$ mm (29.9 inches).* Slight variations from this value occur, depending on weather conditions and altitude.

Figure 11.12 shows another kind of pressure gauge, the open-tube manometer. The phrase "open-tube" refers to the fact that one side of the U-tube is open to atmospheric pressure. The tube contains a liquid, often mercury, and its other side is connected to the container whose pressure $P_2$ is to be measured. When the pressure in the container is equal to the atmospheric pressure, the liquid levels in both sides of the U-tube are the same. When the pressure in the container is greater than atmospheric pressure, as in Figure 11.12, the liquid in the tube is pushed downward on the left side and upward on the right side. The relation $P_2 = P_1 + \rho gh$ can be used to determine the container pressure. Atmospheric pressure exists at the top of the right column, so that $P_1 = P_{atm}$. The pressure $P_2$ is the same at points $A$ and $B$, so we find that $P_2 = P_{atm} + \rho gh$, or

$$P_2 - P_{atm} = \rho gh$$

The height $h$ is proportional to $P_2 - P_{atm}$, which is called the **gauge pressure**. The gauge pressure is the amount by which the container pressure differs from atmospheric pressure. The actual value for $P_2$ is called the **absolute pressure**.

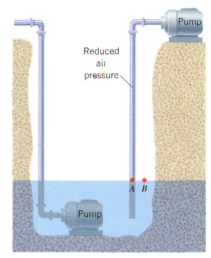

**Figure 11.11** A mercury barometer.

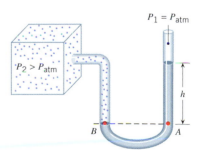

**Figure 11.12** The U-shaped tube is called an open-tube manometer and can be used to measure the pressure $P_2$ in a container.

---

* A pressure of one millimeter of mercury is sometimes referred to as one *torr*, to honor the inventor of the barometer, Evangelista Torricelli (1608–1647). Thus, one atmosphere of pressure is 760 torr.

**Problem solving insight**
When solving problems that deal with pressure, be sure to note the distinction between gauge pressure and absolute pressure.

**The physics of** measuring blood pressure.

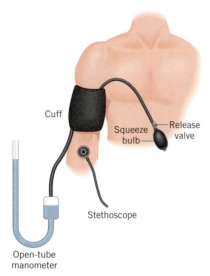

**Figure 11.13** A sphygmomanometer is used to measure blood pressure.

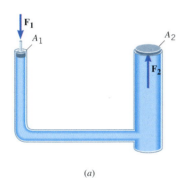

(a)

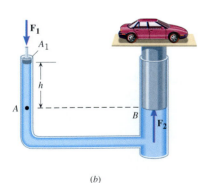

(b)

**Figure 11.14** (a) An external force $\mathbf{F_1}$ is applied to the piston on the left. As a result, a force $\mathbf{F_2}$ is exerted on the cap on the chamber on the right. (b) The familiar hydraulic car lift.

**The physics of** a hydraulic car lift.

The sphygmomanometer is a familiar device for measuring blood pressure. As Figure 11.13 illustrates, a squeeze bulb can be used to inflate the cuff with air, which cuts off the flow of blood through the artery below the cuff. When the release valve is opened, the cuff pressure drops. Blood begins to flow again when the pressure created by the heart at the peak of its beating cycle exceeds the cuff pressure. Using a stethoscope to listen for the initial flow, the operator can measure the corresponding cuff gauge pressure with, for example, an open-tube manometer. This cuff gauge pressure is called the *systolic* pressure. Eventually, there comes a point when even the pressure created by the heart at the low point of its beating cycle is sufficient to cause blood to flow. Identifying this point with the stethoscope, the operator can measure the corresponding cuff gauge pressure, which is referred to as the *diastolic* pressure. The systolic and diastolic pressures are reported in millimeters of mercury, and values of 120 and 80, respectively, are typical of a young, healthy heart.

## 11.5 Pascal's Principle

As we have seen, the pressure in a fluid increases with depth, due to the weight of the fluid above the point of interest. A completely enclosed fluid may be subjected to an additional pressure by the application of an external force. For example, Figure 11.14a shows two interconnected cylindrical chambers. The chambers have different diameters and, together with the connecting tube, are completely filled with a liquid. The larger chamber is sealed at the top with a cap, while the smaller one is fitted with a movable piston. We now begin by asking what determines the pressure $P_1$ at a point immediately beneath the piston. According to the definition of pressure, it is the magnitude $F_1$ of the external force divided by the area $A_1$ of the piston: $P_1 = F_1/A_1$. If it is necessary to know the pressure $P_2$ *at any deeper place in the liquid,* we just add to the value of $P_1$ the increment $\rho gh$, which takes into account the depth $h$ below the piston: $P_2 = P_1 + \rho gh$. The important feature here is this: The pressure $P_1$ adds to the pressure $\rho gh$ due to the depth of the liquid at any point, whether that point is in the smaller chamber, the connecting tube, or the larger chamber. Therefore, if the applied pressure $P_1$ is increased or decreased, the pressure at any other point within the confined liquid changes correspondingly. This behavior is described by *Pascal's principle.*

> ■ **PASCAL'S PRINCIPLE**
>
> Any change in the pressure applied to a completely enclosed fluid is transmitted undiminished to all parts of the fluid and the enclosing walls.

The usefulness of the arrangement in Figure 11.14a becomes apparent when we calculate the force $\mathbf{F_2}$ applied by the liquid to the cap on the right side. The area of the cap is $A_2$ and the pressure there is $P_2$. As long as the tops of the left and right chambers are at the same level, the pressure increment $\rho gh$ is zero, so that the relation $P_2 = P_1 + \rho gh$ becomes $P_2 = P_1$. Consequently, $F_2/A_2 = F_1/A_1$, and

$$F_2 = F_1\left(\frac{A_2}{A_1}\right) \tag{11.5}$$

If area $A_2$ is larger than area $A_1$, a large force $\mathbf{F_2}$ can be applied to the cap on the right chamber, starting with a smaller force $\mathbf{F_1}$ on the left. Depending on the ratio of the areas $A_2/A_1$, the force $\mathbf{F_2}$ can be large indeed, as in the familiar hydraulic car lift shown in part *b*. In this device the force $\mathbf{F_2}$ is not applied to a cap that seals the larger chamber, but, rather, to a movable plunger that lifts a car. Example 7 deals with a hydraulic car lift.

## Example 7 A Car Lift

In a hydraulic car lift, the input piston has a radius of $r_1 = 0.0120$ m and a negligible weight. The output plunger has a radius of $r_2 = 0.150$ m. The combined weight of the car and the plunger is $F_2 = 20\,500$ N. The lift uses hydraulic oil that has a density of $8.00 \times 10^2$ kg/m$^3$.

What input force $F_1$ is needed to support the car and the output plunger when the bottom surfaces of the piston and plunger are at (a) the same level and (b) the levels shown in Figure 11.14b with $h = 1.10$ m?

**Reasoning** When the bottom surfaces of the piston and plunger are at the same levels, as in part (a), Equation 11.5 applies. However, this equation does not apply in part (b), where the bottom surface of the output plunger is $h = 1.10$ m below the input piston. Therefore, our solution in part (b) will take into account the pressure increment $\rho gh$. In either case, we will see that the input force is less than the combined weight of the plunger and car.

**Solution**

(a) Using $A = \pi r^2$ for the circular areas of the piston and plunger, we rearrange Equation 11.5 and find that

$$F_1 = F_2\left(\frac{A_1}{A_2}\right) = F_2\left(\frac{\pi r_1^2}{\pi r_2^2}\right) = (20\ 500\ \text{N})\frac{(0.0120\ \text{m})^2}{(0.150\ \text{m})^2} = \boxed{131\ \text{N}}$$

(b) In Figure 11.14b, the bottom surface of the plunger at point $B$ is at the same level as point $A$, which is at a depth $h$ beneath the input piston. Therefore, we can apply Equation 11.4, $P_2 = P_1 + \rho gh$, with $P_2 = F_2/(\pi r_2^2)$ and $P_1 = F_1/(\pi r_1^2)$:

$$\frac{F_2}{\pi r_2^2} = \frac{F_1}{\pi r_1^2} + \rho gh$$

Solving for $F_1$ gives

$$F_1 = F_2\left(\frac{r_1^2}{r_2^2}\right) - \rho gh(\pi r_1^2)$$

$$= (20\ 500\ \text{N})\frac{(0.0120\ \text{m})^2}{(0.150\ \text{m})^2} - (8.00 \times 10^2\ \text{kg/m}^3)$$

$$\times (9.80\ \text{m/s}^2)(1.10\ \text{m})\ \pi\ (0.0120\ \text{m})^2 = \boxed{127\ \text{N}}$$

The answer here is less than in part (a) because the weight of the 1.10-m column of hydraulic oil provides some of the input force to support the car.

▲

**Problem solving insight**
Note that the relation $F_1 = F_2(A_1/A_2)$, which results from Pascal's principle, applies only when the points 1 and 2 lie at the same depth ($h = 0$ m) in the fluid.

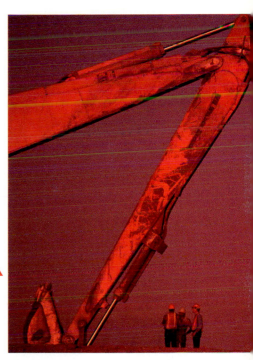

In a device such as a hydraulic car lift, the same amount of work is done by both the input and output forces in the absence of friction. The larger output force $\mathbf{F}_2$ moves through a smaller distance, while the smaller input force $\mathbf{F}_1$ moves through a larger distance. The work, being the product of the magnitude of the force and the distance, is the same in either case since mechanical energy is conserved.

An enormous variety of clever devices use hydraulic fluids, just as the car lift does. In a backhoe, for instance, the fluids multiply a small input force into the large output force required for digging (see Figure 11.15).

**Figure 11.15** A backhoe uses a hydraulic fluid to generate a large output force, starting with a small input force. The output force is far greater than anything generated by humans. (© Richard Hamilton Smith/Corbis Images)

## 11.6 *Archimedes' Principle*

Anyone who has tried to push a beach ball under the water has felt how the water pushes back with a strong upward force. This upward force is called the **buoyant force,** and all fluids apply such a force to objects that are immersed in them. The buoyant force exists because fluid pressure is larger at greater depths.

In Figure 11.16 a cylinder of height $h$ is being held under the surface of a liquid. The pressure $P_1$ on the top face generates the downward force $P_1A$, where $A$ is the area of the face. Similarly, the pressure $P_2$ on the bottom face generates the upward force $P_2A$. Since the pressure is greater at greater depths, the upward force exceeds the downward force. Consequently, the liquid applies to the cylinder a net upward force, or buoyant force, whose magnitude $F_B$ is

$$F_B = P_2A - P_1A = (P_2 - P_1)A = \rho ghA$$

We have substituted $P_2 - P_1 = \rho gh$ from Equation 11.4 into this result. In so doing, we find that the buoyant force equals $\rho ghA$. The quantity $hA$ is the volume of liquid that the cylinder moves aside or displaces in being submerged, and $\rho$ denotes the density of the

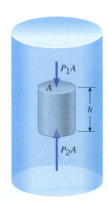

**Figure 11.16** The fluid applies a downward force $P_1A$ to the top face of the submerged cylinder and an upward force $P_2A$ to the bottom face.

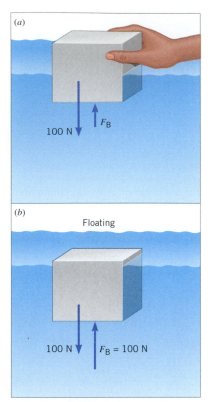

**Figure 11.17** (*a*) An object of weight 100 N is being immersed in a liquid. The deeper the object is, the more liquid it displaces, and the stronger the buoyant force is. (*b*) The buoyant force matches the 100-N weight, so the object floats.

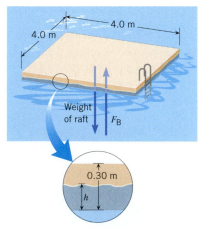

**Figure 11.18** A raft floating with a distance *h* beneath the surface of the water.

**Problem solving insight**
When using Archimedes' principle to find the buoyant force $F_B$ that acts on an object, be sure to use the density of the displaced fluid, not the density of the object.

liquid, not the density of the material from which the cylinder is made. Therefore, $\rho h A$ gives the mass $m$ of the displaced fluid, so that the magnitude of the buoyant force equals $mg$, the weight of the displaced fluid. The phrase "weight of the displaced fluid" refers to the weight of the fluid that would spill out, if the container were filled to the brim before the cylinder is inserted into the liquid. The "buoyant force" is not a new type of force. It is just the name given to the net upward force exerted by the fluid on the object.

The shape of the object in Figure 11.16 is not important. No matter what its shape, the buoyant force pushes it upward in accord with ***Archimedes' principle.*** It was an impressive accomplishment that the Greek scientist Archimedes (ca. 287–212 B.C.) discovered the essence of this principle so long ago.

■ **ARCHIMEDES' PRINCIPLE**

Any fluid applies a buoyant force to an object that is partially or completely immersed in it; the magnitude of the buoyant force equals the weight of the fluid that the object displaces:

$$\underbrace{F_B}_{\substack{\text{Magnitude of}\\\text{buoyant force}}} = \underbrace{W_{\text{fluid}}}_{\substack{\text{Weight of}\\\text{displaced fluid}}} \tag{11.6}$$

The effect that the buoyant force has depends on its strength compared with the strengths of the other forces that are acting. For example, if the buoyant force is strong enough to balance the force of gravity, an object will float in a fluid. Figure 11.17 explores this possibility. In part *a*, a block that weighs 100 N displaces some liquid, and the liquid applies a buoyant force $F_B$ to the block, according to Archimedes' principle. Nevertheless, if the block were released, it would fall further into the liquid because the buoyant force is not sufficiently strong to balance the weight of the block. In part *b*, however, enough of the block is submerged to provide a buoyant force that can balance the 100-N weight, so the block is in equilibrium and floats when released. If the buoyant force were not large enough to balance the weight, even with the block completely submerged, the block would sink. Even if an object sinks, there is still a buoyant force acting on it; it's just that the buoyant force is not large enough to balance the weight. Example 8 provides additional insight into what determines whether an object floats or sinks in a fluid.

### Example 8  A Swimming Raft

A solid, square pinewood raft measures 4.0 m on a side and is 0.30 m thick. (a) Determine whether the raft floats in water, and (b) if so, how much of the raft is beneath the surface (see the distance *h* in Figure 11.18).

**Reasoning** To determine whether the raft floats, we will compare the weight of the raft to the maximum possible buoyant force and see whether there could be enough buoyant force to balance the weight. If so, then the value of the distance *h* can be obtained by utilizing the fact that the floating raft is in equilibrium, with the magnitude of the buoyant force equaling the raft's weight.

**Solution**

(a) The weight of the raft can be calculated from the density $\rho_{\text{pine}} = 550 \text{ kg/m}^3$ (Table 11.1), the volume of the wood, and the acceleration due to gravity. The volume of the wood is $V_{\text{pine}} = 4.0 \text{ m} \times 4.0 \text{ m} \times 0.30 \text{ m} = 4.8 \text{ m}^3$, so that

$$\substack{\text{Weight}\\\text{of raft}} = (\rho_{\text{pine}} V_{\text{pine}})g = (550 \text{ kg/m}^3)(4.8 \text{ m}^3)(9.80 \text{ m/s}^2) = 26\,000 \text{ N}$$

The maximum possible buoyant force occurs when the entire raft is under the surface, displacing 4.8 m³ of water. According to Archimedes' principle, the weight of this volume of water is the maximum buoyant force $F_B^{\text{max}}$. It can be obtained using the density of water:

$$F_B^{\text{max}} = \rho_{\text{water}}V_{\text{pine}} g = (1.000 \times 10^3 \text{ kg/m}^3)(4.8 \text{ m}^3)(9.80 \text{ m/s}^2) = 47\,000 \text{ N}$$

Since the maximum possible buoyant force exceeds the 26 000-N weight of the raft, the raft will float only partially submerged at a distance *h* beneath the water.

**(b)** We now find the value of $h$. The buoyant force balances the raft's weight, so $F_B = 26\ 000$ N. But according to Equation 11.6, the magnitude of the buoyant force is also the weight of the displaced fluid, so $F_B = 26\ 000$ N $= W_{fluid}$. Using the density of water, we can also express the weight of the displaced water as $W_{fluid} = \rho_{water} V_{water} g$, where the volume is $V_{water} = 4.0$ m $\times$ 4.0 m $\times h$. As a result,

$$26\ 000\ \text{N} = W_{fluid} = \rho_{water}(4.0\ \text{m} \times 4.0\ \text{m} \times h)g$$

$$h = \frac{26\ 000\ \text{N}}{\rho_{water}(4.0\ \text{m} \times 4.0\ \text{m})g}$$

$$= \frac{26\ 000\ \text{N}}{(1.000 \times 10^3\ \text{kg/m}^3)(4.0\ \text{m} \times 4.0\ \text{m})(9.80\ \text{m/s}^2)} = \boxed{0.17\ \text{m}}$$

In deciding whether the raft floats in part (a) of Example 8, we compared the raft's weight $(\rho_{pine} V_{pine})g$ to the maximum possible buoyant force $(\rho_{water} V_{pine})g$. The comparison depends only on the densities $\rho_{pine}$ and $\rho_{water}$. The take-home message is this: Any object that is *solid throughout* will float in a liquid if the density of the object is less than or equal to the density of the liquid. For instance, at 0 °C ice has a density of 917 kg/m$^3$, whereas water has a density of 1000 kg/m$^3$. Therefore, ice floats in water.

Although a solid piece of a high-density material like steel will sink in water, such materials can, nonetheless, be used to make floating objects. A supertanker, for example, floats because it is *not* solid metal. It contains enormous amounts of empty space and, because of its shape, displaces enough water to balance its own large weight. Conceptual Example 9 focuses on the reason ships float.

### Conceptual Example 9    How Much Water Is Needed to Float a Ship?

A ship floating in the ocean is a familiar sight. But is all that water really necessary? Can an ocean vessel float in the amount of water that a swimming pool contains, for instance?

**Reasoning and Solution**  In principle, a ship can float in much less than the amount of water that a swimming pool contains. To see why, look at Figure 11.19. Part *a* shows the ship floating in the ocean because it contains empty space within its hull and displaces enough water to balance its own weight. Part *b* shows the water that the ship displaces, which, according to Archimedes' principle, has a weight that equals the ship's weight. Although this wedge-shaped portion of water represents the water *displaced* by the ship, it is not the amount that must be present to float the ship, as part *c* illustrates. This part of the drawing shows a canal, the cross section of which matches the shape in part *b*. *All that is needed, in principle, is a thin section of water that separates the hull of the floating ship from the sides of the canal.* This thin section of water could have a very small volume indeed.

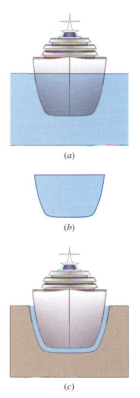

**Figure 11.19**  (*a*) A ship floating in the ocean. (*b*) This is the water that the ship displaces. (*c*) The ship floats here in a canal that has a cross section similar in shape to that in part *b*.

A useful application of Archimedes' principle can be found in car batteries. To alert the owner that recharging is necessary, some batteries include a state-of-charge indicator, such as the one illustrated in Figure 11.20 on p. 218. The battery includes a viewing port that looks down through a plastic rod, which extends into the battery acid. Attached to the end of this rod is a "cage," containing a green ball. The cage has holes in it that allow the acid to enter. When the battery is charged, the density of the acid is great enough that its buoyant force makes the ball rise to the top of the cage, to just beneath the plastic rod. The viewing port shows a green dot. As the battery discharges, the density of the acid decreases. Since the buoyant force is the weight of the acid displaced by the ball, the buoyant force also decreases. As a result, the ball sinks into one of the two chambers oriented at an angle beneath it (see Figure 11.20). With the ball no longer visible, the viewing port shows a dark or black dot, warning that the battery charge is low.

Archimedes' principle has allowed us to determine how an object can float in a liquid. This principle also applies to gases, as the next example illustrates.

**The physics of**
**a state-of-charge battery indicator.**

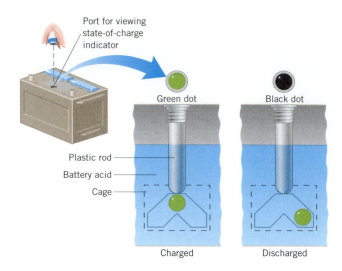

Port for viewing
state-of-charge
indicator

Green dot

Black dot

Plastic rod

Battery acid

Cage

Charged

Discharged

**Figure 11.20** A state-of-charge indicator for a car battery.

### Example 10  A Goodyear Airship

**The physics of**
**a Goodyear airship.**

Normally, a Goodyear airship, such as that in Figure 11.21, contains about $5.40 \times 10^3$ m$^3$ of helium (He) whose density is 0.179 kg/m$^3$. Find the weight of the load $W_L$ that the airship can carry in equilibrium at an altitude where the density of air is 1.20 kg/m$^3$.

**Reasoning**  The airship and its load are in equilibrium. Thus, the buoyant force $F_B$ applied to the airship by the surrounding air balances the weight $W_{He}$ of the helium and the weight $W_L$ of the load, including the solid parts of the airship. The free-body diagram in Figure 11.21$b$ shows these forces.

**Solution**  Because the forces in the free-body diagram balance, we have

$$W_{He} + W_L = F_B \quad \text{or} \quad W_L = F_B - W_{He}$$

According to Archimedes' principle, the buoyant force is the weight of the displaced air: $F_B = W_{air} = \rho_{air} V_{ship} g$. The weight of the helium is $W_{He} = \rho_{He} V_{ship} g$, where we have assumed that the volume of the helium and the volume of the airship have nearly the same value of $V_{ship} = 5.40 \times 10^3$ m$^3$. Thus, we find that

$$W_L = \rho_{air} V_{ship} g - \rho_{He} V_{ship} g = (\rho_{air} - \rho_{He}) V_{ship} g$$
$$W_L = (1.20 \text{ kg/m}^3 - 0.179 \text{ kg/m}^3)(5.40 \times 10^3 \text{ m}^3) \times (9.80 \text{ m/s}^2) = \boxed{5.40 \times 10^4 \text{ N}}$$

**Figure 11.21**  ($a$) A helium-filled Goodyear airship soars over the Sydney Olympic stadium in Australia at the summer 2000 Olympic games. The airship has been painted with the words 'G'day' on one side and 'Good Luck' on the other. ($b$) The free-body diagram of the airship, including the load weight $W_L$. (©AP/Wide World Photos)

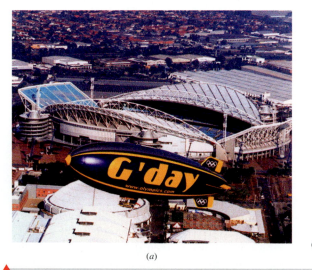

($a$)

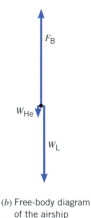

$F_B$

$W_{He}$

$W_L$

($b$) Free-body diagram
of the airship

## 11.7  *Fluids in Motion*

Fluids can move or flow in many ways. Water may flow smoothly and slowly in a quiet stream or violently over a waterfall. The air may form a gentle breeze or a raging tornado. To deal with such diversity, it helps to identify some of the basic types of fluid flow.

**Fluid flow can be steady or unsteady.** In *steady flow* the velocity of the fluid particles at any point is constant as time passes. For instance, in Figure 11.22 a fluid particle flows with a velocity of $\mathbf{v_1} = +2$ m/s past point 1. In steady flow every particle passing through this point has this same velocity. At another location the velocity may be different, as in a river, which usually flows fastest near its center and slowest near its banks. Thus, at point 2 in the figure, the fluid velocity is $\mathbf{v_2} = +0.5$ m/s, and if the flow is steady, all particles passing through this point have a velocity of $+0.5$ m/s. *Unsteady flow* exists whenever the velocity at a point in the fluid changes as time passes. *Turbulent flow* is an extreme kind of unsteady flow and occurs when there are sharp obstacles or bends in the path of a fast-moving fluid, as in the rapids in Figure 11.23. In turbulent flow, the velocity at a point changes erratically from moment to moment, both in magnitude and direction.

**Fluid flow can be compressible or incompressible.** Most liquids are nearly incompressible; that is, the density of a liquid remains almost constant as the pressure changes. To a good approximation, then, liquids flow in an incompressible manner. In contrast, gases are highly compressible. However, there are situations in which the density of a flowing gas remains constant enough that the flow can be considered incompressible.

**Fluid flow can be viscous or nonviscous.** A viscous fluid, such as honey, does not flow readily and is said to have a large viscosity. In contrast, water is less viscous and flows more readily; water has a smaller viscosity than honey. The flow of a viscous fluid is an energy-dissipating process. The viscosity hinders neighboring layers of fluid from sliding freely past one another. A fluid with zero viscosity flows in an unhindered manner with no dissipation of energy. Although no real fluid has zero viscosity at normal temperatures, some fluids have negligibly small viscosities. An incompressible, nonviscous fluid is called an *ideal fluid.*

When the flow is steady, *streamlines* are often used to represent the trajectories of the fluid particles. A streamline is a line drawn in the fluid such that a tangent to the streamline at any point is parallel to the fluid velocity at that point. Figure 11.24 shows the velocity vectors at three points along a streamline. The fluid velocity can vary (in both magnitude and direction) from point to point along a streamline, but at any given point, the velocity is constant in time, as required by the condition of steady flow. In fact, steady flow is often called *streamline flow.*

Figure 11.25*a* illustrates a method for making streamlines visible by using small tubes to release a colored dye into the moving liquid. The dye does not immediately mix with the liquid and is carried along a streamline. In the case of a flowing gas, such as that in a wind tunnel, streamlines are often revealed by smoke streamers, as part *b* of the figure shows.

In steady flow, the pattern of streamlines is steady in time, and, as Figure 11.25*a* indicates, no two streamlines cross one another. If they did cross, every particle arriving at the crossing point could go one way or the other. This would mean that the velocity at the crossing point would change from moment to moment, a condition that does not exist in steady flow.

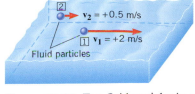

**Figure 11.22** Two fluid particles in a stream. At different locations in the stream the particle velocities may be different, as indicated by $\mathbf{v_1}$ and $\mathbf{v_2}$.

**Figure 11.23** The flow of water in white-water rapids is an example of turbulent flow. (© Michael Kevin Daly/Corbis Images)

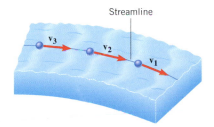

**Figure 11.24** At any point along a streamline, the velocity vector of the fluid particle at that point is tangent to the streamline.

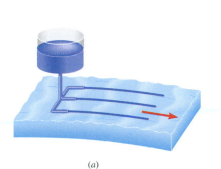

(*a*)

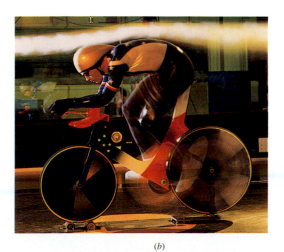

(*b*)

**Figure 11.25** (*a*) In the steady flow of a liquid, a colored dye reveals the streamlines. (*b*) A smoke streamer reveals a streamline pattern for the air flowing around this pursuit cyclist, as he tests his bike for wind resistance in a wind tunnel. (© Nicholas Pinturas/Stone/ Getty Images)

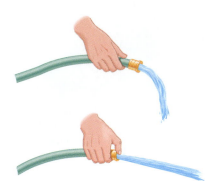

**Figure 11.26** When the end of a hose is partially closed off, thus reducing its cross-sectional area, the fluid velocity increases.

# 11.8   *The Equation of Continuity*

Have you ever used your thumb to control the water flowing from the end of a hose, as in Figure 11.26? If so, you have seen that the water velocity increases when your thumb reduces the cross-sectional area of the hose opening. This kind of fluid behavior is described by the **equation of continuity.** This equation expresses the following simple idea: If a fluid enters one end of a pipe at a certain rate (e.g., 5 kilograms per second), then fluid must also leave at the same rate, assuming that there are no places between the entry and exit points to add or remove fluid. The mass of fluid per second (e.g., 5 kg/s) that flows through a tube is called the **mass flow rate.**

Figure 11.27 shows a small mass of fluid or fluid element (dark blue) moving along a tube. Upstream at position 2, where the tube has a cross-sectional area $A_2$, the fluid has a speed $v_2$ and a density $\rho_2$. Downstream at location 1, the corresponding quantities are $A_1$, $v_1$, and $\rho_1$. During a small time interval $\Delta t$, the fluid at point 2 moves a distance of $v_2\Delta t$, as the drawing shows. The volume of fluid that has flowed past this point is the cross-sectional area times this distance, or $A_2v_2\Delta t$. The mass $\Delta m_2$ of this fluid element is the product of the density and volume: $\Delta m_2 = \rho_2 A_2 v_2 \Delta t$. Dividing $\Delta m_2$ by $\Delta t$ gives the mass flow rate (the mass per second):

$$\frac{\text{Mass flow rate}}{\text{at position 2}} = \frac{\Delta m_2}{\Delta t} = \rho_2 A_2 v_2 \qquad (11.7a)$$

Similar reasoning leads to the mass flow rate at position 1:

$$\frac{\text{Mass flow rate}}{\text{at position 1}} = \frac{\Delta m_1}{\Delta t} = \rho_1 A_1 v_1 \qquad (11.7b)$$

Since no fluid can cross the sidewalls of the tube, the mass flow rates at positions 1 and 2 must be equal. But these positions were selected arbitrarily, so the mass flow rate has the same value everywhere in the tube, an important result known as the **equation of continuity.** The equation of continuity is an expression of the fact that mass is conserved (i.e., neither created nor destroyed) as the fluid flows.

---

■  **EQUATION OF CONTINUITY**

The mass flow rate ($\rho Av$) has the same value at every position along a tube that has a single entry and a single exit point for fluid flow. For two positions along such a tube

$$\rho_1 A_1 v_1 = \rho_2 A_2 v_2 \qquad (11.8)$$

where $\rho$ = fluid density (kg/m$^3$).
     $A$ = cross-sectional area of tube (m$^2$)
     $v$ = fluid speed (m/s)

**SI Unit of Mass Flow Rate:** kg/s

---

The density of an incompressible fluid does not change during flow, so that $\rho_1 = \rho_2$, and the equation of continuity reduces to

*Incompressible fluid*          $A_1 v_1 = A_2 v_2$          (11.9)

**Figure 11.27** In general, a fluid flowing in a tube that has different cross-sectional areas $A_1$ and $A_2$ at positions 1 and 2 also has different velocities $\mathbf{v_1}$ and $\mathbf{v_2}$ at these positions.

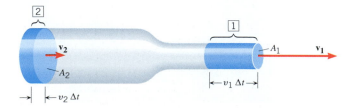

The quantity $Av$ represents the volume of fluid per second that passes through the tube and is referred to as the **volume flow rate Q:**

$$Q = \text{Volume flow rate} = Av \qquad (11.10)$$

Equation 11.9 shows that where the tube's cross-sectional area is large, the fluid speed is small, and, conversely, where the tube's cross-sectional area is small, the speed is large. Example 11 explores this behavior in more detail for the hose in Figure 11.26.

### *Example 11*   A Garden Hose

A garden hose has an unobstructed opening with a cross-sectional area of $2.85 \times 10^{-4} \text{ m}^2$, from which water fills a bucket in 30.0 s. The volume of the bucket is $8.00 \times 10^{-3} \text{ m}^3$ (about two gallons). Find the speed of the water that leaves the hose through (a) the unobstructed opening and (b) an obstructed opening with half as much area.

**Reasoning** If we can determine the volume flow rate $Q$, the speed of the water can be obtained from Equation 11.10 as $v = Q/A$, since the area $A$ is given. The volume flow rate can be found from the volume of the bucket and its fill time.

**Solution**

(a) The volume flow rate $Q$ is equal to the volume of the bucket divided by the fill time. Therefore, the speed of the water is

$$v = \frac{Q}{A} = \frac{(8.00 \times 10^{-3} \text{ m}^3)/(30.0 \text{ s})}{2.85 \times 10^{-4} \text{ m}^2} = \boxed{0.936 \text{ m/s}} \qquad (11.10)$$

(b) Water can be considered incompressible, so the equation of continuity can be applied in the form $A_1 v_1 = A_2 v_2$. Since $A_2 = \frac{1}{2} A_1$, we find that

$$v_2 = \left(\frac{A_1}{A_2}\right) v_1 = \left(\frac{A_1}{\frac{1}{2} A_1}\right)(0.936 \text{ m/s}) = \boxed{1.87 \text{ m/s}} \qquad (11.9)$$

The next example applies the equation of continuity to the flow of blood.

### *Example 12*   A Clogged Artery

In the condition known as atherosclerosis, a deposit or atheroma forms on the arterial wall and reduces the opening through which blood can flow. In the carotid artery in the neck, blood flows three times faster through a partially blocked region than it does through an unobstructed region. Determine the ratio of the effective radii of the artery at the two places.

 *The physics of* a clogged artery.

**Reasoning** Blood, like most liquids, is incompressible, and the equation of continuity in the form of $A_1 v_1 = A_2 v_2$ (Equation 11.9) can be applied. In applying this equation, we use the fact that the area of a circle is $\pi r^2$.

**Solution** From Equation 11.9, it follows that

$$\underbrace{(\pi r_U^2) v_U}_{\substack{\text{Unobstructed} \\ \text{volume flow rate}}} = \underbrace{(\pi r_O^2) v_O}_{\substack{\text{Obstructed} \\ \text{volume flow rate}}}$$

The ratio of the radii is

$$\frac{r_U}{r_O} = \sqrt{\frac{v_O}{v_U}} = \sqrt{3} = \boxed{1.7}$$

**Problem solving insight**
The equation of continuity in the form $A_1 v_1 = A_2 v_2$ applies only when the density of the fluid is constant. If the density is not constant, the equation of continuity is $\rho_1 A_1 v_1 = \rho_2 A_2 v_2$.

## 11.9  *Bernoulli's Equation*

For *steady flow,* the speed, pressure, and elevation of an *incompressible and nonviscous* fluid are related by an equation discovered by Daniel Bernoulli (1700–1782). To derive **Bernoulli's equation,** we will use the work–energy theorem. This theorem, which is introduced in Chapter 6, states that the net work $W_{nc}$ done on an object by external nonconservative forces is equal to the change in the total mechanical energy of the object (see Equation 6.8). As mentioned earlier, the pressure within a fluid is caused by collisional

**Figure 11.28** (*a*) In this horizontal pipe, the pressure in region 2 is greater than that in region 1. The difference in pressures leads to the net force that accelerates the fluid to the right. (*b*) When the fluid changes elevation, the pressure at the bottom is greater than that at the top, assuming the cross-sectional area of the pipe is constant.

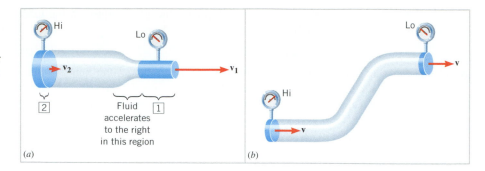

forces, which are nonconservative. Therefore, when a fluid is accelerated because of a difference in pressures, work is being done by nonconservative forces ($W_{nc} \neq 0$ J), and this work changes the total mechanical energy of the fluid from an initial value of $E_0$ to a final value of $E_f$. The total mechanical energy is not conserved. We will now see how the work–energy theorem leads directly to Bernoulli's equation.

To begin with, let us make two observations about a moving fluid. First, whenever a fluid is flowing in a horizontal pipe and encounters a region of reduced cross-sectional area, the pressure of the fluid drops, as the pressure gauges in Figure 11.28*a* indicate. The reason for this follows from Newton's second law. When moving from the wider region 2 to the narrower region 1, the fluid speeds up or accelerates, consistent with the conservation of mass (as expressed by the equation of continuity). According to the second law, the accelerating fluid must be subjected to an unbalanced force. But there can be an unbalanced force only if the pressure in region 2 exceeds the pressure in region 1. We will see that the difference in pressures is given by Bernoulli's equation. The second observation is that if the fluid moves to a higher elevation, the pressure at the lower level is greater than the pressure at the higher level, as in Figure 11.28*b*. The basis for this observation is our previous study of static fluids, and Bernoulli's equation will confirm it, provided that the cross-sectional area of the pipe does not change.

To derive Bernoulli's equation, consider Figure 11.29*a*. This drawing shows a fluid element of mass $m$, upstream in region 2 of a pipe. Both the cross-sectional area and the elevation are different at different places along the pipe. The speed, pressure, and elevation in this region are $v_2$, $P_2$, and $y_2$, respectively. Downstream in region 1 these variables have the values $v_1$, $P_1$, and $y_1$. As Chapter 6 discusses, an object moving under the influence of gravity has a total mechanical energy $E$ that is the sum of the kinetic energy KE and the gravitational potential energy PE: $E = \text{KE} + \text{PE} = \frac{1}{2}mv^2 + mgy$. When work $W_{nc}$ is done on the fluid element by external nonconservative forces, the total mechanical energy changes. According to the work–energy theorem, the work equals the change in the total mechanical energy:

$$W_{nc} = E_1 - E_2 = \underbrace{(\tfrac{1}{2}mv_1^2 + mgy_1)}_{\substack{\text{Total mechanical} \\ \text{energy in region 1}}} - \underbrace{(\tfrac{1}{2}mv_2^2 + mgy_2)}_{\substack{\text{Total mechanical} \\ \text{energy in region 2}}} \qquad (6.8)$$

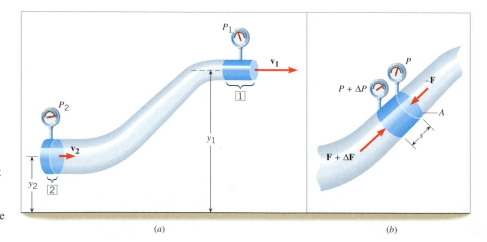

**Figure 11.29** (*a*) A fluid element (dark blue) moving through a pipe whose cross-sectional area and elevation change. (*b*) The fluid element experiences a force $-\mathbf{F}$ on its top surface due to the fluid above it, and a force $\mathbf{F} + \Delta\mathbf{F}$ on its bottom surface due to the fluid below it.

Figure 11.29b helps us understand how the work $W_{nc}$ arises. On the top surface of the fluid element, the surrounding fluid exerts a pressure $P$. This pressure gives rise to a force of magnitude $F = PA$, where $A$ is the cross-sectional area. On the bottom surface, the surrounding fluid exerts a slightly greater pressure, $P + \Delta P$, where $\Delta P$ is the pressure difference between the ends of the element. As a result, the force on the bottom surface has a magnitude of $F + \Delta F = (P + \Delta P)A$. The magnitude of the *net* force pushing the fluid element up the pipe is $\Delta F = (\Delta P)A$. When the fluid element moves through its own length $s$, the work done is the product of the magnitude of the net force and the distance: Work $= (\Delta F)s = (\Delta P)As$. The quantity $As$ is the volume $V$ of the element, so the work is $(\Delta P)V$. The total work done on the fluid element in moving it from region 2 to region 1 is the sum of the small increments of work $(\Delta P)V$ done as the element moves along the pipe. This sum amounts to $W_{nc} = (P_2 - P_1)V$, where $P_2 - P_1$ is the pressure difference between the two regions. With this expression for $W_{nc}$, the work–energy theorem becomes

$$W_{nc} = (P_2 - P_1)V = (\tfrac{1}{2}mv_1^2 + mgy_1) - (\tfrac{1}{2}mv_2^2 + mgy_2)$$

By dividing both sides of this result by the volume $V$, recognizing that $m/V$ is the density $\rho$ of the fluid, and rearranging terms, we obtain Bernoulli's equation.

■ **BERNOULLI'S EQUATION**

In the steady flow of a nonviscous, incompressible fluid of density $\rho$, the pressure $P$, the fluid speed $v$, and the elevation $y$ at any two points (1 and 2) are related by

$$P_1 + \tfrac{1}{2}\rho v_1^2 + \rho g y_1 = P_2 + \tfrac{1}{2}\rho v_2^2 + \rho g y_2 \qquad (11.11)$$

Since the points 1 and 2 were selected arbitrarily, the term $P + \tfrac{1}{2}\rho v^2 + \rho g y$ has a constant value at all positions in the flow. For this reason, Bernoulli's equation is sometimes expressed as $P + \tfrac{1}{2}\rho v^2 + \rho g y = $ constant.

Equation 11.11 can be regarded as an extension of the earlier result that specifies how the pressure varies with depth in a static fluid ($P_2 = P_1 + \rho g h$), the terms $\tfrac{1}{2}\rho v_1^2$ and $\tfrac{1}{2}\rho v_2^2$ accounting for the effects of fluid speed. Bernoulli's equation reduces to the result for static fluids when the speed of the fluid is the same everywhere ($v_1 = v_2$), as it is when the cross-sectional area remains constant. Under such conditions, Bernoulli's equation is $P_1 + \rho g y_1 = P_2 + \rho g y_2$. After rearrangement, this result becomes

$$P_2 = P_1 + \rho g (y_1 - y_2) = P_1 + \rho g h$$

which is the result (Equation 11.4) for static fluids.

## 11.10 Applications of Bernoulli's Equation

When a moving fluid is contained in a horizontal pipe, all parts of it have the same elevation ($y_1 = y_2$), and Bernoulli's equation simplifies to

$$P_1 + \tfrac{1}{2}\rho v_1^2 = P_2 + \tfrac{1}{2}\rho v_2^2 \qquad (11.12)$$

Thus, the quantity $P + \tfrac{1}{2}\rho v^2$ remains constant throughout a horizontal pipe; if $v$ increases, $P$ decreases and vice versa. This is exactly the result that we deduced qualitatively from Newton's second law at the beginning of Section 11.9, and Conceptual Example 13 illustrates it.

### Conceptual Example 13  Tarpaulins and Bernoulli's Equation

A tarpaulin is a piece of canvas that is used to cover a cargo, like that pulled by the truck in Figure 11.30. When the truck is stationary the tarpaulin lies flat, but it bulges outward when the truck is speeding down the highway. Account for this behavior.

**Reasoning and Solution**  The behavior is a direct consequence of the pressure changes that Bernoulli's equation describes for flowing fluids. When the truck is stationary, the air outside

Tarpaulin is flat

Stationary

Tarpaulin bulges outward

Moving

**Figure 11.30**  The tarpaulin that covers the cargo is flat when the truck is stationary but bulges outward when the truck is moving.

and inside the cargo area is stationary, so the air pressure is the same in both places. This pressure applies the same force to the outer and inner surfaces of the canvas, with the result that the tarpaulin lies flat. When the truck is moving, the outside air rushes over the top surface of the canvas. In accord with Bernoulli's equation (Equation 11.12), the moving air has a lower pressure than the stationary air within the cargo area. *The greater inside pressure generates a greater force on the inner surface of the canvas, and the tarpaulin bulges outward.*

**Related Homework:** *Problem 57*

Example 14 applies Equation 11.12 to a dangerous physiological condition known as an aneurysm.

### *Example 14*   An Enlarged Blood Vessel

**The physics of an aneurysm.**

An aneurysm is an abnormal enlargement of a blood vessel such as the aorta. Suppose that, because of an aneurysm, the cross-sectional area $A_1$ of the aorta increases to a value $A_2 = 1.7A_1$. The speed of the blood ($\rho = 1060$ kg/m³) through a normal portion of the aorta is $v_1 = 0.40$ m/s. Assuming that the aorta is horizontal (the person is lying down), determine the amount by which the pressure $P_2$ in the enlarged region exceeds the pressure $P_1$ in the normal region.

**Reasoning**  Bernoulli's equation (Equation 11.12) may be used to find the pressure difference between two points in a fluid moving horizontally. However, in order to use this relation we need to know the speed of the blood in the enlarged region of the artery, as well as the speed in the normal section. The speed in the enlarged region can be obtained from the equation of continuity (Equation 11.9), which relates the blood speeds in the enlarged and normal regions to the cross-sectional areas of these regions.

**Solution**  For purposes of using Bernoulli's equation, we define region 2 to be the aneurysm. In the form for horizontal flow, this equation is $P_1 + \frac{1}{2}\rho v_1^2 = P_2 + \frac{1}{2}\rho v_2^2$, the difference in pressure being $P_2 - P_1 = \frac{1}{2}\rho(v_1^2 - v_2^2)$. From the equation of continuity, the speed $v_2$ of the blood in the aneurysm is given by $v_2 = (A_1/A_2)v_1$, where $A_2$ is the cross-sectional area of the enlarged artery, and $A_1$ and $v_1$ are the cross-sectional area and blood speed in the normal artery. Substituting this value for $v_2$ into Bernoulli's equation yields

$$P_2 - P_1 = \tfrac{1}{2}\rho(v_1^2 - v_2^2) = \tfrac{1}{2}\rho v_1^2\left[1 - \left(\frac{A_1}{A_2}\right)^2\right]$$

$$= \tfrac{1}{2}(1060 \text{ kg/m}^3)(0.40 \text{ m/s})^2\left[1 - \frac{A_1^2}{(1.7A_1)^2}\right] = \boxed{55 \text{ Pa}}$$

**The physics of household plumbing (see below).**

This result is positive, indicating that $P_2$ is greater than $P_1$. The excess pressure puts added stress on the already weakened tissue of the arterial wall at the aneurysm.

**Figure 11.31** In a household plumbing system, a vent is necessary to equalize the pressures at points $A$ and $B$, thus preventing the trap from being emptied. An empty trap allows sewer gas to enter the house.

The impact of fluid flow on pressure is widespread. Figure 11.31, for instance, illustrates how household plumbing takes into account the implications of Bernoulli's equation. The U-shaped section of pipe beneath the sink is called a "trap," because

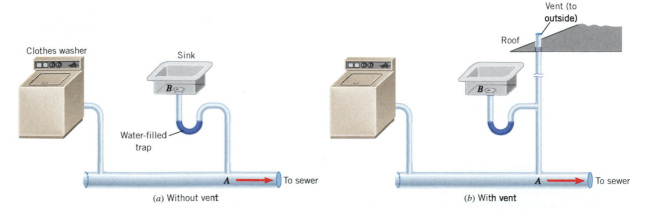

(a) Without vent                    (b) With **vent**

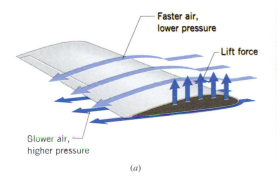

*(a)*  *(b)*

**Figure 11.32** (*a*) Air flowing around an airplane wing. The wing is moving to the right. (*b*) The end of this wing has roughly the shape indicated in part *a*. (© Gary Gladstone/The Image Bank/Getty Images)

it traps water, which serves as a barrier to prevent sewer gas from leaking into the house. Part *a* of the drawing shows poor plumbing. When water from the clothes washer rushes through the sewer pipe, the high-speed flow causes the pressure at point *A* to drop. The pressure at point *B* in the sink, however, remains at the higher atmospheric pressure. As a result of this pressure difference, the water is pushed out of the trap and into the sewer line, leaving no protection against sewer gas. A correctly designed system is vented to the outside of the house, as in Figure 11.31*b*. The vent ensures that the pressure at *A* remains the same as that at *B* (atmospheric pressure), even when water from the clothes washer is rushing through the pipe. Thus, the purpose of the vent is to prevent the trap from being emptied, not to provide an escape route for sewer gas.

One of the most spectacular examples of how fluid flow affects pressure is the dynamic lift on airplane wings. Figure 11.32*a* shows a wing (in cross section) moving to the right, with the air flowing leftward past the wing. Because of the shape of the wing, the air travels faster over the curved upper surface than it does over the flatter lower surface. According to Bernoulli's equation, the pressure above the wing is lower (faster moving air), while the pressure below the wing is higher (slower moving air). Thus, the wing is lifted upward. Part *b* of the figure shows the wing of an airplane.

*The physics of* **airplane wings.**

The curveball, one of the main weapons in the arsenal of a baseball pitcher, is another illustration of the effects of fluid flow. Figure 11.33*a* shows a baseball moving to the right with no spin. The view is from above, looking down toward the ground. In this situation, air flows with the same speed around both sides of the ball, and the pressure is the same on both sides. No net force exists to make the ball curve to either side. However, when the ball is given a spin, the air close to its surface is dragged around with it; the air on one-half of the ball is speeded up (lower pressure), while that on the other half is slowed down (higher pressure). Part *b* of the picture illustrates the effects of a counterclockwise spin. The baseball experiences a net deflection force and curves on its way from the pitcher's mound to the plate, as part *c* shows.*

*The physics of* **a curveball.**

As a final application of Bernoulli's equation, Figure 11.34*a* shows a large tank from which water is emerging through a small pipe near the bottom. Bernoulli's equation can

**Figure 11.33** These views of a baseball are from above, looking down toward the ground, with the ball moving to the right. (*a*) Without spin, the ball does not curve to either side. (*b*) A spinning ball curves in the direction of the deflection force. (*c*) The spin in part *b* causes the ball to curve as shown here.

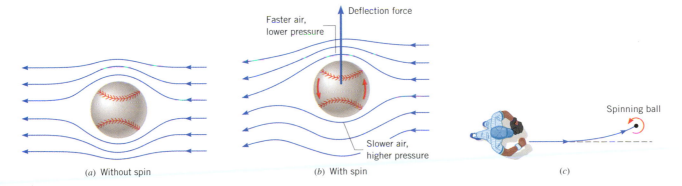

*(a)* Without spin    *(b)* With spin    *(c)*

* In the jargon used in baseball, the pitch shown in Figure 11.33 is called a "slider."

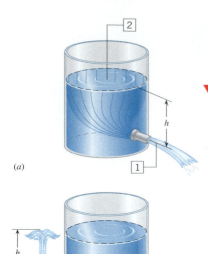

*(a)*

*(b)*

**Figure 11.34** (*a*) Bernoulli's equation can be used to determine the speed of the liquid leaving the small pipe. (*b*) An ideal fluid (no viscosity) will rise to the fluid level in the tank after leaving a vertical outlet nozzle.

be used to determine the speed (called the efflux speed) at which the water leaves the pipe, as the next example shows.

### Example 15 Efflux Speed

The tank in Figure 11.34*a* is open to the atmosphere at the top. Find an expression for the speed of the liquid leaving the pipe at the bottom.

**Reasoning** We assume that the liquid behaves as an ideal fluid. Therefore, we can apply Bernoulli's equation, and in preparation for doing so, we locate two points in the liquid in Figure 11.34*a*. Point 1 is just outside the efflux pipe, and point 2 is at the top surface of the liquid. The pressure at each of these points is equal to the atmospheric pressure, a fact that will be used to simplify Bernoulli's equation.

**Solution** Since the pressures at points 1 and 2 are the same, we have $P_1 = P_2$, and Bernoulli's equation becomes $\frac{1}{2}\rho v_1^2 + \rho g y_1 = \frac{1}{2}\rho v_2^2 + \rho g y_2$. The density $\rho$ can be eliminated algebraically from this result, which can then be solved for the square of the efflux speed $v_1$:

$$v_1^2 = v_2^2 + 2g(y_2 - y_1) = v_2^2 + 2gh$$

We have substituted $h = y_2 - y_1$ for the height of the liquid above the efflux tube. If the tank is very large, the liquid level changes only slowly, and the speed at point 2 can be set equal to zero, so that $\boxed{v_1 = \sqrt{2gh}}$.

In Example 15 the liquid is assumed to be an ideal fluid, and the speed with which it leaves the pipe is the same as if the liquid had freely fallen through a height $h$ (see Equation 2.9 with $x = h$ and $a = g$). This result is known as **Torricelli's theorem.** If the outlet pipe were pointed directly upward, as in Figure 11.34*b*, the liquid would rise to a height $h$ equal to the fluid level above the pipe. However, if the liquid is not an ideal fluid, its viscosity cannot be neglected. Then, the efflux speed would be less than that given by Bernoulli's equation, and the liquid would rise to a height less than $h$.

## Concept Summary

This summary presents an abridged version of the chapter, including the important equations and all available learning aids. For convenient reference, the learning aids (including the text's examples) are placed next to or immediately after the relevant equation or discussion. The following learning aids may be found on-line at **www.wiley.com/college/cutnell**:

| | |
|---|---|
| **Interactive LearningWare** examples are solved according to a five-step interactive format that is designed to help you develop problem-solving skills. | **Concept Simulations** are animated versions of text figures or animations that illustrate important concepts. You can control parameters that affect the display, and we encourage you to experiment. |
| **Interactive Solutions** offer specific models for certain types of problems in the chapter homework. The calculations are carried out interactively. | **Self-Assessment Tests** include both qualitative and quantitative questions. Extensive feedback is provided for both incorrect and correct answers, to help you evaluate your understanding of the material. |

| *Topic* | *Discussion* | *Learning Aids* |
|---|---|---|

### 11.1 Mass Density

Fluids are materials that can flow, and they include gases and liquids.

The mass density $\rho$ of a substance is its mass $m$ divided by its volume $V$:

**Mass Density**

$$\rho = \frac{m}{V} \qquad (11.1) \quad \textbf{Example 1}$$

The specific gravity of a substance is its mass density divided by the density of water at 4 °C ($1.000 \times 10^3$ kg/m³):

**Specific gravity**

$$\text{Specific gravity} = \frac{\text{Density of substance}}{1.000 \times 10^3 \text{ kg/m}^3} \qquad (11.2)$$

| Topic | Discussion | Learning Aids |
|---|---|---|

### 11.2 Pressure

The pressure $P$ exerted by a fluid is the magnitude $F$ of the force acting perpendicular to a surface embedded in the fluid divided by the area $A$ over which the force acts:

**Pressure**

$$P = \frac{F}{A} \qquad (11.3)$$

**Example 2**

**Interactive Solution 11.15**

The SI unit for measuring pressure is the pascal (Pa); $1\ \text{Pa} = 1\ \text{N/m}^2$.

**Atmospheric pressure**  One atmosphere of pressure is $1.013 \times 10^5$ Pa or 14.7 lb/in.$^2$

### 11.3 Pressure and Depth in a Static Fluid

In the presence of gravity, the upper layers of a fluid push downward on the layers beneath, with the result that fluid pressure is related to depth. In an incompressible static fluid whose density is $\rho$, the relation is

$$P_2 = P_1 + \rho g h \qquad (11.4)$$

**Examples 3–6**

where $P_1$ is the pressure at one level, $P_2$ is the pressure at a level that is $h$ meters deeper, and $g$ is the magnitude of the acceleration due to gravity.

### 11.4 Pressure Gauges

Two basic types of pressure gauges are the mercury barometer and the open-tube manometer.

**Gauge pressure and absolute pressure**  The gauge pressure is the amount by which a pressure $P$ differs from atmospheric pressure. The absolute pressure is the actual value for $P$.

### 11.5 Pascal's Principle

Pascal's principle states that any change in the pressure applied to a completely enclosed fluid is transmitted undiminished to all parts of the fluid and the enclosing walls.

**Example 7**

**Interactive Solution 11.33**

### 11.6 Archimedes' Principle

The buoyant force is the upward force that a fluid applies to an object that is partially or completely immersed in it.

Archimedes' principle states that the magnitude of the buoyant force equals the weight of the fluid that the partially or completely immersed object displaces:

**Archimedes' principle**

$$\underbrace{F_{\text{B}}}_{\substack{\text{Magnitude of} \\ \text{buoyant force}}} = \underbrace{W_{\text{fluid}}}_{\substack{\text{Weight of} \\ \text{displaced fluid}}} \qquad (11.6)$$

**Examples 8, 9, 10**

**Interactive LearningWare 11.1**

Use **Self-Assessment Test 11.1** to evaluate your understanding of Sections 11.1–11.6.

### 11.7 Fluids in Motion

### 11.8 The Equation of Continuity

**Steady flow**  In steady flow, the velocity of the fluid particles at any point is constant as time passes.

**Ideal fluid**  An incompressible, nonviscous fluid is known as an ideal fluid.

The mass flow rate of a fluid with a density $\rho$, flowing with a speed $v$ in a pipe of cross-sectional area $A$, is the mass per second (kg/s) flowing past a point and is given by

**Mass flow rate**

$$\text{Mass flow rate} = \rho A v \qquad (11.7)$$

The equation of continuity expresses the fact that mass is conserved: what flows into one end of a pipe flows out the other end, assuming there are no additional entry or exit points in between. Expressed in terms of the mass flow rate, the equation of continuity is

**Equation of continuity**

$$\rho_1 A_1 v_1 = \rho_2 A_2 v_2 \qquad (11.8)$$

where the subscripts 1 and 2 denote two points along the pipe.

| Topic | Discussion | Learning Aids |
|---|---|---|

**Incompressible fluid**

If a fluid is incompressible, the density at any two points is the same, $\rho_1 = \rho_2$. For an incompressible fluid, the equation of continuity becomes

$$A_1 v_1 = A_2 v_2 \qquad (11.9)$$

**Examples 11, 12**
**Concept Simulation 11.1**

The product $Av$ is known as the volume flow rate $Q$:

**Volume flow rate**

$$Q = \text{Volume flow rate} = Av \qquad (11.10)$$

### 11.9 Bernoulli's Equation
### 11.10 Applications of Bernoulli's Equation

In the steady flow of an ideal fluid whose density is $\rho$, the pressure $P$, the fluid speed $v$, and the elevation $y$ at any two points (1 and 2) in the fluid are related by Bernoulli's equation:

**Examples 13, 14, 15**

**Interactive LearningWare 11.2**

**Bernoulli's equation**

$$P_1 + \tfrac{1}{2}\rho v_1^2 + \rho g y_1 = P_2 + \tfrac{1}{2}\rho v_2^2 + \rho g y_2 \qquad (11.11)$$

**Interactive Solution 11.63**

When the flow is horizontal ($y_1 = y_2$), Bernoulli's equation indicates that higher fluid speeds are associated with lower fluid pressures.

 Use **Self-Assessment Test 11.2** to evaluate your understanding of Sections 11.7–11.10.

# Problems

## Section 11.1 Mass Density

**1. ssm** Accomplished silver workers in India can pound silver into incredibly thin sheets, as thin as $3.00 \times 10^{-7}$ m (about one-hundredth of the thickness of this sheet of paper). Find the area of such a sheet that can be formed from 1.00 kg of silver.

**2.** The *karat* is a dimensionless unit that is used to indicate the proportion of gold in a gold-containing alloy. An alloy that is one karat gold contains a weight of pure gold that is one part in twenty-four. What is the volume of gold in a 14.0-karat gold necklace whose weight is 1.27 N?

**3.** A pirate in a movie is carrying a chest (0.30 m × 0.30 m × 0.20 m) that is supposed to be filled with gold. To see how ridiculous this is, determine the weight (in newtons) of the gold. To judge how large this weight is, remember that 1 N = 0.225 lb.

**4.** A water bed has dimensions of 1.83 m × 2.13 m × 0.229 m. The floor of the bedroom will tolerate an additional weight of no more than 6660 N. Find the weight of the water in the bed and determine whether it should be purchased.

**5. ssm** The ice on a lake is 0.010 m thick. The lake is circular, with a radius of 480 m. Find the mass of the ice.

**\* 6.** An irregularly shaped chunk of concrete has a hollow spherical cavity inside. The mass of the chunk is 33 kg, and the volume enclosed by the outside surface of the chunk is 0.025 m³. What is the radius of the spherical cavity?

**\* 7.** A bar of gold measures 0.15 m × 0.050 m × 0.050 m. How many gallons of water have the same mass as this bar?

**\* 8.** A hypothetical spherical planet consists entirely of iron. What is the period of a satellite that orbits this planet just above its surface? Consult Table 11.1 as necessary.

**\*\* 9. ssm www** An antifreeze solution is made by mixing ethylene glycol ($\rho = 1116$ kg/m³) with water. Suppose the specific gravity of such a solution is 1.0730. Assuming that the total volume of the solution is the sum of its parts, determine the volume percentage of ethylene glycol in the solution.

## Section 11.2 Pressure

**10.** A glass bottle of soda is sealed with a screw cap. The absolute pressure of the carbon dioxide inside the bottle is $1.80 \times 10^5$ Pa. Assuming that the top and bottom surfaces of the cap each have an area of $4.10 \times 10^{-4}$ m², obtain the magnitude of the force that the screw thread exerts on the cap in order to keep it on the bottle. The air pressure outside the bottle is one atmosphere.

**11.** An airtight box has a removable lid of area $1.3 \times 10^{-2}$ m² and negligible weight. The box is taken up a mountain where the air pressure outside the box is $0.85 \times 10^5$ Pa. The inside of the box is completely evacuated. What is the magnitude of the force required to pull the lid off the box?

**12.** United States currency is printed using intaglio presses that generate a printing pressure of $8.0 \times 10^4$ lb/in.² A $20 bill is 6.1 in. by 2.6 in. Calculate the magnitude of the force that the printing press applies to one side of the bill.

**13. ssm** High-heeled shoes can cause tremendous pressure to be applied to a floor. Suppose the radius of a heel is $6.00 \times 10^{-3}$ m. At times during a normal walking motion, nearly the entire body weight acts perpendicular to the surface of such a heel. Find the pressure that is applied to the floor under the heel because of the weight of a 50.0-kg woman.

**14.** A person who weighs 625 N is riding a 98-N mountain bike. Suppose the entire weight of the rider and bike is supported equally by the two tires. If the gauge pressure in each tire is $7.60 \times 10^5$ Pa, what is the area of contact between each tire and the ground?

**15. Interactive Solution 11.15** at www.wiley.com/college/cutnell presents a model for solving this problem. A solid concrete block weighs 169 N and is resting on the ground. Its dimensions are 0.400 m × 0.200 m × 0.100 m. A number of identical blocks are stacked on top of this one. What is the smallest number of whole blocks (including the one on the ground) that can be stacked so that their weight creates a pressure of at least two atmospheres on the ground beneath the first block?

\* **16.** A cylinder is fitted with a piston, beneath which is a spring, as in the drawing. The cylinder is open at the top. Friction is absent. The spring constant of the spring is 3600 N/m. The piston has a negligible mass and a radius of 0.025 m. (a) When air beneath the piston is completely pumped out, how much does the atmospheric pressure cause the spring to compress? (b) How much work does the atmospheric pressure do in compressing the spring?

\* **17. ssm** A cylinder (with circular ends) and a hemisphere are solid throughout and made from the same material. They are resting on the ground, the cylinder on one of its ends and the hemisphere on its flat side. The weight of each causes the same pressure to act on the ground. The cylinder is 0.500 m high. What is the radius of the hemisphere?

\*\* **18.** A house has a roof (colored gray) with the dimensions shown in the drawing. Determine the magnitude and direction of the net force that the atmosphere applies to the roof when the outside pressure rises suddenly by 10.0 mm of mercury, before the pressure in the attic can adjust.

14.5 m    4.21 m

30.0°

30.0°

### Section 11.3 Pressure and Depth in a Static Fluid, Section 11.4 Pressure Gauges

**19. ssm** Some researchers believe that the dinosaur Barosaurus held its head erect on a long neck, much as a giraffe does. If so, fossil remains indicate that its heart would have been about 12 m below its brain. Assume that the blood has the density of water, and calculate the amount by which the blood pressure in the heart would have exceeded that in the brain. Size estimates for the single heart needed to withstand such a pressure range up to two tons. Alternatively, Barosaurus may have had a number of smaller hearts.

**20.** At a given instant, the blood pressure in the heart is $1.6 \times 10^4$ Pa. If an artery in the brain is 0.45 m above the heart, what is the pressure in the artery? Ignore any pressure changes due to blood flow.

**21.** The Mariana trench is located in the Pacific Ocean at a depth of about 11 000 m below the surface of the water. The density of seawater is 1025 kg/m³. (a) If an underwater vehicle were to explore such a depth, what force would the water exert on the vehicle's observation window (radius = 0.10 m)? (b) For comparison, determine the weight of a jetliner whose mass is $1.2 \times 10^5$ kg.

**22.** Measured along the surface of the water, a rectangular swimming pool has a length of 15 m. Along this length, the flat bottom of the pool slopes downward at an angle of 11° below the horizontal, from one end to the other. By how much does the pressure at the

bottom of the deep end exceed the pressure at the bottom of the shallow end?

**23. ssm** A water tower is a familiar sight in many towns. The purpose of such a tower is to provide storage capacity and to provide sufficient pressure in the pipes that deliver the water to customers. The drawing shows a spherical reservoir that contains $5.25 \times 10^5$ kg of water when full. The reservoir is vented to the atmosphere at the top. For a full reservoir, find the gauge pressure that the water has at the faucet in (a) house A and (b) house B. Ignore the diameter of the delivery pipes.

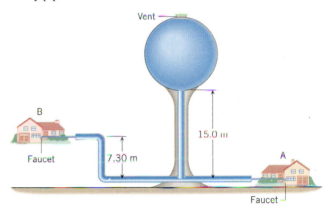

Vent

B

15.0 m

Faucet

7.30 m

A

Faucet

**24.** The human lungs can function satisfactorily up to a limit where the pressure difference between the outside and inside of the lungs is one-twentieth of an atmosphere. If a diver uses a snorkel for breathing, how far below the water can she swim? Assume the diver is in salt water whose density is 1025 kg/m³.

**25.** As background for this problem, review Conceptual Example 6. A submersible pump is put under the water at the bottom of a well and is used to push water up through a pipe. What minimum output gauge pressure must the pump generate to make the water reach the nozzle at ground level, 71 m above the pump?

\* **26.** A mercury barometer reads 747.0 mm on the roof of a building and 760.0 mm on the ground. Assuming a constant value of 1.29 kg/m³ for the density of air, determine the height of the building.

\* **27. ssm www** Mercury is poured into a tall glass. Ethyl alcohol is then poured on top of the mercury until the height of the ethyl alcohol itself is 110 cm. The two fluids do not mix, and the air pressure at the top of the ethyl alcohol is one atmosphere. What is the absolute pressure at a point that is 7.10 cm below the ethyl alcohol–mercury interface?

\* **28.** Figure 11.11 shows a mercury barometer. Consider two barometers, one using mercury and another using an unknown liquid. Suppose that the pressure above the liquid in each tube is maintained at the same value $P$, between zero and atmospheric pressure. The height of the unknown liquid is 16 times greater than the height of the mercury. Find the density of the unknown liquid.

\* **29.** A 1.00-m-tall container is filled to the brim, partway with mercury and the rest of the way with water. The container is open to the atmosphere. What must be the depth of the mercury so that the absolute pressure on the bottom of the container is twice the atmospheric pressure?

\*\* **30.** As the drawing illustrates, a pond has the shape of an inverted cone with the tip sliced off and has a depth of 5.00 m. The atmospheric pressure above the pond is $1.01 \times 10^5$ Pa. The circular top surface (radius = $R_2$) and circular bottom surface (radius = $R_1$)

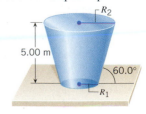

$R_2$

5.00 m

60.0°

$R_1$

of the pond are both parallel to the ground. The magnitude of the force acting on the top surface is the same as the magnitude of the force acting on the bottom surface. Obtain (a) $R_2$ and (b) $R_1$.

### Section 11.5 Pascal's Principle

**31. ssm** The atmospheric pressure above a swimming pool changes from 755 to 765 mm of mercury. The bottom of the pool is a 12-m × 24-m rectangle. By how much does the force on the bottom of the pool increase?

**32.** In the hydraulic press used in a trash compactor, the radii of the input piston and the output plunger are $6.4 \times 10^{-3}$ m and $5.1 \times 10^{-2}$ m, respectively. The height difference between the input piston and the output plunger can be neglected. What force is applied to the trash when the input force is 330 N?

**33. Interactive Solution 11.33** at **www.wiley.com/college/cutnell** presents a model for solving this problem. You can also prepare for this problem by reviewing Example 7. The hydraulic oil in a car lift has a density of $8.30 \times 10^2$ kg/m³. The weight of the input piston is negligible. The radii of the input piston and output plunger are $7.70 \times 10^{-3}$ m and 0.125 m, respectively. What input force $F$ is needed to support the 24 500-N combined weight of a car and the output plunger, when (a) the bottom surfaces of the piston and plunger are at the same level, and (b) the bottom surface of the output plunger is 1.30 m *above* that of the input piston?

**34.** A dentist's chair with a patient in it weighs 2100 N. The output plunger of a hydraulic system begins to lift the chair when the dentist's foot applies a force of 55 N to the input piston. Neglect any height difference between the plunger and the piston. What is the ratio of the radius of the plunger to the radius of the piston?

**\*35. ssm** A dump truck uses a hydraulic cylinder, as the drawing illustrates. When activated by the operator, a pump injects hydraulic oil into the cylinder at an absolute pressure of $3.54 \times 10^6$ Pa and drives the output plunger, which has a radius of 0.150 m. Assuming the plunger remains perpendicular to the floor of the load bed, find the torque that the plunger creates about the axis identified in the drawing.

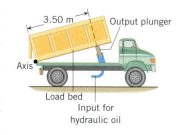

**\*36.** The drawing shows a hydraulic system used with disc brakes. The force **F** is applied perpendicularly to the brake pedal. The pedal rotates about the axis shown in the drawing and causes a force to be applied perpendicularly to the input piston (radius = $9.50 \times 10^{-3}$ m) in the master cylinder. The resulting pressure is transmitted by the brake fluid to the output plungers (radii = $1.90 \times 10^{-2}$ m), which are covered with the brake linings. The linings are pressed against both sides of a disc attached to the rotating wheel. Suppose that the magnitude of **F** is 9.00 N. Assume that the input piston and the output plungers are at the same vertical level, and find the force applied to each side of the rotating disc.

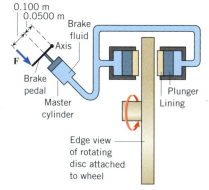

**\*37.** The drawing shows a hydraulic chamber in which a spring (spring constant = 1600 N/m) is attached to the input piston, and a

rock of mass 40.0 kg rests on the output plunger. The piston and plunger are nearly at the same height, and each has a negligible mass. By how much is the spring compressed from its unstrained position?

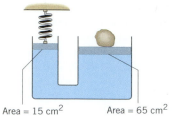

Area = 15 cm²     Area = 65 cm²

### Section 11.6 Archimedes' Principle

**38.** A duck is floating on a lake with 25% of its volume beneath the water. What is the average density of the duck?

**39. ssm** What is the radius of a hydrogen-filled balloon that would carry a load of 5750 N (in addition to the weight of the hydrogen) when the density of air is 1.29 kg/m³?

**40.** Only a small part of an iceberg protrudes above the water, while the bulk lies below the surface. The density of ice is 917 kg/m³ and that of seawater is 1025 kg/m³. Find the percentage of the iceberg's volume that lies below the surface.

**41.** A paperweight, when weighed in air, has a weight of $W = 6.9$ N. When completely immersed in water, however, it has a weight of $W_{\text{in water}} = 4.3$ N. Find the volume of the paperweight.

**42.** What is the total mass of swimmers that the raft in Example 8 can carry and float with its top surface at water level?

**43. ssm www** A person can change the volume of his body by taking air into his lungs. The amount of change can be determined by weighing the person under water. Suppose that under water a person weighs 20.0 N with partially full lungs and 40.0 N with empty lungs. Find the change in body volume.

**\*44.** What is the smallest number of whole logs ($\rho = 725$ kg/m³, radius = 0.0800 m, length = 3.00 m) that can be used to build a raft that will carry four people, each of whom has a mass of 80.0 kg?

**\*45.** A hollow cubical box is 0.30 m on an edge. This box is floating in a lake with one-third of its height beneath the surface. The walls of the box have a negligible thickness. Water is poured into the box. What is the depth of the water in the box at the instant the box begins to sink?

**\*46.** An object is solid throughout. When the object is completely submerged in ethyl alcohol, its apparent weight is 15.2 N. When completely submerged in water, its apparent weight is 13.7 N. What is the volume of the object?

**\*\*47. ssm** A solid cylinder (radius = 0.150 m, height = 0.120 m) has a mass of 7.00 kg. This cylinder is floating in water. Then oil ($\rho = 725$ kg/m³) is poured on top of the water until the situation shown in the drawing results. How much of the height of the cylinder is in the oil?

**\*\*48. Interactive LearningWare 11.1** at **www.wiley.com/college/cutnell** provides a review of the concepts that are important in this problem. A spring is attached to the bottom of an empty swimming pool, with the axis of the spring oriented vertically. An 8.00-kg block of wood ($\rho = 840$ kg/m³) is fixed to the top of the spring and compresses it. Then the pool is filled with water, completely covering the block. The spring is now observed to be stretched twice as much as it had been compressed. Determine the percentage of the block's total volume that is hollow. Ignore any air in the hollow space.

**\*\*49.** One kilogram of glass ($\rho = 2.60 \times 10^3$ kg/m³) is shaped into a hollow spherical shell that just barely floats in water. What are the inner and outer radii of the shell? Do not assume the shell is thin.

## Section 11.8 The Equation of Continuity

**50.** Oil is flowing with a speed of 1.22 m/s through a pipeline with a radius of 0.305 m. How many gallons of oil (1 gal = $3.79 \times 10^{-3}$ m³) flow in one day?

**51. ssm** Water flows with a volume flow rate of 1.50 m³/s in a pipe. Find the water speed where the pipe radius is 0.500 m.

**52.** **Concept Simulation 11.1** at **www.wiley.com/college/ cutnell** reviews the concept that plays the central role in this problem. (a) The volume flow rate in an artery supplying the brain is $3.6 \times 10^{-6}$ m³/s. If the radius of the artery is 5.2 mm, determine the average blood speed. (b) Find the average blood speed at a constriction in the artery if the constriction reduces the radius by a factor of 3. Assume that the volume flow rate is the same as that in part (a).

**53.** A room has a volume of 120 m³. An air-conditioning system is to replace the air in this room every twenty minutes, using ducts that have a square cross section. Assuming that air can be treated as an incompressible fluid, find the length of a side of the square if the air speed within the ducts is (a) 3.0 m/s and (b) 5.0 m/s.

**\*54.** Three fire hoses are connected to a fire hydrant. Each hose has a radius of 0.020 m. Water enters the hydrant through an underground pipe of radius 0.080 m. In this pipe the water has a speed of 3.0 m/s. (a) How many kilograms of water are poured onto a fire in one hour? (b) Find the water speed in each hose.

**\*55. ssm** A water line with an internal radius of $6.5 \times 10^{-3}$ m is connected to a shower head that has 12 holes. The speed of the water in the line is 1.2 m/s. (a) What is the volume flow rate in the line? (b) At what speed does the water leave one of the holes (effective hole radius = $4.6 \times 10^{-4}$ m) in the head?

## Section 11.9 Bernoulli's Equation,
## Section 11.10 Applications of Bernoulli's Equation

**56.** One way to administer an inoculation is with a "gun" that shoots the vaccine through a narrow opening. No needle is necessary, for the vaccine emerges with sufficient speed to pass directly into the tissue beneath the skin. The speed is high, because the vaccine ($\rho = 1100$ kg/m³) is held in a reservoir where a high pressure pushes it out. The pressure on the surface of the vaccine in one gun is $4.1 \times 10^6$ Pa above the atmospheric pressure outside the narrow opening. The dosage is small enough that the vaccine's surface in the reservoir is nearly stationary during an inoculation. The vertical height between the vaccine's surface in the reservoir and the opening can be ignored. Find the speed at which the vaccine emerges.

**57. ssm** Review Conceptual Example 13 as an aid in understanding this problem. Suppose that a 15-m/s wind is blowing across the roof of your house. The density of air is 1.29 kg/m³. (a) Determine the reduction in pressure (below atmospheric pressure of stationary air) that accompanies this wind. (b) Explain why some roofs are "blown outward" during high winds.

**58.** The blood speed in a normal segment of a horizontal artery is 0.11 m/s. An abnormal segment of the artery is narrowed down by an arteriosclerotic plaque to one-fourth the normal cross-sectional area. What is the difference in blood pressures between the normal and constricted segments of the artery?

**59. ssm** An airplane wing is designed so that the speed of the air across the top of the wing is 251 m/s when the speed of the air below the wing is 225 m/s. The density of the air is 1.29 kg/m³. What is the lifting force on a wing of area 24.0 m²?

**60. Interactive LearningWare 11.2** at **www.wiley.com/college/ cutnell** reviews the approach taken in problems such as this one. A

small crack occurs at the base of a 15.0-m-high dam. The effective crack area through which water leaves is $1.30 \times 10^{-3}$ m². (a) Ignoring viscous losses, what is the speed of water flowing through the crack? (b) How many cubic meters of water per second leave the dam?

**61.** The water tower in the drawing is drained by a pipe that extends to the ground. The flow is nonviscous. The top surface of the water at point 2 is at atmospheric pressure. (a) What is the absolute pressure at point 1 if the valve is *closed*? (b) What is the absolute pressure at point 1 when the valve is opened and the water is flowing? Assume that the water speed at point 2 is negligible. (c) Assuming the effective cross-sectional area of the valve opening is $2.00 \times 10^{-2}$ m², find the volume flow rate at point 1.

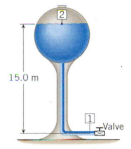

**\*62.** In a very large closed tank, the absolute pressure of the air above the water is $6.01 \times 10^5$ Pa. The water leaves the bottom of the tank through a nozzle that is directed straight upward. The opening of the nozzle is 4.00 m below the surface of the water. (a) Find the speed at which the water leaves the nozzle. (b) Ignoring air resistance and viscous effects, determine the height to which the water rises.

**\*63. Interactive Solution 11.63** at **www.wiley.com/college/cutnell** presents one method for modeling this problem. The construction of a flat rectangular roof (5.0 m × 6.3 m) allows it to withstand a maximum net outward force of 22 000 N. The density of the air is 1.29 kg/m³. At what wind speed will this roof blow outward?

**\*64.** A pump and its horizontal intake pipe are located 12 m beneath the surface of a reservoir. The speed of the water in the intake pipe causes the pressure there to decrease, in accord with Bernoulli's principle. Assuming nonviscous flow, what is the maximum speed with which water can flow through the intake pipe?

**\*65. ssm** A Venturi meter is a device for measuring the speed of a fluid within a pipe. The drawing shows a gas flowing at speed $v_2$ through a horizontal section of pipe whose cross-sectional area is $A_2 = 0.0700$ m². The gas has a density of $\rho = 1.30$ kg/m³. The Venturi meter has a cross-sectional area of $A_1 = 0.0500$ m² and has been substituted for a section of the larger pipe. The pressure difference between the two sections is $P_2 - P_1 = 120$ Pa. Find (a) the speed $v_2$ of the gas in the larger original pipe and (b) the volume flow rate $Q$ of the gas.

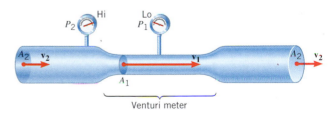

Venturi meter

**\*66.** A liquid is flowing through a horizontal pipe whose radius is 0.0200 m. The pipe bends straight upward through a height of 10.0 m and joins another horizontal pipe whose radius is 0.0400 m. What volume flow rate will keep the pressures in the two horizontal pipes the same?

**\*67.** An airplane has an effective wing surface area of 16 m² that is generating the lift force. In level flight the air speed over the top of the wings is 62.0 m/s, while the air speed beneath the wings is 54.0 m/s. What is the weight of the plane?

** **68.** A siphon tube is useful for removing liquid from a tank. The siphon tube
is first filled with liquid, and then one
end is inserted into the tank. Liquid then
drains out the other end, as the drawing
illustrates. (a) Using reasoning similar
to that employed in obtaining Torricelli's theorem, derive an expression for
the speed $v$ of the fluid emerging from
the tube. This expression should give $v$
in terms of the vertical height $y$ and the
acceleration due to gravity $g$. (Note that
this speed does not depend on the depth
$d$ of the tube below the surface of the liquid.) (b) At what value of the
vertical distance $y$ will the siphon stop working? (c) Derive an expression for the absolute pressure at the highest point in the siphon (point
$A$) in terms of the atmospheric pressure $P_0$, the fluid density $\rho$, $g$, and
the heights $h$ and $y$. (Note that the fluid speed at point $A$ is the same as
the speed of the fluid emerging from the tube, because the cross-sectional area of the tube is the same everywhere.)

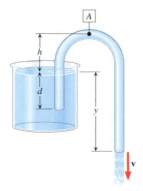

** **69. ssm** A uniform rectangular plate is hanging vertically downward from a hinge that passes along its left edge. By blowing air at
11.0 m/s *over the top of the plate only,* it is possible to keep the
plate in a horizontal position, as illustrated in part *a* of the drawing.
To what value should the air speed be reduced so that the plate is
kept at a 30.0° angle with respect to the vertical, as in part *b* of the
drawing? (*Hint: Apply Bernoulli's equation in the form of Equation
11.12.*)

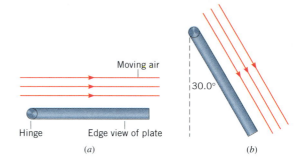

Moving air

Hinge

Edge view of plate

(a)

30.0°

(b)

# Chapter 12 Temperature and Heat

## 12.1 Common Temperature Scales

To measure temperature we use a thermometer. Many thermometers make use of the fact that materials usually expand with increasing temperature. For example, Figure 12.1 shows the common mercury-in-glass thermometer, which consists of a mercury-filled glass bulb connected to a capillary tube. When the mercury is heated, it expands into the capillary tube, the amount of expansion being proportional to the change in temperature. The outside of the glass is marked with an appropriate scale for reading the temperature.

A number of different temperature scales have been devised, two popular choices being the *Celsius* (formerly, centigrade) and *Fahrenheit scales.* Figure 12.1 illustrates these scales. Historically,* both scales were defined by assigning two temperature points on the scale and then dividing the distance between them into a number of equally spaced intervals. One point was chosen to be the temperature at which ice melts under one atmosphere of pressure (the "ice point"), and the other was the temperature at which water boils under one atmosphere of pressure (the "steam point"). On the Celsius scale, an ice point of 0 °C (0 degrees Celsius) and a steam point of 100 °C were selected. On the Fahrenheit scale, an ice point of 32 °F (32 degrees Fahrenheit) and a steam point of 212 °F were chosen. The Celsius scale is used worldwide, while the Fahrenheit scale is used mostly in the United States, often in home medical thermometers.

There is a subtle difference in the way the temperature of an object is reported, as compared to a *change* in its temperature. For example, the temperature of the human body is about 37 °C, where the symbol °C stands for "degrees Celsius." However, the *change* between two temperatures is specified in "Celsius degrees" (C°)—not in "degrees Celsius." Thus, if the body temperature rises to 39 °C, the change in temperature is 2 Celsius degrees or 2 C°, not 2 °C.

As Figure 12.1 indicates, the separation between the ice and steam points on the Celsius scale is divided into 100 Celsius degrees, while on the Fahrenheit scale the separation is divided into 180 Fahrenheit degrees. Therefore, the size of the Celsius degree is larger than that of the Fahrenheit degree by a factor of $\frac{180}{100}$, or $\frac{9}{5}$. Examples 1 and 2 illustrate how to convert between the Celsius and Fahrenheit scales using this factor.

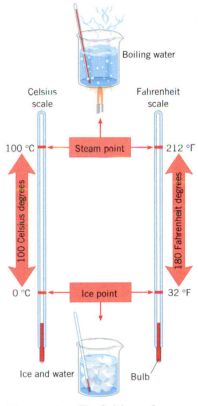

**Figure 12.1** The Celsius and Fahrenheit temperature scales.

### Example 1  Converting from a Fahrenheit to a Celsius Temperature

A healthy person has an oral temperature of 98.6 °F. What would this reading be on the Celsius scale?

**Reasoning and Solution** A temperature of 98.6 °F is 66.6 Fahrenheit degrees above the ice point of 32.0 °F. Since 1 C° = $\frac{9}{5}$ F°, the difference of 66.6 F° is equivalent to

$$(66.6 \text{ F}°)\left(\frac{1 \text{ C}°}{\frac{9}{5} \text{ F}°}\right) = 37.0 \text{ C}°$$

Thus, the person's temperature is 37.0 Celsius degrees above the ice point. Adding 37.0 Celsius degrees to the ice point of 0 °C on the Celsius scale gives a Celsius temperature of $\boxed{37.0 \text{ °C}}$.

### Example 2  Converting from a Celsius to a Fahrenheit Temperature

A time and temperature sign on a bank indicates that the outdoor temperature is −20.0 °C. Find the corresponding temperature on the Fahrenheit scale.

---

* Today, the Celsius and Fahrenheit scales are defined in terms of the Kelvin temperature scale; Section 12.2 discusses the Kelvin scale.

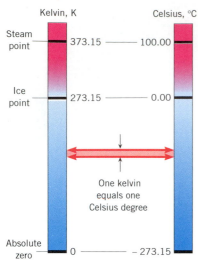

**Figure 12.2** A comparison of the Kelvin and Celsius temperature scales.

**Reasoning and Solution** The temperature of −20.0 °C is 20.0 Celsius degrees *below* the ice point of 0 °C. This number of Celsius degrees corresponds to

$$(20.0 \text{ C}°)\left(\frac{\frac{9}{5} \text{ F}°}{1 \text{ C}°}\right) = 36.0 \text{ F}°$$

The temperature, then, is 36.0 Fahrenheit degrees below the ice point. Subtracting 36.0 Fahrenheit degrees from the ice point of 32.0 °F on the Fahrenheit scale gives a Fahrenheit temperature of $\boxed{-4.0 \text{ °F}}$.

## 12.2   *The Kelvin Temperature Scale*

Although the Celsius and Fahrenheit scales are widely used, the ***Kelvin temperature scale*** has greater scientific significance. It was introduced by the Scottish physicist William Thompson (Lord Kelvin, 1824–1907), and in his honor each degree on the scale is called a kelvin (K). By international agreement, the symbol K is not written with a degree sign (°), nor is the word "degrees" used when quoting temperatures. For example, a temperature of 300 K (not 300 °K) is read as "three hundred kelvins," not "three hundred degrees kelvin." The kelvin is the SI base unit for temperature.

Figure 12.2 compares the Kelvin and Celsius scales. The size of one kelvin is identical to that of one Celsius degree because there are one hundred divisions between the ice and steam points on both scales. As we will discuss shortly, experiments have shown that there exists a lowest possible temperature, below which no substance can be cooled. This lowest temperature is defined to be the zero point on the Kelvin scale and is referred to as absolute zero.

The ice point (0 °C) occurs at 273.15 K on the Kelvin scale. Thus, the Kelvin temperature $T$ and the Celsius temperature $T_c$ are related by

$$T = T_c + 273.15 \tag{12.1}$$

The number 273.15 in Equation 12.1 is an experimental result, obtained in studies that utilize a gas-based thermometer. When a gas confined to a fixed volume is heated, its pressure increases. Conversely, when the gas is cooled, its pressure decreases. For example, the air pressure in automobile tires can rise by as much as 20% after the car has been driven and the tires have become warm. The change in gas pressure with temperature is the basis for the ***constant-volume gas thermometer.***

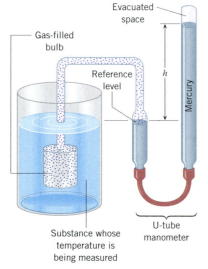

**Figure 12.3** A constant-volume gas thermometer.

A constant-volume gas thermometer consists of a gas-filled bulb to which a pressure gauge is attached, as in Figure 12.3. The gas is often hydrogen or helium at a low density, and the pressure gauge can be a U-tube manometer filled with mercury. The bulb is placed in thermal contact with the substance whose temperature is being measured. The volume of the gas is held constant by raising or lowering the *right* column of the U-tube manometer in order to keep the mercury level in the *left* column at the same reference level. The absolute pressure of the gas is proportional to the height $h$ of the mercury on the right. As the temperature changes, the pressure changes and can be used to indicate the temperature, once the constant-volume gas thermometer has been calibrated.

Suppose the absolute pressure of the gas in Figure 12.3 is measured at different temperatures. If the results are plotted on a pressure-versus-temperature graph, a straight line is obtained, as in Figure 12.4. If the straight line is extended or extrapolated to lower and lower temperatures, the line crosses the temperature axis at −273.15 °C. In reality, no gas can be cooled to this temperature, because all gases liquify before reaching it. However, helium and hydrogen liquify at such low temperatures that they are often used in the thermometer. This kind of graph can be obtained for different amounts and types of low-density gases. In all cases, it is found that the straight line extrapolates back to −273.15° C on the temperature axis, which suggests that the value of −273.15° C has fundamental significance. The significance of this number is that it is the ***absolute zero point*** for temperature measurement. The phrase "absolute zero" means that temperatures lower than −273.15 °C cannot be reached by continually cooling a gas or any other substance. If lower temperatures could be reached, then further extrapolation of the straight line in Figure 12.4 would suggest that negative absolute gas pressures could exist. Such a situation

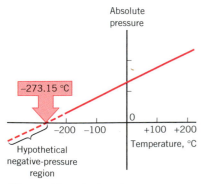

**Figure 12.4** A plot of absolute pressure versus temperature for a low-density gas at constant volume. The graph is a straight line and, when extrapolated (dashed line), crosses the temperature axis at −273.15 °C.

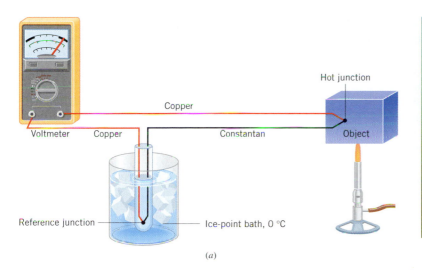

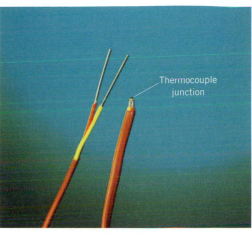

(a)

(b)

**Figure 12.5** (*a*) A thermocouple is made from two different types of wires, copper and constantan in this case. (*b*) A thermocouple junction between two different wires. (Courtesy Omega Engineering, Inc.)

would be impossible, because a negative absolute gas pressure has no meaning. Thus, the Kelvin scale is chosen so that its zero temperature point is the lowest temperature attainable.

## 12.3 Thermometers

All thermometers make use of the change in some physical property with temperature. A property that changes with temperature is called a *thermometric property*. For example, the thermometric property of the mercury thermometer is the length of the mercury column, while in the constant-volume gas thermometer it is the pressure of the gas. Several other thermometers and their thermometric properties will now be discussed.

The *thermocouple* is a thermometer used extensively in scientific laboratories. It consists of thin wires of different metals, welded together at the ends to form two junctions, as Figure 12.5 illustrates. Often the metals are copper and constantan (a copper–nickel alloy). One of the junctions, called the "hot" junction, is placed in thermal contact with the object whose temperature is being measured. The other junction, termed the "reference" junction, is kept at a known constant temperature (usually an ice–water mixture at 0 °C). The thermocouple generates a voltage that depends on the *difference in temperature* between the two junctions. This voltage is the thermometric property and is measured by a voltmeter, as the drawing indicates. With the aid of calibration tables, the temperature of the hot junction can be obtained from the voltage. Thermocouples are used to measure temperatures as high as 2300 °C or as low as −270 °C.

Most substances offer resistance to the flow of electricity. Because this electrical resistance changes with temperature, electrical resistance is another thermometric property. *Electrical resistance thermometers* are often made from platinum wire, because platinum has excellent mechanical and electrical properties in the temperature range from −270 °C to +700 °C. The electrical resistance of platinum wire is known as a function of temperature. Thus, the temperature of a substance can be determined by placing the resistance thermometer in thermal contact with the substance and measuring the resistance of the platinum wire.

Radiation emitted by an object can also be used to indicate temperature. At low to moderate temperatures, the predominant radiation emitted is infrared. As the temperature is raised, the intensity of the radiation increases substantially. In one interesting application, an infrared camera registers the intensity of the infrared radiation produced at different locations on the human body. The camera is connected to a color monitor that displays the different infrared intensities as different colors. This "thermal painting" is called a *thermograph* or *thermogram*. Thermography is an important diagnostic tool in medicine. For example, breast cancer may be indicated in a thermograph by the elevated temperatures associated with malignant tissue. Figure 12.6 shows a thermograph used to di-

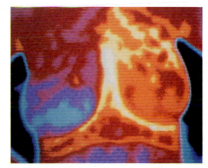

**Figure 12.6** Invasive carcinoma (cancer) of the breast registers colors from red to yellow/white in this thermograph, indicating markedly elevated temperatures. (© Science Photo Library/Photo Researchers)

 **The physics of thermography.**

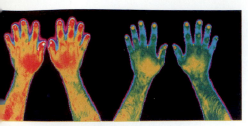

**Figure 12.7** Thermogram showing a smoker's forearms before *(left)* and 5 minutes after *(right)* he has smoked a cigarette. Temperatures range from over 34 °C (white) to about 28 °C (blue). (© Dr. Arthur Tucker/Science Photo Library/Photo Researchers)

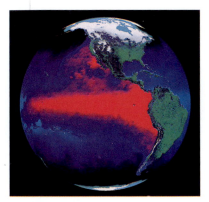

**Figure 12.8** A thermogram of the 1997/98 El Niño (red), a large region of abnormally high temperatures in the Pacific Ocean. (Courtesy NOAA)

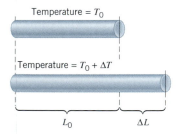

**Figure 12.9** When the temperature of a rod is raised by $\Delta T$, the length of the rod increases by $\Delta L$.

agnose breast cancer. Figure 12.7 shows thermographic images of a smoker's forearms before (left) and 5 minutes after (right) he has smoked a cigarette. After smoking, the forearms are cooler due to the effect of nicotine, which causes vasoconstriction (narrowing of the blood vessels) and reduces blood flow, a result that can lead to a higher risk from blood clotting. Temperatures in these images range from over 34 °C to about 28 °C and are indicated in decreasing order by the colors white, red, yellow, green, and blue.

Oceanographers and meteorologists use thermographs extensively to map the temperature distribution on the surface of the earth. For example, Figure 12.8 shows a satellite image of the sea-surface temperature of the Pacific Ocean. The region depicted in red is the 1997/98 El Niño, a large area of the ocean, approximately twice the width of the United States, where temperatures reached abnormally high values. This El Niño caused major weather changes in certain regions of the earth.

## 12.4 *Linear Thermal Expansion*

### NORMAL SOLIDS

Have you ever found the metal lid on a glass jar too tight to open? One solution is to run hot water over the lid, which loosens because the metal expands more than the glass does. To varying extents, most materials expand when heated and contract when cooled. The increase in any one dimension of a solid is called *linear expansion,* linear in the sense that the expansion occurs along a line. Figure 12.9 illustrates the linear expansion of a rod whose length is $L_0$ when the temperature is $T_0$. When the temperature increases to $T_0 + \Delta T$, the length becomes $L_0 + \Delta L$, where $\Delta T$ and $\Delta L$ are the magnitudes of the changes in temperature and length, respectively. Conversely, when the temperature decreases to $T_0 - \Delta T$, the length decreases to $L_0 - \Delta L$.

For modest temperature changes, experiments show that the change in length is directly proportional to the change in temperature ($\Delta L \propto \Delta T$). In addition, the change in length is proportional to the initial length of the rod, a fact that can be understood with the aid of Figure 12.10. Part *a* of the drawing shows two identical rods. Each rod has a length $L_0$ and expands by $\Delta L$ when the temperature increases by $\Delta T$. Part *b* shows the two heated rods combined into a single rod, for which the total expansion is the sum of the expansions of each part—namely, $\Delta L + \Delta L = 2\Delta L$. Clearly, the amount of expansion doubles if the rod is twice as long to begin with. In other words, the change in length is directly proportional to the original length ($\Delta L \propto L_0$). Equation 12.2 expresses the fact that $\Delta L$ is proportional to both $L_0$ and $\Delta T$ ($\Delta L \propto L_0 \Delta T$) by using a proportionality constant $\alpha$, which is called the *coefficient of linear expansion.*

■ **LINEAR THERMAL EXPANSION OF A SOLID**

The length $L_0$ of an object changes by an amount $\Delta L$ when its temperature changes by an amount $\Delta T$:

$$\Delta L = \alpha L_0 \Delta T \qquad (12.2)$$

where $\alpha$ is the coefficient of linear expansion.

*Common Unit for the Coefficient of Linear Expansion:* $\dfrac{1}{\text{C}°} = (\text{C}°)^{-1}$

Solving Equation 12.2 for $\alpha$ shows that $\alpha = \Delta L/(L_0 \Delta T)$. Since the length units of $\Delta L$ and $L_0$ algebraically cancel, the coefficient of linear expansion $\alpha$ has the unit of $(\text{C}°)^{-1}$ when the temperature difference $\Delta T$ is expressed in Celsius degrees (C°). Different materials with the same initial length expand and contract by different amounts as the temperature changes, so the value of $\alpha$ depends on the nature of the material. Table 12.1 shows some typical values. Coefficients of linear expansion also vary somewhat depending on the range of temperatures involved, but the values in Table 12.1 are adequate approximations. Example 3 deals with a situation where a dramatic effect due to thermal expansion can be observed, even though the change in temperature is small.

**Table 12.1** *Coefficients of Thermal Expansion for Solids and Liquids*[a]

| Substance | Coefficient of Thermal Expansion $(C°)^{-1}$ | |
| --- | --- | --- |
| | Linear ($\alpha$) | Volumetric ($\beta$) |
| **Solids** | | |
| Aluminum | $23 \times 10^{-6}$ | $69 \times 10^{-6}$ |
| Brass | $19 \times 10^{-6}$ | $57 \times 10^{-6}$ |
| Concrete | $12 \times 10^{-6}$ | $36 \times 10^{-6}$ |
| Copper | $17 \times 10^{-6}$ | $51 \times 10^{-6}$ |
| Glass (common) | $8.5 \times 10^{-6}$ | $26 \times 10^{-6}$ |
| Glass (Pyrex) | $3.3 \times 10^{-6}$ | $9.9 \times 10^{-6}$ |
| Gold | $14 \times 10^{-6}$ | $42 \times 10^{-6}$ |
| Iron or steel | $12 \times 10^{-6}$ | $36 \times 10^{-6}$ |
| Lead | $29 \times 10^{-6}$ | $87 \times 10^{-6}$ |
| Nickel | $13 \times 10^{-6}$ | $39 \times 10^{-6}$ |
| Quartz (fused) | $0.50 \times 10^{-6}$ | $1.5 \times 10^{-6}$ |
| Silver | $19 \times 10^{-6}$ | $57 \times 10^{-6}$ |
| **Liquids**[b] | | |
| Benzene | — | $1240 \times 10^{-6}$ |
| Carbon tetrachloride | — | $1240 \times 10^{-6}$ |
| Ethyl alcohol | — | $1120 \times 10^{-6}$ |
| Gasoline | — | $950 \times 10^{-6}$ |
| Mercury | — | $182 \times 10^{-6}$ |
| Methyl alcohol | — | $1200 \times 10^{-6}$ |
| Water | — | $207 \times 10^{-6}$ |

[a] The values for $\alpha$ and $\beta$ pertain to a temperature near 20 °C.
[b] Since liquids do not have fixed shapes, the coefficient of linear expansion is not defined for them.

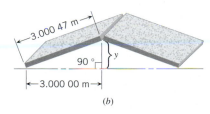

**Figure 12.10** (*a*) Each of two identical rods expands by $\Delta L$ when heated. (*b*) When the rods are combined into a single rod of length $2L_0$, the "combined" rod expands by $2 \Delta L$.

## Example 3 Buckling of a Sidewalk

A concrete sidewalk is constructed between two buildings on a day when the temperature is 25 °C. The sidewalk consists of two slabs, each three meters in length and of negligible thickness (Figure 12.11*a*). As the temperature rises to 38 °C, the slabs expand, but no space is provided for thermal expansion. The buildings do not move, so the slabs buckle upward. Determine the vertical distance *y* in part *b* of the drawing.

**Reasoning** The expanded length of each slab is equal to its original length plus the change in length $\Delta L$ due to the rise in temperature. We know the original length, and Equation 12.2 can be used to find the change in length. Once the expanded length has been determined, the Pythagorean theorem can be employed to find the vertical distance *y* in Figure 12.11*b*.

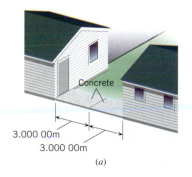

**Figure 12.11** (*a*) Two concrete slabs completely fill the space between the buildings. (*b*) When the temperature increases, each slab expands, causing the sidewalk to buckle.

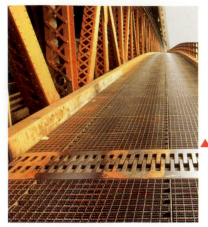

**Figure 12.12** An expansion joint in a bridge. (© Richard Choy/Peter Arnold, Inc.)

**The physics of**
**an antiscalding device.**

Movable plunger    Actuator spring

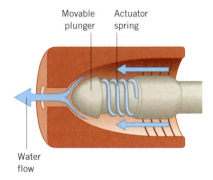

Water flow

**Figure 12.13** An antiscalding device.

**The physics of**
**thermal stress.**

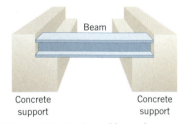
Beam

Concrete support          Concrete support

**Figure 12.14** A steel beam is mounted between concrete supports with no room provided for thermal expansion of the beam.

**Solution** The change in temperature is $\Delta T = 38\ °C - 25\ °C = 13\ C°$, and the coefficient of linear expansion for concrete is given in Table 12.1. The change in length of each slab associated with this temperature change is

$$\Delta L = \alpha L_0 \Delta T = [12 \times 10^{-6}\ (C°)^{-1}](3.0\ m)(13\ C°) = 0.000\ 47\ m \qquad (12.2)$$

The expanded length of each slab is, thus, 3.000 47 m. The vertical distance $y$ can be obtained by applying the Pythagorean theorem to the right triangle in Figure 12.11*b*:

$$y = \sqrt{(3.000\ 47\ m)^2 - (3.000\ 00\ m)^2} = \boxed{0.053\ m}$$

The buckling of a sidewalk is one consequence of not providing sufficient room for thermal expansion. To eliminate such problems, engineers incorporate expansion joints or spaces at intervals along bridge roadbeds, as Figure 12.12 shows.

Although Example 3 shows how thermal expansion can cause problems, there are also times when it can be useful. For instance, each year thousands of children are taken to emergency rooms suffering from burns caused by scalding tap water. Such accidents can be reduced with the aid of the antiscalding device shown in Figure 12.13. This device screws onto the end of a faucet and quickly shuts off the flow of water when it becomes too hot. As the water temperature rises, the actuator spring expands and pushes the plunger forward, shutting off the flow. When the water cools, the spring contracts and the water flow resumes.

## THERMAL STRESS

If the concrete slabs in Figure 12.11 had not buckled upward, they would have been subjected to immense forces from the buildings. The forces needed to prevent a solid object from expanding must be strong enough to counteract any change in length that would occur due to a change in temperature. Although the change in temperature may be small, the forces—and hence the stresses—can be enormous. They can, in fact, lead to serious structural damage. Example 4 illustrates just how large the stresses can be.

**Example 4** **The Stress on a Steel Beam**

A steel beam is used in the roadbed of a bridge. The beam is mounted between two concrete supports when the temperature is 23 °C, with no room provided for thermal expansion (Figure 12.14). What compressional stress must the concrete supports apply to each end of the beam, if they are to keep the beam from expanding when the temperature rises to 42 °C?

**Reasoning** Recall from Section 10.8 that the stress (force per unit cross-sectional area or $F/A$) required to change the length $L_0$ of an object by an amount $\Delta L$ is

$$\text{Stress} = \frac{F}{A} = Y\frac{\Delta L}{L_0} \qquad (10.17)$$

where $Y$ is Young's modulus. If the steel beam were free to expand because of the change in temperature, the length would change by $\Delta L = \alpha L_0 \Delta T$. Because the concrete supports do not permit any expansion, they must supply a stress to compress the beam by an amount $\Delta L$. Thus,

$$\text{Stress} = Y\frac{\Delta L}{L_0} = Y\frac{\alpha L_0 \Delta T}{L_0} = Y\alpha\Delta T$$

**Solution** For steel, the values of Young's modulus and the coefficient of linear expansion are $Y = 2.0 \times 10^{11}\ N/m^2$ (Table 10.1) and $\alpha = 12 \times 10^{-6}\ (C°)^{-1}$ (Table 12.1), respectively. The change in temperature from 23 to 42 °C is $\Delta T = 19\ C°$. The thermal stress is

$$\text{Stress} = Y\alpha\Delta T = (2.0 \times 10^{11}\ N/m^2)[12 \times 10^{-6}\ (C°)^{-1}](19\ C°) = \boxed{4.6 \times 10^7\ N/m^2}$$

This stress is huge. If the beam has a cross-sectional area of $A = 0.10\ m^2$, the force applied to each end by a concrete support is $F = (\text{Stress})A = 4.6 \times 10^6\ N$ (over 1 million lb).

# THE BIMETALLIC STRIP

A *bimetallic strip* is made from two thin strips of metal that have *different* coefficients of linear expansion, as Figure 12.15a shows. Often brass [$\alpha = 19 \times 10^{-6}\,(\text{C}°)^{-1}$] and steel [$\alpha = 12 \times 10^{-6}\,(\text{C}°)^{-1}$] are selected. The two pieces are welded or riveted together. When the bimetallic strip is heated, the brass, having the larger value of $\alpha$, expands more than the steel. Since the two metals are bonded together, the bimetallic strip bends into an arc as in part b, with the longer brass piece having a larger radius than the steel piece. When the strip is cooled, the bimetallic strip bends in the opposite direction, as in part c.

Bimetallic strips are frequently used as adjustable automatic switches in electrical appliances. Figure 12.16 shows an automatic coffee maker that turns off when the coffee is brewed to the selected strength. In part a, while the brewing cycle is on, electricity passes through the heating coil that heats the water. The electricity can flow because the contact mounted on the bimetallic strip touches the contact mounted on the "strength" adjustment knob, thus providing a continuous path for the electricity. When the bimetallic strip gets hot enough to bend away, as in part b of the drawing, the contacts separate. The electricity stops because it no longer has a continuous path along which to flow, and the brewing cycle is shut off. Turning the "strength" knob adjusts the brewing time by adjusting the distance through which the bimetallic strip must bend for the contact points to separate.

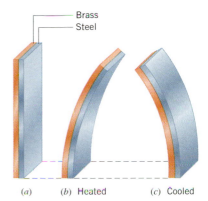

(a)  (b) Heated  (c) Cooled

**Figure 12.15** (a) A bimetallic strip and how it behaves when (b) heated and (c) cooled.

**The physics of an automatic coffee maker.**

**Figure 12.16** A bimetallic strip controls whether this coffee pot is (a) "on" (strip cool, straight) or (b) "off" (strip hot, bent).

Heating coil

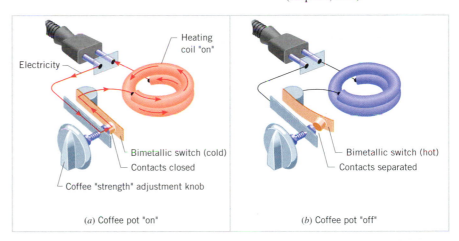

Electricity
Heating coil "on"
Bimetallic switch (cold)
Contacts closed
Coffee "strength" adjustment knob
(a) Coffee pot "on"

Bimetallic switch (hot)
Contacts separated
(b) Coffee pot "off"

# THE EXPANSION OF HOLES

An interesting example of linear expansion occurs when there is a hole in a piece of solid material. We know that the material itself expands when heated. But what about the hole? Does it expand, contract, or remain the same? Conceptual Example 5 provides some insight into the answer to this question.

## Conceptual Example 5
### Do Holes Expand or Contract When the Temperature Increases?

Figure 12.17a shows eight square tiles that are arranged to form a square pattern with a hole in the center. If the tiles are heated, what happens to the size of the hole?

**Reasoning and Solution** We can analyze this problem by disassembling the pattern into separate tiles, heating them, and then reassembling the pattern. Because each tile expands upon heating, it is evident from Figure 12.17b that the heated pattern expands and so does the hole in the center. In fact, if we had a ninth tile that was identical to and also heated like the others, it would fit exactly into the center hole, as Figure 12.17c indicates. Thus, not only does the hole in the pattern expand, but it also expands exactly as much as one of the tiles. Since the ninth tile is made of the same material as the others, we see that *the hole expands just as if it were made of the material of the surrounding tiles.* The thermal expansion of the hole

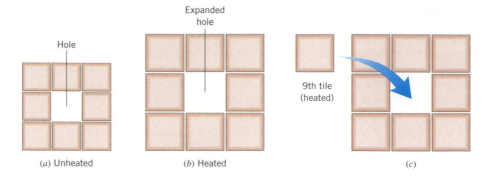

**Figure 12.17** (*a*) The tiles are arranged to form a square pattern with a hole in the center. (*b*) When the tiles are heated, the hole in the center, as well as the pattern, expands. (*c*) The expanded hole is the same size as a heated tile.

(*a*) Unheated          (*b*) Heated          (*c*)

and the surrounding material is analogous to a photographic enlargement; in both situations everything is enlarged, including holes.

**Related Homework:** *Problems 12, 18*

---

*Problem solving insight*

Instead of the separate tiles in Example 5, we could have used a square plate with a square hole in the center. The hole in the plate would have expanded just like the hole in the pattern of tiles. Furthermore, the same conclusion applies to a hole of any shape. Thus, it follows that *a hole in a piece of solid material expands when heated and contracts when cooled, just as if it were filled with the material that surrounds it.* If the hole is circular, the equation $\Delta L = \alpha L_0 \Delta T$ can be used to find the change in any linear dimension of the hole, such as its radius or diameter. Example 6 illustrates this type of linear expansion.

### Example 6   A Heated Engagement Ring

A gold engagement ring has an inner diameter of $1.5 \times 10^{-2}$ m and a temperature of 27 °C. The ring falls into a sink of hot water whose temperature is 49 °C. What is the change in the diameter of the hole in the ring?

**Reasoning** The hole expands as if it were filled with gold, so the change in the diameter is given by $\Delta L = \alpha L_0 \Delta T$, where $\alpha = 14 \times 10^{-6}\,(\text{C}°)^{-1}$ is the coefficient of linear expansion for gold (Table 12.1), $L_0$ is the original diameter, and $\Delta T$ is the change in temperature.

**Solution** The change in the ring's diameter is

$$\Delta L = \alpha L_0 \Delta T$$
$$= [14 \times 10^{-6}\,(\text{C}°)^{-1}](1.5 \times 10^{-2}\text{ m})(49\,°\text{C} - 27\,°\text{C}) = \boxed{4.6 \times 10^{-6}\text{ m}}$$

The previous two examples illustrate that holes expand like the surrounding material when heated. Therefore, holes in materials with larger coefficients of linear expansion expand more than those in materials with smaller coefficients of linear expansion. Conceptual Example 7 explores this aspect of thermal expansion.

### Conceptual Example 7   Expanding Cylinders

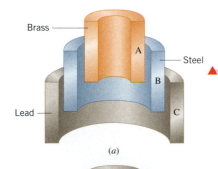

Figure 12.18 shows a cross-sectional view of three cylinders, A, B, and C. Each is made from a different material: one is lead, one is brass, and one is steel. All three have the same temperature, and they barely fit inside each other. As the cylinders are heated to the same, but higher, temperature, cylinder C falls off, while cylinder A becomes tightly wedged to cylinder B. Which cylinder is made from which material?

**Reasoning and Solution** We need to consider how the outer and inner diameters of each cylinder change as the temperature is raised. With respect to the inner diameter, we will be guided by the fact that a hole expands as if it were filled with the surrounding material. According to Table 12.1, lead has the greatest coefficient of linear expansion, followed by brass, and then by steel. These data indicate that the outer and inner diameters of the lead cylinder change the most, while those of the steel cylinder change the least.

**Figure 12.18** Conceptual Example 7 discusses the arrangements of the three cylinders shown in cutaway views in parts *a* and *b*.

Since the steel cylinder expands the least, it cannot be the outer one, for if it were, the greater expansion of the middle cylinder would prevent the steel cylinder from falling off. The steel cylinder also cannot be the inner one, because then the greater expansion of the middle cylinder would allow the steel cylinder to fall out, contrary to what is observed. The only place left for the steel cylinder is in the middle, which leads to the two possibilities in Figure 12.18. In part *a*, lead is on the outside and will fall off as the temperature is raised, since lead expands more than steel. On the other hand, the inner brass cylinder expands more than the steel that surrounds it and becomes tightly wedged, as observed. Thus, one possibility is *A = brass, B = steel, and C = lead.*

In part *b* of the drawing, brass is on the outside. As the temperature is raised, brass expands more than steel, so the outer cylinder will again fall off. The inner lead cylinder has the greatest expansion and will be wedged against the middle steel cylinder. A second possible answer, then, is *A = lead, B = steel, and C = brass.*

# 12.5 *Volume Thermal Expansion*

The volume of a normal material increases as the temperature increases. Most solids and liquids behave in this fashion. By analogy with linear thermal expansion, the change in volume $\Delta V$ is proportional to the change in temperature $\Delta T$ and to the initial volume $V_0$, provided the change in temperature is not too large. These two proportionalities can be converted into Equation 12.3 with the aid of a proportionality constant $\beta$, known as the *coefficient of volume expansion.* The algebraic form of this equation is similar to that for linear expansion, $\Delta L = \alpha L_0 \Delta T$.

---

■ **VOLUME THERMAL EXPANSION**

The volume $V_0$ of an object changes by an amount $\Delta V$ when its temperature changes by an amount $\Delta T$:

$$\Delta V = \beta V_0 \Delta T \qquad (12.3)$$

where $\beta$ is the coefficient of volume expansion.

*Common Unit for the Coefficient of Volume Expansion:* $(C°)^{-1}$

---

The unit for $\beta$, like that for $\alpha$, is $(C°)^{-1}$. Values for $\beta$ depend on the nature of the material, and Table 12.1 lists some examples measured near 20 °C. The values of $\beta$ for liquids are substantially larger than those for solids, because liquids typically expand more than solids, given the same initial volumes and temperature changes. Table 12.1 also shows that, for most solids, the coefficient of volume expansion is three times as much as the coefficient of linear expansion: $\beta = 3\alpha$.

If a cavity exists within a solid object, the volume of the cavity increases when the object expands, just as if the cavity were filled with the surrounding material. The expansion of the cavity is analogous to the expansion of a hole in a sheet of material. Accordingly, the change in volume of a cavity can be found using the relation $\Delta V = \beta V_0 \Delta T$, where $\beta$ is the coefficient of volume expansion of the material that surrounds the cavity. Example 8 illustrates this point.

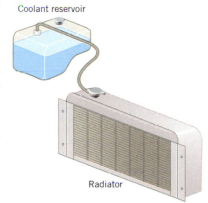

**Figure 12.19** An automobile radiator and a coolant reservoir for catching the overflow from the radiator.

## *Example 8* An Automobile Radiator

A small plastic container, called the coolant reservoir, catches the radiator fluid that overflows when an automobile engine becomes hot (see Figure 12.19). The radiator is made of copper, and the coolant has a coefficient of volume expansion of $\beta = 4.10 \times 10^{-4} \, (C°)^{-1}$. If the radiator is filled to its 15-quart capacity when the engine is cold (6.0 °C), how much overflow from the radiator will spill into the reservoir when the coolant reaches its operating temperature of 92 °C?

**Reasoning** When the temperature increases, both the coolant and radiator expand. If they were to expand by the same amount, there would be no overflow. However, the liquid coolant

**The physics of** the overflow of an automobile radiator.

*Problem solving insight*
The way in which the level of a liquid in a container changes with temperature depends on the change in volume of both the liquid and the container.

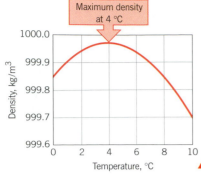

**Figure 12.20** The density of water in the temperature range from 0 to 10 °C. At 4 °C water has a maximum density of 999.973 kg/m³. (This value is equivalent to the often-quoted density of 1.000 00 gram per milliliter.)

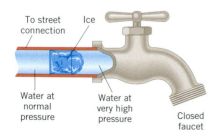

**Figure 12.21** As water freezes and expands, enormous pressure is applied to the liquid water between the ice and the faucet.

**The physics of** ice formation and the survival of aquatic life.

**The physics of** bursting water pipes.

expands more than the radiator, and the overflow volume is the amount of coolant expansion *minus* the amount of the radiator cavity expansion.

**Solution** When the temperature increases by 86 C°, the coolant expands by an amount

$$\Delta V = \beta V_0 \Delta T = [4.10 \times 10^{-4}\ (C°)^{-1}](15\ \text{quarts})(86\ C°) = 0.53\ \text{quart} \qquad (12.3)$$

The radiator cavity expands as if it were filled with copper [$\beta = 51 \times 10^{-6}\ (C°)^{-1}$; see Table 12.1]. The expansion of the radiator cavity is

$$\Delta V = \beta V_0 \Delta T = [51 \times 10^{-6}\ (C°)^{-1}](15\ \text{quarts})(86\ C°) = 0.066\ \text{quart}$$

The overflow volume is 0.53 quart − 0.066 quart = $\boxed{0.46\ \text{quart}}$.

▲

Although most substances expand when heated, a few do not. For instance, if water at 0 °C is heated, its volume *decreases* until the temperature reaches 4 °C. Above 4 °C water behaves normally, and its volume increases as the temperature increases. Because a given mass of water has a minimum volume at 4 °C, the density (mass per unit volume) of water is greatest at 4 °C, as Figure 12.20 shows.

The fact that water has its greatest density at 4 °C, rather than at 0 °C, has important consequences for the way in which a lake freezes. When the air temperature drops, the surface layer of water is chilled. As the temperature of the surface layer drops toward 4 °C, this layer becomes more dense than the warmer water below. The denser water sinks and pushes up the deeper and warmer water, which in turn is chilled at the surface. This process continues until the temperature of the entire lake reaches 4 °C. Further cooling of the surface water below 4 °C makes it *less dense* than the deeper layers; consequently, the surface layer does not sink but stays on top. Continued cooling of the top layer to 0 °C leads to the formation of ice that floats on the water, because ice has a smaller density than water at any temperature. Below the ice, however, the water temperature remains above 0 °C. The sheet of ice acts as an insulator that reduces the loss of heat from the lake, especially if the ice is covered with a blanket of snow, which is also an insulator. As a result, lakes usually do not freeze solid, even during prolonged cold spells, so fish and other aquatic life can survive.

The fact that the density of ice is smaller than the density of water has an important consequence for home owners, who have to contend with the possibility of bursting water pipes during severe winters. Water often freezes in a section of pipe exposed to unusually cold temperatures. The ice can form an immovable plug that prevents the subsequent flow of water, as Figure 12.21 illustrates. When water (larger density) turns to ice (smaller density), its volume expands by 8.3%. Therefore, when more water freezes at the left side of the plug, the expanding ice pushes liquid back into the pipe leading to the street connection, and no damage is done. However, when ice forms on the right side of the plug, the expanding ice pushes liquid to the right. But it has nowhere to go if the faucet is closed. As ice continues to form and expand, the water pressure between the plug and faucet rises. Even a small increase in the amount of ice produces a large increase in the pressure. This situation is analogous to the thermal stress discussed in Example 4, where a small change in the length of the steel beam produces a large stress on the concrete supports. The entire section of pipe to the right of the blockage experiences the same elevated pressure, according to Pascal's principle (Section 11.5). Therefore, the pipe can burst at any point where it is structurally weak, even within the heated space of the building. If you should lose heat during the winter, there is a simple way to prevent pipes from bursting. Simply open the faucet, so it drips a little. The excessive pressure will be relieved.

## 12.6 *Heat and Internal Energy*

An object with a high temperature is said to be hot, and the word "hot" brings to mind the word "heat." *Heat* flows from a hotter object to a cooler object when the two are placed in contact. It is for this reason that a cup of hot coffee feels hot to the touch, while a glass of ice water feels cold. When the person in Figure 12.22a touches the coffee cup, heat flows from the hotter cup into the cooler hand. When the person touches the glass in part *b* of the drawing, heat again flows from hot to cold, in this case from the warmer hand into the

(a) 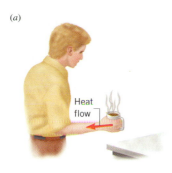   (b)

Heat flow

Heat flow

**Figure 12.22** Heat is energy in transit from hot to cold. (*a*) Heat flows from the hotter coffee cup to the colder hand. (*b*) Heat flows from the warmer hand to the colder glass of ice water.

colder glass. The response of the nerves in the hand to the arrival or departure of heat prompts the brain to identify the coffee cup as being hot and the glass as being cold.

But just what is heat? As the following definition indicates, heat is a form of energy, energy in transit from hot to cold.

---

■ **DEFINITION OF HEAT**

Heat is energy that flows from a higher-temperature object to a lower-temperature object because of the difference in temperatures.

*SI Unit of Heat:* joule (J)

---

Being a kind of energy, heat is measured in the same units used for work, kinetic energy, and potential energy. Thus, the SI unit for heat is the joule.

The heat that flows from hot to cold in Figure 12.22 originates in the ***internal energy*** of the hot substance. The internal energy of a substance is the sum of the molecular kinetic energy (due to the random motion of the molecules), the molecular potential energy (due to forces that act between the atoms of a molecule and between molecules), and other kinds of molecular energy. When heat flows in circumstances where the work done is negligible, the internal energy of the hot substance decreases and the internal energy of the cold substance increases. Although heat may originate in the internal energy supply of a substance, *it is not correct to say that a substance contains heat*. The substance has internal energy, not heat. The word "heat" is used only when referring to the energy actually in transit from hot to cold.

# 12.7 *Heat and Temperature Change: Specific Heat Capacity*

## SOLIDS AND LIQUIDS

Greater amounts of heat are needed to raise the temperature of solids or liquids to higher values. A greater amount of heat is also required to raise the temperature of a greater mass of material. Similar comments apply when the temperature is lowered, except that heat must be removed. For limited temperature ranges, experiment shows that the amount of heat $Q$ is directly proportional to the change in temperature $\Delta T$ and to the mass $m$. These two proportionalities are expressed below in Equation 12.4, with the help of a proportionality constant $c$ that is referred to as the ***specific heat capacity*** of the material.

---

■ **HEAT SUPPLIED OR REMOVED IN CHANGING THE TEMPERATURE OF A SUBSTANCE**

The heat $Q$ that must be supplied or removed to change the temperature of a substance of mass $m$ by an amount $\Delta T$ is

$$Q = cm\Delta T \qquad (12.4)$$

where $c$ is the specific heat capacity of the substance.

*Common Unit for Specific Heat Capacity:* $J/(kg \cdot C°)$

---

**Table 12.2** *Specific Heat Capacities[a] of Some Solids and Liquids*

| Substance | Specific Heat Capacity, $c$ J/(kg·C°) |
|---|---|
| **Solids** | |
| Aluminum | $9.00 \times 10^2$ |
| Copper | 387 |
| Glass | 840 |
| Human body | 3500 |
| (37 °C, average) | |
| Ice (−15 °C) | $2.00 \times 10^3$ |
| Iron or steel | 452 |
| Lead | 128 |
| Silver | 235 |
| **Liquids** | |
| Benzene | 1740 |
| Ethyl alcohol | 2450 |
| Glycerin | 2410 |
| Mercury | 139 |
| Water (15 °C) | 4186 |

[a] Except as noted, the values are for 25 °C and 1 atm of pressure.

Solving Equation 12.4 for the specific heat capacity shows that $c = Q/(m\Delta T)$, so the unit for specific heat capacity is J/(kg·C°). Table 12.2 reveals that the value of the specific heat capacity depends on the nature of the material. Examples 9 and 10 illustrate the use of Equation 12.4.

## Example 9    A Hot Jogger

In a half hour, a 65-kg jogger can generate $8.0 \times 10^5$ J of heat. This heat is removed from the jogger's body by a variety of means, including the body's own temperature-regulating mechanisms. If the heat were not removed, how much would the body temperature increase?

**Reasoning** The increase in body temperature depends on the amount of heat $Q$ generated by the jogger, her mass $m$, and the specific heat capacity $c$ of the human body. Since numerical values are known for these three variables, we can determine the potential rise in temperature by using Equation 12.4.

**Solution** Table 12.2 gives the average specific heat capacity of the human body as 3500 J/(kg·C°). With this value, Equation 12.4 shows that

$$\Delta T = \frac{Q}{cm} = \frac{8.0 \times 10^5 \text{ J}}{[3500 \text{ J/(kg·C°)}](65 \text{ kg})} = \boxed{3.5 \text{ C°}}$$

An increase in body temperature of 3.5 °C could be life-threatening. One way in which the jogger's body prevents it from occurring is to remove excess heat by perspiring. In contrast, dogs, such as the one in Figure 12.23, do not perspire but often pant to remove excess heat.

## Example 10    Taking a Hot Shower

Cold water at a temperature of 15 °C enters a heater, and the resulting hot water has a temperature of 61 °C. A person uses 120 kg of hot water in taking a shower. (a) Find the energy needed to heat the water. (b) Assuming that the utility company charges $0.10 per kilowatt·hour for electrical energy, determine the cost of heating the water.

**Reasoning** The amount $Q$ of heat needed to raise the water temperature can be found from the relation $Q = cm\Delta T$, since the specific heat capacity, mass, and temperature change of the water are known. To determine the cost of this energy, we multiply the cost per unit of energy ($0.10 per kilowatt·hour) by the amount of energy used, expressed in energy units of kilowatt·hours.

**Solution**

**(a)** The amount of heat needed to heat the water is

$$Q = cm\Delta T = [4186 \text{ J/(kg·C°)}](120 \text{ kg})(61 \text{ °C} - 15 \text{ °C}) = \boxed{2.3 \times 10^7 \text{ J}} \quad (12.4)$$

**(b)** The kilowatt·hour (kWh) is the unit of energy that utility companies use in your electric bill. To calculate the cost, we need to determine the number of joules in one kilowatt·hour. Recall that 1 kilowatt is 1000 watts (1 kW = 1000 W), 1 watt is 1 joule per second (1 W = 1 J/s; see Equation 6.10b), and 1 hour is equal to 3600 seconds (1 h = 3600 s). Thus,

$$1 \text{ kWh} = (1 \text{ kWh})\left(\frac{1000 \text{ W}}{1 \text{ kW}}\right)\left(\frac{1 \text{ J/s}}{1 \text{ W}}\right)\left(\frac{3600 \text{ s}}{1 \text{ h}}\right) = 3.60 \times 10^6 \text{ J}$$

The number of kilowatt·hours of energy used to heat the water is

$$(2.3 \times 10^7 \text{ J})\left(\frac{1 \text{ kWh}}{3.60 \times 10^6 \text{ J}}\right) = 6.4 \text{ kWh}$$

At a cost of $0.10 per kWh, the bill for the heat is $\boxed{\$0.64}$ or 64 cents.

## GASES

As we will see in Section 15.6, the value of the specific heat capacity depends on whether the pressure or volume is held constant while energy in the form of heat is added to or removed from a substance. The distinction between constant pressure and constant volume

**Figure 12.23** Dogs often pant to get rid of excess heat. (© James L. Stanfield/ National Geographic/Getty Images)

is usually not important for solids and liquids but is significant for gases. As we will see in Section 15.6, a greater value for the specific heat capacity is obtained for a gas at constant pressure than for a gas at constant volume.

## HEAT UNITS OTHER THAN THE JOULE

There are three heat units other than the joule in common use. One kilocalorie (1 kcal) was defined historically as the amount of heat needed to raise the temperature of one kilogram of water by one Celsius degree.* With $Q = 1.00$ kcal, $m = 1.00$ kg, and $\Delta T = 1.00$ C°, the equation $Q = cm\Delta T$ shows that such a definition is equivalent to a specific heat capacity for water of $c = 1.00$ kcal/(kg·C°). Similarly, one calorie (1 cal) was defined as the amount of heat needed to raise the temperature of one gram of water by one Celsius degree, which yields a value of $c = 1.00$ cal/(g·C°). (Nutritionists use the word "Calorie," with a capital C, to specify the energy content of foods; this use is unfortunate, since 1 Calorie = 1000 calories = 1 kcal.) The British thermal unit (Btu) is the other commonly used heat unit and was defined historically as the amount of heat needed to raise the temperature of one pound of water by one Fahrenheit degree.

It was not until the time of James Joule (1818–1889) that the relationship between energy in the form of work (in units of joules) and energy in the form of heat (in units of kilocalories) was firmly established. Joule's experiments revealed that the performance of mechanical work, like rubbing your hands together, can make the temperature of a substance rise, just as the absorption of heat can. His experiments and those of later workers have shown that

$$1 \text{ kcal} = 4186 \text{ joules} \quad \text{or} \quad 1 \text{ cal} = 4.186 \text{ joules}$$

Because of its historical significance, this conversion factor is known as the **mechanical equivalent of heat.**

## CALORIMETRY

In Section 6.8 we encountered the principle of conservation of energy, which states that energy can be neither created nor destroyed, but can only be converted from one form to another. There we dealt with kinetic and potential energies. In this chapter we have expanded our concept of energy to include heat, which is energy that flows from a higher-temperature object to a lower-temperature object because of the difference in temperature. No matter what its form, whether kinetic energy, potential energy, or heat, energy can be neither created nor destroyed. This fact governs the way objects at different temperatures come to an equilibrium temperature when they are placed in contact. If there is no heat loss to the external surroundings, the heat lost by the hotter objects equals the heat gained by the cooler ones, a process that is consistent with the conservation of energy. Just this kind of process occurs within a thermos. A perfect thermos would prevent any heat from leaking out or in. However, energy in the form of heat can flow *between* materials inside the thermos to the extent that they have different temperatures; for example, between ice cubes and warm tea. The transfer of energy continues until a common temperature is reached at thermal equilibrium.

The kind of heat transfer that occurs within a thermos of iced tea also occurs within a calorimeter, which is the experimental apparatus used in a technique known as *calorimetry.* Figure 12.24 shows that, like a thermos, a calorimeter is essentially an insulated container. It can be used to determine the specific heat capacity of a substance, as the next example illustrates.

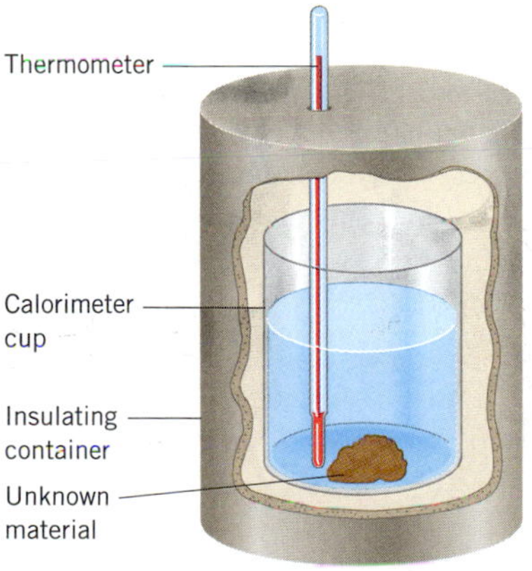

**Figure 12.24** A calorimeter can be used to measure the specific heat capacity of an unknown material.

### *Example 11*  Measuring the Specific Heat Capacity

The calorimeter cup in Figure 12.24 is made from 0.15 kg of aluminum and contains 0.20 kg of water. Initially, the water and the cup have a common temperature of 18.0 °C. A 0.040-kg mass of unknown material is heated to a temperature of 97.0 °C and then added to the water. The temperature of the water, the cup, and the unknown material is 22.0 °C after thermal equi-

* From 14.5 to 15.5 °C.

librium is reestablished. Ignoring the small amount of heat gained by the thermometer, find the specific heat capacity of the unknown material.

**Reasoning** Since energy is conserved and there is negligible heat flow between the calorimeter and the outside surroundings, the heat gained by the cold water and the aluminum cup as they warm up is equal to the heat lost by the unknown material as it cools down. Each quantity of heat can be calculated using the relation $Q = cm\Delta T$, where we always write the change in temperature $\Delta T$ as the higher temperature minus the lower temperature. The equation "Heat gained = Heat lost" will then contain a single unknown quantity, the desired specific heat capacity.

**Solution**

$$\underbrace{(cm\Delta T)_{\text{Al}} + (cm\Delta T)_{\text{water}}}_{\substack{\text{Heat gained by} \\ \text{aluminum and water}}} = \underbrace{(cm\Delta T)_{\text{unknown}}}_{\substack{\text{Heat lost by} \\ \text{unknown material}}}$$

$$c_{\text{unknown}} = \frac{c_{\text{Al}} m_{\text{Al}} \Delta T_{\text{Al}} + c_{\text{water}} m_{\text{water}} \Delta T_{\text{water}}}{m_{\text{unknown}} \Delta T_{\text{unknown}}}$$

The changes in temperature for the three substances are $\Delta T_{\text{Al}} = \Delta T_{\text{water}} = 22.0\ °\text{C} - 18.0\ °\text{C} = 4.0\ \text{C}°$, and $\Delta T_{\text{unknown}} = 97.0\ °\text{C} - 22.0\ °\text{C} = 75.0\ \text{C}°$. Table 12.2 contains values for the specific heat capacities of aluminum and water. Substituting these data into the equation above, we find that

$$c_{\text{unknown}} = \frac{[9.00 \times 10^2\ \text{J/(kg} \cdot \text{C}°)](0.15\ \text{kg})(4.0\ \text{C}°) + [4186\ \text{J/(kg} \cdot \text{C}°)](0.20\ \text{kg})(4.0\ \text{C}°)}{(0.040\ \text{kg})(75.0\ \text{C}°)}$$

$$= \boxed{1300\ \text{J/(kg} \cdot \text{C}°)}$$

## 12.8  Heat and Phase Change: Latent Heat

Surprisingly, there are situations in which the addition or removal of heat does not cause a temperature change. Consider a well-stirred glass of iced tea that has come to thermal equilibrium. Even though heat enters the glass from the warmer room, the temperature of the tea does not rise above 0 °C as long as ice cubes are present. Apparently the heat is being used for some purpose other than raising the temperature. In fact, the heat is being used to melt the ice, and only when all of it is melted will the temperature of the liquid begin to rise.

An important point illustrated by the iced tea example is that there is more than one type or phase of matter. For instance, some of the water in the glass is in the solid phase (ice) and some in the liquid phase. The gas or vapor phase is the third familiar phase of matter. In the gas phase, water is referred to as water vapor or steam. All three phases of water are present in the scene depicted in Figure 12.25.

Matter can change from one phase to another, and heat plays a role in the change. Figure 12.26 summarizes the various possibilities. A solid can *melt* or *fuse* into a liquid if heat is added, while the liquid can *freeze* into a solid if heat is removed. Similarly, a liquid can *evaporate* into a gas if heat is supplied, while the gas can *condense* into a liquid if heat is taken away. Rapid evaporation, with the formation of vapor bubbles within the liquid, is called *boiling*. Finally, a solid can sometimes change directly into a gas if heat is provided. We say that the solid *sublimes* into a gas. Examples of sublimation are (1) solid carbon dioxide, $CO_2$ (dry ice), turning into gaseous $CO_2$ and (2) solid naphthalene (moth balls) turning into naphthalene fumes. Conversely, if heat is removed under the right conditions, the gas will condense directly into a solid.

Figure 12.27 displays a graph that indicates what typically happens when heat is added to a material that changes phase. The graph records temperature versus heat added and refers to water at the normal atmospheric pressure of $1.01 \times 10^5$ Pa. The water starts off as ice at the subfreezing temperature of $-30$ °C. As heat is added, the temperature of the ice increases, in accord with the specific heat capacity of ice [2000 J/(kg·C°)]. Not until the temperature reaches the normal melting/freezing point of 0 °C does the water begin to change phase. Then, when heat is added, the solid changes into the liquid, the temperature staying at 0 °C until *all the ice has melted*. Once all the material is in the liquid

**Figure 12.25** The three phases of water: ice is floating in liquid water, and water vapor (invisible) is present in the air. (© Paul Souders/Stone/Getty Images)

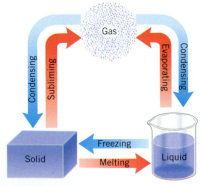

**Figure 12.26** Three familiar phases of matter—solid, liquid, and gas—and the phase changes that can occur between any two of them.

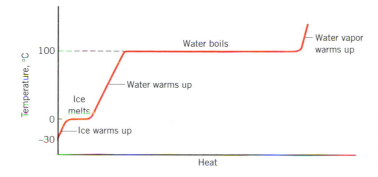

**Figure 12.27** The graph shows the way the temperature of water changes as heat is added, starting with ice at $-30\ °C$. The pressure is atmospheric pressure.

phase, additional heat causes the temperature to increase again, now in accord with the specific heat capacity of liquid water [4186 J/(kg·C°)]. When the temperature reaches the normal boiling/condensing point of 100 °C, the water begins to change from the liquid to the gas phase and continues to do so as long as heat is added. The temperature remains at 100 °C *until all liquid is gone.* When all of the material is in the gas phase, additional heat once again causes the temperature to rise, this time according to the specific heat capacity of water vapor at constant atmospheric pressure [2020 J/(kg·C°)]. Conceptual Example 12 applies the information in Figure 12.27 to a familiar situation.

## Conceptual Example 12  Saving Energy

Suppose you are cooking spaghetti for dinner, and the instructions say "boil the pasta in water for ten minutes." To cook spaghetti in an open pot with the least amount of energy, should you turn up the burner to its fullest so the water vigorously boils, or should you turn down the burner so the water barely boils?

**Reasoning and Solution** The spaghetti needs to cook at a temperature of 100 °C for ten minutes. It doesn't matter whether the water is vigorously boiling or barely boiling, because its temperature is still 100 °C. Remember, as long as there is boiling water in the pot, and the pot is open to one atmosphere of pressure, no amount of additional heat will cause the water temperature to rise above 100 °C. Additional heat only vaporizes the water and produces steam, which is of no use in cooking the spaghetti. So, save your money and turn down the heat, because *the least amount of energy is expended when the water barely boils.*

When a substance changes from one phase to another, the amount of heat that must be added or removed depends on the type of material and the nature of the phase change. The heat per kilogram associated with a phase change is referred to as *latent heat:*

■ **HEAT SUPPLIED OR REMOVED IN CHANGING THE PHASE OF A SUBSTANCE**

The heat $Q$ that must be supplied or removed to change the phase of a mass $m$ of a substance is

$$Q = mL \qquad (12.5)$$

where $L$ is the latent heat of the substance.

*SI Unit of Latent Heat:* J/kg

The *latent heat of fusion* $L_f$ refers to the change between solid and liquid phases, the *latent heat of vaporization* $L_v$ applies to the change between liquid and gas phases, and the *latent heat of sublimation* $L_s$ refers to the change between solid and gas phases.

Table 12.3 gives some typical values of latent heats of fusion and vaporization. For instance, the latent heat of fusion for water is $L_f = 3.35 \times 10^5$ J/kg. Thus, $3.35 \times 10^5$ J of heat must be supplied to melt one kilogram of ice at 0 °C into liquid water at 0 °C; conversely, this amount of heat must be removed from one kilogram of liquid water at 0 °C to freeze the liquid into ice at 0 °C. In comparison, the latent heat of vaporization for

**Table 12.3** *Latent Heats[a] of Fusion and Vaporization*

| Substance | Melting Point (°C) | Latent Heat of Fusion, $L_f$ (J/kg) | Boiling Point (°C) | Latent Heat of Vaporization, $L_v$ (J/kg) |
|---|---|---|---|---|
| Ammonia | −77.8 | $33.2 \times 10^4$ | −33.4 | $13.7 \times 10^5$ |
| Benzene | 5.5 | $12.6 \times 10^4$ | 80.1 | $3.94 \times 10^5$ |
| Copper | 1083 | $20.7 \times 10^4$ | 2566 | $47.3 \times 10^5$ |
| Ethyl alcohol | −114.4 | $10.8 \times 10^4$ | 78.3 | $8.55 \times 10^5$ |
| Gold | 1063 | $6.28 \times 10^4$ | 2808 | $17.2 \times 10^5$ |
| Lead | 327.3 | $2.32 \times 10^4$ | 1750 | $8.59 \times 10^5$ |
| Mercury | −38.9 | $1.14 \times 10^4$ | 356.6 | $2.96 \times 10^5$ |
| Nitrogen | −210.0 | $2.57 \times 10^4$ | −195.8 | $2.00 \times 10^5$ |
| Oxygen | −218.8 | $1.39 \times 10^4$ | −183.0 | $2.13 \times 10^5$ |
| Water | 0.0 | $33.5 \times 10^4$ | 100.0 | $22.6 \times 10^5$ |

[a] The values pertain to 1 atm pressure.

**The physics of steam burns.**

water has the much larger value of $L_v = 22.6 \times 10^5$ J/kg. When water boils at 100 °C, $22.6 \times 10^5$ J of heat must be supplied for each kilogram of liquid turned into steam. And when steam condenses at 100 °C, this amount of heat is released from each kilogram of steam that changes back into liquid. Liquid water at 100 °C is hot enough by itself to cause a bad burn, and the additional effect of the large latent heat can cause severe tissue damage if condensation occurs on the skin.

**The physics of high-tech clothing.**

By taking advantage of the latent heat of fusion, designers can now engineer clothing that can absorb or release heat to help maintain a comfortable and approximately constant temperature close to your body. As the photograph in Figure 12.28 shows, the fabric in this type of clothing is coated with microscopic balls of heat-resistant plastic that contain a substance known as a "phase-change material" (PCM). When you are enjoying your favorite winter sport, for example, it is easy to become overheated. The PCM prevents this by melting, absorbing excess body heat in the process. When you are taking a break and cooling down, however, the PCM freezes and releases heat to keep you warm. The temperature range over which the PCM can maintain a comfort zone is related to its melting/freezing temperature, which is determined by its chemical composition.

Examples 13 and 14 illustrate how to take into account the effect of latent heat when using the conservation-of-energy principle.

### Example 13　Ice-cold Lemonade

Ice at 0 °C is placed in a Styrofoam cup containing 0.32 kg of lemonade at 27 °C. The specific heat capacity of lemonade is virtually the same as that of water; that is, $c = 4186$ J/(kg·C°). After the ice and lemonade reach an equilibrium temperature, some ice still remains. The latent heat of fusion for water is $L_f = 3.35 \times 10^5$ J/kg. Assume that the mass of the cup is so small that it absorbs a negligible amount of heat, and ignore any heat lost to the surroundings. Determine the mass of ice that has melted.

**Reasoning** According to the principle of energy conservation, the heat gained by the melting ice equals the heat lost by the cooling lemonade. According to Equation 12.5, the heat gained by the melting ice is $Q = mL_f$, where $m$ is the mass of the melted ice, and $L_f$ is the latent heat of fusion for water. The heat lost by the lemonade is given by $Q = cm\Delta T$, where $\Delta T$ is the higher temperature of 27 °C minus the lower equilibrium temperature. The equilibrium temperature is 0 °C, because there is some ice remaining, and ice is in equilibrium with liquid water when the temperature is 0 °C.

**Solution**

$$\underbrace{(mL_f)_{ice}}_{\substack{\text{Heat gained} \\ \text{by ice}}} = \underbrace{(cm\Delta T)_{lemonade}}_{\substack{\text{Heat lost} \\ \text{by lemonade}}}$$

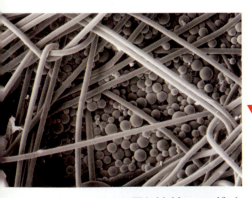
**Figure 12.28** This highly magnified image shows a fabric that has been coated with microscopic balls of heat-resistant plastic. The balls contain a substance known as a "phase-change material," the melting and freezing of which absorbs and releases heat. Clothing made from such fabrics can automatically adjust itself in reaction to your body heat and help maintain a constant temperature next to your skin. (Courtesy Outlast Technologies, Boulder, CO.)

The mass $m_{\text{ice}}$ of ice that has melted is

$$m_{\text{ice}} = \frac{(cm\Delta T)_{\text{lemonade}}}{L_{\text{f}}} = \frac{[4186 \text{ J/(kg} \cdot \text{C}°)](0.32 \text{ kg})(27 \text{ °C} - 0 \text{ °C})}{3.35 \times 10^5 \text{ J/kg}} = \boxed{0.11 \text{ kg}}$$

## Example 14  Getting Ready for a Party

A 7.00-kg glass bowl [$c = 840$ J/(kg·C°)] contains 16.0 kg of punch at 25.0 °C. Two-and-a-half kilograms of ice [$c = 2.00 \times 10^3$ J/(kg·C°)] are added to the punch. The ice has an initial temperature of −20.0 °C, having been kept in a very cold freezer. The punch may be treated as if it were water [$c = 4186$ J/(kg·C°)], and it may be assumed that there is no heat flow between the punch bowl and the external environment. The latent heat of fusion for water is $3.35 \times 10^5$ J/kg. When thermal equilibrium is reached, all the ice has melted, and the final temperature of the mixture is above 0 °C. Determine this temperature.

**Reasoning**  The final temperature can be found by using the conservation of energy: the heat gained is equal to the heat lost. Heat is gained (a) by the ice in warming up to the melting point, (b) by the ice in changing phase from a solid to a liquid, and (c) by the liquid that results from the ice warming up to the final temperature; heat is lost (d) by the punch and (e) by the bowl in cooling down. The heat gained or lost by each component in changing temperature can be determined from the relation $Q = cm\Delta T$. The heat gained when water changes phase from a solid to a liquid at 0 °C is $Q = mL_{\text{f}}$, where $m$ is the mass of water and $L_{\text{f}}$ is the latent heat of fusion.

**Solution**  The heat gained or lost by each component is listed as follows:

**(a)**  Heat gained when
ice warms to 0.0 °C        $= [2.00 \times 10^3 \text{ J/(kg·C°)}](2.50 \text{ kg})[0.0 \text{ °C} - (-20.0 \text{ °C})]$

**(b)**  Heat gained when
ice melts at 0.0 °C        $= (2.50 \text{ kg})(3.35 \times 10^5 \text{ J/kg})$

**(c)**  Heat gained when melted
ice (liquid) warms        $= [4186 \text{ J/(kg·C°)}](2.50 \text{ kg})(T - 0.0 \text{ °C})$
to temperature $T$

**(d)**  Heat lost when
punch cools to        $= [4186 \text{ J/(kg·C°)}](16.0 \text{ kg})(25.0 \text{ °C} - T)$
temperature $T$

**(e)**  Heat lost when
bowl cools to        $= [840 \text{ J/(kg·C°)}](7.00 \text{ kg})(25.0 \text{ °C} - T)$
temperature $T$

Setting the heat gained equal to the heat lost gives:

$$\underbrace{(a) + (b) + (c)}_{\text{Heat gained}} = \underbrace{(d) + (e)}_{\text{Heat lost}}$$

This equation can be solved to show that $\boxed{T = 11 \text{ °C}}$.

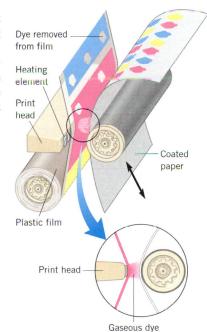

Dye removed from film

Heating element

Print head

Coated paper

Plastic film

Print head

Gaseous dye

**Figure 12.29**  A dye-sublimation printer. As the plastic film passes in front of the print head, the heat from a given heating element causes one of three pigments or dyes on the film to sublime from a solid to a gas. The gaseous dye is absorbed onto the coated paper as a dot of color. The size of the dots on the paper has been exaggerated for clarity.

An interesting application of the phase change between a solid and a gas is found in one kind of color printer used with computers. A dye-sublimation printer uses a thin plastic film coated with separate panels of cyan (blue), yellow, and magenta pigment or dye. A full spectrum of colors is produced by using combinations of tiny spots of these dyes. As Figure 12.29 shows, the coated film passes in front of a print head that extends across the width of the paper and contains 2400 heating elements. When a heating element is turned on, the dye in front of it absorbs heat and goes from a solid to a gas—it sublimes—with no liquid phase in between. A coating on the paper absorbs the gaseous dye on contact, producing a small spot of color. The intensity of the spot is controlled by the heating element, since each element can produce 256 different temperatures; the hotter the element, the greater the amount of dye transferred to the paper. The paper makes three separate passes across the print head, once for each of the dyes. The final result is an image of near-photographic quality.

**The physics of**
**a dye-sublimation color printer.**

# Concept Summary

This summary presents an abridged version of the chapter, including the important equations and all available learning aids. For convenient reference, the learning aids (including the text's examples) are placed next to or immediately after the relevant equation or discussion. The following learning aids may be found on-line at **www.wiley.com/college/cutnell:**

**Interactive LearningWare** examples are solved according to a five-step interactive format that is designed to help you develop problem-solving skills.

**Concept Simulations** are animated versions of text figures or animations that illustrate important concepts. You can control parameters that affect the display, and we encourage you to experiment.

**Interactive Solutions** offer specific models for certain types of problems in the chapter homework. The calculations are carried out interactively.

**Self-Assessment Tests** include both qualitative and quantitative questions. Extensive feedback is provided for both incorrect and correct answers, to help you evaluate your understanding of the material.

| Topic | Discussion | Learning Aids |
|---|---|---|
| | **12.1  Common Temperature Scales** | |
| **Celsius temperature scale** **Fahrenheit temperature scale** | On the Celsius temperature scale, there are 100 equal divisions between the ice point (0 °C) and the steam point (100 °C). On the Fahrenheit temperature scale, there are 180 equal divisions between the ice point (32 °F) and the steam point (212 °F). | **Examples 1, 2** |
| | **12.2  The Kelvin Temperature Scale** | |
| **Kelvin temperature scale** | For scientific work, the Kelvin temperature scale is the scale of choice. One kelvin (K) is equal in size to one Celsius degree. However, the temperature $T$ on the Kelvin scale differs from the temperature $T_c$ on the Celsius scale by an additive constant of 273.15: $$T = T_c + 273.15 \qquad (12.1)$$ | |
| **Absolute zero** | The lower limit of temperature is called absolute zero and is designated as 0 K on the Kelvin scale. | |
| | **12.3  Thermometers** | |
| **Thermometric property** | The operation of any thermometer is based on the change in some physical property with temperature; this physical property is called a thermometric property. Examples of thermometric properties are the length of a column of mercury, electrical voltage, and electrical resistance. | |
| | **12.4  Linear Thermal Expansion** | |
| | Most substances expand when heated. For linear expansion, an object of length $L_0$ experiences a change $\Delta L$ in length when the temperature changes by $\Delta T$: | |
| **Linear thermal expansion** | $$\Delta L = \alpha L_0 \Delta T \qquad (12.2)$$ where $\alpha$ is the coefficient of linear expansion. | **Example 3** |
| **Thermal stress** | For an object held rigidly in place, a thermal stress can occur when the object attempts to expand or contract. The stress can be large, even for small temperature changes. | **Example 4** |
| **How a hole in a plate expands or contracts** | When the temperature changes, a hole in a plate of solid material expands or contracts as if the hole were filled with the surrounding material. | **Examples 5, 6, 7** |
| | **12.5  Volume Thermal Expansion** | |
| | For volume expansion, the change $\Delta V$ in the volume of an object of volume $V_0$ is given by | |
| **Volume thermal expansion** | $$\Delta V = \beta V_0 \Delta T \qquad (12.3)$$ where $\beta$ is the coefficient of volume expansion. | **Example 8** **Interactive LearningWare 12.1** |
| **How a cavity expands or contracts** | When the temperature changes, a cavity in a piece of solid material expands or contracts as if the cavity were filled with the surrounding material. | **Interactive Solution 12.27** |

 *Use Self-Assessment Test 12.1 to evaluate your understanding of Sections 12.1–12.5.*

| Topic | Discussion | Learning Aids |
|---|---|---|

### 12.6  Heat and Internal Energy

**Internal energy**
**Heat**

The internal energy of a substance is the sum of the kinetic, potential, and other kinds of energy that the molecules of the substance have. Heat is energy that flows from a higher-temperature object to a lower-temperature object because of the difference in temperatures. The SI unit for heat is the joule (J).

### 12.7  Heat and Temperature Change: Specific Heat Capacity

The amount of heat $Q$ that must be supplied or removed to change the temperature of a substance of mass $m$ by an amount $\Delta T$ is

**Heat needed to change the temperature**

$$Q = cm\Delta T \qquad (12.4)$$

**Examples 9, 10, 11**

where $c$ is a constant known as the specific heat capacity.

**Interactive Solutions 12.43, 12.89**

**Energy conservation and heat**  When materials are placed in thermal contact within a perfectly insulated container, the principle of energy conservation requires that heat lost by warmer materials equals heat gained by cooler materials.

Heat is sometimes measured with a unit called the kilocalorie (kcal). The conversion factor between kilocalories and joules is known as the mechanical equivalent of heat:

**Mechanical equivalent of heat**

$$1 \text{ kcal} = 4186 \text{ joules}$$

**Interactive LearningWare 12.2**

### 12.8  Heat and Phase Change: Latent Heat

Heat must be supplied or removed to make a material change from one phase to another. The heat $Q$ that must be supplied or removed to change the phase of a mass $m$ of a substance is

**Heat needed to change the phase**

$$Q = mL \qquad (12.5)$$

**Examples 12, 13, 14**

where $L$ is the latent heat of the substance and has SI units of J/kg. The latent heats of fusion, vaporization, and sublimation refer, respectively, to the solid/liquid, the liquid/vapor, and the solid/vapor phase changes.

**Interactive Solution 12.63**

Use **Self-Assessment Test 12.2** to evaluate your understanding of Sections 12.6–12.8.

# Problems

*Note: For problems in this set, use the values of α and β given in Table 12.1, and the values of c, L$_f$, and L$_v$ given in Tables 12.2 and 12.3, unless stated otherwise.*

 **ssm**  Solution is in the Student Solutions Manual.     **www**  Solution is available on the World Wide Web at www.wiley.com/college/cutnell
This icon represents a biomedical application.

**Section 12.1 Common Temperature Scales,**
**Section 12.2 The Kelvin Temperature Scale,**
**Section 12.3 Thermometers**

**1. ssm**  What's your normal body temperature? It may not be 98.6 °F, the oft-quoted average that was determined in the nineteenth century. A more recent study has reported an average temperature of 98.2 °F. What is the *difference* between these averages, expressed in Celsius degrees?

**2.** On the moon the surface temperature ranges from 375 K during the day to $1.00 \times 10^2$ K at night. What are these temperatures on the (a) Celsius and (b) Fahrenheit scales?

**3.** A personal computer is designed to operate over the temperature range from 50.0 to 104 °F. To what do these temperatures correspond (a) on the Celsius scale and (b) on the Kelvin scale?

**4.**  Dermatologists often remove small precancerous skin lesions by freezing them quickly with liquid nitrogen, which

has a temperature of 77 K. What is this temperature on the (a) Celsius and (b) Fahrenheit scales?

**5. ssm**  A temperature of absolute zero occurs at −273.15 °C. What is this temperature on the Fahrenheit scale?

* **6.** A copper–constantan thermocouple generates a voltage of $4.75 \times 10^{-3}$ volts when the temperature of the hot junction is 110.0 °C and the reference junction is kept at 0.0 °C. If the voltage is proportional to the difference in temperature between the junctions, what is the temperature of the hot junction when the voltage is $1.90 \times 10^{-3}$ volts?

* **7. ssm**  A constant-volume gas thermometer (see Figures 12.3 and 12.4) has a pressure of $5.00 \times 10^3$ Pa when the gas temperature is 0.00 °C. What is the temperature (in °C) when the pressure is $2.00 \times 10^3$ Pa?

* **8.** Space invaders land on earth. On the invaders' temperature scale, the ice point is at 25 °I (I = invader), and the steam point is at 156 °I. The invaders' thermometer shows the temperature on earth to

be 58 °I. Using logic similar to that in Example 1 in the text, what would this temperature be on the Celsius scale?

### Section 12.4 Linear Thermal Expansion

**9.** A steel aircraft carrier is 370 m long when moving through the icy North Atlantic at a temperature of 2.0 °C. By how much does the carrier lengthen when it is traveling in the warm Mediterranean Sea at a temperature of 21 °C?

**10.** The Concorde is 62 m long when its temperature is 23 °C. In flight, the outer skin of this supersonic aircraft can reach 105 °C due to air friction. The coefficient of linear expansion of the skin is $2.0 \times 10^{-5}$ (C °)$^{-1}$. Find the amount by which the Concorde expands.

**11. ssm www** Find the approximate length of the Golden Gate bridge if it is known that the steel in the roadbed expands by 0.53 m when the temperature changes from +2 to +32 °C.

**12.** Conceptual Example 5 provides background for this problem. A hole is drilled through a copper plate whose temperature is 11 °C. (a) When the temperature of the plate is increased, will the radius of the hole be larger or smaller than the radius at 11 °C? Why? (b) When the plate is heated to 110 °C, by what fraction $\Delta r/r_0$ will the radius of the hole change?

**13.** A commonly used method of fastening one part to another part is called "shrink fitting." A steel rod has a diameter of 2.0026 cm, and a flat plate contains a hole whose diameter is 2.0000 cm. The rod is cooled so that it just fits into the hole. When the rod warms up, the enormous thermal stress exerted by the plate holds the rod securely to the plate. By how many Celsius degrees should the rod be cooled?

**14.** A thin rod consists of two parts joined together. One-third of it is silver and two-thirds is gold. The temperature decreases by 26 C°. Determine the fractional decrease $\dfrac{\Delta L}{L_{0,\,\text{Silver}} + L_{0,\,\text{Gold}}}$ in the rod's length, where $L_{0,\,\text{Silver}}$ and $L_{0,\,\text{Gold}}$ are the initial lengths of the silver and gold rods.

**15. ssm** A rod made from a particular alloy is heated from 25.0 °C to the boiling point of water. Its length increases by $8.47 \times 10^{-4}$ m. The rod is then cooled from 25.0 °C to the freezing point of water. By how much does the rod shrink?

**\*16.** As the drawing shows, two thin strips of metal are bolted together at one end and have the same temperature. One is steel, and the other is aluminum. The steel strip is 0.10% longer than the aluminum strip. By how much should the temperature of the strips be increased, so that the strips have the same length?

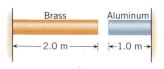

**\*17.** The brass bar and the aluminum bar in the drawing are each attached to an immovable wall. At 28 °C the air gap between the rods is $1.3 \times 10^{-3}$ m. At what temperature will the gap be closed?

**\*18.** Consult Conceptual Example 5 for background pertinent to this problem. A lead sphere has a diameter that is 0.050% larger than the inner diameter of a steel ring when each has a temperature of 70.0 °C. Thus, the ring will not slip over the sphere. At what common temperature will the ring just slip over the sphere?

**\*19. ssm www** A simple pendulum consists of a ball connected to one end of a thin brass wire. The period of the pendulum is 2.0000 s. The temperature rises by 140 C°, and the length of the wire increases. Determine the period of the heated pendulum.

**\*20.** Concrete sidewalks are always laid in sections, with gaps between each section. For example, the drawing shows three identical 2.4-m sections, the outer two of which are against immovable walls. The two identical gaps between the sections are provided so that thermal expansion will not create the thermal stress that could lead to cracks. What is the minimum gap width necessary to account for an increase in temperature of 32 C°?

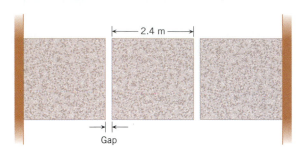

**\*21.** A steel ruler is accurate when the temperature is 25 °C. When the temperature drops to −15 °C, the ruler no longer reads correctly, but it can be made to read correctly if a stress is applied to each end of the ruler. (a) Should the stress be a compression or a tension? Why? (b) What is the magnitude of the necessary stress?

**\*\*22.** A steel ruler is calibrated to read true at 20.0 °C. A draftsman uses the ruler at 40.0 °C to draw a line on a 40.0 °C copper plate. As indicated on the warm ruler, the length of the line is 0.50 m. To what temperature should the plate be cooled, such that the length of the line truly becomes 0.50 m?

**\*\*23. ssm** A wire is made by attaching two segments together, end to end. One segment is made of aluminum and the other is steel. The effective coefficient of linear expansion of the two-segment wire is $19 \times 10^{-6}$ (C°)$^{-1}$. What fraction of the length is aluminum?

**\*\*24.** An 85.0-N backpack is hung from the middle of an aluminum wire, as the drawing shows. The temperature of the wire then drops by 20.0 C°. Find the tension in the wire at the lower temperature. Assume that the distance between the supports does not change, and ignore any thermal stress.

### Section 12.5 Volume Thermal Expansion

**25.** A copper kettle contains water at 24 °C. When the water is heated to its boiling point, the volume of the kettle expands by $1.2 \times 10^{-5}$ m$^3$. Determine the volume of the kettle at 24 °C.

**26.** A swimming pool contains 110 m$^3$ of water. The sun heats the water from 17 to 27 °C. What is the change in the volume of the water?

**27. Interactive Solution 12.27** at **www.wiley.com/college/cutnell** presents a model for solving problems of this type. A thin spherical shell of silver has an inner radius of $2.0 \times 10^{-2}$ m when the temperature is 18 °C. The shell is heated to 147 °C. Find the change in the interior volume of the shell.

**28.** A lead object and a quartz object each have the same initial volume. The volume of each increases by the same amount, because the temperature increases. If the temperature of the lead object increases by 4.0 C°, by how much does the temperature of the quartz object increase?

**29. ssm** When heated at constant pressure, a given volume of gas expands much more than an equal volume of liquid. The coefficient of volume expansion for air is about $3.7 \times 10^{-3}$ (C°)$^{-1}$ at room temperature and one atmosphere of pressure. Find the ratio of the change in volume of air to the change in volume of water, assuming the same initial volumes and temperature changes.

**30. Interactive LearningWare 12.1** at www.wiley.com/college/cutnell provides some useful background for this problem. Many hot-water heating systems have a reservoir tank connected directly to the pipeline, so as to allow for expansion when the water becomes hot. The heating system of a house has 76 m of copper pipe whose inside radius is $9.5 \times 10^{-3}$ m. When the water and pipe are heated from 24 to 78 °C, what must be the minimum volume of the reservoir tank to hold the overflow of water?

**31. ssm** Suppose that the steel gas tank in your car is completely filled when the temperature is 17 °C. How many gallons will spill out of the twenty-gallon tank when the temperature rises to 35 °C?

**32.** Consult **Interactive LearningWare 12.1** at **www.wiley.com/college/cutnell** for help in solving this problem. During an all-night cram session, a student heats up a one-half liter ($0.50 \times 10^{-3}$ m$^3$) glass (Pyrex) beaker of cold coffee. Initially, the temperature is 18 °C, and the beaker is filled to the brim. A short time later when the student returns, the temperature has risen to 92 °C. The coefficient of volume expansion of coffee is the same as that of water. How much coffee (in cubic meters) has spilled out of the beaker?

*** 33.** At the bottom of an old mercury-in-glass thermometer is a 45-mm$^3$ reservoir filled with mercury. When the thermometer was placed under your tongue, the warmed mercury would expand into a very narrow cylindrical channel, called a capillary, whose radius was $1.7 \times 10^{-2}$ mm. Marks were placed along the capillary that indicated the temperature. Ignore the thermal expansion of the glass and determine how far (in mm) the mercury would expand into the capillary when the temperature changed by 1.0 C°.

*** 34.** A spherical brass shell has an interior volume of $1.60 \times 10^{-3}$ m$^3$. Within this interior volume is a solid steel ball that has a volume of $0.70 \times 10^{-3}$ m$^3$. The space between the steel ball and the inner surface of the brass shell is filled completely with mercury. A small hole is drilled through the brass, and the temperature of the arrangement is increased by 12 C°. What is the volume of the mercury that spills out of the hole?

*** 35. ssm** The bulk modulus of water is $B = 2.2 \times 10^9$ N/m$^2$. What change in pressure $\Delta P$ (in atmospheres) is required to keep water from expanding when it is heated from 15 to 25 °C?

*** 36.** A solid aluminum sphere has a radius of 0.50 m and a temperature of 75 °C. The sphere is then completely immersed in a pool of water whose temperature is 25 °C. The sphere cools, while the water temperature remains nearly at 25 °C, because the pool is very large. The sphere is weighed in the water immediately after being submerged (before it begins to cool) and then again after cooling to 25 °C. (a) Which weight is larger? Why? (b) Use Archimedes' principle to find the magnitude of the *difference* between the weights.

**** 37. ssm www** Two identical thermometers made of Pyrex glass contain, respectively, identical volumes of mercury and methyl alcohol. If the expansion of the glass is taken into account, how many times greater is the distance between the degree marks on the methyl alcohol thermometer than that on the mercury thermometer?

**** 38.** The column of mercury in a barometer (see Figure 11.11) has a height of 0.760 m when the pressure is one atmosphere and the tem-

perature is 0.0 °C. Ignoring any change in the glass containing the mercury, what will be the height of the mercury column for the same one atmosphere of pressure when the temperature rises to 38.0 °C on a hot day? (*Hint: The pressure in the barometer is given by Pressure = $\rho g h$, and the density $\rho$ of the mercury changes when the temperature changes.*)

### Section 12.6 Heat and Internal Energy, Section 12.7 Heat and Temperature Change: Specific Heat Capacity

**39.** Blood can carry excess energy from the interior to the surface of the body, where the energy is dispersed in a number of ways. While a person is exercising, 0.6 kg of blood flows to the surface of the body and releases 2000 J of energy. The blood arriving at the surface has the temperature of the body interior, 37.0 °C. Assuming that blood has the same specific heat capacity as water, determine the temperature of the blood that leaves the surface and returns to the interior.

**40.** A piece of glass has a temperature of 83.0 °C. Liquid that has a temperature of 43.0 °C is poured over the glass, completely covering it, and the temperature at equilibrium is 53.0 °C. The mass of the glass and the liquid is the same. Ignoring the container that holds the glass and liquid and assuming that the heat lost to or gained from the surroundings is negligible, determine the specific heat capacity of the liquid.

**41. ssm** If the price of electrical energy is $0.10 per kilowatt·hour, what is the cost of using electrical energy to heat the water in a swimming pool (12.0 m × 9.00 m × 1.5 m) from 15 to 27 °C?

**42.** Ideally, when a thermometer is used to measure the temperature of an object, the temperature of the object itself should not change. However, if a significant amount of heat flows from the object to the thermometer, the temperature will change. A thermometer has a mass of 31.0 g, a specific heat capacity of $c = 815$ J/(kg·C°), and a temperature of 12.0 °C. It is immersed in 119 g of water, and the final temperature of the water and thermometer is 41.5 °C. What was the temperature of the water before the insertion of the thermometer?

**43.** Review **Interactive Solution 12.43** at **www.wiley.com/college/cutnell** for help in approaching this problem. When resting, a person has a metabolic rate of about $3.0 \times 10^5$ joules per hour. The person is submerged neck-deep into a tub containing $1.2 \times 10^3$ kg of water at 21.00 °C. If the heat from the person goes only into the water, find the water temperature after half an hour.

**44.** When you take a bath, how many kilograms of hot water (49.0 °C) must you mix with cold water (13.0 °C) so that the temperature of the bath is 36.0 °C? The total mass of water (hot plus cold) is 191 kg. Ignore any heat flow between the water and its external surroundings.

**45. ssm** At a fabrication plant, a hot metal forging has a mass of 75 kg and a specific heat capacity of 430 J/(kg·C°). To harden it, the forging is immersed in 710 kg of oil that has a temperature of 32 °C and a specific heat capacity of 2700 J/(kg·C°). The final temperature of the oil and forging at thermal equilibrium is 47 °C. Assuming that heat flows only between the forging and the oil, determine the initial temperature of the forging.

*** 46.** Refer to **Interactive LearningWare 12.2** at **www.wiley.com/college/cutnell** for a review of the concepts that play roles in this problem. The box of a well-known breakfast cereal states that one ounce of the cereal contains 110 Calories (1 food Calorie = 4186 J). If 2.0% of this energy could be converted by a weight lifter's body into work done in lifting a barbell, what is the heaviest barbell that could be lifted a distance of 2.1 m?

\* **47.** In a passive solar house, the sun heats water stored in barrels to a temperature of 38 °C. The stored energy is then used to heat the house on cloudy days. Suppose that $2.4 \times 10^8$ J of heat is needed to maintain the inside of the house at 21 °C. How many barrels (1 barrel = 0.16 $m^3$) of water are needed?

\* **48.** The water in a swimming pool absorbs $2.00 \times 10^9$ J of heat from the sun. What is the change in volume of the water?

\* **49.** **ssm** A rock of mass 0.20 kg falls from rest from a height of 15 m into a pail containing 0.35 kg of water. The rock and water have the same initial temperature. The specific heat capacity of the rock is 1840 J/(kg·C°). Ignore the heat absorbed by the pail itself, and determine the rise in the temperature of the rock and water.

\*\* **50.** A steel rod ($\rho = 7860$ kg/$m^3$) has a length of 2.0 m. It is bolted at both ends between immobile supports. Initially there is no tension in the rod, because the rod just fits between the supports. Find the tension that develops when the rod loses 3300 J of heat.

## Section 12.8 Heat and Phase Change: Latent Heat

**51.** **ssm** How much heat must be added to 0.45 kg of aluminum to change it from a solid at 130 °C to a liquid at 660 °C (its melting point)? The latent heat of fusion for aluminum is $4.0 \times 10^5$ J/kg.

**52.** Assume that the pressure is one atmosphere and determine the heat required to produce 2.00 kg of water vapor at 100.0 °C, starting with (a) 2.00 kg of water at 100.0 °C and (b) 2.00 kg of liquid water at 0.0 °C.

**53.** Suppose that the amount of heat removed when 3.0 kg of water freezes at 0 °C were removed from ethyl alcohol at its freezing/melting point of −114 °C. How many kilograms of ethyl alcohol would freeze?

**54.** A person eats a container of yogurt. The Nutritional Facts label states that it contains 240 Calories (1 Calorie = 4186 J). What mass of perspiration would one have to lose to get rid of this energy? At body temperature, the latent heat of vaporization of water is $2.42 \times 10^6$ J/kg.

**55.** **ssm** Find the mass of water that vaporizes when 2.10 kg of mercury at 205 °C is added to 0.110 kg of water at 80.0 °C.

**56.** The latent heat of vaporization of $H_2O$ at body temperature (37.0 °C) is $2.42 \times 10^6$ J/kg. To cool the body of a 75-kg jogger [average specific heat capacity = 3500 J/(kg·C°)] by 1.5 C°, how many kilograms of water in the form of sweat have to be evaporated?

**57.** A woman finds the front windshield of her car covered with ice at −12.0 °C. The ice has a thickness of $4.50 \times 10^{-4}$ m, and the windshield has an area of 1.25 $m^2$. The density of ice is 917 kg/$m^3$. How much heat is required to melt the ice?

**58.** A 0.200-kg piece of aluminum that has a temperature of −155 °C is added to 1.5 kg of water that has a temperature of 3.0 °C. At equilibrium the temperature is 0.0 °C. Ignoring the container and assuming that the heat exchanged with the surroundings is negligible, determine the mass of water that has been frozen into ice.

\* **59.** Occasionally, huge icebergs are found floating on the ocean's currents. Suppose one such iceberg is 120 km long, 35 km wide, and

230 m thick. (a) How much heat would be required to melt this iceberg (assumed to be at 0 °C) into liquid water at 0 °C? The density of ice is 917 kg/$m^3$. (b) The annual energy consumption by the United States in 1994 was $9.3 \times 10^{19}$ J. If this energy were delivered to the iceberg every year, how many years would it take before the ice melted?

\* **60.** To help keep his barn warm on cold days, a farmer stores 840 kg of solar-heated water ($L_f = 3.35 \times 10^5$ J/kg) in barrels. For how many hours would a 2.0-kW electric space heater have to operate to provide the same amount of heat as the water does when it cools from 10.0 to 0.0 °C and completely freezes?

\* **61.** **ssm www** An unknown material has a normal melting/freezing point of −25.0 °C, and the liquid phase has a specific heat capacity of 160 J/(kg·C°). One-tenth of a kilogram of the solid at −25.0 °C is put into a 0.150-kg aluminum calorimeter cup that contains 0.100 kg of glycerin. The temperature of the cup and the glycerin is initially 27.0 °C. All the unknown material melts, and the final temperature at equilibrium is 20.0 °C. The calorimeter neither loses energy to nor gains energy from the external environment. What is the latent heat of fusion of the unknown material?

\* **62.** Two grams of liquid water are at 0 °C, and another two grams are at 100 °C. Heat is removed from the water at 0 °C, completely freezing it at 0 °C. This heat is then used to vaporize some of the water at 100 °C. What is the mass (in grams) of the liquid water that remains?

\* **63.** **Interactive Solution 12.63** at **www.wiley.com/college/cutnell** provides a model for solving problems such as this. A 42-kg block of ice at 0 °C is sliding on a horizontal surface. The initial speed of the ice is 7.3 m/s and the final speed is 3.5 m/s. Assume that the part of the block that melts has a very small mass and that all the heat generated by kinetic friction goes into the block of ice, and determine the mass of ice that melts into water at 0 °C.

\* **64.** Equal masses of two different liquids have the same temperature of 25.0 °C. Liquid A has a freezing point of −68.0 °C and a specific heat capacity of 1850 J/(kg·C°). Liquid B has a freezing point of −96.0 °C and a specific heat capacity of 2670 J/(kg·C°). The same amount of heat must be removed from each liquid in order to freeze it into a solid at its respective freezing point. Determine the difference $L_{f, A} - L_{f, B}$ between the latent heats of fusion for these liquids.

\* **65.** **ssm** It is claimed that if a lead bullet goes fast enough, it can melt completely when it comes to a halt suddenly, and all its kinetic energy is converted into heat via friction. Find the minimum speed of a lead bullet (initial temperature = 30.0 °C) for such an event to happen.

\*\* **66.** A locomotive wheel is 1.00 m in diameter. A 25.0-kg steel band has a temperature of 20.0 °C and a diameter that is $6.00 \times 10^{-4}$ m less than that of the wheel. What is the smallest mass of water vapor at 100 °C that can be condensed on the steel band to heat it, so that it will fit onto the wheel? Do not ignore the water that results from the condensation.

# Chapter 13 The Transfer of Heat

## 13.1 Convection

When heat is transferred to or from a substance, the internal energy of the substance can change, as we saw in Chapter 12. This change in internal energy is accompanied by a change in temperature or a change in phase. The transfer of heat affects us in many ways. For instance, within our homes furnaces distribute heat on cold days, and air conditioners remove it on hot days. Our bodies constantly transfer heat in one direction or another, to prevent the adverse effects of hypo- and hyperthermia. And virtually all our energy originates in the sun and is transferred to us over a distance of 150 million kilometers through the void of space. Today's sunlight provides the energy to drive photosynthesis in the plants that provide our food and, hence, metabolic energy. Ancient sunlight nurtured the organic matter that became the fossil fuels of oil, natural gas, and coal. This chapter examines the three processes by which heat is transferred: convection, conduction, and radiation. We will see how these three methods of heat transfer are related to our previous study of heat and internal energy.

When part of a fluid is warmed, such as the air above a fire, the volume of the fluid expands, and the density decreases. According to Archimedes' principle (see Section 11.6), the surrounding cooler and denser fluid exerts a buoyant force on the warmer fluid and pushes it upward. As warmer fluid rises, the surrounding cooler fluid replaces it. This cooler fluid, in turn, is warmed and pushed upward. Thus, a continuous flow is established, which carries along heat. Whenever heat is transferred by the bulk movement of a gas or a liquid, the heat is said to be transferred by *convection*. The fluid flow itself is called a *convection current.*

**Figure 13.1** During a volcanic eruption, smoke at the top of the plume rises thousands of meters because of convection. (© Art Wolfe/The Image Bank/Getty Images)

> ■ **CONVECTION**
>
> Convection is the process in which heat is carried from place to place by the bulk movement of a fluid.

The smoke rising from the volcanic eruption in Figure 13.1 is one visible result of convection. Figure 13.2 shows the less visible example of convection currents in a pan of water being heated on a gas burner. The currents distribute the heat from the burning gas to all parts of the water. Conceptual Example 1 deals with some of the important roles that convection plays in the home.

**Figure 13.2** Convection currents are set up when a pan of water is heated.

### Conceptual Example 1
### Hot Water Baseboard Heating and Refrigerators

Hot water baseboard heating units are frequently used in homes, where they are mounted on the wall next to the floor, as in Figure 13.3a. In contrast, the cooling coil in a refrigerator is mounted near the top of the refrigerator, as in part *b* of the drawing. The locations for these heating and cooling devices are different, yet each location is designed to maximize the production of convection currents. Explain how.

**The physics of** heating and cooling by convection.

**Reasoning and Solution** An important goal for a heating system is to distribute heat throughout a room. The analogous goal for the cooling coil is to remove heat from all of the space within a refrigerator. In each case, the heating or cooling device is positioned so that convection makes the goal achievable. The air above the baseboard unit is heated, like the air above a fire. Buoyant forces from the surrounding cooler air push the warm air upward. Cooler air near the ceiling is displaced downward and then warmed by the baseboard heating unit, leading to the convection current illustrated in Figure 13.3a. Had the heating unit been located near the ceiling, the warm air would have remained there, with very little convection to distribute the heat.

Within the refrigerator, the air in contact with the top-mounted coil is cooled, its volume decreases, and its density increases. The surrounding warmer and less dense air cannot provide sufficient buoyant force to support the colder air, which sinks downward. In the process,

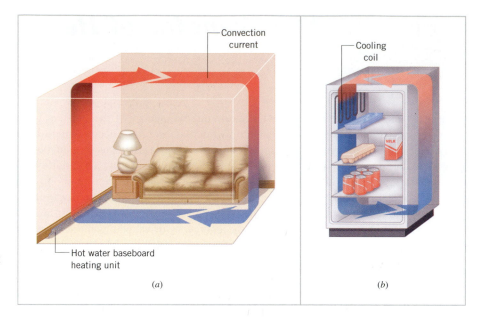

**Figure 13.3** (*a*) Air warmed by the baseboard heating unit is pushed to the top of the room by the cooler and denser air. (*b*) Air cooled by the cooling coil sinks to the bottom of the refrigerator. In both (*a*) and (*b*) a convection current is established.

warmer air near the bottom is displaced upward and is then cooled by the coil, establishing the convection current shown in Figure 13.3*b*. Had the cooling coil been placed at the bottom of the refrigerator, stagnant, cool air would have collected there, with little convection to carry the heat from other parts of the refrigerator to the coil for removal.

Another example of convection occurs when the ground, heated by the sun's rays, warms the neighboring air. Surrounding cooler and denser air pushes the heated air upward. The resulting updraft or "thermal" can be quite strong, depending on the amount of heat that the ground can supply. As Figure 13.4 illustrates, these thermals can be used by glider pilots to gain considerable altitude. Birds such as eagles utilize thermals in a similar fashion.

It is usual for air temperature to decrease with increasing altitude, and the resulting upward convection currents are important for dispersing pollutants from industrial sources and automobile exhaust systems. Sometimes, however, meteorological conditions cause a layer to form in the atmosphere where the temperature increases with increasing altitude. Such a layer is called an ***inversion layer*** because its temperature profile is inverted compared to the usual situation. An inversion layer arrests the normal upward convection currents, causing a stagnant-air condition in which the concentration of pollutants increases substantially. This leads to a smog layer that can often be seen hovering over large cities (Figure 13.5).

We have been discussing ***natural convection,*** in which a temperature difference causes the density at one place in a fluid to be different from that at another. Sometimes, natural convection is inadequate to transfer sufficient amounts of heat. In such cases ***forced convection*** is often used, and an external device such as a pump or a fan mixes the warmer and cooler portions of the fluid. Figure 13.6 shows an example of an automobile engine, where forced convection occurs in two ways. First, a pump circulates radiator fluid (water and antifreeze) through the engine to remove excess heat from the combustion process. Second, a radiator fan draws air through the radiator. Heat is transferred from the hotter radiator fluid to the cooler air, thereby cooling the fluid.

**The physics of**
**"thermals."**

**The physics of**
**an inversion layer.**

**The physics of**
**cooling by forced convection.**

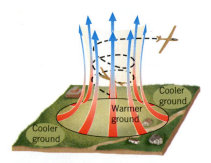

**Figure 13.4** Updrafts, or thermals, are caused by the convective movement of air that the ground has warmed.

## 13.2 *Conduction*

Anyone who has fried a hamburger in an all-metal skillet knows that the metal handle becomes hot. Somehow, heat is transferred from the burner to the handle. Clearly, heat is not being transferred by the bulk movement of the metal or the surrounding air, so convection can be ruled out. Instead, heat is transferred directly through the metal by a process called ***conduction.***

### ■ CONDUCTION

Conduction is the process whereby heat is transferred directly through a material, with any bulk motion of the material playing no role in the transfer.

One mechanism for conduction occurs when the atoms or molecules in a hotter part of the material vibrate or move with greater energy than those in a cooler part. By means of collisions, the more energetic molecules pass on some of their energy to their less energetic neighbors. For example, imagine a gas filling the space between two walls that face each other and are maintained at different temperatures. Molecules strike the hotter wall, absorb energy from it, and rebound with a greater kinetic energy than when they arrived. As these more energetic molecules collide with their less energetic neighbors, they transfer some of their energy to them. Eventually, this energy is passed on until it reaches the molecules next to the cooler wall. These molecules, in turn, collide with the wall, giving up some of their energy to it in the process. Through such molecular collisions, heat is conducted from the hotter to the cooler wall.

A similar mechanism for the conduction of heat occurs in metals. Metals are different from most substances in having a pool of electrons that are more or less free to wander throughout the metal. These free electrons can transport energy and allow metals to transfer heat very well. The free electrons are also responsible for the excellent electrical conductivity that metals have.

Those materials that conduct heat well are called *thermal conductors,* and those that conduct heat poorly are known as *thermal insulators.* Most metals are excellent thermal conductors; wood, glass, and most plastics are common thermal insulators. Thermal insulators have many important applications. Virtually all new housing construction incorporates thermal insulation in attics and walls to reduce heating and cooling costs. And the wooden or plastic handles on many pots and pans reduce the flow of heat to the cook's hand.

To illustrate the factors that influence the conduction of heat, Figure 13.7 displays a rectangular bar. The ends of the bar are in thermal contact with two bodies, one of which is kept at a constant higher temperature, while the other is kept at a constant lower temperature. Although not shown for the sake of clarity, the sides of the bar are insulated, so the heat lost through them is negligible. The amount of heat $Q$ conducted through the bar from the warmer end to the cooler end depends on a number of factors:

1. $Q$ is proportional to the time $t$ during which conduction takes place ($Q \propto t$). More heat flows in longer time periods.

2. $Q$ is proportional to the temperature difference $\Delta T$ between the ends of the bar ($Q \propto \Delta T$). A larger difference causes more heat to flow. No heat flows when both ends have the same temperature, so that $\Delta T = 0$ C°.

3. $Q$ is proportional to the cross-sectional area $A$ of the bar ($Q \propto A$). Figure 13.8 helps to explain this fact by showing two identical bars (insulated sides not shown) placed between the warmer and cooler bodies. Clearly, twice as much heat flows through two bars as through one, since the cross-sectional area has been doubled.

4. $Q$ is inversely proportional to the length $L$ of the bar ($Q \propto 1/L$). Greater lengths of material conduct less heat. To experience this effect, put two insulated mittens (the pot holders that cooks keep around the stove) on the *same hand*. Then,

**Figure 13.5** In the absence of upward convection currents in the air, pollutants accumulate and form the smog layer that is easily visible near the top of this photograph of Los Angeles. (© Deborah Davis/Stone/Getty Images)

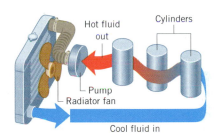

**Figure 13.6** The forced convection generated by a pump circulates radiator fluid through an automobile engine to remove excess heat.

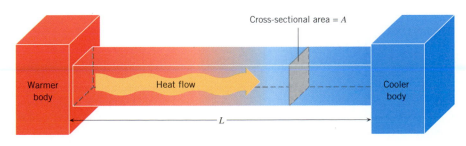

**Figure 13.7** Heat is conducted through the bar when the ends of the bar are maintained at different temperatures. The heat flows from the warmer to the cooler end.

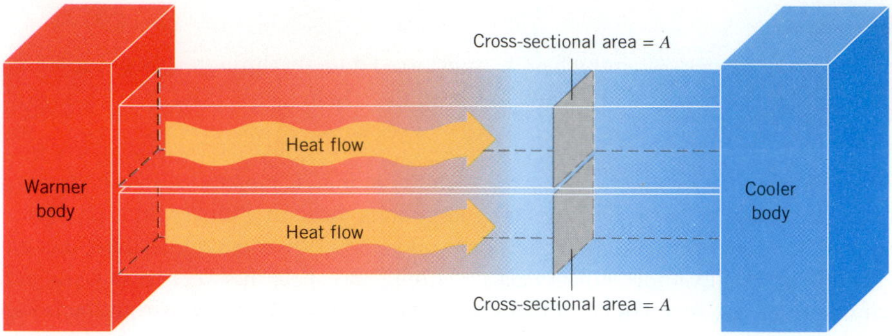

Cross-sectional area = A

Heat flow

Warmer body

Cooler body

Heat flow

Cross-sectional area = A

**Figure 13.8** Twice as much heat flows through two identical bars as through one.

**Table 13.1** *Thermal Conductivities*[a] *of Selected Materials*

| Substance | Thermal Conductivity, $k$ [J/(s·m·C°)] |
|---|---|
| **Metals** | |
| Aluminum | 240 |
| Brass | 110 |
| Copper | 390 |
| Iron | 79 |
| Lead | 35 |
| Silver | 420 |
| Steel (stainless) | 14 |
| **Gases** | |
| Air | 0.0256 |
| Hydrogen ($H_2$) | 0.180 |
| Nitrogen ($N_2$) | 0.0258 |
| Oxygen ($O_2$) | 0.0265 |
| **Other Materials** | |
| Asbestos | 0.090 |
| Body fat | 0.20 |
| Concrete | 1.1 |
| Diamond | 2450 |
| Glass | 0.80 |
| Goose down | 0.025 |
| Ice (0 °C) | 2.2 |
| Styrofoam | 0.010 |
| Water | 0.60 |
| Wood (oak) | 0.15 |
| Wool | 0.040 |

[a] Except as noted, the values pertain to temperatures near 20 °C.

**The physics of**
**dressing warm.**

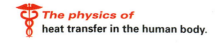

**The physics of**
**heat transfer in the human body.**

touch a hot pot and notice that it feels cooler than when you wear only one mitten, signifying that less heat passes through the greater thickness ("length") of material.

These proportionalities can be stated together as $Q \propto (A\Delta T)t/L$. Equation 13.1 expresses this result with the aid of a proportionality constant $k$, which is called the **thermal conductivity.**

■ **CONDUCTION OF HEAT THROUGH A MATERIAL**

The heat $Q$ conducted during a time $t$ through a bar of length $L$ and cross-sectional area $A$ is

$$Q = \frac{(kA\Delta T)t}{L} \tag{13.1}$$

where $\Delta T$ is the temperature difference between the ends of the bar and $k$ is the thermal conductivity of the material.

**SI Unit of Thermal Conductivity:** J/(s·m·C°)

Since $k = QL/(tA\Delta T)$, the SI unit for thermal conductivity is J·m/(s·m²·C°) or J/(s·m·C°). The SI unit of power is the joule per second (J/s) or watt (W), so the thermal conductivity is also given in units of W/(m·C°).

Different materials have different thermal conductivities, and Table 13.1 gives some representative values. Because metals are such good thermal conductors, they have large thermal conductivities. In comparison, liquids and gases generally have small thermal conductivities. In fact, in most fluids the heat transferred by conduction is negligible compared to that transferred by convection when there are strong convection currents. Air, for instance, with its small thermal conductivity, is an excellent thermal insulator when confined to small spaces where no appreciable convection currents can be established. Goose down, Styrofoam, and wool derive their fine insulating properties in part from the small dead-air spaces within them, as Figure 13.9 illustrates. We also take advantage of dead-air spaces when we dress "in layers" during very cold weather and put on several layers of relatively thin clothing rather than one thick layer. The air trapped between the layers acts as an excellent insulator.

Example 2 deals with the role that conduction through body fat plays in regulating body temperature.

**Example 2**  **Heat Transfer in the Human Body**

When excessive heat is produced within the body, it must be transferred to the skin and dispersed if the temperature at the body interior is to be maintained at the normal value of 37.0 °C. One possible mechanism for transfer is conduction through body fat. Suppose that heat travels through 0.030 m of fat in reaching the skin, which has a total surface area of 1.7 m² and a temperature of 34.0 °C. Find the amount of heat that reaches the skin in half an hour (1800 s).

**Reasoning and Solution** In Table 13.1 the thermal conductivity of body fat is given as $k = 0.20 \text{ J/(s·m·C°)}$. According to Equation 13.1,

$$Q = \frac{(kA\Delta T)t}{L}$$

$$Q = \frac{[0.20 \text{ J/(s·m·C°)}](1.7 \text{ m}^2)(37.0 \text{ °C} - 34.0 \text{ °C})(1800 \text{ s})}{0.030 \text{ m}} = \boxed{6.1 \times 10^4 \text{ J}}.$$

For comparison, a jogger can generate over ten times this amount of heat in a half hour. Thus, conduction through body fat is not a particularly effective way of removing excess heat. Heat transfer via blood flow to the skin is more effective and has the added advantage that the body can vary the blood flow as needed (see Problem 7).

Virtually all homes contain insulation in the walls to reduce heat loss. Example 3 illustrates how to determine this loss with and without insulation.

## Example 3 Layered Insulation

One wall of a house consists of 0.019-m-thick plywood backed by 0.076-m-thick insulation, as Figure 13.10 shows. The temperature at the inside surface is 25.0 °C, while the temperature at the outside surface is 4.0 °C, both being constant. The thermal conductivities of the insulation and the plywood are, respectively, 0.030 and 0.080 J/(s·m·C°), and the area of the wall is 35 m². Find the heat conducted through the wall in one hour (a) with the insulation and (b) without the insulation.

**Reasoning** The temperature $T$ at the insulation–plywood interface (see Figure 13.10) must be determined before the heat conducted through the wall can be obtained. In calculating this temperature, we use the fact that no heat is accumulating in the wall because the inner and outer temperatures are constant. Therefore, the heat conducted through the insulation must equal the heat conducted through the plywood during the same time; that is, $Q_{\text{insulation}} = Q_{\text{plywood}}$. Each of the $Q$ values can be expressed as $Q = (kA\Delta T)t/L$, according to Equation 13.1, leading to an expression that can be solved for the interface temperature. Once a value for $T$ is available, Equation 13.1 can be used to obtain the heat conducted through the wall.

**Solution**

**(a)** Using Equation 13.1 and the fact that $Q_{\text{insulation}} = Q_{\text{plywood}}$, we find that

$$\left[\frac{(kA\Delta T)t}{L}\right]_{\text{insulation}} = \left[\frac{(kA\Delta T)t}{L}\right]_{\text{plywood}}$$

$$\frac{[0.030 \text{ J/(s·m·C°)}]A(25.0 \text{ °C} - T)t}{0.076 \text{ m}} = \frac{[0.080 \text{ J/(s·m·C°)}]A(T - 4.0 \text{ °C})t}{0.019 \text{ m}}$$

Eliminating the area $A$ and time $t$ algebraically and solving this equation for $T$ reveals that the temperature at the insulation–plywood interface is $T = 5.8$ °C.

The heat conducted through the wall is either $Q_{\text{insulation}}$ or $Q_{\text{plywood}}$, since the two quantities are equal. Choosing $Q_{\text{insulation}}$ and using $T = 5.8$ °C in Equation 13.1, we find that

$$Q_{\text{insulation}} = \frac{[0.030 \text{ J/(s·m·C°)}](35 \text{ m}^2)(25.0 \text{ °C} - 5.8 \text{ °C})(3600 \text{ s})}{0.076 \text{ m}}$$

$$= \boxed{9.5 \times 10^5 \text{ J}}$$

**(b)** It is straightforward to use Equation 13.1 to calculate the amount of heat that would flow through the plywood in one hour if the insulation were absent:

$$Q_{\text{plywood}} = \frac{[0.080 \text{ J/(s·m·C°)}](35 \text{ m}^2)(25.0 \text{ °C} - 4.0 \text{ °C})(3600 \text{ s})}{0.019 \text{ m}}$$

$$= \boxed{110 \times 10^5 \text{ J}}$$

Without insulation, the heat loss is increased by a factor of about 12.

Fruit growers sometimes protect their crops by spraying them with water when overnight temperatures are expected to drop below freezing. Some fruit crops, like the

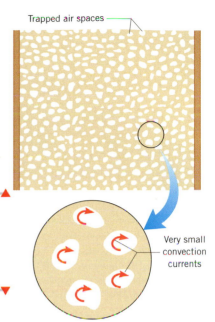

Trapped air spaces

Very small convection currents

**Figure 13.9** Styrofoam is an excellent thermal insulator because it contains many small, dead-air spaces. These small spaces inhibit heat transfer by convection currents, and air itself has a very low thermal conductivity.

**The physics of** layered insulation.

**Problem solving insight**
When heat is conducted through a multi-layered material, and the high and low temperatures are constant, the heat conducted through each layer is the same.

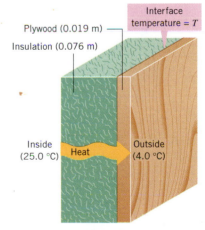

Plywood (0.019 m)
Insulation (0.076 m)

Interface temperature = $T$

Inside (25.0 °C)    Heat    Outside (4.0 °C)

**Figure 13.10** Heat flows through the insulation and plywood from the warmer inside to the cooler outside. The temperature of the insulation–plywood interface is $T$.

**The physics of**
**protecting fruit plants from freezing.**

**Figure 13.11** After a subfreezing night, this blueberry crop is being checked for freeze damage. The plants were sprayed the previous evening by sprinklers that put a coating of ice on them, thereby insulating the plants against the subfreezing temperatures. (© AP/Wide World Photos)

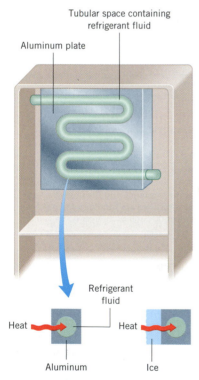

**Figure 13.12** In a refrigerator, cooling is accomplished by a cold refrigerant fluid that circulates through a tubular space embedded within an aluminum plate. The arrangement works less well when the plate is coated with a layer of ice.

blueberry plants in Figure 13.11, can withstand temperatures down to freezing (0 °C). However, as the temperature falls below freezing, the risk of damage rises significantly. When water is sprayed on the plants, it can freeze and form a covering of ice. When the water freezes it releases heat (see Section 12.8), some of which goes into warming the plant. In addition, both water and ice have relatively small thermal conductivities, as Table 13.1 indicates. Thus, they also protect the crop by acting as thermal insulators that reduce heat loss from the plants.

Although a layer of ice may be beneficial to blueberry plants, it is not so desirable inside a refrigerator, as Conceptual Example 4 discusses.

### Conceptual Example 4    An Iced-up Refrigerator

In a refrigerator, heat is removed by a cold refrigerant fluid that circulates within a tubular space embedded within a metal plate, as Figure 13.12 illustrates. A good refrigerator cools food as quickly as possible. Decide whether the plate should be made from aluminum or stainless steel and whether the arrangement works better or worse when it becomes coated with a layer of ice.

**Reasoning and Solution** Figure 13.12 (see blowups) shows the metal cooling plate with and without a layer of ice. Without ice, heat passes by conduction through the metal to the refrigerant fluid within. For a given temperature difference across the thickness of metal, heat is transferred more quickly through the metal with the largest thermal conductivity. Table 13.1 indicates that the thermal conductivity of aluminum is more than 17 times larger than that of stainless steel. Therefore, *the plate should be made from aluminum.* When the plate becomes coated with ice, any heat that is removed by the refrigerant fluid must first be transferred by conduction through the ice before it encounters the aluminum plate. But the conduction of heat through ice occurs much less readily than through aluminum, because, as Table 13.1 indicates, ice has a much smaller thermal conductivity. Moreover, Equation 13.1 indicates that the heat conducted per unit time ($Q/t$) is inversely proportional to the thickness $L$ of the ice. Thus, as ice builds up, the heat removed per unit time by the cooling plate decreases. *When covered with ice, the cooling plate works less well.*

## 13.3  *Radiation*

Energy from the sun is brought to earth by large amounts of visible light waves, as well as by substantial amounts of infrared and ultraviolet waves. These waves belong to a class of waves known as electromagnetic waves, a class that also includes the microwaves used for cooking and the radio waves used for AM and FM broadcasts. The sunbather in Figure 13.13 feels hot because her body absorbs energy from the sun's electromagnetic waves. And anyone who has stood by a roaring fire or put a hand near an incandescent light bulb has experienced a similar effect. Thus, fires and light bulbs also emit electromagnetic waves, and when the energy of such waves is absorbed, it can have the same effect as heat.

The process of transferring energy via electromagnetic waves is called **radiation,** and, unlike convection or conduction, it does not require a material medium. Electromagnetic waves from the sun, for example, travel through the void of space during their journey to earth.

**Figure 13.13** Suntans are produced by ultraviolet rays. (© Angelo Cavalli/The Image Bank/Getty Images)

■ **RADIATION**

Radiation is the process in which energy is transferred by means of electromagnetic waves.

All bodies continuously radiate energy in the form of electromagnetic waves. Even an ice cube radiates energy, although so little of it is in the form of visible light that an ice cube cannot be seen in the dark. Likewise, the human body emits insufficient visible light to be seen in the dark. However, as Figures 12.6 and 12.7 illustrate, the infrared waves radiating from the body can be detected in the dark by electronic cameras. Generally, an object does not emit much visible light until the temperature of the object exceeds about 1000 K. Then a characteristic red glow appears, like that of a heating coil on an electric stove. When its temperature reaches about 1700 K, an object begins to glow white-hot, like the tungsten filament in an incandescent light bulb.

In the transfer of energy by radiation, the absorption of electromagnetic waves is just as important as their emission. The surface of an object plays a significant role in determining how much radiant energy the object will absorb or emit. The two blocks in sunlight in Figure 13.14, for example, are identical, except that one has a rough surface coated with lampblack (a fine black soot), while the other has a highly polished silver surface. As the thermometers indicate, the temperature of the black block rises at a much faster rate than that of the silvery block. This is because lampblack absorbs about 97% of the incident radiant energy, while the silvery surface absorbs only about 10%. The remaining part of the incident energy is reflected in each case. We see the lampblack as black in color because it reflects so little of the light falling on it, while the silvery surface looks like a mirror because it reflects so much light. Since the color black is associated with nearly complete absorption of visible light, the term **perfect blackbody** or, simply, **blackbody** is used when referring to an object that absorbs all the electromagnetic waves falling on it.

All objects emit and absorb electromagnetic waves simultaneously. When a body has the same constant temperature as its surroundings, the amount of radiant energy being absorbed must balance the amount being emitted in a given interval of time. The block coated with lampblack absorbs and emits the same amount of radiant energy, and the silvery block does too. In either case, if absorption were greater than emission, the block would experience a net gain in energy. As a result, the temperature of the block would rise and not be constant. Similarly, if emission were greater than absorption, the temperature would fall. Since absorption and emission are balanced, **a material that is a good absorber, like lampblack, is also a good emitter, and a material that is a poor absorber, like polished silver, is also a poor emitter.** A perfect blackbody, being a perfect absorber, is also a perfect emitter.

The fact that a black surface is both a good absorber and a good emitter is the reason people are uncomfortable wearing dark clothes during the summer. Dark clothes absorb a large fraction of the sun's radiation and then reemit it in all directions. About one-half of the emitted radiation is directed inward toward the body and creates the sensation of warmth. Light-colored clothes, in contrast, are cooler to wear, since they absorb and reemit relatively little of the incident radiation.

The use of light colors for comfort also occurs in nature. Most lemurs, for instance, are nocturnal and have dark fur like the one shown in Figure 13.15a. Since they are active at night, the dark fur poses no disadvantage in absorbing excessive sunlight. Figure 13.15b

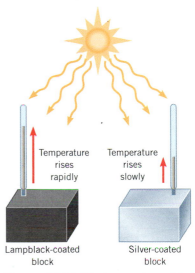

**Figure 13.14** The temperature of the block coated with lampblack rises faster than the temperature of the block coated with silver because the black surface absorbs radiant energy from the sun at the greater rate.

*The physics of* **summer clothing.**

(a)

(b)

**Figure 13.15** (a) Most lemurs, like this one, are nocturnal and have dark fur. (© Wolfgang Kaehler/Corbis) (b) The species of lemur called the white sifaka, however, is active during the day and has white fur. (© Nigel Dennis/ Wildlife Pictures/Peter Arnold)

shows a species of lemur called the white sifaka, which lives in semiarid regions where there is little shade. The white color of the fur may help in thermoregulation, by reflecting sunlight during the hot part of the day. However, during the cool mornings, reflection of sunlight would be a hindrance in warming up. It is interesting to note that these lemurs have black skin and only sparse fur on their bellies, and that to warm up in the morning, they turn their dark bellies toward the sun. The dark color enhances the absorption of sunlight.

The amount of radiant energy $Q$ emitted by a perfect blackbody is proportional to the radiation time interval $t$ ($Q \propto t$). The longer the time, the greater is the amount of energy radiated. Experiment shows that $Q$ is also proportional to the surface area $A$ ($Q \propto A$). An object with a large surface area radiates more energy than one with a small surface area, other things being equal. Finally, experiment reveals that $Q$ is proportional to the *fourth power of the Kelvin temperature* $T$ ($Q \propto T^4$), so the emitted energy increases markedly with increasing temperature. If, for example, the Kelvin temperature of an object doubles, the object emits $2^4$ or 16 times more energy. Combining these factors into a single proportionality, we see that $Q \propto T^4At$. This proportionality is converted into an equation by inserting a proportionality constant $\sigma$, known as the *Stefan–Boltzmann constant*. It has been found experimentally that $\sigma = 5.67 \times 10^{-8}$ J/(s·m²·K⁴):

$$Q = \sigma T^4 At$$

The relationship above holds only for a perfect emitter. Most objects are not perfect emitters, however. Suppose that an object radiates only about 80% of the visible light energy that a perfect emitter would radiate, so $Q$ (for the object) $= (0.80)\sigma T^4At$. The factor such as the 0.80 in this equation is called the **emissivity e** and is a dimensionless number between zero and one. The emissivity is the ratio of the energy an object actually radiates to the energy the object would radiate if it were a perfect emitter. For visible light, the value of $e$ for the human body, for instance, varies between about 0.65 and 0.80, the smaller values pertaining to lighter skin colors. For infrared radiation, $e$ is nearly one for all skin colors. For a perfect blackbody emitter, $e = 1$. Including the factor $e$ on the right side of the expression $Q = \sigma T^4At$ leads to the **Stefan–Boltzmann law of radiation**.

■ **THE STEFAN–BOLTZMANN LAW OF RADIATION**

The radiant energy $Q$, emitted in a time $t$ by an object that has a Kelvin temperature $T$, a surface area $A$, and an emissivity $e$, is given by

$$Q = e\sigma T^4 At \tag{13.2}$$

where $\sigma$ is the Stefan–Boltzmann constant and has a value of $5.67 \times 10^{-8}$ J/(s·m²·K⁴).

In Equation 13.2, the Stefan–Boltzmann constant $\sigma$ is a universal constant in the sense that its value is the same for all bodies, regardless of the nature of their surfaces. The emissivity $e$, however, depends on the condition of the surface.

Example 5 shows how the Stefan–Boltzmann law can be used to determine the size of a star.

### Example 5   A Supergiant Star

The supergiant star Betelgeuse has a surface temperature of about 2900 K (about one-half that of our sun) and emits a radiant power (in joules per second, or watts) of approximately $4 \times 10^{30}$ W (about 10 000 times as great as that of our sun). Assuming that Betelgeuse is a perfect emitter (emissivity $e = 1$) and spherical, find its radius.

**Reasoning** According to the Stefan–Boltzmann law, the power emitted is $Q/t = e\sigma T^4A$. A star with a relatively small temperature $T$ can have a relatively large radiant power $Q/t$ only if the area $A$ is large. As we will see, Betelgeuse has a very large surface area, so its radius is enormous.

**Solution** Solving the Stefan–Boltzmann law for the area, we find

$$A = \frac{Q/t}{e\sigma T^4}$$

But the surface area of a sphere is $A = 4\pi r^2$, so $r = \sqrt{A/4\pi}$. Therefore, we have

$$r = \sqrt{\frac{Q/t}{4\pi e\sigma T^4}} = \sqrt{\frac{4 \times 10^{30}\ \text{W}}{4\pi(1)[5.67 \times 10^{-8}\ \text{J/(s·m}^2\text{·K}^4)](2900\ \text{K})^4}}$$

$$= \boxed{3 \times 10^{11}\ \text{m}}$$

**Problem solving insight**
First solve an equation for the unknown in terms of the known variables. Then substitute numbers for the known variables, as this example shows.

For comparison, Mars orbits the sun at a distance of $2.28 \times 10^{11}$ m. Betelgeuse is certainly a "supergiant."

The next example explains how to apply the Stefan–Boltzmann law when an object, such as a wood stove, simultaneously emits and absorbs radiant energy.

## Example 6 A Wood-Burning Stove

A wood-burning stove stands unused in a room where the temperature is 18 °C (291 K). A fire is started inside the stove. Eventually, the temperature of the stove surface reaches a constant 198 °C (471 K), and the room warms to a constant 29 °C (302 K). The stove has an emissivity of 0.900 and a surface area of 3.50 m². Determine the *net* radiant power generated by the stove when the stove (a) is unheated and has a temperature equal to room temperature and (b) has a temperature of 198 °C.

*The physics of a wood-burning stove.*

**Reasoning** The stove emits more radiant power when heated than when unheated. In both cases, however, the Stefan–Boltzmann law can be used to determine the amount of power emitted. Power is the change in energy per unit time (Equation 6.10b), or $Q/t$. But in this problem we need to find the *net* power produced by the stove. The net power is the power the stove emits minus the power the stove absorbs. The power the stove absorbs comes from the walls, ceiling, and floor of the room, all of which emit radiation.

**Solution**

**(a)** Remembering that temperature must be expressed in kelvins when using the Stefan–Boltzmann law, we find that

Power emitted by unheated stove at 18 °C $= \dfrac{Q}{t} = e\sigma T^4 A$ $\qquad$ (13.2)

**Problem solving insight**
In the Stefan–Boltzmann law of radiation, the temperature $T$ must be expressed in kelvins, not in degrees Celsius or degrees Fahrenheit.

$$= (0.900)[5.67 \times 10^{-8}\ \text{J/(s·m}^2\text{·K}^4)](291\ \text{K})^4(3.50\ \text{m}^2) = 1280\ \text{W}$$

The fact that the unheated stove emits 1280 W of power and yet maintains a constant temperature means that the stove also absorbs 1280 W of radiant power from its surroundings. Thus, the *net* power generated by the unheated stove is zero:

Net power generated by stove at 18 °C $= \underbrace{1280\ \text{W}}_{\substack{\text{Power emitted} \\ \text{by stove at} \\ 18\ °\text{C}}} - \underbrace{1280\ \text{W}}_{\substack{\text{Power emitted by} \\ \text{room at 18 °C and} \\ \text{absorbed by stove}}} = \boxed{0\ \text{W}}$

**(b)** The hot stove (198 °C or 471 K) emits more radiant power than it absorbs from the cooler room. The radiant power the stove emits is

Power emitted by stove at 198 °C $= \dfrac{Q}{t} = e\sigma T^4 A$

$$= (0.900)[5.67 \times 10^{-8}\ \text{J/(s·m}^2\text{·K}^4)](471\ \text{K})^4(3.50\ \text{m}^2) = 8790\ \text{W}$$

The radiant power the stove absorbs from the room is identical to the power that the stove would emit at the constant room temperature of 29 °C (302 K). The reasoning here is exactly like that in part (a):

Power emitted by room at 29 °C and absorbed by stove $= \dfrac{Q}{t} = e\sigma T^4 A$

$$= (0.900)[5.67 \times 10^{-8}\ \text{J/(s·m}^2\text{·K}^4)](302\ \text{K})^4(3.50\ \text{m}^2) = 1490\ \text{W}$$

The *net* radiant power the stove produces from the fuel it burns is

$$
\begin{array}{c}
\text{Net power} \\
\text{generated by} \\
\text{stove at 198 °C}
\end{array}
=
\underbrace{8790 \text{ W}}_{\substack{\text{Power emitted} \\ \text{by stove at} \\ \text{198 °C}}}
-
\underbrace{1490 \text{ W}}_{\substack{\text{Power emitted by} \\ \text{room at 29 °C and} \\ \text{absorbed by stove}}}
= \boxed{7300 \text{ W}}
$$

**Figure 13.16** Highly reflective metal foil covering this satellite minimizes temperature changes. (Courtesy NASA.)

Example 6 illustrates that when an object has a higher temperature than its surroundings, the object emits a net radiant power $P_{net} = (Q/t)_{net}$. The net power is the power the object emits minus the power it absorbs. Applying the Stefan–Boltzmann law as in Example 6 leads to the following expression for $P_{net}$ when the temperature of the object is $T$ and the temperature of the environment is $T_0$:

$$P_{net} = e\sigma A(T^4 - T_0^4) \tag{13.3}$$

## 13.4  Applications

To keep heating and air-conditioning bills to a minimum, it pays to use good thermal insulation in your home. Insulation inhibits convection between inner and outer walls and minimizes heat transfer by conduction. With respect to conduction, the logic behind home insulation ratings comes directly from Equation 13.1. According to this equation, the heat per unit time $Q/t$ flowing through a thickness of material is $Q/t = kA\Delta T/L$. Keeping the value for $Q/t$ to a minimum means using materials that have small thermal conductivities $k$ and large thicknesses $L$. Construction engineers, however, prefer to use Equation 13.1 in the slightly different form shown below:

$$\frac{Q}{t} = \frac{A\Delta T}{L/k}$$

**The physics of**
**rating thermal insulation by R values.**

The term $L/k$ in the denominator is called the $R$ value of the insulation. For a building material, it is convenient to talk about an $R$ value because it expresses in a single number the combined effects of thermal conductivity and thickness. Larger $R$ values reduce the heat per unit time flowing through the material and, therefore, mean better insulation. It is also convenient to use $R$ values to describe layered slabs formed by sandwiching together a number of materials with different thermal conductivities and different thicknesses. The $R$ values for the individual layers can be added to give a single $R$ value for the entire slab. It should be noted, however, that $R$ values are expressed using units of feet, hours, F°, and Btu for thickness, time, temperature, and heat, respectively.

**The physics of**
**regulating the temperature of an**
**orbiting satellite.**

When it is in the earth's shadow, an orbiting satellite is shielded from the intense electromagnetic waves emitted by the sun. But when a satellite moves out of the earth's shadow, the satellite experiences the full effect of these waves. As a result, the temperature within a satellite would decrease and increase sharply during an orbital period and sensitive electronic circuitry would suffer, unless precautions are taken. To minimize temperature fluctuations, satellites are often covered with a highly reflecting and, hence, poorly absorbing metal foil, as Figure 13.16 shows. By reflecting much of the sunlight, the foil minimizes temperature rises. Being a poor absorber, the foil is also a poor emitter and reduces radiant energy losses. Reducing these losses keeps the temperature from falling excessively when the satellite is in the earth's shadow.

**The physics of**
**a thermos bottle   and**
**a halogen cooktop stove.**

A thermos bottle, sometimes referred to as a Dewar flask, reduces the rate at which hot liquids cool down or cold liquids warm up. A thermos usually consists of a double-walled glass vessel with silvered inner walls (see Figure 13.17) and accomplishes its job by minimizing heat transfer via convection, conduction, and radiation. The space between the walls is evacuated to minimize energy losses due to conduction and convection. The silvered surfaces reflect most of the radiant energy that would otherwise enter or leave the liquid in the thermos. Finally, little heat is lost through the glass or the rubberlike gaskets and stopper, since these materials have relatively small thermal conductivities.

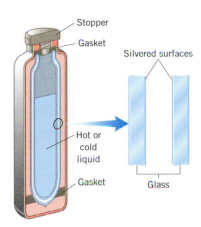

**Figure 13.17** A thermos bottle minimizes energy transfer due to convection, conduction, and radiation.

**Figure 13.18** In a halogen cooktop, quartz–iodine lamps emit a large amount of electromagnetic energy that is absorbed directly by a pot or pan.

Halogen cooktops use radiant energy to heat pots and pans. A halogen cooktop uses several quartz–iodine lamps, like the ones used for ultrabright automobile headlights. These lamps are electrically powered and are mounted below a ceramic top. (See Figure 13.18.) They radiate a great deal of electromagnetic energy, which passes through the ceramic top and is absorbed directly by the bottom of the pot. Consequently, the pot heats up very quickly, rivaling the time of a pot on an open gas burner.

# Concept Summary

This summary presents an abridged version of the chapter, including the important equations and all available learning aids. For convenient reference, the learning aids (including the text's examples) are placed next to or immediately after the relevant equation or discussion. The following learning aids may be found on-line at **www.wiley.com/college/cutnell**:

| | |
|---|---|
| **Interactive LearningWare** examples are solved according to a five-step interactive format that is designed to help you develop problem-solving skills. | **Concept Simulations** are animated versions of text figures or animations that illustrate important concepts. You can control parameters that affect the display, and we encourage you to experiment. |
| **Interactive Solutions** offer specific models for certain types of problems in the chapter homework. The calculations are carried out interactively. | **Self-Assessment Tests** include both qualitative and quantitative questions. Extensive feedback is provided for both incorrect and correct answers, to help you evaluate your understanding of the material. |

| Topic | Discussion | Learning Aids |
|---|---|---|
| | **13.1 Convection** | |
| | Convection is the process in which heat is carried from place to place by the bulk movement of a fluid. | |
| Natural convection | During natural convection, the warmer, less dense part of a fluid is pushed upward by the buoyant force provided by the surrounding cooler and denser part. | Example 1 |
| Forced convection | Forced convection occurs when an external device, such as a fan or a pump, causes the fluid to move. | |
| | **13.2 Conduction** | |
| | Conduction is the process whereby heat is transferred directly through a material, with any bulk motion of the material playing no role in the transfer. | |

| Topic | Discussion | Learning Aids |
|---|---|---|
| Thermal conductors and thermal insulators | Materials that conduct heat well, such as most metals, are known as thermal conductors. Materials that conduct heat poorly, such as wood, glass, and most plastics, are referred to as thermal insulators. | |
| Conduction of heat through a material | The heat $Q$ conducted during a time $t$ through a bar of length $L$ and cross-sectional area $A$ is $$Q = \frac{(kA\Delta T)t}{L} \qquad (13.1)$$ where $\Delta T$ is the temperature difference between the ends of the bar and $k$ is the thermal conductivity of the material. | Examples 2, 3, 4<br><br>**Concept Simulation 13.1**<br><br>**Interactive LearningWare 13.1**<br><br>**Interactive Solutions 13.13, 13.33** |

### 13.3 Radiation

Radiation is the process in which energy is transferred by means of electromagnetic waves.

| | | |
|---|---|---|
| Absorbers and emitters | All objects, regardless of their temperature, simultaneously absorb and emit electromagnetic waves. Objects that are good absorbers of radiant energy are also good emitters, and objects that are poor absorbers are also poor emitters. | |
| A perfect blackbody | An object that absorbs all the radiation incident upon it is called a perfect blackbody. A perfect blackbody, being a perfect absorber, is also a perfect emitter. | |
| Stefan–Boltzmann law of radiation | The radiant energy $Q$ emitted during a time $t$ by an object whose surface area is $A$ and whose Kelvin temperature is $T$ is given by the Stefan–Boltzmann law of radiation: $$Q = e\sigma T^4 A t \qquad (13.2)$$ | Example 5 |
| Emissivity | where $\sigma = 5.67 \times 10^{-8}$ J/(s·m²·K⁴) is the Stefan–Boltzmann constant and $e$ is the emissivity, a dimensionless number characterizing the surface of the object. The emissivity lies between 0 and 1, being zero for a nonemitting surface and one for a perfect blackbody. | |
| Net radiant power | The net radiant power is the power an object emits minus the power it absorbs. The net radiant power $P_{net}$ emitted by an object of temperature $T$ located in an environment of temperature $T_0$ is $$P_{net} = e\sigma A(T^4 - T_0^4) \qquad (13.3)$$ | Example 6<br><br>**Interactive LearningWare 13.2** |

**Use Self-Assessment Test 13.1 to evaluate your understanding of Sections 13.1–13.3.**

# Problems

*Note: For problems in this set, use the values for thermal conductivities given in Table 13.1 unless stated otherwise.*

**ssm** Solution is in the Student Solutions Manual. **www** Solution is available on the World Wide Web at www.wiley.com/college/cutnell
 This icon represents a biomedical application.

### Section 13.2 Conduction

**1. ssm Concept Simulation 13.1** at www.wiley.com/college/cutnell illustrates the concepts pertinent to this problem. A person's body is covered with 1.6 m² of wool clothing. The thickness of the wool is $2.0 \times 10^{-3}$ m. The temperature at the outside surface of the wool is 11 °C, and the skin temperature is 36 °C. How much heat per second does the person lose due to conduction?

**2.** The temperature in an electric oven is 160 °C. The temperature at the outer surface in the kitchen is 50 °C. The oven (surface area = 1.6 m²) is insulated with material that has a thickness of 0.020 m and a thermal conductivity of 0.045 J/(s·m·C°). (a) How much energy is used to operate the oven for six hours? (b) At a price of

$0.10 per kilowatt·hour for electrical energy, what is the cost of operating the oven?

**3.** In an electrically heated home, the temperature of the ground in contact with a concrete basement wall is 12.8 °C. The temperature at the inside surface of the wall is 20.0 °C. The wall is 0.10 m thick and has an area of 9.0 m². Assume that one kilowatt·hour of electrical energy costs $0.10. How many hours are required for one dollar's worth of energy to be conducted through the wall? $14 h$

**4.** The amount of heat per second conducted from the blood capillaries beneath the skin to the surface is 240 J/s. The energy is transferred a distance of $2.0 \times 10^{-3}$ m through a body whose surface area is 1.6 m². Assuming that the thermal conductivity is that of body fat, determine the temperature difference between the capillaries and the surface of the skin.

**5. ssm** Due to a temperature difference $\Delta T$, heat is conducted through an aluminum plate that is 0.035 m thick. The plate is then replaced by a stainless steel plate that has the same temperature difference and cross-sectional area. How thick should the steel plate be so that the same amount of heat per second is conducted through it?

**6.** A skier wears a jacket filled with goose down that is 15 mm thick. Another skier wears a wool sweater that is 5.0 mm thick. Both have the same surface area. Assuming that the temperature difference between the inner and outer surfaces of each garment is the same, calculate the ratio (wool/goose down) of the heat lost due to conduction during the same time interval.

**7. ssm www** In the conduction equation $Q = (kA\Delta T)t/L$, the combination of factors $kA/L$ is called the *conductance*. The human body has the ability to vary the conductance of the tissue beneath the skin by means of vasoconstriction and vasodilation, in which the flow of blood to the veins and capillaries underlying the skin is decreased and increased, respectively. The conductance can be adjusted over a range such that the tissue beneath the skin is equivalent to a thickness of 0.080 mm of Styrofoam or 3.5 mm of air. By what factor can the body adjust the conductance?

**8. Interactive LearningWare 13.1** at **www.wiley.com/college/cutnell** explores the approach taken in problems such as this one. A composite rod is made from stainless steel and iron and has a length of 0.50 m. The cross section of this composite rod is shown in the drawing and consists of a square within a circle. The square cross section of the steel is 1.0 cm on a side. The temperature at one end of the rod is 78 °C, while it is 18 °C at the other end. Assuming that no heat exits through the cylindrical outer surface, find the total amount of heat conducted through the rod in two minutes.

Iron

Stainless steel

**9.** Three building materials, plasterboard [$k = 0.30$ J/(s·m·C°)], brick [$k = 0.60$ J/(s·m·C°)], and wood [$k = 0.10$ J/(s·m·C°)], are sandwiched together as the drawing illustrates. The temperatures at the inside and outside surfaces are 27 °C and 0 °C, respectively. Each material has the same thickness and cross-sectional area. Find the temperature (a) at the plasterboard–brick interface and (b) at the brick–wood interface.

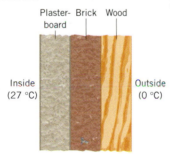
Plaster-board  Brick  Wood

Inside (27 °C)   Outside (0 °C)

**10.** In a house the temperature at the surface of a window is 25 °C. The temperature outside at the window surface is 5.0 °C. Heat is lost through the window via conduction, and the heat lost per second has a certain value. The temperature outside begins to fall, while the conditions inside the house remain the same. As a result, the heat lost per second increases. What is the temperature at the outside window surface when the heat lost per second doubles?

**11. ssm** Two rods, one of aluminum and the other of copper, are joined end to end. The cross-sectional area of each is $4.0 \times 10^{-4}$ m², and the length of each is 0.040 m. The free end of the aluminum rod is kept at 302 °C, while the free end of the copper rod is kept at 25 °C. The loss of heat through the sides of the rods may be ignored. (a) What is the temperature at the aluminum–copper interface? (b) How much heat is conducted through the unit in 2.0 s? (c) What is the temperature in the aluminum rod at a distance of 0.015 m from the hot end?

**12.** A copper rod has a length of 1.5 m and a cross-sectional area of $4.0 \times 10^{-4}$ m². One end of the rod is in contact with boiling water and the other with a mixture of ice and water. What is the mass of ice per second that melts? Assume that no heat is lost through the side surface of the rod.

**13.** Consult **Interactive Solution 13.13** at **www.wiley.com/college/cutnell** to explore a model for solving this problem. One end of a brass bar is maintained at 306 °C, while the other end is kept at a constant, but lower temperature. The cross-sectional area of the bar is $2.6 \times 10^{-4}$ m². Because of insulation, there is negligible heat loss through the sides of the bar. Heat flows through the bar, however, at the rate of 3.6 J/s. What is the temperature of the bar at a point 0.15 m from the hot end?

**14.** A 0.30-m-thick sheet of ice covers a lake. The air temperature at the ice surface is −15 °C. In five minutes, the ice thickens by a small amount. Assume that no heat flows from the ground below into the water and that the added ice is very thin compared to 0.30 m. Find the number of millimeters by which the ice thickens.

**15. ssm www** Two cylindrical rods have the same mass. One is made of silver (density = 10 500 kg/m³), and one is made of iron (density = 7860 kg/m³). Both rods conduct the same amount of heat per second when the same temperature difference is maintained across their ends. What is the ratio (silver-to-iron) of (a) the lengths and (b) the radii of these rods?

## Section 13.3 Radiation

**16.** A person is standing outdoors in the shade where the temperature is 28 °C. (a) What is the radiant energy absorbed per second by his head when it is covered with hair? The surface area of the hair (assumed to be flat) is 160 cm² and its emissivity is 0.85. (b) What would be the radiant energy absorbed per second by the same person if he were bald and the emissivity of his head were 0.65?

**17. ssm www** How many days does it take for a perfect blackbody cube (0.0100 m on a side, 30.0 °C) to radiate the same amount of energy that a one-hundred-watt light bulb uses in one hour?

**18.** The filament of a light bulb has a temperature of $3.0 \times 10^3$ °C and radiates sixty watts of power. The emissivity of the filament is 0.36. Find the surface area of the filament.

**19.** An object emits 30 W of radiant power. If it were a perfect blackbody, other things being equal, it would emit 90 W of radiant power. What is the emissivity of the object?

**20.** The amount of radiant power produced by the sun is approximately $3.9 \times 10^{26}$ W. Assuming the sun to be a perfect blackbody sphere with a radius of $6.96 \times 10^8$ m, find its surface temperature (in kelvins).

**21. ssm** A car parked in the sun absorbs energy at a rate of 560 watts per square meter of surface area. The car reaches a temperature at which it radiates energy at this same rate. Treating the car as a perfect radiator ($e = 1$), find the temperature.

**22. Interactive LearningWare 13.2** at **www.wiley.com/college/cutnell** reviews the concepts that are involved in this problem. Suppose the skin temperature of a naked person is 34 °C when the person is standing inside a room whose temperature is 25 °C. The skin area of the individual is 1.5 m². (a) Assuming the emissivity is 0.80, find the net loss of radiant power from the body. (b) Determine the number of food Calories of energy (1 food Calorie = 4186 J) that is lost in one hour due to the net loss rate obtained in part (a). Metabolic conversion of food into energy replaces this loss.

**23.** The concrete wall of a building is 0.10 m thick. The temperature inside the building is 20.0 °C, while the temperature outside is 0.0 °C. Heat is conducted through the wall. When the building is unheated, the inside temperature falls to 0.0 °C, and heat conduction ceases. However, the wall does emit radiant energy when its temperature is 0.0 °C. The radiant energy emitted per second per square meter is the same as the heat lost per second per square meter due to conduction. What is the emissivity of the wall?

* **24.** A solid sphere has a temperature of 773 K. The sphere is melted down and recast into a cube that has the same emissivity and emits the same radiant power as the sphere. What is the cube's temperature?

** **25.** **ssm** A solid cylinder is radiating power. It has a length that is ten times its radius. It is cut into a number of smaller cylinders, each of which has the same length. Each small cylinder has the same temperature as the original cylinder. The total radiant power emitted by the pieces is twice that emitted by the original cylinder. How many smaller cylinders are there?

** **26.** A small sphere (emissivity = 0.90, radius = $r_1$) is located at the center of a spherical asbestos shell (thickness = 1.0 cm, outer radius = $r_2$). The thickness of the shell is small compared to the inner and outer radii of the shell. The temperature of the small sphere is 800.0 °C, while the temperature of the inner surface of the shell is 600.0 °C, both temperatures remaining constant. Assuming that $r_2/r_1 = 10.0$ and ignoring any air inside the shell, find the temperature of the outer surface of the shell.

# Chapter 14 The Ideal Gas Law and Kinetic Theory

## 14.1 Molecular Mass, the Mole, and Avogadro's Number

Often, we wish to compare the mass of one atom with another. To facilitate the comparison, a mass scale known as the **atomic mass scale** has been established. To set up this scale, a reference value (along with a unit) is chosen for one of the elements. The unit is called the **atomic mass unit** (symbol: u). By international agreement, the reference element is chosen to be the most abundant type or isotope* of carbon, which is called carbon-12. Its atomic mass† is defined to be exactly twelve atomic mass units, or 12 u. The relationship between the atomic mass unit and the kilogram is

$$1 \text{ u} = 1.6605 \times 10^{-27} \text{ kg}$$

The atomic masses of all the elements are listed in the periodic table, part of which is shown in Figure 14.1. The complete periodic table is given on the inside of the back cover. In general, the masses listed are average values and take into account the various isotopes of an element that exist naturally. For brevity, the unit "u" is often omitted from the table. For example, a magnesium atom (Mg) has an average atomic mass of 24.305 u, while the corresponding average value for the lithium atom (Li) is 6.941 u; thus, atomic magnesium is more massive than atomic lithium by a factor of (24.305 u)/(6.941 u) = 3.502. In the periodic table, the atomic mass of carbon (C) is given as 12.011 u, rather than exactly 12 u, because a small amount (about 1%) of the naturally occurring material is an isotope called carbon-13. The value of 12.011 u is an average that reflects the small contribution of carbon-13.

The molecular mass of a molecule is the sum of the atomic masses of its atoms. For instance, hydrogen and oxygen have atomic masses of 1.007 94 u and 15.9994 u, respectively, so the molecular mass of a water molecule ($H_2O$) is 2(1.007 94 u) + 15.9994 u = 18.0153 u.

Macroscopic amounts of materials contain large numbers of atoms or molecules. Even in a small volume of gas, 1 $cm^3$, for example, the number is enormous. It is convenient to express such large numbers in terms of a single unit, the **gram-mole**, or simply the **mole** (symbol: mol). *One gram-mole of a substance contains as many particles (atoms or molecules) as there are atoms in 12 grams of the isotope carbon-12.* Experiment shows that 12 grams of carbon-12 contain $6.022 \times 10^{23}$ atoms. The number of atoms per mole is known as **Avogadro's number $N_A$**, after the Italian scientist Amedeo Avogadro (1776–1856):

$$N_A = 6.022 \times 10^{23} \text{ mol}^{-1}$$

Thus, the number of moles $n$ contained in any sample is the number of particles $N$ in the sample divided by the number of particles per mole $N_A$ (Avogadro's number):

$$n = \frac{N}{N_A}$$

Although defined in terms of carbon atoms, the concept of a mole can be applied to any collection of objects by noting that one mole contains Avogadro's number of objects. Thus, one mole of atomic sulfur contains $6.022 \times 10^{23}$ sulfur atoms, one mole of water contains $6.022 \times 10^{23}$ $H_2O$ molecules, and one mole of golf balls contains $6.022 \times 10^{23}$ golf balls. Just as the meter is the SI base unit for length, the mole is the SI base unit for expressing "the amount of a substance."

* Isotopes are discussed in Section 31.1.
† In chemistry the expression "atomic weight" is frequently used in place of "atomic mass."

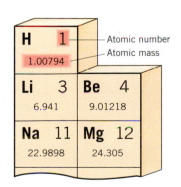

**Figure 14.1** A portion of the periodic table showing the atomic number and atomic mass of each element. In the periodic table it is customary to omit the symbol "u" denoting the atomic mass unit.

The number $n$ of moles contained in a sample can also be found from its mass. To see how, multiply and divide the right-hand side of the previous equation by the mass $m_{particle}$ of a single particle, expressed in grams:

$$n = \frac{m_{particle}N}{m_{particle}N_A} = \frac{m}{\text{Mass per mole}}$$

The numerator $m_{particle}N$ is the mass of a particle times the number of particles in the sample, which is the mass $m$ of the sample. The denominator $m_{particle}N_A$ is the mass of a particle times the number of particles per mole, which is the mass per mole, expressed in grams per mole. The mass per mole of carbon-12 is 12 g/mol, since, by definition, 12 grams of carbon-12 contain one mole of atoms. On the other hand, the mass per mole of sodium (Na) is 22.9898 g/mol for the following reason: as indicated in Figure 14.1, a sodium atom is more massive than a carbon-12 atom by the ratio of their atomic masses, (22.9898 u)/(12 u) = 1.915 82. Therefore, the mass per mole of sodium is 1.915 82 times greater than that of carbon-12, or (1.915 82)(12 g/mol) = 22.9898 g/mol. Note that the numerical value of the mass per mole of sodium (22.9898) is the same as the numerical value of its atomic mass. This is true in general, so *the mass per mole (in g/mol) of a substance has the same numerical value as the atomic or molecular mass of the substance (in atomic mass units).*

*Problem solving insight*

Since one gram-mole of a substance contains Avogadro's number of particles (atoms or molecules), the mass $m_{particle}$ of a particle (in grams) can be obtained by dividing the mass per mole (in g/mol) by Avogadro's number:

$$m_{particle} = \frac{\text{Mass per mole}}{N_A}$$

*The physics of* gemstones.

Example 1 illustrates how to use the concepts of the mole, atomic mass, and Avogadro's number to determine the number of atoms and molecules present in two famous gemstones.

### Example 1   The Hope Diamond and the Rosser Reeves Ruby

Figure 14.2a shows the Hope diamond (44.5 carats), which is almost pure carbon. Figure 14.2b shows the Rosser Reeves ruby (138 carats), which is primarily aluminum oxide ($Al_2O_3$). One carat is equivalent to a mass of 0.200 g. Determine (a) the number of carbon atoms in the diamond and (b) the number of $Al_2O_3$ molecules in the ruby.

**Reasoning** The number $N$ of atoms (or molecules) in a sample is the number of moles $n$ times the number of atoms per mole $N_A$ (Avogadro's number); $N = nN_A$. We can determine the number of moles by dividing the mass of the sample $m$ by the mass per mole of the substance.

**Solution**

(a) The mass of the Hope diamond is $m$ = (44.5 carats)[(0.200 g)/(1 carat)] = 8.90 g. Since the average atomic mass of naturally occurring carbon is 12.011 u (see the periodic table on the inside of the back cover), the mass per mole of this substance is 12.011 g/mol. The number of moles of carbon in the Hope diamond is

$$n = \frac{m}{\text{Mass per mole}} = \frac{8.90 \text{ g}}{12.011 \text{ g/mol}} = 0.741 \text{ mol}$$

The number of carbon atoms in the Hope diamond is

$$N = nN_A = (0.741 \text{ mol})(6.022 \times 10^{23} \text{ atoms/mol}) = \boxed{4.46 \times 10^{23} \text{ atoms}}$$

(b) The mass of the Rosser Reeves ruby is $m$ = (138 carats)[(0.200 g)/(1 carat)] = 27.6 g. The molecular mass of an aluminum oxide molecule ($Al_2O_3$) is the sum of the atomic masses of its atoms, which are 26.9815 u for aluminum and 15.9994 u for oxygen (see the periodic table on the inside of the back cover):

$$\text{Molecular mass} = \underbrace{2(26.9815 \text{ u})}_{\substack{\text{Mass of 2} \\ \text{aluminum} \\ \text{atoms}}} + \underbrace{3(15.9994 \text{ u})}_{\substack{\text{Mass of 3} \\ \text{oxygen} \\ \text{atoms}}} = 101.9612 \text{ u}$$

**Figure 14.2** (a) The Hope diamond surrounded by 16 smaller diamonds. (b) The Rosser Reeves ruby. Both gems are on display at the Smithsonian Institution in Washington, D.C. (Courtesy National Museum of Natural History. © 2004 Smithsonian Institution)

Thus, the mass per mole of $Al_2O_3$ is 101.9612 g/mol. Calculations like those in part (a) reveal that the Rosser Reeves ruby contains 0.271 mol or $\boxed{1.63 \times 10^{23} \text{ molecules of } Al_2O_3}$.

## 14.2 The Ideal Gas Law

An *ideal gas* is an idealized model for real gases that have sufficiently low densities. The condition of low density means that the molecules of the gas are so far apart that they do not interact (except during collisions that are effectively elastic). The ideal gas law expresses the relationship between the absolute pressure, the Kelvin temperature, the volume, and the number of moles of the gas.

In discussing the constant-volume gas thermometer, Section 12.2 has already explained the relationship between the absolute pressure and Kelvin temperature of a low-density gas. This thermometer utilizes a small amount of gas (e.g., hydrogen or helium) placed inside a bulb and kept at a constant volume. Since the density is low, the gas behaves as an ideal gas. Experiment reveals that a plot of gas pressure versus temperature is a straight line, as in Figure 12.4. This plot is redrawn in Figure 14.3, with the change that the temperature axis is now labeled in kelvins rather than in degrees Celsius. The graph indicates that the absolute pressure $P$ is directly proportional to the Kelvin temperature $T$ ($P \propto T$), for a fixed volume and a fixed number of molecules.

The relation between absolute pressure and the number of molecules of an ideal gas is simple. Experience indicates that it is possible to increase the pressure of a gas by adding more molecules; this is exactly what happens when a tire is pumped up. When the volume and temperature of a low-density gas are kept constant, doubling the number of molecules doubles the pressure. Thus, the absolute pressure of an ideal gas is proportional to the number of molecules or, equivalently, to the number of moles $n$ of the gas ($P \propto n$).

To see how the absolute pressure of a gas depends on the volume of the gas, look at the partially filled balloon in Figure 14.4a. This balloon is "soft," because the pressure of the air is low. However, if all the air in the balloon is squeezed into a smaller "bubble," as in part b of the figure, the "bubble" has a very tight feel. This tightness indicates that the pressure in the smaller volume is high enough to stretch the rubber substantially. Thus, it is possible to increase the pressure of a gas by reducing its volume, and if the number of molecules and the temperature are kept constant, the absolute pressure of an ideal gas is inversely proportional to its volume $V$ ($P \propto 1/V$).

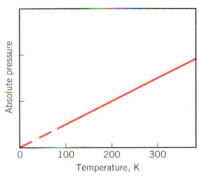

**Figure 14.3** The pressure inside a constant-volume gas thermometer is directly proportional to the Kelvin temperature, which is characteristic of an ideal gas.

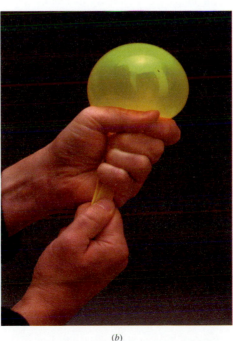

(a)    (b)

**Figure 14.4** (a) The air pressure in the partially filled balloon can be increased by decreasing the volume of the balloon, as illustrated in (b). (© Andy Washnik)

The three relations just discussed for the absolute pressure of an ideal gas can be expressed as a single proportionality, $P \propto nT/V$. This proportionality can be written as an equation by inserting a proportionality constant $R$, called the ***universal gas constant.*** The value of $R$ has been determined experimentally to be 8.31 J/(mol · K) for any real gas with a density sufficiently low to ensure ideal gas behavior. The resulting equation is known as the ***ideal gas law.***

■ **IDEAL GAS LAW**

The absolute pressure $P$ of an ideal gas is directly proportional to the Kelvin temperature $T$ and the number of moles $n$ of the gas and is inversely proportional to the volume $V$ of the gas: $P = R(nT/V)$. In other words,

$$PV = nRT \qquad (14.1)$$

where $R$ is the universal gas constant and has the value of 8.31 J/(mol · K).

Sometimes, it is convenient to express the ideal gas law in terms of the total number of particles $N$, instead of the number of moles $n$. To obtain such an expression, we multiply and divide by Avogadro's number $N_A = 6.022 \times 10^{23}$ particles/mol* on the right in Equation 14.1 and recognize that the product $nN_A$ is equal to the total number $N$ of particles:

$$PV = nRT = nN_A\left(\frac{R}{N_A}\right)T = N\left(\frac{R}{N_A}\right)T$$

The constant term $R/N_A$ is referred to as ***Boltzmann's constant,*** in honor of the Austrian physicist Ludwig Boltzmann (1844–1906), and is represented by the symbol $k$:

$$k = \frac{R}{N_A} = \frac{8.31 \text{ J/(mol·K)}}{6.022 \times 10^{23} \text{ mol}^{-1}} = 1.38 \times 10^{-23} \text{ J/K}$$

With this substitution, the ideal gas law becomes

$$PV = NkT \qquad (14.2)$$

Example 2 presents an application of the ideal gas law.

### Example 2   Oxygen in the Lungs

**The physics of oxygen in the lungs.**

In the lungs, the respiratory membrane separates tiny sacs of air (absolute pressure = $1.00 \times 10^5$ Pa) from the blood in the capillaries. These sacs are called alveoli, and it is from them that oxygen enters the blood. The average radius of the alveoli is 0.125 mm, and the air inside contains 14% oxygen. Assuming that the air behaves as an ideal gas at body temperature (310 K), find the number of oxygen molecules in one of the sacs.

**Reasoning**   The pressure and temperature of the air inside an alveolus are known, and its volume can be determined since we know the radius. Thus, the ideal gas law in the form $PV = NkT$ can be used directly to find the number $N$ of air particles inside one of the sacs. The number of oxygen molecules is 14% of the number of air particles.

**Solution**   The volume of a spherical sac is $V = \frac{4}{3}\pi r^3$. Solving Equation 14.2 for the number of air particles, we have

**Problem solving insight**
In the ideal gas law, the temperature $T$ must be expressed on the Kelvin scale. The Celsius and Fahrenheit scales cannot be used.

$$N = \frac{PV}{kT} = \frac{(1.00 \times 10^5 \text{ Pa})[\frac{4}{3}\pi(0.125 \times 10^{-3} \text{ m})^3]}{(1.38 \times 10^{-23} \text{ J/K})(310 \text{ K})} = 1.9 \times 10^{14}$$

The number of oxygen molecules is 14% of this value, or $0.14N = \boxed{2.7 \times 10^{13}}$.

With the aid of the ideal gas law, it can be shown that one mole of an ideal gas occupies a volume of 22.4 liters at a temperature of 273 K (0 °C) and a pressure of one atmosphere (1.013 × 10⁵ Pa). These conditions of temperature and pressure are known as ***stan-***

---

* "Particles" is not an SI unit and is often omitted. Then, particles/mol = 1/mol = mol⁻¹.

*dard temperature and pressure (STP).* Conceptual Example 3 discusses another interesting application of the ideal gas law.

## Conceptual Example 3   Beer Bubbles on the Rise

The next time you get a chance, watch the bubbles rise in a glass of beer (see Figure 14.5). If you look carefully, you'll see them grow in size as they move upward, often doubling in volume by the time they reach the surface. Why does a bubble grow as it ascends?

**Reasoning and Solution**  Beer bubbles contain mostly carbon dioxide ($CO_2$), a gas that is in the beer because of the fermentation process. The volume $V$ of gas in a bubble is related to its temperature $T$, pressure $P$, and the number $n$ of moles of $CO_2$ by the ideal gas law: $V = nRT/P$. Thus, one or more of these variables must be responsible for the growth of a bubble. Temperature can be eliminated immediately, since it is constant throughout the beer. What about the pressure? As a bubble rises, its depth decreases, and so does the fluid pressure. Since the volume is inversely proportional to pressure, part of the bubble growth is due to the decreasing pressure of the surrounding beer. However, some bubbles double in volume on the way up. To account for the doubling, there would have to be two atmospheres of pressure at the bottom of the glass, compared to the one atmosphere at the top. The pressure increment due to depth is $\rho g h$ according to Equation 11.4, so the extra pressure of one atmosphere at the bottom would mean $1.01 \times 10^5$ Pa $= \rho g h$. Solving for $h$ with $\rho$ equal to the density of water reveals that $h = 10.3$ m. Since most beer glasses are only about 0.2 m tall, we can rule out a change in pressure as the major cause of the change in volume. We are left with only one variable, the number of moles of $CO_2$ gas in the bubble. In fact, the number of moles does increase as the bubble rises. Each bubble acts as a nucleation site for $CO_2$ molecules, so as a bubble moves upward, it accumulates carbon dioxide from the surrounding beer and grows larger.

**Related Homework:** *Problem 22*

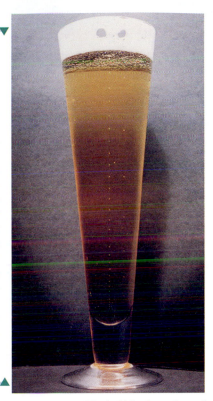

**Figure 14.5**  The bubbles in a glass of beer grow larger as they move upward. (Courtesy Richard Zare, Stanford University)

Historically, the work of several investigators led to the formulation of the ideal gas law. The Irish scientist Robert Boyle (1627–1691) discovered that at a constant temperature, the absolute pressure of a fixed mass (fixed number of moles) of a low-density gas is inversely proportional to its volume ($P \propto 1/V$). This fact is often called Boyle's law and can be derived from the ideal gas law by noting that $P = nRT/V = $ constant$/V$ when $n$ and $T$ are constants. Alternatively, if an ideal gas changes from an initial pressure and volume ($P_i, V_i$) to a final pressure and volume ($P_f, V_f$), it is possible to write $P_i V_i = nRT$ and $P_f V_f = nRT$. Since the right sides of these equations are equal, we may equate the left sides to give the following concise way of expressing **Boyle's law:**

**Constant T, constant n**          $$P_i V_i = P_f V_f \qquad (14.3)$$

Figure 14.6 illustrates how pressure and volume change according to Boyle's law for a fixed number of moles of an ideal gas at a constant temperature of 100 K. The gas begins with an initial pressure and volume of $P_i$ and $V_i$ and is compressed. The pressure increases as the volume decreases, according to $P = nRT/V$, until the final pressure and volume of $P_f$ and $V_f$ are reached. The curve that passes through the initial and final points is called an **isotherm,** meaning "same temperature." If the temperature had been 300 K, rather than 100 K, the compression would have occurred along the 300-K isotherm. Different isotherms do not intersect. Example 4 deals with an application of Boyle's law to scuba diving.

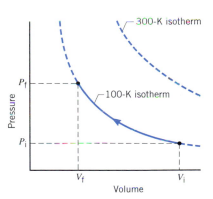

**Figure 14.6**  A pressure-versus-volume plot for a gas at a constant temperature is called an isotherm. For an ideal gas, each isotherm is a plot of the equation $P = nRT/V = $ constant$/V$.

## Example 4   Scuba* Diving

In scuba diving, a greater water pressure acts on a diver at greater depths. The air pressure inside the body cavities (e.g., lungs, sinuses) must be maintained at the same pressure as that of the surrounding water; otherwise they would collapse. A special valve automatically adjusts the pressure of the air breathed from a scuba tank to ensure that the air pressure equals the water pressure at all times. The scuba gear in Figure 14.7 consists of a 0.0150-m³ tank filled with compressed air at an absolute pressure of $2.02 \times 10^7$ Pa. Assuming that air is consumed at a

* The word is an acronym for self-contained underwater breathing apparatus.

**Figure 14.7** The air pressure inside the body cavities of a scuba diver must be maintained at the same pressure as that of the surrounding water. (© Chris McLaughlin/Corbis Images)

rate of 0.0300 m³ per minute and that the temperature is the same at all depths, determine how long the diver can stay under seawater at a depth of (a) 10.0 m and (b) 30.0 m.

**Reasoning** The time (in minutes) that a scuba diver can remain under water is equal to the volume of air that is available divided by the volume per minute consumed by the diver. The available volume is the volume of air at the pressure $P_2$ breathed by the diver. This pressure is determined by the depth $h$ beneath the surface, according to $P_2 = P_1 + \rho g h$ (Equation 11.4), where $P_1 = 1.01 \times 10^5$ Pa is the atmospheric pressure at the surface. Since we know the pressure and volume of air in the scuba tank, and since the temperature is constant, we can use Boyle's law to find the volume of air available at the pressure $P_2$.

**Solution**

(a) Using $\rho = 1025$ kg/m³ for the density of seawater, we find that the absolute pressure $P_2$ at the depth of $h = 10.0$ m is

$$P_2 = P_1 + \rho g h = 1.01 \times 10^5 \text{ Pa} + (1025 \text{ kg/m}^3)(9.80 \text{ m/s}^2)(10.0 \text{ m})$$
$$= 2.01 \times 10^5 \text{ Pa}$$

The pressure and volume of the air in the tank are $P_i = 2.02 \times 10^7$ Pa and $V_i = 0.0150$ m³, respectively. According to Boyle's law, the volume of air $V_f$ available at a pressure of $P_f = 2.01 \times 10^5$ Pa is

$$V_f = \frac{P_i V_i}{P_f} = \frac{(2.02 \times 10^7 \text{ Pa})(0.0150 \text{ m}^3)}{2.01 \times 10^5 \text{ Pa}} = 1.51 \text{ m}^3 \qquad (14.3)$$

Of this volume, only 1.51 m³ − 0.0150 m³ = 1.50 m³ is available for breathing, because 0.0150 m³ of air always remains in the tank. At a consumption rate of 0.0300 m³/min, the compressed air will last for

$$t = \frac{1.50 \text{ m}^3}{0.0300 \text{ m}^3/\text{min}} = \boxed{50.0 \text{ min}}$$

(b) The calculation here is like that in part (a). Equation 11.4 indicates that at a depth of 30.0 m, the absolute water pressure is $4.02 \times 10^5$ Pa. Because this pressure is twice that at the 10.0-m depth, Boyle's law reveals that the volume of air provided by the tank is now only $V_f = 0.754$ m³. The air available for use is 0.754 m³ − 0.0150 m³ = 0.739 m³. At a consumption rate of 0.0300 m³/min, the air will last for $\boxed{t = 24.6 \text{ min}}$, so the deeper dive must have a shorter duration.

▲

**Problem solving insight**
When using the ideal gas law, either directly or in the form of Boyle's law, remember that the pressure $P$ must be the absolute pressure, not the gauge pressure.

Another investigator whose work contributed to the formulation of the ideal gas law was the Frenchman Jacques Charles (1746–1823). He discovered that at a constant pressure, the volume of a fixed mass (fixed number of moles) of a low-density gas is directly proportional to the Kelvin temperature ($V \propto T$). This relationship is known as Charles' law and can be obtained from the ideal gas law by noting that $V = nRT/P = (\text{constant})T$, if $n$ and $P$ are constant. Equivalently, when an ideal gas changes from an initial volume and temperature ($V_i, T_i$) to a final volume and temperature ($V_f, T_f$), it is possible to write $V_i/T_i = nR/P$ and $V_f/T_f = nR/P$. Thus, one way of stating **Charles' law** is

*Constant P, constant n*

$$\frac{V_i}{T_i} = \frac{V_f}{T_f} \qquad (14.4)$$

## 14.3   *Kinetic Theory of Gases*

As useful as it is, the ideal gas law provides no insight as to how pressure and temperature are related to properties of the molecules themselves, such as their masses and speeds. To show how such microscopic properties are related to the pressure and temperature of an ideal gas, this section examines the dynamics of molecular motion. The pressure that a gas exerts on the walls of a container is due to the force exerted by the gas molecules when they collide with the walls. Therefore, we will begin by combining the notion of collisional forces exerted by a fluid (Section 11.2) with Newton's second and third laws of motion (Sections 4.3 and 4.5). These concepts will allow us to obtain an

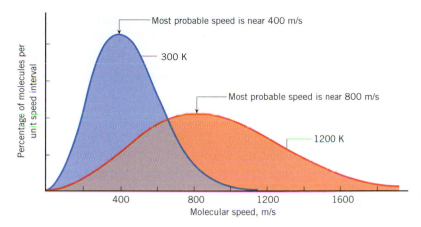

**Figure 14.8** The Maxwell distribution curves for molecular speeds in oxygen gas at temperatures of 300 and 1200 K.

expression for the pressure in terms of microscopic properties. We will then combine this with the ideal gas law to show that the average translational kinetic energy $\overline{KE}$ of a particle in an ideal gas is $\overline{KE} = \frac{3}{2}kT$, where $k$ is Boltzmann's constant and $T$ is the Kelvin temperature. In the process, we will also see that the internal energy $U$ of a monatomic ideal gas is $U = \frac{3}{2}nRT$, where $n$ is the number of moles and $R$ is the universal gas constant.

## THE DISTRIBUTION OF MOLECULAR SPEEDS

A macroscopic container filled with a gas at standard temperature and pressure contains a large number of particles (atoms or molecules). These particles are in constant, random motion, colliding with each other and with the walls of the container. In the course of one second, a particle undergoes many collisions, and each one changes the particle's speed and direction of motion. As a result, the atoms or molecules have different speeds. It is possible, however, to speak about an average particle speed. At any given instant, some particles have speeds less than, some near, and some greater than the average. For conditions of low gas density, the distribution of speeds within a large collection of molecules at a constant temperature was calculated by the Scottish physicist James Clerk Maxwell (1831–1879). Figure 14.8 displays the Maxwell speed distribution curves for $O_2$ gas at two different temperatures. When the temperature is 300 K, the maximum in the curve indicates that the most probable speed is about 400 m/s. At a temperature of 1200 K, the distribution curve shifts to the right, and the most probable speed increases to about 800 m/s.

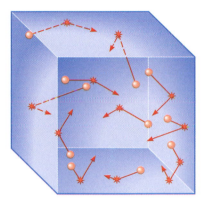

**Figure 14.9** The pressure that a gas exerts is caused by the collisions of its molecules with the walls of the container.

## KINETIC THEORY

If a ball is thrown against a wall, it exerts a force on the wall. As Figure 14.9 suggests, gas particles do the same thing, except that their masses are smaller and their speeds are greater. The number of particles is so great and they strike the wall so often that the effect of their individual impacts appears as a continuous force. Dividing the magnitude of this force by the area of the wall gives the pressure exerted by the gas.

To calculate the force, consider an ideal gas composed of $N$ identical particles in a cubical container whose sides have length $L$. Except for elastic* collisions, these particles do not interact. Figure 14.10 focuses attention on one particle of mass $m$ as it strikes the right wall perpendicularly and rebounds elastically. While approaching the wall, the particle has a velocity $+v$ and linear momentum $+mv$ (see Section 7.1 for a review of linear momentum). The particle rebounds with a velocity $-v$ and momentum $-mv$, travels to the left wall, rebounds again, and heads back toward the right. The time $t$ between collisions with the right wall is the round-trip distance $2L$ divided by the speed of the particle; that is, $t = 2L/v$. According to Newton's second law of motion, in the form of the

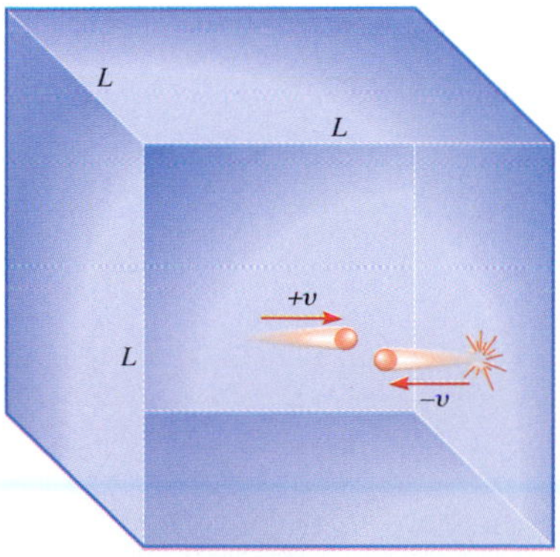

**Figure 14.10** A gas particle is shown colliding elastically with the right wall of the container and rebounding from it.

* The term "elastic" is used here to mean that *on the average,* in a large number of particles, there is no gain or loss of translational kinetic energy because of collisions.

impulse–momentum theorem, the average force exerted on the particle by the wall is given by the change in the particle's momentum per unit time:

$$\text{Average force} = \frac{\text{Final momentum} - \text{Initial momentum}}{\text{Time between successive collisions}} \qquad (7.4)$$

$$= \frac{(-mv) - (+mv)}{2L/v} = \frac{-mv^2}{L}$$

According to Newton's law of action–reaction, the force applied to the wall by the particle is equal in magnitude to this value, but oppositely directed (i.e., $+mv^2/L$). The magnitude $F$ of the *total* force exerted on the right wall is equal to the number of particles that collide with the wall during the time $t$ multiplied by the average force exerted by each particle. Since the $N$ particles move randomly in three dimensions, one-third of them on the average strike the right wall during the time $t$. Therefore, the total force is

$$F = \left(\frac{N}{3}\right)\left(\frac{m\overline{v^2}}{L}\right)$$

In this result $v^2$ has been replaced by $\overline{v^2}$, *the average value* of the squared speed. The collection of particles possesses a Maxwell distribution of speeds, so an average value for $v^2$ must be used, rather than a value for any individual particle. The square root of the quantity $\overline{v^2}$ is called the **root-mean-square speed,** or, for short, the *rms speed;* $v_{rms} = \sqrt{\overline{v^2}}$. With this substitution, the total force becomes

$$F = \left(\frac{N}{3}\right)\left(\frac{mv_{rms}^2}{L}\right)$$

Pressure is force per unit area, so the pressure $P$ acting on a wall of area $L^2$ is

$$P = \frac{F}{L^2} = \left(\frac{N}{3}\right)\left(\frac{mv_{rms}^2}{L^3}\right)$$

Since the volume of the box is $V = L^3$, the equation above can be written as

$$PV = \tfrac{2}{3}N(\tfrac{1}{2}mv_{rms}^2) \qquad (14.5)$$

Equation 14.5 relates the macroscopic properties of the gas, its pressure and volume, to the microscopic properties of the constituent particles, their mass and speed. Since the term $\tfrac{1}{2}mv_{rms}^2$ is the average translational kinetic energy $\overline{KE}$ of an individual particle, it follows that

$$PV = \tfrac{2}{3}N(\overline{KE})$$

This result is similar to the ideal gas law, $PV = NkT$. Both equations have identical terms on the left, so the terms on the right must be equal: $\tfrac{2}{3}N(\overline{KE}) = NkT$. Therefore,

$$\overline{KE} = \tfrac{1}{2}mv_{rms}^2 = \tfrac{3}{2}kT \qquad (14.6)$$

Equation 14.6 is significant, because it allows us to interpret temperature in terms of the motion of gas particles. This equation indicates that the Kelvin temperature is directly proportional to the average translational kinetic energy per particle in an ideal gas, no matter what the pressure and volume are. On the average, the particles have greater kinetic energies when the gas is hotter than when it is cooler. Conceptual Example 5 discusses a common misconception about the relation between kinetic energy and temperature.

### *Conceptual Example 5*   Does a Single Particle Have a Temperature?

Each particle in a gas has kinetic energy. Furthermore, the equation $\tfrac{1}{2}mv_{rms}^2 = \tfrac{3}{2}kT$ establishes the relationship between the average kinetic energy per particle and the temperature of an ideal gas. Is it valid, then, to conclude that a single particle has a temperature?

**Reasoning and Solution** We know that a gas contains an enormous number of particles that are traveling with a distribution of speeds, such as those indicated by the graphs in Figure 14.8. Therefore, the particles do not all have the same kinetic energy, but possess a distribution of kinetic energies ranging from very nearly zero to extremely large values. If each particle had a temperature that was associated with its kinetic energy, there would be a whole range of different temperatures within the gas. This is not so, for a gas at thermal equilibrium has only one temperature (see Section 15.2), a temperature that would be registered by a thermometer placed in the gas. Thus, temperature is a property that characterizes the gas as a whole, a fact that is inherent in the relation $\frac{1}{2}mv_{\text{rms}}^2 = \frac{3}{2}kT$. The term $v_{\text{rms}}$ is a kind of *average particle speed*. Therefore, $\frac{1}{2}mv_{\text{rms}}^2$ is the *average kinetic energy* per particle and is characteristic of the gas as a whole. Since the Kelvin temperature is proportional to $\frac{1}{2}mv_{\text{rms}}^2$, it is also a characteristic of the gas as a whole and cannot be ascribed to each gas particle individually. Thus, **a single gas particle does not have a temperature.**

If two ideal gases have the same temperature, the relation $\frac{1}{2}mv_{\text{rms}}^2 = \frac{3}{2}kT$ indicates that the average kinetic energy of each kind of gas particle is the same. In general, however, the rms speeds of the different particles are not the same, for the masses may be different. The next example illustrates these facts and shows how rapidly gas particles move at normal temperatures.

## *Example 6*  The Speed of Molecules in Air

Air is primarily a mixture of nitrogen $N_2$ (molecular mass = 28.0 u) and oxygen $O_2$ (molecular mass = 32.0 u). Assume that each behaves as an ideal gas and determine the rms speeds of the nitrogen and oxygen molecules when the temperature of the air is 293 K.

**Reasoning** The rms speed can be obtained from Equation 14.6 as $v_{\text{rms}} = \sqrt{2\overline{\text{KE}}/m_{\text{particle}}}$, where $m_{\text{particle}}$ is the mass of a single particle (nitrogen or oxygen molecule). In this expression, the average kinetic energy is $\overline{\text{KE}} = \frac{3}{2}kT$ and is the same for both types of molecules at the same temperature. However, the masses of the nitrogen and oxygen molecules are different. The mass of each particle is equal to the mass per mole divided by the number $N_A$ of particles per mole (Avogadro's number).

**Solution** The average kinetic energy per particle for both nitrogen and oxygen is

$$\overline{\text{KE}} = \tfrac{3}{2}kT = \tfrac{3}{2}(1.38 \times 10^{-23} \text{ J/K})(293 \text{ K}) = 6.07 \times 10^{-21} \text{ J} \qquad (14.6)$$

Since the molecular mass of nitrogen is 28.0 u, its mass per mole is 28.0 g/mol. Therefore, the mass of a nitrogen molecule is

$$m_{\text{particle}} = \frac{\text{Mass per mole}}{N_A} = \frac{28.0 \text{ g/mol}}{6.022 \times 10^{23} \text{ mol}^{-1}} = 4.65 \times 10^{-23} \text{ g} = 4.65 \times 10^{-26} \text{ kg}$$

Similarly, we find the particle mass for oxygen to be $5.31 \times 10^{-26}$ kg. The calculations of the rms speeds are shown below:

*Nitrogen* $\qquad v_{\text{rms}} = \sqrt{\dfrac{2(\overline{\text{KE}})}{m_{\text{particle}}}} = \sqrt{\dfrac{2(6.07 \times 10^{-21} \text{ J})}{4.65 \times 10^{-26} \text{ kg}}} = \boxed{511 \text{ m/s}}$

*Oxygen* $\qquad v_{\text{rms}} = \sqrt{\dfrac{2(\overline{\text{KE}})}{m_{\text{particle}}}} = \sqrt{\dfrac{2(6.07 \times 10^{-21} \text{ J})}{5.31 \times 10^{-26} \text{ kg}}} = \boxed{478 \text{ m/s}}$

For comparison, the speed of sound at a temperature of 293 K is 343 m/s (767 mi/h).

*Problem solving insight*
The average translational kinetic energy is the same for all ideal-gas molecules at the same temperature, regardless of their masses. The rms translational speed of the molecules is not the same, however, but depends on the mass.

The equation $\overline{\text{KE}} = \frac{3}{2}kT$ has also been applied to particles much larger than atoms or molecules. The English botanist Robert Brown (1773–1858) observed through a microscope that pollen grains suspended in water move on very irregular, zigzag paths. This Brownian motion can also be observed with other particle suspensions, such as fine smoke particles in air. In 1905, Albert Einstein (1879–1955) showed that Brownian motion could be explained as a response of the large suspended particles to impacts from the moving molecules of the fluid medium (e.g., water or air). As a result of the impacts, the suspended particles have the same average translational kinetic energy as the fluid molecules—namely, $\overline{\text{KE}} = \frac{3}{2}kT$. But unlike the molecules, the particles are large enough to be seen through a microscope and, because of their relatively large mass, have a comparatively small average speed.

## THE INTERNAL ENERGY OF A MONATOMIC IDEAL GAS

Chapter 15 deals with the science of thermodynamics, in which the concept of internal energy plays an important role. Using the results just developed for the average translational kinetic energy, we conclude this section by expressing the internal energy of a monatomic ideal gas in a form that is suitable for use later on.

The internal energy of a substance is the sum of the various kinds of energy that the atoms or molecules of the substance possess. A monatomic ideal gas is composed of single atoms. These atoms are assumed to be so small that the mass is concentrated at a point, with the result that the moment of inertia $I$ about the center of mass is negligible. Thus, the rotational kinetic energy $\frac{1}{2}I\omega^2$ is also negligible. Vibrational kinetic and potential energies are absent, because the atoms are not connected by chemical bonds and, except for elastic collisions, do not interact. As a result, the internal energy $U$ is the total translational kinetic energy of the $N$ atoms that constitute the gas: $U = N(\frac{1}{2}mv_{rms}^2)$. Since $\frac{1}{2}mv_{rms}^2 = \frac{3}{2}kT$ according to Equation 14.6, the internal energy can be written in terms of the Kelvin temperature as

$$U = N(\tfrac{3}{2}kT)$$

Usually, $U$ is expressed in terms of the number of moles $n$, rather than the number of atoms $N$. Using the fact that Boltzmann's constant is $k = R/N_A$, where $R$ is the universal gas constant and $N_A$ is Avogadro's number, and realizing that $N/N_A = n$, we find that

*Monatomic ideal gas* $\qquad\qquad U = \tfrac{3}{2}nRT \qquad\qquad$ (14.7)

Thus, the internal energy depends on the number of moles and Kelvin temperature of the gas. In fact, it can be shown that the internal energy is proportional to the Kelvin temperature for *any type* of ideal gas (e.g., monatomic, diatomic, etc.). For example, when hot-air balloonists turn on the burner, they increase the temperature, and hence the internal energy per mole, of the air inside the balloon (see Figure 14.11).

**Figure 14.11** Since air behaves approximately as an ideal gas, the internal energy per mole inside a hot-air balloon increases as the temperature rises. (© Donovan Reese/Stone/Getty Images)

## Concept Summary

This summary presents an abridged version of the chapter, including the important equations and all available learning aids. For convenient reference, the learning aids (including the text's examples) are placed next to or immediately after the relevant equation or discussion. The following learning aids may be found on-line at **www.wiley.com/college/cutnell**:

| | |
|---|---|
| **Interactive LearningWare** examples are solved according to a five-step interactive format that is designed to help you develop problem-solving skills. | **Concept Simulations** are animated versions of text figures or animations that illustrate important concepts. You can control parameters that affect the display, and we encourage you to experiment. |
| **Interactive Solutions** offer specific models for certain types of problems in the chapter homework. The calculations are carried out interactively. | **Self-Assessment Tests** include both qualitative and quantitative questions. Extensive feedback is provided for both incorrect and correct answers, to help you evaluate your understanding of the material. |

| Topic | Discussion | Learning Aids |
|---|---|---|

### 14.1 Molecular Mass, the Mole, and Avogadro's Number

Atomic mass unit

Each element in the periodic table is assigned an atomic mass. One atomic mass unit (u) is exactly one-twelfth the mass of an atom of carbon-12. The molecular mass of a molecule is the sum of the atomic masses of its atoms.

The number of moles $n$ contained in a sample is equal to the number of particles $N$ (atoms or molecules) in the sample divided by the number of particles per mole $N_A$

Number of moles

$$n = \frac{N}{N_A}$$

Avogadro's number

where $N_A$ is called Avogadro's number and has a value of $N_A = 6.022 \times 10^{23}$ particles per mole. The number of moles is also equal to the mass $m$ of the sam-

| Topic | Discussion | Learning Aids |
|-------|-----------|---------------|

ple (expressed in grams) divided by the mass per mole (expressed in grams per mole):

**Number of moles**

$$n = \frac{m}{\text{Mass per mole}}$$

Example 1

**Mass per mole**

The mass per mole (in g/mol) of a substance has the same numerical value as the atomic or molecular mass of one of its particles (in atomic mass units).

**Mass of a particle**

The mass $m_{\text{particle}}$ of a particle (in grams) can be obtained by dividing the mass per mole (in g/mol) by Avogadro's number:

$$m_{\text{particle}} = \frac{\text{Mass per mole}}{N_A}$$

### 14.2 The Ideal Gas Law

The ideal gas law relates the absolute pressure $P$, the volume $V$, the number $n$ of moles, and the Kelvin temperature $T$ of an ideal gas according to

**Ideal gas law**

$$PV = nRT \qquad (14.1)$$

**Universal gas constant**

where $R = 8.31$ J/(mol · K) is the universal gas constant. An alternative form of the ideal gas law is

**Ideal gas law**

$$PV = NkT \qquad (14.2)$$

Examples 2, 3

**Boltzmann's constant**

where $N$ is the number of particles and $k = \dfrac{R}{N_A}$ is Boltzmann's constant. A real gas behaves as an ideal gas when its density is low enough that its particles do not interact, except via elastic collisions.

Interactive Solutions 14.20, 14.21

A form of the ideal gas law that applies when the number of moles and the temperature are constant is known as Boyle's law. Using the subscripts "i" and "f" to denote, respectively, initial and final conditions, we can write Boyle's law as

**Boyle's law**

$$P_i V_i = P_f V_f \qquad (14.3)$$

Example 4

A form of the ideal gas law that applies when the number of moles and the pressure are constant is called Charles' law:

**Charles' law**

$$\frac{V_i}{T_i} = \frac{V_f}{T_f} \qquad (14.4)$$

### 14.3 Kinetic Theory of Gases

**Maxwell speed distribution**

The distribution of particle speeds in an ideal gas at constant temperature is the Maxwell speed distribution (see Figure 14.8). The kinetic theory of gases indicates that the Kelvin temperature $T$ of an ideal gas is related to the average translational kinetic energy $\overline{\text{KE}}$ of a particle according to

**Average translational kinetic energy**

$$\overline{\text{KE}} = \tfrac{1}{2}mv_{\text{rms}}^2 = \tfrac{3}{2}kT \qquad (14.6)$$

Examples 5, 6

**Root-mean-square speed**

where $v_{\text{rms}}$ is the root-mean-square speed of the particles.

Interactive Solutions 14.35, 14.51

The internal energy $U$ of $n$ moles of a monatomic ideal gas is

**Internal energy**

$$U = \tfrac{3}{2}nRT \qquad (14.7)$$

The internal energy of any type of ideal gas (e.g., monatomic, diatomic) is proportional to its Kelvin temperature.

 **Use Self-Assessment Test 14.1 to evaluate your understanding of Sections 14.1 – 14.3.**

# Problems

*Note: The pressures referred to in these problems are absolute pressures, unless indicated otherwise.*

**ssm** Solution is in the Student Solutions Manual.   **www** Solution is available on the World Wide Web at www.wiley.com/college/cutnell
  This icon represents a biomedical application.

### Section 14.1 Molecular Mass, the Mole, and Avogadro's Number

**1. ssm** A mass of 135 g of an element is known to contain $30.1 \times 10^{23}$ atoms. What is the element?

**2.** Manufacturers of headache remedies routinely claim that their own brands are more potent pain relievers than the competing brands. The best way of making the comparison is to compare the number of molecules in the standard dosage. Tylenol uses 325 mg of acetaminophen ($C_8H_9NO_2$) as the standard dose, while Advil uses $2.00 \times 10^2$ mg of ibuprofen ($C_{13}H_{18}O_2$). Find the number of molecules of pain reliever in the standard doses of (a) Tylenol and (b) Advil.

**3.** The artificial sweetener NutraSweet is a chemical called aspartame ($C_{14}H_{18}N_2O_5$). What is (a) its molecular mass (in atomic mass units) and (b) the mass (in kg) of an aspartame molecule?

**4.** The active ingredient in the allergy medication Claritin contains carbon (C), hydrogen (H), chlorine (Cl), nitrogen (N), and oxygen (O). Its molecular formula is $C_{22}H_{23}ClN_2O_2$. The standard adult dosage utilizes $1.572 \times 10^{19}$ molecules of this species. Determine the mass (in grams) of the active ingredient in the standard dosage.

**5. ssm** Hemoglobin has a molecular mass of 64 500 u. Find the mass (in kg) of one molecule of hemoglobin.

**6.** A cylindrical glass of water ($H_2O$) has a radius of 4.50 cm and a height of 12.0 cm. The density of water is 1.00 g/cm$^3$. How many moles of water molecules are contained in the glass?

**7. ssm** Estimate the spacing between the centers of neighboring atoms in a piece of solid aluminum, based on a knowledge of the density (2700 kg/m$^3$) and atomic mass (26.9815 u) of aluminum. *(Hint: Assume that the volume of the solid is filled with many small cubes, with one atom at the center of each.)*

**8.** A sample of ethyl alcohol ($C_2H_5OH$) has a density of 806 kg/m$^3$ and a volume of $2.00 \times 10^{-3}$ m$^3$. (a) Determine the mass (in kg) of a molecule of ethyl alcohol, and (b) find the number of molecules in the liquid.

### Section 14.2 The Ideal Gas Law   $R = 8.317$

**9.** A Goodyear blimp typically contains 5400 m$^3$ of helium (He) at an absolute pressure of $1.1 \times 10^5$ Pa. The temperature of the helium is 280 K. What is the mass (in kg) of the helium in the blimp?

**10.** At the start of a trip, a driver adjusts the absolute pressure in her tires to be $2.81 \times 10^5$ Pa when the outdoor temperature is 284 K. At the end of the trip she measures the pressure to be $3.01 \times 10^5$ Pa. Ignoring the expansion of the tires, find the air temperature inside the tires at the end of the trip.

**11. ssm** A bicycle tire whose volume is $4.1 \times 10^{-4}$ m$^3$ has a temperature of 296 K and an absolute pressure of $4.8 \times 10^5$ Pa. A cyclist brings the pressure up to $6.2 \times 10^5$ Pa without changing the temperature or volume. How many moles of air must have been pumped into the tire?

**12.** A clown at a birthday party has brought along a helium cylinder, with which he intends to fill balloons. When full, each balloon contains 0.034 m$^3$ of helium at an absolute pressure of $1.2 \times 10^5$ Pa. The cylinder contains helium at an absolute pressure of $1.6 \times$ $10^7$ Pa and has a volume of 0.0031 m$^3$. The temperature of the helium in the tank and in the balloons is the same and remains constant. What is the maximum number of people who will get a balloon?

**13.** An ideal gas at 15.5 °C and a pressure of $1.72 \times 10^5$ Pa occupies a volume of 2.81 m$^3$. (a) How many moles of gas are present? (b) If the volume is raised to 4.16 m$^3$ and the temperature raised to 28.2 °C, what will be the pressure of the gas?

**14.** Oxygen for hospital patients is kept in special tanks, where the oxygen has a pressure of 65.0 atmospheres and a temperature of 288 K. The tanks are stored in a separate room, and the oxygen is pumped to the patient's room, where it is administered at a pressure of 1.00 atmosphere and a temperature of 297 K. What volume does 1.00 m$^3$ of oxygen in the tanks occupy at the conditions in the patient's room?

**15.** **ssm** A young male adult takes in about $5.0 \times 10^{-4}$ m$^3$ of fresh air during a normal breath. Fresh air contains approximately 21% oxygen. Assuming that the pressure in the lungs is $1.0 \times 10^5$ Pa and air is an ideal gas at a temperature of 310 K, find the number of oxygen molecules in a normal breath.

**16.** A frictionless gas-filled cylinder is fitted with a movable piston, as the drawing shows. The block resting on the top of the piston determines the constant pressure that the gas has. The height $h$ is 0.120 m when the temperature is 273 K and increases as the temperature increases. What is the value of $h$ when the temperature reaches 318 K?

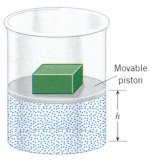

Movable piston

$h$

**17.** Two ideal gases have the same mass density and the same absolute pressure. One of the gases is helium (He), and its temperature is 175 K. The other gas is neon (Ne). What is the temperature of the neon?

**18.** On the sunlit surface of Venus, the atmospheric pressure is $9.0 \times 10^6$ Pa, and the temperature is 740 K. On the earth's surface the atmospheric pressure is $1.0 \times 10^5$ Pa, while the surface temperature can reach 320 K. These data imply that Venus has a "thicker" atmosphere at its surface than does the earth, which means that the number of molecules per unit volume ($N/V$) is greater on the surface of Venus than on the earth. Find the ratio $(N/V)_{\text{Venus}}/(N/V)_{\text{Earth}}$.

**19.** In a diesel engine, the piston compresses air at 305 K to a volume that is one-sixteenth of the original volume and a pressure that is 48.5 times the original pressure. What is the temperature of the air after the compression?

**20. Interactive Solution 14.20** at **www.wiley.com/college/cutnell** offers one approach to this problem. One assumption of the ideal gas law is that the atoms or molecules themselves occupy a negligible volume. Verify that this assumption is reasonable by considering gaseous xenon (Xe). Xenon has an atomic radius of $2.0 \times 10^{-10}$ m. For STP conditions, calculate the percentage of the total volume occupied by the atoms.

* **21.** Refer to **Interactive Solution 14.21** at **www.wiley.com/college/cutnell** for help with problems like this one. An apartment has a living room whose dimensions are 2.5 m × 4.0 m × 5.0 m. Assume that the air in the room is composed of 79% nitrogen ($N_2$) and 21% oxygen ($O_2$). At a temperature of 22 °C and a pressure of $1.01 \times 10^5$ Pa, what is the mass (in grams) of the air?

* **22.** Review Conceptual Example 3 before starting this problem. A bubble, located 0.200 m beneath the surface of the beer, rises to the top. The air pressure at the top is $1.01 \times 10^5$ Pa. Assume that the density of beer is the same as that of fresh water. If the temperature and number of moles of $CO_2$ remain constant as the bubble rises, find the ratio of its volume at the top to that at the bottom.

* **23.** **ssm www** A primitive diving bell consists of a cylindrical tank with one end open and one end closed. The tank is lowered into a freshwater lake, open end downward. Water rises into the tank, compressing the trapped air, whose temperature remains constant during the descent. The tank is brought to a halt when the distance between the surface of the water in the tank and the surface of the lake is 40.0 m. Atmospheric pressure at the surface of the lake is $1.01 \times 10^5$ Pa. Find the fraction of the tank's volume that is filled with air.

* **24.** The drawing shows two thermally insulated tanks. They are connected by a valve that is initially closed. Each tank contains neon gas at the pressure, temperature, and volume indicated in the drawing. When the valve is opened, the contents of the two tanks mix, and the pressure becomes constant throughout. (a) What is the final temperature? Ignore any change in temperature of the tanks themselves. *(Hint: The heat gained by the gas in one tank is equal to that lost by the other.)* (b) What is the final pressure?

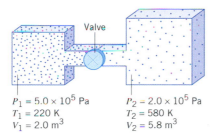

$P_1 = 5.0 \times 10^5$ Pa
$T_1 = 220$ K
$V_1 = 2.0$ m$^3$

$P_2 = 2.0 \times 10^5$ Pa
$T_2 = 580$ K
$V_2 = 5.8$ m$^3$

** **25.** A spherical balloon is made from a material whose mass is 3.00 kg. The thickness of the material is negligible compared to the 1.50-m radius of the balloon. The balloon is filled with helium (He) at a temperature of 305 K and just floats in air, neither rising nor falling. The density of the surrounding air is 1.19 kg/m$^3$. Find the absolute pressure of the helium gas.

** **26.** A gas fills the right portion of a horizontal cylinder whose radius is 5.00 cm. The initial pressure of the gas is $1.01 \times 10^5$ Pa. A frictionless movable piston separates the gas from the left portion of the

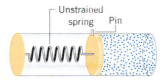

cylinder, which is evacuated and contains an ideal spring, as the drawing shows. The piston is initially held in place by a pin. The spring is initially unstrained, and the length of the gas-filled portion is 20.0 cm. When the pin is removed and the gas is allowed to expand, the length of the gas-filled chamber doubles. The initial and final temperatures are equal. Determine the spring constant of the spring.

** **27.** **ssm** A cylindrical glass beaker of height 1.520 m rests on a table. The bottom half of the beaker is filled with a gas, and the top half is filled with liquid mercury that is exposed to the atmosphere. The gas and mercury do not mix because they are separated by a frictionless, movable piston of negligible mass and thickness. The initial

temperature is 273 K. The temperature is increased until a value is reached when one-half of the mercury has spilled out. Ignore the thermal expansion of the glass and mercury, and find this temperature.

### Section 14.3 Kinetic Theory of Gases

**28.** If the translational rms speed of the water vapor molecules ($H_2O$) in air is 648 m/s, what is the translational rms speed of the carbon dioxide molecules ($CO_2$) in the same air? Both gases are at the same temperature.

**29.** **ssm** The average value of the squared speed $\overline{v^2}$ does not equal the square of the average speed $(\overline{v})^2$. To verify this fact, consider three particles with the following speeds: $v_1 = 3.0$ m/s, $v_2 = 7.0$ m/s, and $v_3 = 9.0$ m/s. Calculate (a) $\overline{v^2} = \frac{1}{3}(v_1^2 + v_2^2 + v_3^2)$ and (b) $(\overline{v})^2 = [\frac{1}{3}(v_1 + v_2 + v_3)]^2$.

**30.** Near the surface of Venus, the rms speed of carbon dioxide molecules ($CO_2$) is 650 m/s. What is the temperature (in kelvins) of the atmosphere at that point?

**31.** The surface of the sun has a temperature of about $6.0 \times 10^3$ K. This hot gas contains hydrogen atoms ($m = 1.67 \times 10^{-27}$ kg). Find the rms speed of these atoms.

**32.** Two gas cylinders are identical. One contains the monatomic gas argon (Ar), and the other contains an equal mass of the monatomic gas krypton (Kr). The pressures in the cylinders are the same, but the temperatures are different. Determine the ratio $\dfrac{\overline{KE}_{Krypton}}{\overline{KE}_{Argon}}$ of the average kinetic energy of a krypton atom to the average kinetic energy of an argon atom.

**33.** **ssm** Suppose that a tank contains 680 m$^3$ of neon at an absolute pressure of $1.01 \times 10^5$ Pa. The temperature is changed from 293.2 to 294.3 K. What is the increase in the internal energy of the neon?

**34.** Initially, the translational rms speed of a molecule of an ideal gas is 463 m/s. The pressure and volume of this gas are kept constant, while the number of molecules is doubled. What is the final translational rms speed of the molecules?

* **35.** **Interactive Solution 14.35** at **www.wiley.com/college/cutnell** provides a model for problems of this type. The temperature near the surface of the earth is 291 K. A xenon atom (atomic mass = 131.29 u) has a kinetic energy equal to the average translational kinetic energy and is moving straight up. If the atom does not collide with any other atoms or molecules, how high up would it go before coming to rest? Assume that the acceleration due to gravity is constant throughout the ascent.

* **36.** Helium (He), a monatomic gas, fills a 0.010-m$^3$ container. The pressure of the gas is $6.2 \times 10^5$ Pa. How long would a 0.25-hp engine have to run (1 hp = 746 W) to produce an amount of energy equal to the internal energy of this gas?

** **37.** **ssm www** In a TV, electrons with a speed of $8.4 \times 10^7$ m/s strike the screen from behind, causing it to glow. The electrons come to a halt after striking the screen. Each electron has a mass of $9.11 \times 10^{-31}$ kg, and there are $6.2 \times 10^{16}$ electrons per second hitting the screen over an area of $1.2 \times 10^{-7}$ m$^2$. What is the pressure that the electrons exert on the screen?

** **38.** When perspiration on the human body absorbs heat, some of the perspiration turns into water vapor. The latent heat of vaporization at body temperature (37 °C) is $2.42 \times 10^6$ J/kg. The heat absorbed is approximately equal to the average energy $\overline{E}$ given to a single water molecule ($H_2O$) times the number of water molecules that are vaporized. What is $\overline{E}$?

# Chapter 15 Thermodynamics

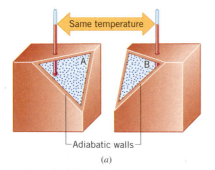

**Figure 15.1** The hot air in a balloon is one example of a thermodynamic system. (© Chase Swift/Corbis Images)

## 15.1 Thermodynamic Systems and Their Surroundings

We have studied heat (Chapter 12) and work (Chapter 6) as separate topics. Often, however, they occur simultaneously. In an automobile engine, for instance, fuel is burned at a relatively high temperature, some of its internal energy is used for doing the work of driving the pistons up and down, and the excess heat is removed by the cooling system to prevent overheating. *Thermodynamics* is the branch of physics that is built upon the fundamental laws that heat and work obey.

In thermodynamics the collection of objects on which attention is being focused is called the *system,* while everything else in the environment is called the *surroundings.* For example, the system in an automobile engine could be the burning gasoline, while the surroundings would then include the pistons, the exhaust system, the radiator, and the outside air. The system and its surroundings are separated by walls of some kind. Walls that permit heat to flow through them, such as those of the engine block, are called *diathermal walls.* Perfectly insulating walls that do not permit heat to flow between the system and its surroundings are known as *adiabatic walls.*

To understand what the laws of thermodynamics have to say about the relationship between heat and work, it is necessary to describe the physical condition or *state of a system.* We might be interested, for instance, in the hot air within the balloon in Figure 15.1. The hot air itself would be the system, and the skin of the balloon provides the walls that separate this system from the surrounding cooler air. The state of the system would be specified by giving values for the pressure, volume, temperature, and mass of the hot air.

As this chapter discusses, there are four laws of thermodynamics. We begin with the one known as the zeroth law and then consider the remaining three.

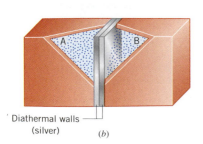

**Figure 15.2** (*a*) Systems A and B are surrounded by adiabatic walls and register the same temperature on the thermometer. (*b*) When A is put into thermal contact with B through diathermal walls, no net flow of heat occurs between the systems.

## 15.2 The Zeroth Law of Thermodynamics

The zeroth law of thermodynamics deals with the concept of *thermal equilibrium.* Two systems are said to be in thermal equilibrium if there is no net flow of heat between them when they are brought into thermal contact. For instance, you are *not* in thermal equilibrium with the water in Lake Michigan in January. Just dive into it, and you will find out how quickly your body loses heat to the frigid water. To help explain the central idea of the zeroth law of thermodynamics, Figure 15.2*a* shows two systems labeled A and B. Each is within a container whose adiabatic walls are made from insulation that prevents the flow of heat, and each has the same temperature, as indicated by the thermometers. In part *b*, one wall of each container is replaced by a thin silver sheet, and the two sheets are touched together. Silver has a large thermal conductivity, so heat flows through it readily and the silver sheets behave as diathermal walls. Even though the diathermal walls would permit it, no net flow of heat occurs in part *b*, indicating that the two systems are in thermal equilibrium. There is no net flow of heat because the two systems have the same temperature. We see, then, that *temperature is the indicator of thermal equilibrium in the sense that there is no net flow of heat between two systems in thermal contact that have the same temperature.*

In Figure 15.2 the thermometer plays an important role. System A is in equilibrium with the thermometer, and so is system B. In each case, the thermometer registers the same temperature, thereby indicating that the two systems are equally hot. Consequently, systems A and B are found to be in thermal equilibrium with each other. In effect, the thermometer is a third system. The fact that system A and system B are each in thermal equilibrium with this third system at the same temperature means that they are in thermal equilibrium with each other. This finding is an example of the *zeroth law of thermodynamics.*

■ **THE ZEROTH LAW OF THERMODYNAMICS**

Two systems individually in thermal equilibrium with a third system* are in thermal equilibrium with each other.

The zeroth law establishes temperature as the indicator of thermal equilibrium and implies that all parts of a system must be in thermal equilibrium if the system is to have a definable single temperature. In other words, there can be no flow of heat within a system that is in thermal equilibrium.

## 15.3 The First Law of Thermodynamics

In Chapter 6 we saw that forces can do work and that work can change the kinetic and potential energy of an object. For example, the atoms and molecules of a substance exert forces on one another. As a result, they have kinetic and potential energy. These and other kinds of molecular energy constitute the internal energy of a substance. When a substance participates in a process involving energy in the form of work and heat, the internal energy of the substance can change. The relationship between work, heat, and changes in the internal energy is known as the first law of thermodynamics. We will now see that the first law of thermodynamics is an expression of the conservation of energy.

Suppose that a system gains heat $Q$ and that this is the only effect occurring. Consistent with the law of conservation of energy, the internal energy of the system increases from an initial value of $U_i$ to a final value of $U_f$, the change being $\Delta U = U_f - U_i = Q$. In writing this equation, we use the convention that *heat $Q$ is positive when the system gains heat and negative when the system loses heat.* The internal energy of a system can also change because of work. If a system does work $W$ on its surroundings and there is no heat flow, energy conservation indicates that the internal energy of the system decreases from $U_i$ to $U_f$, the change now being $\Delta U = U_f - U_i = -W$. The minus sign is included because we follow the convention that *work is positive when it is done by the system and negative when it is done on the system.* A system can gain or lose energy simultaneously in the form of heat $Q$ and work $W$. The change in internal energy due to both factors is given by Equation 15.1. Thus, the *first law of thermodynamics* is just the conservation of energy principle applied to heat, work, and the change in the internal energy.

*Problem solving insight*

*Problem solving insight*

■ **THE FIRST LAW OF THERMODYNAMICS**

The internal energy of a system changes from an initial value $U_i$ to a final value of $U_f$ due to heat $Q$ and work $W$:

$$\Delta U = U_f - U_i = Q - W \qquad (15.1)$$

$Q$ is positive when the system gains heat and negative when it loses heat. $W$ is positive when work is done by the system and negative when work is done on the system.

*Problem solving insight*
When using the first law of thermodynamics, as expressed by Equation 15.1, be careful to follow the proper sign conventions for the heat $Q$ and the work $W$.

Example 1 illustrates the use of Equation 15.1 and the sign conventions for $Q$ and $W$.

### Example 1  Positive and Negative Work

▼

Figure 15.3 illustrates a system and its surroundings. In part *a*, the system gains 1500 J of heat from its surroundings, and 2200 J of work is done *by* the system on the surroundings. In part *b*, the system also gains 1500 J of heat, but 2200 J of work is done *on* the system by the surroundings. In each case, determine the change in the internal energy of the system.

**Reasoning** In Figure 15.3*a* the system loses more energy in doing work than it gains in the form of heat, so the internal energy of the system decreases. Thus, we expect the change in the

*The state of the third system is the same when it is in thermal equilibrium with either of the two systems. In Figure 15.2, for example, the mercury is at the same position in the thermometer in either system.

**Figure 15.3** (*a*) The system gains energy in the form of heat but loses energy because work is done by the system. (*b*) The system gains energy in the form of heat and also gains energy because work is done on the system.

internal energy, $\Delta U = U_\text{f} - U_\text{i}$, to be negative. In part *b* of the drawing, the system gains energy in the form of both heat and work. The internal energy of the system increases, and we expect $\Delta U$ to be positive.

**Solution**

(a) The heat is positive, $Q = +1500$ J, since it is gained by the system. The work is positive, $W = +2200$ J, since it is done *by* the system. According to the first law of thermodynamics

$$\Delta U = Q - W = (+1500 \text{ J}) - (+2200 \text{ J}) = \boxed{-700 \text{ J}} \tag{15.1}$$

The minus sign for $\Delta U$ indicates that the internal energy has decreased, as expected.

(b) The heat is positive, $Q = +1500$ J, since it is gained by the system. But the work is negative, $W = -2200$ J, since it is done *on* the system. Thus,

$$\Delta U = Q - W = (+1500 \text{ J}) - (-2200 \text{ J}) = \boxed{+3700 \text{ J}} \tag{15.1}$$

The plus sign for $\Delta U$ indicates that the internal energy has increased, as expected.

In the first law of thermodynamics, the internal energy $U$, heat $Q$, and work $W$ are energy quantities, and each is expressed in energy units such as joules. However, there is a fundamental difference between $U$, on the one hand, and $Q$ and $W$ on the other. The next example sets the stage for explaining this difference.

### *Example 2*  **An Ideal Gas**

The temperature of three moles of a monatomic ideal gas is reduced from $T_\text{i} = 540$ K to $T_\text{f} = 350$ K by two different methods. In the first method 5500 J of heat flows into the gas, while in the second, 1500 J of heat flows into it. In each case find (a) the change in the internal energy and (b) the work done by the gas.

**Reasoning** Since the internal energy of a monatomic ideal gas is $U = \frac{3}{2}nRT$ (Equation 14.7) and since the number of moles $n$ is fixed, only a change in temperature $T$ can alter the internal energy. Because the change in $T$ is the same in both methods, the change in $U$ is also the same. From the given temperatures, the change $\Delta U$ in internal energy can be determined. Then, the first law of thermodynamics can be used with $\Delta U$ and the given heat values to calculate the work for each of the methods.

**Solution**

(a) Using Equation 14.7 for the internal energy of a monatomic ideal gas, we find for each method of adding heat that

$$\Delta U = \tfrac{3}{2}nR(T_\text{f} - T_\text{i}) = \tfrac{3}{2}(3.0 \text{ mol})[8.31 \text{ J/(mol·K)}](350 \text{ K} - 540 \text{ K}) = \boxed{-7100 \text{ J}}$$

(b) Since $\Delta U$ is now known and the heat is given in each method, Equation 15.1 can be used to determine the work:

*1st method*   $W = Q - \Delta U = 5500 \text{ J} - (-7100 \text{ J}) = \boxed{12\,600 \text{ J}}$

*2nd method*   $W = Q - \Delta U = 1500 \text{ J} - (-7100 \text{ J}) = \boxed{8600 \text{ J}}$

In each method the gas does work, but it does more in the first method.

*Problem solving insight*

To understand the difference between $U$ and either $Q$ or $W$, consider the value for $\Delta U$ in Example 2. In both methods $\Delta U$ is the same. Its value is determined once the initial and final temperatures are specified because the internal energy of an ideal gas depends only on the temperature. Temperature is one of the variables (along with pressure and volume) that define the state of a system. *The internal energy depends only on the state of a system, not on the method by which the system arrives at a given state.* In recognition of this characteristic, internal energy is referred to as a *function of state.** In contrast, heat and work are not functions of state because they have different values for each different method used to make the system change from one state to another, as in Example 2.

*The fact that an ideal gas is used in Example 2 does not restrict our conclusion. Had a real (nonideal) gas or other material been used, the only difference would have been that the expression for the internal energy would have been more complicated. It might have involved the volume $V$, as well as the temperature $T$, for instance.

# 15.4 Thermal Processes

A system can interact with its surroundings in many ways, and the heat and work that come into play always obey the first law of thermodynamics. This section introduces four common thermal processes. In each case, the process is assumed to be *quasi-static,* which means that it occurs slowly enough that a uniform pressure and temperature exist throughout all regions of the system at all times.

*An isobaric process is one that occurs at constant pressure.* For instance, Figure 15.4 shows a substance (solid, liquid, or gas) contained in a chamber fitted with a frictionless piston. The pressure $P$ experienced by the substance is always the same and is determined by the external atmosphere and the weight of the piston and the block resting on it. Heating the substance makes it expand and do work $W$ in lifting the piston and block through the displacement $\mathbf{s}$. The work can be calculated from $W = Fs$ (Equation 6.1), where $F$ is the magnitude of the force and $s$ is the magnitude of the displacement. The force is generated by the pressure $P$ acting on the bottom surface of the piston (area $= A$), according to $F = PA$ (Equation 10.19). With this substitution for $F$, the work becomes $W = (PA)s$. But the product $A \cdot s$ is the change in volume of the material, $\Delta V = V_f - V_i$, where $V_f$ and $V_i$ are the final and initial volumes, respectively. Thus, the expression for the work is

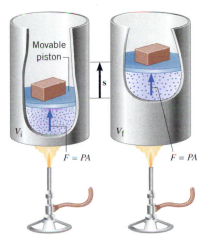

**Figure 15.4** The substance in the chamber is expanding isobarically because the pressure is held constant by the external atmosphere and the weight of the piston and the block.

**Isobaric process**
$$W = P\,\Delta V = P(V_f - V_i) \tag{15.2}$$

Consistent with our sign convention, this result predicts a positive value for the work done *by a system* when it expands isobarically ($V_f$ exceeds $V_i$). Equation 15.2 also applies to an isobaric compression ($V_f$ less than $V_i$). Then, the work is negative, since work must be done *on the system* to compress it. Example 3 emphasizes that $W = P\,\Delta V$ applies to any system, solid, liquid, or gas, as long as the pressure remains constant while the volume changes.

## Example 3 Isobaric Expansion of Water

One gram of water is placed in the cylinder in Figure 15.4, and the pressure is maintained at $2.0 \times 10^5$ Pa. The temperature of the water is raised by 31 C°. In one case, the water is in the liquid phase and expands by the small amount of $1.0 \times 10^{-8}$ m³. In another case, the water is in the gas phase and expands by the much greater amount of $7.1 \times 10^{-5}$ m³. For the water in each case, find (a) the work done and (b) the change in the internal energy.

**Reasoning** In both cases the material expands, so work is done by the system (water) on the surroundings and is, therefore, positive. The work is done at a constant pressure, so it can be found from $W = P\,\Delta V$. Once the work is known, the first law of thermodynamics, $\Delta U = Q - W$, can be used to find the change in internal energy, provided a value for the heat $Q$ can be found. The heat needed to raise the temperature can be obtained from $Q = cm\Delta T$ (Equation 12.4), where the specific heat capacity of liquid water is $c = 4186$ J/(kg·C°) (see Table 12.2) and the specific heat capacity of water vapor at constant pressure is $c_P = 2020$ J/(kg·C°).

**Solution**

(a) In both cases the process is isobaric, so the work is given by Equation 15.2:

$$W_{\text{liquid}} = P\,\Delta V = (2.0 \times 10^5 \text{ Pa})(1.0 \times 10^{-8} \text{ m}^3) = \boxed{0.0020 \text{ J}}$$
$$W_{\text{gas}} = P\,\Delta V = (2.0 \times 10^5 \text{ Pa})(7.1 \times 10^{-5} \text{ m}^3) = \boxed{14 \text{ J}}$$

Under comparable conditions, liquids (and solids) change volume much less than gases do, which leads to a much smaller work of expansion or compression for liquids than for gases.

(b) Using $\Delta U = Q - W$ (Equation 15.1) and $Q = cm\Delta T$ (Equation 12.4), we find that $\Delta U = cm\Delta T - W$. Therefore,

$$\Delta U_{\text{liquid}} = [4186 \text{ J/(kg·C°)}](0.0010 \text{ kg})(31 \text{ C°}) - 0.0020 \text{ J}$$
$$= 130 \text{ J} - 0.0020 \text{ J} = \boxed{130 \text{ J}}$$
$$\Delta U_{\text{gas}} = [2020 \text{ J/(kg·C°)}](0.0010 \text{ kg})(31 \text{ C°}) - 14 \text{ J}$$
$$= 63 \text{ J} - 14 \text{ J} = \boxed{49 \text{ J}}$$

For the liquid, virtually all the 130 J of heat serves to change the internal energy, since the volume change and the corresponding work of expansion are so small. In contrast, a significant fraction of the 63 J of heat added to the vapor causes work of expansion to be done, so that only 49 J is left to change the internal energy.

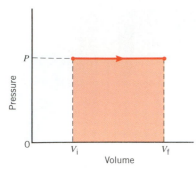

**Figure 15.5** For an isobaric process, a pressure-versus-volume plot is a horizontal straight line, and the work done [$W = P(V_f - V_i)$] is the colored rectangular area under the graph.

**Problem solving insight**

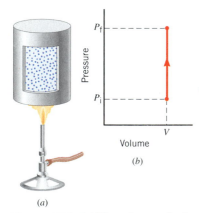

**Figure 15.6** (a) The substance in the chamber is being heated isochorically because the rigid chamber keeps the volume constant. (b) The pressure–volume plot for an isochoric process is a vertical straight line. The area under the graph is zero, indicating that no work is done.

It is often convenient to display thermal processes graphically. For instance, Figure 15.5 shows a plot of pressure versus volume for an isobaric expansion. Since the pressure is constant, the graph is a horizontal straight line, beginning at the initial volume $V_i$ and ending at the final volume $V_f$. In terms of such a plot, the work $W = P(V_f - V_i)$ is the area under the graph, which is the shaded rectangle of height $P$ and width $V_f - V_i$.

Another common thermal process is an *isochoric process, one that occurs at constant volume.* Figure 15.6a illustrates an isochoric process in which a substance (solid, liquid, or gas) is heated. The substance would expand if it could, but the rigid container keeps the volume constant, so the pressure–volume plot shown in Figure 15.6b is a vertical straight line. Because the volume is constant, the pressure inside rises, and the substance exerts more and more force on the walls. Although enormous forces can be generated in the closed container, no work is done, since the walls do not move. Consistent with zero work being done, the area under the vertical straight line in Figure 15.6b is zero. Since no work is done, the first law of thermodynamics indicates that the heat in an isochoric process serves only to change the internal energy: $\Delta U = Q - W = Q$.

A third important thermal process is an *isothermal process, one that takes place at constant temperature.* The next section illustrates the important features of an isothermal process when the system is an ideal gas.

Last, there is the *adiabatic process, one that occurs without the transfer of heat.* Since there is no heat transfer, $Q$ equals zero, and the first law indicates that $\Delta U = Q - W = -W$. Thus, when work is done by a system adiabatically, $W$ is positive and the internal energy of the system decreases by exactly the amount of the work done. When work is done on a system adiabatically, $W$ is negative and the internal energy increases correspondingly. The next section discusses an adiabatic process for an ideal gas.

A process may be complex enough that it is not recognizable as one of the four just discussed. For instance, Figure 15.7 shows a process for a gas in which the pressure, volume, and temperature are changed along the straight line from $X$ to $Y$. With the aid of integral calculus, it can be shown that *the area under a pressure–volume graph is the work for any kind of process,* so the area representing the work has been colored in the drawing. The volume increases, so that work is done by the gas. This work is positive by convention, as is the area. In contrast, if a process reduces the volume, work is done on the gas, and this work is negative by convention. Correspondingly, the area under the pressure–volume graph would be assigned a negative value. In Example 4, we determine the work for the case shown in Figure 15.7.

### Example 4  Work and the Area Under a Pressure–Volume Graph

Determine the work for the process in which the pressure, volume, and temperature of a gas are changed along the straight line from $X$ to $Y$ in Figure 15.7.

**Reasoning** The work is given by the area (in color) under the straight line between $X$ and $Y$. Since the volume increases, work is done by the gas on the surroundings, so the work is positive. The area can be found by counting squares in Figure 15.7 and multiplying by the area per square.

**Solution** We estimate that there are 8.9 colored squares in the drawing. The area of one square is $(2.0 \times 10^5 \text{ Pa})(1.0 \times 10^{-4} \text{ m}^3) = 2.0 \times 10^1 \text{ J}$, so the work is

$$W = +(8.9 \text{ squares})(2.0 \times 10^1 \text{ J/square}) = \boxed{+180 \text{ J}}$$

## 15.5  *Thermal Processes Using an Ideal Gas*

### ISOTHERMAL EXPANSION OR COMPRESSION

When a system performs work isothermally, the temperature remains constant. In Figure 15.8a, for instance, a metal cylinder contains $n$ moles of an ideal gas, and the large mass of hot water maintains the cylinder and gas at a constant Kelvin temperature $T$. The piston is held in place initially so the volume of the gas is $V_i$. As the external force applied to the piston is reduced quasi-statically, the gas expands to the final volume $V_f$. Figure 15.8b gives a plot of pressure ($P = nRT/V$) versus volume for the process. The solid red line in the graph is called an isotherm (meaning "constant temperature") and represents the relation between pressure and volume when the temperature is held constant. The work $W$ done by the gas is *not* given by $W = P \Delta V = P(V_f - V_i)$ because the pressure is not constant. Nevertheless, the work is equal to the area under the graph. The techniques of integral calculus lead to the following result* for $W$:

*Isothermal expansion or compression of an ideal gas*

$$W = nRT \ln\left(\frac{V_f}{V_i}\right) \tag{15.3}$$

Where does the energy for this work originate? Since the internal energy of any ideal gas is proportional to the Kelvin temperature ($U = \frac{3}{2}nRT$ for a monatomic ideal gas, for example), the internal energy remains constant throughout an isothermal process, and the change in internal energy is zero. The first law of thermodynamics becomes $\Delta U = 0 = Q - W$. In other words, $Q = W$, and the energy for the work originates in the hot water. Heat flows into the gas from the water, as Figure 15.8a illustrates. If the gas is compressed isothermally, Equation 15.3 still applies, and heat flows out of the gas into the water. The following example deals with the isothermal expansion of an ideal gas.

### Example 5 Isothermal Expansion of an Ideal Gas

Two moles of the monatomic gas argon expand isothermally at 298 K, from an initial volume of $V_i = 0.025 \text{ m}^3$ to a final volume of $V_f = 0.050 \text{ m}^3$. Assuming that argon is an ideal gas, find (a) the work done by the gas, (b) the change in the internal energy of the gas, and (c) the heat supplied to the gas.

**Reasoning and Solution**

(a) The work done by the gas can be found from Equation 15.3:

$$W = nRT \ln\left(\frac{V_f}{V_i}\right) = (2.0 \text{ mol})[8.31 \text{ J/(mol·K)}](298 \text{ K}) \ln\left(\frac{0.050 \text{ m}^3}{0.025 \text{ m}^3}\right) = \boxed{+3400 \text{ J}}$$

(b) The internal energy of a monatomic ideal gas is $U = \frac{3}{2}nRT$ (Equation 14.7) and does not change when the temperature is constant. Therefore, $\boxed{\Delta U = 0 \text{ J}}$.

(c) The heat $Q$ supplied can be determined from the first law of thermodynamics:

$$Q = \Delta U + W = 0 \text{ J} + 3400 \text{ J} = \boxed{+3400 \text{ J}} \tag{15.1}$$

## ADIABATIC EXPANSION OR COMPRESSION

When a system performs work adiabatically, no heat flows into or out of the system. Figure 15.9a shows an arrangement in which $n$ moles of an ideal gas do work under adiabatic conditions, expanding quasi-statically from an initial volume $V_i$ to a final volume $V_f$. The arrangement is similar to that in Figure 15.8 for isothermal expansion. However, a different amount of work is done here, because the cylinder is now surrounded by insulating material that prevents the flow of heat, so $Q = 0$ J. According to the first law of thermodynamics, the change in internal energy is $\Delta U = Q - W = -W$. Since the internal energy of an ideal monatomic gas is $U = \frac{3}{2}nRT$ (Equation 14.7), it follows that $\Delta U = U_f - U_i = \frac{3}{2}nR(T_f - T_i)$, where $T_i$ and $T_f$ are the initial and final Kelvin temperatures. With this substitution, the relation $\Delta U = -W$ becomes

*Adiabatic expansion or compression of a monatomic ideal gas*

$$W = \frac{3}{2}nR(T_i - T_f) \tag{15.4}$$

*In this result, "ln" denotes the natural logarithm to the base $e = 2.71828$. The natural logarithm is related to the common logarithm to the base ten by $\ln(V_f/V_i) = 2.303 \log(V_f/V_i)$.

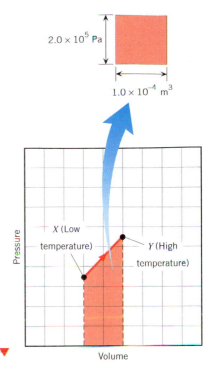

**Figure 15.7** The colored area gives the work done by the gas for the process from $X$ to $Y$.

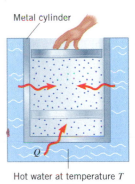

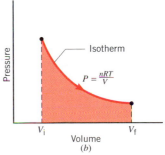

**Figure 15.8** (a) The ideal gas in the cylinder is expanding isothermally at temperature $T$. The force holding the piston in place is reduced slowly, so the expansion occurs quasi-statically. (b) The work done by the gas is given by the colored area.

Metal cylinder

Insulating material
*(a)*

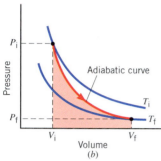

*(b)*

**Figure 15.9** *(a)* The ideal gas in the cylinder is expanding adiabatically. The force holding the piston in place is reduced slowly, so the expansion occurs quasi-statically. *(b)* A plot of pressure versus volume yields the adiabatic curve shown in red, which intersects the isotherms (blue) at the initial temperature $T_i$ and the final temperature $T_f$. The work done by the gas is given by the colored area.

**Table 15.1    *Summary of Thermal Processes***

| Type of Thermal Process | Work Done | First Law of Thermodynamics ($\Delta U = Q - W$) |
|---|---|---|
| Isobaric (constant pressure) | $W = P(V_f - V_i)$ | $\Delta U = Q - \underbrace{P(V_f - V_i)}_{W}$ |
| Isochoric (constant volume) | $W = 0 \ \text{J}$ | $\Delta U = Q - \underbrace{0 \ \text{J}}_{W}$ |
| Isothermal (constant temperature) | $W = nRT \ln\left(\dfrac{V_f}{V_i}\right)$ (for an ideal gas) | $\underbrace{0 \ \text{J}}_{\substack{\Delta U \text{ for an}\\ \text{ideal gas}}} = Q - \underbrace{nRT \ln\left(\dfrac{V_f}{V_i}\right)}_{W}$ |
| Adiabatic (no heat flow) | $W = \frac{3}{2}nR(T_f - T_i)$ (for a monatomic ideal gas) | $\Delta U = \underbrace{0 \ \text{J}}_{Q} - \underbrace{\frac{3}{2}nR(T_i - T_f)}_{W}$ |

When an ideal gas expands adiabatically, it does positive work, so $W$ is positive in Equation 15.4. Therefore, the term $T_i - T_f$ is also positive, so the final temperature of the gas must be less than the initial temperature. The internal energy of the gas is reduced to provide the necessary energy to do the work, and because the internal energy is proportional to the Kelvin temperature, the temperature decreases. Figure 15.9*b* shows a plot of pressure versus volume for an adiabatic process. The adiabatic curve (red) intersects the isotherms (blue) at the higher initial temperature $[T_i = P_iV_i/(nR)]$ and the lower final temperature $[T_f = P_fV_f/(nR)]$. The colored area under the adiabatic curve represents the work done.

The reverse of an adiabatic expansion is an adiabatic compression ($W$ is negative), and Equation 15.4 indicates that the final temperature exceeds the initial temperature. The energy provided by the agent doing the work increases the internal energy of the gas. As a result, the gas becomes hotter.

The equation that gives the adiabatic curve (red) between the initial pressure and volume ($P_i$, $V_i$) and the final pressure and volume ($P_f$, $V_f$) in Figure 15.9*b* can be derived using integral calculus. The result is

**Adiabatic expansion or compression of an ideal gas**

$$P_iV_i^{\gamma} = P_fV_f^{\gamma} \tag{15.5}$$

where the exponent $\gamma$ (Greek gamma) is the ratio of the specific heat capacities at constant pressure and constant volume, $\gamma = c_P/c_V$. Equation 15.5 applies in conjunction with the ideal gas law, for *each point* on the adiabatic curve satisfies the relation $PV = nRT$.

Table 15.1 summarizes the work done in the four types of thermal processes that we have been considering. For each process it also shows how the first law of thermodynamics depends on the work and other variables.

## 15.6  *Specific Heat Capacities*

In this section the first law of thermodynamics is used to gain an understanding of the factors that determine the specific heat capacity of a material. Remember, when the temperature of a substance changes as a result of heat flow, the change in temperature $\Delta T$ and the amount of heat $Q$ are related according to $Q = cm\Delta T$ (Equation 12.4). In this expression $c$ denotes the specific heat capacity in units of J/(kg·C°), and $m$ is the mass in kilograms. It is more convenient now to express the amount of material as the number of moles $n$, rather than the number of kilograms. Therefore, we replace the expression $Q = cm\Delta T$ with the following analogous expression:

$$Q = Cn\Delta T \tag{15.6}$$

where the capital letter $C$ (as opposed to the lowercase $c$) refers to the ***molar specific heat capacity*** in units of J/(mol·K). In addition, the unit for measuring the temperature change $\Delta T$ is the Kelvin (K) rather than the Celsius degree (C°), and $\Delta T = T_f - T_i$, where $T_f$ and $T_i$ are the final and initial temperatures. For gases it is necessary to distinguish between the molar specific heat capacities $C_P$ and $C_V$, which apply, respectively, to conditions of constant pressure and constant volume. With the help of the first law of thermodynamics and an ideal gas as an example, it is possible to see why $C_P$ and $C_V$ differ.

To determine the molar specific heat capacities, we must first calculate the heat $Q$ needed to raise the temperature of an ideal gas from $T_i$ to $T_f$. According to the first law, $Q = \Delta U + W$. We also know that the internal energy of a monatomic ideal gas is $U = \frac{3}{2}nRT$ (Equation 14.7). As a result, $\Delta U = U_f - U_i = \frac{3}{2}nR(T_f - T_i)$. When the heating process occurs at constant pressure, the work done is given by Equation 15.2: $W = P\,\Delta V = P(V_f - V_i)$. For an ideal gas, $PV = nRT$, so the work becomes $W = nR(T_f - T_i)$. On the other hand, when the volume is constant, $\Delta V = 0$ m³, and the work done is zero. The calculation of the heat is summarized below:

$$Q = \Delta U + W$$

$$Q_{\text{constant pressure}} = \tfrac{3}{2}nR(T_f - T_i) + nR(T_f - T_i) = \tfrac{5}{2}nR(T_f - T_i)$$

$$Q_{\text{constant volume}} = \tfrac{3}{2}nR(T_f - T_i) + 0$$

The molar specific heat capacities can now be determined, since Equation 15.6 indicates that $C = Q/[n(T_f - T_i)]$:

*Constant pressure for a monatomic ideal gas*
$$C_P = \frac{Q_{\text{constant pressure}}}{n(T_f - T_i)} = \tfrac{5}{2}R \qquad (15.7)$$

*Constant volume for a monatomic ideal gas*
$$C_V = \frac{Q_{\text{constant volume}}}{n(T_f - T_i)} = \tfrac{3}{2}R \qquad (15.8)$$

The ratio $\gamma$ of the specific heats is

*Monatomic ideal gas*
$$\gamma = \frac{C_P}{C_V} = \frac{\tfrac{5}{2}R}{\tfrac{3}{2}R} = \frac{5}{3} \qquad (15.9)$$

For real monatomic gases near room temperature, experimental values of $C_P$ and $C_V$ give ratios very close to the theoretical value of $\frac{5}{3}$.

The difference between $C_P$ and $C_V$ arises because work is done when the gas expands in response to the addition of heat under conditions of constant pressure, whereas no work is done under conditions of constant volume. For a monatomic ideal gas, $C_P$ exceeds $C_V$ by an amount equal to $R$, the ideal gas constant:

$$C_P - C_V = R \qquad (15.10)$$

In fact, it can be shown that Equation 15.10 applies to any kind of ideal gas—monatomic, diatomic, etc.

# 15.7 *The Second Law of Thermodynamics*

Ice cream melts when left out on a warm day. A cold can of soda warms up on a hot day at a picnic. Ice cream and soda never become colder when left in a hot environment, for heat always flows spontaneously from hot to cold, and never from cold to hot. The spontaneous flow of heat is the focus of one of the most profound laws in all of science, the ***second law of thermodynamics.***

■ **THE SECOND LAW OF THERMODYNAMICS: THE HEAT FLOW STATEMENT**

Heat flows spontaneously from a substance at a higher temperature to a substance at a lower temperature and does not flow spontaneously in the reverse direction.

It is important to realize that the second law of thermodynamics deals with a different aspect of nature than does the first law of thermodynamics. The second law is a statement about the natural tendency of heat to flow from hot to cold, whereas the first law deals with energy conservation and focuses on both heat and work. A number of important devices depend on heat and work in their operation, and to understand such devices both laws are needed. For instance, an automobile engine is a type of heat engine because it uses heat to produce work. In discussing heat engines, Sections 15.8 and 15.9 will bring together the first and second laws to analyze engine efficiency. Then, in Section 15.10 we will see that refrigerators, air conditioners, and heat pumps also utilize heat and work and are closely related to heat engines. The way in which these three appliances operate also depends on both the first and second laws of thermodynamics.

## 15.8 Heat Engines

**The physics of a heat engine.**

A *heat engine* is any device that uses heat to perform work. It has three essential features:

1. Heat is supplied to the engine at a relatively high input temperature from a place called the *hot reservoir*.

2. Part of the input heat is used to perform work by the *working substance* of the engine, which is the material within the engine that actually does the work (e.g., the gasoline–air mixture in an automobile engine).

3. The remainder of the input heat is rejected to a place called the *cold reservoir*, which has a temperature lower than the input temperature.

Figure 15.10 emphasizes these features in a schematic fashion. The symbol $Q_H$ refers to the magnitude of the input heat, and the subscript H indicates the hot reservoir. The symbol $Q_C$ refers to the magnitude of the rejected heat, and the subscript C denotes the cold reservoir. The symbol $W$ stands for the magnitude of the work done. ***These three symbols refer to magnitudes only, without reference to algebraic signs. Therefore, when these symbols appear in an equation, they do not have negative values assigned to them.***

**Problem solving insight**

To be highly efficient, a heat engine must produce a relatively large amount of work from as little input heat as possible. Thus, the ***efficiency e*** of a heat engine is defined as the ratio of the work $W$ done by the engine to the input heat $Q_H$:

$$e = \frac{\text{Work done}}{\text{Input heat}} = \frac{W}{Q_H} \qquad (15.11)$$

If the input heat were converted entirely into work, the engine would have an efficiency of 1.00, since $W = Q_H$; such an engine would be 100% efficient. ***Efficiencies are often quoted as percentages obtained by multiplying the ratio $W/Q_H$ by a factor of 100.*** Thus, an efficiency of 68% would mean that a value of 0.68 is used for the efficiency in Equation 15.11.

An engine, like any device, must obey the principle of conservation of energy. Some of the engine's input heat $Q_H$ is converted into work $W$, and the remainder $Q_C$ is rejected to the cold reservoir. If there are no other losses in the engine, the principle of energy conservation requires that

$$Q_H = W + Q_C \qquad (15.12)$$

Solving this equation for $W$ and substituting the result into Equation 15.11 leads to the following alternative expression for the efficiency $e$ of a heat engine:

$$e = \frac{Q_H - Q_C}{Q_H} = 1 - \frac{Q_C}{Q_H} \qquad (15.13)$$

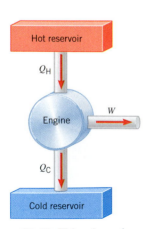

**Figure 15.10** This schematic representation of a heat engine shows the input heat (magnitude = $Q_H$) that originates from the hot reservoir, the work (magnitude = $W$) that the engine does, and the heat (magnitude = $Q_C$) that the engine rejects to the cold reservoir.

Example 6 illustrates how the concepts of efficiency and energy conservation are applied to a heat engine.

### Example 6 An Automobile Engine

An automobile engine has an efficiency of 22.0% and produces 2510 J of work. How much heat is rejected by the engine?

**Reasoning** Energy conservation indicates that the amount of heat rejected to the cold reservoir is the part of the input heat that is not converted into work. The amount $Q_C$ rejected is $Q_C = Q_H - W$, according to Equation 15.12. To use this equation, however, we need a value for the input heat $Q_H$, so we begin our solution by finding this value.

**Solution** From Equation 15.11 for the efficiency $e$, we find that $Q_H = W/e$. Substituting this result into Equation 15.12, we see that the rejected heat is

$$Q_C = Q_H - W = \frac{W}{e} - W = (2510 \text{ J}) \left( \frac{1}{0.220} - 1 \right) = \boxed{8900 \text{ J}}$$

In Example 6, less than one-quarter of the input heat is converted into work because the efficiency of the automobile engine is only 22.0%. If the engine were 100% efficient, all the input heat would be converted into work. Unfortunately, nature does not permit 100% efficient heat engines to exist, as the next section discusses.

## 15.9 Carnot's Principle and the Carnot Engine

What is it that allows a heat engine to operate with maximum efficiency? The French engineer Sadi Carnot (1796–1832) proposed that a heat engine has maximum efficiency when the processes within the engine are reversible. *A reversible process is one in which both the system and its environment can be returned to exactly the states they were in before the process occurred.*

In a reversible process, *both* the system and its environment can be returned to their initial states. Therefore, a process that involves an energy-dissipating mechanism, such as friction, cannot be reversible because the energy wasted due to friction would alter the system or the environment or both. There are also reasons other than friction why a process may not be reversible. For instance, the spontaneous flow of heat from a hot substance to a cold substance is irreversible, even though friction is not present. For heat to flow in the reverse direction, work must be done, as we will see in Section 15.10. The agent doing such work must be located in the environment of the hot and cold substances, and, therefore, the environment must change while the heat is moved back from cold to hot. Since the system and the environment cannot *both* be returned to their initial states, the process of spontaneous heat flow is irreversible. In fact, all spontaneous processes are irreversible, such as the explosion of an unstable chemical or the bursting of a bubble. When the word "reversible" is used in connection with engines, it does not just mean a gear that allows the engine to operate a device in reverse. All cars have a reverse gear, for instance, but no automobile engine is thermodynamically reversible, since friction exists no matter which way the car moves.

Today, the idea that the efficiency of a heat engine is a maximum when the engine operates reversibly is referred to as *Carnot's principle.*

■ **CARNOT'S PRINCIPLE: AN ALTERNATIVE STATEMENT OF THE SECOND LAW OF THERMODYNAMICS**

No irreversible engine operating between two reservoirs at constant temperatures can have a greater efficiency than a reversible engine operating between the same temperatures. Furthermore, all reversible engines operating between the same temperatures have the same efficiency.

Carnot's principle is quite remarkable, for no mention is made of the working substance of the engine. It does not matter whether the working substance is a gas, a liquid, or a solid. As long as the process is reversible, the efficiency of the engine is a maximum. Furthermore, Carnot's principle does *not* state, or even imply, that a reversible engine has an efficiency of 100%.

It can be shown that if Carnot's principle were not valid, it would be possible for heat to flow spontaneously from a cold substance to a hot substance, in violation of the second

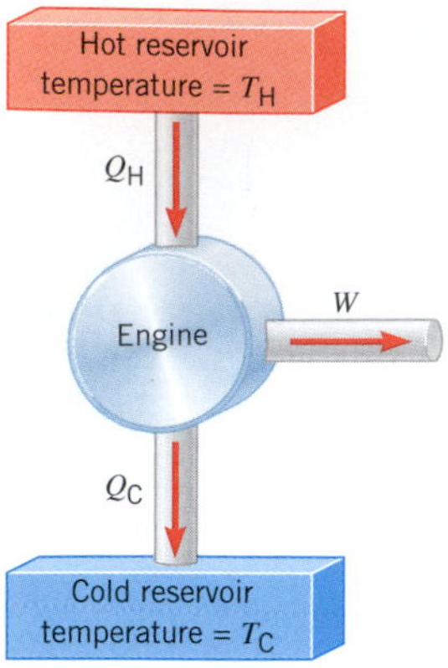

**Figure 15.11** A Carnot engine is a reversible engine in which all input heat $Q_H$ originates from a hot reservoir at a single temperature $T_H$, and all rejected heat $Q_C$ goes into a cold reservoir at a single temperature $T_C$. The work done by the engine is $W$.

law of thermodynamics. In effect, then, Carnot's principle is another way of expressing the second law.

No real engine operates reversibly. Nonetheless, the idea of a reversible engine provides a useful standard for judging the performance of real engines. Figure 15.11 shows a reversible engine, called a ***Carnot engine,*** that is particularly useful as an idealized model. An important feature of the Carnot engine is that all input heat $Q_H$ originates from a hot reservoir at a *single temperature* $T_H$ and all rejected heat $Q_C$ goes into a cold reservoir *at a single temperature* $T_C$.

Carnot's principle implies that the efficiency of a reversible engine is independent of the working substance of the engine, and therefore can depend only on the temperatures of the hot and cold reservoirs. Since efficiency $e = 1 - Q_C/Q_H$ according to Equation 15.13, the ratio $Q_C/Q_H$ can depend only on the reservoir temperatures. This observation led Lord Kelvin to propose a ***thermodynamic temperature scale.*** He proposed that the thermodynamic temperatures of the cold and hot reservoirs be defined such that their ratio is equal to $Q_C/Q_H$. Thus, the thermodynamic temperature scale is related to the heats absorbed and rejected by a Carnot engine, and is independent of the working substance. If a reference temperature is properly chosen, it can be shown that the thermodynamic temperature scale is identical to the Kelvin scale introduced in Section 12.2 and used in the ideal gas law. As a result, the ratio of the rejected heat $Q_C$ to the input heat $Q_H$ is

$$\frac{Q_C}{Q_H} = \frac{T_C}{T_H} \tag{15.14}$$

where the temperatures $T_C$ and $T_H$ *must be expressed in kelvins.*

The efficiency $e_{\text{Carnot}}$ of a Carnot engine can be written in a particularly useful way by substituting Equation 15.14 into Equation 15.13 for the efficiency, $e = 1 - Q_C/Q_H$:

$$\begin{matrix}\text{Efficiency of a}\\\text{Carnot engine}\end{matrix} = e_{\text{Carnot}} = 1 - \frac{T_C}{T_H} \tag{15.15}$$

This relation gives the *maximum possible efficiency* for a heat engine operating between two Kelvin temperatures $T_C$ and $T_H$, and the next example illustrates its application.

### Example 7    A Tropical Ocean as a Heat Engine

**The physics of extracting work from a warm ocean.**

Water near the surface of a tropical ocean has a temperature of 298.2 K (25.0 °C), whereas water 700 m beneath the surface has a temperature of 280.2 K (7.0 °C). It has been proposed that the warm water be used as the hot reservoir and the cool water as the cold reservoir of a heat engine. Find the maximum possible efficiency for such an engine.

**Reasoning** The maximum possible efficiency is the efficiency that a Carnot engine would have (Equation 15.15) operating between temperatures of $T_H = 298.2$ K and $T_C = 280.2$ K.

**Solution** Using $T_H = 298.2$ K and $T_C = 280.2$ K in Equation 15.15, we find that

**Problem solving insight**
When determining the efficiency of a Carnot engine, be sure the temperatures $T_C$ and $T_H$ of the cold and hot reservoirs are expressed in kelvins; degrees Celsius or degrees Fahrenheit will not do.

$$e_{\text{Carnot}} = 1 - \frac{T_C}{T_H} = 1 - \frac{280.2 \text{ K}}{298.2 \text{ K}} = \boxed{0.060 \ (6.0\%)}$$

In Example 7 the maximum possible efficiency is only 6.0%. The small efficiency arises because the Kelvin temperatures of the hot and cold reservoirs are so close. A greater efficiency is possible only when there is a greater difference between the reservoir temperatures. However, there are limits on how large the efficiency of a heat engine can be, as Conceptual Example 8 discusses.

### Conceptual Example 8
### Natural Limits on the Efficiency of a Heat Engine

Consider a hypothetical engine that receives 1000 J of heat as input from a hot reservoir and delivers 1000 J of work, rejecting no heat to a cold reservoir whose temperature is above 0 K. Decide whether this engine violates the first or the second law of thermodynamics, or both.

**Reasoning and Solution**  The first law of thermodynamics is an expression of energy conservation. From the point of view of energy conservation, nothing is wrong with an engine that converts 1000 J of heat into 1000 J of work. Energy has been neither created nor destroyed; it has only been transformed from one form (heat) to another (work). This engine does, however, violate the second law of thermodynamics. Since all of the input heat is converted into work, the efficiency of the engine is 1, or 100%. But Equation 15.15, which is based on the second law, indicates that the maximum possible efficiency is $1 - T_C/T_H$, where $T_C$ and $T_H$ are the temperatures of the cold and hot reservoirs, respectively. Since we know that $T_C$ is above 0 K, it is clear that the ratio $T_C/T_H$ is greater than zero, so the maximum possible efficiency is less than 1, or less than 100%. It is important to understand that the first and second laws of thermodynamics address different aspects of nature. *It is the second law, not the first law, that limits the efficiencies of heat engines to values less than 100%.*

Example 8 has emphasized that *even a perfect heat engine has an efficiency that is less than 1.0 or 100%.* In this regard, we note that the maximum possible efficiency, as given by Equation 15.15, approaches 1.0 when $T_C$ approaches absolute zero (0 K). However, experiments have shown that it is not possible to cool a substance to absolute zero (see Section 15.12), so nature does not permit the existence of a 100%-efficient heat engine. As a result, there will always be heat rejected to a cold reservoir whenever a heat engine is used to do work, even if friction and other irreversible processes are eliminated completely. This rejected heat is a form of thermal pollution. Thus, the second law of thermodynamics requires that at least some thermal pollution be generated whenever heat engines are used to perform work. This kind of thermal pollution can be reduced only if society reduces its dependence on heat engines to do work.

*The physics of* **thermal pollution.**

## 15.10  *Refrigerators, Air Conditioners, and Heat Pumps*

The natural tendency of heat is to flow from hot to cold, as indicated by the second law of thermodynamics. However, if work is used, heat can be *made* to flow from cold to hot, against its natural tendency. Refrigerators, air conditioners, and heat pumps are, in fact, devices that do just that. As Figure 15.12 *(left)* illustrates, these devices use work $W$ to extract an amount of heat $Q_C$ from the cold reservoir and deposit an amount of heat $Q_H$ into the hot reservoir. Generally speaking, such a process is called a *refrigeration process.* A comparison of this drawing with Figure 15.11 shows that the directions of the arrows symbolizing heat and work in a refrigeration process are opposite to those in an engine process. Nonetheless, energy is conserved during a refrigeration process, just as it is in an engine process, so $Q_H = W + Q_C$. Moreover, if the process occurs reversibly, we have ideal devices that are called Carnot refrigerators, Carnot air conditioners, and Carnot heat pumps. For these ideal devices, the relation $Q_C/Q_H = T_C/T_H$ (Equation 15.14) applies, just as it does for the Carnot engine.

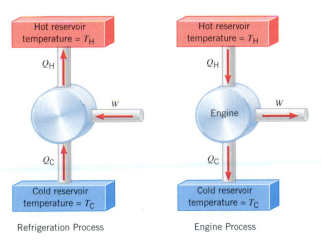

**Figure 15.12**  In the refrigeration process on the left, work $W$ is used to remove heat $Q_C$ from the cold reservoir and deposit heat $Q_H$ into the hot reservoir. Compare this with the engine process on the right.

In a *refrigerator,* the interior of the unit is the cold reservoir, while the warmer exterior is the hot reservoir. As Figure 15.13 illustrates, the refrigerator takes heat from the food inside and deposits it into the kitchen, along with the energy needed to do the work of making the heat flow from cold to hot. For this reason, the outside surfaces (usually the sides and back) of most refrigerators are warm to the touch while the units operate. Thus, a refrigerator warms the kitchen.

An *air conditioner* is like a refrigerator, except that the room itself is the cold reservoir and the outdoors is the hot reservoir. Figure 15.14 shows a window unit, which cools a room by removing heat and depositing it outside, along with the work used to make the heat flow from cold to hot. Conceptual Example 9 considers a common misconception about refrigerators and air conditioners.

## Conceptual Example 9
### You Can't Beat the Second Law of Thermodynamics

Is it possible to cool your kitchen by leaving the refrigerator door open or cool your bedroom by putting a window air conditioner on the floor by the bed?

**Reasoning and Solution** Whatever heat $Q_C$ is removed from the air directly in front of the open refrigerator is deposited back into the kitchen at the rear of the unit. Moreover, according to the second law, work $W$ is needed to move that heat from cold to hot, and the energy from this work is also deposited into the kitchen as additional heat. Thus, the open refrigerator puts into the kitchen an amount of heat $Q_H = Q_C + W$, which is more than it removes. *Rather than cooling the kitchen, the open refrigerator warms it up.* Putting a window air conditioner on the floor to cool your bedroom is similarly a no-win game. The heat pumped out the back of the air conditioner and into the bedroom is greater than the heat pulled into the front of the unit. Consequently, *the air conditioner actually warms the bedroom.*

**Related Homework:** *Problem 63*

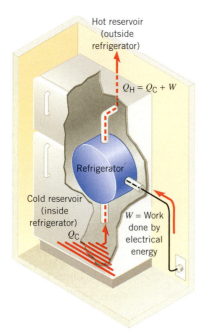

**Figure 15.13** A refrigerator.

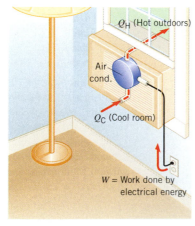

**Figure 15.14** A window air conditioner removes heat from a room, which is the cold reservoir, and deposits heat outdoors, which is the hot reservoir.

The quality of a refrigerator or air conditioner is rated according to its coefficient of performance. Such appliances perform well when they remove a relatively large amount of heat $Q_C$ from the cold reservoir with as little work $W$ as possible. Therefore, the coefficient of performance is defined as the ratio of $Q_C$ to $W$, and the greater this ratio is, the better the performance is:

| *Refrigerator or air conditioner* | $\dfrac{\text{Coefficient of}}{\text{performance}} = \dfrac{Q_C}{W}$ | (15.16) |
|---|---|---|

Commercially available refrigerators and air conditioners have coefficients of performance in the range 2 to 6, depending on the temperatures involved. The coefficients of performance for these real devices are less than those for ideal, or Carnot, refrigerators and air conditioners.

In a sense, refrigerators and air conditioners operate like pumps. They pump heat "uphill" from a lower temperature to a higher temperature, just as a water pump forces water uphill from a lower elevation to a higher elevation. It would be appropriate to call them heat pumps. However, the name "heat pump" is reserved for the device illustrated in Figure 15.15, which is a home heating appliance. The *heat pump* uses work $W$ to make heat $Q_C$ from the wintry outdoors (the cold reservoir) flow up the temperature "hill" into a warm house (the hot reservoir). According to the conservation of energy, the heat pump deposits inside the house an amount of heat $Q_H = Q_C + W$. The air conditioner and the heat pump do closely related jobs. The air conditioner refrigerates the inside of the house and heats up the outdoors, while the heat pump refrigerates the outdoors and heats up the inside. These jobs are so closely related that most heat pump systems serve in a dual capacity, being equipped with a switch that converts them from heaters in the winter into air conditioners in the summer.

Heat pumps are popular for home heating in today's energy-conscious world, and it is easy to understand why. Suppose 1000 J of energy is available to use for home heating. Figure 15.16 shows that a conventional electric heating system uses this 1000 J to heat a coil of wire, just as in a toaster. A fan blows air across the hot coil, and forced convection carries the 1000 J of heat into the house. In contrast, the heat pump in Figure 15.15 does not use the 1000 J directly as heat. Instead, it uses the 1000 J to do the work $W$ of pumping heat $Q_C$ from the cooler outdoors into the warmer house. The heat pump delivers to the house an amount of energy $Q_H = Q_C + W$. With $W = 1000$ J, this becomes $Q_H = Q_C + 1000$ J, so that the heat pump delivers more than 1000 J of heat into the house, whereas the conventional electric heating system delivers only 1000 J. The next example shows how the basic relations $Q_H = W + Q_C$ and $Q_C/Q_H = T_C/T_H$ are used with heat pumps.

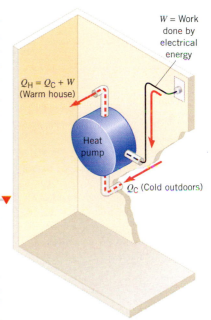

### Example 10   A Heat Pump

An ideal or Carnot heat pump is used to heat a house to a temperature of $T_H = 294$ K (21 °C). How much work must be done by the pump to deliver $Q_H = 3350$ J of heat into the house when the outdoor temperature $T_C$ is (a) 273 K (0 °C) and (b) 252 K (−21 °C)?

**Reasoning** The conservation of energy ($Q_H = W + Q_C$) applies to the heat pump. Thus, the work can be determined from $W = Q_H - Q_C$, provided we can obtain a value for $Q_C$, the heat taken by the pump from the outside. To determine $Q_C$, we use the fact that the pump is a Carnot heat pump and operates reversibly. Therefore, the relation $Q_C/Q_H = T_C/T_H$ (Equation 15.14) applies. Solving it for $Q_C$, we obtain $Q_C = Q_H(T_C/T_H)$. Using this result, we find that

**Figure 15.15** In a heat pump the cold reservoir is the wintry outdoors, and the hot reservoir is the inside of the house.

$$W = Q_H - Q_C = Q_H - Q_H\left(\frac{T_C}{T_H}\right) = Q_H\left(1 - \frac{T_C}{T_H}\right)$$

### Solution

(a) At an indoor temperature of $T_H = 294$ K and an outdoor temperature of $T_C = 273$ K, the work needed is

$$W = Q_H\left(1 - \frac{T_C}{T_H}\right) = (3350 \text{ J})\left(1 - \frac{273 \text{ K}}{294 \text{ K}}\right) = \boxed{240 \text{ J}}$$

(b) This solution is identical to that in part (a), except that it is now cooler outside, so $T_C = 252$ K. The necessary work is $W = \boxed{479 \text{ J}}$, which is more than in part (a). More work must be done because the heat is pumped up a greater temperature "hill" when the outside is colder than when it is warmer.

**Problem solving insight**
When applying Equation 15.14 ($Q_C/Q_H = T_C/T_H$) to heat pumps, refrigerators, or air conditioners, be sure the temperatures $T_C$ and $T_H$ are expressed in kelvins; degrees Celsius or degrees Fahrenheit will not do.

It is also possible to specify a coefficient of performance for heat pumps. However, unlike refrigerators and air conditioners, the job of a heat pump is to heat, not to cool. As a result, the coefficient of performance of a heat pump is the ratio of the heat $Q_H$ delivered into the house to the work $W$ required to deliver it:

**Heat pump**    $$\frac{\text{Coefficient of}}{\text{performance}} = \frac{Q_H}{W}$$    (15.17)

The coefficient of performance depends on the indoor and outdoor temperatures. Commercial units have coefficients of about 3 to 4 under favorable conditions.

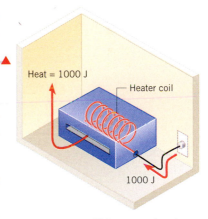

**Figure 15.16** This conventional electric heating system is delivering 1000 J of heat to the living room.

## 15.11   Entropy

A Carnot engine has the maximum possible efficiency for its operating conditions because the processes occurring within it are reversible. Irreversible processes, such as friction, cause real engines to operate at less than maximum efficiency, for they reduce our ability to use heat to perform work. As an extreme example, imagine that a hot object is placed in thermal contact with a cold object, so heat flows spontaneously, and hence irreversibly, from hot to cold. Eventually both objects reach the same temperature, and $T_C = T_H$. A Carnot engine using these two objects as heat reservoirs is unable to do work, because the efficiency of the engine is zero [$e_{\text{Carnot}} = 1 - (T_C/T_H) = 0$].

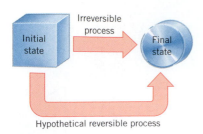

$\Delta S$ for irreversible　=　$\Delta S$ for hypothetical
　　　process　　　　　　　reversible process

**Figure 15.17** Although the relation $\Delta S = (Q/T)_R$ applies to reversible processes, it can be used as part of an indirect procedure to find the entropy change for an irreversible process. This drawing illustrates the procedure discussed in the text.

In general, irreversible processes cause us to lose some, but not necessarily all, of the ability to perform work. This partial loss can be expressed in terms of a concept called *entropy*.

To introduce the idea of entropy we recall the relation $Q_C/Q_H = T_C/T_H$ that applies to a Carnot engine. This equation can be rearranged as $Q_C/T_C = Q_H/T_H$, which focuses attention on the heat $Q$ divided by the Kelvin temperature $T$. The quantity $Q/T$ is called the change in the entropy $\Delta S$:

$$\Delta S = \left(\frac{Q}{T}\right)_R \tag{15.18}$$

In this expression the temperature $T$ must be in kelvins, and the subscript R refers to the word "reversible." It can be shown that Equation 15.18 applies to any process in which heat $Q$ enters or leaves a system reversibly at a constant temperature. Such is the case for the heat that flows into and out of the reservoirs of a Carnot engine. Equation 15.18 indicates that the SI unit for entropy is a joule per kelvin (J/K).

Entropy, like internal energy, is a function of the state or condition of the system. Only the state of a system determines the entropy $S$ that a system has. Therefore, the change in entropy $\Delta S$ is equal to the entropy of the final state of the system minus the entropy of the initial state.

We can now describe what happens to the entropy of a Carnot engine. As the engine operates, the entropy of the hot reservoir decreases, since heat $Q_H$ departs at a Kelvin temperature $T_H$. The change in the entropy of the hot reservoir is $\Delta S_H = -Q_H/T_H$, where the minus sign is needed to indicate a decrease in entropy, since the symbol $Q_H$ denotes only the magnitude of the heat. In contrast, the entropy of the cold reservoir increases by an amount $\Delta S_C = +Q_C/T_C$, for the rejected heat enters the cold reservoir at a Kelvin temperature $T_C$. The total change in entropy is

$$\Delta S_C + \Delta S_H = \frac{Q_C}{T_C} - \frac{Q_H}{T_H} = 0$$

because $Q_C/T_C = Q_H/T_H$ according to Equation 15.14.

The fact that the total change in entropy is zero for a Carnot engine is a specific illustration of a general result. It can be proved that when *any* reversible process occurs, the change in the entropy of the universe is zero; $\Delta S_{universe} = 0$ J/K for a reversible process. The word "universe" means that $\Delta S_{universe}$ takes into account the entropy changes of all parts of the system and all parts of the environment. ***Reversible processes do not alter the total entropy of the universe.*** To be sure, the entropy of one part of the universe may change because of a reversible process, but if so, the entropy of another part changes in the opposite way by the same amount.

What happens to the entropy of the universe when an *irreversible* process occurs is more complex, because the expression $\Delta S = (Q/T)_R$ does not apply directly. However, if a system changes irreversibly from an initial state to a final state, this expression can be used to calculate $\Delta S$ indirectly, as Figure 15.17 indicates. We imagine a hypothetical reversible process that causes the system to change between *the same initial and final states* and then find $\Delta S$ for this reversible process. The value obtained for $\Delta S$ also applies to the irreversible process that actually occurs, since only the nature of the initial and final states, and not the path between them, determines $\Delta S$. Example 11 illustrates this indirect method and shows that spontaneous (irreversible) processes increase the entropy of the universe.

## Example 11　The Entropy of the Universe Increases

Figure 15.18 shows 1200 J of heat flowing spontaneously through a copper rod from a hot reservoir at 650 K to a cold reservoir at 350 K. Determine the amount by which this irreversible process changes the entropy of the universe, assuming that no other changes occur.

**Reasoning** The hot-to-cold heat flow is irreversible, so the relation $\Delta S = (Q/T)_R$ is applied to a hypothetical process whereby the 1200 J of heat is taken reversibly from the hot reservoir and added reversibly to the cold reservoir.

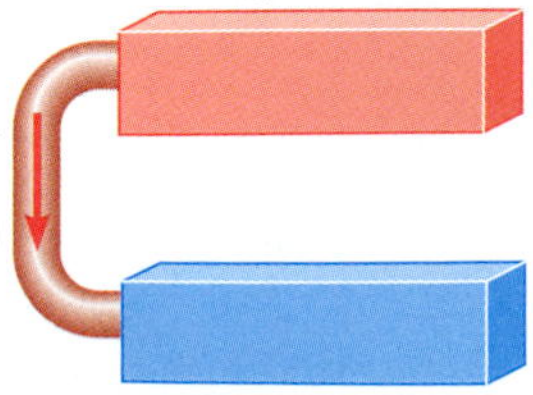

**Figure 15.18** Heat flows spontaneously from a hot reservoir to a cold reservoir.

**Solution** The total entropy change of the universe is the algebraic sum of the entropy changes for each reservoir:

$$\Delta S_{\text{universe}} = -\frac{1200 \text{ J}}{650 \text{ K}} + \frac{1200 \text{ J}}{350 \text{ K}} = \boxed{+1.6 \text{ J/K}}$$

Entropy lost by hot reservoir    Entropy gained by cold reservoir

The irreversible process causes the entropy of the universe to increase by 1.6 J/K.

Example 11 is a specific illustration of a general result: ***Any irreversible process increases the entropy of the universe.*** In other words, $\Delta S_{\text{universe}} > 0$ J/K for an irreversible process. Reversible processes do not alter the entropy of the universe, whereas irreversible processes cause the entropy to increase. Therefore, the entropy of the universe continually increases, like time itself, and entropy is sometimes called "time's arrow." It can be shown that this behavior of the entropy of the universe provides a completely general statement of the second law of thermodynamics, which applies not only to heat flow but also to all kinds of other processes.

■ **THE SECOND LAW OF THERMODYNAMICS STATED IN TERMS OF ENTROPY**

The total entropy of the universe does not change when a reversible process occurs ($\Delta S_{\text{universe}} = 0$ J/K) and does increase when an irreversible process occurs ($\Delta S_{\text{universe}} > 0$ J/K).

When an irreversible process occurs and the entropy of the universe increases, the energy available for doing work decreases, as the next example illustrates.

## Example 12   Energy Unavailable for Doing Work

Suppose that 1200 J of heat is used as input for an engine under two different conditions. In Figure 15.19a the heat is supplied by a hot reservoir whose temperature is 650 K. In part b of the drawing, the heat flows irreversibly through a copper rod into a second reservoir whose temperature is 350 K and then enters the engine. In either case, a 150-K reservoir is used as the cold reservoir. For each case, determine the maximum amount of work that can be obtained from the 1200 J of heat.

**Reasoning** According to Equation 15.11, the work $W$ obtained from the engine is the product of its efficiency $e$ and the input heat $Q_H$: $W = eQ_H = e(1200 \text{ J})$. The maximum amount of work is obtained when the efficiency is a maximum—that is, when the engine is a Carnot engine. The efficiency of a Carnot engine is given by Equation 15.15 as $e_{\text{Carnot}} = 1 - T_C/T_H$. Therefore, the efficiency may be determined from the Kelvin temperatures of the hot and cold reservoirs.

**Solution**

*Before irreversible heat flow*

$$e_{\text{Carnot}} = 1 - \frac{T_C}{T_H} = 1 - \frac{150 \text{ K}}{650 \text{ K}} = 0.77$$

$$W = (e_{\text{Carnot}})(1200 \text{ J}) = (0.77)(1200 \text{ J}) = \boxed{920 \text{ J}}$$

*After irreversible heat flow*

$$e_{\text{Carnot}} = 1 - \frac{T_C}{T_H} = 1 - \frac{150 \text{ K}}{350 \text{ K}} = 0.57$$

$$W = (e_{\text{Carnot}})(1200 \text{ J}) = (0.57)(1200 \text{ J}) = \boxed{680 \text{ J}}$$

When the 1200 J of input heat is taken from the 350-K reservoir instead of the 650-K reservoir, the efficiency of the Carnot engine is smaller. As a result, less work (680 J versus 920 J) can be extracted from the input heat.

Example 12 shows that 240 J less work (920 J − 680 J) can be performed when the input heat is obtained from the hot reservoir with the smaller temperature. In other words, the irreversible process of heat flow through the copper rod causes energy to become unavailable

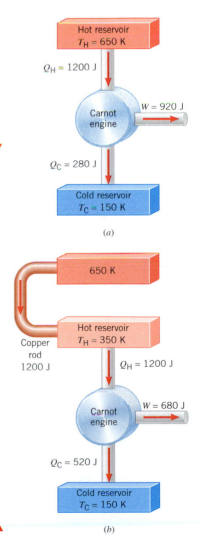

**Figure 15.19** Heat in the amount of $Q_H = 1200$ J is used as input for an engine under two different conditions in parts a and b.

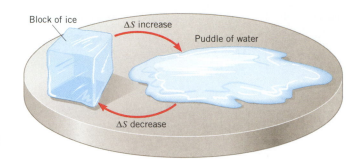

**Figure 15.20** A block of ice is an example of an ordered system relative to a puddle of water.

for doing work in the amount of $W_{\text{unavailable}} = 240$ J. Example 11 shows that this irreversible process also causes the entropy of the universe to increase by an amount $\Delta S_{\text{universe}} = +1.6$ J/K. These values for $W_{\text{unavailable}}$ and $\Delta S_{\text{universe}}$ are in fact related. If you multiply $\Delta S_{\text{universe}}$ by 150 K, which is the lowest Kelvin temperature in Example 12, you obtain $W_{\text{unavailable}} = (150 \text{ K}) \times (1.6 \text{ J/K}) = 240$ J. This is one illustration of the following general result:

$$W_{\text{unavailable}} = T_0 \Delta S_{\text{universe}} \tag{15.19}$$

where $T_0$ is the Kelvin temperature of the coldest heat reservoir. Since irreversible processes cause the entropy of the universe to increase, they cause energy to be degraded, in the sense that part of the energy becomes unavailable for the performance of work. In contrast, there is no penalty when reversible processes occur, because for them $\Delta S_{\text{universe}} = 0$ J/K, and there is no loss of work.

Entropy can also be interpreted in terms of order and disorder. As an example, consider a block of ice (Figure 15.20) with each of its $H_2O$ molecules fixed rigidly in place in a highly structured and ordered arrangement. In comparison, the puddle of water into which the ice melts is disordered and unorganized, because the molecules in a liquid are free to move from place to place. Heat is required to melt the ice and produce the disorder. Moreover, heat flow into a system increases the entropy of the system, according to $\Delta S = (Q/T)_{\text{R}}$. We associate an increase in entropy, then, with an increase in disorder. Conversely, we associate a decrease in entropy with a decrease in disorder or a greater degree of order. Example 13 illustrates an order-to-disorder change and the increase of entropy that accompanies it.

### Example 13 Order to Disorder

Find the change in entropy that results when a 2.3-kg block of ice melts slowly (reversibly) at 273 K (0 °C).

**Reasoning** Since the phase change occurs reversibly at a constant temperature, the change in entropy can be found by using Equation 15.18, $\Delta S = (Q/T)_{\text{R}}$, where $Q$ is the heat absorbed by the melting ice. This heat can be determined by using the relation $Q = mL_f$ (Equation 12.5), where $m$ is the mass and $L_f = 3.35 \times 10^5$ J/kg is the latent heat of fusion of water.

**Solution** Using Equation 15.18 and Equation 12.5, we find that the change in entropy is

$$\Delta S = \left(\frac{Q}{T}\right)_{\text{R}} = \frac{mL_f}{T} = \frac{(2.3 \text{ kg})(3.35 \times 10^5 \text{ J/kg})}{273 \text{ K}} = \boxed{+2.8 \times 10^3 \text{ J/K}}$$

a result that is positive, since the ice absorbs heat as it melts.

Figure 15.21 shows another order-to-disorder change that can be described in terms of entropy.

**Figure 15.21** With the aid of 100 pounds of dynamite, demolition experts caused this hotel-casino in Las Vegas to go from an ordered state (lower entropy) to a disordered state (higher entropy). (© Reuters/Aaron Mayes/Archive Photos/Getty Images)

## 15.12 The Third Law of Thermodynamics

To the zeroth, first, and second laws of thermodynamics we add the third (and last) law. The *third law of thermodynamics* indicates that it is impossible to reach a temperature of absolute zero.

■ **THE THIRD LAW OF THERMODYNAMICS**

It is not possible to lower the temperature of any system to absolute zero ($T = 0$ K) in a finite number of steps.

This law, like the second law, can be expressed in a number of ways, but a discussion of them is beyond the scope of this text. The third law is needed to explain a number of experimental observations that cannot be explained by the other laws of thermodynamics.

# Concept Summary

This summary presents an abridged version of the chapter, including the important equations and all available learning aids. For convenient reference, the learning aids (including the text's examples) are placed next to or immediately after the relevant equation or discussion. The following learning aids may be found on-line at **www.wiley.com/college/cutnell**:

| | |
|---|---|
| **Interactive LearningWare** examples are solved according to a five-step interactive format that is designed to help you develop problem-solving skills. | **Concept Simulations** are animated versions of text figures or animations that illustrate important concepts. You can control parameters that affect the display, and we encourage you to experiment. |
| **Interactive Solutions** offer specific models for certain types of problems in the chapter homework. The calculations are carried out interactively. | **Self-Assessment Tests** include both qualitative and quantitative questions. Extensive feedback is provided for both incorrect and correct answers, to help you evaluate your understanding of the material. |

| *Topic* | *Discussion* | *Learning Aids* |
|---|---|---|
| | **15.1  Thermodynamic Systems and Their Surroundings** | |
| | A thermodynamic system is the collection of objects on which attention is being focused, and the surroundings are everything else in the environment. The state of a system is the physical condition of the system, as described by values for physical parameters, often pressure, volume, and temperature. | |
| | **15.2  The Zeroth Law of Thermodynamics** | |
| | Two systems are in thermal equilibrium if there is no net flow of heat between them when they are brought into thermal contact. | |
| **Thermal equilibrium and temperature** | Temperature is the indicator of thermal equilibrium in the sense that there is no net flow of heat between two systems in thermal contact that have the same temperature. | |
| **Zeroth law of thermodynamics** | The zeroth law of thermodynamics states that two systems individually in thermal equilibrium with a third system are in thermal equilibrium with each other. | |
| | **15.3  The First Law of Thermodynamics** | |
| | The first law of thermodynamics states that due to heat $Q$ and work $W$, the internal energy of a system changes from its initial value of $U_i$ to a final value of $U_f$ according to | |
| | $$\Delta U = U_f - U_i = Q - W \qquad (15.1)$$ | **Example 1** |
| **Sign convention for Q and W** | $Q$ is positive when the system gains heat and negative when it loses heat. $W$ is positive when work is done by the system and negative when work is done on the system. | |
| | The first law of thermodynamics is the conservation-of-energy principle applied to heat, work, and the change in the internal energy. | |
| **Function of state** | The internal energy is called a function of state because it depends only on the state of the system and not on the method by which the system came to be in a given state. | **Example 2** |
| | **15.4  Thermal Processes** | |
| **Quasi-static process** | A thermal process is quasi-static when it occurs slowly enough that a uniform pressure and temperature exist throughout the system at all times. | |

| Topic | Discussion | Learning Aids |
|---|---|---|
| **Isobaric process** | An isobaric process is one that occurs at constant pressure. The work $W$ done when a system changes at a constant pressure $P$ from an initial volume $V_i$ to a final volume $V_f$ is<br><br>$$W = P\,\Delta V = P(V_f - V_i) \qquad (15.2)$$ | **Example 3** |
| **Isochoric process** | An isochoric process is one that takes place at constant volume, and no work is done in such a process. | |
| **Isothermal process** | An isothermal process is one that takes place at constant temperature. | |
| **Adiabatic process** | An adiabatic process is one that takes place without the transfer of heat. | |
| **Work done as the area under a pressure–volume graph** | The work done in any kind of quasi-static process is given by the area under the corresponding pressure-versus-volume graph. | **Example 4** |

### 15.5 Thermal Processes Using an Ideal Gas

When $n$ moles of an ideal gas change quasi-statically from an initial volume $V_i$ to a final volume $V_f$ at a constant Kelvin temperature $T$, the work done is

| | | |
|---|---|---|
| **Work done during an isothermal process** | $$W = nRT \ln\left(\frac{V_f}{V_i}\right) \qquad (15.3)$$ | **Example 5**<br><br>*Interactive LearningWare 15.1* |

When $n$ moles of a monatomic ideal gas change quasi-statically and adiabatically from an initial temperature $T_i$ to a final temperature $T_f$, the work done is

| **Work done during an adiabatic process** | $$W = \tfrac{3}{2}nR(T_i - T_f) \qquad (15.4)$$ | |
|---|---|---|

During an adiabatic process, and in addition to the ideal gas law, an ideal gas obeys the relation

| **Adiabatic change in pressure and volume** | $$P_i V_i^{\gamma} = P_f V_f^{\gamma} \qquad (15.5)$$ | *Interactive Solution 15.27* |
|---|---|---|

where $\gamma = c_P/c_V$ is the ratio of the specific heat capacities at constant pressure and constant volume.

### 15.6 Specific Heat Capacities

The molar specific heat capacity $C$ of a substance determines how much heat $Q$ is added or removed when the temperature of $n$ moles of the substance changes by an amount $\Delta T$:

| | $$Q = Cn\,\Delta T \qquad (15.6)$$ | *Interactive Solution 15.85* |
|---|---|---|

For a monatomic ideal gas, the molar specific heat capacities at constant pressure and constant volume are, respectively,

| **Specific heat capacities of a monatomic ideal gas** | $$C_P = \tfrac{5}{2}R \qquad (15.7)$$<br>$$C_V = \tfrac{3}{2}R \qquad (15.8)$$ | |
|---|---|---|

where $R$ is the ideal gas constant. For any type of ideal gas, the difference between $C_P$ and $C_V$ is

$$C_P - C_V = R \qquad (15.10)$$

 **Use Self-Assessment Test 15.1 to evaluate your understanding of Sections 15.1–15.6.**

### 15.7 The Second Law of Thermodynamics

The second law of thermodynamics can be stated in a number of equivalent forms. In terms of heat flow, the second law declares that heat flows spontaneously from a substance at a higher temperature to a substance at a lower temperature and does not flow spontaneously in the reverse direction.

### 15.8 Heat Engines

A heat engine produces work $W$ from input heat $Q_H$ that is extracted from a heat reservoir at a relatively high temperature. The engine rejects heat $Q_C$ into a reservoir at a relatively low temperature.

The efficiency $e$ of a heat engine is

| **Efficiency of a heat engine** | $$e = \frac{\text{Work done}}{\text{Input heat}} = \frac{W}{Q_H} \qquad (15.11)$$ |
|---|---|

| Topic | Discussion | Learning Aids |
|---|---|---|

**Conservation of energy for a heat engine**

The conservation of energy requires that the input heat of magnitude $Q_H$ must be equal to the work $W$ done by the engine plus the heat of magnitude $Q_C$ rejected to the cold reservoir:

$$Q_H = W + Q_C \qquad (15.12)$$

**Example 6**

By combining Equation 15.12 with Equation 15.11, the efficiency of a heat engine can also be written as

$$e = 1 - \frac{Q_C}{Q_H} \qquad (15.13)$$

### 15.9 Carnot's Principle and the Carnot Engine

**Reversible process**

A reversible process is one in which both the system and its environment can be returned to exactly the states they were in before the process occurred.

**Carnot's principle**

Carnot's principle is an alternative statement of the second law of thermodynamics. It states that no irreversible engine operating between two reservoirs at constant temperatures can have a greater efficiency than a reversible engine operating between the same temperatures. Furthermore, all reversible engines operating between the same temperatures have the same efficiency.

**A Carnot engine**

A Carnot engine is a reversible engine in which all input heat $Q_H$ originates from a hot reservoir at a single Kelvin temperature $T_H$ and all rejected heat $Q_C$ goes into a cold reservoir at a single Kelvin temperature $T_C$. For a Carnot engine

$$\frac{Q_C}{Q_H} = \frac{T_C}{T_H} \qquad (15.14)$$

The efficiency $e_{Carnot}$ of a Carnot engine is the maximum efficiency that an engine operating between two fixed temperatures can have:

**Examples 7, 8**

**Efficiency of a Carnot engine**

$$e_{Carnot} = 1 - \frac{T_C}{T_H} \qquad (15.15)$$

**Concept Simulation 15.1**

### 15.10 Refrigerators, Air Conditioners, and Heat Pumps

Refrigerators, air conditioners, and heat pumps are devices that utilize work $W$ to make heat of magnitude $Q_C$ flow from a lower Kelvin temperature $T_C$ to a higher Kelvin temperature $T_H$. In the process (the refrigeration process) they deposit heat of magnitude $Q_H$ at the higher temperature. The principle of the conservation of energy requires that $Q_H = W + Q_C$.

**Example 9**
**Interactive LearningWare 15.2**

If the refrigeration process is ideal, in the sense that it occurs reversibly, the devices are called Carnot devices and the relation $Q_C/Q_H = T_C/T_H$ (Equation 15.14) holds.

**Example 10**

The coefficient of performance of a refrigerator or an air conditioner is

**Coefficient of performance (refrigerator or air conditioner)**

$$\frac{\text{Coefficient of}}{\text{performance}} = \frac{Q_C}{W} \qquad (15.16)$$

The coefficient of performance of a heat pump is

**Coefficient of performance (heat pump)**

$$\frac{\text{Coefficient of}}{\text{performance}} = \frac{Q_H}{W} \qquad (15.17)$$

### 15.11 Entropy

The change in entropy $\Delta S$ for a process in which heat $Q$ enters or leaves a system reversibly at a constant Kelvin temperature $T$ is

**Change in entropy**

$$\Delta S = \left(\frac{Q}{T}\right)_R \qquad (15.18)$$

**Examples 11, 13**

where the subscript R stands for "reversible."

**The second law of thermodynamics (entropy statement)**

The second law of thermodynamics can be stated in a number of equivalent forms. In terms of entropy, the second law states that the total entropy of the universe does not change when a reversible process occurs ($\Delta S_{universe} = 0$ J/K) and increases when an irreversible process occurs ($\Delta S_{universe} > 0$ J/K).

**Interactive Solution 15.73**

| Topic | Discussion | Learning Aids |
|-------|-----------|---------------|
| | Irreversible processes cause energy to be degraded in the sense that part of the energy becomes unavailable for the performance of work. The energy $W_{unavailable}$ that is unavailable for doing work because of an irreversible process is | **Example 12** |
| **Unavailable work** | $$W_{unavailable} = T_0 \, \Delta S_{universe} \qquad (15.19)$$ where $\Delta S_{universe}$ is the total entropy change of the universe and $T_0$ is the Kelvin temperature of the coldest reservoir into which heat can be rejected. | |
| **Entropy and disorder** | Increased entropy is associated with a greater degree of disorder and decreased entropy with a lesser degree of disorder (more order). | |

Use **Self-Assessment Test 15.2** to evaluate your understanding of Sections **15.7–15.11.**

### 15.12 The Third Law of Thermodynamics

The third law of thermodynamics states that it is not possible to lower the temperature of any system to absolute zero ($T = 0$ K) in a finite number of steps.

# Problems

**ssm** Solution is in the Student Solutions Manual. **www** Solution is available on the World Wide Web at www.wiley.com/college/cutnell
This icon represents a biomedical application.

### Section 15.3 The First Law of Thermodynamics

**1. ssm** The internal energy of a system changes because the system gains 165 J of heat and performs 312 J of work. In returning to its initial state, the system loses 114 J of heat. During this return process, (a) how much work is involved, and (b) is work done by the system or is work done on the system?

**2.** In moving out of a dormitory at the end of the semester, a student does $1.6 \times 10^4$ J of work. In the process, his internal energy decreases by $4.2 \times 10^4$ J. Determine each of the following quantities (including the algebraic sign): (a) $W$, (b) $\Delta U$, and (c) $Q$.

**3.** A system does 164 J of work on its environment and gains 77 J of heat in the process. Find the change in the internal energy of (a) the system and (b) the environment.

**4.** Three moles of an ideal monatomic gas are at a temperature of 345 K. Then, 2438 J of heat are added to the gas, and 962 J of work are done on it. What is the final temperature of the gas?

**5. ssm** When one gallon of gasoline is burned in a car engine, $1.19 \times 10^8$ J of internal energy is released. Suppose that $1.00 \times 10^8$ J of this energy flows directly into the surroundings (engine block and exhaust system) in the form of heat. If $6.0 \times 10^5$ J of work is required to make the car go one mile, how many miles can the car travel on one gallon of gas?

**\*6.** In exercising, a weight lifter loses 0.150 kg of water through evaporation, the heat required to evaporate the water coming from the weight lifter's body. The work done in lifting weights is $1.40 \times 10^5$ J. (a) Assuming that the latent heat of vaporization of perspiration is $2.42 \times 10^6$ J/kg, find the change in the internal energy of the weight lifter. (b) Determine the minimum number of nutritional Calories of food (1 nutritional Calorie = 4186 J) that must be consumed to replace the loss of internal energy.

### Section 15.4 Thermal Processes

**7.** The specific heat capacity of a material is 1100 J/(kg·C°). The temperature of 2.0 kg of this solid material is raised by 6.0 C°. Ig-

noring the work that corresponds to the small change in the volume of the material, determine the change in the internal energy of the material.

**8.** A gas, while expanding under isobaric conditions, does 480 J of work. The pressure of the gas is $1.6 \times 10^5$ Pa, and its initial volume is $1.5 \times 10^{-3}$ m³. What is the final volume of the gas?

**9. ssm** The volume of a gas is changed along the curved line between $A$ and $B$ in the drawing. Do not assume that the curved line is an isotherm or that the gas is ideal. (a) Find the magnitude of the work for the process, and (b) determine whether the work is positive or negative.

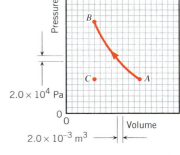

**10.** The pressure and volume of a gas are changed along the path $ABCA$. Determine the work done (including the algebraic sign) in each segment of the path: (a) $A$ to $B$, (b) $B$ to $C$, and (c) $C$ to $A$.

**11.** A gas is contained in a chamber such as that in Figure 15.4. Suppose the region outside the chamber is evacuated and the total mass of the block and the movable piston is 135 kg. When 2050 J of heat flows into the gas, the internal energy of the gas increases by 1730 J. What is the distance $s$ through which the piston rises?

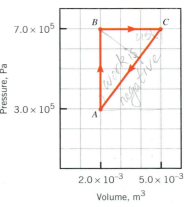

*Problem 10*

**12.** (a) Using the data presented in the accompanying pressure-versus-volume graph, estimate the magnitude of the work done when the system changes from $A$ to $B$ to $C$ along the path shown. (b) Determine whether the work is done by the system or on the system and, hence, whether the work is positive or negative.

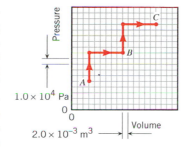

**13.** **ssm** A system gains 1500 J of heat, while the internal energy of the system increases by 4500 J and the volume decreases by 0.010 m$^3$. Assume the pressure is constant and find its value.

* **14.** Refer to the drawing in problem 9, where the curve between $A$ and $B$ is now an isotherm. An ideal gas begins at $A$ and is changed along the horizontal line from $A$ to $C$ and then along the vertical line from $C$ to $B$. (a) Find the heat for the process $ACB$ and (b) determine whether it flows into or out of the gas.

* **15.** **ssm** **www** A monatomic ideal gas expands isobarically. Using the first law of thermodynamics, prove that the heat $Q$ is positive, so that it is impossible for heat to flow out of the gas.

* **16.** Refer to the drawing that accompanies problem 12. When a system changes from $A$ to $B$ along the path shown on the pressure-versus-volume graph, it gains 2700 J of heat. What is the change in the internal energy of the system?

** **17.** Water is heated in an open pan where the air pressure is one atmosphere. The water remains a liquid, which expands by a small amount as it is heated. Determine the ratio of the work done by the water to the heat absorbed by the water.

**Section 15.5 Thermal Processes Using an Ideal Gas**

**18.** The temperature of a monatomic ideal gas remains constant during a process in which 4700 J of heat flows out of the gas. How much work (including the proper + or − sign) is done?

**19.** **ssm** Three moles of an ideal gas are compressed from $5.5 \times 10^{-2}$ to $2.5 \times 10^{-2}$ m$^3$. During the compression, $6.1 \times 10^3$ J of work is done on the gas, and heat is removed to keep the temperature of the gas constant at all times. Find (a) $\Delta U$, (b) $Q$, and (c) the temperature of the gas.

**20.** Five moles of a monatomic ideal gas expand adiabatically, and its temperature decreases from 370 to 290 K. Determine (a) the work done (including the algebraic sign) by the gas, and (b) the change in its internal energy.

**21.** The pressure of a monatomic ideal gas ($\gamma = \frac{5}{3}$) doubles during an adiabatic compression. What is the ratio of the final volume to the initial volume?

**22.** One-half mole of a monatomic ideal gas expands adiabatically and does 610 J of work. By how many kelvins does its temperature change? Specify whether the change is an increase or a decrease.

**23.** **ssm** A monatomic ideal gas has an initial temperature of 405 K. This gas expands and does the same amount of work whether the expansion is adiabatic or isothermal. When the expansion is adiabatic, the final temperature of the gas is 245 K. What is the ratio of the final to the initial volume when the expansion is isothermal?

* **24.** The drawing at the top of the right column refers to one mole of a monatomic ideal gas and shows a process that has four steps, two isobaric ($A$ to $B$, $C$ to $D$) and two isochoric ($B$ to $C$, $D$ to $A$). Complete the following table by calculating $\Delta U$, $W$, and $Q$ (including the algebraic signs) for each of the four steps.

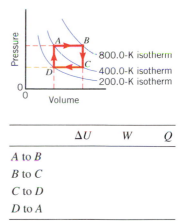

| | $\Delta U$ | $W$ | $Q$ |
|---|---|---|---|
| $A$ to $B$ | | | |
| $B$ to $C$ | | | |
| $C$ to $D$ | | | |
| $D$ to $A$ | | | |

* **25.** The pressure and volume of an ideal monatomic gas change from $A$ to $B$ to $C$, as the drawing shows. The curved line between $A$ and $C$ is an isotherm. (a) Determine the total heat for the process and (b) state whether the flow of heat is into or out of the gas.

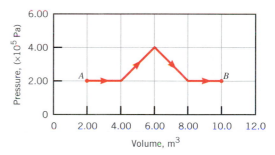

* **26.** A monatomic ideal gas expands from point $A$ to point $B$ along the path shown in the drawing. (a) Determine the work done by the gas. (b) The temperature of the gas at point $A$ is 185 K. What is its temperature at point $B$? (c) How much heat has been added to or removed from the gas during the process?

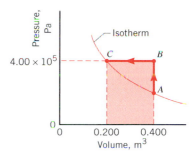

* **27.** Refer to **Interactive Solution 15.27** at **www.wiley.com/college/cutnell** for help in solving this problem. A diesel engine does not use spark plugs to ignite the fuel and air in the cylinders. Instead, the temperature required to ignite the fuel occurs because the pistons compress the air in the cylinders. Suppose air at an initial temperature of 21 °C is compressed adiabatically to a temperature of 688 °C. Assume the air to be an ideal gas for which $\gamma = \frac{7}{5}$. Find the compression ratio, which is the ratio of the initial volume to the final volume.

** **28.** One mole of a monatomic ideal gas has an initial pressure, volume, and temperature of $P_0$, $V_0$, and 438 K, respectively. It undergoes an isothermal expansion that triples the volume of the gas. Then, the gas undergoes an isobaric compression back to its original volume. Finally, the gas undergoes an isochoric increase in pressure, so that the final pressure, volume, and temperature are $P_0$, $V_0$, and 438 K, respectively. Find the total heat for this three-step process, and state whether it is absorbed by or given off by the gas.

** **29.** **ssm** The work done by one mole of a monatomic ideal gas ($\gamma = \frac{5}{3}$) in expanding adiabatically is 825 J. The initial temperature and volume of the gas are 393 K and 0.100 m$^3$. Obtain (a) the final temperature and (b) the final volume of the gas.

### Section 15.6 Specific Heat Capacities

**30.** Heat is added to two identical samples of a monatomic ideal gas. In the first sample the heat is added while the volume of the gas is kept constant, and the heat causes the temperature to rise by 75 K. In the second sample, an identical amount of heat is added while the pressure (but not the volume) of the gas is kept constant. By how much does the temperature of this sample increase?

**31. ssm** How much heat is required to change the temperature of 1.5 mol of a monatomic ideal gas by 77 K if the pressure is held constant?

**32.** Argon is a monatomic gas whose atomic mass is 39.9 u. The temperature of eight grams of argon is raised by 75 K under conditions of constant pressure. Assuming that argon is an ideal gas, how much heat is required?

**33. ssm** The temperature of 2.5 mol of a monatomic ideal gas is 350 K. The internal energy of this gas is doubled by the addition of heat. How much heat is needed when it is added at (a) constant volume and (b) constant pressure?

**34.** The temperature of 2.5 mol of helium (a monatomic gas) is lowered by 35 K under conditions of constant volume. Assuming that helium behaves as an ideal gas, how much heat is removed from the gas?

**35.** Heat $Q$ is added to a monatomic ideal gas at constant pressure. As a result, the gas does work $W$. Find the ratio $Q/W$.

* **36.** ⚕ Even at rest, the human body generates heat. The heat arises because of the body's metabolism—that is, the chemical reactions that are always occurring in the body to generate energy. In rooms designed for use by large groups, adequate ventilation or air conditioning must be provided to remove this heat. Consider a classroom containing 200 students. Assume that the metabolic rate of generating heat is 130 W for each student and that the heat accumulates during a fifty-minute lecture. In addition, assume that the air has a molar specific heat of $C_V = \frac{5}{2}R$ and that the room (volume = 1200 m³, initial pressure = $1.01 \times 10^5$ Pa, and initial temperature = 21 °C) is sealed shut. If all the heat generated by the students were absorbed by the air, by how much would the air temperature rise during a lecture?

* **37. ssm** A monatomic ideal gas expands at constant pressure. (a) What percentage of the heat being supplied to the gas is used to increase the internal energy of the gas? (b) What percentage is used for doing the work of expansion?

* **38.** Suppose a monatomic ideal gas is contained within a vertical cylinder that is fitted with a movable piston. The piston is frictionless and has a negligible mass. The area of the piston is $3.14 \times 10^{-2}$ m², and the pressure outside the cylinder is $1.01 \times 10^5$ Pa. Heat (2093 J) is removed from the gas. Through what distance does the piston drop?

** **39.** One mole of neon, a monatomic gas, starts out at conditions of standard temperature and pressure. The gas is heated at constant volume until its pressure is tripled, then further heated at constant pressure until its volume is doubled. Assume that neon behaves as an ideal gas. For the entire process, find the heat added to the gas.

### Section 15.8 Heat Engines

**40.** Due to a tune-up, the efficiency of an automobile engine increases by 5.0%. For an input heat of 1300 J, how much more work does the engine produce after the tune-up than before?

**41. ssm** In doing 16 600 J of work, an engine rejects 9700 J of heat. What is the efficiency of the engine?

**42.** Engine A discards 72% of its input heat into a cold reservoir. Engine B has twice the efficiency of engine A. What percentage of its input heat does engine B discard?

**43.** An engine has an efficiency of 64% and produces 5500 J of work. Determine (a) the input heat and (b) the rejected heat.

* **44.** Due to design changes, the efficiency of an engine increases from 0.23 to 0.42. For the same input heat $Q_H$, these changes increase the work done by the more efficient engine and reduce the amount of heat rejected to the cold reservoir. Find the ratio of the heat rejected to the cold reservoir for the improved engine to that for the original engine.

** **45. ssm** An engine has an efficiency $e_1$. The engine takes input heat $Q_H$ from a hot reservoir and delivers work $W_1$. The heat rejected by this engine is used as input heat for a second engine, which has an efficiency $e_2$ and delivers work $W_2$. The overall efficiency of this two-engine device is the total work delivered ($W_1 + W_2$) divided by the input heat $Q_H$. Find an expression for the overall efficiency $e$ in terms of $e_1$ and $e_2$.

### Section 15.9 Carnot's Principle and the Carnot Engine

**46.** Five thousand joules of heat is put into a Carnot engine whose hot and cold reservoirs have temperatures of 500 and 200 K, respectively. How much heat is converted into work?

**47.** An engine has a hot reservoir temperature of 950 K and a cold reservoir temperature of 620 K. The engine operates at three-fifths maximum efficiency. What is the efficiency of the engine?

**48.** A Carnot engine operates with an efficiency of 27.0% when the temperature of its cold reservoir is 275 K. Assuming that the temperature of the hot reservoir remains the same, what must be the temperature of the cold reservoir in order to increase the efficiency to 32.0%?

**49. ssm** A Carnot engine has an efficiency of 0.700, and the temperature of its cold reservoir is 378 K. (a) Determine the temperature of its hot reservoir. (b) If 5230 J of heat is rejected to the cold reservoir, what amount of heat is put into the engine?

**50.** An engine does 18 500 J of work and rejects 6550 J of heat into a cold reservoir whose temperature is 285 K. What would be the smallest possible temperature of the hot reservoir?

**51. Concept Simulation 15.1** at **www.wiley.com/college/cutnell** illustrates the concepts pertinent to this problem. A Carnot engine operates between temperatures of 650 and 350 K. To improve the efficiency of the engine, it is decided either to raise the temperature of the hot reservoir by 40 K or to lower the temperature of the cold reservoir by 40 K. Which change gives the greatest improvement? Justify your answer by calculating the efficiency in each case.

* **52.** From a hot reservoir at a temperature of $T_1$, Carnot engine A takes an input heat of 5550 J, delivers 1750 J of work, and rejects heat to a cold reservoir that has a temperature of 503 K. This cold reservoir at 503 K also serves as the hot reservoir for Carnot engine B, which uses the rejected heat of the first engine as input heat. Engine B also delivers 1750 J of work, while rejecting heat to an even colder reservoir that has a temperature of $T_2$. Find the temperatures (a) $T_1$ and (b) $T_2$.

* **53. ssm** A power plant taps steam superheated by geothermal energy to 505 K (the temperature of the hot reservoir) and uses the steam to do work in turning the turbine of an electric generator. The steam is then converted back into water in a condenser at 323 K (the temperature of the cold reservoir), after which the water is pumped back down into the earth where it is heated again. The output power (work per unit time) of the plant is 84 000 kilowatts. Determine (a) the maximum efficiency at which this plant can operate and (b) the minimum amount of rejected heat that must be removed from the condenser every twenty-four hours. $e = 1 - \frac{323}{505} =$

$e = \frac{W}{Q_h} \Rightarrow Q_h = \frac{W}{e} = \frac{84000 \times 3600}{e}$

**54.** Suppose the gasoline in a car engine burns at 631 °C, while the exhaust temperature (the temperature of the cold reservoir) is 139 °C and the outdoor temperature is 27 °C. Assume that the engine can be treated as a Carnot engine (a gross oversimplification). In an attempt to increase mileage performance, an inventor builds a second engine that functions between the exhaust and outdoor temperatures and uses the exhaust heat to produce additional work. Assume that the inventor's engine can also be treated as a Carnot engine. Determine the ratio of the total work produced by both engines to that produced by the first engine alone.

**55. ssm** A nuclear-fueled electric power plant utilizes a so-called "boiling water reactor." In this type of reactor, nuclear energy causes water under pressure to boil at 285 °C (the temperature of the hot reservoir). After the steam does the work of turning the turbine of an electric generator, the steam is converted back into water in a condenser at 40 °C (the temperature of the cold reservoir). To keep the condenser at 40 °C, the rejected heat must be carried away by some means—for example, by water from a river. The plant operates at three-fourths of its Carnot efficiency, and the electrical output power of the plant is $1.2 \times 10^9$ watts. A river with a water flow rate of $1.0 \times 10^5$ kg/s is available to remove the rejected heat from the plant. Find the number of Celsius degrees by which the temperature of the river rises.

**56.** The hot and cold reservoirs of a Carnot engine have temperatures of 845 and 395 K, respectively. The engine does the work of lifting a 15.0-kg block straight up from rest, so that at a height of 5.00 m the block has a speed of 8.50 m/s. How much heat must be put into the engine?

**Section 15.10 Refrigerators, Air Conditioners, and Heat Pumps**

**57.** A Carnot refrigerator maintains the food inside it at 276 K, while the temperature of the kitchen is 298 K. The refrigerator removes $3.00 \times 10^4$ J of heat from the food. How much heat is delivered to the kitchen?

**58.** The inside of a Carnot refrigerator is maintained at a temperature of 277 K, while the temperature in the kitchen is 299 K. Using 2500 J of work, how much heat can this refrigerator remove from its inside compartment?

**59. ssm www** The temperatures indoors and outdoors are 299 and 312 K, respectively. A Carnot air conditioner deposits $6.12 \times 10^5$ J of heat outdoors. How much heat is removed from the house?

**60.** A refrigerator operates between temperatures of 296 and 275 K. What would be its maximum coefficient of performance?

**61.** A Carnot heat pump operates between an outdoor temperature of 265 K and an indoor temperature of 298 K. Find its coefficient of performance.

**62.** A Carnot air conditioner maintains the temperature in a house at 297 K on a day when the temperature outside is 311 K. What is the coefficient of performance of the air conditioner?

**63.** Review Conceptual Example 9 before attempting this problem. A window air conditioner has an average coefficient of performance of 2.0. This unit has been placed on the floor by the bed, in a futile attempt to cool the bedroom. During this attempt $7.6 \times 10^4$ J of heat is pulled in the front of the unit. The room is sealed and contains 3800 mol of air. Assuming that the molar specific heat capacity of the air is $C_V = \frac{5}{2}R$, determine the rise in temperature caused by operating the air conditioner in this manner.

**64.** Two kilograms of liquid water at 0 °C is put into the freezer compartment of a Carnot refrigerator. The temperature of the compartment is −15 °C, and the temperature of the kitchen is 27 °C. If

the cost of electrical energy is ten cents per kilowatt·hour, how much does it cost to make two kilograms of ice at 0 °C?

**65. ssm www** A Carnot refrigerator transfers heat from its inside (6.0 °C) to the room air outside (20.0 °C). (a) Find the coefficient of performance of the refrigerator. (b) Determine the magnitude of the minimum work needed to cool 5.00 kg of water from 20.0 to 6.0 °C when it is placed in the refrigerator.

**66.** An air conditioner keeps the inside of a house at a temperature of 19.0 °C when the outdoor temperature is 33.0 °C. Heat, leaking into the house at the rate of 10 500 joules per second, is removed by the air conditioner. Assuming that the air conditioner is a Carnot air conditioner, what is the work per second that must be done by the electrical energy in order to keep the inside temperature constant?

**67. ssm** A Carnot engine uses hot and cold reservoirs that have temperatures of 1684 and 842 K, respectively. The input heat for this engine is $Q_H$. The work delivered by the engine is used to operate a Carnot heat pump. The pump removes heat from the 842-K reservoir and puts it into a hot reservoir at a temperature $T'$. The amount of heat removed from the 842-K reservoir is also $Q_H$. Find the temperature $T'$.

**Section 15.11 Entropy**

**68.** Four kilograms of carbon dioxide sublimes from solid dry ice to a gas at a pressure of one atmosphere and a temperature of 194.7 K. The latent heat of sublimation is $5.77 \times 10^5$ J/kg. Find the change in entropy of the carbon dioxide.

**69.** On a cold day, 24 500 J of heat leaks out of a house. The inside temperature is 21 °C, and the outside temperature is −15 °C. What is the increase in the entropy of the universe that this heat loss produces?

**70.** Heat $Q$ flows spontaneously from a reservoir at 394 K into a reservoir that has a lower temperature $T$. Because of the spontaneous flow, thirty percent of $Q$ is rendered unavailable for work when a Carnot engine operates between the reservoir at temperature $T$ and a reservoir at 248 K. Find the temperature $T$.

**71. ssm** Find the change in entropy of the $H_2O$ molecules when (a) three kilograms of ice melts into water at 273 K and (b) three kilograms of water changes into steam at 373 K. (c) On the basis of the answers to parts (a) and (b), discuss which change creates more disorder in the collection of $H_2O$ molecules.

**72.** (a) Find the equilibrium temperature that results when one kilogram of liquid water at 373 K is added to two kilograms of liquid water at 283 K in a perfectly insulated container. (b) When heat is added to or removed from a solid or liquid of mass $m$ and specific heat capacity $c$, the change in entropy can be shown to be $\Delta S = mc \ln(T_f/T_i)$, where $T_i$ and $T_f$ are the initial and final Kelvin temperatures. Use this equation to calculate the entropy change for each amount of water. Then combine the two entropy changes algebraically to obtain the total entropy change of the universe. Note that the process is irreversible, so the total entropy change of the universe is greater than zero. (c) Assuming that the coldest reservoir at hand has a temperature of 273 K, determine the amount of energy that becomes unavailable for doing work because of the irreversible process.

**73.** Refer to **Interactive Solution 15.73** at **www.wiley.com/college/cutnell** to review a method by which this problem can be solved. (a) After 6.00 kg of water at 85.0 °C is mixed in a perfect thermos with 3.00 kg of ice at 0.0 °C, the mixture is allowed to reach equilibrium. Using the expression $\Delta S = mc \ln(T_f/T_i)$ [see problem 72] and the change in entropy for melting, find the change in entropy that occurs. (b) Should the entropy of the universe increase or decrease as a result of the mixing process? Give your reasoning and state whether your answer in part (a) is consistent with your answer here.

# Chapter 16  Waves and Sound

## 16.1  The Nature of Waves

Water waves have two features common to all waves:

1. A wave is a traveling disturbance.
2. A wave carries energy from place to place.

In Figure 16.1 the wave created by the motorboat travels across the lake and disturbs the fisherman. However, *there is no bulk flow of water* outward from the motorboat. The wave is not a bulk movement of water such as a river, but, rather, a disturbance traveling on the surface of the lake. Part of the wave's energy in Figure 16.1 is transferred to the fisherman and his boat.

We will consider two basic types of waves, transverse and longitudinal. Figure 16.2 illustrates how a transverse wave can be generated using a Slinky, a remarkable toy in the form of a long, loosely coiled spring. If one end of the Slinky is jerked up and down, as in part *a*, an upward pulse is sent traveling toward the right. If the end is then jerked down and up, as in part *b*, a downward pulse is generated and also moves to the right. If the end is continually moved up and down in simple harmonic motion, an entire wave is produced. As part *c* illustrates, the wave consists of a series of alternating upward and downward sections that propagate to the right, disturbing the vertical position of the Slinky in the process. To focus attention on the disturbance, a colored dot is attached to the Slinky in part *c* of the drawing. As the wave advances, the dot is displaced up and down in simple harmonic motion. The motion of the dot occurs perpendicular, or transverse, to the direction in which the wave travels. Thus, *a transverse wave is one in which the disturbance occurs perpendicular to the direction of travel of the wave.* Radio waves, light waves, and microwaves are transverse waves. Transverse waves also travel on the strings of instruments such as guitars and banjos.

A longitudinal wave can also be generated with a Slinky, and Figure 16.3 demonstrates how. When one end of the Slinky is pushed forward along its length (i.e., longitudinally) and then pulled back to its starting point, as in part *a*, a region where the coils are squeezed together or compressed is sent traveling to the right. If the end is pulled backward and then pushed forward to its starting point, as in part *b*, a region where the coils are pulled apart or stretched is formed and also moves to the right. If the end is continually moved back and forth in simple harmonic motion, an entire wave is created. As part *c* shows, the wave consists of a series of alternating compressed and stretched regions that travel to the right and disturb the separation between adjacent coils. A colored dot is once

**Figure 16.1** The wave created by the motorboat travels across the lake and disturbs the fisherman.

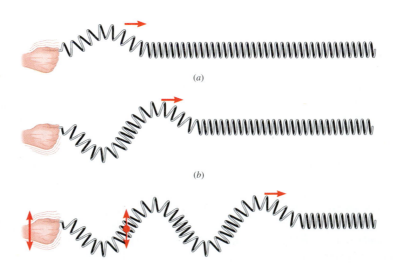

**Figure 16.2** (*a*) An upward pulse moves to the right, followed by (*b*) a downward pulse. (*c*) When the end of the Slinky is moved up and down continuously, a transverse wave is produced.

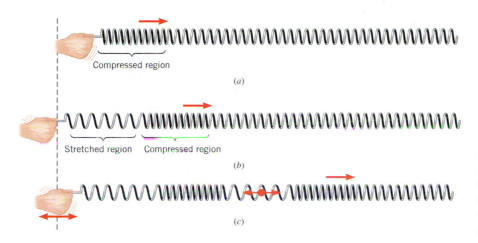

Compressed region

(a)

Stretched region    Compressed region

(b)

(c)

**Figure 16.3** (*a*) A compressed region moves to the right, followed by (*b*) a stretched region. (*c*) When the end of the Slinky is moved back and forth continuously, a longitudinal wave is produced.

again attached to the Slinky to emphasize the vibratory nature of the disturbance. In response to the wave, the dot moves back and forth in simple harmonic motion along the line of travel of the wave. Thus, *a longitudinal wave is one in which the disturbance occurs parallel to the line of travel of the wave.* A sound wave is a longitudinal wave.

Some waves are neither transverse nor longitudinal. For instance, in a water wave the motion of the water particles is not strictly perpendicular or strictly parallel to the line along which the wave travels. Instead, the motion includes both transverse and longitudinal components, since the water particles at the surface move on nearly circular paths, as Figure 16.4 indicates.

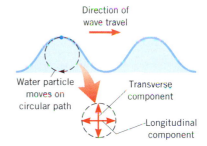

Direction of wave travel

Water particle moves on circular path

Transverse component

Longitudinal component

**Figure 16.4** A water wave is neither transverse nor longitudinal, since water particles at the surface move clockwise on nearly circular paths as the wave moves from left to right.

## 16.2 Periodic Waves

The transverse and longitudinal waves that we have been discussing are called *periodic waves* because they consist of cycles or patterns that are produced over and over again by the source. In Figures 16.2 and 16.3 the repetitive patterns occur as a result of the simple harmonic motion of the left end of the Slinky. Every segment of the Slinky vibrates in simple harmonic motion. Sections 10.1 and 10.2 discuss the simple harmonic motion of an object on a spring and introduce the concepts of cycle, amplitude, period, and frequency. This same terminology is used to describe periodic waves, such as the sound waves we hear and the light waves we see (discussed in Chapter 24).

Figure 16.5 uses a graphical representation of a transverse wave to review this terminology. One *cycle* of a wave is shaded in color in both parts of the drawing. A wave is a series of many cycles. In part *a* the vertical position of the Slinky is plotted on the vertical axis, and the corresponding distance along the length of the Slinky is plotted on the horizontal axis. Such a graph is equivalent to a photograph of the wave taken at one instant in time and shows the disturbance that exists at each point along the Slinky's length. As marked on this graph, the *amplitude A* is the maximum excursion of a particle of the medium from the particle's undisturbed position. The amplitude is the distance between a crest, or highest point on the wave pattern, and the undisturbed position; it is also the distance between a trough, or lowest point on the wave pattern, and the undisturbed position.

**Figure 16.5** One cycle of the wave is shaded in color, and the amplitude of the wave is denoted as *A*.

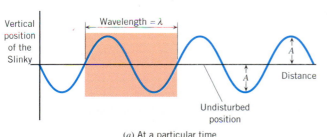

Vertical position of the Slinky

Wavelength = λ

A

Distance

A

Undisturbed position

(a) At a particular time

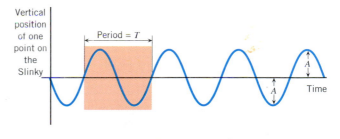

Vertical position of one point on the Slinky

Period = T

A

Time

A

(b) At a particular location

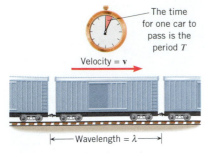

**Figure 16.6** A train moving at a constant speed serves as an analogy for a traveling wave.

The *wavelength* λ is the horizontal length of one cycle of the wave, as shown in Figure 16.5*a*. The wavelength is also the horizontal distance between two successive crests, two successive troughs, or any two successive equivalent points on the wave.

Part *b* of Figure 16.5 shows a graph in which time, rather than distance, is plotted on the horizontal axis. This graph is obtained by observing a single point on the Slinky. As the wave passes, the point under observation oscillates up and down in simple harmonic motion. As indicated on the graph, the *period T* is the time required for one complete up/down cycle, just as it is for an object vibrating on a spring. Equivalently, the period is the time required for the wave to travel a distance of one wavelength. The period *T* is related to the *frequency f*, just as it is for any example of simple harmonic motion:

$$f = \frac{1}{T} \tag{10.5}$$

The period is commonly measured in seconds, and frequency is measured in cycles per second or hertz (Hz). If, for instance, one cycle of a wave takes one-tenth of a second to pass an observer, then ten cycles pass per second, as Equation 10.5 indicates [$f = 1/(0.1\text{ s}) = 10$ cycles/s $= 10$ Hz].

A simple relation exists between the period, the wavelength, and the speed of a wave, a relation that Figure 16.6 helps to introduce. Imagine waiting at a railroad crossing, while a freight train moves by at a constant speed *v*. The train consists of a long line of identical boxcars, each of which has a length λ and requires a time *T* to pass, so the speed is $v = \lambda/T$. This same equation applies for a wave and relates the speed of the wave to the wavelength λ and the period *T*. Since the frequency of a wave is $f = 1/T$, the expression for the speed is

$$v = \frac{\lambda}{T} = f\lambda \tag{16.1}$$

The terminology just discussed and the fundamental relations $f = 1/T$ and $v = f\lambda$ apply to longitudinal as well as to transverse waves. Example 1 illustrates how the wavelength of a wave is determined by the wave speed and the frequency established by the source.

### Example 1  The Wavelengths of Radio Waves

AM and FM radio waves are transverse waves that consist of electric and magnetic disturbances. These waves travel at a speed of $3.00 \times 10^8$ m/s. A station broadcasts an AM radio wave whose frequency is $1230 \times 10^3$ Hz (1230 kHz on the dial) and an FM radio wave whose frequency is $91.9 \times 10^6$ Hz (91.9 MHz on the dial). Find the distance between adjacent crests in each wave.

**Reasoning** The distance between adjacent crests is the wavelength λ. Since the speed of each wave is $v = 3.00 \times 10^8$ m/s and the frequencies are known, the relation $v = f\lambda$ can be used to determine the wavelengths.

**Solution**

**Problem solving insight**
The equation $v = f\lambda$ applies to any kind of periodic wave.

*AM*  $\qquad \lambda = \dfrac{v}{f} = \dfrac{3.00 \times 10^8 \text{ m/s}}{1230 \times 10^3 \text{ Hz}} = \boxed{244 \text{ m}}$

*FM*  $\qquad \lambda = \dfrac{v}{f} = \dfrac{3.00 \times 10^8 \text{ m/s}}{91.9 \times 10^6 \text{ Hz}} = \boxed{3.26 \text{ m}}$

Notice that the wavelength of an AM radio wave is longer than two and one-half football fields!

## 16.3  *The Speed of a Wave on a String*

The properties of the material* or medium through which a wave travels determine the speed of the wave. For example, Figure 16.7 shows a transverse wave on a string and draws attention to four string particles that have been drawn as colored dots. As the

---

* Electromagnetic waves (discussed in Chapter 24) can move through a vacuum, as well as through materials such as glass and water.

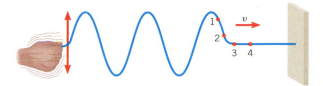

**Figure 16.7** As a transverse wave moves to the right with speed $v$, each string particle is displaced, one after the other, from its undisturbed position.

wave moves to the right, each particle is displaced, one after the other, from its undisturbed position. In the drawing, particles 1 and 2 have already been displaced upward, while particles 3 and 4 are not yet affected by the wave. Particle 3 will be next to move because the section of string immediately to its left (i.e., particle 2) will pull it upward.

Figure 16.7 leads us to conclude that the speed with which the wave moves to the right depends on how quickly one particle of the string is accelerated upward in response to the net pulling force exerted by its adjacent neighbors. In accord with Newton's second law, a stronger net force results in a greater acceleration, and, thus, a faster-moving wave. The ability of one particle to pull on its neighbors depends on how tightly the string is stretched—that is, on the tension (see Section 4.10 for a review of tension). The greater the tension, the greater the pulling force the particles exert on each other, and the faster the wave travels, other things being equal. Along with the tension, a second factor influences the wave speed. According to Newton's second law, the inertia or mass of particle 3 in Figure 16.7 also affects how quickly it responds to the upward pull of particle 2. For a given net pulling force, a smaller mass has a greater acceleration than a larger mass. Therefore, other things being equal, a wave travels faster on a string whose particles have a small mass, or, as it turns out, on a string that has a small mass per unit length. The mass per unit length is called the *linear density* of the string. It is the mass $m$ of the string divided by its length $L$, or $m/L$. The effects of the tension $F$ and the mass per unit length are evident in the following expression for the speed $v$ of a small-amplitude wave on a string:

$$v = \sqrt{\frac{F}{m/L}} \tag{16.2}$$

The motion of transverse waves along a string is important in the operation of musical instruments, such as the guitar, the violin, and the piano. In these instruments, the strings are either plucked, bowed, or struck to produce transverse waves. Example 2 discusses the speed of the waves on the strings of a guitar.

## Example 2   Waves Traveling on Guitar Strings

Transverse waves travel on the strings of an electric guitar after the strings are plucked (see Figure 16.8). The length of each string between its two fixed ends is 0.628 m, and the mass is 0.208 g for the highest pitched E string and 3.32 g for the lowest pitched E string. Each string is under a tension of 226 N. Find the speeds of the waves on the two strings.

*The physics of waves on guitar strings.*

**Reasoning** The speed of a wave on a guitar string, as expressed by Equation 16.2, depends on the tension $F$ in the string and its linear density $m/L$. Since the tension is the same for both strings, and smaller linear densities give rise to greater speeds, we expect the wave speed to be greatest on the string with the smallest linear density.

**Solution** The speeds of the waves are given by Equation 16.2 as

*High-pitched E*     $v = \sqrt{\dfrac{F}{m/L}} = \sqrt{\dfrac{226\ \text{N}}{(0.208 \times 10^{-3}\ \text{kg})/(0.628\ \text{m})}} = \boxed{826\ \text{m/s}}$

*Low-pitched E*     $v = \sqrt{\dfrac{F}{m/L}} = \sqrt{\dfrac{226\ \text{N}}{(3.32 \times 10^{-3}\ \text{kg})/(0.628\ \text{m})}} = \boxed{207\ \text{m/s}}$

Notice how fast the waves move; the speeds correspond to 1850 and 463 mi/h.

Conceptual Example 3 offers additional insight into the nature of a wave as a traveling disturbance.

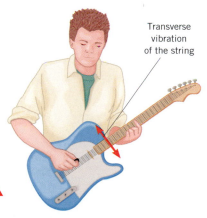

Transverse vibration of the string

**Figure 16.8** Plucking a guitar string generates transverse waves.

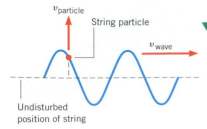

**Figure 16.9** A transverse wave on a string is moving to the right with a constant speed $v_{wave}$. A string particle moves up and down in simple harmonic motion about the undisturbed position of the string. The speed of the particle $v_{particle}$ changes from moment to moment as the wave passes.

## Conceptual Example 3  Wave Speed Versus Particle Speed

Is the speed of a transverse wave on a string the same as the speed at which a particle on the string moves (see Figure 16.9)?

**Reasoning and Solution** The particle speed $v_{particle}$ specifies how fast the particle is moving as it oscillates up and down, and it is different from the wave speed. If the source of the wave (e.g., the hand in Figure 16.2c) vibrates in simple harmonic motion, each string particle vibrates in a like manner, with the same amplitude and frequency as the source. Moreover, the particle speed, unlike the wave speed, is not constant. As for any object in simple harmonic motion, the particle speed is greatest when the particle is passing through the undisturbed position of the string and zero when the particle is at its maximum displacement. According to Equation 10.7, the particle speed depends on the amplitude $A$, the angular frequency $\omega$, and the time $t$ through the relation $v_{particle} = A\omega \sin \omega t$, where the minus sign has been omitted because we are interested only in the speed, which is the magnitude of the particle's velocity. Thus, the speed of a string particle is determined by the *properties of the source* creating the wave and not by the properties of the string itself. In contrast, the speed of the wave is determined by the properties of the string—that is, the tension $F$ and the mass per unit length $m/L$, according to Equation 16.2. We see, then, that *the two speeds, $v_{wave}$ and $v_{particle}$, are not the same.*

**Related Homework:** *Problem 18*

## 16.4  *The Mathematical Description of a Wave*

When a wave travels through a medium, it displaces the particles of the medium from their undisturbed positions. Suppose a particle is located at a distance $x$ from a coordinate origin. We would like to know the displacement $y$ of this particle from its undisturbed position at any time $t$ as the wave passes. For periodic waves that result from simple harmonic motion of the source, the expression for the displacement involves a sine or cosine, a fact that is not surprising. After all, in Chapter 10 simple harmonic motion is described using sinusoidal equations, and the graphs for a wave in Figure 16.5 look like a plot of displacement versus time for an object oscillating on a spring (see Figure 10.5).

Our tack will be to present the expression for the displacement and then show graphically that it gives a correct description. Equation 16.3 represents the displacement of a particle caused by a wave traveling in the $+x$ direction (to the right), with an amplitude $A$, frequency $f$, and wavelength $\lambda$. Equation 16.4 applies to a wave moving in the $-x$ direction (to the left).

*Wave motion toward $+x$*
$$y = A \sin \left( 2\pi ft - \frac{2\pi x}{\lambda} \right) \qquad (16.3)$$

*Wave motion toward $-x$*
$$y = A \sin \left( 2\pi ft + \frac{2\pi x}{\lambda} \right) \qquad (16.4)$$

These equations apply to transverse or longitudinal waves and assume that $y = 0$ m when $x = 0$ m and $t = 0$ s.

Consider a transverse wave moving in the $+x$ direction along a string. The term $(2\pi ft - 2\pi x/\lambda)$ in Equation 16.3 is called the *phase angle* of the wave. A string particle located at the origin ($x = 0$ m) exhibits simple harmonic motion with a phase angle of $2\pi ft$; that is, its displacement as a function of time is $y = A \sin (2\pi ft)$. A particle located at a distance $x$ also exhibits simple harmonic motion, but its phase angle is

$$2\pi ft - \frac{2\pi x}{\lambda} = 2\pi f \left( t - \frac{x}{f\lambda} \right) = 2\pi f \left( t - \frac{x}{v} \right)$$

The quantity $x/v$ is the time needed for the wave to travel the distance $x$. In other words, the simple harmonic motion that occurs at $x$ is delayed by the time interval $x/v$ compared to the motion at the origin.

Figure 16.10 shows the displacement $y$ plotted as a function of position $x$ along the string at a series of time intervals separated by one-fourth of the period $T$ ($t = 0$ s, $\frac{1}{4}T$, $\frac{2}{4}T$, $\frac{3}{4}T$, $T$). These graphs are constructed by substituting the corresponding value for $t$ into Equation 16.3, remembering that $f = 1/T$, and then calculating $y$ at a series of values for $x$. The graphs are like photographs taken at various times as the wave moves to the right. For reference, the colored square on each graph marks the place on the wave that is located at $x = 0$ m when $t = 0$ s. As time passes, the colored square moves to the right, along with the wave. In a similar manner, it can be shown that Equation 16.4 represents a wave moving in the $-x$ direction. Note that the phase angles ($2\pi ft - 2\pi x/\lambda$) in Equation 16.3 and ($2\pi ft + 2\pi x/\lambda$) in Equation 16.4 are measured in *radians*, not degrees. **When a calculator is used to evaluate the functions sin ($2\pi ft - 2\pi x/\lambda$) or sin ($2\pi ft + 2\pi x/\lambda$), it must be set to its radian mode.**

*Problem solving insight*

# 16.5 The Nature of Sound

## LONGITUDINAL SOUND WAVES

Sound is a longitudinal wave that is created by a vibrating object, such as a guitar string, the human vocal cords, or the diaphragm of a loudspeaker. Moreover, sound can be created or transmitted only in a medium, such as a gas, liquid, or solid. As we will see, the particles of the medium must be present for the disturbance of the wave to move from place to place. Sound cannot exist in a vacuum.

To see how sound waves are produced and why they are longitudinal, consider the vibrating diaphragm of a loudspeaker. When the diaphragm moves outward, it compresses the air directly in front of it, as in Figure 16.11a. This compression causes the air pressure to rise slightly. The region of increased pressure is called a *condensation,* and it travels away from the speaker at the speed of sound. The condensation is analogous to the compressed region of coils in a longitudinal wave on a Slinky, which is included in Figure 16.11a for comparison. After producing a condensation, the diaphragm reverses its motion and moves inward, as in part *b* of the drawing. The inward motion produces a region known as a *rarefaction,* where the air pressure is slightly less than normal. The rarefaction is similar to the stretched region of coils in a longitudinal Slinky wave. Following immediately behind the condensation, the rarefaction also travels away from the speaker at the speed of sound. Figure 16.12 on p. 312 further emphasizes the similarity between a sound wave and a longitudinal Slinky wave. As the wave passes, the colored dots attached both to the Slinky and to an air molecule execute simple harmonic motion about their undisturbed positions. The colored arrows on either side of the dots indicate that the simple harmonic motion occurs parallel to the line of travel. The drawing also shows that the wavelength $\lambda$ is the distance between the centers of two successive condensations; $\lambda$ is also the distance between the centers of two successive rarefactions.

Figure 16.13 illustrates a sound wave spreading out in space after being produced by a loudspeaker. When the condensations and rarefactions arrive at the ear, they force the eardrum to vibrate at the same frequency as the speaker diaphragm. The vibratory motion

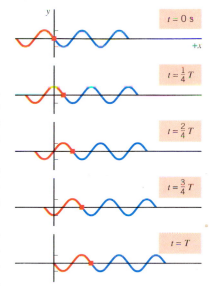

**Figure 16.10** Equation 16.3 is plotted here at a series of times separated by one-fourth of the period $T$. The colored square in each graph marks the place on the wave that is located at $x = 0$ m when $t = 0$ s. As time passes, the wave moves to the right.

*The physics of*
*a loudspeaker diaphragm.*

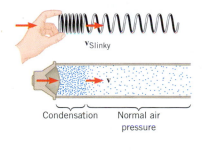

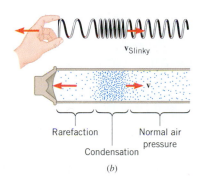

| Condensation | Normal air pressure |
|---|---|

(*a*)

| Rarefaction | Normal air pressure |
|---|---|

Condensation

(*b*)

**Figure 16.11** (*a*) When the speaker diaphragm moves outward, it creates a condensation. (*b*) When the diaphragm moves inward, it creates a rarefaction. The condensation and rarefaction on the Slinky are included for comparison. In reality, the velocity of the wave on the Slinky $v_{Slinky}$ is much smaller than the velocity of sound in air $v$. For simplicity, the two waves are shown here to have the same velocity.

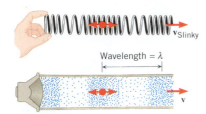

**Figure 16.12** Both the wave on the Slinky and the sound wave are longitudinal. The colored dots attached to the Slinky and to an air molecule vibrate back and forth parallel to the line of travel of the wave.

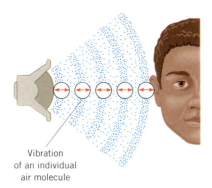

**Figure 16.13** Condensations and rarefactions travel from the speaker to the listener, but the individual air molecules do not move with the wave. A given molecule vibrates back and forth about a fixed location.

**The physics of**
**push-button telephones.**

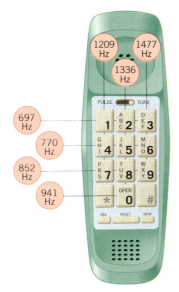

**Figure 16.14** A push-button telephone and a schematic showing the two pure tones produced when each button is pressed.

of the eardrum is interpreted by the brain as sound. It should be emphasized that sound is not a mass movement of air, like the wind. As the condensations and rarefactions of the sound wave travel outward from the vibrating diaphragm in Figure 16.13, the individual air molecules are not carried along with the wave. Rather, each molecule executes simple harmonic motion about a fixed location. In doing so, one molecule collides with its neighbor and passes the condensations and rarefactions forward. The neighbor, in turn, repeats the process.

## THE FREQUENCY OF A SOUND WAVE

Each cycle of a sound wave includes one condensation and one rarefaction, and the *frequency* is the number of cycles per second that passes by a given location. For example, if the diaphragm of a speaker vibrates back and forth in simple harmonic motion at a frequency of 1000 Hz, then 1000 condensations, each followed by a rarefaction, are generated every second, thus forming a sound wave whose frequency is also 1000 Hz. A sound with a single frequency is called a *pure tone.* Experiments have shown that a healthy young person hears all sound frequencies from approximately 20 to 20 000 Hz (20 kHz). The ability to hear the high frequencies decreases with age, however, and a normal middle-aged adult hears frequencies only up to 12–14 kHz.

Pure tones are used in push-button telephones, such as that shown in Figure 16.14. These phones simultaneously produce two pure tones when each button is pressed, a different pair of tones for each different button. The tones are transmitted electronically to the central telephone office, where they activate switching circuits that complete the call. For example, the drawing indicates that pressing the "5" button produces pure tones of 770 and 1336 Hz simultaneously, while the "9" button generates tones of 852 and 1477 Hz.

Sound can be generated whose frequency lies below 20 Hz or above 20 kHz, although humans normally do not hear it. Sound waves with frequencies below 20 Hz are said to be *infrasonic,* while those with frequencies above 20 kHz are referred to as *ultrasonic.* Rhinoceroses use infrasonic frequencies as low as 5 Hz to call one another (Figure 16.15), while bats use ultrasonic frequencies up to 100 kHz for locating their food sources and navigating (Figure 16.16).

Frequency is an objective property of a sound wave because frequency can be measured with an electronic frequency counter. A listener's perception of frequency, however, is subjective. The brain interprets the frequency detected by the ear primarily in terms of the subjective quality called *pitch.* A pure tone with a large (high) frequency is interpreted as a high-pitched sound, while a pure tone with a small (low) frequency is interpreted as a low-pitched sound. A piccolo produces high-pitched sounds, and a tuba produces low-pitched sounds.

**Figure 16.15** Rhinoceroses call to one another using infrasonic sound waves. (Photo by Ron Garrison, courtesy Zoological Society of San Diego.)

**Figure 16.16** Bats use ultrasonic sound waves for navigating and locating food sources. (© Merlin D. Tuttle/Bat Conservation International/ Photo Researchers)

## THE PRESSURE AMPLITUDE OF A SOUND WAVE

Figure 16.17 illustrates a pure-tone sound wave traveling in a tube. Attached to the tube is a series of gauges that indicate the pressure variations along the wave. The graph shows that the air pressure varies sinusoidally along the length of the tube. Although this graph has the appearance of a transverse wave, remember that the sound itself is a longitudinal wave. The graph also shows the ***pressure amplitude*** of the wave, which is the magnitude of the maximum change in pressure, measured relative to the undisturbed or atmospheric pressure. The pressure fluctuations in a sound wave are normally very small. For instance, in a typical conversation between two people the pressure amplitude is about $3 \times 10^{-2}$ Pa, certainly a small amount compared with the atmospheric pressure of $1.01 \times 10^{+5}$ Pa. The ear is remarkable in being able to detect such small changes.

*Loudness* is an attribute of sound that depends primarily on the amplitude of the wave: the larger the amplitude, the louder the sound. The pressure amplitude is an objective property of a sound wave, since it can be measured. Loudness, on the other hand, is subjective. Each individual determines what is loud, depending on the acuteness of his or her hearing.

## 16.6 *The Speed of Sound*

### GASES

Sound travels through gases, liquids, and solids at considerably different speeds, as Table 16.1 reveals. Near room temperature, the speed of sound in air is 343 m/s (767 mi/h) and is markedly greater in liquids and solids. For example, sound travels more than four times

**Table 16.1** *Speed of Sound in Gases, Liquids, and Solids*

| Substance | Speed (m/s) |
|---|---|
| *Gases* | |
| Air (0 °C) | 331 |
| Air (20 °C) | 343 |
| Carbon dioxide (0 °C) | 259 |
| Oxygen (0 °C) | 316 |
| Helium (0 °C) | 965 |
| *Liquids* | |
| Chloroform (20 °C) | 1004 |
| Ethyl alcohol (20 °C) | 1162 |
| Mercury (20 °C) | 1450 |
| Fresh water (20 °C) | 1482 |
| Seawater (20 °C) | 1522 |
| *Solids* | |
| Copper | 5010 |
| Glass (Pyrex) | 5640 |
| Lead | 1960 |
| Steel | 5960 |

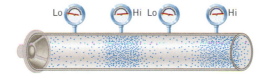

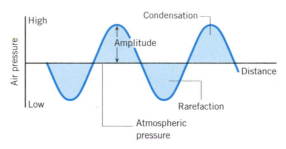

**Figure 16.17** A sound wave is a series of alternating condensations and rarefactions. The graph shows that the condensations are regions of higher than normal air pressure, and the rarefactions are regions of lower than normal air pressure.

faster in water and more than seventeen times faster in steel than it does in air. In general, sound travels slowest in gases, faster in liquids, and fastest in solids.

Like the speed of a wave on a guitar string, the speed of sound depends on the properties of the medium. In a gas, it is only when molecules collide that the condensations and rarefactions of a sound wave can move from place to place. It is reasonable, then, to expect the speed of sound in a gas to have the same order of magnitude as the average molecular speed between collisions. For an ideal gas this average speed is the translational rms speed given by Equation 14.6: $v_{rms} = \sqrt{3kT/m}$, where $T$ is the Kelvin temperature, $m$ is the mass of a molecule, and $k$ is Boltzmann's constant. Although the expression for $v_{rms}$ overestimates the speed of sound, it does give the correct dependence on Kelvin temperature and particle mass. Careful analysis shows that the speed of sound in an ideal gas is given by

**Ideal gas**
$$v = \sqrt{\frac{\gamma kT}{m}}$$
(16.5)

where $\gamma = c_P/c_V$ is the ratio of the specific heat capacity at constant pressure $c_P$ to the specific heat capacity at constant volume $c_V$.

The factor $\gamma$ is introduced in Section 15.5, where the adiabatic compression and expansion of an ideal gas is discussed. It appears in Equation 16.5 because the condensations and rarefactions of a sound wave are formed by adiabatic compressions and expansions of the gas. The regions that are compressed (the condensations) become slightly warmed, and the regions that are expanded (the rarefactions) become slightly cooled. However, no appreciable heat flows from a condensation to an adjacent rarefaction because the distance between the two (half a wavelength) is relatively large for most audible sound waves and a gas is a poor thermal conductor. Thus, the compression and expansion process is adiabatic. Example 4 illustrates the use of Equation 16.5.

### Example 4    An Ultrasonic Ruler

**The physics of an ultrasonic ruler.**

Figure 16.18 shows an ultrasonic ruler that is used to measure the distance between itself and a target, such as a wall. To initiate the measurement, the ruler generates a pulse of ultrasonic sound that travels to the wall and, like an echo, reflects from it. The reflected pulse returns to the ruler, which measures the time it takes for the round-trip. Using a preset value for the speed of sound, the unit determines the distance to the wall and displays it on a digital readout. Suppose the round-trip travel time is 20.0 ms on a day when the air temperature is 23 °C. Assuming that air is an ideal gas for which $\gamma = 1.40$ and that the average molecular mass of air is 28.9 u, find the distance $x$ to the wall.

**Reasoning** The distance between the ruler and the wall is $x = vt$, where $v$ is the speed of sound and $t$ is the time for the sound pulse to reach the wall. The time $t$ is one-half the round-trip time, so $t = 10.0$ ms. The speed of sound in air can be obtained directly from Equation 16.5, provided the temperature and mass are expressed in the SI units of kelvins and kilograms, respectively.

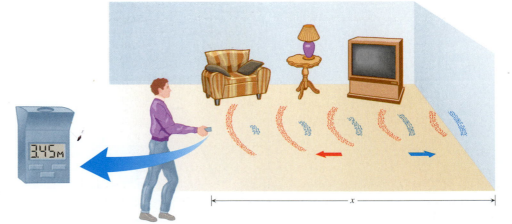

**Figure 16.18** An ultrasonic ruler uses sound with a frequency greater than 20 kHz to measure the distance $x$ to the wall. The blue arcs and blue arrow denote the outgoing sound wave, and the red arcs and red arrow denote the wave reflected from the wall.

**Solution** To convert the air temperature of 23 °C to the Kelvin temperature scale, we add 23 to 273.15 (see Equation 12.1): $T = 23 + 273.15 = 296$ K. The mass of a molecule (in kilograms) can be obtained from the conversion relation between atomic mass units and kilograms (see Section 14.1), $1 \text{ u} = 1.6605 \times 10^{-27}$ kg:

$$m = (28.9 \text{ u}) \left( \frac{1.6605 \times 10^{-27} \text{ kg}}{1 \text{ u}} \right) = 4.80 \times 10^{-26} \text{ kg}$$

For the speed of sound, we find

$$v = \sqrt{\frac{\gamma k T}{m}} = \sqrt{\frac{(1.40)(1.38 \times 10^{-23} \text{ J/K})(296 \text{ K})}{4.80 \times 10^{-26} \text{ kg}}} = 345 \text{ m/s} \qquad (16.5)$$

The distance to the wall is

$$x = vt = (345 \text{ m/s})(10.0 \times 10^{-3} \text{ s}) = \boxed{3.45 \text{ m}}$$

▲

**Problem solving insight**
When using equation $v = \sqrt{\gamma k T/m}$ to calculate the speed of sound in an ideal gas, be sure to express the temperature $T$ in kelvins and not in degrees Celsius or Fahrenheit.

Sonar (**so**und **na**vigation **r**anging) is a technique for determining water depth and locating underwater objects, such as reefs, submarines, and schools of fish. The core of a sonar unit consists of an ultrasonic transmitter and receiver mounted on the bottom of a ship. The transmitter emits a short pulse of ultrasonic sound, and at a later time the reflected pulse returns and is detected by the receiver. The water depth is determined from the electronically measured round-trip time of the pulse and a knowledge of the speed of sound in water; the depth registers automatically on an appropriate meter. Such a depth measurement is similar to the distance measurement discussed for the ultrasonic ruler in Example 4.

*The physics of* sonar.

Conceptual Example 5 illustrates how the speed of sound in air can be used to estimate the distance to a thunderstorm, using a handy rule of thumb.

## *Conceptual Example 5* Lightning, Thunder, and a Rule of Thumb

▼

There is a rule of thumb for estimating how far away a thunderstorm is. After you see a flash of lightning, count off the seconds until the thunder is heard. Divide the number of seconds by five. The result gives the approximate distance (in miles) to the thunderstorm. Why does this rule work?

**Reasoning and Solution** Figure 16.19 shows a lightning bolt and a person who is standing one mile ($1.6 \times 10^3$ m) away. When the lightning occurs, light and sound (thunder) are produced very nearly at the same instant. Light travels so rapidly ($v_{\text{light}} = 3.0 \times 10^8$ m/s) that it reaches the observer almost instantaneously. Its travel time is only $(1.6 \times 10^3 \text{ m})/(3.0 \times 10^8 \text{ m/s}) = 5.3 \times 10^{-6}$ s. In comparison, sound travels very slowly ($v_{\text{sound}} = 343$ m/s). The time for the thunder to reach the person is $(1.6 \times 10^3 \text{ m})/(343 \text{ m/s}) = 5$ s. Thus, the time interval between seeing the flash and hearing the thunder is about 5 seconds for every mile of travel. *This rule of thumb works because the speed of light is so much greater than the speed of sound* that the time needed for the light to reach the observer is negligible compared to the time needed for the sound.

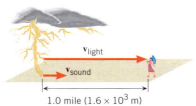

$v_{\text{light}}$

$v_{\text{sound}}$

1.0 mile ($1.6 \times 10^3$ m)

**Figure 16.19** A lightning bolt from a thunderstorm generates a flash of light and sound (thunder). The speed of light is much greater than the speed of sound. Therefore, the light reaches the person first, followed about 5 seconds later by the sound.

▲

## LIQUIDS

In a liquid, the speed of sound depends on the density $\rho$ and the *adiabatic* bulk modulus $B_{\text{ad}}$ of the liquid:

*Liquid*
$$v = \sqrt{\frac{B_{\text{ad}}}{\rho}} \qquad (16.6)$$

The bulk modulus is introduced in Section 10.7 in a discussion of the volume deformation of liquids and solids. There it is tacitly assumed that the temperature remains constant while the volume of the material changes; that is, the compression or expansion is isothermal. However, the condensations and rarefactions in a sound wave occur under *adiabatic* rather than isothermal conditions. Thus, the adiabatic bulk modulus $B_{\text{ad}}$ must be used when calculating the speed of sound in liquids. Values of $B_{\text{ad}}$ will be provided as needed in this text.

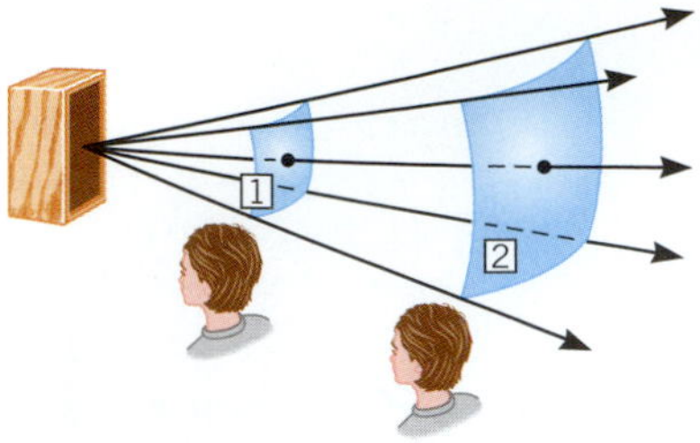

**Figure 16.20** The power carried by a sound wave spreads out after leaving a source, such as a loudspeaker. Thus, the power passes perpendicularly through surface 1 and then through surface 2, which has the larger area.

## SOLID BARS

When sound travels through a long slender solid bar, the speed of the sound depends on the properties of the medium according to

*Long slender solid bar*
$$v = \sqrt{\frac{Y}{\rho}} \qquad (16.7)$$

where $Y$ is Young's modulus (defined in Section 10.7) and $\rho$ is the density.

## 16.7 *Sound Intensity*

Sound waves carry energy that can be used to do work, like forcing the eardrum to vibrate. In an extreme case such as a sonic boom, the energy can be sufficient to cause damage to windows and buildings. The amount of energy transported per second by a sound wave is called the *power* of the wave and is measured in SI units of joules per second (J/s) or watts (W).

When a sound wave leaves a source, such as the loudspeaker in Figure 16.20, the power spreads out and passes through imaginary surfaces that have increasingly larger areas. For instance, the same sound power passes through the surfaces labeled 1 and 2 in the drawing. However, the power is spread out over a greater area in surface 2 and we have to take this spreading-out effect into account. We will bring together the ideas of sound power and the area through which the power passes and, in the process, formulate the concept of sound intensity. The idea of wave intensity is not confined to sound waves. It will recur, for example, in Chapter 24 when we discuss another important type of waves, electromagnetic waves.

The *sound intensity I* is defined as the sound power $P$ that passes perpendicularly through a surface divided by the area $A$ of that surface:

$$I = \frac{P}{A} \qquad (16.8)$$

*Problem solving insight*
Sound intensity $I$ and sound power $P$ are different concepts. They are related, however, since intensity equals power per unit area.

The unit of sound intensity is power per unit area, or $W/m^2$. The next example illustrates how the sound intensity changes as the distance from a loudspeaker changes.

### Example 6  Sound Intensities

In Figure 16.20, $12 \times 10^{-5}$ W of sound power passes perpendicularly through the surfaces labeled 1 and 2. These surfaces have areas of $A_1 = 4.0$ m$^2$ and $A_2 = 12$ m$^2$. Determine the sound intensity at each surface and discuss why listener 2 hears a quieter sound than listener 1.

**Reasoning** The sound intensity $I$ is the sound power $P$ passing perpendicularly through a surface divided by the area $A$ of that surface. Since the same sound power passes through both surfaces and surface 2 has the greater area, the sound intensity is less at surface 2.

**Solution** The sound intensity at each surface follows from Equation 16.8:

*Surface 1*
$$I_1 = \frac{P}{A_1} = \frac{12 \times 10^{-5} \text{ W}}{4.0 \text{ m}^2} = \boxed{3.0 \times 10^{-5} \text{ W/m}^2}$$

*Surface 2*
$$I_2 = \frac{P}{A_2} = \frac{12 \times 10^{-5} \text{ W}}{12 \text{ m}^2} = \boxed{1.0 \times 10^{-5} \text{ W/m}^2}$$

The sound intensity is less at the more distant surface, where the same power passes through a threefold greater area. The ear of a listener, with its fixed area, intercepts less power where the intensity, or power per unit area, is smaller. Thus, listener 2 intercepts less of the sound power than listener 1. With less power striking the ear, the sound is quieter.

For a 1000-Hz tone, the smallest sound intensity that the human ear can detect is about $1 \times 10^{-12}$ W/m$^2$; this intensity is called the *threshold of hearing.* On the other extreme, continuous exposure to intensities greater than 1 W/m$^2$ can be painful and result in permanent hearing damage. The human ear is remarkable for the wide range of intensities to which it is sensitive.

If a source emits sound *uniformly in all directions,* the intensity depends on distance in a simple way. Figure 16.21 shows such a source at the center of an imaginary sphere (for clarity only a hemisphere is shown). The radius of the sphere is $r$. Since all the radiated sound power $P$ passes through the spherical surface of area $A = 4\pi r^2$, the intensity at a distance $r$ is

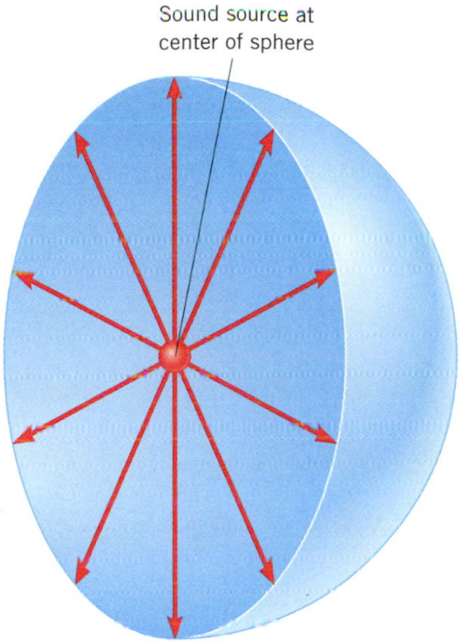

**Spherically uniform radiation**
$$I = \frac{P}{4\pi r^2} \qquad (16.9)$$

From this we see that the intensity of a source that radiates sound uniformly in all directions varies as $1/r^2$. For example, if the distance increases by a factor of two, the sound intensity decreases by a factor of $2^2 = 4$. Example 7 illustrates the effect of the $1/r^2$ dependence of intensity on distance.

### Example 7  Fireworks

During a fireworks display, a rocket explodes high in the air, as Figure 16.22 illustrates. Assume that the sound spreads out uniformly in all directions and that reflections from the ground can be ignored. When the sound reaches listener 2, who is $r_2 = 640$ m away from the explosion, the sound has an intensity of $I_2 = 0.10$ W/m$^2$. What is the sound intensity detected by listener 1, who is $r_1 = 160$ m away from the explosion?

**Reasoning** Listener 1 is four times closer to the explosion than listener 2. Therefore, the sound intensity detected by listener 1 is $4^2 = 16$ times greater than that detected by listener 2.

**Solution** The ratio of the sound intensities can be found using Equation 16.9:

$$\frac{I_1}{I_2} = \frac{\dfrac{P}{4\pi r_1^2}}{\dfrac{P}{4\pi r_2^2}} = \frac{r_2^2}{r_1^2} = \frac{(640 \text{ m})^2}{(160 \text{ m})^2} = 16$$

As a result, $I_1 = (16)I_2 = (16)(0.10 \text{ W/m}^2) = \boxed{1.6 \text{ W/m}^2}$.

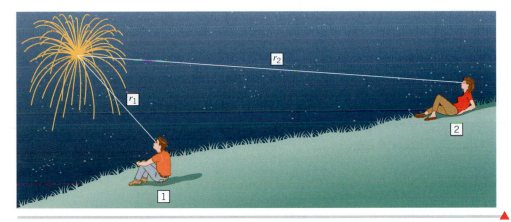

Equation 16.9 is valid only when no walls, ceilings, floors, etc. are present to reflect the sound and cause it to pass through the same surface more than once. Conceptual Example 8 demonstrates why this is so.

### Conceptual Example 8  Reflected Sound and Sound Intensity

Suppose the person singing in the shower in Figure 16.23 produces a sound power $P$. Sound reflects from the surrounding shower stall. At a distance $r$ in front of the person, does Equation 16.9, $I = P/(4\pi r^2)$, underestimate, overestimate, or give the correct sound intensity?

**Reasoning and Solution** In arriving at Equation 16.9, it was assumed that the sound spreads out uniformly from the source and passes only once through the imaginary surface that surrounds it (see Figure 16.21). Only part of this imaginary surface is shown in Figure 16.23. The drawing illustrates three paths by which the sound passes through the surface. The "direct"

Sound source at center of sphere

**Figure 16.21** The sound source at the center of the sphere emits sound uniformly in all directions. In this drawing, only a hemisphere is shown for clarity.

**Problem solving insight**
Equation 16.9 can be used only when the sound spreads out uniformly in all directions and there are no reflections of the sound waves.

**Figure 16.22** If an explosion in a fireworks display radiates sound uniformly in all directions, the intensity at any distance $r$ is $I = P/(4\pi r^2)$, where $P$ is the sound power of the explosion.

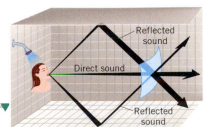

**Figure 16.23** When someone sings in the shower, the sound power passing through part of an imaginary spherical surface (shown in blue) is the sum of the direct sound power and the reflected sound power.

sound travels directly along a path from its source to the surface. It is the intensity of the direct sound that is given by $I = P/(4\pi r^2)$. The remaining paths are two of the many that characterize the sound reflected from the shower stall. The total sound power that passes through the surface is the sum of the direct and reflected powers. Thus, the total sound intensity at a distance $r$ from the source is greater than that of the direct sound alone, and *the relation $I = P/(4\pi r^2)$ underestimates the sound intensity from the singing because it does not take into account the reflected sound.* People like to sing in the shower because their voices sound so much louder due to the enhanced intensity caused by the reflected sound.

**Related Homework:** *Problems 52, 64*

## 16.8 *Decibels*

The *decibel* (dB) is a measurement unit used when comparing two sound intensities. The simplest method of comparison would be to compute the ratio of the intensities. For instance, we could compare $I = 8 \times 10^{-12}$ W/m² to $I_0 = 1 \times 10^{-12}$ W/m² by computing $I/I_0 = 8$ and stating that $I$ is eight times greater than $I_0$. However, because of the way in which the human hearing mechanism responds to intensity, it is more appropriate to use a logarithmic scale for the comparison. For this purpose, the *intensity level $\beta$* (expressed in decibels) is defined as follows:

$$\beta = (10 \text{ dB}) \log\left(\frac{I}{I_0}\right) \tag{16.10}$$

where "log" denotes the logarithm to the base ten. $I_0$ is the intensity of the reference level to which $I$ is being compared and is often the threshold of hearing, $I_0 = 1.00 \times 10^{-12}$ W/m². With the aid of a calculator, the intensity level can be evaluated for the values of $I$ and $I_0$ given above:

$$\beta = (10 \text{ dB}) \log\left(\frac{8 \times 10^{-12} \text{ W/m}^2}{1 \times 10^{-12} \text{ W/m}^2}\right) = (10 \text{ dB}) \log 8 = (10 \text{ dB})(0.9) = 9 \text{ dB}$$

This result indicates that $I$ is 9 decibels greater than $I_0$. Although $\beta$ is called the "intensity level," it is *not* an intensity and does *not* have intensity units of W/m². In fact, the decibel, like the radian, is dimensionless.

Notice that if both $I$ and $I_0$ are at the threshold of hearing, then $I = I_0$, and the intensity level is 0 dB according to Equation 16.10:

$$\beta = (10 \text{ dB}) \log\left(\frac{I_0}{I_0}\right) = (10 \text{ dB}) \log 1 = 0$$

since $\log 1 = 0$. Thus, *an intensity level of zero decibels does not mean that the sound intensity $I$ is zero; it means that $I = I_0$.*

Intensity levels can be measured with a sound level meter, such as the one in Figure 16.24. The intensity level $\beta$ is displayed on its scale, assuming that the threshold of hearing is 0 dB. Table 16.2 lists the intensities $I$ and the associated intensity levels $\beta$ for some common sounds, using the threshold of hearing as the reference level.

When a sound wave reaches a listener's ear, the sound is interpreted by the brain as loud or soft, depending on the intensity of the wave. Greater intensities give rise to louder sounds. However, the relation between intensity and loudness is not a simple proportionality, because doubling the intensity does *not* double the loudness, as we will now see.

Suppose you are sitting in front of a stereo system that is producing an intensity level of 90 dB. If the volume control on the amplifier is turned up slightly to produce a 91-dB level, you would just barely notice the change in loudness. *Hearing tests have revealed that a one-decibel (1-dB) change in the intensity level corresponds to approximately the smallest change in loudness that an average listener with normal hearing can detect.* Since 1 dB is the smallest perceivable increment in loudness, a change of 3 dB—say, from 90 to 93 dB—is still a rather small change in loudness. Example 9 determines the factor by which the sound intensity must be increased to achieve such a change.

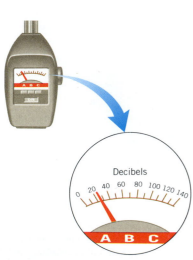

**Figure 16.24** A sound level meter and a close-up view of its decibel scale.

**Table 16.2**   *Typical Sound Intensities and Intensity Levels Relative to the Threshold of Hearing*

|  | Intensity $I$ (W/m$^2$) | Intensity Level $\beta$ (dB) |
|---|---|---|
| Threshold of hearing | $1.0 \times 10^{-12}$ | 0 |
| Rustling leaves | $1.0 \times 10^{-11}$ | 10 |
| Whisper | $1.0 \times 10^{-10}$ | 20 |
| Normal conversation (1 meter) | $3.2 \times 10^{-6}$ | 65 |
| Inside car in city traffic | $1.0 \times 10^{-4}$ | 80 |
| Car without muffler | $1.0 \times 10^{-2}$ | 100 |
| Live rock concert | 1.0 | 120 |
| Threshold of pain | 10 | 130 |

### Example 9   Comparing Sound Intensities

Audio system 1 produces an intensity level of $\beta_1 = 90.0$ dB, and system 2 produces an intensity level of $\beta_2 = 93.0$ dB. The corresponding intensities (in W/m$^2$) are $I_1$ and $I_2$. Determine the ratio $I_2/I_1$.

**Reasoning** Intensity levels are related to intensities by logarithms (see Equation 16.10), and it is a property of logarithms (see Appendix D) that $\log A - \log B = \log (A/B)$. Subtracting the two intensity levels and using this property, we find that

$$\beta_2 - \beta_1 = (10 \text{ dB}) \log \left(\frac{I_2}{I_0}\right) - (10 \text{ dB}) \log \left(\frac{I_1}{I_0}\right) = (10 \text{ dB}) \log \left(\frac{I_2/I_0}{I_1/I_0}\right)$$

$$= (10 \text{ dB}) \log \left(\frac{I_2}{I_1}\right)$$

**Solution** Using the result just obtained, we find

$$93.0 \text{ dB} - 90.0 \text{ dB} = (10 \text{ dB}) \log \left(\frac{I_2}{I_1}\right)$$

$$0.30 = \log \left(\frac{I_2}{I_1}\right) \quad \text{or} \quad \frac{I_2}{I_1} = 10^{0.30} = \boxed{2.0}$$

Doubling the intensity changes the loudness by only a small amount (3 dB) and does not double it, so there is no simple proportionality between intensity and loudness.

To double the loudness of a sound, the intensity must be increased by more than a factor of two. *Experiment shows that if the intensity level increases by 10 dB, the new sound seems approximately twice as loud as the original sound.* For instance, a 70-dB intensity level sounds about twice as loud as a 60-dB level, and an 80-dB intensity level sounds about twice as loud as a 70-dB level. The factor by which the sound intensity must be increased to double the loudness can be determined by the method used in Example 9:

$$\beta_2 - \beta_1 = 10.0 \text{ dB} = (10 \text{ dB}) \left[ \log \left(\frac{I_2}{I_0}\right) - \log \left(\frac{I_1}{I_0}\right) \right]$$

Solving this equation reveals that $I_2/I_1 = 10.0$. Thus, increasing the sound intensity by a factor of ten will double the perceived loudness. Consequently, with both audio systems in Figure 16.25 set at maximum volume, the 200-watt system will sound only twice as loud as the much cheaper 20-watt system.

## 16.9  The Doppler Effect

Have you ever heard an approaching fire truck and noticed the distinct change in the sound of the siren as the truck passes? The effect is similar to what you get when you put together the two syllables "eee" and "yow" to produce "eee-yow." While the truck ap-

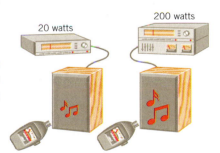

**Figure 16.25** In spite of its tenfold greater power, the 200-watt audio system has only about double the loudness of the 20-watt system, when both are set for maximum volume.

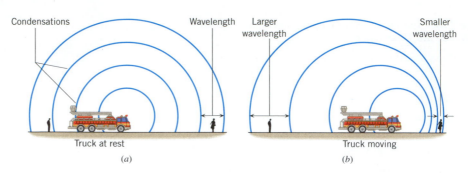

**Figure 16.26** (*a*) When the truck is stationary, the wavelength of the sound is the same in front of and behind the truck. (*b*) When the truck is moving, the wavelength in front of the truck becomes smaller, while the wavelength behind the truck becomes larger.

proaches, the pitch of the siren is relatively high ("eee"), but as the truck passes and moves away, the pitch suddenly drops ("yow"). Something similar, but less familiar, occurs when an observer moves toward or away from a stationary source of sound. Such phenomena were first identified in 1842 by the Austrian physicist Christian Doppler (1803–1853) and are collectively referred to as the Doppler effect.

To explain why the Doppler effect occurs, we will bring together concepts that we have discussed previously—namely, the velocity of an object and the wavelength and frequency of a sound wave (Section 16.5). We will combine the effects of the velocities of the source and observer of the sound with the definitions of wavelength and frequency. In so doing, we will learn that the ***Doppler effect*** is the change in frequency or pitch of the sound detected by an observer because the sound source and the observer have different velocities with respect to the medium of sound propagation.

## MOVING SOURCE

To see how the Doppler effect arises, consider the sound emitted by a siren on the stationary fire truck in Figure 16.26*a*. Like the truck, the air is assumed to be stationary with respect to the earth. Each solid blue arc in the drawing represents a condensation of the sound wave. Since the sound pattern is symmetrical, listeners standing in front of or behind the truck detect the same number of condensations per second and, consequently, hear the same frequency. Once the truck begins to move, the situation changes, as part *b* of the picture illustrates. Ahead of the truck, the condensations are now closer together, resulting in a decrease in the wavelength of the sound. This "bunching-up" occurs because the moving truck "gains ground" on a previously emitted condensation before emitting the next one. Since the condensations are closer together, the observer standing in front of the truck senses more of them arriving per second than she does when the truck is stationary. The increased rate of arrival corresponds to a greater sound frequency, which the observer hears as a higher pitch. Behind the moving truck, the condensations are farther apart than they are when the truck is stationary. This increase in the wavelength occurs because the truck pulls away from condensations emitted toward the rear. Consequently, fewer condensations per second arrive at the ear of an observer behind the truck, corresponding to a smaller sound frequency or lower pitch.

If the stationary siren in Figure 16.26*a* emits a condensation at the time $t = 0$ s, it will emit the next one at time $T$, where $T$ is the period of the wave. The distance between these two condensations is the wavelength $\lambda$ of the sound produced by the stationary source, as Figure 16.27*a* indicates. When the truck is moving with a speed $v_s$ (the subscript "s" stands for the "source" of sound) toward a stationary observer, the siren also emits condensations at $t = 0$ s and at time $T$. However, prior to emitting the second condensation, the truck moves closer to the observer by a distance $v_s T$, as Figure 16.27*b* shows. As a result, the distance between successive condensations is no longer the wavelength $\lambda$ created by the stationary siren, but, rather, a wavelength $\lambda'$ that is shortened by the amount $v_s T$:

$$\lambda' = \lambda - v_s T$$

Let's denote the frequency perceived by the stationary observer as $f_o$, where the subscript "o" stands for "observer." According to Equation 16.1, $f_o$ is equal to the speed of sound $v$ divided by the shortened wavelength $\lambda'$:

$$f_o = \frac{v}{\lambda'} = \frac{v}{\lambda - v_s T}$$

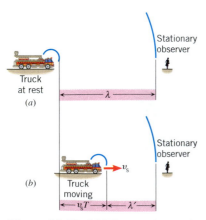

**Figure 16.27** (*a*) When the fire truck is stationary, the distance between successive condensations is one wavelength $\lambda$. (*b*) When the truck moves with a speed $v_s$, the wavelength of the sound in front of the truck is shortened to $\lambda'$.

But for the stationary siren, we have $\lambda = v/f_s$ and $T = 1/f_s$, where $f_s$ is the frequency of the sound emitted by the source (not the frequency $f_o$ perceived by the observer). With the aid of these substitutions for $\lambda$ and $T$, the expression for $f_o$ can be arranged to give the following result:

**Source moving toward stationary observer**

$$f_o = f_s \left( \frac{1}{1 - \dfrac{v_s}{v}} \right)$$

(16.11)

Since the term $1 - v_s/v$ is in the denominator in Equation 16.11 and is less than one, the frequency $f_o$ heard by the observer is *greater* than the frequency $f_s$ emitted by the source. The difference between these two frequencies, $f_o - f_s$, is called the **Doppler shift,** and its magnitude depends on the ratio of the speed of the source $v_s$ to the speed of sound $v$.

When the siren moves away from, rather than toward, the observer, the wavelength $\lambda'$ becomes *greater* than $\lambda$ according to

$$\lambda' = \lambda + v_s T$$

Notice the presence of the "+" sign in this equation, in contrast to the "−" sign that appeared earlier. The same reasoning that led to Equation 16.11 can be used to obtain an expression for the observed frequency $f_o$:

**Source moving away from stationary observer**

$$f_o = f_s \left( \frac{1}{1 + \dfrac{v_s}{v}} \right)$$

(16.12)

The denominator $1 + v_s/v$ in Equation 16.12 is greater than one, so the frequency $f_o$ heard by the observer is *less* than the frequency $f_s$ emitted by the source. The next example illustrates how large the Doppler shift is in a familiar situation.

## Example 10 The Sound of a Passing Train

A high-speed train is traveling at a speed of 44.7 m/s (100 mi/h) when the engineer sounds the 415-Hz warning horn. The speed of sound is 343 m/s. What are the frequency and wavelength of the sound, as perceived by a person standing at a crossing, when the train is (a) approaching and (b) leaving the crossing?

**Reasoning** When the train approaches, the person at the crossing hears a sound whose frequency is greater than 415 Hz because of the Doppler effect. As the train moves away, the person hears a frequency that is less than 415 Hz. We may use Equations 16.11 and 16.12, respectively, to determine these frequencies. In either case, the observed wavelength can be obtained according to Equation 16.1 as the speed of sound divided by the observed frequency.

**Solution**

(a) When the train approaches, the observed frequency is

$$f_o = f_s \left( \frac{1}{1 - \dfrac{v_s}{v}} \right) = (415 \text{ Hz}) \left( \frac{1}{1 - \dfrac{44.7 \text{ m/s}}{343 \text{ m/s}}} \right) = \boxed{477 \text{ Hz}}$$

(16.11)

The observed wavelength is

$$\lambda' = \frac{v}{f_o} = \frac{343 \text{ m/s}}{477 \text{ Hz}} = \boxed{0.719 \text{ m}}$$

(16.1)

(b) When the train leaves the crossing, the observed frequency is

$$f_o = f_s \left( \frac{1}{1 + \dfrac{v_s}{v}} \right) = (415 \text{ Hz}) \left( \frac{1}{1 + \dfrac{44.7 \text{ m/s}}{343 \text{ m/s}}} \right) = \boxed{367 \text{ Hz}}$$

(16.12)

In this case, the observed wavelength is

$$\lambda' = \frac{v}{f_o} = \frac{343 \text{ m/s}}{367 \text{ Hz}} = \boxed{0.935 \text{ m}}$$

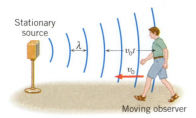

Stationary source

Moving observer

**Figure 16.28** An observer moving with a speed $v_o$ toward the stationary source intercepts more wave condensations per unit of time than does a stationary observer.

## MOVING OBSERVER

Figure 16.28 shows how the Doppler effect arises when the sound source is stationary and the observer moves, again assuming the air is stationary. The observer moves with a speed $v_o$ ("o" stands for "observer") toward the stationary source and covers a distance $v_o t$ in a time $t$. During this time, the moving observer encounters all the condensations that he would if he were stationary, **plus an additional number.** The additional number of condensations encountered is the distance $v_o t$ divided by the distance $\lambda$ between successive condensations, or $v_o t/\lambda$. Thus, the additional number of condensations encountered per second is $v_o/\lambda$. Since a stationary observer would hear a frequency $f_s$ emitted by the source, the moving observer hears a higher frequency $f_o$ given by

$$f_o = f_s + \frac{v_o}{\lambda} = f_s \left( 1 + \frac{v_o}{f_s \lambda} \right)$$

Using the fact that $v = f_s \lambda$, we find that

**Observer moving toward stationary source**
$$f_o = f_s \left( 1 + \frac{v_o}{v} \right) \tag{16.13}$$

An observer moving *away from* a stationary source moves in the same direction as the sound wave and, as a result, intercepts *fewer* condensations per second than a stationary observer does. In this case, the moving observer hears a smaller frequency $f_o$ that is given by

**Observer moving away from stationary source**
$$f_o = f_s \left( 1 - \frac{v_o}{v} \right) \tag{16.14}$$

The physical mechanism producing the Doppler effect in the case of the moving observer is different from that in the case of the moving source. When the source moves and the observer is stationary, the wavelength $\lambda$ in Figure 16.27b changes, giving rise to the frequency $f_o$ heard by the observer. On the other hand, when the observer moves and the source is stationary, the *wavelength $\lambda$ in Figure 16.28 does not change.* Instead, a moving observer intercepts a different number of wave condensations per second than does a stationary observer and, therefore, detects a different frequency $f_o$.

## GENERAL CASE

It is possible for *both* the sound source and the observer to move with respect to the medium of sound propagation. If the medium is stationary, Equations 16.11–16.14 may be combined to give the observed frequency $f_o$ as

**Source and observer both moving**
$$f_o = f_s \left( \frac{1 \pm \dfrac{v_o}{v}}{1 \mp \dfrac{v_s}{v}} \right) \tag{16.15}$$

In the numerator, the plus sign applies when the observer moves toward the source, and the minus sign applies when the observer moves away from the source. In the denominator, the minus sign is used when the source moves toward the observer, and the plus sign is used when the source moves away from the observer. The symbols $v_o$, $v_s$, and $v$ denote numbers without an algebraic sign because the direction of travel has been taken into account by the plus and minus signs that appear directly in this equation.

## NEXRAD

**The physics of**
**Next Generation Weather Radar.**

NEXRAD stands for **Nex**t Generation Weather **Rad**ar and is a nationwide system used by the National Weather Service to provide dramatically improved early warning of severe storms, such as the tornado in Figure 16.29. The system is based on radar waves, which are a type of electromagnetic wave (see Chapter 24) and, like sound waves, can exhibit the Doppler effect. The Doppler effect is at the heart of NEXRAD. As the drawing illustrates, a tornado is a swirling mass of air and water droplets. Radar pulses are sent out by a NEXRAD unit, whose protective covering is shaped like a soccer ball. The waves reflect from the water droplets and return to the unit, where the frequency is observed and compared to the outgoing frequency. For instance, droplets at point A in the drawing are

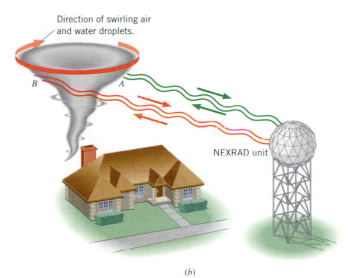

(a)   (b)

moving toward the unit, and the radar waves reflected from them have their frequency Doppler-shifted to higher values. Droplets at point *B*, however, are moving away from the unit. The frequency of the waves reflected from these droplets is Doppler-shifted to lower values. Computer processing of the Doppler frequency shifts leads to color-enhanced views on display screens (see Figure 16.30). These views reveal the direction and magnitude of the wind velocity and can identify, from distances up to 140 mi, the swirling air masses that are likely to spawn tornadoes. The equations that specify the Doppler frequency shifts are different from those given for sound waves by Equations 16.11–16.15. The reason for the difference is that radar waves propagate from one place to another by a different mechanism than that of sound waves (see Section 24.5).

**Figure 16.29** (*a*) A tornado is one of nature's most dangerous storms. (*b*) The National Weather Service uses the NEXRAD system, which is based on Doppler-shifted radar, to identify the storms that are likely to spawn tornadoes. (© Paul and Lindamarie Ambrose/Taxi/Getty Images)

**The physics of** ultrasonic imaging.

# 16.10 *Applications of Sound in Medicine*

When ultrasonic waves are used in medicine for diagnostic purposes, high-frequency sound pulses are produced by a transmitter and directed into the body. As in sonar, reflections occur. They occur each time a pulse encounters a boundary between two tissues that have different densities or a boundary between a tissue and the adjacent fluid. By scanning ultrasonic waves across the body and detecting the echoes generated from various internal locations, it is possible to obtain an image or sonogram of the inner anatomy. Ultrasonic imaging is employed extensively in obstetrics to examine the developing fetus (Figure 16.31 on p. 324). The fetus, surrounded by the amniotic sac, can be distinguished from other anatomical features so that fetal size, position, and possible abnormalities can be detected.

Ultrasound is also used in other medically related areas. For instance, malignancies in the liver, kidney, brain, and pancreas can be detected with ultrasound. Yet another application involves monitoring the real-time movement of pulsating structures, such as heart valves ("echocardiography") and large blood vessels.

When ultrasound is used to form images of internal anatomical features or foreign objects in the body, the wavelength of the sound wave must be about the same size as, or smaller than, the object to be located. Therefore, high frequencies in the range from 1 to 15 MHz (1 MHz = 1 megahertz = $1 \times 10^6$ Hz) are the norm. For instance, the wavelength of 5-MHz ultrasound is $\lambda = v/f = 0.3$ mm, if a value of 1540 m/s is used for the speed of sound through tissue. A sound wave with a frequency higher than 5 MHz and a correspondingly shorter wavelength is required for locating objects smaller than 0.3 mm.

Ultrasound also has applications other than imaging. Neurosurgeons use a device called a **c**avitron **u**ltra**s**onic **s**urgical **a**spirator (CUSA) to remove brain tumors once thought to be inoperable. Ultrasonic sound waves cause the slender tip of the CUSA probe (see Figure 16.32) to vibrate at approximately 23 kHz. The probe shatters any section of the tumor that it touches, and the fragments are flushed out of the brain with a saline solution. Because the tip of the probe is small, the surgeon can selectively remove small bits of malignant tissue without damaging the surrounding healthy tissue.

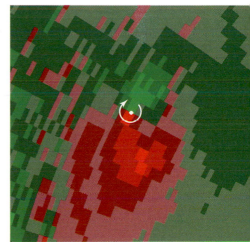

**Figure 16.30** This color-enhanced NEXRAD view of a tornado shows winds moving toward (green) and away from (red) a NEXRAD station, which is below and to the right of the figure. The white dot and arrow indicate the the storm center and direction of wind circulation. (Courtesy Kurt Hondl, National Severe Storms Laboratory, Norman, OK.)

**The physics of** the cavitron ultrasonic surgical aspirator.

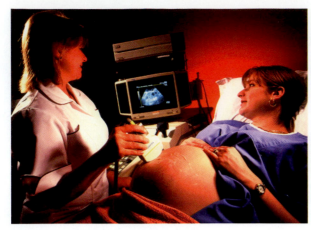

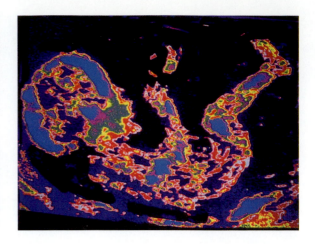

**Figure 16.31** An ultrasonic scanner can be used to produce an image of the fetus as it develops in the uterus. (*Left,* © Deep Light Productions/Science Photo Library/Photo Researchers; *right,* © Howard Sochurek)

**The physics of** bloodless surgery with HIFU.

**The physics of** the Doppler flow meter.

Another application of ultrasound is in a new type of bloodless surgery, which can eliminate abnormal cells, such as those in benign hyperplasia of the prostate gland. Still in the experimental phase, this technique is known as HIFU (**h**igh-**i**ntensity **f**ocused **u**ltrasound). It is analogous to focusing the sun's electromagnetic waves by using a magnifying glass and producing a small region where the energy carried by the waves can cause localized heating. Ultrasonic waves can be used in a similar fashion. The waves enter directly through the skin and come into focus inside the body over a region that is sufficiently well defined to be surgically useful. Within this region the energy of the waves causes localized heating, leading to a temperature of about 56 °C (normal body temperature is 37 °C), which is sufficient to kill abnormal cells. The killed cells are eventually removed by the body's natural processes.

The Doppler flow meter is a particularly interesting medical application of the Doppler effect. This device measures the speed of blood flow, using transmitting and receiving elements that are placed directly on the skin, as in Figure 16.33. The transmitter emits a continuous sound whose frequency is typically about 5 MHz. When the sound is reflected from the red blood cells, its frequency is changed in a kind of Doppler effect because the cells are moving. The receiving element detects the reflected sound, and an electronic counter measures its frequency, which is Doppler-shifted relative to the transmitter frequency. From the change in frequency the speed of the blood flow can be determined. Typically, the change in frequency is around 600 Hz for flow speeds of about 0.1 m/s. The Doppler flow meter can be used to locate regions where blood vessels have narrowed, since greater flow speeds occur in the narrowed regions, according to the equation of continuity (see Section 11.8). In addition, the Doppler flow meter can be used to detect the motion of a fetal heart as early as 8–10 weeks after conception.

**Figure 16.32** Neurosurgeons use a cavitron ultrasonic surgical aspirator (CUSA) to "cut out" brain tumors without adversely affecting the surrounding healthy tissue.

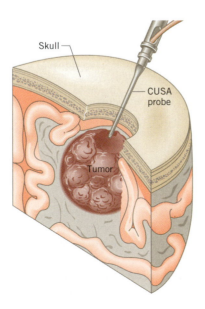

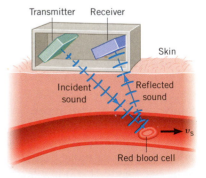

**Figure 16.33** A Doppler flow meter measures the speed of red blood cells.

# Concept Summary

This summary presents an abridged version of the chapter, including the important equations and all available learning aids. For convenient reference, the learning aids (including the text's examples) are placed next to or immediately after the relevant equation or discussion. The following learning aids may be found on-line at **www.wiley.com/college/cutnell**:

| | |
|---|---|
| **Interactive LearningWare** examples are solved according to a five-step interactive format that is designed to help you develop problem-solving skills. | **Concept Simulations** are animated versions of text figures or animations that illustrate important concepts. You can control parameters that affect the display, and we encourage you to experiment. |
| **Interactive Solutions** offer specific models for certain types of problems in the chapter homework. The calculations are carried out interactively. | **Self-Assessment Tests** include both qualitative and quantitative questions. Extensive feedback is provided for both incorrect and correct answers, to help you evaluate your understanding of the material. |

| Topic | Discussion | Learning Aids |
|---|---|---|
| | **16.1 The Nature of Waves** | |
| Transverse wave<br>Longitudinal wave | A wave is a traveling disturbance and carries energy from place to place. In a transverse wave, the disturbance occurs perpendicular to the direction of travel of the wave. In a longitudinal wave, the disturbance occurs parallel to the line along which the wave travels. | |
| | **16.2 Periodic Waves** | |
| Cycle<br>Amplitude<br><br>Wavelength<br>Period<br>Frequency | A periodic wave consists of cycles or patterns that are produced over and over again by the source of the wave. The amplitude of the wave is the maximum excursion of a particle of the medium from the particle's undisturbed position. The wavelength $\lambda$ is the distance along the length of the wave between two successive equivalent points, such as two crests or two troughs. The period $T$ is the time required for the wave to travel a distance of one wavelength. The frequency $f$ (in hertz) is the number of wave cycles per second that passes an observer and is the reciprocal of the period (in seconds): | |
| Relation between frequency and period | $$f = \frac{1}{T} \qquad (10.5)$$ | |
| Relation between speed, frequency, and wavelength | The speed $v$ of a wave is related to its wavelength and frequency according to $$v = f\lambda \qquad (16.1)$$ | **Example 1** |
| | **16.3 The Speed of a Wave on a String** | |
| | The speed of a wave depends on the properties of the medium in which the wave travels. For a transverse wave on a string that has a tension $F$ and a mass per unit length $m/L$, the wave speed is | **Examples 2, 3** |
| Speed of a wave on a string | $$v = \sqrt{\frac{F}{m/L}} \qquad (16.2)$$ | **Concept Simulation 16.1** |
| Linear density | The mass per unit length is also called the linear density. | **Interactive Solution 16.17** |
| | **16.4 The Mathematical Description of a Wave** | |
| | When a wave of amplitude $A$, frequency $f$, and wavelength $\lambda$ moves in the $+x$ direction through a medium, the wave causes a displacement $y$ of a particle at position $x$ according to | |
| | $$y = A \sin\left(2\pi ft - \frac{2\pi x}{\lambda}\right) \qquad (16.3)$$ | |
| | For a wave moving in the $-x$ direction, the expression is | |
| | $$y = A \sin\left(2\pi ft + \frac{2\pi x}{\lambda}\right) \qquad (16.4)$$ | |

  ***Use Self-Assessment Test 16.1 to evaluate your understanding of Sections 16.1–16.4.***

| Topic | Discussion | Learning Aids |
|---|---|---|
| | **16.5 The Nature of Sound** | |
| Condensation | Sound is a longitudinal wave that can be created only in a medium; it cannot exist in a vacuum. Each cycle of a sound wave includes one condensation (a re- | |

| Topic | Discussion | Learning Aids |
|---|---|---|
| Rarefaction | gion of greater than normal pressure) and one rarefaction (a region of less than normal pressure). | |
| Infrasonic frequency<br>Ultrasonic frequency<br><br>Pitch | A sound wave with a single frequency is called a pure tone. Frequencies less than 20 Hz are called infrasonic. Frequencies greater than 20 kHz are called ultrasonic. The brain interprets the frequency detected by the ear primarily in terms of the subjective quality known as pitch. A high-pitched sound is one with a large frequency (e.g., piccolo). A low-pitched sound is one with a small frequency (e.g., tuba). | |
| Pressure amplitude<br><br>Loudness | The pressure amplitude of a sound wave is the magnitude of the maximum change in pressure, measured relative to the undisturbed pressure. The pressure amplitude is associated with the subjective quality of loudness. The larger the pressure amplitude, the louder the sound. | |

### 16.6 The Speed of Sound

The speed of sound $v$ depends on the properties of the medium. In an ideal gas, the speed of sound is

| Topic | Discussion | Learning Aids |
|---|---|---|
| Speed of sound<br>in an ideal gas | $$v = \sqrt{\frac{\gamma k T}{m}} \qquad (16.5)$$ | **Examples 4, 5**<br><br>**Interactive LearningWare 16.1** |

where $\gamma = c_P/c_V$ is the ratio of the specific heat capacities at constant pressure and constant volume, $k$ is Boltzmann's constant, $T$ is the Kelvin temperature, and $m$ is the mass of a molecule of the gas. In a liquid, the speed of sound is

| Topic | Discussion | Learning Aids |
|---|---|---|
| Speed of sound in a liquid | $$v = \sqrt{\frac{B_{ad}}{\rho}} \qquad (16.6)$$ | |

where $B_{ad}$ is the adiabatic bulk modulus and $\rho$ is the mass density. In a solid that has a Young's modulus of $Y$ and the shape of a long slender bar, the speed of sound is

| Topic | Discussion | Learning Aids |
|---|---|---|
| Speed of sound in solid bar | $$v = \sqrt{\frac{Y}{\rho}} \qquad (16.7)$$ | |

### 16.7 Sound Intensity

The intensity $I$ of a sound wave is the power $P$ that passes perpendicularly through a surface divided by the area $A$ of the surface

| Topic | Discussion | Learning Aids |
|---|---|---|
| Intensity | $$I = \frac{P}{A} \qquad (16.8)$$ | **Example 6**<br><br>**Interactive Solution 16.55** |
| Threshold of hearing | The SI unit for intensity is watts per square meter (W/m²). The smallest sound intensity that the human ear can detect is known as the threshold of hearing and is about $1 \times 10^{-12}$ W/m² for a 1-kHz sound. When a source radiates sound uniformly in all directions and no reflections are present, the intensity of the sound is inversely proportional to the square of the distance from the source, according to | |
| Spherically uniform radiation | $$I = \frac{P}{4\pi r^2} \qquad (16.9)$$ | **Examples 7, 8** |

### 16.8 Decibels

The intensity level $\beta$ (in decibels) is used to compare a sound intensity $I$ to the sound intensity $I_0$ of a reference level:

| Topic | Discussion | Learning Aids |
|---|---|---|
| Intensity level in decibels | $$\beta = (10 \text{ dB}) \log\left(\frac{I}{I_0}\right) \qquad (16.10)$$ | **Example 9** |

The decibel, like the radian, is dimensionless. An intensity level of zero decibels means that $I = I_0$. One decibel is approximately the smallest change in loudness that an average listener with healthy hearing can detect. An increase of ten decibels in the intensity level corresponds approximately to a doubling of the loudness of the sound.

### 16.9 The Doppler Effect

The Doppler effect is the change in frequency detected by an observer because the sound source and the observer have different velocities with respect to the

| Topic | Discussion | Learning Aids |
|-------|-----------|---------------|

medium of sound propagation. If the observer and source move with speeds $v_o$ and $v_s$, respectively, and if the medium is stationary, the frequency $f_o$ detected by the observer is

**The Doppler effect**

$$f_o = f_s \left( \frac{1 \pm \dfrac{v_o}{v}}{1 \mp \dfrac{v_s}{v}} \right)$$  (16.15)

**Example 10**

Interactive LearningWare 16.2

Interactive Solution 16.77

where $f_s$ is the frequency of the sound emitted by the source and $v$ is the speed of sound. In the numerator, the plus sign applies when the observer moves toward the source, and the minus sign applies when the observer moves away from the source. In the denominator, the minus sign is used when the source moves toward the observer, and the plus sign is used when the source moves away from the observer.

 **Use Self-Assessment Test 16.2 to evaluate your understanding of Sections 16.5–16.9.**

# Problems

## Section 16.1 The Nature of Waves, Section 16.2 Periodic Waves

**1. ssm** A person standing in the ocean notices that after a wave crest passes by, ten more crests pass in a time of 120 s. What is the frequency of the wave?

**2.** Light is an electromagnetic wave and travels at a speed of $3.00 \times 10^8$ m/s. The human eye is most sensitive to yellow-green light, which has a wavelength of $5.45 \times 10^{-7}$ m. What is the frequency of this light?

**3.** A longitudinal wave with a frequency of 3.0 Hz takes 1.7 s to travel the length of a 2.5-m Slinky (see Figure 16.3). Determine the wavelength of the wave.

**4.** Consider the freight train in Figure 16.6. Suppose 15 boxcars pass by in a time of 12.0 s and each has a length of 14.0 m. (a) What is the frequency at which each boxcar passes? (b) What is the speed of the train?

**5. ssm** In Figure 16.2c the hand moves the end of the Slinky up and down through two complete cycles in one second. The wave moves along the Slinky at a speed of 0.50 m/s. Find the distance between two adjacent crests on the wave.

**6.** A person lying on an air mattress in the ocean rises and falls through one complete cycle every five seconds. The crests of the wave causing the motion are 20.0 m apart. Determine (a) the frequency and (b) the speed of the wave.

**7. ssm** Suppose the amplitude and frequency of the transverse wave in Figure 16.2c are, respectively, 1.3 cm and 5.0 Hz. Find the *total vertical distance* (in cm) through which the colored dot moves in 3.0 s.

**\* 8.** A jetskier is moving at 8.4 m/s in the direction in which the waves on a lake are moving. Each time he passes over a crest, he feels a bump. The bumping frequency is 1.2 Hz, and the crests are separated by 5.8 m. What is the wave speed?

**\* 9.** The speed of a transverse wave on a string is 450 m/s, and the wavelength is 0.18 m. The amplitude of the wave is 2.0 mm. How much time is required for a particle of the string to move through a total distance of 1.0 km?

**\* 10.** A 3.49-rad/s ($33\frac{1}{3}$ rpm) record has a 5.00-kHz tone cut in the groove. If the groove is located 0.100 m from the center of the record (see drawing), what is the wavelength in the groove?

**\*\* 11.** A water-skier is moving at a speed of 12.0 m/s. When she skis in the same direction as a traveling wave, she springs upward every 0.600 s because of the wave crests. When she skis in the direction opposite to that in which the wave moves, she springs upward every 0.500 s in response to the crests. The speed of the skier is greater than the speed of the wave. Determine (a) the speed and (b) the wavelength of the wave.

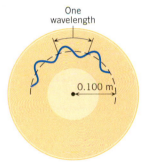

*Problem 10*

## Section 16.3 The Speed of a Wave on a String

**12.** The mass of a string is $5.0 \times 10^{-3}$ kg, and it is stretched so that the tension in it is 180 N. A transverse wave traveling on this string has a frequency of 260 Hz and a wavelength of 0.60 m. What is the length of the string?

**13. ssm** The linear density of the A string on a violin is $7.8 \times 10^{-4}$ kg/m. A wave on the string has a frequency of 440 Hz and a wavelength of 65 cm. What is the tension in the string?

**14.** A vibrator moves one end of a rope up and down to generate a wave. The tension in the rope is 58 N. The frequency is then doubled. To what value must the tension be adjusted, so the new wave has the same wavelength as the old one?

**15.** A transverse wave is traveling with a speed of 300 m/s on a horizontal string. If the tension in the string is increased by a factor of four, what is the speed of the wave?

**16.** Two wires are parallel, and one is directly above the other. Each has a length of 50.0 m and a mass per unit length of 0.020 kg/m. However, the tension in wire A is $6.00 \times 10^2$ N, and the tension in wire B is $3.00 \times 10^2$ N. Transverse wave pulses are generated simultaneously, one at the left end of wire A and one at the right end of wire B. The pulses travel toward each other. How much time does it take until the pulses pass each other?

**17.** Consult **Interactive Solution 16.17** at **www.wiley.com/college/cutnell** in order to review a model for solving this problem. To measure the acceleration due to gravity on a distant planet, an astronaut hangs a 0.055-kg ball from the end of a wire. The wire has a length of 0.95 m and a linear density of $1.2 \times 10^{-4}$ kg/m. Using electronic equipment, the astronaut measures the time for a transverse pulse to travel the length of the wire and obtains a value of 0.016 s. The mass of the wire is negligible compared to the mass of the ball. Determine the acceleration due to gravity.

* **18.** Review Conceptual Example 3 before starting this problem. The amplitude of a transverse wave on a string is 4.5 cm. The ratio of the maximum particle speed to the speed of the wave is 3.1. What is the wavelength (in cm) of the wave?

* **19. ssm www** The drawing at right shows a frictionless incline and pulley. The two blocks are connected by a wire (mass per unit length = 0.0250 kg/m) and remain stationary. A transverse wave on the wire has a speed of

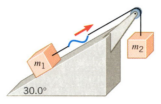

75.0 m/s. Neglecting the weight of the wire relative to the tension in the wire, find the masses $m_1$ and $m_2$ of the blocks.

** **20.** A copper wire, whose cross-sectional area is $1.1 \times 10^{-6}$ m², has a linear density of $7.0 \times 10^{-3}$ kg/m and is strung between two walls. At the ambient temperature, a transverse wave travels with a speed of 46 m/s on this wire. The coefficient of linear expansion for copper is $17 \times 10^{-6}$ (C°)$^{-1}$, and Young's modulus for copper is $1.1 \times 10^{11}$ N/m². What will be the speed of the wave when the temperature is lowered by 14 C°? Ignore any change in the linear density caused by the change in temperature.

** **21.** The drawing shows a 15.0-kg ball being whirled in a circular path on the end of a string. The motion occurs on a frictionless, horizontal table. The angular speed of the ball is $\omega = 12.0$ rad/s. The string has a mass of 0.0230 kg. How much time does it take for a wave on the string to travel from the center of the circle to the ball?

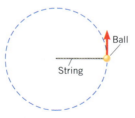

## Section 16.4 The Mathematical Description of a Wave

*(Note: The phase angles $(2\pi ft - 2\pi x/\lambda)$ and $(2\pi ft + 2\pi x/\lambda)$ are measured in radians, not degrees.)*

**22.** A wave traveling in the $+x$ direction has an amplitude of 0.35 m, a speed of 5.2 m/s, and a frequency of 14 Hz. Write the equation of the wave in the form given by either Equation 16.3 or 16.4.

**23. ssm** A wave has the following properties: amplitude = 0.37 m, period = 0.77 s, wave speed = 12 m/s. The wave is traveling in the $-x$ direction. What is the mathematical expression (similar to Equation 16.3 or 16.4) for the wave?

**24.** The drawing at the top of the right column shows two graphs that represent a transverse wave on a string. The wave is moving in the $+x$ direction. Using the information contained in these graphs, write the mathematical expression (similar to Equation 16.3 or 16.4) for the wave.

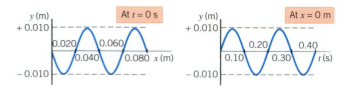

**25.** A wave causes a displacement $y$ that is given in meters according to $y = (0.45) \sin (8.0\pi t + \pi x)$, where $t$ and $x$ are expressed in seconds and meters, respectively. (a) Find the amplitude, the frequency, the wavelength, and the speed of the wave. (b) Is this wave traveling in the $+x$ or $-x$ direction?

* **26.** The tension in a string is 15 N, and its linear density is 0.85 kg/m. A wave on the string travels toward the $-x$ direction; it has an amplitude of 3.6 cm and a frequency of 12 Hz. What are the (a) speed and (b) wavelength of the wave? (c) Write down a mathematical expression (like Equation 16.3 or 16.4) for the wave, substituting numbers for the variables $A$, $f$, and $\lambda$.

* **27. ssm** A transverse wave is traveling on a string. The displacement $y$ of a particle from its equilibrium position is given by $y = (0.021$ m$) \sin (25t - 2.0x)$. Note that the phase angle $25t - 2.0x$ is in radians, $t$ is in seconds, and $x$ is in meters. The linear density of the string is $1.6 \times 10^{-2}$ kg/m. What is the tension in the string?

** **28.** A transverse wave on a string has an amplitude of 0.20 m and a frequency of 175 Hz. Consider the particle of the string at $x = 0$ m. It begins with a displacement of $y = 0$ m when $t = 0$ s, according to Equation 16.3 or 16.4. How much time passes between the first two instants when this particle has a displacement of $y = 0.10$ m?

## Section 16.5 The Nature of Sound,
## Section 16.6 The Speed of Sound

**29. ssm** The speed of a sound in a container of hydrogen at 201 K is 1220 m/s. What would be the speed of sound if the temperature were raised to 405 K? Assume that hydrogen behaves like an ideal gas.

**30.** For research purposes a sonic buoy is tethered to the ocean floor and emits an infrasonic pulse of sound. The period of this sound is 71 ms. Determine the wavelength of the sound.

**31.** The distance between a loudspeaker and the left ear of a listener is 2.70 m. (a) Calculate the time required for sound to travel this distance if the air temperature is 20 °C. (b) Assuming that the sound frequency is 523 Hz, how many wavelengths of sound are contained in this distance?

**32.** Have you ever listened for an approaching train by kneeling next to a railroad track and putting your ear to the rail? Young's modulus for steel is $Y = 2.0 \times 10^{11}$ N/m², and the density of steel is $\rho = 7860$ kg/m³. On a day when the temperature is 20 °C, how many times greater is the speed of sound in the rail than in the air?

**33. ssm** At 20 °C the densities of fresh water and ethyl alcohol are, respectively, 998 and 789 kg/m³. Find the ratio of the adiabatic bulk modulus of fresh water to the adiabatic bulk modulus of ethyl alcohol at 20 °C.

**34.** The wavelength of a sound wave in air is 2.74 m at 20 °C. What is the wavelength of this sound wave in fresh water at 20 °C? *(Hint: The frequency of the sound is the same in both media.)*

**35.** An explosion occurs at the end of a pier. The sound reaches the other end of the pier by traveling through three media: air, fresh water, and a slender metal handrail. The speeds of sound in air, water, and the handrail are 343, 1482, and 5040 m/s, respectively. The sound travels a distance of 125 m in each medium. (a) Through which medium does the sound arrive first, second, and third? (b) After the first sound arrives, how much later do the second and third sounds arrive?

**36.** Consult **Interactive LearningWare 16.1** at **www.wiley.com/college/cutnell** for insight into this problem. At what temperature is the speed of sound in helium (ideal gas, $\gamma = 1.67$, atomic mass = 4.003 u) the same as its speed in oxygen at 0 °C?

**37.** As the drawing illustrates, a siren can be made by blowing a jet of air through 20 equally spaced holes in a rotating disk. The time it takes for successive holes to move past the air jet is the period of the sound. The siren is to produce a 2200-Hz tone. What must be the angular speed $\omega$ (in rad/s) of the disk?

* **38.** As the drawing shows, one microphone is located at the origin, and a second microphone is located on the $+y$ axis. The microphones are separated by a distance of $D = 1.50$ m. A source of sound is located on the $+x$ axis, its distances from microphones 1 and 2 being $L_1$ and $L_2$, respectively. The speed of sound is 343 m/s. The sound reaches microphone 1 first, and then, 1.46 ms later, it reaches microphone 2. Find the distances $L_1$ and $L_2$.

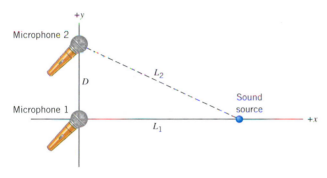

* **39. ssm** A long slender bar is made from an unknown material. The length of the bar is 0.83 m, its cross-sectional area is $1.3 \times 10^{-4}$ m², and its mass is 2.1 kg. A sound wave travels from one end of the bar to the other end in $1.9 \times 10^{-4}$ s. From which one of the materials listed in Table 10.1 is the bar most likely to be made?

* **40.** When an earthquake occurs, two types of sound waves are generated and travel through the earth. The primary, or P, wave has a speed of about 8.0 km/s and the secondary, or S, wave has a speed of about 4.5 km/s. A seismograph, located some distance away, records the arrival of the P wave and then, 78 s later, records the arrival of the S wave. Assuming that the waves travel in a straight line, how far is the seismograph from the earthquake?

* **41. ssm** A hunter is standing on flat ground between two vertical cliffs that are directly opposite one another. He is closer to one cliff than to the other. He fires a gun and, after a while, hears three echoes. The second echo arrives 1.6 s after the first, and the third echo arrives 1.1 s after the second. Assuming that the speed of sound is 343 m/s and that there are no reflections of sound from the ground, find the distance between the cliffs.

* **42.** A sound wave travels twice as far in neon (Ne) as it does in krypton (Kr) in the same time interval. Both neon and krypton can be treated as monatomic ideal gases. The atomic mass of neon is 20.2 u, and that of krypton is 83.8 u. The temperature of the krypton is 293 K. What is the temperature of the neon?

* **43.** A monatomic ideal gas ($\gamma = 1.67$) is contained within a box whose volume is 2.5 m³. The pressure of the gas is $3.5 \times 10^5$ Pa. The total mass of the gas is 2.3 kg. Find the speed of sound in the gas.

* **44.** At a height of ten meters above the surface of a freshwater lake, a sound pulse is generated. The echo from the bottom of the lake re-

turns to the point of origin 0.140 s later. The air and water temperatures are 20 °C. How deep is the lake?

** **45.** A jet is flying horizontally, as the drawing shows. When the plane is directly overhead at $B$, a person on the ground hears the sound coming from $A$ in the drawing. The average temperature of the air is 20 °C. If the speed of the plane at $A$ is 164 m/s, what is its speed at $B$, assuming that it has a constant acceleration?

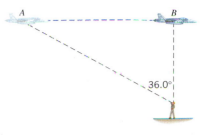

** **46.** The sonar unit on a boat is designed to measure the depth of fresh water ($\rho = 1.00 \times 10^3$ kg/m³, $B_{ad} = 2.20 \times 10^9$ Pa). When the boat moves into salt water ($\rho = 1025$ kg/m³, $B_{ad} = 2.37 \times 10^9$ Pa), the sonar unit is no longer calibrated properly. In salt water, the sonar unit indicates the water depth to be 10.0 m. What is the actual depth of the water?

** **47. ssm www** As a prank, someone drops a water-filled balloon out of a window. The balloon is released from rest at a height of 10.0 m above the ears of a man who is the target. Then, because of a guilty conscience, the prankster shouts a warning after the balloon is released. The warning will do no good, however, if shouted after the balloon reaches a certain point, even if the man could react infinitely quickly. Assuming that the air temperature is 20 °C and ignoring the effect of air resistance on the balloon, determine how far above the man's ears this point is.

**Section 16.7 Sound Intensity**

**48.** A typical adult ear has a surface area of $2.1 \times 10^{-3}$ m². The sound intensity during a normal conversation is about $3.2 \times 10^{-6}$ W/m² at the listener's ear. Assume that the sound strikes the surface of the ear perpendicularly. How much power is intercepted by the ear?

**49. ssm** A loudspeaker has a circular opening with a radius of 0.0950 m. The electrical power needed to operate the speaker is 25.0 W. The average sound intensity at the opening is 17.5 W/m². What percentage of the electrical power is converted by the speaker into sound power?

**50.** The average sound intensity inside a busy restaurant is $3.2 \times 10^{-5}$ W/m². How much energy goes into each ear (area = $2.1 \times 10^{-3}$ m²) during a one-hour meal?

**51.** Suppose that sound is emitted uniformly in all directions by a public address system. The intensity at a location 22 m away from the sound source is $3.0 \times 10^{-4}$ W/m². What is the intensity at a spot that is 78 m away?

**52.** Suppose in Conceptual Example 8 (see Figure 16.23) that the person is producing 1.1 mW of sound power. Some of the sound is reflected from the floor and ceiling. The intensity of this reflected sound at a distance of 3.0 m from the source is $4.4 \times 10^{-6}$ W/m². What is the total sound intensity due to both the direct and reflected sounds, at this point?

**53. ssm www** At a distance of 3.8 m from a siren, the sound intensity is $3.6 \times 10^{-2}$ W/m². Assuming that the siren radiates sound uniformly in all directions, find the total power radiated.

* **54.** Deep ultrasonic heating is used to promote healing of torn tendons. It is produced by applying ultrasonic sound to the body. The sound transducer (generator) is circular with a radius of 1.8 cm, and it produces a sound intensity of $5.9 \times 10^3$ W/m². How much time is required for the transducer to emit 4800 J of sound energy?

* **55.** Review **Interactive Solution 16.55** at **www.wiley.com/college/cutnell** for one approach to this problem. A dish of lasagna is being heated in a microwave oven. The effective area of the lasagna that is exposed to the microwaves is $2.2 \times 10^{-2}$ m$^2$. The mass of the lasagna is 0.35 kg, and its specific heat capacity is 3200 J/(kg·C°). The temperature rises by 72 C° in 8.0 minutes. What is the intensity of the microwaves in the oven?

* **56.** When a helicopter is hovering 1450 m directly overhead, an observer on the ground measures a sound intensity $I$. Assume that sound is radiated uniformly from the helicopter and that ground reflections are negligible. How far must the helicopter fly in a straight line parallel to the ground before the observer measures a sound intensity of $\frac{1}{4}I$?

** **57.** **ssm** A rocket, starting from rest, travels straight up with an acceleration of 58.0 m/s$^2$. When the rocket is at a height of 562 m, it produces sound that eventually reaches a ground-based monitoring station directly below. The sound is emitted uniformly in all directions. The monitoring station measures a sound intensity $I$. Later, the station measures an intensity $\frac{1}{3}I$. Assuming that the speed of sound is 343 m/s, find the time that has elapsed between the two measurements.

## Section 16.8 Decibels

**58.** A middle-aged man typically has poorer hearing than a middle-aged woman. In one case a woman can just begin to hear a musical tone, while a man can just begin to hear the tone only when its intensity level is increased by 7.8 dB relative to that for the woman. What is the ratio of the sound intensity just detected by the man to that just detected by the woman?

**59.** The bellow of a territorial bull hippopotamus has been measured at 115 dB above the threshold of hearing. What is the sound intensity?

**60.** A recording engineer works in a soundproofed room that is 44.0 dB quieter than the outside. If the sound intensity in the room is $1.20 \times 10^{-10}$ W/m$^2$, what is the intensity outside?

**61.** **ssm** Humans can detect a difference in sound intensity levels as small as 1.0 dB. What is the ratio of the sound intensities?

**62.** The equation $\beta = (10 \text{ dB}) \log (I/I_0)$, which defines the decibel, is sometimes written in terms of power $P$ (in watts) rather than intensity $I$ (in watts/meter$^2$). The form $\beta = (10 \text{ dB}) \log (P/P_0)$ can be used to compare two power levels in terms of decibels. Suppose that stereo amplifier A is rated at $P = 250$ watts per channel, and amplifier B has a rating of $P_0 = 45$ watts per channel. (a) Expressed in decibels, how much more powerful is A compared to B? (b) Will A sound more than twice as loud as B? Justify your answer.

**63.** For information, read Problem 62 before working this problem. Stereo manufacturers express the power output of a stereo amplifier using the decibel, abbreviated as dBW, where the "W" indicates that a reference power level of $P_0 = 1.00$ W has been used. If an amplifier has a power rated at 17.5 dBW, how many watts of power can this amplifier deliver?

* **64.** **ssm** Review Conceptual Example 8 as background for this problem. A loudspeaker is generating sound in a room. At a certain point, the sound waves coming directly from the speaker (without reflecting from the walls) create an intensity level of 75.0 dB. The waves reflected from the walls create, by themselves, an intensity level of 72.0 dB at the same point. What is the total intensity level? (*Hint: The answer is not 147.0 dB.*)

* **65.** In a discussion person A is talking 1.5 dB louder than person B, and person C is talking 2.7 dB louder than person A. What is the ratio of the sound intensity of person C to the sound intensity of person B?

* **66.** The sound intensity level of a person speaking normally is about 65 dB above the threshold of hearing. What is the minimum number of people speaking simultaneously, each with this intensity level, that is necessary to produce a sound intensity level at least 78 dB above the threshold of hearing?

* **67.** A portable radio is sitting at the edge of a balcony 5.1 m above the ground. The unit is emitting sound uniformly in all directions. By accident, it falls from rest off the balcony and continues to play on the way down. A gardener is working in a flower bed directly below the falling unit. From the instant the unit begins to fall, how much time is required for the sound intensity level heard by the gardener to increase by 10.0 dB?

** **68.** A source emits sound uniformly in all directions. A radial line is drawn from this source. On this line, determine the positions of two points, 1.00 m apart, such that the intensity level at one point is 2.00 dB greater than that at the other.

** **69.** **ssm** Suppose that when a certain sound intensity level (in dB) triples, the sound intensity (in W/m$^2$) also triples. Determine this sound intensity level.

## Section 16.9 The Doppler Effect

**70.** You are riding your bicycle directly away from a stationary source of sound and hear a frequency that is 1.0% lower than the emitted frequency. The speed of sound is 343 m/s. What is your speed?

**71.** The security alarm on a parked car goes off and produces a frequency of 960 Hz. The speed of sound is 343 m/s. As you drive toward this parked car, pass it, and drive away, you observe the frequency to change by 95 Hz. At what speed are you driving?

**72.** Suppose you are stopped for a traffic light, and an ambulance approaches you from behind with a speed of 18 m/s. The siren on the ambulance produces sound with a frequency of 955 Hz. The speed of sound in air is 343 m/s. What is the wavelength of the sound reaching your ears?

**73.** **ssm** A speeder looks in his rearview mirror. He notices that a police car has pulled behind him and is matching his speed of 38 m/s. The siren on the police car has a frequency of 860 Hz when the police car and the listener are stationary. The speed of sound is 343 m/s. What frequency does the speeder hear when the siren is turned on in the moving police car?

**74.** A bird is flying directly toward a stationary bird-watcher and emits a frequency of 1250 Hz. The bird-watcher, however, hears a frequency of 1290 Hz. What is the speed of the bird, expressed as a percentage of the speed of sound?

* **75.** **ssm** An aircraft carrier has a speed of 13.0 m/s relative to the water. A jet is catapulted from the deck and has a speed of 67.0 m/s relative to the water. The engines produce a 1550-Hz whine, and the speed of sound is 343 m/s. What is the frequency of the sound heard by the crew on the ship?

* **76.** A bungee jumper jumps from rest and screams with a frequency of 589 Hz. The air temperature is 20 °C. What is the frequency heard by the people on the ground below when she has fallen a distance of 11.0 m? Assume that the bungee cord has not yet taken effect, so she is in free-fall.

* **77.** Refer to **Interactive Solution 16.77** at **www.wiley.com/college/cutnell** for one approach to this type of problem. Two trucks travel at the same speed. They are far apart on adjacent lanes and approach each other essentially head-on. One driver hears the horn of the other truck at a frequency that is 1.14 times the frequency he hears

when the trucks are stationary. The speed of sound is 343 m/s. At what speed is each truck moving?

*78. Two submarines are underwater and approaching each other head-on. Sub A has a speed of 12 m/s and sub B has a speed of 8 m/s. Sub A sends out a 1550-Hz sonar wave that travels at a speed of 1522 m/s. (a) What is the frequency detected by sub B? (b) Part of the sonar wave is reflected from B and returns to A. What frequency does A detect for this reflected wave?

*79. **ssm** A motorcycle starts from rest and accelerates along a straight line at 2.81 m/s². The speed of sound is 343 m/s. A siren at the starting point remains stationary. How far has the motorcycle gone when the driver hears the frequency of the siren at 90.0% of the value it has when the motorcycle is stationary?

**80. A microphone is attached to a spring that is suspended from the ceiling, as the drawing indicates. Directly below on the floor is a stationary 440-Hz source of sound. The microphone vibrates up and down in simple harmonic motion with a period of 2.0 s. The difference between the maximum and minimum sound frequencies detected by the microphone is 2.1 Hz. Ignoring any reflections of sound in the room and using 343 m/s for the speed of sound, determine the amplitude of the simple harmonic motion.

Sound source

# Chapter 17 The Principle of Linear Superposition and Interference Phenomena

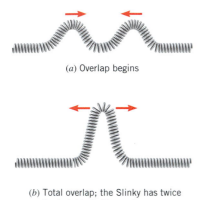

*(a)* Overlap begins

*(b)* Total overlap; the Slinky has twice the height of either pulse

*(c)* The receding pulses

**Figure 17.1** Two transverse "up" pulses passing through each other.

## 17.1 The Principle of Linear Superposition

Often, two or more sound waves are present at the same place at the same time, as is the case with sound waves when everyone is talking at a party or when music plays from the speakers of a stereo system. To illustrate what happens when several waves pass simultaneously through the same region, let's first consider Figures 17.1 and 17.2, which show two transverse pulses of equal heights moving toward each other along a Slinky. In Figure 17.1 both pulses are "up," while in Figure 17.2 one is "up" and the other is "down." Part *a* of each drawing shows the two pulses beginning to overlap. The pulses merge, and the Slinky assumes a shape that is *the sum of the shapes of the individual pulses.* Thus, when the two "up" pulses overlap completely, as in Figure 17.1*b*, the Slinky has a pulse height that is twice the height of an individual pulse. Likewise, when the "up" pulse and the "down" pulse overlap exactly, as in Figure 17.2*b*, they momentarily cancel, and the Slinky becomes straight. In either case, the two pulses move apart after overlapping, and the Slinky once again conforms to the shapes of the individual pulses.

The adding together of individual pulses to form a resultant pulse is an example of a more general concept called the **principle of linear superposition.**

> ■ **THE PRINCIPLE OF LINEAR SUPERPOSITION**
>
> When two or more waves are present simultaneously at the same place, the resultant disturbance is the sum of the disturbances from the individual waves.

This principle can be applied to all types of waves, including sound waves, water waves, and electromagnetic waves such as light. It embodies one of the most important concepts in physics, and the remainder of this chapter deals with examples related to it.

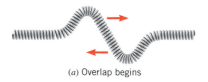

*(a)* Overlap begins

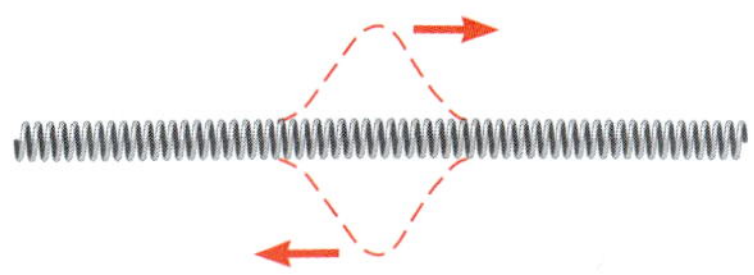

*(b)* Total overlap

*(c)* The receding pulses

**Figure 17.2** Two transverse pulses, one "up" and one "down," passing through each other.

## 17.2 Constructive and Destructive Interference of Sound Waves

Suppose that the sounds from two speakers overlap in the middle of a listening area, as in Figure 17.3, and that each speaker produces a sound wave of the same amplitude and fre-

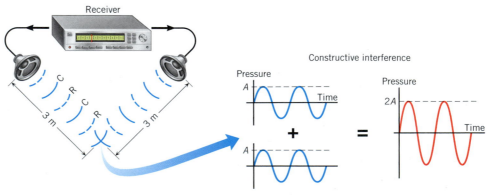

**Figure 17.3** As a result of constructive interference between the two sound waves (amplitude = *A*), a loud sound (amplitude = 2*A*) is heard at an overlap point located equally distant from two in-phase speakers (C, condensation; R, rarefaction).

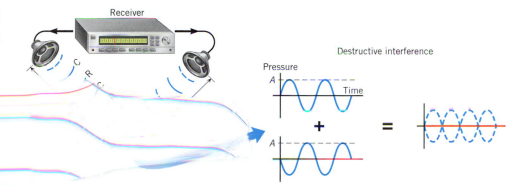

Destructive interference

Pressure

$A$

Time

$+$

$A$

$=$

**Figure 17.4** The speakers in this drawing vibrate in phase. However, the left speaker is one-half of a wavelength ($\frac{1}{2}$ m) farther from the overlap point than the right speaker. Because of destructive interference, no sound is heard at the overlap point (C, condensation; R, rarefaction).

quency. For convenience, the wavelength of the sound is chosen to be $\lambda = 1$ m. In addition, assume the diaphragms of the speakers vibrate in phase; that is, they move outward together and inward together. If the distance of each speaker from the overlap point is the same (3 m in the drawing), the condensations (C) of one wave always meet the condensations of the other when the waves come together; similarly, rarefactions (R) always meet rarefactions. According to the principle of linear superposition, the combined pattern is the sum of the individual patterns. As a result, the pressure fluctuations at the overlap point have twice the amplitude $A$ that the individual waves have, and a listener at this spot hears a louder sound than that coming from either speaker alone. When two waves always meet condensation-to-condensation and rarefaction-to-rarefaction (or crest-to-crest and trough-to-trough), they are said to be ***exactly in phase*** and to exhibit ***constructive interference.***

Now consider what happens if one of the speakers is moved. The result is surprising. In Figure 17.4, the left speaker is moved away* from the overlap point by a distance equal to one-half of the wavelength, or 0.5 m. Therefore, at the overlap point, a condensation arriving from the left meets a rarefaction arriving from the right. Likewise, a rarefaction arriving from the left meets a condensation arriving from the right. According to the principle of linear superposition, the net effect is a mutual cancellation of the two waves. The condensations from one wave offset the rarefactions from the other, leaving only a *constant air pressure.* A constant air pressure, devoid of condensations and rarefactions, means that a listener detects no sound. When two waves always meet condensation-to-rarefaction (or crest-to-trough), they are said to be ***exactly out of phase*** and to exhibit ***destructive interference.***

When two waves meet, they interfere constructively if they always meet exactly in phase and destructively if they always meet exactly out of phase. In either case, this means that the wave patterns do not shift relative to one another as time passes. Sources that produce waves in this fashion are called ***coherent sources.***

Destructive interference is the basis of a useful technique for reducing the loudness of undesirable sounds. For instance, Figure 17.5 shows a pair of noise-canceling head-

**Figure 17.5** Noise-canceling headphones utilize destructive interference.

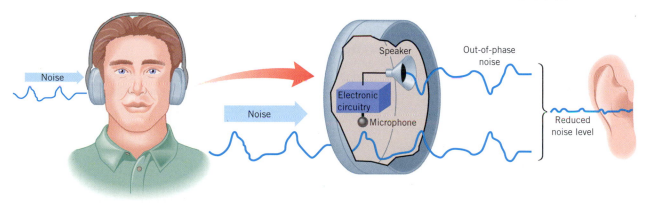

* When the left speaker is moved back, its sound intensity and, hence, its pressure amplitude decrease at the overlap point. In this chapter assume that the power delivered to the left speaker by the receiver is increased slightly to keep the amplitudes equal at the overlap point.

phones. Small microphones are mounted inside the headphones and detect noise such as the engine noise that an airplane pilot would hear. The headphones also contain circuitry to process the electronic signals from the microphones and reproduce the noise in a form that is exactly out of phase compared to the original. This out-of-phase version is played back through the headphone speakers and, because of destructive interference, combines with the original noise to produce a quieter background.

It should be apparent that if the left speaker in Figure 17.4 were moved away from the overlap point by *another* one-half wavelength ($3\frac{1}{2}$ m + $\frac{1}{2}$ m = 4 m), the two waves would again be in phase, and constructive interference would occur. The listener would hear a loud sound because the left wave travels one whole wavelength ($\lambda = 1$ m) farther than the right wave and, at the overlap point, condensation meets condensation and rarefaction meets rarefaction. In general, the important issue is the *difference* in the distances traveled by each wave in reaching the overlap point:

**Problem solving insight**

> *For two wave sources vibrating in phase, a difference in path lengths that is zero or an integer number (1, 2, 3, . . .) of wavelengths leads to constructive interference; a difference in path lengths that is a half-integer number ($\frac{1}{2}$, $1\frac{1}{2}$, $2\frac{1}{2}$, . . .) of wavelengths leads to destructive interference.*

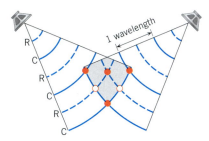

**Figure 17.6** Two sound waves overlap in the shaded region. The solid lines denote the middle of the condensations (C), and the dashed lines denote the middle of the rarefactions (R). Constructive interference occurs at each solid dot (●) and destructive interference at each open dot (○).

Interference effects can also be detected if the two speakers are fixed in position and the listener moves about the room. Consider Figure 17.6, where the sound waves spread outward from each speaker, as indicated by the concentric circular arcs. Each solid arc represents the middle of a condensation, and each dashed arc represents the middle of a rarefaction. Where the two waves overlap, there are places of constructive interference and places of destructive interference. Constructive interference occurs at any spot where two condensations or two rarefactions intersect, and the drawing shows four such places as solid dots. A listener stationed at any one of these locations hears a loud sound. On the other hand, destructive interference occurs at any place where a condensation and a rarefaction intersect, such as the two open dots in the picture. A listener situated at a point of destructive interference hears no sound. At locations where neither constructive nor destructive interference occurs, the two waves partially reinforce or partially cancel, depending on the position relative to the speakers. Thus, it is possible for a listener to walk about the overlap region and hear marked variations in loudness.

The individual sound waves from the speakers in Figure 17.6 carry energy, and the energy delivered to the overlap region is the sum of the energies of the individual waves. This fact is consistent with the principle of conservation of energy, which we first encountered in Section 6.8. This principle states that energy can neither be created nor destroyed, but can only be converted from one form to another. One of the interesting consequences of interference is that the energy is redistributed, so there are places within the overlap region where the sound is loud and other places where there is no sound at all. Interference, so to speak, "robs Peter to pay Paul," but energy is always conserved in the process. Example 1 illustrates how to decide what a listener hears.

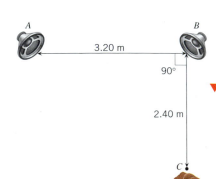

**Figure 17.7** Example 1 discusses whether this setup leads to constructive or destructive interference at point C for 214-Hz sound waves.

## Example 1   What Does a Listener Hear?

In Figure 17.7 two in-phase loudspeakers, A and B, are separated by 3.20 m. A listener is stationed at point C, which is 2.40 m in front of speaker B. The triangle ABC is a right triangle. Both speakers are playing identical 214-Hz tones, and the speed of sound is 343 m/s. Does the listener hear a loud sound or no sound?

**Reasoning** The listener will hear either a loud sound or no sound, depending on whether the interference occurring at point C is constructive or destructive. To determine which it is, we need to find the difference in the distances traveled by the two sound waves that reach point C and see whether the difference is an integer or half-integer number of wavelengths. In either event, the wavelength can be found from the relation $\lambda = v/f$ (Equation 16.1).

**Solution** Since the triangle ABC is a right triangle, the distance AC is given by the Pythagorean theorem as $\sqrt{(3.20 \text{ m})^2 + (2.40 \text{ m})^2} = 4.00$ m. The distance BC is given as 2.40 m. Thus, the difference in the travel distances for the waves is 4.00 m − 2.40 m = 1.60 m.

The wavelength of the sound is

$$\lambda = \frac{v}{f} = \frac{343 \text{ m/s}}{214 \text{ Hz}} = 1.60 \text{ m} \qquad (16.1)$$

Since the difference in the distances is one wavelength, constructive interference occurs at point *C*, and the *listener hears a loud sound.*

Up to this point, we have been assuming that the speaker diaphragms vibrate synchronously, or in phase; that is, they move outward together and inward together. This may not be the case, however, and Conceptual Example 2 considers what happens then.

### Conceptual Example 2  Out-of-Phase Speakers

To make a speaker operate, two wires must be connected between the speaker and the receiver (amplifier), as in Figure 17.8. To ensure that the diaphragms of two speakers vibrate in phase, it is necessary to make these connections in exactly the same way. If the wires for one speaker are not connected just as they are for the other speaker, the two diaphragms will vibrate out of phase. Whenever one diaphragm moves outward, the other will move inward, and vice versa. Suppose that in Figures 17.3 and 17.4 the connections are made so that the speaker diaphragms vibrate out of phase, everything else remaining the same. In each case, what kind of interference would result at the overlap point?

**Reasoning and Solution** Since the diaphragms are vibrating out of phase, one of them is now moving exactly opposite to the way it was moving originally. Let us assume that it is the one on the left in the drawings. The effect of this change is that every condensation originating from the left speaker becomes a rarefaction and every rarefaction becomes a condensation. As a result, in Figure 17.3 a rarefaction from the left now meets a condensation from the right at the overlap point, and *destructive interference results.* In Figure 17.4, a condensation from the left now meets a condensation from the right at the overlap point, *leading to constructive interference.* These results are opposite to those shown in the figures.

Instructions for connecting stereo systems specifically warn users to avoid out-of-phase vibration of the speaker diaphragms. One way to check your system is to play some music that is rich in low-frequency bass tones. Use the monaural mode on your receiver so the same sound comes from each speaker, and slide the speakers toward each other. If the diaphragms are moving in phase, the bass sound will either remain the same or slightly improve as the speakers come together. If the diaphragms are moving out of phase, the bass sound will decrease noticeably because of destructive interference when the speakers are right next to each other. In this event, simply interchange the wires to the terminals on one (but not both) of the speakers.

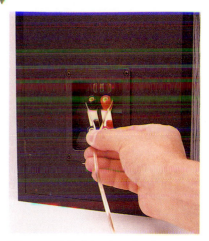

**Figure 17.8** A loudspeaker is connected to a receiver (amplifier) by two wires. (© Andy Washnik)

**The physics of**
**wiring stereo speakers.**

The phenomena of constructive and destructive interference are exhibited by all types of waves, not just sound waves. We will encounter interference effects again in Chapter 27, in connection with light waves.

## 17.3  Diffraction

Section 16.5 discusses the fact that sound is a pressure wave created by a vibrating object, such as a loudspeaker. The previous two sections of this chapter have examined what happens when two sound waves are present simultaneously at the same place; according to the principle of linear superposition, a resultant disturbance is formed from the sum of the individual waves. This principle reveals that overlapping sound waves exhibit interference effects, whereby the sound energy is redistributed within the overlap region. We will now use the principle of linear superposition to explore another interference effect, that of diffraction.

When a wave encounters an obstacle or the edges of an opening, it bends around them. For instance, a sound wave produced by a stereo system bends around the edges of an open doorway, as Figure 17.9a illustrates. If such bending did not occur, sound could be heard outside the room only at locations directly in front of the doorway, as part *b* of the drawing suggests. (It is assumed that no sound is transmitted directly through the

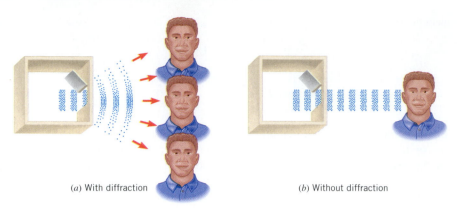

**Figure 17.9** (*a*) The bending of a sound wave around the edges of the doorway is an example of diffraction. The source of the sound within the room is not shown. (*b*) If diffraction did not occur, the sound wave would not bend as it passes through the doorway.

(*a*) With diffraction     (*b*) Without diffraction

walls.) The bending of a wave around an obstacle or the edges of an opening is called *diffraction*. All kinds of waves exhibit diffraction.

To demonstrate how the bending of waves arises, Figure 17.10 shows an expanded view of Figure 17.9a. When the sound wave reaches the doorway, the air in the doorway is set into longitudinal vibration. In effect, each molecule of the air in the doorway becomes a source of a sound wave in its own right, and, for purposes of illustration, the drawing shows two of the molecules. Each produces a sound wave that expands outward in three dimensions, much like a water wave does in two dimensions when a stone is dropped into a pond. The sound waves generated by all the molecules in the doorway must be added together to obtain the total sound wave at any location outside the room, in accord with the principle of linear superposition. However, even considering only the waves from the two molecules in the picture, it is clear that the expanding wave patterns reach locations off to either side of the doorway. The net effect is a "bending," or diffraction, of the sound around the edges of the opening. Further insight into the origin of diffraction can be obtained with the aid of Huygens' principle (see Section 27.5).

When the sound waves generated by every molecule in the doorway are added together, it is found that there are places where the intensity is a maximum and places where it is zero, in a fashion similar to that discussed in the previous section. Analysis shows that at a great distance from the doorway the intensity is a maximum directly opposite the center of the opening. As the distance to either side of the center increases, the intensity decreases and reaches zero, then rises again to a maximum, falls again to zero, rises back to a maximum, and so on. Only the maximum at the center is a strong one. The other maxima are weak and become progressively weaker at greater distances from the center. In Figure 17.10 the angle $\theta$ defines the location of the first minimum intensity point on either side of the center. Equation 17.1 gives $\theta$ in terms of the wavelength $\lambda$ and the width $D$ of the doorway and assumes that the doorway can be treated like a slit whose height is very large compared to its width:

*Single slit—first minimum*     $$\sin\theta = \frac{\lambda}{D}$$     (17.1)

Waves also bend around the edges of openings other than single slits. Particularly important is the diffraction of sound by a circular opening, such as that in a loudspeaker. In this case, the angle $\theta$ is related to the wavelength $\lambda$ and the diameter $D$ of the opening by

*Circular opening —first minimum*     $$\sin\theta = 1.22\frac{\lambda}{D}$$     (17.2)

An important point to remember about Equations 17.1 and 17.2 is that the extent of the diffraction depends on the ratio of the wavelength to the size of the opening. If the ratio $\lambda/D$ is small, then $\theta$ is small and little diffraction occurs. The waves are beamed in the forward direction as they leave an opening, much like the light from a flashlight. Such sound waves are said to have "narrow dispersion." Since high-frequency sound has a relatively small wavelength, it tends to have a narrow dispersion. On the other hand, for

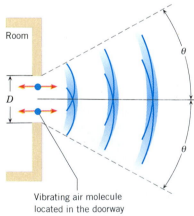

Room

$D$

Vibrating air molecule located in the doorway

**Figure 17.10** Each vibrating molecule of the air in the doorway generates a sound wave that expands outward and bends, or diffracts, around the edges of the doorway. Because of interference effects among the sound waves produced by all the molecules, the sound intensity is mostly confined to the region defined by the angle $\theta$ on either side of the doorway.

larger values of the ratio $\lambda/D$, the angle $\theta$ is larger. The waves spread out over a larger region and are said to have a "wide dispersion." Low-frequency sound, with its relatively large wavelength, typically has a wide dispersion.

In a stereo loudspeaker, a wide dispersion of the sound is desirable. Example 3 illustrates, however, that there are limitations to the dispersion that can be achieved, depending on the loudspeaker design.

### Example 3  Designing a Loudspeaker for Wide Dispersion

A 1500-Hz sound and a 8500-Hz sound each emerges from a loudspeaker through a circular opening whose diameter is 0.30 m (see Figure 17.11). Assuming that the speed of sound in air is 343 m/s, find the diffraction angle $\theta$ for each sound.

**Reasoning** The diffraction angle $\theta$ for each sound wave is given by $\sin\theta = 1.22(\lambda/D)$. However, it will first be necessary to calculate the wavelengths of the sounds from $\lambda = v/f$ (Equation 16.1).

**Solution** The wavelengths of the two sounds are

$$\lambda_{1500} = \frac{343 \text{ m/s}}{1500 \text{ Hz}} = 0.23 \text{ m} \quad \text{and} \quad \lambda_{8500} = \frac{343 \text{ m/s}}{8500 \text{ Hz}} = 0.040 \text{ m}$$

The diffraction angles can now be determined:

**1500-Hz sound**
$$\sin\theta = 1.22 \frac{\lambda_{1500}}{D} = 1.22\left(\frac{0.23 \text{ m}}{0.30 \text{ m}}\right) = 0.94 \qquad (17.2)$$

$$\theta = \sin^{-1} 0.94 = \boxed{70°}$$

**8500-Hz sound**
$$\sin\theta = 1.22 \frac{\lambda_{8500}}{D} = 1.22\left(\frac{0.040 \text{ m}}{0.30 \text{ m}}\right) = 0.16 \qquad (17.2)$$

$$\theta = \sin^{-1} 0.16 = \boxed{9.2°}$$

Figure 17.11 illustrates these results. With a 0.30-m opening, the dispersion of the higher-frequency sound is limited to only 9.2°. To increase the dispersion, a smaller opening is needed. It is for this reason that loudspeaker designers use a small-diameter speaker called a *tweeter* to generate the high-frequency sound, as Figure 17.12 indicates.

As we have seen, diffraction is an interference effect, one in which some of the wave's energy is directed into regions that would otherwise not be accessible. Energy, of course, is conserved during this process, because energy is only redistributed during diffraction; no energy is created or destroyed.

## 17.4  Beats

In situations where waves with the *same frequency* overlap, we have seen how the principle of linear superposition leads to constructive and destructive interference and how it explains diffraction. We will see in this section that two overlapping waves with *slightly different frequencies* give rise to the phenomenon of beats. However, the principle of linear superposition again provides an explanation of what happens when the waves overlap.

A tuning fork has the property of producing a single-frequency sound wave when struck with a sharp blow. Figure 17.13 shows sound waves coming from two tuning forks placed side by side. The tuning forks in the drawing are identical, and each is designed to produce a 440-Hz tone. However, a small piece of putty has been attached to one fork, whose frequency is lowered to 438 Hz because of the added mass. When the forks are sounded simultaneously, the loudness of the resulting sound rises and falls periodically — faint, then loud, then faint, then loud, and so on. The periodic variations in loudness are called *beats* and result from the interference between two sound waves with slightly different frequencies.

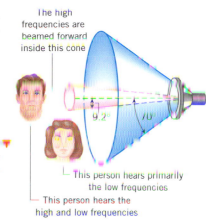

The high frequencies are beamed forward inside this cone

This person hears primarily the low frequencies

This person hears the high and low frequencies

**Figure 17.11** Because the dispersion of high frequencies is less than that of low frequencies, you should be directly in front of the speaker to hear both the high and low frequencies equally well.

**Problem solving insight**
When a wave passes through an opening, the extent of diffraction is greater when the ratio $\lambda/D$ is greater, where $\lambda$ is the wavelength of the wave and $D$ is the width or diameter of the opening.

Tweeter

**Figure 17.12** Small-diameter speakers, called tweeters, are used to produce high-frequency sound. The small diameter helps to promote a wider dispersion of the sound. (© Christopher Gould/The Image Bank/Getty Images)

**The physics of**
tweeter loudspeakers. (See above.)

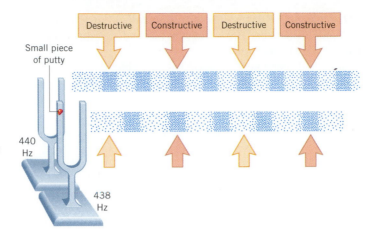

**Figure 17.13** Two tuning forks have slightly different frequencies of 440 and 438 Hz. The phenomenon of beats occurs when the forks are sounded simultaneously. The sound waves are not drawn to scale.

For clarity, Figure 17.13 shows the condensations and rarefactions of the sound waves separately. In reality, however, the waves spread out and overlap. In accord with the principle of linear superposition, the ear detects the combined total of the two. Notice that there are places where the waves interfere constructively and places where they interfere destructively. When a region of constructive interference reaches the ear, a loud sound is heard. When a region of destructive interference arrives, the sound intensity drops to zero (assuming each of the waves has the same amplitude). The number of times per second that the loudness rises and falls is the **beat frequency** and is the **difference** between the two sound frequencies. Thus, in the situation illustrated in Figure 17.13, an observer hears the sound loudness rise and fall at the rate of 2 times per second (440 Hz − 438 Hz).

Figure 17.14 helps to explain why the beat frequency is the difference between the two frequencies. The drawing displays graphical representations of the pressure patterns of a 10-Hz wave and a 12-Hz wave, along with the pressure pattern that results when the two overlap. These frequencies have been chosen for convenience, even though they lie below the audio range and are inaudible. Audible sound waves behave in exactly the same way. The top two drawings, in blue, show the pressure variations in a one-second interval of each wave. The third drawing, in red, shows the result of adding together the blue patterns according to the principle of linear superposition. Notice that the amplitude in the red drawing is not constant, as it is in the individual waves. Instead, the amplitude changes from a minimum to a maximum, back to a minimum, and so on. When such pressure variations reach the ear and occur in the audible frequency range, they produce a loud sound when the amplitude is a maximum and a faint sound when the amplitude is a minimum. Two loud–faint cycles, or beats, occur in the one-second interval shown in the

**Figure 17.14** A 10-Hz sound wave and a 12-Hz sound wave, when added together, produce a wave with a beat frequency of 2 Hz. The drawings show the pressure patterns (in blue) of the individual waves and the pressure pattern (in red) that results when the two overlap. The time interval shown is one second.

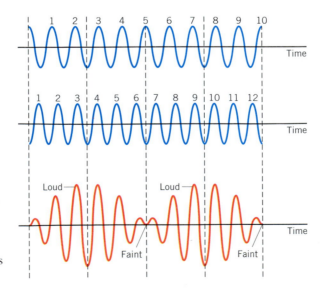

drawing, corresponding to a beat frequency of 2 Hz. Thus, the beat frequency is the difference between the frequencies of the individual waves, or 12 Hz − 10 Hz = 2 Hz.

Musicians often tune their instruments by listening to a beat frequency. For instance, a guitar player plucks an out-of-tune string along with a tone from a source known to have the correct frequency. The guitarist adjusts the tension in the string until the beats vanish, ensuring that the string is vibrating at the correct frequency.

**The physics of** tuning a musical instrument.

## 17.5 Transverse Standing Waves

A standing wave is another interference effect that can occur when two waves overlap. Standing waves can arise with transverse waves, such as those on a guitar string, and also with longitudinal sound waves, such as those in a flute. In any case, the principle of linear superposition provides an explanation of the effect, just as it does for diffraction and beats.

Figure 17.15 shows some of the essential features of transverse standing waves. In this figure the left end of each string is vibrated back and forth, while the right end is attached to a wall. Regions of the string move so fast that they appear only as a blur in the photographs. Each of the patterns shown is called a *transverse standing wave pattern.* Notice that the patterns include special places called nodes and antinodes. The *nodes* are places that do not vibrate at all, and the *antinodes* are places where maximum vibration occurs. To the right of each photograph is a series of superimposed drawings that help us to visualize the motion of the string as it vibrates in a standing wave pattern. These drawings freeze the shape of the string at various times and emphasize the maximum vibration that occurs at an antinode with the aid of a red dot attached to the string.

Each standing wave pattern is produced at a unique frequency of vibration. These frequencies form a series, the smallest frequency $f_1$ corresponding to the one-loop pattern and the larger frequencies being integer multiples of $f_1$, as Figure 17.15 indicates.

**Figure 17.15** Vibrating a string at certain unique frequencies sets up transverse standing wave patterns, such as the three shown in the photographs on the left. Each drawing on the right shows the various shapes that the string assumes at various times as it vibrates. The red dots attached to the strings focus attention on the maximum vibration that occurs at an antinode. In each of the drawings, one-half of a wave cycle is outlined in red. (© Richard Megna/Fundamental Photographs)

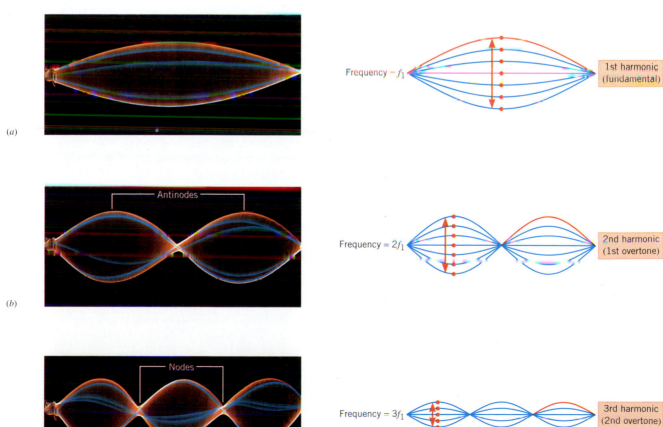

(a) Frequency = $f_1$ — 1st harmonic (fundamental)

(b) Antinodes — Frequency = $2f_1$ — 2nd harmonic (1st overtone)

(c) Nodes — Frequency = $3f_1$ — 3rd harmonic (2nd overtone)

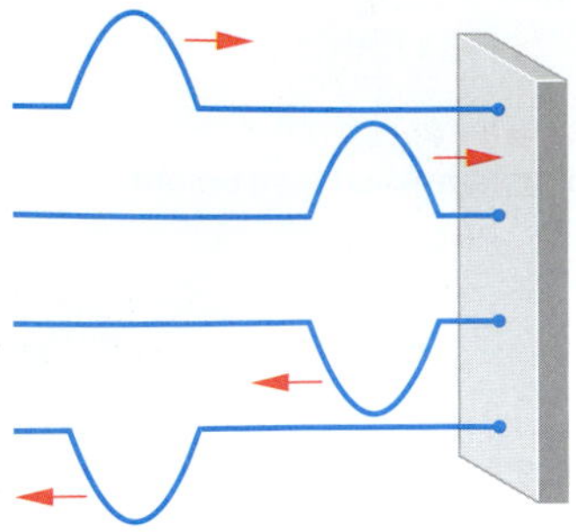

**Figure 17.16** In reflecting from the wall, a forward-traveling half-cycle becomes a backward-traveling half-cycle that is inverted.

Thus, if $f_1$ is 10 Hz, the frequency needed to establish the 2-loop pattern is $2f_1$ or 20 Hz, while that needed to create the 3-loop pattern is $3f_1$ or 30 Hz, and so on. The frequencies in this series ($f_1$, $2f_1$, $3f_1$, etc.) are called **harmonics.** The lowest frequency $f_1$ is called the first harmonic, and the higher frequencies are designated as: the second harmonic ($2f_1$), the third harmonic ($3f_1$), and so forth. The harmonic number (1st, 2nd, 3rd, etc.) corresponds to the number of loops in the standing wave pattern. The frequencies in this series are also referred to as the fundamental frequency, the first overtone, the second overtone, and so on. Thus, frequencies above the fundamental are **overtones** (see Figure 17.15).

Standing waves arise because identical waves travel on the string in *opposite directions* and combine in accord with the principle of linear superposition. A standing wave is said to be *standing* because it does not travel in one direction or the other, as do the individual waves that produce it. Figure 17.16 shows why there are waves traveling in both directions on the string. At the top of the picture, one-half of a wave cycle (the remainder of the wave is omitted for clarity) is moving toward the wall on the right. When the half-cycle reaches the wall, it causes the string to pull upward on the wall. Consistent with Newton's action–reaction law, the wall pulls downward on the string, and a downward-pointing half-cycle is sent back toward the left. Thus, the wave reflects from the wall. Upon arriving back at the point of origin, the wave reflects again, this time from the hand vibrating the string. For small vibration amplitudes, the hand is essentially fixed and behaves as the wall does in causing reflections. Repeated reflections at both ends of the string create a multitude of wave cycles traveling in both directions.

As each new cycle is formed by the vibrating hand, previous cycles that have reflected from the wall arrive and reflect again from the hand. Unless the timing is right, however, the new cycles and the reflected cycles tend to offset one another, and the formation of a standing wave is inhibited. Think about pushing someone on a swing and timing your pushes so that the effect of one push reinforces that of another. Such reinforcement in the case of the wave cycles leads to a large-amplitude standing wave. Suppose the string has a length $L$ and its left end is being vibrated at a frequency $f_1$. The time required to create a new wave cycle is the period $T$ of the wave, where $T = 1/f_1$ (Equation 10.5). On the other hand, the time needed for a cycle to travel from the hand to the wall and back, a distance of $2L$, is $2L/v$, where $v$ is the wave speed. Reinforcement between new and reflected cycles occurs if these two times are equal; that is, if $1/f_1 = 2L/v$. Thus, a standing wave is established when the string is vibrated with a frequency of $f_1 = v/(2L)$.

Repeated reinforcement between newly created and reflected cycles causes a large-amplitude standing wave to develop on the string, *even when the hand itself vibrates with only a small amplitude*. Thus, the motion of the string is a resonance effect, analogous to that discussed in Section 10.6 for an object attached to a spring. The frequency $f_1$ at which resonance occurs is sometimes called a **natural frequency** of the string, similar to the frequency at which an object oscillates on a spring.

There is a difference between the resonance of the string and the resonance of a spring system, however. An object on a spring has only a single natural frequency, whereas the string has a *series* of natural frequencies. The series arises because a reflected wave cycle need not return to its point of origin in time to reinforce *every* newly created cycle. Reinforcement can occur, for instance, on *every other* new cycle, as it does if the string is vibrated at twice the frequency $f_1$, or $f_2 = 2f_1$. Likewise, if the vibration frequency is $f_3 = 3f_1$, reinforcement occurs on *every third* new cycle. Similar arguments apply for any frequency $f_n = nf_1$, where $n$ is an integer. As a result, the series of natural frequencies that lead to standing waves on a string fixed at both ends is given by

**String fixed at both ends** $$f_n = n \left( \frac{v}{2L} \right) \qquad n = 1, 2, 3, 4, \ldots \qquad (17.3)$$

**Problem solving insight**
The distance between two successive nodes (or between two successive antinodes) of a standing wave is equal to one-half of a wavelength.

It is also possible to obtain Equation 17.3 in another way. In Figure 17.15, one-half of a wave cycle is outlined in red for each of the harmonics, to show that each loop in a standing wave pattern corresponds to one-half a wavelength. Since the two fixed ends of the string are nodes, the length $L$ of the string must contain an integer number $n$ of half-wavelengths: $L = n(\frac{1}{2}\lambda_n)$ or $\lambda_n = 2L/n$. Using this result for the wavelength in the relation $f_n \lambda_n = v$ shows that $f_n(2L/n) = v$, which can be rearranged to give Equation 17.3.

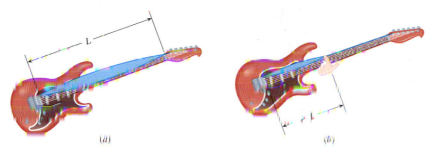

Standing waves on a string play an important role in the way many musical instruments produce sound. For instance, a guitar string is stretched between two supports and, when plucked, vibrates according to the series of natural frequencies given by Equation 17.3. The next two examples deal with how this series of frequencies governs the playing and the design of a guitar.

### Example 1   Playing a Guitar

The heaviest string on an electric guitar has a linear density of $m/L = 5.28 \times 10^{-3}$ kg/m and is stretched with a tension of $F = 226$ N. This string produces the musical note E when vibrating along its entire length in a standing wave at the fundamental frequency of 164.8 Hz. (a) Find the length $L$ of the string between its two fixed ends (see Figure 17.17a). (b) A guitar player wants the string to vibrate at a fundamental frequency of $2 \times 164.8$ Hz $= 329.6$ Hz, as it must if the musical note E is to be sounded one octave higher in pitch. To accomplish this, he presses the string against the proper fret and then plucks the string (see part b of the drawing). Find the distance $L$ between the fret and the bridge of the guitar.

**Reasoning** The fundamental frequency $f_1$ is given by Equation 17.3 with $n = 1$: $f_1 = v/(2L)$. Since $f_1$ is known in both parts (a) and (b), the length $L$ in each case can be calculated directly from this expression, once the speed $v$ is known. The speed, in turn, is related to the tension $F$ and the linear density $m/L$ according to Equation 16.2.

**Solution**

(a) The speed is

$$v = \sqrt{\frac{F}{m/L}} = \sqrt{\frac{226 \text{ N}}{5.28 \times 10^{-3} \text{ kg/m}}} = 207 \text{ m/s} \qquad (16.2)$$

According to $f_1 = v/(2L)$, the length of the string is

$$L = \frac{v}{2f_1} = \frac{207 \text{ m/s}}{2(164.8 \text{ Hz})} = \boxed{0.628 \text{ m}}$$

(b) The distance $L$ that locates the fret can be determined exactly as in part (a) by using the wave speed $v = 207$ m/s and noting that the frequency is now $f_1 = 329.6$ Hz: $\boxed{L = 0.314 \text{ m}}$. This length is exactly half that determined in part (a) because the frequencies have a ratio of $2:1$.

### Conceptual Example 5   The Frets on a Guitar

Figure 17.18 shows the frets on the neck of a guitar. They allow the player to produce a complete sequence of musical notes using a single string. Starting with the fret at the top of the neck, each successive fret indicates where the player should press to get the next note in the sequence. Musicians call the sequence the chromatic scale, and every thirteenth note in it corresponds to one octave, or a doubling of the sound frequency. The spacing between the frets is greatest at the top of the neck and decreases with each additional fret further on down. The spacing eventually becomes smaller than the width of a finger, limiting the number of frets that can be used. Why does the spacing between the frets decrease going down the neck?

*The physics of the frets on a guitar.*

**Reasoning and Solution** Our reasoning is based on Equation 17.3, with $n = 1$ [$f_1 = v/(2L)$]. The value of $n$ is 1 because a string vibrates mainly at its fundamental frequency

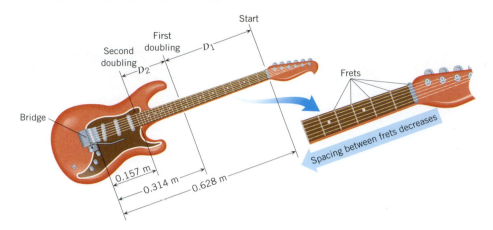

**Figure 17.18** The spacing between the frets on the neck of a guitar decreases going down the neck toward the bridge.

when plucked, as mentioned in Example 4. This equation indicates that the fundamental frequency $f_1$ is inversely proportional to the length $L$ between a given fret and the bridge of the guitar. Thus, in Example 4 we found that the E-string had a length of 0.628 m, corresponding to a frequency of $f_1 = 164.8$ Hz. We also found that the length between the bridge and the fret that must be pressed to double this frequency to 329.6 Hz is one-half of 0.628 m, or 0.314 m. To understand why the spacing between frets decreases going down the neck, consider the fret that must be pressed to double the frequency again, from 329.6 Hz to 659.2 Hz. The length between the bridge and this fret would be one-half of 0.314 m, or 0.157 m. The three lengths of 0.628 m, 0.314 m, and 0.157 m are shown in Figure 17.18. The distances $D_1$ and $D_2$ are also shown and give the spacings between the starting fret at the top of the neck, the fret for the first doubling, and the fret for the second doubling. Clearly, $D_1$ is greater than $D_2$, and the frets near the top of the neck have more space between them than those further down.

**Related Homework:** *Problem 32*

## 17.6 *Longitudinal Standing Waves*

Standing wave patterns can also be formed from longitudinal waves. For example, when sound reflects from a wall, the forward- and backward-going waves can produce a standing wave. Figure 17.19 illustrates the vibrational motion in a longitudinal standing wave on a Slinky. As in a transverse standing wave, there are nodes and antinodes. At the nodes the coils of the Slinky do not vibrate at all; that is, they have no displacement. At the antinodes the coils vibrate with maximum amplitude. The red dots in Figure 17.19 indicate the lack of vibration at a node and the maximum vibration at an antinode. At an antinode the coils have a maximum displacement. The vibration occurs along the line of travel of the individual waves, as is to be expected for longitudinal waves. In a standing wave of sound, the molecules or atoms of the medium behave as the red dots do.

Musical instruments in the wind family depend on longitudinal standing waves in producing sound. Since wind instruments (trumpet, flute, clarinet, pipe organ, etc.) are modified tubes or columns of air, it is useful to examine the standing waves that can be set up in such tubes. Figure 17.20 shows two cylindrical columns of air that are open at both ends. Sound waves, originating from a tuning fork, travel up and down within each tube, since they reflect from the ends of the tubes, even though the ends are open. If the frequency $f$ of the tuning fork matches one of the natural frequencies of the air column, the downward- and upward-traveling waves combine to form a standing wave, and the

**Figure 17.19** A longitudinal standing wave on a Slinky showing the displacement nodes (N) and antinodes (A).

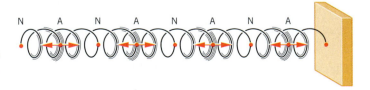

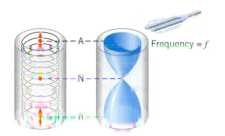
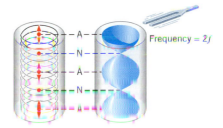

sound of the tuning fork becomes markedly louder. To emphasize the longitudinal nature of the standing wave patterns, the left side of each pair of drawings in Figure 17.20 replaces the air in the tubes with Slinkies, on which the nodes and antinodes are indicated with red dots. As an additional aid in visualizing the standing waves, the right side of each pair of drawings shows blurred blue patterns within each tube. These patterns symbolize the amplitude of the vibrating air molecules at various locations. Wherever the pattern is widest, the amplitude of vibration is greatest (a displacement antinode), and wherever the pattern is narrowest there is no vibration (a displacement node).

To determine the natural frequencies of the air columns in Figure 17.20, notice that there is a displacement antinode at each end of the open tube because the air molecules there are free to move.* As in a transverse standing wave, the distance between two successive antinodes is one-half of a wavelength, so the length $L$ of the tube must be an integer number $n$ of half-wavelengths: $L = n(\frac{1}{2}\lambda_n)$ or $\lambda_n = 2L/n$. Using this wavelength in the relation $f_n = v/\lambda_n$ shows that the natural frequencies $f_n$ of the tube are

**Tube open at both ends**
$$f_n = n\left(\frac{v}{2L}\right) \qquad n = 1, 2, 3, 4, \ldots \qquad (17.4)$$

At these frequencies, large-amplitude standing waves develop within the tube due to resonance. Example 6 illustrates how Equation 17.4 is involved when a flute is played.

### Example 6  Playing a Flute

When all the holes are closed on one type of flute, the lowest note it can sound is a middle C, whose fundamental frequency is 261.6 Hz. (a) The air temperature is 293 K, and the speed of sound is 343 m/s. Assuming the flute is a cylindrical tube open at both ends, determine the distance $L$ in Figure 17.21—that is, the distance from the mouthpiece to the end of the tube. (This distance is only approximate, since the antinode does not occur exactly at the mouthpiece.) (b) A flautist can alter the length of the flute by adjusting the extent to which the head joint is inserted into the main stem of the instrument. If the air temperature rises to 305 K, to what length must the flute be adjusted to play a middle C?

*The physics of a flute.*

**Reasoning** The fundamental frequency $f_1$ is given by Equation 17.4 with $n = 1$; $f_1 = v/(2L)$. This expression can be used to calculate the length as $L = v/(2 f_1)$. When the speed of sound $v$ changes, as it does when the temperature changes, the length of the flute must be changed. The effect of temperature on the speed of sound in air is given by $v = \sqrt{\gamma kT/m}$ (Equation 16.5), assuming air behaves as an ideal gas. Thus, the speed is proportional to the square root of the Kelvin temperature ($v \propto \sqrt{T}$), a fact that we can use to find the speed at the higher temperature.

**Solution**

(a) At a temperature of 293 K, when the speed of sound is $v = 343$ m/s, the length of the flute is

$$L = \frac{v}{2f_1} = \frac{343 \text{ m/s}}{2(261.6 \text{ Hz})} = \boxed{0.656 \text{ m}}$$

(b) Since $v \propto \sqrt{T}$, it follows that

$$\frac{v_{305 \text{ K}}}{v_{293 \text{ K}}} = \frac{\sqrt{305 \text{ K}}}{\sqrt{293 \text{ K}}} = 1.02$$

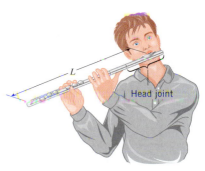

**Figure 17.21** The length $L$ of a flute between the mouthpiece and the end of the instrument determines the fundamental frequency of the lowest playable note.

* In reality, the antinode does not occur exactly at the open end. However, if the tube's diameter is small compared to its length, little error is made in assuming that the antinode is located right at the end.

As a result, $v_{305 \text{ K}} = 1.02(v_{293 \text{ K}}) = 1.02(343 \text{ m/s}) = 3.50 \times 10^2$ m/s. The adjusted flute length is

$$L = \frac{v}{2f_1} = \frac{3.50 \times 10^2 \text{ m/s}}{2(261.6 \text{ Hz})} = \boxed{0.669 \text{ m}}$$

Thus, to play in tune at the higher temperature, a flautist must lengthen the flute by 0.013 m.

Standing waves can also exist in a tube with only one end open, as the patterns in Figure 17.22 indicate. Note the difference between these patterns and those in Figure 17.20. Here the standing waves have a displacement antinode at the open end and a displacement node at the closed end, where the air molecules are not free to move. Since the distance between a node and an adjacent antinode is one-fourth of a wavelength, the length $L$ of the tube must be an odd number of quarter-wavelengths: $L = 1(\frac{1}{4}\lambda)$ and $L = 3(\frac{1}{4}\lambda)$ for the two standing wave patterns in Figure 17.22. In general, then, $L = n(\frac{1}{4}\lambda_n)$, where $n$ is any odd integer ($n = 1, 3, 5, \ldots$). From this result it follows that $\lambda_n = 4L/n$, and the natural frequencies $f_n$ can be obtained from the relation $f_n = v/\lambda_n$:

**Tube open at only one end**    $$f_n = n\left(\frac{v}{4L}\right) \qquad n = 1, 3, 5, \ldots \qquad (17.5)$$

A tube open at only one end can develop standing waves only at the odd harmonic frequencies $f_1, f_3, f_5$, etc. In contrast, a tube open at both ends can develop standing waves at all harmonic frequencies $f_1, f_2, f_3$, etc. Moreover, the fundamental frequency $f_1$ of a tube open at only one end (Equation 17.5) is one-half that of a tube open at both ends (Equation 17.4). In other words, a tube open only at one end needs to be only one-half as long as a tube open at both ends in order to produce the *same* fundamental frequency.

Energy is also conserved when a standing wave is produced, either on a string or in a tube of air. The energy of the standing wave is the sum of the energies of the individual waves that comprise the standing wave. Once again, interference redistributes the energy of the individual waves to create locations of greatest energy (displacement antinodes) and locations of no energy (displacement nodes).

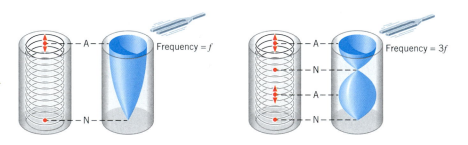

**Figure 17.22**  A pictorial representation of the longitudinal standing waves on a Slinky (left side of each pair) and in a tube of air (right side of each pair) that is open only at one end (A, antinode; N, node).

# Concept Summary

This summary presents an abridged version of the chapter, including the important equations and all available learning aids. For convenient reference, the learning aids (including the text's examples) are placed next to or immediately after the relevant equation or discussion. The following learning aids may be found on-line at **www.wiley.com/college/cutnell**:

| | |
|---|---|
| **Interactive LearningWare** examples are solved according to a five-step interactive format that is designed to help you develop problem-solving skills. | **Concept Simulations** are animated versions of text figures or animations that illustrate important concepts. You can control parameters that affect the display, and we encourage you to experiment. |
| **Interactive Solutions** offer specific models for certain types of problems in the chapter homework. The calculations are carried out interactively. | **Self-Assessment Tests** include both qualitative and quantitative questions. Extensive feedback is provided for both incorrect and correct answers, to help you evaluate your understanding of the material. |

| Topic | Discussion | Learning Aids |
|---|---|---|

### 17.1 The Principle of Linear Superposition

**Principle of linear superposition**

The principle of linear superposition states that when two or more waves are present simultaneously at the same place, the resultant disturbance is the sum of the disturbances from the individual waves.

**Concept Simulation 17.1**

### 17.2 Constructive and Destructive Interference of Sound Waves

**Constructive and destructive interference**

Constructive interference occurs at a point when two waves meet there crest-to-crest and trough-to-trough, thus reinforcing each other. Destructive interference occurs when the waves meet crest-to-trough and cancel each other.

**In-phase and out-of-phase waves**

When waves meet crest-to-crest and trough-to-trough, they are exactly in phase. When they meet crest-to-trough, they are exactly out of phase.

**Conditions for constructive and destructive interference**

For two wave sources vibrating in phase, a difference in path lengths that is zero or an integer number $(1, 2, 3, \ldots)$ of wavelengths leads to constructive interference; a difference in path lengths that is a half-integer number $(\frac{1}{2}, 1\frac{1}{2}, 2\frac{1}{2}, \ldots)$ of wavelengths leads to destructive interference.

**Examples 1, 2**

**Interactive Solution 17.53**

### 17.3 Diffraction

**Diffraction**

Diffraction is the bending of a wave around an obstacle or the edges of an opening. The angle through which the wave bends depends on the ratio of the wavelength $\lambda$ of the wave to the width $D$ of the opening; the greater the ratio $\lambda/D$, the greater the angle.

When a sound wave of wavelength $\lambda$ passes through an opening, the place where the intensity of the sound is a minimum relative to the center of the opening is specified by the angle $\theta$. If the opening is a rectangular slit of width $D$, such as a doorway, the angle is

**Single slit—first minimum**

$$\sin \theta = \frac{\lambda}{D} \qquad (17.1)$$

If the opening is a circular opening of diameter $D$, such as that in a loudspeaker, the angle is

**Circular opening—first minimum**

$$\sin \theta = 1.22 \frac{\lambda}{D} \qquad (17.2)$$

**Example 3**

### 17.4 Beats

**Beats**

**Beat frequency**

Beats are the periodic variations in amplitude that arise from the linear superposition of two waves that have slightly different frequencies. When the waves are sound waves, the variations in amplitude cause the loudness to vary at the beat frequency, which is the difference between the frequencies of the waves.

**Concept Simulation 17.2**

**Interactive LearningWare 17.1**

---

*Use Self-Assessment Test 17.1 to evaluate your understanding of Sections 17.1–17.4.*

---

### 17.5 Transverse Standing Waves

**Standing waves**

A standing wave is the pattern of disturbance that results when oppositely traveling waves of the same frequency and amplitude pass through each other. A standing wave has places of minimum and maximum vibration called, respectively, nodes and antinodes.

**Nodes and antinodes**

**Natural frequencies**

**Harmonics**

Under resonance conditions, standing waves can be established only at certain natural frequencies. The frequencies in this series $(f_1, 2f_1, 3f_1, \text{etc.})$ are called harmonics. The lowest frequency $f_1$ is called the first harmonic, the next frequency $2f_1$ is the second harmonic, and so on.

For a string that is fixed at both ends and has a length $L$, the natural frequencies are

**String fixed at both ends**

$$f_n = n\left(\frac{v}{2L}\right) \qquad n = 1, 2, 3, 4, \ldots \qquad (17.3)$$

where $v$ is the speed of the wave on the string and $n$ is a positive integer.

**Examples 4, 5**

**Interactive LearningWare 17.2**

| Topic | Discussion | Learning Aids |
|---|---|---|
| | **17.6 Longitudinal Standing Waves** | |
| | For a gas in a cylindrical tube open at both ends, the natural frequencies of vibration are | |
| **Tube open at both ends** | $$f_n = n\left(\frac{v}{2L}\right) \qquad n = 1, 2, 3, 4, \ldots \qquad (17.4)$$ | **Example 6** |
| | where $v$ is the speed of sound in the gas and $L$ is the length of the tube. | |
| | For a gas in a cylindrical tube open at only one end, the natural frequencies of vibration are | |
| **Tube open at only one end** | $$f_n = n\left(\frac{v}{4L}\right) \qquad n = 1, 3, 5, 7, \ldots \qquad (17.5)$$ | **Interactive Solution 17.49** |

 **Use Self-Assessment Test 17.2 to evaluate your understanding of Sections 17.5 and 17.6.**

# Problems

ssm  Solution is in the Student Solutions Manual.    www  Solution is available on the World Wide Web at www.wiley.com/college/cutnell
  This icon represents a biomedical application.

**Section 17.1 The Principle of Linear Superposition, Section 17.2 Constructive and Destructive Interference of Sound Waves**

**1.** In **Concept Simulation 17.1** at **www.wiley.com/college/cutnell** you can explore how two pulses, traveling in opposite directions, combine to form a resultant pulse. Two pulses are traveling toward each other, each having a speed of 1 cm/s. At $t = 0$ s, their positions are shown in the drawing. When $t = 1$ s, what is the height of the resultant pulse at (a) $x = 3$ cm and at (b) $x = 4$ cm?

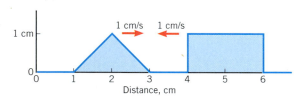

**2.** Two speakers, one directly behind the other, are each generating a 245-Hz sound wave. What is the smallest separation distance between the speakers that will produce destructive interference at a listener standing in front of them? The speed of sound is 343 m/s.

**3. ssm** The drawing graphs a string on which two rectangular pulses are traveling at a constant speed of 1 cm/s at time $t = 0$ s. Using the principle of linear superposition, draw the shape of the string's pulses at $t = 1$ s, 2 s, 3 s, and 4 s.

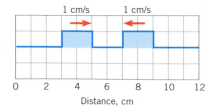

**4.** In Figure 17.7, suppose that the separation between speakers $A$ and $B$ is 5.00 m and the speakers are vibrating in phase. They are playing identical 125-Hz tones, and the speed of sound is 343 m/s. What is the largest possible distance between speaker $B$ and the observer at $C$, such that he observes destructive interference?

**5. ssm** Two loudspeakers are vibrating in phase. They are set up as in Figure 17.7, and point $C$ is located as shown there. The speed of sound is 343 m/s. The speakers play the same tone. What is the smallest frequency that will produce destructive interference at point $C$?

**6.** The sound produced by the loudspeaker in the drawing has a frequency of 12 000 Hz and arrives at the microphone via two different paths. The sound travels through the left tube $LXM$, which has a fixed length. Simultaneously, the sound travels through the right tube $LYM$, the length of which can be changed by moving the sliding section. At $M$, the sound waves coming from the two paths interfere. As the length of the path $LYM$ is changed, the sound loudness detected by the microphone changes. When the sliding section is pulled out by 0.020 m, the loudness changes from a maximum to a minimum. Find the speed at which sound travels through the gas in the tube.

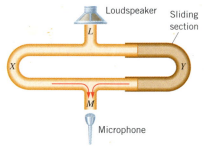

**7. ssm www** The drawing shows a loudspeaker $A$ and point $C$, where a listener is positioned. A second loudspeaker $B$ is located somewhere to the right of $A$. Both speakers vibrate in phase and are playing a 68.6-Hz tone. The speed of sound is 343 m/s. What is the closest to speaker $A$ that speaker $B$ can be located, so that the listener hears no sound?

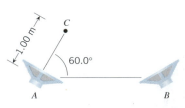

**\*8.** The two speakers in the drawing are vibrating in phase, and a listener is standing at point $P$. Does constructive or destructive interference occur at $P$ when the speakers produce sound waves whose frequency is (a) 1466 Hz and (b) 977 Hz? Justify your answers with appropriate calculations. Take the speed of sound to be 343 m/s.

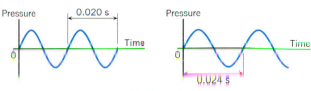

**Problem 17**

**Problem 8**

**\*\*9.** Speakers A and B are vibrating in phase. They are directly facing each other, are 7.80 m apart, and are each playing a 73.0-Hz tone. The speed of sound is 343 m/s. On the line between the speakers there are three points where constructive interference occurs. What are the distances of these three points from speaker A?

### Section 17.3 Diffraction

**10.** The entrance to a large lecture room consists of two side-by-side doors, one hinged on the left and the other hinged on the right. Each door is 0.700 m wide. Sound of frequency 607 Hz is coming through the entrance from within the room. The speed of sound is 343 m/s. What is the diffraction angle $\theta$ of the sound after it passes through the doorway when (a) one door is open and (b) both doors are open?

**11. ssm** A speaker has a diameter of 0.30 m. (a) Assuming that the speed of sound is 343 m/s, find the diffraction angle $\theta$ for a 2.0-kHz tone. (b) What speaker diameter D should be used to generate a 6.0-kHz tone whose diffraction angle is as wide as that for the 2.0-kHz tone in part (a)?

**12.** Sound emerges through a doorway, as in Figure 17.10. The width of the doorway is 77 cm, and the speed of sound is 343 m/s. Find the diffraction angle $\theta$ when the frequency of the sound is (a) 5.0 kHz and (b) $5.0 \times 10^2$ Hz.

**13.** Sound exits a diffraction horn loudspeaker through a rectangular opening like a small doorway. A person is sitting at an angle $\theta$ off to the side of a diffraction horn that has a width D of 0.060 m. The speed of sound is 343 m/s. This individual does not hear a sound wave that has a frequency of 8100 Hz. When she is sitting at an angle $\theta/2$, there is a different frequency that she does not hear. What is it?

**\*14.** A row of seats is parallel to a stage at a distance of 8.7 m from it. At the center and front of the stage is a diffraction horn loudspeaker. This speaker sends out its sound through an opening that is like a small doorway with a width D of 7.5 cm. The speaker is playing a tone that has a frequency of $1.0 \times 10^4$ Hz. The speed of sound is 343 m/s. What is the distance between two seats, located near the center of the row, at which the tone cannot be heard?

**\*15.** A 3.00-kHz tone is being produced by a speaker with a diameter of 0.175 m. The air temperature changes from 0 to 29 °C. Assuming air to be an ideal gas, find the *change* in the diffraction angle $\theta$.

### Section 17.4 Beats

**16.** Two out-of-tune flutes play the same note. One produces a tone that has a frequency of 262 Hz, while the other produces 266 Hz. When a tuning fork is sounded together with the 262-Hz tone, a beat frequency of 1 Hz is produced. When the same tuning fork is sounded together with the 266-Hz tone, a beat frequency of 3 Hz is produced. What is the frequency of the tuning fork?

**17. ssm** Two pure tones are sounded together. The drawing shows the pressure variations of the two sound waves, measured with respect to atmospheric pressure. What is the beat frequency?

**18.** In Concept Simulation 17.2 at **www.wiley.com/college/cutnell** you can explore the concepts that are important in this problem. A 440.0-Hz tuning fork is sounded together with an out-of-tune guitar string, and a beat frequency of 3 Hz is heard. When the string is tightened, the frequency at which it vibrates increases, and the beat frequency is heard to decrease. What was the original frequency of the guitar string?

**19. ssm** Two ultrasonic sound waves combine and form a beat frequency that is in the range of human hearing. The frequency of one of the ultrasonic waves is 70 kHz. What is (a) the smallest possible and (b) the largest possible value for the frequency of the other ultrasonic wave?

**20.** A tuning fork vibrates at a frequency of 524 Hz. An out-of-tune piano string vibrates at 529 Hz. How much time separates successive beats?

**\*21.** A sound wave is traveling in seawater, where the adiabatic bulk modulus and density are $2.31 \times 10^9$ Pa and 1025 kg/m$^3$, respectively. The wavelength of the sound is 3.35 m. A tuning fork is struck underwater and vibrates at 440.0 Hz. What would be the beat frequency heard by an underwater swimmer?

**\*\*22.** Two loudspeakers are mounted on a merry-go-round whose radius is 9.01 m. When stationary, the speakers both play a tone whose frequency is 100.0 Hz. As the drawing illustrates, they are situated at opposite ends of a diameter. The speed of sound is 343.00 m/s, and the merry-go-round revolves once every 20.0 s. What is the beat frequency that is detected by the listener when the merry-go-round is near the position shown?

### Section 17.5 Transverse Standing Waves

**23. ssm** The A string on a string bass is tuned to vibrate at a fundamental frequency of 55.0 Hz. If the tension in the string were increased by a factor of four, what would be the new fundamental frequency?

**24.** A string of length 0.28 m is fixed at both ends. The string is plucked and a standing wave is set up that is vibrating at its second harmonic. The traveling waves that make up the standing wave have a speed of 140 m/s. What is the frequency of vibration?

**25.** The G string on a guitar has a fundamental frequency of 196 Hz and a length of 0.62 m. This string is pressed against the proper fret to produce the note C, whose fundamental frequency is 262 Hz. What is the distance L between the fret and the end of the string at the bridge of the guitar (see Figure 17.17b)?

**26.** A 41-cm length of wire has a mass of 6.0 g. It is stretched between two fixed supports and is under a tension of 160 N. What is the fundamental frequency of this wire?

**27. ssm** On a cello, the string with the largest linear density ($1.56 \times 10^{-2}$ kg/m) is the C string. This string produces a fundamental frequency of 65.4 Hz and has a length of 0.800 m between the two fixed ends. Find the tension in the string.

**28.** The fundamental frequency of a string fixed at both ends is 256 Hz. How long does it take for a wave to travel the length of this string?

**29.** A string has a linear density of $8.5 \times 10^{-3}$ kg/m and is under a tension of 280 N. The string is 1.8 m long, is fixed at both ends, and is vibrating in the standing wave pattern shown in the drawing. Determine the (a) speed, (b) wavelength, and (c) frequency of the traveling waves that make up the standing wave.

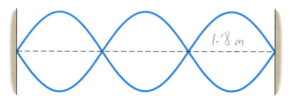

**30.** Two strings have different lengths and linear densities, as the drawing shows. They are joined together and stretched so that the tension in each string is 190.0 N. The free ends of the joined string are fixed in place. Find the lowest frequency that permits standing waves in both strings with a node at the junction. The standing wave pattern in each string may have a different number of loops.

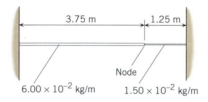

**31.** **ssm** The E string on an electric bass guitar has a length of 0.628 m and, when producing the note E, vibrates at a fundamental frequency of 41.2 Hz. Players sometimes add to their instruments a device called a "D-tuner." This device allows the E string to be used to produce the note D, which has a fundamental frequency of 36.7 Hz. The D-tuner works by extending the length of the string, keeping all other factors the same. By how much does a D-tuner extend the length of the E string?

**32.** Review Conceptual Example 5 before attempting this problem. As the drawing shows, the length of a guitar string is 0.628 m. The frets are numbered for convenience. A performer can play a musical scale on a single string because the spacing *between the frets* is designed according to the following rule: When the string is pushed against any fret $j$, the fundamental frequency of the shortened string is larger by a factor of the twelfth root of two ($\sqrt[12]{2}$) than it is when the string is pushed against the fret $j - 1$. Assuming that the tension in the string is the same for any note, find the spacing (a) between fret 1 and fret 0 and (b) between fret 7 and fret 6.

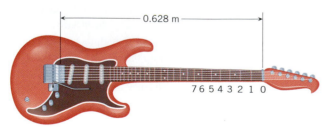

**33.** **ssm www** The drawing shows an arrangement in which a block (mass = 15.0 kg) is held in position on a frictionless incline by a cord (length = 0.600 m). The mass per unit length of the cord is $1.20 \times 10^{-2}$ kg/m,

so the mass of the cord is negligible compared to the mass of the block. The cord is being vibrated at a frequency of 165 Hz (vibration source not shown in the drawing). What are the values of the angle $\theta$ between 15.0° and 90.0° at which a standing wave exists on the cord?

**Section 17.6 Longitudinal Standing Waves**

**34.** An organ pipe is open at both ends. It is producing sound at its third harmonic, the frequency of which is 262 Hz. The speed of sound is 343 m/s. What is the length of the pipe?

**35.** The fundamental frequency of a vibrating system is 400 Hz. For each of the following systems, give the three lowest frequencies (excluding the fundamental) at which standing waves can occur: (a) a string fixed at both ends, (b) a cylindrical pipe with both ends open, and (c) a cylindrical pipe with only one end open.

**36.** ☤ Sound enters the ear, travels through the auditory canal, and reaches the eardrum. The auditory canal is approximately a tube open at only one end. The other end is closed by the eardrum. A typical length for the auditory canal in an adult is about 2.9 cm. The speed of sound is 343 m/s. What is the fundamental frequency of the canal? (Interestingly, the fundamental frequency is in the frequency range where human hearing is most sensitive.)

**37.** **ssm** A tube of air is open at only one end and has a length of 1.5 m. This tube sustains a standing wave at its third harmonic. What is the distance between one node and the adjacent antinode?

**38.** A tube is open only at one end. A certain harmonic produced by the tube has a frequency of 450 Hz. The next higher harmonic has a frequency of 750 Hz. The speed of sound in air is 343 m/s. (a) What is the integer $n$ that describes the harmonic whose frequency is 450 Hz? (b) What is the length of the tube?

**39.** **ssm** Both neon (Ne) and helium (He) are monatomic gases and can be assumed to be ideal gases. The fundamental frequency of a tube of neon is 268 Hz. What is the fundamental frequency of the tube if the tube is filled with helium, all other factors remaining the same?

**40.** Two loudspeakers face each other, vibrate in phase, and produce identical 440-Hz tones. A listener walks from one speaker toward the other at a constant speed and hears the loudness change (loud–soft–loud) at a frequency of 3.0 Hz. The speed of sound is 343 m/s. What is the walking speed?

**41.** A person hums into the top of a well and finds that standing waves are established at frequencies of 42, 70.0, and 98 Hz. The frequency of 42 Hz is not necessarily the fundamental frequency. The speed of sound is 343 m/s. How deep is the well?

**42.** A tube, open at both ends, contains an unknown ideal gas for which $\gamma = 1.40$. At 293 K, the shortest tube in which a standing wave can be set up with a 294-Hz tuning fork has a length of 0.248 m. Find the mass of a gas molecule.

**43.** **ssm www** A vertical tube is closed at one end and open to air at the other end. The air pressure is $1.01 \times 10^5$ Pa. The tube has a length of 0.75 m. Mercury (mass density = 13 600 kg/m³) is poured into it to shorten the effective length for standing waves. What is the absolute pressure at the bottom of the mercury column, when the fundamental frequency of the shortened, air-filled tube is equal to the third harmonic of the original tube?

**44.** A tube, open at only one end, is cut into two shorter (nonequal) lengths. The piece that is open at both ends has a fundamental frequency of 425 Hz, while the piece open only at one end has a fundamental frequency of 675 Hz. What is the fundamental frequency of the original tube?

# Chapter 18 Electric Forces and Electric Fields

## 18.1 The Origin of Electricity

The electrical nature of matter is inherent in atomic structure. An atom consists of a small, relatively massive nucleus that contains particles called protons and neutrons. A proton has a mass of $1.673 \times 10^{-27}$ kg, and a neutron has a slightly greater mass of $1.675 \times 10^{-27}$ kg. Surrounding the nucleus is a diffuse cloud of orbiting particles called electrons, as Figure 18.1 suggests. An electron has a mass of $9.11 \times 10^{-31}$ kg. Like mass, *electric charge* is an intrinsic property of protons and electrons, and only two types of charge have been discovered, positive and negative. A proton has a positive charge, and an electron has a negative charge. A neutron has no net electric charge.

Experiment reveals that the magnitude of the charge on the proton *exactly equals* the magnitude of the charge on the electron; the proton carries a charge $+e$, and the electron carries a charge $-e$. The SI unit for measuring the magnitude of an electric charge is the *coulomb** (C), and $e$ has been determined experimentally to have the value

$$e = 1.60 \times 10^{-19} \text{ C}$$

The symbol $e$ represents only the magnitude of the charge on a proton or an electron and does not include the algebraic sign that indicates whether the charge is positive or negative. In nature, atoms are normally found with equal numbers of protons and electrons. Usually, then, an atom carries no net charge because the algebraic sum of the positive charge of the nucleus and the negative charge of the electrons is zero. When an atom, or any object, carries no net charge, the object is said to be *electrically neutral*. The neutrons in the nucleus are electrically neutral particles.

The charge on an electron or a proton is the *smallest* amount of free charge that has been discovered. Charges of larger magnitude are built up on an object by adding or removing electrons. Thus, any charge of magnitude $q$ is an integer multiple of $e$; that is, $q = Ne$, where $N$ is an integer. Because any electric charge $q$ occurs in integer multiples of elementary, indivisible charges of magnitude $e$, electric charge is said to be *quantized*. Example 1 emphasizes the quantized nature of electric charge.

- electron
- proton
- neutron

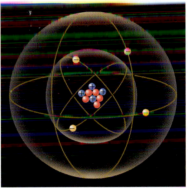

**Figure 18.1** An atom contains a small, positively charged nucleus, about which the negatively charged electrons move. The closed-loop paths shown here are symbolic only. In reality, the electrons do not follow discrete paths, as Section 30.5 discusses.

### Example 1   A Lot of Electrons

How many electrons are there in one coulomb of negative charge?

**Reasoning** The negative charge is due to the presence of excess electrons, since they carry negative charge. Because an electron has a charge whose magnitude is $e = 1.60 \times 10^{-19}$ C, the number of electrons is equal to the charge $q$ divided by the charge $e$ on each electron.

**Solution** The number $N$ of electrons is

$$N = \frac{q}{e} = \frac{1.00 \text{ C}}{1.60 \times 10^{-19} \text{ C}} = \boxed{6.25 \times 10^{18}}$$

## 18.2 Charged Objects and the Electric Force

Electricity has many useful applications, and they are related to the fact that it is possible to transfer electric charge from one object to another. Usually electrons are transferred, and the body that gains electrons acquires an excess of negative charge. The body that loses electrons has an excess of positive charge. Such separation of charge occurs often when

---

*The definition of the coulomb depends on electric currents and magnetic fields, concepts that will be discussed later. Therefore, we postpone its definition until Section 21.7.

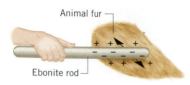

**Figure 18.2** When an ebonite rod is rubbed against animal fur, electrons from atoms of the fur are transferred to the rod. This transfer gives the rod a negative charge (−) and leaves a positive charge (+) on the fur.

two unlike materials are rubbed together. For example, when an ebonite (hard, black rubber) rod is rubbed against animal fur, some of the electrons from atoms of the fur are transferred to the rod. The ebonite becomes negatively charged, and the fur becomes positively charged, as Figure 18.2 indicates. Similarly, if a glass rod is rubbed with a silk cloth, some of the electrons are removed from the atoms of the glass and deposited on the silk, leaving the silk negatively charged and the glass positively charged. There are many familiar examples of charge separation, as when you walk across a nylon rug or run a comb through dry hair. In each case, objects become "electrified" as surfaces rub against one another.

When an ebonite rod is rubbed with animal fur, the rubbing process serves only to separate electrons and protons already present in the materials. No electrons or protons are created or destroyed. Whenever an electron is transferred to the rod, a proton is left behind on the fur. Since the charges on the electron and proton have identical magnitudes but opposite signs, the algebraic sum of the two charges is zero, and the transfer does not change the net charge of the fur/rod system. If each material contains an equal number of protons and electrons to begin with, the net charge of the system is zero initially and remains zero at all times during the rubbing process.

Electric charges play a role in many situations other than rubbing two surfaces together. They are involved, for instance, in chemical reactions, electric circuits, and radioactive decay. A great number of experiments have verified that in any situation, the *law of conservation of electric charge* is obeyed.

■ **LAW OF CONSERVATION OF ELECTRIC CHARGE**

During any process, the net electric charge of an isolated system remains constant (is conserved).

It is easy to demonstrate that two electrically charged objects exert a force on one another. Consider Figure 18.3a, which shows two small balls that have been *oppositely charged* and are light and free to move. The balls attract each other. On the other hand, balls with the *same* type of charge, either both positive or both negative, repel each other, as parts b and c of the drawing indicate. The behavior depicted in Figure 18.3 illustrates the following fundamental characteristic of electric charges:

*Like charges repel and unlike charges attract each other.*

Like other forces that we have encountered, the electric force (also called the electrostatic force) can alter the motion of an object. It can do so by contributing to the net external force ΣF that acts on the object. Newton's second law, ΣF = ma, specifies the acceleration a that arises because of the net external force. Any external electric force that acts on an object must be included when determining the net external force to be used in the second law.

A new technology based on the electric force may revolutionize the way books and other printed matter are made. This technology, called electronic ink, allows letters and graphics on a page to be changed instantly, much like the symbols displayed on a computer monitor. Figure 18.4a illustrates the essential features of electronic ink. It consists of millions of clear microcapsules, each having the diameter of a human hair and filled with a dark inky liquid. Inside each microcapsule are several dozen extremely tiny white beads that carry a slightly negative charge. The microcapsules are sandwiched between two sheets, an opaque base layer and a transparent top layer, at which the reader looks. When a positive charge is applied to a small region of the base layer, as shown in part b of the drawing, the negatively charged white beads are drawn to it, leaving dark ink at the top layer. Thus, a viewer sees only the dark liquid. When a negative charge is applied to a region of the base layer, the negatively charged white beads are repelled from it and are forced to the top of the microcapsules; now a viewer sees a white area due to the beads. Thus, electronic ink is based on the principle that like charges repel and unlike charges attract each other; a positive charge causes one color to appear, and a negative charge causes another color to appear. Each small region, whether dark or light, is known as a pixel (short for "picture element"). Computer chips provide the instructions to produce the negative and positive charges on the base layer of each pixel. Letters and graphics are produced by the patterns generated with the two colors.

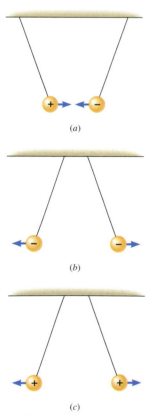

*(a)*

*(b)*

*(c)*

**Figure 18.3** (a) A positive charge (+) and a negative charge (−) attract each other. (b) Two negative charges repel each other. (c) Two positive charges repel each other.

*The physics of...*

**electronic ink.**

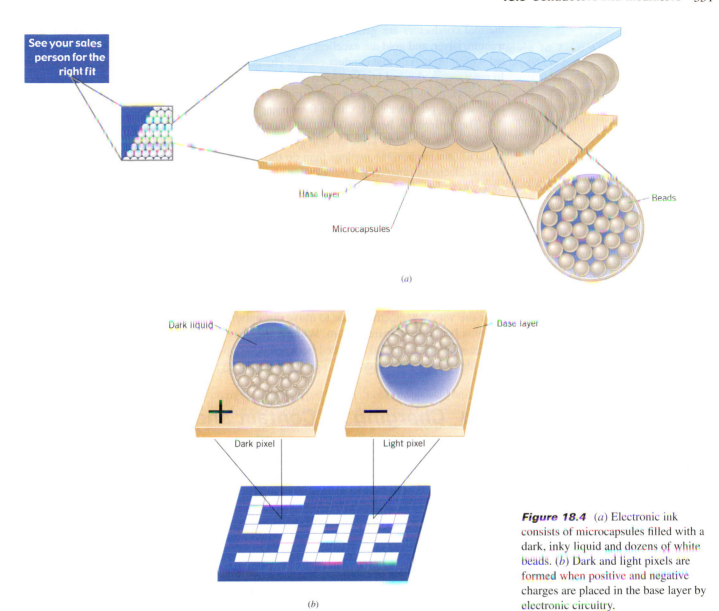

See your sales person for the right fit

Base layer

Microcapsules

Beads

(a)

Dark liquid

Base layer

+

Dark pixel

−

Light pixel

(b)

**Figure 18.4** (a) Electronic ink consists of microcapsules filled with a dark, inky liquid and dozens of white beads. (b) Dark and light pixels are formed when positive and negative charges are placed in the base layer by electronic circuitry.

# 18.3  *Conductors and Insulators*

Not only can electric charge exist *on an object,* but it can also move *through an object.* However, materials differ vastly in their abilities to allow electric charge to move or be conducted through them. To help illustrate such differences in conductivity, Figure 18.5a recalls the conduction of heat through a bar of material whose ends are maintained at different temperatures. As Section 13.2 discusses, metals conduct heat readily and, therefore, are known as thermal conductors. On the other hand, substances that conduct heat poorly are referred to as thermal insulators.

A situation analogous to the conduction of heat arises when a metal bar is placed between two charged objects, as in Figure 18.5b. Electrons are conducted through the bar from the negatively charged object toward the positively charged object. Substances that readily conduct electric charge are called *electrical conductors.* Although there are exceptions, good

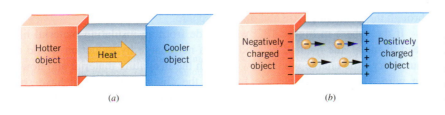

Hotter object

Heat

Cooler object

(a)

Negatively charged object

Positively charged object

(b)

**Figure 18.5** (a) Heat is conducted from the hotter end of the metal bar to the cooler end. (b) Electrons are conducted from the negatively charged end of the metal bar to the positively charged end.

It is common practice to express $k$ in terms of another constant $\epsilon_0$, by writing $k = 1/(4\pi\epsilon_0)$; $\epsilon_0$ is called the **permittivity of free space** and has a value of $\epsilon_0 = 1/(4\pi k) = 8.85 \times 10^{-12}$ C$^2$/(N·m$^2$). Equation 18.1 gives the magnitude of the electrostatic force that each point charge exerts on the other. When using this equation, then, it is important to remember to substitute only the charge magnitudes (without algebraic signs) for $|q_1|$ and $|q_2|$, as Example 2 illustrates.

### Example 2 A Large Attractive Force

Two objects, whose charges are $+1.0$ and $-1.0$ C, are separated by 1.0 km. Compared to 1.0 km, the sizes of the objects are small. Find the magnitude of the attractive force that either charge exerts on the other.

**Reasoning** Considering that the sizes of the objects are small compared to the separation distance, we can treat the charges as point charges. Coulomb's law may then be used to find the magnitude of the attractive force, provided that only the *magnitudes of the charges* are used for the symbols $|q_1|$ and $|q_2|$ that appear in the law.

**Solution** The magnitude of the force is

$$F = k\frac{|q_1||q_2|}{r^2} = \frac{(8.99 \times 10^9 \text{ N·m}^2/\text{C}^2)(1.0 \text{ C})(1.0 \text{ C})}{(1.0 \times 10^3 \text{ m})^2} = \boxed{9.0 \times 10^3 \text{ N}} \quad (18.1)$$

The force calculated in Example 2 corresponds to about 2000 pounds and is so large because charges of $\pm 1.0$ C are enormous. Such large charges are encountered only in the most severe conditions, as in a lightning bolt, where as much as 25 coulombs can be transferred between the cloud and the ground. The typical charges produced in the laboratory are much smaller and are measured conveniently in microcoulombs (1 microcoulomb = 1 $\mu$C = $10^{-6}$ C).

Coulomb's law has a form remarkably similar to Newton's law of gravitation ($F = Gm_1m_2/r^2$). The force in both laws depends on the inverse square ($1/r^2$) of the distance between the two objects and is directed along the line between them. In addition, the force is proportional to the product of an intrinsic property of each of the objects, the magnitudes of the charges $|q_1|$ and $|q_2|$ in Coulomb's law and the masses $m_1$ and $m_2$ in the gravitation law. But there is a major difference between the two laws. The electrostatic force can be either repulsive or attractive, depending on whether or not the charges have the same sign; in contrast, the gravitational force is always an attractive force.

Section 5.5 discusses how the gravitational attraction between the earth and a satellite provides the centripetal force that keeps a satellite in orbit. Example 3 illustrates that the electrostatic force of attraction plays a similar role in a famous model of the atom created by the Danish physicist Niels Bohr (1885–1962).

### Example 3 A Model of the Hydrogen Atom

In the Bohr model of the hydrogen atom, the electron ($-e$) is in orbit about the nuclear proton ($+e$) at a radius of $r = 5.29 \times 10^{-11}$ m, as Figure 18.10 shows. Determine the speed of the electron, assuming the orbit to be circular.

**Reasoning** Recall from Section 5.2 that any object moving with speed $v$ on a circular path of radius $r$ has a centripetal acceleration of $a_c = v^2/r$. This acceleration is directed toward the center of the circle. Newton's second law specifies that the net force $\Sigma F$ needed to create this acceleration is $\Sigma F = ma_c = mv^2/r$, where $m$ is the mass of the object. This equation can be solved for the speed: $v = \sqrt{(\Sigma F)r/m}$. Since the mass of the electron is $m = 9.11 \times 10^{-31}$ kg and the radius is given, we can calculate the speed, once a value for the net force is available. For the electron in the hydrogen atom, the net force is provided almost exclusively by the electrostatic force, as given by Coulomb's law. This force points toward the center of the circle, since the electron and the proton have opposite signs. The electron is also pulled toward the proton by the gravitational force. However, the gravitational force is negligible in comparison to the electrostatic force.

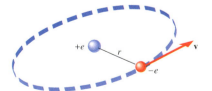

**Figure 18.10** In the Bohr model of the hydrogen atom, the electron ($-e$) orbits the proton ($+e$) at a distance of $r = 5.29 \times 10^{-11}$ m. The velocity of the electron is **v**.

**Solution** The electron experiences an electrostatic force of attraction because of the proton, and the magnitude of this force is

$$F = k\frac{|q_1||q_2|}{r^2} = \frac{(8.99 \times 10^9 \text{ N}\cdot\text{m}^2/\text{C}^2)(1.60 \times 10^{-19} \text{ C})(1.60 \times 10^{-19} \text{ C})}{(5.29 \times 10^{-11} \text{ m})^2}$$

$$= 8.22 \times 10^{-8} \text{ N}$$

Using this value for the net force, we find

$$v = \sqrt{\frac{(\Sigma F)r}{m}} = \sqrt{\frac{(8.22 \times 10^{-8} \text{ N})(5.29 \times 10^{-11} \text{ m})}{9.11 \times 10^{-31} \text{ kg}}} = \boxed{2.18 \times 10^6 \text{ m/s}}$$

This orbital speed is almost five million miles per hour.

**Problem solving insight**
When using Coulomb's law ($F = k|q_1||q_2|/r^2$), remember that the symbols $|q_1|$ and $|q_2|$ stand for the charge magnitudes. Do not substitute negative numbers for these symbols.

Since the electrostatic force depends on the inverse square of the distance between the charges, it becomes larger for smaller distances, such as those involved when a strip of adhesive tape is stuck to a smooth surface. Electrons shift over the small distances between the tape and the surface. As a result, the materials become oppositely charged. Since the distance between the charges is relatively small, the electrostatic force of attraction is large enough to contribute to the adhesive bond. Figure 18.11 shows an image of the sticky surface of a piece of tape after it has been pulled off a metal surface. The image was obtained using an atomic-force microscope and reveals the tiny pits left behind when microscopic portions of the adhesive remain stuck to the metal because of the strong adhesive bonding forces.

*The physics of...*
adhesion.

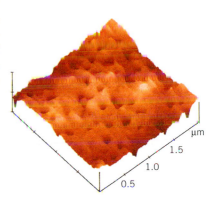

**Figure 18.11** After a strip of adhesive tape has been pulled off a metal surface, there are tiny pits (approximately one ten-millionth of a meter in diameter) in the sticky surface of the tape, as this image shows. It was obtained using an atomic-force microscope. (Courtesy Louis Scudiero and J. Thomas Dickinson, Washington State University.)

## THE FORCE ON A POINT CHARGE DUE TO TWO OR MORE OTHER POINT CHARGES

Up to now, we have been discussing the electrostatic force on a point charge (magnitude $|q_1|$) due to another point charge (magnitude $|q_2|$). Suppose that a third point charge (magnitude $|q_3|$) is also present. What would be the net force on $q_1$ due to both $q_2$ and $q_3$? It is convenient to deal with such a problem in parts. First, find the magnitude and direction of the force exerted on $q_1$ by $q_2$ (ignoring $q_3$). Then, determine the force exerted on $q_1$ by $q_3$ (ignoring $q_2$). The *net force* on $q_1$ is the *vector sum* of these forces. Examples 4 and 5 illustrate this approach when the charges lie along a straight line and on a plane, respectively.

### *Example 4* Three Charges on a Line

Figure 18.12a shows three point charges that lie along the x axis in a vacuum. Determine the magnitude and direction of the net electrostatic force on $q_1$.

**Reasoning** Part b of the drawing shows a free-body diagram of the forces that act on $q_1$. Since $q_1$ and $q_2$ have opposite signs, they attract one another. Thus, the force exerted on $q_1$ by $q_2$ is $\mathbf{F}_{12}$, and it points to the left. Similarly, the force exerted on $q_1$ by $q_3$ is $\mathbf{F}_{13}$ and is also an attractive force. It points to the right in Figure 18.12b. The magnitudes of these forces can be obtained from Coulomb's law. The net force is the vector sum of $\mathbf{F}_{12}$ and $\mathbf{F}_{13}$.

**Solution** The magnitudes of the forces are

$$F_{12} = k\frac{|q_1||q_2|}{r_{12}^2} = \frac{(8.99 \times 10^9 \text{ N}\cdot\text{m}^2/\text{C}^2)(3.0 \times 10^{-6} \text{ C})(4.0 \times 10^{-6} \text{ C})}{(0.20 \text{ m})^2} = 2.7 \text{ N}$$

$$F_{13} = k\frac{|q_1||q_3|}{r_{13}^2} = \frac{(8.99 \times 10^9 \text{ N}\cdot\text{m}^2/\text{C}^2)(3.0 \times 10^{-6} \text{ C})(7.0 \times 10^{-6} \text{ C})}{(0.15 \text{ m})^2} = 8.4 \text{ N}$$

Since $\mathbf{F}_{12}$ points in the negative x direction, and $\mathbf{F}_{13}$ points in the positive x direction, the net force $\mathbf{F}$ is

$$\mathbf{F} = \mathbf{F}_{12} + \mathbf{F}_{13} = (-2.7 \text{ N}) + (8.4 \text{ N}) = \boxed{+5.7 \text{ N}}$$

The plus sign in the answer indicates that the net force points to the right in the drawing.

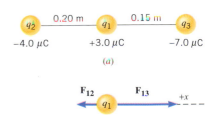

(b) Free-body diagram for $q_1$

**Figure 18.12** (a) Three charges lying along the x axis. (b) The force exerted on $q_1$ by $q_2$ is $\mathbf{F}_{12}$, while the force exerted on $q_1$ by $q_3$ is $\mathbf{F}_{13}$.

### Example 5    Three Charges in a Plane

Figure 18.13a shows three point charges that lie in the x, y plane in a vacuum. Find the magnitude and direction of the net electrostatic force on $q_1$.

**Reasoning** The force exerted on $q_1$ by $q_2$ is $\mathbf{F_{12}}$ and is an attractive force because the two charges have opposite signs. It points along the line between the charges. The force exerted on $q_1$ by $q_3$ is $\mathbf{F_{13}}$ and is also an attractive force. It points along the line between $q_1$ and $q_3$. Coulomb's law specifies the magnitudes of these forces. Since the forces point in different directions (see Figure 18.13b), we will use vector components to find the net force.

**Solution** The magnitudes of the forces are

$$F_{12} = k\frac{|q_1||q_2|}{r_{12}^2} = \frac{(8.99 \times 10^9 \, \text{N} \cdot \text{m}^2/\text{C}^2)(4.0 \times 10^{-6} \, \text{C})(6.0 \times 10^{-6} \, \text{C})}{(0.15 \, \text{m})^2} = 9.6 \, \text{N}$$

$$F_{13} = k\frac{|q_1||q_3|}{r_{13}^2} = \frac{(8.99 \times 10^9 \, \text{N} \cdot \text{m}^2/\text{C}^2)(4.0 \times 10^{-6} \, \text{C})(5.0 \times 10^{-6} \, \text{C})}{(0.10 \, \text{m})^2} = 18 \, \text{N}$$

The net force $\mathbf{F}$ is the vector sum of $\mathbf{F_{12}}$ and $\mathbf{F_{13}}$, as part b of the drawing shows. The components of $\mathbf{F}$ that lie in the x and y directions are $\mathbf{F_x}$ and $\mathbf{F_y}$, respectively. Our approach to finding $\mathbf{F}$ is the same as that used in Chapters 1 and 4. The forces $\mathbf{F_{12}}$ and $\mathbf{F_{13}}$ are resolved into x and y components. Then, the x components are combined to give $\mathbf{F_x}$, and the y components are combined to give $\mathbf{F_y}$. Once $\mathbf{F_x}$ and $\mathbf{F_y}$ are known, the magnitude and direction of $\mathbf{F}$ can be determined using trigonometry.

| Force | x component | y component |
|---|---|---|
| $\mathbf{F_{12}}$ | +(9.6 N) cos 73° = +2.8 N | +(9.6 N) sin 73° = +9.2 N |
| $\mathbf{F_{13}}$ | +18 N | 0 N |
| $\mathbf{F}$ | $\mathbf{F_x}$ = +21 N | $\mathbf{F_y}$ = +9.2 N |

**Problem solving insight**
The electrostatic force is a vector and has a direction as well as a magnitude. When adding electrostatic forces, take into account the directions of all forces, using vector components as needed.

The magnitude F and the angle $\theta$ of the net force are

$$F = \sqrt{F_x^2 + F_y^2} = \sqrt{(21 \, \text{N})^2 + (9.2 \, \text{N})^2} = \boxed{23 \, \text{N}}$$

$$\theta = \tan^{-1}\left(\frac{F_y}{F_x}\right) = \tan^{-1}\left(\frac{9.2 \, \text{N}}{21 \, \text{N}}\right) = \boxed{24°}$$

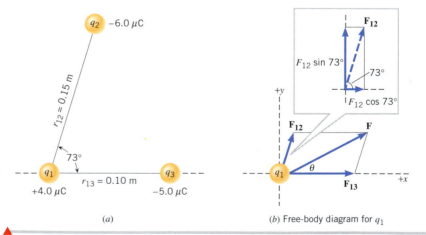

**Figure 18.13** (a) Three charges lying in a plane. (b) The net force acting on $q_1$ is $\mathbf{F} = \mathbf{F_{12}} + \mathbf{F_{13}}$. The angle that $\mathbf{F}$ makes with the +x axis is $\theta$.

## 18.6  *The Electric Field*

### DEFINITION

As we know, a charge can experience an electrostatic force due to the presence of other charges. For instance, the positive charge $q_0$ in Figure 18.14 experiences a force $\mathbf{F}$, which is the vector sum of the forces exerted by the charges on the rod and the two spheres. It is useful to think of $q_0$ as a **test charge** for determining the extent to which the surrounding charges generate a force. However, in using a test charge, we must be careful to select one with a very small magnitude, so that it does not alter the locations of the other charges. The next example illustrates how the concept of a test charge is applied.

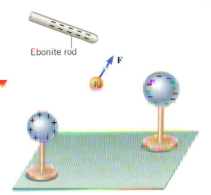

Ebonite rod

### Example 6  A Test Charge

The positive test charge in Figure 18.14 is $q_0 = +3.0 \times 10^{-8}$ C and experiences a force $F = 6.0 \times 10^{-8}$ N in the direction shown in the drawing. (a) Find the *force per coulomb* that the test charge experiences. (b) Using the result of part (a), predict the force that a charge of $+12 \times 10^{-8}$ C would experience if it replaced $q_0$.

**Reasoning** The charges in the environment apply a force **F** to the test charge $q_0$. The force per coulomb experienced by the test charge is $F/q_0$. If $q_0$ is replaced by a new charge $q$, then the force on this new charge is the force per coulomb times $q$.

**Figure 18.14** A positive charge $q_0$ experiences an electrostatic force **F** due to the surrounding charges on the ebonite rod and the two spheres.

#### Solution

**(a)** The force per coulomb of charge is

$$\frac{F}{q_0} = \frac{6.0 \times 10^{-8}\,\text{N}}{3.0 \times 10^{-8}\,\text{C}} = \boxed{2.0\,\text{N/C}}$$

**(b)** The result from part (a) indicates that the surrounding charges can exert 2.0 newtons of force per coulomb of charge. Thus, a charge of $+12 \times 10^{-8}$ C would experience a force whose magnitude is

$$F = (2.0\,\text{N/C})(12 \times 10^{-8}\,\text{C}) = \boxed{24 \times 10^{-8}\,\text{N}}$$

The direction of this force would be the same as that experienced by the test charge, since both have the same positive sign.

The electric force per coulomb, $F/q_0$, calculated in Example 6(a) is one illustration of an idea that is very important in the study of electricity. The idea is called the *electric field;* the notion of an electric field emerges when we combine the electrostatic force with the idea of a test charge. Equation 18.2 presents the definition of the electric field.

---

■ **DEFINITION OF THE ELECTRIC FIELD**

The electric field **E** that exists at a point is the electrostatic force **F** experienced by a small test charge* $q_0$ placed at that point divided by the charge itself:

$$\mathbf{E} = \frac{\mathbf{F}}{q_0} \qquad (18.2)$$

The electric field is a vector, and its direction is the same as the direction of the force **F** on a positive test charge.

*SI Unit of Electric Field:* newton per coulomb (N/C)

---

Equation 18.2 indicates that the unit for the electric field is that of force divided by charge, which is a newton/coulomb (N/C) in SI units.

It is the surrounding charges that create an electric field at a given point. Any positive or negative charge placed at the point interacts with the field and, as a result, experiences a force, as the next example indicates.

### Example 7  An Electric Field Leads to a Force

In Figure 18.15a the charges on the two metal spheres and the ebonite rod create an electric field **E** at the spot indicated. This field has a magnitude of 2.0 N/C and is directed as in the

---

*As long as the test charge is small enough that it does not disturb the surrounding charges, it may be either positive or negative. Compared to a positive test charge, a negative test charge of the same magnitude experiences a force of the same magnitude that points in the opposite direction. However, the same electric field is given by Equation 18.2, in which **F** is replaced by $-\mathbf{F}$ and $q_0$ is replaced by $-q_0$.

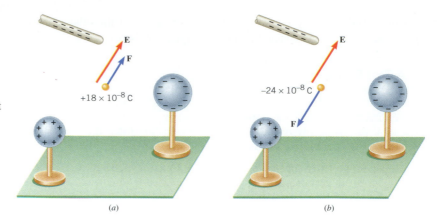

**Figure 18.15** The electric field **E** that exists at a given spot can exert a variety of forces. The force exerted depends on the magnitude and sign of the charge placed at that spot. (*a*) The force on a positive charge points in the same direction as **E**, while (*b*) the force on a negative charge points opposite to **E**.

drawing. Determine the force on a charge placed at that spot, if the charge has a value of (a) $q_0 = +18 \times 10^{-8}$ C and (b) $q_0 = -24 \times 10^{-8}$ C.

**Reasoning** The electric field at a given spot can exert a variety of forces, depending on the magnitude and sign of the charge placed there. The charge is assumed to be small enough that it does not alter the locations of the surrounding charges that create the field.

**Solution**

(a) The magnitude of the force is the product of the magnitudes of $q_0$ and **E**:

$$F = |q_0| E = (18 \times 10^{-8} \text{ C})(2.0 \text{ N/C}) = \boxed{36 \times 10^{-8} \text{ N}} \qquad (18.2)$$

Since $q_0$ is positive, the force points in the same direction as the electric field, as part *a* of the drawing indicates.

(b) In this case, the magnitude of the force is

$$F = |q_0| E = (24 \times 10^{-8} \text{ C})(2.0 \text{ N/C}) = \boxed{48 \times 10^{-8} \text{ N}} \qquad (18.2)$$

The force on the negative charge points in the direction *opposite* to the force on the positive charge—that is, opposite to the electric field (see part *b* of the drawing).

---

At a particular point in space, each of the surrounding charges contributes to the net electric field that exists there. To determine the net field, it is necessary to obtain the various contributions separately and then find the vector sum of them all. Such an approach is an illustration of the principle of linear superposition, as applied to electric fields. (This principle is introduced in Section 17.1, in connection with waves.) Example 8 emphasizes the vector nature of the electric field.

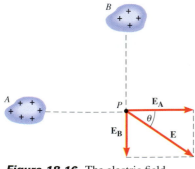

**Figure 18.16** The electric field contributions **E**$_\text{A}$ and **E**$_\text{B}$, which come from the two charge distributions, are added vectorially to obtain the net field **E** at point *P*.

### Example 8  Electric Fields Add as Vectors Do

Figure 18.16 shows two charged objects, *A* and *B*. Each contributes as follows to the net electric field at point *P*: $\mathbf{E}_A = 3.00$ N/C directed to the right, and $\mathbf{E}_B = 2.00$ N/C directed downward. Thus, $\mathbf{E}_A$ and $\mathbf{E}_B$ are perpendicular. What is the net electric field at *P*?

**Reasoning** The net electric field **E** is the vector sum of $\mathbf{E}_A$ and $\mathbf{E}_B$: $\mathbf{E} = \mathbf{E}_A + \mathbf{E}_B$. As illustrated in Figure 18.16, $\mathbf{E}_A$ and $\mathbf{E}_B$ are perpendicular, so **E** is the diagonal of the rectangle shown in the drawing. Thus, we can use the Pythagorean theorem to find the magnitude of **E** and trigonometry to find the directional angle $\theta$.

**Solution** The magnitude of the net electric field is

$$E = \sqrt{E_A{}^2 + E_B{}^2} = \sqrt{(3.00 \text{ N/C})^2 + (2.00 \text{ N/C})^2} = \boxed{3.61 \text{ N/C}}$$

The direction of **E** is given by the angle $\theta$ in the drawing:

$$\theta = \tan^{-1}\left(\frac{E_B}{E_A}\right) = \tan^{-1}\left(\frac{2.00 \text{ N/C}}{3.00 \text{ N/C}}\right) = \boxed{33.7°}$$

## POINT CHARGES

A more complete understanding of the electric field concept can be gained by considering the field created by a point charge, as in the following example.

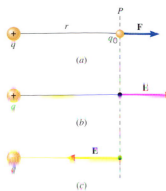

(a)

(b)

(c)

**Figure 18.17** (a) At location P, a positive test charge $q_0$ experiences a repulsive force **F** due to the positive point charge q. (b) At P the electric field **E** is directed to the right. (c) If the charge q were negative rather than positive, the electric field would have the same magnitude as in (b) but point to the left.

### Example 9  The Electric Field of a Point Charge

There is an isolated point charge of $q = +15 \, \mu C$ in a vacuum at the left in Figure 18.17a. Using a test charge of $q_0 = +0.80 \, \mu C$, determine the electric field at point P, which is 0.20 m away.

**Reasoning** Following the definition of the electric field, we place the test charge $q_0$ at point P, determine the force acting on the test charge, and then divide the force by the test charge.

**Solution** Coulomb's law (Equation 18.1), gives the magnitude of the force:

$$F = k \frac{|q_0||q|}{r^2} = \frac{(8.99 \times 10^9 \, \text{N} \cdot \text{m}^2/\text{C}^2)(0.80 \times 10^{-6} \, \text{C})(15 \times 10^{-6} \, \text{C})}{(0.20 \, \text{m})^2} = 2.7 \, \text{N}$$

Equation 18.2 gives the magnitude of the electric field:

$$E = \frac{F}{|q_0|} = \frac{2.7 \, \text{N}}{0.80 \times 10^{-6} \, \text{C}} = \boxed{3.4 \times 10^6 \, \text{N/C}}$$

The electric field **E** points in the *same direction* as the force **F** on the positive test charge. Since the test charge experiences a force of repulsion directed to the right, the electric field vector also points to the right, as Figure 18.17b shows.

The electric field produced by a point charge q can be obtained in general terms from Coulomb's law. First, note that the magnitude of the force exerted by the charge q on a test charge $q_0$ is $F = k|q||q_0|/r^2$. Then, divide this value by $|q_0|$ to obtain the magnitude of the field. Since $|q_0|$ is eliminated algebraically from the result, *the electric field does not depend on the test charge:*

**Point charge q** 
$$E = \frac{k|q|}{r^2} \qquad (18.3)$$

As in Coulomb's law, the symbol $|q|$ denotes the magnitude of q in Equation 18.3, without regard to whether q is positive or negative. If q is positive, then **E** is directed away from q, as in Figure 18.17b. On the other hand, if q is negative, then **E** is directed toward q, since a negative charge attracts a positive test charge. For instance, Figure 18.17c shows the electric field that would exist at P if there were a charge of $-q$ instead of $+q$ at the left of the drawing. Example 10 reemphasizes the fact that all the surrounding charges make a contribution to the electric field that exists at a given place.

### Example 10  The Electric Fields from Separate Charges May Cancel

Two positive point charges, $q_1 = +16 \, \mu C$ and $q_2 = +4.0 \, \mu C$, are separated in a vacuum by a distance of 3.0 m, as Figure 18.18 illustrates. Find the spot on the line between the charges where the net electric field is zero.

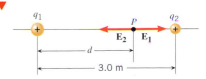

**Figure 18.18** The two point charges $q_1$ and $q_2$ create electric fields **E**$_1$ and **E**$_2$ that cancel at a location P on the line between the charges.

**Reasoning** Between the charges the two field contributions have opposite directions, and the net electric field is zero at the place where the magnitude of **E**$_1$ equals that of **E**$_2$. However, since $q_2$ is smaller than $q_1$, this location must be *closer* to $q_2$, in order that the field of the smaller charge can balance the field of the larger charge. In the drawing, the cancellation spot is labeled P, and its distance from $q_1$ is d.

**Solution** At P, $E_1 = E_2$, and using the expression $E = k|q|/r^2$, we have

$$\frac{k(16 \times 10^{-6} \, \text{C})}{d^2} = \frac{k(4.0 \times 10^{-6} \, \text{C})}{(3.0 \, \text{m} - d)^2}$$

Rearranging this expression shows that $4.0(3.0 \, \text{m} - d)^2 = d^2$. Taking the square root of each side of this equation reveals that

$$2.0(3.0 \, \text{m} - d) = \pm d$$

**Problem solving insight**
Equation 18.3 gives only the magnitude of the electric field produced by a point charge. Therefore, do not use negative numbers for the symbol $|q|$ in this equation.

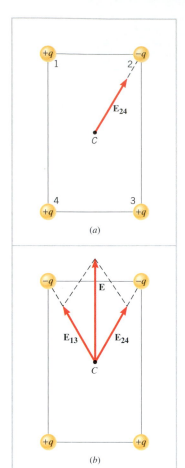

(a)

(b)

**Figure 18.19** Charges of identical magnitude, but different signs, are placed at the corners of a rectangle. The charges give rise to different electric fields at the center *C* of the rectangle, depending on the signs of the charges.

The plus and minus signs on the right occur because either the positive or negative root can be taken. Therefore, there are two possible values for $d$: $+2.0$ m and $+6.0$ m. The value $+6.0$ m corresponds to a location off to the right of both charges, where the magnitudes of $\mathbf{E}_1$ and $\mathbf{E}_2$ are equal, but where the directions are the same. Thus, $\mathbf{E}_1$ and $\mathbf{E}_2$ do not cancel at this spot. The other value for $d$ corresponds to the location shown in the drawing and is the zero-field location: $\boxed{d = +2.0 \text{ m}}$.

When point charges are arranged in a symmetrical fashion, it is often possible to deduce useful information about the magnitude and direction of the electric field by taking advantage of the symmetry. Conceptual Example 11 illustrates the use of this technique.

### *Conceptual Example 11* **Symmetry and the Electric Field**

Figure 18.19 shows point charges fixed to the corners of a rectangle in two different ways. The charges all have the same magnitudes, but they have different signs. Consider the net electric field at the center *C* of the rectangle in each case. Which field is larger?

**Reasoning and Solution** In Figure 18.19*a*, the charges at corners 1 and 3 are both $+q$. The positive charge at corner 1 produces an electric field at *C* that points toward corner 3. In contrast, the positive charge at corner 3 produces an electric field at *C* that points toward corner 1. Thus, the two fields have opposite directions. The magnitudes of the fields are identical because the charges have the same magnitude and are equally far from the center. Therefore, the fields from the two positive charges cancel.

Now, let's look at the electric field produced by the charges on corners 2 and 4 in Figure 18.19*a*. The electric field due to the negative charge at corner 2 points toward corner 2, and the field due to the positive charge at corner 4 points the same way. Furthermore, the magnitudes of these fields are equal because each charge has the same magnitude and is located at the same distance from the center of the rectangle. As a result, the two fields combine to give the net electric field $\mathbf{E}_{24}$ shown in the drawing.

In Figure 18.19*b*, the charges on corners 2 and 4 are identical to those in part *a* of the drawing, so this pair gives rise to the electric field labeled $\mathbf{E}_{24}$. The charges on corners 1 and 3 are identical to those on corners 2 and 4, so they give rise to an electric field labeled $\mathbf{E}_{13}$, which has the same magnitude as $\mathbf{E}_{24}$. The net electric field $\mathbf{E}$ is the vector sum of $\mathbf{E}_{24}$ and $\mathbf{E}_{13}$ and is also shown in the drawing. Clearly, this sum is greater than $\mathbf{E}_{24}$ alone. Therefore, ***the net field in part b is larger than that in part a.***

**Related Homework:** *Problem 34*

### THE PARALLEL PLATE CAPACITOR

Equation 18.3, which gives the electric field of a point charge, is a very useful result. With the aid of integral calculus, this equation can be applied in a variety of situations where point charges are distributed over one or more surfaces. One such example that has considerable practical importance is the ***parallel plate capacitor.*** As Figure 18.20 shows, this device consists of two parallel metal plates, each with area $A$. A charge $+q$ is spread uniformly over one plate, while a charge $-q$ is spread uniformly over the other plate. In the region between the plates and away from the edges, the electric field points from the positive plate toward the negative plate and is perpendicular to both. Using Gauss' law (see Section 18.9), it can be shown that the electric field has a magnitude of

*Parallel plate capacitor*
$$E = \frac{q}{\epsilon_0 A} = \frac{\sigma}{\epsilon_0} \qquad (18.4)$$

where $\epsilon_0$ is the permittivity of free space. In this expression the Greek symbol sigma ($\sigma$) denotes the charge per unit area ($\sigma = q/A$) and is sometimes called the charge density. Except in the region near the edges, the field has the same value at all places between the plates. The field does *not* depend on the distance from the charges, in distinct contrast to the field created by an isolated point charge.

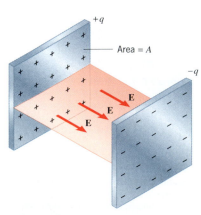

**Figure 18.20** A parallel plate capacitor.

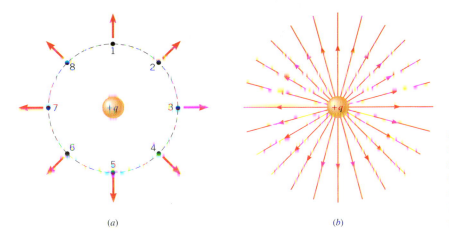

(a)

(b)

**Figure 18.21** (a) At any of the eight marked spots around a positive point charge $+q$, a positive test charge would experience a repulsive force directed radially outward. (b) The electric field lines are directed radially outward from a positive point charge $+q$.

# 18.7 Electric Field Lines

As we have seen, electric charges create an electric field in the space surrounding them. It is useful to have a kind of "map" that gives the direction and indicates the strength of the field at various places. The great English physicist Michael Faraday (1791–1867) proposed an idea that provides such a "map," the idea of *electric field lines*. Since the electric field is the electric force per unit charge, the electric field lines are sometimes called *lines of force*.

To introduce the electric field line concept, Figure 18.21a shows a positive point charge $+q$. At the locations numbered 1–8, a positive test charge would experience a repulsive force, as the arrows in the drawing indicate. Therefore, the electric field created by the charge $+q$ is directed radially outward. The electric field lines are lines drawn to show this direction, as part b of the drawing illustrates. They begin on the charge $+q$ and point radially *outward*. Figure 18.22 shows the field lines in the vicinity of a negative charge $-q$. In this case they are directed radially *inward* because the force on a positive test charge is one of attraction, indicating that the electric field points inward. In general, *electric field lines are always directed away from positive charges and toward negative charges.*

The electric field lines in Figures 18.21 and 18.22 are drawn in only two dimensions, as a matter of convenience. Field lines radiate from the charges in three dimensions, and an infinite number of lines could be drawn. However, for clarity only a small number is ever included in pictures. The number is chosen to be proportional to the magnitude of the charge; thus, five times as many lines would emerge from a $+5q$ charge as from a $+q$ charge.

The pattern of electric field lines also provides information about the magnitude or strength of the field. Notice in Figures 18.21 and 18.22 that near the charges, where the electric field is stronger, the lines are closer together. At distances far from the charges, where the electric field is weaker, the lines are more spread out. It is true in general that the electric field is stronger in regions where the field lines are closer together. In fact, no matter how many charges are present, the number of lines per unit area passing perpendicularly through a surface is proportional to the magnitude of the electric field.

In regions where the electric field lines are equally spaced, there is the same number of lines per unit area everywhere, and the electric field has the same strength at all points. For example, Figure 18.23 shows that the field lines between the plates of a parallel plate capacitor are parallel and equally spaced, except near the edges where they bulge outward. The equally spaced, parallel lines indicate that the electric field has the same magnitude and direction at all points in the central region of the capacitor.

Often, electric field lines are curved, as in the case of an *electric dipole.* An electric dipole consists of two separated point charges that have the same magnitude but opposite signs. The electric field of a dipole is proportional to the product of the magnitude of one of the charges and the distance between the charges. This product is called the *dipole*

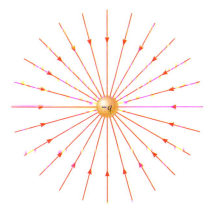

**Figure 18.22** The electric field lines are directed radially inward toward a negative point charge $-q$.

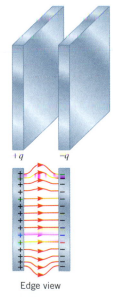

Edge view

**Figure 18.23** In the central region of a parallel plate capacitor, the electric field lines are parallel and evenly spaced, indicating that the electric field there has the same magnitude and direction at all points.

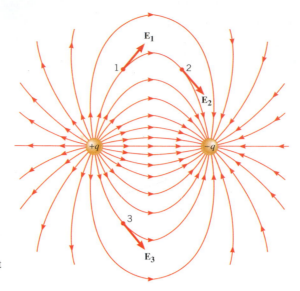

**Figure 18.24** The electric field lines of an electric dipole are curved and extend from the positive to the negative charge. At any point, such as 1, 2, or 3, the field created by the dipole is tangent to the line through the point.

*moment.* Many molecules, such as $H_2O$ and HCl, have dipole moments. Figure 18.24 depicts the field lines in the vicinity of a dipole. For a curved field line, the electric field vector at a point is *tangent* to the line at that point (see points 1, 2, and 3 in the drawing). The pattern of the lines for the dipole indicates that the electric field is greatest in the region between and immediately surrounding the two charges, since the lines are closest together there.

Notice in Figure 18.24 that any given field line starts on the positive charge and ends on the negative charge. In general, *electric field lines always begin on a positive charge and end on a negative charge and do not start or stop in midspace. Furthermore, the number of lines leaving a positive charge or entering a negative charge is proportional to the magnitude of the charge.* This means, for example, that if 100 lines are drawn leaving a $+4 \ \mu C$ charge, then 75 lines would have to end on a $-3 \ \mu C$ charge and 25 lines on a $-1 \ \mu C$ charge. Thus, 100 lines leave the charge of $+4 \ \mu C$ and end on a total charge of $-4 \ \mu C$, so the lines begin and end on equal amounts of total charge.

The electric field lines are also curved in the vicinity of two identical charges. Figure 18.25 shows the pattern associated with two positive point charges and reveals that there is an absence of lines in the region between the charges. The absence of lines indicates that the electric field is relatively weak between the charges.

Some of the important properties of electric field lines are reexamined in Conceptual Example 12.

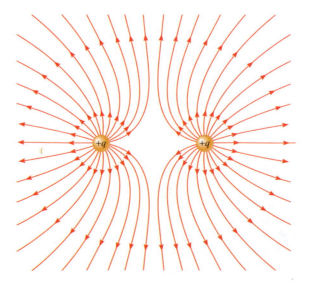

**Figure 18.25** The electric field lines for two identical positive point charges. If the charges were both negative, the directions of the lines would be reversed.

## Conceptual Example 12 Drawing Electric Field Lines

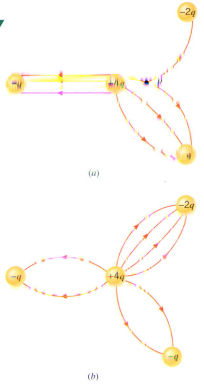

Figure 18.26a shows three negative point charges ($-q$, $-q$, and $-2q$) and one positive point charge ($+4q$), along with some electric field lines drawn between the charges. There are three things wrong with this drawing. What are they?

**Reasoning and Solution** One aspect of Figure 18.26a that is incorrect is that electric field lines cross at point $P$. Field lines can never cross, and here's why. An electric charge placed at $P$ experiences a single net force due to the presence of the other charges in its environment. Therefore, there is only one value for the electric field (which is the force per unit charge) at that point. If two field lines intersected, there would be two electric fields at the point of intersection, one associated with each line. Since there can be only one value of the electric field at any point, there can be only one electric field line passing through that point.

Another mistake in Figure 18.26a is the number of electric field lines that end on the negative charges. Remember that the number of field lines leaving a positive charge or entering a negative charge is proportional to the magnitude of the charge. The $-2q$ charge has half the magnitude of the $+4q$ charge. Therefore, since 8 lines leave the $+4q$ charge, 4 of them (one-half of them) must enter the $-2q$ charge. Of the remaining 4 lines that leave the positive charge, 2 enter each of the $-q$ charges, according to a similar line of reasoning.

The third error in Figure 18.26a is the way in which the electric field lines are drawn between the $+4q$ charge and the $-q$ charge at the left of the drawing. As drawn, the lines are parallel and evenly spaced. This would indicate that the electric field everywhere in this region has a constant magnitude and direction, as is the case in the central region of a parallel plate capacitor. But the electric field between the $+4q$ and $-q$ charges is not constant everywhere. It certainly is stronger in places close to the $+4q$ or $-q$ charge than it is midway between them. The field lines, therefore, should be drawn with a curved nature, similar (but not identical) to those that surround a dipole. Figure 18.26b shows more nearly correct representations of the field lines for the four charges.

**Related Homework:** *Problem 26*

**Figure 18.26** (*a*) Incorrectly and (*b*) correctly drawn electric field lines.

## 18.8 The Electric Field Inside a Conductor: Shielding

In conducting materials such as copper, electric charges move readily in response to the forces that electric fields exert. This property of conducting materials has a major effect on the electric field that can exist within and around them. Suppose that a piece of copper carries a number of excess electrons somewhere within it, as in Figure 18.27a. Each electron would experience a force of repulsion because of the electric field of its neighbors. And, since copper is a conductor, the excess electrons move readily in response to that force. In fact, as a consequence of the $1/r^2$ dependence on distance in Coulomb's law, they rush to the surface of the copper. Once static equilibrium is established with all of the excess charge on the surface, no further movement of charge occurs, as part *b* of the drawing indicates. Similarly, excess positive charge also moves to the surface of a conductor. In general, *at equilibrium under electrostatic conditions, any excess charge resides on the surface of a conductor.*

Now consider the interior of the copper in Figure 18.27b. The interior is electrically neutral, although there are still free electrons that can move under the influence of an electric field. The absence of a net movement of these free electrons indicates that there is no net electric field present within the conductor. In fact, the excess charges arrange themselves on the conductor surface precisely in the manner needed to make the electric field zero within the material. Thus, *at equilibrium under electrostatic conditions, the electric field is zero at any point within a conducting material.* This fact has some fascinating implications.

Figure 18.28a shows an uncharged, solid, cylindrical conductor at equilibrium in the central region of a parallel plate capacitor. Induced charges on the surface of the cylinder alter the electric field lines of the capacitor. Since an electric field cannot exist within the conductor under these conditions, the electric field lines do not penetrate the cylinder. Instead, they end or begin on the induced charges. Consequently, a test charge placed *inside* the

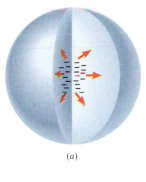

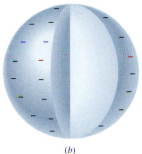

**Figure 18.27** (*a*) Excess charge within a conductor (copper) moves quickly (*b*) to the surface.

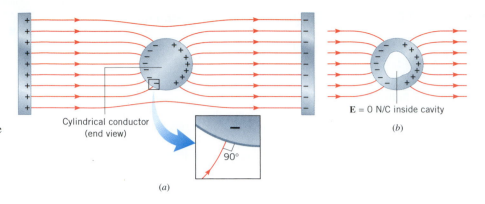

**Figure 18.28** (*a*) A cylindrical conductor (shown as an end view) is placed between the oppositely charged plates of a capacitor. The electric field lines do not penetrate the conductor. The blowup shows that, just outside the conductor, the electric field lines are perpendicular to its surface. (*b*) The electric field is zero in a cavity within the conductor.

conductor would feel no force due to the presence of the charges on the capacitor. In other words, *the conductor shields any charge within it from electric fields created outside the conductor.* The shielding results from the induced charges on the conductor surface.

**The physics of...**
**shielding electronic circuits.**

Since the electric field is zero inside the conductor, nothing is disturbed if a cavity is cut from the interior of the material, as in part *b* of the drawing. Thus, the interior of the cavity is also shielded from external electric fields, a fact that has important applications, particularly for shielding electronic circuits. "Stray" electric fields are produced by various electrical appliances (e.g., hair dryers, blenders, and vacuum cleaners), and these fields can interfere with the operation of sensitive electronic circuits, such as those in stereo amplifiers, televisions, and computers. To eliminate such interference, circuits are often enclosed within metal boxes that provide shielding from external fields.

The blowup in Figure 18.28*a* shows another aspect of how conductors alter the electric field lines created by external charges. The lines are altered because *the electric field just outside the surface of a conductor is perpendicular to the surface at equilibrium under electrostatic conditions.* If the field were not perpendicular, there would be a component of the field parallel to the surface. Since the free electrons on the surface of the conductor can move, they would do so under the force exerted by that parallel component. In reality, however, no electron flow occurs at equilibrium. Therefore, there can be no parallel component, and the electric field is perpendicular to the surface.

The preceding discussion deals with features of the electric field within and around a conductor at equilibrium under electrostatic conditions. These features are related to the fact that conductors contain mobile free electrons and *do not apply to insulators,* which contain very few free electrons. Example 13 further explores the behavior of a conducting material in the presence of an electric field.

### Conceptual Example 13 A Conductor in an Electric Field

A charge $+q$ is suspended at the center of a hollow, electrically neutral, spherical conductor, as Figure 18.29 illustrates. Show that this charge induces (a) a charge of $-q$ on the interior surface and (b) a charge of $+q$ on the exterior surface of the conductor.

**Reasoning and Solution**

(a) Electric field lines emanate from the positive charge $+q$. Since the electric field inside the metal conductor must be zero at equilibrium under electrostatic conditions, each field line ends when it reaches the conductor, as the picture shows. Since field lines terminate only on negative charges, there must be an induced *negative* charge on the interior surface of the conductor. Furthermore, the lines begin and end on equal amounts of charge, so the magnitude of the total induced charge is the same as the magnitude of the charge at the center. Thus, the total induced charge on the interior surface is $-q$.

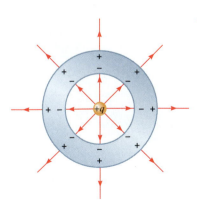

**Figure 18.29** A positive charge $+q$ is suspended at the center of a hollow spherical conductor that is electrically neutral. Induced charges appear on the inner and outer surfaces of the conductor. The electric field within the conductor itself is zero.

(b) Before the charge $+q$ is introduced, the conductor is electrically neutral. Therefore, it carries no net charge. We have also seen that there can be no excess charge within the metal. Thus, since an induced charge of $-q$ appears on the interior surface, a charge of $+q$ must be induced on the outer surface. The positive charge on the outer surface generates field lines that radiate outward (see the drawing) as if they originated from the central charge and the conductor were absent. The conductor does not shield the outside from the field produced by the charge on the inside.

# 18.9 Gauss' Law

Section 18.6 discusses how a point charge creates an electric field in the space around the charge. There are also many situations in which an electric field is produced by charges that are spread out over a region, rather than by a single point charge. Such an extended collection of charges is called a charge distribution. For example, the electric field within the parallel plate capacitor in Figure 18.20 is produced by positive charges spread uniformly over one plate and an equal number of negative charges spread over the other plate. As we will see, Gauss' law describes the relationship between a charge distribution and the electric field it produces. This law was formulated by the German mathematician and physicist Carl Friedrich Gauss (1777–1855).

In presenting Gauss' law, it will be necessary to introduce a new idea called electric flux. The idea of flux involves both the electric field and the surface through which it passes. By bringing together the electric field and the surface through which it passes, we will be able to define electric flux and then present Gauss' law.

We begin by developing a version of Gauss' law that applies only to a point charge, which we assume to be positive. The electric field lines for a positive point charge radiate outward in all directions from the charge, as Figure 18.21b indicates. The magnitude $E$ of the electric field at a distance $r$ from the charge is $E = kq/r^2$, according to Equation 18.3, in which we have replaced the symbol $|q|$ with the symbol $q$ since we are assuming that the charge is positive. As mentioned in Section 18.5, the constant $k$ can be expressed as $k = 1/(4\pi\epsilon_0)$, where $\epsilon_0$ is the permittivity of free space. With this substitution, the magnitude of the electric field becomes $E = q/(4\pi\epsilon_0 r^2)$. We now place this point charge at the center of an imaginary spherical surface of radius $r$, as Figure 18.30 shows. Such a hypothetical closed surface is called a **Gaussian surface,** although in general it need not be spherical. The surface area $A$ of a sphere is $A = 4\pi r^2$, and the magnitude of the electric field can be written in terms of this area as $E = q/(A\epsilon_0)$, or

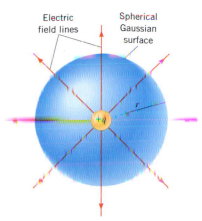

**Figure 18.30** A positive point charge is located at the center of an imaginary spherical surface of radius $r$. Such a surface is one example of a Gaussian surface. Here the electric field is perpendicular to the surface and has the same magnitude everywhere on it.

*Gauss' law for a point charge*

$$\underbrace{EA}_{\text{Electric flux, } \Psi_E} = \frac{q}{\epsilon_0} \qquad (18.5)$$

The left side of Equation 18.5 is the product of the magnitude $E$ of the electric field at any point on the Gaussian surface and the area $A$ of the surface. In Gauss' law this product is especially important and is called the **electric flux,** $\Psi_E$: $\Psi_E = EA$. (It will be necessary to modify this definition of flux when we consider the general case of a Gaussian surface with an arbitrary shape.)

Equation 18.5 is the result we have been seeking, for it is the form of Gauss' law that applies to a point charge. This result indicates that, aside from the constant $\epsilon_0$, the electric flux $\Phi_E$ depends only on the charge $q$ within the Gaussian surface and is independent of the radius $r$ of the surface. We will now see how to generalize Equation 18.5 to account for distributions of charges and Gaussian surfaces with arbitrary shapes.

Figure 18.31 shows a charge distribution whose net charge is labeled $Q$. The charge distribution is surrounded by a Gaussian surface—that is, an imaginary closed surface. The surface can have *any arbitrary shape* (it need not be spherical), but *it must be closed* (an open surface would be like that of half an eggshell). The direction of the electric field is not necessarily perpendicular to the Gaussian surface. Furthermore, the magnitude of the electric field need not be constant on the surface, but can vary from point to point.

To determine the electric flux through such a surface, we divide the surface into many tiny sections with areas $\Delta A_1$, $\Delta A_2$, and so on. Each section is so small that it is essentially flat and the electric field **E** is a constant (both in magnitude and direction) over it. For reference, a dashed line called the "normal" is drawn perpendicular to each section on the outside of the surface. To determine the electric flux for each of the sections, we use only the component of **E** that is perpendicular to the surface—that is, the component of the electric field that passes through the surface. From the drawing it can be seen that this component has a magnitude of $E \cos \phi$, where $\phi$ is the angle between the electric field and the normal. The electric flux through any one section is then $(E \cos \phi)\Delta A$. The

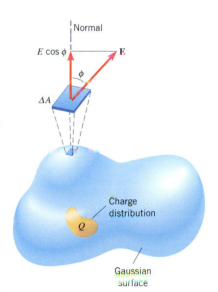

**Figure 18.31** The charge distribution $Q$ is surrounded by an arbitrarily shaped Gaussian surface. The electric flux $\Phi$ through any tiny segment of the surface is the product of $E \cos \phi$ and the area $\Delta A$ of the segment: $\Phi = (E \cos \phi) \Delta A$. The angle $\phi$ is the angle between the electric field and the normal to the surface.

electric flux $\Phi_E$ that passes through the entire Gaussian surface is the sum of all of these individual fluxes: $\Phi_E = (E_1 \cos \phi_1)\Delta A_1 + (E_2 \cos \phi_2)\Delta A_2 + \cdots$, or

$$\Phi_E = \Sigma(E \cos \phi)\Delta A \tag{18.6}$$

where, as usual, the symbol $\Sigma$ means "the sum of." Gauss' law relates the electric flux $\Phi_E$ to the net charge $Q$ enclosed by the arbitrarily shaped Gaussian surface.

■ **GAUSS' LAW**

The electric flux $\Phi_E$ through a Gaussian surface is equal to the net charge $Q$ enclosed by the surface divided by $\epsilon_0$, the permittivity of free space:

$$\underbrace{\sum (E \cos \phi)\Delta A}_{\text{Electric flux, } \Phi_E} = \frac{Q}{\epsilon_0} \tag{18.7}$$

**SI Unit of Electric Flux:** $\text{N} \cdot \text{m}^2/\text{C}$

Gauss' law is often used to find the magnitude of the electric field produced by a distribution of charges. The law is most useful when the distribution is uniform and symmetrical. In the next two examples we will see how to apply Gauss' law in such situations.

### Example 14   The Electric Field of a Charged Thin Spherical Shell

Figure 18.32 shows a thin spherical shell of radius $R$. A positive charge $q$ is spread uniformly over the shell. Find the magnitude of the electric field at any point (a) outside the shell and (b) inside the shell.

**Reasoning** Because the charge is distributed uniformly over the spherical shell, the electric field is symmetrical. This means that the electric field is directed radially outward in all directions, and its magnitude is the same at all points that are equidistant from the shell. All such points lie on a sphere, so the symmetry is called *spherical symmetry*. With this symmetry in mind, we will use a spherical Gaussian surface to evaluate the electric flux $\Phi_E$. We will then use Gauss' law to determine the magnitude of the electric field.

**Solution**

**(a)** To find the magnitude of the electric field outside the charged shell, we evaluate the electric flux $\Phi_E = \Sigma(E \cos \phi)\Delta A$ by using a spherical Gaussian surface of radius $r$ ($r > R$) that is concentric with the shell. See the surface labeled S in Figure 18.32. Since the electric field **E** is everywhere perpendicular to the Gaussian surface, $\phi = 0°$ and $\cos \phi = 1$. In addition, $E$ has the same value at all points on the surface, since they are equidistant from the charged shell. Being constant over the surface, $E$ can be factored outside the summation, with the result that

$$\Phi_E = \Sigma(E \cos 0°)\Delta A = E\,(\underbrace{\Sigma \Delta A}_{\substack{\text{Area of}\\\text{Gaussian}\\\text{surface}}}) = E\,(\underbrace{4\pi r^2}_{\substack{\text{Surface area}\\\text{of sphere}}})$$

The term $\Sigma \Delta A$ is just the sum of the tiny areas that make up the Gaussian surface. Since the area of a spherical surface is $4\pi r^2$, we have $\Sigma \Delta A = 4\pi r^2$. Setting the electric flux equal to $Q/\epsilon_0$, as specified by Gauss' law, yields $E(4\pi r^2) = Q/\epsilon_0$. Since the only charge within the Gaussian surface is the charge $q$ on the shell, it follows that the net charge within the Gaussian surface is $Q = q$. Thus, we can solve for $E$ and find that

$$\boxed{E = \frac{q}{4\pi\epsilon_0 r^2} \qquad (\text{for } r > R)}$$

This is a surprising result, for it is the same as that for a point charge (see Equation 18.3 with $|q| = q$). Thus, the electric field outside a uniformly charged spherical shell is the same as if all the charge $q$ were concentrated as a point charge at the center of the shell.

**(b)** To find the magnitude of the electric field inside the charged shell, we select a spherical Gaussian surface that lies inside the shell and is concentric with it. See the surface labeled $S_1$ in Figure 18.32. Inside the charged shell, the electric field (if it exists) must also have spherical

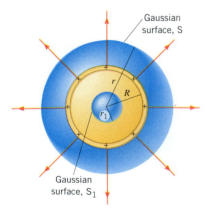

**Figure 18.32** A uniform distribution of positive charge resides on a thin spherical shell of radius $R$. The spherical Gaussian surfaces S and $S_1$ are used in Example 14 to evaluate the electric flux outside and inside the shell, respectively.

symmetry. Therefore, using reasoning like that in part (a), the electric flux through the Gaussian surface is $\Phi_E = \Sigma(E\cos\phi)\Delta A = E(4\pi r_1^2)$. In accord with Gauss' law, the electric flux must be equal to $Q/\epsilon_0$, where $Q$ is the net charge *inside* the Gaussian surface. But now $Q = 0$ C, since all the charge lies on the shell that is *outside* the surface $S_1$. Consequently, we have $E(4\pi r_1^2) = Q/\epsilon_0 = 0$, or

$$\boxed{E = 0\ \text{N/C} \qquad (\text{for } r < R)}$$

Gauss' law allows us to deduce that there is no electric field inside a uniform spherical shell of charge. An electric field exists only on the outside.

### Example 15 The Electric Field Inside a Parallel Plate Capacitor

According to Equation 18.4, the electric field inside a parallel plate capacitor, and away from the edges, is constant and has a magnitude of $E = \sigma/\epsilon_0$, where $\sigma$ is the charge density (the charge per unit area) on a plate. Use Gauss' law to prove this result.

**Reasoning** Figure 18.33a shows the electric field inside a parallel plate capacitor. Because the positive and negative charges are distributed uniformly over the surfaces of the plates, symmetry requires that the electric field be perpendicular to the plates. We will take advantage of this symmetry by choosing our Gaussian surface to be a small cylinder whose axis is perpendicular to the plates (see part b of the figure). With this choice, we will be able to evaluate the electric flux and then, with the aid of Gauss' law, determine $E$.

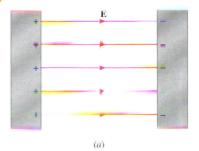

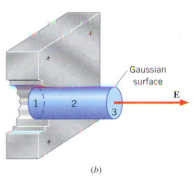

(a)

Gaussian surface

E

(b)

**Figure 18.33** (a) A side view of a parallel plate capacitor, showing some of the electric field lines. (b) The Gaussian surface is a cylinder oriented so its axis is perpendicular to the positive plate and its left end is inside the plate.

**Solution** Figure 18.33b shows that we have placed our Gaussian cylinder so that its left end is inside the positive metal plate, and the right end is in the space between the plates. To determine the electric flux through this Gaussian surface, we evaluate the flux through each of the three parts—labeled 1, 2, and 3 in the drawing—that make up the total surface of the cylinder and then add up the fluxes.

Surface 1—the flat left end of the cylinder—is embedded inside the positive metal plate. As discussed in Section 18.8, the electric field is zero everywhere inside a conductor that is in equilibrium under electrostatic conditions. Since $E = 0$ N/C, the electric flux through this surface is also zero:

$$\Phi_1 = \Sigma(E\cos\phi)\Delta A = \Sigma[(0\ \text{N/C})\cos\phi]\Delta A = 0$$

Surface 2—the curved wall of the cylinder—is everywhere parallel to the electric field between the plates, so that $\cos\phi = \cos 90° = 0$. Therefore, the electric flux through this surface is also zero:

$$\Phi_2 = \Sigma(E\cos\phi)\Delta A = \Sigma(E\cos 90°)\Delta A = 0$$

Surface 3—the flat right end of the cylinder—is perpendicular to the electric field between the plates, so $\cos\phi = \cos 0° = 1$. The electric field is constant over this surface, so $E$ can be taken outside the summation in Equation 18.6. Noting that $\Sigma\Delta A = A$ is the area of surface 3, we find that the electric flux through this surface is

$$\Phi_3 = \Sigma(E\cos\phi)\Delta A = \Sigma(E\cos 0°)\Delta A = E(\Sigma\Delta A) = EA$$

The electric flux through the entire Gaussian cylinder is the sum of the three fluxes determined above:

$$\Phi_E = \Phi_1 + \Phi_2 + \Phi_3 = 0 + 0 + EA = EA$$

According to Gauss' law, we set the electric flux equal to $Q/\epsilon_0$, where $Q$ is the net charge *inside* the Gaussian cylinder: $EA = Q/\epsilon_0$. But $Q/A$ is the charge per unit area, $\sigma$, on the plate. Therefore, we arrive at the value of the electric field inside a parallel plate capacitor: $\boxed{E = \sigma/\epsilon_0}$. Notice that the distance of the right end of the Gaussian cylinder from the positive plate does not appear in this result, indicating that the electric field has the same value everywhere between the plates.

# Concept Summary

This summary presents an abridged version of the chapter, including the important equations and all available learning aids. For convenient reference, the learning aids (including the text's examples) are placed next to or immediately after the relevant equation or discussion. The following learning aids may be found on-line at **www.wiley.com/college/cutnell**:

| | |
|---|---|
| **Interactive LearningWare** examples are solved according to a five-step interactive format that is designed to help you develop problem-solving skills. | **Concept Simulations** are animated versions of text figures or animations that illustrate important concepts. You can control parameters that affect the display, and we encourage you to experiment. |
| **Interactive Solutions** offer specific models for certain types of problems in the chapter homework. The calculations are carried out interactively. | **Self-Assessment Tests** include both qualitative and quantitative questions. Extensive feedback is provided for both incorrect and correct answers, to help you evaluate your understanding of the material. |

| Topic | Discussion | Learning Aids |
|---|---|---|
| | **18.1 The Origin of Electricity** | |
| The coulomb (C) | There are two kinds of electric charge: positive and negative. The SI unit of electric charge is the coulomb (C). The magnitude of the charge on an electron or a proton is | |
| Magnitude of charge on electron or proton | $$e = 1.60 \times 10^{-19}\ \text{C}$$ | |
| | Since the symbol $e$ denotes a magnitude, it has no algebraic sign. Thus, the electron carries a charge of $-e$, and the proton carries a charge of $+e$. | |
| Charge is quantized | The charge on any object, whether positive or negative, is quantized, in the sense that the charge consists of an integer number of protons or electrons. | **Example 1** |
| | **18.2 Charged Objects and the Electric Force** | |
| Law of conservation of electric charge | The law of conservation of electric charge states that the net electric charge of an isolated system remains constant during any process. | |
| Electrical repulsion and attraction | Like charges repel and unlike charges attract each other. | |
| | **18.3 Conductors and Insulators** | |
| Conductor | An electrical conductor is a material, such as copper, that conducts electric charge readily. | |
| Insulator | An electrical insulator is a material, such as rubber, that conducts electric charge poorly. | |
| | **18.4 Charging by Contact and by Induction** | |
| Charging by contact | Charging by contact is the process of giving one object a net electric charge by placing it in contact with an object that is already charged. | |
| Charging by induction | Charging by induction is the process of giving an object a net electric charge without touching it to a charged object. | |
| | **18.5 Coulomb's Law** | |
| Point charge | A point charge is a charge that occupies so little space that it can be regarded as a mathematical point. | |
| | Coulomb's law gives the magnitude $F$ of the electric force that two point charges $q_1$ and $q_2$ exert on each other: | **Examples 2–5**<br>**Interactive LearningWare 18.1** |
| Coulomb's law | $$F = k\frac{|q_1||q_2|}{r^2} \qquad (18.1)$$ | **Concept Simulation 18.1**<br>**Interactive LearningWare 18.2** |
| | where $|q_1|$ and $|q_2|$ are the magnitudes of the charges and have no algebraic sign. The term $k$ is a constant and has the value $k = 8.99 \times 10^9\ \text{N}\cdot\text{m}^2/\text{C}^2$. The force specified by Equation 18.1 acts along the line between the two charges. | **Interactive Solution 18.15** |
| Permittivity of free space | The permittivity of free space $\epsilon_0$ is defined by the relation | |
| | $$k = \frac{1}{4\pi\epsilon_0}$$ | |

| Topic | Discussion | Learning Aids |
|---|---|---|

 **Use Self-Assessment Test 18.1 to evaluate your understanding of Sections 18.1–18.5.**

### 18.6 The Electric Field

The electric field **E** at a given spot is a vector and is the electrostatic force **F** experienced by a very small test charge $q_0$ placed at that spot divided by the charge itself:

**Electric field**

$$E = \frac{F}{q_0} \qquad (18.2)$$

Examples 6, 7, 8
Interactive Solution 18.65

The direction of the electric field is the same as the direction of the force on a positive test charge. The SI unit for the electric field is the Newton per coulomb (N/C). The source of the electric field at any spot is the charged objects surrounding that spot.

The magnitude of the electric field created by a point charge $q$ is

**Electric field of a point charge**

$$E = \frac{k|q|}{r^2} \qquad (18.3)$$

Examples 9, 10, 11
Interactive LearningWare 18.3
Interactive Solution 18.41

where $|q|$ is the magnitude of the charge and has no algebraic sign and $r$ is the distance from the charge.

For a parallel plate capacitor that has a charge per unit area of $\sigma$ on each plate, the magnitude of the electric field between the plates is

**Electric field of a parallel plate capacitor**

$$E = \frac{\sigma}{\epsilon_0} \qquad (18.4)$$

### 18.7 Electric Field Lines

**Electric field lines**

Electric field lines are lines that can be thought of as a "map," insofar as the lines provide information about the direction and strength of the electric field. The lines are directed away from positive charges and toward negative charges.

Example 12

**Direction of electric field**

The direction of the lines gives the direction of the electric field, since the electric field vector at a point is tangent to the line at that point. The electric field is

Concept Simulation 18.2

**Strength of electric field**

strongest in regions where the number of lines per unit area passing perpendicularly through a surface is the greatest—that is, where the lines are packed together most tightly.

### 18.8 The Electric Field Inside a Conductor: Shielding

Example 13

**Excess charge carried by a conductor at equilibrium**

Excess negative or positive charge resides on the surface of a conductor at equilibrium under electrostatic conditions. In such a situation, the electric field at any point within the conducting material is zero, and the electric field just outside the surface of the conductor is perpendicular to the surface.

### 18.9 Gauss' Law

The electric flux $\Phi_E$ through a surface is related to the magnitude $E$ of the electric field, the area $A$ of the surface, and the angle $\phi$ that specifies the direction of the field relative to the normal to the surface:

**Electric flux**

$$\Phi_E = \Sigma(E \cos \phi)\Delta A \qquad (18.6)$$

Gauss' law states that the electric flux through a closed surface (a Gaussian surface) is equal to the net charge $Q$ enclosed by the surface divided by $\epsilon_0$, the permittivity of free space:

**Gauss' law**

$$\Phi_E = \Sigma(E \cos \phi)\Delta A = \frac{Q}{\epsilon_0} \qquad (18.7)$$

Examples 14, 15
Concept Simulation 18.3
Interactive Solution 18.53

 **Use Self-Assessment Test 18.2 to evaluate your understanding of Sections 18.6–18.9.**

# Problems

*Problems that are not marked with a star are considered the easiest to solve. Problems that are marked with a single star (\*) are more difficult, while those marked with a double star (\*\*) are the most difficult. Note: All charges are point charges, unless specified otherwise.*

**ssm**  Solution is in the Student Solutions Manual.    **www**  Solution is available on the World Wide Web at www.wiley.com/college/cutnell

☤  This icon represents a biomedical application.

### Section 18.1 The Origin of Electricity,
### Section 18.2 Charged Objects and the Electric Force,
### Section 18.3 Conductors and Insulators,
### Section 18.4 Charging by Contact and by Induction

**1. ssm** How many electrons must be removed from an electrically neutral silver dollar to give it a charge of $+2.4\ \mu C$?

**2.** A metal sphere has a charge of $+8.0\ \mu C$. What is the net charge after $6.0 \times 10^{13}$ electrons have been placed on it?

**3.** A plate carries a charge of $-3.0\ \mu C$, while a rod carries a charge of $+2.0\ \mu C$. How many electrons must be transferred from the plate to the rod, so that both objects have the same charge?

**4.** Object A is metallic and electrically neutral. It is charged by induction so that it acquires a charge of $-3.0 \times 10^{-6}$ C. Object B is identical to object A and is also electrically neutral. It is charged by induction so that it acquires a charge of $+3.0 \times 10^{-6}$ C. Find the *difference* in mass between the charged objects and state which has the greater mass.

**5. ssm** Consider three identical metal spheres, A, B, and C. Sphere A carries a charge of $+5q$. Sphere B carries a charge of $-q$. Sphere C carries no net charge. Spheres A and B are touched together and then separated. Sphere C is then touched to sphere A and separated from it. Last, sphere C is touched to sphere B and separated from it. (a) How much charge ends up on sphere C? What is the total charge on the three spheres (b) before they are allowed to touch each other and (c) after they have touched?

\* **6.** Water has a mass per mole of 18.0 g/mol, and each water molecule ($H_2O$) has 10 electrons. (a) How many electrons are there in one liter ($1.00 \times 10^{-3}\ m^3$) of water? (b) What is the net charge of all these electrons?

### Section 18.5 Coulomb's Law

**7. ssm** Two charges attract each other with a force of 1.5 N. What will be the force if the distance between them is reduced to one-ninth of its original value?

**8.** The nucleus of the helium atom contains two protons that are separated by about $3.0 \times 10^{-15}$ m. Find the magnitude of the electrostatic force that each proton exerts on the other. (The protons remain together in the nucleus because the repulsive electrostatic force is balanced by an attractive force called the strong nuclear force.)

**9.** The force of repulsion that two like charges exert on each other is 3.5 N. What will be the force if the distance between the charges is increased to five times its original value?

**10.** In a vacuum, two particles have charges of $q_1$ and $q_2$, where $q_1 = +3.5\ \mu C$. They are separated by a distance of 0.26 m, and particle 1 experiences an attractive force of 3.4 N. What is $q_2$ (magnitude and sign)?

**11. ssm  Interactive LearningWare 18.1** at www.wiley.com/college/cutnell offers some perspective on this problem. Two tiny spheres have the same mass and carry charges of the same magnitude. The mass of each sphere is $2.0 \times 10^{-6}$ kg. The gravitational force that each sphere exerts on the other is balanced by the electric force. (a) What algebraic signs can the charges have? (b) Determine the charge magnitude.

**12.** Two tiny conducting spheres are identical and carry charges of $-20.0\ \mu C$ and $+50.0\ \mu C$. They are separated by a distance of 2.50 cm. (a) What is the magnitude of the force that each sphere experiences, and is the force attractive or repulsive? (b) The spheres are brought into contact and then separated to a distance of 2.50 cm. Determine the magnitude of the force that each sphere now experiences, and state whether the force is attractive or repulsive.

**13. ssm  www** Two particles, with identical positive charges and a separation of $2.60 \times 10^{-2}$ m, are released from rest. Immediately after the release, particle 1 has an acceleration $\mathbf{a_1}$ whose magnitude is $4.60 \times 10^3\ m/s^2$, while particle 2 has an acceleration $\mathbf{a_2}$ whose magnitude is $8.50 \times 10^3\ m/s^2$. Particle 1 has a mass of $6.00 \times 10^{-6}$ kg. Find (a) the charge on each particle and (b) the mass of particle 2.

**14.** A charge of $-3.00\ \mu C$ is fixed at the center of a compass. Two additional charges are fixed on the circle of the compass (radius = 0.100 m). The charges on the circle are $-4.00\ \mu C$ at the position due north and $+5.00\ \mu C$ at the position due east. What is the magnitude and direction of the net electrostatic force acting on the charge at the center? Specify the direction relative to due east.

**15. Interactive Solution 18.15** at www.wiley.com/college/cutnell provides a model for solving this type of problem. Two small objects, A and B, are fixed in place and separated by 3.00 cm in a vacuum. Object A has a charge of $+2.00\ \mu C$, and object B has a charge of $-2.00\ \mu C$. How many electrons must be removed from A and put onto B to make the electrostatic force that acts on each object an attractive force whose magnitude is 68.0 N?

\* **16.** The drawing shows an equilateral triangle, each side of which has a length of 2.00 cm. Point charges are fixed to each corner, as shown. The $4.00\ \mu C$ charge experiences a net force due to the charges $q_A$ and $q_B$. This net force points vertically downward in the drawing and has a magnitude of 405 N. Determine the magnitudes and algebraic signs of the charges $q_A$ and $q_B$.

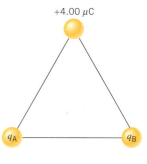

\* **17.** A point charge of $-0.70\ \mu C$ is fixed to one corner of a square. An identical charge is fixed to the diagonally opposite corner. A point charge $q$ is fixed to each of the remaining corners. The net force acting on either of the charges $q$ is zero. Find the magnitude and algebraic sign of $q$.

\* **18. Interactive LearningWare 18.2** at www.wiley.com/college/cutnell provides one approach to solving problems such as this one. The drawing shows three point charges fixed in place. The charge at the coordinate origin has a value of $q_1 = +8.00\ \mu C$; the other two have identical magnitudes, but opposite signs: $q_2 = -5.00\ \mu C$ and $q_3 = +5.00\ \mu C$. (a) Determine the net force (magnitude and direction) exerted on $q_1$ by the other two charges. (b) If $q_1$ had a mass of 1.50 g and it were free to move, what would be its acceleration?

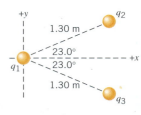

* **19. ssm** In the rectangle in the drawing, a charge is to be placed at the empty corner to make the net force on the charge at corner A point along the vertical direction. What charge (magnitude and algebraic sign) must be placed at the empty corner?

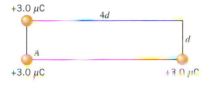

+3.0 μC

4d

d

A

+3.0 μC

+3.0 μC

* **20.** Four point charges have equal magnitudes. Three are positive, and one is negative, as the drawing shows. They are fixed in place on the same straight line, and adjacent charges are equally separated by a distance of $d$. Consider the net electrostatic force acting on each charge. Calculate the ratio of the largest to the smallest net force.

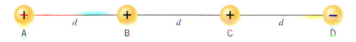

A    $d$    B    $d$    C    $d$    D

* **21.** An electrically neutral model airplane is flying in a horizontal circle on a 3.0-m guideline, which is nearly parallel to the ground. The line breaks when the kinetic energy of the plane is 50.0 J. Reconsider the same situation, except that now there is a point charge of $+q$ on the plane and a point charge of $-q$ at the other end of the guideline. In this case, the line breaks when the kinetic energy of the plane is 51.8 J. Find the magnitude of the charges.

** **22.** Two objects are identical and small enough that their sizes can be ignored relative to the distance between them, which is 0.200 m. In a vacuum, each object carries a different charge, and they attract each other with a force of 1.20 N. The objects are brought into contact, so the net charge is shared equally, and then they are returned to their initial positions. Now it is found that the objects repel one another with a force whose magnitude is equal to that of the initial attractive force. What is the initial charge on each object? Note that there are two answers.

** **23. ssm** A small spherical insulator of mass $8.00 \times 10^{-2}$ kg and charge $+0.600$ μC is hung by a thin wire of negligible mass. A charge of $-0.900$ μC is held 0.150 m away from the sphere and directly to the right of it, so the wire makes an angle $\theta$ with the vertical (see the drawing). Find (a) the angle $\theta$ and (b) the tension in the wire.

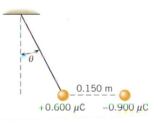

$\theta$

0.150 m

+0.600 μC    −0.900 μC

** **24.** There are four charges, each with a magnitude of 2.0 μC. Two are positive and two are negative. The charges are fixed to the corners of a 0.30-m square, one to a corner, in such a way that the net force on any charge is directed toward the center of the square. Find the magnitude of the net electrostatic force experienced by any charge.

**Section 18.6 The Electric Field,**
**Section 18.7 Electric Field Lines,**
**Section 18.8 The Electric Field Inside a Conductor: Shielding**

**25.** An electric field of 260 000 N/C points due west at a certain spot. What are the magnitude and direction of the force that acts on a charge of −7.0 μC at this spot?

**26.** Review Conceptual Example 12 as an aid in working this problem. Charges of $-4q$ are fixed to diagonally opposite corners of a square. A charge of $+5q$ is fixed to one of the remaining corners, and a charge of $+3q$ is fixed to the last corner. Assuming that ten electric field lines emerge from the $+5q$ charge, sketch the field lines in the vicinity of the four charges.

**27. ssm www** A tiny ball (mass = 0.012 kg) carries a charge of −18 μC. What electric field (magnitude and direction) is needed to cause the ball to float above the ground?

**28.** At a distance $r_1$ from a point charge, the magnitude of the electric field created by the charge is 248 N/C. At a distance $r_2$ from the charge, the field has a magnitude of 132 N/C. Find the ratio $r_2/r_1$.

**29.** Two charges are placed on the $x$ axis. One charge ($q_1 = +8.5$ μC) is at $x_1 = +3.0$ cm and the other ($q_2 = -21$ μC) is at $r_1 = +9.0$ cm. Find the net electric field (magnitude and direction) at (a) $x = 0$ cm and (b) $x = +6.0$ cm.

**30.** Two charges, −16 and +4.0 μC, are fixed in place and separated by 3.0 m. (a) At what spot along a line through the charges is the net electric field zero? Locate this spot relative to the positive charge. (*Hint: The spot does not necessarily lie between the two charges.*) (b) What would be the force on a charge of +14 μC placed at this spot?

**31. ssm** Background pertinent to this problem is available in **Interactive LearningWare 18.3** at **www.wiley.com/college/cutnell**. A 3.0-μC point charge is placed in an external uniform electric field of $1.6 \times 10^4$ N/C. At what distance from the charge is the net electric field zero?

**32.** A charge of $q = +7.50$ μC is located in an electric field. The $x$ and $y$ components of the electric field are $E_x = 6.00 \times 10^3$ N/C and $E_y = 8.00 \times 10^3$ N/C, respectively. (a) What is the magnitude of the force on the charge? (b) Determine the angle that the force makes with the $+x$ axis.

**33. ssm** A small drop of water is suspended motionless in air by a uniform electric field that is directed upward and has a magnitude of 8480 N/C. The mass of the water drop is $3.50 \times 10^{-9}$ kg. (a) Is the excess charge on the water drop positive or negative? Why? (b) How many excess electrons or protons reside on the drop?

**34.** Review Conceptual Example 11 before attempting this problem. The magnitude of each of the charges in Figure 18.19 is $8.60 \times 10^{-12}$ C. The lengths of the sides of the rectangles are 3.00 cm and 5.00 cm. Find the magnitude of the electric field at the center of the rectangle in Figures 18.19a and b.

* **35.** Two charges are located on the $x$ axis: $q_1 = +6.0$ μC at $x_1 = +4.0$ cm, and $q_2 = +6.0$ μC at $x_2 = -4.0$ cm. Two other charges are located on the $y$ axis: $q_3 = +3.0$ μC at $y_3 = +5.0$ cm, and $q_4 = -8.0$ μC at $y_4 = +7.0$ cm. Find the net electric field (magnitude and direction) at the origin.

* **36.** Two parallel plate capacitors have circular plates. The magnitude of the charge on these plates is the same. However, the electric field between the plates of the first capacitor is $2.2 \times 10^5$ N/C, while the field within the second capacitor is $3.8 \times 10^5$ N/C. Determine the ratio $r_2/r_1$ of the plate radius for the second capacitor to the plate radius for the first capacitor.

* **37.** A proton is moving parallel to a uniform electric field. The electric field accelerates the proton and increases its linear momentum to $5.0 \times 10^{-23}$ kg·m/s from $1.5 \times 10^{-23}$ kg·m/s in a time of $6.3 \times 10^{-6}$ s. What is the magnitude of the electric field?

* **38.** An electron is released from rest at the negative plate of a parallel plate capacitor. The charge per unit area on each plate is $\sigma = 1.8 \times 10^{-7}$ C/m², and the plates are separated by a distance of $1.5 \times 10^{-2}$ m. How fast is the electron moving just before it reaches the positive plate?

* **39. ssm www** A rectangle has a length of $2d$ and a height of $d$. Each of the following three charges is located at a corner of the rec-

tangle: $+q_1$ (upper left corner), $+q_2$ (lower right corner), and $-q$ (lower left corner). The net electric field at the (empty) upper right corner is zero. Find the magnitudes of $q_1$ and $q_2$. Express your answers in terms of $q$.

*  **40.** A uniform electric field has a magnitude of $2.3 \times 10^3$ N/C. In a vacuum, a proton begins with a speed of $2.5 \times 10^4$ m/s and moves in the direction of this field. Find the speed of the proton after it has moved a distance of 2.0 mm.

*  **41.** Review **Interactive Solution 18.41** at **www.wiley.com/college/cutnell** for help with this problem. The drawing shows two positive charges $q_1$ and $q_2$ fixed to a circle. At the center of the circle they produce a net electric field that is directed upward along the vertical axis. Determine the ratio $|q_2|/|q_1|$ of the charge magnitudes.

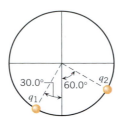

**  **42.** The drawing shows an electron entering the lower left side of a parallel plate capacitor and exiting at the upper right side. The initial speed of the electron is $7.00 \times 10^6$ m/s. The capacitor is

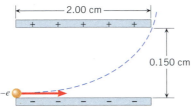

2.00 cm long, and its plates are separated by 0.150 cm. Assume that the electric field between the plates is uniform everywhere and find its magnitude.

**  **43.** A small plastic ball of mass $6.50 \times 10^{-3}$ kg and charge $+0.150$ $\mu$C is suspended from an insulating thread and hangs between the plates of a capacitor (see the drawing). The ball is in equilibrium, with the thread making an angle of $30.0°$ with respect to the vertical. The area of each plate is $0.0150$ m$^2$. What is the magnitude of the charge on each plate?

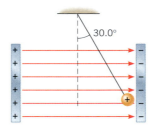

**  **44.** The magnitude of the electric field between the plates of a parallel plate capacitor is 480 N/C. A silver dollar is placed between the plates and oriented parallel to the plates. (a) Ignoring the edges of the coin, find the induced charge density $\sigma$ on each face of the coin. (b) Assuming the coin has a radius of 1.9 cm, find the magnitude of the total charge on each face of the coin.

**  **45. ssm** Two point charges of the same magnitude but opposite signs are fixed to either end of the base of an isosceles triangle, as the drawing shows. The electric field at the midpoint $M$ between the charges has a magnitude $E_M$. The field directly above the midpoint at point $P$ has a magnitude $E_P$. The ratio of these two field magnitudes is $E_M/E_P = 9.0$. Find the angle $\alpha$ in the drawing.

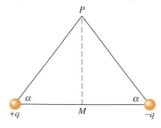

### Section 18.9 Gauss' Law

**46.** A rectangular surface ($0.16$ m $\times$ $0.38$ m) is oriented in a uniform electric field of 580 N/C. What is the maximum possible electric flux through the surface?

**47. ssm** The drawing shows an edge-on view of two planar surfaces that intersect and are mutually perpendicular. Surface 1

has an area of $1.7$ m$^2$, while surface 2 has an area of $3.2$ m$^2$. The electric field **E** in the drawing is uniform and has a magnitude of 250 N/C. Find the electric flux through (a) surface 1 and (b) surface 2.

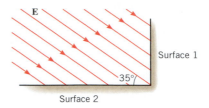

**48.** A spherical surface completely surrounds a collection of charges. Find the electric flux through the surface if the collection consists of (a) a single $+3.5 \times 10^{-6}$ C charge, (b) a single $-2.3 \times 10^{-6}$ C charge, and (c) both of the charges in (a) and (b).

**49.** A surface completely surrounds a $+2.0 \times 10^{-6}$ C charge. Find the electric flux through this surface when the surface is (a) a sphere with a radius of 0.50 m, (b) a sphere with a radius of 0.25 m, and (c) a cube with edges that are 0.25 m long.

**50.** A vertical wall ($5.9$ m $\times$ $2.5$ m) in a house faces due east. A uniform electric field has a magnitude of 150 N/C. This field is parallel to the ground and points $35°$ north of east. What is the electric flux through the wall?

*  **51. ssm** A cube is located with one corner at the origin of an $x$, $y$, $z$ coordinate system. One of the cube's faces lies in the $x$, $y$ plane, another in the $y$, $z$ plane, and another in the $x$, $z$ plane. In other words, the cube is in the first octant of the coordinate system. The edges of the cube are 0.20 m long. A uniform electric field is parallel to the $x$, $y$ plane and points in the direction of the $+y$ axis. The magnitude of the field is 1500 N/C. (a) Find the electric flux through each of the six faces of the cube. (b) Add the six values obtained in part (a) to show that the electric flux through the cubical surface is zero, as Gauss' law predicts, since there is no net charge within the cube.

*  **52.** Refer to **Concept Simulation 18.3** at **www.wiley.com/college/cutnell** for a perspective that is useful in solving this problem. Two spherical shells have a common center. A $-1.6 \times 10^{-6}$ C charge is spread uniformly over the inner shell, which has a radius of 0.050 m. A $+5.1 \times 10^{-6}$ C charge is spread uniformly over the outer shell, which has a radius of 0.15 m. Find the magnitude and direction of the electric field at a distance (measured from the common center) of (a) 0.20 m, (b) 0.10 m, and (c) 0.025 m.

*  **53. Interactive Solution 18.53** at **www.wiley.com/college/cutnell** offers help with this problem in an interactive environment. A solid nonconducting sphere has a positive charge $q$ spread uniformly throughout its volume. The charge density or charge per unit volume, therefore, is $\dfrac{q}{\frac{4}{3}\pi R^3}$. Use Gauss' law to show that the electric field at a point within the sphere at a radius $r$ has a magnitude of $\dfrac{qr}{4\pi\epsilon_0 R^3}$.

*(Hint: For a Gaussian surface, use a sphere of radius $r$ centered within the solid sphere. Note that the net charge within any volume is the charge density times the volume.)*

**  **54.** A long, thin, straight wire of length $L$ has a positive charge $Q$ distributed uniformly along it. Use Gauss' law to show that the electric field created by this wire at a radial distance $r$ has a magnitude of $E = \lambda/(2\pi\epsilon_0 r)$, where $\lambda = Q/L$. *(Hint: For a Gaussian surface, use a cylinder aligned with its axis along the wire and note that the cylinder has a flat surface at either end, as well as a curved surface.)*

# Chapter 19 Electric Potential Energy and the Electric Potential

## 19.1 Potential Energy

In Chapter 18 we discussed the electrostatic force that two point charges exert on each other, the magnitude of which is $F = k|q_1||q_2|/r^2$. The form of this equation is similar to that for the gravitational force that two particles exert on each other, which is $F = Gm_1m_2/r^2$, according to Newton's law of universal gravitation (see Section 4.7). Both of these forces are conservative and, as Section 6.4 explains, a potential energy can be associated with a conservative force. Thus, an electric potential energy exists that is analogous to the gravitational potential energy. To set the stage for a discussion of the electric potential energy, let's review some of the important aspects of the gravitational counterpart.

Figure 19.1, which is essentially Figure 6.9, shows a basketball of mass $m$ falling from point $A$ to point $B$. The gravitational force, $m\mathbf{g}$, is the only force acting on the ball, where $g$ is the magnitude of the acceleration due to gravity. As Section 6.3 discusses, the work $W_{AB}$ done by the gravitational force when the ball falls from a height of $h_A$ to a height of $h_B$ is

$$W_{AB} = \underbrace{mgh_A}_{\substack{\text{Initial} \\ \text{gravitational} \\ \text{potential energy,} \\ \text{GPE}_A}} - \underbrace{mgh_B}_{\substack{\text{Final} \\ \text{gravitational} \\ \text{potential energy,} \\ \text{GPE}_B}} = \text{GPE}_A - \text{GPE}_B \qquad (6.4)$$

Recall that the quantity $mgh$ is the gravitational potential energy* of the ball, $\text{GPE} = mgh$ (Equation 6.5), and represents the energy that the ball has by virtue of its position relative to the surface of the earth. Thus, the work done by the gravitational force equals the initial gravitational potential energy minus the final gravitational potential energy.

Figure 19.2 clarifies the analogy between electric and gravitational potential energies. In this drawing a positive test charge $+q_0$ is situated at point $A$ between two oppositely charged plates. Because of the charges on the plates, an electric field $\mathbf{E}$ exists in the region between them. Consequently, the test charge experiences an electric force, $\mathbf{F} = q_0\mathbf{E}$, that is directed downward, toward the lower plate. (The gravitational force is being neglected here.) As the charge moves from $A$ to $B$, work is done by this force, in a fashion analogous to the work done by the gravitational force in Figure 19.1. The work $W_{AB}$ done by the electric force equals the difference between the electric potential energy EPE at $A$ and that at $B$:

$$W_{AB} = \text{EPE}_A - \text{EPE}_B \qquad (19.1)$$

This expression is similar to Equation 6.4. The path along which the test charge moves from $A$ to $B$ is of no consequence because the electric force is a conservative force, and so the work $W_{AB}$ is the same for all paths.

## 19.2 The Electric Potential Difference

Since the electric force is $\mathbf{F} = q_0\mathbf{E}$, the work that it does as the charge moves from $A$ to $B$ in Figure 19.2 depends on the charge $q_0$. It is useful, therefore, to express this work on a per-unit-charge basis, by dividing both sides of Equation 19.1 by the charge:

$$\frac{W_{AB}}{q_0} = \frac{\text{EPE}_A}{q_0} - \frac{\text{EPE}_B}{q_0} \qquad (19.2)$$

Notice that the right-hand side of this equation is the difference between two terms, each of which is an electric potential energy divided by the test charge, $\text{EPE}/q_0$. The quantity

*The gravitational potential energy is now being denoted by GPE to distinguish it from the electric potential energy EPE.

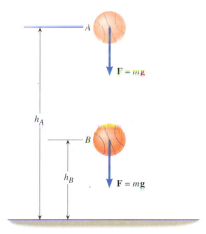

**Figure 19.1** Gravity exerts a force, $\mathbf{F} = m\mathbf{g}$, on the basketball of mass $m$. Work is done by the gravitational force as the ball falls from $A$ to $B$.

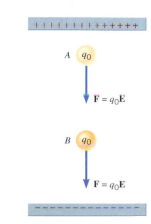

**Figure 19.2** Because of the electric field $\mathbf{E}$, an electric force, $\mathbf{F} = q_0\mathbf{E}$, is exerted on a positive test charge $+q_0$. Work is done by the force as the charge moves from $A$ to $B$.

$EPE/q_0$ is the electric potential energy per unit charge and is an important concept in electricity. It is called the *electric potential* or, simply, the *potential* and is referred to with the symbol $V$, as in Equation 19.3. The electric potential is a concept that incorporates both the electric potential energy and the idea of a test charge.

---

■ **DEFINITION OF ELECTRIC POTENTIAL**

The electric potential $V$ at a given point is the electric potential energy EPE of a small test charge $q_0$ situated at that point divided by the charge itself:

$$V = \frac{EPE}{q_0} \qquad (19.3)$$

*SI Unit of Electric Potential:* joule/coulomb = volt (V)

---

The SI unit of electric potential is a joule per coulomb, a quantity known as a *volt*. The name honors Alessandro Volta (1745–1827), who invented the voltaic pile, the forerunner of the battery. In spite of the similarity in names, the electric potential energy EPE and the electric potential $V$ are *not* the same. The electric potential energy, as its name implies, is an *energy* and, therefore, is measured in joules. In contrast, the electric potential is an *energy per unit charge* and is measured in joules per coulomb, or volts.

We can now relate the work $W_{AB}$ done by the electric force when a charge $q_0$ moves from $A$ to $B$ to the potential difference $V_B - V_A$ between the points. Combining Equations 19.2 and 19.3, we have:

$$V_B - V_A = \frac{EPE_B}{q_0} - \frac{EPE_A}{q_0} = \frac{-W_{AB}}{q_0} \qquad (19.4)$$

Often the "delta" notation is used to express the difference (final value minus initial value) in potentials and in potential energies: $\Delta V = V_B - V_A$ and $\Delta(EPE) = EPE_B - EPE_A$. In terms of this notation, Equation 19.4 takes the following more compact form:

$$\Delta V = \frac{\Delta(EPE)}{q_0} = \frac{-W_{AB}}{q_0} \qquad (19.4)$$

Neither the potential $V$ nor the potential energy EPE can be determined in an absolute sense, because only the *differences* $\Delta V$ and $\Delta(EPE)$ are measurable in terms of the work $W_{AB}$. The gravitational potential energy has this same characteristic, since only the value at one height relative to that at some reference height has any significance. Example 1 emphasizes the relative nature of the electric potential.

### *Example 1*  Work, Electric Potential Energy, and Electric Potential

▼

In Figure 19.2, the work done by the electric force as the test charge ($q_0 = +2.0 \times 10^{-6}$ C) moves from $A$ to $B$ is $W_{AB} = +5.0 \times 10^{-5}$ J. (a) Find the difference, $\Delta(EPE) = EPE_B - EPE_A$, in the electric potential energies of the charge between these points. (b) Determine the potential difference, $\Delta V = V_B - V_A$, between the points.

**Reasoning**  The work done by the electric force when the charge moves from $A$ to $B$ is $W_{AB} = EPE_A - EPE_B$, according to Equation 19.1. Therefore, the difference in the electric potential energies (final value minus initial value) is $\Delta(EPE) = EPE_B - EPE_A = -W_{AB}$. The potential difference, $\Delta V = V_B - V_A$, is the difference in the electric potential energies divided by the charge $q_0$, according to Equation 19.4.

**Solution**

(a)  The difference in the electric potential energies of the charge at points $A$ and $B$ is

$$\underbrace{EPE_B - EPE_A}_{= \Delta(EPE)} = -W_{AB} = \boxed{-5.0 \times 10^{-5} \text{ J}} \qquad (19.1)$$

Thus the charge has a higher electric potential energy at $A$ than at $B$.

**(b)** The potential difference $\Delta V$ between $A$ and $B$ is

$$\underbrace{V_B - V_A}_{= \Delta V} = \frac{EPE_B - EPE_A}{q_0} = \frac{-5.0 \times 10^{-5}\,\text{J}}{2.0 \times 10^{-6}\,\text{C}} = \boxed{-25\,\text{V}} \qquad (19.4)$$

We can see, then, that the electric potential is higher at $A$ than at $B$.

In Figure 19.1 the speed of the basketball increases as it falls from $A$ to $B$. Since point $A$ has a greater gravitational potential energy than point $B$, we see that an object of mass $m$ accelerates when it moves from a region of higher potential energy toward a region of lower potential energy. Likewise, the positive charge in Figure 19.2 accelerates as it moves from $A$ to $B$ because of the electric repulsion from the upper plate and the attraction to the lower plate. Since point $A$ has a higher electric potential than point $B$, we conclude that *a positive charge accelerates from a region of higher electric potential toward a region of lower electric potential.* On the other hand, a negative charge placed between the plates in Figure 19.2 behaves in the opposite fashion, since the electric force acting on the negative charge is directed opposite to that on the positive charge. *A negative charge accelerates from a region of lower potential toward a region of higher potential.* The next example illustrates the way positive and negative charges behave.

*Problem solving insight*

*Problem solving insight*

## Conceptual Example 2
### The Accelerations of Positive and Negative Charges

Three points, $A$, $B$, and $C$, are located along a horizontal line, as Figure 19.3 illustrates. A positive test charge is released from rest at $A$ and accelerates toward $B$. Upon reaching $B$, the test charge continues to accelerate toward $C$. Assuming that only motion along the line is possible, what will a negative test charge do when it is released from rest at $B$?

**Reasoning and Solution** A negative charge will accelerate from a region of lower potential toward a region of higher potential. Therefore, before we can decide what the negative test charge will do, it is necessary to know how the electric potentials compare at points $A$, $B$, and $C$. This information can be deduced from the behavior of the positive test charge, which accelerates from a region of higher toward a region of lower potential. Since the positive test charge accelerates from $A$ to $B$, the potential at $A$ must exceed that at $B$. And since the positive test charge accelerates from $B$ to $C$, the potential at $B$ must exceed that at $C$. The potential at point $B$, then, must lie between that at points $A$ and $C$, as Figure 19.3 illustrates. When the negative test charge is released from rest at $B$, it will accelerate toward the region of higher potential. In other words, *it will begin moving toward $A$.*

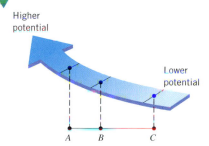

**Figure 19.3** The electric potentials at points $A$, $B$, and $C$ are different. Under the influence of these potentials, positive and negative charges accelerate in opposite directions.

As a familiar application of electric potential energy and electric potential, Figure 19.4 shows a 12-V automobile battery with a headlight connected between its terminals. The positive terminal, point $A$, has a potential that is 12 V higher than the potential at the negative terminal, point $B$; in other words, $V_A - V_B = 12\,\text{V}$. Positive charges are repelled from the positive terminal and travel through the wires and headlight toward the negative terminal.* As the charges pass through the headlight, virtually all their potential energy is converted into heat, which causes the filament to glow "white hot" and emit light. When the charges reach the negative terminal, they no longer have any potential energy. The battery then gives the charges an additional "shot" of potential energy by moving them to the higher-potential positive terminal, and the cycle is repeated. In raising the potential energy of the charges, the battery does work on them and draws from its reserve of chemical energy to do so. Example 3 illustrates the concepts of electric potential energy and electric potential as applied to a battery.

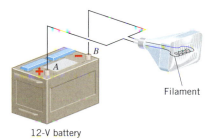

**Figure 19.4** A headlight connected to a 12-V battery.

---

*Historically, it was believed that positive charges flow in the wires of an electric circuit. Today, it is known that negative charges flow in wires from the negative toward the positive terminal. Here, however, we follow the customary practice of describing the flow of negative charges by specifying the opposite but equivalent flow of positive charges. This hypothetical flow of positive charges is called the "conventional electric current," as we will see in Section 20.1.

### *Example 3*   Operating a Headlight

▼

Determine the number of particles, each carrying a charge of $1.60 \times 10^{-19}$ C (the magnitude of the charge on an electron), that pass between the terminals of a 12-V car battery when a 60.0-W headlight burns for one hour.

**Reasoning**  To obtain the number of charged particles, we determine the total charge needed to provide the energy consumed by the headlight in one hour. Dividing the total charge by the charge on each particle gives the number of particles.

**Solution**  Using energy at a rate of 60.0 joules per second (60.0 watts) for one hour, the headlight uses a total energy of

$$\text{Energy} = \text{Power} \times \text{Time} = (60.0 \text{ W})(3600 \text{ s}) = 2.2 \times 10^5 \text{ J} \qquad (6.10\text{b})$$

Since this energy comes from the battery, it represents the difference between the electrical potential energy of the charges at point $A$ and that at point $B$: $2.2 \times 10^5 \text{ J} = \text{EPE}_A - \text{EPE}_B$. According to Equation 19.4, we have

$$\frac{\text{EPE}_A - \text{EPE}_B}{q_0} = V_A - V_B$$

$$q_0 = \frac{\text{EPE}_A - \text{EPE}_B}{V_A - V_B} = \frac{2.2 \times 10^5 \text{ J}}{12 \text{ V}} = 1.8 \times 10^4 \text{ C}$$

The number of particles whose individual charges combine to provide this total charge is $(1.8 \times 10^4 \text{ C})/(1.60 \times 10^{-19} \text{ C}) = \boxed{1.1 \times 10^{23}}$.

▲

As used in connection with batteries, the volt is a familiar unit for measuring electric potential difference. The word "volt" also appears in another context, as part of a unit that is used to measure energy, particularly the energy of an atomic particle, such as an electron or a proton. This energy unit is called the *electron volt* (eV). ***One electron volt is the magnitude of the amount by which the potential energy of an electron changes when the electron moves through a potential difference of one volt.*** Since the magnitude of the change in potential energy is $|q_0 \Delta V| = |(-1.60 \times 10^{-19} \text{ C}) \times (1.00 \text{ V})| = 1.60 \times 10^{-19}$ J, it follows that

$$1 \text{ eV} = 1.60 \times 10^{-19} \text{ J}$$

One million ($10^{+6}$) electron volts of energy is referred to as one MeV, and one billion ($10^{+9}$) electron volts of energy is one GeV, where the "G" stands for the prefix "giga" (pronounced "jig′a").

The total energy of an object, which is the sum of its kinetic and potential energies, is an important concept. Its significance lies in the fact that the total energy remains the same (is conserved) during the object's motion, provided that nonconservative forces, such as friction, are either absent or do no net work. While the sum of the energies at each instant remains constant, energy may be converted from one form to another; for example, gravitational potential energy is converted into kinetic energy as a ball falls. We now include the electric potential energy EPE as part of the total energy that an object can have:

$$E = \underbrace{\tfrac{1}{2}mv^2}_{\substack{\text{Total} \\ \text{energy}}} \quad + \quad \underbrace{\tfrac{1}{2}I\omega^2}_{\substack{\text{Translational} \\ \text{kinetic energy}}} \quad + \quad \underbrace{mgh}_{\substack{\text{Rotational} \\ \text{kinetic energy}}} \quad + \quad \underbrace{\tfrac{1}{2}kx^2}_{\substack{\text{Gravitational} \\ \text{potential} \\ \text{energy}}} \quad + \quad \underbrace{\text{EPE}}_{\substack{\text{Electric} \\ \text{potential} \\ \text{energy}}}$$

| Total energy | Translational kinetic energy | Rotational kinetic energy | Gravitational potential energy | Elastic potential energy | Electric potential energy |
|---|---|---|---|---|---|

If the total energy is conserved as the object moves, then its final energy $E_\text{f}$ is equal to its initial energy $E_0$, or $E_\text{f} = E_0$. Example 4 illustrates how the conservation of energy is applied to a charge moving in an electric field.

### *Example 4*   The Conservation of Energy

▼

(a) A particle has a mass of $m = 1.8 \times 10^{-5}$ kg and a positive charge of $q_0 = +3.0 \times 10^{-5}$ C. It is released from rest at point $A$ and accelerates horizontally until it reaches point $B$, as Figure

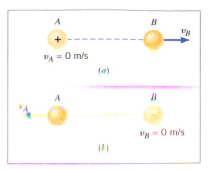

19.5*a* shows. During transit, the particle does not rotate. The only force acting on the particle is an electric force (not shown in the drawing), and the electric potential at *A* is 25 V greater than that at *B*. In other words, $V_A - V_B = 25$ V. What is the speed $v_B$ of the particle when it reaches point *B*? (b) If the same particle had a *negative* charge and were released from rest at *B* (see Figure 19.5b), what would be its speed $v_A$ at *A*?

**Reasoning** Since the only force acting on the particle is the conservative electric force, the total energy of the particle is conserved. This means that the total energy is the same at points *B* and *A*:

$$\tfrac{1}{2}mv_B^2 + \underbrace{\tfrac{1}{2}I\omega_B^2}_{\substack{=\,0\,\text{J, since}\\ \omega_B\,=\,0\text{ rad/s}}} + mgh_B + \underbrace{\tfrac{1}{2}kx_B^2}_{\substack{=\,0\,\text{J, since no}\\ \text{elastic forces}\\ \text{are present}}} + EPE_B$$

$$= \tfrac{1}{2}mv_A^2 + \underbrace{\tfrac{1}{2}I\omega_A^2}_{\substack{=\,0\,\text{J, since}\\ \omega_A\,=\,0\text{ rad/s}}} + mgh_A + \underbrace{\tfrac{1}{2}kx_A^2}_{\substack{=\,0\,\text{J, since no}\\ \text{elastic forces}\\ \text{are present}}} + EPE_A$$

In this expression the angular speeds $\omega_A$ and $\omega_B$ are zero because the particle does not rotate. Moreover, we set $h_A = h_B$ (the particle moves horizontally) and note that $EPE_A - EPE_B = q_0(V_A - V_B)$; the conservation of energy equation then reduces to

$$\tfrac{1}{2}mv_B^2 = \tfrac{1}{2}mv_A^2 + q_0(V_A - V_B)$$

We will use this equation to determine the final speeds.

**Solution**

**(a)** The positively charged particle starts from rest at *A* ($v_A = 0$ m/s). Its speed $v_B$ at *B* can be found from the equation above:

$$v_B = \sqrt{\frac{2q_0(V_A - V_B)}{m}}$$

$$= \sqrt{\frac{2(+3.0 \times 10^{-5}\text{ C})(25\text{ V})}{1.8 \times 10^{-5}\text{ kg}}} = \boxed{9.1\text{ m/s}}$$

**(b)** The negatively charged particle starts from rest at *B* ($v_B = 0$ m/s). Its speed $v_A$ at *A* can also be found from the last equation in the reasoning section:

$$v_A = \sqrt{\frac{-2q_0(V_A - V_B)}{m}}$$

$$= \sqrt{\frac{-2(-3.0 \times 10^{-5}\text{ C})(25\text{ V})}{1.8 \times 10^{-5}\text{ kg}}} = \boxed{9.1\text{ m/s}}$$

**Figure 19.5** (a) A positive charge starts from rest at point *A* and accelerates toward point *B*. (b) A negative charge starts from rest at *B* and accelerates toward *A*.

**Problem solving insight**
A positive charge accelerates from a region of higher potential toward a region of lower potential. In contrast, a negative charge accelerates from a region of lower potential toward a region of higher potential.

## 19.3 The Electric Potential Difference Created by Point Charges

A positive point charge $+q$ creates an electric potential in a way that Figure 19.6 helps explain. This picture shows two locations *A* and *B*, at distances $r_A$ and $r_B$ from the charge. At any position between *A* and *B* an electrostatic force of repulsion **F** acts on a positive test charge $+q_0$. The magnitude of the force is given by Coulomb's law as $F = kq_0q/r^2$, where we assume for convenience that $q_0$ and $q$ are positive, so that $|q_0| = q_0$ and $|q| = q$. When the test charge moves from *A* to *B*, work is done by this force. Since *r* varies between $r_A$ and $r_B$, the force *F* also varies, and the work is not the product of the force and the distance between the points. (Recall from Section 6.1 that work is force times distance only if the force is constant.) However, the work $W_{AB}$ can be found with the methods of integral calculus. The result is

$$W_{AB} = \frac{kqq_0}{r_A} - \frac{kqq_0}{r_B}$$

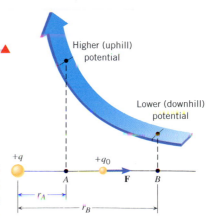

**Figure 19.6** The positive test charge $+q_0$ experiences a repulsive force **F** due to the positive point charge $+q$. As a result, work is done by this force when the test charge moves from *A* to *B*. Consequently, the electric potential is higher (uphill) at *A* and lower (downhill) at *B*.

This result is valid whether $q$ is positive or negative, and whether $q_0$ is positive or negative. The potential difference, $V_B - V_A$, between $A$ and $B$ can now be determined by substituting this expression for $W_{AB}$ into Equation 19.4:

$$V_B - V_A = \frac{-W_{AB}}{q_0} = \frac{kq}{r_B} - \frac{kq}{r_A} \qquad (19.5)$$

As point $B$ is located farther and farther from the charge $q$, $r_B$ becomes larger and larger. In the limit that $r_B$ is infinitely large, the term $kq/r_B$ becomes zero, and it is customary to set $V_B$ equal to zero also. In this limit, Equation 19.5 becomes $V_A = kq/r_A$, and it is standard convention to omit the subscripts and write the potential in the following form:

*Potential of a point charge* $\qquad\qquad V = \dfrac{kq}{r} \qquad (19.6)$

The symbol $V$ in this equation does not refer to the potential in any absolute sense. Rather, $V = kq/r$ stands for the amount by which the potential at a distance $r$ from a point charge differs from the potential at an infinite distance away. In other words, $V$ refers to a potential difference with the arbitrary assumption that the potential at infinity is zero.

With the aid of Equation 19.6, we can describe the effect that a point charge $q$ has on the surrounding space. When $q$ is positive, the value of $V = kq/r$ is also positive, indicating that the positive charge has everywhere raised the potential above the zero reference value. Conversely, when $q$ is negative, the potential $V$ is also negative, indicating that the negative charge has everywhere decreased the potential below the zero reference value. The next example deals with these effects quantitatively.

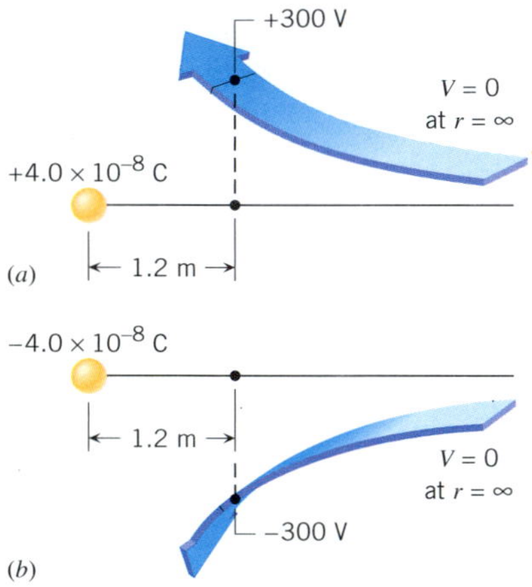

### Example 5   The Potential of a Point Charge

Using a zero reference potential at infinity, determine the amount by which a point charge of $4.0 \times 10^{-8}$ C alters the electric potential at a spot 1.2 m away when the charge is (a) positive and (b) negative.

**Reasoning** A point charge $q$ alters the potential at every location in the surrounding space. In the expression $V = kq/r$, the effect of the charge in increasing or decreasing the potential is conveyed by the algebraic sign for the value of $q$.

**Solution**

(a) Figure 19.7a shows the potential when the charge is positive:

$$V = \frac{kq}{r} = \frac{(8.99 \times 10^9 \text{ N}\cdot\text{m}^2/\text{C}^2)(+4.0 \times 10^{-8} \text{ C})}{1.2 \text{ m}} = \boxed{+300 \text{ V}} \qquad (19.6)$$

(b) Part $b$ of the drawing illustrates the results when the charge is negative. A calculation similar to the one in part (a) shows that the potential is now negative: $\boxed{-300 \text{ V}}$.

**Figure 19.7** A point charge of $4.0 \times 10^{-8}$ C alters the potential at a spot 1.2 m away. The potential is (a) increased by 300 V when the charge is positive and (b) decreased by 300 V when the charge is negative, relative to a zero reference potential at infinity.

A single point charge raises or lowers the potential at a given location, depending on whether the charge is positive or negative. ***When two or more charges are present, the potential due to all the charges is obtained by adding together the individual potentials,*** as the next two examples show.

**Problem solving insight**

### Example 6   The Total Electric Potential

At locations $A$ and $B$ in Figure 19.8, find the total electric potential due to the two point charges.

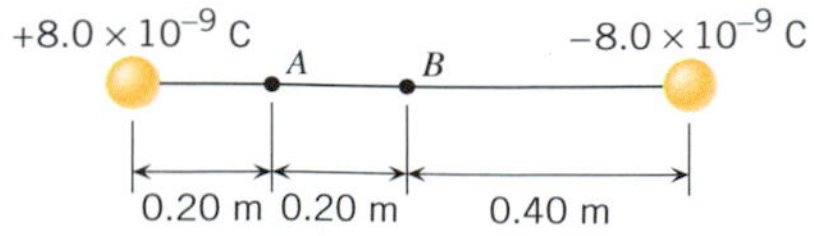

**Reasoning** At each location, each charge contributes to the total electric potential. We obtain the individual contributions by using $V = kq/r$ and find the total potential by adding the individual contributions algebraically. The two charges have the same magnitude, but different signs. Thus, at $A$ the total potential is positive because this spot is closer to the positive charge, whose effect dominates over that of the more distant negative charge. At $B$, midway between the charges, the total potential is zero, since the potential of one charge exactly offsets that of the other.

**Figure 19.8** Both the positive and negative charges affect the electric potential at locations $A$ and $B$.

**Solution**

| Location | Contribution from + Charge | | Contribution from − Charge | Total Potential |
|---|---|---|---|---|
| A | $\dfrac{(8.99 \times 10^9 \text{ N} \cdot \text{m}^2/\text{C}^2)(+8.0 \times 10^{-9} \text{ C})}{0.20 \text{ m}}$ | + | $\dfrac{(8.99 \times 10^9 \text{ N} \cdot \text{m}^2/\text{C}^2)(-8.0 \times 10^{-9} \text{ C})}{0.60 \text{ m}}$ = | +240 V |
| B | $\dfrac{(8.99 \times 10^9 \text{ N} \cdot \text{m}^2/\text{C}^2)(+8.0 \times 10^{-9} \text{ C})}{0.40 \text{ m}}$ | + | $\dfrac{(8.99 \times 10^9 \text{ N} \cdot \text{m}^2/\text{C}^2)(-8.0 \times 10^{-9} \text{ C})}{0.40 \text{ m}}$ = | 0 V |

## Conceptual Example 7  Where Is the Potential Zero?

Two point charges are fixed in place, as in Figure 19.9. The positive charge is $+2q$ and has twice the magnitude of the negative charge, which is $-q$. On the line that passes through the charges, how many places are there at which the total potential is zero?

**Reasoning and Solution** The total potential is the algebraic sum of the individual potentials created by each charge. It will be zero if the potential due to the positive charge is exactly offset by the potential due to the negative charge. The potential of a point charge is directly proportional to the charge and inversely proportional to the distance from the charge. The positive charge has the larger magnitude and, in the region to its left, is closer to any spot than the negative charge is. As a result, the potential of the positive charge here always dominates over that of the negative charge. Therefore, the total potential cannot be zero anywhere to the left of the positive charge.

*Between the charges there is a location where the individual potentials do balance.* We saw a similar situation in Example 6, where the balance occurred at the midpoint between two charges that had equal magnitudes. Now the charges have unequal magnitudes, so the balance does not occur at the midpoint. Instead, it occurs at a location that is closer to the charge with the smaller magnitude—namely, the negative charge. Then, since the potential of a point charge is inversely proportional to the distance from the charge, the effect of the smaller charge will be able to offset that of the more distant larger charge.

*To the right of the negative charge, there is another location at which the individual potentials exactly offset one another.* All places on this section of the line are closer to the negative than to the positive charge. In this region, therefore, a location exists at which the potential of the smaller negative charge again exactly offsets that of the more distant and larger positive charge.

**Related Homework:** *Problem 18*

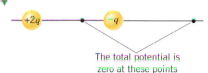

The total potential is zero at these points

**Figure 19.9** Two point charges, one positive and one negative. The positive charge, $+2q$, has twice the magnitude of the negative charge, $-q$.

**Problem solving insight**
At any location, the total electric potential is the algebraic sum of the individual potentials created by each point charge that is present.

In Example 6 we determined the total potential at a spot due to several point charges. In Example 8 we now extend this technique to find the total potential energy of three charges.

## Example 8  The Potential Energy of a Group of Charges

Figure 19.10 shows three point charges; initially, they are infinitely far apart. They are then brought together and placed at the corners of an equilateral triangle. Each side of the triangle has a length of 0.50 m. Determine the electric potential energy of the triangular group. In other words, determine the amount by which the electric potential energy of the group differs from that of the three charges in their initial, infinitely separated locations.

**Reasoning** We will proceed in steps by adding charges to the triangle, one at a time, and then determining the electric potential energy at each step. According to Equation 19.3, EPE = $q_0V$, the electric potential energy is the product of the charge and the electric potential at the spot where the charge is placed. The total electric potential energy of the triangular group is the sum of the energies of each step in assembling the group.

**Solution** The order in which the charges are put on the triangle does not matter; we begin with the charge of $+5.0 \ \mu$C. When this charge is placed at a corner of the triangle, it has no electric potential energy, according to EPE = $q_0V$. This is because the total potential $V$ produced by the other two charges is zero at this corner, since they are infinitely far away. Once the charge is in place, the potential it creates at either empty corner ($r$ = 0.50 m) is

$$V = \frac{kq}{r} = \frac{(8.99 \times 10^9 \text{ N} \cdot \text{m}^2/\text{C}^2)(+5.0 \times 10^{-6} \text{ C})}{0.50 \text{ m}} = +9.0 \times 10^4 \text{ V} \quad (19.6)$$

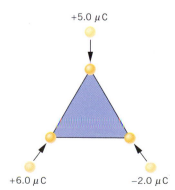

+5.0 $\mu$C

+6.0 $\mu$C          −2.0 $\mu$C

**Figure 19.10** Three point charges are placed on the corners of an equilateral triangle. Example 8 illustrates how to determine the total electric potential energy of this group of charges.

**Problem solving insight**
Be careful to distinguish between the concepts of potential $V$ and electric potential energy EPE. Potential is electric potential energy per unit charge: $V = \text{EPE}/q$.

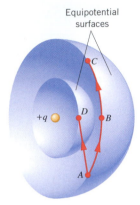

Equipotential surfaces

**Figure 19.11** The equipotential surfaces that surround the point charge $+q$ are spherical. The electric force does no work as a charge moves on a path that lies on an equipotential surface, such as the path $ABC$. However, work is done by the electric force when a charge moves between two equipotential surfaces, as along the path $AD$.

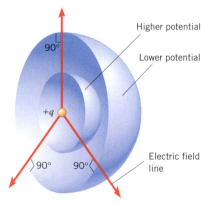

Higher potential

Lower potential

Electric field line

**Figure 19.12** The radially directed electric field of a point charge is perpendicular to the spherical equipotential surfaces that surround the charge. The electric field points in the direction of *decreasing* potential.

**Problem solving insight**

Therefore, when the $+6.0$-$\mu$C charge is placed at the second corner of the triangle, its electric potential energy is

$$\text{EPE} = qV = (+6.0 \times 10^{-6} \text{ C})(+9.0 \times 10^4 \text{ V}) = +0.54 \text{ J} \qquad (19.3)$$

The electric potential at the remaining empty corner is the sum of the potentials due to the two charges that are already in place:

$$V = \frac{(8.99 \times 10^9 \text{ N} \cdot \text{m}^2/\text{C}^2)(+5.0 \times 10^{-6} \text{ C})}{0.50 \text{ m}}$$
$$+ \frac{(8.99 \times 10^9 \text{ N} \cdot \text{m}^2/\text{C}^2)(+6.0 \times 10^{-6} \text{ C})}{0.50 \text{ m}} = +2.0 \times 10^5 \text{ V}$$

When the third charge, $-2.0$ $\mu$C, is placed at the remaining empty corner, its electric potential energy is

$$\text{EPE} = qV = (-2.0 \times 10^{-6} \text{ C})(+2.0 \times 10^5 \text{ V}) = -0.40 \text{ J} \qquad (19.3)$$

The total potential energy of the triangular group differs from that of the widely separated charges by an amount that is the sum of the potential energies calculated above:

$$\text{Total potential energy} = 0 \text{ J} + 0.54 \text{ J} - 0.40 \text{ J} = \boxed{+0.14 \text{ J}}$$

This energy originates in the work done to bring the charges together.

## 19.4 Equipotential Surfaces and Their Relation to the Electric Field

An *equipotential surface* is a surface on which the electric potential is the same everywhere. The easiest equipotential surfaces to visualize are those that surround an isolated point charge. According to Equation 19.6, the potential at a distance $r$ from a point charge $q$ is $V = kq/r$. Thus, wherever $r$ is the same, the potential is the same, and the equipotential surfaces are spherical surfaces centered on the charge. There are an infinite number of such surfaces, one for every value of $r$, and Figure 19.11 illustrates two of them. The larger the distance $r$, the smaller is the potential of the equipotential surface.

*The net electric force does no work as a charge moves on an equipotential surface.* This important characteristic arises because when an electric force does work $W_{AB}$ as a charge moves from $A$ to $B$, the potential changes according to Equation 19.4, $V_B - V_A = -W_{AB}/q_0$. Since the potential remains the same on an equipotential surface, $V_A = V_B$, and we see that $W_{AB} = 0$ J. In Figure 19.11, for instance, the electric force does no work as a test charge moves along the circular arc $ABC$, which lies on an equipotential surface. In contrast, the electric force does work when a charge moves *between* equipotential surfaces, as from $A$ to $D$ in the picture.

The spherical equipotential surfaces that surround an isolated point charge illustrate another characteristic of such surfaces. Figure 19.12 shows two of the surfaces around a positive point charge, along with some electric field lines. The electric field lines give the direction of the electric field, and for a positive point charge, the electric field is directed radially outward. Therefore, at each location on an equipotential sphere the electric field is perpendicular to the surface and points outward in the direction of decreasing potential, as the drawing emphasizes. This perpendicular relation is valid whether or not the equipotential surfaces result from a positive charge or have a spherical shape; *the electric field created by any charge or group of charges is everywhere perpendicular to the associated equipotential surfaces and points in the direction of decreasing potential.* For example, Figure 19.13 shows the electric field lines around an electric dipole, along with some equipotential surfaces (shown in cross section). Since the field lines are not simply radial, the equipotential surfaces are no longer spherical but, instead, have the shape necessary to be everywhere perpendicular to the field lines.

To see why an equipotential surface must be perpendicular to the electric field, consider Figure 19.14, which shows a hypothetical situation in which the perpendicular relation does *not* hold. If **E** were not perpendicular to the equipotential surface, there would be a component of **E** parallel to the surface. This field component would exert an electric force on a test charge placed on the surface. As the charge moved along the sur-

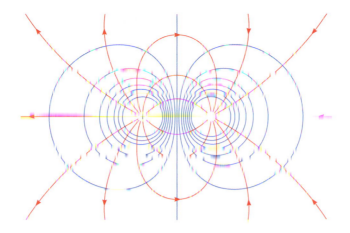

**Figure 19.13** A cross-sectional view of the equipotential surfaces (in blue) of an electric dipole. The surfaces are drawn so that at every point they are perpendicular to the electric field lines (in red) of the dipole.

face, work would be done by this component of the electric force. The work, according to Equation 19.4, would cause the potential to change, and, thus, the surface could not be an equipotential surface as assumed. The only way out of the dilemma is for the electric field to be perpendicular to the surface, so there is no component of the field parallel to the surface.

We have already encountered one equipotential surface. In Section 18.8, we found that the direction of the electric field just outside an electrical conductor is perpendicular to the conductor's surface, when the conductor is at equilibrium under electrostatic conditions. Thus, the surface of any conductor is an equipotential surface under such conditions. In fact, since the electric field is zero everywhere inside a conductor whose charges are in equilibrium, the entire conductor can be regarded as an equipotential volume.

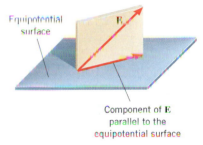

**Figure 19.14** In this hypothetical situation, the electric field **E** is not perpendicular to the equipotential surface. As a result, there is a component of **E** parallel to the surface.

There is a quantitative relation between the electric field and the equipotential surfaces. One example that illustrates this relation is the parallel plate capacitor in Figure 19.15. As Section 18.6 discusses, the electric field **E** between the metal plates is perpendicular to them and is the same everywhere, ignoring fringe fields at the edges. To be perpendicular to the electric field, the equipotential surfaces must be planes that are parallel to the plates, which themselves are equipotential surfaces. The potential difference between the plates is given by Equation 19.4 as $\Delta V = V_B - V_A = -W_{AB}/q_0$, where $A$ is a point on the positive plate and $B$ is a point on the negative plate. The work done by the electric force as a positive test charge $q_0$ moves from $A$ to $B$ is $W_{AB} = F\Delta s$, where $F$ refers to the electric force and $\Delta s$ to the displacement along a line perpendicular to the plates. The force equals the product of the charge and the electric field $E$ ($F = q_0E$), so the work becomes $W_{AB} = F\Delta s = q_0E\Delta s$. Therefore, the potential difference between the capacitor plates can be written in terms of the electric field as $\Delta V = -W_{AB}/q_0 = -q_0E\Delta s/q_0$, or

$$E = -\frac{\Delta V}{\Delta s} \tag{19.7}$$

The quantity $\Delta V/\Delta s$ is referred to as the *potential gradient* and has units of volts per meter. In general, the relation $E = -\Delta V/\Delta s$ gives only the component of the electric field along the displacement $\Delta s$; it does not give the perpendicular component. The next example deals further with the equipotential surfaces between the plates of a capacitor.

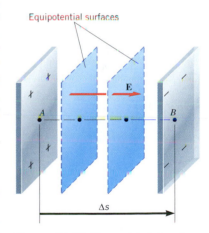

**Figure 19.15** The metal plates of a parallel plate capacitor are equipotential surfaces. Two additional equipotential surfaces are shown between the plates. These two equipotential surfaces are parallel to the plates and are perpendicular to the electric field **E** between the plates.

### Example 9   The Electric Field and Potential Are Related

The plates of the capacitor in Figure 19.15 are separated by a distance of 0.032 m, and the potential difference between them is $\Delta V = V_B - V_A = -64$ V. Between the two equipotential surfaces shown in color there is a potential difference of $-3.0$ V. Find the spacing between the two colored surfaces.

**Reasoning** The electric field is $E = -\Delta V/\Delta s$. To find the spacing between the two colored equipotential surfaces, we solve this equation for $\Delta s$, with $\Delta V = -3.0$ V and $E$ equal to the electric field between the plates of the capacitor. A value for $E$ can be obtained by using the values given for the distance and potential difference between the plates.

**Solution** The electric field between the capacitor plates is

$$E = -\frac{\Delta V}{\Delta s} = -\frac{-64 \text{ V}}{0.032 \text{ m}} = 2.0 \times 10^3 \text{ V/m} \tag{19.7}$$

The spacing between the colored equipotential surfaces can now be determined:

$$\Delta s = -\frac{\Delta V}{E} = -\frac{-3.0 \text{ V}}{2.0 \times 10^3 \text{ V/m}} = \boxed{1.5 \times 10^{-3} \text{ m}}$$

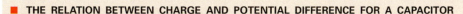

# 19.5 *Capacitors and Dielectrics*

## THE CAPACITANCE OF A CAPACITOR

In Section 18.6 we saw that a parallel plate capacitor consists of two parallel metal plates placed near one another but not touching. This type of capacitor is only one among many. In general, a *capacitor* consists of two conductors of any shape placed near one another without touching. For a reason that will become clear later on, it is common practice to fill the region between the conductors or plates with an electrically insulating material called a *dielectric*, as Figure 19.16 illustrates.

A capacitor stores electric charge. Each capacitor plate carries a charge of the *same magnitude,* one positive and the other negative. Because of the charges, the electric potential of the positive plate exceeds that of the negative plate by an amount V, as Figure 19.16 indicates. Experiment shows that when the magnitude q of the charge on each plate is doubled, the electric potential difference V is also doubled, so q is proportional to V: q ∝ V. Equation 19.8 expresses this proportionality with the aid of a proportionality constant C, which is the *capacitance* of the capacitor.

> ■ **THE RELATION BETWEEN CHARGE AND POTENTIAL DIFFERENCE FOR A CAPACITOR**
>
> The magnitude q of the charge on each plate of a capacitor is directly proportional to the magnitude V of the potential difference between the plates:
>
> $$q = CV \tag{19.8}$$
>
> where C is the capacitance.
>
> *SI Unit of Capacitance:* coulomb/volt = farad (F)

Equation 19.8 shows that the SI unit of capacitance is the coulomb per volt (C/V). This unit is called the *farad* (F), named after the English scientist Michael Faraday (1791–1867). One farad is an enormous capacitance. Usually smaller amounts, such as a microfarad (1 $\mu$F = $10^{-6}$ F) or a picofarad (1 pF = $10^{-12}$ F), are used in electric circuits. The capacitance reflects the ability of the capacitor to store charge, in the sense that a larger capacitance C allows more charge q to be put onto the plates for a given value of the potential difference V.

The ability of a capacitor to store charge lies at the heart of the random-access memory (RAM) chips used in computers, where information is stored in the form of the "ones" and "zeros" that comprise binary numbers. Figure 19.17 illustrates the role of a capacitor in a RAM chip. The capacitor is connected to a transistor switch, to which two lines are connected, an address line and a data line. A single RAM chip often contains millions of such transistor–capacitor units. The address line is used by the computer to locate a particular transistor–capacitor combination, and the data line carries the data to be stored. A pulse on the address line turns on the transistor switch. With the switch turned on, a pulse coming in on the data line can cause the capacitor to charge. A charged capacitor means that a "one" has been stored, whereas an uncharged capacitor means that a "zero" has been stored.

## THE DIELECTRIC CONSTANT

If a dielectric is inserted between the plates of a capacitor, the capacitance can increase markedly because of the way in which the dielectric alters the electric field between the

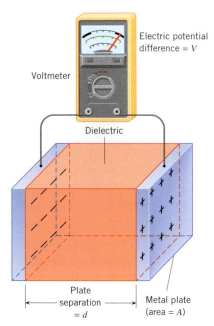

**Figure 19.16** A parallel plate capacitor consists of two metal plates, one carrying a charge +q and the other a charge −q. The potential of the positive plate exceeds that of the negative plate by an amount V. The region between the plates is filled with a dielectric.

**The physics of**
**random-access memory (RAM) chips.**

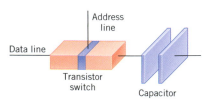

**Figure 19.17** A transistor–capacitor combination is part of a RAM chip used in computer memories.

plates. Figure 19.18 shows how this effect comes about. In part *a*, the region between the charged plates is empty. The field lines point from the positive toward the negative plate. In part *b*, a dielectric has been inserted between the plates. Since the capacitor is not connected to anything, the charge on the plates remains constant as the dielectric is inserted. In many materials (e.g., water) the molecules possess permanent dipole moments, even though the molecules are electrically neutral. The dipole moment exists because one end of a molecule has a slight excess of negative charge while the other end has a slight excess of positive charge. When such molecules are placed between the charged plates of the capacitor, the negative ends are attracted to the positive plate and the positive ends are attracted to the negative plate. As a result, the dipolar molecules tend to orient themselves end to end, as in part *b*. Whether or not a molecule has a permanent dipole moment, the electric field can cause the electrons to shift position within a molecule, making one end slightly negative and the opposite end slightly positive. Because of the end-to-end orientation, the left surface of the dielectric becomes positively charged, and the right surface becomes negatively charged. The surface charges are shown in red in the picture.

Because of the surface charges on the dielectric, not all the electric field lines generated by the charges on the plates pass through the dielectric. As Figure 19.18c shows, some of the field lines end on the negative surface charges and begin again on the positive surface charges. Thus, the electric field inside the dielectric is less strong than the electric field inside the empty capacitor, assuming the charge on the plates remains constant. This reduction in the electric field is described by the ***dielectric constant*** $\kappa$, which is the ratio of the field magnitude $E_0$ without the dielectric to the field magnitude $E$ inside the dielectric:

$$\kappa = \frac{E_0}{E} \tag{19.9}$$

Being a ratio of two field strengths, the dielectric constant is a number without units. Moreover, since the field $\mathbf{E_0}$ without the dielectric is greater than the field $\mathbf{E}$ inside the dielectric, the dielectric constant is greater than unity. The value of $\kappa$ depends on the nature of the dielectric material, as Table 19.1 indicates.

## THE CAPACITANCE OF A PARALLEL PLATE CAPACITOR

The capacitance of a capacitor is affected by the geometry of the plates and the dielectric constant of the material between them. For example, Figure 19.16 shows a parallel plate capacitor in which the area of each plate is $A$ and the separation between the plates is $d$. The magnitude of the electric field inside the dielectric is given by Equation 19.7 (without the minus sign) as $E = V/d$, where $V$ is the magnitude of the potential difference between the plates. If the charge on each plate is kept fixed, the electric field inside the dielectric is related to the electric field in the absence of the dielectric via Equation 19.9. Therefore,

$$E = \frac{E_0}{\kappa} = \frac{V}{d}$$

Since the electric field within an empty capacitor is $E_0 = q/(\epsilon_0 A)$ (see Equation 18.4), it follows that $q/(\kappa \epsilon_0 A) = V/d$, which can be solved for $q$ to give

$$q = \left( \frac{\kappa \epsilon_0 A}{d} \right) V$$

A comparison of this expression with $q = CV$ (Equation 19.8) reveals that the capacitance $C$ is

***Parallel plate capacitor filled with a dielectric***
$$C = \frac{\kappa \epsilon_0 A}{d} \tag{19.10}$$

Notice that only the geometry of the plates ($A$ and $d$) and the dielectric constant $\kappa$ affect the capacitance. With $C_0$ representing the capacitance of the empty capacitor ($\kappa = 1$), Equation 19.10 shows that $C = \kappa C_0$. In other words, the capacitance with the dielectric

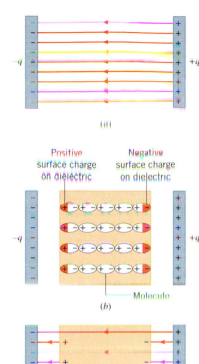

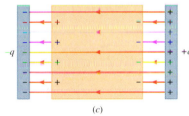

**Figure 19.18** (*a*) The electric field lines inside an empty capacitor. (*b*) The electric field produced by the charges on the plates aligns the molecular dipoles within the dielectric end to end. (*c*) The surface charges on the dielectric reduce the electric field inside the dielectric. The space between the dielectric and the plates is added for clarity. In reality, the dielectric fills the region between the plates.

**Table 19.1 Dielectric Constants of Some Common Substances[a]**

| Substance | Dielectric Constant, $\kappa$ |
|---|---|
| Vacuum | 1 |
| Air | 1.000 54 |
| Teflon | 2.1 |
| Benzene | 2.28 |
| Paper (royal gray) | 3.3 |
| Ruby mica | 5.4 |
| Neoprene rubber | 6.7 |
| Methyl alcohol | 33.6 |
| Water | 80.4 |

[a]Near room temperature.

present is increased by a factor of $\kappa$ over the capacitance without the dielectric. It can be shown that the relation $C = \kappa C_0$ applies to any capacitor, not just to a parallel plate capacitor. One reason, then, that capacitors are filled with dielectric materials is to increase the capacitance. Example 10 illustrates the effect that increasing the capacitance has on the charge stored by a capacitor.

### Example 10  Storing Electric Charge

The capacitance of an empty capacitor is 1.2 $\mu$F. The capacitor is connected to a 12-V battery and charged up. With the capacitor connected to the battery, a slab of dielectric material is inserted between the plates. As a result, $2.6 \times 10^{-5}$ C of *additional* charge flows from one plate, through the battery, and on to the other plate. What is the dielectric constant $\kappa$ of the material?

**Reasoning** The charge stored by a capacitor is $q = CV$, according to Equation 19.8. The battery maintains a constant potential difference of $V = 12$ volts between the plates of the capacitor while the dielectric is inserted. Inserting the dielectric causes the capacitance $C$ to increase, so that with $V$ held constant, the charge $q$ must increase. Thus, additional charge flows onto the plates. To find the dielectric constant, we apply Equation 19.8 to the empty capacitor and then to the capacitor filled with the dielectric material.

**Solution** The empty capacitor has a capacitance of $C_0 = 1.2$ $\mu$F and, according to Equation 19.8, stores an amount of charge $q_0 = C_0 V$. With the dielectric material in place, the capacitor has a capacitance $C = \kappa C_0$ and stores an amount of charge $q = (\kappa C_0)V$. Thus, the additional charge that the battery supplies is

$$q - q_0 = (\kappa C_0)V - C_0 V$$

Solving for the dielectric constant, we find that

$$\kappa = \frac{q - q_0}{C_0 V} + 1 = \frac{2.6 \times 10^{-5}\ \text{C}}{(1.2 \times 10^{-6}\ \text{F})(12\ \text{V})} + 1 = \boxed{2.8}$$

In Example 10 the capacitor remains connected to the battery while the dielectric is inserted between the plates. The next example discusses what happens if the capacitor is disconnected from the battery before the dielectric is inserted.

### Conceptual Example 11
### The Effect of a Dielectric When a Capacitor Has a Constant Charge

An empty capacitor is connected to a battery and charged up. The capacitor is then disconnected from the battery, and a slab of dielectric material is inserted between the plates. Does the voltage across the plates increase, remain the same, or decrease?

**Reasoning and Solution** Our reasoning is guided by the following fact: once the capacitor is disconnected from the battery, the charge on its plates remains constant, for there is no longer any way for charge to be added or removed. According to Equation 19.8, the charge $q$ stored by the capacitor is $q = CV$. Inserting the dielectric causes the capacitance $C$ to increase. Therefore, *the voltage V across the plates must decrease in order for q to remain unchanged.* The amount by which the voltage decreases from the value initially established by the battery depends on the dielectric constant of the slab.

**Related Homework:** *Problem 46*

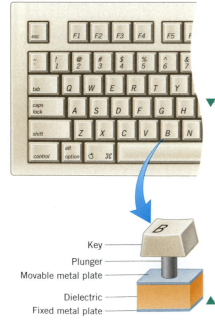

Key
Plunger
Movable metal plate
Dielectric
Fixed metal plate

**Figure 19.19** In one kind of computer keyboard, each key, when pressed, changes the separation between the plates of a capacitor.

**The physics of a computer keyboard.**

Capacitors are used often in electronic devices, and Example 12 deals with one familiar application.

### Example 12  A Computer Keyboard

One common kind of computer keyboard is based on the idea of capacitance. Each key is mounted on one end of a plunger, the other end being attached to a movable metal plate (see Figure 19.19). The movable plate is separated from a fixed plate, the two plates forming a capacitor. When the key is pressed, the movable plate is pushed closer to the fixed plate, and the capacitance increases. Electronic circuitry enables the computer to detect the *change* in capacitance, thereby recognizing which key has been pressed. The separation of the plates is normally $5.00 \times 10^{-3}$ m but decreases to $0.150 \times 10^{-3}$ m when a key is pressed. The plate area is $9.50 \times 10^{-5}$ m² and the capacitor is filled with a material whose dielectric constant is 3.50. Determine the change in capacitance that is detected by the computer.

**Reasoning** We can use Equation 19.10 directly to find the capacitance of the key, since the dielectric constant $\kappa$, the plate area $A$, and the plate separation $d$ are known. We will use this relation twice, once to find the capacitance when the key is pressed and once when it is not pressed. The change in capacitance will be the difference between these two values.

**Solution** When the key is pressed, the capacitance is

$$C = \frac{\kappa \epsilon_0 A}{d} = \frac{(3.50)[8.85 \times 10^{-12}\ \text{C}^2/(\text{N} \cdot \text{m}^2)](9.50 \times 10^{-5}\ \text{m}^2)}{0.150 \times 10^{-3}\ \text{m}}$$

$$= 19.6 \times 10^{-12}\ \text{F}\quad (19.6\ \text{pF}) \tag{19.10}$$

A similar calculation reveals that when the key is *not* pressed, the capacitance has a value of $0.589 \times 10^{-12}$ F (0.589 pF). The *change* in capacitance is an increase of $\boxed{19.0 \times 10^{-12}\ \text{F}\ (19.0\ \text{pF})}$. The *change* in the capacitance is greater with the dielectric present, which makes it easier for the circuitry within the computer to detect it.

# ENERGY STORAGE IN A CAPACITOR

When a capacitor stores charge, it also stores energy. In charging up a capacitor, for example, a battery does work in transferring an increment of charge from one plate of the capacitor to the other plate. The work done is equal to the product of the charge increment and the potential difference between the plates. However, as each increment of charge is moved, the potential difference increases slightly, and a larger amount of work is needed to move the next increment. The total work $W$ done in completely charging the capacitor is the product of the total charge $q$ transferred and the average potential difference $\overline{V}$; $W = q\overline{V}$. Since the average potential difference is one-half the final potential $V$, or $\overline{V} = \frac{1}{2}V$, the total work done by the battery is $W = \frac{1}{2}qV$. This work does not disappear but is stored as electric potential energy in the capacitor, so that EPE $= \frac{1}{2}qV$. Since $q = CV$, the energy stored becomes

$$\text{Energy} = \tfrac{1}{2}(CV)V = \tfrac{1}{2}CV^2 \tag{19.11}$$

It is also possible to regard the energy as being stored in the electric field between the plates. The relation between energy and field strength can be obtained for a parallel plate capacitor by substituting $V = Ed$ (Equation 19.7 without the minus sign) and $C = \kappa \epsilon_0 A/d$ (Equation 19.10) into Equation 19.11:

$$\text{Energy} = \frac{1}{2}\left(\frac{\kappa \epsilon_0 A}{d}\right)(Ed)^2$$

Since the area $A$ times the separation $d$ is the volume between the plates, the energy per unit volume or ***energy density*** is

$$\text{Energy density} = \frac{\text{Energy}}{\text{Volume}} = \tfrac{1}{2}\kappa \epsilon_0 E^2 \tag{19.12}$$

It can be shown that this expression is valid for any electric field strength, not just that between the plates of a capacitor.

The energy-storing capability of a capacitor is often put to good use in electronic circuits. For example, in an electronic flash attachment for a camera, energy from the battery pack is stored in a capacitor. The capacitor is then discharged between the electrodes of the flash tube, which converts the energy into light. Flash duration times range from 1/200 to 1/1 000 000 second or less, with the shortest flashes being used in high-speed photography (see Figure 19.20). Some flash attachments automatically control the flash duration by monitoring the light reflected from the photographic subject and quickly stop-

**Figure 19.20** This time-lapse photo of a bike rider was obtained using a camera with an electronic flash attachment. The energy for each "flash" comes from the electrical energy stored in a capacitor. (© Charlie Samuels/Corbis Images)

**Figure 19.21** A defibrillator uses the electrical energy stored in a capacitor to deliver a controlled electric current that can restore normal heart rhythm in a heart attack victim. Airplane personnel are now being trained to use portable units such as the one shown here to save lives in the air. (Courtesy Heartstream Operation/Philips Medical Systems)

**The physics of**
**an electronic flash attachment for a camera.**

ping or quenching the capacitor discharge when the reflected light reaches a predetermined level.

During a heart attack, the heart produces a rapid, unregulated pattern of beats, a condition known as cardiac fibrillation. Cardiac fibrillation can often be stopped by sending a very fast discharge of electrical energy through the heart. Emergency medical personnel use defibrillators, such as the one shown in Figure 19.21 (on p. 385). A paddle is connected to each plate of a large capacitor, and the paddles are placed on the chest near the heart. The capacitor is charged to a potential difference of about a thousand volts. The capacitor is then discharged in a few thousandths of a second; the discharge current passes through a paddle, the heart, and the other paddle. Within a few seconds, the heart often returns to its normal beating pattern.

# Concept Summary

This summary presents an abridged version of the chapter, including the important equations and all available learning aids. For convenient reference, the learning aids (including the text's examples) are placed next to or immediately after the relevant equation or discussion. The following learning aids may be found on-line at **www.wiley.com/college/cutnell**:

| | |
|---|---|
| **Interactive LearningWare** examples are solved according to a five-step interactive format that is designed to help you develop problem-solving skills. | **Concept Simulations** are animated versions of text figures or animations that illustrate important concepts. You can control parameters that affect the display, and we encourage you to experiment. |
| **Interactive Solutions** offer specific models for certain types of problems in the chapter homework. The calculations are carried out interactively. | **Self-Assessment Tests** include both qualitative and quantitative questions. Extensive feedback is provided for both incorrect and correct answers, to help you evaluate your understanding of the material. |

| *Topic* | *Discussion* | *Learning Aids* |
|---|---|---|

### 19.1 *Potential Energy*

**Work and electric potential energy**

When a positive test charge $+q_0$ moves from point $A$ to point $B$ in an electric field, work $W_{AB}$ is done by the electric force. The work equals the electric potential energy (EPE) at $A$ minus that at $B$:

$$W_{AB} = \text{EPE}_A - \text{EPE}_B \qquad (19.1)$$

**Path independence**

The electric force is a conservative force, so the path along which the test charge moves from $A$ to $B$ is of no consequence, for the work $W_{AB}$ is the same for all paths.

### 19.2 *The Electric Potential Difference*

**Electric potential**

The electric potential $V$ at a given point is the electric potential energy of a small test charge $q_0$ situated at that point divided by the charge itself:

$$V = \frac{\text{EPE}}{q_0} \qquad (19.3)$$

The SI unit of electric potential is the joule per coulomb (J/C) or volt (V).

The electric potential difference between two points $A$ and $B$ is

**Electric potential difference**

$$V_B - V_A = \frac{\text{EPE}_B}{q_0} - \frac{\text{EPE}_A}{q_0} = \frac{-W_{AB}}{q_0} \qquad (19.4)$$

Examples 1, 3
**Interactive LearningWare 19.1**
**Interactive Solution 19.7**

**Acceleration of positive and negative charges**

A positive charge accelerates from a region of higher potential toward a region of lower potential. Conversely, a negative charge accelerates from a region of lower potential toward a region of higher potential.

Example 2

**Electron volt**

An electron volt (eV) is a unit of energy. The relation between electron volts and joules is $1\ \text{eV} = 1.60 \times 10^{-19}\ \text{J}$.

The total energy $E$ of a system is the sum of its translational ($\frac{1}{2}mv^2$) and rotational ($\frac{1}{2}I\omega^2$) kinetic energies, gravitational potential energy ($mgh$), elastic potential energy ($\frac{1}{2}kx^2$), and electric potential energy (EPE):

**Total energy**

$$E = \tfrac{1}{2}mv^2 + \tfrac{1}{2}I\omega^2 + mgh + \tfrac{1}{2}kx^2 + \text{EPE}$$

| Topic | Discussion | Learning Aids |
|---|---|---|
| Conservation of energy | If external nonconservative forces like friction do no net work, the total energy of the system is conserved. That is, the final total energy $E_f$ is equal to the initial total energy $E_0$; $E_f = E_0$. | Example 4 |

**19.3 The Electric Potential Difference Created by Point Charges**

The electric potential $V$ at a distance $r$ from a point charge $q$ is

| | | |
|---|---|---|
| Electric potential of a point charge | $$V = \frac{kq}{r} \qquad (19.6)$$ where $k = 8.99 \times 10^9 \ \text{N} \cdot \text{m}^2/\text{C}^2$. This expression for $V$ assumes that the electric potential is zero at an infinite distance away from the charge. | Example 5 Concept Simulation 19.1 Interactive Solution 19.21 |
| Total electric potential | The total electric potential at a given location due to two or more charges is the algebraic sum of the potentials due to each charge. | Examples 6, 7 |
| Total potential energy of a group of charges | The total potential energy of a group of charges is the amount by which the electric potential energy of the group differs from its initial value when the charges are infinitely far apart. It is also equal to the work required to assemble the group, one charge at a time, starting with the charges infinitely far apart. | Example 8 |

**Use Self-Assessment Test 19.1 to evaluate your understanding of Sections 19.1–19.3.**

**19.4 Equipotential Surfaces and Their Relation to the Electric Field**

| | | |
|---|---|---|
| Equipotential surface | An equipotential surface is a surface on which the electric potential is the same everywhere. The electric force does no work as a charge moves on an equipotential surface, because the force is always perpendicular to the displacement of the charge. | |
| | The electric field created by any group of charges is everywhere perpendicular to the associated equipotential surfaces and points in the direction of decreasing potential. | |
| | The electric field is related to two equipotential surfaces by | |
| Relation between the electric field and the potential gradient | $$E = -\frac{\Delta V}{\Delta s} \qquad (19.7)$$ where $\Delta V$ is the potential difference between the surfaces and $\Delta s$ is the displacement. The term $\Delta V/\Delta s$ is called the potential gradient. | Example 9 Interactive Solution 19.33 |

**19.5 Capacitors and Dielectrics**

| | | |
|---|---|---|
| A capacitor | A capacitor is a device that stores charge and energy. It consists of two conductors or plates that are near one another, but not touching. The magnitude $q$ of the charge on each plate is given by | |
| Relation between charge and potential difference | $$q = CV \qquad (19.8)$$ where $V$ is the magnitude of the potential difference between the plates and $C$ is the capacitance. The SI unit for capacitance is the coulomb per volt (C/V) or farad (F). | |
| | The insulating material included between the plates of a capacitor is called a dielectric. The dielectric constant $\kappa$ of the material is defined as | |
| Dielectric constant | $$\kappa = \frac{E_0}{E} \qquad (19.9)$$ where $E_0$ and $E$ are, respectively, the magnitudes of the electric fields between the plates without and with a dielectric, assuming the charge on the plates is kept fixed. | |
| | The capacitance of a parallel plate capacitor filled with a dielectric is | |
| Capacitance of a parallel plate capacitor | $$C = \frac{\kappa \epsilon_0 A}{d} \qquad (19.10)$$ where $\epsilon_0 = 8.85 \times 10^{-12} \ \text{C}^2 / (\text{N} \cdot \text{m}^2)$ is the permittivity of free space, $A$ is the area of each plate, and $d$ is the distance between the plates. | Examples 10, 11, 12 Interactive Solution 19.57 |

| Topic | Discussion | Learning Aids |
|---|---|---|
| **Energy stored in a capacitor** | The electric potential energy stored in a capacitor is<br><br>$$\text{Energy} = \tfrac{1}{2}CV^2 \qquad (19.11)$$ | **Interactive LearningWare 19.2**<br>**Interactive Solution 19.41** |
| **Energy density** | The energy density is the energy stored per unit volume and is related to the magnitude $E$ of the electric field as follows:<br><br>$$\text{Energy density} = \tfrac{1}{2}\kappa\epsilon_0 E^2 \qquad (19.12)$$ | |

 **Use Self-Assessment Test 19.2 to evaluate your understanding of Sections 19.4 and 19.5.**

# Problems

*Note: All charges are assumed to be point charges unless specified otherwise.*

**ssm**  Solution is in the Student Solutions Manual.      **www**  Solution is available on the World Wide Web at www.wiley.com/college/cutnell

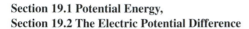 This icon represents a biomedical application.

## Section 19.1 Potential Energy,
## Section 19.2 The Electric Potential Difference

**1.** **ssm** Suppose that the electric potential outside a living cell is higher than that inside the cell by 0.070 V. How much work is done by the electric force when a sodium ion (charge $= +e$) moves from the outside to the inside?

**2.** An electric force moves a charge of $+1.80 \times 10^{-4}$ C from point $A$ to point $B$ and performs $5.80 \times 10^{-3}$ J of work on the charge. (a) What is the difference (EPE$_A$ − EPE$_B$) between the electric potential energies of the charge at the two points? (b) Determine the potential difference $(V_A - V_B)$ between the two points. (c) State which point is at the higher potential.

**3.** The anode (positive terminal) of an X-ray tube is at a potential of $+125\ 000$ V with respect to the cathode (negative terminal). (a) How much work (in joules) is done by the electric force when an electron is accelerated from the cathode to the anode? (b) If the electron is initially at rest, what kinetic energy does the electron have when it arrives at the anode?

**4.** During a particular thunderstorm, the electric potential difference between a cloud and the ground is $V_{\text{cloud}} - V_{\text{ground}} = 1.3 \times 10^8$ V, with the cloud being at the higher potential. What is the change in an electron's electric potential energy when the electron moves from the ground to the cloud?

**5.** **ssm** In a television picture tube, electrons strike the screen after being accelerated from rest through a potential difference of 25 000 V. The speeds of the electrons are quite large, and for accurate calculations of the speeds, the effects of special relativity must be taken into account. Ignoring such effects, find the electron speed just before the electron strikes the screen.

**6.** Point $A$ is at a potential of $+250$ V, and point $B$ is at a potential of $-150$ V. An $\alpha$-particle is a helium nucleus that contains two protons and two neutrons; the neutrons are electrically neutral. An $\alpha$-particle starts from rest at $A$ and accelerates toward $B$. When the $\alpha$-particle arrives at $B$, what kinetic energy (in electron volts) does it have?

**7.** Consult **Interactive Solution 19.7** at **www.wiley.com/college/cutnell** to review one method for modeling this problem. An electric car accelerates for 7.0 s by drawing energy from its 290-V battery pack. During this time, 1200 C of charge pass through the battery pack. Find the minimum horsepower rating of the car.

**\*8.** A typical 12-V car battery can deliver about $7.5 \times 10^5$ C of charge before dying. This is not very much. To get a feel for this, calculate the maximum number of kilograms of water (100 °C) that could be boiled into steam (100 °C) using energy from this battery.

**\*9.** **ssm** **www** The potential at location $A$ is 452 V. A positively charged particle is released there from rest and arrives at location $B$ with a speed $v_B$. The potential at location $C$ is 791 V, and when released from rest from this spot, the particle arrives at $B$ with twice the speed it previously had, or $2v_B$. Find the potential at $B$.

**\*\*10.** A particle is uncharged and is thrown vertically upward from ground level with a speed of 25.0 m/s. As a result, it attains a maximum height $h$. The particle is then given a positive charge $+q$ and reaches the same maximum height $h$ when thrown vertically upward with a speed of 30.0 m/s. The electric potential at the height $h$ exceeds the electric potential at ground level. Finally, the particle is given a negative charge $-q$. Ignoring air resistance, determine the speed with which the negatively charged particle must be thrown vertically upward, so that it attains exactly the maximum height $h$. In all three situations, be sure to include the effect of gravity.

## Section 19.3 The Electric Potential Difference Created by Point Charges

**11.** **ssm** There is an electric potential of $+130$ V at a spot that is 0.25 m away from a charge. Find the magnitude and sign of the charge.

**12.** Two point charges, $+3.40$ $\mu$C and $-6.10$ $\mu$C, are separated by 1.20 m. What is the electric potential midway between them?

**13.** An electron and a proton are initially very far apart (effectively an infinite distance apart). They are then brought together to form a hydrogen atom, in which the electron orbits the proton at an average distance of $5.29 \times 10^{-11}$ m. What is EPE$_{\text{final}}$ − EPE$_{\text{initial}}$, which is the change in the electric potential energy?

**14.** Location $A$ is 3.00 m to the right of a point charge $q$. Location $B$ lies on the same line and is 4.00 m to the right of the charge. The potential difference between the two locations is $V_B - V_A = 45.0$ V. What is the magnitude and sign of the charge?

**15.** **ssm** **www** Two identical point charges are fixed to diagonally opposite corners of a square that is 0.500 m on a side. Each charge is $+3.0 \times 10^{-6}$ C. How much work is done by the electric force as one of the charges moves to an empty corner?

**16.** The drawing shows four point charges. The value of $q$ is 2.0 $\mu$C, and the distance $d$ is 0.96 m. Find the total potential at the location $P$. Assume that the potential of a point charge is zero at infinity.

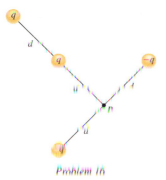

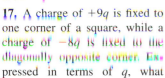

**Problem 16**

**17.** A charge of $+9q$ is fixed to one corner of a square, while a charge of $-8q$ is fixed to the diagonally opposite corner. Expressed in terms of $q$, what charge should be fixed to the center of the square, so the potential is zero at each of the two empty corners?

**18.** Review Conceptual Example 7 as background for this problem. Two charges are fixed in place with a separation $d$. One charge is positive and has twice the magnitude of the other charge, which is negative. The positive charge lies to the left of the negative charge, as in Figure 19.9. Relative to the negative charge, locate the two spots on the line through the charges where the total potential is zero.

**\* 19. ssm** Determine the electric potential energy for the array of three charges shown in the drawing, relative to its value when the charges are infinitely far away.

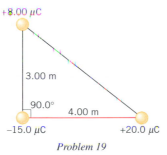

Problem 19

**\* 20.** Four identical charges ($+2.0$ $\mu$C each) are brought from infinity and fixed to a straight line. The charges are located 0.40 m apart. Determine the electric potential energy of this group.

**\* 21.** Refer to **Interactive Solution 19.21** at **www.wiley.com/college/ cutnell** to review one way in which this problem can be solved. Two protons are moving directly toward one another. When they are very far apart, their initial speeds are $3.00 \times 10^6$ m/s. What is the distance of closest approach?

**\* 22.** Identical point charges of $+1.7$ $\mu$C are fixed to diagonally opposite corners of a square. A third charge is then fixed at the center of the square, such that it causes the potentials at the empty corners to change signs without changing magnitudes. Find the sign and magnitude of the third charge.

**\* 23. ssm** A charge of $-3.00$ $\mu$C is fixed in place. From a horizontal distance of 0.0450 m, a particle of mass $7.20 \times 10^{-3}$ kg and charge $-8.00$ $\mu$C is fired with an initial speed of 65.0 m/s directly toward the fixed charge. How far does the particle travel before its speed is zero?

**\*\* 24.** A positive charge of $+q_1$ is located 3.00 m to the left of a negative charge $-q_2$. The charges have different magnitudes. On the line through the charges, the net *electric field* is zero at a spot 1.00 m to the right of the negative charge. On this line there are also two spots where the potential is zero. Locate these two spots relative to the negative charge.

**\*\* 25.** Charges $q_1$ and $q_2$ are fixed in place, $q_2$ being located at a distance $d$ to the right of $q_1$. A third charge $q_3$ is then fixed to the line joining $q_1$ and $q_2$ at a distance $d$ to the right of $q_2$. The third charge is chosen so the potential energy of the group is zero; that is, the potential energy has the same value as that of the three charges when they are widely separated. Determine $q_3$, assuming that (a) $q_1 =$

$q_2 = q$ and (b) $q_1 = q$ and $q_2 = -q$. Express your answers in terms of $q$.

**\*\* 26.** One particle has a mass of $3.00 \times 10^{-3}$ kg and a charge of $+8.00$ $\mu$C. A second particle has a mass of $6.00 \times 10^{-3}$ kg and the same charge. The two particles are initially held in place and then released. The particles fly apart, and when the separation between them is 0.100 m, the speed of the $3.00 \times 10^{-3}$-kg particle is 125 m/s. Find the initial separation between the particles.

**Section 19.4 Equipotential Surfaces and Their Relation to the Electric Field**

**27. ssm** An equipotential surface that surrounds a $+3.0 \times 10^{-7}$-C point charge has a radius of 0.15 m. What is the potential of this surface?

**28.** Two equipotential surfaces surround a $+1.50 \times 10^{-8}$-C point charge. How far is the 190-V surface from the 75.0-V surface?

**29.** A spark plug in an automobile engine consists of two metal conductors that are separated by a distance of 0.75 mm. When an electric spark jumps between them, the magnitude of the electric field is $4.7 \times 10^7$ V/m. What is the magnitude of the potential difference $\Delta V$ between the conductors?

**30.** The inner and outer surfaces of a cell membrane carry a negative and positive charge, respectively. Because of these charges, a potential difference of about 0.070 V exists across the membrane. The thickness of the membrane is $8.0 \times 10^{-9}$ m. What is the magnitude of the electric field in the membrane?

**31. ssm www** When you walk across a rug on a dry day, your body can become electrified, and its electric potential can change. When the potential becomes large enough, a spark of negative charges can jump between your hand and a metal surface. A spark occurs when the electric field strength created by the charges on your body reaches the dielectric strength of the air. The dielectric strength of the air is $3.0 \times 10^6$ N/C and is the electric field strength at which the air suffers electrical breakdown. Suppose a spark 3.0 mm long jumps between your hand and a metal doorknob. Assuming that the electric field is uniform, find the potential difference ($V_{knob} - V_{hand}$) between your hand and the doorknob.

**\* 32.** The drawing shows a uniform electric field that points in the negative $y$ direction; the magnitude of the field is 3600 N/C. Determine the electric potential difference (a) $V_B - V_A$ between points $A$ and $B$, (b) $V_C - V_B$ between points $B$ and $C$, and (c) $V_A - V_C$ between points $C$ and $A$.

**\* 33.** Refer to **Interactive Solution 19.33** at **www.wiley.com/ college/cutnell** to review a method by which this problem can be solved. The electric field has a constant value of $4.0 \times 10^3$ V/m and is directed downward. The field is the same everywhere. The potential at a point $P$ within this region is 155 V. Find the potential at the following points: (a) $6.0 \times 10^{-3}$ m directly above $P$, (b) $3.0 \times 10^{-3}$ m directly below $P$, (c) $8.0 \times 10^{-3}$ m directly to the right of $P$.

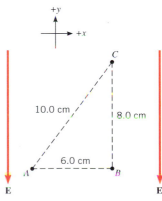

Problem 32

* **34.** The drawing shows the electric potential as a function of distance along the $x$ axis. Determine the magnitude of the electric field in the region (a) $A$ to $B$, (b) $B$ to $C$, and (c) $C$ to $D$.

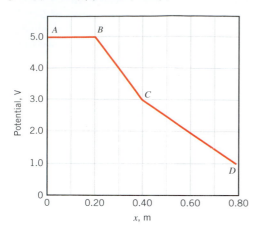

* **35.** **ssm** Equipotential surface $A$ has a potential of 5650 V, while equipotential surface $B$ has a potential of 7850 V. A particle has a mass of $5.00 \times 10^{-2}$ kg and a charge of $+4.00 \times 10^{-5}$ C. The particle has a speed of 2.00 m/s on surface $A$. An outside force is applied to the particle, and it moves to surface $B$, arriving there with a speed of 3.00 m/s. How much work is done by the outside force in moving the particle from $A$ to $B$?

### Section 19.5 Capacitors and Dielectrics

**36.** What voltage is required to store $7.2 \times 10^{-5}$ C of charge on the plates of a 6.0-$\mu$F capacitor?

**37.** **ssm** The electric potential energy stored in the capacitor of a defibrillator is 73 J, and the capacitance is 120 $\mu$F. What is the potential difference across the capacitor plates?

**38.** An axon is the relatively long tail-like part of a neuron, or nerve cell. The outer surface of the axon membrane (dielectric constant = 5, thickness = $1 \times 10^{-8}$ m) is charged positively, and the inner portion is charged negatively. Thus, the membrane is a kind of capacitor. Assuming that an axon can be treated like a parallel plate capacitor with a plate area of $5 \times 10^{-6}$ m$^2$, what is its capacitance?

**39.** **ssm** A parallel plate capacitor has a capacitance of 7.0 $\mu$F when filled with a dielectric. The area of each plate is 1.5 m$^2$ and the separation between the plates is $1.0 \times 10^{-5}$ m. What is the dielectric constant of the dielectric?

**40.** A capacitor has a capacitance of $2.5 \times 10^{-8}$ F. In the charging process, electrons are removed from one plate and placed on the other plate. When the potential difference between the plates is 450 V, how many electrons have been transferred?

**41.** Refer to **Interactive Solution 19.41** at **www.wiley.com/college/cutnell** for one approach to this problem. The electronic flash at-

tachment for a camera contains a capacitor for storing the energy used to produce the flash. In one such unit, the potential difference between the plates of an 850-$\mu$F capacitor is 280 V. (a) Determine the energy that is used to produce the flash in this unit. (b) Assuming that the flash lasts for $3.9 \times 10^{-3}$ s, find the effective power or "wattage" of the flash.

**42.** Two capacitors are identical, except that one is empty and the other is filled with a dielectric ($\kappa = 4.50$). The empty capacitor is connected to a 12.0-V battery. What must be the potential difference across the plates of the capacitor filled with a dielectric such that it stores the same amount of electrical energy as the empty capacitor?

* **43.** **ssm Interactive LearningWare 19.2** at **www.wiley.com/college/cutnell** reviews the concepts pertinent to this problem. What is the potential difference between the plates of a 3.3-F capacitor that stores sufficient energy to operate a 75-W light bulb for one minute?

* **44.** The dielectric strength of an insulating material is the maximum electric field strength to which the material can be subjected without electrical breakdown occurring. Suppose a parallel plate capacitor is filled with a material whose dielectric constant is 3.5 and whose dielectric strength is $1.4 \times 10^7$ N/C. If this capacitor is to store $1.7 \times 10^{-7}$ C of charge on each plate without suffering breakdown, what must be the radius of its circular plates?

* **45.** Two hollow metal spheres are concentric with each other. The inner sphere has a radius of 0.1500 m and a potential of 85.0 V. The radius of the outer sphere is 0.1520 m and its potential is 82.0 V. If the region between the spheres is filled with Teflon, find the electric energy contained in this space.

* **46.** Review Conceptual Example 11 before attempting this problem. An empty capacitor is connected to a 12.0-V battery and charged up. The capacitor is then disconnected from the battery, and a slab of dielectric material ($\kappa = 2.8$) is inserted between the plates. Find the amount by which the potential difference across the plates changes. Specify whether the change is an increase or a decrease.

** **47.** **ssm** The drawing shows a parallel plate capacitor. One-half of the region between the plates is filled with a material that has a dielectric constant $\kappa_1$. The other half is filled with a material that has a dielectric constant $\kappa_2$. The area of each plate is $A$, and the plate separation is $d$. The potential difference across the plates is $V$. Note especially that the charge stored by the capacitor is $q_1 + q_2 = CV$, where $q_1$ and $q_2$ are the charges on the area of the plates in contact with materials 1 and 2, respectively. Show that $C = \epsilon_0 A(\kappa_1 + \kappa_2)/(2d)$.

** **48.** The plate separation of a charged capacitor is 0.0800 m. A proton and an electron are released from rest at the midpoint between the plates. Ignore the attraction between the two particles, and determine how far the proton has traveled by the time the electron strikes the positive plate.

# Chapter 20  Electric Circuits

## 20.1  Electromotive Force and Current

Look around you. Chances are that there is an electrical device nearby—a radio, a hair dryer, a computer—something that uses electrical energy to operate. The energy needed to run a portable CD player, for instance, comes from batteries, as Figure 20.1 illustrates. The transfer of energy takes place via an electric circuit, in which the energy source (the battery pack) and the energy consuming device (the CD player) are connected by conducting wires, through which electric charges move.

**Figure 20.1** In an electric circuit, energy is transferred from a source (the battery pack) to a device (the CD player) by charges that move through a conducting wire.

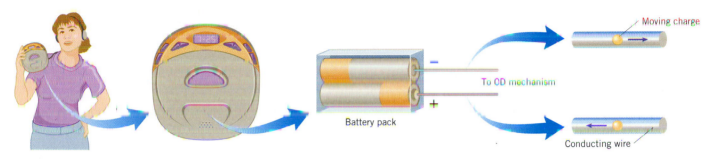

Battery pack — To CD mechanism — Moving charge — Conducting wire

Within a battery, a chemical reaction occurs that transfers electrons from one terminal (leaving it positively charged) to another terminal (leaving it negatively charged). Figure 20.2 shows the two terminals of a car battery and a flashlight battery. The drawing also illustrates the symbol $(\overset{+}{\dashv}\overset{-}{\vdash})$ used to represent a battery in circuit drawings. Because of the positive and negative charges on the battery terminals, an electric potential difference exists between them. The maximum potential difference is called the *electromotive force* (emf) of the battery, for which the symbol $\mathscr{E}$ is used. In a typical car battery, the chemical reaction maintains the potential of the positive terminal at a maximum of 12 volts (12 joules/coulomb) higher than the potential of the negative terminal, so the emf is $\mathscr{E} = 12$ V. Thus, one coulomb of charge emerging from the battery and entering a circuit has at most 12 joules of energy. In a typical flashlight battery the emf is 1.5 V. In reality, the potential difference between the terminals of a battery is somewhat less than the maximum value indicated by the emf, for reasons that Section 20.9 discusses.

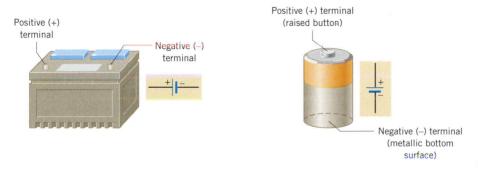

Positive (+) terminal — Negative (−) terminal

Positive (+) terminal (raised button) — Negative (−) terminal (metallic bottom surface)

**Figure 20.2** Typical batteries and the symbol $(\overset{+}{\dashv}\overset{-}{\vdash})$ used to represent them in electric circuits.

In a circuit such as that in Figure 20.1, the battery creates an electric field within and parallel to the wire, directed from the positive toward the negative terminal. The electric field exerts a force on the free electrons in the wire, and they respond by moving. Figure 20.3 shows charges moving inside a wire and crossing an imaginary surface that is per-

---

* The word "force" appears in this context for historical reasons, even though it is incorrect. As we have seen in Section 19.2, electric potential is energy per unit charge, which is not force.

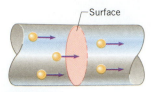

**Figure 20.3** The electric current is the amount of charge per unit time that passes through a surface that is perpendicular to the motion of the charges.

pendicular to their motion. This flow of charge is known as an *electric current.* We will formulate this new concept of electric current by bringing together the amount of charge $\Delta q$ that passes through the surface and the time interval $\Delta t$ during which it does so.

The electric current $I$ is defined as the amount of charge per unit time that crosses the imaginary surface in Figure 20.3, in much the same sense that a river current is the amount of water flowing per unit of time. If the rate is constant, the current is

$$I = \frac{\Delta q}{\Delta t} \tag{20.1}$$

If the rate of flow is not constant, then Equation 20.1 gives the average current. Since the units for charge and time are the coulomb (C) and the second (s), the SI unit for current is a coulomb per second (C/s). One coulomb per second is referred to as an *ampere* (A), after the French mathematician André-Marie Ampère (1775–1836).

If the charges move around a circuit in the same direction at all times, the current is said to be *direct current (dc),* which is the kind produced by batteries. In contrast, the current is said to be *alternating current (ac)* when the charges move first one way and then the opposite way, changing direction from moment to moment. Many energy sources produce alternating current—for example, generators at power companies and microphones. Example 1 deals with direct current.

### Example 1 A Pocket Calculator

The current from the 3.0-V battery pack of a pocket calculator is 0.17 mA. In one hour of operation, (a) how much charge flows in the circuit and (b) how much energy does the battery deliver to the calculator circuit?

**Reasoning** Since current is defined as charge per unit time, the charge that flows in one hour is the product of the current and the time (3600 s). The charge that leaves the 3.0-V battery pack has 3.0 joules of energy per coulomb of charge. Thus, the total energy delivered to the calculator circuit is the charge (in coulombs) times the energy per unit charge (in volts or joules/coulomb).

**Solution**

(a) The charge that flows in one hour can be determined from Equation 20.1:

$$\Delta q = I(\Delta t) = (0.17 \times 10^{-3}\ \text{A})(3600\ \text{s}) = \boxed{0.61\ \text{C}}$$

(b) The energy delivered to the calculator circuit is

$$\text{Energy} = \text{Charge} \times \underbrace{\frac{\text{Energy}}{\text{Charge}}}_{\text{Battery emf}} = (0.61\ \text{C})(3.0\ \text{V}) = \boxed{1.8\ \text{J}}$$

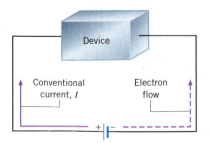

**Figure 20.4** In a circuit, electrons actually flow through the metal wires. However, it is customary to use a conventional current $I$ to describe the flow of charges.

Today, it is known that electrons flow in metal wires. Figure 20.4 shows the negative electrons emerging from the negative terminal of the battery and moving around the circuit toward the positive terminal. It is customary, however, *not* to use the flow of electrons when discussing circuits. Instead, a so-called *conventional current* is used, for reasons that date to the time when it was believed that positive charges moved through metal wires. Conventional current is the hypothetical flow of positive charges that would have the same effect in the circuit as the movement of negative charges that actually does occur. In Figure 20.4, negative electrons leave the negative terminal of the battery, pass through the device, and arrive at the positive terminal. The same effect would have been achieved if an equivalent amount of positive charge had left the positive terminal, passed through the device, and arrived at the negative terminal. Therefore, the drawing shows the conventional current originating from the positive terminal. A conventional current of hypothetical positive charges is consistent with our earlier use of a positive test charge for defining electric fields and potentials. The direction of conventional current is always from a point of higher potential toward a point of lower potential—that is, from the positive toward the negative terminal. In this text, the symbol $I$ stands for conventional current.

# 20.2 Ohm's Law

The current that a battery can push through a wire is analogous to the water flow that a pump can push through a pipe. Greater pump pressures lead to larger water flow rates, and, similarly, greater battery voltages[1] lead to larger electric currents. In the simplest case, the current $I$ is directly proportional to the voltage $V$; that is, $I \propto V$. Thus, a 12-V battery leads to twice as much current as a 6-V battery, when each is connected to the same circuit.

In a water pipe, the flow rate is not only determined by the pump pressure but is also affected by the length and diameter of the pipe. Longer and narrower pipes offer higher resistance to the moving water and lead to smaller flow rates for a given pump pressure. A similar situation exists in electric circuits, and to deal with it we introduce the concept of electrical resistance. Electrical resistance is defined in terms of two ideas that have already been discussed—the electric potential difference, or voltage (see Section 19.2), and the electric current (see Section 20.1).

The **resistance (R)** is defined as the ratio of the voltage $V$ applied across a piece of material to the current $I$ through the material, or $R = V/I$. When only a small current results from a large voltage, there is a high resistance to the moving charge. For many materials (e.g., metals), the ratio $V/I$ is the same for a given piece of material over a wide range of voltages and currents. In such a case, the resistance is a constant. Then, the relation $R = V/I$ is referred to as **Ohm's law**, after the German physicist Georg Simon Ohm (1789–1854), who discovered it.

■ **OHM'S LAW**

The ratio $V/I$ is a constant, where $V$ is the voltage applied across a piece of material (such as a wire) and $I$ is the current through the material:

$$\frac{V}{I} = R = \text{constant} \quad \text{or} \quad V = IR \qquad (20.2)$$

$R$ is the resistance of the piece of material.

**SI Unit of Resistance:** volt/ampere (V/A) = ohm (Ω)

The SI unit for resistance is a volt per ampere, which is called an *ohm* and is represented by the Greek capital letter omega (Ω). Ohm's law is not a fundamental law of nature like Newton's laws of motion. It is only a statement of the way certain materials behave in electric circuits.

To the extent that a wire or an electrical device offers resistance to the flow of charges, it is called a **resistor.** The resistance can have a wide range of values. The copper wires in a television set, for instance, have a very small resistance. On the other hand, commercial resistors can have resistances up to many kilohms (1 kΩ = $10^3$ Ω) or megohms (1 MΩ = $10^6$ Ω). Such resistors play an important role in electric circuits, where they are used to limit the amount of current and establish proper voltage levels.

In drawing electric circuits we follow the usual conventions: (1) a zigzag line (—WW—) represents a resistor and (2) a straight line (———) represents an ideal conducting wire, or one with a negligible resistance. Example 2 illustrates an application of Ohm's law to the circuit in a flashlight.

## Example 2  A Flashlight

The filament in a light bulb is a resistor in the form of a thin piece of wire. The wire becomes hot enough to emit light because of the current in it. Figure 20.5 shows a flashlight that uses two 1.5-V batteries (effectively a single 3.0-V battery) to provide a current of 0.40 A in the filament. Determine the resistance of the glowing filament.

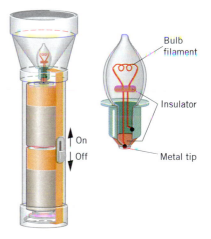

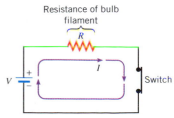

**Figure 20.5** The circuit in this flashlight consists of a resistor (the filament of the light bulb) connected to a 3.0-V battery (two 1.5-V batteries).

---

[1] The potential difference between two points, such as the terminals of a battery, is commonly called the voltage between the points.

**Reasoning** The filament resistance is assumed to be the only resistance in the circuit. The potential difference applied across the filament is that of the 3.0-V battery. The resistance, given by Equation 20.2, is equal to this potential difference divided by the current.

**Solution** The resistance of the filament is

$$R = \frac{V}{I} = \frac{3.0 \text{ V}}{0.40 \text{ A}} = \boxed{7.5 \text{ }\Omega} \tag{20.2}$$

## 20.3 Resistance and Resistivity

In a water pipe, the length and cross-sectional area of the pipe determine the resistance that the pipe offers to the flow of water. Longer pipes with smaller cross-sectional areas offer greater resistance. Analogous effects are found in the electrical case. For a wide range of materials, the resistance of a piece of material of length $L$ and cross-sectional area $A$ is

$$R = \rho \frac{L}{A} \tag{20.3}$$

where $\rho$ is a proportionality constant known as the **resistivity** of the material. It can be seen from Equation 20.3 that the unit for resistivity is the ohm $\cdot$ meter ($\Omega \cdot$ m), and Table 20.1 lists values for various materials. All the conductors in the table are metals and have small resistivities. Insulators such as rubber have large resistivities. Materials like germanium and silicon have intermediate resistivity values and are, accordingly, called *semiconductors*.

   Resistivity is an inherent property of a material, inherent in the same sense that density is an inherent property. Resistance, on the other hand, depends on both the resistivity and the geometry of the material. Thus, two wires can be made from copper, which has a resistivity of $1.72 \times 10^{-8} \text{ }\Omega \cdot$ m, but Equation 20.3 indicates that a short wire with a large cross-sectional area has a smaller resistance than does a long, thin wire. Wires that carry large currents, such as main power cables, are thick rather than thin so that the resistance of the wires is kept as small as possible. Similarly, electric tools that are to be used far away from wall sockets require thicker extension cords, as Example 3 illustrates.

**Table 20.1  Resistivities[a] of Various Materials**

| Material | Resistivity $\rho$ ($\Omega \cdot$ m) | Material | Resistivity $\rho$ ($\Omega \cdot$ m) |
|---|---|---|---|
| *Conductors* | | *Semiconductors* | |
| Aluminum | $2.82 \times 10^{-8}$ | Carbon | $3.5 \times 10^{-5}$ |
| Copper | $1.72 \times 10^{-8}$ | Germanium | $0.5^{b}$ |
| Gold | $2.44 \times 10^{-8}$ | Silicon | $20 – 2300^{b}$ |
| Iron | $9.7 \times 10^{-8}$ | *Insulators* | |
| Mercury | $95.8 \times 10^{-8}$ | Mica | $10^{11} – 10^{15}$ |
| Nichrome (alloy) | $100 \times 10^{-8}$ | Rubber (hard) | $10^{13} – 10^{16}$ |
| Silver | $1.59 \times 10^{-8}$ | Teflon | $10^{16}$ |
| Tungsten | $5.6 \times 10^{-8}$ | Wood (maple) | $3 \times 10^{10}$ |

[a] The values pertain to temperatures near 20 °C.
[b] Depending on purity.

## Example 3  Longer Extension Cords

**The physics of electrical extension cords.**

The instructions for an electric lawn mower suggest that a 20-gauge extension cord can be used for distances up to 35 m, but a thicker 16-gauge cord should be used for longer distances, to keep the resistance of the wire as small as possible. The cross-sectional area of 20-gauge wire is $5.2 \times 10^{-7}$ m$^2$, while that of 16-gauge wire is $13 \times 10^{-7}$ m$^2$. Determine the resistance of (a) 35 m of 20-gauge copper wire and (b) 75 m of 16-gauge copper wire.

**Reasoning** According to Equation 20.3, the resistance of a copper wire depends on the resistivity of copper and the length and cross-sectional area of the wire. The resistivity can be obtained from Table 20.1, while the length and cross-sectional area are given in the problem statement.

**Solution** According to Table 20.1 the resistivity of copper is $1.72 \times 10^{-8} \ \Omega \cdot m$. The resistance of the wires can be found using Equation 20.3:

20-gauge wire
$$R = \frac{\rho L}{A} = \frac{(1.72 \times 10^{-8} \ \Omega \cdot m)(35 \ m)}{5.2 \times 10^{-7} \ m^2} = \boxed{1.2 \ \Omega}$$

16-gauge wire
$$R = \frac{\rho L}{A} = \frac{(1.72 \times 10^{-8} \ \Omega \cdot m)(75 \ m)}{13 \times 10^{-7} \ m^2} = \boxed{0.99 \ \Omega}$$

Even though it is more than twice as long, the thicker 16-gauge wire has less resistance than the thinner 20-gauge wire. It is necessary to keep the resistance as low as possible to minimize heating of the wire, thereby reducing the possibility of a fire, as Conceptual Example 7 in Section 20.5 emphasizes.

▲

Equation 20.3 provides the basis for an important medical diagnostic technique known as impedance (or resistance) plethysmography. Figure 20.6 shows how the technique is applied to diagnose blood clotting in the veins (deep venous thrombosis) near the knee. A pressure cuff, like that used in blood pressure measurements, is placed around the midthigh, while electrodes are attached around the calf. The two outer electrodes are connected to a source of a small amount of ac current. The two inner electrodes are separated by a distance $L$, and the voltage between them is measured. The voltage divided by the current gives the resistance. The key to this technique is the fact that resistance can be related to the volume $V_{calf}$ of the calf between the inner electrodes. The volume is the product of the length $L$ and the cross-sectional area $A$ of the calf, or $V_{calf} = LA$. Solving for $A$ and substituting in Equation 20.3 shows that

 ***The physics of impedance plethysmography.***

$$R = \rho \frac{L}{A} = \rho \frac{L}{V_{calf}/L} = \rho \frac{L^2}{V_{calf}}$$

Thus, resistance is inversely proportional to volume, a fact that is exploited in diagnosing deep venous thrombosis. Blood flows from the heart into the calf through arteries in the leg and returns through the system of veins. The pressure cuff in Figure 20.6 is inflated to the point where it cuts off the venous flow, but not the arterial flow. As a result, more blood enters than leaves the calf, the volume of the calf increases, and the electrical resistance decreases. When the cuff pressure is removed suddenly, the volume returns to a normal value, and so does the electrical resistance. With healthy (unclotted) veins, there is a rapid return to normal values. A slow return, however, reveals the presence of clotting.

The resistivity of a material depends on temperature. In metals, the resistivity increases with increasing temperature, whereas in semiconductors the reverse is true. For many materials and limited temperature ranges, it is possible to express the temperature dependence of the resistivity as follows:

$$\rho = \rho_0[1 + \alpha(T - T_0)] \qquad (20.4)$$

In this expression $\rho$ and $\rho_0$ are the resistivities at temperatures $T$ and $T_0$, respectively. The term $\alpha$ has the unit of reciprocal temperature and is the ***temperature coefficient of resistivity***. When the resistivity increases with increasing temperature, $\alpha$ is positive, as it is for metals. When the resistivity decreases with increasing temperature, $\alpha$ is negative, as it is for the semiconductors carbon, germanium, and silicon. Since resistance is given by $R = \rho L/A$, both sides of Equation 20.4 can be multiplied by $L/A$ to show that resistance depends on temperature according to

$$R = R_0[1 + \alpha(T - T_0)] \qquad (20.5)$$

The next example illustrates the role of the resistivity and its temperature coefficient in determining the electrical resistance of a piece of material.

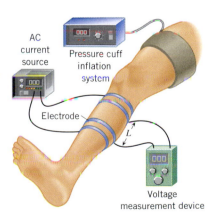

**Figure 20.6** Using the technique of impedance plethysmography, the electrical resistance of the calf can be measured to diagnose deep venous thrombosis (blood clotting in the veins).

## Example 4    The Heating Element of an Electric Stove

Figure 20.7a shows a cherry-red heating element on an electric stove. The element contains a wire (length = 1.1 m, cross-sectional area = $3.1 \times 10^{-6}$ m²) through which electric charge flows. As Figure 20.7b shows, this wire is embedded within an electrically insulating material that is contained within a metal casing. The wire becomes hot in response to the flowing charge and heats the casing. The material of the wire has a resistivity of $\rho_0 = 6.8 \times 10^{-5}$ $\Omega \cdot$m at $T_0 = 320$ °C and a temperature coefficient of resistivity of $\alpha = 2.0 \times 10^{-3}$ $(C°)^{-1}$. Determine the resistance of the heater wire at an operating temperature of 420 °C.

**Reasoning** Equation 20.3 $(R = \rho L/A)$ can be used to find the resistance of the wire at 420 °C, once the resistivity $\rho$ is determined at this temperature. Since the resistivity at 320 °C is given, Equation 20.4 can be employed to find the resistivity at 420 °C.

**Solution** At the operating temperature of 420 °C, the material of the wire has a resistivity of

$$\rho = \rho_0[1 + \alpha(T - T_0)]$$
$$\rho = (6.8 \times 10^{-5} \ \Omega \cdot m)\{1 + [2.0 \times 10^{-3} \ (C°)^{-1}](420 \ °C - 320 \ °C)\}$$
$$= 8.2 \times 10^{-5} \ \Omega \cdot m$$

This value of the resistivity can be used along with the given length and cross-sectional area to find the resistance of the heater wire:

$$R = \frac{\rho L}{A} = \frac{(8.2 \times 10^{-5} \ \Omega \cdot m)(1.1 \ m)}{3.1 \times 10^{-6} \ m^2} = \boxed{29 \ \Omega} \tag{20.3}$$

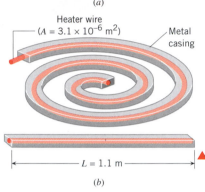

Heater wire
$(A = 3.1 \times 10^{-6}$ m²$)$    Metal casing

$L = 1.1$ m

(b)

**Figure 20.7** A heating element from an electric stove. (© David Chasey/PhotoDisc, Inc./Getty Images)

**The physics of**
**a heating element on an electric stove. (See above.)**

There is an important class of materials whose resistivity suddenly goes to zero below a certain temperature $T_c$, which is called the *critical temperature* and is commonly a few degrees above absolute zero. Below this temperature, such materials are called *superconductors*. The name derives from the fact that with zero resistivity, these materials offer no resistance to electric current and are, therefore, perfect conductors. One of the remarkable properties of zero resistivity is that once a current is established in a superconducting ring, it continues indefinitely without the need for an emf. Currents have persisted in superconductors for many years without measurable decay. In contrast, the current in a nonsuperconducting material drops to zero almost immediately after the emf is removed.

Many metals become superconductors only at very low temperatures, such as aluminum $(T_c = 1.18$ K$)$, tin $(T_c = 3.72$ K$)$, lead $(T_c = 7.20$ K$)$, and niobium $(T_c = 9.25$ K$)$. Materials involving copper oxide complexes have been made that undergo the transition to the superconducting state at 175 K. Superconductors have many technological applications, including magnetic resonance imaging (Section 21.7), magnetic levitation of trains (Section 21.9), cheaper transmission of electric power, powerful (yet small) electric motors, and faster computer chips.

## 20.4    Electric Power

Suppose that an amount of charge $\Delta q$ emerges from a battery in a time $\Delta t$ and that the potential difference between the battery terminals is $V$. According to the definition of electric potential given in Equation 19.3, the energy of this charge is the product of the charge and the potential difference, or $(\Delta q)V$. Since the change in energy per unit time is the power $P$ (see Equation 6.10b), the electric power provided to the circuit is

$$P = \frac{(\Delta q)V}{\Delta t} = \underbrace{\frac{\Delta q}{\Delta t}}_{\text{Current } I} V$$

The term $\Delta q/\Delta t$ is the charge per unit time, or the current $I$ in the circuit, according to Equation 20.1. Thus, the electric power is the product of current and voltage.

■ **ELECTRIC POWER**

When there is a current $I$ in a circuit as a result of a voltage $V$, the electric power $P$ delivered to the circuit is

$$P = IV \qquad (20.6)$$

*SI Unit of Power:* watt (W)

Power is measured in watts, and Equation 20.6 indicates that the product of an ampere and a volt is equal to a watt.

Many electrical devices are essentially resistors that become hot when provided with sufficient electric power: toasters, irons, space heaters, heating elements on electric stoves, and incandescent light bulbs, to name a few. In such cases, it is possible to obtain two additional, but equivalent, expressions for the power. These two expressions follow directly upon substituting $V = IR$, or equivalently $I = V/R$, into the relation $P = IV$:

$$P = IV \qquad (20.6a)$$

$$P = I(IR) = I^2R \qquad (20.6b)$$

$$P = \left(\frac{V}{R}\right)V = \frac{V^2}{R} \qquad (20.6c)$$

Example 5 deals with the electric power delivered to the bulb of a flashlight.

### Example 5   The Power and Energy Used in a Flashlight

In the flashlight in Figure 20.5, the current is 0.40 A, and the voltage is 3.0 V. Find (a) the power delivered to the bulb and (b) the energy dissipated in the bulb in 5.5 minutes of operation.

**Reasoning** The electric power delivered to the bulb is the product of the current and voltage. Since power is energy per unit time, the energy delivered to the bulb is the product of the power and time.

**Solution**

(a) The power is
$$P = IV = (0.40 \text{ A})(3.0 \text{ V}) = \boxed{1.2 \text{ W}} \qquad (20.6a)$$

The "wattage" rating of this bulb would, therefore, be 1.2 W.

(b) The energy consumed in 5.5 minutes (330 s) follows from the definition of power as energy per unit time:
$$\text{Energy} = P \, \Delta t = (1.2 \text{ W})(330 \text{ s}) = \boxed{4.0 \times 10^2 \text{ J}}$$

Monthly electric bills specify the cost for the energy consumed during the month. Energy is the product of power and time, and electric companies compute your energy consumption by expressing power in kilowatts and time in hours. Therefore, a commonly used unit for energy is the *kilowatt·hour* (kWh). For instance, if you used an average power of 1440 watts (1.44 kW) for 30 days (720 h), your energy consumption would be $(1.44 \text{ kW})(720 \text{ h}) = 1040$ kWh. At a cost of $0.10 per kWh, your monthly bill would be $104. As shown in Example 10 in Chapter 12, 1 kWh $= 3.60 \times 10^6$ J of energy.

## 20.5  *Alternating Current*

Many electric circuits use batteries and involve direct current (dc). However, there are considerably more circuits that operate with alternating current (ac), in which the charge flow reverses direction periodically. The common generators that create ac electricity depend on magnetic forces for their operation and are discussed in Chapter 22. In an ac circuit, these generators serve the same purpose as a battery serves in a dc circuit; that is, they give energy to the moving electric charges. This section deals with ac circuits that contain only resistance.

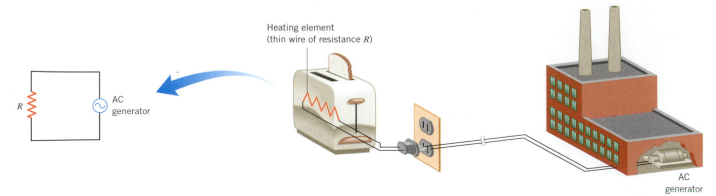

Heating element
(thin wire of resistance $R$)

$R$

AC generator

AC generator

**Figure 20.8** This circuit consists of a toaster (resistance = $R$) and an ac generator at the electric power company.

Since the electrical outlets in a house provide alternating current, we all use ac circuits routinely. Figure 20.8 shows the ac circuit that is formed when a toaster is plugged into a wall socket. The heating element of a toaster is essentially a thin wire of resistance $R$ and becomes red hot when electrical energy is dissipated in it. The circuit schematic in the picture introduces the symbol $\odot$ that is used to represent the generator. In this case, the generator is located at the electric power company.

Figure 20.9 shows a graph that records the voltage $V$ produced between the terminals of the ac generator in Figure 20.8 at each moment of time $t$. This is the most common type of ac voltage. It fluctuates sinusoidally between positive and negative values as a function of time:

$$V = V_0 \sin 2\pi f t \qquad (20.7)$$

where $V_0$ is the maximum or peak value of the voltage, and $f$ is the frequency (in cycles/s or Hz) at which the voltage oscillates. The angle $2\pi f t$ in Equation 20.7 is expressed in radians, so a calculator must be set to its radian mode before the sine of this angle is evaluated. In the United States, the voltage at most home wall outlets has a *peak value* of approximately $V_0 = 170$ volts and oscillates with a frequency of $f = 60$ Hz. Thus, the period of each cycle is $\frac{1}{60}$ s, and the polarity of the generator terminals reverses twice during each cycle, as Figure 20.9 indicates.

The current in an ac circuit also oscillates. In circuits that contain only resistance, the current reverses direction each time the polarity of the generator terminals reverses. Thus, the current in a circuit like that in Figure 20.8 would have a frequency of 60 Hz and would change direction twice during each cycle. Substituting $V = V_0 \sin 2\pi f t$ into $V = IR$ shows that the current can be represented as

$$I = \frac{V_0}{R} \sin 2\pi f t = I_0 \sin 2\pi f t \qquad (20.8)$$

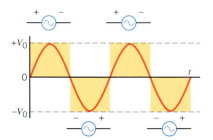

**Figure 20.9** In the most common case, the ac voltage is a sinusoidal function of time. The relative polarity of the generator terminals during the positive and negative parts of the sine wave is indicated.

The peak current is given by $I_0 = V_0/R$, so it can be determined if the peak voltage and the resistance are known.

The power delivered to an ac circuit by the generator is given by $P = IV$, just as it is in a dc circuit. However, since both $I$ and $V$ depend on time, the power fluctuates as time passes. Substituting Equations 20.7 and 20.8 for $V$ and $I$ into $P = IV$ gives

$$P = I_0 V_0 \sin^2 2\pi f t \qquad (20.9)$$

This expression is plotted in Figure 20.10.

Since the power fluctuates in an ac circuit, it is customary to consider the average power $\overline{P}$, which is one-half the peak power, as Figure 20.10 indicates:

$$\overline{P} = \tfrac{1}{2} I_0 V_0 \qquad (20.10)$$

On the basis of this expression, a kind of average current and average voltage can be introduced that are very useful when discussing ac circuits. A slight rearrangement of Equation 20.10 reveals that

$$\overline{P} = \left(\frac{I_0}{\sqrt{2}}\right)\left(\frac{V_0}{\sqrt{2}}\right) = I_{\text{rms}} V_{\text{rms}} \qquad (20.11)$$

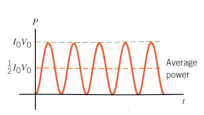

**Figure 20.10** In an ac circuit, the power $P$ delivered to a resistor oscillates between zero and a peak value of $I_0 V_0$, where $I_0$ and $V_0$ are the peak current and voltage, respectively.

$I_{rms}$ and $V_{rms}$ are called the *root mean square (rms)* current and voltage, respectively, and may be calculated from the peak values by dividing them by $\sqrt{2}$*:

$$I_{rms} = \frac{I_0}{\sqrt{2}} \qquad (20.12)$$

$$V_{rms} = \frac{V_0}{\sqrt{2}} \qquad (20.13)$$

Since the maximum ac voltage at a home wall socket in the United States is typically $V_0 = 170$ volts, the corresponding rms voltage is $V_{rms} = (170 \text{ volts})/\sqrt{2} = 120$ volts. Thus, when the instructions for an electrical appliance specify 120 V, it is an rms voltage that is meant. Similarly, when we specify an ac voltage or current in this text, it is an rms value, unless indicated otherwise. Likewise, when we specify ac power, it is an average power, unless stated otherwise.

Except for dealing with average quantities, the relation $\overline{P} = I_{rms} V_{rms}$ has the same form as Equation 20.6a ($P = IV$). Moreover, Ohm's law can be written conveniently in terms of rms quantities:

$$V_{rms} = I_{rms} R \qquad (20.14)$$

Substituting Equation 20.14 into $\overline{P} = I_{rms} V_{rms}$ shows that the average power can be expressed in the following ways:

$$\overline{P} = I_{rms} V_{rms} \qquad (20.15a)$$
$$\overline{P} = I_{rms}^2 R \qquad (20.15b)$$
$$\overline{P} = \frac{V_{rms}^2}{R} \qquad (20.15c)$$

These expressions are completely analogous to $P = IV = I^2 R = V^2/R$ for dc circuits. Example 6 deals with the average power in one familiar ac circuit.

### Example 6 Electrical Power Sent to a Loudspeaker

A stereo receiver applies a peak ac voltage of 34 V to a speaker. The speaker behaves approximately† as if it had a resistance of 8.0 Ω, as the circuit in Figure 20.11 indicates. Determine (a) the rms voltage, (b) the rms current, and (c) the average power for this circuit.

**Reasoning** The rms voltage is, by definition, equal to the peak voltage divided by $\sqrt{2}$. Furthermore, we are assuming that the circuit contains only a resistor. Therefore, we can use Ohm's law (Equation 20.14) to calculate the rms current as the rms voltage divided by the resistance, and we can then determine the average power as the rms current times the rms voltage (Equation 20.15a).

**Solution**

(a) The peak value of the voltage is $V_0 = 34$ V, so the corresponding rms value is

$$V_{rms} = \frac{V_0}{\sqrt{2}} = \frac{34 \text{ V}}{\sqrt{2}} = \boxed{24 \text{ V}} \qquad (20.13)$$

(b) The rms current can be obtained from Ohm's law:

$$I_{rms} = \frac{V_{rms}}{R} = \frac{24 \text{ V}}{8.0 \text{ }\Omega} = \boxed{3.0 \text{ A}} \qquad (20.14)$$

(c) The average power is

$$\overline{P} = I_{rms} V_{rms} = (3.0 \text{ A})(24 \text{ V}) = \boxed{72 \text{ W}} \qquad (20.15a)$$

The electric power dissipated in a resistor causes it to heat up. Excessive power can lead to a potential fire hazard, as Conceptual Example 7 discusses.

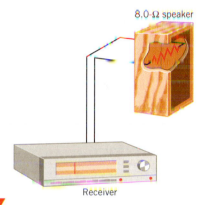

8.0-Ω speaker

Receiver

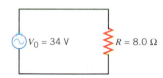

$V_0 = 34$ V $\qquad R = 8.0$ Ω

**Figure 20.11** A receiver applies an ac voltage (peak value = 34 V) to an 8.0-Ω speaker.

*Problem solving insight*
The rms values of the ac voltage and the ac current, $V_{rms}$ and $I_{rms}$, respectively, are not the same as the peak values $V_0$ and $I_0$. The rms values are always smaller than the peak values by a factor of $\sqrt{2}$.

* Equations 20.12 and 20.13 apply only for sinusoidal current and voltage.
† Other factors besides resistance can affect the current and voltage in ac circuits; they are discussed in Chapter 23.

**Figure 20.12** When an extension cord is used with a space heater, the cord must have a resistance that is sufficiently small to prevent overheating of the cord.

## Conceptual Example 7 Extension Cords and a Potential Fire Hazard

During the winter, many people use portable electric space heaters to keep warm. Sometimes, however, the heater must be located far from a 120-V wall receptacle, so an extension cord must be used (see Figure 20.12). However, manufacturers often warn against using an extension cord. If one must be used, they recommend a certain wire gauge, or smaller. Why the warning, and why are smaller-gauge wires better than larger-gauge wires?

**Reasoning and Solution** An electric space heater contains a heater element that is a piece of wire of resistance $R$, which is heated to a high temperature. The heating occurs because of the power $I_{rms}^2 R$ dissipated in the heater element. A typical heater uses a relatively large current $I_{rms}$ of about 12 A. On its way to the heater, this current passes through the wires of the extension cord. Since these additional wires offer resistance to the current, the extension cord can also heat up, just as the heater element does. The extent of the heating depends on the resistance of the extension cord. If the heating becomes excessive, the insulation around the wire can melt, and a fire can result. Thus, *manufacturers of space heaters warn about extension cords in order to prevent a fire hazard.* To keep the heating of the extension cord to a safe level, the resistance of the wire must be kept small. Recall from Section 20.3 that the resistance of a wire depends inversely on its cross-sectional area, so a large cross-sectional area gives rise to a small resistance. A wire with a large cross-sectional area is one that has a small gauge number, as discussed in Example 3. *Smaller-gauge wires are better because they offer less resistance than larger-gauge wires.*

**Related Homework:** *Problem 34*

## 20.6 Series Wiring

Thus far, we have dealt with circuits that include only a single device, such as a light bulb or a loudspeaker. There are, however, many circuits in which more than one device is connected to a voltage source. This section introduces one method by which such connections may be made—namely, series wiring. *Series wiring means that the devices are connected in such a way that there is the same electric current through each device.* Figure 20.13 shows a circuit in which two different devices, represented by resistors $R_1$ and $R_2$, are connected in series with a battery. Note that if the current in one resistor is interrupted, the current in the other is too. This could occur, for example, if two light bulbs were connected in series and the filament of one bulb broke. Because of the series wiring, the voltage $V$ supplied by the battery is divided between the two resistors. The drawing indicates that the portion of the voltage across $R_1$ is $V_1$, while the portion across $R_2$ is $V_2$, so $V = V_1 + V_2$. For each resistance, the definition of resistance is $R = V/I$, which may be solved for the corresponding voltage $V = IR$. Therefore, we have

$$V = V_1 + V_2 = IR_1 + IR_2 = I(R_1 + R_2) = IR_S$$

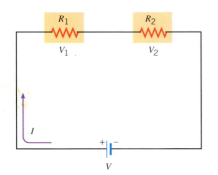

**Figure 20.13** When two resistors are connected in series, the same current $I$ is in both of them.

where $R_S$ is called the *equivalent resistance* of the series circuit. Thus, two resistors in series are equivalent to a single resistor whose resistance is $R_S = R_1 + R_2$, in the sense that there is the same current through $R_S$ as there is through the series combination of $R_1$ and $R_2$. This line of reasoning can be extended to any number of resistors in series, with the result that

*Series resistors*          $R_S = R_1 + R_2 + R_3 + \cdots$          (20.16)

Example 8 illustrates the concept of equivalent resistance in a series circuit.

### Example 8 Resistors in a Series Circuit

A 6.00-$\Omega$ resistor and a 3.00-$\Omega$ resistor are connected in series with a 12.0-V battery, as Figure 20.14 indicates. Assuming that the battery contributes no resistance to the circuit, find (a) the current, (b) the power dissipated in each resistor, and (c) the total power delivered to the resistors by the battery.

**Reasoning** The current $I$ can be obtained from Ohm's law as $I = V/R_S$, where $R_S = R_1 + R_2$ is the equivalent resistance of the two resistors. The power delivered to each resistor is given by Equation 20.6b as $P = I^2 R$, where $R$ is the resistance of the resistor being considered and $I$

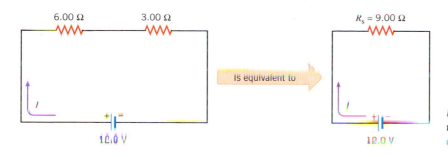

**Figure 20.14** A 6.00-Ω and a 3.00-Ω resistor connected in series are equivalent to a single 9.00-Ω resistor.

is the current through it. The total power delivered by the battery is the power delivered to the 6.00-Ω resistor plus the power delivered to the 3.00-Ω resistor.

**Solution**

**(a)** The equivalent resistance is

$$R_S = 6.00 \ \Omega + 3.00 \ \Omega = 9.00 \ \Omega \tag{20.16}$$

Ohm's law indicates that the current is

$$I = \frac{V}{R_s} = \frac{12.0 \ V}{9.00 \ \Omega} = \boxed{1.33 \ A} \tag{20.2}$$

**(b)** Now that the current is known, the power dissipated in each resistor can be obtained from $P = I^2 R$.

**6.00-Ω resistor**  $P = I^2 R = (1.33 \ A)^2 (6.00 \ \Omega) = \boxed{10.6 \ W}$  (20.6b)

**3.00-Ω resistor**  $P = I^2 R = (1.33 \ A)^2 (3.00 \ \Omega) = \boxed{5.31 \ W}$

**(c)** The total power delivered by the battery is the sum of the contributions in part (b): $P = 10.6 \ W + 5.31 \ W = 15.9 \ W$. Alternatively, the total power can be obtained directly by using the equivalent resistance $R_S = 9.00 \ \Omega$ and the current from part (a):

$$P = I^2 R_S = (1.33 \ A)^2 (9.00 \ \Omega) = \boxed{15.9 \ W}$$

In general, *the total power delivered to any number of resistors in series is equal to the power delivered to the equivalent resistor.*

*Problem solving insight*

Pressure-sensitive pads form the heart of computer input devices that function as personal electronic assistants or personal digital assistants, and they offer an interesting application of series resistors. These devices are simple to use. You write directly on the pad with a plastic stylus that itself contains no electronics (see Figure 20.15). The writing appears as the stylus is moved, and recognition software interprets it as input for the built-in computer. The pad utilizes two transparent conductive layers that are separated by a small distance, except where pressure from the stylus brings them into contact (see point P in the drawing). Current I enters the positive side of the top layer, flows into the bottom layer through point P, and exits that layer through its negative side. Each layer provides resistance to the current, the amounts depending on where the point P is located. As the

*The physics of* **personal electronic assistants.**

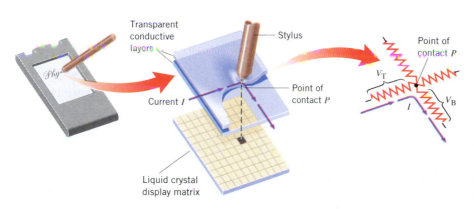

**Figure 20.15** The pressure pad on which the user writes in a personal electronic assistant is based on the use of resistances that are connected in series.

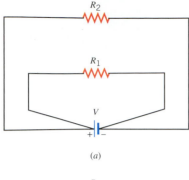

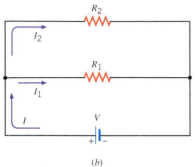

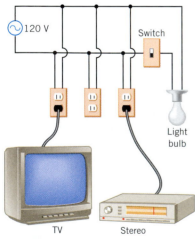

**Figure 20.16** (*a*) When two resistors are connected in parallel, the same voltage *V* is applied across each resistor. (*b*) This circuit drawing is equivalent to that in part *a*. $I_1$ and $I_2$ are, respectively, the currents in $R_1$ and $R_2$.

**Figure 20.17** This drawing shows some of the parallel connections found in a typical home. Each wall socket provides 120 V to the appliance connected to it. In addition, 120 V is applied to the light bulb when the switch is turned on.

right side of the drawing indicates, the resistances from the layers are in series, since the same current flows through both of them. The voltage across the top-layer resistance is $V_T$, and the voltage across the bottom-layer resistance is $V_B$. In each case, the voltage is the current times the resistance. These two voltages are used to locate the point *P* and to activate (darken) one of the elements or pixels in a liquid crystal display matrix that lies beneath the transparent layers (see Section 24.6 for a discussion of liquid crystal displays). As the stylus is moved, the writing becomes visible as one element after another in the display matrix is activated.

## 20.7 *Parallel Wiring*

Parallel wiring is another method of connecting electrical devices. *Parallel wiring means that the devices are connected in such a way that the same voltage is applied across each device.* Figure 20.16 shows two resistors connected in parallel between the terminals of a battery. Part *a* of the picture is drawn so as to emphasize that the entire voltage of the battery is applied across each resistor. Actually, parallel connections are rarely drawn in this manner; instead they are drawn as in part *b*, where the dots indicate the points where the wires for the two branches are joined together. Parts *a* and *b* are equivalent representations of the same circuit.

Parallel wiring is very common. For example, when an electrical appliance is plugged into a wall socket, the appliance is connected in parallel with other appliances, as in Figure 20.17, where the entire voltage of 120 V is applied across the television, the stereo, and the light bulb. The presence of the unused socket or other devices that are turned off does not affect the operation of those devices that are turned on. Moreover, if the current in one device is interrupted (perhaps by an opened switch or a broken wire), the current in the other devices is not interrupted. In contrast, if household appliances were connected in series, there would be no current through any appliance if the current in the circuit were halted at any point.

When two resistors $R_1$ and $R_2$ are connected as in Figure 20.16, each receives current from the battery as if the other were not present. Therefore, $R_1$ and $R_2$ together draw more current from the battery than does either resistor alone. According to the definition of resistance, $R = V/I$, a larger current implies a smaller resistance. Thus, the two parallel resistors behave as a single equivalent resistance that is *smaller* than either $R_1$ or $R_2$. Figure 20.18 returns to the water-flow analogy to provide additional insight into this important feature of parallel wiring. In part *a*, two sections of pipe that have the same length are connected in parallel with a pump. In part *b* these two sections have been replaced with a single pipe of the same length, whose cross-sectional area equals the combined cross-sectional areas of section 1 and section 2. The pump can push more water per second through the wider pipe in part *b* than it can through *either* of the narrower pipes in part *a*. In effect, the wider pipe offers less resistance to the flow than either of the narrower pipes offers individually.

As in a series circuit, it is possible to replace a parallel combination of resistors with an equivalent resistor that results in the same total current and power for a given voltage as the original combination. To determine the equivalent resistance for the two resistors in Figure 20.16*b*, note that the total current *I* from the battery is the sum of $I_1$ and $I_2$, where $I_1$ is the current in resistor $R_1$ and $I_2$ is the current in resistor $R_2$: $I = I_1 + I_2$. Since the same voltage *V* is applied across each resistor, the definition of resistance indicates that $I_1 = V/R_1$ and $I_2 = V/R_2$. Therefore,

$$I = I_1 + I_2 = \frac{V}{R_1} + \frac{V}{R_2} = V\left(\frac{1}{R_1} + \frac{1}{R_2}\right) = V\left(\frac{1}{R_P}\right)$$

where $R_P$ is the equivalent resistance. Hence, when two resistors are connected in parallel, they are equivalent to a single resistor whose resistance $R_P$ can be obtained from $1/R_P = 1/R_1 + 1/R_2$. For any number of resistors in parallel, a similar line of reasoning reveals that

*Parallel resistors*
$$\frac{1}{R_P} = \frac{1}{R_1} + \frac{1}{R_2} + \frac{1}{R_3} + \cdots \qquad (20.17)$$

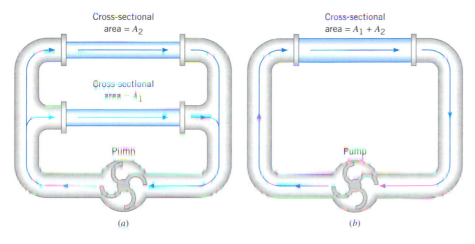

**Figure 20.18** (*a*) Two equally long pipe sections, with cross-sectional areas $A_1$ and $A_2$, are connected in parallel to a water pump. (*b*) The two parallel pipe sections in part *a* are equivalent to a single pipe of the same length whose cross-sectional area is $A_1 + A_2$.

The next example deals with a parallel combination of resistors that occurs in a stereo system.

### Example 9 Main and Remote Stereo Speakers

Most receivers allow the user to connect "remote" speakers (to play music in another room, for instance) in addition to the main speakers. Figure 20.19 shows that the remote speaker and the main speaker for the right stereo channel are connected to the receiver in parallel (for clarity, the speakers for the left channel are not shown). At the instant represented in the picture, the ac voltage across the speakers is 6.00 V. The main-speaker resistance is 8.00 Ω, and the remote-speaker resistance is 4.00 Ω.* Determine (a) the equivalent resistance of the two speakers, (b) the total current supplied by the receiver, (c) the current in each speaker, (d) the power dissipated in each speaker, and (e) the total power delivered by the receiver.

*The physics of* **main and remote stereo speakers.**

**Reasoning** The total current supplied to the two speakers by the receiver can be calculated as $I_{\text{rms}} = V_{\text{rms}}/R_{\text{P}}$, where $R_{\text{P}}$ is the equivalent resistance of the two speakers in parallel and can be obtained from $1/R_{\text{P}} = 1/R_1 + 1/R_2$. The current in each speaker is different, however, since the speakers have different resistances. The average power delivered to a given speaker is the product of its current and voltage. In the parallel connection the same voltage is applied to each speaker.

**Figure 20.19** (*a*) The main and remote speakers in a stereo system are connected in parallel to the receiver. (*b*) The circuit schematic shows the situation when the ac voltage across the speakers is 6.00 V.

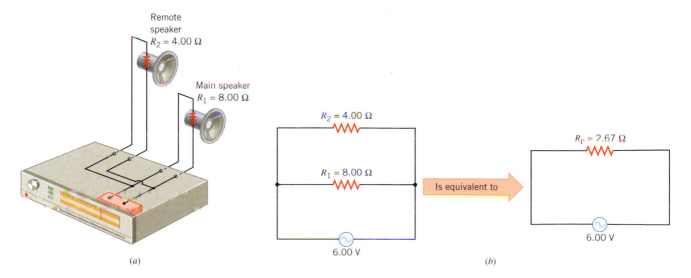

* In reality, frequency-dependent characteristics of the speaker (see Chapter 23) play a role in the operation of a loudspeaker. We assume here, however, that the frequency of the sound is low enough that the speakers behave as pure resistances.

**Solution**

**(a)** According to Equation 20.17, the equivalent resistance of the two speakers is given by

$$\frac{1}{R_P} = \frac{1}{8.00\ \Omega} + \frac{1}{4.00\ \Omega} = \frac{3}{8.00\ \Omega} \quad \text{or} \quad R_P = \frac{8.00\ \Omega}{3} = \boxed{2.67\ \Omega}$$

This result is illustrated in part *b* of the drawing.

**(b)** Using the equivalent resistance in Ohm's law shows that the total current is

$$I_{\text{rms}} = \frac{V_{\text{rms}}}{R_P} = \frac{6.00\ \text{V}}{2.67\ \Omega} = \boxed{2.25\ \text{A}} \tag{20.14}$$

**(c)** Applying Ohm's law to each speaker gives the individual speaker currents:

*8.00-$\Omega$ speaker* $\qquad\qquad I_{\text{rms}} = \dfrac{V_{\text{rms}}}{R} = \dfrac{6.00\ \text{V}}{8.00\ \Omega} = \boxed{0.750\ \text{A}}$

*4.00-$\Omega$ speaker* $\qquad\qquad I_{\text{rms}} = \dfrac{V_{\text{rms}}}{R} = \dfrac{6.00\ \text{V}}{4.00\ \Omega} = \boxed{1.50\ \text{A}}$

The sum of these currents is equal to the total current obtained in part (b).

**(d)** The average power dissipated in each speaker can be calculated using $\overline{P} = I_{\text{rms}} V_{\text{rms}}$ with the individual currents obtained in part (c):

*8.00-$\Omega$ speaker* $\qquad\qquad \overline{P} = (0.750\ \text{A})(6.00\ \text{V}) = \boxed{4.50\ \text{W}} \tag{20.15a}$

*4.00-$\Omega$ speaker* $\qquad\qquad \overline{P} = (1.50\ \text{A})(6.00\ \text{V}) = \boxed{9.00\ \text{W}} \tag{20.15a}$

**(e)** The total power delivered by the receiver is the sum of the individual values that were found in part (d), $\overline{P} = 4.50\ \text{W} + 9.00\ \text{W} = 13.5\ \text{W}$. Alternatively, the total power can be obtained from the equivalent resistance $R_P = 2.67\ \Omega$ and the total current in part (b):

$$\overline{P} = I_{\text{rms}}^2 R_P = (2.25\ \text{A})^2 (2.67\ \Omega) = \boxed{13.5\ \text{W}} \tag{20.15b}$$

In general, *the total power delivered to any number of resistors in parallel is equal to the power delivered to the equivalent resistor.*

▲

In a parallel combination of resistances, it is the *smallest* resistance that has the largest impact in determining the equivalent resistance. In fact, if one resistance approaches zero, then according to Equation 20.17, the equivalent resistance also approaches zero. In such a case, the near-zero resistance is said to *short out* the other resistances by providing a near-zero resistance path for the current to follow as a shortcut around the other resistances.

An interesting application of parallel wiring occurs in a three-way light bulb, as Conceptual Example 10 discusses.

### *Conceptual Example 10*  A Three-Way Light Bulb and Parallel Wiring

Three-way light bulbs are popular because they can provide three levels of illumination (e.g., 50 W, 100 W, and 150 W) using a 120-V socket. The socket, however, must be equipped with a special three-way switch that enables one to select the illumination level. This switch does not select different voltages, because a three-way bulb uses a single voltage of 120 V. Within the bulb are two separate filaments. When one of these burns out (i.e., vaporizes), the bulb can produce only one level of illumination (either its lowest or its intermediate level) but not the highest level. How are the two filaments connected, in series or in parallel? And how can two filaments be used to produce three different illumination levels?

**Reasoning and Solution** If the filaments were wired in series and one of them burned out, no current would pass through the bulb and none of the illumination levels would be available, contrary to what is observed. Therefore, *the filaments must be connected in parallel,* as in Figure 20.20.

As to how two filaments can be used to produce three illumination levels, we note that the power dissipated in a resistance $R$ is $\overline{P} = V_{\text{rms}}^2 / R$, according to Equation 20.15c. With a single value of 120 V for the voltage $V_{\text{rms}}$, three different wattage ratings for the bulb can be obtained only if three different values for the resistance $R$ can occur. In a 50-W/100-W/150-W three-way bulb, for example, one resistance $R_{50}$ is provided by the 50-W filament, and the second re-

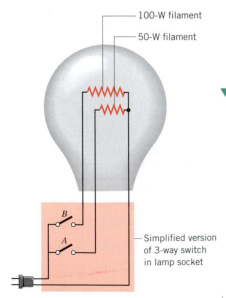

**Figure 20.20** A three-way light bulb uses two filaments connected in parallel. The filaments can be turned on one at a time or both together.

100-W filament
50-W filament

*B*
*A*
Simplified version of 3-way switch in lamp socket

sistance $R_{100}$ comes from the 100-W filament. The third resistance $R_{150}$ is the parallel combination of the other two and can be obtained from $1/R_{150} = 1/R_{50} + 1/R_{100}$. Figure 20.20 illustrates a simplified version of how the three-way switch operates to provide these three alternatives. In the first position of the switch, contact $A$ is closed and contact $B$ is open, so current passes only through the 50-W filament. In the second position, contact $A$ is open and contact $B$ is closed, so current passes only through the 100-W filament. In the third position, both contacts $A$ and $B$ are closed, and both filaments light up, giving 150 W of illumination.

**Related Homework:** Problem 50

# 20.8 *Circuits Wired Partially in Series and Partially in Parallel*

Often an electric circuit is wired partially in series and partially in parallel. The key to determining the current, voltage, and power in such a case is to deal with the circuit in parts, with the resistances in each part being either in series or in parallel with each other. Example 11 shows how such an analysis is carried out.

### Example 11 A Four-Resistor Circuit

Figure 20.21 shows a circuit composed of a 24-V battery and four resistors, whose resistances are 110, 180, 220, and 250 $\Omega$. Find (a) the total current supplied by the battery and (b) the voltage between points $A$ and $B$ in the circuit.

**Reasoning** The total current supplied by the battery can be obtained from Ohm's law, $I = V/R$, where $R$ is the equivalent resistance of the four resistors. The equivalent resistance can be calculated by dealing with the circuit in parts. The voltage $V_{AB}$ between the points $A$ and $B$ is

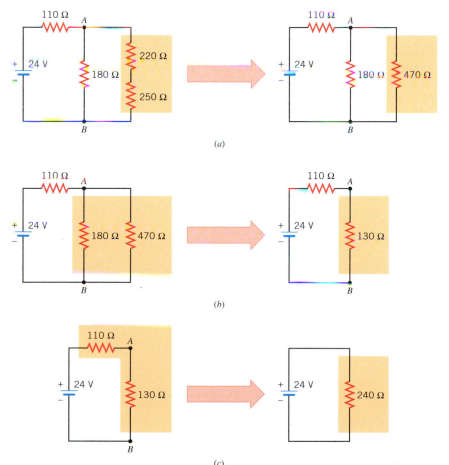

**Figure 20.21** The circuits shown in this picture are equivalent.

also given by Ohm's law, $V_{AB} = IR_{AB}$, where $I$ is the current and $R_{AB}$ is the equivalent resistance between the two points.

**Solution**

(a) The 220-$\Omega$ resistor and the 250-$\Omega$ resistor are in series, so they are equivalent to a single resistor whose resistance is 220 $\Omega$ + 250 $\Omega$ = 470 $\Omega$ (see Figure 20.21a). The 470-$\Omega$ resistor is in parallel with the 180-$\Omega$ resistor. Their equivalent resistance can be obtained from Equation 20.17:

$$\frac{1}{R_{AB}} = \frac{1}{470 \ \Omega} + \frac{1}{180 \ \Omega} = 0.0077 \ \Omega^{-1} \quad \text{or} \quad R_{AB} = \frac{1}{0.0077 \ \Omega^{-1}} = 130 \ \Omega$$

The circuit is now equivalent to a circuit containing a 110-$\Omega$ resistor in series with a 130-$\Omega$ resistor (see Figure 20.21b). This combination acts like a single resistor whose resistance is $R = 110 \ \Omega + 130 \ \Omega = 240 \ \Omega$ (see Figure 20.21c). The total current from the battery is, then,

$$I = \frac{V}{R} = \frac{24 \text{ V}}{240 \ \Omega} = \boxed{0.10 \text{ A}}$$

(b) The current $I = 0.10$ A passes through the resistance between points $A$ and $B$. Therefore, Ohm's law indicates that the voltage across the 130-$\Omega$ resistor between points $A$ and $B$ is

$$V_{AB} = IR_{AB} = (0.10 \text{ A})(130 \ \Omega) = \boxed{13 \text{ V}}$$

# 20.9 *Internal Resistance*

So far, the circuits we have considered include batteries or generators that contribute only their emfs to a circuit. In reality, however, such devices also add some resistance. This resistance is called the ***internal resistance*** of the battery or generator because it is located inside the device. In a battery, the internal resistance is due to the chemicals within the battery. In a generator, the internal resistance is the resistance of wires and other components within the generator.

Figure 20.22 shows a schematic representation of the internal resistance $r$ of a battery. The drawing emphasizes that when an external resistance $R$ is connected to the battery, the resistance is connected *in series* with the internal resistance. The internal resistance of a functioning battery is typically small (several thousandths of an ohm for a new car battery). Nevertheless, the effect of the internal resistance may not be negligible. Example 12 illustrates that when current is drawn from a battery, the internal resistance causes the voltage between the terminals to drop below the maximum value specified by the battery's emf. The actual voltage between the terminals of a battery is known as the ***terminal voltage.***

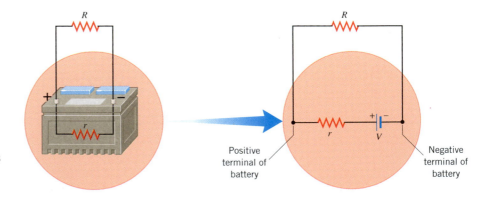

*Positive terminal of battery*

*Negative terminal of battery*

**Figure 20.22** When an external resistance $R$ is connected between the terminals of a battery, the resistance is connected in series with the internal resistance $r$ of the battery.

## Example 12 The Terminal Voltage of a Battery

*The physics of*
**automobile batteries.**

Figure 20.23 shows a car battery whose emf is 12.0 V and whose internal resistance is 0.010 $\Omega$. This resistance is relatively large because the battery is old and the terminals are corroded. What is the terminal voltage when the current $I$ drawn from the battery is (a) 10.0 A and (b) 100.0 A?

**Reasoning** The voltage between the terminals is not the entire 12.0-V emf, because part of the emf is needed to make the current go through the internal resistance. The amount of voltage needed can be determined from Ohm's law as the current $I$ through the battery times the internal resistance $r$. For larger currents, a larger amount of voltage is needed, leaving less of the emf between the terminals.

**Solution**

**(a)** The voltage needed to make a current of $I = 10.0$ A go through an internal resistance of $r = 0.010 \ \Omega$ is $V = Ir = (10.0 \ \text{A})(0.010 \ \Omega) = 0.10$ V. To find the terminal voltage, remember that the direction of conventional current is always from a higher toward a lower potential. To emphasize this fact in the drawing, plus and minus signs have been included at the right and left ends, respectively, of the resistance $r$. The terminal voltage can be calculated by starting at the negative terminal and keeping track of how the voltage increases and decreases as we move toward the positive terminal. The voltage rises by 12.0 V due to the battery's emf. However, the voltage drops by 0.10 V because of the potential difference across the internal resistance. Therefore, the terminal voltage is 12.0 V − 0.10 V = $\boxed{11.9 \ \text{V}}$.

**(b)** When the current through the battery is 100.0 A, the voltage needed to make the current go through the internal resistance is $V = (100.0 \ \text{A})(0.010 \ \Omega) = 1.0$ V. The terminal voltage now decreases to 12.0 V − 1.0 V = $\boxed{11.0 \ \text{V}}$.

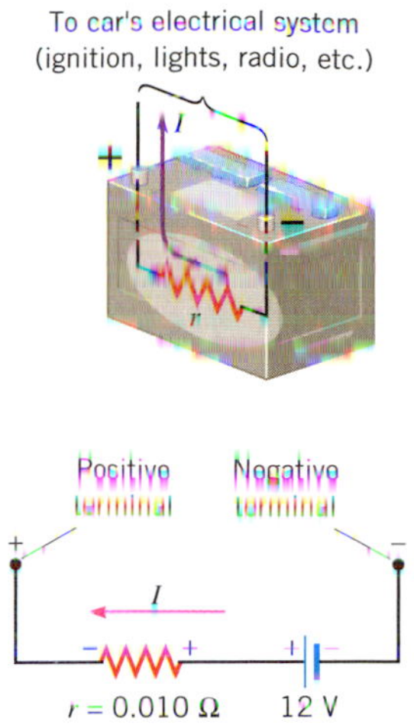

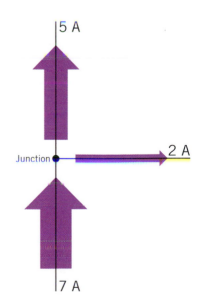

To car's electrical system (ignition, lights, radio, etc.)

Positive terminal  Negative terminal

$r = 0.010 \ \Omega$   12 V

**Figure 20.23** A car battery whose emf is 12 V and whose internal resistance is $r$.

---

Example 12 indicates that the terminal voltage of a battery is smaller when the current drawn from the battery is larger, an effect that any car owner can demonstrate. Turn the headlights on before starting your car, so the current through the battery is about 10 A, as in part (a) of Example 12. Then start the car. The starter motor draws a large amount of additional current from the battery, momentarily increasing the total current by an appreciable amount. Consequently, the terminal voltage of the battery decreases, causing the headlights to dim.

# 20.10 *Kirchhoff's Rules*

Electric circuits that contain a number of resistors can often be analyzed by combining individual groups of resistors in series and parallel, as Section 20.8 discusses. However, there are many circuits in which no two resistors are in series or in parallel. To deal with such circuits it is necessary to employ methods other than the series–parallel method. One alternative is to take advantage of Kirchhoff's rules, named after their developer Gustav Kirchhoff (1824–1887).

There are two rules, the *junction rule* and the *loop rule.* These rules arise from principles and ideas that we have encountered earlier. The junction rule is an application of the law of conservation of electric charge (see Section 18.2) to the electric current in a circuit. The loop rule is an application of the principle of conservation of energy (see Section 6.8) to the electric potential (see Section 19.2) that exists at various places in a circuit.

Figure 20.24 illustrates in greater detail the basic idea behind Kirchhoff's junction rule. The picture shows a junction where several wires are connected together. As Section 18.2 discusses, electric charge is conserved. Therefore, since there is no accumulation of charges at the junction itself, the total charge per second flowing into the junction must equal the total charge per second flowing out of it. In other words, *the junction rule states that the total current directed into a junction must equal the total current directed out of the junction,* or 7 A = 5 A + 2 A for the specific case shown in the picture.

To help explain Kirchhoff's loop rule, Figure 20.25 shows a circuit in which a 12-V battery is connected to a series combination of a 5-Ω and a 1-Ω resistor. The plus and minus signs associated with each resistor remind us that, outside a battery, conventional current is directed from a higher toward a lower potential. From left to right, there is a potential drop of 10 V across the first resistor and another drop of 2 V across the second resistor. Keeping in mind that potential is the electric potential energy per unit charge, let us follow a positive test charge clockwise*

5 A

Junction

2 A

7 A

**Figure 20.24** A junction is a point in a circuit where a number of wires are connected together. If 7 A of current is directed into the junction, then a total of 7 A (5 A + 2 A) of current must be directed out of the junction.

* The choice of the clockwise direction is arbitrary.

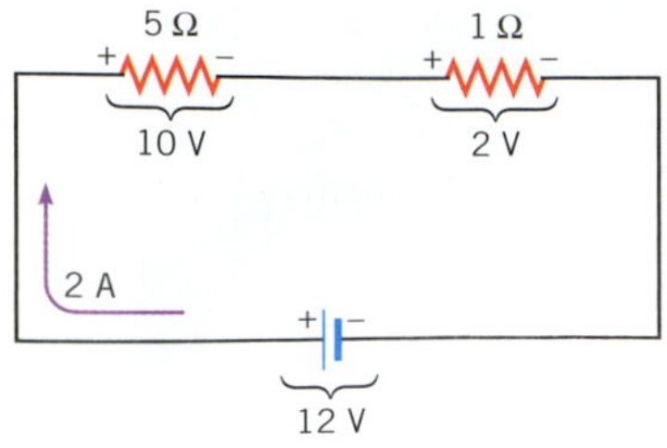

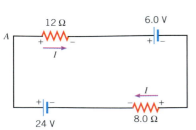

**Figure 20.25** Following a positive test charge clockwise around the circuit, we see that the total voltage drop of 10 V + 2 V across the two resistors equals the voltage rise of 12 V due to the battery. The plus and minus signs on the resistors emphasize that, outside a battery, conventional current is directed from a higher potential (+) toward a lower potential (−).

around the circuit. Starting at the negative terminal of the battery, we see that the test charge gains potential energy because of the 12-V rise in potential due to the battery. The test charge then loses potential energy because of the 10-V and 2-V drops in potential across the resistors, ultimately arriving back at the negative terminal. In traversing the closed-circuit loop, the test charge is like a skier gaining gravitational potential energy in going up a hill on a chair lift and then losing it to friction in coming down and stopping. When the skier returns to the starting point, the gain equals the loss, so there is no net change in potential energy. Similarly, when the test charge arrives back at its starting point, there is no net change in electric potential energy, the gains matching the losses. *The loop rule expresses this example of energy conservation in terms of the electric potential and states that for a closed-circuit loop, the total of all the potential rises is the same as the total of all the potential drops,* or 12 V = 10 V + 2 V for the specific case in Figure 20.25.

Kirchhoff's rules can be applied to any circuit, even when the resistors are not in series or in parallel. The two rules are summarized below, and Examples 13 and 14 illustrate how to use them.

■ **KIRCHHOFF'S RULES**

*Junction rule.* The sum of the magnitudes of the currents directed into a junction equals the sum of the magnitudes of the currents directed out of the junction.

*Loop rule.* Around any closed-circuit loop, the sum of the potential drops equals the sum of the potential rises.

## *Example 13*  Using Kirchhoff's Loop Rule

Figure 20.26 shows a circuit that contains two batteries and two resistors. Determine the current *I* in the circuit.

**Reasoning** The first step is to draw the current, which we have chosen to be clockwise around the circuit. The choice of direction is *arbitrary,* and if it is incorrect, *I* will turn out to be negative.

*Problem solving insight*

The second step is to mark the resistors with plus and minus signs, which serve as an aid in identifying the potential drops and rises for Kirchhoff's loop rule. *Remember that, outside a battery, conventional current is always directed from a higher potential (+) toward a lower potential (−).* Thus, we *must* mark the resistors as indicated in Figure 20.26, to be consistent with the clockwise direction chosen for the current.

We may now apply Kirchhoff's loop rule to the circuit, starting at corner *A*, proceeding clockwise around the loop, and identifying the potential drops and rises as we go. The potential across each resistor is given by Ohm's law as *V = IR*. The clockwise direction is arbitrary, and the same result is obtained with a counterclockwise path.

**Solution** Starting at corner *A* and moving clockwise around the loop, there is

**1.** A potential drop (+ to −) of *IR* = *I*(12 Ω) across the 12-Ω resistor

**2.** A potential drop (+ to −) of 6.0 V across the 6.0-V battery

**3.** A potential drop (+ to −) of *IR* = *I*(8.0 Ω) across the 8.0-Ω resistor

**4.** A potential rise (− to +) of 24 V across the 24-V battery

**Figure 20.26** A single-loop circuit that contains two batteries and two resistors.

Setting the sum of the potential drops equal to the sum of the potential rises, as required by Kirchhoff's loop rule, gives

$$\underbrace{I(12\ \Omega) + 6.0\ \text{V} + I(8.0\ \Omega)}_{\text{Potential drops}} = \underbrace{24\ \text{V}}_{\text{Potential rises}}$$

Solving this equation for the current yields $\boxed{I = 0.90\ \text{A}}$. The current is a positive number, indicating that our initial choice for the direction (clockwise) of the current was correct.

## Example 14 The Electrical System of a Car

In a car, the headlights are connected to the battery and would discharge the battery if it were not for the alternator, which is run by the engine. Figure 20.27 indicates how the car battery, headlights, and alternator are connected. The circuit includes an internal resistance of 0.0100 Ω for the car battery and its leads and a resistance of 1.20 Ω for the headlights. For the sake of simplicity, the alternator is approximated as an additional 14.00-V battery with an internal resistance of 0.100 Ω. Determine the currents through the car battery ($I_B$), the headlights ($I_H$), and the alternator ($I_A$).

**Reasoning** The drawing shows the directions chosen arbitrarily for the currents $I_B$, $I_H$, and $I_A$. If any direction is incorrect, the analysis will reveal a negative value for the corresponding current.

Next, we mark the resistors with the plus and minus signs that serve as an aid in identifying the potential drops and rises for the loop rule, recalling that, outside a battery, conventional current is always directed from a higher potential (+) toward a lower potential (−). Thus, given the directions selected for $I_B$, $I_H$, and $I_A$, the plus and minus signs *must* be those indicated in Figure 20.27.

Kirchhoff's junction and loop rules can now be used.

*The physics of an automobile electrical system.*

**Solution** The junction rule can be applied to junction B or junction E. In either case, the same equation results:

***Junction rule applied at B***

$$\underbrace{I_A + I_B}_{\substack{\text{Into} \\ \text{junction}}} = \underbrace{I_H}_{\substack{\text{Out of} \\ \text{junction}}}$$

In applying the loop rule to the lower loop *BEFA*, we start at point *B*, move clockwise around the loop, and identify the potential drops and rises. There is a potential rise (− to +) of $I_B$ (0.0100 Ω) across the 0.0100-Ω resistor and then a drop (+ to −) of 12.00 V due to the car battery. Continuing around the loop, we find a 14.00-V rise (− to +) across the alternator, followed by a potential drop (+ to −) of $I_A$ (0.100 Ω) across the 0.100-Ω resistor. Setting the sum of the potential drops equal to the sum of the potential rises gives the following result:

***Loop rule: BEFA***  $\underbrace{I_A(0.100\ \Omega) + 12.00\ \text{V}}_{\text{Potential drops}} = \underbrace{I_B(0.0100\ \Omega) + 14.00\ \text{V}}_{\text{Potential rises}}$

Since there are three unknown variables in this problem, $I_B$, $I_H$, and $I_A$, a third equation is needed for a solution. To obtain the third equation, we apply the loop rule to the upper loop *CDEB*, choosing a clockwise path for convenience. The result is

***Loop rule: CDEB***  $\underbrace{I_B(0.0100\ \Omega) + I_H(1.20\ \Omega)}_{\substack{\text{Potential} \\ \text{drops}}} = \underbrace{12.00\ \text{V}}_{\text{Potential rises}}$

These three equations can be solved simultaneously to show that

$$\boxed{I_B = -9.0\ \text{A},\ I_H = 10.1\ \text{A},\ I_A = 19.1\ \text{A}}$$

The negative answer for $I_B$ indicates that the current through the battery is not directed from right to left, as drawn in Figure 20.27. Instead, the 9.0-A current is directed from left to right, opposite to the way current would be directed if the alternator were not connected. It is the left-to-right current created by the alternator that keeps the battery charged.

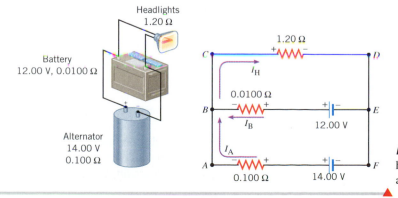

**Figure 20.27** A circuit showing the headlight(s), battery, and alternator of a car.

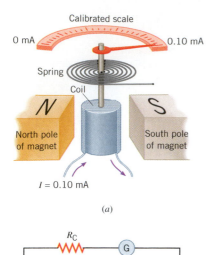

Calibrated scale

0 mA    0.10 mA

Spring

Coil

N
North pole
of magnet

S
South pole
of magnet

$I = 0.10$ mA

(a)

$R_C$

G

$I$    $I$

(b)

**Figure 20.28** (a) A dc galvanometer. The coil of wire and pointer rotate when there is a current in the wire. (b) A galvanometer with a coil resistance of $R_C$ is represented in a circuit diagram as shown here.

**The physics of**
**an ammeter.**

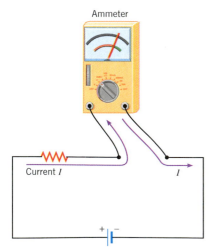

Ammeter

Current $I$    $I$

**Figure 20.29** An ammeter must be inserted into a circuit so that the current passes directly through it.

**The physics of**
**a voltmeter.**

# 20.11 The Measurement of Current and Voltage

Current and voltage can be measured with devices known, respectively, as ammeters and voltmeters. There are two types of such devices: those that use digital electronics and those that do not. The essential feature of nondigital devices is the dc *galvanometer.* As Figure 20.28a illustrates, a galvanometer consists of a magnet, a coil of wire, a spring, a pointer, and a calibrated scale. The coil is mounted in such a way that it can rotate, which causes the pointer to move in relation to the scale. As Section 21.6 will discuss, the coil rotates in response to the torque applied by the magnet when there is a current in the coil. The coil stops rotating when this torque is balanced by the torque of the spring.

A galvanometer has two important characteristics that must be considered when it is used as part of a measurement device. First, the amount of dc current that causes a full-scale deflection of the pointer indicates the sensitivity of the galvanometer. For instance, Figure 20.28a shows an instrument that deflects full scale when the current in the coil is 0.10 mA. The second important characteristic is the resistance $R_C$ of the wire in the coil. Figure 20.28b shows how a galvanometer with a coil resistance of $R_C$ is represented in a circuit diagram.

Since an ***ammeter*** is an instrument that measures current, it must be inserted in the circuit so the current passes directly through it, as Figure 20.29 shows. An ammeter includes a galvanometer and one or more *shunt resistors,* which are connected in parallel with the galvanometer and provide a bypass for current in excess of the galvanometer's full-scale limit. The bypass allows the ammeter to be used to measure a current exceeding the full-scale limit. In Figure 20.30, for instance, a current of 60.0 mA enters terminal A of an ammeter, which uses a galvanometer with a full-scale current of 0.100 mA. The shunt resistor R can be selected so that 0.100 mA of current enters the galvanometer, while 59.9 mA bypasses it. In such a case, the measurement scale on the ammeter would be labeled 0 to 60.0 mA. To determine the shunt resistance, it is necessary to know the coil resistance $R_C$ (see problem 80).

When an ammeter is inserted into a circuit, the equivalent resistance of the coil and the shunt resistor adds to the circuit resistance. Any increase in circuit resistance causes a reduction in current, and this is a problem, because an ammeter should only measure the current, not change it. Therefore, an *ideal* ammeter would have zero resistance. In practice, a good ammeter is designed with an equivalent resistance small enough so there is only a negligible reduction of the current in the circuit when the ammeter is inserted.

A ***voltmeter*** is an instrument that measures the voltage between two points, A and B, in a circuit. Figure 20.31 shows that the voltmeter must be connected between the points and is *not* inserted into the circuit as an ammeter is. A voltmeter includes a galvanometer whose scale is calibrated in volts. Suppose, for instance, that the galvanometer in Figure 20.32 has a full-scale current of 0.1 mA and a coil resistance of 50 $\Omega$. Under full-scale conditions, the voltage across the coil would be $V = IR_C = (0.1 \times 10^{-3} \text{ A})(50 \ \Omega) = 0.005$ V. Thus, this galvanometer could be used to register voltages in the range $0 - 0.005$ V. A voltmeter, then, is a galvanometer used in this fashion, along with some provision for adjusting the range of voltages to be measured. To adjust the range, an additional resistance R is connected in series with the coil resistance $R_C$. The value of R needed to extend the measurement range to a full-scale voltage V can be determined from the full-scale current $I$ by applying Ohm's law in the form $V = I(R + R_C)$ (see problem 81).

Ideally, the voltage registered by a voltmeter should be the same as the voltage that exists when the voltmeter is not connected. However, a voltmeter takes some current from a circuit and, thus, alters the circuit voltage to some extent. An *ideal* voltmeter would have infinite resistance and draw away only an infinitesimal amount of current. In reality, a good voltmeter is designed with a resistance that is large enough so the unit does not appreciably alter the voltage in the circuit to which it is connected.

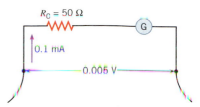

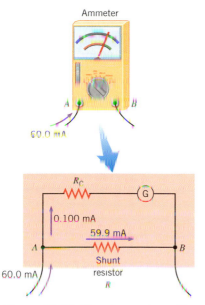

**Figure 20.30** If a galvanometer with a full-scale limit of 0.100 mA is to be used to measure a current of 60.0 mA, a shunt resistance R must be used, so the excess current of 59.9 mA can detour around the galvanometer coil.

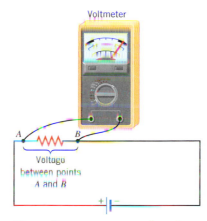

**Figure 20.31** To measure the voltage between two points A and B in a circuit, a voltmeter is connected between the points.

# 20.12 Capacitors in Series and Parallel

Figure 20.33 shows two different capacitors connected in parallel to a battery. Since the capacitors are in parallel, they have the same voltage V across their plates. However, the capacitors *contain different amounts of charge*. The charge stored by a capacitor is $q = CV$ (Equation 19.8), so $q_1 = C_1V$ and $q_2 = C_2V$.

As with resistors, it is always possible to replace a parallel combination of capacitors with an *equivalent capacitor* that stores the same charge and energy for a given voltage as the combination does. To determine the equivalent capacitance $C_P$, note that the total charge $q$ stored by the two capacitors is

$$q = q_1 + q_2 = C_1V + C_2V = (C_1 + C_2)V = C_PV$$

This result indicates that two capacitors in parallel can be replaced by an equivalent capacitor whose capacitance is $C_P = C_1 + C_2$. For any number of capacitors in parallel, the equivalent capacitance is

**Parallel capacitors** $$C_P = C_1 + C_2 + C_3 + \cdots \tag{20.18}$$

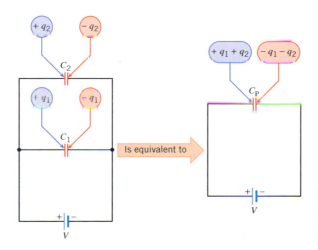

**Figure 20.33** In a parallel combination of capacitances $C_1$ and $C_2$, the voltage V across each capacitor is the same, but the charges $q_1$ and $q_2$ on each capacitor are different.

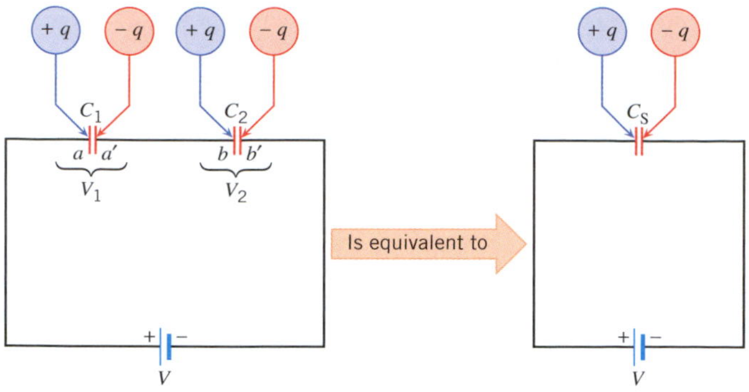

**Figure 20.34** In a series combination of capacitances $C_1$ and $C_2$, the same amount of charge $q$ is on the plates of each capacitor, but the voltages $V_1$ and $V_2$ across each capacitor are different.

Is equivalent to

Capacitances in parallel simply add together to give an equivalent capacitance. This behavior contrasts with that of resistors in parallel, which combine as reciprocals, according to Equation 20.17.

The equivalent capacitor not only stores the same amount of charge as the parallel combination of capacitors, but also stores the same amount of energy. For instance, the energy stored in a single capacitor is $\frac{1}{2}CV^2$ (Equation 19.11), so the total energy stored by two capacitors in parallel is

$$\text{Total energy} = \tfrac{1}{2}C_1V^2 + \tfrac{1}{2}C_2V^2 = \tfrac{1}{2}(C_1 + C_2)V^2 = \tfrac{1}{2}C_PV^2$$

which is equal to the energy stored in the equivalent capacitor $C_P$.

When capacitors are connected in series, the equivalent capacitance is different than when they are in parallel. As an example, Figure 20.34 shows two capacitors in series and reveals the following important fact: **_All capacitors in series, regardless of their capacitances, contain charges of the same magnitude, $+q$ and $-q$, on their plates._** The battery places a charge of $+q$ on plate $a$ of capacitor $C_1$, and this charge induces a charge of $+q$ to depart from the opposite plate $a'$, leaving behind a charge $-q$. The $+q$ charge that leaves plate $a'$ is deposited on plate $b$ of capacitor $C_2$ (since these two plates are connected by a wire), where it induces a $+q$ charge to move away from the opposite plate $b'$, leaving behind a charge of $-q$. Thus, all capacitors in series contain charges of the same magnitude on their plates. Note the difference between charging capacitors in parallel and in series. When charging parallel capacitors, the battery moves a charge $q$ that is the sum of the charges moved for each of the capacitors: $q = q_1 + q_2 + q_3 + \cdots$. In contrast, when charging a series combination of $n$ capacitors, the battery only moves a charge $q$, not $nq$, because the charge $q$ passes by induction from one capacitor directly to the next one in line.

The equivalent capacitance $C_S$ for the series connection in Figure 20.34 can be determined by observing that the battery voltage $V$ is shared by the two capacitors. The drawing indicates that the voltages across $C_1$ and $C_2$ are $V_1$ and $V_2$, so that $V = V_1 + V_2$. Since the voltages across the capacitors are $V_1 = q/C_1$ and $V_2 = q/C_2$, we find that

$$V = V_1 + V_2 = \frac{q}{C_1} + \frac{q}{C_2} = q\left(\frac{1}{C_1} + \frac{1}{C_2}\right) = q\left(\frac{1}{C_S}\right)$$

Thus, two capacitors in series can be replaced by an equivalent capacitor whose capacitance $C_S$ can be obtained from $1/C_S = 1/C_1 + 1/C_2$. For any number of capacitors connected in series the equivalent capacitance is given by

**_Series capacitors_**
$$\frac{1}{C_S} = \frac{1}{C_1} + \frac{1}{C_2} + \frac{1}{C_3} + \cdots \tag{20.19}$$

Equation 20.19 indicates that capacitances in series combine as reciprocals and do not simply add together as resistors in series do. It is left as an exercise (problem 91) to show that the equivalent series capacitance stores the same electrostatic energy as the sum of the energies of the individual capacitors.

It is possible to simplify circuits containing a number of capacitors in the same general fashion as that outlined for resistors in Example 11 and Figure 20.21. The capacitors

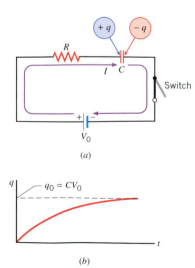

(a)

$q_0 = CV_0$

(b)

**Figure 20.35** Charging a capacitor.

in a parallel grouping can be combined according to Equation 20.18, and those in a series grouping can be combined according to Equation 20.19.

## 20.13 RC Circuits

Many electric circuits contain both resistors and capacitors. Figure 20.35 illustrates an example of a resistor–capacitor or *RC* circuit. Part *a* of the drawing shows the circuit at a time *t* after the switch has been closed and the battery has begun to charge up the capacitor plates. The charge on the plates builds up gradually to its equilibrium value of $q_0 = CV_0$, where $V_0$ is the voltage of the battery. Assuming that the capacitor is uncharged at time $t = 0$ s when the switch is closed, it can be shown that the magnitude $q$ of the charge on the plates at time $t$ is

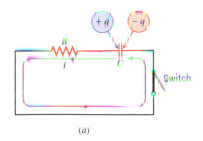

*Capacitor charging*    $$q = q_0[1 - e^{-t/(RC)}] \qquad (20.20)$$

where the exponential $e$ has the value of 2.718, . . . Part *b* of the drawing shows a graph of this expression, which indicates that the charge is $q = 0$ C when $t = 0$ s and increases gradually toward the equilibrium value of $q_0 = CV_0$. The voltage $V$ across the capacitor at any time can be obtained from Equation 20.20 by dividing the charges $q$ and $q_0$ by the capacitance $C$, since $V = q/C$ and $V_0 = q_0/C$.

The term $RC$ in the exponent in Equation 20.20 is called the ***time constant*** $\tau$ of the circuit:

$$\tau = RC \qquad (20.21)$$

The time constant is measured in seconds; verification of the fact that an ohm times a farad is equivalent to a second is left to the reader. The time constant is the amount of time required for the capacitor to accumulate 63.2% of its equilibrium charge, as can be seen by substituting $t = \tau = RC$ in Equation 20.20; $q_0(1 - e^{-1}) = q_0(0.632)$. The charge approaches its equilibrium value rapidly when the time constant is small and slowly when the time constant is large.

Figure 20.36 shows a circuit at a time $t$ after a switch is closed to allow a charged capacitor to begin discharging. There is no battery in this circuit, so the charge $+q$ on the left plate of the capacitor can flow counterclockwise through the resistor and neutralize the charge $-q$ on the right plate. Assuming that the capacitor has a charge $q_0$ at time $t = 0$ s when the switch is closed, it can be shown that

*Capacitor discharging*    $$q = q_0 e^{-t/(RC)} \qquad (20.22)$$

where $q$ is the amount of charge remaining on either plate at time $t$. The graph of this expression in part *b* of the drawing shows that the charge begins at $q_0$ when $t = 0$ s and decreases gradually toward zero. Smaller values of the time constant $RC$ lead to a more rapid discharge. Equation 20.22 indicates that when $t = \tau = RC$, the magnitude of the charge remaining on each plate is $q_0 e^{-1} = q_0(0.368)$. Therefore, the time constant is also the amount of time required for a charged capacitor to *lose* 63.2% of its charge.

The charging/discharging of a capacitor has many applications. Heart pacemakers, for instance, incorporate *RC* circuits to control the timing of voltage pulses that are delivered to a malfunctioning heart to regulate its beating cycle. The pulses are delivered by electrodes attached externally to the chest or located internally near the heart when the pacemaker is implanted surgically (see Figure 20.37). A voltage pulse is delivered when the capacitor discharges to a preset level, following which the capacitor is recharged rapidly and the cycle repeats. The value of the time constant $RC$ controls the pulsing rate, which is about once per second.

The charging/discharging of a capacitor is also used in automobiles that have windshield wipers equipped for intermittent operation during a light drizzle. In this mode of operation, the wipers remain off for a while and then turn on briefly. The timing of the on–off cycle is determined by the time constant of a resistor–capacitor combination.

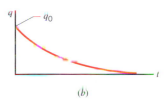
(b)

**Figure 20.36** Discharging a capacitor.

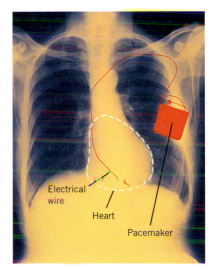

**Figure 20.37** This X-ray photograph shows a heart pacemaker that has been implanted surgically. (© RNHRD NHS Trust/Stone/Getty Images)

**The physics of heart pacemakers.**

**The physics of windshield wipers.**

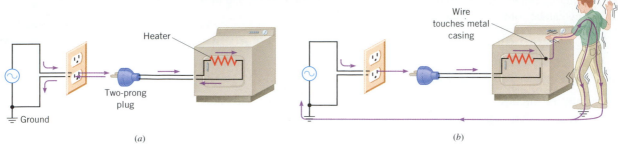

*(a)*                    *(b)*

**Figure 20.38** (*a*) A normally operating clothes dryer that is connected to a wall socket via a two-prong plug. (*b*) An internal wire accidentally touches the metal casing, and a person who touches the casing receives an electrical shock.

## 20.14 Safety and the Physiological Effects of Current

Electric circuits, although very useful, can also be hazardous. To reduce the danger inherent in using circuits, proper *electrical grounding* is necessary. The next two figures help to illustrate what electrical grounding means and how it is achieved.

Figure 20.38*a* shows a clothes dryer connected to a wall socket via a two-prong plug. The dryer is operating normally; that is, the wires inside are insulated from the metal casing of the dryer, so no charge flows through the casing itself. Notice that one terminal of the ac generator is customarily connected to ground ( ⏚ ) by the electric power company. Part *b* of the drawing shows the hazardous result that occurs if a wire comes loose and contacts the metal casing. A person touching it receives a shock, since electric charge flows through the casing, the person's body, and the ground on the way back to the generator.

Figure 20.39 shows the same appliance connected to a wall socket via a three-prong plug that provides safe electrical grounding. The third prong connects the metal casing directly to a copper rod driven into the ground or to a copper water pipe that is in the ground. This arrangement protects against electrical shock in the event that a broken wire touches the metal casing. In this event, charge would flow through the casing, through the third prong, and into the ground, returning eventually to the generator. No charge would flow through the person's body, because the copper rod provides much less electrical resistance than does the body.

Serious and sometimes fatal injuries can result from electrical shock. The severity of the injury depends on the magnitude of the current and the parts of the body through which the moving charges pass. The amount of current that causes a mild tingling sensation is about 0.001 A. Currents on the order of 0.01–0.02 A can lead to muscle spasms, in which a person "can't let go" of the object causing the shock. Currents of approximately 0.2 A are potentially fatal because they can make the heart fibrillate, or beat in an uncontrolled manner. Substantially larger currents stop the heart completely. However, since the heart often begins beating normally again after the current ceases, the larger currents can be less dangerous than the smaller currents that cause fibrillation.

**The physics of** safe electrical grounding.

**The physics of** the physiological effects of current.

**Figure 20.39** When the dryer malfunctions, a person touching it receives no shock, since electric charge flows through the third prong and into the ground via a copper rod, rather than through the person's body.

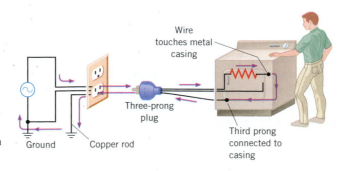

# Concept Summary

This summary presents an abridged version of the chapter, including the important equations and all available learning aids. For convenient reference, the learning aids (including the text's examples) are placed next to or immediately after the relevant equation or discussion. The following learning aids may be found on-line at www.wiley.com/college/cutnell:

**Interactive LearningWare** examples are solved according to a five-step interactive format that is designed to help you develop problem-solving skills.

**Concept Simulations** are animated versions of text figures or animations that illustrate important concepts. You can control parameters that affect the display, and we encourage you to experiment.

**Interactive Solutions** offer specific models for certain types of problems in the chapter homework. The calculations are carried out interactively.

**Self-Assessment Tests** include both qualitative and quantitative questions. Extensive feedback is provided for both incorrect and correct answers, to help you evaluate your understanding of the material.

| Topic | Discussion | Learning Aids |
|-------|-----------|---------------|

### 20.1 Electromotive Force and Current

**Electromotive force**

There must be at least one source or generator of electrical energy in an electric circuit. The electromotive force (emf) of a generator, such as a battery, is the maximum potential difference (in volts) that exists between the terminals of the generator.

The rate of flow of charge is called the electric current. If the rate is constant, the current $I$ is given by

**Current**

$$I = \frac{\Delta q}{\Delta t}$$

(20.1)  **Example 1**

where $\Delta q$ is the magnitude of the charge crossing a surface in a time $\Delta t$, the surface being perpendicular to the motion of the charge. The SI unit for current is the coulomb per second (C/s), which is referred to as an ampere (A).

**The ampere**

When the charges flow only in one direction around a circuit, the current is called direct current (dc). When the direction of charge flow changes from moment to moment, the current is known as alternating current (ac).

**Direct current**
**Alternating current**

**Conventional current**

Conventional current is the hypothetical flow of positive charges that would have the same effect in a circuit as the movement of negative charges that actually does occur.

### 20.2 Ohm's Law

**Resistance**

The definition of electrical resistance $R$ is $R = V/I$, where $V$ (in volts) is the voltage applied across a piece of material and $I$ (in amperes) is the current through the material. Resistance is measured in volts per ampere, a unit called an ohm ($\Omega$).

**The ohm**

If the ratio of the voltage to the current is constant for all values of voltage and current, the resistance is constant. In this event, the definition of resistance becomes Ohm's law as follows:

**Example 2**

**Ohm's law**

$$\frac{V}{I} = R = \text{constant} \quad \text{or} \quad V = IR$$

(20.2)

**Concept Simulation 20.1**

### 20.3 Resistance and Resistivity

The resistance of a piece of material of length $L$ and cross-sectional area $A$ is

**Resistance in terms of length and area**

$$R = \rho \frac{L}{A}$$

(20.3)  **Example 3**

**Resistivity**

where $\rho$ is the resistivity of the material.

The resistivity of a material depends on the temperature. For many materials and limited temperature ranges, the temperature dependence is given by

**Temperature dependence of resistivity**

$$\rho = \rho_0[1 + \alpha(T - T_0)]$$

(20.4)  **Example 4**

**Temperature coefficient of resistivity**

where $\rho$ and $\rho_0$ are the resistivities at temperatures $T$ and $T_0$, respectively, and $\alpha$ is the temperature coefficient of resistivity.

| Topic | Discussion | Learning Aids |
|-------|-----------|---------------|
| | The temperature dependence of the resistance $R$ is given by | |
| **Temperature dependence of resistance** | $$R = R_0[1 + \alpha(T - T_0)] \tag{20.5}$$ where $R$ and $R_0$ are the resistances at temperatures $T$ and $T_0$, respectively. | |

## 20.4 Electric Power

In a circuit in which a current $I$ results from a voltage $V$, the electric power delivered to the circuit is

**Electric power**

$$P = IV \tag{20.6a}$$ **Example 5**

For a resistor, Ohm's law applies, and it follows that the power dissipated in the resistance is also given by either of the following two equations:

$$P = I^2 R \tag{20.6b}$$

**Power dissipated in a resistor**

$$P = \frac{V^2}{R} \tag{20.6c}$$

## 20.5 Alternating Current

The alternating voltage between the terminals of an ac generator can be represented by

**Alternating voltage**

$$V = V_0 \sin 2\pi f t \tag{20.7}$$

where $V_0$ is the peak value of the voltage, $t$ is the time, and $f$ is the frequency (in Hertz) at which the voltage oscillates.

Correspondingly, in a circuit containing only resistance, the ac current is

**Alternating current**

$$I = I_0 \sin 2\pi f t \tag{20.8}$$

where $I_0$ is the peak value of the current and is related to the peak voltage via $I_0 = V_0/R$.

The root mean square (rms) current and voltage are related to the peak values according to the following equations:

**Root mean square current**

$$I_{\text{rms}} = \frac{I_0}{\sqrt{2}} \tag{20.12}$$

**Examples 6, 7**

**Root mean square voltage**

$$V_{\text{rms}} = \frac{V_0}{\sqrt{2}} \tag{20.13}$$

The power in an ac circuit is the product of the current and the voltage and oscillates in time. The average power is

**Average ac power**

$$\overline{P} = I_{\text{rms}} V_{\text{rms}} \tag{20.15a}$$

For a resistor, Ohm's law applies, so that $V_{\text{rms}} = I_{\text{rms}} R$ and the average power dissipated in the resistor is also given by the following two equations:

$$\overline{P} = I_{\text{rms}}^2 R \tag{20.15b}$$

**Average ac power dissipated in a resistor**

$$\overline{P} = \frac{V_{\text{rms}}^2}{R} \tag{20.15c}$$

## 20.6 Series Wiring

When devices are connected in series, there is the same current through each device. The equivalent resistance $R_S$ of a series combination of resistances ($R_1$, $R_2$, $R_3$, etc.) is

**Interactive Solution 20.45**

**Equivalent series resistance**

$$R_S = R_1 + R_2 + R_3 + \cdots \tag{20.16}$$

**Example 8**

The equivalent resistance dissipates the same total power as the series combination.

## 20.7 Parallel Wiring

When devices are connected in parallel, the same voltage is applied across each device. In general, devices wired in parallel carry different currents. The reciprocal of the equivalent resistance $R_P$ of a parallel combination of resistances ($R_1$, $R_2$, $R_3$, etc.) is

**Interactive Solution 20.49**

**Equivalent parallel resistance**

$$\frac{1}{R_P} = \frac{1}{R_1} + \frac{1}{R_2} + \frac{1}{R_3} + \cdots \tag{20.17}$$

**Examples 9, 10**

| Topic | Discussion | Learning Aids |
|---|---|---|

The equivalent resistance dissipates the same total power as the parallel combination.

 Use Self-Assessment Test 20.1 to evaluate your understanding of Sections 20.1–20.7.

### 20.8 Circuits Wired Partially in Series and Partially in Parallel

Sometimes, one section of a circuit is wired in series, while another is wired in parallel. In such cases the circuit can be analyzed in parts, according to the respective series and parallel equivalent resistances of the various sections.

**Example 11**

**Concept Simulations 20.2, 20.5**

### 20.9 Internal Resistance

**Internal resistance**
**Terminal voltage**

The internal resistance of a battery or generator is the resistance within the battery or generator. The terminal voltage is the voltage between the terminals of a battery or generator and is equal to the emf only when there is no current through the device. When there is a current $I$, the internal resistance $r$ causes the terminal voltage to be less than the emf by an amount $Ir$.

**Example 12**

### 20.10 Kirchhoff's Rules

**The junction rule**

Kirchhoff's junction rule states that the sum of the magnitudes of the currents directed into a junction equals the sum of the magnitudes of the currents directed out of the junction.

**Example 13**

**Interactive LearningWare 20.1**

**The loop rule**

Kirchhoff's loop rule states that, around any closed-circuit loop, the sum of the potential drops equals the sum of the potential rises.

**Example 14**

**Concept Simulation 20.3**

### 20.11 The Measurement of Current and Voltage

**Galvanometer**
**Ammeter**

**Voltmeter**

A galvanometer is a device that responds to electric current and is used in nondigital ammeters and voltmeters. An ammeter is an instrument that measures current and must be inserted into a circuit in such a way that the current passes directly through the ammeter. A voltmeter is an instrument for measuring the voltage between two points in a circuit. A voltmeter must be connected between the two points and is not inserted into a circuit as an ammeter is.

### 20.12 Capacitors in Series and Parallel

**Equivalent parallel capacitance**

The equivalent capacitance $C_P$ for a parallel combination of capacitances ($C_1$, $C_2$, $C_3$, etc.) is

$$C_P = C_1 + C_2 + C_3 + \cdots \qquad (20.18)$$

In general, each capacitor in a parallel combination carries a different amount of charge. The equivalent capacitor carries the same total charge and stores the same total energy as the parallel combination.

**Equivalent series capacitance**

The reciprocal of the equivalent capacitance $C_S$ for a series combination ($C_1$, $C_2$, $C_3$, etc.) of capacitances is

$$\frac{1}{C_S} = \frac{1}{C_1} + \frac{1}{C_2} + \frac{1}{C_3} + \cdots \qquad (20.19)$$

The equivalent capacitor carries the same amount of charge as *any one* of the capacitors in the combination and stores the same total energy as the series combination.

### 20.13 RC Circuits

The charging or discharging of a capacitor in a dc series circuit (resistance $R$, capacitance $C$) does not occur instantaneously. The charge on a capacitor builds up gradually, as described by the following equation:

**Charging a capacitor**

$$q = q_0[1 - e^{-t/(RC)}] \qquad (20.20)$$

**Interactive Solution 20.99**

| Topic | Discussion | Learning Aids |
|-------|-----------|---------------|
| | where $q$ is the charge on the capacitor at time $t$ and $q_0$ is the equilibrium value of the charge. The time constant $\tau$ of the circuit is | |
| Time constant | $$\tau = RC$$ | (20.21) **Concept Simulation 20.4** |
| | The discharging of a capacitor through a resistor is described as follows: | |
| Discharging a capacitor | $$q = q_0 e^{-t/(RC)}$$ | (20.22) |
| | where $q_0$ is the charge on the capacitor at time $t = 0$ s. | |

 **Use Self-Assessment Test 20.2 to evaluate your understanding of Sections 20.8–20.13.**

# Problems

*Note: For problems that involve ac conditions, the current and voltage are rms values and the power is an average value, unless indicated otherwise.*

**ssm**    Solution is in the Student Solutions Manual.        **www**    Solution is available on the World Wide Web at www.wiley.com/college/cutnell
☤    This icon represents a biomedical application.

### Section 20.1 Electromotive Force and Current,
### Section 20.2 Ohm's Law

**1. ssm** A fax machine uses 0.110 A of current in its normal mode of operation, but only 0.067 A in the standby mode. The machine uses a potential difference of 120 V. (a) In one minute, how much more charge passes through the machine in the normal mode versus the standby mode, and (b) how much more energy is used?

**2.** A defibrillator is used during a heart attack to restore the heart to its normal beating pattern (see Section 19.5). A defibrillator passes 18 A of current through the torso of a person in 2.0 ms. (a) How much charge moves during this time? (b) How many electrons pass through the wires connected to the patient?

**3.** The filament of a light bulb has a resistance of 580 Ω. A voltage of 120 V is connected across the filament. How much current is in the filament?

**4.** A battery charger is connected to a dead battery and delivers a current of 6.0 A for 5.0 hours, keeping the voltage across the battery terminals at 12 V in the process. How much energy is delivered to the battery?

**5. ssm** In the Arctic, electric socks are useful. A pair of socks uses a 9.0-V battery pack for each sock. A current of 0.11 A is drawn from each battery pack by wire woven into the socks. Find the resistance of the wire in one sock.

*  **6.** A car battery has a rating of 220 ampere · hours (A · h). This rating is one indication of the *total charge* that the battery can provide to a circuit before failing. (a) What is the total charge (in coulombs) that this battery can provide? (b) Determine the maximum current that the battery can provide for 38 minutes.

*  **7.** The resistance of a bagel toaster is 14 Ω. To prepare a bagel, the toaster is operated for one minute from a 120-V outlet. How much energy is delivered to the toaster?

*  **8.** A resistor is connected across the terminals of a 9.0-V battery, which delivers $1.1 \times 10^5$ J of energy to the resistor in six hours. What is the resistance of the resistor?

** **9. ssm** A beam of protons is moving toward a target in a particle accelerator. This beam constitutes a current whose value is 0.50 µA. (a) How many protons strike the target in 15 seconds? (b) Each proton has

a kinetic energy of $4.9 \times 10^{-12}$ J. Suppose the target is a 15-gram block of aluminum, and all the kinetic energy of the protons goes into heating it up. What is the change in temperature of the block at the end of 15 s?

### Section 20.3 Resistance and Resistivity

**10.** High-voltage power lines are a familiar sight throughout the country. The aluminum wire used for some of these lines has a cross-sectional area of $4.9 \times 10^{-4}$ m². What is the resistance of ten kilometers of this wire?

**11. ssm www** Two wires have the same length and the same resistance. One is made from aluminum and the other from copper. Obtain the ratio of the cross-sectional area of the aluminum wire to that of the copper wire.

**12.** Two wires are identical, except that one is aluminum and one is copper. The aluminum wire has a resistance of 0.20 Ω. What is the resistance of the copper wire?

**13.** A cylindrical copper cable carries a current of 1200 A. There is a potential difference of $1.6 \times 10^{-2}$ V between two points on the cable that are 0.24 m apart. What is the radius of the cable?

**14.** A coil of wire has a resistance of 38.0 Ω at 25 °C and 43.7 Ω at 55 °C. What is the temperature coefficient of resistivity?

**15.** In Section 12.3 it was mentioned that temperatures are often measured with electrical resistance thermometers made of platinum wire. Suppose that the resistance of a platinum resistance thermometer is 125 Ω when its temperature is 20.0 °C. The wire is then immersed in boiling chlorine, and the resistance drops to 99.6 Ω. The temperature coefficient of resistivity of platinum is $\alpha = 3.72 \times 10^{-3}$ (C°)⁻¹. What is the temperature of the boiling chlorine?

*  **16.** The temperature coefficient of resistivity for gold is 0.0034 (C°)⁻¹, and for tungsten it is 0.0045 (C°)⁻¹. The resistance of a gold wire increases by 7.0% due to an increase in temperature. For the same increase in temperature, what is the percentage increase in the resistance of a tungsten wire?

*  **17. ssm www** A wire has a resistance of 21.0 Ω. It is melted down, and from the same volume of metal a new wire is made that is three times longer than the original wire. What is the resistance of the new wire?

* **18.** An iron wire has a resistance of 5.90 Ω at 20.0 °C, and a gold wire has a resistance of 6.70 Ω at the same temperature. The temperature coefficient of resistivity for iron is 0.0050 (C°)$^{-1}$, and for gold it is 0.0034 (C°)$^{-1}$. At what temperature do the wires have the same resistance?

* **19. ssm www** Two wires have the same cross-sectional area and are joined end to end to form a single wire. One is tungsten, which has a temperature coefficient of resistivity of $\alpha = 0.0045$ (C°)$^{-1}$. The other is carbon, for which $\alpha = -0.0005$ (C°)$^{-1}$. The total resistance of the composite wire is the sum of the resistances of the pieces. The total resistance of the composite *does not change with temperature*. What is the ratio of the lengths of the tungsten and carbon sections? Ignore any changes in length due to thermal expansion.

** **20.** An aluminum wire is hung between two towers and has a length of 175 m. A current of 125 A exists in the wire, and the potential difference between the ends of the wire is 0.300 V. The density of aluminum is 2700 kg/m³. Find the mass of the wire.

## Section 20.4 Electric Power

**21.** The heating element in an iron has a resistance of 24 Ω. The iron is plugged into a 120-V outlet. What is the power dissipated by the iron?

**22.** A cigarette lighter in a car is a resistor that, when activated, is connected across the 12-V battery. Suppose a lighter dissipates 33 W of power. Find (a) the resistance of the lighter and (b) the current that the battery delivers to the lighter.

**23. ssm** In doing a load of clothes, a clothes dryer uses 16 A of current at 240 V for 45 min. A personal computer, in contrast, uses 2.7 A of current at 120 V. With the energy used by the clothes dryer, how long (in hours) could you use this computer to "surf" the Internet?

**24.** There are approximately 110 million TVs in the United States. Each uses, on average, 75 W of power and is turned on for 6.0 hours a day. If electrical energy costs $0.10 per KWh, how much money is spent every day in keeping the TVs turned on?

**25.** A blow-dryer and a vacuum cleaner each operate with a voltage of 120 V. The current rating of the blow-dryer is 11 A, and that of the vacuum cleaner is 4.0 A. Determine the power consumed by (a) the blow-dryer and (b) the vacuum cleaner. (c) Determine the ratio of the energy used by the blow-dryer in 15 minutes to the energy used by the vacuum cleaner in one-half hour.

* **26.** An electric heater is used to boil small amounts of water and consists of a 15-Ω coil that is immersed directly in the water. It operates from a 120-V socket. How much time is required for this heater to raise the temperature of 0.50 kg of water from 13 °C to the normal boiling point?

* **27. ssm** Tungsten has a temperature coefficient of resistivity of 0.0045 (C°)$^{-1}$. A tungsten wire is connected to a source of constant voltage via a switch. At the instant the switch is closed, the temperature of the wire is 28 °C, and the initial power dissipated in the wire is $P_0$. At what wire temperature has the power dissipated in the wire decreased to $\frac{1}{2}P_0$?

* **28.** A piece of Nichrome wire has a radius of $6.5 \times 10^{-4}$ m. It is used in a laboratory to make a heater that dissipates $4.00 \times 10^2$ W of power when connected to a voltage source of 120 V. Ignoring the effect of temperature on resistance, estimate the necessary length of wire.

** **29.** An iron wire has a resistance of 12 Ω at 20.0 °C and a mass of $1.3 \times 10^{-3}$ kg. A current of 0.10 A is sent through the wire for

one minute and causes the wire to become hot. Assuming that all the electrical energy is dissipated in the wire and remains there, find the final temperature of the wire. *[Hint: Use the average resistance of the wire during the heating process, and see Table 12.2 for the specific heat capacity of iron. Note that $\alpha = 0.0050$ (C°)$^{-1}$.]*

## Section 20.5 Alternating Current

**30.** According to Equation 20.7, an ac voltage $V$ is given as a function of time $t$ by $V = V_0 \sin 2\pi f t$, where $V_0$ is the peak voltage and $f$ is the frequency (in hertz). For a frequency of 60.0 Hz, what is the smallest value of the time at which the voltage equals one-half of the peak value?

**31.** The current in a circuit is ac and has a peak value of 2.50 A. Determine the rms current.

**32.** The average power dissipated in a stereo speaker is 55 W. Assuming that the speaker can be treated as a 4.0-Ω resistance, find the peak value of the ac voltage applied to the speaker.

**33. ssm** The heating element in an iron has a resistance of 16 Ω and is connected to a 120-V wall socket. (a) What is the average power consumed by the iron, and (b) the peak power?

**34.** Review Conceptual Example 7 as an aid in solving this problem. A portable electric heater uses 18 A of current. The manufacturer recommends that an extension cord attached to the heater dissipate no more than 2.0 W of power per meter of length. What is the smallest radius of copper wire that can be used in the extension cord? *(Note: An extension cord contains two wires.)*

**35.** An electric furnace runs nine hours a day to heat a house during January (31 days). The heating element has a resistance of 5.3 Ω and carries a current of 25 A. The cost of electricity is $0.10/kWh. Find the cost of running the furnace for the month of January.

* **36.** A light bulb is connected to a 120.0-V wall socket. The current in the bulb depends on the time $t$ according to the relation $I = (0.707 \text{ A})\sin[(314 \text{ Hz})t]$. (a) What is the frequency of the alternating current? (b) Determine the resistance of the bulb's filament. (c) What is the average power consumed by the light bulb?

* **37. ssm** The *recovery time* of a hot water heater is the time required to heat all the water in the unit to the desired temperature. Suppose that a 52-gal (1.00 gal = $3.79 \times 10^{-3}$ m³) unit starts with cold water at 11 °C and delivers hot water at 53 °C. The unit is electric and utilizes a resistance heater (120 V ac, 3.0 Ω) to heat the water. Assuming that no heat is lost to the environment, determine the recovery time (in hours) of the unit.

** **38.** To save on heating costs, the owner of a greenhouse keeps 660 kg of water around in barrels. During a winter day, the water is heated by the sun to 10.0 °C. During the night the water freezes into ice at 0.0 °C in nine hours. What is the minimum ampere rating of an electric heating system (240 V) that would provide the same heating effect as the water does?

## Section 20.6 Series Wiring

**39. ssm** The current in a series circuit is 15.0 A. When an additional 8.00-Ω resistor is inserted in series, the current drops to 12.0 A. What is the resistance in the original circuit?

**40.** The current in a 47-Ω resistor is 0.12 A. This resistor is in series with a 28-Ω resistor, and the series combination is connected across a battery. What is the battery voltage?

**41.** A 36.0-Ω resistor and an 18.0-Ω resistor are connected in series across a 15.0-V battery. What is the voltage across (a) the 36.0-Ω resistor and (b) the 18.0-Ω resistor?

**42.** A 60.0-W lamp is placed in series with a resistor and a 120.0-V source. If the voltage across the lamp is 25 V, what is the resistance $R$ of the resistor?

**43. ssm** Three resistors, 25, 45, and 75 Ω, are connected in series, and a 0.51-A current passes through them. What is (a) the equivalent resistance and (b) the potential difference across the three resistors?

**\* 44.** A 47-Ω resistor can dissipate up to 0.25 W of power without burning up. What is the smallest number of such resistors that can be connected in series across a 9.0-V battery without any one of them burning up?

**\* 45. Interactive Solution 20.45** at **www.wiley.com/college/cutnell** provides one approach to problems like this one. Three resistors are connected in series across a battery. The value of each resistance and its maximum power rating are as follows: 2.0 Ω and 4.0 W, 12.0 Ω and 10.0 W, and 3.0 Ω and 5.0 W. (a) What is the greatest voltage that the battery can have without one of the resistors burning up? (b) How much power does the battery deliver to the circuit in (a)?

**\* 46.** One heater uses 340 W of power when connected by itself to a battery. Another heater uses 240 W of power when connected by itself to the same battery. How much total power do the heaters use when they are both connected in series across the battery?

**\*\* 47.** Two resistances, $R_1$ and $R_2$, are connected in series across a 12-V battery. The current increases by 0.20 A when $R_2$ is removed, leaving $R_1$ connected across the battery. However, the current increases by just 0.10 A when $R_1$ is removed, leaving $R_2$ connected across the battery. Find (a) $R_1$ and (b) $R_2$.

**Section 20.7 Parallel Wiring**

**48.** What resistance must be placed in parallel with a 155-Ω resistor to make the equivalent resistance 115 Ω?

**49.** Help with problems of this kind is available in **Interactive Solution 20.49** at **www.wiley.com/college/cutnell**. A coffee cup heater and a lamp are connected in parallel to the same 120-V outlet. Together, they use a total of 111 W of power. The resistance of the heater is $4.0 \times 10^2$ Ω. Find the resistance of the lamp.

**50.** For the three-way bulb (50 W, 100 W, 150 W) discussed in Conceptual Example 10, find the resistance of each of the two filaments. Assume that the wattage ratings are not limited by significant figures and ignore any heating effects on the resistances.

**51. ssm** A 16-Ω loudspeaker and an 8.0-Ω loudspeaker are connected in parallel across the terminals of an amplifier. Assuming the speakers behave as resistors, determine the equivalent resistance of the two speakers.

**52.** A wire whose resistance is $R$ is cut into three equally long pieces, which are then connected in parallel. In terms of $R$, what is the resistance of the parallel combination?

**53. ssm** Two resistors, 42.0 and 64.0 Ω, are connected in parallel. The current through the 64.0-Ω resistor is 3.00 A. (a) Determine the current in the other resistor. (b) What is the total power consumed by the two resistors?

**\* 54.** Two resistors have resistances $R_1$ and $R_2$. When the resistors are connected in series to a 12.0-V battery, the current from the battery is 2.00 A. When the resistors are connected in parallel to the battery, the total current from the battery is 9.00 A. Determine $R_1$ and $R_2$.

**\* 55. ssm** The total current delivered to a number of devices connected in parallel is the sum of the individual currents in each device. Circuit breakers are resettable automatic switches that protect against a dangerously large total current by "opening" to stop the

current at a specified safe value. A 1650-W toaster, a 1090-W iron, and a 1250-W microwave oven are turned on in a kitchen. As the drawing shows, they are all connected through a 20-A circuit breaker to an ac voltage of 120 V. (a) Find the equivalent resistance of the three devices. (b) Obtain the total current delivered by the source and determine whether the breaker will "open" to prevent an accident.

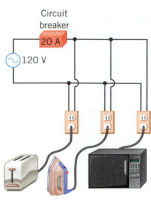

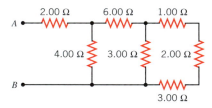

**\* 56.** A resistor (resistance $= R$) is connected first in parallel and then in series with a 2.00-Ω resistor. A battery delivers five times as much current to the parallel combination as it does to the series combination. Determine the two possible values for $R$.

**\*\* 57.** The rear window defogger of a car consists of thirteen thin wires (resistivity $= 88.0 \times 10^{-8}$ Ω·m) embedded in the glass. The wires are connected in parallel to the 12.0-V battery, and each has a length of 1.30 m. The defogger can melt $2.10 \times 10^{-2}$ kg of ice at 0 °C into water at 0 °C in two minutes. Assume that all the power dissipated in the wires is used immediately to melt the ice. Find the cross-sectional area of each wire.

**Section 20.8 Circuits Wired Partially in Series and Partially in Parallel**

**58.** Find the equivalent resistance between points $A$ and $B$ in the drawing.

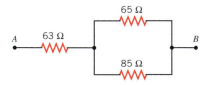

**59. ssm** For the combination of resistors shown in the drawing, determine the equivalent resistance between points $A$ and $B$.

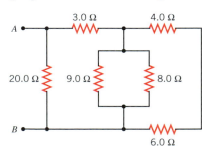

**60.** A 14-Ω coffee maker and a 16-Ω frying pan are connected in series across a 120-V source of voltage. A 23-Ω bread maker is also connected across the 120-V source and is in parallel with the series combination. Find the total current supplied by the source of voltage.

**61. ssm** Determine the equivalent resistance between the points $A$ and $B$ for the group of resistors in the drawing.

**62.** A 60.0-Ω resistor is connected in parallel with a 120.0-Ω resistor. This parallel group is connected in series with a 20.0-Ω resistor. The total combination is connected across a 15.0-V battery. Find (a) the current and (b) the power dissipated in the 120.0-Ω resistor.

**63.** The circuit in the drawing contains five identical resistors. The 45-V battery delivers 58 W of power to the circuit. What is the resistance R of each resistor?

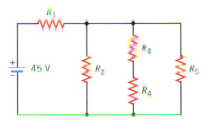

**\* 64. Concept Simulation 20.5** at **www.wiley.com/college/cutnell** provides some background pertinent to this problem. Determine the power dissipated in each of the resistors in the drawing.

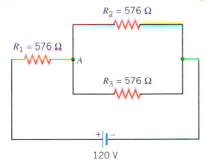

**\* 65. ssm** Eight different values of resistance can be obtained by connecting together three resistors (1.00, 2.00, and 3.00 Ω) in all possible ways. What are they?

**\*\* 66.** Three identical resistors are connected in parallel. The equivalent resistance increases by 700 Ω when one resistor is removed and connected in series with the remaining two, which are still in parallel. Find the resistance of each resistor.

**Section 20.9 Internal Resistance**

**67.** A new "D" battery has an emf of 1.5 V. When a wire of negligible resistance is connected between the terminals of the battery, a current of 28 A is produced. Find the internal resistance of the battery.

**68.** A battery has an internal resistance of 0.012 Ω and an emf of 9.00 V. What is the maximum current that can be drawn from the battery without the terminal voltage dropping below 8.90 V?

**69. ssm** A battery has an internal resistance of 0.50 Ω. A number of identical light bulbs, each with a resistance of 15 Ω, are connected in parallel across the battery terminals. The terminal voltage of the battery is observed to be one-half the emf of the battery. How many bulbs are connected?

**70.** A battery has an emf of 12.0 V and an internal resistance of 0.15 Ω. What is the terminal voltage when the battery is connected to a 1.50-Ω resistor?

**\* 71.** A battery delivering a current of 55.0 A to a circuit has a terminal voltage of 23.4 V. The electric power being dissipated by the internal resistance of the battery is 34.0 W. Find the emf of the battery.

**72.** A resistor has a resistance R, and a battery has an internal resistance r. When the resistor is connected across the battery, ten percent less power is dissipated in R than would be dissipated if the battery had no internal resistance. Find the ratio r/R.

**Section 20.10 Kirchhoff's Rules**

**73. ssm** Consider the circuit in the drawing. Determine (a) the magnitude of the current in the circuit and (b) the magnitude of the voltage between the points labeled A and B. (c) State which point, A or B, is at the higher potential.

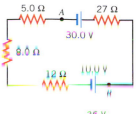

**74.** A current of 2.0 A exists in the partial circuit shown in the drawing. What is the magnitude of the potential difference between the points (a) A and B, and (b) A and C?

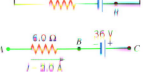

**75.** Two batteries, each with an internal resistance of 0.015 Ω, are connected as in the drawing. In effect, the 9.0-V battery is being used to charge the 8.0-V battery. What is the current in the circuit?

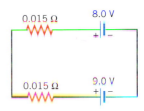

**76. Interactive LearningWare 20.1** at **www.wiley.com/college/cutnell** provides background for this problem. Find the magnitude and direction of the current in the 2.0-Ω resistor in the drawing.

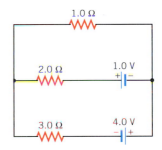

**\* 77. ssm** Determine the voltage across the 5.0-Ω resistor in the drawing. Which end of the resistor is at the higher potential?

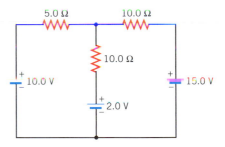

**\* 78. Concept Simulation 20.3** at **www.wiley.com/college/cutnell** allows you to verify your answer for this problem. Find the current in the 4.00-Ω resistor in the drawing. Specify the direction of the current.

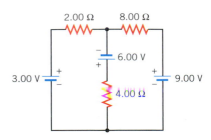

**\*\*79.** The circuit in the drawing is known as a Wheatstone bridge circuit. Find the voltage between points *B* and *D*, and state which point is at the higher potential.

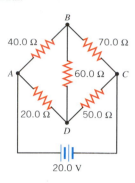

### Section 20.11 The Measurement of Current and Voltage

**80.** A galvanometer has a full-scale current of 0.100 mA and a coil resistance of 50.0 $\Omega$. This instrument is used with a shunt resistor to form an ammeter that will register full scale for a current of 60.0 mA. Determine the resistance of the shunt resistor.

**81. ssm** A voltmeter utilizes a galvanometer that has a 180-$\Omega$ coil resistance and a full-scale current of 8.30 mA. The voltmeter measures voltages up to 30.0 V. Determine the resistance that is connected in series with the galvanometer.

**82.** Voltmeter A has an equivalent resistance of $2.40 \times 10^5$ $\Omega$ and a full-scale voltage of 50.0 V. Voltmeter B, using the same galvanometer as voltmeter A, has an equivalent resistance of $1.44 \times 10^5$ $\Omega$. What is its full-scale voltage?

**83.** A galvanometer with a coil resistance of 12.0 $\Omega$ and a full-scale current of 0.150 mA is used with a shunt resistor to make an ammeter. The ammeter registers a maximum current of 4.00 mA. Find the equivalent resistance of the ammeter.

**\*84.** Two scales on a voltmeter measure voltages up to 20.0 and 30.0 V, respectively. The resistance connected in series with the galvanometer is 1680 $\Omega$ for the 20.0-V scale and 2930 $\Omega$ for the 30.0-V scale. Determine the coil resistance and the full-scale current of the galvanometer that is used in the voltmeter.

**\*\*85. ssm** In measuring a voltage, a voltmeter uses some current from the circuit. Consequently, the voltage measured is only an approximation to the voltage present when the voltmeter is not connected. Consider a circuit consisting of two 1550-$\Omega$ resistors connected in series across a 60.0-V battery. (a) Find the voltage across one of the resistors. (b) A voltmeter has a full-scale voltage of 60.0 V and uses a galvanometer with a full-scale deflection of 5.00 mA. Determine the voltage that this voltmeter registers when it is connected across the resistor used in part (a).

### Section 20.12 Capacitors in Series and Parallel

**86.** Two capacitors are connected in parallel across the terminals of a battery. One has a capacitance of 2.0 $\mu$F and the other a capacitance of 4.0 $\mu$F. These two capacitors together store $5.4 \times 10^{-5}$ C of charge. What is the voltage of the battery?

**87.** Determine the equivalent capacitance between *A* and *B* for the group of capacitors in the drawing.

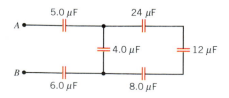

**88.** Three capacitors (4.0, 6.0, and 12.0 $\mu$F) are connected in series across a 50.0-V battery. Find the voltage across the 4.0-$\mu$F capacitor.

**89. ssm** Three capacitors (3.0, 7.0, and 9.0 $\mu$F) are connected in series. What is their equivalent capacitance?

**90.** Three capacitors have identical geometries. One is filled with a material whose dielectric constant is 2.50. Another is filled with a material whose dielectric constant is 4.00. The third capacitor is filled with a material whose dielectric constant $\kappa$ is such that this single capacitor has the same capacitance as the series combination of the other two. Determine $\kappa$.

**91.** Suppose two capacitors ($C_1$ and $C_2$) are connected in series. Show that the sum of the energies stored in these capacitors is equal to the energy stored in the equivalent capacitor. [*Hint: The energy stored in a capacitor can be expressed as* $q^2/(2C)$.]

**\*92.** A 3.00-$\mu$F and a 5.00-$\mu$F capacitor are connected in series across a 30.0-V battery. A 7.00-$\mu$F capacitor is then connected in parallel across the 3.00-$\mu$F capacitor. Determine the voltage across the 7.00-$\mu$F capacitor.

**\*93. ssm** A sheet of gold foil (negligible thickness) is placed between the plates of a capacitor and has the same area as each of the plates. The foil is parallel to the plates, at a position one-third of the way from one to the other. Before the foil is inserted, the capacitance is $C_0$. What is the capacitance after the foil is in place? Express your answer in terms of $C_0$.

**\*\*94.** The drawing shows two fully charged capacitors ($C_1 = 2.00$ $\mu$F, $q_1 = 6.00$ $\mu$C; $C_2 = 8.00$ $\mu$F, $q_2 = 12.0$ $\mu$C). The switch is closed, and charge flows until equilibrium is reestablished (i.e., until both capacitors have the same voltage across their plates). Find the resulting voltage across either capacitor.

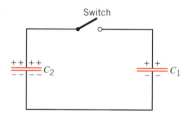

### Section 20.13 *RC* Circuits

**95.** The circuit in the drawing contains two resistors and two capacitors that are connected to a battery via a switch. When the switch is closed, the capacitors begin to charge up. What is the time constant for the charging process?

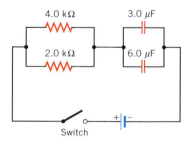

**96.** An electronic flash attachment for a camera produces a flash by using the energy stored in a 750-$\mu$F capacitor. Between flashes, the capacitor recharges through a resistor whose resistance is chosen so the capacitor recharges with a time constant of 3.0 s. Determine the value of the resistance.

**97.** ssm In a heart pacemaker, a pulse is delivered to the heart 81 times per minute. The capacitor that controls this pulsing rate discharges through a resistance of $1.8 \times 10^6 \, \Omega$. One pulse is delivered every time the fully charged capacitor loses 63.2% of its original charge. What is the capacitance of the capacitor?

**98. Concept Simulation 20.4** at www.wiley.com/college/cutnell provides background for this problem and gives you the opportunity to verify your answer graphically. How many time constants must elapse before a capacitor in a series $RC$ circuit is charged to 80.0% of its equilibrium charge?

* **99.** For one approach to problems like this one, consult **Interactive Solution 20.99** at **www.wiley.com/college/cutnell.** Four identical capacitors are connected with a resistor in two different ways. When they are connected as in part $a$ of the drawing, the time constant to charge up this circuit is 0.72 s. What is the time constant when they are connected with the same resistor as in part $b$?

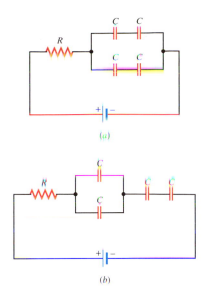

(a)

(b)

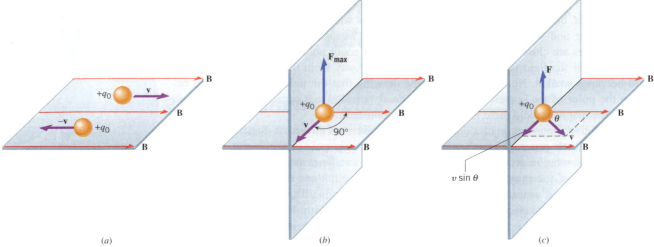

(a)                                      (b)                                      (c)

**Figure 21.6** (a) No magnetic force acts on a charge moving with a velocity **v** that is parallel or antiparallel to a magnetic field **B**. (b) The charge experiences a maximum force **F**$_{max}$ when the charge moves perpendicular to the field. (c) If the charge travels at an angle $\theta$ with respect to **B**, only the velocity component perpendicular to **B** gives rise to a magnetic force **F**, which is smaller than **F**$_{max}$. This component is $v \sin \theta$.

part *a* of the drawing, the charge experiences *no magnetic force.* If, on the other hand, the charge moves *perpendicular* to the field, as in part *b*, the charge experiences the *maximum possible force* **F**$_{max}$. In general, if a charge moves at an angle $\theta$* with respect to the field (see part *c* of the drawing), only the velocity component $v \sin \theta$, which is perpendicular to the field, gives rise to a magnetic force. This force **F** is smaller than the maximum possible force. The component of the velocity that is parallel to the magnetic field yields no force.

Figure 21.6 shows that the direction of the magnetic force **F** is perpendicular to both the velocity **v** and the magnetic field **B**; in other words, **F** is perpendicular to the plane defined by **v** and **B**. As an aid in remembering the direction of the force, it is convenient to use *Right-Hand Rule No. 1 (RHR-1)*, as Figure 21.7 illustrates:

> **Right-Hand Rule No. 1.** Extend the right hand so the fingers point along the direction of the magnetic field **B** and the thumb points along the velocity **v** of the charge. The palm of the hand then faces in the direction of the magnetic force **F** that acts on a positive charge.

It is as if the open palm of the right hand pushes on the positive charge in the direction of the magnetic force. *If the moving charge is negative instead of positive, the direction of the magnetic force is opposite to that predicted by RHR-1.* Thus, there is an easy method for finding the force on a moving negative charge. First, assume that the charge is positive and use RHR-1 to find the direction of the force. Then, reverse this direction to find the direction of the force acting on the negative charge.

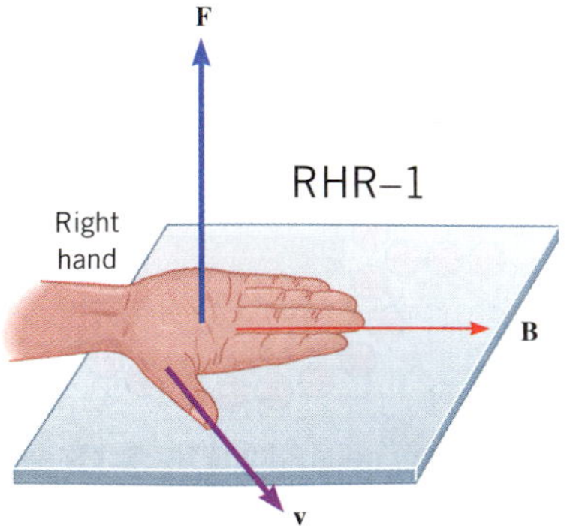

We will now use what we know about the magnetic force to define the magnetic field, in a procedure that is analogous to that used in Section 18.6 to define the electric field. A test charge and the electrostatic force acting on it are brought together to define the electric field. A similar procedure is used in the magnetic case. Recall that the electric field at any point in space is the force per unit charge that acts on a test charge $q_0$ placed at that point. In other words, to determine the electric field **E**, we divide the electrostatic force **F** by the charge $q_0$: **E** = **F**/$q_0$. In the magnetic case, however, the test charge is moving, and the force depends not only on the charge $q_0$, but also on the velocity component $v \sin \theta$ that is perpendicular to the magnetic field. Therefore, to determine the magnitude of the magnetic field, we divide the magnitude of the magnetic force not only by $q_0$, but also by $v \sin \theta$, according to the following definition.

**RHR–1**

Right hand

**Figure 21.7** Right-Hand Rule No. 1 is illustrated. When the right hand is oriented so the fingers point along the magnetic field **B** and the thumb points along the velocity **v** of a positively charged particle, the palm faces in the direction of the magnetic force **F** applied to the particle.

* The angle $\theta$ between the velocity of the charge and the magnetic field is chosen so that it lies in the range $0 \leq \theta \leq 180°$.

■ **DEFINITION OF THE MAGNETIC FIELD**

The magnitude $B$ of the magnetic field at any point in space is defined as

$$B = \frac{F}{q_0(v \sin \theta)} \tag{21.1}$$

where $F$ is the magnitude of the magnetic force on a positive test charge $q_0$ and $v$ is the velocity of the charge and makes an angle $\theta$ ($0 \le \theta \le 180°$) with the direction of the magnetic field. The magnetic field $B$ is a vector, and its direction can be determined by using a small compass needle.

**SI Unit of Magnetic Field:** $\dfrac{\text{newton} \cdot \text{second}}{\text{coulomb} \cdot \text{meter}} = 1$ tesla (T)

The unit of magnetic field strength that follows from Equation 21.1 is the $N \cdot s/(C \cdot m)$. This unit is called the *tesla* (T), a tribute to the Croatian-born American engineer Nikola Tesla (1856–1943). Thus, one tesla is the strength of the magnetic field in which a unit test charge, traveling perpendicular to the magnetic field with a speed of one meter per second, experiences a force of one newton. Because a coulomb per second is an ampere (1 C/s = 1 A, see Section 20.1), the tesla is often written as $1\ T = 1\ N/(A \cdot m)$.

In many situations the magnetic field has a value that is considerably less than one tesla. For example, the strength of the magnetic field near the earth's surface is approximately $10^{-4}$ T. In such circumstances, a magnetic field unit called the *gauss* (G) is sometimes used. Although not an SI unit, the gauss is a convenient size for many applications involving magnetic fields. The relation between the gauss and the tesla is

$$1 \text{ gauss} = 10^{-4} \text{ tesla}$$

Example 1 deals with the magnetic force exerted on a moving proton and on a moving electron.

### Example 1 Magnetic Forces on Charged Particles

A proton in a particle accelerator has a speed of $5.0 \times 10^6$ m/s. The proton encounters a magnetic field whose magnitude is 0.40 T and whose direction makes an angle of $\theta = 30.0°$ with respect to the proton's velocity (see Figure 21.6c). Find (a) the magnitude and direction of the magnetic force on the proton and (b) the acceleration of the proton. (c) What would be the force and acceleration if the particle were an electron instead of a proton?

**Reasoning** For both the proton and the electron, the magnitude of the magnetic force is given by Equation 21.1. The magnetic forces that act on these particles have opposite directions, however, because the charges have opposite signs. In either case, the acceleration is given by Newton's second law, which applies to the magnetic force just as it does to any force. In using the second law, we must take into account the fact that the masses of the proton and the electron are different.

**Solution**

(a) The positive charge on a proton is $1.60 \times 10^{-19}$ C, and according to Equation 21.1, the magnitude of the magnetic force is $F = q_0 vB \sin \theta$. Therefore,

$$F = (1.60 \times 10^{-19} \text{ C})(5.0 \times 10^6 \text{ m/s})(0.40 \text{ T})(\sin 30.0°) = \boxed{1.6 \times 10^{-13} \text{ N}}$$

The direction of the magnetic force is given by RHR-1 and is directed *upward* in Figure 21.6c, with the magnetic field pointing to the right.

(b) The magnitude $a$ of the proton's acceleration follows directly from Newton's second law as the magnitude of the net force divided by the mass $m_p$ of the proton. Since the only force acting on the proton is the magnetic force $F$, it is the net force. Thus,

$$a = \frac{F}{m_p} = \frac{1.6 \times 10^{-13} \text{ N}}{1.67 \times 10^{-27} \text{ kg}} = \boxed{9.6 \times 10^{13} \text{ m/s}^2} \tag{4.1}$$

The direction of the acceleration is the same as the direction of the net force (the magnetic force).

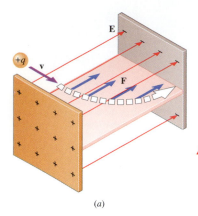

(a)

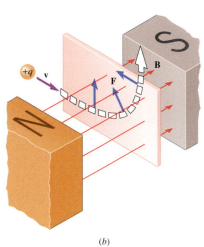

(b)

**Figure 21.8** (a) The electric force **F** that acts on a positive charge is parallel to the electric field **E** and causes the particle's trajectory to bend in a horizontal plane. (b) The magnetic force **F** is perpendicular to both the magnetic field **B** and the velocity **v** and causes the particle's trajectory to bend in a vertical plane.

**The physics of**
**a velocity selector.**

**(c)** The magnitude of the magnetic force on the electron is the same as that on the proton, since both have the same speed and charge magnitude. However, the direction of the force on the electron is opposite to that on the proton, or *downward* in Figure 21.6c, since the electron charge is negative. Furthermore, the electron has a smaller mass $m_e$ and, therefore, experiences a significantly greater acceleration:

$$a = \frac{F}{m_e} = \frac{1.6 \times 10^{-13}\ \text{N}}{9.11 \times 10^{-31}\ \text{kg}} = \boxed{1.8 \times 10^{17}\ \text{m/s}^2}$$

The direction of this acceleration is downward in Figure 21.6c.

## 21.3   The Motion of a Charged Particle in a Magnetic Field

### COMPARING PARTICLE MOTION IN ELECTRIC AND MAGNETIC FIELDS

The motion of a charged particle in an electric field is noticeably different from the motion in a magnetic field. For example, Figure 21.8a shows a positive charge moving between the plates of a parallel plate capacitor. Initially, the charge is moving perpendicular to the direction of the electric field. Since the direction of the electric force on a positive charge is in the same direction as the electric field, the particle is deflected sideways. Part b of the drawing shows the same particle traveling initially at right angles to a magnetic field. An application of RHR-1 shows that when the charge enters the field, the charge is deflected upward (not sideways) by the magnetic force. As the charge moves upward, the direction of the magnetic force changes, always remaining perpendicular to both the magnetic field and the velocity. Conceptual Example 2 focuses on the difference in how electric and magnetic fields apply forces to a moving charge.

### Conceptual Example 2   A Velocity Selector

A velocity selector is a device for measuring the velocity of a charged particle. The device operates by applying electric and magnetic forces to the particle in such a way that these forces balance. Figure 21.9a shows a particle with a positive charge $+q$ and a velocity **v**, which is perpendicular to a constant magnetic field* **B**. How should an electric field **E** be directed so that the force it applies to the particle can balance the magnetic force produced by **B**?

**Reasoning and Solution** If the electric and magnetic forces are to balance, they must have opposite directions. By applying RHR-1, we find that the magnetic force acting on the positively charged particle in Figure 21.9a is directed upward, toward the top of the page. Therefore, the electric force must be directed downward. But the force applied to a positive charge by an electric field has the same direction as the field itself. Therefore, **the electric field must point downward** if the electric force is to balance the magnetic force. In other words, the electric and magnetic fields are perpendicular. Figure 21.9b shows a velocity selector that uses perpendicular electric and magnetic fields. The device is a cylindrical tube located within the magnetic field **B**. Inside the tube is a parallel plate capacitor that produces the electric field **E**. The charged particle enters the left end of the tube perpendicular to both fields. If the strengths of the fields **E** and **B** are adjusted properly, the electric and magnetic forces acting on the particle will cancel each other. With no net force acting on the particle, the velocity remains un-

---

* In many instances it is convenient to orient the magnetic field **B** so its direction is perpendicular to the page. In these cases it is customary to use a dot to symbolize the magnetic field pointing out of the page (toward the reader); this dot symbolizes the tip of the arrow representing the **B** vector. A region where a magnetic field is directed *into the page* is drawn as a series of crosses that indicate the tail feathers of the arrows representing the **B** vectors. Therefore, regions where a magnetic field is directed out of the page or into the page are drawn as shown below:

Out of page          Into page

changed, according to Newton's first law. As a result, the particle moves in a straight line at a constant speed and exits at the right end of the tube. The magnitude of the velocity "selected" can be determined from a knowledge of the strengths of the electric and magnetic fields. Particles with velocities different from the one "selected" are deflected and do not exit at the right end of the tube.

**Related Homework:** *Problems 16, 20*

We have seen that a charged particle traveling in a magnetic field experiences a magnetic force that is always perpendicular to the field. In contrast, the force applied by an electric field is always parallel (or antiparallel) to the field direction. Because of the difference in the way that electric and magnetic fields exert forces, the work done on a charged particle by each field is different, as we now discuss.

## THE WORK DONE ON A CHARGED PARTICLE MOVING THROUGH ELECTRIC AND MAGNETIC FIELDS

In Figure 21.8a an electric field applies a force to a positively charged particle, and, consequently, the path of the particle bends in the direction of the force. Because there is a component of the particle's displacement in the direction of the electric force, the force does work on the particle, according to Equation 6.1. This work increases the kinetic energy and, hence, the speed of the particle, in accord with the work–energy theorem (see Section 6.2). In contrast, the magnetic force in Figure 21.8b always acts in a direction that is perpendicular to the motion of the charge. Consequently, the displacement of the moving charge never has a component in the direction of the magnetic force. As a result, *the magnetic force cannot do work and change the kinetic energy of the charged particle* in Figure 21.8b. Thus, the speed of the particle *does not* change, although the force does alter the direction of the motion.

## THE CIRCULAR TRAJECTORY

To describe the motion of a charged particle in a constant magnetic field more completely, let's discuss the special case in which the velocity of the particle is perpendicular to a uniform magnetic field. As Figure 21.10 illustrates, the magnetic force serves to move the particle in a circular path. To understand why the path is circular, consider two points on the circumference labeled 1 and 2. When the positively charged particle is at point 1, the magnetic force **F** is perpendicular to the velocity **v** and points directly upward in the drawing. This force causes the trajectory to bend upward. When the particle reaches point 2, the magnetic force still remains perpendicular to the velocity but is now directed to the left in the drawing. *[Problem solving insight]* **The magnetic force always remains perpendicular to the velocity and is directed toward the center of the circular path.**

To find the radius of the path in Figure 21.10, we use the concept of centripetal force from Section 5.3. The centripetal force is the net force, directed toward the center of the circle, that is needed to keep a particle moving along a circular path. The magnitude $F_c$ of this force depends on the speed $v$ and mass $m$ of the particle, as well as the radius $r$ of the circle:

$$F_c = \frac{mv^2}{r} \tag{5.3}$$

In the present situation, the magnetic force furnishes the centripetal force. Being perpendicular to the velocity, the magnetic force does no work in keeping the charge $+q$ on the circular path. According to Equation 21.1, the magnitude of the magnetic force is $qvB \sin 90°$, so $qvB = mv^2/r$ or

$$r = \frac{mv}{qB} \tag{21.2}$$

This result shows that the radius of the circle is inversely proportional to the magnitude of the magnetic field, with stronger fields producing "tighter" circular paths. Example 3 illustrates an application of Equation 21.2.

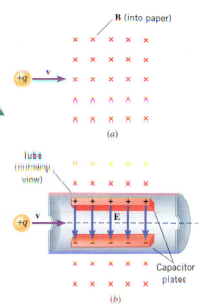

**Figure 21.9** (a) A particle with a positive charge $q$ and velocity **v** moves perpendicularly into a magnetic field **B**. (b) A velocity selector is a tube in which an electric field **E** is perpendicular to a magnetic field, and the field magnitudes are adjusted so that the electric and magnetic forces acting on the particle balance.

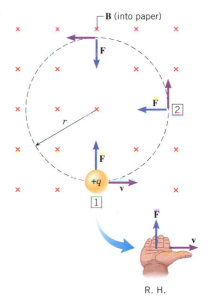

**Figure 21.10** A positively charged particle is moving perpendicular to a constant magnetic field. The magnetic force **F** causes the particle to move on a circular path (R.H. = right hand).

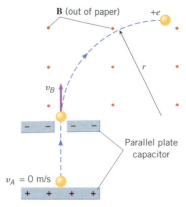

**Figure 21.11** The proton, starting from rest at the positive plate of the capacitor, accelerates toward the negative plate. After leaving the capacitor, the proton enters a magnetic field, where it moves on a circular path of radius *r*.

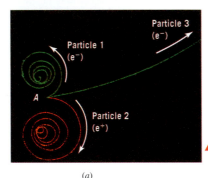

(a)

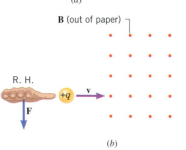

(b)

**Figure 21.12** (a) A photograph of tracks in a bubble chamber. A magnetic field is directed out of the paper. At point *A* a gamma ray (not visible) spontaneously transforms into an electron ($e^-$) and a positron ($e^+$), which produce the spirals. In addition, an electron is knocked forward out of a hydrogen atom in the chamber. (© Lawrence Berkeley Laboratory/Photo Researchers). (b) In accord with RHR-1, the magnetic field applies a downward force to a positively charged particle that moves to the right.

## Example 3 The Motion of a Proton

A proton is released from rest at point *A*, which is located next to the positive plate of a parallel plate capacitor (see Figure 21.11). The proton then accelerates toward the negative plate, leaving the capacitor at point *B* through a small hole in the plate. The electric potential of the positive plate is 2100 V greater than that of the negative plate, so $V_A - V_B = 2100$ V. Once outside the capacitor, the proton travels at a constant velocity until it enters a region of constant magnetic field of magnitude 0.10 T. The velocity is perpendicular to the magnetic field, which is directed out of the page in Figure 21.11. Find (a) the speed $v_B$ of the proton when it leaves the negative plate of the capacitor, and (b) the radius *r* of the circular path on which the proton moves in the magnetic field.

**Reasoning** The only force that acts on the proton (charge = $+e$) while it is between the capacitor plates is the conservative electric force. Thus, we can use the conservation of energy to find the speed of the proton when it leaves the negative plate. The total energy of the proton is the sum of its kinetic energy, $\frac{1}{2}mv^2$, and electric potential energy, EPE. Following Example 4 in Chapter 19, we set the total energy at point *B* equal to the total energy at point *A*:

$$\underbrace{\tfrac{1}{2}mv_B^2 + \text{EPE}_B}_{\text{Total energy at }B} = \underbrace{\tfrac{1}{2}mv_A^2 + \text{EPE}_A}_{\text{Total energy at }A}$$

We note that $v_A = 0$ m/s, since the proton starts from rest, and use Equation 19.3 to set $\text{EPE}_A - \text{EPE}_B = e(V_A - V_B)$. Then the conservation of energy reduces to $\frac{1}{2}mv_B^2 = e(V_A - V_B)$. Solving for $v_B$ gives $v_B = \sqrt{2e(V_A - V_B)/m}$. The proton enters the magnetic field with this speed and moves on a circular path with a radius that is given by Equation 21.2.

**Solution**

**(a)** The speed of the proton is

$$v_B = \sqrt{\frac{2e(V_A - V_B)}{m}} = \sqrt{\frac{2(1.60 \times 10^{-19}\text{ C})(2100\text{ V})}{1.67 \times 10^{-27}\text{ kg}}} = \boxed{6.3 \times 10^5\text{ m/s}}$$

**(b)** When the proton moves in the magnetic field, the radius of the circular path is

$$r = \frac{mv_B}{eB} = \frac{(1.67 \times 10^{-27}\text{ kg})(6.3 \times 10^5\text{ m/s})}{(1.60 \times 10^{-19}\text{ C})(0.10\text{ T})} = \boxed{6.6 \times 10^{-2}\text{ m}} \quad (21.2)$$

One of the exciting areas in physics today is the study of elementary particles, which are the basic building blocks from which all matter is constructed. Important information about an elementary particle can be obtained from its motion in a magnetic field, with the aid of a device known as a bubble chamber. A bubble chamber contains a superheated liquid such as hydrogen, which will boil and form bubbles readily. When an electrically charged particle passes through the chamber, a thin track of bubbles is left in its wake. This track can be photographed to show how a magnetic field affects the particle motion. Conceptual Example 4 illustrates how physicists deduce information from such photographs.

## Conceptual Example 4 Particle Tracks in a Bubble Chamber

Figure 21.12a shows the bubble-chamber tracks resulting from an event that begins at point *A*. At this point a gamma ray (emitted by certain radioactive substances) travels in from the left, spontaneously transforms into two charged particles. There is no track from the gamma ray itself. These particles move away from point *A*, producing the two spiral tracks. A third charged particle is knocked out of a hydrogen atom and moves forward, producing the long track with the slight upward curvature. Each of the three particles has the same mass and carries a charge of the same magnitude. A uniform magnetic field is directed out of the paper toward you. Guided by RHR-1 and Equation 21.2, deduce the sign of the charge carried by each particle, identify which particle is moving most rapidly, and account for the two spiral paths.

**Reasoning and Solution** To help in our reasoning, Figure 21.12b shows a positively charged particle traveling with a velocity **v** that is perpendicular to a magnetic field. The field is directed out of the paper. RHR-1 indicates that the magnetic force points downward. Therefore, downward-curving tracks in the photograph indicate a positive charge, while upward-curving tracks indicate a negative charge. ***Particles 1 and 3, then, must carry a negative charge.*** They are, in fact, electrons (e⁻). In contrast, ***particle 2 must have a positive charge.*** It is called a positron (e⁺), an elementary particle that has the same mass as an electron but an opposite charge.

In Equation 21.2 the mass $m$, the charge magnitude $q$, and the magnetic field strength $B$ are the same for each particle. Therefore, the radius $r$ is proportional to the speed $v$, and a greater radius means a greater speed. The track for particle 3 has the greatest radius, so ***particle 3 has the greatest speed.***

Each spiral indicates that the radius is decreasing as the particle moves. Since the radius is proportional to the speed, the speeds of particles 1 and 2 must be decreasing as these particles move. Correspondingly, the kinetic energies of these particles must be decreasing. ***Particles 1 and 2 are losing energy each time they collide with a hydrogen atom in the bubble chamber.***

**Related Homework:** *Problem 22*

**The physics of** a mass spectrometer. (See below.)

## 21.4 The Mass Spectrometer

Physicists use mass spectrometers for determining the relative masses and abundances of isotopes.* Chemists use these instruments to help identify unknown molecules produced in chemical reactions. Mass spectrometers are also used during surgery, where they give the anesthesiologist information on the gases, including the anesthetic, in the patient's lungs.

In the type of mass spectrometer illustrated in Figure 21.13, the atoms or molecules are first vaporized and then ionized by the ion source. The ionization process removes one electron from the particle, leaving it with a net positive charge of $+e$. The positive ions are then accelerated through the potential difference $V$, which is applied between the ion source and the metal plate. With a speed $v$, the ions pass through a hole in the plate and enter a region of constant magnetic field $\mathbf{B}$, where they are deflected in semicircular paths. Only those ions following a path with the proper radius $r$ strike the detector, which records the number of ions arriving per second.

The mass $m$ of the detected ions can be expressed in terms of $r$, $B$, and $v$ by recalling that the radius of the path followed by a particle of charge $+e$ is $r = mv/(eB)$ (Equation 21.2). In addition, the Reasoning section in Example 3 shows that the ion speed $v$ can be expressed in terms of the potential difference $V$ as $v = \sqrt{2eV/m}$. This expression for $v$ is the same as that used in Example 3, except that, for convenience, we have replaced the potential difference, $V_A - V_B$, by the symbol $V$. Eliminating $v$ from these two equations algebraically and solving for the mass gives

$$m = \left(\frac{er^2}{2V}\right) B^2$$

This result shows that the mass of each ion reaching the detector is proportional to $B^2$. Experimentally changing the value of $B$ and keeping the term in the parentheses constant will allow ions of different masses to enter the detector. A plot of the detector output as a function of $B^2$ then gives an indication of what masses are present and the abundance of each mass.

Figure 21.14 shows a record obtained by a mass spectrometer for naturally occurring neon gas. The results show that the element neon has three isotopes whose atomic mass numbers are 20, 21, and 22. These isotopes occur because neon atoms exist with different numbers of neutrons in the nucleus. Notice that the isotopes have different abundances, with neon-20 being the most abundant.

* Isotopes are atoms that have the same atomic number but different atomic masses due to the presence of different numbers of neutrons in the nucleus. They are discussed in Section 31.1.

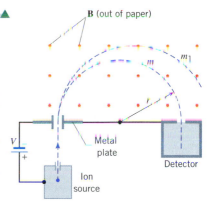

**Figure 21.13** In this mass spectrometer the dashed lines are the paths traveled by ions of different masses. Ions with mass $m$ follow the path of radius $r$ and enter the detector. Ions with the larger mass $m_1$ follow the outer path and miss the detector.

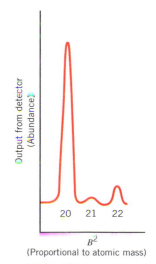

**Figure 21.14** The mass spectrum (not to scale) of naturally occurring neon, showing three isotopes whose atomic mass numbers are 20, 21, and 22. The larger the peak, the more abundant the isotope.

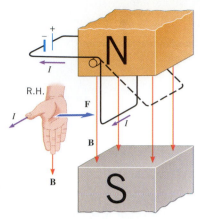

**Figure 21.15** The wire carries a current $I$, and the bottom segment of the wire is oriented perpendicular to a magnetic field **B**. A magnetic force deflects the wire to the right.

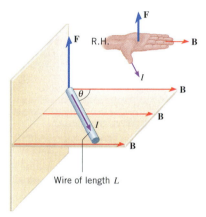

**Figure 21.16** The current $I$ in the wire, oriented at an angle $\theta$ with respect to a magnetic field **B**, is acted upon by a magnetic force **F**.

**The physics of**
a loudspeaker.

# 21.5 The Force on a Current in a Magnetic Field

As we have seen, a charge moving through a magnetic field can experience a magnetic force. Since an electric current is a collection of moving charges, a current in the presence of a magnetic field can also experience a magnetic force. In Figure 21.15, for instance, a current-carrying wire is placed between the poles of a magnet. When the direction of the current $I$ is as shown, the moving charges experience a magnetic force that pushes the wire to the right in the drawing. The direction of the force is determined in the usual manner by using RHR-1, with the minor modification that the direction of the velocity of a positive charge is replaced by the direction of the conventional current $I$. If the current in the drawing were reversed by switching the leads to the battery, the direction of the force would be reversed, and the wire would be pushed to the left.

When a charge moves through a magnetic field, the magnitude of the force that acts on the charge is $F = qvB \sin \theta$ (Equation 21.1). With the aid of Figure 21.16, this expression can be put into a form that is more suitable for use with an electric current. The drawing shows a wire of length $L$ that carries a current $I$. The wire is oriented at an angle $\theta$ with respect to a magnetic field **B**. This picture is similar to Figure 21.6c, except that now the charges move in a wire. The magnetic force exerted on this length of wire is the net force acting on the total amount of charge moving in the wire. Suppose that an amount of charge $\Delta q$ travels the length of the wire in a time interval $\Delta t$. The magnitude of the magnetic force on this amount of charge is given by Equation 21.1 as $F = (\Delta q)vB \sin \theta$. Multiplying and dividing the right side of this equation by $\Delta t$, we find that

$$F = \underbrace{\left(\frac{\Delta q}{\Delta t}\right)}_{I} \underbrace{(v\,\Delta t)}_{L} B \sin \theta$$

According to Equation 20.1, the term $\Delta q/\Delta t$ is the current $I$ in the wire, and the term $v\Delta t$ is the length $L$ of the wire. With these two substitutions, the expression for the magnetic force exerted on a current-carrying wire becomes

**Magnetic force on a current-carrying wire of length $L$**      $F = ILB \sin \theta$      (21.3)

As in the case of a single charge traveling in a magnetic field, the magnetic force on a current-carrying wire is a maximum when the wire is oriented perpendicular to the field ($\theta = 90°$) and vanishes when the current is parallel or antiparallel to the field ($\theta = 0°$ or $180°$). The direction of the magnetic force is given by RHR-1, as Figure 21.16 indicates.

Most loudspeakers operate on the principle that a magnetic field exerts a force on a current-carrying wire. Figure 21.17a shows a speaker design that consists of three basic

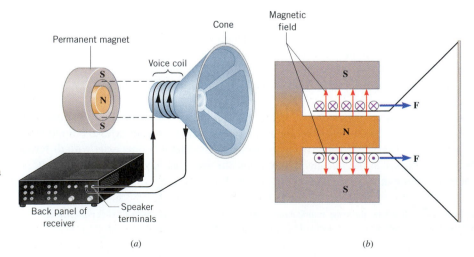

**Figure 21.17** (a) An "exploded" view of one type of speaker design, which shows a cone, a voice coil, and a permanent magnet. (b) Because of the current in the voice coil (shown as ⊗ and ⊙), the magnetic field causes a force **F** to be exerted on the voice coil and cone.

parts: a cone, a voice coil, and a permanent magnet. The cone is mounted so it can vibrate back and forth. When vibrating, it pushes and pulls on the air in front of it, thereby creating sound waves. Attached to the apex of the cone is the voice coil, which is a hollow cylinder around which coils of wire are wound. The voice coil is slipped over one of the poles of the stationary permanent magnet (the north pole in the drawing) and can move freely. The two ends of the voice-coil wire are connected to the speaker terminals on the back panel of a receiver.

The receiver acts as an ac generator, sending an alternating current to the voice coil. The alternating current interacts with the magnetic field to generate an alternating force that pushes and pulls on the voice coil and the attached cone. To see how the magnetic force arises, consider Figure 21.17*b*, which is a cross-sectional view of the voice coil and the magnet. In the cross-sectional view, the current is directed into the page in the upper half of the voice coil (⊗⊗⊗) and out of the page in the lower half (⊙⊙⊙). In both cases the magnetic field is perpendicular to the current, so the maximum possible force is exerted on the wire. An application of RHR-1 to both the upper and lower halves of the voice coil shows that the magnetic force **F** in the drawing is directed to the right, causing the cone to accelerate in that direction. One half of a cycle later when the current is reversed, the direction of the magnetic force is also reversed, and the cone accelerates to the left. If, for example, the alternating current from the receiver has a frequency of 1000 Hz, the alternating magnetic force causes the cone to vibrate back and forth at the same frequency, and a 1000-Hz sound wave is produced. Thus, it is the magnetic force on a current-carrying wire that is responsible for converting an electrical signal into a sound wave. In Example 5 a typical force and acceleration in a loudspeaker are determined.

**Problem solving insight**
Whenever the current in a wire reverses direction, the force exerted on the wire by a given magnetic field also reverses direction.

## Example 5   The Force and Acceleration in a Loudspeaker

The voice coil of a speaker has a diameter of $d = 0.025$ m, contains 55 turns of wire, and is placed in a 0.10-T magnetic field. The current in the voice coil is 2.0 A. (a) Determine the magnetic force that acts on the coil and cone. (b) The voice coil and cone have a combined mass of 0.020 kg. Find their acceleration.

**Reasoning** The magnetic force that acts on the current-carrying voice coil is given by Equation 21.3 as $F = ILB \sin \theta$. The effective length $L$ of the wire in the voice coil is very nearly the number of turns $N$ times the circumference ($\pi d$) of one turn: $L = N\pi d$. The acceleration of the voice coil and cone is given by Newton's second law as the magnetic force divided by the combined mass.

**Solution**

(a) Since the magnetic field acts perpendicular to all parts of the wire, $\theta = 90°$ and the force on the voice coil is

$$F = ILB \sin \theta = (2.0 \text{ A})[55\pi(0.025 \text{ m})](0.10 \text{ T})\sin 90° = \boxed{0.86 \text{ N}} \qquad (21.3)$$

(b) According to Newton's second law, the acceleration of the voice coil and cone is

$$a = \frac{F}{m} = \frac{0.86 \text{ N}}{0.020 \text{ kg}} = \boxed{43 \text{ m/s}^2} \qquad (4.1)$$

This acceleration is more than four times the acceleration due to gravity.

Magnetohydrodynamic (MHD) propulsion is a revolutionary type of propulsion system that does not use propellers to power ships and submarines, but, instead, uses a magnetic force on a current. The method eliminates motors, drive shafts, and gears, as well as propellers, so it promises to be a low-noise system with great reliability at relatively low cost. Figure 21.18*a* shows the first ship to use MHD technology, the *Yamato 1*. In part *b* the schematic side view shows one of the two MHD propulsion units that are mounted underneath the vessel. Seawater enters the front of the unit and is expelled from the rear. A jet engine uses air in a similar fashion, taking it in in the front and pushing it out at the back to propel the plane forward.

Figure 21.18*c* presents an enlarged view of a propulsion unit. An electromagnet (see Section 21.7) within the vessel uses superconducting wire to produce a strong magnetic

**The physics of**
magnetohydrodynamic propulsion.

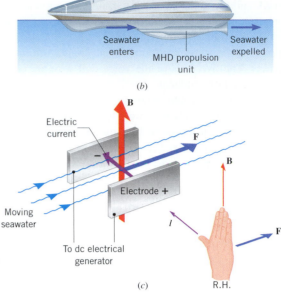

**Figure 21.18** (*a*) The *Yamato 1* is the first ship to use magnetohydrodynamic (MHD) propulsion (© Dennis Budd Gray). (*b*) A schematic side view of the *Yamato 1*. (*c*) In the MHD unit, the magnetic force exerted on the current forces water out the back.

field. Electrodes (metal plates) mounted on either side of the unit are attached to a dc electrical generator. The generator sends an electric current through the seawater, from one electrode to the other, perpendicular to the magnetic field. Consistent with RHR-1, the field applies a force **F** to the current, as the drawing shows. The magnetic force pushes the seawater, similar to the way in which it pushes the wire in Figure 21.15. As a result, a water jet is expelled from the tube. Since the MHD unit exerts a magnetic force on the water, the water exerts a force on the propulsion unit and the vessel attached to it. According to Newton's third law, this "reaction" force is equal in magnitude, but opposite in direction, to the magnetic force, and it is the reaction force that provides the thrust to drive the vessel.

## 21.6 *The Torque on a Current-Carrying Coil*

We have seen that a current-carrying wire can experience a force when placed in a magnetic field. If a loop of wire is suspended properly in a magnetic field, the magnetic force produces a torque that tends to rotate the loop. This torque is responsible for the operation of a widely used type of electric motor.

Figure 21.19*a* shows a rectangular loop of wire attached to a vertical shaft. The shaft is mounted such that it is free to rotate in a uniform magnetic field. When there is a current in the loop, the loop rotates because magnetic forces act on the two vertical sides, labeled 1 and 2 in the drawing. Part *b* shows a top view of the loop and the magnetic forces

**Figure 21.19** (*a*) A current-carrying loop of wire, which can rotate about a vertical shaft, is situated in a magnetic field. (*b*) A top view of the loop. The current in side 1 is directed out of the page (⊙), while that in side 2 is directed into the page (⊗). The current in side 1 experiences a force **F** that is opposite to the force exerted on side 2. The two forces produce a clockwise torque about the shaft.

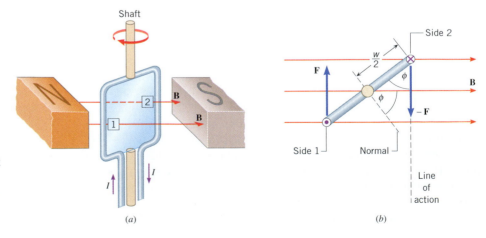

F and −F that act on the two sides. These two forces have the same magnitude, but an application of RHR-1 shows that they point in opposite directions, so the loop experiences no net force. The loop does, however, experience a net torque that tends to rotate it in a clockwise fashion about the vertical shaft. Figure 21.20a shows that the torque is maximum when the normal to the plane of the loop is perpendicular to the field. In contrast, part b shows that the torque is zero when the normal is parallel to the field. **When a current-carrying loop is placed in a magnetic field, the loop tends to rotate such that its normal becomes aligned with the magnetic field.** In this respect, a current loop behaves like a magnet (e.g., a compass needle) suspended in a magnetic field, since a magnet also rotates to align itself with the magnetic field.

It is possible to determine the magnitude of the torque on the loop. From Equation 21.3 the magnetic force on each vertical side has a magnitude of $F = ILB \sin 90°$, where $L$ is the length of side 1 or side 2, and $\theta = 90°$ because the current $I$ always remains perpendicular to the magnetic field as the loop rotates. As Section 9.1 discusses, the torque produced by a force is the product of the magnitude of the force and the lever arm. In Figure 21.19b the lever arm is the perpendicular distance from the line of action of the force to the shaft. This distance is given by $(w/2) \sin \phi$, where $w$ is the width of the loop, and $\phi$ is the angle between the normal to the plane of the loop and the direction of the magnetic field. The net torque is the sum of the torques on the two sides, so

$$\text{Net torque} = \tau = ILB \left(\tfrac{1}{2} w \sin \phi\right) + ILB \left(\tfrac{1}{2} w \sin \phi\right) = IAB \sin \phi$$

In this result the product $Lw$ has been replaced by the area $A$ of the loop. If the wire is wrapped so as to form a coil containing $N$ loops, each of area $A$, the force on each side is $N$ times larger, and the torque becomes proportionally greater:

$$\tau = NIAB \sin \phi \tag{21.4}$$

Equation 21.4 has been derived for a rectangular coil, but it is valid for any shape of flat coil, such as a circular coil. The torque depends on the geometric properties of the coil and the current in it through the quantity $NIA$. This quantity is known as the *magnetic moment* of the coil, and its unit is ampere · meter$^2$. The greater the magnetic moment of a current-carrying coil, the greater the torque that the coil experiences when placed in a magnetic field. Example 6 discusses the torque that a magnetic field applies to such a coil.

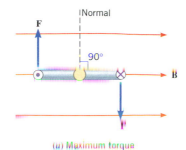

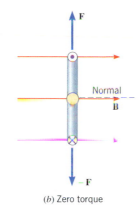

(b) Zero torque

**Figure 21.20** (a) Maximum torque occurs when the normal to the plane of the loop is perpendicular to the magnetic field. (b) The torque is zero when the normal is parallel to the field.

### Example 6   The Torque Exerted on a Current-Carrying Coil

A coil of wire has an area of $2.0 \times 10^{-4}$ m$^2$, consists of 100 loops or turns, and contains a current of 0.045 A. The coil is placed in a uniform magnetic field of magnitude 0.15 T. (a) Determine the magnetic moment of the coil. (b) Find the maximum torque that the magnetic field can exert on the coil.

**Reasoning and Solution**

**(a)** The magnetic moment of the coil is

$$\text{Magnetic moment} = NIA = (100)(0.045 \text{ A})(2.0 \times 10^{-4} \text{ m}^2) = \boxed{9.0 \times 10^{-4} \text{ A} \cdot \text{m}^2}$$

**(b)** According to Equation 21.4, the torque is the product of the magnetic moment $NIA$ and $B \sin \phi$. However, the maximum torque occurs when $\phi = 90°$, so

$$\tau = \underbrace{(NIA)}_{\substack{\text{Magnetic} \\ \text{moment}}}(B \sin 90°) = (9.0 \times 10^{-4} \text{ A} \cdot \text{m}^2)(0.15 \text{ T}) = \boxed{1.4 \times 10^{-4} \text{ N} \cdot \text{m}}$$

The electric motor is found in many devices, such as CD players, tape decks, automobiles, washing machines, and air conditioners. Figure 21.21 shows that a direct-current (dc) motor consists of a coil of wire placed in a magnetic field and free to rotate about a vertical shaft. The coil of wire contains many turns and is wrapped around an iron cylinder that rotates with the coil, although these features have been omitted to simplify the drawing. The coil and iron cylinder assembly is known as the armature. Each end of the

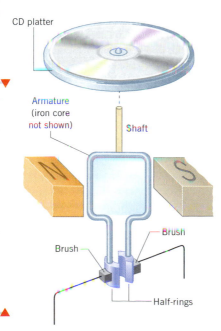

**Figure 21.21** The basic components of a dc motor. A CD platter is shown as it might be attached to the motor.

**The physics of**
**a direct-current electric motor.**

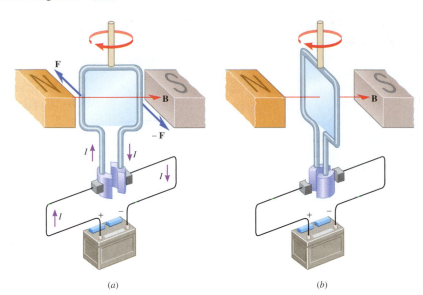

(a)                    (b)

**Figure 21.22** (a) When a current exists in the coil, the coil experiences a torque. (b) Because of its inertia, the coil continues to rotate when there is no current.

wire coil is attached to a metallic half-ring. Rubbing against each of the half-rings is a graphite contact called a brush. While the half-rings rotate with the coil, the graphite brushes remain stationary. The two half-rings and the associated brushes are referred to as a split-ring commutator, the purpose of which will be explained shortly.

The operation of a motor can be understood by considering Figure 21.22. In part *a* the current from the battery enters the coil through the left brush and half-ring, goes around the coil, and then leaves through the right half-ring and brush. Consistent with RHR-1, the directions of the magnetic forces **F** and **−F** on the two sides of the coil are as shown in the drawing. These forces produce the torque that turns the coil. Eventually the coil reaches the position shown in part *b* of the drawing. In this position the half-rings momentarily lose electrical contact with the brushes, so that there is no current in the coil and no applied torque. However, like any moving object, the rotating coil does not stop immediately, for its inertia carries it onward. When the half-rings reestablish contact with the brushes, there again is a current in the coil, and a magnetic torque again rotates the coil in the same direction. The split-ring commutator ensures that the current is always in the proper direction to yield a torque that produces a continuous rotation of the coil.

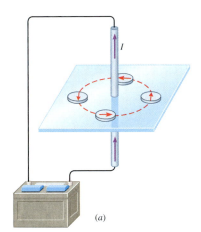

(a)

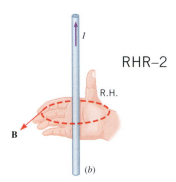

RHR–2

R.H.

**B**

(b)

**Figure 21.23** (a) A very long, straight current-carrying wire produces magnetic field lines that are circular about the wire. One such circular line is indicated by the compass needles. (b) If the thumb of the right hand (R.H.) is pointed in the direction of the current *I*, the curled fingers point in the direction of the magnetic field, according to RHR-2.

## 21.7 *Magnetic Fields Produced by Currents*

We have seen that a current-carrying wire can experience a magnetic force when placed in a magnetic field that is produced by an external source, such as a permanent magnet. *A current-carrying wire also produces a magnetic field of its own,* as we will see in this section. Hans Christian Oersted (1777–1851) first discovered this effect in 1820 when he observed that a current-carrying wire influences the orientation of a nearby compass needle. The compass needle aligns itself with the net magnetic field produced by the current and the magnetic field of the earth. Oersted's discovery, which linked the motion of electric charges with the creation of a magnetic field, marked the beginning of an important discipline called *electromagnetism.*

### A LONG, STRAIGHT WIRE

Figure 21.23*a* illustrates Oersted's discovery with a very long, straight wire. When a current is present, the compass needles point in a circular pattern about the wire. The pattern indicates that the magnetic field lines produced by the current are circles centered on the wire. If the direction of the current is reversed, the needles also reverse their directions, indicating that the direction of the magnetic field has reversed. The direction of the field can be obtained by using *Right-Hand Rule No. 2 (RHR-2),* as part *b* of the drawing indicates:

***Right-Hand Rule No. 2.*** Curl the fingers of the right hand into the shape of a half-circle. Point the thumb in the direction of the conventional current $I$, and the tips of the fingers will point in the direction of the magnetic field **B**.

Experimentally, it is found that the magnitude $B$ of the magnetic field produced by an infinitely long, straight wire is directly proportional to the current $I$ and inversely proportional to the radial distance $r$ from the wire, $B \propto I/r$. As usual, this proportionality is converted into an equation by introducing a proportionality constant, which, in this instance, is written as $\mu_0/(2\pi)$. Thus, the magnitude of the magnetic field is

*Infinitely long, straight wire*
$$B = \frac{\mu_0 I}{2\pi r}$$
(21.5)

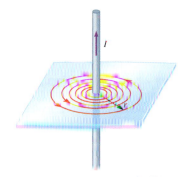

**Figure 21.24** The magnetic field becomes stronger as the radial distance $r$ decreases, so the field lines are closer together near the wire.

The constant $\mu_0$ is known as the ***permeability of free space,*** and its value is $\mu_0 = 4\pi \times 10^{-7}$ T·m/A. The magnetic field becomes stronger nearer the wire, where $r$ is smaller. Therefore, the field lines near the wire are closer together than those located farther away, where the field is weaker. Figure 21.24 shows the pattern of field lines.

The magnetic field that surrounds a current-carrying wire can exert a force on a moving charge, as the next example illustrates.

### Example 7   A Current Exerts a Magnetic Force on a Moving Charge

Figure 21.25 shows a very long, straight wire carrying a current of $I = 3.0$ A. A particle of charge $q_0 = +6.5 \times 10^{-6}$ C is moving parallel to the wire at a distance of $r = 0.050$ m; the speed of the particle is $v = 280$ m/s. Determine the magnitude and direction of the magnetic force exerted on the moving charge by the current in the wire.

**Reasoning** The current generates a magnetic field in the space around the wire. A charge moving through this field experiences a magnetic force **F** whose magnitude is given by Equation 21.1 as $F = q_0 v B \sin \theta$, where $\theta$ is the angle between the magnetic field and the velocity of the charge. The magnitude of the magnetic field follows from Equation 21.5 as $B = \mu_0 I/(2\pi r)$. Thus, the magnitude of the magnetic force can be expressed as

$$F = q_0 v B \sin \theta = q_0 v \left( \frac{\mu_0 I}{2\pi r} \right) \sin \theta$$

The direction of the magnetic force is predicted by RHR-1 (see Section 21.2).

**Solution** Figure 21.25 shows that the magnetic field **B** lies in the plane that is perpendicular to both the wire and velocity **v** of the particle. Thus, the angle between **B** and **v** is $\theta = 90°$, and the magnitude of the magnetic force is

$$F = q_0 v \left( \frac{\mu_0 I}{2\pi r} \right) \sin 90°$$

$$F = (6.5 \times 10^{-6} \text{ C})(280 \text{ m/s}) \left[ \frac{(4\pi \times 10^{-7} \text{ T·m/A})(3.0 \text{ A})}{2\pi(0.050 \text{ m})} \right]$$

$$= \boxed{2.2 \times 10^{-8} \text{ N}}$$

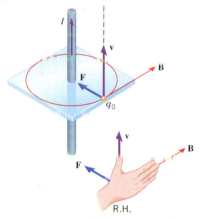

**Figure 21.25** The moving positive charge $q_0$ experiences a magnetic force **F** because of the magnetic field **B** produced by the current in the wire.

The direction of the magnetic force is predicted by RHR-1 and, as the drawing shows, is radially inward toward the wire.

We have now seen that an electric current can create a magnetic field of its own. Earlier, we have also seen that an electric current can experience a force created by another magnetic field. Therefore, the magnetic field that one current creates can exert a force on another nearby current. Examples 8 and 9 deal with this magnetic interaction between currents.

### Example 8
### Two Current-Carrying Wires Exert Magnetic Forces on One Another

Figure 21.26 shows two parallel straight wires that are very long. The wires are separated by a distance of $r = 0.065$ m and carry currents of $I_1 = 15$ A and $I_2 = 7.0$ A. Find the magnitude and direction of the force that the magnetic field of wire 1 applies to a 1.5-m length of wire 2 when the currents are (a) in opposite directions and (b) in the same direction.

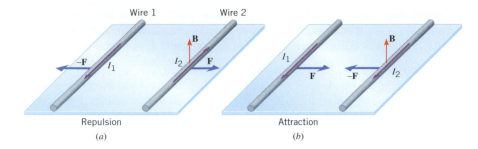

**Figure 21.26** (*a*) Two long, parallel wires carrying currents $I_1$ and $I_2$ in opposite directions repel each other. (*b*) The wires attract each other when the currents are in the same direction.

**Reasoning** The current $I_2$ in wire 2 is situated in the magnetic field produced by the current in wire 1. The magnitude $F$ of the magnetic force experienced by a length $L$ of wire 2 is given by Equation 21.3 as $F = I_2 LB \sin \theta$. Here $B$ is the magnitude of the magnetic field produced by wire 1 and is given by Equation 21.5 as $B = \mu_0 I_1/(2\pi r)$. The direction of the magnetic force can be determined by using RHR-1.

**Solution**

(**a**) At wire 2, the magnitude of the magnetic field created by wire 1 is

$$B = \frac{\mu_0 I_1}{2\pi r} = \frac{(4\pi \times 10^{-7}\ \text{T·m/A})(15\ \text{A})}{2\pi(0.065\ \text{m})} = 4.6 \times 10^{-5}\ \text{T} \quad (21.5)$$

The direction of this field is upward at the location of wire 2, as part *a* of the figure shows. The direction can be obtained using RHR-2 (thumb of right hand along $I_1$, curled fingers point upward at wire 2 and indicate the direction of **B**). The magnetic field is perpendicular to wire 2 ($\theta = 90°$), so the magnitude of the force on a 1.5-m section of its length is

$$F = I_2 LB \sin \theta = (7.0\ \text{A})(1.5\ \text{m})(4.6 \times 10^{-5}\ \text{T}) \sin 90° = \boxed{4.8 \times 10^{-4}\ \text{N}} \quad (21.3)$$

The direction of the magnetic force on wire 2 is away from wire 1, as part *a* of the drawing indicates; the force direction is found by using RHR-1 (fingers of the right hand extended upward along **B**, thumb points along $I_2$, palm pushes in the direction of the force **F**).

In a like manner, the current in wire 2 also creates a magnetic field that produces a force on wire 1. Reasoning similar to that above shows that a 1.5-m length of wire 1 is repelled from wire 2 with a force that also has a magnitude of $4.8 \times 10^{-4}$ N. Thus, each wire generates a force on the other, and, if the currents are in *opposite* directions, the wires *repel* each other. The fact that the two wires exert equal but oppositely directed forces on each other is consistent with Newton's third law, the action–reaction law.*

(**b**) If the current in wire 2 is reversed, as part *b* of the drawing indicates, wire 2 is attracted to wire 1 because the direction of the magnetic force is reversed. However, the magnitude of the force is the same as that calculated in part *a* above. Likewise, wire 1 is attracted to wire 2. Two parallel wires carrying currents in the *same* direction *attract* each other.

▲

## Conceptual Example 9
### The Net Force That a Current-Carrying Wire Exerts on a Current-Carrying Coil

▼

Figure 21.27 shows a very long, straight wire carrying a current $I_1$ and a rectangular coil carrying a current $I_2$. The wire and the coil lie in the same plane, with the wire parallel to the long sides of the rectangle. Is the coil attracted to or repelled from the wire?

**Reasoning and Solution** The current in the straight wire exerts a force on each of the four sides of the coil. The net force acting on the coil is the vector sum of these four forces. To decide whether the net force is attractive or repulsive, we need to consider the directions and magnitudes of the individual forces. In the long side of the coil near the wire, the current $I_2$

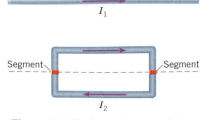

**Figure 21.27** A very long, straight wire carries a current $I_1$, and a rectangular coil carries a current $I_2$. The dashed line is parallel to the wire and locates a small segment on each short side of the coil.

---

* In this example the currents in the two wires and the distance between them are known; therefore, the magnetic force that one wire exerts on the other can be calculated. If, instead, the force and the distance were known and the wires carried the same current, that current could be calculated. This is, in fact, the procedure used to define the ampere, which is the unit for electric current. This procedure is chosen because force and distance are quantities that can be measured with a high degree of precision. One ampere is the amount of electric current in each of two long, parallel wires that gives rise to a magnetic force per unit length of $2 \times 10^{-7}$ N/m on each wire when the wires are separated by one meter. With the ampere defined in terms of force and distance, the coulomb is defined as the quantity of electrical charge that passes a given point in one second when the current is one ampere or one coulomb per second.

has the same direction as the current $I_1$, and we have just seen in Example 8 that two such currents attract each other. In the long side of the coil farthest from the wire, $I_2$ has a direction opposite to that of $I_1$ and, according to Example 8, they repel one another. However, the attractive force is stronger than the repulsive force because the magnetic field produced by the current $I_1$ is stronger at shorter distances than at greater distances. Consequently, we reach the preliminary conclusion that the coil is attracted to the wire.

But what about the forces that act on the two short sides? Consider a small segment of each of the short sides, located at the same distance from the straight wire, as indicated by the dashed line in Figure 21.27. Each of these segments experiences the same magnetic field from the current $I_1$. RHR-2 shows that this field is directed downward into the plane of the paper, so that it is perpendicular to the current $I_2$ in each segment. But the directions of $I_2$ in the segments are opposite. As a result, RHR-1 reveals that the magnetic field from the straight wire applies a force to one segment that is opposite to that applied to the other segment. Thus, the forces acting on the two short sides of the coil cancel, and our preliminary conclusion is valid; *the coil is attracted to the wire.*

**Related Homework:** *Problem 54*

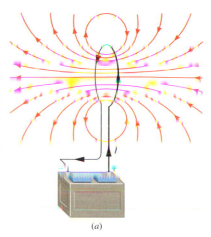

## A LOOP OF WIRE

If a current-carrying wire is bent into a circular loop, the magnetic field lines around the loop have the pattern shown in Figure 21.28a. At the *center* of a loop of radius $R$, the magnetic field is perpendicular to the plane of the loop and has the value $B = \mu_0 I/(2R)$, where $I$ is the current in the loop. Often, the loop consists of $N$ turns of wire that are wound sufficiently close together that they form a flat coil with a single radius. In this case, the magnetic fields of the individual turns add together to give a net field that is $N$ times greater than that of a single loop. For such a coil the magnetic field at the center is

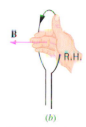

*Center of a circular loop* 
$$B = N\frac{\mu_0 I}{2R}$$
(21.6)

**Figure 21.28** (a) The magnetic field lines in the vicinity of a current-carrying circular loop. (b) The direction of the magnetic field at the center of the loop is given by RHR-2.

The direction of the magnetic field at the center of the loop can be determined with the help of RHR-2. If the thumb of the right hand is pointed in the direction of the current and the curled fingers are placed at the center of the loop, as in Figure 21.28b, the fingers indicate that the magnetic field points from right to left.

Example 10 shows how the magnetic fields produced by the current in a loop of wire and the current in a long, straight wire combine to form a net magnetic field.

### *Example 10* Finding the Net Magnetic Field

A long, straight wire carries a current of $I_1 = 8.0$ A. As Figure 21.29a illustrates, a circular loop of wire lies immediately to the right of the straight wire. The loop has a radius of $R = 0.030$ m and carries a current of $I_2 = 2.0$ A. Assuming that the thickness of the wires is negligible, find the magnitude and direction of the net magnetic field at the center $C$ of the loop.

**Reasoning** The net magnetic field at the point $C$ is the sum of two contributions: (1) the field $\mathbf{B_1}$ produced by the long, straight wire, and (2) the field $\mathbf{B_2}$ produced by the circular loop. An application of RHR-2 shows that at point $C$ the field $\mathbf{B_1}$ is directed upward, perpendicular to the plane containing the straight wire and the loop (see part b of the drawing). Similarly, RHR-2 shows that the magnetic field $\mathbf{B_2}$ is directed downward, opposite to the direction of $\mathbf{B_1}$.

**Solution** If we choose the upward direction in Figure 21.29b as positive, the net magnetic field at the point $C$ is

$$B = \underbrace{\frac{\mu_0 I_1}{2\pi r}}_{\substack{\text{Long, straight}\\\text{wire}}} - \underbrace{\frac{\mu_0 I_2}{2R}}_{\substack{\text{Center of a}\\\text{circular loop}}} = \frac{\mu_0}{2}\left(\frac{I_1}{\pi r} - \frac{I_2}{R}\right)$$

$$B = \frac{(4\pi \times 10^{-7}\ \text{T·m/A})}{2}\left[\frac{8.0\ \text{A}}{\pi(0.030\ \text{m})} - \frac{2.0\ \text{A}}{0.030\ \text{m}}\right] = \boxed{1.1 \times 10^{-5}\ \text{T}}$$

The net field is positive, so it is directed upward, perpendicular to the plane.

**Problem solving insight**
Do not confuse the formula for the magnetic field produced at the center of a circular loop with that of a very long, straight wire. The formulas are similar, differing only by a factor of $\pi$ in the denominator.

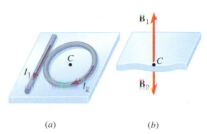

**Figure 21.29** (a) A long, straight wire carrying a current $I_1$ lies next to a circular loop that carries a current $I_2$. (b) The magnetic fields at the center $C$ of the loop produced by the straight wire ($\mathbf{B_1}$) and the loop ($\mathbf{B_2}$).

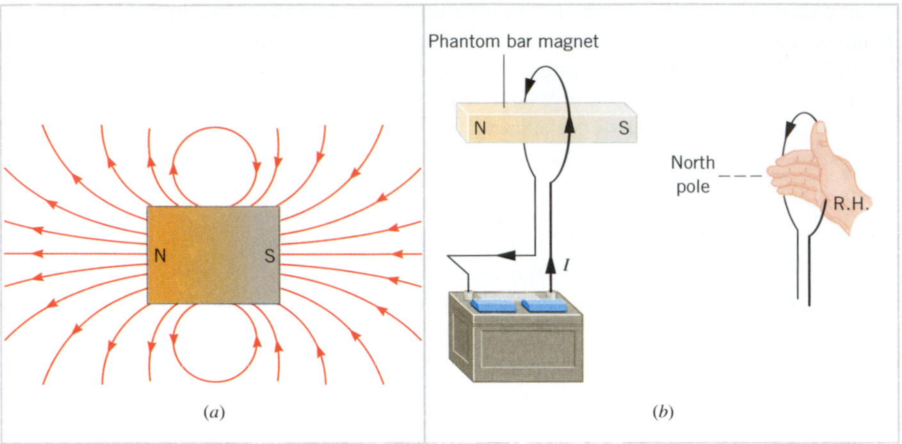

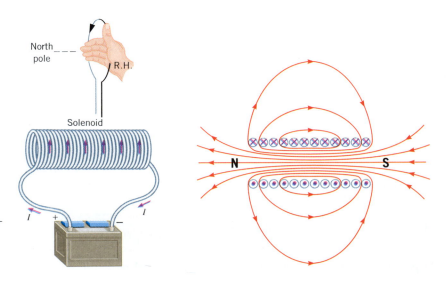

Phantom bar magnet

**Figure 21.30** (*a*) The field lines around the bar magnet resemble those around the loop in Figure 21.28*a*. (*b*) The current loop can be imagined to be a phantom bar magnet with a north pole and a south pole.

A comparison of the magnetic field lines around the current loop in Figure 21.28*a* with those in the vicinity of the short bar magnet in Figure 21.30*a* shows that the two patterns are quite similar. Not only are the patterns similar, but the loop itself behaves as a bar magnet with a "north pole" on one side and a "south pole" on the other side. To emphasize that the loop may be imagined to be a bar magnet, Figure 21.30*b* includes a "phantom" bar magnet at the center of the loop. The side of the loop that acts like a north pole can be determined with the aid of RHR-2; curl the fingers of the right hand into the shape of a half-circle, point the thumb along the current *I*, and place the curled fingers at the center of the loop. The fingers not only point in the direction of **B**, but they also point toward the north pole.

Because a current-carrying loop acts like a bar magnet, two adjacent loops can be either attracted to or repelled from each other, depending on the relative directions of the currents. Figure 21.31 includes a "phantom" magnet for each loop and shows that the loops are attracted to each other when the currents are in the same direction and repelled from each other when the currents are in opposite directions. This behavior is analogous to that of the two long, straight wires discussed in Example 8 (see Figure 21.26).

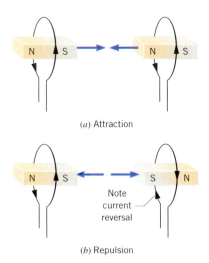

(*a*) Attraction

(*b*) Repulsion

**Figure 21.31** (*a*) The two current loops attract each other if the directions of the currents are the same and (*b*) repel each other if the directions are opposite. The "phantom" magnets help explain the attraction and repulsion.

## A SOLENOID

A solenoid is a long coil of wire in the shape of a helix (see Figure 21.32). If the wire is wound so the turns are packed close to each other and the solenoid is long compared to its diameter, the magnetic field lines have the appearance shown in the drawing. Notice that the field inside the solenoid and away from its ends is nearly constant in magnitude and directed parallel to the axis. The direction of the field inside the solenoid is given by

**Figure 21.32** A solenoid and a cross-sectional view of it, showing the magnetic field lines and the north and south poles.

RHR-2, just as it is for a circular current loop. The magnitude of the magnetic field in the interior of a long solenoid is

***Interior of a long solenoid***
$$B = \mu_0 n I \qquad (21.7)$$

where $n$ is the number of turns per unit length of the solenoid and $I$ is the current. If, for example, the solenoid contains 100 turns and has a length of 0.05 m, the number of turns per unit length is $n = (100 \text{ turns})/(0.05 \text{ m}) = 2000 \text{ turns/m}$. The magnetic field outside the solenoid is not constant and is much weaker than the interior field. In fact, the magnetic field outside is nearly zero if the length of the solenoid is much greater than its diameter.

As with a single loop of wire, a solenoid can also be imagined to be a bar magnet, for the solenoid is just an array of connected current loops. And, as with a circular current loop, the location of the north pole can be determined with RHR-2. Figure 21.32 shows that the left end of the solenoid acts as a north pole, and the right end behaves as a south pole. Solenoids are often referred to as *electromagnets,* and they have several advantages over permanent magnets. For one thing, the strength of the magnetic field can be altered by changing the current and/or the number of turns per unit length. Furthermore, the north and south poles of an electromagnet can be readily switched by reversing the current.

Applications of the magnetic field produced by a current-carrying solenoid are widespread. An exciting medical application is in the technique of magnetic resonance imaging (MRI). With this technique, detailed pictures of the internal parts of the body can be obtained in a noninvasive way that involves none of the risks inherent in the use of X-rays. Figure 21.33 shows a patient who has just been removed from a magnetic resonance imaging machine. The opening behind the patient provides access to the interior of a solenoid, which is typically made from superconducting wire. The superconducting wire facilitates the use of large currents to produce a strong magnetic field. In the presence of this field, the nuclei of certain atoms can be made to behave as tiny radio transmitters and emit radio waves similar to those used by FM stations. The hydrogen atom, which is so prevalent in the human body, can be made to behave in this fashion. The strength of the magnetic field determines where a given collection of hydrogen atoms will "broadcast" on an imaginary FM dial. With a magnetic field that has a slightly different strength at different places, it is possible to associate the location on this imaginary FM dial with a physical location within the body. Computer processing of these locations produces the magnetic resonance image. When hydrogen atoms are used in this way, the image is essentially a map showing their distribution within the body. Remarkably detailed magnetic resonance images can now be obtained, such as those shown in Figure 21.34. They provide doctors with a powerful diagnostic tool that complements those available from X-ray and other techniques. Surgeons can now perform operations more accurately by stepping inside specially designed MRI scanners along with the patient and seeing live MRI images of the area into which they are cutting.

Television sets and computer display monitors use electromagnets (solenoids) to produce images by exerting magnetic forces on moving electrons. An evacuated glass tube, called a cathode-ray tube (CRT), contains an electron gun that sends a narrow beam of high-speed electrons toward the screen of the tube, as Figure 21.35*a* illustrates. The inner surface of the screen is covered with a phosphor coating, and when the electrons strike it, they generate a spot of visible light. This spot is called a pixel (a contraction of "picture element").

To create a black-and-white picture, the electron beam is scanned rapidly from left to right across the screen. As the beam makes each horizontal scan, the number of electrons per second striking the screen is changed by electronics controlling the electron gun, making the scan line brighter in some places and darker in others. When the beam reaches the right side of the screen, it is turned off and returned to the left side slightly below where it started (see part *b* of the figure). The beam is then scanned across the next line, and so on. In current TV sets, a complete picture consists of 525 scan lines (or 625 in Europe) and is formed in $\frac{1}{30}$ of a second. High-definition TV sets have about 1100 scan lines, giving a much sharper, more detailed picture.

The electron beam is deflected by a pair of electromagnets placed around the neck of the tube, between the electron gun and the screen. One electromagnet is responsible for producing the horizontal deflection of the beam and the other for the vertical deflection. For clarity, Figure 21.35*a* shows the net magnetic field generated by the electromagnets at

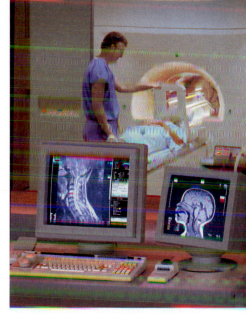

**Figure 21.33** A magnetic resonance imaging (MRI) machine. The patient has just been removed from the large semicircular opening in the background, which is the interior of a solenoid. MRI scans of the spinal cord and head can be seen on the monitors. (© Lester Lefkowitz/Taxi/Getty Images)

***The physics of*** **magnetic resonance imaging (MRI).**

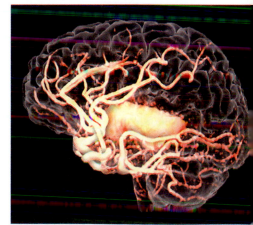

**Figure 21.34** Magnetic resonance imaging provides one way to diagnose brain disorders. This three-dimensional magnetic resonance angiogram scan shows a human brain after a stroke. Major arteries are white. The central region (yellow) is an area of bleeding. (© Zephyr/Photo Researchers)

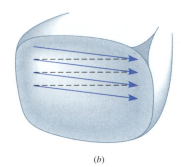
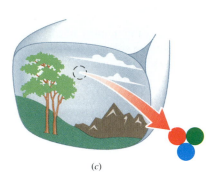

**Figure 21.35** (*a*) A cathode-ray tube contains an electron gun, a magnetic field for deflecting the electron beam, and a phosphor-coated screen. A color TV actually uses three guns, although only one is shown here for clarity. (*b*) The image is formed by scanning the electron beam across the screen. (*c*) The red, green, and blue phosphors of a color TV.

**The physics of** television screens and computer display monitors.

one instant, and not the electromagnets themselves. The electric current in the electromagnets produces a net magnetic field that exerts a force on the moving electrons, causing their trajectories to bend and reach different points on the screen. Changing the current changes the field, so the electrons can be deflected to any point on the screen.

A color TV operates with three electron guns instead of one. And the single phosphor of a black-and-white TV is replaced by a large number of three-dot clusters of phosphors that glow red, green, and blue when struck by an electron beam, as indicated in Figure 21.35*c*. Each red, green, and blue color in a cluster is produced when electrons from one of the three guns strike the corresponding phosphor dot. The three dots are so close together that, from a normal viewing distance, they cannot be separately distinguished. Red, green, and blue are primary colors, so virtually all other colors can be created by varying the intensities of the three beams focused on a cluster.

## 21.8 *Ampere's Law*

We have seen that an electric current creates a magnetic field. However, the magnitude and direction of the field at any point in space depends on the specific geometry of the wire carrying the current. For instance, distinctly different magnetic fields surround a long, straight wire, a circular loop of wire, and a solenoid. Although different, each of these fields can be obtained from a general law known as ***Ampere's law,*** which is valid for a wire of any geometrical shape. Ampere's law specifies the relationship between a current and its associated magnetic field.

To see how Ampere's law is stated, consider Figure 21.36, which shows two wires carrying currents $I_1$ and $I_2$. In general, there may be any number of currents. Around the wires we construct an arbitrarily shaped but closed path. This path encloses a surface and is constructed from a large number of short segments, each of length $\Delta\ell$. Ampere's law deals with the product of $\Delta\ell$ and $B_\parallel$ for each segment, where $B_\parallel$ is the component of the magnetic field that is *parallel* to $\Delta\ell$ (see the blow-up view in the drawing). For magnetic fields that do not change as time passes, the law states that the sum of all the $B_\parallel \Delta\ell$ terms is proportional to the net current $I$ passing through the surface enclosed by the path. For the specific example in Figure 21.36, we see that $I = I_1 + I_2$. Ampere's law is stated in equation form as follows:

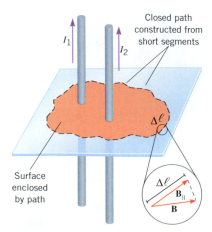

**Figure 21.36** This setup is used in the text to explain Ampere's law.

■ **AMPERE'S LAW FOR STATIC MAGNETIC FIELDS**

For any current geometry that produces a magnetic field that does not change in time,

$$\Sigma B_\parallel \Delta\ell = \mu_0 I \qquad (21.8)$$

where $\Delta\ell$ is a small segment of length along a closed path of arbitrary shape around the current, $B_\parallel$ is the component of the magnetic field parallel to $\Delta\ell$, $I$ is the net current passing through the surface bounded by the path, and $\mu_0$ is the permeability of free space. The symbol $\Sigma$ indicates that the sum of all $B_\parallel \Delta\ell$ terms must be taken around the closed path.

To illustrate the use of Ampere's law, we apply it in Example 11 to the special case of the current in a long, straight wire and show that it leads to the proper expression for the magnetic field.

### Example 11 | An Infinitely Long, Straight, Current-Carrying Wire

Use Ampere's law to obtain the magnetic field produced by the current in an infinitely long, straight wire.

**Reasoning** Figure 21.23a shows that compass needles point in a circular pattern around the wire, so we know that the magnetic field lines are circular. Therefore, it is convenient to use a circular path of radius $r$ when applying Ampere's law, as Figure 21.37 indicates.

**Solution** Along the circular path in Figure 21.37, the magnetic field is everywhere parallel to $\Delta \ell$ and has a constant magnitude, since each point is at the same distance from the wire. Thus, $B_\parallel = B$ and, according to Ampere's law, we have

$$\Sigma B_\parallel \Delta \ell = B(\Sigma \Delta \ell) = \mu_0 I$$

But $\Sigma \Delta \ell$ is just the circumference $2\pi r$ of the circle, so Ampere's law reduces to

$$B(\Sigma \Delta \ell) = B(2\pi r) = \mu_0 I$$

Dividing both sides by $2\pi r$ shows that $\boxed{B = \mu_0 I/(2\pi r)}$, as given earlier in Equation 21.5.

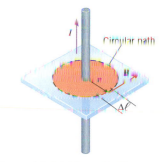

**Figure 21.37** Example 11 uses Ampere's law to find the magnetic field in the vicinity of this long, straight, current-carrying wire.

## 21.9 Magnetic Materials

### FERROMAGNETISM

The similarity between the magnetic field lines in the neighborhood of a bar magnet and those around a current loop suggests that the magnetism in each case arises from a common cause. The field that surrounds the loop is created by the charges moving in the wire. The magnetic field around a bar magnet is also due to the motion of charges, but the motion is not that of a bulk current through the magnetic material. Instead, the motion responsible for the magnetism is that of the electrons within the atoms of the material.

The magnetism produced by electrons within an atom can arise from two motions. First, each electron orbiting the nucleus behaves like an atomic-sized loop of current that generates a small magnetic field, similar to the field created by the current loop in Figure 21.28a. Second, each electron possesses a spin that also gives rise to a magnetic field. The net magnetic field created by the electrons within an atom is due to the combined fields created by their orbital and spin motions.

In most substances the magnetism produced at the atomic level tends to cancel out, with the result that the substance is nonmagnetic overall. However, there are some materials, known as *ferromagnetic materials,* in which the cancellation does not occur for groups of approximately $10^{16} - 10^{19}$ neighboring atoms, because they have electron spins that are naturally aligned parallel to each other. This alignment results from a special type of quantum mechanical* interaction between the spins. The result of the interaction is a small but highly magnetized region of about 0.01 to 0.1 mm in size, depending on the nature of the material; this region is called a *magnetic domain.* Each domain behaves as a small magnet with its own north and south poles. Common ferromagnetic materials are iron, nickel, cobalt, chromium dioxide, and alnico (an *al*uminum–*nickel–co*balt alloy).

### INDUCED MAGNETISM

Often the magnetic domains in a ferromagnetic material are arranged randomly, as Figure 21.38a illustrates for a piece of iron. In such a situation, the magnetic fields of the domains cancel each other, so the iron displays little, if any, overall magnetism. However, an unmagnetized piece of iron can be magnetized by placing it in an external magnetic field

---

* The branch of physics called quantum mechanics is mentioned in Section 29.5, although a detailed discussion of it is beyond the scope of this book.

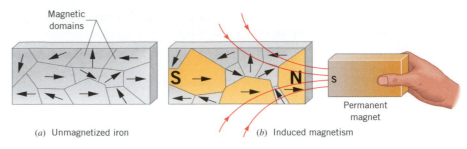

(a) Unmagnetized iron    (b) Induced magnetism

**Figure 21.38** (*a*) Each magnetic domain is a highly magnetized region that behaves like a small magnet (represented by an arrow whose head indicates a north pole). An unmagnetized piece of iron consists of many domains that are randomly aligned. The size of each domain is exaggerated for clarity. (*b*) The external magnetic field of the permanent magnet causes those domains that are parallel or nearly parallel to the field to grow in size (shown in gold).

provided by a permanent magnet or an electromagnet. The external magnetic field penetrates the unmagnetized iron and *induces* (or brings about) a state of magnetism in the iron by causing two effects. Those domains whose magnetism is parallel or nearly parallel to the external magnetic field grow in size at the expense of other domains that are not so oriented. Part *b* of the drawing shows the growing domains in gold. In addition, the magnetic alignment of some domains may rotate and become more oriented in the direction of the external field. The resulting preferred alignment of the domains gives the iron an overall magnetism, so the iron behaves like a magnet with associated north and south poles. In some types of ferromagnetic materials, such as the chromium dioxide used in cassette tapes, the domains remain aligned for the most part when the external magnetic field is removed, and the material thus becomes permanently magnetized.

The magnetism induced in a ferromagnetic material can be surprisingly large, even in the presence of a weak external field. For instance, it is not unusual for the induced magnetic field to be a hundred to a thousand times stronger than the external field that causes the alignment. For this reason, high-field electromagnets are constructed by wrapping the current-carrying wire around a solid core made from iron or other ferromagnetic material.

Induced magnetism explains why a permanent magnet sticks to a refrigerator door and why an electromagnet can pick up scrap iron at a junkyard. The photo in Figure 21.39 illustrates yet another example of induced magnetism. Notice in Figure 21.38*b* that there is a north pole at the end of the iron that is closest to the south pole of the permanent magnet. The net result is that the two opposite poles give rise to an attraction between the iron and the permanent magnet. Conversely, the north pole of the permanent magnet would also attract the piece of iron by inducing a south pole in the nearest side of the iron. In nonferromagnetic materials, such as aluminum and copper, the formation of

**Figure 21.39** Because of induced magnetism, powerful magnetic grippers enable this worker to climb the metal bulkhead of a ship. (© Trident Technologies. Reproduced with permission.)

magnetic domains does not occur, so magnetism cannot be induced into these substances. Consequently, magnets do not stick to aluminum cans or to copper pennies.

## MAGNETIC TAPE RECORDING

The process of magnetic tape recording uses induced magnetism, as Figure 21.40 illustrates. The weak electrical signal from a microphone is routed to an amplifier where it is amplified. The current from the output of the amplifier is then sent to the recording head, which is a coil of wire wrapped around an iron core. The iron core has the approximate shape of a horseshoe with a small gap between the two ends. The ferromagnetic iron substantially enhances the magnetic field produced by the current in the wire.

When there is a current in the coil, the recording head becomes an electromagnet with a north pole at one end and a south pole at the other end. The magnetic field lines pass through the iron core and cross the gap. Within the gap, the lines are directed from the north pole to the south pole. Some of the field lines in the gap "bow outward," as Figure 21.40 indicates, the bowed region of magnetic field being called the *fringe field*. The fringe field penetrates the magnetic coating on the tape and induces magnetism in the coating. This induced magnetism is retained when the tape leaves the vicinity of the recording head and, thus, provides a means for storing audio information. Audio information is stored, because at any instant in time the way in which the tape is magnetized depends on the amount and direction of current in the recording head. The current, in turn, depends on the sound picked up by the microphone, so that changes in the sound that occur from moment to moment are preserved as changes in the tape's induced magnetism.

## MAGLEV TRAINS

A magnetically levitated train—maglev, for short—uses forces that arise from induced magnetism to levitate or float above a guideway. Since it rides a few centimeters above the guideway, a maglev does not need wheels. Freed from friction with the guideway, the train can achieve significantly greater speeds than do conventional trains. For example, the Transrapid maglev in Figure 21.41 has achieved speeds of 110 m/s (250 mph).

Figure 21.41*a* shows that the Transrapid maglev achieves levitation with electromagnets mounted on arms that extend around and under the guideway. When a current is sent to an electromagnet, the resulting magnetic field creates induced magnetism in a rail mounted in the guideway. The upward attractive force from the induced magnetism is balanced by the weight of the train, so the train moves without touching the rail or the guideway.

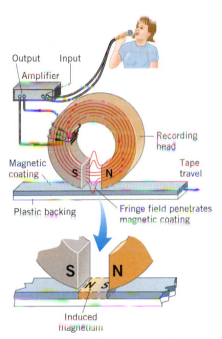

*The physics of magnetic tape recording.*

**Figure 21.40** The magnetic fringe field of the recording head penetrates the magnetic coating on the tape and magnetizes it.

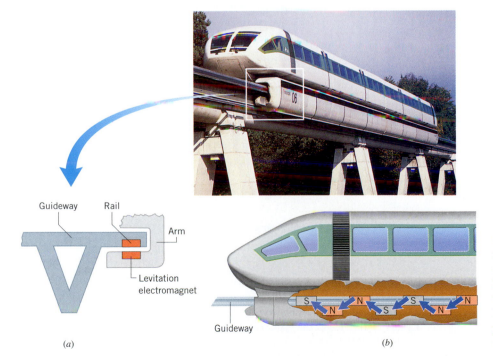

Guideway
Rail
Arm
Levitation electromagnet
Guideway

(a)

(b)

**Figure 21.41** (*a*) The Transrapid maglev (a German train) has achieved speeds of 110 m/s (250 mph). The levitation electromagnets are drawn up toward the rail in the guideway, levitating the train. (*b*) The magnetic propulsion system. (© Courtesy Deutsche Bundesbahn/Transrapid)

**The physics of
a magnetically levitated train.**

Magnetic levitation only lifts the train and does not move it forward. Figure 21.41*b* illustrates how magnetic propulsion is achieved. In addition to the levitation electromagnets, propulsion electromagnets are also placed along the guideway. By controlling the direction of the currents in the train and guideway electromagnets, it is possible to create an unlike pole in the guideway just ahead of each electromagnet on the train and a like pole just behind. Each electromagnet on the train is thus both pulled and pushed forward by electromagnets in the guideway. By adjusting the timing of the like and unlike poles in the guideway, the speed of the train can be adjusted. Reversing the poles in the guideway electromagnets serves to brake the train.

# Concept Summary

This summary presents an abridged version of the chapter, including the important equations and all available learning aids. For convenient reference, the learning aids (including the text's examples) are placed next to or immediately after the relevant equation or discussion. The following learning aids may be found on-line at **www.wiley.com/college/cutnell**:

| | |
|---|---|
| **Interactive LearningWare** examples are solved according to a five-step interactive format that is designed to help you develop problem-solving skills. | **Concept Simulations** are animated versions of text figures or animations that illustrate important concepts. You can control parameters that affect the display, and we encourage you to experiment. |
| **Interactive Solutions** offer specific models for certain types of problems in the chapter homework. The calculations are carried out interactively. | **Self-Assessment Tests** include both qualitative and quantitative questions. Extensive feedback is provided for both incorrect and correct answers, to help you evaluate your understanding of the material. |

| Topic | Discussion | Learning Aids |
|---|---|---|
| | **21.1 Magnetic Fields** | |
| North and south poles | A magnet has a north pole and a south pole. The north pole is the end that points toward the north magnetic pole of the earth when the magnet is freely suspended. Like magnetic poles repel each other, and unlike poles attract each other. | |
| Direction of the magnetic field | A magnetic field exists in the space around a magnet. The magnetic field is a vector whose direction at any point is the direction indicated by the north pole of a small compass needle placed at that point. | |
| Magnetic field lines | As an aid in visualizing the magnetic field, magnetic field lines are drawn in the vicinity of a magnet. The lines appear to originate from the north pole and end on the south pole. The magnetic field at any point in space is tangent to the magnetic field line at the point. Furthermore, the strength of the magnetic field is proportional to the number of lines per unit area that passes through a surface oriented perpendicular to the lines. | |
| | **21.2 The Force That a Magnetic Field Exerts on a Moving Charge** | |
| Magnetic force (direction) | The direction of the magnetic force acting on a charge moving with a velocity **v** in a magnetic field **B** is perpendicular to both **v** and **B**. For a positive charge the direction can be determined with the aid of Right-Hand Rule No. 1 (see below). The magnetic force on a moving negative charge is opposite to the force on a moving positive charge. | |
| Right-Hand Rule No. 1 | Extend the right hand so the fingers point along the direction of the magnetic field **B** and the thumb points along the velocity **v** of the charge. The palm of the hand then faces in the direction of the magnetic force **F** that acts on a positive charge. | |
| | The magnitude $B$ of the magnetic field at any point in space is defined as | |
| Magnetic field (magnitude) | $$B = \frac{F}{q_0 v \sin \theta}$$ (21.1) | **Example 1** |
| The tesla and the gauss | where $F$ is the magnitude of the magnetic force that acts on a positive charge $q_0$ whose velocity **v** makes an angle $\theta$ with respect to the magnetic field. The SI unit for the magnetic field is the tesla (T). Another, smaller unit for the magnetic field is the gauss; 1 gauss = $10^{-4}$ tesla. The gauss is not an SI unit. | |

| Topic | Discussion | Learning Aids |
|-------|-----------|---------------|

**21.3 The Motion of a Charged Particle in a Magnetic Field**

When a charged particle moves in a region that contains both magnetic and electric fields, the net force on the particle is the vector sum of the magnetic and electric forces.

**Example 7**

**A constant magnetic force does no work**

A magnetic force does no work on a particle, because the direction of the force is always perpendicular to the motion of the particle. Being unable to do work, the magnetic force cannot change the kinetic energy, and hence the speed, of the particle; however, the magnetic force does change the direction in which the particle moves.

When a particle of charge $q$ and mass $m$ moves with speed $v$ perpendicular to a uniform magnetic field of magnitude $B$, the magnetic force causes the particle to move on a circular path of radius

**Radius of circular path**

$$r = \frac{mv}{qB} \qquad (21.2)$$

Concept Simulation 21.1
**Examples 3, 4**
Interactive LearningWare 21.1
Interactive Solution 21.21

**21.4 The Mass Spectrometer**

The mass spectrometer is an instrument for measuring the abundance of ionized atoms or molecules that have different masses. The atoms or molecules are ionized ($+e$), accelerated to a speed $v$ by a potential difference $V$, and sent into a uniform magnetic field of magnitude $B$. The magnetic field causes the particles (each with a mass $m$) to move on a circular path of radius $r$. The relation between $m$ and $B$ is

**Relation between mass and magnetic field**

$$m = \left(\frac{er^2}{2V}\right) B^2$$

---

*Use Self-Assessment Test 21.1 to evaluate your understanding of Sections 21.1–21.4.*

---

**21.5 The Force on a Current in a Magnetic Field**

An electric current, being composed of moving charges, can experience a magnetic force when placed in a magnetic field of magnitude $B$. For a straight wire that has a length $L$ and carries a current $I$, the magnetic force has a magnitude of

**Force on a current (magnitude)**

$$F = ILB \sin \theta \qquad (21.3)$$

**Example 5**
Interactive LearningWare 21.2
Interactive Solution 21.34

where $\theta$ is the angle between the directions of the current and the magnetic field. The direction of the force is perpendicular to both the current and the magnetic field and is given by Right-Hand Rule No. 1.

**Force on a current (direction)**

**21.6 The Torque on a Current-Carrying Coil**

Magnetic forces can exert a torque on a current-carrying loop of wire and thus cause the loop to rotate. When a current $I$ exists in a coil of wire with $N$ turns, each of area $A$, in the presence of a magnetic field of magnitude $B$, the coil experiences a net torque of magnitude

**Torque on a current-carrying coil**

$$\tau = NIAB \sin \phi \qquad (21.4)$$

**Example 6**
Interactive Solution 21.43

where $\phi$ is the angle between the direction of the magnetic field and the normal to the plane of the coil.

**Magnetic moment**

The quantity $NIA$ is known as the magnetic moment of the coil.

**21.7 Magnetic Fields Produced by Currents**

An electric current produces a magnetic field, with different current geometries giving rise to different field patterns. For an infinitely long, straight wire, the magnetic field lines are circles centered on the wire, and their direction is given by Right-Hand Rule No. 2 (see page 448). The magnitude of the magnetic field at a radial distance $r$ from the wire is

| Topic | Discussion | Learning Aids |
|---|---|---|
| **Long, straight wire** | $$B = \frac{\mu_0 I}{2\pi r} \qquad (21.5)$$ | **Examples 7, 8, 9**<br>**Interactive LearningWare 21.3**<br>**Concept Simulation 21.2** |
| **Permeability of free space** | where $I$ is the current in the wire and $\mu_0$ is a constant known as the permeability of free space ($\mu_0 = 4\pi \times 10^{-7}$ T·m/A). | |
| **Right-Hand Rule No. 2** | Curl the fingers of the right hand into the shape of a half-circle. Point the thumb in the direction of the conventional current $I$, and the tips of the fingers will point in the direction of the magnetic field **B**. | |
| | The magnitude of the magnetic field at the center of a flat circular loop consisting of $N$ turns, each of radius $R$, is | |
| **Center of a circular loop** | $$B = \frac{N \mu_0 I}{2R} \qquad (21.6)$$ | **Concept Simulation 21.3**<br>**Example 10**<br>**Interactive Solution 21.55** |
| | The loop has associated with it a north pole on one side and a south pole on the other side. The side of the loop that behaves like a north pole can be predicted by using Right-Hand Rule No. 2. | |
| | A solenoid is a coil of wire wound in the shape of a helix. Inside a long solenoid the magnetic field is nearly constant and has a magnitude of | |
| **Interior of a solenoid** | $$B = \mu_0 n I \qquad (21.7)$$ | **Concept Simulation 21.4** |
| | where $n$ is the number of turns per unit length of the solenoid. One end of the solenoid behaves like a north pole, and the other end like a south pole. The end that is the north pole can be predicted by using Right-Hand Rule No. 2. | |

### 21.8 *Ampere's Law*

Ampere's law specifies the relationship between a current and its associated magnetic field. For any current geometry that produces a magnetic field that does not change in time, Ampere's law states that

| | | |
|---|---|---|
| **Ampere's law** | $$\Sigma B_\parallel \Delta \ell = \mu_0 I \qquad (21.8)$$ | **Example 11**<br>**Interactive Solution 21.62** |

where $\Delta \ell$ is a small segment of length along a closed path of arbitrary shape around the current, $B_\parallel$ is the component of the magnetic field parallel to $\Delta \ell$, $I$ is the net current passing through the surface bounded by the path, and $\mu_0$ is the permeability of free space. The symbol $\Sigma$ indicates that the sum of all $B_\parallel \Delta \ell$ terms must be taken around the closed path.

 **Use Self-Assessment Test 21.2 to evaluate your understanding of Sections 21.5–21.8.**

### 21.9 *Magnetic Materials*

| | |
|---|---|
| **Ferromagnetic materials** | Ferromagnetic materials, such as iron, are made up of tiny regions called magnetic domains, each of which behaves as a small magnet. In an unmagnetized ferromagnetic material, the domains are randomly aligned. In a permanent magnet, many of the domains are aligned, and a high degree of magnetism results. An unmagnetized ferromagnetic material can be induced into becoming magnetized by placing it in an external magnetic field. |

# Problems

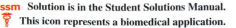

**ssm** Solution is in the Student Solutions Manual.    **www** Solution is available on the World Wide Web at www.wiley.com/college/cutnell

This icon represents a biomedical application.

**Section 21.1 Magnetic Fields,**
**Section 21.2 The Force That a Magnetic Field Exerts on a Moving Charge**

**1. ssm** Due to friction with the air, an airplane has acquired a net charge of $1.70 \times 10^{-5}$ C. The plane moves with a speed of $2.80 \times 10^2$ m/s at an angle $\theta$ with respect to the earth's magnetic field, the magnitude of which is $5.00 \times 10^{-5}$ T. The magnetic force on the airplane has a magnitude of $2.30 \times 10^{-7}$ N. Find the angle $\theta$. (There are two possible angles.)

**2.** A particle with a charge of $+8.4$ $\mu$C and a speed of 45 m/s enters a uniform magnetic field whose magnitude is 0.30 T. For each of the cases in the drawing, find the magnitude and direction of the magnetic force on the particle.

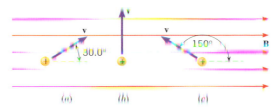

**3.** In a television set, electrons are accelerated from rest through a potential difference of 19 kV. The electrons then pass through a 0.28-T magnetic field that deflects them to the appropriate spot on the screen. Find the magnitude of the maximum magnetic force that an electron can experience.

**4.** Two charged particles move in the same direction with respect to the same magnetic field. Particle 1 travels three times faster than particle 2. However, each particle experiences a magnetic force of the same magnitude. Find the ratio $q_1/q_2$ of the magnitudes of the charges.

**5.** **ssm** At a certain location, the horizontal component of the earth's magnetic field is $2.5 \times 10^{-5}$ T, due north. A proton moves eastward with just the right speed, so the magnetic force on it balances its weight. Find the speed of the proton.

**6.** In New England, the horizontal component of the earth's magnetic field has a magnitude of $1.6 \times 10^{-5}$ T. An electron is shot vertically straight up from the ground with a speed of $2.1 \times 10^6$ m/s. What is the magnitude of the acceleration caused by the magnetic force? Ignore the gravitational force acting on the electron.

**7.** An electron is moving through a magnetic field whose magnitude is $8.70 \times 10^{-4}$ T. The electron experiences only a magnetic force and has an acceleration of magnitude $3.50 \times 10^{14}$ m/s². At a certain instant, it has a speed of $6.80 \times 10^6$ m/s. Determine the angle $\theta$ (less than 90°) between the electron's velocity and the magnetic field.

**\*8.** The drawing shows a parallel plate capacitor that is moving with a speed of 32 m/s through a 3.6-T magnetic field. The velocity **v** is perpendicular to the magnetic field. The electric field within the capacitor has a value of 170 N/C, and each plate has an area of $7.5 \times 10^{-4}$ m². What is the magnetic force (magnitude and direction) exerted on the positive plate of the capacitor?

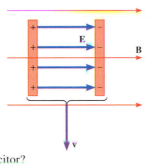

**\*9.** **ssm www** The electrons in the beam of a television tube have a kinetic energy of $2.40 \times 10^{-15}$ J. Initially, the electrons move horizontally from west to east. The vertical component of the earth's magnetic field points down, toward the surface of the earth, and has a magnitude of $2.00 \times 10^{-5}$ T. (a) In what direction are the electrons deflected by this field component? (b) What is the acceleration of an electron in part (a)?

**Section 21.3 The Motion of a Charged Particle in a Magnetic Field,**
**Section 21.4 The Mass Spectrometer**

**10.** A magnetic field has a magnitude of $1.2 \times 10^{-3}$ T, and an electric field has a magnitude of $4.6 \times 10^3$ N/C. Both fields point

in the same direction. A positive 1.8-$\mu$C charge moves at a speed of $3.1 \times 10^6$ m/s in a direction that is perpendicular to both fields. Determine the magnitude of the net force that acts on the charge.

**11.** An electron moves at a speed of $6.0 \times 10^6$ m/s perpendicular to a constant magnetic field. The path is a circle of radius $1.3 \times 10^{-3}$ m. (a) Draw a sketch showing the magnetic field and the electron's path. (b) What is the magnitude of the field? (c) Find the magnitude of the electron's acceleration.

**12.** A charged particle enters a uniform magnetic field and follows the circular path shown in the drawing. (a) Is the particle positively or negatively charged? Why? (b) The particle's speed is 140 m/s, the magnitude of the magnetic field is 0.48 T, and the radius of the path is 960 m. Determine the mass of the particle, given that its charge has a magnitude of $8.2 \times 10^{-4}$ C.

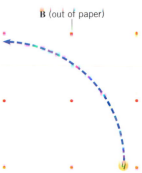

**13.** **ssm www** A charged particle with a charge-to-mass ratio of $q/m = 5.7 \times 10^8$ C/kg travels on a circular path that is perpendicular to a magnetic field whose magnitude is 0.72 T. How much time does it take for the particle to complete one revolution?

**14.** The solar wind is a thin, hot gas given off by the sun. Charged particles in this gas enter the magnetic field of the earth and can experience a magnetic force. Suppose a charged particle traveling with a speed of $9.0 \times 10^6$ m/s encounters the earth's magnetic field at an altitude where the field has a magnitude of $1.2 \times 10^{-7}$ T. Assuming that the particle's velocity is perpendicular to the magnetic field, find the radius of the circular path on which the particle would move if it were (a) an electron and (b) a proton.

**15.** **ssm** A beam of protons moves in a circle of radius 0.25 m. The protons move perpendicular to a 0.30-T magnetic field. (a) What is the speed of each proton? (b) Determine the magnitude of the centripetal force that acts on each proton.

**16.** Review Conceptual Example 2 before attempting this problem. Derive an expression for the magnitude $v$ of the velocity "selected" by the velocity selector. This expression should give $v$ in terms of the strengths $E$ and $B$ of the electric and magnetic fields, respectively.

**17.** Two isotopes of carbon, carbon-12 and carbon-13, have masses of $19.93 \times 10^{-27}$ kg and $21.59 \times 10^{-27}$ kg, respectively. These two isotopes are singly ionized ($+e$) and each is given a speed of $6.667 \times 10^5$ m/s. The ions then enter the bending region of a mass spectrometer where the magnetic field is 0.8500 T. Determine the spatial separation between the two isotopes after they have traveled through a half-circle.

**\*18.** The ion source in a mass spectrometer produces both singly and doubly ionized species, $X^+$ and $X^{2+}$. The difference in mass between these species is too small to be detected. Both species are accelerated through the same electric potential difference, and both experience the same magnetic field, which causes them to move on circular paths. The radius of the path for the species $X^+$ is $r_1$, while the radius for species $X^{2+}$ is $r_2$. Find the ratio $r_1/r_2$ of the radii.

**\*19.** **ssm** A particle of charge $+7.3$ $\mu$C and mass $3.8 \times 10^{-8}$ kg is traveling perpendicular to a 1.6-T magnetic field, as the drawing shows. The speed of the particle is 44 m/s. (a) What is the value of the angle $\theta$, such that the particle's subsequent path will intersect the $y$ axis at the greatest possible value of $y$? (b) Determine this value of $y$.

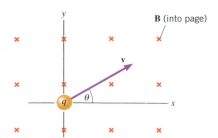

*20. Review Conceptual Example 2 as background for this problem. A charged particle moves through a velocity selector at a constant speed in a straight line. The electric field of the velocity selector is $3.80 \times 10^3$ N/C, while the magnetic field is 0.360 T. When the electric field is turned off, the charged particle travels on a circular path whose radius is 4.30 cm. Find the charge-to-mass ratio of the particle.

*21. Consult **Interactive Solution 21.21** at **www.wiley.com/college/cutnell** to review a model for solving this problem. A proton with a speed of $3.5 \times 10^6$ m/s is shot into a region between two plates that are separated by a distance of 0.23 m. As the drawing shows, a magnetic field exists between the plates, and it is perpendicular to the velocity of the proton. What must be the magnitude of the magnetic field, so the proton just misses colliding with the opposite plate?

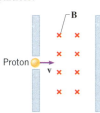

*22. Conceptual Example 4 provides background pertinent to this problem. An electron has a kinetic energy of $2.0 \times 10^{-17}$ J. It moves on a circular path that is perpendicular to a uniform magnetic field of magnitude $5.3 \times 10^{-5}$ T. Determine the radius of the path.

*23. **ssm www** A particle of mass $6.0 \times 10^{-8}$ kg and charge $+7.2$ $\mu$C is traveling due east. It enters perpendicularly a magnetic field whose magnitude is 3.0 T. After entering the field, the particle completes one-half of a circle and exits the field traveling due west. How much time does the particle spend in the magnetic field?

*24. A positively charged particle of mass $7.2 \times 10^{-8}$ kg is traveling due east with a speed of 85 m/s and enters a 0.31-T uniform magnetic field. The particle moves through one-quarter of a circle in a time of $2.2 \times 10^{-3}$ s, at which time it leaves the field heading due south. All during the motion the particle moves perpendicular to the magnetic field. (a) What is the magnitude of the magnetic force acting on the particle? (b) Determine the magnitude of its charge.

**25. The drawing shows a proton at the coordinate origin, as well as a target located at the coordinates $x = -0.10$ m and $y = -0.10$ m. A uniform magnetic field of magnitude of 0.010 T is directed perpendicularly into the plane of the paper. The proton can be projected in the plane of the paper only, along the positive or negative $x$ or $y$ axis. Thus, there are four possible directions for the initial velocity of the proton. The proton can be made to hit the target for only two of the four directions. The speed at which the proton is projected is the same for either of the two paths leading to the target. Find the speed.

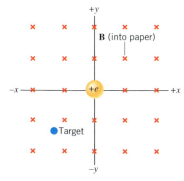

### Section 21.5 The Force on a Current in a Magnetic Field

26. A 45-m length of wire is stretched horizontally between two vertical posts. The wire carries a current of 75 A and experiences a

magnetic force of 0.15 N. Find the magnitude of the earth's magnetic field at the location of the wire, assuming the field makes an angle of $60.0°$ with respect to the wire.

27. **ssm** An electric power line carries a current of 1400 A in a location where the earth's magnetic field is $5.0 \times 10^{-5}$ T. The line makes an angle of $75°$ with respect to the field. Determine the magnitude of the magnetic force on a 120-m length of line.

28. A wire carries a current of 0.66 A. This wire makes an angle of $58°$ with respect to a magnetic field of magnitude $4.7 \times 10^{-5}$ T. The wire experiences a magnetic force of magnitude $7.1 \times 10^{-5}$ N. What is the length of the wire?

29. A square coil of wire containing a single turn is placed in a uniform 0.25-T magnetic field, as the drawing shows. Each side has a length of 0.32 m, and the current in the coil is 12 A. Determine the magnitude of the magnetic force on each of the four sides.

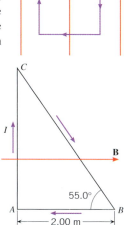

30. The triangular loop of wire shown in the drawing carries a current of $I = 4.70$ A. A uniform magnetic field is directed parallel to side $AB$ of the triangle and has a magnitude of 1.80 T. (a) Find the magnitude and direction of the magnetic force exerted on each side of the triangle. (b) Determine the magnitude of the net force exerted on the triangle.

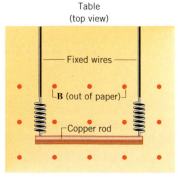

31. **ssm** A wire of length 0.655 m carries a current of 21.0 A. In the presence of a 0.470-T magnetic field, the wire experiences a force of 5.46 N. What is the angle (less than $90°$) between the wire and the magnetic field?

32. Two insulated wires, each 2.40 m long, are taped together to form a two-wire unit that is 2.40 m long. One wire carries a current of 7.00 A; the other carries a smaller current $I$ in the opposite direction. The two-wire unit is placed at an angle of $65.0°$ relative to a magnetic field whose magnitude is 0.360 T. The magnitude of the net magnetic force experienced by the two-wire unit is 3.13 N. What is the current $I$?

*33. **ssm** A copper rod of length 0.85 m is lying on a frictionless table (see the drawing). Each end of the rod is attached to a fixed wire by an unstretched spring whose spring constant is $k = 75$ N/m. A magnetic field with a strength of 0.16 T is oriented perpendicular to the surface of the table. (a) What must be the direction of the current in the copper rod that causes the springs to stretch? (b) If the current is 12 A, by how much does each spring stretch?

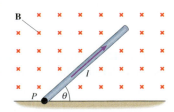

*34. Consult **Interactive Solution 21.34** at **www.wiley.com/college/cutnell** to explore a model for solving this problem. The drawing shows a thin, uniform rod, which has a length of 0.45 m and a mass of 0.094 kg. This rod lies in the plane of the paper and is attached to the floor by a hinge at point $P$. A uniform magnetic field of 0.36 T is directed perpendicularly into the plane of the paper. There is a current $I = 4.1$ A in the rod, which does not rotate clockwise or counterclockwise. Find the angle $\theta$. (Hint: The magnetic force may be taken to act at the center of gravity.)

†† **35.** The two conducting rails in the drawing are tilted upward so they each make an angle of 30.0° with respect to the ground. The vertical magnetic field has a magnitude of 0.050 T. The 0.20-kg aluminum rod (length = 1.6 m) slides without friction down the rails at a constant velocity. How much current flows through the bar?

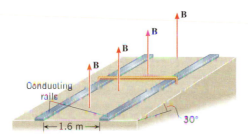

Conducting rails

|← 1.6 m →|

30°

### Section 21.6 The Torque on a Current-Carrying Coil

**36.** The 1200-turn coil in a dc motor has an area per turn of $1.1 \times 10^{-2}$ m². The design for the motor specifies that the magnitude of the maximum torque is 5.8 N·m when the coil is placed in a 0.20-T magnetic field. What is the current in the coil?

**37. ssm** A circular coil of wire has a radius of 0.10 m. The coil has 50 turns and a current of 15 A, and is placed in a magnetic field whose magnitude is 0.20 T. (a) Determine the magnetic moment of the coil. (b) What is the maximum torque the coil can experience in this field?

**38.** Two coils have the same number of circular turns and carry the same current. Each rotates in a magnetic field as in Figure 21.19. Coil 1 has a radius of 5.0 cm and rotates in a 0.18-T field. Coil 2 rotates in a 0.42-T field. Each coil experiences the same maximum torque. What is the radius (in cm) of coil 2?

**39.** The maximum torque experienced by a coil in a 0.75-T magnetic field is $8.4 \times 10^{-4}$ N·m. The coil is circular and consists of only one turn. The current in the coil is 3.7A. What is the length of the wire from which the coil is made?

**40.** Two pieces of the same wire have the same length. From one piece, a square coil containing a single loop is made. From the other, a circular coil containing a single loop is made. The coils carry different currents. When placed in the same magnetic field with the same orientation, they experience the same torque. What is the ratio $I_{square}/I_{circle}$ of the current in the square coil to that in the circular coil?

**41. ssm www** The rectangular loop in the drawing consists of 75 turns and carries a current of $I = 4.4$ A. A 1.8-T magnetic field is directed along the $+y$ axis. The loop is free to rotate about the $z$ axis. (a) Determine the magnitude of the net torque exerted on the loop and (b) state whether the 35° angle will increase or decrease.

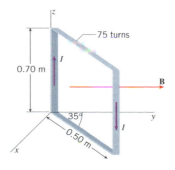

$z$

75 turns

$I$

0.70 m

$B$

35°

$y$

0.50 m

$I$

$x$

**42.** A coil carries a current and experiences a torque due to a magnetic field. The value of the torque is 80.0% of the maximum possi-

ble torque. (a) What is the smallest angle between the magnetic field and the normal to the plane of the coil? (b) Make a drawing, showing how this coil would be oriented relative to the magnetic field. Be sure to include the angle in the drawing.

* **43.** Consult **Interactive Solution** 21.43 at www.wiley.com/college/cutnell to see how this problem can be solved. The coil in Figure 21.70a contains 410 turns and has an area per turn of $3.1 \times 10^{-3}$ m². The magnetic field is 0.23 T, and the current in the coil is 0.26 A. A brake shoe is pressed perpendicularly against the shaft to keep the coil from turning. The coefficient of static friction between the shaft and the brake shoe is 0.76. The radius of the shaft is 0.012 m. What is the magnitude of the minimum normal force that the brake shoe exerts on the shaft?

* **44.** A square coil and a rectangular coil are each made from the same length of wire. Each contains a single turn. The long sides of the rectangle are twice as long as the short sides. Find the ratio $\tau_{square}/\tau_{rectangle}$ of the maximum torques that these coils experience in the same magnetic field when they contain the same current.

** **45. ssm** A charge of $4.0 \times 10^{-6}$ C is placed on a small conducting sphere that is located at the end of a thin insulating rod whose length is 0.20 m. The rod rotates with an angular speed of $\omega = 150$ rad/s about an axis that passes perpendicularly through its other end. Find the magnetic moment of the rotating charge. (Hint: The charge travels around a circle in a time equal to the period of the motion.)

### Section 21.7 Magnetic Fields Produced by Currents

**46.** A long, straight wire carries a current of 48 A. The magnetic field produced by this current at a certain point is $8.0 \times 10^{-5}$ T. How far is the point from the wire?

**47.** In a lightning bolt, 15 C of charge flows in a time of $1.5 \times 10^{-3}$ s. Assuming that the lightning bolt can be represented as a long, straight line of current, what is the magnitude of the magnetic field at a distance of 25 m from the bolt?

**48.** What must be the radius of a circular loop of wire so the magnetic field at its center is $1.8 \times 10^{-4}$ T when the loop carries a current of 12 A?

**49. ssm** A long solenoid has 1400 turns per meter of length, and it carries a current of 3.5 A. A small circular coil of wire is placed inside the solenoid with the normal to the coil oriented at an angle of 90.0° with respect to the axis of the solenoid. The coil consists of 50 turns, has an area of $1.2 \times 10^{-3}$ m², and carries a current of 0.50 A. Find the torque exerted on the coil.

**50.** A +6.00 μC charge is moving with a speed of $7.50 \times 10^4$ m/s parallel to a very long, straight wire. The wire is 5.00 cm from the charge and carries a current of 67.0 A in a direction opposite to that of the moving charge. Find the magnitude and direction of the force on the charge.

**51. ssm** Two rigid rods are oriented parallel to each other and to the ground. The rods carry the same current in the same direction. The length of each rod is 0.85 m, while the mass of each is 0.073 kg. One rod is held in place above the ground, and the other floats beneath it at a distance of $8.2 \times 10^{-3}$ m. Determine the current in the rods.

**52.** Two circular loops of wire, each containing a single turn, have the same radius of 4.0 cm and a common center. The planes of the loops are perpendicular. Each carries a current of 1.7 A. What is the magnitude of the net magnetic field at the common center?

**53.** A circular loop of wire and a long, straight wire carry currents of $I_1$ and $I_2$ (see the drawing), where $I_2 = 6.6I_1$. The loop and the straight wire lie in the same plane. The net magnetic field at the center of the loop is zero. Find the distance $H$, expressing your answer in terms of $R$, the radius of the loop.

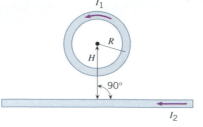

* **54.** As background for this problem, review Conceptual Example 9. A rectangular current loop is located near a long, straight wire that carries a current of 12 A (see the drawing). The current in the loop is 25 A. Determine the magnitude of the net magnetic force that acts on the loop.

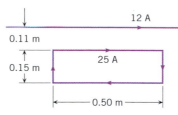

* **55.** Review **Interactive Solution 21.55** at **www.wiley.com/college/cutnell** for one approach to this problem. Two circular coils are concentric and lie in the same plane. The inner coil contains 140 turns of wire, has a radius of 0.015 m, and carries a current of 7.2 A. The outer coil contains 180 turns and has a radius of 0.023 m. What must be the magnitude and direction (relative to the current in the inner coil) of the current in the outer coil, such that the net magnetic field at the common center of the two coils is zero?

* **56.** Two long, straight, parallel wires $A$ and $B$ are separated by a distance of one meter. They carry currents in opposite directions, and the current in wire $A$ is one-third of that in wire $B$. On a line drawn perpendicular to the wires, find the point where the net magnetic field is zero. Determine this point relative to wire $A$.

* **57.** **ssm** A piece of copper wire has a resistance per unit length of $5.90 \times 10^{-3}$ $\Omega$/m. The wire is wound into a thin, flat coil of many turns that has a radius of 0.140 m. The ends of the wire are connected to a 12.0-V battery. Find the magnetic field strength at the center of the coil.

** **58.** The drawing shows two wires that carry the same current of $I = 85.0$ A and are oriented perpendicular to the plane of the paper. The current in one wire is directed out of the paper, while the current in the other is directed into the paper. Find the magnitude and direction of the net magnetic field at point $P$.

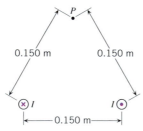

** **59.** The drawing shows an end-on view of three wires. They are long, straight, and perpendicular to the plane of the paper. Their cross sections lie at the corners of a square. The currents in wires 1 and 2 are $I_1 = I_2 = I$ and are directed into the paper. What is the direction of the current in wire 3, and what is the ratio $I_3/I$, such that the net magnetic field at the empty corner is zero?

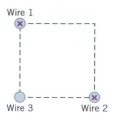

### Section 21.8 Ampere's Law

**60.** Suppose a uniform magnetic field is everywhere perpendicular to this page. The field points directly upward toward you. A circular path is drawn on the page. Use Ampere's law to show that there can be no net current passing through the circular surface.

**61.** **ssm** The wire in Figure 21.37 carries a current of 12 A. Suppose that a second long, straight wire is placed right next to this wire. The current in the second wire is 28 A. Use Ampere's law to find the magnitude of the magnetic field at a distance of $r = 0.72$ m from the wires when the currents are (a) in the same direction and (b) in opposite directions.

* **62.** Refer to **Interactive Solution 21.62** at **www.wiley.com/college/cutnell** for help with problems like this one. A very long, hollow cylinder is formed by rolling up a thin sheet of copper. Electric charges flow along the copper sheet parallel to the axis of the cylinder. The arrangement is, in effect, a hollow tube of current $I$. Use Ampere's law to show that the magnetic field (a) is $\mu_0 I/(2\pi r)$ outside the cylinder at a distance $r$ from the axis and (b) is zero at any point within the hollow interior of the cylinder. (*Hint: For closed paths, use circles perpendicular to and centered on the axis of the cylinder.*)

** **63.** A long, cylindrical conductor is solid throughout and has a radius $R$. Electric charges flow parallel to the axis of the cylinder and pass uniformly through the entire cross section. The arrangement is, in effect, a solid tube of current $I_0$. The current per unit cross-sectional area (i.e., the current density) is $I_0/(\pi R^2)$. Use Ampere's law to show that the magnetic field inside the conductor at a distance $r$ from the axis is $\mu_0 I_0 r/(2\pi R^2)$. (*Hint: For a closed path, use a circle of radius $r$ perpendicular to and centered on the axis. Note that the current through any surface is the area of the surface times the current density.*)

# Chapter 22 Electromagnetic Induction

## 22.1 Induced Emf and Induced Current

There are a number of ways a magnetic field can be used to generate an electric current, and Figure 22.1 illustrates one of them. This drawing shows a bar magnet and a helical coil of wire to which an ammeter is connected. When there is no *relative* motion between the magnet and the coil, as in part *a* of the drawing, the ammeter reads zero, indicating that no current exists. However, when the magnet moves toward the coil, as in part *b*, a current *I* appears. As the magnet approaches, the magnetic field that it creates at the location of the coil becomes stronger and stronger, and it is this *changing* field that produces the current. When the magnet moves away from the coil, as in part *c*, a current is also produced, but with a reversed direction. Now the magnetic field at the coil becomes weaker as the magnet moves away, and once again it is the *changing* field that generates the current.

A current would also be created in Figure 22.1 if the magnet were held stationary and the coil were moved, because the magnetic field at the coil would be changing as the coil approached or receded from the magnet. Only relative motion between the magnet and the coil is needed to generate a current; it does not matter which one moves.

The current in the coil is called an ***induced current*** because it is brought about (or "induced") by a changing magnetic field. Since a source of emf is always needed to produce a current, the coil itself behaves as if it were a source of emf. This emf is known as an ***induced emf.*** Thus, a changing magnetic field induces an emf in the coil, and the emf leads to an induced current.

**The physics of an automobile cruise control device.**

Induced emf and induced current are frequently used in the cruise control devices found in many cars. Figure 22.2 (on p. 454) illustrates how a cruise control device operates. Usually two magnets are mounted on opposite sides of the vehicle's drive shaft, with a stationary sensing coil positioned nearby. As the shaft turns, the magnets pass by the coil and cause an induced emf and current to appear in it. A microprocessor (the "brain" of a computer) counts the pulses of current and, with the aid of its internal clock and a knowledge of the shaft's radius, determines the rotational speed of the drive shaft. The rotational speed, in turn, is related to the car's speed. Thus, once the driver sets the desired cruising speed with the speed control switch (mounted near the steering wheel), the microprocessor can compare it with the measured speed. To the extent that the selected cruising speed and the measured speed differ, a signal is sent to a servo or control mechanism, which causes the throttle/fuel injector to send more or less fuel to the engine. The car speeds up or slows down accordingly, until the desired cruising speed is reached.

Figure 22.3 shows another way to induce an emf and a current in a coil. An emf can be induced by *changing the area* of a coil in a constant magnetic field. Here the shape of

**Figure 22.1** (*a*) When there is no relative motion between the coil of wire and the bar magnet, there is no current in the coil. (*b*) A current is created in the coil when the magnet moves toward the coil. (*c*) A current also exists when the magnet moves away from the coil, but the direction of the current is opposite to that in (*b*).

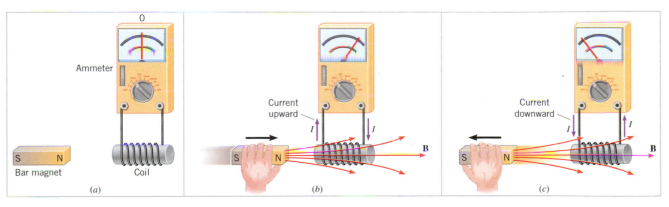

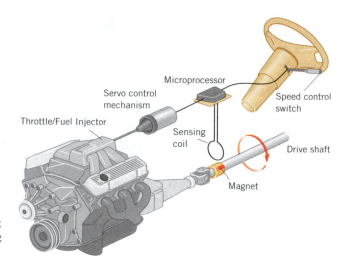

**Figure 22.2** Induced emf lies at the heart of an automobile cruise control device. The emf is induced in a sensing coil by magnets attached to the rotating drive shaft.

the coil is being distorted so as to reduce the area. As long as the area is changing, an induced emf and current exist; they vanish when the area is no longer changing. If the distorted coil is returned to its original shape, thereby increasing the area, an oppositely directed current is generated while the area is changing.

In each of the previous examples, both an emf and a current are induced in the coil because the coil is part of a complete, or closed, circuit. If the circuit were open — perhaps because of an open switch — there would be no induced current. However, an emf would still be induced in the coil, whether the current exists or not.

Changing a magnetic field and changing the area of a coil are methods that can be used to create an induced emf. The phenomenon of producing an induced emf with the aid of a magnetic field is called *electromagnetic induction.* The next section discusses yet another method by which an induced emf can be created.

## 22.2  *Motional Emf*

### THE EMF INDUCED IN A MOVING CONDUCTOR

When a conducting rod moves through a constant magnetic field, an emf is induced in the rod. This special case of electromagnetic induction arises as a result of the magnetic force (see Section 21.2) that acts on a moving charge. Consider the metal rod of length $L$ moving to the right in Figure 22.4a. The velocity **v** of the rod is constant and is perpendicular to a uniform magnetic field **B**. Each charge $q$ within the rod also moves with a velocity **v** and experiences a magnetic force of magnitude $F = qvB$, according to Equation 21.1. By using RHR-1, it can be seen that the mobile, free electrons are driven to the bottom of the rod, leaving behind an equal amount of positive charge at the top. (Remember to reverse the direction of the force that RHR-1 predicts, since the electrons have a negative charge. See Section 21.2.) The positive and negative charges accumulate until the attractive electric force that they exert on each other becomes equal in magnitude to the magnetic force. When the two forces balance, equilibrium is reached and no further charge separation occurs.

The separated charges on the ends of the moving conductor give rise to an induced emf, called a *motional emf* because it originates from the motion of charges through a magnetic field. The emf exists as long as the rod moves. If the rod is brought to a halt, the magnetic force vanishes, with the result that the attractive electric force reunites the positive and negative charges and the emf disappears. The emf of the moving rod is analogous to that between the terminals of a battery. However, the emf of a battery is produced by chemical reactions, whereas the motional emf is created by the agent that moves the rod through the magnetic field (like the hand in Figure 22.4b.)

The fact that the electric and magnetic forces balance at equilibrium in Figure 22.4a can be used to determine the magnitude of the motional emf $\mathcal{E}$. Acccording to Equation

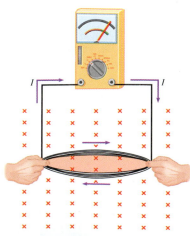

**Figure 22.3** While the area of the coil is changing, an induced emf and current are generated.

18.2, the magnitude of the electric force acting on the positive charge $q$ at the top of the rod is $Eq$, where $E$ is the magnitude of the electric field due to the separated charges. And according to Equation 19.7 (without the minus sign), the electric field magnitude is given by the voltage between the ends of the rod (the emf $\mathscr{E}$) divided by the length $L$ of the rod. Thus, the electric force is $Eq = (\mathscr{E}/L)q$. The magnetic force is $qvB$, according to Equation 21.1, because the charge $q$ moves perpendicular to the magnetic field. Since these two forces balance, it follows that $(\mathscr{E}/L)q = qvB$. The emf, then, is

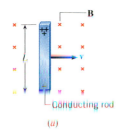

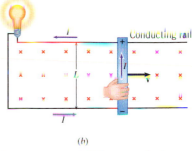

Motional emf when v, B,
and L are mutually
perpendicular

$$\mathscr{E} = vBL \qquad (22.1)$$

As expected, $\mathscr{E} = 0$ V when $v = 0$ m/s, because no motional emf is developed in a stationary rod. Greater speeds and stronger magnetic fields lead to greater emfs for a given length $L$. As with batteries, $\mathscr{E}$ is expressed in volts. In Figure 22.4b the rod is sliding on conducting rails that form part of a closed circuit, and $L$ is the length of the rod between the rails. Due to the emf, electrons flow in a clockwise direction around the circuit. Positive charge would flow in the direction opposite to the electron flow, so the conventional current $I$ is drawn counterclockwise in the picture. Example 1 illustrates how to determine the electrical energy that the motional emf delivers to a device such as the light bulb in the drawing.

**Figure 22.4** (*a*) When a conducting rod moves at right angles to a constant magnetic field, the magnetic force causes opposite charges to appear at the ends of the rod, giving rise to an induced emf. (*b*) The induced emf causes an induced current $I$ to appear in the circuit.

## Example 1  Operating a Light Bulb with Motional Emf

Suppose the rod in Figure 22.4b is moving at a speed of 5.0 m/s in a direction perpendicular to a 0.80-T magnetic field. The rod has a length of 1.6 m and a negligible electrical resistance. The rails also have negligible resistance. The light bulb, however, has a resistance of 96 $\Omega$. Find (*a*) the emf produced by the rod, (*b*) the current induced in the circuit, (*c*) the electric power delivered to the bulb, and (*d*) the energy used by the bulb in 60.0 s.

**Reasoning** The moving rod acts like an imaginary battery and supplies a motional emf of $vBL$ to the circuit. The induced current can be determined from Ohm's law as the motional emf divided by the resistance of the bulb. The electric power delivered to the bulb is the product of the induced current and the potential difference across the bulb (which, in this case, is the motional emf). The energy used is the product of the power and the time.

**Solution**

(*a*) The motional emf is given by Equation 22.1 as

$$\mathscr{E} = vBL = (5.0 \text{ m/s})(0.80 \text{ T})(1.6 \text{ m}) = \boxed{6.4 \text{ V}}$$

(*b*) According to Ohm's law, the induced current is equal to the motional emf divided by the resistance of the circuit:

$$I = \frac{\mathscr{E}}{R} = \frac{6.4 \text{ V}}{96 \ \Omega} = \boxed{0.067 \text{ A}} \qquad (20.2)$$

(*c*) The electric power $P$ delivered to the light bulb is the product of the current $I$ and the potential difference across the bulb:

$$P = I\mathscr{E} = (0.067 \text{ A})(6.4 \text{ V}) = \boxed{0.43 \text{ W}} \qquad (20.6a)$$

(*d*) Since power is energy per unit time, the energy $E$ used in 60.0 s is the product of the power and the time:

$$E = Pt = (0.43 \text{ W})(60.0 \text{ s}) = \boxed{26 \text{ J}} \qquad (6.10b)$$

## MOTIONAL EMF AND ELECTRICAL ENERGY

Motional emf arises because a magnetic force acts on the charges in a conductor that is moving through a magnetic field. Whenever this emf causes a current, a second magnetic force enters the picture. In Figure 22.4b, for instance, the second force arises because the current $I$ in the rod is perpendicular to the magnetic field. The current, and hence

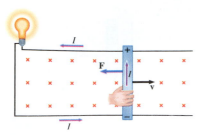

**Figure 22.5** A magnetic force **F** is exerted on the current $I$ in the moving rod and is opposite to the velocity **v**. The rod will slow down unless a counterbalancing force is applied by the hand.

the rod, experiences a magnetic force **F** whose magnitude is given by Equation 21.3 as $F = ILB \sin 90°$. Using the values of $I$, $L$, and $B$ from Example 1, we see that $F = (0.067 \text{ A})(1.6 \text{ m})(0.80 \text{ T}) = 0.086 \text{ N}$. The direction of **F** is specified by RHR-1 and is *opposite* to the velocity **v** of the rod, as Figure 22.5 shows. By itself, **F** would *slow down* the rod, and here lies the crux of the matter. To keep the rod moving to the right with a constant velocity, a counterbalancing force must be applied to the rod by an external agent, such as the hand in the picture. The counterbalancing force must have a magnitude of 0.086 N and must be directed opposite to the magnetic force **F**. If the counterbalancing force were removed, the rod would decelerate under the influence of **F** and eventually come to rest. During the deceleration, the motional emf would decrease and the light bulb would eventually go out.

We can now answer an important question—Who or what provides the 26 J of electrical energy that the light bulb in Example 1 uses in sixty seconds? The provider is the external agent that applies the 0.086-N counterbalancing force needed to keep the rod moving. This agent does work, and Example 2 shows that the work done is equal to the electrical energy used by the bulb.

### Example 2    The Work Needed to Keep the Light Bulb Burning

In Example 1, an external agent (the hand in Figures 22.4*b* and 22.5) supplies a 0.086-N force that keeps the rod moving at a constant speed of 5.0 m/s. Determine the work done in 60.0 s by the external agent.

**Reasoning** The work $W$ done by the hand in Figure 22.5 is given by Equation 6.1 as $W = (F \cos \theta)x$. In this equation $F$ is the magnitude of the external force, $\theta$ is the angle between the external force and the displacement of the rod ($\theta = 0°$), and $x$ is the distance that the rod moves in a time $t$. The distance is the product of the speed of the rod and the time, $x = vt$, so the work can be expressed as $W = (F \cos 0°)vt = Fvt$.

**Solution** The work done by the external agent is

$$W = Fvt = (0.086 \text{ N})(5.0 \text{ m/s})(60.0 \text{ s}) = \boxed{26 \text{ J}}$$

The 26 J of work done on the rod by the external agent is the same as the 26 J of energy used by the bulb. Hence, the moving rod and the magnetic force convert mechanical work into electrical energy, much as a battery converts chemical energy into electrical energy.

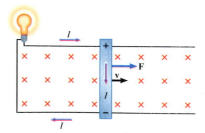

**Figure 22.6** The current cannot be directed clockwise in this circuit, because the magnetic force **F** exerted on the rod would then be in the same direction as the velocity **v**. The rod would accelerate to the right and create energy on its own, violating the principle of conservation of energy.

It is important to realize that the direction of the current in Figure 22.5 is consistent with the principle of conservation of energy. Consider what would happen if the direction of the current were reversed, as in Figure 22.6. With the direction of the current reversed, the direction of the magnetic force **F** would also be reversed and point in the direction of the velocity **v** of the rod. As a result, the force would cause the rod to accelerate rather than decelerate. The rod would accelerate without the need for an external force (like that provided by the hand in Figure 22.5) and create a motional emf that supplies energy to the light bulb. Thus, this hypothetical generator would produce energy out of nothing, since there is no external agent. Such a device cannot exist because it violates the principle of conservation of energy, which states that energy cannot be created or destroyed, but can only be converted from one form to another. Therefore, the current cannot be directed clockwise around the circuit, as in Figure 22.6. In situations such as that in Examples 1 and 2, when a motional emf leads to an induced current, a magnetic force always appears that opposes the motion, in accord with the principle of conservation of energy. Conceptual Example 3 deals further with the important issue of energy conservation.

### Conceptual Example 3    Conservation of Energy

Figure 22.7*a* illustrates a conducting rod that is free to slide down between two vertical copper tracks. There is no kinetic friction between the rod and the tracks, although the rod maintains electrical contact with the tracks during its fall. A constant magnetic field **B** is directed perpendicular to the motion of the rod, as the drawing shows. Because there is no friction, the only force that acts on the rod is its weight **W**, so the rod falls with an acceleration equal to the acceleration due to gravity, $a = 9.8$ m/s². Suppose that a resistance $R$ is connected between the tops of the tracks, as in part $b$ of the drawing. (a) Does the rod now fall with the acceleration due to gravity? (b) How does the principle of conservation of energy apply to what happens in Figure 22.7b?

### Reasoning and Solution

(a) As the rod falls perpendicular to the magnetic field, a motional emf is induced between its ends. This emf is induced whether or not the resistance $R$ is attached between the tracks. However, when $R$ is present, a complete circuit is formed, and the emf produces an induced current $I$ that is perpendicular to the field. The direction of this current is such that the rod experiences an upward magnetic force **F**, opposite to the falling motion and the weight of the rod (see part $b$ of the drawing and use RHR-1). The net force acting downward on the rod is $W + F$, which is *less* than the weight, since **F** points upward and the weight **W** points downward. In accord with Newton's second law of motion, the downward acceleration is proportional to the net force, so **the rod falls with an acceleration that is less than the acceleration due to gravity.** In fact, as the speed of the rod in Figure 22.7b increases during the descent, the magnetic force becomes larger and larger. There will be a time when the magnitude of the magnetic force becomes equal to the magnitude of the rod's weight. When this occurs, the net force on the rod will be zero, and the rod's acceleration will also be zero. From then on, the rod will fall at a constant velocity.

(b) For the freely falling rod in Figure 22.7a, gravitational potential energy (GPE) is converted into kinetic energy (KE) as the rod picks up speed on the way down. In Figure 22.7b, however, the rod always has a smaller speed than that of a freely falling rod at the same place. This is because the acceleration is less than the acceleration due to gravity. In terms of energy conservation, the smaller speed arises because only part of the GPE is converted into KE, with part also being dissipated as heat in the resistance $R$. In fact, when the rod reaches the point when it falls with a constant velocity, none of the GPE is converted into KE because all of it is dissipated as heat in the resistance.

**Related Homework:** *Problem 9*

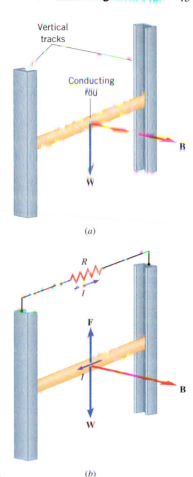

**Figure 22.7** (*a*) Because there is no kinetic friction between the falling rod and the tracks, the only force acting on the rod is its weight **W**. (*b*) When an induced current $I$ exists in the circuit, a magnetic force **F** also acts on the rod.

## 22.3 *Magnetic Flux*

### MOTIONAL EMF AND MAGNETIC FLUX

Motional emf, as well as any other type of induced emf, can be described in terms of a concept called *magnetic flux*. Magnetic flux is analogous to electric flux, which deals with the electric field and the surface through which it passes (see Section 18.9). Magnetic flux is defined in a similar way. A magnetic field is defined in terms of the magnetic force that acts on a moving test charge. By bringing together the magnetic field and the surface through which it passes, we will see how the concept of magnetic flux arises.

Now let us see how the motional emf given in Equation 22.1 ($\mathcal{E} = vBL$) can be written in terms of the magnetic flux. Figure 22.8a shows the rod used to derive

**Figure 22.8** (*a*) In a time $t_0$, the moving rod sweeps out an area $A_0 = x_0 L$. (*b*) The area swept out in a time $t$ is $A = xL$. In both parts of the figure the areas are shaded in color.

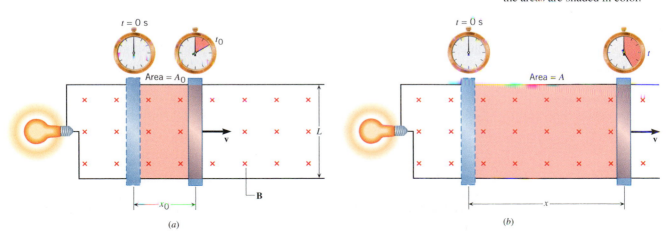

(a)                                                                 (b)

Equation 22.1; it is moving through a magnetic field between a time $t = 0$ s and a later time $t_0$. During this interval, it has moved a distance $x_0$ to the right. At a later time $t$, the rod has moved an even greater distance $x$, as part $b$ of the drawing indicates. The speed $v$ of the rod is the distance traveled divided by the elapsed time: $v = (x - x_0)/(t - t_0)$. Substituting this expression for $v$ into $\mathscr{E} = vBL$ gives

$$\mathscr{E} = \left( \frac{x - x_0}{t - t_0} \right) BL = \left( \frac{xL - x_0L}{t - t_0} \right) B$$

As the drawing indicates, the term $x_0L$ is the area $A_0$ swept out by the rod in moving a distance $x_0$, while $xL$ is the area $A$ swept out in moving a distance $x$. In terms of these areas, the emf is

$$\mathscr{E} = \left( \frac{A - A_0}{t - t_0} \right) B = \frac{(BA) - (BA)_0}{t - t_0}$$

Notice how the product $BA$ of the magnetic field strength and the area appears in the numerator of this expression. The quantity $BA$ is given the name ***magnetic flux*** and is represented by the symbol $\Phi$ (Greek capital letter phi); thus $\Phi = BA$. The magnitude of the induced emf is the *change* in flux $\Delta\Phi = \Phi - \Phi_0$ divided by the time interval $\Delta t = t - t_0$ during which the change occurs:

$$\mathscr{E} = \frac{\Phi - \Phi_0}{t - t_0} = \frac{\Delta\Phi}{\Delta t}$$

In other words, the induced emf equals the time rate of change of the magnetic flux.

You will almost always see the previous equation written with a minus sign—namely, $\mathscr{E} = -\Delta\Phi/\Delta t$. The minus sign is introduced for the following reason: The direction of the current induced in the circuit is such that the magnetic force **F** acts on the rod to *oppose* its motion, thereby tending to slow down the rod (see Figure 22.5). The minus sign is a reminder that the polarity of the induced emf sends the induced current in the proper direction so as to give rise to this opposing magnetic force.

The advantage of writing the induced emf as $\mathscr{E} = -\Delta\Phi/\Delta t$ is that this relation is far more general than our present discussion suggests. In Section 22.4 we will see that $\mathscr{E} = -\Delta\Phi/\Delta t$ can be applied to *all possible ways of generating induced emfs.*

## A GENERAL EXPRESSION FOR MAGNETIC FLUX

In Figure 22.8 the direction of the magnetic field **B** is perpendicular to the surface swept out by the moving rod. In general, however, **B** may not be perpendicular to the surface. For instance, in Figure 22.9 the direction perpendicular to the surface is indicated by the normal to the surface, but the magnetic field is inclined at an angle $\phi$ with respect to this direction. In such a case the flux is computed using only the component of the field that is perpendicular to the surface, $B \cos \phi$. The general expression for magnetic flux is

$$\Phi = (B \cos \phi)A = BA \cos \phi \qquad (22.2)$$

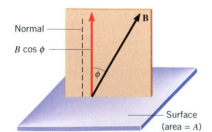

**Figure 22.9** When computing the magnetic flux, the component of the magnetic field that is perpendicular to the surface must be used; this component is $B \cos \phi$.

If either the magnitude $B$ of the magnetic field or the angle $\phi$ is not constant over the surface, an average value of the product $B \cos \phi$ must be used to compute the flux. Equation 22.2 shows that the unit of magnetic flux is the tesla·meter² (T·m²). This unit is called a *weber* (Wb), after the German physicist Wilhelm Weber (1804–1891): 1 Wb = 1 T·m². Example 4 illustrates how to determine the magnetic flux for three different orientations of the surface of a coil relative to the magnetic field.

### *Example 4*  Magnetic Flux

A rectangular coil of wire is situated in a constant magnetic field whose magnitude is 0.50 T. The coil has an area of 2.0 m². Determine the magnetic flux for the three orientations, $\phi = 0°$, $60.0°$, and $90.0°$, shown in Figure 22.10.

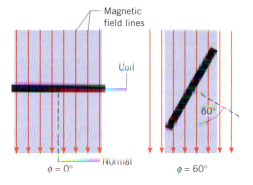

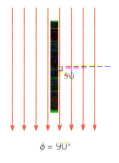

Figure 22.10 Three orientations of a rectangular coil (drawn as an edge view) relative to the magnetic field lines. The magnetic field lines that pass through the coil are those in the regions shaded in blue.

**Reasoning** The magnetic flux $\Phi$ is defined as $\Phi = BA \cos \phi$, where $B$ is the magnitude of the magnetic field, $A$ is the area of the surface through which the magnetic field passes, and $\phi$ is the angle between the magnetic field and the normal to the surface.

**Solution** The magnetic flux for the three cases is:

$\phi = 0°$      $\Phi = (0.50 \text{ T})(2.0 \text{ m}^2) \cos 0° = \boxed{1.0 \text{ Wb}}$

$\phi = 60.0°$      $\Phi = (0.50 \text{ T})(2.0 \text{ m}^2) \cos 60.0° = \boxed{0.50 \text{ Wb}}$

$\phi = 90.0°$      $\Phi = (0.50 \text{ T})(2.0 \text{ m}^2) \cos 90.0° = \boxed{0 \text{ Wb}}$

**Problem solving insight**
The magnetic flux $\Phi$ is determined by more than just the magnitude $B$ of the magnetic field and the area $A$. It also depends on the angle $\phi$ (see Figure 22.9 and Equation 22.2).

## GRAPHICAL INTERPRETATION OF MAGNETIC FLUX

It is possible to interpret the magnetic flux graphically because the magnitude of the magnetic field **B** is proportional to the number of field lines per unit area that pass through a surface perpendicular to the lines (see Section 21.1). For instance, the magnitude of **B** in Figure 22.11a is three times larger than it is in part b of the drawing, since the number of field lines drawn through the identical surfaces is in the ratio of 3:1. Because $\Phi$ is directly proportional to $B$ for a given area, the flux in part a is also three times larger than that in part b. Therefore, *the magnetic flux is proportional to the number of field lines that pass through a surface.*

The graphical interpretation of flux also applies when the surface is oriented at an angle with respect to **B**. For example, as the coil in Figure 22.10 is rotated from $\phi = 0°$ to 60° to 90°, the number of magnetic field lines passing through the surface (see the field lines in the regions shaded in blue) changes in the ratio of 8:4:0 or 2:1:0. The results of Example 4 show that the flux in the three orientations changes by the same ratio. Because the magnetic flux is proportional to the number of field lines passing through a surface, we often use phrases such as "the flux that passes through a surface bounded by a loop of wire."

# 22.4 *Faraday's Law of Electromagnetic Induction*

Two scientists are given credit for the discovery of electromagnetic induction: the Englishman Michael Faraday (1791–1867) and the American Joseph Henry (1797–1878). Although Henry was the first to observe electromagnetic induction, Faraday investigated it in more detail and published his findings first. Consequently, the law that describes the phenomenon bears his name.

Faraday discovered that whenever there is a *change in flux* through a loop of wire, an emf is induced in the loop. In this context, the word "change" refers to a change as time passes. A flux that is constant in time creates no emf. *Faraday's law of electromagnetic induction* is expressed by bringing together the idea of magnetic flux and the time interval during which it changes. In fact, Faraday found that the magnitude of the induced emf is equal to the time rate of change of the magnetic flux. This is consistent with the relation we obtained in Section 22.3 for the specific case of motional emf: $\mathcal{E} = -\Delta\Phi/\Delta t$.

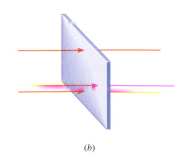

Figure 22.11 The magnitude of the magnetic field in (a) is three times as great as that in (b), because the number of magnetic field lines crossing the surfaces is in the ratio of 3:1.

Often the magnetic flux passes through a coil of wire containing more than one loop (or turn). If the coil consists of $N$ loops, and if the same flux passes through each loop, it is found experimentally that the total induced emf is $N$ times that induced in a single loop. An analogous situation occurs in a flashlight when two 1.5-V batteries are stacked in series on top of one another to give a total emf of 3.0 volts. For the general case of $N$ loops, the total induced emf is described by Faraday's law of electromagnetic induction in the following manner:

---

■ **FARADAY'S LAW OF ELECTROMAGNETIC INDUCTION**

The average emf $\mathcal{E}$ induced in a coil of $N$ loops is

$$\mathcal{E} = -N\left(\frac{\Phi - \Phi_0}{t - t_0}\right) = -N\frac{\Delta\Phi}{\Delta t} \tag{22.3}$$

where $\Delta\Phi$ is the change in magnetic flux through one loop and $\Delta t$ is the time interval during which the change occurs. The term $\Delta\Phi/\Delta t$ is the average time rate of change of the flux that passes through one loop.

*SI Unit of Induced Emf:* volt (V)

---

Faraday's law states that an emf is generated if the magnetic flux changes for any reason. Since the flux is given by Equation 22.2 as $\Phi = BA\cos\phi$, it depends on three factors, $B$, $A$, and $\phi$, any of which may change. Example 5 considers a change in $B$.

### *Example 5* The Emf Induced by a Changing Magnetic Field

A coil of wire consists of 20 turns, or loops, each of which has an area of $1.5 \times 10^{-3}$ m². A magnetic field is perpendicular to the surface of each loop at all times, so that $\phi = \phi_0 = 0°$. At time $t_0 = 0$ s, the magnitude of the field at the location of the coil is $B_0 = 0.050$ T. At a later time $t = 0.10$ s, the magnitude of the field at the coil has increased to $B = 0.060$ T. (a) Find the average emf induced in the coil during this time. (b) What would be the value of the average induced emf if the magnitude of the magnetic field decreased from 0.060 T to 0.050 T in 0.10 s?

**Reasoning** To find the induced emf, we use Faraday's law of electromagnetic induction (Equation 22.3), combining it with the definition of magnetic flux from Equation 22.2. We note that only the magnitude of the magnetic field changes in time. All other factors remain constant.

**Solution**

(a) Since $\phi = \phi_0$, the induced emf is

*Problem solving insight*
The change in any quantity is the final value minus the initial value: e.g., the change in flux is $\Delta\Phi = \Phi - \Phi_0$ and the change in time is $\Delta t = t - t_0$.

$$\mathcal{E} = -N\left(\frac{\Phi - \Phi_0}{t - t_0}\right) = -N\left(\frac{BA\cos\phi - B_0A\cos\phi_0}{t - t_0}\right)$$

$$= -NA\cos\phi\left(\frac{B - B_0}{t - t_0}\right)$$

$$\mathcal{E} = -(20)(1.5 \times 10^{-3}\text{ m}^2)(\cos 0°)\left(\frac{0.060\text{ T} - 0.050\text{ T}}{0.10\text{ s} - 0\text{ s}}\right) = \boxed{-3.0 \times 10^{-3}\text{ V}}$$

(b) The calculation here is similar to that in part (a), except the initial and final values of $B$ are interchanged. This interchange reverses the sign of the emf, so $\boxed{\mathcal{E} = +3.0 \times 10^{-3}\text{ V}}$. Because the algebraic sign or polarity of the emf is reversed, the direction of the induced current would be opposite to that in part (a).

---

The next example demonstrates that an emf can be created when a coil is rotated in a magnetic field.

### *Example 6* The Emf Induced in a Rotating Coil

A flat coil of wire has an area of 0.020 m² and consists of 50 turns. At $t_0 = 0$ s the coil is oriented so the normal to its surface is parallel ($\phi_0 = 0°$) to a constant magnetic field of magnitude 0.18 T. The coil is then rotated through an angle of $\phi = 30.0°$ in a time of 0.10 s (see Figure 22.10). (a) Determine the average induced emf. (b) What would be the induced emf if the coil were returned to its initial orientation in the same time of 0.10 s?

**Reasoning** As in Example 5 we can determine the induced emf by using Faraday's law of electromagnetic induction, along with the definition of magnetic flux. In the present case, however, only $\phi$ (the angle between the normal to the surface of the coil and the magnetic field) changes in time. All other factors remain constant.

**Solution**

(a) Faraday's law yields

$$\mathscr{E} = -N\left(\frac{\Phi - \Phi_0}{t - t_0}\right) = -N\left(\frac{BA\cos\phi - BA\cos\phi_0}{t - t_0}\right)$$

$$= -NBA\left(\frac{\cos\phi - \cos\phi_0}{t - t_0}\right)$$

$$\mathscr{E} = -(50)(0.18\text{ T})(0.020\text{ m}^2)\left(\frac{\cos 30.0° - \cos 0°}{0.10\text{ s} - 0\text{ s}}\right) = \boxed{+0.24\text{ V}}$$

(b) When the coil is rotated back to its initial orientation in a time of 0.10 s, the initial and final values of $\phi$ are interchanged. As a result, the induced emf has the same magnitude, but opposite polarity, so $\boxed{\mathscr{E} = -0.24\text{ V}}$.

One application of Faraday's law that is found in the home is a safety device called a ground fault interrupter. This device protects against electrical shock from an appliance, such as a clothes dryer. It plugs directly into a wall socket, as in Figure 22.12 or, in new home construction, replaces the socket entirely. The interrupter consists of a circuit breaker that can be triggered to stop the current to the dryer, depending on whether an induced voltage appears across a sensing coil. This coil is wrapped around an iron ring, through which the current-carrying wires pass. In the drawing, the current going to the dryer is shown in red, and the returning current is shown in green. Each of the currents creates a magnetic field that encircles the corresponding wire, according to RHR-2 (see Section 21.7). However, the field lines have opposite directions since the currents have opposite directions. As the drawing shows, the iron ring guides the field lines through the sensing coil. Since the current is ac, the fields from the red and green current are changing, but the red and green field lines always have opposite directions and the opposing fields cancel at all times. As a result, the net flux through the coil remains zero, and no induced emf appears in the coil. Thus, when the dryer operates normally, the circuit breaker is not triggered and does not shut down the current. The picture changes if the dryer malfunctions, as when a wire inside the unit breaks and accidentally contacts the metal case.

**The physics of a ground fault interrupter.**

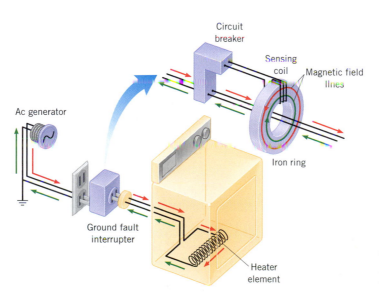

**Figure 22.12** The clothes dryer is connected to the wall socket through a ground fault interrupter. The dryer is operating normally.

When someone touches the case, some of the current begins to pass through the person's body and into the ground, returning to the ac generator *without using the return wire that passes through the ground fault interrupter*. Under this condition, the net magnetic field through the sensing coil is no longer zero and changes with time, since the current is ac. The changing flux causes an induced voltage to appear in the sensing coil, which triggers the circuit breaker to stop the current. Ground fault interrupters work very fast (in less than a millisecond) and turn off the current before it reaches a dangerous level.

Conceptual Example 7 discusses another application of electromagnetic induction—namely, how a stove can cook food without getting hot.

### Conceptual Example 7 An Induction Stove

*The physics of* **an induction stove**.

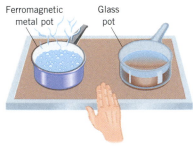

Ferromagnetic metal pot  Glass pot

**Figure 22.13** The water in the ferromagnetic metal pot is boiling. Yet the water in the glass pot is not boiling, and the stove top is cool to the touch. The stove operates in this way by using electromagnetic induction.

Figure 22.13 shows two pots of water that were placed on an induction stove at the same time. There are two interesting features in this drawing. First, the stove itself is cool to the touch. Second, the water in the ferromagnetic metal pot is boiling while that in the glass pot is not. How can such a "cool" stove boil water, and why isn't the water in the glass pot boiling?

**Reasoning and Solution** The key to this puzzle is related to the fact that one pot is made from a ferromagnetic metal and one from glass. We know that metals are good conductors, while glass is an insulator. Perhaps the stove causes electricity to flow directly in the metal pot. This is exactly what happens. The stove is called an *induction stove* because it operates by using electromagnetic induction. Just beneath the cooking surface is a metal coil that carries an ac current (frequency about 25 kHz). This current produces an alternating magnetic field that extends outward to the location of the metal pot. As the changing field crosses the pot's bottom surface, an emf is induced in it. Because the pot is metallic, an induced current is generated by the induced emf. The metal has a finite resistance to the induced current, however, and heats up as energy is dissipated in this resistance. The fact that the metal is ferromagnetic is important. Ferromagnetic materials contain magnetic domains (see Section 21.9), and the boundaries between them move extremely rapidly in response to the external magnetic field, thus enhancing the induction effect. A normal aluminum cooking pot, in contrast, is not ferromagnetic, so this enhancement is absent and such cookware is not used with induction stoves. An emf is also induced in the glass pot and the cooking surface of the stove. However, these materials are insulators, so very little induced current exists within them. Thus, they do not heat up very much and remain cool to the touch.

## 22.5 Lenz's Law

An induced emf drives current around a circuit just as the emf of a battery does. With a battery, conventional current is directed out of the positive terminal, through the attached device, and into the negative terminal. The same is true for an induced emf, although the locations of the positive and negative terminals are generally not as obvious. Therefore, a method is needed for determining the polarity or algebraic sign of the induced emf, so the terminals can be identified. As we discuss this method, it will be helpful to keep in mind that the net magnetic field penetrating a coil of wire results from two contributions. One is the original magnetic field that produces the changing flux that leads to the induced emf. The other arises because of the induced current, which, like any current, creates its own magnetic field. The field created by the induced current is called the ***induced magnetic field.***

To determine the polarity of the induced emf, we will use a method based on a discovery made by the Russian physicist Heinrich Lenz (1804–1865). This discovery is known as ***Lenz's law.***

■ **LENZ'S LAW**

The induced emf resulting from a changing magnetic flux has a polarity that leads to an induced current whose direction is such that the induced magnetic field opposes the original flux change.

The next two examples illustrate Lenz's law.

## Conceptual Example 8  The Emf Produced by a Moving Magnet

Figure 22.14a shows a permanent magnet approaching a loop of wire. The external circuit attached to the loop consists of the resistance R, which could be the resistance of the filament in a light bulb, for instance. Find the direction of the induced current and the polarity of the induced emf.

**Reasoning and Solution** We apply Lenz's law, the essence of which is that the change in magnetic flux must be opposed by the induced magnetic field. The magnetic flux through the loop is increasing, since the magnitude of the magnetic field at the loop is increasing as the magnet approaches. To oppose the increase in the flux, the direction of the induced magnetic field must be opposite to the field of the bar magnet. Since the field of the bar magnet passes through the loop from left to right in part a of the drawing, the induced field must pass through the loop from right to left, as in part b. To create such an induced field, *the induced current must be directed counterclockwise around the loop, when viewed from the side nearest the magnet.* (See the application of RHR-2 in the drawing.) The loop behaves as a source of emf, just like a battery. Since conventional current is directed into the external circuit from the positive terminal, *point A in Figure 22.14b must be the positive terminal, and point B must be the negative terminal.*

In Conceptual Example 8 the direction of the induced magnetic field is opposite to the direction of the external field of the bar magnet. However, *[Problem solving insight]* the induced field is not always opposite to the external field, because Lenz's law requires only that it must oppose the change in the flux that generates the emf. Conceptual Example 9 illustrates this point.

## Conceptual Example 9  The Emf Produced by a Moving Copper Ring

In Figure 22.15 there is a constant magnetic field in a rectangular region of space. This field is directed perpendicularly into the page. Outside this region there is no magnetic field. A copper ring slides through the region, from position 1 to position 5. For each of the five positions, determine whether an induced current exists in the ring and, if so, find its direction.

**Reasoning and Solution** Lenz's law requires that the induced magnetic field must oppose the change in flux. Sometimes this means that the induced field will be opposite to the external field, as in Example 8, but not always. We will see that the induced field must sometimes have the same direction as the external field in order to oppose the flux change.

**Position 1:** No flux passes through the ring because the magnetic field is zero outside the rectangular region. Consequently, *there is no change in flux and no induced emf or current.*

**Position 2:** As the ring moves into the field region, the flux increases and there is an induced emf and current. To determine the direction of the current, we require that the induced field point out of the plane of the paper, opposite to the external field. Only then can the induced field oppose the increase in the flux, in accord with Lenz's law. To produce such an induced field, RHR-2 indicates that *the direction of the induced current must be counterclockwise,* as the drawing shows.

**Position 3:** Even though a flux passes through the moving ring, *there is no induced emf or current* because the flux remains constant within the rectangular region. To induce an emf, it is not sufficient just to have a flux; the flux must change as time passes.

**Position 4:** As the ring leaves the field region, the flux decreases, and the induced field must oppose this change. Since the change is a decrease in flux, the induced field must point in the *same* direction as the external field—namely, into the paper. With this orientation, the induced field will increase the net magnetic field through the ring and thereby increase the flux. To produce an induced field pointing into the paper, RHR-2 indicates that *the induced current must be clockwise,* opposite to what it is in position 2. By comparing the results for positions 2 and 4, it should be clear that the induced magnetic field is not always opposite to the external field.

**Position 5:** As in position 1, *there is no induced current,* since the magnetic field is everywhere zero.

**Related Homework:** *Problem 31*

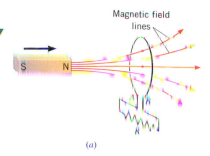

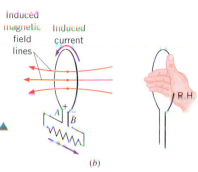

(a)

(b)

**Figure 22.14** (a) As the magnet moves to the right, the magnetic flux through the loop increases. The external circuit attached to the loop has a resistance R. (b) The polarity of the induced emf is indicated by the + and − symbols.

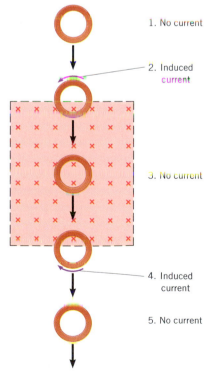

**Figure 22.15** A copper ring passes through a rectangular region where a constant magnetic field is directed into the page. The picture shows that a current is induced in the ring at locations 2 and 4.

Lenz's law should not be thought of as an independent law, because it is a consequence of the law of conservation of energy. The connection between energy conservation and induced emf has already been discussed in Section 22.2 for the specific case of motional emf. However, the connection is valid for any type of induced emf. In fact, the polarity of the induced emf, as specified by Lenz's law, ensures that energy is conserved.

## 22.6 *The Electric Generator*

### HOW A GENERATOR PRODUCES AN EMF

**The physics of** an electric generator.

Electric generators, such as those in Figure 22.16, produce virtually all of the world's electrical energy. A generator produces electrical energy from mechanical work, just the opposite of what a motor does. In a motor, an *input* electric current causes a coil to rotate, thereby doing mechanical work on any object attached to the shaft of the motor. In a generator, the shaft is rotated by some mechanical means, such as an engine or a turbine, and an emf is induced in a coil. If the generator is connected to an external circuit, an electric current is the *output* of the generator.

In its simplest form, an ac generator consists of a coil of wire that is rotated in a uniform magnetic field, as Figure 22.17a indicates. Although not shown in the picture, the wire is usually wound around an iron core. As in an electric motor, the coil/core combination is called the *armature*. Each end of the wire forming the coil is connected to the external circuit by means of a metal ring that rotates with the coil. Each ring slides against a stationary carbon brush, to which the external circuit (the lamp in the drawing) is connected.

To see how current is produced by the generator, consider the two vertical sides of the coil in Figure 22.17b. Since each is moving in a magnetic field **B**, the magnetic force exerted on the charges in the wire causes them to flow, thus creating a current. With the aid of RHR-1 (fingers of extended right hand point along **B**, thumb along the velocity **v**, palm pushes in the direction of the force on a positive charge), it can be seen that the direction of the current is from bottom to top in the left side and from top to bottom in the right side. Thus, charge flows around the loop. The upper and lower segments of the loop are also moving. However, these segments can be ignored because the magnetic force on the charges within them points toward the sides of the wire and not along the length.

The magnitude of the motional emf developed in a conductor moving through a magnetic field is given by Equation 22.1. To apply this expression to the left side of the coil,

**Figure 22.16** Electric generators such as this one supply electric power by producing an induced emf according to Faraday's law of electromagnetic induction. (© Peter Bowater/Age Fotostock America, Inc.)

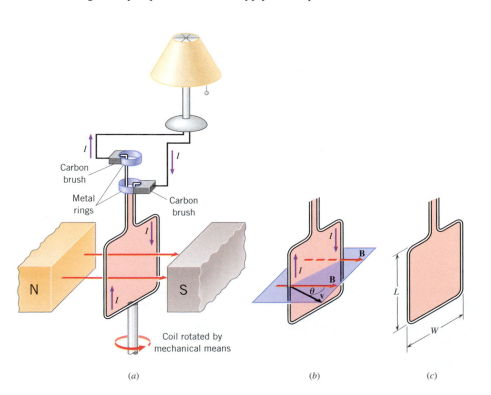

**Figure 22.17** (*a*) This electric generator consists of a coil (only one loop is shown) of wire that is rotated in a magnetic field **B** by some mechanical means. (*b*) The current *I* arises because of the magnetic force exerted on the charges in the moving wire. (*c*) The dimensions of the coil.

whose length is $L$ (see Figure 22.17c), we need to use the velocity component $v_\perp$ that is perpendicular to **B**. Letting $\theta$ be the angle between **v** and **B**, it follows that $v_\perp = v \sin \theta$, and, with the aid of Equation 22.1, the emf can be written as

$$\mathcal{E} = BLv_\perp = BLv \sin \theta$$

The emf induced in the right side has the same magnitude as that in the left side. Since the emfs from both sides drive current in the same direction around the loop, the emf for the complete loop is $\mathcal{E} = 2BLv \sin \theta$. If the coil consists of $N$ loops, the net emf is $N$ times as great as that of one loop, so

$$\mathcal{E} = N(2BLv \sin \theta)$$

It is convenient to express the variables $v$ and $\theta$ in terms of the angular speed $\omega$ at which the coil rotates. Equation 8.2 shows that the angle $\theta$ is the product of the angular speed and the time, $\theta = \omega t$, if it is assumed that $\theta = 0$ rad when $t = 0$ s. Furthermore, any point on each vertical side moves on a circular path of radius $r = W/2$, where $W$ is the width of the coil (see Figure 22.17c). Thus, the tangential speed $v$ of each side is related to the angular speed $\omega$ via Equation 8.9 as $v = r\omega = (W/2)\omega$. Substituting these expressions for $\theta$ and $v$ in the previous equation for $\mathcal{E}$, and recognizing that the product $LW$ is the area $A$ of the coil, we can write the induced emf as

**Emf induced in a rotating planar coil**
$$\mathcal{E} = NAB\omega \sin \omega t = \mathcal{E}_0 \sin \omega t \qquad \text{where } \omega = 2\pi f \qquad (22.4)$$

In this result, the angular speed $\omega$ is in radians per second and is related to the frequency $f$ [in cycles per second or hertz (Hz)] according to $\omega = 2\pi f$ (Equation 10.6).

Although Equation 22.4 was derived for a rectangular coil, the result is valid for any planar shape of area $A$ (e.g., circular) and shows that the emf varies sinusoidally with time. The peak, or maximum, emf $\mathcal{E}_0$ occurs when $\sin \omega t = 1$ and has the value $\mathcal{E}_0 = NAB\omega$. Figure 22.18 shows a plot of Equation 22.4 and reveals that the emf changes polarity as the coil rotates. This changing polarity is exactly the same as that discussed for an ac voltage in Section 20.5 and illustrated in Figure 20.9. If the external circuit connected to the generator is a closed circuit, an alternating current results that changes direction at the same frequency $f$ as the emf changes polarity. Therefore, this electric generator is also called an *alternating current (ac) generator*. The next two examples show how Equation 22.4 is applied.

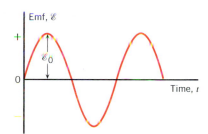

**Figure 22.18** An ac generator produces this alternating emf $\mathcal{E}$ according to $\mathcal{E} = \mathcal{E}_0 \sin \omega t$.

## Example 10 An AC Generator

In Figure 22.17 the coil of the ac generator rotates at a frequency of $f = 60.0$ Hz and develops an emf of 120 V (rms). The coil has an area of $A = 3.0 \times 10^{-3}$ m$^2$ and consists of $N = 500$ turns. Find the magnitude of the magnetic field in which the coil rotates.

**Reasoning** The magnetic field can be found from the relation $\mathcal{E}_0 = NAB\omega$. However, in using this equation we must remember that $\mathcal{E}_0$ is the peak emf, whereas the given value of 120 V is not a peak value but an rms value. The peak emf is related to the rms emf by $\mathcal{E}_0 = \sqrt{2}\mathcal{E}_{rms}$, according to Equation 20.13.

**Solution** The peak emf is $\mathcal{E}_0 = \sqrt{2}\mathcal{E}_{rms} = \sqrt{2}(120 \text{ V}) = 170$ V. Solving $\mathcal{E}_0 = NAB\omega$ for $B$ and using the fact that $\omega = 2\pi f$, we find that the magnitude of the magnetic field is

$$B = \frac{\mathcal{E}_0}{NA\omega} = \frac{\mathcal{E}_0}{NA2\pi f} = \frac{170 \text{ V}}{(500)(3.0 \times 10^{-3} \text{ m}^2)2\pi(60.0 \text{ Hz})} = \boxed{0.30 \text{ T}}$$

**Problem solving insight**
When using the equation $\mathcal{E}_0 = NAB\omega$, remember that the angular frequency $\omega$ must be in rad/s and is related to the frequency $f$ (in Hz) according to $\omega = 2\pi f$ (Equation 10.6).

## Example 11 A Bike Generator

A generator is mounted on a bicycle to power a headlight. A small wheel on the shaft of the generator is pressed against the bike tire and turns the armature 44 times for each revolution of the tire. The tire has a radius of 0.33 m. The armature has 75 turns, each with an area of $2.6 \times 10^{-3}$ m$^2$, and rotates in a 0.10-T magnetic field. When the peak emf being generated is 6.0 V, what is the linear speed of the bike?

**The physics of a bike generator.**

**Reasoning**  The relation $\mathscr{E}_0 = NAB\omega$ provides a solution to this problem, if we take advantage of two additional facts. The first is that the angular speed $\omega$ of the armature is 44 times larger than the angular speed $\omega_{\text{tire}}$ of the bike tire. The second is that the tire is rolling, so that $\omega_{\text{tire}}$ is related to the linear speed $v$ of the bike according to $\omega_{\text{tire}} = v/r$ (Equation 8.12), where $r$ is the radius of the tire.

**Solution**  Using $\omega = 44\omega_{\text{tire}} = 44(v/r)$, we find that the peak emf is $\mathscr{E}_0 = NAB\,[44(v/r)]$. Solving for the linear speed $v$ gives

$$v = \frac{\mathscr{E}_0 r}{44NAB} = \frac{(6.0\ \text{V})(0.33\ \text{m})}{44(75)(2.6 \times 10^{-3}\ \text{m}^2)(0.10\ \text{T})} = \boxed{2.3\ \text{m/s}}$$

## THE ELECTRICAL ENERGY DELIVERED BY A GENERATOR AND THE COUNTERTORQUE

Some power-generating stations burn fossil fuel (coal, gas, or oil) to heat water and produce pressurized steam for turning the blades of a turbine, whose shaft is linked to the generator. Others use nuclear fuel or falling water as a source of energy. As the turbine rotates, the generator coil also rotates and mechanical work is transformed into electrical energy.

The devices to which the generator supplies electricity are known collectively as the "load," because they place a burden or load on the generator by taking electrical energy from it. If all the devices are switched off, the generator runs under a no-load condition, because there is no current in the external circuit and the generator does not supply electrical energy. Then, work needs to be done on the turbine only to overcome friction and other mechanical losses within the generator itself, and fuel consumption is at a minimum.

Figure 22.19 illustrates a situation in which a load is connected to a generator. Because there is now a current $I = I_1 + I_2$ in the coil of the generator and the coil is situated in a magnetic field, the current experiences a magnetic force $\mathbf{F}$. Figure 22.20 shows the magnetic force acting on the left side of the coil, the direction of $\mathbf{F}$ being given by RHR-1. A force of equal magnitude but opposite direction acts on the right side of the coil, although this force is not shown in the drawing. The magnetic force $\mathbf{F}$ gives rise to a *countertorque* that opposes the rotational motion. The greater the current drawn from the generator, the greater the countertorque, and the harder it is for the turbine to turn the coil. To compensate for this countertorque and keep the coil rotating at a constant angular speed, work must be done by the turbine, which means more fuel must be burned. This is another example of the law of conservation of energy, since the electrical energy consumed by the load must ultimately come from the energy source used to drive the turbine.

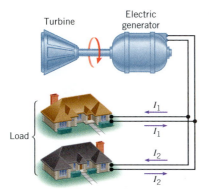

**Figure 22.19**  The generator supplies a total current of $I = I_1 + I_2$ to the load.

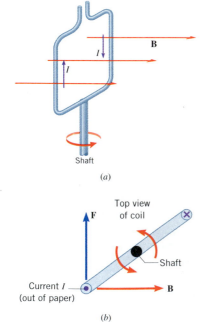

**Figure 22.20**  (*a*) A current $I$ exists in the rotating coil of the generator. (*b*) A top view of the coil, showing the magnetic force $\mathbf{F}$ exerted on the left side of the coil.

## THE BACK EMF GENERATED BY AN ELECTRIC MOTOR

A generator converts mechanical work into electrical energy; in contrast, an electric motor converts electrical energy into mechanical work. Both devices are similar and consist of a coil of wire that rotates in a magnetic field. In fact, as the armature of a motor rotates, the magnetic flux passing through the coil changes and an emf is induced in the coil. Thus, when a motor is operating, two sources of emf are present: (1) the applied emf $V$ that provides current to drive the motor (e.g., from a 120-V outlet), and (2) the emf $\mathscr{E}$ induced by the generator-like action of the rotating coil. The circuit diagram in Figure 22.21 shows these two emfs.

Consistent with Lenz's law, the induced emf $\mathscr{E}$ acts to oppose the applied emf $V$ and is called the **back emf** or the **counter emf** of the motor. The greater the speed of the motor, the greater the flux change through the coil, and the greater is the back emf. Because $V$ and $\mathscr{E}$ have opposite polarities, the net emf in the circuit is $V - \mathscr{E}$. In Figure 22.21, $R$ is the resistance of the wire in the coil, and the current $I$ drawn by the motor is determined from Ohm's law as the net emf divided by the resistance:

$$I = \frac{V - \mathscr{E}}{R} \tag{22.5}$$

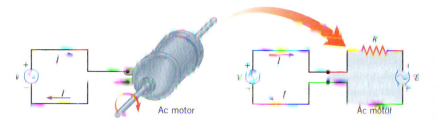

*Figure 22.21* The applied emf *V* supplies the current *I* to drive the motor. The circuit on the right shows *V* along with the electrical equivalent of the motor, including the resistance *R* of its coil and the back emf $\mathcal{E}$.

The next example uses this result to illustrate that the current in a motor depends on both the applied emf *V* and the back emf $\mathcal{E}$.

## Example 12  Operating a Motor

The coil of an ac motor has a resistance of $R = 4.1\ \Omega$. The motor is plugged into an outlet where $V = 120.0$ volts (rms), and the coil develops a back emf of $\mathcal{E} = 118.0$ volts (rms) when rotating at normal speed. The motor is turning a wheel. Find (a) the current when the motor first starts up and (b) the current when the motor is operating at normal speed.

**Reasoning** Once normal operating speed is attained, the motor need only work to compensate for frictional losses. But in bringing the wheel up to speed from rest, the motor must also do work to increase the wheel's rotational kinetic energy. Thus, bringing the wheel up to speed requires more work, and hence more current, than maintaining the normal operating speed. We expect our answers to parts (a) and (b) to reflect this fact.

**Solution**

(a) When the motor just starts up, the coil is not rotating, so there is no back emf induced in the coil and $\mathcal{E} = 0$ V. The start-up current drawn by the motor is

$$I = \frac{V - \mathcal{E}}{R}$$

$$= \frac{120\ \text{V} - 0\ \text{V}}{4.1\ \Omega} = \boxed{29\ \text{A}} \tag{22.5}$$

(b) At normal speed, the motor develops a back emf of $\mathcal{E} = 118.0$ volts, so the current is

$$I = \frac{V - \mathcal{E}}{R}$$

$$= \frac{120.0\ \text{V} - 118.0\ \text{V}}{4.1\ \Omega} = \boxed{0.49\ \text{A}}$$

*Problem solving insight*
The current in an electric motor depends on both the applied emf *V* and any back emf $\mathcal{E}$ developed because the coil of the motor is rotating.

Example 12 illustrates that when a motor is just starting, there is little back emf, and, consequently, a relatively large current exists in the coil. As the motor speeds up, the back emf increases until it reaches a maximum value when the motor is rotating at normal speed. The back emf becomes almost equal to the applied emf, and the current is reduced to a relatively small value, which is sufficient to provide the torque on the coil needed to overcome frictional and other losses in the motor and to drive the load (e.g., a fan).

# 22.7  *Mutual Inductance and Self-Inductance*

## MUTUAL INDUCTANCE

We have seen that an emf can be induced in a coil by keeping the coil stationary and moving a magnet nearby, or by moving the coil near a stationary magnet. Figure 22.22 illustrates another important method of inducing an emf. Here, two coils of wire are placed close to each other, the *primary coil* and the *secondary coil*. The primary coil is the one connected to an ac generator, which sends an alternating current $I_p$ through it. The sec-

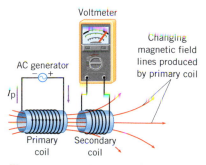

*Figure 22.22* An alternating current $I_p$ in the primary coil creates an alternating magnetic field. This changing field induces an emf in the secondary coil.

ondary coil is not attached to a generator, although a voltmeter is connected across it to register any induced emf.

The current-carrying primary coil is an electromagnet and creates a magnetic field in the surrounding region. If the two coils are close to each other, a significant fraction of this magnetic field penetrates the secondary coil and produces a magnetic flux. The flux is changing, since the current in the primary coil and its associated magnetic field are changing. Because of the change in flux, an emf is induced in the secondary coil.

The effect in which a changing current in one circuit induces an emf in another circuit is called *mutual induction.* According to Faraday's law of electromagnetic induction, the average emf $\mathcal{E}_s$ induced in the secondary coil is proportional to the change in flux $\Delta\Phi_s$ passing through it. However, $\Delta\Phi_s$ is produced by the change in current $\Delta I_p$ in the primary coil. Therefore, it is convenient to recast Faraday's law into a form that relates $\mathcal{E}_s$ to $\Delta I_p$. To see how this recasting is accomplished, note that the net magnetic flux passing through the secondary coil is $N_s\Phi_s$, where $N_s$ is the number of loops in the secondary coil and $\Phi_s$ is the flux through one loop (assumed to be the same for all loops). The net flux is proportional to the magnetic field, which, in turn, is proportional to the current $I_p$ in the primary. Thus, we can write $N_s\Phi_s \propto I_p$. This proportionality can be converted into an equation in the usual manner by introducing a proportionality constant $M$, known as the *mutual inductance:*

$$N_s\Phi_s = MI_p \quad \text{or} \quad M = \frac{N_s\Phi_s}{I_p} \tag{22.6}$$

Substituting this equation into Faraday's law, we find that

$$\mathcal{E}_s = -N_s\frac{\Delta\Phi_s}{\Delta t} = -\frac{\Delta(N_s\Phi_s)}{\Delta t} = -\frac{\Delta(MI_p)}{\Delta t} = -M\frac{\Delta I_p}{\Delta t}$$

***Emf due to mutual induction***

$$\mathcal{E}_s = -M\frac{\Delta I_p}{\Delta t} \tag{22.7}$$

Writing Faraday's law in this manner makes it clear that the average emf $\mathcal{E}_s$ induced in the secondary coil is due to the change in the current $\Delta I_p$ in the primary coil.

Equation 22.7 shows that the measurement unit for the mutual inductance $M$ is $V \cdot s/A$, which is called a henry (H) in honor of Joseph Henry: $1\ V \cdot s/A = 1\ H$. The mutual inductance depends on the geometry of the coils and the nature of any ferromagnetic core material that is present. Although $M$ can be calculated for some highly symmetrical arrangements, it is usually measured experimentally. In most situations, values of $M$ are less than 1 H and are often on the order of millihenries ($1\ mH = 1 \times 10^{-3}\ H$) or microhenries ($1\ \mu H = 1 \times 10^{-6}\ H$).

**The physics of transcranial magnetic stimulation (TMS)**

A new technique that shows promise for the treatment of psychiatric disorders such as depression is based on mutual induction. This technique is called transcranial magnetic stimulation (TMS) and is a type of indirect and gentler electric shock therapy. In traditional electric shock therapy, electric current is delivered directly through the skull and penetrates the brain, disrupting its electrical circuitry and in the process alleviating the symptoms of the psychiatric disorder. The treatment is not gentle and requires an anesthetic, because relatively large electric currents must be used to penetrate the skull. In contrast, TMS produces its electric current by using a time-varying magnetic field. A primary coil is held over the part of the brain to be treated (see Figure 22.23), and a time-varying current is applied to this coil. The arrangement is analogous to that in Figure 22.22, except that the brain and the electrically conductive pathways within it take the place of the secondary coil. The magnetic field produced by the primary coil penetrates the brain and, since the field is changing in time, it induces an emf in the brain. This induced emf causes an electric current to flow in the conductive brain tissue, with therapeutic results similar to those of conventional electric shock treatment. The current delivered to the brain, however, is much smaller than that in the conventional treatment, so that patients receive TMS treatments without anesthetic and severe after-effects such as headaches and memory loss. TMS remains in the experimental stage, however, and the optimal protocol for applying the technique has not yet been determined.

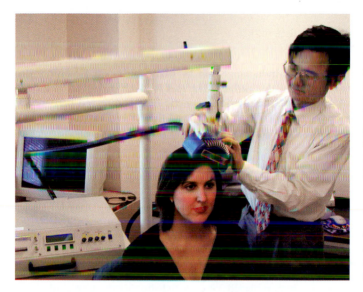

**Figure 22.23** In the technique of transcranial magnetic stimulation (TMS), a time-varying electric current is applied to a primary coil, which is held over a region of the brain, as this photograph illustrates. The time-varying magnetic field produced by the coil penetrates the brain and creates an induced emf within it. This induced emf leads to an induced current that disrupts the electric circuits of the brain, thereby relieving some of the symptoms of psychiatric disorders such as depression. (Courtesy Dr. Mark S. George, Medical University of South Carolina.)

## SELF-INDUCTANCE

In all the examples of induced emfs presented so far, the magnetic field has been produced by an external source, such as a permanent magnet or an electromagnet. However, the magnetic field need not arise from an external source. An emf can be induced in a current-carrying coil by a change in the magnetic field that the current itself produces. For instance, Figure 22.24 shows a coil connected to an ac generator. The alternating current creates an alternating magnetic field that, in turn, creates a changing flux through the coil. The change in flux induces an emf in the coil, in accord with Faraday's law. The effect in which a changing current in a circuit induces an emf in the same circuit is referred to as *self-induction.*

When dealing with self-induction, as with mutual induction, it is customary to recast Faraday's law into a form in which the induced emf is proportional to the change in current in the coil rather than to the change in flux. If $\Phi$ is the magnetic flux that passes through one turn of the coil, then $N\Phi$ is the net flux through a coil of $N$ turns. Since $\Phi$ is proportional to the magnetic field, and the magnetic field is proportional to the current $I$, it follows that $N\Phi \propto I$. By inserting a constant $L$, called the *self-inductance* or simply the *inductance* of the coil, we can convert this proportionality into Equation 22.8:

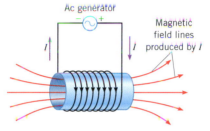

**Figure 22.24** The alternating current in the coil generates an alternating magnetic field that induces an emf in the coil.

$$N\Phi = LI \quad \text{or} \quad L = \frac{N\Phi}{I} \qquad (22.8)$$

Faraday's law of induction now gives the average induced emf as

$$\mathscr{E} = -N\frac{\Delta\Phi}{\Delta t} = -\frac{\Delta(N\Phi)}{\Delta t} = -\frac{\Delta(LI)}{\Delta t} = -L\frac{\Delta I}{\Delta t}$$

**Emf due to self-induction**

$$\mathscr{E} = -L\frac{\Delta I}{\Delta t} \qquad (22.9)$$

Like mutual inductance, $L$ is measured in henries. The magnitude of $L$ depends on the geometry of the coil and on the core material. By wrapping the coil around a ferromagnetic (iron) core, the magnetic flux—and therefore the inductance—can be increased substantially relative to that for an air core. Because of their self-inductance, coils are known as *inductors* and are widely used in electronics. Inductors come in all sizes, typically in the range between millihenries and microhenries. Example 13 shows how to determine the inductance of a solenoid.

### Example 13 The Self-Inductance of a Long Solenoid

A long solenoid of length $\ell = 8.0 \times 10^{-2}$ m and cross-sectional area $A = 5.0 \times 10^{-5}$ m$^2$ contains $n = 6500$ turns per meter. (a) Find the self-inductance of the solenoid, assuming the core is air. (b) Determine the emf induced in the solenoid when the current increases from 0 to 1.5 A in a time of 0.20 s.

### Reasoning

**(a)** The self-inductance can be found by using Equation 22.8 ($L = N\Phi/I$) provided the flux $\Phi$ can be determined. The flux is given by Equation 22.2 as $\Phi = BA \cos \phi$. In the case of a solenoid, the interior magnetic field is directed perpendicular to the plane of the loops (see Section 21.7), so $\phi = 0°$ and $\Phi = BA$. The magnetic field inside a long solenoid has the value $B = \mu_0 nI$, according to Equation 21.7, where $n$ is the number of turns per unit length.

**(b)** The emf induced in the solenoid can be obtained from the relation $\mathcal{E} = -L(\Delta I/\Delta t)$, since $L$ is determined in part (a) and $\Delta I$ and $\Delta t$ are given.

### Solution

**(a)** The self-inductance of the solenoid is

$$L = \frac{N\Phi}{I} = \frac{N(BA)}{I} = \frac{N(\mu_0 nI)A}{I} = \mu_0 nNA = \mu_0 n^2 A\ell$$

where we have replaced $N$ by $n\ell$. Substituting the given values into this result yields

$$L = \mu_0 n^2 A\ell = (4\pi \times 10^{-7}\,\text{T}\cdot\text{m/A})(6500\,\text{turns/m})^2$$
$$\times (5.0 \times 10^{-5}\,\text{m}^2)(8.0 \times 10^{-2}\,\text{m}) = \boxed{2.1 \times 10^{-4}\,\text{H}}$$

**(b)** The induced emf that results from the increasing current is

$$\mathcal{E} = -L\frac{\Delta I}{\Delta t} = -(2.1 \times 10^{-4}\,\text{H})\left(\frac{1.5\,\text{A} - 0\,\text{A}}{0.20\,\text{s}}\right) = \boxed{-1.6 \times 10^{-3}\,\text{V}} \qquad (22.9)$$

The negative sign reminds us that the induced emf opposes the increasing current.

## THE ENERGY STORED IN AN INDUCTOR

An inductor, like a capacitor, can store energy. This stored energy arises because a generator does work to establish a current in an inductor. Suppose an inductor is connected to a generator whose terminal voltage can be varied continuously from zero to some final value. As the voltage is increased, the current $I$ in the circuit rises continuously from zero to its final value. While the current is rising, an induced emf $\mathcal{E} = -L(\Delta I/\Delta t)$ appears across the inductor. Conforming to Lenz's law, the polarity of the induced emf $\mathcal{E}$ is opposite to that of the generator voltage, so as to oppose the increase in the current. Thus, the generator must do work to push the charges through the inductor against this induced emf. The increment of work $\Delta W$ done by the generator in moving a small amount of charge $\Delta Q$ through the inductor is $\Delta W = -(\Delta Q)\mathcal{E} = -(\Delta Q)[-L(\Delta I/\Delta t)]$, according to Equation 19.4. Since $\Delta Q/\Delta t$ is the current $I$, the work done is

$$\Delta W = LI(\Delta I)$$

In this expression $\Delta W$ represents the work done by the generator to increase the current in the inductor by an amount $\Delta I$. To determine the total work $W$ done while the current is changed from zero to its final value, all the small increments $\Delta W$ must be added together. This summation is left as an exercise (see Problem 52). The result is $W = \frac{1}{2}LI^2$, where $I$ represents the final current in the inductor. This work is stored as energy in the inductor, so that

*Energy stored in an inductor*

$$\text{Energy} = \tfrac{1}{2}LI^2 \qquad (22.10)$$

It is possible to regard the energy in an inductor as being stored in its magnetic field. For the special case of a long solenoid, Example 13 shows that the self-inductance is $L = \mu_0 n^2 A\ell$, where $n$ is the number of turns per unit length, $A$ is the cross-sectional area, and $\ell$ is the length of the solenoid. As a result, the energy stored in a long solenoid is

$$\text{Energy} = \tfrac{1}{2}LI^2 = \tfrac{1}{2}\mu_0 n^2 A\ell I^2$$

Since $B = \mu_0 nI$ at the interior of a long solenoid (Equation 21.7), this energy can be expressed as

$$\text{Energy} = \frac{1}{2\mu_0} B^2 A\ell$$

The term $A\ell$ is the volume inside the solenoid where the magnetic field exists, so the energy per unit volume or **energy density** is

$$\text{Energy density} = \frac{\text{Energy}}{\text{Volume}} = \frac{1}{2\mu_0} B^2 \tag{22.11}$$

Although this result was obtained for the special case of a long solenoid, it is quite general and is valid for any point where a magnetic field exists in air or vacuum or in a nonmagnetic material. Thus, energy can be stored in a magnetic field, just as it can in an electric field.

## 22.8 Transformers

One of the most important applications of mutual induction and self-induction takes place in a transformer. A **transformer** is a device for increasing or decreasing an ac voltage. For instance, whenever cordless appliances (e.g., a hand-held vacuum cleaner) are plugged into a wall receptacle to recharge the batteries, a transformer plays a role in reducing the 120-V ac voltage to a much smaller value. Typically, between 3 and 9 V are needed to energize batteries. In another example, a picture tube in a television set needs about 15 000 V to accelerate the electron beam, and a transformer is used to obtain this high voltage from the 120 V at a wall socket.

*The physics of transformers.*

Figure 22.25 shows a drawing of a transformer. The transformer consists of an iron core on which two coils are wound: a primary coil with $N_p$ turns and a secondary coil with $N_s$ turns. The primary coil is the one connected to the ac generator. For the moment, suppose the switch in the secondary circuit is open, so there is no current in this circuit.

The alternating current in the primary coil establishes a changing magnetic field in the iron core. Because iron is easily magnetized, it greatly enhances the magnetic field relative to that in an air core and guides the field lines to the secondary coil. In a well-designed core, nearly all the magnetic flux $\Phi$ that passes through each turn of the primary also goes through each turn of the secondary. Since the magnetic field is changing, the flux through the primary and secondary coils is also changing, and consequently an emf is induced in both coils. In the secondary coil the induced emf $\mathcal{E}_s$ arises from mutual induction and is given by Faraday's law as

$$\mathcal{E}_s = -N_s \frac{\Delta\Phi}{\Delta t}$$

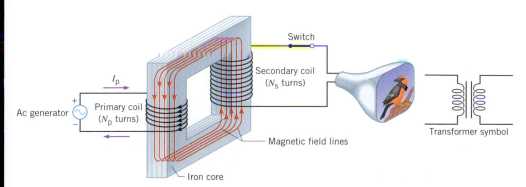

**Figure 22.25** A transformer consists of a primary coil and a secondary coil, both wound on an iron core. The changing magnetic flux produced by the current in the primary coil induces an emf in the secondary coil. At the far right is the symbol for a transformer.

In the primary coil the induced emf $\mathcal{E}_p$ is due to self-induction and is specified by Faraday's law as

$$\mathcal{E}_p = -N_p \frac{\Delta\Phi}{\Delta t}$$

The term $\Delta\Phi/\Delta t$ is the same in both of these equations, since the same flux penetrates each turn of both coils. Dividing the two equations shows that

$$\frac{\mathcal{E}_s}{\mathcal{E}_p} = \frac{N_s}{N_p}$$

In a high-quality transformer the resistances of the coils are negligible, so the magnitudes of the emfs, $\mathcal{E}_s$ and $\mathcal{E}_p$, are nearly equal to the terminal voltages, $V_s$ and $V_p$, across the coils (see Section 20.9 for a discussion of terminal voltage). The relation $\mathcal{E}_s/\mathcal{E}_p = N_s/N_p$ is called the **transformer equation** and is usually written in terms of the terminal voltages:

*Transformer*
*equation*
$$\frac{V_s}{V_p} = \frac{N_s}{N_p} \tag{22.12}$$

According to the transformer equation, if $N_s$ is greater than $N_p$, the secondary (output) voltage is greater than the primary (input) voltage. In this case we have a *step-up* transformer. On the other hand, if $N_s$ is less than $N_p$, the secondary voltage is less than the primary voltage, and we have a *step-down* transformer. The ratio $N_s/N_p$ is referred to as the *turns ratio* of the transformer. A turns ratio of 8/1 (often written as 8:1) means, for example, that the secondary coil has eight times more turns than does the primary coil. Conversely, a turns ratio of 1:8 implies that the secondary coil has one-eighth as many turns as the primary coil.

A transformer operates with ac electricity and not with dc. A steady direct current in the primary coil produces a flux that does not change in time, and thus no emf is induced in the secondary coil. The ease with which transformers can change voltages from one value to another is a principal reason why ac is preferred over dc.

With the switch in the secondary circuit of Figure 22.25 closed, a current $I_s$ exists in the circuit and electrical energy is fed to the TV tube. This energy comes from the ac generator connected to the primary coil. Although the secondary voltage $V_s$ may be larger or smaller than the primary voltage $V_p$, energy is not being created or destroyed by the transformer. Energy conservation requires that the energy delivered to the secondary coil must be the same as the energy delivered to the primary coil, provided no energy is dissipated in heating these coils or is otherwise lost. In a well-designed transformer, less than 1% of the input energy is lost in the form of heat. Noting that power is energy per unit time, and assuming 100% energy transfer, the average power $\overline{P}_p$ delivered to the primary coil is equal to the average power $\overline{P}_s$ delivered to the secondary coil: $\overline{P}_p = \overline{P}_s$. But $\overline{P} = IV$ (Equation 20.15a), so $I_p V_p = I_s V_s$, or

$$\frac{I_s}{I_p} = \frac{V_p}{V_s} = \frac{N_p}{N_s} \tag{22.13}$$

*Problem solving insight*

Observe that $V_s/V_p$ is equal to the turns ratio $N_s/N_p$, while $I_s/I_p$ is equal to the inverse turns ratio $N_p/N_s$. Consequently, *a transformer that steps up the voltage simultaneously steps down the current, and a transformer that steps down the voltage steps up the current.* However, the power is neither stepped up nor stepped down, since $\overline{P}_p = \overline{P}_s$. Example 14 emphasizes this fact.

### Example 14    A Step-Down Transformer

A step-down transformer inside a stereo receiver has 330 turns in the primary coil and 25 turns in the secondary coil. The plug connects the primary coil to a 120-V wall socket, and there is a current of 0.83 A in the primary coil while the receiver is turned on. Connected to the secondary coil are the transistor circuits of the receiver. Find (a) the voltage across the secondary coil, (b) the current in the secondary coil, and (c) the average electric power delivered to the transistor circuits.

**Reasoning** The transformer equation, Equation 22.12, states that the secondary voltage $V_s$ is equal to the product of the primary voltage $V_p$ and the turns ratio $N_s/N_p$. On the other hand, the secondary current $I_s$ is equal to the product of the primary current $I_p$ and the inverse turns ratio $N_p/N_s$. The average power delivered to the transistor circuits is the product of the secondary current and the secondary voltage.

**Solution**

(a) The voltage across the secondary coil can be found from the transformer equation:

$$V_s = V_p \frac{N_s}{N_p} = (120 \ \text{V})\left(\frac{25}{330}\right) = \boxed{9.1 \ \text{V}} \tag{22.12}$$

(b) The current in the secondary coil is

$$I_s = I_p \frac{N_p}{N_s} = (0.83 \ \text{A})\left(\frac{330}{25}\right) = \boxed{11 \ \text{A}} \tag{22.13}$$

(c) The average power $\overline{P}_s$ delivered to the secondary is the product of $I_s$ and $V_s$:

$$\overline{P}_s = I_s V_s = (11 \ \text{A})(9.1 \ \text{V}) = \boxed{1.0 \times 10^2 \ \text{W}} \tag{20.15a}$$

As a check on our calculation, we verify that the power delivered to the secondary coil is the same as that sent to the primary coil from the wall receptacle: $\overline{P}_p = I_p V_p = (0.83 \ \text{A})(120 \ \text{V}) = 1.0 \times 10^2 \ \text{W}$.

Transformers play an important role in the transmission of power between electrical generating plants and the communities they serve. Whenever electricity is transmitted, there is always some loss of power in the transmission lines themselves due to resistive heating. Since the resistance of the wires is proportional to their length, the longer the wires the greater is the power loss. Power companies reduce this loss by using transformers that step up the voltage to high levels while reducing the current. A smaller current means less power loss, since $P = I^2R$, where $R$ is the resistance of the transmission wires (see problem 59). Figure 22.26 shows one possible way of transmitting power. The power plant produces a voltage of 12 000 V. This voltage is then raised to 240 000 V by a 20:1 step-up transformer. The high-voltage power is sent over the long-distance transmission line. Upon arrival at the city, the voltage is reduced to about 8000 V at a substation using a 1:30 step-down transformer. However, before any domestic use, the voltage is further reduced to 240 V (or possibly 120 V) by another step-down transformer that is often mounted on a utility pole. The power is then distributed to consumers.

**Figure 22.26** Transformers play a key role in the transmission of electric power.

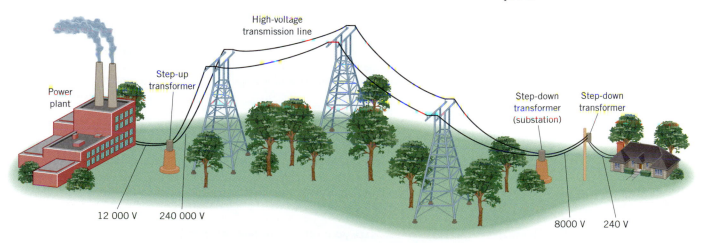

# Concept Summary

This summary presents an abridged version of the chapter, including the important equations and all available learning aids. For convenient reference, the learning aids (including the text's examples) are placed next to or immediately after the relevant equation or discussion. The following learning aids may be found on-line at **www.wiley.com/college/cutnell**:

| | |
|---|---|
| **Interactive LearningWare** examples are solved according to a five-step interactive format that is designed to help you develop problem-solving skills. | **Concept Simulations** are animated versions of text figures or animations that illustrate important concepts. You can control parameters that affect the display, and we encourage you to experiment. |
| **Interactive Solutions** offer specific models for certain types of problems in the chapter homework. The calculations are carried out interactively. | **Self-Assessment Tests** include both qualitative and quantitative questions. Extensive feedback is provided for both incorrect and correct answers, to help you evaluate your understanding of the material. |

| *Topic* | *Discussion* | *Learning Aids* |
|---|---|---|

### 22.1 Induced Emf and Induced Current

**Electromagnetic induction**

Electromagnetic induction is the phenomenon in which an emf is induced in a piece of wire or a coil of wire with the aid of a magnetic field. The emf is called an induced emf, and any current that results from the emf is called an induced current.

### 22.2 Motional Emf

An emf $\mathcal{E}$ is induced in a conducting rod of length $L$ when the rod moves with a speed $v$ in a magnetic field of magnitude $B$, according to

**Motional emf**

$$\mathcal{E} = vBL \qquad (22.1)$$

**Example 1**
**Interactive LearningWare 22.1**

Equation 22.1 applies when the velocity of the rod, the length of the rod, and the magnetic field are mutually perpendicular.

**Conservation of energy**

When the motional emf is used to operate an electrical device, such as a light bulb, the energy delivered to the device originates in the work done to move the rod, and the law of conservation of energy applies.

**Examples 2, 3**

### 22.3 Magnetic Flux

The magnetic flux $\Phi$ that passes through a surface is

**Magnetic flux**

$$\Phi = BA \cos \phi \qquad (22.2)$$

**Example 4**

where $B$ is the magnitude of the magnetic field, $A$ is the area of the surface, and $\phi$ is the angle between the field and the normal to the surface.

The magnetic flux is proportional to the number of magnetic field lines that pass through the surface.

### 22.4 Faraday's Law of Electromagnetic Induction

Faraday's law of electromagnetic induction states that the average emf $\mathcal{E}$ induced in a coil of $N$ loops is

**Faraday's law**

$$\mathcal{E} = -N\left(\frac{\Phi - \Phi_0}{t - t_0}\right) = -N\frac{\Delta\Phi}{\Delta t} \qquad (22.3)$$

**Example 5**
**Concept Simulation 22.1**
**Example 6**
**Interactive LearningWare 22.2**
**Example 7**
**Interactive Solution 22.69**

where $\Delta\Phi$ is the change in magnetic flux through one loop and $\Delta t$ is the time interval during which the change occurs. Motional emf is a special case of induced emf.

### 22.5 Lenz's Law

**Lenz's law**

Lenz's law provides a way to determine the polarity of an induced emf. Lenz's law is stated as follows: The induced emf resulting from a changing magnetic flux has a polarity that leads to an induced current whose direction is such that the induced magnetic field opposes the original flux change. This statement is a consequence of the law of conservation of energy.

**Examples 8, 9**
**Interactive Solution 22.33**

 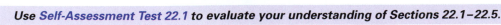 **Use *Self-Assessment Test 22.1* to evaluate your understanding of Sections 22.1–22.5.**

| Topic | Discussion | Learning Aids |
|---|---|---|

### 22.6 The Electric Generator

In its simplest form, an electric generator consists of a coil of $N$ loops that rotates in a uniform magnetic field **B**. The emf produced by this generator is

**Emf of a generator**

$$\mathscr{E} = NAB\omega \sin \omega t = \mathscr{E}_0 \sin \omega t \qquad (22.4)$$

**Examples 10, 11**
**Interactive LearningWare 22.3**
**Interactive Solution 22.63**

where $A$ is the area of the coil, $\omega$ is the angular speed (in rad/s) of the coil, and $\mathscr{E}_0$ is the peak emf. The angular speed in rad/s is related to the frequency $f$ in cycles/s, or Hz, according to $\omega = 2\pi f$.

When an electric motor is running, it exhibits a generator-like behavior by producing an induced emf, called the back emf. The current $I$ needed to keep the motor running at a constant speed is

**Back emf of a motor**

$$I = \frac{V - \mathscr{E}}{R} \qquad (22.5)$$

**Example 12**

where $V$ is the emf applied to the motor by an external source, $\mathscr{E}$ is the back emf, and $R$ is the resistance of the motor coil.

### 22.7 Mutual Inductance and Self Inductance

Mutual induction is the effect in which a changing current in the primary coil induces an emf in the secondary coil. The average emf $\mathscr{E}_s$ induced in the secondary coil by a change in current $\Delta I_p$ in the primary coil is

**Emf due to mutual induction**

$$\mathscr{E}_s = -M \frac{\Delta I_p}{\Delta t} \qquad (22.7)$$

**Mutual inductance**

where $\Delta t$ is the time interval during which the change occurs. The constant $M$ is the mutual inductance between the two coils and is measured in henries (H).

Self-induction is the effect in which a change in current $\Delta I$ in a coil induces an average emf $\mathscr{E}$ in the same coil, according to

**Emf due to self-induction**

$$\mathscr{E} = -L \frac{\Delta I}{\Delta t} \qquad (22.9)$$

**Example 13**

**Self-inductance**

The constant $L$ is the self-inductance, or inductance, of the coil and is measured in henries.

To establish a current $I$ in an inductor, work must be done by an external agent. This work is stored as energy in the inductor, the amount being

**Energy stored in an inductor**

$$\text{Energy} = \tfrac{1}{2}LI^2 \qquad (22.10)$$

The energy stored in an inductor can be regarded as being stored in its magnetic field. At any point in air or vacuum or in a nonmagnetic material where a magnetic field **B** exists, the energy density, or the energy stored per unit volume, is

**Energy density of a magnetic field**

$$\text{Energy density} = \frac{1}{2\mu_0} B^2 \qquad (22.11)$$

### 22.8 Transformers

A transformer consists of a primary coil of $N_p$ turns and a secondary coil of $N_s$ turns. If the resistances of the coils are negligible, the voltage $V_p$ across the primary and the voltage $V_s$ across the secondary are related according to the transformer equation:

**Transformer equation**

$$\frac{V_s}{V_p} = \frac{N_s}{N_p} \qquad (22.12)$$

**Example 14**
**Interactive Solution 22.55**

**Turns ratio**

where the ratio $N_s/N_p$ is called the turns ratio of the transformer.

A transformer functions with ac electricity, not with dc. If the transformer is 100% efficient in transferring power from the primary to the secondary coil, the ratio of the secondary current $I_s$ to the primary current $I_p$ is

$$\frac{I_s}{I_p} = \frac{N_p}{N_s} \qquad (22.13)$$

 **Use Self-Assessment Test 22.2 to evaluate your understanding of Sections 22.6–22.8.**

# Problems

ssm   Solution is in the Student Solutions Manual.     www   Solution is available on the World Wide Web at www.wiley.com/college/cutnell

🜊   This icon represents a biomedical application.

### Section 22.2 Motional Emf

**1. ssm** A spark can jump between two nontouching conductors if the potential difference between them is sufficiently large. A potential difference of approximately 940 V is required to produce a spark in an air gap of $1.0 \times 10^{-4}$ m. Suppose the light bulb in Figure 22.4b is replaced by such a gap. How fast would a 1.3-m rod have to be moving in a magnetic field of 4.8 T to cause a spark to jump across the gap?

**2.** 🜊 The drawing shows a type of flow meter that can be used to measure the speed of blood in situations when a blood vessel is sufficiently exposed (e.g., during surgery). Blood is conductive enough that it can be treated as a moving conductor. When it flows perpendicularly with respect to a magnetic field, as in the drawing, electrodes can be used to measure the small voltage that develops across the vessel. Suppose the speed of the blood is 0.30 m/s and the diameter of the vessel is 5.6 mm. In a 0.60-T magnetic field what is the magnitude of the voltage that is measured with the electrodes in the drawing?

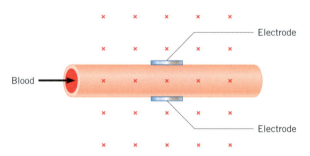

**3.** The wingspan (tip to tip) of a Boeing 747 jetliner is 59 m. The plane is flying horizontally at a speed of 220 m/s. The vertical component of the earth's magnetic field is $5.0 \times 10^{-6}$ T. Find the emf induced between the wing tips.

**4.** In 1996, NASA performed an experiment called the Tethered Satellite experiment. In this experiment a $2.0 \times 10^4$-m length of wire was let out by the space shuttle *Atlantis* to generate a motional emf. The shuttle had an orbital speed of $7.6 \times 10^3$ m/s, and the magnitude of the earth's magnetic field at the location of the wire was $5.1 \times 10^{-5}$ T. If the wire had moved perpendicular to the earth's magnetic field, what would have been the motional emf generated between the ends of the wire?

**5. ssm www** The drawing shows three identical rods (A, B, and C) moving in different planes. A constant magnetic field of magnitude 0.45 T is directed along the +y axis. The length of each rod is $L = 1.3$ m, and the speeds are the same, $v_A = v_B = v_C = 2.7$ m/s. For each rod, find the magnitude of the motional emf, and indicate which end (1 or 2) of the rod is positive.

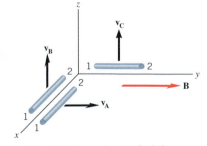

**\*6.** Suppose the light bulb in Figure 22.4b is a 60.0-W bulb with a resistance of 240 Ω. The magnetic field has a magnitude of 0.40 T,

and the length of the rod is 0.60 m. The only resistance in the circuit is that due to the bulb. Minimally, how long would the rails on which the moving rod slides have to be, in order that the bulb can remain lit for one-half second?

**\*7. ssm** Suppose the light bulb in Figure 22.4b is replaced by a 6.0-Ω electric heater that consumes 15 W of power. The conducting bar moves to the right at a constant speed, the field strength is 2.4 T, and the length of the bar between the rails is 1.2 m. (a) How fast is the bar moving? (b) What force must be applied to the bar to keep it moving to the right at the constant speed found in part (a)?

**\*8.** Suppose that the voltage of the battery in the circuit in the drawing is 3.0 V, the magnitude of the magnetic field (directed perpendicularly into the plane of the paper) is 0.60 T, and the length of the rod between the rails is 0.20 m. Assuming that the rails are very long and have negligible resistance, find the maximum speed attained by the rod after the switch is closed.

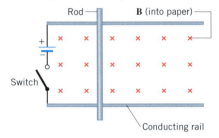

**\*\*9.** Review Conceptual Example 3 and Figure 22.7b as an aid in solving this problem. A conducting rod slides down between two frictionless vertical copper tracks at a constant speed of 4.0 m/s perpendicular to a 0.50-T magnetic field. The resistance of the rod and tracks is negligible. The rod maintains electrical contact with the tracks at all times and has a length of 1.3 m. A 0.75-Ω resistor is attached between the tops of the tracks. (a) What is the mass of the rod? (b) Find the change in the gravitational potential energy that occurs in a time of 0.20 s. (c) Find the electrical energy dissipated in the resistor in 0.20 s.

### Section 22.3 Magnetic Flux

*For problems in this set, assume that the magnetic flux is a positive quantity.*

**10.** The drawing shows two surfaces that have the same area. A uniform magnetic field **B** fills the space occupied by these surfaces and is oriented parallel to the *yz* plane as shown. Find the ratio $\Phi_{xz}/\Phi_{xy}$ of the magnetic fluxes that pass through the surfaces.

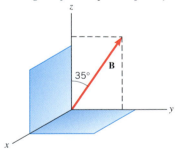

**11. ssm** A standard door into a house rotates about a vertical axis through one side, as defined by the door's hinges. A uniform magnetic field is parallel to the ground and perpendicular to this axis. Through what angle must the door rotate so that the magnetic flux that passes through it decreases from its maximum value to one-third of its maximum value?

**12.** A house has a floor area of 112 m² and an outside wall that has an area of 28 m². The earth's magnetic field here has a horizontal component of $2.6 \times 10^{-5}$ T that points due north and a vertical component of $4.2 \times 10^{-5}$ T that points straight down, toward the earth. Determine the magnetic flux through the wall if the wall faces (a) north and (b) east. (c) Calculate the magnetic flux that passes through the floor.

**13.** The drawing shows three square surfaces, one lying in the *xy* plane, one in the *xz* plane, and one in the *yz* plane. The sides of each square have lengths of $2.0 \times 10^{-2}$ m. A uniform magnetic field exists in this region, and its components are: $B_x = 0.50$ T, $B_y = 0.80$ T, and $B_z = 0.30$ T. Determine the magnetic flux that passes through the surface that lies in (a) the *xy* plane, (b) the *xz* plane, and (c) the *yz* plane.

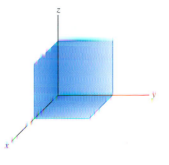

**14.** A loop of wire has the shape shown in the drawing. The top part of the wire is bent into a semicircle of radius $r = 0.20$ m. The normal to the plane of the loop is parallel to a constant magnetic field ($\phi = 0°$) of magnitude 0.75 T. What is the change $\Delta\Phi$ in the magnetic flux that passes through the loop when, starting with the position shown in the drawing, the semicircle is rotated through half a revolution?

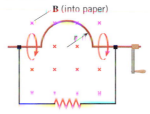

B (into paper)

**15. ssm www** A five-sided object, whose dimensions are shown in the drawing, is placed in a uniform magnetic field. The magnetic field has a magnitude of 0.25 T and points along the positive *y* direction. Determine the magnetic flux through each of the five sides.

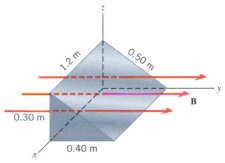

**16.** A long and narrow rectangular loop of wire is moving toward the bottom of the page with a speed of 0.020 m/s (see the drawing). The loop is leaving a region in which a 2.4-T magnetic field exists; the magnetic field outside this region is zero. During a time of 2.0 s, what is the magnitude of the *change* in the magnetic flux?

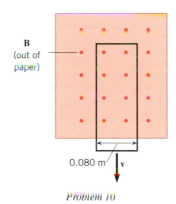

B (out of paper)

0.080 m    v

Problem 16

**Section 22.4 Faraday's Law of Electromagnetic Induction**

**17.** A planar coil of wire has a single turn. The normal to this coil is parallel to a uniform and constant (in time) magnetic field of 1.7 T. An emf that has a magnitude of 2.6 V is induced in this coil because the coil's area A is shrinking. What is the magnitude of $\Delta A/\Delta t$, which is the rate (in m²/s) at which the area changes?

**18.** In each of two coils the rate of change of the magnetic flux in a single loop is the same. The emf induced in coil 1, which has 184 loops, is 2.82 V. The emf induced in coil 2 is 4.23 V. How many loops does coil 2 have?

**19. ssm** A circular coil (950 turns, radius = 0.060 m) is rotating in a uniform magnetic field. At $t = 0$ s, the normal to the coil is perpendicular to the magnetic field. At $t = 0.010$ s, the normal makes an angle of $\phi = 45°$ with the field because the coil has made one-

eighth of a revolution. An average emf of magnitude 0.065 V is induced in the coil. Find the magnitude of the magnetic field at the location of the coil.

**20.** A magnetic field is perpendicular to the plane of a single-turn circular coil. The magnitude of the field is changing, so that an emf of 0.80 V and a current of 3.2 A are induced in the coil. The wire is then re-formed into a single-turn square coil, which is used in the same magnetic field (again perpendicular to the plane of the coil and with a magnitude changing at the same rate). What emf and current are induced in the square coil?

**21.** Magnetic resonance imaging (MRI) is a medical technique for producing pictures of the interior of the body. The patient is placed within a strong magnetic field. One safety concern is what would happen to the positively and negatively charged particles in the body fluids if an equipment failure caused the magnetic field to be shut off suddenly. An induced emf could cause these particles to flow, producing an electric current within the body. Suppose the largest surface of the body through which flux passes has an area of 0.032 m² and a normal that is parallel to a magnetic field of 1.5 T. Determine the smallest time period during which the field can be allowed to vanish if the magnitude of the average induced emf is to be kept less than 0.010 V.

**22. Interactive LearningWare 22.2** at **www.wiley.com/college/cutnell** reviews the fundamental approach in problems such as this. A constant magnetic field passes through a single rectangular loop whose dimensions are 0.35 m × 0.55 m. The magnetic field has a magnitude of 2.1 T and is inclined at an angle of 65° with respect to the normal to the plane of the loop. (a) If the magnetic field decreases to zero in a time of 0.45 s, what is the magnitude of the average emf induced in the loop? (b) If the magnetic field remains constant at its initial value of 2.1 T, what is the magnitude of the rate $\Delta A/\Delta t$ at which the area should change so that the average emf has the same magnitude as in part (a)?

**\* 23.** A piece of copper wire is formed into a single circular loop of radius 12 cm. A magnetic field is oriented parallel to the normal to the loop, and it increases from 0 to 0.60 T in a time of 0.45 s. The wire has a resistance per unit length of $3.3 \times 10^{-3}$ Ω/m. What is the average electrical energy dissipated in the resistance of the wire?

**\* 24.** Parts *a* and *b* of the drawing show the same uniform and constant (in time) magnetic field **B** directed perpendicularly into the paper over a rectangular region. Outside this region, there is no field. Also shown is a rectangular coil (one turn), which lies in the plane of the paper. In part *a* the long side of the coil (length = $L$) is just at the edge of the field region, while in part *b* the short side (width = $W$) is just at the edge. It is known that $L/W = 3.0$. In both parts of the drawing the coil is pushed into the field with the same velocity **v** until it is completely within the field region. The magnitude of the average emf induced in the coil in part *a* is 0.15 V. What is its magnitude in part *b*?

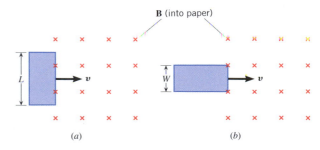

(a)    (b)

**\* 25. ssm** A magnetic field is passing through a loop of wire whose area is 0.018 m². The direction of the magnetic field is parallel to the normal to the loop, and the magnitude of the field is increasing at the rate of 0.20 T/s. (a) Determine the magnitude of the emf induced in the loop. (b) Suppose the area of the loop can be enlarged or shrunk. If the magnetic field is increasing as in part (a), at what rate (in m²/s) should the area be changed at the instant when $B = 1.8$ T if the induced emf is to be zero? Explain whether the area is to be enlarged or shrunk.

* **26.** The drawing shows a copper wire (negligible resistance) bent into a circular shape with a radius of 0.50 m. The radial section $BC$ is fixed in place, while the copper bar $AC$ sweeps around at an angular speed of 15 rad/s. The bar makes electrical contact with the wire at all times. The wire and the bar have negligible resistance. A uniform magnetic field exists everywhere, is perpendicular to the plane of the circle, and has a magnitude of $3.8 \times 10^{-3}$ T. Find the magnitude of the current induced in the loop $ABC$.

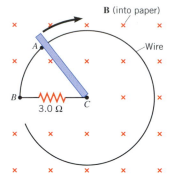

** **27.** Two 0.68-m-long conducting rods are rotating at the same speed in opposite directions, and both are perpendicular to a 4.7-T magnetic field. As the drawing shows, the ends of these rods come to within 1.0 mm of each other as they rotate. Moreover, the fixed ends about which the rods are rotating are connected by a wire, so these ends are at the same electric potential. If a potential difference of 4.5 $\times 10^3$ V is required to cause a 1.0-mm spark in air, what is the angular speed (in rad/s) of the rods when a spark jumps across the gap?

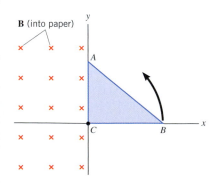

### Section 22.5 Lenz's Law

**28.** The drawing shows that a uniform magnetic field is directed perpendicularly into the plane of the paper and fills the entire region to the left of the $y$ axis. There is no magnetic field to the right of the $y$ axis. A rigid right triangle $ABC$ is made of copper wire. The triangle rotates counterclockwise about the origin at point $C$. What is the direction (clockwise or counterclockwise) of the induced current when the triangle is crossing (a) the $+y$ axis, (b) the $-x$ axis, (c) the $-y$ axis, and (d) the $+x$ axis? For each case, justify your answer.

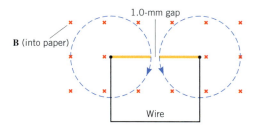

**29.** **ssm** Suppose in Figure 22.1 that the bar magnet is held stationary, but the coil of wire is free to move. Which way will current be directed through the ammeter, left to right or right to left, when the coil is moved (a) to the left and (b) to the right? Explain.

**30.** Review the drawing that accompanies Problem 14. The semicircular piece of wire rotates through half a revolution in the direction shown, starting from the position indicated in the drawing. Which end of the resistor, the left or the right end, is positive? Explain your reasoning.

**31.** **ssm** Review Conceptual Example 9 as an aid in understanding this problem. A long, straight wire lies on a table and carries a current $I$. As the drawing shows, a small circular loop of wire is pushed across the top of the table from position 1 to position 2. Determine the direction of the induced current, clockwise or counterclockwise, as the loop moves past (a) position 1 and (b) position 2. Justify your answers.

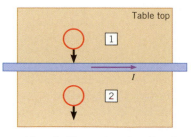

* **32.** Indicate the direction of the electric field between the plates of the parallel plate capacitor shown in the drawing if the magnetic field is decreasing in time. Give your reasoning.

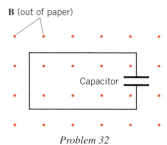

*Problem 32*

* **33.** Consult **Interactive Solution 22.33** at **www.wiley.com/college/cutnell** for one approach to this problem. A circular loop of wire rests on a table. A long, straight wire lies on this loop, directly over its center, as the drawing illustrates. The current $I$ in the straight wire is decreasing. In what direction is the induced current, if any, in the loop? Give your reasoning.

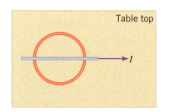

** **34.** A wire loop is suspended from a string that is attached to point $P$ in the drawing. When released, the loop swings downward, from left to right, through a uniform magnetic field, with the plane of the loop remaining perpendicular to the plane of the paper at all times. (a) Determine the direction of the current induced in the loop as it swings past the locations labeled I and II. Specify the direction of the current in terms of the points $x$, $y$, and $z$ on the loop (e.g., $x \rightarrow y \rightarrow z$ or $z \rightarrow y \rightarrow x$). The points $x$, $y$, and $z$ lie behind the plane of the paper. (b) What is the direction of the induced current at the locations II and I when the loop swings back, from right to left? Provide reasons for your answers.

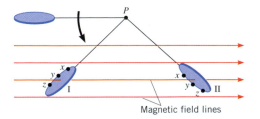

### Section 22.6 The Electric Generator

**35.** **ssm** The coil of an ac generator has an area per turn of $1.2 \times 10^{-2}$ m² and consists of 500 turns. The coil is situated in a 0.13-T magnetic field and is rotating at an angular speed of 34 rad/s. What is the emf induced in the coil at the instant when the normal to the loop makes an angle of 27° with respect to the direction of the magnetic field?

**36.** A vacuum cleaner is plugged into a 120.0-V socket and uses 3.0 A of current in normal operation when the back emf generated by the electric motor is 72.0 V. Find the coil resistance of the motor.

**37.** One generator uses a magnetic field of 0.10 T and has a coil area per turn of 0.045 m². A second generator has a coil area per turn of 0.015 m². The generator coils have the same number of turns and rotate at the same angular speed. What magnetic field should be used in the second generator so that its peak emf is the same as that of the first generator?

**38.** A 120.0-volt motor draws a current of 7.00 A when running at normal speed. The resistance of the armature wire is 0.720 Ω. (a) Determine the back emf generated by the motor. (b) What is the current at the instant when the motor is just turned on and has not begun to rotate? (c) What series resistance must be added to limit the starting current to 15.0 A?

**39. ssm www** The drawing shows a plot of the output emf of a generator as a function of time $t$. The coil of this device has a cross-sectional area per turn of 0.020 m² and contains 150 turns. Find (a) the frequency $f$ of the generator in hertz, (b) the angular speed $\omega$ in rad/s, and (c) the magnitude of the magnetic field.

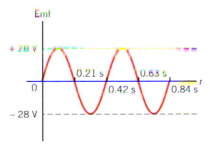

**40.** A generator uses a coil that has 100 turns and a 0.50-T magnetic field. The frequency of this generator is 60.0 Hz, and its emf has an rms value of 120 V. Assuming that each turn of the coil is a square (an approximation), determine the length of the wire from which the coil is made.

**41. ssm** At its normal operating speed, an electric fan motor draws only 15.0% of the current it draws when it just begins to turn the fan blade. The fan is plugged into a 120.0-V socket. What back emf does the motor generate at its normal operating speed?

**42.** The coil of a generator has a radius of 0.14 m. When this coil is unwound, the wire from which it is made has a length of 5.7 m. The magnetic field of the generator is 0.20 T, and the coil rotates at an angular speed of 25 rad/s. What is the peak emf of this generator?

**43.** The armature of an electric drill motor has a resistance of 15.0 Ω. When connected to a 120.0-V outlet, the motor rotates at its normal speed and develops a back emf of 108 V. (a) What is the current through the motor? (b) If the armature freezes up due to a lack of lubrication in the bearings and can no longer rotate, what is the current in the stationary armature? (c) What is the current when the motor runs at only half speed?

### Section 22.7 Mutual Inductance and Self-Inductance

**44.** The average emf induced in the secondary coil is 0.12 V when the current in the primary coil changes from 3.4 to 1.6 A in 0.14 s. What is the mutual inductance of the coils?

**45. ssm** The earth's magnetic field, like any magnetic field, stores energy. The maximum strength of the earth's field is about $7.0 \times 10^{-5}$ T. Find the maximum magnetic energy stored in the space above a city if the space occupies an area of $5.0 \times 10^8$ m² and has a height of 1500 m.

**46.** The current through a 3.2-mH inductor varies with time according to the graph shown in the drawing. What is the average induced emf during the time intervals (a) 0–2.0 ms, (b) 2.0–5.0 ms, and (c) 5.0–9.0 ms?

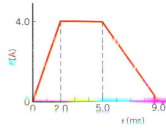

*Problem 46*

**47.** Mutual induction can be used as the basis for a metal detector. A typical setup uses two large coils that are parallel to each other and have a common axis. Because of mutual induction, the ac generator connected to the primary coil causes an emf of 0.46 V to be induced in the secondary coil. When someone without metal objects walks through the coils, the mutual inductance and, thus, the induced emf do not change much. But when a person carrying a handgun walks through, the mutual inductance increases. The change in emf can be used to trigger an alarm. If the mutual inductance increases by a factor of three, find the new value of the induced emf.

**48.** During a 72-ms interval, a change in the current in a primary coil occurs. This change leads to the appearance of a 6.0-mA current in a nearby secondary coil. The secondary coil is part of a circuit in which the resistance is 12 Ω. The mutual inductance between the two coils is 3.2 mH. What is the change in the primary current?

**49. ssm** Suppose you wish to make a solenoid whose self-inductance is 1.4 mH. The inductor is to have a cross-sectional area of $1.2 \times 10^{-3}$ m² and a length of 0.052 m. How many turns of wire are needed?

**50.** A magnetic field has a magnitude of 12 T. What is the magnitude of an electric field that stores the same energy per unit volume as this magnetic field?

**51.** A long, current-carrying solenoid with an air core has 1750 turns per meter of length and a radius of 0.0180 m. A coil of 125 turns is wrapped tightly around the outside of the solenoid. What is the mutual inductance of this system?

**52.** The purpose of this problem is to show that the work $W$ needed to establish a final current $I_f$ in an inductor is $W = \frac{1}{2}LI_f^2$ (Equation 22.10). In Section 22.7 we saw that the amount of work $\Delta W$ needed to change the current through an inductor by an amount $\Delta I$ is $\Delta W = LI(\Delta I)$, where $L$ is the inductance. The drawing shows a graph of $LI$ versus $I$. Notice that $LI(\Delta I)$ is the area of the shaded vertical rectangle whose height is $LI$ and whose width is $\Delta I$. Use this fact to show that the total work $W$ needed to establish a current $I_f$ is $W = \frac{1}{2}LI_f^2$.

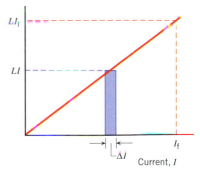

**53.** Coil 1 is a flat circular coil that has $N_1$ turns and a radius $R_1$. At its center is a much smaller flat, circular coil that has $N_2$ turns and radius $R_2$. The planes of the coils are parallel. Assume that coil 2 is so small that the magnetic field due to coil 1 has nearly the same value at all points covered by the area of coil 2. Determine an expression for the mutual inductance between these two coils in terms of $\mu_0$, $N_1$, $R_1$, $N_2$, and $R_2$.

## Section 22.8 Transformers

**54.** A neon sign requires 12 000 V for its operation. It operates from a 220-V receptacle. (a) What type of transformer, step-up or step-down, is needed? (b) What must be the turns ratio $N_s/N_p$ of the transformer?

**55.** Interactive Solution 22.55 at **www.wiley.com/college/cutnell** offers one approach to problems such as this one. The secondary coil of a step-up transformer provides the voltage that operates an electrostatic air filter. The turns ratio of the transformer is 50:1. The primary coil is plugged into a standard 120-V outlet. The current in the secondary coil is $1.7 \times 10^{-3}$ A. Find the power consumed by the air filter.

**56.** The batteries in a portable CD player are recharged by a unit that plugs into a wall socket. Inside the unit is a step-down transformer with a turns ratio of 1:13. The wall socket provides 120 V. What voltage does the secondary coil of the transformer provide?

**57.** ssm Electric doorbells found in many homes require 10.0 V to operate. To obtain this voltage from the standard 120-V supply, a transformer is used. Is a step-up or a step-down transformer needed, and what is its turns ratio $N_s/N_p$?

**58.** In some places, insect "zappers," with their blue lights, are a familiar sight on a summer's night. These devices use a high voltage to electrocute insects. One such device uses an ac voltage of 4320 V, which is obtained from a standard 120.0-V outlet by means of a transformer. If the primary coil has 21 turns, how many turns are in the secondary coil?

*  **59.** ssm A generating station is producing $1.2 \times 10^6$ W of power that is to be sent to a small town located 7.0 km away. Each of the two wires that comprise the transmission line has a resistance per kilometer of length of $5.0 \times 10^{-2}$ $\Omega$/km. (a) Find the power lost in heating the wires if the power is transmitted at 1200 V. (b) A 100:1 step-up transformer is used to raise the voltage before the power is transmitted. How much power is now lost in heating the wires?

*  **60.** Suppose there are two transformers between your house and the high-voltage transmission line that distributes the power. In addition, assume that your house is the only one using electric power. At a substation the primary of a step-down transformer (turns ratio = 1:29) receives the voltage from the high-voltage transmission line. Because of your usage, a current of 48 mA exists in the primary of this transformer. The secondary is connected to the primary of another step-down transformer (turns ratio = 1:32) somewhere near your house, perhaps up on a telephone pole. The secondary of this transformer delivers a 240-V emf to your house. How much power is your house using? Remember that the current and voltage given in this problem are rms values.

** **61.** A generator is connected across the primary coil ($N_p$ turns) of a transformer, while a resistance $R_2$ is connected across the secondary coil ($N_s$ turns). This circuit is equivalent to a circuit in which a single resistance $R_1$ is connected directly across the generator, without the transformer. Show that $R_1 = (N_p/N_s)^2 R_2$, by starting with Ohm's law as applied to the secondary coil.

# Chapter 23 Alternating Current Circuits

## 23.1 Capacitors and Capacitive Reactance

Our experience with capacitors so far has been in dc circuits. As we have seen in Section 20.13, charge flows in a dc circuit only for the brief period after the battery voltage is applied across the capacitor. In other words, charge flows only while the capacitor is charging up. After the capacitor becomes fully charged, no more charge leaves the battery. However, suppose the battery connections to the fully charged capacitor were suddenly reversed. Then charge would flow again, but in the reverse direction, until the battery recharges the capacitor according to the new connections. What happens in an ac circuit is similar. The polarity of the voltage applied to the capacitor continually switches back and forth, and, in response, charges flow first one way around the circuit and then the other way. This flow of charge, surging back and forth, constitutes an alternating current. Thus, charge flows continuously in an ac circuit containing a capacitor.

To help set the stage for the present discussion, recall that, for a purely resistive ac circuit, the rms voltage $V_{rms}$ across the resistor is related to the rms current $I_{rms}$ through it by $V_{rms} = I_{rms}R$ (Equation 20.14). The resistance $R$ has the same value for any frequency of the ac voltage or current. Figure 23.1 emphasizes this fact by showing that a graph of resistance versus frequency is a horizontal straight line.

For the rms voltage across a capacitor the following expression applies, which is analogous to $V_{rms} = I_{rms}R$:

$$V_{rms} = I_{rms}X_C \tag{23.1}$$

The term $X_C$ appears in place of the resistance $R$ and is called the **capacitive reactance.** The capacitive reactance, like resistance, is measured in *ohms* and determines how much rms current exists in a capacitor in response to a given rms voltage across the capacitor. It is found experimentally that the capacitive reactance $X_C$ is inversely proportional to both the frequency $f$ and the capacitance $C$, according to the following equation:

$$X_C = \frac{1}{2\pi fC} \tag{23.2}$$

For a fixed value of the capacitance $C$, Figure 23.2 gives a plot of $X_C$ versus frequency, according to Equation 23.2. A comparison of this drawing with Figure 23.1 reveals that a capacitor and a resistor behave differently. As the frequency becomes very large, Figure 23.2 shows that $X_C$ approaches zero, signifying that a capacitor offers only a negligibly small opposition to the alternating current. In contrast, in the limit of zero frequency (i.e., direct current), $X_C$ becomes infinitely large, and a capacitor provides so much opposition to the motion of charges that there is no current. Example 1 illustrates how frequency and capacitance determine the amount of current in an ac circuit.

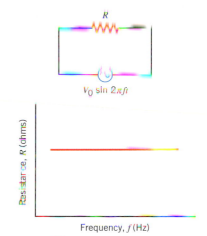

**Figure 23.1** The resistance in a purely resistive circuit has the same value at all frequencies. The maximum emf of the generator is $V_0$.

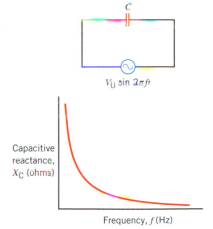

**Figure 23.2** The capacitive reactance $X_C$ is inversely proportional to the frequency $f$ according to $X_C = 1/(2\pi fC)$.

## Example 1 A Capacitor in an AC Circuit

For the circuit in Figure 23.2, the capacitance of the capacitor is 1.50 $\mu$F, and the rms voltage of the generator is 25.0 V. What is the rms current in the circuit when the frequency of the generator is (a) $1.00 \times 10^2$ Hz and (b) $5.00 \times 10^3$ Hz?

**Reasoning** The current can be found from $I_{rms} = V_{rms}/X_C$, once the capacitive reactance $X_C$ is determined. The values for the capacitive reactance will reflect the fact that the capacitor provides more opposition to the current when the frequency is smaller.

**Problem solving insight**
The capacitive reactance $X_C$ is inversely proportional to the frequency $f$ of the voltage. If the frequency increases by a factor of 50, for example, the capacitive reactance decreases by a factor of 50.

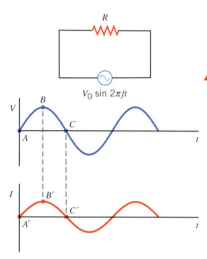

**Figure 23.3** The instantaneous voltage $V$ and current $I$ in a purely resistive circuit are *in phase,* which means that they increase and decrease in step with one another.

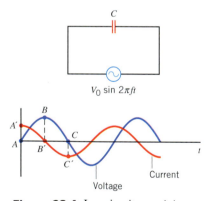

**Figure 23.4** In a circuit containing only a capacitor, the instantaneous voltage and current are not in phase. Instead, the current *leads* the voltage by one-quarter of a cycle or by a phase angle of 90°.

**Problem solving insight**

**Solution**

(a) At a frequency of $1.00 \times 10^2$ Hz, we find

$$X_C = \frac{1}{2\pi fC} = \frac{1}{2\pi (1.00 \times 10^2 \text{ Hz})(1.50 \times 10^{-6} \text{ F})} = 1060 \text{ } \Omega \qquad (23.2)$$

$$I_{rms} = \frac{V_{rms}}{X_C} = \frac{25.0 \text{ V}}{1060 \text{ } \Omega} = \boxed{0.0236 \text{ A}} \qquad (23.1)$$

(b) When the frequency is $5.00 \times 10^3$ Hz, the calculations are similar:

$$X_C = \frac{1}{2\pi fC} = \frac{1}{2\pi (5.00 \times 10^3 \text{ Hz})(1.50 \times 10^{-6} \text{ F})} = 21.2 \text{ } \Omega \qquad (23.2)$$

$$I_{rms} = \frac{V_{rms}}{X_C} = \frac{25.0 \text{ V}}{21.2 \text{ } \Omega} = \boxed{1.18 \text{ A}} \qquad (23.1)$$

We now consider the behavior of the instantaneous (not rms) voltage and current. For comparison, Figure 23.3 shows graphs of voltage and current versus time in a resistive circuit. These graphs indicate that when only resistance is present, the voltage and current are proportional to each other at every moment. For example, when the voltage increases from $A$ to $B$ on the graph, the current follows along in step, increasing from $A'$ to $B'$ during the same time interval. Likewise, when the voltage decreases from $B$ to $C$, the current decreases from $B'$ to $C'$. For this reason, the current in a resistance $R$ is said to be *in phase* with the voltage across the resistance.

For a capacitor, this in-phase relation between instantaneous voltage and current does *not* exist. Figure 23.4 shows graphs of the ac voltage and current versus time for a circuit that contains only a capacitor. As the voltage increases from $A$ to $B$, the charge on the capacitor increases and reaches its full value at $B$. The current, however, is not the same thing as the charge. The current is the rate of flow of charge and has a maximum positive value at the start of the charging process at $A'$. It is a maximum because there is no charge on the capacitor at the start and hence no capacitor voltage to oppose the generator voltage. When the capacitor is fully charged at $B$, the capacitor voltage has a magnitude equal to that of the generator and completely opposes the generator voltage. The result is that the current decreases to zero at $B'$. While the capacitor voltage decreases from $B$ to $C$, the charges flow out of the capacitor in a direction opposite to that of the charging current, as indicated by the negative current from $B'$ to $C'$. Thus, voltage and current are not in phase but are, in fact, one-quarter wave cycle out of step, or out of phase. More specifically, assuming that the voltage fluctuates as $V_0 \sin(2\pi ft)$, the current varies as $I_0 \sin(2\pi ft + \pi/2) = I_0 \cos(2\pi ft)$. Since $\pi/2$ radians correspond to 90° and since the current reaches its maximum value *before* the voltage does, it is said that the current in a capacitor *leads* the voltage across the capacitor by a phase angle of 90°.

The fact that the current and voltage for a capacitor are 90° out of phase has an important consequence from the point of view of electric power, since power is the product of current and voltage. For the time interval between points $A$ and $B$ (or $A'$ and $B'$) in Figure 23.4, both current and voltage are positive. Therefore, the instantaneous power is also positive, meaning that the generator is delivering energy to the capacitor. However, during the period between $B$ and $C$ (or $B'$ and $C'$), the current is negative while the voltage remains positive, and the power, being the product of the two, is negative. During this period, the capacitor is returning energy to the generator. Thus, the power alternates between positive and negative values for equal periods of time. In other words, the capacitor alternately absorbs and releases energy. Consequently, ***the average power (and, hence, the average energy) used by a capacitor in an ac circuit is zero.***

It will prove useful later on to use a model for the voltage and current when analyzing ac circuits. In this model, voltage and current are represented by rotating arrows, often called ***phasors,*** whose lengths correspond to the maximum voltage $V_0$ and maximum current $I_0$, as Figure 23.5 indicates. These phasors rotate counterclockwise at a frequency $f$. For a resistor, the phasors are co-linear as they rotate (see part *a* of the drawing) because voltage and current are in phase. For a capacitor (see part *b*), the phasors remain perpendicular while rotating because the phase angle between the current and the voltage is 90°. Since current leads voltage for a capacitor, the current phasor is ahead of the voltage phasor in the direction of rotation.

Note from the two circuit drawings in Figure 23.5 that the instantaneous voltage across the resistor or the capacitor is $V_0 \sin(2\pi f t)$. We can find this instantaneous value of the voltage directly from the phasor diagram. Imagine that the voltage phasor $V_0$ in this diagram represents the hypotenuse of a right triangle. Then, the *vertical component* of the phasor would be $V_0 \sin(2\pi f t)$. In a similar manner, the instantaneous current can be found as the vertical component of the current phasor.

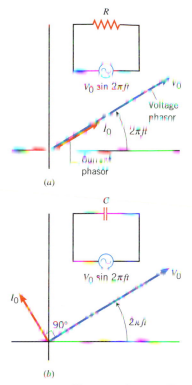

# 23.2 Inductors and Inductive Reactance

As Section 22.7 discusses, an inductor is usually a coil of wire, and the basis of its operation is Faraday's law of electromagnetic induction. According to Faraday's law, an inductor develops a voltage that opposes a change in the current. This voltage $V$ is given by $V = -L(\Delta I/\Delta t)$ (see Equation 22.9*), where $\Delta I/\Delta t$ is the rate at which the current changes and $L$ is the inductance of the inductor. In an ac circuit the current is always changing, and Faraday's law can be used to show that the rms voltage across an inductor is

$$V_{rms} = I_{rms} X_L \qquad (23.3)$$

Equation 23.3 is analogous to $V_{rms} = I_{rms} R$, with the term $X_L$ appearing in place of the resistance $R$ and being called the **inductive reactance**. The inductive reactance is measured in ohms and determines how much rms current exists in an inductor for a given rms voltage across the inductor. It is found experimentally that the inductive reactance $X_L$ is directly proportional to the frequency $f$ and the inductance $L$, as indicated by the following equation:

$$X_L = 2\pi f L \qquad (23.4)$$

This relation indicates that the larger the inductance, the larger is the inductive reactance. Note that the inductive reactance is directly proportional to the frequency ($X_L \propto f$), whereas the capacitive reactance is inversely proportional to the frequency ($X_C \propto 1/f$).

Figure 23.6 shows a graph of the inductive reactance versus frequency for a fixed value of the inductance, according to Equation 23.4. As the frequency becomes very large, $X_L$ also becomes very large. In such a situation, an inductor provides a large opposition to the alternating current. In the limit of zero frequency (i.e., direct current), $X_L$ becomes zero, indicating that an inductor does not oppose direct current at all. The next example demonstrates the effect of inductive reactance on the current in an ac circuit.

**Figure 23.5** These rotating-arrow models represent the voltage and the current in ac circuits that contain (a) only a resistor and (b) only a capacitor.

## Example 2 An Inductor in an AC Circuit

The circuit in Figure 23.6 contains a 3.60-mH inductor. The rms voltage of the generator is 25.0 V. Find the rms current in the circuit when the generator frequency is (a) $1.00 \times 10^2$ Hz and (b) $5.00 \times 10^3$ Hz.

**Reasoning** The current can be calculated from $I_{rms} = V_{rms}/X_L$, provided the inductive reactance is obtained first. The inductor offers more opposition to the changing current when the frequency is larger, and the values for the inductive reactance will reflect this fact.

**Solution**

**(a)** At a frequency of $1.00 \times 10^2$ Hz, we find

$$X_L = 2\pi f L = 2\pi(1.00 \times 10^2 \text{ Hz})(3.60 \times 10^{-3} \text{H}) = 2.26 \ \Omega \qquad (23.4)$$

$$I_{rms} = \frac{V_{rms}}{X_L} = \frac{25.0 \text{ V}}{2.26 \ \Omega} = \boxed{11.1 \text{ A}} \qquad (23.3)$$

**(b)** The calculation is similar when the frequency is $5.00 \times 10^3$ Hz:

$$X_L = 2\pi f L = 2\pi(5.00 \times 10^3 \text{ Hz})(3.60 \times 10^{-3} \text{ H}) = 113 \ \Omega \qquad (23.4)$$

$$I_{rms} = \frac{V_{rms}}{X_L} = \frac{25.0 \text{ V}}{113 \ \Omega} = \boxed{0.221 \text{ A}} \qquad (23.3)$$

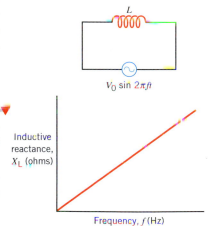

Inductive reactance, $X_L$ (ohms)

Frequency, $f$ (Hz)

**Figure 23.6** In an ac circuit the inductive reactance $X_L$ is directly proportional to the frequency $f$, according to $X_L = 2\pi f L$.

**Problem solving insight**
The inductive reactance $X_L$ is directly proportional to the frequency $f$ of the voltage. If the frequency increases by a factor of 50, for example, the inductive reactance also increases by a factor of 50.

---

\* When an inductor is used in a circuit, the notation is simplified if we designate the potential difference across the inductor as the voltage $V$, rather than the emf $\mathcal{E}$.

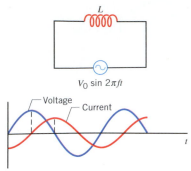

**Figure 23.7** The instantaneous voltage and current in a circuit containing only an inductor are not in phase. The current *lags behind* the voltage by one-quarter of a cycle or by a phase angle of 90°.

**Problem solving insight**

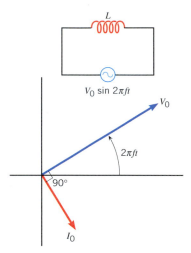

**Figure 23.8** This phasor model represents the voltage and current in a circuit that contains only an inductor.

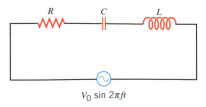

**Figure 23.9** A series RCL circuit contains a resistor, a capacitor, and an inductor.

By virtue of its inductive reactance, an inductor affects the amount of current in an ac circuit. The inductor also influences the current in another way, as Figure 23.7 shows. This figure displays graphs of voltage and current versus time for a circuit containing only an inductor. At a maximum or minimum on the current graph, the current does not change much with time, so the voltage generated by the inductor to oppose a change in the current is zero. At the points on the current graph where the current is zero, the graph is at its steepest, and the current has the largest rate of increase or decrease. Correspondingly, the voltage generated by the inductor to oppose a change in the current has the largest positive or negative value. Thus, current and voltage are not in phase but are one-quarter of a wave cycle out of phase. If the voltage varies as $V_0 \sin (2\pi ft)$, the current fluctuates as $I_0 \sin (2\pi ft - \pi/2) = -I_0 \cos (2\pi ft)$. The current reaches its maximum *after* the voltage does, and it is said that the current *lags behind* the voltage by a phase angle of 90° ($\pi/2$ radians). In a purely capacitive circuit, in contrast, the current leads the voltage by 90° (see Figure 23.4).

In an inductor the 90° phase difference between current and voltage leads to the same result for average power that it does in a capacitor. An inductor alternately absorbs and releases energy for equal periods of time, so **the average power (and, hence, the average energy) used by an inductor in an ac circuit is zero.**

As an alternative to the graphs in Figure 23.7, Figure 23.8 uses phasors to describe the instantaneous voltage and current in a circuit containing only an inductor. The voltage and current phasors remain perpendicular as they rotate, because there is a 90° phase angle between them. The current phasor lags behind the voltage phasor, relative to the direction of rotation, in contrast to the equivalent picture in Figure 23.5b for a capacitor. Once again, the instantaneous values are given by the vertical components of the phasors.

## 23.3 *Circuits Containing Resistance, Capacitance, and Inductance*

Capacitors and inductors can be combined along with resistors in a single circuit. The simplest combination contains a resistor, a capacitor, and an inductor in series, as Figure 23.9 shows. In a series RCL circuit the total opposition to the flow of charge is called the **impedance** of the circuit and comes partially from (1) the resistance $R$, (2) the capacitive reactance $X_C$, and (3) the inductive reactance $X_L$. It is tempting to follow the analogy of a series combination of resistors and calculate the impedance by simply adding together $R$, $X_C$, and $X_L$. However, such a procedure is not correct. Instead, the phasors shown in Figure 23.10 must be used. The lengths of the voltage phasors in this drawing represent the maximum voltages $V_R$, $V_C$, and $V_L$ across the resistor, the capacitor, and the inductor, respectively. The current is the same for each device, since the circuit is wired in series. The length of the current phasor represents the maximum current $I_0$. Notice that the drawing shows the current phasor to be (1) in phase with the voltage phasor for the resistor, (2) ahead of the voltage phasor for the capacitor by 90°, and (3) behind the voltage phasor for the inductor by 90°. These three facts are consistent with our earlier discussion in Sections 23.1 and 23.2.

The basis for dealing with the voltage phasors in Figure 23.10 is Kirchhoff's loop rule. In an ac circuit this rule applies to the *instantaneous* voltages across each circuit component and the generator. Therefore, it is necessary to take into account the fact that these voltages do not have the same phase; that is, the phasors $V_R$, $V_C$, and $V_L$ point in different directions in the drawing. Kirchhoff's loop rule indicates that the phasors add together to give the total voltage $V_0$ that is supplied to the circuit by the generator. The addition, however, must be like a vector addition, to take into account the different directions. Since $V_L$ and $V_C$ point in opposite directions, they combine to give a resultant phasor of $V_L - V_C$, as Figure 23.11 shows. In this drawing the resultant $V_L - V_C$ is perpendicular to $V_R$ and may be combined with it to give the total voltage $V_0$. Using the Pythagorean theorem, we find

$$V_0^2 = V_R^2 + (V_L - V_C)^2$$

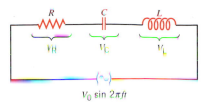

In this equation each of the symbols stands for a maximum voltage and when divided by $\sqrt{2}$ gives the corresponding rms voltage. Therefore, it is possible to divide both sides of the equation by $(\sqrt{2})^2$ and obtain a result for $V_{rms} = V_0/\sqrt{2}$. This result has exactly the same form as that above, but involves the rms voltages $V_{R\text{-}rms}$, $V_{C\text{-}rms}$, and $V_{L\text{-}rms}$. However, to avoid such awkward symbols, we simply interpret $V_R$, $V_C$, and $V_L$ as rms quantities in the following expression:

$$V_{rms}^2 = V_R^2 + (V_L - V_C)^2 \qquad (23.5)$$

The last step in determining the impedance of the circuit is to remember that $V_R = I_{rms}R$, $V_C = I_{rms}X_C$, and $V_L = I_{rms}X_L$. With these substitutions Equation 23.5 can be written as

$$V_{rms} = I_{rms}\sqrt{R^2 + (X_L - X_C)^2}$$

Therefore, for the entire RCL circuit, it follows that

$$V_{rms} = I_{rms}Z \qquad (23.6)$$

where the impedance $Z$ of the circuit is defined as

***Series RCL*** *combination*

$$Z = \sqrt{R^2 + (X_L - X_C)^2} \qquad (23.7)$$

The impedance of the circuit, like $R$, $X_C$, and $X_L$, is measured in ohms. In Equation 23.7, $X_L = 2\pi fL$ and $X_C = 1/(2\pi fC)$.

The phase angle between the current in and the voltage across a series RCL combination is the angle $\phi$ between the current phasor $I_0$ and the voltage phasor $V_0$ in Figure 23.11. According to the drawing, the tangent of this angle is

$$\tan\phi = \frac{V_L - V_C}{V_R} = \frac{I_{rms}X_L - I_{rms}X_C}{I_{rms}R}$$

***Series RCL*** *combination*

$$\tan\phi = \frac{X_L - X_C}{R} \qquad (23.8)$$

The phase angle $\phi$ is important because it has a major effect on the average power $\overline{P}$ dissipated by the circuit. Remember that, on the average, only the resistance consumes power; that is, $\overline{P} = I_{rms}^2 R$ (Equation 20.15b). According to Figure 23.11, $\cos\phi = V_R/V_0 = (I_{rms}R)/(I_{rms}Z) = R/Z$, so that $R = Z\cos\phi$. Therefore,

$$\overline{P} = I_{rms}^2 Z\cos\phi = I_{rms}(I_{rms}Z)\cos\phi$$
$$\overline{P} = I_{rms}V_{rms}\cos\phi \qquad (23.9)$$

where $V_{rms}$ is the rms voltage of the generator. The term $\cos\phi$ is called the ***power factor*** of the circuit. As a check on the validity of Equation 23.9, note that if no resistance is present, $R = 0\ \Omega$, and $\cos\phi = R/Z = 0$. Consequently, $\overline{P} = I_{rms}V_{rms}\cos\phi = 0$, a result that is expected since, on the average, neither a capacitor nor an inductor consumes energy. Conversely, if only resistance is present, $Z = \sqrt{R^2 + (X_L - X_C)^2} = R$, and $\cos\phi = R/Z = 1$. In this case, $\overline{P} = I_{rms}V_{rms}\cos\phi = I_{rms}V_{rms}$, which is the expression for the average power dissipated in a resistor. Example 3 deals with the current, voltages, and power for a series RCL circuit.

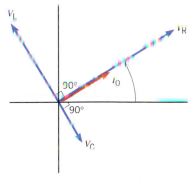

**Figure 23.10** The three voltage phasors ($V_R$, $V_C$, and $V_L$) and the current phasor ($I_0$) for a series RCL circuit.

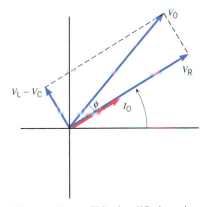

**Figure 23.11** This simplified version of Figure 23.10 results when the phasors $V_L$ and $V_C$, which point in opposite directions, are combined to give a resultant of $V_L - V_C$.

### Example 3   Current, Voltages, and Power in a Series RCL Circuit

A series RCL circuit contains a 148-$\Omega$ resistor, a 1.50-$\mu$F capacitor, and a 35.7-mH inductor. The generator has a frequency of 512 Hz and an rms voltage of 35.0 V. Obtain (a) the rms voltage across each circuit element and (b) the average electric power consumed by the circuit.

**Reasoning** The rms voltages across each circuit element can be determined from $V_R = I_{rms}R$, $V_C = I_{rms}X_C$, and $V_L = I_{rms}X_L$, as soon as the rms current and the reactances $X_C$ and $X_L$ are known. Since the rms current can be found from $I_{rms} = V_{rms}/Z$ and $V_{rms} = 35.0$ V, the first step in the solution is to find the impedance $Z$ from the individual reactances. The average power consumed is given by $\overline{P} = I_{rms}V_{rms}\cos\phi$, where the phase angle $\phi$ can be obtained from $\tan\phi = (X_L - X_C)/R$.

**Solution**

**(a)** The individual reactances are

$$X_C = \frac{1}{2\pi fC} = \frac{1}{2\pi(512 \text{ Hz})(1.50 \times 10^{-6} \text{ F})} = 207 \text{ }\Omega \qquad (23.2)$$

$$X_L = 2\pi fL = 2\pi(512 \text{ Hz})(35.7 \times 10^{-3} \text{ H}) = 115 \text{ }\Omega \qquad (23.4)$$

The impedance of the circuit is

$$Z = \sqrt{R^2 + (X_L - X_C)^2} = \sqrt{(148 \text{ }\Omega)^2 + (115 \text{ }\Omega - 207 \text{ }\Omega)^2} = 174 \text{ }\Omega \qquad (23.7)$$

The rms current in each circuit element is

$$I_{rms} = \frac{V_{rms}}{Z} = \frac{35.0 \text{ V}}{174 \text{ }\Omega} = 0.201 \text{ A} \qquad (23.6)$$

The rms voltage across each circuit element now follows immediately:

$$V_R = I_{rms}R = (0.201 \text{ A})(148 \text{ }\Omega) = \boxed{29.7 \text{ V}} \qquad (20.14)$$

$$V_C = I_{rms}X_C = (0.201 \text{ A})(207 \text{ }\Omega) = \boxed{41.6 \text{ V}} \qquad (23.1)$$

$$V_L = I_{rms}X_L = (0.201 \text{ A})(115 \text{ }\Omega) = \boxed{23.1 \text{ V}} \qquad (23.3)$$

Observe that these three rms voltages do not add up to give the generator's rms voltage, which is 35.0 V. Instead, the rms voltages satisfy Equation 23.5. It is the sum of the *instantaneous* voltages across R, C, and L, rather than the sum of the rms voltages, that equals the generator's *instantaneous* voltage, according to Kirchhoff's loop rule.

**(b)** The average power consumed by the circuit is $\overline{P} = I_{rms}V_{rms}\cos\phi$. Therefore, a value for the phase angle $\phi$ is needed and can be obtained from $\tan\phi = (X_L - X_C)/R$ as follows:

$$\phi = \tan^{-1}\left(\frac{X_L - X_C}{R}\right) = \tan^{-1}\left(\frac{115 \text{ }\Omega - 207 \text{ }\Omega}{148 \text{ }\Omega}\right) = -32° \qquad (23.8)$$

The phase angle is negative since the circuit is more capacitive than inductive ($X_C$ is greater than $X_L$), and the current leads the voltage. The average power consumed is

$$\overline{P} = I_{rms}V_{rms}\cos\phi = (0.201 \text{ A})(35.0 \text{ V})\cos(-32°) = \boxed{6.0 \text{ W}} \qquad (23.9)$$

In addition to the series RCL circuit, there are many different ways to connect resistors, capacitors, and inductors. In analyzing these additional possibilities, it helps to keep in mind the behavior of capacitors and inductors at the extreme limits of the frequency. When the frequency approaches zero (i.e., dc conditions), the reactance of a capacitor becomes so large that no charge can flow through the capacitor. It is as if the capacitor were cut out of the circuit, leaving an open gap in the connecting wire. In the limit of zero frequency the reactance of an inductor is vanishingly small. The inductor offers no opposition to a dc current. It is as if the inductor were replaced with a wire of zero resistance. In the limit of very large frequency, the behaviors of a capacitor and an inductor are reversed. The capacitor has a very small reactance and offers little opposition to the current, as if it were replaced by a wire with zero resistance. In contrast, the inductor has a very large reactance when the frequency is very large. The inductor offers so much opposition to the current that it might as well be cut out of the circuit, leaving an open gap in the connecting wire. Conceptual Example 4 illustrates how to gain insight into more complicated circuits using the limiting behaviors of capacitors and inductors.

## Conceptual Example 4
### The Limiting Behavior of Capacitors and Inductors

Figure 23.12*a* shows two circuits. The rms voltage of the generator is the same in each case. The values of the resistance R, the capacitance C, and the inductance L in these circuits are the same. The frequency of the ac generator is very near zero. In which circuit does the generator supply more rms current?

**Reasoning and Solution** According to Equation 23.6, the rms current is given by $I_{rms} = V_{rms}/Z$. However, the impedance Z is not given by Equation 23.7, since the circuits in Figure

**Problem solving insight**
In a series RCL circuit, the rms voltages across the resistor, capacitor, and inductor do not add up to equal the rms voltage across the generator.

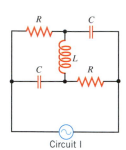

Circuit I

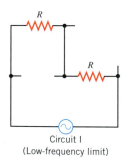

Circuit II

(a)

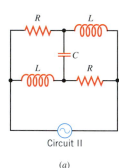

Circuit I
(Low-frequency limit)

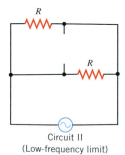

Circuit II
(Low-frequency limit)

(b)

**Figure 23.12** (*a*) These circuits are discussed in the limit of very small or low frequency in Conceptual Example 4. (*b*) For a frequency very near zero, the circuits in part *a* behave as if they were as shown here.

23.12*a* are not series RCL circuits. Since $V_{rms}$ is the same in each case, the greater current is delivered to the circuit with the smaller impedance Z. In the limit of very small frequency, the capacitors behave as if they were cut out of the circuit, leaving gaps in the connecting wires. In this limit, however, the inductors behave as if they were replaced by wires with zero resistance. Figure 23.12*b* shows the circuits as they would appear according to these changes. Circuit I behaves as if it contained only two identical resistors in series, with a total impedance of $Z = 2R$. In contrast, circuit II behaves as if it contained two identical resistors in parallel, in which case the total impedance is given by $1/Z = 1/R + 1/R$, or $Z = R/2$. At a frequency very near zero, then, *circuit II has the smaller impedance and its generator supplies the greater rms current.*

**Related Homework:** *Problem 22*

## 23.4 Resonance in Electric Circuits

The behavior of the current and voltage in a series RCL circuit can give rise to a condition of *resonance.* Resonance occurs when the frequency of a vibrating force exactly matches a natural (resonant) frequency of the object to which the force is applied. When resonance occurs, the force can transmit a large amount of energy to the object, leading to a large-amplitude vibratory motion. We have already encountered several examples of resonance. First, resonance can occur when a vibrating force is applied to an object of mass *m* that is attached to a spring whose spring constant is *k* (Section 10.6). In this case there is one natural frequency $f_0$, whose value is $f_0 = [1/(2\pi)]\sqrt{k/m}$. Second, resonance occurs when standing waves are set up on a string (Section 17.5) or in a tube of air (Section 17.6). The string and tube of air have many natural frequencies, one for each allowed standing wave. As we will now see, a condition of resonance can also be established in a series RCL circuit. In this case there is only one natural frequency, and the vibrating force is provided by the oscillating electric field that is related to the voltage of the generator.

Figure 23.13 helps us to understand why there is a resonant frequency for an ac circuit. This drawing presents an analogy between the electrical case (ignoring resistance) and the mechanical case of an object attached to a horizontal spring (ignoring friction). Part *a* shows a fully stretched spring that has just been released, with the initial speed *v* of the object being zero. All the energy is stored in the form of elastic potential energy. When the object begins to move, it gradually loses potential energy and picks up kinetic energy. In part *b*, the object moves with maximum kinetic energy through the position where the spring is unstretched (zero potential energy). Because of its inertia, the moving object coasts through this position and eventually comes to a halt in part *c* when the spring is fully compressed and all kinetic energy has been converted back into elastic potential energy. Part *d* of the picture is like part *b*, except the direction of motion is reversed. The resonant frequency $f_0$ of the object on the spring is the natural frequency at which the object vibrates and is given as $f_0 = [1/(2\pi)]\sqrt{k/m}$ according to Equations 10.6 and 10.11. In this expression, *m* is the mass of the object, and *k* is the spring constant.

**Figure 23.13** The oscillation of an object on a spring is analogous to the oscillation of the electric and magnetic fields that occur, respectively, in a capacitor and in an inductor.

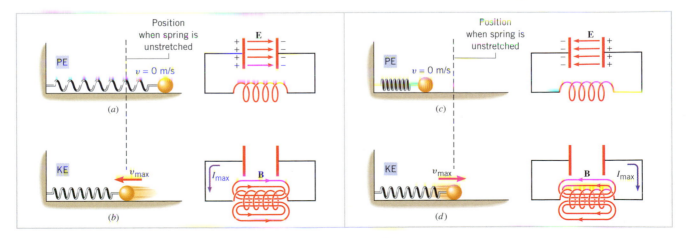

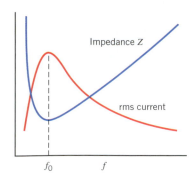

**Figure 23.14** In a series RCL circuit the impedance is a minimum, and the current is a maximum, when the frequency $f$ equals the resonant frequency $f_0$ of the circuit.

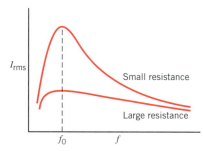

**Figure 23.15** The effect of resistance on the current in a series RCL circuit.

**Figure 23.16** A heterodyne metal detector can be used to locate buried metal objects, although in this case the objects are just rusty cans. (© American Images, Inc./Taxi/Getty Images)

**The physics of**
**a heterodyne metal detector.**

In the electrical case, Figure 23.13*a* begins with a fully charged capacitor that has just been connected to an inductor. At this instant the energy is stored in the electric field between the capacitor plates. As the capacitor discharges, the electric field **E** between the plates decreases, while a magnetic field **B** builds up around the inductor because of the increasing current in the circuit. The maximum current and the maximum magnetic field exist at the instant when the capacitor is completely discharged, as in part *b* of the figure. Energy is now stored entirely in the magnetic field of the inductor. The voltage induced in the inductor keeps the charges flowing until the capacitor again becomes fully charged, but now with reversed polarity, as in part *c*. Once again, the energy is stored in the electric field between the plates, and no energy resides in the magnetic field of the inductor. Part *d* of the cycle repeats part *b*, but with reversed directions of current and magnetic field. Thus, an ac circuit can have a resonant frequency because there is a natural tendency for energy to shuttle back and forth between the electric field of the capacitor and the magnetic field of the inductor.

To determine the resonant frequency at which energy shuttles back and forth between the capacitor and the inductor, we note that the current in a series RCL circuit is $I_{rms} = V_{rms}/Z$ (Equation 23.6). In this expression $Z$ is the impedance of the circuit and is given by $Z = \sqrt{R^2 + (X_L - X_C)^2}$ (Equation 23.7). As Figure 23.14 illustrates, the rms current is a maximum when the impedance is a minimum, assuming a given generator voltage. The minimum impedance of $Z = R$ occurs when the frequency is $f_0$, such that $X_L = X_C$ or $2\pi f_0 L = 1/(2\pi f_0 C)$. This result can be solved for $f_0$, which is the resonant frequency:

$$f_0 = \frac{1}{2\pi\sqrt{LC}} \tag{23.10}$$

The resonant frequency is determined by the inductance and the capacitance, but not the resistance.

The effect of resistance on electrical resonance is to make the "sharpness" of the circuit response less pronounced, as Figure 23.15 indicates. When the resistance is small, the current-versus-frequency graph falls off suddenly on either side of the maximum current. When the resistance is large, the falloff is more gradual, and there is less current at the maximum.

The following example deals with one application of resonance in electric circuits. In this example the focus is on the oscillation of energy between a capacitor and an inductor. Once a capacitor/inductor combination is energized, the energy will oscillate indefinitely as in Figure 23.13, provided there is some provision to replace any dissipative losses that occur because of resistance. Circuits that include this type of provision are called oscillator circuits.

### *Example 5* A Heterodyne Metal Detector

Figure 23.16 shows a heterodyne metal detector being used. As Figure 23.17 illustrates, this device utilizes two capacitor/inductor oscillator circuits, A and B. Each circuit produces its own resonant frequency, $f_{0A} = 1/(2\pi\sqrt{L_A C})$ and $f_{0B} = 1/(2\pi\sqrt{L_B C})$. Any difference between these two frequencies is detected through earphones as a beat frequency $f_{0B} - f_{0A}$, similar to the beat frequency that two musical tones produce. In the absence of any nearby metal object, the inductances $L_A$ and $L_B$ are the same, and $f_{0A}$ and $f_{0B}$ are identical. There is no beat frequency. When inductor B (the search coil) comes near a piece of metal, the inductance $L_B$ decreases, the corresponding oscillator frequency $f_{0B}$ increases, and a beat frequency is heard. Suppose that initially each inductor is adjusted so $L_B = L_A$, and each oscillator has a resonant frequency of 855.5 kHz. Assuming that the inductance of search coil B decreases by 1.00% due to a nearby piece of metal, determine the beat frequency heard through the earphones.

**Reasoning** To find the beat frequency $f_{0B} - f_{0A}$, we need to determine the amount by which the resonant frequency $f_{0B}$ changes because of a 1.00% decrease in the inductance $L_B$.

**Solution** We begin by obtaining the ratio of $f_{0B}$ to $f_{0A}$:

$$\frac{f_{0B}}{f_{0A}} = \frac{\dfrac{1}{2\pi\sqrt{L_B C}}}{\dfrac{1}{2\pi\sqrt{L_A C}}} = \sqrt{\frac{L_A}{L_B}}$$

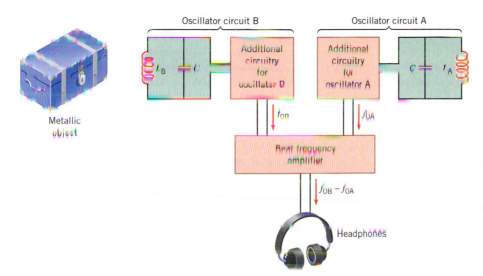

**Figure 23.17** A heterodyne metal detector uses two electrical oscillators, A and B, in its operation. When the resonant frequency of oscillator B is changed due to the proximity of a metallic object, a beat frequency, whose value is $f_{0B} - f_{0A}$, is heard in the headphones.

Due to the 1.00% decrease in the inductance of coil B, $L_B = 0.9900L_A$, so that

$$\frac{f_{0B}}{f_{0A}} = \sqrt{\frac{L_A}{0.9900L_A}} = 1.005$$

Therefore, the new value for $f_{0B}$ is $f_{0B} = 1.005f_{0A} = 1.005 \times (855.5 \text{ kHz}) = 859.8 \text{ kHz}$. As a result, the detected beat frequency is

$$f_{0B} - f_{0A} = 859.8 \text{ kHz} - 855.5 \text{ kHz} = \boxed{4.3 \text{ kHz}}$$

# 23.5 *Semiconductor Devices*

Semiconductor devices such as diodes and transistors are widely used in modern electronics, and Figure 23.18 illustrates one application. The drawing shows an audio system in which small ac voltages (originating in a compact disc player, an FM tuner, or a cassette deck) are amplified so they can drive the speaker(s). The electric circuits that accomplish the amplification do so with the aid of a dc voltage provided by the power supply. In portable units the power supply is simply a battery. In nonportable units, however, the power supply is a separate electric circuit containing diodes, along with other elements. As we will see, the diodes convert the 60-Hz ac voltage present at a wall outlet into the dc voltage needed by the amplifier, which, in turn, performs its job of amplification with the aid of transistors.

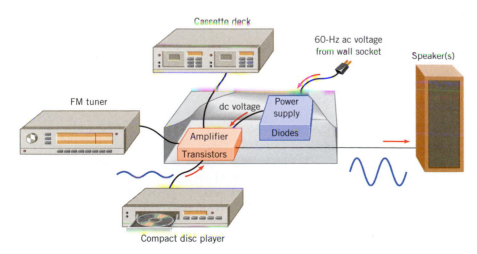

**Figure 23.18** In a typical audio system, diodes are used in the power supply to create a dc voltage from the ac voltage present at the wall socket. This dc voltage is necessary so the transistors in the amplifier can perform their task of enlarging the small ac voltages originating in the compact disc player, etc.

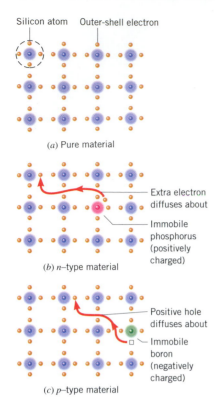

(a) Pure material

(b) n–type material

— Extra electron diffuses about

— Immobile phosphorus (positively charged)

(c) p–type material

— Positive hole diffuses about

— Immobile boron (negatively charged)

**Figure 23.19** A silicon crystal that is (a) undoped, or pure, (b) doped with phosphorus to produce an *n*-type material, and (c) doped with boron to produce a *p*-type material.

# *n*-TYPE AND *p*-TYPE SEMICONDUCTORS

The materials used in diodes and transistors are semiconductors, such as silicon and germanium. However, they are not pure materials because small amounts of "impurity" atoms (about one part in a million) have been added to them to change their conductive properties. For instance, Figure 23.19*a* shows an array of atoms that symbolizes the crystal structure in pure silicon. Each silicon atom has four outer-shell* electrons, and each electron participates with electrons from neighboring atoms in forming the bonds that hold the crystal together. Since they participate in forming bonds, these electrons generally do not move throughout the crystal. Consequently, pure silicon and germanium are not good conductors of electricity. It is possible, however, to increase their conductivities by adding tiny amounts of impurity atoms, such as phosphorus or arsenic, whose atoms have five outer-shell electrons. For example, when a phosphorus atom replaces a silicon atom in the crystal, only four of the five outer-shell electrons of phosphorus fit into the crystal structure. The extra fifth electron does not fit in and is relatively free to diffuse throughout the crystal, as part *b* of the drawing suggests. A semiconductor containing small amounts of phosphorus can, therefore, be envisioned as containing immobile, positively charged phosphorus atoms and a pool of electrons that are free to wander throughout the material. These mobile electrons allow the semiconductor to conduct electricity.

The process of adding impurity atoms is called *doping*. A semiconductor doped with an impurity that contributes mobile electrons is called an ***n*-type semiconductor,** since the mobile charge carriers have a **n**egative charge. Note that an *n*-type semiconductor is overall electrically neutral, since it contains equal numbers of positive and negative charges.

It is also possible to dope a silicon crystal with an impurity whose atoms have only three outer-shell electrons (e.g., boron or gallium). Because of the missing fourth electron, there is a "hole" in the lattice structure at the boron atom, as Figure 23.19*c* illustrates. An electron from a neighboring silicon atom can move into this hole, in which event the region around the boron atom, having acquired the electron, becomes negatively charged. Of course, when a nearby electron does move, it leaves behind a hole. This hole is positively charged, since it results from the removal of an electron from the vicinity of a neutral silicon atom. The vast majority of atoms in the lattice are silicon, so the hole is almost always next to another silicon atom. Consequently, an electron from one of these adjacent atoms can move into the hole, with the result that the hole moves to yet another location. In this fashion, a positively charged hole can wander through the crystal. This type of semiconductor can, therefore, be viewed as containing immobile, negatively charged boron atoms and an equal number of positively charged, mobile holes. Because of the mobile holes, the semiconductor can conduct electricity. In this case the charge carriers are positive. A semiconductor doped with an impurity that introduces mobile **p**ositive holes is called a ***p*-type semiconductor.**

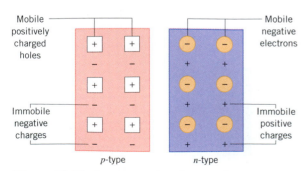

**Figure 23.20** A *p*-type semiconductor and an *n*-type semiconductor.

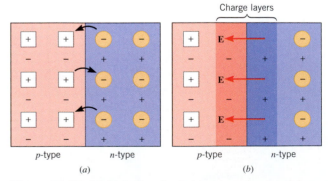

**Figure 23.21** (*a*) At the junction between *n* and *p* materials, mobile electrons and holes combine and (*b*) create positive and negative charge layers. The electric field produced by the charge layers is **E**.

* Section 30.6 discusses the electronic structure of the atom in terms of "shells."

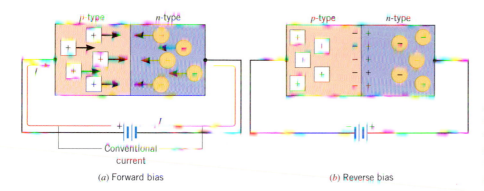

*(a)* Forward bias                    *(b)* Reverse bias

**Figure 23.22** *(a)* There is an appreciable current through the diode when the diode is forward biased. *(b)* Under a reverse bias condition, there is almost no current through the diode.

## THE SEMICONDUCTOR DIODE

A *p-n junction diode* is a device formed from a *p*-type semiconductor and an *n*-type semiconductor. The *p-n* junction between the two materials is of fundamental importance to the operation of diodes and transistors. Figure 23.20 shows separate *p*-type and *n*-type semiconductors, each electrically neutral. Figure 23.21*a* shows them joined together to form a diode. Mobile electrons from the *n*-type semiconductor and mobile holes from the *p*-type semiconductor flow across the junction and combine. This process leaves the *n*-type material with a positive charge layer and the *p*-type material with a negative charge layer, as part *b* of the drawing indicates. The positive and negative charge layers on the two sides of the junction set up an electric field **E**, much like that in a parallel plate capacitor. This electric field tends to prevent any further movement of charge across the junction, and all charge flow quickly stops.

Suppose now that a battery is connected across the *p-n* junction, as in Figure 23.22*a*, where the negative terminal of the battery is attached to the *n*-material, and the positive terminal is attached to the *p*-material. In this situation the junction is said to be in a condition of *forward bias,* and, as a result, there is a current in the circuit. The negative terminal of the battery repels the mobile electrons in the *n*-type material, and they move toward the junction. Likewise, the positive terminal repels the positive holes in the *p*-type material, and they also move toward the junction. At the junction the electrons fill the holes. In the meantime, the negative terminal provides a fresh supply of electrons to the *n*-material, and the positive terminal pulls off electrons from the *p*-material, forming new holes in the process. Consequently, a continual flow of charge, and hence a current, is maintained.

In Figure 23.22*b* the battery polarity has been reversed, and the *p-n* junction is in a condition known as *reverse bias.* The battery forces electrons in the *n*-material and holes in the *p*-material away from the junction. As a result, the potential across the junction builds up until it opposes the battery potential, and very little current can be sustained through the diode. The diode, then, is a unidirectional device, for it allows current to pass only in one direction.

The graph in Figure 23.23 shows the dependence of the current on the magnitude and polarity of the voltage applied across a *p-n* junction diode. The exact values of the current depend on the nature of the semiconductor and the extent of the doping. Also shown in the drawing is the symbol used for a diode (–▶–). The direction of the arrowhead in the symbol indicates the direction of the conventional current in the diode under a forward bias condition. In a forward bias condition, the side of the symbol that contains the arrowhead has a positive potential relative to the other side.

A special kind of diode is called an **LED**, which stands for **light-emitting diode.** You can see LEDs in the form of small bright red, green, and yellow lights that appear on many electronic devices, such as computers, TV sets, and stereo systems. These diodes, like others, carry current in only one direction. Imagine a forward biased diode, like that shown in Figure 23.22*a*, in which a current exists. An LED emits light whenever electrons and holes combine, the light coming from the *p-n* junction. Commercial LEDs are often made from gallium, suitably doped with arsenic and phosphorus atoms.

**The physics of a semiconductor diode.**

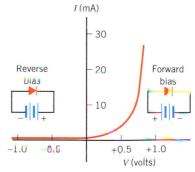

**Figure 23.23** The current-versus-voltage characteristics of a typical *p-n* junction diode.

**The physics of light-emitting diodes (LEDs).**

Sensor

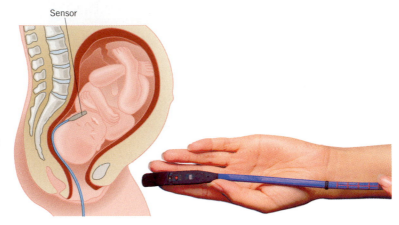

**Figure 23.24** A fetal oxygen monitor uses a sensor that contains LEDs to measure the level of oxygen in the fetus's blood. (Reprinted by permission of Nellcor Puritan Bennett, Inc., Pleasanton, California.)

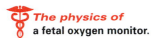

**The physics of**
**a fetal oxygen monitor.**

**The physics of**
**rectifier circuits.**

**The physics of**
**solar cells.**

**Figure 23.25** A half-wave rectifier circuit, together with a capacitor and a transformer (not shown), constitutes a dc power supply because the rectifier converts an ac voltage into a dc voltage.

A **fetal oxygen monitor** uses LEDs to measure the level of oxygen in a fetus's blood. A sensor is inserted into the mother's uterus and positioned against the cheek of the fetus, as indicated in Figure 23.24. Two light-emitting diodes are located within the sensor, and each shines light of a different wavelength (or color) into the fetal tissue. The light is reflected by the oxygen-carrying red blood cells and is detected by an adjacent photodetector. Light from one of the LEDs is used to measure the level of oxyhemoglobin in the blood, and light from the other LED is used to measure the level of deoxyhemoglobin. From a comparison of these two levels, the oxygen saturation in the blood is determined.

Because diodes are unidirectional devices, they are commonly used in *rectifier circuits*, which convert an ac voltage into a dc voltage. For instance, Figure 23.25 shows a circuit in which charges flow through the resistance $R$ only while the ac generator biases the diode in the forward direction. Since current occurs only during one-half of every generator voltage cycle, the circuit is called a half-wave rectifier. A plot of the output voltage across the resistor reveals that only the positive halves of each cycle are present. If a capacitor is added in parallel with the resistor, as indicated in the drawing, the capacitor charges up and keeps the voltage from dropping to zero between each positive half-cycle. It is also possible to construct full-wave rectifier circuits, in which both halves of every cycle of the generator voltage drive current through the load resistor in the same direction.

When a circuit such as that in Figure 23.25 includes a capacitor and also a transformer to establish the desired voltage level, the circuit is called a power supply. In the audio system in Figure 23.18, the power supply receives the 60-Hz ac voltage from a wall socket and produces a dc output voltage that is used for the transistors within the amplifier. Power supplies using diodes are also found in virtually all electronic appliances, such as televisions and microwave ovens.

## SOLAR CELLS

Solar cells use *p-n* junctions to convert sunlight directly into electricity, as Figure 23.26 illustrates. The solar cell in this drawing consists of a *p*-type semiconductor surrounding an *n*-type semiconductor. As discussed earlier, charge layers form at the junction between

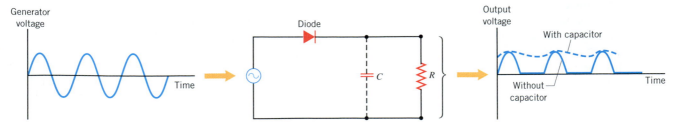

the two types of semiconductors, leading to an electric field **E** pointing from the *n*-type toward the *p*-type layer. The outer covering of *p*-type material is so thin that sunlight penetrates into the charge layers and ionizes some of the atoms there. In the process of ionization, the energy of the sunlight causes a negative electron to be ejected from the atom, leaving behind a positive hole. As the drawing indicates, the electric field in the charge layers causes the electron and the hole to move away from the junction. The electron moves into the *n*-type material, and the hole moves into the *p*-type material. As a result, the sunlight causes the solar cell to develop negative and positive terminals, much like the terminals of a battery. The current that a single solar cell can provide is small, so applications of solar cells often use many of them mounted to form large panels, as Figure 23.27 illustrates.

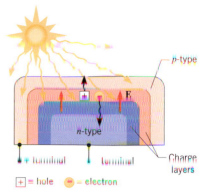

# TRANSISTORS

A number of different kinds of transistors are in use today. One type is the **bipolar junction transistor,** which consists of two *p-n* junctions formed by three layers of doped semiconductors. As Figure 23.28 indicates, there are *pnp* and *npn* transistors. In either case, the middle region is made very thin compared to the outer regions.

A transistor is useful because it can be used in circuits that amplify a smaller voltage into a larger one. A transistor plays the same kind of role in an amplifier circuit that a valve does when it controls the flow of water through a pipe. A small change in the valve setting produces a large change in the amount of water per second that flows through the pipe. In other words, a small change in the voltage input to a transistor produces a large change in the output from the transistor.

Figure 23.29 shows a *pnp* transistor connected to two batteries, labeled $V_E$ and $V_C$. The voltages $V_E$ and $V_C$ are applied in such a way that the *p-n* junction on the left has a forward bias, while the *p-n* junction on the right has a reverse bias. Moreover, the voltage $V_C$ is usually much larger than $V_E$ for a reason to be discussed shortly. The drawing also shows the standard symbol and nomenclature for the three sections of the transistor—namely, the *emitter,* the *base,* and the *collector.* The arrowhead in the symbol points in the direction of the conventional current through the emitter.

The positive terminal of $V_E$ pushes the mobile positive holes in the *p*-type material of the emitter toward the emitter/base junction. Since this junction has a forward bias, the holes enter the base region readily. Once in the base region, the holes come under the strong influence of $V_C$ and are attracted to its negative terminal. Since the base is so thin (about $10^{-6}$ m or so), approximately 98% of the holes are drawn through the base and into the collector. The remaining 2% of the holes combine with free electrons in the base region, thereby giving rise to a small base current $I_B$. As the

**Figure 23.26** A solar cell formed from a *p-n* junction. When sunlight strikes it, the solar cell acts like a battery, with + and − terminals.

**The physics of transistors.**

**Figure 23.27** The Helios Prototype flying wing is propelled by solar energy. The solar cells are mounted on the top of the wing. (© Gamma Press, Inc.)

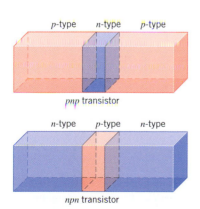

**Figure 23.28** There are two kinds of bipolar junction transistors, *pnp* and *npn*.

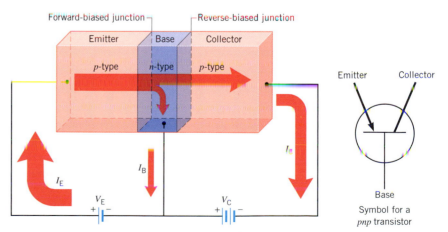

**Figure 23.29** A *pnp* transistor, along with its bias voltages $V_E$ and $V_C$. On the symbol for the *pnp* transistor, the emitter is marked with an arrowhead that denotes the direction of conventional current through the emitter.

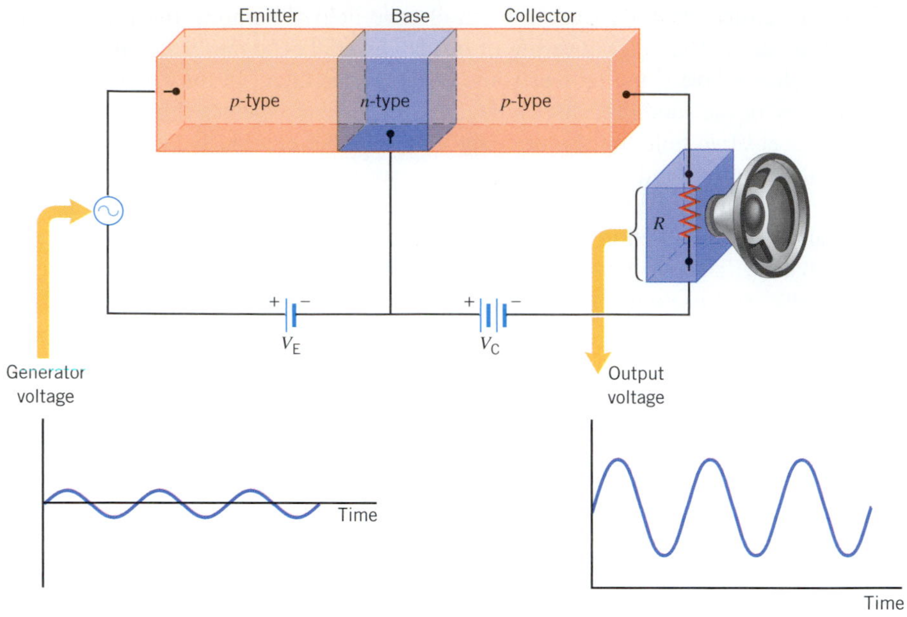

Emitter   Base   Collector

p-type    n-type    p-type

$V_E$

$V_C$

Generator voltage

Output voltage

Time

Time

**Figure 23.30** The basic *pnp* transistor amplifier in this drawing amplifies a small generator voltage to produce an enlarged voltage across the resistance *R*.

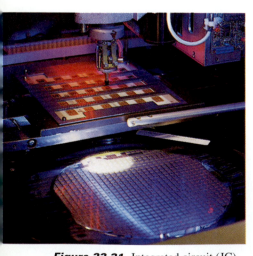

**Figure 23.31** Integrated circuit (IC) chips are manufactured on wafers of semiconductor material. Shown here is one wafer containing many chips. Some of the so-called smart cards in which the chips are used are also shown. (Courtesy ORGA Card Systems, Inc.)

drawing shows, the moving holes in the emitter and collector constitute currents that are labeled $I_E$ and $I_C$, respectively. From Kirchhoff's junction rule it follows that $I_C = I_E - I_B$.

Because the base current $I_B$ is small, the collector current is determined primarily by current from the emitter ($I_C = I_E - I_B \approx I_E$). This means that a change in $I_E$ will cause a change in $I_C$ of nearly the same amount. Furthermore, a substantial change in $I_E$ can be caused by only a small change in the forward bias voltage $V_E$. To see that this is the case, look back at Figure 23.23 and notice how steep the current-versus-voltage curve is for a *p-n* junction; small changes in the forward bias voltage give rise to large changes in the current.

With the help of Figure 23.30 we can now appreciate what was meant by the earlier statement that a small change in the voltage input to a transistor leads to a large change in the output. This picture shows an ac generator connected in series with the battery $V_E$ and a resistance *R* connected in series with the collector. The generator voltage could originate from many sources, such as an electric guitar pickup or a compact disc player, while the resistance *R* could represent a loudspeaker. The generator introduces small voltage changes in the forward bias across the emitter/base junction and, thus, causes large corresponding changes in the current $I_C$ leaving the collector and passing through the resistance *R*. As a result, the output voltage across *R* is an enlarged or amplified version of the input voltage of the generator. The operation of an *npn* transistor is similar to that of a *pnp* transistor. The main difference is that the bias voltages and current directions are reversed.

It is important to realize that the increased power available at the output of a transistor amplifier does *not* come from the transistor itself. Rather, it comes from the power provided by the voltage source $V_C$. The transistor, acting like an automatic valve, merely allows the small, weak signals from the input generator to control the power taken from the source $V_C$ and delivered to the resistance *R*.

Today it is possible to combine arrays of tens of thousands of transistors, diodes, resistors, and capacitors on a tiny chip of silicon that usually measures less than a centimeter on a side. These arrays are called *integrated circuits* (ICs) and can be designed to perform almost any desired electronic function. Integrated circuits, such as the type in Figure 23.31, have revolutionized the electronics industry and lie at the heart of computers, cellular phones, digital watches, and programmable appliances.

# Concept Summary

This summary presents an abridged version of the chapter, including the important equations and all available learning aids. For convenient reference, the learning aids (including the text's examples) are placed next to or immediately after the relevant equation or discussion. The following learning aids may be found on-line at **www.wiley.com/college/cutnell**:

**Interactive LearningWare** examples are solved according to a five-step interactive format that is designed to help you develop problem-solving skills.

**Concept Simulations** are animated versions of text figures or animations that illustrate important concepts. You can control parameters that affect the display, and we encourage you to experiment.

**Interactive Solutions** offer specific models for certain types of problems in the chapter homework. The calculations are carried out interactively.

**Self-Assessment Tests** include both qualitative and quantitative questions. Extensive feedback is provided for both incorrect and correct answers, to help you evaluate your understanding of the material.

| Topic | Discussion | Learning Aids |
|---|---|---|

### 23.1 Capacitors and Capacitive Reactance

In an ac circuit the rms voltage $V_{rms}$ across a capacitor is related to the rms current $I_{rms}$ by

**Relation between voltage and current**

$$V_{rms} = I_{rms} X_C \qquad (23.1)$$

where $X_C$ is the capacitive reactance. The capacitive reactance is measured in ohms ($\Omega$) and is given by

**Capacitive reactance**

$$X_C = \frac{1}{2\pi f C} \qquad (23.2)$$  **Example 1**
**Concept Simulation 23.1**

where $f$ is the frequency and $C$ is the capacitance.

**Phase angle between current and voltage**

The ac current in a capacitor leads the voltage across the capacitor by a phase angle of 90° or $\pi/2$ radians. As a result, a capacitor consumes no power, on average.

**Voltage and current phasors**

The phasor model is useful for analyzing the voltage and current in an ac circuit. In this model, the voltage and current are represented by rotating arrows, called phasors.

The length of the voltage phasor represents the maximum voltage $V_0$, and the length of the current phasor represents the maximum current $I_0$. The phasors rotate in a counterclockwise direction at a frequency $f$. Since the current leads the voltage by 90° in a capacitor, the current phasor is ahead of the voltage phasor by 90° in the direction of rotation.

The instantaneous values of the voltage and current are equal to the vertical components of the corresponding phasors.

### 23.2 Inductors and Inductive Reactance

In an ac circuit the rms voltage $V_{rms}$ across an inductor is related to the rms current $I_{rms}$ by

**Relation between voltage and current**

$$V_{rms} = I_{rms} X_L \qquad (23.3)$$

where $X_L$ is the inductive reactance. The inductive reactance is measured in ohms ($\Omega$) and is given by

**Inductive reactance**

$$X_L = 2\pi f L \qquad (23.4)$$  **Example 2**
**Concept Simulation 23.1**

where $f$ is the frequency and $L$ is the inductance.

**Phase angle between current and voltage**

The ac current in an inductor lags behind the voltage across the inductor by a phase angle of 90° or $\pi/2$ radians. Consequently, an inductor, like a capacitor, consumes no power, on average.

**Voltage and current phasors**

The voltage and current phasors in a circuit containing only an inductor also rotate in a counterclockwise direction at a frequency $f$. However, since the current lags the voltage by 90° in an inductor, the current phasor is behind the voltage phasor by 90° in the direction of rotation.

The instantaneous values of the voltage and current are equal to the vertical components of the corresponding phasors.

 **Use Self-Assessment Test 23.1 to evaluate your understanding of Sections 23.1 and 23.2.**

| Topic | Discussion | Learning Aids |
|---|---|---|

### 23.3 Circuits Containing Resistance, Capacitance, and Inductance

When a resistor, a capacitor, and an inductor are connected in series, the rms voltage across the combination is related to the rms current according to

**Relation between voltage and current**

$$V_{rms} = I_{rms} Z \qquad (23.6)$$

where $Z$ is the impedance of the combination. The impedance is measured in ohms ($\Omega$) and is given by

**Impedance**

$$Z = \sqrt{R^2 + (X_L - X_C)^2} \qquad (23.7)$$

**Interactive Solution 23.23**
**Concept Simulation 23.2**

where $R$ is the resistance, and $X_L$ and $X_C$ are, respectively, the inductive and capacitive reactances.

The tangent of the phase angle $\phi$ between current and voltage in a series RCL circuit is

**Phase angle between current and voltage**

$$\tan \phi = \frac{X_L - X_C}{R} \qquad (23.8)$$

Only the resistor in the RCL combination dissipates power, on average. The average power $\overline{P}$ dissipated in the circuit is

**Average power dissipated**

$$\overline{P} = I_{rms} V_{rms} \cos \phi \qquad (23.9)$$   **Examples 3, 4**

**Power factor**

where $\cos \phi$ is called the power factor of the circuit.

### 23.4 Resonance in Electric Circuits

A series RCL circuit has a resonant frequency $f_0$ that is given by

**Resonant frequency**

$$f_0 = \frac{1}{2\pi\sqrt{LC}} \qquad (23.10)$$   **Example 5**

where $L$ is the inductance and $C$ is the capacitance. At resonance, the impedance of the circuit has a minimum value equal to the resistance $R$, and the rms current has a maximum value.

 **Use Self-Assessment Test 23.2 to evaluate your understanding of Sections 23.3 and 23.4.**

### 23.5 Semiconductor Devices

**n-type semiconductor**

In an *n*-type semiconductor, mobile negative electrons carry the current. An *n*-type material is produced by doping a semiconductor such as silicon with a small amount of impurity atoms such as phosphorus.

**p-type semiconductor**

In a *p*-type semiconductor, mobile positive "holes" in the crystal structure carry the current. A *p*-type material is produced by doping a semiconductor with a small amount of impurity atoms such as boron.

These two types of semiconductors are used in *p-n* junction diodes, light-emitting diodes, and solar cells, and in *pnp* and *npn* bipolar junction transistors.

# Problems

*Note: For problems in this set, the ac current and voltage are rms values, and the power is an average value, unless indicated otherwise.*

### Section 23.1 Capacitors and Capacitive Reactance

**1. ssm** At what frequency does a 7.50-$\mu$F capacitor have a reactance of 168 $\Omega$?

**2.** What voltage is needed to create a current of 35 mA in a circuit containing only a 0.86-$\mu$F capacitor, when the frequency is 3.4 kHz?

**3.** A capacitor is connected across the terminals of an ac generator that has a frequency of 440 Hz and supplies a voltage of 24 V. When a second capacitor is connected in parallel with the first one, the current from the generator increases by 0.18 A. Find the capacitance of the second capacitor.

**4.** Two identical capacitors are connected in parallel to an ac generator that has a frequency of 610 Hz and produces a voltage of 24 V. The current in the circuit is 0.16 A. What is the capacitance of each capacitor?

**5. ssm** A capacitor is attached to a 5.00-Hz generator. The instantaneous current is observed to reach a maximum value at a certain time. What is the least amount of time that passes before the instantaneous voltage across the capacitor reaches its maximum value?

*** 6.** A capacitor is connected across an ac generator whose frequency is 750 Hz and whose *peak* output voltage is 140 V. The rms current in the circuit is 3.0 A. (a) What is the capacitance of the capacitor? (b) What is the magnitude of the *maximum* charge on one plate of the capacitor?

**** 7.** A capacitor (capacitance $C_1$) is connected across the terminals of an ac generator. Without changing the voltage or frequency of the generator, a second capacitor (capacitance $C_2$) is added in series with the first one. As a result, the current delivered by the generator decreases by a factor of three. Suppose the second capacitor had been added in parallel with the first one, instead of in series. By what factor would the current delivered by the generator have increased?

### Section 23.2 Inductors and Inductive Reactance

**8.** The current in an inductor is 0.20 A, and the frequency is 750 Hz. If the inductance is 0.080 H, what is the voltage across the inductor?

**9. ssm** At what frequency (in Hz) are the reactances of a 52-mH inductor and a 76-$\mu$F capacitor equal?

**10.** Two ac generators supply the same voltage. However, the first generator has a frequency of 1.5 kHz, and the second has a frequency of 6.0 kHz. When an inductor is connected across the terminals of the first generator, the current delivered is 0.30 A. How much current is delivered when this inductor is connected across the terminals of the second generator?

**11. ssm www** A 40.0-$\mu$F capacitor is connected across a 60.0-Hz generator. An inductor is then connected in parallel with the capacitor. What is the value of the inductance if the rms currents in the inductor and capacitor are equal?

**12.** A 30.0-mH inductor has a reactance of 2.10 k$\Omega$. (a) What is the frequency of the ac current that passes through the inductor? (b) What is the capacitance of a capacitor that has the same reactance at this frequency? The frequency is tripled, so that the reactances of the inductor and capacitor are no longer equal. What are the new reactances of (c) the inductor and (d) the capacitor?

*** 13.** The rms current in a solenoid is 0.036 A when the solenoid is connected to an 18-kHz generator. The solenoid has a cross-sectional area of $3.1 \times 10^{-5}$ m$^2$ and a length of 2.5 cm. The solenoid has 135 turns. Determine the *peak voltage* of the generator.

**** 14.** Two inductors are connected in parallel across the terminals of a generator. One has an inductance of $L_1 = 0.030$ H, and the other has an inductance of $L_2 = 0.060$ H. A single inductor, with an inductance $L$, is connected across the terminals of a second generator that has the same frequency and voltage as the first one. The current delivered by the second generator is equal to the *total* current delivered by the first generator. Find $L$.

### Section 23.3 Circuits Containing Resistance, Capacitance, and Inductance

**15. ssm** A series RCL circuit includes a resistance of 275 $\Omega$, an inductive reactance of 648 $\Omega$, and a capacitive reactance of 415 $\Omega$. The current in the circuit is 0.233 A. What is the voltage of the generator?

**16.** A light bulb has a resistance of 240 $\Omega$. It is connected to a standard wall socket (120 V, 60.0 Hz). (a) Determine the current in the bulb. (b) Determine the current in the bulb after a 10.0-$\mu$F capacitor is added in series in the circuit. (c) It is possible to return the current in the bulb to the value calculated in part (a) by adding an inductor in series with the bulb and the capacitor. What is the value of the inductance of this inductor?

**17.** Suppose that the inductance is zero ($L = 0$ H) in the series RCL circuit shown in Figure 23.10. The rms voltages across the generator and the resistor are 45 and 24 V, respectively. What is the rms voltage across the capacitor?

**18.** A 2700-$\Omega$ resistor and a 1.1-$\mu$F capacitor are connected in series across a generator (60.0 Hz, 120 V). Determine the power dissipated in the circuit.

**19. ssm** A circuit consists of a 215-$\Omega$ resistor and a 0.200-H inductor. These two elements are connected in series across a generator that has a frequency of 106 Hz and a voltage of 234 V. (a) What is the current in the circuit? (b) Determine the phase angle between the current and the voltage of the generator.

**20.** In a series circuit, a generator (1350 Hz, 15.0 V) is connected to a 16.0-$\Omega$ resistor, a 4.10-$\mu$F capacitor, and a 5.30-mH inductor. Find the voltage across each circuit element.

*** 21. ssm www** A circuit consists of an 85-$\Omega$ resistor in series with a 4.0-$\mu$F capacitor, the two being connected between the terminals of an ac generator. The voltage of the generator is fixed. At what frequency is the current in the circuit one-half the value that exists when the frequency is very large?

*** 22.** Refer to Conceptual Example 4 as background for this problem. For the circuit shown in the drawing, find the current provided by the generator when the frequency is (a) very large and (b) very small.

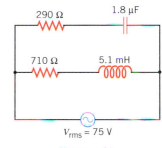

*Problem 22*

*** 23.** Refer to **Interactive Solution 23.23** at **www.wiley.com/college/ cutnell** for help with problems like this one. A series RCL circuit contains only a capacitor ($C = 6.60$ $\mu$F), an inductor ($L = 7.20$ mH), and a generator (*peak* voltage = 32.0 V, frequency = $1.50 \times 10^3$ Hz). When $t = 0$ s, the instantaneous value of the voltage is zero, and it rises to a maximum one-quarter of a period later. (a) Find the *instantaneous* value of the voltage across the capacitor/inductor combination when $t = 1.20 \times 10^{-4}$ s. (b) What is the *instantaneous* value of the current when $t = 1.20 \times 10^{-4}$ s? (*Hint: The instantaneous values of the voltage and current are, respectively, the vertical components of the voltage and current phasors.*)

*** 24.** A series circuit contains only a resistor and an inductor. The voltage $V$ of the generator is fixed. If $R = 16$ $\Omega$ and $L = 4.0$ mH, find the frequency at which the current is one-half its value at zero frequency.

**** 25.** When a resistor is connected by itself to an ac generator, the average power dissipated in the resistor is 1.000 W. When a capacitor is added in series with the resistor, the power dissipated is 0.500 W. When an inductor is added in series with the resistor (without the capacitor), the power dissipated is 0.250 W. Determine the power dissipated when both the capacitor and the inductor are added in series with the resistor.

### Section 23.4 Resonance in Electric Circuits

**26.** A series RCL circuit has a resonant frequency of 690 kHz. If the value of the capacitance is $2.0 \times 10^{-9}$ F, what is the value of the inductance?

**27. ssm** A 10.0-$\Omega$ resistor, a 12.0-$\mu$F capacitor, and a 17.0-mH inductor are connected in series with a 155-V generator. (a) At what frequency is the current a maximum? (b) What is the maximum value of the rms current?

**28.** A series RCL circuit is at resonance and contains a variable resistor that is set to 175 $\Omega$. The power dissipated in the circuit is 2.6 W. Assuming that the voltage remains constant, how much power is dissipated when the variable resistor is set to 562 $\Omega$?

**29. ssm** The resonant frequency of a series RCL circuit is 9.3 kHz. The inductance and capacitance of the circuit are each tripled. What is the new resonant frequency?

**30.** A series RCL circuit has a resonant frequency of 1500 Hz. When operating at a frequency other than 1500 Hz, the circuit has a capacitive reactance of 5.0 $\Omega$ and an inductive reactance of 30.0 $\Omega$. What are the values of (a) $L$ and (b) $C$?

**31.** The resonant frequency of an RCL circuit is 1.3 kHz, and the value of the inductance is 7.0 mH. What is the resonant frequency (in kHz) when the value of the inductance is 1.5 mH?

* **32.** The ratio of the inductive reactance to the capacitive reactance is observed to be 5.36 in a series RCL circuit. The resonant frequency of the circuit is 225 Hz. What is the frequency (nonresonant) of the generator that is connected to the circuit?

* **33. ssm www** A series RCL circuit contains a 5.10-$\mu$F capacitor and a generator whose voltage is 11.0 V. At a resonant frequency of 1.30 kHz the power dissipated in the circuit is 25.0 W. Find the values of (a) the inductance and (b) the resistance. (c) Calculate the power factor when the generator frequency is 2.31 kHz.

** **34.** A 108-$\Omega$ resistor, a 0.200-$\mu$F capacitor, and a 5.42-mH inductor are connected in series to a generator whose voltage is 26.0 V. The current in the circuit is 0.141 A. Because of the shape of the current-versus-frequency graph (see Figure 23.14), there are two possible values for the frequency that correspond to this current. Obtain these two values.

** **35.** When the frequency is twice the resonant frequency, the impedance of a series RCL circuit is twice the value of the impedance at resonance. Obtain the ratios of the inductive and capacitive reactances to the resistance; that is, obtain (a) $X_L/R$ and (b) $X_C/R$ when the frequency is twice the resonant frequency.

# Chapter 24 Electromagnetic Waves

## 24.1 The Nature of Electromagnetic Waves

In Section 13.3 we saw that energy is transported to us from the sun via a class of waves known as electromagnetic waves. This class includes the familiar visible, ultraviolet, and infrared waves. In Sections 18.6, 21.1, and 21.2 we studied the concepts of electric and magnetic fields. It was the great Scottish physicist James Clerk Maxwell (1831–1879) who showed that these two fields fluctuating together can form a propagating *electromagnetic wave*. We will now bring together our knowledge of electric and magnetic fields in order to understand this important type of wave.

Figure 24.1 illustrates one way to create an electromagnetic wave. The setup consists of two straight metal wires that are connected to the terminals of an ac generator and serve as an antenna. The potential difference between the terminals changes sinusoidally with time $t$ and has a period $T$. Part $a$ shows the instant $t = 0$ s, when there is no charge at the ends of either wire. Since there is no charge, there is no electric field at the point $P$ just to the right of the antenna. As time passes, the top wire becomes positively charged and the bottom wire negatively charged. One-quarter of a cycle later ($t = \frac{1}{4}T$), the charges have attained their maximum values, as part $b$ of the drawing indicates. The corresponding electric field $E$ at point $P$ is represented by the red arrow and has increased to its maximum strength in the downward direction.* Part $b$ also shows that the electric field created at earlier times (see the black arrow in the picture) has not disappeared but has moved to the right. Here lies the crux of the matter: At distant points, the electric field of the charges is not felt immediately. Instead, the field is created first near the wires and then, like the effect of a pebble dropped into a pond, moves outward as a wave in all directions. Only the field moving to the right is shown in the picture for the sake of clarity.

Parts $c$–$e$ of Figure 24.1 show the creation of the electric field at point $P$ (red arrow) at later times during the generator cycle. In each part, the fields produced earlier in the sequence (black arrows) continue propagating toward the right. Part $d$ shows the charges on the wires when the polarity of the generator has reversed, so the top wire is negative and the bottom wire is positive. As a result, the electric field at $P$ has reversed its direction and points upward. In part $e$ of the sequence, a complete sine wave has been drawn through the tips of the electric field vectors to emphasize that the field changes sinusoidally.

Along with the electric field in Figure 24.1, a magnetic field $B$ is also created, because the charges flowing in the antenna constitute an electric current, which produces a magnetic field. Figure 24.2 illustrates the field direction at point $P$ at the instant when the current in the antenna wire is upward. With the aid of Right-Hand Rule No. 2 (thumb of right hand points along the current $I$, fingers curl in the direction of $B$), the magnetic field at $P$ can be seen to point into the page. As the oscillating current changes, the magnetic field changes accordingly. The magnetic fields created at earlier times propagate outward as a wave, just as the electric fields do.

Notice that the magnetic field in Figure 24.2 is perpendicular to the page, whereas the electric field in Figure 24.1 lies in the plane of the page. Thus, the electric and magnetic fields created by the antenna are mutually perpendicular and remain so as they move outward. Moreover, both fields are perpendicular to the direction of travel. These perpendicular electric and magnetic fields, moving together, constitute an electromagnetic wave.

The electric and magnetic fields in Figures 24.1 and 24.2 decrease to zero rapidly with increasing distance from the antenna. Therefore, they exist mainly near the antenna and together are called the *near field*. Electric and magnetic fields do form a wave at large distances from the antenna, however. These fields arise from an effect that is different from that which produces the near field and are referred to as the *radiation field*.

---

* The direction of the electric field can be obtained by imagining a positive test charge at $P$ and determining the direction in which it would be pushed because of the charges on the wires.

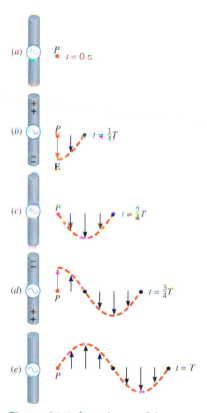

**Figure 24.1** In each part of the drawing, the red arrow represents the electric field $E$ produced at point $P$ by the oscillating charges on the antenna at the indicated time. The black arrows represent the electric fields created at earlier times. For simplicity, only the fields propagating to the right are shown.

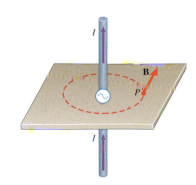

**Figure 24.2** The oscillating current $I$ in the antenna wires creates a magnetic field $B$ at point $P$ that is tangent to a circle centered on the wires. The field is directed into the page when the current is upward and out of the page when the current is downward.

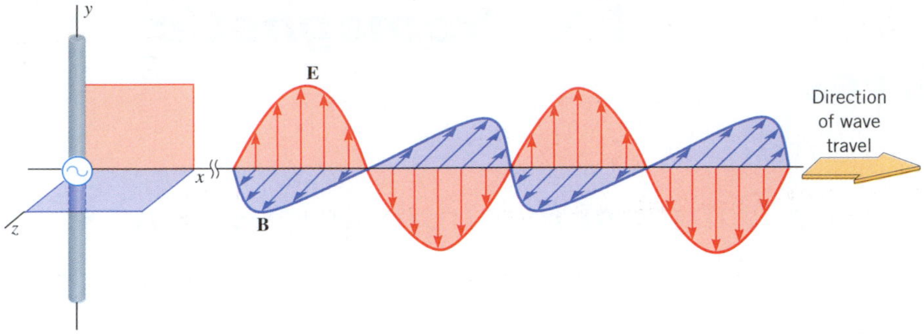

**Figure 24.3** This picture shows the wave of the radiation field far from the antenna. Observe that **E** and **B** are perpendicular to each other, and both are perpendicular to the direction of travel.

Faraday's law of induction provides part of the basis for the radiation field. As Section 22.4 discusses, this law describes the emf or potential difference produced by a changing magnetic field. And, as Section 19.4 explains, a potential difference can be related to an electric field. Thus, a changing magnetic field produces an electric field. Maxwell predicted that the reverse effect also occurs—namely, that a changing electric field produces a magnetic field. The radiation field arises because the changing magnetic field creates an electric field that fluctuates in time and the changing electric field creates the magnetic field.

Figure 24.3 shows the electromagnetic wave of the radiation field far from the antenna. The picture shows only the part of the wave traveling along the $+x$ axis. The parts traveling in the other directions have been omitted for clarity. It should be clear from the drawing that *an electromagnetic wave is a transverse wave* because the electric and magnetic fields are both perpendicular to the direction in which the wave travels. Moreover, an electromagnetic wave, unlike a wave on a string or a sound wave, does not require a medium in which to propagate. *Electromagnetic waves can travel through a vacuum or a material substance,* since electric and magnetic fields can exist in either one.

Electromagnetic waves can be produced in situations that do not involve a wire antenna. In general, any electric charge that is accelerating emits an electromagnetic wave, whether the charge is inside a wire or not. In an alternating current, an electron oscillates in simple harmonic motion along the length of the wire and is one example of an accelerating charge.

All electromagnetic waves move through a vacuum at the same speed, and the symbol $c$ is used to denote its value. This speed is called the *speed of light in a vacuum* and is $c = 3.00 \times 10^8$ m/s. In air, electromagnetic waves travel at nearly the same speed as they do in a vacuum, but, in general, they move through a substance such as glass at a speed that is substantially less than $c$.

The frequency of an electromagnetic wave is determined by the oscillation frequency of the electric charges at the source of the wave. In Figures 24.1–24.3 the wave frequency would equal the frequency of the ac generator. Suppose, for example, that the antenna is broadcasting electromagnetic waves known as radio waves. The frequencies of AM radio waves lie between 545 and 1605 kHz, these numbers corresponding to the limits of the AM broadcast band on the radio dial. The frequencies of FM radio waves lie between 88 and 108 MHz on the dial. Television channels 2–6, on the other hand, utilize electromagnetic waves with frequencies between 54 and 88 MHz, and channels 7–13 use frequencies between 174 and 216 MHz.

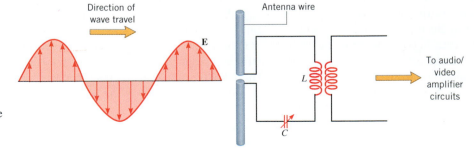

**Figure 24.4** A radio wave can be detected with a receiving antenna wire that is parallel to the electric field of the wave. The magnetic field of the radio wave has been omitted for simplicity.

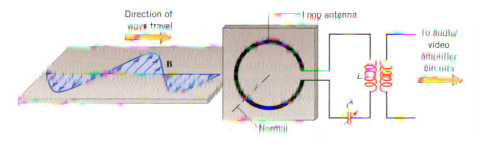

**Figure 24.5** With a receiving antenna in the form of a loop, the magnetic field of a broadcasted radio wave can be detected. The normal to the plane of the loop should be parallel to the magnetic field for best reception. For clarity, the electric field of the radio wave has been omitted.

*The physics of* radio and television reception.

Radio and television reception involves a process that is the reverse of that outlined earlier for the creation of electromagnetic waves. When broadcasted waves reach a receiving antenna, they interact with the electric charges in the antenna wires. Either the electric field or the magnetic field of the waves can be used. To take full advantage of the electric field, the wires of the receiving antenna must be parallel to the electric field, as Figure 24.4 indicates. The electric field acts on the electrons in the wire, forcing them to oscillate back and forth along the length of the wire. Consequently, an ac current exists in the antenna and the circuit connected to it. The variable-capacitor $C$ ($\dashv\vdash$) and the inductor $L$ in the circuit provide one way to select the frequency of the desired electromagnetic wave. By adjusting the value of the capacitance, it is possible to adjust the corresponding resonant frequency $f_0$ of the circuit [$f_0 = 1/(2\pi\sqrt{LC})$, Equation 23.10] to match the frequency of the wave. Under the condition of resonance there will be a maximum oscillating current in the inductor. Because of mutual inductance, this current creates a maximum voltage in the second coil in the drawing, and this voltage can then be amplified and processed by the remaining radio or television circuitry.

To detect the magnetic field of a broadcasted radio wave, a receiving antenna in the form of a loop can be used, as Figure 24.5 shows. For best reception, the normal to the plane of the wire loop is oriented parallel to the magnetic field. Then, as the wave sweeps by, the magnetic field penetrates the loop, and the changing magnetic flux induces a voltage and a current in the loop, in accord with Faraday's law. Once again, the resonant frequency of a capacitor/inductor combination can be adjusted to match the frequency of the desired electromagnetic wave. Both straight wire and loop antennas can be seen on the ship in Figure 24.6.

Cochlear implants use the broadcasting and receiving of radio waves to provide assistance to hearing-impaired people who have auditory nerves that are at least partially intact. These implants utilize radio waves to bypass the damaged part of the hearing mechanism and access the auditory nerve directly, as Figure 24.7 illustrates. An external microphone (often set into an ear mold) detects sound waves and sends a corresponding

**Figure 24.6** This cruise ship uses both straight wire and loop antennas to communicate with other vessels and on-shore stations. (© Harvey Lloyd.)

 *The physics of* cochlear implants.

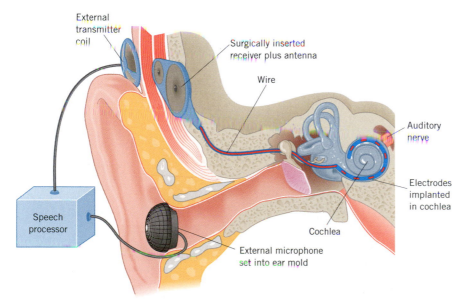

**Figure 24.7** Hearing-impaired people can sometimes recover part of their hearing with the help of a cochlear implant. The broadcasting and receiving of electromagnetic waves lies at the heart of this device.

**Figure 24.8** This wireless capsule endoscope is designed to be swallowed. As it passes through a patient's intestines, it broadcasts video images of the interior of the intestines. (Courtesy Given Imaging, Ltd.)

**The physics of** wireless capsule endoscopy (see above).

electrical signal to a speech processor small enough to be carried in a pocket. The speech processor encodes these signals into a radio wave, which is broadcast from an external transmitter coil placed over the site of a miniature receiver (and its receiving antenna) that has been surgically inserted beneath the skin. The receiver acts much like a radio. It detects the broadcasted wave and from the encoded audio information produces electrical signals that represent the sound wave. These signals are sent along a wire to electrodes that are implanted in the cochlea of the inner ear. The electrodes stimulate the auditory nerves that feed directly between structures within the cochlea and the brain. To the extent that the nerves are intact, a person can learn to recognize sounds.

The broadcasting and receiving of radio waves is also now being used in the practice of endoscopy. In this medical diagnostic technique a device called an endoscope is used to peer inside the body. For example, to examine the interior of the colon for signs of cancer, a conventional endoscope (known as a colonoscope) is inserted through the rectum. (See Section 26.3.) The wireless capsule endoscope shown in Figure 24.8 bypasses this invasive procedure completely. With a size of about $11 \times 26$ mm, this capsule can be swallowed and carried through the gastrointestinal tract by the involuntary contractions of the walls of the intestines (peristalsis). The capsule is self-contained and uses no external wires. A marvel of miniaturization, it contains a radio transmitter and its associated antenna, batteries, a white-light-emitting diode (see Section 23.5) for illumination, and an optical system to capture the digital images. As the capsule moves through the intestine, the transmitter broadcasts the images to an array of small receiving antennas attrached to the patient's body. These receiving antennas also are used to determine the position of the capsule within the body. The radio waves that are used lie in the ultrahigh frequency, or UHF band, from $3 \times 10^8$ to $3 \times 10^9$ Hz.

Radio waves are only one part of the broad spectrum of electromagnetic waves that has been discovered. The next section discusses the entire spectrum.

## 24.2 *The Electromagnetic Spectrum*

An electromagnetic wave, like any periodic wave, has a frequency $f$ and a wavelength $\lambda$ that are related to the speed $v$ of the wave by $v = f\lambda$ (Equation 16.1). For electromagnetic waves traveling through a vacuum or, to a good approximation, through air, the speed is $v = c$, so $c = f\lambda$.

As Figure 24.9 shows, electromagnetic waves exist with an enormous range of frequencies, from values less than $10^4$ Hz to greater than $10^{24}$ Hz. Since all these waves travel through a vacuum at the same speed of $c = 3.00 \times 10^8$ m/s, Equation 16.1 can be used to find the correspondingly wide range of wavelengths that the picture also displays. The ordered series of electromagnetic wave frequencies or wavelengths in Figure 24.9 is called the *electromagnetic spectrum.* Historically, regions of the spectrum have been

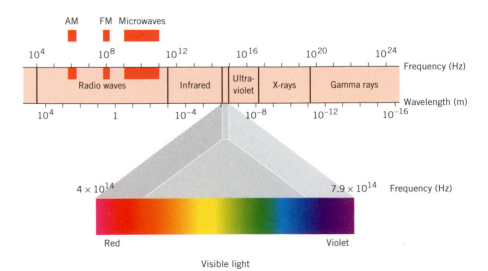

**Figure 24.9** The electromagnetic spectrum.

given names such as radio waves and infrared waves. Although the boundary between adjacent regions is shown as a sharp line in the drawing, the boundary is not so well defined in practice, and the regions often overlap.

Beginning on the left in Figure 24.9, we find radio waves. Lower-frequency radio waves are generally produced by electrical oscillator circuits, while higher-frequency radio waves (called microwaves) are usually generated using electron tubes called klystrons. Infrared radiation, sometimes loosely called heat waves, originates with the vibration and rotation of molecules within a material. Visible light is emitted by hot objects, such as the sun, a burning log, or the filament of an incandescent light bulb, when the temperature is high enough to excite the electrons within an atom. Ultraviolet frequencies can be produced from the discharge of an electric arc. X-rays are produced by the sudden deceleration of high-speed electrons. And, finally, gamma rays are radiation from nuclear decay.

The human body, like any object, radiates infrared radiation, and the amount emitted depends on the temperature of the body. Although infrared radiation cannot be seen by the human eye, it can be detected by sensors. An ear thermometer, like the pyroelectric thermometer shown in Figure 24.10, determines the body's temperature by measuring the amount of infrared radiation that emanates from the eardrum and surrounding tissue. The ear is one of the best places to measure body temperature because it is close to the hypothalamus, an area at the bottom of the brain that controls body temperature. The ear is also not cooled or warmed by eating, drinking, or breathing. When the probe of the thermometer is inserted into the ear canal, infrared radiation travels down the barrel of the probe and strikes the sensor. The absorption of infrared radiation warms the sensor, and, as a result, its electrical conductivity changes. The change in electrical conductivity is measured by an electronic circuit. The output from the circuit is sent to a microprocessor, which calculates the body temperature and displays the result on a digital readout.

Of all the frequency ranges in the electromagnetic spectrum, the most familiar is that of visible light, although it is the most narrow (see Figure 24.9). Only waves with frequencies between about $4.0 \times 10^{14}$ Hz and $7.9 \times 10^{14}$ Hz are perceived by the human eye as visible light. Usually visible light is discussed in terms of wavelengths (in vacuum) rather than frequencies. As Example 1 indicates, the wavelengths of visible light are extremely small and, therefore, are normally expressed in *nanometers* (nm); $1 \text{ nm} = 10^{-9}$ m. An obsolete (non-SI) unit occasionally used for wavelengths is the *angstrom* (Å); $1 \text{ Å} = 10^{-10}$ m.

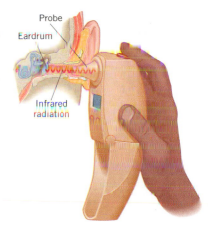

**Figure 24.10** A pyroelectric thermometer measures body temperature by determining the amount of infrared radiation emitted by the eardrum and surrounding tissue.

*The physics of* a pyroelectric ear thermometer.

## Example 1 The Wavelengths of Visible Light

Find the range in wavelengths (in vacuum) for visible light in the frequency range between $4.0 \times 10^{14}$ Hz (red light) and $7.9 \times 10^{14}$ Hz (violet light). Express the answers in nanometers.

**Reasoning** According to Equation 16.1, the wavelength (in vacuum) $\lambda$ of a light wave is equal to the speed of light $c$ in a vacuum divided by the frequency $f$ of the wave, $\lambda = c/f$.

**Solution** The wavelength corresponding to a frequency of $4.0 \times 10^{14}$ Hz is

$$\lambda = \frac{c}{f} = \frac{3.00 \times 10^8 \text{ m/s}}{4.0 \times 10^{14} \text{ Hz}} = 7.5 \times 10^{-7} \text{ m}$$

Since $1 \text{ nm} = 10^{-9}$ m, it follows that

$$\lambda = (7.5 \times 10^{-7} \text{ m}) \left( \frac{1 \text{ nm}}{10^{-9} \text{ m}} \right) = \boxed{750 \text{ nm}}$$

The calculation for a frequency of $7.9 \times 10^{14}$ Hz is similar:

$$\lambda = \frac{c}{f} = \frac{3.00 \times 10^8 \text{ m/s}}{7.9 \times 10^{14} \text{ Hz}} = 3.8 \times 10^{-7} \text{ m} \quad \text{or} \quad \boxed{\lambda = 380 \text{ nm}}$$

The eye/brain recognizes light of different wavelengths as different colors. A wavelength of 750 nm (in vacuum) is approximately the longest wavelength of red light, whereas 380 nm (in vacuum) is approximately the shortest wavelength of violet light. Between these limits are found the other familiar colors, as Figure 24.9 indicates.

The association between color and wavelength in the visible part of the electromagnetic spectrum is well known. The wavelength also plays a central role in governing the behavior and use of electromagnetic waves in all regions of the spectrum. For instance, Conceptual Example 2 considers the influence of the wavelength on diffraction.

### Conceptual Example 2   The Diffraction of AM and FM Radio Waves

As we have seen in Section 17.3, diffraction is the ability of a wave to bend around an obstacle or the edges of an opening. Based on the discussion in that chapter, would you expect AM or FM radio waves to bend more readily around an obstacle such as a building?

**Reasoning and Solution** Section 17.3 points out that, other things being equal, sound waves exhibit diffraction to a greater extent when the wavelength is longer than when it is shorter. Based on this information, we expect that longer-wavelength electromagnetic waves will bend more readily around obstacles than will shorter-wavelength waves. Figure 24.9 shows that AM radio waves have considerably longer wavelengths than FM waves. Therefore, **AM waves have a greater ability to bend around buildings than do FM waves.** The inability of FM waves to diffract around obstacles is a major reason why FM stations broadcast their signals essentially in a "line-of-sight" fashion.

The picture of light as a wave is supported by experiments that will be discussed in Chapter 27. However, there are also experiments indicating that light can behave as if it were composed of discrete particles rather than waves. These experiments will be discussed in Chapter 29. Wave theories and particle theories of light have been around for hundreds of years, and it is now widely accepted that light, as well as other electromagnetic radiation, exhibits a dual nature. Either wave-like or particle-like behavior can be observed, depending on the kind of experiment being performed.

## 24.3   *The Speed of Light*

At a speed of $3.00 \times 10^8$ m/s, light travels from the earth to the moon in a little over a second, so the time required for light to travel between two places on earth is very short. Therefore, the earliest attempts at measuring the speed of light had only limited success. One of the first accurate measurements employed a rotating mirror, and Figure 24.11 shows a simplified version of the setup. It was used first by the French scientist Jean Foucault (1819–1868) and later in a more refined version by the American physicist Albert Michelson (1852–1931). If the angular speed of the rotating eight-sided mirror in Figure 24.11 is adjusted correctly, light reflected from one side travels to the fixed mirror, reflects, and can be detected after reflecting from another side that has rotated into place at just the right time. The minimum angular speed must be such that one side of the mirror rotates one-eighth of a revolution during the time it takes for the light to make the round trip between the mirrors. For one of his experiments, Michelson placed mirrors on Mt. San Antonio and Mt. Wilson in California, a distance of 35 km apart. From a value of the minimum angular speed in such experiments, he obtained the value of $c = (2.997\,96 \pm 0.000\,04) \times 10^8$ m/s in 1926.

Today, the speed of light has been determined with such high accuracy that it is used to define the meter. As discussed in Section 1.2, the speed of light is now *defined* to be

| | |
|---|---|
| ***Speed of light in a vacuum*** | $c = 299\,792\,458$ m/s |

(although a value of $3.00 \times 10^8$ m/s is adequate for most calculations). The second is defined in terms of a cesium clock, and the meter is then defined as the distance light travels in a vacuum during a time of $1/(299\,792\,458)$ s. Although the speed of light in a vacuum is large, it is finite, so it takes a finite amount of time for light to travel from one place to another. The travel time is especially long for light traveling between astronomical objects, as Conceptual Example 3 discusses.

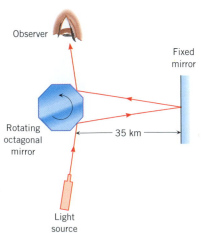

**Figure 24.11** Between 1878 and 1931, Michelson used a rotating eight-sided mirror to measure the speed of light. This is a simplified version of the setup.

## Conceptual Example 3  Looking Back in Time

A supernova is a violent explosion that occurs at the death of certain stars. For a few days after the explosion, the intensity of the emitted light can become a billion times greater than that of our own sun. But after several years, the intensity usually returns to zero. Supernovae are relatively rare events in the universe, for only six have been observed in our galaxy within the past 400 years. One of them was recorded in 1987. It occurred in a neighboring galaxy, approximately $1.66 \times 10^{21}$ m away. Figure 24.12 shows a photograph of the sky (a) before and (b) a few hours after the explosion. Why do astronomers say that viewing an event like the supernova is like looking back in time?

**Reasoning and Solution** The light from the supernova traveled to earth at a speed of $c = 3.00 \times 10^8$ m/s. Even at such a speed, the travel time is large because the distance $d = 1.66 \times 10^{21}$ m is enormous. The time $t$ is $t = d/c = (1.66 \times 10^{21}$ m$)/(3.00 \times 10^8$ m/s$)$ $= 5.53 \times 10^{12}$ s. This corresponds to 175 000 years. So when astronomers saw the explosion in 1987, they were actually seeing the light that left the supernova 175 000 years earlier; in other words, they were "looking back in time." In fact, whenever we view any celestial object, such as a star, we are seeing it as it was a long time ago. The farther the object is from earth, the longer it takes for the light to reach us, and the further back in time we are looking.

**Related Homework:** *Problem 14*

**Figure 24.12** A view of the sky (a) before and (b) a few hours after the 1987 supernova. (© Courtesy Anglo-Australian Telescope Board)

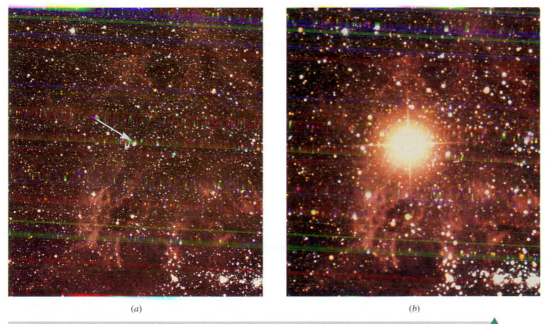

(a)                    (b)

In 1865, Maxwell determined theoretically that electromagnetic waves propagate through a vacuum at a speed given by

$$c = \frac{1}{\sqrt{\epsilon_0 \mu_0}} \qquad (24.1)$$

where $\epsilon_0 = 8.85 \times 10^{-12}$ C$^2$/(N·m$^2$) is the (electric) permittivity of free space and $\mu_0 = 4\pi \times 10^{-7}$ T·m/A is the (magnetic) permeability of free space. Originally $\epsilon_0$ was introduced in Section 18.5 as an alternative way of writing the proportionality constant $k$ in Coulomb's law [$k = 1/(4\pi\epsilon_0)$] and, hence, plays a basic role in determining the strengths of the electric fields created by point charges. The role of $\mu_0$ is similar for magnetic fields; it was introduced in Section 21.7 as part of a proportionality constant in the expression for the magnetic field created by the current in a long, straight wire. Substituting the values for $\epsilon_0$ and $\mu_0$ into Equation 24.1 shows that

$$c = \frac{1}{\sqrt{[8.85 \times 10^{-12}\ \text{C}^2/(\text{N·m}^2)][4\pi \times 10^{-7}\ \text{T·m/A}]}} = 3.00 \times 10^8\ \text{m/s}$$

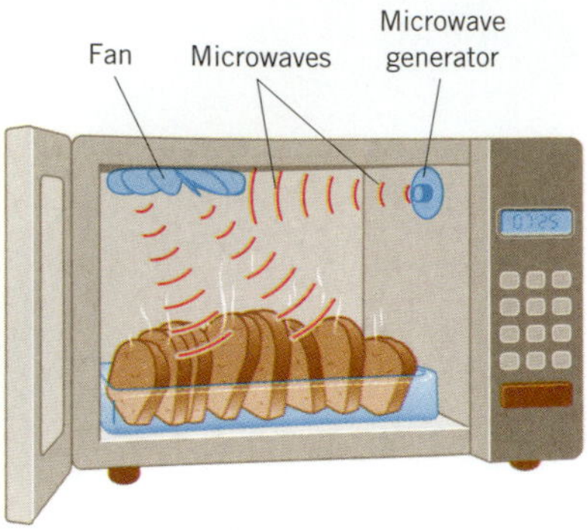

**Figure 24.13** A microwave oven. The rotating fan blades reflect the microwaves to all parts of the oven.

**The physics of**
a microwave oven.

**The physics of**
the greenhouse effect.

The experimental and theoretical values for $c$ agree. Maxwell's success in predicting $c$ provided a basis for inferring that light behaves as a wave consisting of oscillating electric and magnetic fields.

## 24.4 *The Energy Carried by Electromagnetic Waves*

Electromagnetic waves, like water waves or sound waves, carry energy. The energy is carried by the electric and magnetic fields that comprise the wave. In a microwave oven, for example, microwaves penetrate food and deliver their energy to it, as Figure 24.13 illustrates. The electric field of the microwaves is largely responsible for delivering the energy, and water molecules in the food absorb it. The absorption occurs because each water molecule has a permanent dipole moment; that is, one end of a molecule has a slight positive charge, and the other end has a negative charge of equal magnitude. As a result, the positive and negative ends of different molecules can form a bond. However, the electric field of the microwaves exerts forces on the positive and negative ends of a molecule, causing it to spin. Because the field is oscillating rapidly—about $2.4 \times 10^9$ times a second—the water molecules are kept spinning at a high rate. In the process, the energy of the microwaves is used to break bonds between neighboring water molecules and ultimately is converted into internal energy. As the internal energy increases, the temperature of the water increases, and the food cooks.

The energy carried by electromagnetic waves in the infrared and visible regions of the spectrum plays the key role in the greenhouse effect that is a contributing factor to global warming. The infrared waves from the sun are largely prevented from reaching the earth's surface by carbon dioxide and water in the atmosphere, which reflect them back into space. The visible waves do reach the earth's surface, however, and the energy they carry heats the earth. Heat also flows to the surface from the interior of the earth. The heated surface in turn radiates infrared waves outward, which, if they could, would carry their energy into space. However, the atmospheric carbon dioxide and water reflect these infrared waves back toward the earth, just as they reflect the infrared waves from the sun. Thus, their energy is trapped, and the earth becomes warmer, like plants in a greenhouse. In a greenhouse, however, energy is trapped mainly for a different reason—namely, the lack of effective convection currents to carry warm air past the cold glass walls.

A measure of the energy stored in the electric field **E** of an electromagnetic wave, such as a microwave, is provided by the electric energy density. As we saw in Section 19.5, this density is the electric energy per unit volume of space in which the electric field exists:

$$\text{Electric energy density} = \frac{\text{Electric energy}}{\text{Volume}} = \frac{1}{2}\kappa\epsilon_0 E^2 = \frac{1}{2}\epsilon_0 E^2 \qquad (19.12)$$

In this equation, the dielectric constant $\kappa$ has been set equal to unity, since we are dealing with an electric field in a vacuum (or air). From Section 22.7, the analogous expression for the magnetic energy density is

$$\text{Magnetic energy density} = \frac{\text{Magnetic energy}}{\text{Volume}} = \frac{1}{2\mu_0} B^2 \qquad (22.11)$$

The **total energy density** $u$ of an electromagnetic wave in a vacuum is the sum of these two energy densities:

$$u = \frac{\text{Total energy}}{\text{Volume}} = \frac{1}{2}\epsilon_0 E^2 + \frac{1}{2\mu_0} B^2 \qquad (24.2)$$

In an electromagnetic wave propagating through a vacuum or air, the electric field and the magnetic field carry equal amounts of energy per unit volume of space. Since $\frac{1}{2}\epsilon_0 E^2 = \frac{1}{2}(B^2/\mu_0)$, it is possible to rewrite Equation 24.2 for the total energy density in two additional, but equivalent, forms:

$$u = \frac{1}{2}\epsilon_0 E^2 + \frac{1}{2\mu_0} B^2 \qquad (24.2a)$$

$$u = \epsilon_0 E^2 \qquad (24.2b)$$

$$u = \frac{1}{\mu_0} B^2 \qquad (24.2c)$$

The fact that the two energy densities are equal implies that the electric and magnetic fields are related. To see how, we set the electric energy density equal to the magnetic energy density and obtain

$$\frac{1}{2}\epsilon_0 E^2 = \frac{1}{2\mu_0} B^2 \quad \text{or} \quad E^2 = \frac{1}{\epsilon_0 \mu_0} B^2$$

But according to Equation 24.1, $c = 1/\sqrt{\epsilon_0 \mu_0}$, so it follows that $E^2 = c^2 B^2$. Taking the square root of both sides of this result shows that the relation between the magnitudes of the electric and magnetic fields in an electromagnetic wave is

$$E = cB \tag{24.3}$$

In an electromagnetic wave, the electric and magnetic fields fluctuate sinusoidally in time, so Equations 24.2a–c give the energy density of the wave at any instant in time. If an average value $\bar{u}$ for the total energy density is desired, average values are needed for $E^2$ and $B^2$. In Section 20.5 we faced a similar situation for alternating currents and voltages and introduced rms (root mean square) quantities. Using an analogous procedure here, we find that the rms values for the electric and magnetic fields, $E_{rms}$ and $B_{rms}$, are related to the maximum values of these fields, $E_0$ and $B_0$, by

$$E_{rms} = \frac{1}{\sqrt{2}} E_0 \quad \text{and} \quad B_{rms} = \frac{1}{\sqrt{2}} B_0$$

Equations 24.2a–c can now be interpreted as giving the average energy density $\bar{u}$, provided the symbols $E$ and $B$ are interpreted to mean the rms values given above. The average density of the sunlight reaching the earth is determined in the next example.

## Example 4  The Average Energy Density of Sunlight

Sunlight enters the top of the earth's atmosphere with an electric field whose rms value is $E_{rms} = 720$ N/C. Find (a) the average total energy density of this electromagnetic wave and (b) the rms value of the sunlight's magnetic field.

**Reasoning** The average total energy density $\bar{u}$ of the sunlight can be obtained from Equation 24.2b, provided the rms value is used for the electric field. Since the magnitudes of the magnetic and electric fields are related according to Equation 24.3, the rms value of the magnetic field is $B_{rms} = E_{rms}/c$.

**Solution**

(a) According to Equation 24.2b, the average total energy density is

$$\bar{u} = \epsilon_0 E_{rms}^2 = [8.85 \times 10^{-12} \text{ C}^2/(\text{N}\cdot\text{m}^2)](720 \text{ N/C})^2 = \boxed{4.6 \times 10^{-6} \text{ J/m}^3}$$

(b) Using Equation 24.3, we find that the rms magnetic field is

$$B_{rms} = \frac{E_{rms}}{c} = \frac{720 \text{ N/C}}{3.0 \times 10^8 \text{ m/s}} = \boxed{2.4 \times 10^{-6} \text{ T}}$$

As an electromagnetic wave moves through space, it carries energy from one region to another. This energy transport is characterized by the *intensity* of the wave. We have encountered the concept of intensity before, in connection with sound waves in Section 16.7. The intensity of a wave is formulated by bringing together the power of the wave and the area through which it passes. The sound intensity is the sound power that passes perpendicularly through a surface divided by the area of the surface. The intensity of an electromagnetic wave is defined similarly. For an electromagnetic wave, the intensity is the electromagnetic power divided by the area of the surface.

Using this definition of intensity, we can show that the electromagnetic intensity $S$ is related to the energy density $u$. According to Equation 16.8 the intensity is the power $P$ that passes perpendicularly through a surface divided by the area $A$ of that surface, or $S = P/A$. Furthermore, the power is equal to the total energy passing through the surface divided by the elapsed time $t$ (Equation 6.10b), so that $P =$ (Total energy)/$t$. Combining these two relations gives

$$S = \frac{P}{A} = \frac{\text{Total energy}}{tA}$$

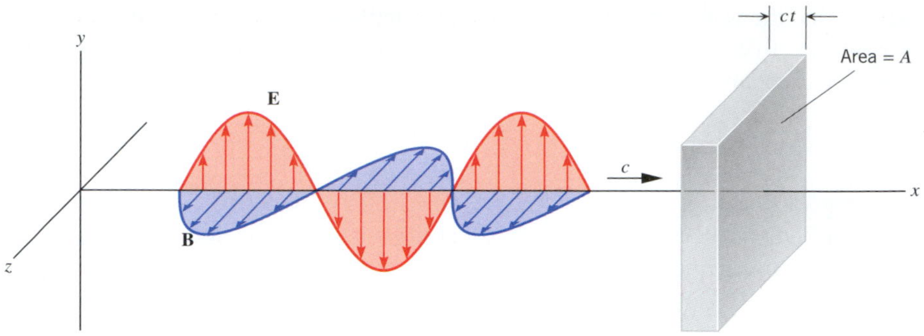

**Figure 24.14** In a time $t$, an electromagnetic wave moves a distance $ct$ along the $x$ axis and passes through a surface of area $A$.

Now, consider Figure 24.14, which shows an electromagnetic wave traveling in a vacuum along the $x$ axis. In a time $t$ the wave travels the distance $ct$, passing through the surface of area $A$. Consequently, the volume of space through which the wave passes is $ctA$. The total (electric and magnetic) energy in this volume is

$$\text{Total energy} = (\text{Total energy density}) \times \text{Volume} = u(ctA)$$

Using this result in the expression for the intensity, we obtain

$$S = \frac{\text{Total energy}}{tA} = \frac{uctA}{tA} = cu \qquad (24.4)$$

Thus, the intensity and the energy density are related by the speed of light, $c$. Substituting Equations 24.2a–c, one at a time, into Equation 24.4 shows that the intensity of an electromagnetic wave depends on the electric and magnetic fields according to the following equivalent relations:

$$S = cu = \frac{1}{2}c\epsilon_0 E^2 + \frac{c}{2\mu_0}B^2 \qquad (24.5a)$$

$$S = c\epsilon_0 E^2 \qquad (24.5b)$$

$$S = \frac{c}{\mu_0}B^2 \qquad (24.5c)$$

If the rms values for the electric and magnetic fields are used in Equations 24.5a–c, the intensity becomes an average intensity, $\overline{S}$, as Example 5 illustrates.

### Example 5   A Neodymium–Glass Laser

A neodymium–glass laser emits short pulses of high-intensity electromagnetic waves. The electric field has an rms value of $E_{rms} = 2.0 \times 10^9$ N/C. Find the average power of each pulse that passes through a $1.6 \times 10^{-5}$-m$^2$ surface that is perpendicular to the laser beam.

**Reasoning** Since the intensity of a wave is the power per unit area that passes perpendicularly through a surface, the average power $\overline{P}$ of the wave is the product of the average intensity $\overline{S}$ and the area $A$.

**Problem solving insight**
The concepts of power and intensity are similar, but they are not the same. Intensity is the power that passes perpendicularly through a surface divided by the area of the surface.

**Solution** The average power is $\overline{P} = \overline{S}A$. But from Equation 24.5b, $\overline{S} = c\epsilon_0 E_{rms}^2$ so that

$$\overline{P} = c\epsilon_0 E_{rms}^2 A$$

$$\overline{P} = (3.0 \times 10^8 \text{ m/s})[8.85 \times 10^{-12} \text{ C}^2/(\text{N} \cdot \text{m}^2)](2.0 \times 10^9 \text{ N/C})^2(1.6 \times 10^{-5} \text{ m}^2)$$

$$= \boxed{1.7 \times 10^{11} \text{ W}}$$

## 24.5 *The Doppler Effect and Electromagnetic Waves*

Section 16.9 presents a discussion of the Doppler effect that sound waves exhibit when either the source of a sound wave, the observer of the wave, or both are moving with respect to the medium of propagation (e.g., air). This effect is one in which the observed sound frequency is greater or smaller than the frequency emitted by the source. A different Doppler effect arises when the source moves than when the observer moves.

Electromagnetic waves also can exhibit a Doppler effect, but it differs from that for sound waves for two reasons. First, sound waves require a medium such as air in which to propagate. In the Doppler effect for sound, it is the motion (of the source, the observer, and the waves themselves) relative to this medium that is important. In the Doppler effect for electromagnetic waves, motion relative to a medium plays no role, because the waves do not require a medium in which to propagate. They can travel in a vacuum. Second, in the equations for the Doppler effect in Section 16.9, the speed of sound plays an important role, and it depends on the reference frame relative to which it is measured. For example, the speed of sound with respect to moving air is different than it is with respect to stationary air. As we will see in Section 28.2, electromagnetic waves behave in a different way. The speed at which they travel has the same value, whether it is measured relative to a stationary observer or relative to one moving at a constant velocity. For these two reasons, the same Doppler effect arises for electromagnetic waves when either the source or the observer of the waves moves; only the relative motion of the source and the observer with respect to one another is important.

When electromagnetic waves and the source and the observer of the waves all travel along the same line in a vacuum (or in air, to a good degree of approximation), the single equation that specifies the Doppler effect is

$$f_o = f_s \left( 1 \pm \frac{v_{rel}}{c} \right) \qquad \text{if } v_{rel} \ll c \qquad (24.6)$$

In this expression, $f_o$ is the observed frequency, and $f_s$ is the frequency emitted by the source. The symbol $v_{rel}$ stands for the speed of the source and the observer relative to one another, and $c$ is the speed of light in a vacuum. Equation 24.6 applies only if $v_{rel}$ is very small compared to $c$—that is, if $v_{rel} \ll c$. The plus sign in Equation 24.6 applies when the source and observer are moving toward one another, and the minus sign applies when they are moving apart.

It is essential to realize that $v_{rel}$ is the **relative** speed of the source and the observer. Thus, if the source is moving due east at a speed of 28 m/s with respect to the earth, while the observer is moving due east at a speed of 22 m/s, the value for $v_{rel}$ is 28 m/s − 22 m/s = 6 m/s. Because $v_{rel}$ is the relative **speed,** it has no algebraic sign. The direction of the relative motion is taken into account by choosing the plus or minus sign in Equation 24.6. The plus sign is used when the source and the observer come together, and the minus sign is used when they move apart. Example 6 illustrates one familiar use of the Doppler effect for electromagnetic waves.

## Example 6 Radar Guns and Speed Traps

**The physics of radar speed traps.**

Police use radar guns and the Doppler effect to catch speeders. One such gun emits an electromagnetic wave with a frequency of $f_s = 8.0 \times 10^9$ Hz, as Figure 24.15 illustrates. In this picture, a car is approaching a police car parked on the side of the road. The direction of approach is essentially head-on. The wave from the radar gun reflects from the speeding car and returns to the police car, where on-board equipment measures its frequency to be greater than that of the emitted wave by 2100 Hz. Find the speed of the car with respect to the highway.

**Reasoning** The Doppler effect depends on the relative speed $v_{rel}$ between the speeding car and the police car. We will find this speed and then relate it to the speed of the moving car

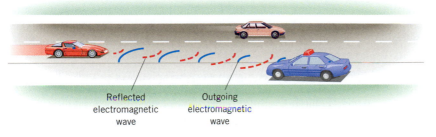

Reflected electromagnetic wave

Outgoing electromagnetic wave

**Figure 24.15** The radar gun used by police emits an electromagnetic wave in the radio frequency region of the spectrum. The Doppler effect exhibited by the wave after it reflects from a moving vehicle is used to determine the vehicle's speed (see Example 6).

with respect to the highway by using the fact that the police car is at rest. There are two Doppler frequency changes in this situation. First, the speeder's car "observes" the wave frequency coming from the radar gun to have a frequency $f_o$ that is different from the emitted frequency $f_s$. According to Equation 24.6 (with the plus sign, since the two cars are coming together), $f_o - f_s = f_s(v_{rel}/c)$. Then, the wave reflects and returns to the police car, where it is observed to have a frequency $f_o'$ that is different than its frequency $f_o$ at the instant of reflection. Again using Equation 24.6, we find that $f_o' - f_o = f_o(v_{rel}/c)$. Adding the two previous equations gives the following result for the total Doppler change in frequency:

$$(f_o' - f_o) + (f_o - f_s) = f_o' - f_s = f_o \left( \frac{v_{rel}}{c} \right) + f_s \left( \frac{v_{rel}}{c} \right) \approx 2f_s \left( \frac{v_{rel}}{c} \right)$$

where we have assumed that $f_o$ and $f_s$ differ by only a negligibly small amount, since $v_{rel}$ is small compared to the speed of light $c$. Solving for the relative speed $v_{rel}$ gives

$$v_{rel} \approx \left( \frac{f_o' - f_s}{2f_s} \right) c$$

**Solution** The speed of the moving car relative to the police car is

$$v_{rel} \approx \left( \frac{f_o' - f_s}{2f_s} \right) c = \left[ \frac{2100 \text{ Hz}}{2(8.0 \times 10^9 \text{ Hz})} \right] (3.0 \times 10^8 \text{ m/s}) = 39 \text{ m/s}$$

Since the police car is at rest, its speed is $v_P = 0$ m/s, and the speeder has a speed of $v_{rel} = v - v_P = v = \boxed{39 \text{ m/s}}$.

The Doppler effect of electromagnetic waves provides a powerful tool for astronomers. For instance, Example 10 in Chapter 5 discusses how astronomers have identified a supermassive black hole at the center of galaxy M87 by using the Hubble space telescope. They focused the telescope on regions to either side of the center of the galaxy (see Figure 5.14). From the light emitted by these two regions, they were able to use the Doppler effect to determine that one side is moving away from the earth, while the other side is moving toward the earth. In other words, the galaxy is rotating. The speeds of recession and approach enabled astronomers to determine the rotational speed of the galaxy, and Example 10 in Chapter 5 shows how the value for this speed leads to the identification of the black hole. Astronomers routinely study the Doppler effect of the light that reaches the earth from distant parts of the universe. From such studies, they have determined the speeds at which distant light-emitting objects are receding from the earth.

## 24.6 *Polarization*

### POLARIZED ELECTROMAGNETIC WAVES

One of the essential features of electromagnetic waves is that they are transverse waves, and because of this feature they can be polarized. Figure 24.16 illustrates the idea of polarization by showing a transverse wave as it travels along a rope toward a slit. The wave is said to be *linearly polarized*, which means that its vibrations always occur along one direction. This direction is called the direction of polarization. In part *a* of the picture, the direction of polarization is vertical, parallel to the slit. Consequently, the wave passes through easily. However, when the slit is turned perpendicular to the direction of polarization, as in part *b*, the wave cannot pass, because the slit prevents the rope from oscillating. For longitudinal waves, such as sound waves, the notion of polarization has no meaning. In a longitudinal wave the direction of vibration is along the direction of travel, and the orientation of the slit would have no effect on the wave.

In an electromagnetic wave such as that in Figure 24.3, the electric field oscillates along the $y$ axis. Similarly, the magnetic field oscillates along the $z$ axis. Therefore, the wave is linearly polarized, with the direction of polarization taken arbitrarily to be that along which the electric field oscillates. If the wave is a radio wave generated by a

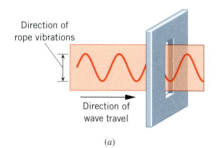

Direction of rope vibrations

Direction of wave travel

*(a)*

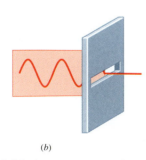

*(b)*

**Figure 24.16** A transverse wave is linearly polarized when its vibrations always occur along one direction. (*a*) A linearly polarized wave on a rope can pass through a slit that is parallel to the direction of the rope vibrations, but (*b*) cannot pass through a slit that is perpendicular to the vibrations.

straight-wire antenna, the direction of polarization is determined by the orientation of the antenna. In comparison, the visible light given off by an incandescent light bulb consists of electromagnetic waves that are completely unpolarized. In this case the waves are emitted by a large number of atoms in the hot filament of the bulb. When an electron in an atom oscillates, the atom behaves as a miniature antenna that broadcasts light for brief periods of time, about $10^{-8}$ seconds. However, the directions of these atomic antennas change randomly as a result of collisions. Unpolarized light, then, consists of many individual waves, emitted in short bursts by many "atomic antennas," each with its own direction of polarization. Figure 24.17 compares polarized and unpolarized light. In the unpolarized case, the arrows shown around the direction of wave travel symbolize the random directions of polarization of the individual waves that comprise the light.

Linearly polarized light can be produced from unpolarized light with the aid of certain materials. One commercially available material goes under the name of Polaroid. Such materials allow only the component of the electric field along one direction to pass through, while absorbing the field component perpendicular to this direction. As Figure 24.18 indicates, the direction of polarization that a polarizing material allows through is called the *transmission axis*. No matter how this axis is oriented, the intensity of the transmitted polarized light is one-half that of the incident unpolarized light. The reason for this is that the unpolarized light contains all polarization directions to an equal extent. Moreover, the electric field for each direction can be resolved into components perpendicular and parallel to the transmission axis, with the result that the average components perpendicular and parallel to the axis are equal. As a result, the polarizing material absorbs as much of the electric (and magnetic) field strength as it transmits.

## MALUS' LAW

Once polarized light has been produced with a piece of polarizing material, it is possible to use a second piece to change the polarization direction and simultaneously adjust the intensity of the light. Figure 24.19 shows how. As in this picture the first piece of polarizing material is called the *polarizer*, and the second piece is referred to as the *analyzer*. The transmission axis of the analyzer is oriented at an angle $\theta$ relative to the transmission axis of the polarizer. If the electric field strength of the polarized light incident on the analyzer is $E$, the field strength passing through is the component parallel to the transmission axis, or $E \cos\theta$. According to Equation 24.5b, the intensity is proportional to the square of the electric field strength. Consequently, the average intensity of polarized light passing through the analyzer is proportional to $\cos^2\theta$. Thus, both the polarization direction and the intensity of the light can be adjusted by rotating the transmission axis of the analyzer relative to that of the polarizer. The average intensity $\overline{S}$ of the light leaving the analyzer, then, is

*Malus' law*
$$\overline{S} = \overline{S}_0 \cos^2\theta \qquad (24.7)$$

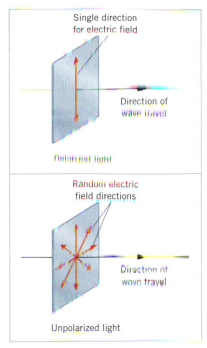

**Figure 24.17** In polarized light, the electric field of the electromagnetic wave fluctuates along a single direction. Unpolarized light consists of short bursts of electromagnetic waves emitted by many different atoms. The electric field directions of these bursts are perpendicular to the direction of wave travel but are distributed randomly about it.

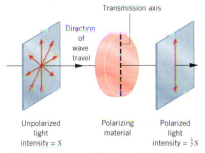

**Figure 24.18** With the aid of a piece of polarizing material, polarized light may be produced from unpolarized light. The transmission axis of the material is the direction of polarization of the light that passes through the material.

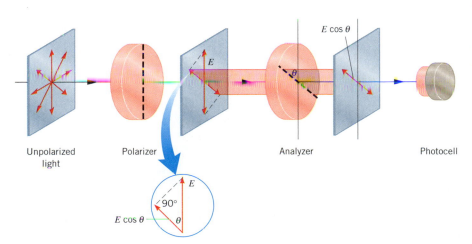

Unpolarized light | Polarizer | Analyzer | Photocell

**Figure 24.19** Two sheets of polarizing material, called the polarizer and the analyzer, may be used to adjust the polarization direction and intensity of the light reaching the photocell. This can be done by changing the angle $\theta$ between the transmission axes of the polarizer and analyzer.

where $\overline{S}_0$ is the average intensity of the light entering the analyzer. Equation 24.7 is sometimes called *Malus' law,* for it was discovered by the French engineer Etienne-Louis Malus (1775–1812). Example 7 illustrates the use of Malus' law.

### *Example 7*  Using Polarizers and Analyzers

**Problem solving insight**
Remember that when unpolarized light strikes a polarizer, only one half of the incident light is transmitted, the other half being absorbed by the polarizer.

What value of $\theta$ should be used in Figure 24.19, so the average intensity of the polarized light reaching the photocell is one-tenth the average intensity of the unpolarized light?

**Reasoning**  Both the polarizer and the analyzer reduce the intensity of the light. The polarizer reduces the intensity by a factor of one-half, as discussed earlier. Therefore, if the average intensity of the unpolarized light is $\overline{I}$, the average intensity of the polarized light leaving the polarizer and striking the analyzer is $\overline{S}_0 = \overline{I}/2$. The angle $\theta$ must now be selected so the average intensity of the light leaving the analyzer is $\overline{S} = \overline{I}/10$. Malus' law provides the solution.

**Solution**  Using $\overline{S}_0 = \overline{I}/2$ and $\overline{S} = \overline{I}/10$ in Malus' law, we find

$$\tfrac{1}{10}\overline{I} = \tfrac{1}{2}\overline{I}\cos^2\theta$$

$$\tfrac{1}{5} = \cos^2\theta \quad \text{or} \quad \theta = \cos^{-1}\left(\frac{1}{\sqrt{5}}\right) = \boxed{63.4°}$$

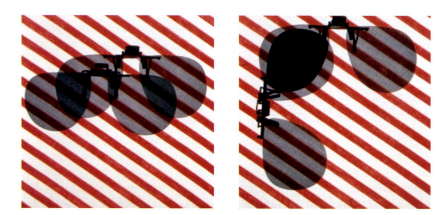

**Figure 24.20**  When Polaroid sunglasses are uncrossed (left photograph), the transmitted light is dimmed due to the extra thickness of tinted plastic. However, when they are crossed (right photograph), the intensity of the transmitted light is reduced to zero because of the effects of polarization. (© Diane Schiumo/ Fundamental Photographs)

When $\theta = 90°$ in Figure 24.19, the polarizer and analyzer are said to be *crossed,* and no light is transmitted by the polarizer/analyzer combination. As an illustration of this effect, Figure 24.20 shows two pairs of Polaroid sunglasses in uncrossed and crossed configurations.

**The physics of**
**IMAX 3-D films.**

An exciting application of crossed polarizers is used in viewing IMAX 3-D movies. These movies are recorded on two separate rolls of film, using a camera that provides images from the two different perspectives that correspond to what is observed by human eyes and allow us to see in three dimensions. The camera has two apertures or openings located at roughly the spacing between our eyes. The films are projected using a projector with two lenses, as Figure 24.21 indicates. Each lens has its own polarizer, and the two polarizers are crossed (see the drawing). In one type of theater, viewers watch the action

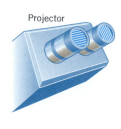

**Figure 24.21**  In an IMAX 3-D film, two separate rolls of film are projected using a projector with two lenses, each with its own polarizer. The two polarizers are crossed. Viewers watch the action on-screen through glasses that have corresponding crossed polarizers for each eye. The result is a 3-D moving picture, as the text discusses.

Projector

on-screen using glasses with corresponding polarizers for the left and right eyes, as the drawing shows. Because of the crossed polarizers the left eye sees only the image from the left lens of the projector, and the right eye sees only the image from the right lens. Since the two images have the approximate perspectives that the left and right eyes would see in reality, the brain combines the images to produce a realistic 3-D effect.

Conceptual Example 8 illustrates an interesting result that occurs when a piece of polarizing material is inserted between a crossed polarizer and analyzer.

## Conceptual Example 8
### How Can a Crossed Polarizer and Analyzer Transmit Light?

As explained earlier, no light reaches the photocell in Figure 24.19 when the polarizer and analyzer are crossed. Suppose that a third piece of polarizing material is inserted between the polarizer and analyzer, as in Figure 24.22a. Does light now reach the photocell?

**Reasoning and Solution** Surprisingly, the answer is yes, even though the polarizer and analyzer are crossed. To show why, we note that if any light is to pass through the analyzer, it must have an electric field component parallel to the transmission axis of the analyzer. Without the insert, there is no such component. But with the insert, there is, as parts *b* and *c* of the drawing illustrate. Part *b* shows that the electric field $E$ of the light leaving the polarizer makes an angle $\theta$ with respect to the transmission axis of the insert. The component of the field that is parallel to the insert's transmission axis is $E \cos \theta$, and this component passes through the insert. Part *c* of the drawing illustrates that the field ($E \cos \theta$) incident on the analyzer has a component parallel to the transmission axis of the analyzer—namely, ($E \cos \theta$) $\sin \theta$. This component passes through the analyzer, so that light now reaches the photocell.

**Light will reach the photocell as long as the angle $\theta$ is between 0 and 90°.** If the angle is 0 or 90°, however, no light reaches the photocell. Can you explain, without using any equations, why no light reaches the photocell under either of these conditions?

**Related Homework:** *Problem 37*

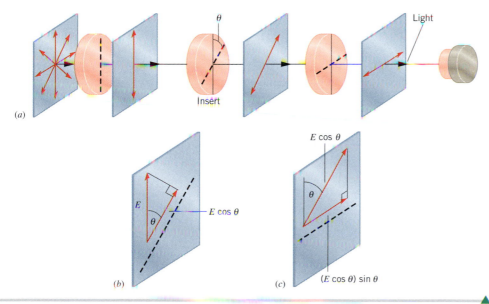

(a)

(b)     (c)

*Figure 24.22* (*a*) Light reaches the photocell when a piece of polarizing material is inserted between a crossed polarizer and analyzer. (*b*) The component of the electric field parallel to the transmission axis of the insert is $E \cos \theta$. (*c*) Light incident on the analyzer has a component ($E \cos \theta$) $\sin \theta$ parallel to its transmission axis.

**The physics of**
**a liquid crystal display. (See below.)**

*Figure 24.23* Liquid crystal displays use liquid crystal segments to form the numbers.

An application of a crossed polarizer/analyzer combination occurs in one kind of liquid crystal display (LCD). LCDs are widely used in pocket calculators and digital watches. The display usually consists of blackened numbers and letters set against a light gray background. As Figure 24.23 indicates, each number or letter is formed from a combination of liquid crystal segments that have been turned on and appear black. The liquid crystal part of an LCD segment consists of the liquid crystal material sandwiched between two transparent electrodes, as in Figure 24.24. When a voltage is applied between the electrodes, the liquid crystal is said to be "on." Part *a* of the picture shows that linearly polarized incident light passes through the "on" material without having its

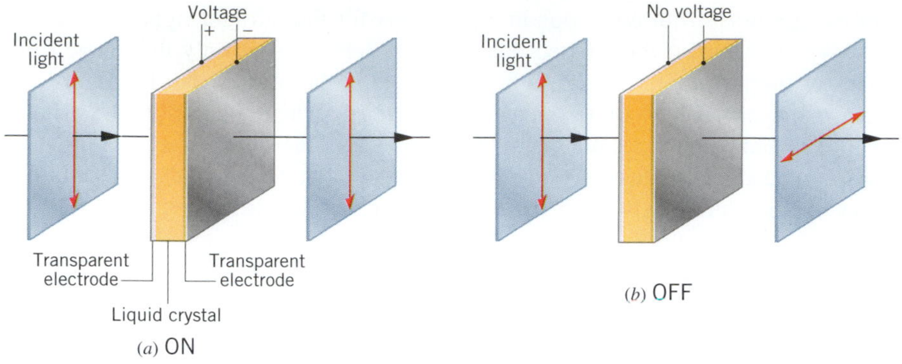

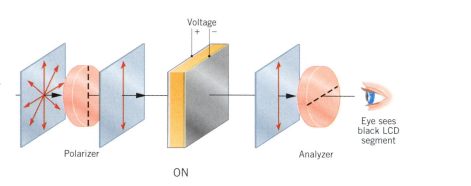

**Figure 24.24** A liquid crystal in its (*a*) "on" state and (*b*) "off" state.

**Figure 24.25** A liquid crystal display (LCD) incorporates a crossed polarizer/analyzer combination. When the LCD segment is turned on (voltage applied), no light is transmitted through the analyzer, and the observer sees a black segment.

**Figure 24.26** This personal entertainment organizer utilizes a color LCD display screen because it is lightweight and space-efficient. (© AP/Wide World Photos)

**The physics of**
**Polaroid sunglasses.**

direction of polarization affected. When the voltage is removed, as in part *b*, the liquid crystal is said to be "off" and now rotates the direction of polarization by 90°. A complete LCD segment also includes a crossed polarizer/analyzer combination, as Figure 24.25 illustrates. The polarizer, analyzer, electrodes, and liquid crystal material are packaged as a single unit. The polarizer produces polarized light from incident unpolarized light. With the display segment turned on, as in Figure 24.25, the polarized light emerges from the liquid crystal only to be absorbed by the analyzer, since the light is polarized perpendicular to the transmission axis of the analyzer. Since no light emerges from the analyzer, an observer sees a black segment against a light gray background, as in Figure 24.23. On the other hand, the segment is turned off when the voltage is removed, in which case the liquid crystal rotates the direction of polarization by 90° to coincide with the axis of the analyzer. The light now passes through the analyzer and enters the eye of the observer. However, the light coming from the segment has been designed to have the same color and shade (light gray) as the background of the display, so the segment becomes indistinguishable from the background.

Color LCD display screens and computer monitors are popular because they occupy less space and weigh less than traditional units do. An LCD display screen, such as that in Figure 24.26, uses thousands of LCD segments arranged like the squares on graph paper. To produce color, three segments are grouped together to form a tiny picture element (or "pixel"). Color filters are used to enable one segment in the pixel to produce red light, one to produce green, and one to produce blue. The eye blends the colors from each pixel into a composite color. By varying the intensity of the red, green, and blue colors, the pixel can generate an entire spectrum of colors.

## THE OCCURRENCE OF POLARIZED LIGHT IN NATURE

Polaroid is a familiar material because of its widespread use in sunglasses. Such sunglasses are designed so that the transmission axis of the Polaroid is oriented vertically when the glasses are worn in the usual fashion. Thus, the glasses prevent any light that is polarized horizontally from reaching the eye. Light from the sun is unpolarized, but a considerable amount of horizontally polarized sunlight originates by reflection from horizontal surfaces such as that of a lake. Section 26.4 discusses this effect. Polaroid sunglasses reduce glare by preventing the horizontally polarized reflected light from reaching the eyes.

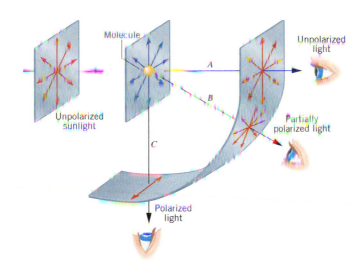

**Figure 24.27** In the process of being scattered from atmospheric molecules, unpolarized light from the sun becomes partially polarized.

Polarized sunlight also originates from the scattering of light by molecules in the atmosphere. Figure 24.27 shows light being scattered by a single atmospheric molecule. The electric fields in the unpolarized sunlight cause the electrons in the molecule to vibrate perpendicular to the direction in which the light is traveling. The electrons, in turn, reradiate the electromagnetic waves in different directions, as the drawing illustrates. The light radiated straight ahead in direction A is unpolarized, just like the incident light. But light radiated perpendicular to the incident light in direction C is polarized. Light radiated in the intermediate direction B is partially polarized. There is experimental evidence that some bird species use such polarized light as a navigational aid.

# Concept Summary

This summary presents an abridged version of the chapter, including the important equations and all available learning aids. For convenient reference, the learning aids (including the text's examples) are placed next to or immediately after the relevant equation or discussion. The following learning aids may be found on-line at **www.wiley.com/college/cutnell:**

| | |
|---|---|
| **Interactive LearningWare** examples are solved according to a five-step interactive format that is designed to help you develop problem-solving skills. | **Concept Simulations** are animated versions of text figures or animations that illustrate important concepts. You can control parameters that affect the display, and we encourage you to experiment. |
| **Interactive Solutions** offer specific models for certain types of problems in the chapter homework. The calculations are carried out interactively. | **Self-Assessment Tests** include both qualitative and quantitative questions. Extensive feedback is provided for both incorrect and correct answers, to help you evaluate your understanding of the material. |

| Topic | Discussion | Learning Aids |
|---|---|---|
| | **24.1 The Nature of Electromagnetic Waves** | |
| Electromagnetic wave | An electromagnetic wave consists of mutually perpendicular and oscillating electric and magnetic fields. The wave is a transverse wave, since the fields are perpendicular to the direction in which the wave travels. Electromagnetic waves can travel through a vacuum or a material substance. All electromagnetic waves travel through a vacuum at the same speed, which is known as the speed of | |
| Speed of light | light $c$ ($c = 3.00 \times 10^8$ m/s). | |
| | **24.2 The Electromagnetic Spectrum** | |
| | The frequency $f$ and wavelength $\lambda$ of an electromagnetic wave in a vacuum are related to its speed $c$ through the relation | |
| Relation between frequency, wavelength, and speed of light in a vacuum | $$c = f\lambda$$ | Examples 1, 2 |
| Electromagnetic spectrum | The series of electromagnetic waves, arranged in order of their frequencies or wavelengths, is called the electromagnetic spectrum. In increasing order of frequency (decreasing order of wavelength), the spectrum includes radio waves, infrared | |

| Topic | Discussion | Learning Aids |
|---|---|---|

**Visible light**

radiation, visible light, ultraviolet radiation, X-rays, and gamma rays. Visible light has frequencies between about $4.0 \times 10^{14}$ and $7.9 \times 10^{14}$ Hz. The human eye and brain perceive different frequencies or wavelengths as different colors.

### 24.3 The Speed of Light

James Clerk Maxwell showed that the speed of light in a vacuum is

**Speed of light**

$$c = \frac{1}{\sqrt{\epsilon_0 \mu_0}}$$  (24.1)  **Example 3**

where $\epsilon_0$ is the (electric) permittivity of free space and $\mu_0$ is the (magnetic) permeability of free space.

### 24.4 The Energy Carried by Electromagnetic Waves

The total energy density $u$ of an electromagnetic wave is the total energy per unit volume of the wave and, in a vacuum, is given by

**Total energy density**

$$u = \frac{1}{2} \epsilon_0 E^2 + \frac{1}{2\mu_0} B^2$$  (24.2a)

where $E$ and $B$, respectively, are the magnitudes of the electric and magnetic fields of the wave. Since the electric and magnetic parts of the total energy density are equal, the following two equations are equivalent to Equation 24.2a:

$$u = \epsilon_0 E^2$$  (24.2b)

$$u = \frac{1}{\mu_0} B^2$$  (24.2c)

In a vacuum, $E$ and $B$ are related according to

**Relation between electric and magnetic fields**

$$E = cB$$  (24.3)

Equations 24.2a–c can be used to determine the average total energy density, if the rms average values $E_{rms}$ and $B_{rms}$ are used in place of the symbols $E$ and $B$. **Example 4** The rms values are related to the peak values $E_0$ and $B_0$ in the usual way:

**Root mean square fields**

$$E_{rms} = \frac{1}{\sqrt{2}} E_0 \quad \text{and} \quad B_{rms} = \frac{1}{\sqrt{2}} B_0$$

**Intensity**

The intensity of an electromagnetic wave is the power that the wave carries per- **Example 5** pendicularly through a surface divided by the area of the surface. In a vacuum, **Interactive Solution 24.27** the intensity $S$ is related to the total energy density $u$ according to

**Relation between intensity and total energy density**

$$S = cu$$  (24.4)

### 24.5 The Doppler Effect and Electromagnetic Waves

When electromagnetic waves and the source and observer of the waves all travel along the same line in a vacuum, the Doppler effect is given by

**Doppler effect**

$$f_o = f_s\left(1 \pm \frac{v_{rel}}{c}\right) \quad \text{if } v_{rel} \ll c$$  (24.6)  **Example 6**
**Interactive Solution 24.31**

where $f_o$ and $f_s$ are, respectively, the observed and emitted wave frequencies and $v_{rel}$ is the relative speed of the source and the observer. The plus sign is used when the source and the observer come together, and the minus sign is used when they move apart.

### 24.6 Polarization

**Linear polarized electromagnetic wave**

A linearly polarized electromagnetic wave is one in which the oscillation of the electric field occurs only along one direction, which is taken to be the direction of polarization. The magnetic field also oscillates along only one direction, which is perpendicular to the electric field direction. In an unpolarized wave such as the light from an incandescent bulb, the direction of polarization does not remain fixed, but fluctuates randomly in time.

**Unpolarized electromagnetic wave**

Polarizing materials allow only the component of the wave's electric field along one direction (and the associated magnetic field component) to pass through

| Topic | Discussion | Learning Aids |
|---|---|---|
| Transmission axis | them. The preferred transmission direction for the electric field is called the transmission axis of the material. | |
| | When unpolarized light is incident on a piece of polarizing material, the transmitted polarized light has an intensity that is one-half that of the incident light. | |
| Polarizer and analyzer | When two pieces of polarizing material are used one after the other, the first is called the polarizer, and the second is referred to as the analyzer. If the average intensity of polarized light falling on the analyzer is $\overline{S}_0$, the average intensity $\overline{S}$ of the light leaving the analyzer is given by Malus' law as | Examples 7, 8 |
| Malus' law | $$\overline{S} = \overline{S}_0 \cos^2 \theta \qquad (24.7)$$ | Interactive LearningWare 24.1 |
| Crossed polarizer and analyzer | where $\theta$ is the angle between the transmission axes of the polarizer and analyzer. When $\theta = 90°$, the polarizer and the analyzer are said to be "crossed," and no light passes through the analyzer. | |

Use Self-Assessment Test 24.1 to evaluate your understanding of Sections 24.1–24.6.

# Problems

## Section 24.1 The Nature of Electromagnetic Waves

1. ssm www  The distance between earth and the moon can be determined from the time it takes for a laser beam to travel from earth to a reflector on the moon and back. If the round-trip time can be measured to an accuracy of one-tenth of a nanosecond (1 ns = $10^{-9}$ s), what is the corresponding error in the earth–moon distance?

2. (a) Neil A. Armstrong was the first person to walk on the moon. The distance between the earth and the moon is $3.85 \times 10^8$ m. Find the time it took for his voice to reach earth via radio waves. (b) Someday a person will walk on Mars, which is $5.6 \times 10^{10}$ m from earth at the point of closest approach. Determine the minimum time that will be required for that person's voice to reach earth.

3. An AM station is broadcasting a radio wave whose frequency is 1400 kHz. The value of the capacitance in Figure 24.4 is $8.4 \times 10^{-11}$ F. What must be the value of the inductance in order that this station can be tuned in by the radio?

4. FM radio stations use radio waves with frequencies from 88.0 to 108 MHz to broadcast their signals. Assuming that the inductance in Figure 24.4 has a value of $6.00 \times 10^{-7}$ H, determine the range of capacitance values that are needed so the antenna can pick up all the radio waves broadcasted by FM stations.

* 5. ssm  Equation 16.3, $y = A \sin (2\pi f t - 2\pi x/\lambda)$, gives the mathematical representation of a wave oscillating in the $y$ direction and traveling in the positive $x$ direction. Let $y$ in this equation equal the electric field of an electromagnetic wave traveling in a vacuum. The maximum electric field is $A = 156$ N/C, and the frequency is $f = 1.50 \times 10^8$ Hz. Plot a graph of the electric field strength versus position, using for $x$ the following values: 0, 0.50, 1.00, 1.50, and 2.00 m. Plot this graph for (a) a time $t = 0$ s and (b) a time $t$ that is one-fourth of the wave's period.

** 6. A flat coil of wire is used with an LC-tuned circuit as a receiving antenna. The coil has a radius of 0.25 m and consists of 450 turns. The transmitted radio wave has a frequency of 1.2 MHz. The mag-

netic field of the wave is parallel to the normal to the coil and has a maximum value of $2.0 \times 10^{-13}$ T. Using Faraday's law of electromagnetic induction and the fact that the magnetic field changes from zero to its maximum value in one-quarter of a wave period, find the magnitude of the average emf induced in the antenna during this time.

## Section 24.2 The Electromagnetic Spectrum

7. Some of the X-rays produced in an X-ray machine have a wavelength of 2.1 nm. What is the frequency of these electromagnetic waves?

8. TV channel 3 (VHF) broadcasts at a frequency of 63.0 MHz. TV channel 23 (UHF) broadcasts at a frequency of 527 MHz. Find the ratio (VHF/UHF) of the wavelengths for these channels.

9. ssm  The human eye is most sensitive to light having a frequency of about $5.5 \times 10^{14}$ Hz, which is in the yellow-green region of the electromagnetic spectrum. How many wavelengths of this light can fit across the width of your thumb, a distance of about 2.0 cm?

10. Magnetic resonance imaging, or MRI (see Section 21.7), and positron emission tomography, or PET scanning (see Section 32.6), are two medical diagnostic techniques. Both employ electromagnetic waves. For these waves, find the ratio of the MRI wavelength (frequency = $6.38 \times 10^7$ Hz) to the PET scanning wavelength (frequency = $1.23 \times 10^{20}$ Hz).

11. ssm www  At one time television sets used "rabbit-ears" antennas. Such an antenna consists of a pair of metal rods. The length of each rod can be adjusted to be one-quarter of a wavelength of an electromagnetic wave whose frequency is 60.0 MHz. How long is each rod?

12. Two radio waves are used in the operation of a cellular telephone. To receive a call, the phone detects the wave emitted at one frequency by the transmitter station or base unit. To send your message to the base unit, your phone emits its own wave at a

different frequency. The difference between these two frequencies is fixed for all channels of cell phone operation. Suppose the wavelength of the wave emitted by the base unit is 0.34339 m and the wavelength of the wave emitted by the phone is 0.36205 m. Using a value of $2.9979 \times 10^8$ m/s for the speed of light, determine the difference between the two frequencies used in the operation of a cell phone.

* **13.** Section 17.5 deals with transverse standing waves on a string. Electromagnetic waves also can form standing waves. In a standing wave pattern formed from microwaves, the distance between a node and an adjacent antinode is 0.50 cm. What is the microwave frequency?

## Section 24.3 The Speed of Light

**14.** Review Conceptual Example 3 before attempting this problem. The brightest star in the night sky is Sirius, which is at a distance of $8.3 \times 10^{16}$ m. When we look at this star, how far back in time are we seeing it? Express your answer in years. (There are $365\frac{1}{4}$ days in one year.)

**15. ssm** Two astronauts are 1.5 m apart in their spaceship. One speaks to the other. The conversation is transmitted to earth via electromagnetic waves. The time it takes for sound waves to travel at 343 m/s through the air between the astronauts equals the time it takes for the electromagnetic waves to travel to the earth. How far away from the earth is the spaceship?

**16.** A communications satellite is in a synchronous orbit that is $3.6 \times 10^7$ m directly above the equator. The satellite is located midway between Quito, Equador, and Belém, Brazil, two cities almost on the equator that are separated by a distance of $3.5 \times 10^6$ m. Find the time it takes for a telephone call to go by way of satellite between these cities. Ignore the curvature of the earth.

**17.** Figure 24.11 illustrates Michelson's setup for measuring the speed of light with the mirrors placed on Mt. San Antonio and Mt. Wilson in California, which are 35 km apart. Using a value of $3.00 \times 10^8$ m/s for the speed of light, find the minimum angular speed (in rev/s) for the rotating mirror.

**18.** A lidar (laser radar) gun is an alternative to the standard radar gun that uses the Doppler effect to catch speeders. A lidar gun uses an infrared laser and emits a precisely timed series of pulses of infrared electromagnetic waves. The time for each pulse to travel to the speeding vehicle and return to the gun is measured. In one situation a lidar gun in a stationary police car observes a difference of $1.27 \times 10^{-7}$ s in round-trip travel times for two pulses that are emitted 0.450 s apart. Assuming that the speeding vehicle is approaching the police car essentially head-on, determine the speed of the vehicle.

* **19. ssm** A mirror faces a cliff located some distance away. Mounted on the cliff is a second mirror, directly opposite the first mirror and facing toward it. A gun is fired very close to the first mirror. The speed of sound is 343 m/s. How many times does the flash of the gunshot travel the round-trip distance between the mirrors before the echo of the gunshot is heard?

* **20.** A celebrity holds a press conference, which is televised live. A television viewer hears the sound picked up by a microphone directly in front of the celebrity. This viewer is seated 2.3 m from the television set. A reporter at the press conference is located 4.1 m from the microphone and hears the words directly *at the very same instant* that the television viewer hears them. Using a value of 343 m/s for the speed of sound, determine the maximum distance between the television viewer and the celebrity.

## Section 24.4 The Energy Carried by Electromagnetic Waves

**21.** A laser emits a narrow beam of light. The radius of the beam is $1.0 \times 10^{-3}$ m, and the power is $1.2 \times 10^{-3}$ W. What is the intensity of the laser beam?

**22.** The maximum strength of the magnetic field in an electromagnetic wave is $3.3 \times 10^{-6}$ T. What is the maximum strength of the wave's electric field?

**23. ssm** The microwave radiation left over from the Big Bang explosion of the universe has an average energy density of $4 \times 10^{-14}$ J/m³. What is the rms value of the electric field of this radiation?

**24.** On a cloudless day, the sunlight that reaches the surface of the earth has an average intensity of about $1.0 \times 10^3$ W/m². What is the average electromagnetic energy contained in 5.5 m³ of space just above the earth's surface?

**25. ssm** A future space station in orbit about the earth is being powered by an electromagnetic beam from the earth. The beam has a cross-sectional area of 135 m² and transmits an average power of $1.20 \times 10^4$ W. What are the rms values of the (a) electric and (b) magnetic fields?

* **26.** The intensity of sunlight at the top of the earth's atmosphere is about 1390 W/m². The distance between the sun and earth is $1.50 \times 10^{11}$ m, while that between the sun and Mars is $2.28 \times 10^{11}$ m. What is the intensity of sunlight at the surface of Mars?

* **27. Interactive Solution 24.27** at **www.wiley.com/college/cutnell** provides one model for problems like this one. The drawing shows an edge-on view of the solar panels on a communications satellite. The dashed line specifies the normal to the panels. Sunlight strikes the panels at an angle $\theta$ with respect to the normal. If the solar power impinging on the panels is 2600 W when $\theta = 65°$, what is it when $\theta = 25°$?

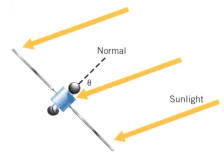

* **28.** A heat lamp emits infrared radiation whose rms electric field is $E_{rms} = 2800$ N/C. (a) What is the average intensity of the radiation? (b) The radiation is focused on a person's leg over a circular area of radius 4.0 cm. What is the average power delivered to the leg? (c) The portion of the leg being radiated has a mass of 0.28 kg and a specific heat capacity of 3500 J/(kg·C°). How long does it take to raise its temperature by 2.0 C°? Assume that there is no other heat transfer into or out of the portion of the leg being heated.

* **29. ssm** The mean distance between earth and the sun is $1.50 \times 10^{11}$ m. The average intensity of solar radiation incident on the upper atmosphere of the earth is 1390 W/m². Assuming the sun emits radiation uniformly in all directions, determine the total power radiated by the sun.

** **30.** The average intensity of sunlight reaching the earth is 1390 W/m². A charge of $2.6 \times 10^{-8}$ C is placed in the path of this electromagnetic wave. (a) What is the magnitude of the maximum electric force that the charge experiences? (b) If the charge is moving at a speed of $3.7 \times 10^4$ m/s, what is the magnitude of the maximum magnetic force that the charge could experience?

## Section 24.5 The Doppler Effect and Electromagnetic Waves

**31.** Review **Interactive Solution 24.31** at **www.wiley.com/college/cutnell** to see one model for solving this problem. A distant galaxy emits light that has a wavelength of 434.1 nm. On earth, the wavelength of this light is measured to be 438.6 nm. (a) Decide whether this galaxy is approaching or receding from the earth. Give your reasoning. (b) Find the speed of the galaxy relative to the earth.

**32.** Suppose that the police car in Example 6 is moving to the right at 27 m/s, while the speeder is coming up from behind at a speed of 39 m/s, both speeds being with respect to the ground. Assume that the electromagnetic wave emitted by the radar gun has a frequency of $8.0 \times 10^9$ Hz. (a) Find the magnitude of the difference between the frequency of the emitted wave and the wave that returns to the police car after reflecting from the speeder's car. (b) Which wave has the greater frequency? Why?

*  **33. ssm** A distant galaxy is simultaneously rotating and receding from the earth. As the drawing shows, the galactic center is receding from the earth at a relative speed of $u_G = 1.6 \times 10^6$ m/s. Relative to the center, the tangential speed is $v_T = 0.4 \times 10^6$ m/s for locations $A$ and $B$, which are equidistant from the center. When the frequencies of the light coming from regions $A$ and $B$ are measured on earth, they are not the same and each is different from the emitted frequency of $6.200 \times 10^{14}$ Hz. Find the measured frequency for the light from (a) region $A$ and (b) region $B$.

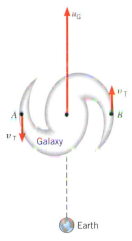

### Section 24.6 Polarization

**34.** Unpolarized light whose intensity is 1.10 W/m² is incident on the polarizer in Figure 24.19. (a) What is the intensity of the light leaving the polarizer? (b) If the analyzer is set at an angle of $\theta = 75°$ with respect to the polarizer, what is the intensity of the light that reaches the photocell?

**35. ssm** In the polarizer/analyzer combination in Figure 24.19, 90.0% of the light intensity falling on the analyzer is absorbed. Determine the angle between the transmission axes of the polarizer and the analyzer.

**36.** For one approach to this problem, consult **Interactive LearningWare 24.1** at **www.wiley.com/college/cutnell.** For each of the three sheets of polarizing material shown in the drawing, the orientation of the transmission axis is labeled relative to the vertical. The incident beam of light is unpolarized and has an intensity of 1260.0 W/m². What is the intensity of the beam transmitted through the three sheets when $\theta_1 = 19.0°$, $\theta_2 = 55.0°$, and $\theta_3 = 100.0°$?

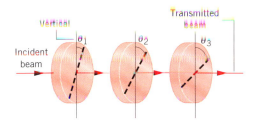

**37.** Review Conceptual Example 8 before solving this problem. Suppose unpolarized light of intensity 150 W/m² falls on the polarizer in Figure 24.22a, and the angle $\theta$ in the drawing is 30.0°. What is the light intensity reaching the photocell?

**38.** Light that is polarized along the vertical direction is incident on a sheet of polarizing material. Only 94% of the intensity of the light passes through the sheet and strikes a second sheet of polarizing material. No light passes through the second sheet. What angle does the transmission axis of the second sheet make with the vertical?

*  **39. ssm www** More than one analyzer can be used in a setup like that in Figure 24.19, each analyzer following the previous one. Suppose that the transmission axis of the first analyzer is rotated 27° relative to the transmission axis of the polarizer, and that the transmission axis of each additional analyzer is rotated 27° relative to the transmission axis of the previous one. What is the minimum number of analyzers needed, so the light reaching the photocell has an intensity that is reduced by at least a factor of one hundred relative to that striking the first analyzer?

*  **40.** A beam of polarized light has an average intensity of 15 W/m² and is sent through a polarizer. The transmission axis makes an angle of 25° with respect to the direction of polarization. Determine the rms value of the electric field of the transmitted beam.

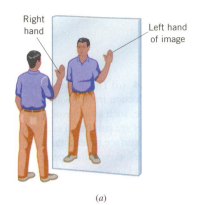

(a)

(b)

**Figure 25.5** (a) The person's right hand becomes the image's left hand when viewed in a plane mirror. (b) Many emergency vehicles are reverse-lettered so the lettering appears normal when viewed through the rearview mirror of a car. (© Mug Shots/Corbis Images)

As Figure 25.5a illustrates, the image of yourself in the mirror is also reversed right to left and left to right. If you wave your *right* hand, it is the *left* hand of the image that waves back. Similarly, letters and words held up to a mirror are reversed. Ambulances and other emergency vehicles are often lettered in reverse, as in Figure 25.5b, so that the letters will appear normal when seen in the rearview mirror of a car.

To illustrate why an image appears to originate from behind a plane mirror, Figure 25.6a shows a light ray leaving the top of an object. This ray reflects from the mirror (angle of reflection equals angle of incidence) and enters the eye. To the eye, it appears that the ray originates from behind the mirror, somewhere back along the dashed line. Actually, rays going in all directions leave each point on the object, but only a small bundle of such rays is intercepted by the eye. Part b of the figure shows a bundle of two rays leaving the top of the object. All the rays that leave a given point on the object, no matter what angle $\theta$ they have when they strike the mirror, appear to originate from a corresponding point on the image behind the mirror (see the dashed lines in part b). For each point on the object, there is a single corresponding point on the image, and it is this fact that makes the image in a plane mirror a sharp and undistorted one.

Although rays of light *seem* to come from the image, it is evident from Figure 25.6b that they do not originate from behind the plane mirror where the image appears to be. Because none of the light rays actually emanate from the image, it is called a **virtual image**. In this text the parts of the light rays that appear to come from a virtual image are represented by dashed lines. *Curved* mirrors, on the other hand, can produce images from which all the light rays actually do emanate. Such images are known as **real images** and are discussed later.

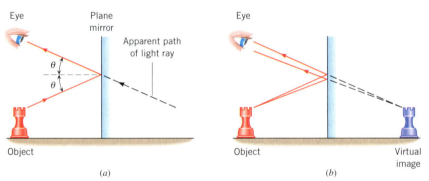

**Figure 25.6** (a) A ray of light from the top of the chess piece reflects from the mirror. To the eye, the ray seems to come from behind the mirror. (b) The bundle of rays from the top of the object appears to originate from the image behind the mirror.

With the aid of the law of reflection, it is possible to show that the image is located as far behind a plane mirror as the object is in front of it. In Figure 25.7 the object distance is $d_o$ and the image distance is $d_i$. A ray of light leaves the base of the object, strikes the mirror at an angle of incidence $\theta$, and is reflected at the same angle. To the eye, this ray appears to come from the base of the image. For the angles $\beta_1$ and $\beta_2$ in the drawing it follows that $\theta + \beta_1 = 90°$ and $\alpha + \beta_2 = 90°$. But the angle $\alpha$ is equal to the angle of reflection $\theta$, since the two are opposite angles formed by intersecting lines. Therefore, $\beta_1 = \beta_2$. As a result, triangles *ABC* and *DBC* are identical (congruent) because they share a common side *BC* and have equal angles ($\beta_1 = \beta_2$) at the top and equal angles (90°) at the base. Thus, the magnitude of the object distance $d_o$ equals the magnitude of the image distance $d_i$.

By starting with a light ray from the top of the object, rather than the bottom, we can extend the line of reasoning given above to show also that the height of the image equals the height of the object.

Conceptual Examples 1 and 2 discuss some interesting features of plane mirrors.

**Figure 25.7** This drawing illustrates the geometry used with a plane mirror to show that the image distance $d_i$ equals the object distance $d_o$.

## Conceptual Example 1 Full-Length Versus Half-Length Mirrors

In Figure 25.8 a woman is standing in front of a plane mirror. What is the minimum mirror height necessary for her to see her full image?

**Reasoning and Solution** The mirror is labeled *ABCD* in the drawing and is the same height as the woman. Light emanating from her body is reflected by the mirror, and some of this light enters her eyes. Consider a ray of light from her foot *F*. This ray strikes the mirror at *B* and enters her eyes at *E*. According to the law of reflection, the angles of incidence and reflection are both *θ*. Any light from her foot that strikes the mirror below *B* is reflected toward a point on her body that is below her eyes. Since light striking the mirror below *B* does not enter her eyes, the part of the mirror between *B* and *A* may be removed. The section *DC* of the mirror that produces the image in one-half the woman's height between *F* and *E*. This follows because the right triangles *FBM* and *EBM* are identical. They are identical because they share a common side *BM* and have two angles, *θ* and 90°, that are the same.

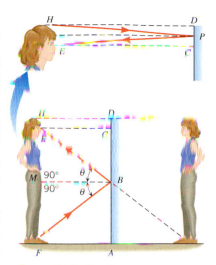

The blowup in Figure 25.8 illustrates a similar line of reasoning, starting with a ray from the woman's head at *H*. This ray is reflected from the mirror at *P* and enters her eyes. The top mirror section *PD* can be removed without disturbing this reflection. The necessary section *CP* is one-half the woman's height between her head at *H* and her eyes at *E*. We find, then, that only the sections *BC* and *CP* are needed for the woman to see her full height. The height of section *BC* plus section *CP* is exactly one-half the woman's height. Thus, *to view one's full length in a mirror, only a half-length mirror is needed.* The conclusions here are valid regardless of how far the person stands from the mirror.

**Related Homework:** *Problem 2*

**Figure 25.8** For the woman to see her full-sized image, only a half-sized mirror is needed.

## Conceptual Example 2 Multiple Reflections

A person is sitting in front of two mirrors that intersect at a 90° angle. As Figure 25.9*a* illustrates, the person sees three images of himself. (The person himself is not shown; only the images are present.) Why are there three images rather than two?

**Reasoning and Solution** Figure 25.9*b* shows a top view of the person in front of the mirrors. It is a straightforward matter to understand two of the images that he sees. These are the images that are normally seen when one stands in front of a mirror. Standing in front of mirror 1, he sees image 1, which is located as far behind that mirror as he is in front of it. He also sees image 2 behind mirror 2, at a distance that matches his distance in front of that mirror. Each of these images arises from light emanating from his body and reflecting from a single mirror. However, it is also possible for light to undergo two reflections in sequence, first from one mirror and then from the other. *When a double reflection occurs, an additional image becomes possible.* Figure 25.9*b* shows two rays of light that strike mirror 1. Each one, according to the law of reflection, has an angle of reflection that equals the angle of incidence. The rays then strike mirror 2, where they again are reflected according to the law of reflection. When the outgoing rays are extended backward (see the dashed lines in the drawing), they intersect and appear to originate from image 3.

**Related Homework:** *Problem 4*

(*a*)

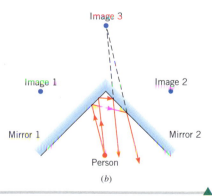

(*b*)

**Figure 25.9** (*a*) These two perpendicular plane mirrors produce three images of the person (not visible) sitting in front of them. (© Dan McCoy/Rainbow) (*b*) A "double" reflection, one from each mirror, gives rise to image 3.

## 25.4 **Spherical Mirrors**

The most common type of curved mirror is a spherical mirror. As Figure 25.10 shows, a spherical mirror has the shape of a section from the surface of a sphere. If the inside surface of the mirror is polished, it is a *concave mirror.* If the outside surface is polished, it

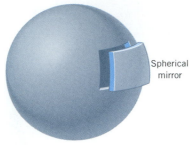

**Figure 25.10** A spherical mirror has the shape of a segment of a spherical surface. The center of curvature is point *C* and the radius is *R*. For a concave mirror, the reflecting surface is the inner one; for a convex mirror it is the outer one.

is a *convex mirror*. The drawing shows both types of mirrors, with a light ray reflecting from the polished surface. The law of reflection applies, just as it does for a plane mirror. For either type of spherical mirror, the normal is drawn perpendicular to the mirror at the point of incidence. For each type, the center of curvature is located at point *C*, and the radius of curvature is *R*. The *principal axis* of the mirror is a straight line drawn through *C* and the midpoint of the mirror.

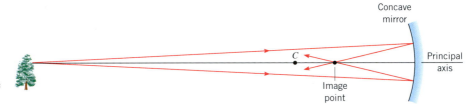

**Figure 25.11** A point on the tree lies on the principal axis of the concave mirror. Rays from this point that are near the principal axis are reflected from the mirror and cross the axis at the image point.

Figure 25.11 shows a tree in front of a concave mirror. A point on this tree lies on the principal axis of the mirror and is beyond the center of curvature *C*. Light rays emanate from this point and reflect from the mirror, consistent with the law of reflection. If the rays are near the principal axis, they cross it at a common point after reflection. This point is called the *image point*. The rays continue to diverge from the image point as if there were an object there. Since light rays actually come from the image point, the image is a real image.

If the tree in Figure 25.11 is infinitely far from the mirror, the rays are parallel to each other and to the principal axis as they approach the mirror. Figure 25.12 shows rays near and parallel to the principal axis, as they reflect from the mirror and pass through an image point. In this special case the image point is referred to as the *focal point F* of the mirror. Therefore, an object infinitely far away on the principal axis gives rise to an image at the focal point of the mirror. The distance between the focal point and the middle of the mirror is the *focal length f* of the mirror.

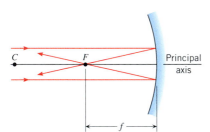

**Figure 25.12** Light rays near and parallel to the principal axis are reflected from a concave mirror and converge at the focal point *F*. The focal length *f* is the distance between *F* and the mirror.

We can show that the focal point *F* lies halfway between the center of curvature *C* and the middle of a concave mirror. In Figure 25.13, a light ray parallel to the principal axis strikes the mirror at point *A*. The line *CA* is the radius of the mirror and, therefore, is the normal to the spherical surface at the point of incidence. The ray reflects from the mirror such that the angle of reflection $\theta$ equals the angle of incidence. Furthermore, the angle *ACF* is also $\theta$ because the radial line *CA* is a transversal of two parallel lines. Since two of its angles are equal, the colored triangle *CAF* is an isosceles triangle; thus, sides *CF* and *FA* are equal. But when the incoming ray lies close to the principal axis, the angle of incidence $\theta$ is small, and the distance *FA* does not differ appreciably from the distance *FB*. Therefore, in the limit that $\theta$ is small, *CF* = *FA* = *FB*, and so the focal point *F* lies halfway between the center of curvature and the mirror. In other words, the focal length *f* is one-half of the radius *R*:

**Focal length of a concave mirror**

$$f = \tfrac{1}{2}R \qquad (25.1)$$

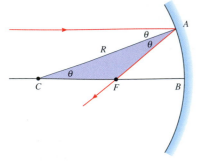

**Figure 25.13** This drawing is used to show that the focal point *F* of a concave mirror is halfway between the center of curvature *C* and the mirror at point *B*.

Rays that lie close to the principal axis are known as *paraxial rays,** and Equation 25.1 is valid only for such rays. Rays that are far from the principal axis do not converge to a single point after reflection from the mirror, as Figure 25.14 shows. The result is a

* Paraxial rays are close to the principal axis but not necessarily parallel to it.

blurred image. The fact that a spherical mirror does not bring all rays parallel to the principal axis to a single image point is known as **spherical aberration.** Spherical aberration can be minimized by using a mirror whose height is small compared to the radius of curvature.

A sharp image point can be obtained with a large mirror, if the mirror is parabolic in shape instead of spherical. The shape of a parabolic mirror is such that all light rays parallel to the principal axis, regardless of their distance from the axis, are reflected through a single image point. However, parabolic mirrors are costly to manufacture and are used where the sharpest images are required, as in research-quality telescopes. Parabolic mirrors are also used in one method of capturing solar energy for commercial purposes. Figure 25.15 shows a long row of concave parabolic mirrors that reflect the sun's rays to the focal point. Located at the focal point and running the length of the row is an oil-filled pipe. The focused rays of the sun heat the oil. In a solar-thermal electric plant, the heat from many such rows is used to generate steam. The steam, in turn, drives a turbine connected to an electric generator. Another application of parabolic mirrors is in automobile headlights. Here, however, the situation is reversed from the operation of a solar collector. In a headlight, a high-intensity light bulb is placed at the focal point of the mirror, and light emerges parallel to the principal axis.

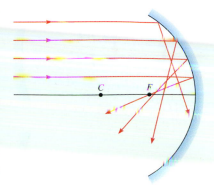

**Figure 25.14** Rays that are farthest from the principal axis have the greatest angle of incidence and miss the focal point *F* after reflection from the mirror.

*The physics of*
**capturing solar energy with mirrors;**
**automobile headlights.**

**Figure 25.15** This long row of parabolic mirrors focuses the sun's rays to heat an oil-filled pipe located at the focal point of each mirror. It is one of many that are used by a solar-thermal electric plant in the Mojave Desert. (© Mastrorillo/Corbis Stock Market)

A convex mirror also has a focal point, and Figure 25.16 illustrates its meaning. In this picture, parallel rays are incident on a convex mirror. Clearly, the rays diverge after being reflected. If the incident parallel rays are paraxial, the reflected rays seem to come from a single point *F* behind the mirror. This point is the focal point of the convex mirror, and its distance from the midpoint of the mirror is the focal length *f*. The focal length of a convex mirror is also one-half of the radius of curvature, just as it is for a concave mirror. However, we assign the focal length of a convex mirror a negative value because it will be convenient later on:

*Focal length of a convex mirror*

$$f = -\tfrac{1}{2}R \qquad (25.2)$$

# 25.5 The Formation of Images by Spherical Mirrors

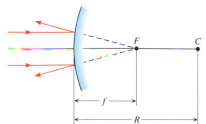

**Figure 25.16** When paraxial light rays that are parallel to the principal axis strike a convex mirror, the reflected rays appear to originate from the focal point *F*. The radius of curvature is *R* and the focal length is *f*.

As we have seen, some of the light rays emitted from an object in front of a mirror strike the mirror, reflect from it, and form an image. We can analyze the image produced by either concave or convex mirrors by using a graphical method called **ray tracing.** This

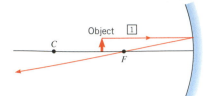

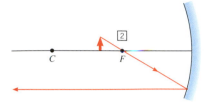

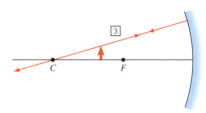

**Figure 25.17** The rays labeled 1, 2, and 3 are useful in locating the image of an object placed in front of a concave spherical mirror. The object is represented as a vertical arrow.

method is based on the law of reflection and the notion that a spherical mirror has a center of curvature $C$ and a focal point $F$. Ray tracing enables us to find the location of the image, as well as its size, by taking advantage of this fact: paraxial rays leave from a point on the object and intersect at a corresponding point on the image after reflection.

## CONCAVE MIRRORS

Three specific paraxial rays are especially convenient to use. Figure 25.17 shows an object in front of a concave mirror, and these three rays leave from a point on the top of the object. The rays are labeled 1, 2, and 3, and when tracing their paths, we use the following conventions.

### Ray Tracing for a Concave Mirror

**Ray 1.** This ray is initially parallel to the principal axis and, therefore, passes through the focal point $F$ after reflection from the mirror.

**Ray 2.** This ray initially passes through the focal point $F$ and is reflected parallel to the principal axis. Ray 2 is analogous to ray 1 except that the reflected, rather than the incident, ray is parallel to the principal axis.

**Ray 3.** This ray travels along a line that passes through the center of curvature $C$ and follows a radius of the spherical mirror; as a result, the ray strikes the mirror perpendicularly and reflects back on itself.

If rays 1, 2, and 3 are superimposed on a scale drawing, they converge at a point on the top of the image, as can be seen in Figure 25.18a.* Although three rays have been used here to locate the image, only two are really needed; the third ray is usually drawn to serve as a check. In a similar fashion, rays from all other points on the object locate corresponding points on the image, and the mirror forms a complete image of the object. If you were to place your eye as shown in the drawing, you would see an image that is *larger* and *inverted* relative to the object. The image is *real* because the light rays actually pass through the image.

If the locations of the object and image in Figure 25.18a are interchanged, the situation in part $b$ of the drawing results. The three rays in part $b$ are the same as those in part $a$, except the directions are reversed. These drawings illustrate the **principle of reversibility,** which states that **if the direction of a light ray is reversed, the light retraces its origi-**

**Problem solving insight**

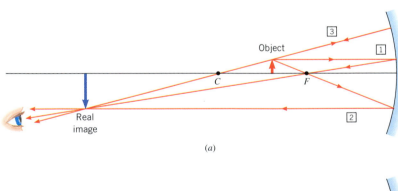

(a)

**Figure 25.18** (a) When an object is placed between the focal point $F$ and the center of curvature $C$ of a concave mirror, a real image is formed. The image is enlarged and inverted relative to the object. (b) When the object is located beyond the center of curvature $C$, a real image is created that is reduced in size and inverted relative to the object.

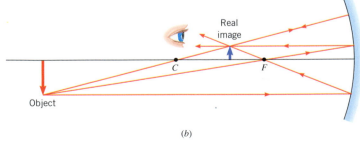

(b)

* In the drawings that follow, we assume that the rays are paraxial, although the distance between the rays and the principal axis is often exaggerated for clarity.

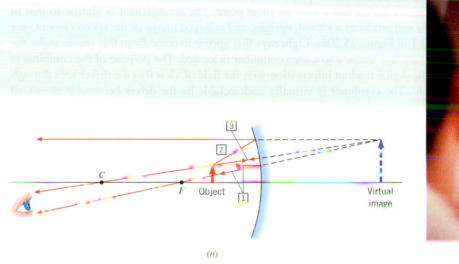

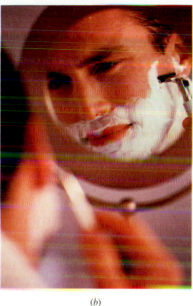

(a)

(b)

**Figure 25.19** (a) When an object is located between a concave mirror and its focal point $F$, an enlarged, upright, and virtual image is produced. (b) A shaving mirror (or makeup mirror) is a concave mirror that can form an enlarged virtual image, as this photograph shows. (© Laurence Monneret/Stone/Getty Images)

*nal path.* This principle is quite general and is not restricted to reflection from mirrors. The image is *real,* and it is *smaller* and *inverted* relative to the object.

When the object is placed between the focal point $F$ and a concave mirror, as in Figure 25.19*a*, three rays can again be drawn to find the image. But now ray 2 does not go through the focal point on its way to the mirror, since the object is inside the focal point. However, when projected backward, ray 2 appears to come from the focal point. Therefore, after reflection, ray 2 is directed parallel to the principal axis. In this case the three reflected rays diverge from each other and do not converge to a common point. However, when projected behind the mirror, the three rays appear to come from a common point; thus, a *virtual* image is formed. This virtual image is *larger* than the object and *upright.* Makeup and shaving mirrors are concave mirrors. When you place your face between the mirror and its focal point, you see an enlarged virtual image of yourself, as part *b* of the drawing shows.

Concave mirrors are also used in one method for displaying the speed of a car. The method presents a digital readout (e.g., "55 mph") that the driver sees when looking directly through the windshield, as in Figure 25.20*a*. The advantage of the method, which is called a

**The physics of** makeup and shaving mirrors.

**The physics of** a head-up display for automobiles.

**Figure 25.20** (a) A head-up display (HUD) presents the driver with a digital readout of the car's speed in the field of view seen through the windshield. (Copyright GM Corporation. All rights reserved.) (b) One version of a HUD uses a concave mirror. (See text for explanation.)

(a)

(b)

head-up display (HUD), is that the driver does not need to take his or her eyes off the road to monitor the speed. Figure 25.20*b* shows how a HUD works. Located below the windshield is a readout device that displays the speed in digital form. This device is located in front of a concave mirror but within its focal point. The arrangement is similar to that in Figure 25.19*a* and produces a virtual, upright, and enlarged image of the speed readout (see virtual image 1 in Figure 25.20*b*). Light rays that appear to come from this image strike the windshield at a place where a so-called combiner is located. The purpose of the combiner is to combine the digital readout information with the field of view that the driver sees through the windshield. The combiner is virtually undetectable by the driver because it allows all colors except one to pass through it unaffected. The one exception is the color produced by the digital readout device. For this color, the combiner behaves as a plane mirror and reflects the light that appears to originate from image 1. Thus, the combiner produces image 2, which is what the driver sees. The location of image 2 is near the front bumper. The driver can then read the speed with his eyes focused just as they are to see the road.

## CONVEX MIRRORS

The procedure for determining the location and size of an image in a convex mirror is similar to that for a concave mirror. The same three rays are used. However, the focal point and center of curvature of a convex mirror lie behind the mirror, not in front of it. Figure 25.21*a* shows the rays. When tracing their paths, we use the following conventions, which take into account the different locations of the focal point and center of curvature.

### Ray Tracing for a Convex Mirror

*Ray 1.* This ray is initially parallel to the principal axis and, therefore, appears to originate from the focal point *F* after reflection from the mirror.

*Ray 2.* This ray heads toward *F*, emerging parallel to the principal axis after reflection. Ray 2 is analogous to ray 1, except that the reflected, rather than the incident, ray is parallel to the principal axis.

*Ray 3.* This ray travels toward the center of curvature *C*; as a result, the ray strikes the mirror perpendicularly and reflects back on itself.

The three rays in Figure 25.21*a* appear to come from a point on a *virtual* image that is behind the mirror. The virtual image is *diminished in size* and *upright,* relative to the object. A convex mirror *always* forms a virtual image of the object, no matter where in

**Figure 25.21** (*a*) An object placed in front of a convex mirror produces a virtual image behind the mirror. The virtual image is reduced in size and upright. (*b*) The sun shield in this pilot's helmet acts as a convex mirror and reflects an image of his plane. (© Chad Slattery/Stone/Getty Images)

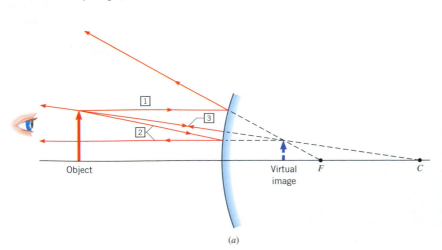

Object   Virtual image   *F*   *C*

(*a*)

(*b*)

front of the mirror the object is placed. Figure 25.21b shows an example of such an image.

Because of its shape, a convex mirror gives a wider field of view than do other types of mirrors. Therefore, they are often used in stores for security purposes. A mirror with a wide field of view is also needed to give a driver a good rear view. Thus, the outside mirror on the passenger side is often a convex mirror. Printed on such a mirror is usually the warning "VEHICLES IN MIRROR ARE CLOSER THAN THEY APPEAR." The reason for the warning is that the virtual image in Figure 25.21a is reduced in size and therefore looks smaller, just as a distant object would appear in a plane mirror. An unwary driver, thinking that the side-view mirror is a plane mirror, might incorrectly deduce from the small size of the image that the car behind is far enough away to ignore.

**The physics of** passenger-side automobile mirrors.

## 25.6 The Mirror Equation and the Magnification Equation

Ray diagrams drawn to scale are useful for determining the location and size of the image formed by a mirror. However, for an accurate description of the image, a more analytical technique is needed. We will now derive two equations, known as the **mirror equation** and the **magnification equation,** that will provide a complete description of the image. Note that these equations are based on the law of reflection and provide relationships between:

$f$ = the focal length of the mirror
$d_o$ = the object distance, which is the distance between the object and the mirror
$d_i$ = the image distance, which is the distance between the image and the mirror
$m$ = the magnification of the mirror, which is the ratio of the height of the image to the height of the object.

### CONCAVE MIRRORS

We begin our derivation of the mirror equation by referring to Figure 25.22a, which shows a ray leaving the top of the object and striking a concave mirror at the point where the principal axis intersects the mirror. Since the principal axis is perpendicular to the mirror, it is also the normal at this point of incidence. Therefore, the ray reflects at an equal angle and passes through the image. The two colored triangles are similar triangles because they have equal angles, so

$$\frac{h_o}{-h_i} = \frac{d_o}{d_i}$$

where $h_o$ is the height of the object and $h_i$ is the height of the image. The minus sign appears on the left in this equation because the image is inverted in Figure 25.22a. In part b another ray leaves the top of the object, this one passing through the focal point $F$, reflecting parallel to the principal axis, and then passing through the image. Provided the ray remains close to the axis, the two colored areas can be considered to be similar triangles, with the result that

$$\frac{h_o}{-h_i} = \frac{d_o - f}{f}$$

Setting the two equations above equal to each other yields $d_o/d_i = (d_o - f)/f$. Rearranging this result gives the **mirror equation:**

**Mirror equation**

$$\frac{1}{d_o} + \frac{1}{d_i} = \frac{1}{f} \tag{25.3}$$

We have derived this equation for a real image formed in front of a concave mirror. In this case, the image distance is a positive quantity, as are the object distance and the focal length. However, we have seen in the last section that a concave mirror can also form a virtual image, if the object is located between the focal point and the mirror.

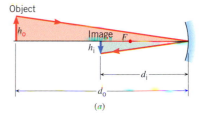

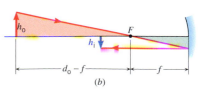

**Figure 25.22** These diagrams are used to derive the mirror equation and the magnification equation. (a) The two colored triangles are similar triangles. (b) If the ray is close to the principal axis, the two colored regions are almost similar triangles.

Equation 25.3 can also be applied to such a situation, provided that we adopt the convention that $d_i$ is negative for an image behind the mirror, as it is for a virtual image.

In deriving the magnification equation, we remember that the **magnification** $m$ of a mirror is the ratio of the image height to the object height: $m = h_i/h_o$. If the image height is less than the object height, the magnitude of $m$ is less than one. Conversely, if the image is larger than the object, the magnitude of $m$ is greater than one. We have already shown that $h_o/(-h_i) = d_o/d_i$, so it follows that

**Magnification equation**

$$m = \frac{\text{Image height, } h_i}{\text{Object height, } h_o} = -\frac{d_i}{d_o} \qquad (25.4)$$

As Examples 3 and 4 show, the value of $m$ is negative if the image is inverted and positive if the image is upright.

## Example 3   A Real Image Formed by a Concave Mirror

▼

A 2.0-cm-high object is placed 7.10 cm from a concave mirror whose radius of curvature is 10.20 cm. Find (a) the location of the image and (b) its size.

**Reasoning**  Since $f = \frac{1}{2}R = \frac{1}{2}(10.20 \text{ cm}) = 5.10$ cm, the object is located between the focal point $F$ and the center of curvature $C$ of the mirror, as in Figure 25.18a. Based on this figure, we expect that the image is real and that, relative to the object, it is farther away from the mirror, inverted, and larger.

**Solution**

**(a)** With $d_o = 7.10$ cm and $f = 5.10$ cm, the mirror equation (Equation 25.3) can be used to find the image distance:

**Problem solving insight**
According to the mirror equation, the image distance $d_i$ has a reciprocal given by $d_i^{-1} = f^{-1} - d_o^{-1}$. After combining the reciprocals $f^{-1}$ and $d_o^{-1}$, do not forget to take the reciprocal of the result to find $d_i$.

$$\frac{1}{d_i} = \frac{1}{f} - \frac{1}{d_o} = \frac{1}{5.10 \text{ cm}} - \frac{1}{7.10 \text{ cm}} = 0.055 \text{ cm}^{-1} \quad \text{or} \quad \boxed{d_i = 18 \text{ cm}}$$

In this calculation, $f$ and $d_o$ are positive numbers, indicating that the focal point and the object are in front of the mirror. The positive answer for $d_i$ means that the image is also in front of the mirror, and the reflected rays actually pass through the image, as Figure 25.18a shows. In other words, the positive value for $d_i$ indicates that the image is a real image.

**(b)** The height of the image can be determined once the magnification $m$ of the mirror is known. The magnification equation (Equation 25.4) can be used to find $m$:

$$m = -\frac{d_i}{d_o} = -\frac{18 \text{ cm}}{7.10 \text{ cm}} = -2.5$$

The image height is $h_i = mh_o = (-2.5)(2.0 \text{ cm}) = \boxed{-5.0 \text{ cm}}$. The image is 2.5 times larger than the object, the negative values for $m$ and $h_i$ indicating that the image is inverted with respect to the object, as in Figure 25.18a.

▲

## Example 4   A Virtual Image Formed by a Concave Mirror

▼

An object is placed 6.00 cm in front of a concave mirror that has a 10.0-cm focal length. (a) Determine the location of the image. (b) The object is 1.2 cm high. Find the image height.

**Reasoning**  The object is located between the focal point and the mirror, as in Figure 25.19a. The setup is analogous to a person using a makeup or shaving mirror. Therefore, we expect that the image is virtual and that, relative to the object, it is upright and larger.

**Solution**

**(a)** Using the mirror equation with $d_o = 6.00$ cm and $f = 10.0$ cm, we have

$$\frac{1}{d_i} = \frac{1}{f} - \frac{1}{d_o} = \frac{1}{10.0 \text{ cm}} - \frac{1}{6.00 \text{ cm}} = -0.067 \text{ cm}^{-1} \quad \text{or} \quad \boxed{d_i = -15 \text{ cm}}$$

The answer for $d_i$ is negative, indicating that the image is *behind* the mirror. Thus, as expected, the image is virtual.

**(b)** The image height $h_i$ can be found from the magnification $m$ and the object height $h_o$:

$$m = -\frac{d_i}{d_o} = -\frac{(-15\text{ cm})}{6.00\text{ cm}} = 2.5$$

The image height is $h_i = mh_o = (2.5)(1.2\text{ cm}) = \boxed{3.0\text{ cm}}$. The image is larger than the object, and the positive values for $m$ and $h_i$ indicate that the image is upright (see Figure 25.19a).

## CONVEX MIRRORS

The mirror equation and the magnification equation can also be used with convex mirrors, provided the focal length $f$ is taken to be a *negative number,* as indicated earlier in Equation 25.2. One way to remember this is to recall that the focal point of a convex mirror lies *behind* the mirror. Example 5 deals with a convex mirror.

### Example 5  A Virtual Image Formed by a Convex Mirror

A convex mirror is used to reflect light from an object placed 66 cm in front of the mirror. The focal length of the mirror is $f = -46$ cm (note the minus sign). Find (a) the location of the image and (b) the magnification.

*Problem solving Insight*
When using the mirror equation, it is useful to construct a ray diagram to guide your thinking and to check your calculation.

**Reasoning** We have seen that a convex mirror always forms a virtual image, as in Figure 25.21a, where the image is upright and smaller than the object. These characteristics should also be indicated by the results of our analysis here.

**Solution**

**(a)** With $d_o = 66$ cm and $f = -46$ cm, the mirror equation gives

$$\frac{1}{d_i} = \frac{1}{f} - \frac{1}{d_o} = \frac{1}{-46\text{ cm}} - \frac{1}{66\text{ cm}} = -0.037\text{ cm}^{-1} \quad \text{or} \quad \boxed{d_i = -27\text{ cm}}$$

The negative sign for $d_i$ indicates that the image is behind the mirror and, therefore, is a virtual image.

**(b)** According to the magnification equation, the magnification is

$$m = -\frac{d_i}{d_o} = -\frac{(-27\text{ cm})}{66\text{ cm}} = \boxed{0.41}$$

The image is smaller ($m$ is less than one) and upright ($m$ is positive) with respect to the object.

Convex mirrors, like plane (flat) mirrors, always produce virtual images behind the mirror. However, the virtual image in a convex mirror is closer to the mirror than it would be if the mirror were planar, as Example 6 illustrates.

### Example 6  A Convex Versus a Plane Mirror

An object is placed 9.00 cm in front of a mirror. The image is 3.00 cm closer to the mirror when the mirror is convex than when it is planar. Find the focal length of the convex mirror.

**Reasoning** For a plane mirror, the image and the object are the same distance on either side of the mirror. Thus, the image would be 9.00 cm behind a plane mirror. If the image in a convex mirror is 3.00 cm closer than this, the image must be located 6.00 cm behind the convex mirror. In other words, when the object distance is $d_o = 9.00$ cm, the image distance for the convex mirror is $d_i = -6.00$ cm (negative because the image is virtual). The mirror equation can be used to find the focal length of the mirror.

**Solution** According to the mirror equation, the reciprocal of the focal length is

$$\frac{1}{f} = \frac{1}{d_o} + \frac{1}{d_i} = \frac{1}{9.00\text{ cm}} + \frac{1}{-6.00\text{ cm}} = -0.056\text{ cm}^{-1} \quad \text{or} \quad \boxed{f = -18\text{ cm}}$$

Below is a summary of the sign conventions that are used with the mirror equation and the magnification equation. These conventions apply to both concave and convex mirrors.

**Summary of Sign Conventions for Spherical Mirrors**

*Focal length*

$f$ is $+$ for a concave mirror.

$f$ is $-$ for a convex mirror.

*Object distance*

$d_o$ is $+$ if the object is in front of the mirror (real object).

$d_o$ is $-$ if the object is behind the mirror (virtual object).*

*Image distance*

$d_i$ is $+$ if the image is in front of the mirror (real image).

$d_i$ is $-$ if the image is behind the mirror (virtual image).

*Magnification*

$m$ is $+$ for an image that is upright with respect to the object.

$m$ is $-$ for an image that is inverted with respect to the object.

* Sometimes optical systems use two (or more) mirrors, and the image formed by the first mirror serves as the object for the second mirror. Occasionally, such an object falls *behind* the second mirror. In this case the object distance is negative, and the object is said to be a virtual object.

# Concept Summary

This summary presents an abridged version of the chapter, including the important equations and all available learning aids. For convenient reference, the learning aids (including the text's examples) are placed next to or immediately after the relevant equation or discussion. The following learning aids may be found on-line at **www.wiley.com/college/cutnell**:

| | |
|---|---|
| **Interactive LearningWare** examples are solved according to a five-step interactive format that is designed to help you develop problem-solving skills. | **Concept Simulations** are animated versions of text figures or animations that illustrate important concepts. You can control parameters that affect the display, and we encourage you to experiment. |
| **Interactive Solutions** offer specific models for certain types of problems in the chapter homework. The calculations are carried out interactively. | **Self-Assessment Tests** include both qualitative and quantitative questions. Extensive feedback is provided for both incorrect and correct answers, to help you evaluate your understanding of the material. |

| Topic | Discussion | Learning Aids |
|---|---|---|
| | **25.1 Wave Fronts and Rays** | |
| Wave fronts<br>Plane waves | Wave fronts are surfaces on which all points of a wave are in the same phase of motion. Waves whose wave fronts are flat surfaces are known as plane waves. | |
| Rays | Rays are lines that are perpendicular to the wave fronts and point in the direction of the velocity of the wave. | |
| | **25.2 The Reflection of Light** | |
| Law of reflection | When light reflects from a smooth surface, the reflected light obeys the law of reflection:<br><br>a. The incident ray, the reflected ray, and the normal to the surface all lie in the same plane.<br>b. The angle of reflection $\theta_r$ equals the angle of incidence $\theta_i$; $\theta_r = \theta_i$. | |
| | **25.3 The Formation of Images by a Plane Mirror** | |
| Virtual image | A virtual image is one from which all the rays of light do not actually come, but only appear to do so. | |
| Real image | A real image is one from which all the rays of light actually do emanate. | |
| Plane mirror | A plane mirror forms an upright, virtual image that is located as far behind the mirror as the object is in front of it. In addition, the heights of the image and the object are equal. | Concept Simulation 25.1<br>Examples 1, 2 |

 **Use *Self Assessment Test 25.1* to evaluate your understanding of Sections 25.1–25.3.**

| Topic | Discussion | Learning Aids |
|---|---|---|

### 25.4 Spherical Mirrors

**Concave and convex mirrors**

A spherical mirror has the shape of a section from the surface of a sphere. If the inside surface of the mirror is polished, it is a concave mirror. If the outside surface is polished, it is a convex mirror.

**Principal axis**

The principal axis of a mirror is a straight line drawn through the center of curvature and the middle of the mirror's surface. Rays that are close to the princi-

**Paraxial rays**

pal axis are known as paraxial rays. Paraxial rays are not necessarily parallel to the principal axis.

**Radius of curvature**

The radius of curvature $R$ of a mirror is the distance from the center of curvature to the mirror.

**Focal point of a concave mirror**

The focal point of a concave spherical mirror is a point on the principal axis, in front of the mirror. Incident paraxial rays that are parallel to the principal axis converge to the focal point after being reflected from the concave mirror.

**Spherical aberration**

The fact that a spherical mirror does not bring all rays parallel to the principal axis to a single image point after reflection is known as spherical aberration.

**Focal point of a convex mirror**

The focal point of a convex spherical mirror is a point on the principal axis, behind the mirror. For a convex mirror, incident paraxial rays that are parallel to the principal axis diverge after reflecting from the mirror. These rays seem to originate from the focal point.

**Focal length**

The focal length $f$ of a mirror is the distance along the principal axis between the focal point and the mirror. The focal length and the radius of curvature $R$ are related by

**Concave mirror**

$$f = \tfrac{1}{2}R \qquad (25.1)$$

**Convex mirror**

$$f = -\tfrac{1}{2}R \qquad (25.2)$$

### 25.5 The Formation of Images by Spherical Mirrors

**Ray tracing**

The image produced by a mirror can be located by a graphical method known as ray tracing.

For a concave mirror, the following paraxial rays are especially useful for ray tracing (see Figure 25.17): **Concept Simulation 25.2**

**Three rays for a concave mirror**

**Ray 1.** This ray leaves the object traveling parallel to the principal axis. The ray reflects from the mirror and passes through the focal point.

**Ray 2.** This ray leaves the object and passes through the focal point. The ray reflects from the mirror and travels parallel to the principal axis.

**Ray 3.** This ray leaves the object and travels along a line that passes through the center of curvature. The ray strikes the mirror perpendicularly and reflects back on itself.

For a convex mirror, these paraxial rays are useful for ray tracing (see Figure **Concept Simulation 25.3** 25.21a):

**Three rays for a convex mirror**

**Ray 1.** This ray leaves the object traveling parallel to the principal axis. After reflection from the mirror, the ray appears to originate from the focal point of the mirror.

**Ray 2.** This ray leaves the object and heads toward the focal point. After reflection, the ray travels parallel to the principal axis.

**Ray 3.** This ray leaves the object and travels toward the center of curvature. The ray strikes the mirror perpendicularly and reflects back on itself.

### 25.6 The Mirror Equation and the Magnification Equation

The mirror equation specifies the relation between the object distance $d_o$, the image distance $d_i$, and the focal length $f$ of the mirror:

**Examples 3, 4**

**Mirror equation**

$$\frac{1}{d_o} + \frac{1}{d_i} = \frac{1}{f} \qquad (25.3)$$

**Interactive LearningWare 25.1**

The mirror equation can be used with either concave or convex mirrors.

| Topic | Discussion | Learning Aids |
|---|---|---|
| **Magnification** | The magnification $m$ of a mirror is the ratio of the image height $h_i$ to the object height $h_o$: $$m = \frac{h_i}{h_o}$$ The magnification is also related to $d_i$ and $d_o$ by the magnification equation: | Examples **5, 6**<br><br>**Interactive Solution 25.19**<br><br>**Interactive Solution 25.25** |
| **Magnification equation** | $$m = -\frac{d_i}{d_o} \qquad (25.4)$$ The algebraic sign conventions for the variables appearing in these equations are summarized at the end of Section 25.6. | |

 **Use Self-Assessment Test 25.2 to evaluate your understanding of Sections 25.4–25.6.**

# Problems

**Section 25.2 The Reflection of Light,**
**Section 25.3 The Formation of Images by a Plane Mirror**

**1. ssm** Two plane mirrors are separated by 120°, as the drawing illustrates. If a ray strikes mirror $M_1$ at a 65° angle of incidence, at what angle $\theta$ does it leave mirror $M_2$?

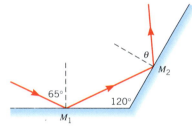

*Problem 1*

**2.** Review Conceptual Example 1 before attempting this problem. A person whose eyes are 1.70 m above the floor stands in front of a plane mirror. The top of her head is 0.12 m above her eyes. (a) What is the height of the shortest mirror in which she can see her entire image? (b) How far above the floor should the bottom edge of the mirror be placed?

**3.** A person stands 3.6 m in front of a wall that is covered floor-to-ceiling with a plane mirror. His eyes are 1.8 m above the floor. He holds a flashlight between his feet and manages to point it at the mirror. At what angle of incidence must the light strike the mirror so the light will reach his eyes?

**4.** Review Conceptual Example 2. Suppose that in Figure 25.9*b* the two perpendicular plane mirrors are represented by the $-x$ and $-y$ axes of an $x, y$ coordinate system. An object is in front of these mirrors at a point whose coordinates are $x = -2.0$ m and $y = -1.0$ m. Find the coordinates that locate each of the three images.

**5. ssm www** Two diverging light rays, originating from the same point, have an angle of 10° between them. After the rays reflect from a plane mirror, what is the angle between them? Construct one possible ray diagram that supports your answer.

**6.** You are trying to photograph a bird sitting on a tree branch, but a tall hedge is blocking your view. However, as the drawing shows, a plane mirror reflects light from the bird into your camera. For what distance must you set the focus of the camera lens in order to snap a sharp picture of the bird's image?

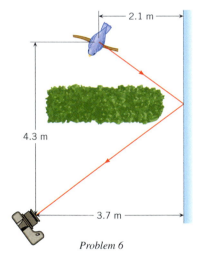

*Problem 6*

**\*7.** The drawing shows a top view of a square room. One wall is missing, and the other three are each mirrors. From point $P$ in the center of the open side, a laser is fired, with the intent of hitting a small target located at the center of one wall. Identify six directions in which the laser can be fired and score a hit, assuming that the light does not strike any mirror more than once. Draw the rays to confirm your choices.

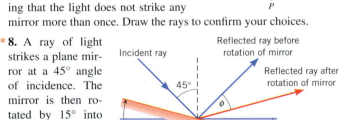

**\*8.** A ray of light strikes a plane mirror at a 45° angle of incidence. The mirror is then rotated by 15° into the position shown in red in the drawing, while the incident ray is kept fixed. (a) Through what angle $\phi$ does the reflected

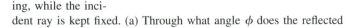

ray rotate? (b) What is the answer to part (a) if the angle of incidence is 60° instead of 45°?

** **9.** In the drawing for problem 7, a laser is fired from point $P$ in the center of the open side of the square room. The laser is pointed at the mirrored wall on the right. At what angle of incidence must the light strike the right-hand wall so that, after being reflected, the light hits the left corner of the back wall?

## Section 25.4 Spherical Mirrors,
## Section 25.5 The Formation of Images by Spherical Mirrors

**10. Concept Simulation 25.2** at **www.wiley.com/college/cutnell** illustrates the concepts pertinent to this problem. A 2.0-cm-high object is situated 15.0 cm in front of a concave mirror that has a radius of curvature of 10.0 cm. Using a ray diagram drawn to scale, measure (a) the location and (b) the height of the image. The mirror must be drawn to scale.

**11.** Repeat problem 10 for a concave mirror with a focal length of 20.0 cm, an object distance of 12.0 cm, and a 2.0 cm-high object.

**12.** The image of a very distant car is located 12 cm behind a convex mirror. (a) What is the radius of curvature of the mirror? (b) Draw a ray diagram to scale showing this situation.

**13. ssm** Repeat problem 10 for a convex mirror with a radius of curvature of $1.00 \times 10^2$ cm, an object distance of 25.0 cm, and a 10.0-cm-high object.

**14.** An object is placed in front of a convex mirror. Draw a convex mirror (radius of curvature = 15 cm) to scale, and place an object 25 cm in front of it. Make the object height 4 cm. Using a ray diagram, locate the image and measure its height. Now move the object closer to the mirror, so the object distance is 5 cm. Again, locate its image using a ray diagram. As the object moves closer to the mirror, (a) does the magnitude of the image distance become larger or smaller, and (b) does the magnitude of the image height become larger or smaller? (c) What is the ratio of the image height when the object distance is 5 cm to that when the object distance is 25 cm? Give your answer to one significant figure.

* **15.** A plane mirror and a concave mirror ($f = 8.0$ cm) are facing each other and are separated by a distance of 20.0 cm. An object is placed 10.0 cm in front of the plane mirror. Consider the light from the object that reflects first from the plane mirror and then from the concave mirror. Using a ray diagram drawn to scale, find the location of the image that this light produces in the concave mirror. Specify this distance relative to the concave mirror.

## Section 25.6 The Mirror Equation and
## the Magnification Equation

**16.** A coin is placed 8.0 cm in front of a concave mirror. The mirror produces a real image that has a diameter 4.0 times larger than that of the coin. What is the image distance?

**17. ssm** A concave mirror has a focal length of 42 cm. The image formed by this mirror is 97 cm in front of the mirror. What is the object distance?

**18.** The focal length of a concave mirror is 17 cm. An object is located 38 cm in front of this mirror. Where is the image located?

**19.** Refer to **Interactive Solution 25.19** at **www.wiley.com/college/cutnell** for help with problems like this one. A concave mirror ($R = 56.0$ cm) is used to project a transparent slide onto a wall. The slide is located at a distance of 31.0 cm from the mirror, and a small flashlight shines light through the slide and onto the mirror. The setup is similar to that in Figure 25.18a. (a) How far from the wall should the mirror be located? (b) The height of the object on the slide is 0.95 cm. What is the height of the image? (c) How should the slide be oriented, so that the picture on the wall looks normal?

**20.** An object that is 25 cm in front of a convex mirror has an image located 17 cm behind the mirror. How far behind the mirror is the image located when the object is 19 cm in front of the mirror?

**21. ssm** The image produced by a concave mirror is located 26 cm in front of the mirror. The focal length of the mirror is 12 cm. How far in front of the mirror is the object located?

**22.** A concave mirror ($f = 45$ cm) produces an image whose distance from the mirror is one-third the object distance. Determine (a) the object distance and (b) the (positive) image distance.

**23. ssm www** When viewed in a spherical mirror, the image of a setting sun is a virtual image. The image lies 12.0 cm behind the mirror. (a) Is the mirror concave or convex? Why? (b) What is the radius of curvature of the mirror?

* **24.** A candle is placed 15.0 cm in front of a convex mirror. When the convex mirror is replaced with a plane mirror, the image moves 7.0 cm farther away from the mirror. Find the focal length of the convex mirror.

* **25.** Consult **Interactive Solution 25.25** at **www.wiley.com/college/cutnell** for insight into this problem. An object is placed in front of a convex mirror, and the size of the image is one-fourth that of the object. What is the ratio $d_o/f$ of the object distance to the focal length of the mirror?

* **26.** The same object is located at the same distance from two spherical mirrors, A and B. The magnifications produced by the mirrors are $m_A = 4.0$ and $m_B = 2.0$. Find the ratio $f_A/f_B$ of the focal lengths of the mirrors.

* **27. ssm** An object is located 14.0 cm in front of a convex mirror, the image being 7.00 cm behind the mirror. A second object, twice as tall as the first one, is placed in front of the mirror, but at a different location. The image of this second object has the same height as the other image. How far in front of the mirror is the second object located?

** **28.** A spherical mirror is polished on both sides. When the convex side is used as a mirror, the magnification is $+1/4$. What is the magnification when the concave side is used as a mirror, the object remaining the same distance from the mirror?

** **29.** Using the mirror equation and the magnification equation, show that for a convex mirror the image is always (a) virtual (i.e., $d_i$ is always negative) and (b) upright and smaller, relative to the object (i.e., $m$ is positive and less than one).

# Chapter 26 The Refraction of Light: Lenses and Optical Instruments

**Table 26.1** *Index of Refraction*[a] *for Various Substances*

| Substance | Index of Refraction, $n$ |
|---|---|
| **Solids at 20 °C** | |
| Diamond | 2.419 |
| Glass, crown | 1.523 |
| Ice (0 °C) | 1.309 |
| Sodium chloride | 1.544 |
| Quartz | |
|    Crystalline | 1.544 |
|    Fused | 1.458 |
| **Liquids at 20 °C** | |
| Benzene | 1.501 |
| Carbon disulfide | 1.632 |
| Carbon tetrachloride | 1.461 |
| Ethyl alcohol | 1.362 |
| Water | 1.333 |
| **Gases at 0 °C, 1 atm** | |
| Air | 1.000 293 |
| Carbon dioxide | 1.000 45 |
| Oxygen, $O_2$ | 1.000 271 |
| Hydrogen, $H_2$ | 1.000 139 |

[a] Measured with light whose wavelength in a vacuum is 589 nm.

## 26.1 The Index of Refraction

As Section 24.3 discusses, light travels through a vacuum at a speed $c = 3.00 \times 10^8$ m/s. It can also travel through many materials, such as air, water, and glass. Atoms in the material absorb, reemit, and scatter the light, however. Therefore, light travels through the material at a speed that is less than $c$, the actual speed depending on the nature of the material. In general, we will see that the change in speed as a ray of light goes from one material to another causes the ray to deviate from its incident direction. This change in direction is called *refraction,* and it is governed by Snell's law of refraction, which will be discussed in the next section.

To describe the extent to which the speed of light in a material medium differs from that in a vacuum, we use a parameter called the *index of refraction* (or *refractive index*). The index of refraction incorporates both the speed of light $c$ in a vacuum and the speed $v$ in the material. The index of refraction is an important parameter because it appears in Snell's law of refraction, the basis of all the phenomena discussed in this chapter.

> ■ **DEFINITION OF THE INDEX OF REFRACTION**
>
> The index of refraction $n$ of a material is the ratio of the speed $c$ of light in a vacuum to the speed $v$ of light in the material:
>
> $$n = \frac{\text{Speed of light in a vacuum}}{\text{Speed of light in the material}} = \frac{c}{v} \qquad (26.1)$$

Table 26.1 lists the refractive indices for some common substances. The values of $n$ are greater than unity because the speed of light in a material medium is less than it is in a vacuum. For example, the index of refraction for diamond is $n = 2.419$, so the speed of light in diamond is $v = c/n = (3.00 \times 10^8 \text{ m/s})/2.419 = 1.24 \times 10^8$ m/s. In contrast, the index of refraction for air (and also for other gases) is so close to unity that $n_{\text{air}} = 1$ for most purposes. The index of refraction depends slightly on the wavelength of the light, and the values in Table 26.1 correspond to a wavelength of $\lambda = 589$ nm in a vacuum.

## 26.2 Snell's Law and the Refraction of Light

### SNELL'S LAW

When light strikes the interface between two transparent materials, such as air and water, the light generally divides into two parts, as Figure 26.1a illustrates. Part of the light is reflected, with the angle of reflection equaling the angle of incidence. The remainder is transmitted across the interface. If the incident ray does not strike the interface at normal incidence, the transmitted ray has a different direction than the incident ray. The ray that enters the second material is said to be refracted.

In Figure 26.1a the light travels from a medium where the refractive index is smaller (air) into a medium where it is larger (water), and the refracted ray is bent *toward* the normal. Both the incident and refracted rays obey the principle of reversibility, so their directions can be reversed to give a situation like that in part b of the drawing. Here light travels from a material with a greater refractive index (water) into one with a smaller refractive index (air), and the refracted ray is bent *away* from the normal. In this case the reflected ray lies in the water rather than in the air. In both parts of the drawing the angles of incidence, refraction, and reflection are measured relative to the normal. Note that the index of refraction of air is labeled $n_1$ in part *a*, whereas it is $n_2$ in part *b*, because *we label all variables associated with the incident (and reflected) ray with subscript 1 and all variables associated with the refracted ray with subscript 2.*

The angle of refraction $\theta_2$ depends on the angle of incidence $\theta_1$ and on the indices of refraction, $n_2$ and $n_1$, of the two media. The relation between these quantities is known as **Snell's law of refraction,** after the Dutch mathematician Willebrord Snell (1591–1626) who discovered it experimentally. At the end of this section is a proof of Snell's law.

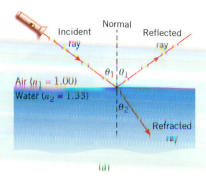

■ **SNELL'S LAW OF REFRACTION**

When light travels from a material with refractive index $n_1$ into a material with refractive index $n_2$, the refracted ray, the incident ray, and the normal to the interface between the materials all lie in the same plane. The angle of refraction $\theta_2$ is related to the angle of incidence $\theta_1$ by

$$n_1 \sin \theta_1 = n_2 \sin \theta_2 \qquad (26.2)$$

Example 1 illustrates the use of Snell's law.

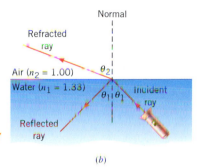

(b)

**Figure 26.1** (*a*) When a ray of light is directed from air into water, part of the light is reflected at the interface and the remainder is refracted into the water. The refracted ray is bent *toward* the normal ($\theta_2 < \theta_1$). (*b*) When a ray of light is directed from water into air, the refracted ray in air is bent *away* from the normal ($\theta_2 > \theta_1$).

## Example 1  Determining the Angle of Refraction

A light ray strikes an air/water surface at an angle of 46° with respect to the normal. The refractive index for water is 1.33. Find the angle of refraction when the direction of the ray is (a) from air to water and (b) from water to air.

**Reasoning** Snell's law of refraction applies to both part (a) and part (b). However, in part (a) the incident ray is in air, whereas in part (b) it is in water. We keep track of this difference by always labeling the variables associated with the incident ray with a subscript 1 and the variables associated with the refracted ray with a subscript 2.

**Solution**

(a) The incident ray is in air, so $\theta_1 = 46°$ and $n_1 = 1.00$. The refracted ray is in water, so $n_2 = 1.33$. Snell's law can be used to find the angle of refraction $\theta_2$:

$$\sin \theta_2 = \frac{n_1 \sin \theta_1}{n_2} = \frac{(1.00) \sin 46°}{1.33} = 0.54 \qquad (26.2)$$

$$\theta_2 = \sin^{-1}(0.54) = \boxed{33°}$$

Since $\theta_2$ is less than $\theta_1$, the refracted ray is bent *toward* the normal, as Figure 26.1*a* shows.

(b) With the incident ray in water, we find that

$$\sin \theta_2 = \frac{n_1 \sin \theta_1}{n_2} = \frac{(1.33) \sin 46°}{1.00} = 0.96$$

$$\theta_2 = \sin^{-1}(0.96) = \boxed{74°}$$

Since $\theta_2$ is greater than $\theta_1$, the refracted ray is bent *away* from the normal, as in Figure 26.1*b*.

**Problem solving insight**
The angle of incidence $\theta_1$ and the angle of refraction $\theta_2$ that appear in Snell's law are measured with respect to the normal to the surface, and not with respect to the surface itself.

We have seen that reflection and refraction of light waves occur simultaneously at the interface between two transparent materials. It is important to keep in mind that light waves are composed of electric and magnetic fields, which carry energy. The principle of conservation of energy (see Chapter 6) indicates that the energy reflected plus the energy refracted must add up to equal the energy carried by the incident light, provided that none of the energy is absorbed by the materials. The percentage of incident energy that appears as reflected versus refracted light depends on the angle of incidence and the refractive indices of the materials on either side of the interface. For instance, when light travels from air toward water at perpendicular incidence, most of its energy is refracted and little is reflected. But when the angle of incidence is nearly 90° and the light barely grazes the water surface, most of its energy is reflected, with only a small amount refracted into the water. On a rainy night, you probably have experienced the annoying glare that results when light from an oncoming car just grazes the wet road. Under such conditions, most of the light energy reflects into your eyes.

The simultaneous reflection and refraction of light have applications in a number of devices. For instance, interior rearview mirrors in cars often have adjustment levers. One

**The physics of**
rearview mirrors that have a day–night adjustment.

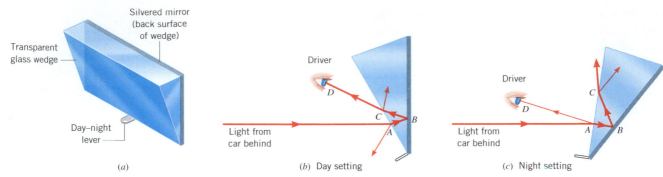

(a)    (b) Day setting    (c) Night setting

**Figure 26.2** An interior rearview mirror with a day–night adjustment lever.

position of the lever is for day driving, while the other is for night driving and reduces glare from the headlights of the car behind. As Figure 26.2a indicates, this kind of mirror is a glass wedge, the back side of which is silvered and highly reflecting. Part b of the picture shows the day setting. Light from the car behind follows path *ABCD* in reaching the driver's eye. At points *A* and *C*, where the light strikes the air–glass surface, there are both reflected and refracted rays. The reflected rays are drawn as thin lines, the thinness denoting that only a small percentage (about 10%) of the light during the day is reflected at *A* and *C*. The weak reflected rays at *A* and *C* do not reach the driver's eye. In contrast, almost all the light reaching the silvered back surface at *B* is reflected toward the driver. Since most of the light follows path *ABCD*, the driver sees a bright image of the car behind. During the night, the adjustment lever can be used to rotate the top of the mirror away from the driver (see part c of the drawing). Now, most of the light from the headlights behind follows path *ABC* and does not reach the driver. Only the light that is weakly reflected from the front surface along path *AD* is seen, and the result is significantly less glare.

## APPARENT DEPTH

One interesting consequence of refraction is that an object lying under water appears to be closer to the surface than it actually is. Example 2 sets the stage for explaining why, by showing what must be done to shine a light on such an object.

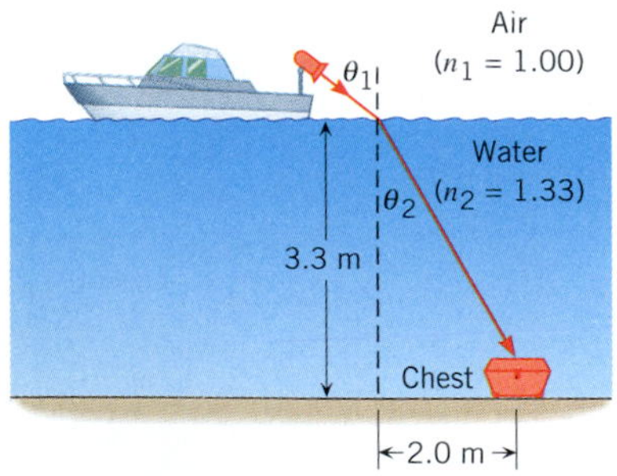

**Figure 26.3** The beam from the searchlight is refracted when it enters the water.

### Example 2   Finding a Sunken Chest

A searchlight on a yacht is being used at night to illuminate a sunken chest, as in Figure 26.3. At what angle of incidence $\theta_1$ should the light be aimed?

**Reasoning** The angle of incidence $\theta_1$ must be such that, after refraction, the light strikes the chest. The angle of incidence can be obtained from Snell's law, once the angle of refraction $\theta_2$ is determined. This angle can be found using the data in Figure 26.3 and trigonometry. The light travels from a region of lower into a region of higher refractive index, so the light is bent toward the normal and we expect $\theta_1$ to be greater than $\theta_2$.

**Solution** From the data in the drawing it follows that $\tan \theta_2 = (2.0 \text{ m})/(3.3 \text{ m})$, so $\theta_2 = 31°$. With $n_1 = 1.00$ for air and $n_2 = 1.33$ for water, Snell's law gives

$$\sin \theta_1 = \frac{n_2 \sin \theta_2}{n_1} = \frac{(1.33) \sin 31°}{1.00} = 0.69$$

$$\theta_1 = \sin^{-1}(0.69) = \boxed{44°}$$

As expected, $\theta_1$ is greater than $\theta_2$.

**Problem solving insight**
Remember that the refractive indices are written as $n_1$ for the medium in which the incident light travels and $n_2$ for the medium in which the refracted light travels.

When the sunken chest in Example 2 is viewed from the boat (Figure 26.4a), light rays from the chest pass upward through the water, refract away from the normal when they enter the air, and then travel to the observer. This picture is similar to Figure 26.3, except the direction of the rays is reversed and the searchlight is replaced by an observer. When the rays entering the air are extended back into the water (see the dashed lines), they indicate that the observer sees a virtual image of the chest at an *apparent depth* that is less than the actual depth. The image is virtual because light rays do not actually pass through it. For the situation shown in Figure 26.4a, it is difficult to determine the apparent depth, A case that is much simpler is shown in part b of the drawing, where the observer is *directly above* the submerged object, and the apparent depth $d'$ is related to the actual depth $d$ by

**Apparent depth,**
**observer directly**
**above object**

$$d' = d\left(\frac{n_2}{n_1}\right) \qquad (26.3)$$

In this result, $n_1$ is the refractive index of the medium associated with the incident ray (the medium in which the object is located), and $n_2$ refers to the medium associated with the refracted ray (the medium in which the observer is situated). The proof of Equation 26.3 is the focus of problem 19 at the end of the chapter. Example 3 illustrates that the effect of apparent depth is quite noticeable in water.

### Example 3  The Apparent Depth of a Swimming Pool

A swimmer is treading water (with her head above the water) at the surface of a 3.00-m-deep pool. She sees a coin on the bottom directly below. How deep does the coin appear to be?

**Reasoning** Equation 26.3 may be used to find the apparent depth, provided we remember that the light rays travel from the coin to the swimmer. Therefore, the incident ray is coming from the coin under the water ($n_1 = 1.33$), while the refracted ray is in the air ($n_2 = 1.00$).

**Solution** The apparent depth $d'$ of the coin is

$$d' = d\left(\frac{n_2}{n_1}\right) = (3.00 \text{ m})\left(\frac{1.00}{1.33}\right) = \boxed{2.26 \text{ m}} \qquad (26.3)$$

In Example 3, a person sees a coin on the bottom of a pool at an apparent depth that is less than the actual depth. Conceptual Example 4 considers the reverse situation—namely, a person looking from under the water at a coin in the air.

### Conceptual Example 4  On the Inside Looking Out

A swimmer is under water and looking up at the surface. Someone holds a coin in the air, directly above the swimmer's eyes. To the swimmer, the coin appears to be at a certain height above the water. Is the apparent height of the coin greater than, less than, or the same as its actual height?

**Reasoning and Solution** Figure 26.5 shows that the apparent height of the coin above the water surface is greater than the actual height. Consider two rays of light leaving a point $P$ on the coin. When the rays enter the water, they are refracted toward the normal because water has a larger index of refraction than air has. By extending the refracted rays backward (see the dashed lines in the drawing), we find that the rays appear to the swimmer as if they originate from a point $P'$ that is farther from the water than point $P$ is. Thus, the swimmer sees the coin at an apparent height $d'$ that is *greater* than the actual height $d$. Equation 26.3 [$d' = d(n_2/n_1)$] reveals the same result, because $n_1$ represents the medium (air) associated with the incident ray and $n_2$ represents the medium (water) associated with the refracted ray. Since $n_2$ for water is greater than $n_1$ for air, the ratio $n_2/n_1$ is greater than one and $d'$ is larger than $d$. This situation is the opposite of that in Figure 26.4b, where an object beneath the water appears to be closer to the surface than it actually is.

**Related Homework:** *Problems 11, 18*

## THE DISPLACEMENT OF LIGHT
## BY A TRANSPARENT SLAB OF MATERIAL

A window pane is an example of a transparent slab of material. It consists of a plate of glass with parallel surfaces. When a ray of light passes through the glass, the emergent ray is parallel to the incident ray but displaced from it, as Figure 26.6 shows. This result

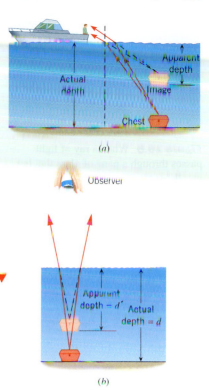

Observer

**Figure 26.4** (a) Because light from the chest is refracted away from the normal when the light enters the air, the apparent depth of the image is less than the actual depth. (b) The observer is viewing the submerged object from directly overhead.

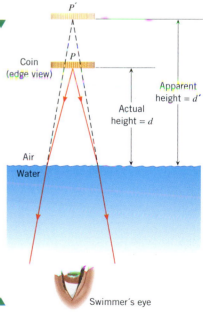

**Figure 26.5** Rays from point $P$ on a coin in the air above the water refract toward the normal as they enter the water. An underwater swimmer perceives the rays as originating from a point $P'$ that is farther above the surface than the actual point $P$.

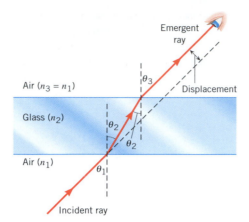

**Figure 26.6** When a ray of light passes through a pane of glass that has parallel surfaces and is surrounded by air, the emergent ray is parallel to the incident ray ($\theta_3 = \theta_1$), but is displaced from it.

can be verified by applying Snell's law to each of the two glass surfaces, with the result that $n_1 \sin \theta_1 = n_2 \sin \theta_2 = n_3 \sin \theta_3$. Since air surrounds the glass, $n_1 = n_3$, and it follows that $\sin \theta_1 = \sin \theta_3$. Therefore, $\theta_1 = \theta_3$, and the emergent and incident rays are parallel. However, as the drawing shows, the emergent ray is displaced laterally relative to the incident ray. The extent of the displacement depends on the angle of incidence, the thickness of the slab, and its refractive index.

## DERIVATION OF SNELL'S LAW

Snell's law can be derived by considering what happens to the wave fronts when the light passes from one medium into another. Figure 26.7a shows light propagating from medium 1, where the speed is relatively large, into medium 2, where the speed is smaller; therefore, $n_1$ is less than $n_2$. The plane wave fronts in this picture are drawn perpendicular to the incident and refracted rays. Since the part of each wave front that penetrates medium 2 slows down, the wave fronts in medium 2 are rotated clockwise relative to those in medium 1. Correspondingly, the refracted ray in medium 2 is bent toward the normal, as the drawing shows.

Although the incident and refracted waves have different speeds, *they have the same frequency f.* The fact that the frequency does not change can be understood in terms of the atomic mechanism underlying the generation of the refracted wave. When the electromagnetic wave strikes the surface, the oscillating electric field forces the electrons in the molecules of medium 2 to oscillate at the same frequency as the wave. The accelerating electrons behave like atomic antennas that radiate "extra" electromagnetic waves, which combine with the original wave. The net electromagnetic wave within medium 2 is a superposition of the original wave plus the extra radiated waves, and it is this superposition that constitutes the refracted wave. Since the extra waves are radiated at the same

**Figure 26.7** (*a*) The wave fronts are refracted as the light passes from medium 1 into medium 2. (*b*) An enlarged view of the incident and refracted wave fronts at the surface.

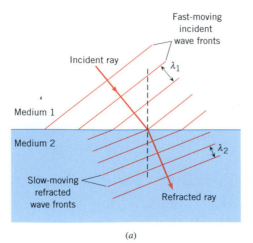

(a)

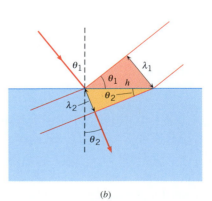

(b)

frequency as the original wave, the refracted wave also has the same frequency as the original wave.

The distance between successive wave fronts in Figure 26.7a has been chosen to be the wavelength $\lambda$. Since the frequencies are the same in both media but the speeds are different, it follows from Equation 16.1 that the wavelengths are different: $\lambda_1 = v_1/f$ and $\lambda_2 = v_2/f$. Since $v_1$ is assumed to be larger than $v_2$, $\lambda_1$ is larger than $\lambda_2$, and the wave fronts are farther apart in medium 1.

Figure 26.7b shows an enlarged view of the incident and refracted wave fronts at the surface. The angles $\theta_1$ and $\theta_2$ within the colored right triangles are, respectively, the angles of incidence and refraction. In addition, the triangles share the same hypotenuse $h$. Therefore,

$$\sin \theta_1 = \frac{\lambda_1}{h} = \frac{v_1/f}{h} = \frac{v_1}{hf}$$

and

$$\sin \theta_2 = \frac{\lambda_2}{h} = \frac{v_2/f}{h} = \frac{v_2}{hf}$$

Combining these two equations into a single equation by eliminating the common term $hf$ gives

$$\frac{\sin \theta_1}{v_1} = \frac{\sin \theta_2}{v_2}$$

By multiplying each side of this result by $c$, the speed of light in a vacuum, and recognizing that the ratio $c/v$ is the index of refraction $n$, we arrive at Snell's law of refraction: $n_1 \sin \theta_1 = n_2 \sin \theta_2$.

## 26.3 Total Internal Reflection

When light passes from a medium of larger refractive index into one of smaller refractive index—for example, from water to air—the refracted ray bends *away* from the normal, as in Figure 26.8a. As the angle of incidence increases, the angle of refraction also increases. When the angle of incidence reaches a certain value, called the ***critical angle, $\theta_c$,*** the angle of refraction is 90°. Then the refracted ray points along the surface; part b illustrates what happens at the critical angle. When the angle of incidence exceeds the critical angle, as in part c of the drawing, there is no refracted light. All the incident light is reflected back into the medium from which it came, a phenomenon called ***total internal reflection.*** Total internal reflection occurs only when light travels from a higher-index medium toward a lower-index medium. It does not occur when light propagates in the reverse direction—for example, from air to water.

An expression for the critical angle $\theta_c$ can be obtained from Snell's law by setting $\theta_1 = \theta_c$ and $\theta_2 = 90°$ (see Figure 26.8b):

$$\sin \theta_c = \frac{n_2 \sin 90°}{n_1}$$

***Critical angle***
$$\sin \theta_c = \frac{n_2}{n_1} \qquad (n_1 > n_2) \qquad (26.4)$$

For instance, the critical angle for light traveling from water ($n_1 = 1.33$) to air ($n_2 = 1.00$) is $\theta_c = \sin^{-1}(1.00/1.33) = 48.8°$. For incident angles greater than 48.8°, Snell's law

**Figure 26.8** (*a*) When light travels from a higher-index medium (water) into a lower-index medium (air), the refracted ray is bent away from the normal. (*b*) When the angle of incidence is equal to the critical angle $\theta_c$, the angle of refraction is 90°. (*c*) If $\theta_1$ is greater than $\theta_c$, there is no refracted ray, and total internal reflection occurs.

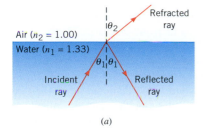

(a)

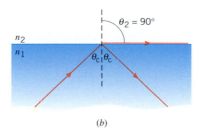

(b)

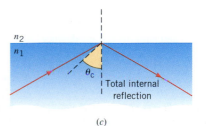

(c)

predicts that $\sin \theta_2$ is greater than unity, a value that is not possible. Thus, light rays with incident angles exceeding 48.8° yield no refracted light, and the light is totally reflected back into the water, as Figure 26.8c indicates.

The next example illustrates how the critical angle changes when the indices of refraction change.

### *Example 5* Total Internal Reflection

A beam of light is propagating through diamond ($n_1 = 2.42$) and strikes a diamond–air interface at an angle of incidence of 28°. (a) Will part of the beam enter the air ($n_2 = 1.00$) or will the beam be totally reflected at the interface? (b) Repeat part (a), assuming that the diamond is surrounded by water ($n_2 = 1.33$).

**Reasoning** Total internal reflection occurs only when the beam of light has an angle of incidence that is greater than the critical angle $\theta_c$. The critical angle is different in parts (a) and (b), since it depends on the ratio $n_2/n_1$ of the refractive indices of the incident ($n_1$) and refracting ($n_2$) media.

**Solution**

(a) The critical angle $\theta_c$ for total internal reflection at the diamond–air interface is given by Equation 26.4 as

$$\theta_c = \sin^{-1}\left(\frac{n_2}{n_1}\right)$$
$$= \sin^{-1}\left(\frac{1.00}{2.42}\right) = 24.4°$$

Because the angle of incidence of 28° is greater than the critical angle, there is no refraction, and the light is totally reflected back into the diamond.

(b) If water, rather than air, surrounds the diamond, the critical angle for total internal reflection becomes larger:

$$\theta_c = \sin^{-1}\left(\frac{n_2}{n_1}\right)$$
$$= \sin^{-1}\left(\frac{1.33}{2.42}\right) = 33.3°$$

Now a beam of light that has an angle of incidence of 28° (less than the critical angle of 33.3°) at the diamond–water interface is refracted into the water.

The critical angle plays an important role in why a diamond sparkles, as Conceptual Example 6 discusses.

### *Conceptual Example 6* The Sparkle of a Diamond

A diamond gemstone is famous for its sparkle because the light coming from it glitters as it is moved about. Why does a diamond exhibit such brilliance? And why does a diamond lose much of its brilliance when placed under water?

**Reasoning and Solution** A diamond sparkles because when it is held a certain way, the intensity of the light coming from it is greatly enhanced. Figure 26.9 helps to explain that this enhancement is related to total internal reflection. Part *a* of the drawing shows a ray of light striking a bottom facet of the diamond at an angle of incidence that is greater than the critical angle. This ray, and all other rays whose angles of incidence exceed the critical angle, are totally reflected back into the diamond, eventually exiting the top surface to give the diamond its sparkle. Many of the rays striking a bottom facet behave in this fashion because diamond has a relatively small critical angle in air. As part (a) of Example 5 shows, the critical angle is 24.4°. The value is so small because the index of refraction of diamond ($n = 2.42$) is large compared to that of air ($n = 1.00$).

Now consider what happens to the same ray of light within the diamond when the diamond is placed in water. Because water has a larger index of refraction than air, the critical angle increases to 33.3°, as part (b) of Example 5 shows. Therefore, this particular ray is no

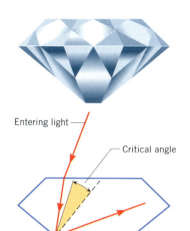

Entering light

Critical angle

*(a)*

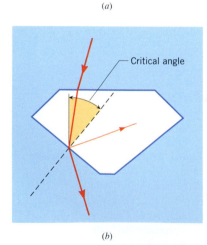

Critical angle

*(b)*

**Figure 26.9** (a) Near the bottom of the diamond, light is totally internally reflected, because the incident angle exceeds the critical angle for diamond and air. (b) When the diamond is in water, the same light is partially reflected and partially refracted, since the incident angle is less than the critical angle for diamond and water.

**The physics of**
**why a diamond sparkles.**

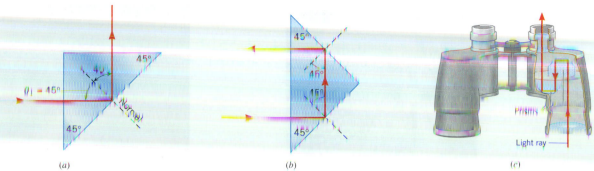

(a)          (b)          (c)

**Figure 26.10** Total internal reflection at a glass–air interface can be used to turn a ray of light through an angle of (a) 90° or (b) 180°. (c) Two prisms, each reflecting the light twice by total internal reflection, are sometimes used in binoculars to produce a lateral displacement of a light ray.

longer totally internally reflected. As Figure 26.9b illustrates, only some of the light is reflected back into the diamond, and the remainder escapes into the water. Consequently, less light exits from the top of the diamond, causing it to lose much of its brilliance.

Many optical instruments, such as binoculars, periscopes, and telescopes, use glass prisms and total internal reflection to turn a beam of light through 90° or 180°. Figure 26.10a shows a light ray entering a 45°–45°–90° glass prism ($n_1 = 1.5$) and striking the hypotenuse of the prism at an angle of incidence of $\theta_1 = 45°$. The critical angle for a glass–air interface is $\theta_c = \sin^{-1}(n_2/n_1) = \sin^{-1}(1.0/1.5) = 42°$. Since the angle of incidence is greater than the critical angle, the light is totally reflected at the hypotenuse and is directed vertically upward in the drawing, having been turned through an angle of 90°. Part b of the picture shows how the same prism can turn the beam through 180° when total internal reflection occurs twice. Prisms can also be used in tandem to produce a lateral displacement of a light ray, while leaving its initial direction unaltered. Figure 26.10c illustrates such an application in binoculars.

An important application of total internal reflection occurs in fiber optics, where hairthin threads of glass or plastic, called optical fibers, "pipe" light from one place to another. Figure 26.11a shows that an optical fiber consists of a cylindrical inner *core* that carries the light and an outer concentric shell, the *cladding*. The core is made from transparent glass or plastic that has a relatively high index of refraction. The cladding is also made of glass, but of a type that has a relatively low index of refraction. Light enters one end of the core, strikes the core/cladding interface at an angle of incidence greater than the critical angle, and, therefore, is reflected back into the core. Light thus travels inside the optical fiber along a zigzag path. In a well-designed fiber, little light is lost as a result of absorption by the core, so light can travel many kilometers before its intensity diminishes appreciably. Fibers are often bundled together to produce cables. Because the fibers themselves are so thin, the cables are relatively small and flexible and can fit into places inaccessible to larger metal wires.

Optical fiber cables are the medium of choice for high-quality telecommunications because the cables are relatively immune to external electrical interference and because a light beam can carry information through an optical fiber just as electricity carries information through copper wires. The information-carrying capacity of light, however, is thousands of times greater than that of electricity. A laser beam traveling through a single optical fiber can carry tens of thousands of telephone conversations and several TV programs simultaneously.

**The physics of fiber optics.**

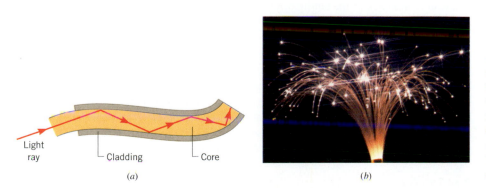

Light ray    Cladding    Core

(a)          (b)

**Figure 26.11** (a) Light can travel with little loss in a curved optical fiber because the light is totally reflected whenever it strikes the core–cladding interface and because the absorption of light by the core itself is small. (b) Light is being transmitted by a bundle of optical fibers. (© Masashiro Sano/Corbis Stock Market)

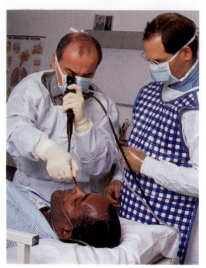

**Figure 26.12** A bronchoscope is being used to look for signs of pulmonary disease. (© Len Lessin/Peter Arnold, Inc.)

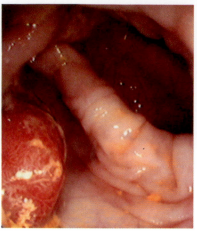

**Figure 26.13** A colonoscope reveals a polyp (red) attached to the wall of the colon. Polyps that turn cancerous or grow large enough to obstruct the colon are surgically removed. (© Dr. Larpent/ CNRI/Photo Researchers)

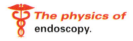

**The physics of** endoscopy.

In the field of medicine, optical fiber cables have had extraordinary impact. In the practice of endoscopy, for instance, a device called an endoscope is used to peer inside the body. Figure 26.12 shows a bronchoscope being used, which is a kind of endoscope that is inserted through the nose or mouth, down the bronchial tubes, and into the lungs. It consists of two optical fiber cables. One provides light to illuminate interior body parts, while the other sends back an image for viewing. A bronchoscope greatly simplifies the diagnosis of pulmonary disease. Tissue samples can even be collected with some bronchoscopes. A colonoscope is another kind of endoscope, and its design is similar to that of the bronchoscope. It is inserted through the rectum and used to examine the interior of the colon (see Figure 26.13). The colonoscope currently offers the best hope for diagnosing colon cancer in its early stages, when it can be treated.

**The physics of** arthroscopic surgery.

The use of optical fibers has also revolutionized surgical techniques. In arthroscopic surgery, a small surgical instrument, several millimeters in diameter, is mounted at the end of an optical fiber cable. The surgeon can insert the instrument and cable into a joint, such as the knee, with only a tiny incision and minimal damage to the surrounding tissue (see Figure 26.14). Consequently, recovery from the procedure is relatively rapid compared to traditional surgical techniques.

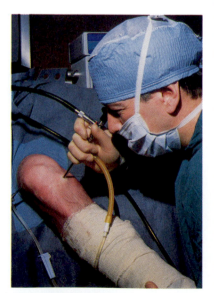

**Figure 26.14** Optical fibers have made arthroscopic surgery possible, such as the repair of a damaged knee shown in this photograph. (© Margaret Rose Orthopaedic Hospital/Photo Researchers)

## 26.4 *Polarization and the Reflection and Refraction of Light*

For incident angles other than 0°, unpolarized light becomes partially polarized in reflecting from a nonmetallic surface, such as water. To demonstrate this fact, rotate a pair of Polaroid sunglasses in the sunlight reflected from a lake. You will see that the light intensity transmitted through the glasses is a minimum when the glasses are oriented as they are normally worn. Since the transmission axis of the glasses is aligned vertically, it follows that the light reflected from the lake is partially polarized in the horizontal direction.

There is one special angle of incidence at which the reflected light is completely polarized parallel to the surface, the refracted ray being only partially polarized. This angle is called the *Brewster angle* $\theta_B$. Figure 26.15 summarizes what happens when unpolarized light strikes a nonmetallic surface at the Brewster angle. The value of $\theta_B$ is given by *Brewster's law,* in which $n_1$ and $n_2$ are, respectively, the refractive indices of the materials in which the incident and refracted rays propagate:

*Brewster's law*
$$\tan \theta_B = \frac{n_2}{n_1}$$
(26.5)

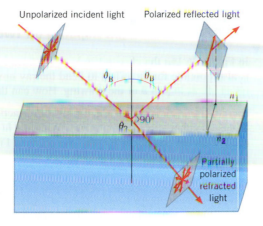

Unpolarized incident light    Polarized reflected light

$\theta_B$    $\theta_B$

$n_1$

$90°$

$\theta_0$

$n_2$

Partially polarized refracted light

**Figure 26.15** When unpolarized light is incident on a nonmetallic surface at the Brewster angle $\theta_B$, the reflected light is 100% polarized in a direction parallel to the surface. The angle between the reflected and refracted rays is 90°.

This relation is named after the Scotsman David Brewster (1781–1868), who discovered it. Figure 26.15 also indicates that the reflected and refracted rays are perpendicular to each other when light strikes the surface at the Brewster angle (see problem 37 at the end of the chapter).

# 26.5 The Dispersion of Light: Prisms and Rainbows

Figure 26.16a shows a ray of monochromatic light passing through a glass prism surrounded by air. When the light enters the prism at the left face, the refracted ray is bent toward the normal, because the refractive index of glass is greater than that of air. When the light leaves the prism at the right face, it is refracted away from the normal. Thus, the net effect of the prism is to change the direction of the ray, causing it to bend downward upon entering the prism, and downward again upon leaving. Because the refractive index of the glass depends on wavelength (see Table 26.2), rays corresponding to different colors are bent by different amounts by the prism and depart traveling in different directions. The greater the index of refraction for a given color, the greater the bending, and part b of the drawing shows the refractions for the colors red and violet, which are at opposite ends of the visible spectrum. If a beam of sunlight, which contains all colors, is sent through the prism, the sunlight is separated into a spectrum of colors, as part c shows. The spreading of light into its color components is called *dispersion*.

In Figure 26.16a the ray of light is refracted twice by a glass prism surrounded by air. Conceptual Example 7 explores what happens to the light when the prism is surrounded by materials other than air.

**Table 26.2** *Indices of Refraction n of Crown Glass at Various Wavelengths*

| Approximate Color | Wavelength in Vacuum (nm) | Index of Refraction, $n$ |
|---|---|---|
| Red | 660 | 1.520 |
| Orange | 610 | 1.522 |
| Yellow | 580 | 1.523 |
| Green | 550 | 1.526 |
| Blue | 470 | 1.531 |
| Violet | 410 | 1.538 |

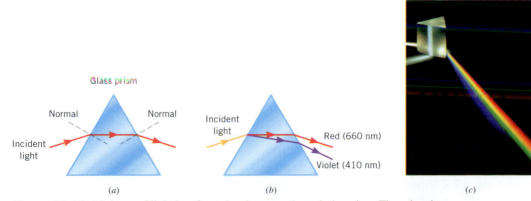

Glass prism

Normal    Normal

Incident light

Incident light

Red (660 nm)

Violet (410 nm)

(a)                    (b)                    (c)

**Figure 26.16** (a) A ray of light is refracted as it passes through the prism. The prism is surrounded by air. (b) Two different colors are refracted by different amounts. For clarity, the amount of refraction has been exaggerated. (c) Sunlight is dispersed into its color components by this prism. (© David Parker/Photo Researchers)

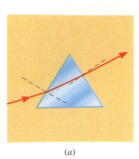

(a)

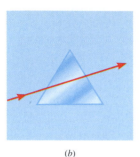

(b)

**Figure 26.17** A ray of light passes through identical prisms, each surrounded by a different fluid. The ray of light is (a) refracted upward and (b) not refracted at all.

**The physics of rainbows.**

### Conceptual Example 7
### The Refraction of Light Depends on Two Refractive Indices

In Figure 26.16a the glass prism is surrounded by air and bends the ray of light downward. It is also possible for the prism to bend the ray upward, as in Figure 26.17a, or to not bend the ray at all, as in part b of the drawing. How can the situations illustrated in Figure 26.17 arise?

**Reasoning and Solution** Snell's law of refraction includes the refractive indices of *both* materials on either side of an interface. With this in mind, we note that the ray bends upward, or away from the normal, as it enters the prism in Figure 26.17a. A ray bends away from the normal when it travels from a medium with a larger refractive index into a medium with a smaller refractive index. When the ray leaves the prism, it again bends upward, which is toward the normal at the point of exit. A ray bends toward the normal when traveling from a smaller toward a larger refractive index. Thus, **the situation in Figure 26.17a could arise if the prism were immersed in a fluid, such as carbon disulfide, which has a larger refractive index than does glass** (see Table 26.1).

We have seen in Figures 26.16a and 26.17a that a glass prism can bend a ray of light either downward or upward, depending on whether the surrounding fluid has a smaller or larger index of refraction than the glass. It seems logical to conclude, then, that **a prism will not bend a ray at all, neither up nor down, if the surrounding fluid has the same index of refraction as the glass**—a condition known as *index matching*. This is exactly what is happening in Figure 26.17b, where the ray proceeds straight through the prism as if it were not even there. If the index of refraction of the surrounding fluid equals that of the glass prism, then $n_1 = n_2$, and Snell's law ($n_1 \sin \theta_1 = n_2 \sin \theta_2$) reduces to $\sin \theta_1 = \sin \theta_2$. Therefore, the angle of refraction equals the angle of incidence, and no bending of the light occurs.

**Related Homework:** *Problem 42*

Another example of dispersion occurs in rainbows, in which refraction by water droplets gives rise to the colors. You can often see a rainbow just as a storm is leaving, if you look at the departing rain with the sun at your back. When light from the sun enters a spherical raindrop, as in Figure 26.18, light of each color is refracted or bent by an amount that depends on the refractive index of water for that wavelength. After reflection from the back surface of the droplet, the different colors are again refracted as they reenter the air. Although each droplet disperses the light into its full spectrum of colors, the observer in Figure 26.19a sees only one color of light coming from any given droplet, since only one color travels in the right direction to reach the observer's eyes. However, all colors are visible in a rainbow (see Figure 26.19b) because each color originates from different droplets at different angles of elevation.

## 26.6  Lenses

The lenses used in optical instruments, such as eyeglasses, cameras, and telescopes, are made from transparent materials that refract light. They refract the light in such a way that an image of the source of the light is formed. Figure 26.20a shows a crude lens formed from two glass prisms. Suppose an object centered on the principal axis is infinitely far from the lens so the rays from the object are parallel to the principal axis. In passing through the prisms, these rays are bent toward the axis because of refraction. Unfortunately, the rays do not all cross the axis at the same place, and, therefore, such a crude lens gives rise to a blurred image of the object.

A better lens can be constructed from a single piece of transparent material with properly curved surfaces, often spherical, as in part b of the drawing. With this improved lens, rays that are near the principal axis (paraxial rays) and parallel to it converge to a single point on the axis after emerging from the lens. This point is called the **focal point F** of the lens. Thus, an object located infinitely far away on the principal axis leads to an image at the focal point of the lens. The distance between the focal point and the lens is the **focal length f.** In what follows, we assume the lens is so thin compared to f that it makes no difference whether f is measured between the focal point and either surface of the lens or the center of the lens. The type of lens in Figure 26.20b is known as a **converging lens** because it causes incident parallel rays to converge at the focal point.

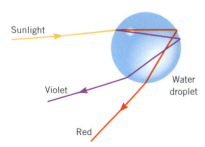

**Figure 26.18** When sunlight emerges from a water droplet, the light is dispersed into its constituent colors, of which only two are shown.

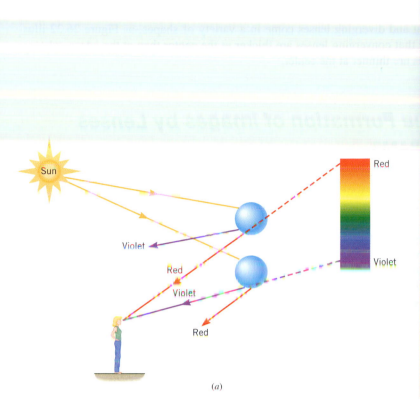

(a)

(b)

**Figure 26.19** The different colors seen in a rainbow originate from water droplets at different angles of elevation. (© David Woodfall/Stone/Getty Images)

Another type of lens found in optical instruments is a **diverging lens,** which causes incident parallel rays to diverge after exiting the lens. Two prisms can also be used to form a crude diverging lens, as in Figure 26.21a. In a properly designed diverging lens, such as that in part b of the picture, paraxial rays that are parallel to the principal axis appear to originate from a single point on the axis after passing through the lens. This point is the focal point F, and its distance f from the lens

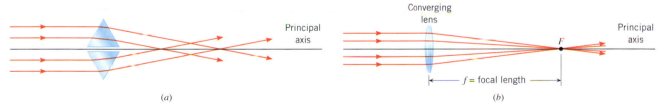

(a)                                                                                          (b)

**Figure 26.20** (a) These two prisms cause rays of light that are parallel to the principal axis to change direction and cross the axis at different points. (b) With a converging lens, paraxial rays that are parallel to the principal axis converge to the focal point F after passing through the lens.

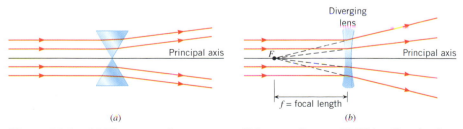

(a)                                                                                          (b)

**Figure 26.21** (a) These two prisms cause parallel rays to diverge. (b) With a diverging lens, paraxial rays that are parallel to the principal axis appear to originate from the focal point F after passing through the lens.

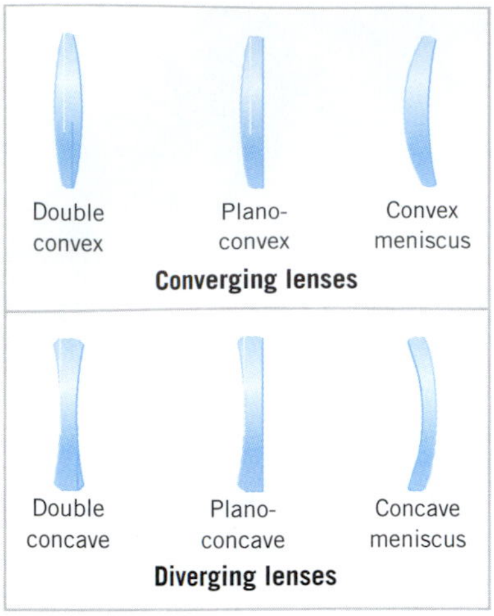

**Converging lenses**

Double convex | Plano-convex | Convex meniscus

**Diverging lenses**

Double concave | Plano-concave | Concave meniscus

*Figure 26.22* Converging and diverging lenses come in a variety of shapes.

is the focal length. Again, we assume that the lens is thin compared to the focal length.

Converging and diverging lenses come in a variety of shapes, as Figure 26.22 illustrates. Observe that converging lenses are thicker at the center than at the edges, whereas diverging lenses are thinner at the center.

## 26.7 *The Formation of Images by Lenses*
### RAY DIAGRAMS

Each point on an object emits light rays in all directions, and when some of these rays pass through a lens, they form an image. As with mirrors, ray diagrams can be drawn to determine the location and size of the image. Lenses differ from mirrors, however, in that light can pass through a lens from left to right or from right to left. Therefore, when constructing ray diagrams, begin by locating a focal point *F* on *each side of the lens;* each point lies on the principal axis at the same distance *f* from the lens. The lens is assumed to be thin, in that its thickness is small compared with the focal length and the distances of the object and the image from the lens. For convenience, it is also assumed that the object is located to the left of the lens and is oriented perpendicular to the principal axis. There are three paraxial rays that leave a point on the top of the object and are especially helpful in drawing ray diagrams. They are labeled 1, 2, and 3 in Figure 26.23. When tracing their paths, we use the following conventions.

**Ray Tracing for Converging and Diverging Lenses**

| **Converging Lens** | **Diverging Lens** |
|---|---|
| *Ray 1* | |
| This ray initially travels parallel to the principal axis. In passing through a converging lens, the ray is refracted toward the axis and travels through the focal point on the right side of the lens, as Figure 26.23*a* shows. | This ray initially travels parallel to the principal axis. In passing through a diverging lens, the ray is refracted away from the axis, and *appears* to have originated from the focal point on the left of the lens. The dashed line in Figure 26.23*d* represents the apparent path of the ray. |
| *Ray 2* | |
| This ray first passes through the focal point on the left and then is refracted by the lens in such a way that it leaves traveling parallel to the axis, as in Figure 26.23*b*. | This ray leaves the object and moves toward the focal point on the right of the lens. Before reaching the focal point, however, the ray is refracted by the lens so as to exit parallel to the axis. See Figure 26.23*e*, where the dashed line indicates the ray's path in the absence of the lens. |
| *Ray 3** | |
| This ray travels directly through the center of the thin lens without any appreciable bending, as in Figure 26.23*c*. | This ray travels directly through the center of the thin lens without any appreciable bending, as in Figure 26.23*f*. |

* Ray 3 does not bend as it proceeds through the lens because the left and right surfaces of each type of lens are nearly parallel at the center. Thus, in either case, the lens behaves as a transparent slab. As Figure 26.6 shows, the rays incident on and exiting from a slab travel in the same direction with only a lateral displacement. If the lens is sufficiently thin, the displacement is negligibly small.

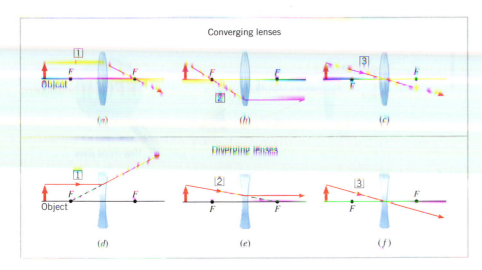

**Figure 26.23** The rays shown here are useful in determining the nature of the images formed by converging and diverging lenses.

# IMAGE FORMATION BY A CONVERGING LENS

Figure 26.24*a* illustrates the formation of a real image by a converging lens. Here the object is located at a distance from the lens that is greater than twice the focal length (beyond the point labeled 2*F*). To locate the image, any two of the three special rays can be drawn from the tip of the object, although all three are shown in the drawing. The point on the right side of the lens where these rays intersect locates the tip of the image. The ray diagram indicates that the image is real, inverted, and smaller than the object. This optical arrangement is similar to that used in a camera, where a piece of film records the image (see part *b* of the drawing).

*The physics of a camera.*

When the object is placed between 2*F* and *F*, as in Figure 26.25*a*, the image is still real and inverted, however, the image is now larger than the object. This optical system is used in a slide or film projector in which a small piece of film is the object and the enlarged image falls on a screen. However, to obtain an image that is right-side up, the film must be placed in the projector upside down.

*The physics of a slide or film projector.*

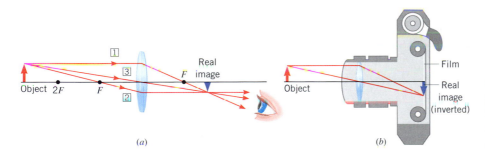

(*a*)

(*b*)

**Figure 26.24** (*a*) When the object is placed to the left of the point labeled 2*F*, a real, inverted, and smaller image is formed. (*b*) The arrangement in part *a* is like that used in a camera.

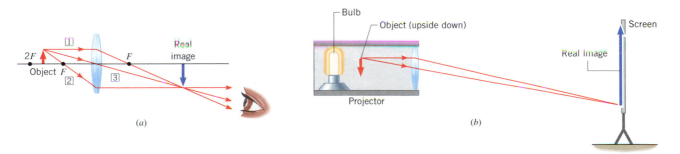

(*a*)

(*b*)

**Figure 26.25** (*a*) When the object is placed between 2*F* and *F*, the image is real, inverted, and larger than the object. (*b*) This arrangement is found in projectors.

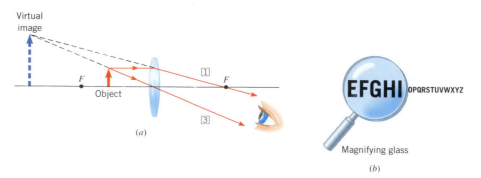

**Figure 26.26** (a) When an object is placed within the focal point F of a converging lens, an upright, enlarged, and virtual image is created. (b) Such an image is seen when looking through a magnifying glass.

When the object is located between the focal point and the lens, as in Figure 26.26, the rays diverge after leaving the lens. To a person viewing the diverging rays, they appear to come from an image behind (to the left of) the lens. Because none of the rays actually come from the image, it is a virtual image. The ray diagram shows that the virtual image is upright and enlarged. A magnifying glass uses this arrangement, as can be seen in part b of the drawing.

### IMAGE FORMATION BY A DIVERGING LENS

Light rays diverge upon leaving a diverging lens, as Figure 26.27 shows, and the ray diagram indicates that a virtual image is formed on the left side of the lens. In fact, regardless of the position of a real object, a diverging lens always forms a virtual image that is upright and smaller relative to the object.

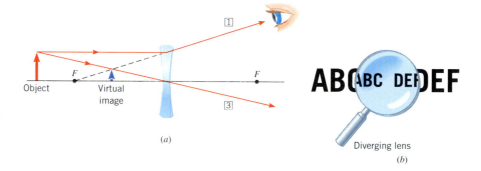

**Figure 26.27** (a) A diverging lens always forms a virtual image of a real object. The image is upright and smaller relative to the object. (b) The image seen through a diverging lens.

## 26.8 *The Thin-Lens Equation and the Magnification Equation*

When an object is placed in front of a spherical mirror, we can determine the location, size, and nature of its image by using the technique of ray tracing or the mirror and magnification equations. Both options are based on the law of reflection. The mirror and magnification equations relate the distances $d_o$ and $d_i$ of the object and image from the mirror to the focal length $f$ and magnification $m$. For an object placed in front of a lens, Snell's law of refraction leads to the technique of ray tracing and to equations that are identical to the mirror and magnification equations.

Snell's law leads to equations that are referred to as the thin-lens equation and the magnification equation:

**Thin-lens equation**
$$\frac{1}{d_o} + \frac{1}{d_i} = \frac{1}{f} \qquad (26.6)$$

**Magnification equation**
$$m = \frac{\text{Image height}}{\text{Object height}} = \frac{h_i}{h_o} = -\frac{d_i}{d_o} \qquad (26.7)$$

Figure 26.28 defines the symbols in these expressions with the aid of a thin converging

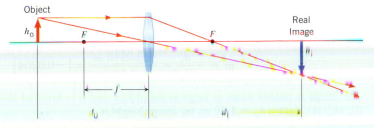

**Figure 26.28** The drawing shows the focal length $f$, the object distance $d_o$, and the image distance $d_i$ for a converging lens. The object and image heights are, respectively, $h_o$ and $h_i$.

lens, but the expressions also apply to a diverging lens, if it is thin. The derivations of these equations are presented at the end of this section.

Certain sign conventions accompany the use of the thin-lens and magnification equations, and the conventions are similar to those used with mirrors in Section 25.6. The issue of real versus virtual images, however, is slightly different with lenses than with mirrors. With a mirror, a real image is formed on the *same side* of the mirror as the object (see Figure 25.18), in which case the image distance $d_i$ is a positive number. With a lens, a positive value for $d_i$ also means the image is real. But, starting with an actual object, a real image is formed on the side of the lens *opposite to* the object (see Figure 26.28). The **sign conventions** listed below apply to light rays traveling from left to right from a real object.

### Summary of Sign Conventions for Lenses

*Focal length*

$f$ is + for a converging lens.

$f$ is − for a diverging lens.

*Object distance*

$d_o$ is + if the object is to the left of the lens (real object), as is usual.

$d_o$ is − if the object is to the right of the lens (virtual object).*

*Image distance*

$d_i$ is + for an image (real) formed to the right of the lens by a real object.

$d_i$ is − for an image (virtual) formed to the left of the lens by a real object.

*Magnification*

$m$ is + for an image that is upright with respect to the object.

$m$ is − for an image that is inverted with respect to the object.

Examples 8 and 9 illustrate the use of the thin-lens and magnification equations.

## Example 8   The Real Image Formed by a Camera Lens

A 1.70-m-tall person is standing 2.50 m in front of a camera. The camera uses a converging lens whose focal length is 0.0500 m. (a) Find the image distance (the distance between the lens and the film) and determine whether the image is real or virtual. (b) Find the magnification and the height of the image on the film.

**Reasoning** This optical arrangement is similar to that in Figure 26.24a, where the object distance is greater than twice the focal length of the lens. Therefore, we expect the image to be real, inverted, and smaller than the object.

### Solution

(a) To find the image distance $d_i$ we use the thin-lens equation with $d_o = 2.50$ m and $f = 0.0500$ m:

$$\frac{1}{d_i} = \frac{1}{f} - \frac{1}{d_o} = \frac{1}{0.0500 \text{ m}} - \frac{1}{2.50 \text{ m}}$$

$$= 19.6 \text{ m}^{-1} \quad \text{or} \quad \boxed{d_i = 0.0510 \text{ m}}$$

**Problem solving insight**
In the thin-lens equation, the reciprocal of the image distance $d_i$ is given by $d_i^{-1} = f^{-1} - d_o^{-1}$, where $f$ is the focal length and $d_o$ is the object distance. After combining the reciprocals $f^{-1}$ and $d_o^{-1}$, do not forget to take the reciprocal of the result to find $d_i$.

* This situation arises in systems containing more than one lens, where the image formed by the first lens becomes the object for the second lens. In such a case, the object of the second lens may lie to the right of that lens, in which event $d_o$ is assigned a negative value and the object is called a virtual object.

Since the image distance is a positive number, a $\boxed{\text{real image}}$ is formed on the film.

**(b)** The magnification follows from the magnification equation:

$$m = -\frac{d_i}{d_o} = -\frac{0.0510 \text{ m}}{2.50 \text{ m}} = \boxed{-0.0204}$$

The image is 0.0204 times as large as the object, and it is inverted since $m$ is negative. Since the object height is $h_o = 1.70$ m, the image height is

$$h_i = mh_o = (-0.0204)(1.70 \text{ m}) = \boxed{-0.0347 \text{ m}}$$

## Example 9 The Virtual Image Formed by a Diverging Lens

An object is placed 7.10 cm to the left of a diverging lens whose focal length is $f = -5.08$ cm (a diverging lens has a negative focal length). (a) Find the image distance and determine whether the image is real or virtual. (b) Obtain the magnification.

**Reasoning** This situation is similar to that in Figure 26.27a. The ray diagram shows that the image is virtual, erect, and smaller than the object.

**Solution**

**(a)** The thin-lens equation can be used to find the image distance $d_i$:

$$\frac{1}{d_i} = \frac{1}{f} - \frac{1}{d_o} = \frac{1}{-5.08 \text{ cm}} - \frac{1}{7.10 \text{ cm}}$$

$$= -0.338 \text{ cm}^{-1} \quad \text{or} \quad \boxed{d_i = -2.96 \text{ cm}}$$

The image distance is negative, indicating that the image is $\boxed{\text{virtual}}$ and located to the left of the lens.

**(b)** Since $d_i$ and $d_o$ are known, the magnification equation shows that

$$m = -\frac{d_i}{d_o} = -\frac{-2.96 \text{ cm}}{7.10 \text{ cm}} = \boxed{0.417}$$

The image is upright ($m$ is $+$) and smaller ($m < 1$) than the object.

The thin-lens and magnification equations can be derived by considering rays 1 and 3 in Figure 26.29a. Ray 1 is shown separately in part $b$ of the drawing, where the angle $\theta$ is the same in each of the two colored triangles. Thus, $\tan \theta$ is the same for each triangle:

$$\tan \theta = \frac{h_o}{f} = \frac{-h_i}{d_i - f}$$

A minus sign has been inserted in the numerator of the ratio $h_i/(d_i - f)$ for the following reason. The angle $\theta$ in Figure 26.29b is assumed to be positive. Since the image is inverted relative to the object, the image height $h_i$ is a negative number. The insertion of the minus sign ensures that the term $-h_i/(d_i - f)$, and hence $\tan \theta$, is a positive quantity.

Ray 3 is shown separately in part $c$ of the drawing, where the angles $\theta'$ are the same. Therefore,

$$\tan \theta' = \frac{h_o}{d_o} = \frac{-h_i}{d_i}$$

A minus sign has been inserted in the numerator of the term $h_i/d_i$ for the same reason that a minus sign was inserted earlier—namely, to ensure that $\tan \theta'$ is a positive quantity. The first equation gives $h_i/h_o = -(d_i - f)/f$, while the second equation yields $h_i/h_o = -d_i/d_o$. Equating these two expressions for $h_i/h_o$ and rearranging the result produces the thin-lens equation, $1/d_o + 1/d_i = 1/f$. The magnification equation follows directly from the equation $h_i/h_o = -d_i/d_o$, if we recognize that $h_i/h_o$ is the magnification $m$ of the lens.

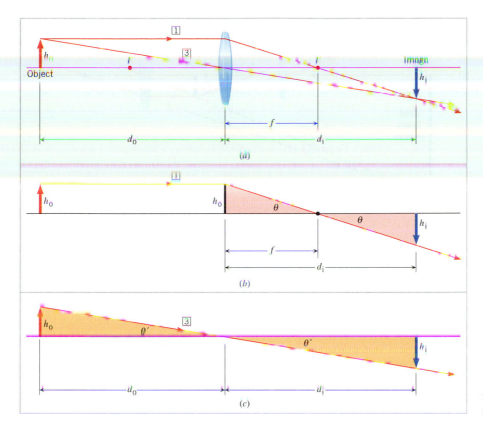

**Figure 26.29** These ray diagrams are used for deriving the thin-lens and magnification equations.

## 26.9 Lenses in Combination

Many optical instruments, such as microscopes and telescopes, use a number of lenses together to produce an image. Among other things, a multiple-lens system can produce an image that is magnified more than is possible with a single lens. For instance, Figure 26.30*a* shows a two-lens system used in a microscope. The first lens, the lens closest to the object, is referred to as the *objective*. The second lens is known as the *eyepiece* (or *ocular*). The object is placed just outside the focal point $F_o$ of the objective. The image formed by the objective—called the "first image" in the drawing—is real, inverted, and enlarged compared to the object. This first image then serves as the object for the eyepiece. Since the first image falls between the eyepiece and its focal point $F_e$, the eyepiece forms an enlarged, virtual, final image, which is what the observer sees.

The location of the final image in a multiple-lens system can be determined by applying the thin-lens equation to each lens separately. The key point to remember in such situations is that *the image produced by one lens serves as the object for the next lens,* as the next example illustrates.

*Problem solving insight*

### Example 10   A Microscope—Two Lenses in Combination

The objective and eyepiece of the compound microscope in Figure 26.30 are both converging lenses and have focal lengths of $f_o = 15.0$ mm and $f_e = 25.5$ mm. A distance of 61.0 mm separates the lenses. The microscope is being used to examine an object placed $d_{o1} = 24.1$ mm in front of the objective. Find the final image distance.

**Reasoning** The thin-lens equation can be used to locate the final image produced by the eyepiece. We know the focal length of the eyepiece, but to determine the final image distance from the thin-lens equation we also need to know the object distance, which is not given. To obtain this distance, we recall that the image produced by one lens (the objective) is the object for the next lens (the eyepiece). We can use the thin-lens equation to locate the image produced by the objective, since the focal length and the object distance for this lens are given. The location of this image relative to the eyepiece will tell us the object distance for the eyepiece.

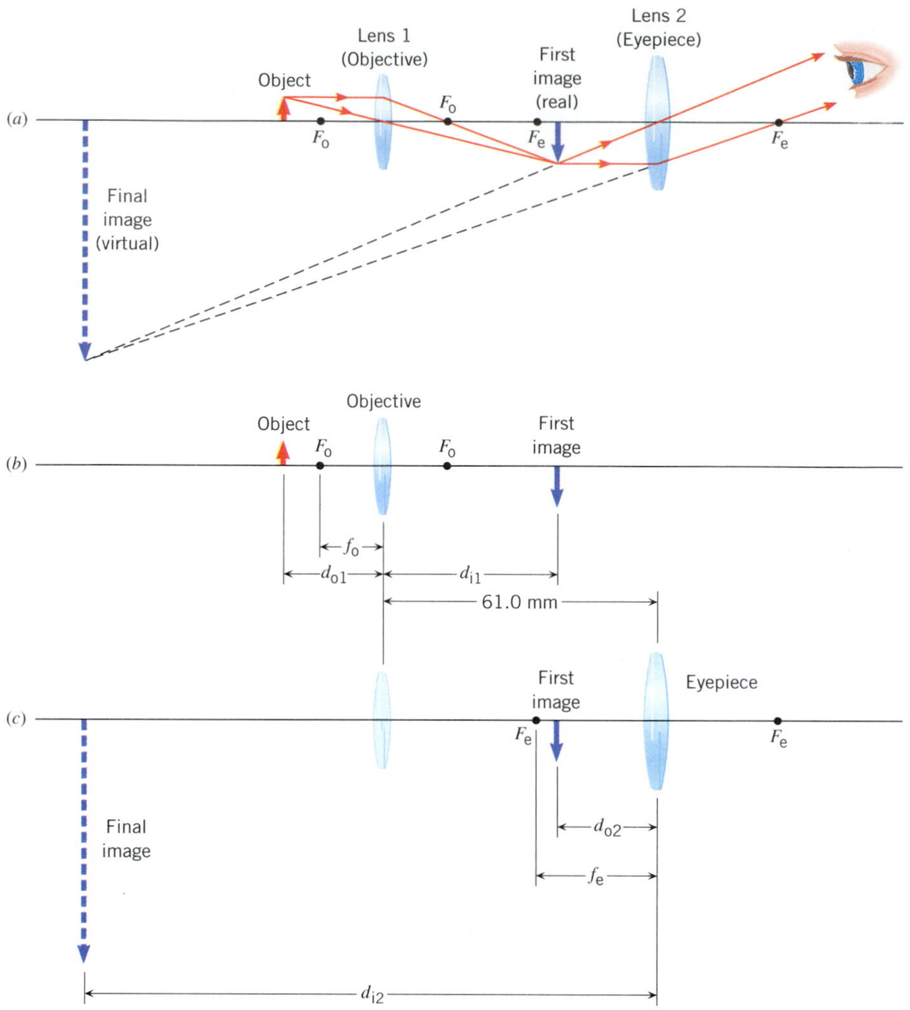

**Figure 26.30** (*a*) This two-lens system can be used as a compound microscope to produce a virtual, enlarged, and inverted final image. (*b*) The objective forms the first image and (*c*) the eyepiece forms the final image.

**Solution** The final image distance relative to the eyepiece is $d_{i2}$, and we can determine it by using the thin-lens equation:

$$\frac{1}{d_{i2}} = \frac{1}{f_e} - \frac{1}{d_{o2}}$$

The focal length $f_e$ of the eyepiece is known, but to obtain a value for the object distance $d_{o2}$ we must locate the first image produced by the objective. The first image distance $d_{i1}$ (see Figure 26.30*b*) can be determined using the thin-lens equation with $d_{o1} = 24.1$ mm and $f_o = 15.0$ mm.

$$\frac{1}{d_{i1}} = \frac{1}{f_o} - \frac{1}{d_{o1}}$$

$$= \frac{1}{15.0 \text{ mm}} - \frac{1}{24.1 \text{ mm}}$$

$$= 0.0252 \text{ mm}^{-1} \quad \text{or} \quad d_{i1} = 39.7 \text{ mm}$$

The first image now becomes the object for the eyepiece (see part *c* of the drawing). Since the distance between the lenses is 61.0 mm, the object distance for the eyepiece is $d_{o2} = 61.0$ mm $- d_{i1} = 61.0$ mm $- 39.7$ mm $= 21.3$ mm. Noting that the focal length of the eyepiece is $f_e = 25.5$ mm, we can determine the final image distance with the aid of the thin-lens equation:

$$\frac{1}{d_{i2}} = \frac{1}{f_e} - \frac{1}{d_{o2}}$$

$$= \frac{1}{25.5 \text{ mm}} - \frac{1}{21.3 \text{ mm}}$$

$$= -0.0077 \text{ mm}^{-1} \quad \text{or} \quad \boxed{d_{i2} = -130 \text{ mm}}$$

The fact that $d_{i2}$ is negative indicates that the final image is virtual. It lies to the left of the eyepiece, as the drawing shows.

# 26.10 The Human Eye

## ANATOMY

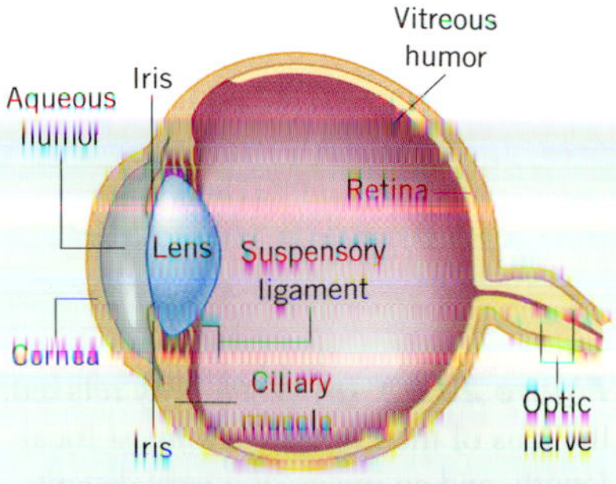

Without doubt, the human eye is the most remarkable of all optical devices. Figure 26.31 shows some of its main anatomical features. The eyeball is approximately spherical with a diameter of about 25 mm. Light enters the eye through a transparent membrane (the *cornea*). This membrane covers a clear liquid region (the *aqueous humor*), behind which are a diaphragm (the *iris*), the *lens*, a region filled with a jelly-like substance (the *vitreous humor*), and, finally, the *retina*. The retina is the light-sensitive part of the eye, consisting of millions of structures called *rods* and *cones*. When stimulated by light, these structures send electrical impulses via the *optic nerve* to the brain, which interprets the image on the retina.

The iris is the colored portion of the eye and controls the amount of light reaching the retina. The iris acts as a controller because it is a muscular diaphragm with a variable opening at its center, through which the light passes. The opening is called the *pupil*. The diameter of the pupil varies from about 2 to 7 mm, decreasing in bright light and increasing (dilating) in dim light.

Of prime importance to the operation of the eye is the fact that the lens is flexible, and its shape can be altered by the action of the *ciliary muscle*. The lens is connected to the ciliary muscle by the *suspensory ligaments* (see the drawing). We will see shortly how the shape-changing ability of the lens affects the focusing ability of the eye.

**Figure 26.31** A cross sectional view of the human eye.

## OPTICS

Optically, the eye and the camera are similar; both have a lens system and a diaphragm with a variable opening or aperture at its center. Moreover, the retina of the eye and the film in the camera serve similar functions, for both record the image formed by the lens system. In the eye, the image formed on the retina is real, inverted, and smaller than the object, just as it is in a camera. Although the image on the retina is inverted, it is interpreted by the brain as being right-side up.

**The physics of the human eye.**

For clear vision, the eye must refract the incoming light rays, so as to form a sharp image on the retina. In reaching the retina, the light travels through five different media, each with a different index of refraction $n$: air ($n = 1.00$), the cornea ($n = 1.38$), the aqueous humor ($n = 1.33$), the lens ($n = 1.40$, on the average), and the vitreous humor ($n = 1.34$). Each time light passes from one medium into another, it is refracted at the boundary. The greatest amount of refraction, about 70% or so, occurs at the air/cornea boundary. According to Snell's law, the large refraction at this interface occurs primarily because the refractive index of air ($n = 1.00$) is so different from that of the cornea ($n = 1.38$). The refraction at all the other boundaries is relatively small because the indices of refraction on either side of these boundaries are nearly equal. The lens itself contributes only about 20–25% of the total refraction, since the surrounding aqueous and vitreous humors have indices of refraction that are nearly the same as that of the lens.

Even though the lens contributes only a quarter of the total refraction or less, its function is an important one. The eye has a fixed image distance; that is, the distance between the lens and the retina is constant. Therefore, the only way that objects located at different distances can produce images on the retina is for the focal length of the lens to be adjustable. And it is the ciliary muscle that adjusts the focal length. When the eye looks at a very distant object, the ciliary muscle is not tensed. The lens has its least curvature and, consequently, its longest focal length. Under this condition the eye is said to be "fully relaxed," and the rays form a sharp image on the retina, as in Figure 26.32a. When the object moves closer to the eye, the ciliary muscle tenses automatically, thereby increasing the curvature of the lens, shortening the focal length, and permitting a sharp image to form again on the retina (Figure 26.32b). When a sharp image of an object is formed on the retina, we say the eye is "focused" on the object. The process in which the lens changes its focal length to focus on objects at different distances is called **accommodation.**

When you hold a book too close, the print is blurred because the lens cannot adjust enough to bring the book into focus. The point nearest the eye at which an object can be

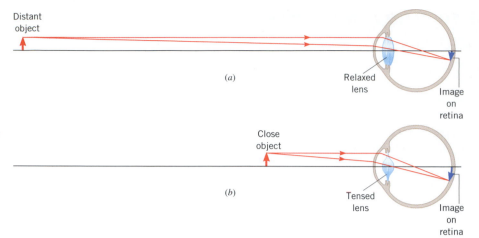

**Figure 26.32** (*a*) When fully relaxed, the lens of the eye has its longest focal length, and an image of a very distant object is formed on the retina. (*b*) When the ciliary muscle is tensed, the lens has a shorter focal length. Consequently, an image of a closer object is also formed on the retina.

placed and still produce a sharp image on the retina is called the **near point** of the eye. The ciliary muscle is fully tensed when an object is placed at the near point. For people in their early twenties with normal vision, the near point is located about 25 cm from the eye. It increases to about 50 cm at age 40 and to roughly 500 cm at age 60. Since most reading material is held at a distance of 25–45 cm from the eye, older adults typically need eyeglasses to overcome the loss of accommodation. The **far point** of the eye is the location of the farthest object on which the fully relaxed eye can focus. A person with normal eyesight can see objects very far away, such as the planets and stars, and thus has a far point located nearly at infinity.

## NEARSIGHTEDNESS

**The physics of nearsightedness.**

A person who is **nearsighted (myopic)** can focus on nearby objects but cannot clearly see objects far away. For such a person, the far point of the eye is not at infinity and may even be as close to the eye as three or four meters. When a nearsighted eye tries to focus on a distant object, the eye is fully relaxed, like a normal eye. However, the nearsighted eye has a focal length that is shorter than it should be, so rays from the distant object form a sharp image in front of the retina, as Figure 26.33*a* shows, and blurred vision results.

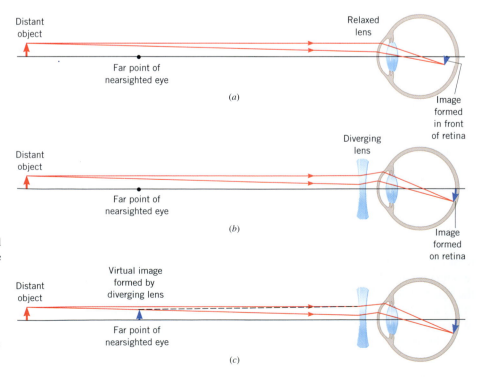

**Figure 26.33** (*a*) When a nearsighted person views a distant object, the image is formed in front of the retina. The result is blurred vision. (*b*) With a diverging lens in front of the eye, the image is moved onto the retina and clear vision results. (*c*) The diverging lens is designed to form a virtual image at the far point of the nearsighted eye.

The nearsighted eye can be corrected with glasses or contacts that use *diverging* lenses, as Figure 26.33b suggests. The rays from the object diverge after leaving the eyeglass lens. Therefore, when they are subsequently refracted toward the principal axis by the eye, a sharp image is formed farther back and falls on the retina. Since the relaxed (but nearsighted) eye can focus on an object at the eye's far point—but not on objects farther away—the diverging lens is designed to transform a very distant object into an image located at the far point. Part c of the drawing shows this transformation, and the next example illustrates how to determine the focal length of the diverging lens that accomplishes it.

### Example 11 Eyeglasses for the Nearsighted Person

A nearsighted person has a far point located only 521 cm from the eye. Assuming that eyeglasses are to be worn 2 cm in front of the eye, find the focal length needed for the diverging lenses of the glasses so the person can see distant objects.

**Reasoning** In Figure 26.33c the far point is 521 cm away from the eye. Since the glasses are worn 2 cm from the eye, the far point is 519 cm to the left of the diverging lens. The image distance, then, is −519 cm, the negative sign indicating that the image is a virtual image formed to the left of the lens. The object is assumed to be infinitely far from the diverging lens. The thin-lens equation can be used to find the focal length of the eyeglasses. We expect the focal length to be negative, since the lens is a diverging lens.

**Problem solving insight**
Eyeglasses are worn about 2 cm from the eyes. Be sure, if necessary, to take this 2 cm into account when determining the object and image distances ($d_o$ and $d_i$) that are used in the thin-lens equation.

**Solution** With $d_i = -519$ cm and $d_o = \infty$, the focal length can be found as follows:

$$\frac{1}{f} = \frac{1}{d_o} + \frac{1}{d_i} = \frac{1}{\infty} + \frac{1}{-519 \text{ cm}} \quad \text{or} \quad \boxed{f = -519 \text{ cm}} \quad (26.6)$$

The value for $f$ is negative, as expected, for a diverging lens.

## FARSIGHTEDNESS

A *farsighted (hyperopic)* person can usually see distant objects clearly, but cannot focus on those nearby. Whereas the near point of a young and normal eye is located about 25 cm from the eye, the near point of a farsighted eye may be considerably farther away than that, perhaps as far as several hundred centimeters. When a farsighted eye tries to focus on a book held closer than the near point, it accommodates and shortens its focal length as much as it can. However, even at its shortest, the focal length is longer than it should be. Therefore, the light rays from the book would form a sharp image behind the retina if they could do so, as Figure 26.34a suggests. In reality, no light passes through the retina, but a blurred image does form on it.

**The physics of** farsightedness.

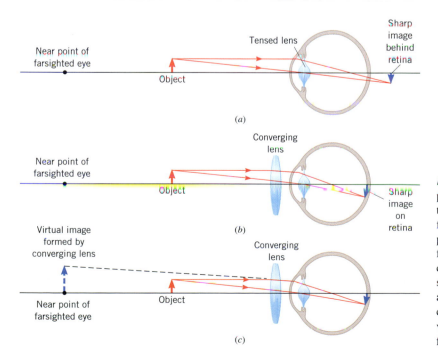

**Figure 26.34** (a) When a farsighted person views an object located inside the near point, a sharp image would be formed behind the retina if light could pass through it. Only a blurred image forms on the retina. (b) With a converging lens in front of the eye, the sharp image is moved onto the retina and clear vision results. (c) The converging lens is designed to form a virtual image at the near point of the farsighted eye.

Figure 26.34*b* shows that farsightedness can be corrected by placing a *converging* lens in front of the eye. The lens refracts the light rays more toward the principal axis before they enter the eye. Consequently, when the rays are refracted even more by the eye, they converge to form an image on the retina. Part *c* of the figure illustrates what the eye sees when it looks through the converging lens. The lens is designed so that the eye perceives the light to be coming from a virtual image located at the near point. Example 12 shows how the focal length of the converging lens is determined to correct for far-sightedness.

### Example 12    Contact Lenses for the Farsighted Person

A farsighted person has a near point located 210 cm from the eyes. Obtain the focal length of the converging lenses in a pair of contacts that can be used to read a book held 25.0 cm from the eyes.

**Reasoning**  A contact lens is placed directly against the eye. Thus, the object distance, which is the distance from the book to the lens, is 25.0 cm. The lens forms an image of the book at the near point of the eye, so the image distance is −210 cm. The minus sign indicates that the image is a virtual image formed to the left of the lens, as in Figure 26.34*c*. The focal length can be obtained from the thin-lens equation.

**Solution**  With $d_o = 25.0$ cm and $d_i = -210$ cm, the focal length can be determined from the thin-lens equation as follows:

$$\frac{1}{f} = \frac{1}{d_o} + \frac{1}{d_i} = \frac{1}{25.0 \text{ cm}} + \frac{1}{-210 \text{ cm}} = 0.0352 \text{ cm}^{-1} \quad \text{or} \quad \boxed{f = 28.4 \text{ cm}}$$

### THE REFRACTIVE POWER OF A LENS—THE DIOPTER

The extent to which rays of light are refracted by a lens depends on its focal length. How-ever, the optometrists who prescribe correctional lenses and the opticians who make the lenses do not specify the focal length directly in prescriptions. Instead, they use the con-cept of *refractive power* to describe the extent to which a lens refracts light:

$$\text{Refractive power of a lens (in diopters)} = \frac{1}{f \text{ (in meters)}} \tag{26.8}$$

The refractive power is measured in units of *diopters*. One diopter is 1 m$^{-1}$.

Equation 26.8 shows that a converging lens has a refractive power of 1 diopter if it focuses parallel light rays to a focal point 1 m beyond the lens. If a lens refracts parallel rays even more and converges them to a focal point only 0.25 m beyond the lens, the lens has four times more refractive power, or 4 diopters. Since a converging lens has a positive focal length and a diverging lens has a negative focal length, the refractive power of a converging lens is positive while that of a diverging lens is negative. For instance, the eyeglasses in Example 11 would be described in a prescription from an optometrist in the following way: Refractive power = 1/(−5.19 m) = −0.193 diopters. The contact lenses in Example 12 would be described in a similar fashion: Refractive power = 1/(0.284 m) = 3.52 diopters.

## 26.11  *Angular Magnification and the Magnifying Glass*

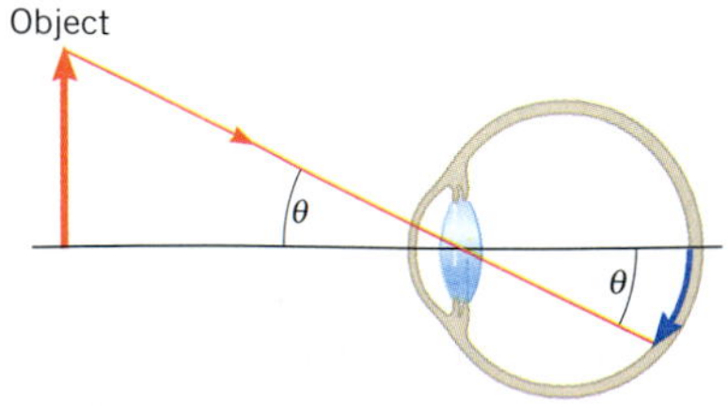

If you hold a penny at arm's length, the penny looks larger than the moon. The reason is that the penny, being so close, forms a larger image on the retina of the eye than does the more distant moon. The brain interprets the larger image of the penny as arising from a larger object. The size of the image on the retina determines how large an object appears to be. However, the size of the image on the retina is difficult to measure. Alternatively, the angle $\theta$ subtended by the image can be used as an indication of the image size. Figure 26.35 shows this alternative, which has the advantage that $\theta$ is also the angle subtended

Object

**Figure 26.35**  The angle $\theta$ is the angular size of both the image and the object.

by the object and, hence, can be measured more easily. The angle $\theta$ is called the *angular size* of both the image and the object. The larger the angular size, the larger the image on the retina, and the larger the object appears to be.

According to Equation 8.1, the angle $\theta$ (measured in radians) is the length of the circular arc that is subtended by the angle divided by the radius of the arc, as Figure 26.36a indicates. Part *b* of the drawing shows the situation for an object of height $h_o$ viewed at a distance $d_o$ from the eye. When $\theta$ is small, $h_o$ is approximately equal to the arc length and $d_o$ is nearly equal to the radius, so that

$$\theta \text{ (in radians)} = \text{Angular size} \approx \frac{h_o}{d_o}$$

This approximation is good to within one percent for angles of 9° or smaller. In the next example the angular size of a penny is compared with that of the moon.

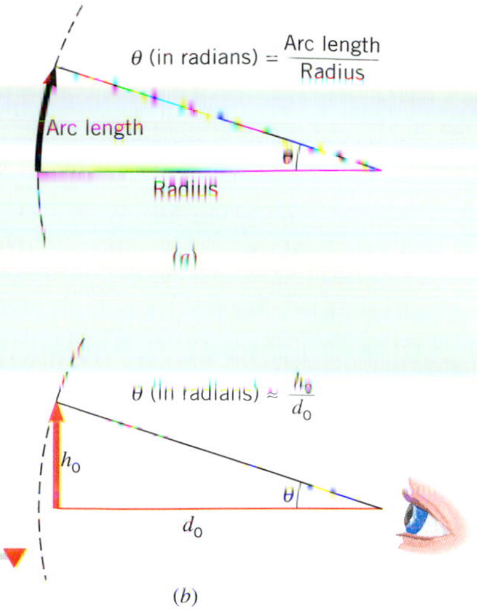

### Example 13  A Penny and the Moon

Compare the angular size of a penny (diameter $= h_o = 1.9$ cm) held at arm's length ($d_o = 71$ cm) with that of the moon (diameter $= h_o = 3.5 \times 10^6$ m, and $d_o = 3.9 \times 10^8$ m).

**Reasoning** The angular size $\theta$ of an object is given approximately by its height $h_o$ divided by its distance $d_o$ from the eye, $\theta \approx h_o/d_o$, provided that the angle involved is less than roughly 9°; this approximation applies here. The "heights" of the penny and the moon are their diameters.

**Solution** The angular sizes of the penny and moon are

*Penny*
$$\theta \approx \frac{h_o}{d_o} = \frac{1.9 \text{ cm}}{71 \text{ cm}} = \boxed{0.027 \text{ rad } (1.5°)}$$

*Moon*
$$\theta \approx \frac{h_o}{d_o} = \frac{3.5 \times 10^6 \text{ m}}{3.9 \times 10^8 \text{ m}} = \boxed{0.0090 \text{ rad } (0.52°)}$$

The penny thus appears to be about three times as large as the moon.

An optical instrument, such as a magnifying glass, allows us to view small or distant objects because it produces a larger image on the retina than would be possible otherwise. In other words, an optical instrument magnifies the angular size of the object. The *angular magnification* (or *magnifying power*) $M$ is the angular size $\theta'$ of the final image produced by the instrument divided by a reference angular size $\theta$. The reference angular size is the angular size of the object when seen without the instrument.

*Angular magnification*
$$M = \frac{\text{Angular size of final image produced by optical instrument}}{\text{Reference angular size of object seen without optical instrument}} = \frac{\theta'}{\theta} \qquad (26.9)$$

A magnifying glass is the simplest device that provides angular magnification. In this case, the reference angular size $\theta$ is chosen to be the angular size of the object when placed at the near point of the eye and seen without the magnifying glass. Since an object cannot be brought closer than the near point and still produce a sharp image on the retina, $\theta$ represents the largest angular size obtainable without the magnifying glass. Figure 26.37a indicates that the reference angular size is $\theta \approx h_o/N$, where $N$ is the distance from the eye to the near point. To compute $\theta'$, recall from Section 26.7 and Figure 26.26 that a magnifying glass is usually a single converging lens, with the object located inside the focal point. In this situation, Figure 26.37b indicates that the lens produces a virtual image that is enlarged and upright with respect to the object. Assuming the eye is next to the magnifying glass, the angular size $\theta'$ seen by the eye is $\theta' \approx h_o/d_o$, where $d_o$ is the object distance. The angular magnification is

$$M = \frac{\theta'}{\theta} \approx \frac{h_o/d_o}{h_o/N} = \frac{N}{d_o}$$

**The physics of a magnifying glass.**

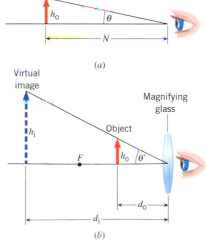

**Figure 26.36** (a) The angle $\theta$, measured in radians, is the arc length divided by the radius. (b) For small angles (less than 9°), $\theta$ is approximately equal to $h_o/d_o$, where $h_o$ and $d_o$ are the object height and distance.

**Figure 26.37** (a) Without a magnifying glass, the largest angular size $\theta$ occurs when the object is placed at the near point, a distance $N$ from the eye. (b) A magnifying glass produces an enlarged, virtual image of an object placed inside the focal point $F$ of the lens. The angular size of both the image and the object is $\theta'$.

According to the thin-lens equation, $d_o$ is related to the image distance $d_i$ and the focal length $f$ of the lens by

$$\frac{1}{d_o} = \frac{1}{f} - \frac{1}{d_i}$$

Substituting this expression for $1/d_o$ into the previous expression for $M$ leads to the following result:

*Angular magnification of a magnifying glass*  $$M = \frac{\theta'}{\theta} \approx \left(\frac{1}{f} - \frac{1}{d_i}\right) N \qquad (26.10)$$

Two special cases of this result are of interest, depending on whether the image is located as close to the eye as possible or as far away as possible. To be seen clearly, the closest the image can be relative to the eye is at the near point, or $d_i = -N$. The minus sign indicates that the image lies to the left of the lens and is virtual. In this event, Equation 26.10 becomes $M \approx (N/f) + 1$. The farthest the image can be from the eye is at infinity ($d_i = -\infty$); this occurs when the object is placed at the focal point of the lens. When the image is at infinity, Equation 26.10 simplifies to $M \approx N/f$. Clearly, the angular magnification is greater when the image is at the near point of the eye rather than at infinity. In either case, however, the greatest magnification is achieved by using a magnifying glass with the shortest possible focal length. Example 14 illustrates how to determine the angular magnification of a magnifying glass that is used in these two ways.

### Example 14  Examining a Diamond with a Magnifying Glass

A jeweler, whose near point is 40.0 cm from his eye and whose far point is at infinity, is using a small magnifying glass (called a loupe) to examine a diamond. The lens of the magnifying glass has a focal length of 5.00 cm, and the image of the gem is $-185$ cm from the lens. The image distance is negative because the image is virtual and is formed on the same side of the lens as the object. (a) Determine the angular magnification of the magnifying glass. (b) Where should the image be located so the jeweler's eye is fully relaxed and has the least strain? What is the angular magnification under this "least strain" condition?

**Reasoning** The angular magnification of the magnifying glass can be determined from Equation 26.10. In part (a) the image distance is $-185$ cm. In part (b) the ciliary muscle of the jeweler's eye is fully relaxed, so the image must be located infinitely far from the eye, at its far point, as Section 26.10 discusses.

**Solution**

(a) With $f = 5.00$ cm, $d_i = -185$ cm, and $N = 40.0$ cm, the angular magnification is

$$M = \left(\frac{1}{f} - \frac{1}{d_i}\right) N = \left(\frac{1}{5.00 \text{ cm}} - \frac{1}{-185 \text{ cm}}\right)(40.0 \text{ cm}) = \boxed{8.22}$$

(b) With $f = 5.00$ cm, $d_i = -\infty$, and $N = 40.0$ cm, the angular magnification is

$$M = \left(\frac{1}{f} - \frac{1}{d_i}\right) N = \left(\frac{1}{5.00 \text{ cm}} - \frac{1}{-\infty}\right)(40.0 \text{ cm}) = \boxed{8.00}$$

Jewelers often prefer to minimize eyestrain when viewing objects, even though it means a slight reduction in angular magnification.

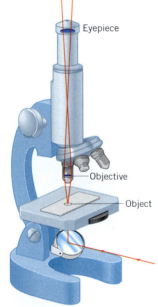

**Figure 26.38** A compound microscope.

**The physics of the compound microscope.**

## 26.12 *The Compound Microscope*

To increase the angular magnification beyond that possible with a magnifying glass, an additional converging lens can be included to "premagnify" the object before the magnifying glass comes into play. The result is an optical instrument known as the *compound microscope* (Figure 26.38). As discussed in Section 26.9, the magnifying glass is called the eyepiece, and the additional lens is called the objective.

The angular magnification of the compound microscope is $M = \theta'/\theta$ (Equation 26.9), where $\theta'$ is the angular size of the final image and $\theta$ is the reference angular size.

As with the magnifying glass in Figure 26.37, the reference angular size is determined by the height $h_o$ of the object when the object is located at the near point of the unaided eye: $\theta \approx h_o/N$, where $N$ is the distance between the eye and the near point. Assuming that the object is placed just outside the focal point $F_o$ of the objective (see Figure 26.30a) and that the final image is very far from the eyepiece (i.e., near infinity, see Figure 26.30c), it can be shown that

**Angular magnification of a compound microscope**

$$M \approx -\frac{(L - f_e)N}{f_o f_e} \qquad (L > f_o + f_e) \qquad (26.11)$$

In Equation 26.11, $f_o$ and $f_e$ are, respectively, the focal lengths of the objective and the eyepiece. The angular magnification is greatest when $f_o$ and $f_e$ are as small as possible (since they are in the denominator in Equation 26.11) and when the distance $L$ between the lenses is as large as possible. Furthermore, $L$ must be greater than the sum of $f_o$ and $f_e$ for this equation to be valid. Example 15 deals with the angular magnification of a compound microscope.

### Example 15 The Angular Magnification of a Compound Microscope

The focal length of the objective of a compound microscope is $f_o = 0.40$ cm, and that of the eyepiece is $f_e = 3.0$ cm. The two lenses are separated by a distance of $L = 20.0$ cm. A person with a near point distance of $N = 25$ cm is using the microscope. (a) Determine the angular magnification of the microscope. (b) Compare the answer in part (a) with the largest angular magnification obtainable by using the eyepiece alone as a magnifying glass.

**Reasoning** The angular magnification of the compound microscope can be obtained directly from Equation 26.11, since all the variables are known. When the eyepiece is used alone as a magnifying glass, as in Figure 26.37b, the largest angular magnification occurs when the image seen through the eyepiece is as close as possible to the eye. The image in this case is at the near point, and according to Equation 26.10, the angular magnification is $M \approx (N/f_e) + 1$.

**Solution**

(a) The angular magnification of the compound microscope is

$$M \approx -\frac{(L - f_e)N}{f_o f_e} = -\frac{(20.0 \text{ cm} - 3.0 \text{ cm})(25 \text{ cm})}{(0.40 \text{ cm})(3.0 \text{ cm})} = \boxed{-350}$$

The minus sign indicates that the final image is inverted relative to the initial object.

(b) The maximum angular magnification of the eyepiece by itself is

$$M \approx \frac{N}{f_e} + 1 = \frac{25 \text{ cm}}{3.0 \text{ cm}} + 1 = \boxed{9.3}$$

The effect of the objective is to increase the angular magnification of the compound microscope by a factor of $350/9.3 = 38$ compared to that of a magnifying glass.

## 26.13 The Telescope

A telescope is an instrument for magnifying distant objects, such as stars and planets. Like a microscope, a telescope consists of an objective and an eyepiece (also called the ocular). Since the object is usually far away, the light rays striking the telescope are nearly parallel, and the "first image" is formed just beyond the focal point $F_o$ of the objective, as Figure 26.39a illustrates. The first image is real and inverted. Unlike that in the compound microscope, however, this image is *smaller* than the object. If, as in part *b* of the drawing, the telescope is constructed so the first image lies just inside the focal point $F_e$ of the eyepiece, the eyepiece acts like a magnifying glass. It forms a final image that is greatly enlarged, virtual, and located near infinity. This final image can then be viewed with a fully relaxed eye.

*The physics of the telescope.*

The angular magnification $M$ of a telescope, like that of a magnifying glass or a microscope, is the angular size $\theta'$ subtended by the final image of the telescope divided by the

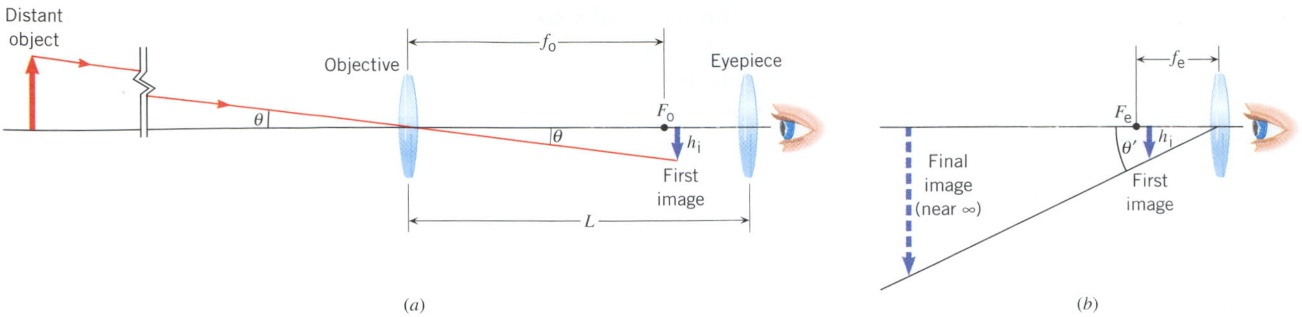

Eyepiece

**Figure 26.39** (*a*) An astronomical telescope is used to view distant objects. (Note the "break" in the principal axis, between the object and the objective.) The objective produces a real, inverted first image. (*b*) The eyepiece magnifies the first image to produce the final image near infinity.

reference angular size $\theta$ of the object. For an astronomical object, such as a planet, it is convenient to use as a reference the angular size of the object seen in the sky with the unaided eye. Since the object is far away, the angular size seen by the unaided eye is nearly the same as the angle $\theta$ subtended at the objective of the telescope in Figure 26.39a. Moreover, $\theta$ is also the angle subtended by the first image, so $\theta \approx -h_i/f_o$, where $h_i$ is the height of the first image and $f_o$ is the focal length of the objective. A minus sign has been inserted into this equation because the first image is inverted relative to the object and the image height $h_i$ is a negative number. The insertion of the minus sign ensures that the term $-h_i/f_o$, and hence $\theta$, is a positive quantity. To obtain an expression for $\theta'$, refer to Figure 26.39b and note that the first image is located very near the focal point $F_e$ of the eyepiece, which has a focal length $f_e$. Therefore, $\theta' \approx h_i/f_e$. The angular magnification of the telescope is approximately

**Angular magnification of an astronomical telescope**

$$M = \frac{\theta'}{\theta} \approx \frac{h_i/f_e}{-h_i/f_o} \approx -\frac{f_o}{f_e} \qquad (26.12)$$

The angular magnification is determined by the ratio of the focal length of the objective to the focal length of the eyepiece. For large angular magnifications, the objective should have a long focal length and the eyepiece a short one. Some of the design features of a telescope are the topic of the next example.

### Example 16 The Angular Magnification of an Astronomical Telescope

The telescope shown in Figure 26.40 has the following specifications: $f_o = 985$ mm and $f_e = 5.00$ mm. From these data, find (a) the angular magnification of the telescope and (b) the approximate length of the telescope.

**Reasoning** The angular magnification of the telescope follows directly from Equation 26.12, since the focal lengths of the objective and eyepiece are known. We can find the length of the telescope by noting that it is approximately equal to the distance $L$ between the objective and eyepiece. Figure 26.39 shows that the first image is located just beyond the focal point $F_o$ of the objective and just inside the focal point $F_e$ of the eyepiece. These two focal points are, therefore, very close together, so the distance $L$ is approximately the sum of the two focal lengths: $L \approx f_o + f_e$.

**Solution**

(a) The angular magnification is approximately

$$M \approx -\frac{f_o}{f_e} = -\frac{985 \text{ mm}}{5.00 \text{ mm}} = \boxed{-197} \qquad (26.12)$$

(b) The approximate length of the telescope is

$$L \approx f_o + f_e = 985 \text{ mm} + 5.00 \text{ mm} = \boxed{990 \text{ mm}}$$

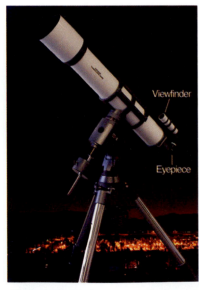

**Figure 26.40** An astronomical telescope. The viewfinder is a separate small telescope with low magnification and serves as an aid in locating the object. Once the object has been found, the viewer looks through the eyepiece to obtain the full magnification of the telescope. (© Tony Freeman/PhotoEdit)

## 26.14 Lens Aberrations

Rather than forming a sharp image, a single lens typically forms an image that is slightly out of focus. This lack of sharpness arises because the rays originating from a single point on the object are not focused to a single point on the image. As a result, each point on the image becomes a small blur. The lack of point-to-point correspondence between object and image is called an *aberration*.

One common type of aberration is **spherical aberration**, and it occurs with converging and diverging lenses made with spherical surfaces. Figure 26.41*a* shows how spherical aberration arises with a converging lens. Ideally, all rays traveling parallel to the principal axis are refracted so they cross the axis at the same point after passing through the lens. However, rays far from the principal axis are refracted more by the lens than those closer in. Consequently, the outer rays cross the axis closer to the lens than do the inner rays, so a lens with spherical aberration does not have a unique focal point. Instead, as the drawing suggests, there is a location along the principal axis where the light converges to the smallest cross-sectional area. This area is circular and is known as the **circle of least confusion.** The circle of least confusion is where the most satisfactory image can be formed by the lens.

Spherical aberration can be reduced substantially by using a variable-aperture diaphragm to allow only those rays close to the principal axis to pass through the lens. Figure 26.41*b* indicates that a reasonably sharp focal point can be achieved by this method, although less light now passes through the lens. Lenses with parabolic surfaces are also used to reduce this type of aberration, but they are difficult and expensive to make.

*Chromatic aberration* also causes blurred images. It arises because the index of refraction of the material from which the lens is made varies with wavelength. Section 26.5 discusses how this variation leads to the phenomenon of dispersion, in which different colors refract by different amounts. Figure 26.42*a* shows sunlight incident on a converging lens, in which the light spreads into its color spectrum because of dispersion. For clarity, however, the picture shows only the colors at the opposite ends of the visible spectrum—red and violet. Violet is refracted more than red, so the violet ray crosses the principal axis closer to the lens than does the red ray. Thus, the focal length of the lens is shorter for violet than for red, with intermediate values of the focal length corresponding to the colors in between. As a result of chromatic aberration, an undesirable color fringe surrounds the image.

Chromatic aberration can be greatly reduced by using a compound lens, such as the combination of a converging lens and a diverging lens shown in Figure 26.42*b*. Each lens is made from a different type of glass. With this lens combination the red and violet rays almost come to a common focus and, thus, chromatic aberration is reduced. A lens combination designed to reduce chromatic aberration is called an *achromatic lens* (from the Greek "achromatos," meaning "without color"). All high-quality cameras use achromatic lenses.

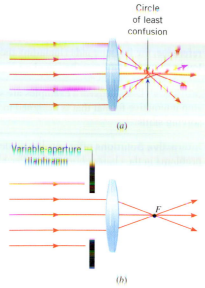

**Figure 26.41** (*a*) In a converging lens, spherical aberration prevents light rays parallel to the principal axis from converging to a common point. (*b*) Spherical aberration can be reduced by allowing only rays near the principal axis to pass through the lens. The refracted rays now converge more nearly to a single focal point *F*.

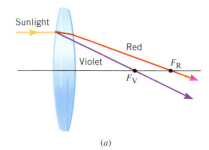

(*a*)

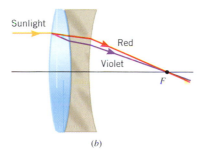

(*b*)

**Figure 26.42** (*a*) Chromatic aberration arises when different colors are focused at different points along the principal axis: $F_V$ = focal point for violet light, $F_R$ = focal point for red light. (*b*) A converging and a diverging lens in tandem can be designed to bring different colors more nearly to the same focal point *F*.

# Concept Summary

This summary presents an abridged version of the chapter, including the important equations and all available learning aids. For convenient reference, the learning aids (including the text's examples) are placed next to or immediately after the relevant equation or discussion. The following learning aids may be found on-line at **www.wiley.com/college/cutnell**:

| | |
|---|---|
| **Interactive LearningWare** examples are solved according to a five-step interactive format that is designed to help you develop problem-solving skills. | **Concept Simulations** are animated versions of text figures or animations that illustrate important concepts. You can control parameters that affect the display, and we encourage you to experiment. |
| **Interactive Solutions** offer specific models for certain types of problems in the chapter homework. The calculations are carried out interactively. | **Self-Assessment Tests** include both qualitative and quantitative questions. Extensive feedback is provided for both incorrect and correct answers, to help you evaluate your understanding of the material. |

| Topic | Discussion | Learning Aids |
|---|---|---|
| | **26.1 The Index of Refraction** | |
| Refraction | The change in speed as a ray of light goes from one material to another causes the ray to deviate from its incident direction. This change in direction is called refraction. The index of refraction $n$ of a material is the ratio of the speed $c$ of light in a vacuum to the speed $v$ of light in the material: | |
| Index of refraction | $$n = \frac{c}{v} \qquad (26.1)$$ | Interactive Solution 26.7 |
| | The values for $n$ are less than unity, because the speed of light in a material medium is less than it is in a vacuum. | |
| | **26.2 Snell's Law and the Refraction of Light** | |
| | The refraction that occurs at the interface between two materials obeys Snell's law of refraction. This law states that (1) the refracted ray, the incident ray, and the normal to the interface all lie in the same plane, and (2) the angle of refraction $\theta_2$ is related to the angle of incidence $\theta_1$ according to | |
| Snell's law of refraction | $$n_1 \sin \theta_1 = n_2 \sin \theta_2 \qquad (26.2)$$ | Example 1 |
| | where $n_1$ and $n_2$ are the indices of refraction of the incident and refracting media, respectively. The angles are measured relative to the normal. | |
| | Because of refraction, a submerged object has an apparent depth that is different from its actual depth. If the observer is directly above (or below) the object, the apparent depth (or height) $d'$ is related to the actual depth (or height) $d$ according to | Example 2 |
| Apparent depth | $$d' = d\left(\frac{n_2}{n_1}\right) \qquad (26.3)$$ | Examples 3, 4 |
| | where $n_1$ and $n_2$ are the refractive indices of the materials in which the object and the observer, respectively, are located. | |
| | **26.3 Total Internal Reflection** | |
| | When light passes from a material with a larger refractive index $n_1$ into a material with a smaller refractive index $n_2$, the refracted ray is bent away from the normal. If the incident ray is at the critical angle $\theta_c$, the angle of refraction is 90°. The critical angle is determined from Snell's law and is given by | Concept Simulation 26.1 |
| Critical angle | $$\sin \theta_c = \frac{n_2}{n_1} \qquad (n_1 > n_2) \qquad (26.4)$$ | Examples 5, 6<br>Interactive LearningWare 26.1<br>Interactive Solution 26.25 |
| Total internal reflection | When the angle of incidence exceeds the critical angle, all the incident light is reflected back into the material from which it came, a phenomenon known as total internal reflection. | |
| | **26.4 Polarization and the Reflection and Refraction of Light** | |
| Brewster angle | When light is incident on a nonmetallic surface at the Brewster angle $\theta_B$, the reflected light is completely polarized parallel to the surface. The Brewster angle is given by | |
| Brewster's law | $$\tan \theta_B = \frac{n_2}{n_1} \qquad (26.5)$$ | |

| Topic | Discussion | Learning Aids |
|-------|------------|---------------|

where $n_1$ and $n_2$ are the refractive indices of the incident and refracting media, respectively. When light is incident at the Brewster angle, the reflected and refracted rays are perpendicular to each other.

### 26.5 The Dispersion of Light: Prisms and Rainbows

**Dispersion**

A glass prism can spread a beam of sunlight into a spectrum of colors because the index of refraction of the glass depends on the wavelength of the light. Thus, a prism bends the refracted rays corresponding to different colors by different amounts. The spreading of light into its color components is known as dispersion. The dispersion of light by water droplets in the air leads to the formation of rainbows.

**Concept Simulation 26.2**

**Index matching**

A prism will not bend a light ray at all, neither up nor down, if the surrounding fluid has the same refractive index as the glass, a condition known as index matching.

**Example 7**

Use Self-Assessment Test 26.1 to evaluate your understanding of Sections 26.1–26.5.

### 26.6 Lenses

### 26.7 The Formation of Images by Lenses

**Focal point and focal length of a converging lens**

**Focal point and focal length of a diverging lens**

**Ray tracing**

Converging lenses and diverging lenses depend on the phenomenon of refraction in forming an image. With a converging lens, paraxial rays that are parallel to the principal axis are focused to a point on the axis by the lens. This point is called the focal point of the lens, and its distance from the lens is the focal length $f$. Paraxial light rays that are parallel to the principal axis of a diverging lens appear to originate from its focal point after passing through the lens. The distance of this point from the lens is the focal length $f$. The image produced by a converging or a diverging lens can be located via a technique known as ray tracing, which utilizes the three rays outlined in the beginning of Section 26.7.

**Image formed by a converging lens**

The nature of the image formed by a converging lens depends on where the object is situated relative to the lens. When the object is located at a distance from the lens that is greater than twice the focal length, the image is real, inverted, and smaller than the object. When the object is located at a distance from the lens that is between the focal length and twice the focal length, the image is real, inverted, and larger than the object. When the object is located within the focal length, the image is virtual, upright, and larger than the object.

**Concept Simulation 26.3**

**Image formed by a diverging lens**

Regardless of the position of a real object, a diverging lens always produces an image that is virtual, upright, and smaller than the object.

**Concept Simulation 26.4**

### 26.8 The Thin-Lens Equation and the Magnification Equation

The thin-lens equation can be used with either converging or diverging lenses that are thin, and it relates the object distance $d_o$, the image distance $d_i$, and the focal length $f$ of the lens:

**Thin-lens equation**

$$\frac{1}{d_o} + \frac{1}{d_i} = \frac{1}{f} \qquad (26.6)$$

**Interactive Solution 26.51**

**Magnification**

The magnification $m$ of a lens is the ratio of the image height $h_i$ to the object height $h_o$ and is also related to $d_o$ and $d_i$ by the magnification equation:

**Examples 8, 9**

**Magnification equation**

$$m = \frac{h_i}{h_o} = -\frac{d_i}{d_o} \qquad (26.7)$$

The algebraic sign conventions for the variables appearing in the thin-lens and magnification equations are summarized in Section 26.8.

### 26.9 Lenses in Combination

When two or more lenses are used in combination, the image produced by one lens serves as the object for the next lens.

**Example 10**

**Interactive LearningWare 26.2**

| *Topic* | *Discussion* | *Learning Aids* |
|---|---|---|

### 26.10 The Human Eye

In the human eye, a real, inverted image is formed on a light-sensitive surface,
called the retina. Accommodation is the process by which the focal length of
the eye is automatically adjusted, so that objects at different distances produce
sharp images on the retina. The near point of the eye is the point nearest the eye
at which an object can be placed and still have a sharp image produced on the
retina. The far point of the eye is the location of the farthest object on which the
fully relaxed eye can focus. For a young and normal eye, the near point is located 25 cm from the eye, and the far point is located at infinity.

**Accommodation**

**Near point**

**Far point**

**Nearsightedness**

**Farsightedness**

A nearsighted (myopic) eye is one that can focus on nearby objects, but not on **Example 11**
distant objects. Nearsightedness can be corrected with eyeglasses or contacts
made from diverging lenses. A farsighted (hyperopic) eye can see distant ob- **Example 12**
jects clearly, but not objects close up. Farsightedness can be corrected with con- **Interactive Solution 26.67**
verging lenses.

The refractive power of a lens is measured in diopters and is given by

**Refractive power**

$$\text{Refractive power (in diopters)} = \frac{1}{f \text{ (in meters)}} \qquad (26.8)$$

where $f$ is the focal length of the lens and must be expressed in meters. A converging lens has a positive refractive power, and a diverging lens has a negative
refractive power.

### 26.11 Angular Magnification and the Magnifying Glass

The angular size of an object is the angle that it subtends at the eye of the
viewer. For small angles, the angular size $\theta$ in radians is

**Angular size**

$$\theta \text{ (in radians)} \approx \frac{h_o}{d_o} \qquad \qquad \textbf{Example 13}$$

where $h_o$ is the height of the object and $d_o$ is the object distance. The angular
magnification $M$ of an optical instrument is the angular size $\theta'$ of the final image produced by the instrument divided by the reference angular size $\theta$ of the
object, which is that seen without the instrument:

**Angular magnification**

$$M = \frac{\theta'}{\theta} \qquad (26.9)$$

A magnifying glass is usually a single converging lens that forms an enlarged,
upright, and virtual image of an object placed at or inside the focal point of the
lens. For a magnifying glass held close to the eye, the angular magnification $M$
is approximately

**Angular magnification
of a magnifying glass**

$$M \approx \left( \frac{1}{f} - \frac{1}{d_i} \right) N \qquad (26.10) \quad \textbf{Example 14}$$

where $f$ is the focal length of the lens, $d_i$ is the image distance, and $N$ is the distance of the viewer's near point from the eye.

### 26.12 The Compound Microscope

A compound microscope usually consists of two lenses, an objective and an
eyepiece. The final image is enlarged, inverted, and virtual. The angular magnification $M$ of such a microscope is approximately

**Angular magnification
of a compound microscope**

$$M \approx -\frac{(L - f_e)N}{f_o f_e} \qquad (L > f_o + f_e) \qquad (26.11) \quad \textbf{Example 15}$$

where $f_o$ and $f_e$ are, respectively, the focal lengths of the objective and eyepiece,
$L$ is the distance between the two lenses, and $N$ is the distance of the viewer's
near point from his or her eye.

### 26.13 The Telescope

An astronomical telescope magnifies distant objects with the aid of an objective
and an eyepiece, and it produces a final image that is inverted and virtual. The

| Topic | Discussion | Learning Aids |
|---|---|---|

angular magnification $M$ of a telescope is approximately

**Angular magnification of an astronomical telescope**

$$M \approx -\frac{f_o}{f_e}$$

(26.12) **Example 16**

where $f_o$ and $f_e$ are, respectively, the focal lengths of the objective and eyepiece

### 26.14 Lens Aberrations

Lens aberrations limit the formation of perfectly focused or sharp images by optical instruments. Spherical aberration occurs because rays that pass through the outer edge of a lens with spherical surfaces are not focused at the same point as those that pass through near the center of the lens.

**Spherical aberration**

**Chromatic aberration**

Chromatic aberration arises because a lens focuses different colors at slightly different points.

 Use *Self-Assessment Test 24.2* to evaluate your understanding of Sections 26.6–26.14.

# Problems

*Unless specified otherwise, use the values given in Table 26.1 for the refractive indices.*

ssm **Solution is in the Student Solutions Manual.**   www **Solution is available on the World Wide Web at www.wiley.com/college/cutnell**
☤ **This icon represents a biomedical application.**

### Section 26.1 The Index of Refraction

1. ssm What is the speed of light in benzene?

2. The refractive indices of materials $A$ and $B$ have a ratio of $n_A/n_B = 1.33$. The speed of light in material $A$ is $1.25 \times 10^8$ m/s. What is the speed of light in material $B$?

3. The frequency of a light wave is the same when the light travels in ethyl alcohol as it is when it travels in carbon disulfide. Find the ratio of the wavelength of the light in ethyl alcohol to that in carbon disulfide.

4. The speed of light is 1.25 times as large in material $A$ as in material $B$. Determine the ratio $n_A/n_B$ of the refractive indices of these materials.

5. ssm www A plate glass window ($n = 1.5$) has a thickness of $4.0 \times 10^3$ m. How long does it take light to pass perpendicularly through the plate?

6. Light has a wavelength of 340.0 nm and a frequency of $5.403 \times 10^{14}$ Hz when traveling through a certain substance. What substance from Table 26.1 could this be?

\* 7. **Interactive Solution 26.7** at **www.wiley.com/college/cutnell** offers one model for problems like this one. In a certain time, light travels 6.20 km in a vacuum. During the same time, light travels only 3.40 km in a liquid. What is the refractive index of the liquid?

\* 8. A flat sheet of ice has a thickness of 2.0 cm. It is on top of a flat sheet of crystalline quartz that has a thickness of 1.1 cm. Light strikes the ice perpendicularly and travels through it and then through the quartz. In the time it takes the light to travel through the two sheets, how far (in cm) would it have traveled in a vacuum?

### Section 26.2 Snell's Law and the Refraction of Light

9. ssm A light ray in air is incident on a water surface at a 43° angle of incidence. Find (a) the angle of reflection and (b) the angle of refraction.

10. A ray of light traveling in material $A$ strikes the interface between materials $A$ and $B$ at an angle of incidence of 72°. The angle of refraction is 56°. Find the ratio $n_A/n_B$ of the refractive indices of the two materials.

11. As an aid in understanding this problem, refer to Conceptual Example 4. A swimmer, who is looking up from under the water, sees a diving board directly above at an apparent height of 4.0 m above the water. What is the actual height of the diving board?

12. A person working on the transmission of a car accidentally drops a bolt into a tray of oil. The oil is 5.00 cm deep. The bolt appears to be 3.40 cm beneath the surface of the oil, when viewed from directly above. What is the index of refraction of the oil?

13. ssm The drawing shows a coin resting on the bottom of a beaker filled with an unknown liquid. A ray of light from the coin travels to the surface of the liquid and is refracted as it enters into the air. A person sees the ray as it skims just above the surface of the liquid. How fast is the light traveling in the liquid?

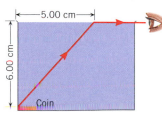

14. A ray of sunlight hits a frozen lake at a 45° angle of incidence. At what angle of refraction does the ray penetrate (a) the ice and (b) the water beneath the ice?

15. Light in a vacuum is incident on a transparent glass slab. The angle of incidence is 35.0°. The slab is then immersed in a pool of liquid. When the angle of incidence for the light striking the slab is 20.3°, the angle of refraction for the light entering the slab is the same as when the slab was in a vacuum. What is the index of refraction of the liquid?

\* 16. A silver medallion is sealed within a transparent block of plastic. An observer in air, viewing the medallion from directly above, sees the medallion at an apparent depth of 1.6 cm beneath the top surface of the block. How far below the top surface would the medallion appear if the observer (not wearing goggles) and the block were under water?

* **17.** **ssm www** In Figure 26.6, suppose that the angle of incidence is $\theta_1 = 30.0°$, the thickness of the glass pane is 6.00 mm, and the refractive index of the glass is $n_2 = 1.52$. Find the amount (in mm) by which the emergent ray is displaced relative to the incident ray.

* **18.** Review Conceptual Example 4 as background for this problem. A man in a boat is looking straight down at a fish in the water directly beneath him. The fish is looking straight up at the man. They are equidistant from the air–water interface. To the man, the fish appears to be 2.0 m beneath his eyes. To the fish, how far above its eyes does the man appear to be?

* **19.** Refer to Figure 26.4a and assume the observer is nearly above the submerged object. For this situation, derive the expression for the apparent depth: $d' = d(n_2/n_1)$, Equation 26.3. *(Hint: Use Snell's law of refraction and the fact that the angles of incidence and refraction are small, so $\tan \theta \approx \sin \theta$.)*

* **20.** The drawing shows a rectangular block of glass $(n = 1.52)$ surrounded by liquid carbon disulfide $(n = 1.63)$. A ray of light is incident on the glass at point A with a 30.0° angle of incidence. At what angle of refraction does the ray leave the glass at point B?

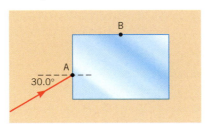

** **21.** **ssm** A small logo is embedded in a thick block of crown glass $(n = 1.52)$, 3.20 cm beneath the top surface of the glass. The block is put under water, so there is 1.50 cm of water above the top surface of the block. The logo is viewed from directly above by an observer in air. How far beneath the top surface of the water does the logo appear to be?

** **22.** A beaker has a height of 30.0 cm. The lower half of the beaker is filled with water, and the upper half is filled with oil $(n = 1.48)$. To a person looking down into the beaker from above, what is the apparent depth of the bottom?

### Section 26.3 Total Internal Reflection

**23.** **ssm** One method of determining the refractive index of a transparent solid is to measure the critical angle when the solid is in air. If $\theta_c$ is found to be 40.5°, what is the index of refraction of the solid?

**24.** What is the critical angle for light emerging from carbon disulfide into air?

**25.** **Interactive Solution 26.25** at **www.wiley.com/college/cutnell** provides one model for solving problems such as this. A glass block $(n = 1.56)$ is immersed in a liquid. A ray of light within the glass hits a glass–liquid surface at a 75.0° angle of incidence. Some of the light enters the liquid. What is the smallest possible refractive index for the liquid?

**26.** A point source of light is submerged 2.2 m below the surface of a lake and emits rays in all directions. On the surface of the lake, directly above the source, the area illuminated is a circle. What is the maximum radius that this circle could have?

**27.** **ssm** The drawing shows a crown glass slab with a rectangular cross section. As illustrated, a laser beam strikes the upper surface at an angle of 60.0°. After reflecting from the upper surface, the beam

reflects from the side and bottom surfaces. (a) If the glass is surrounded by air, determine where part of the beam first exits the glass, at point A, B, or C. (b) Repeat part (a), assuming that the glass is surrounded by water.

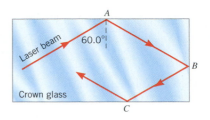

**28.** A layer of liquid B floats on liquid A. A ray of light begins in liquid A and undergoes total internal reflection at the interface between the liquids when the angle of incidence exceeds 36.5°. When liquid B is replaced with liquid C, total internal reflection occurs for angles of incidence greater than 47.0°. Find the ratio $n_B/n_C$ of the refractive indices of liquids B and C.

* **29.** The drawing shows an optical fiber that consists of a core made of flint glass $(n_{flint} = 1.667)$ surrounded by a cladding made of crown glass $(n_{crown} = 1.523)$. A beam of light enters the fiber from air at an angle $\theta_1$ with respect to the normal. What is $\theta_1$ if the light strikes the core–cladding interface at the critical angle $\theta_c$?

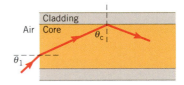

* **30.** **Interactive Learning Ware 26.1** at **www.wiley.com/college/cutnell** provides some helpful background for this problem. The drawing shows a crystalline quartz slab with a rectangular cross section. A ray of light strikes the slab at an incident angle of $\theta_1 = 34°$, enters the quartz, and travels to point P. This slab is surrounded by a fluid with a refractive index $n$. What is the maximum value of $n$ such that total internal reflection occurs at point P?

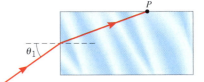

### Section 26.4 Polarization and the Reflection and Refraction of Light

**31.** **ssm www** When light strikes the surface between two materials from above, the Brewster angle is 65.0°. What is the Brewster angle when the light encounters the same surface from below?

**32.** Sunlight strikes a diamond surface. At what angle of incidence is the reflected light completely polarized?

**33.** Light is incident from air onto a beaker of carbon tetrachloride. If the reflected light is 100% polarized, what is the angle of refraction of the light that penetrates into the carbon tetrachloride?

**34.** For light that originates within a liquid and strikes the liquid–air interface, the critical angle is 39°. What is Brewster's angle for this light?

**35.** **ssm** Light is reflected from a glass coffee table. When the angle of incidence is 56.7°, the reflected light is completely polarized parallel to the surface of the glass. What is the index of refraction of the glass?

* **36.** When red light in a vacuum is incident at the Brewster angle on a certain type of glass, the angle of refraction is 29.9°. What are (a) the Brewster angle and (b) the index of refraction of the glass?

* **37.** In Section 26.4 it is mentioned that the reflected and refracted rays are perpendicular to each other when light strikes the surface at the Brewster angle. This is equivalent to saying that the angle of reflection plus the angle of refraction is 90°. Using Snell's law and Brewster's law, prove that the angle of reflection plus the angle of refraction is 90°.

### Section 26.5 The Dispersion of Light: Prisms and Rainbows

**38.** A ray of sunlight is passing from diamond into crown glass; the angle of incidence is 35.00°. The indices of refraction for the blue and red components of the ray are: blue ($n_{diamond}$ = 2.444, $n_{crown glass}$ = 1.531), and red ($n_{diamond}$ = 2.410, $n_{crown glass}$ = 1.520). Determine the angle between the refracted blue and red rays in the crown glass.

**39. ssm** A beam of sunlight encounters a plate of crown glass at a 45.00° angle of incidence. Using the data in Table 26.2, find the angle between the violet ray and the red ray in the glass.

**40.** Red light ($n$ = 1.520) and violet light ($n$ = 1.538) traveling in air are incident on a slab of crown glass. Both colors enter the glass at the same angle of refraction. The red light has an angle of incidence of 30.00°. What is the angle of incidence of the violet light?

**41.** Horizontal rays of red light ($\lambda$ = 660 nm, in vacuum) and violet light ($\lambda$ = 410 nm, in vacuum) are incident on the flint-glass prism shown in the drawing. The indices of refraction for the red and violet light are 1.662 and 1.698, respectively. What is the angle of refraction for each ray as it emerges from the prism?

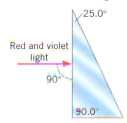

**\* 42.** Refer to Conceptual Example 7 as background material for this problem. The drawing shows a horizontal beam of light that is incident on a prism made from ice. The base of the prism is also horizontal. The prism ($n$ = 1.31) is surrounded by oil whose index of refraction is 1.48. Determine the angle $\theta$ that the exiting light makes with the normal to the right face of the prism.

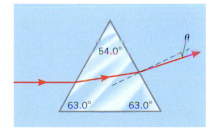

**\* 43. ssm** This problem relates to Figure 26.16 which illustrates the dispersion of light by a prism. The prism is made from glass, and its cross section is an equilateral triangle. The indices of refraction for the red and violet light are 1.662 and 1.698, respectively. The angle of incidence for both the red and violet light is 60.0°. Find the angles of refraction at which the red and violet rays emerge into the air from the prism.

### Section 26.6 Lenses,
### Section 26.7 The Formation of Images by Lenses,
### Section 26.8 The Thin-Lens Equation and the Magnification Equation

*(Note: When drawing ray diagrams, be sure that the object height $h_o$ is much smaller than the focal length $f$ of the lens or mirror.)*

**44.** A diverging lens has a focal length of $-32$ cm. An object is placed 19 cm in front of this lens. Calculate (a) the image distance and (b) the magnification. Is the image (c) real or virtual, (d) upright or inverted, and (e) enlarged or reduced in size?

**45. ssm** An object is located 9.0 cm in front of a converging lens ($f$ = 6.0 cm). Using an accurately drawn ray diagram, determine where the image is located.

**46.** When a diverging lens is held 13 cm above a line of print, as in Figure 26.27, the image is 5.0 cm beneath the lens. What is the focal length of the lens?

**47.** A tourist takes a picture of a mountain 14 km away using a camera that has a lens with a focal length of 50 mm. She then takes a second picture when she is only 5.0 km away. What is the ratio of the height of the mountain's image on the film for the second picture to its height on the film for the first picture?

**48. Concept Simulation 26.3** at **www.wiley.com/college/cutnell** reviews the concepts that play a role in this problem. A converging lens has a focal length of 88.00 cm. A 13.0-cm-tall object is located 155.0 cm in front of this lens. (a) What is the image distance? (b) Is the image real or virtual? (c) What is the image height? Be sure to include the proper algebraic sign.

**49. ssm** An object is located 30.0 cm to the left of a converging lens whose focal length is 50.0 cm. (a) Draw a ray diagram to scale and from it determine the image distance and the magnification. (b) Use the thin-lens and magnification equations to verify your answers to part (a).

**50.** A slide projector has a converging lens whose focal length is 105.00 mm. (a) How far (in meters) from the lens must the screen be located if a slide is placed 108.00 mm from the lens? (b) If the slide measures 24.0 mm × 36.0 mm, what are the dimensions (in mm) of its image?

**51.** Consult **Interactive Solution 26.51** at **www.wiley.com/college/ cutnell** to review the concepts on which this problem depends. A camera is supplied with two interchangeable lenses, whose focal lengths are 35.0 and 150.0 mm. A woman whose height is 1.60 m stands 9.00 m in front of the camera. What is the height (including sign) of her image on the film, as produced by (a) the 35.0-mm lens and (b) the 150.0-mm lens?

**\* 52. Concept Simulation 26.4** at **www.wiley.com/college/cutnell** provides the option of exploring the ray diagram that applies to this problem. The distance between an object and its image formed by a diverging lens is 49.0 cm. The focal length of the lens is $-233.0$ cm. Find (a) the image distance and (b) the object distance.

**\* 53. ssm** An office copier uses a lens to place an image of a document onto a rotating drum. The copy is made from this image. (a) What kind of lens is used? If the document and its copy are to have the same size, but are inverted with respect to one another, (b) how far from the document is the lens located and (c) how far from the lens is the image located? Express your answers in terms of the focal length $f$ of the lens.

**\* 54.** An object is in front of a converging lens ($f$ = 0.30 m). The magnification of the lens is $m$ = 4.0. (a) Relative to the lens, in what direction should the object be moved so that the magnification changes to $m$ = $-4.0$? (b) Through what distance should the object be moved?

**\* 55. ssm** An object is 18 cm in front of a diverging lens that has a focal length of $-12$ cm. How far in front of the lens should the object be placed so that the size of its image is reduced by a factor of 2.0?

**\* 56.** From a distance of 72 m, a photographer uses a telephoto lens ($f$ = 300.0 mm) to take a picture of a charging rhinoceros. How far from the rhinoceros would the photographer have to be to record an image of the same size using a lens whose focal length is 50.0 mm?

**\*\* 57.** A converging lens ($f$ = 25.0 cm) is used to project an image of an object onto a screen. The object and the screen are 125 cm apart, and between them the lens can be placed at either of two locations. Find the two object distances.

**\*\* 58.** The equation

$$\frac{1}{d_o} + \frac{1}{d_i} = \frac{1}{f}$$

is called the *Gaussian* form of the thin-lens equation. The drawing shows the variables $d_o$, $d_i$, and $f$. The drawing also shows the distances $x$ and $x'$, which are, respectively, the distance from the object to the focal point on the left of the lens and the distance

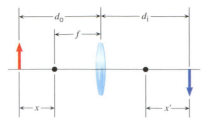

from the focal point on the right of the lens to the image. An equivalent form of the thin-lens equation, involving $x$, $x'$, and $f$, is called the *Newtonian* form. Show that the Newtonian form of the thin-lens equation can be written as $xx' = f^2$.

## Section 26.9 Lenses in Combination

**59.** A converging lens has a focal length of 0.080 m. An object is located 0.040 m to the left of this lens. A second converging lens has the same focal length as the first one and is located 0.120 m to the right of it. Relative to the second lens, where is the final image located?

**60.** Two identical diverging lenses are separated by 16 cm. The focal length of each lens is −8.0 cm. An object is located 4.0 cm to the left of the lens that is on the left. Determine the final image distance relative to the lens on the right.

**61. ssm** A converging lens ($f = 12.0$ cm) is located 30.0 cm to the left of a diverging lens ($f = -6.00$ cm). A postage stamp is placed 36.0 cm to the left of the converging lens. (a) Locate the final image of the stamp relative to the diverging lens. (b) Find the overall magnification. (c) Is the final image real or virtual? With respect to the original object, is the final image (d) upright or inverted, and is it (e) larger or smaller?

**62. Interactive LearningWare 26.2** at **www.wiley.com/college/cutnell** offers a review of the concepts that play roles in this problem. A diverging lens ($f = -10.0$ cm) is located 20.0 cm to the left of a converging lens ($f = 30.0$ cm). A 3.00-cm-tall object stands to the left of the diverging lens, exactly at its focal point. (a) Determine the distance of the final image relative to the converging lens. (b) What is the height of the final image (including the proper algebraic sign)?

**63.** A converging lens ($f_1 = 24.0$ cm) is located 56.0 cm to the left of a diverging lens ($f_2 = -28.0$ cm). An object is placed to the left of the converging lens, and the final image produced by the two-lens combination lies 20.7 cm to the left of the diverging lens. How far is the object from the converging lens?

* **64.** A coin is located 20.0 cm to the left of a converging lens ($f = 16.0$ cm). A second, identical lens is placed to the right of the first lens, such that the image formed by the combination has the same size and orientation as the original coin. Find the separation between the lenses.

* **65. ssm** An object is placed 20.0 cm to the left of a diverging lens ($f = -8.00$ cm). A concave mirror ($f = 12.0$ cm) is placed 30.0 cm to the right of the lens. (a) Find the final image distance, measured relative to the mirror. (b) Is the final image real or virtual? (c) Is the final image upright or inverted with respect to the original object?

** **66.** Two converging lenses ($f_1 = 9.00$ cm and $f_2 = 6.00$ cm) are separated by 18.0 cm. The lens on the left has the longer focal length. An object stands 12.0 cm to the left of the left-hand lens in the combination. (a) Locate the final image relative to the lens on the right. (b) Obtain the overall magnification. (c) Is the final image real or virtual? With respect to the original object, is the final image (d) upright or inverted and is it (e) larger or smaller?

## Section 26.10 The Human Eye

**67. Interactive Solution 26.67** at **www.wiley.com/college/cutnell** illustrates one approach to solving problems such as this one. A farsighted person has a near point that is 67.0 cm from her eyes. She wears eyeglasses that are designed to enable her to read a newspaper held at a distance of 25.0 cm from her eyes. Find the focal length of the eyeglasses, assuming that they are worn (a) 2.2 cm from the eyes and (b) 3.3 cm from the eyes.

**68.** A person holds a book 25 cm in front of the effective lens of her eye; the print in the book is 2.0 mm high. If the effective lens of the eye is located 1.7 cm from the retina, what is the size (including the sign) of the print image on the retina?

**69. ssm** A nearsighted person has a far point located only 220 cm from his eyes. Determine the focal length of contact lenses that will enable him to see distant objects clearly.

**70.** A student is reading a lecture written on a blackboard. The lenses in her eyes have a refractive power of 57.50 diopters, and the lens-to-retina distance is 1.750 cm. (a) How far (in meters) is the blackboard from her eyes? (b) If the writing on the blackboard is 5.00 cm high, what is the size of the image on her retina?

**71.** A nearsighted person cannot read a sign that is more than 5.2 m from his eyes. To deal with this problem, he wears contact lenses that do not correct his vision completely, but do allow him to read signs located up to distances of 12.0 m from his eyes. What is the focal length of the contacts?

**72.** A farsighted woman breaks her current eyeglasses and is using an old pair whose refractive power is 1.660 diopters. Since these eyeglasses do not completely correct her vision, she must hold a newspaper 42.00 cm from her eyes in order to read it. She wears the eyeglasses 2.00 cm from her eyes. How far is her near point from her eyes?

* **73. ssm** At age forty, a man requires contact lenses ($f = 65.0$ cm) to read a book held 25.0 cm from his eyes. At age forty-five, he finds that while wearing these contacts he must now hold a book 29.0 cm from his eyes. (a) By what distance has his near point *changed*? (b) What focal length lenses does he require at age forty-five to read a book at 25.0 cm?

** **74.** The contacts worn by a farsighted person allow her to see objects clearly that are as close as 25.0 cm, even though her uncorrected near point is 79.0 cm from her eyes. When she is looking at a poster, the contacts form an image of the poster at a distance of 217 cm from her eyes. (a) How far away is the poster actually located? (b) If the poster is 0.350 m tall, how tall is the image formed by the contacts?

** **75.** The far point of a nearsighted person is 6.0 m from her eyes, and she wears contacts that enable her to see distant objects clearly. A tree is 18.0 m away and 2.0 m high. (a) When she looks through the contacts at the tree, what is its image distance? (b) How high is the image formed by the contacts?

## Section 26.11 Angular Magnification and the Magnifying Glass

**76.** A jeweler whose near point is 72 cm from his eye uses a magnifying glass as in Figure 26.37b to examine a watch. The watch is held 4.0 cm from the magnifying glass. Find the angular magnification of the magnifying glass.

**77. ssm** A quarter (diameter = 2.4 cm) is held at arm's length (70.0 cm). The sun has a diameter of $1.39 \times 10^9$ m and is $1.50 \times 10^{11}$ m from the earth. What is the ratio of the angular size of the quarter to that of the sun?

**78.** An object has an angular size of 0.0150 rad when placed at the near point (21.0 cm) of an eye. When the eye views this object using a magnifying glass, the largest possible angular size of the image is 0.0380 rad. What is the focal length of the magnifying glass?

**79.** A magnifying glass is held next to the eye, above a magazine. The image formed by the magnifying glass is located at the near point of the eye. The near point is 0.30 m away from the eye, and the angular magnification is 3.4. Find the focal length of the magnifying glass.

**80.** The near point of a naked eye is 32 cm. When an object is placed at the near point and viewed by the naked eye, it has an angular size of 0.060 rad. A magnifying glass has a focal length of 16 cm, and is held next to the eye. The enlarged image that is seen is located 64 cm from the magnifying glass. Determine the angular size of the image.

* **81.** **ssm** A person using a magnifying glass as in Figure 26.37b observes that for clear vision its maximum angular magnification is 1.25 times as large as its minimum angular magnification. Assuming that the person has a near point located 25 cm from her eye, what is the focal length of the magnifying glass?

** **82.** A farsighted person can read printing as close as 25.0 cm when she wears contacts that have a focal length of 45.4 cm. One day, however, she forgets her contacts and uses a magnifying glass, as in Figure 26.37b. It has a maximum angular magnification of 7.50 for a young person with a normal near point of 25.0 cm. What is the maximum angular magnification that the magnifying glass can provide for her?

### Section 26.12 The Compound Microscope

**83.** An insect subtends an angle of only $4.0 \times 10^{-3}$ rad at the unaided eye when placed at the near point. What is the angular size (magnitude only) when the insect is viewed through a microscope whose angular magnification has a magnitude of 160?

**84.** An anatomist is viewing heart muscle cells with a microscope that has two selectable objectives with refracting powers of 100 and 300 diopters. When she uses the 100-diopter objective, the image of a cell subtends an angle of $3 \times 10^{-3}$ rad with the eye. What angle is subtended when she uses the 300-diopter objective?

**85.** **ssm** A compound microscope has a barrel whose length is 16.0 cm and an eyepiece whose focal length is 1.4 cm. The viewer has a near point located 25 cm from his eyes. What focal length must the objective have so that the angular magnification of the microscope will be $-320$?

**86.** The near point of a naked eye is 25 cm. When placed at the near point and viewed by the naked eye, a tiny object would have an angular size of $5.2 \times 10^{-5}$ rad. When viewed through a compound microscope, however, it has an angular size of $-8.8 \times 10^{-3}$ rad. (The minus sign indicates that the image produced by the microscope is inverted.) The objective of the microscope has a focal length of 2.6 cm, and the distance between the objective and the eyepiece is 16 cm. Find the focal length of the eyepiece.

* **87.** The maximum angular magnification of a magnifying glass is 12.0 when a person uses it who has a near point that is 25.0 cm from his eyes. The same person finds that a microscope, using this magni-

fying glass as the eyepiece, has an angular magnification of $-525$. The separation between the eyepiece and the objective of the microscope is 23.0 cm. Obtain the focal length of the objective.

* **88.** In a compound microscope, the focal length of the objective is 3.50 cm and that of the eyepiece is 6.50 cm. The distance between the lenses is 26.0 cm. (a) What is the angular magnification of the microscope if the person using it has a near point of 35.0 cm? (b) If, as usual, the first image lies just inside the focal point of the eyepiece (see Figure 26.30), how far is the object from the objective? (c) What is the magnification (not the angular magnification) of the objective?

### Section 26.13 The Telescope

**89.** **ssm** An astronomical telescope has an angular magnification of $-184$ and uses an objective with a focal length of 48.0 cm. What is the focal length of the eyepiece?

**90.** An astronomical telescope has an angular magnification of $-132$. Its objective has a refractive power of 1.50 diopters. What is the refractive power of its eyepiece?

**91.** Mars subtends an angle of $8.0 \times 10^{-5}$ rad at the unaided eye. An astronomical telescope has an eyepiece with a focal length of 0.032 m. When Mars is viewed using this telescope, it subtends an angle of $2.8 \times 10^{-3}$ rad. Find the focal length of the telescope's objective lens.

**92.** An astronomical telescope for hobbyists has an angular magnification of $-155$. The eyepiece has a focal length of 5.00 mm. (a) Determine the focal length of the objective. (b) About how long is the telescope?

**93.** **ssm** An amateur astronomer decides to build a telescope from a discarded pair of eyeglasses. One of the lenses has a refractive power of 11 diopters, and the other has a refractive power of 1.3 diopters. (a) Which lens should be the objective? (b) How far apart should the lenses be separated? (c) What is the angular magnification of the telescope?

* **94.** The telescope at Yerkes Observatory in Wisconsin has an objective whose focal length is 19.4 m. Its eyepiece has a focal length of 10.0 cm. (a) What is the angular magnification of the telescope? (b) If the telescope is used to look at a lunar crater (diameter = 1500 m), what is the size of the first image, assuming the surface of the moon is $3.77 \times 10^8$ m from the surface of the earth? (c) How close does the crater appear to be when seen through the telescope?

* **95.** The objective and eyepiece of an astronomical telescope are 1.25 m apart, and the eyepiece has a focal length of 5.0 cm. What is the angular magnification of the telescope?

** **96.** An astronomical telescope is being used to examine a relatively close object that is only 114.00 m away from the objective of the telescope. The objective and eyepiece have focal lengths of 1.500 and 0.070 m, respectively. Noting that the expression $M \approx -f_o/f_e$ is no longer applicable because the object is so close, use the thin-lens and magnification equations to find the angular magnification of this telescope. (*Hint: See Figure 26.39 and note that the focal points $F_o$ and $F_e$ are so close together that the distance between them may be ignored.*)

# Chapter 27  Interference and the Wave Nature of Light

## 27.1  THE PRINCIPLE OF LINEAR SUPERPOSITION

Chapter 17 examines what happens when several sound waves are present at the same place at the same time. The pressure disturbance that results is governed by the principle of linear superposition, which states that the resultant disturbance is the sum of the disturbances from the individual waves. Light is also a wave, an electromagnetic wave, and it too obeys the superposition principle. This principle is used to explain interference phenomena associated with light, including Young's double-slit interference, thin-film interference, and the interference effects that occur in the Michelson interferometer. When two or more light waves pass through a given point, their electric fields combine according to the principle of linear superposition and produce a resultant electric field. According to Equation 24.5b, the square of the electric field strength is proportional to the intensity of the light, which, in turn, is related to its brightness. Thus, interference can and does alter the brightness of light, just as it affects the loudness of sound.

Figure 27.1 illustrates what happens when two identical waves (same wavelength $\lambda$ and same amplitude) arrive at the point $P$ in phase—that is, crest-to-crest and trough-to-trough. According to the principle of linear superposition, the waves reinforce each other and *constructive interference* occurs. The resulting total wave at $P$ has an amplitude that is twice the amplitude of either individual wave, and in the case of light waves, the brightness at $P$ is greater than that due to either wave alone. The waves start out in phase and are in phase at $P$ because the distances $\ell_1$ and $\ell_2$ between this spot and the sources of the waves differ by one wavelength $\lambda$. In Figure 27.1, these distances are $\ell_1 = 2\frac{1}{4}$ wavelengths and $\ell_2 = 3\frac{1}{4}$ wavelengths. In general, when the waves start out in phase, constructive interference will result at $P$ whenever the distances are the same or differ by any integer number of wavelengths—in other words, assuming $\ell_2$ is the larger distance, whenever $\ell_2 - \ell_1 = m\lambda$, where $m = 0, 1, 2, 3, \ldots$.

Figure 27.2 shows what occurs when two identical waves arrive at the point $P$ out of phase with one another, or crest-to-trough. Now the waves mutually cancel, according to the principle of linear superposition, and *destructive interference* results. With light waves this would mean that there is no brightness. The waves begin with the same phase but are out of phase at $P$ because the distances through which they travel in reaching this spot differ by one-half of a wavelength ($\ell_1 = 2\frac{3}{4}\lambda$ and $\ell_2 = 3\frac{1}{4}\lambda$ in the drawing). In gen-

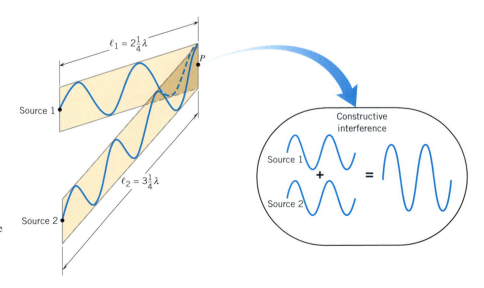

**Figure 27.1** The waves emitted by source 1 and source 2 start out in phase and arrive at point $P$ in phase, leading to constructive interference at that point.

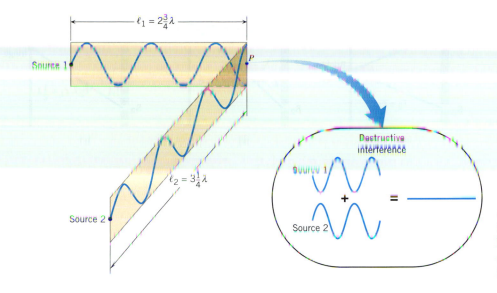

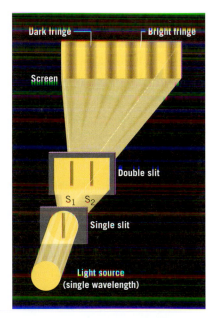

**Figure 27.2** The waves emitted by the two sources have the same phase to begin with, but they arrive at point $P$ out of phase. As a result, destructive interference occurs at $P$.

eral, for waves that start out in phase, destructive interference will take place at $P$ whenever the distances differ by any odd integer number of half-wavelengths—that is, whenever $\ell_2 - \ell_1 = \frac{1}{2}\lambda, \frac{3}{2}\lambda, \frac{5}{2}\lambda \ldots$ , where $\ell_2$ is the larger distance. This is equivalent to $\ell_2 - \ell_1 = (m + \frac{1}{2})\lambda$, where $m = 0, 1, 2, 3, \ldots$ .

If constructive or destructive interference is to continue occurring at a point, the sources of the waves must be **coherent sources.** Two sources are coherent if the waves they emit maintain a constant phase relation. Effectively, this means that the waves do not shift relative to one another as time passes. For instance, suppose that the wave pattern of source 1 in Figure 27.2 shifted forward or backward by random amounts at random moments. Then, on average, neither constructive nor destructive interference would be observed at point $P$ because there would be no stable relation between the two wave patterns. Lasers are coherent sources of light, whereas incandescent light bulbs and fluorescent lamps are incoherent sources.

## 27.2 Young's Double-Slit Experiment

In 1801 the English scientist Thomas Young (1773–1829) performed a historic experiment that demonstrated the wave nature of light by showing that two overlapping light waves interfered with each other. His experiment was particularly important because he was also able to determine the wavelength of the light from his measurements, the first such determination of this important property. Figure 27.3 shows one arrangement of Young's experiment, in which light of a single wavelength (monochromatic light) passes through a single narrow slit and falls on two closely spaced, narrow slits $S_1$ and $S_2$. These two slits act as coherent sources of light waves that interfere constructively and destructively at different points on the screen to produce a pattern of alternating bright and dark fringes. The purpose of the single slit is to ensure that only light from one direction falls on the double slit. Without it, light coming from different points on the light source would strike the double slit from different directions and cause the pattern on the screen to be washed out. The slits $S_1$ and $S_2$ act as coherent sources of light waves because the light from each originates from the same primary source—namely, the single slit.

To help explain the origin of the bright and dark fringes, Figure 27.4 presents three top views of the double slit and the screen. Part $a$ illustrates how a bright fringe arises directly opposite the midpoint between the two slits. In this part of the drawing the waves (identical) from each slit travel to the midpoint on the screen. At this location, the distances $\ell_1$ and $\ell_2$ to the slits are equal, each containing the same number of wavelengths. Therefore, constructive interference results, leading to the bright fringe. Part $b$ indicates that constructive interference produces another bright fringe on one side of the midpoint when the distance $\ell_2$ is larger than $\ell_1$ by exactly one wavelength. A bright fringe also oc-

**Figure 27.3** In Young's double-slit experiment, two slits $S_1$ and $S_2$ act as coherent sources of light. Light waves from these slits interfere constructively and destructively on the screen to produce, respectively, the bright and dark fringes. The slit widths and the distance between the slits have been exaggerated for clarity.

**Figure 27.4** The waves from slits $S_1$ and $S_2$ interfere constructively (parts *a* and *b*) or destructively (part *c*) on the screen, depending on the difference in distances between the slits and the screen. The slit widths and the distance between the slits have been exaggerated for clarity.

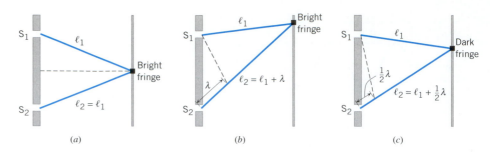

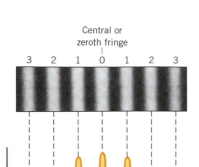

**Figure 27.5** The results of Young's double-slit experiment, showing a photograph of the bright and dark fringes formed on the screen and a graph of the light intensity. The central or zeroth fringe is the brightest fringe (greatest intensity). (From Michel Cagnet, et al., *Atlas of Optical Phenomena*, Springer-Verlag, Berlin.)

curs symmetrically on the other side of the midpoint when the distance $\ell_1$ exceeds $\ell_2$ by one wavelength; for clarity, however, this bright fringe is not shown. Constructive interference produces additional bright fringes (not shown) on both sides of the middle wherever the difference between $\ell_1$ and $\ell_2$ is an integer number of wavelengths: $\lambda$, $2\lambda$, $3\lambda$, and so on. Part *c* shows how the first dark fringe arises. Here the distance $\ell_2$ is larger than $\ell_1$ by exactly one-half a wavelength, so the waves interfere destructively, giving rise to the dark fringe. Destructive interference creates additional dark fringes on both sides of the center wherever the difference between $\ell_1$ and $\ell_2$ equals an odd integer number of half-wavelengths: $1(\frac{\lambda}{2})$, $3(\frac{\lambda}{2})$, $5(\frac{\lambda}{2})$, and so on.

The brightness of the fringes in Young's experiment varies, as the photograph in Figure 27.5 shows. Below the photograph is a graph to suggest the way in which the intensity varies for the fringe pattern. The central fringe is labeled with a zero, and the other bright fringes are numbered in ascending order on either side of the center. It can be seen that the central fringe has the greatest intensity. To either side of the center, the intensities of the other fringes decrease symmetrically in a way that depends on how small the slit widths are relative to the wavelength of the light.

The position of the fringes observed on the screen in Young's experiment can be calculated with the aid of Figure 27.6. If the screen is located far away compared with the separation $d$ of the slits, then the lines labeled $\ell_1$ and $\ell_2$ in part *a* are nearly parallel. Being nearly parallel, these lines make approximately equal angles $\theta$ with the horizontal. The distances $\ell_1$ and $\ell_2$ differ by an amount $\Delta\ell$, which is the length of the short side of the colored triangle in part *b* of the drawing. Since the triangle is a right triangle, it follows that $\Delta\ell = d \sin\theta$. Constructive interference occurs when the distances differ by an integer number $m$ of wavelengths $\lambda$, or $\Delta\ell = d \sin\theta = m\lambda$. Therefore, the angle $\theta$ for the interference maxima can be determined from the following expression:

**Bright fringes of a double slit**
$$\sin\theta = m\frac{\lambda}{d} \qquad m = 0, 1, 2, 3, \ldots \qquad (27.1)$$

The value of $m$ specifies the *order* of the fringe. Thus, $m = 2$ identifies the "second-order" bright fringe. Part *c* of the drawing stresses that the angle $\theta$ given by Equation 27.1 locates bright fringes on either side of the midpoint between the slits. A similar line of reasoning leads to the conclusion that the dark fringes, which lie between the bright fringes, are located according to

**Dark fringes of a double slit**
$$\sin\theta = (m + \tfrac{1}{2})\frac{\lambda}{d} \qquad m = 0, 1, 2, 3, \ldots \qquad (27.2)$$

**Figure 27.6** (a) Rays from slits $S_1$ and $S_2$, which make approximately the same angle $\theta$ with the horizontal, strike a distant screen at the same spot. (b) The difference in the path lengths of the two rays is $\Delta\ell = d \sin\theta$. (c) The angle $\theta$ is the angle at which a bright fringe ($m = 2$, here) occurs on either side of the central bright fringe ($m = 0$).

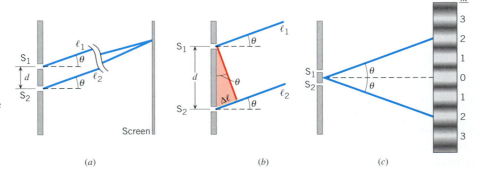

Example 1 illustrates how to determine the distance of a higher-order bright fringe from the central bright fringe with the aid of these expressions.

## Example 1 Young's Double-Slit Experiment

Red light ($\lambda = 664$ nm in vacuum) is used in Young's experiment with the slits separated by a distance $d = 1.20 \times 10^{-4}$ m. The screen in Figure 27.7 is located at a distance of $L = 2.75$ m from the slits. Find the distance $y$ on the screen between the central bright fringe and the third-order bright fringe.

**Reasoning** This problem can be solved by first using Equation 27.1 to determine the value of $\theta$ that locates the third-order ($m = 3$) bright fringe. Then trigonometry can be used to obtain the distance $y$.

**Solution** According to Equation 27.1, we find

$$\theta = \sin^{-1}\left(\frac{m\lambda}{d}\right) = \sin^{-1}\left[\frac{3(664 \times 10^{-9} \text{ m})}{1.20 \times 10^{-4} \text{ m}}\right] = 0.951°$$

According to Figure 27.7, the distance $y$ can be calculated from $\tan \theta = y/L$:

$$y = L \tan \theta = (2.75 \text{ m}) \tan 0.951° = \boxed{0.0456 \text{ m}}$$

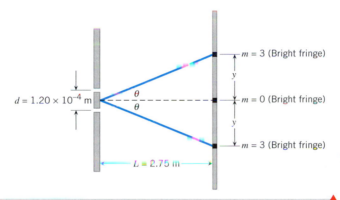

**Figure 27.7** The third-order bright fringe ($m = 3$) is observed on the screen at a distance $y$ from the central bright fringe ($m = 0$).

In the preceding version of Young's experiment, monochromatic light has been used. Light that contains a mixture of wavelengths can also be used. Conceptual Example 2 deals with some of the interesting features of the resulting interference pattern.

## Conceptual Example 2 White Light and Young's Experiment

Figure 27.8 shows a photograph that illustrates the kind of interference fringes that can result when white light, which is a mixture of all colors, is used in Young's experiment. Except for the central fringe, which is white, the bright fringes are a rainbow of colors. Why does Young's experiment separate white light into its constituent colors? In any group of colored fringes, such as the two singled out in Figure 27.8, why is red farther out from the central fringe than green is? And finally, why is the central fringe white rather than colored?

**Reasoning and Solution** To understand how the color separation arises, we need to remember that each color corresponds to a different wavelength $\lambda$ and that constructive and destructive interference depend on the wavelength. According to Equation 27.1 ($\sin \theta = m\lambda/d$), there is a different angle that locates a bright fringe for each value of $\lambda$, and thus for each color. These different angles lead to the separation of colors on the observation screen. In fact, on either side of the central fringe, there is one group of colored fringes for $m = 1$ and another for each value of $m$.

Now, consider what it means that, within any single group of colored fringes, red is farther out from the central fringe than green is. It means that, in the equation $\sin \theta = m\lambda/d$, red light has a larger angle $\theta$ than green light does. Does this make sense? Yes, because red has the larger wavelength (see Table 26.2, where $\lambda_{\text{red}} = 660$ nm and $\lambda_{\text{green}} = 550$ nm).

In Figure 27.8, the central fringe is distinguished from all the other colored fringes by being white. In Equation 27.1, the central fringe is different from the other fringes because

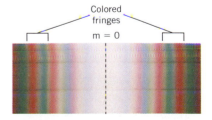

**Figure 27.8** This photograph shows the results observed on the screen in one version of Young's experiment in which white light (a mixture of all colors) is used. (© Andy Washnik)

it is the only one for which $m = 0$. In Equation 27.1, a value of $m = 0$ means that $\sin \theta = m\lambda/d = 0$, which reveals that $\theta = 0°$, no matter what the wavelength $\lambda$ is. In other words, all wavelengths have a zeroth-order bright fringe located at the same place on the screen, so that all colors strike the screen there and mix together to produce the white central fringe.

**Related Homework:** *Problem 6*

Historically, Young's experiment provided strong evidence that light has a wave-like character. If light behaved only as a stream of "tiny particles," as others believed at the time,* then the two slits would deliver the light energy into only two bright fringes located directly opposite the slits on the screen. Instead, Young's experiment shows that wave interference redistributes the energy from the two slits into many bright fringes.

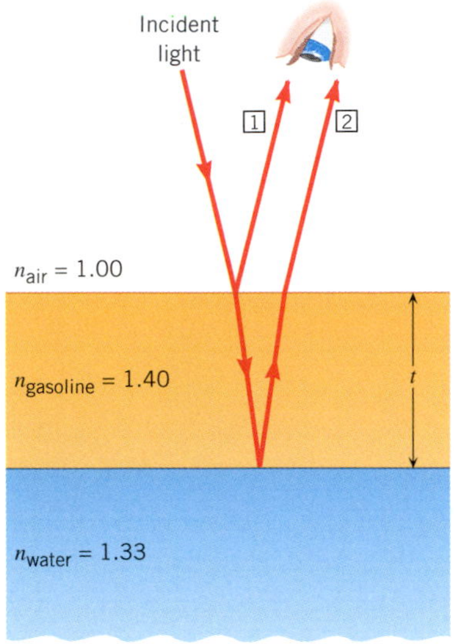

## 27.3 *Thin-Film Interference*

Young's double-slit experiment is one example of interference between light waves. Interference also occurs in more common circumstances. For instance, Figure 27.9 shows a thin film, such as gasoline floating on water. To begin with, let us assume that the film has a constant thickness. Consider what happens when monochromatic light (a single wavelength) strikes the film nearly perpendicularly. At the top surface of the film reflection occurs and produces the light wave represented by ray 1. However, refraction also occurs, and some light enters the film. Part of this light reflects from the bottom surface of the film and passes back up through the film, eventually reentering the air. Thus, a second light wave, which is represented by ray 2, also exists. Moreover, this wave, having traversed the film twice, has traveled farther than wave 1. Because of the extra travel distance, there can be interference between the two waves. If constructive interference occurs, an observer, whose eyes detect the superposition of waves 1 and 2, would see a uniformly bright film. If destructive interference occurs, an observer would see a uniformly dark film. The controlling factor is whether the extra distance for wave 2 is an integer number of whole wavelengths (for constructive interference) or an odd integer number of half-wavelengths (for destructive interference).

In Figure 27.9 the difference in path lengths between waves 1 and 2 occurs inside the thin film. Therefore, *the wavelength that is important for thin-film interference is the wavelength within the film,* not the wavelength in vacuum. The wavelength within the film can be calculated from the wavelength in vacuum by using the index of refraction $n$ for the film. According to Equations 26.1 and 16.1, $n = c/v = (c/f)/(v/f) = \lambda_{\text{vacuum}}/\lambda_{\text{film}}$. In other words,

$$\lambda_{\text{film}} = \frac{\lambda_{\text{vacuum}}}{n} \qquad (27.3)$$

In explaining the interference that can occur in Figure 27.9, we need to add one other important part to the story. Whenever waves reflect at a boundary, it is possible for them to change phase. Figure 27.10, for example, shows that a wave on a string is inverted when it reflects from the end that is tied to a wall (see also Figure 17.16). This inversion is equivalent to a half-cycle of the wave, as if the wave had traveled an additional distance of one-half of a wavelength. In contrast, a phase change does not occur when a wave on a string reflects from the end of a string that is hanging free. When light waves undergo reflection, similar phase changes occur as follows:

1. When light travels through a material with a smaller refractive index toward a material with a larger refractive index (e.g., air to gasoline), reflection at the boundary occurs along with a phase change that is equivalent to one-half of a wavelength in the film.

2. When light travels from a larger toward a smaller refractive index, there is no phase change upon reflection at the boundary.

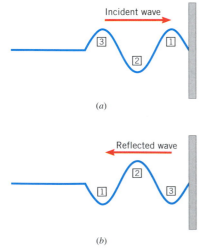

$n_{\text{air}} = 1.00$

$n_{\text{gasoline}} = 1.40$

$n_{\text{water}} = 1.33$

**Figure 27.9** Because of reflection and refraction, two light waves, represented by rays 1 and 2, enter the eye when light shines on a thin film of gasoline floating on a thick layer of water.

Incident wave

(a)

Reflected wave

(b)

**Figure 27.10** When a wave on a string reflects from a wall, the wave undergoes a phase change. Thus, an upward-pointing half-cycle of the wave becomes, after reflection, a downward-pointing half-cycle, and vice versa, as the numbered labels in the drawing indicate.

*It is now known that the particle or corpuscular theory of light, which Isaac Newton promoted, does indeed explain some experiments that the wave theory cannot explain. Today, light is regarded as having both particle and wave characteristics. Chapter 29 discusses this dual nature of light.

The next example indicates how the phase change that can accompany reflection is taken into account when dealing with thin-film interference.

## Example 3  A Colored Thin Film of Gasoline

A thin film of gasoline floats on a puddle of water. Sunlight falls almost perpendicularly on the film and reflects into your eyes. Although sunlight is white since it contains all colors, the film has a yellow hue because destructive interference eliminates the color of blue ($\lambda_{\text{vacuum}} = 469$ nm) from the reflected light. The refractive indices of the blue light in gasoline and water are 1.40 and 1.33, respectively. Determine the minimum nonzero thickness $t$ of the film.

**Reasoning** To solve this problem, we must express the condition for destructive interference in terms of the film thickness $t$ and the wavelength in the gasoline film $\lambda_{\text{film}}$. We must also take into account any phase changes that occur upon reflection.

**Solution** Equation 27.3, with $n = 1.40$, gives the wavelength of blue light in the film as $\lambda_{\text{film}} = (469 \text{ nm})/1.40 = 335$ nm. In Figure 27.9, the phase change for wave 1 is equivalent to one-half of a wavelength, since this light travels from a smaller refractive index ($n_{\text{air}} = 1.00$) toward a larger refractive index ($n_{\text{gasoline}} = 1.40$). In contrast, there is no phase change when wave 2 reflects from the bottom surface of the film, since this light travels from a larger refractive index ($n_{\text{gasoline}} = 1.40$) toward a smaller one ($n_{\text{water}} = 1.33$). The net phase change between waves 1 and 2 due to reflection is, thus, equivalent to one-half of a wavelength, $\frac{1}{2}\lambda_{\text{film}}$. This half-wavelength must be combined with the extra travel distance for wave 2, to determine the condition for destructive interference. For destructive interference, the combined total must be an odd integer number of half-wavelengths. Since wave 2 travels back and forth through the film and since light strikes the film nearly perpendicularly, the extra travel distance is twice the film thickness, or $2t$. Thus, the condition for destructive interference is

*Problem solving insight*
When analyzing thin-film interference effects, remember to use the wavelength of the light in the film ($\lambda_{\text{film}}$) instead of the wavelength in a vacuum ($\lambda_{\text{vacuum}}$).

$$\underbrace{2t}_{\substack{\text{Extra distance} \\ \text{traveled by} \\ \text{wave 2}}} + \underbrace{\tfrac{1}{2}\lambda_{\text{film}}}_{\substack{\text{Half-wavelength} \\ \text{net phase change} \\ \text{due to reflection}}} = \underbrace{\tfrac{1}{2}\lambda_{\text{film}}, \tfrac{3}{2}\lambda_{\text{film}}, \tfrac{5}{2}\lambda_{\text{film}}, \ldots}_{\substack{\text{Condition for} \\ \text{destructive interference}}}$$

After subtracting the term $\frac{1}{2}\lambda_{\text{film}}$ from the left-hand side of this equation and from each term on the right-hand side, we can solve for the thickness $t$ of the film that yields destructive interference:

$$t = \frac{m\lambda_{\text{film}}}{2} \qquad m = 0, 1, 2, 3, \ldots$$

With $m = 1$, the expression above gives the minimum nonzero film thickness for which the blue color is missing in the reflected light:

$$t = \tfrac{1}{2}\lambda_{\text{film}} = \tfrac{1}{2}(335 \text{ nm}) = \boxed{168 \text{ nm}}$$

Example 3 deals with a thin film that has the same yellow color everywhere. In nature, such a uniformly colored thin film would be unusual, and the next example deals with a more realistic situation.

## Conceptual Example 4  Multicolored Thin Films

Under natural conditions, thin films, like gasoline on water or like the soap bubble in Figure 27.11 have a multicolored appearance that often changes while you are watching them. Why are such films multicolored, and what can be inferred from the fact that the colors change in time?

**Reasoning and Solution** In Example 3 we have seen that a thin film can appear yellow if destructive interference removes blue light from the reflected sunlight. The thickness of the film is the key. If the thickness were different, so that destructive interference removed green light from the reflected sunlight, the film would appear magenta. Constructive interference can also cause certain colors to appear brighter than others in the reflected light and give the film a colored appearance. The colors that are enhanced by constructive interference, like those removed by destructive interference, depend on the thickness of the film. Thus, we conclude that

**Figure 27.11** A soap bubble is multicolored when viewed in sunlight because of the effects of thin-film interference. (© Paul A. Souders/Corbis Images)

**Problem solving insight**

**The physics of** nonreflecting lens coatings.

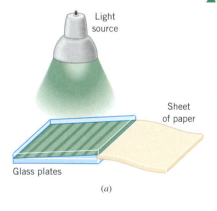

(a)

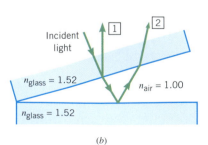

(b)

**Figure 27.12** (a) The wedge of air formed between two flat glass plates causes an interference pattern of alternating dark and bright fringes to appear in reflected light. (b) A side view of the glass plates and the air wedge.

*the different colors in a thin film of gasoline on water or in a soap bubble arise because the thickness is different in different places on the film. Moreover, the fact that the colors change as you watch them indicates that the thickness is changing.* A number of factors can cause the thickness to change, including air currents, temperature fluctuations, and the pull of gravity, which tends to make a vertical film sag, leading to thicker regions at the bottom than at the top.

**Related Homework:** *Problems 12, 14, 16*

---

The colors that you see when sunlight is reflected from a thin film also depend on your viewing angle. At an oblique angle, the light corresponding to ray 2 in Figure 27.9 would travel a greater distance within the film than it does at nearly perpendicular incidence. The greater distance would lead to destructive interference for a different wavelength.

Thin-film interference can be beneficial in optical instruments. For example, some cameras contain six or more lenses. Reflections from all the lens surfaces can reduce considerably the amount of light directly reaching the film. In addition, multiple reflections from the lenses often reach the film indirectly and degrade the quality of the image. To minimize such unwanted reflections, high-quality lenses are often covered with a thin nonreflective coating of magnesium fluoride ($n$ = 1.38). The thickness of the coating is usually chosen to ensure that destructive interference eliminates the reflection of green light, which is in the middle of the visible spectrum. It should be pointed out that the absence of any reflected light does not mean that it has been destroyed by the nonreflective coating. Rather, the "missing" light has been transmitted into the coating and the lens.

Another interesting illustration of thin-film interference is the air wedge. As Figure 27.12a shows, an air wedge is formed when two flat plates of glass are separated along one side, perhaps by a thin sheet of paper. The thickness of this film of air varies between zero, where the plates touch, and the thickness of the paper. When monochromatic light reflects from this arrangement, alternate bright and dark fringes are formed by constructive and destructive interference, as the drawing indicates and Example 5 discusses.

## Example 5 An Air Wedge

(a) Assuming that green light ($\lambda_{\text{vacuum}}$ = 552 nm) strikes the glass plates nearly perpendicularly in Figure 27.12, determine the number of bright fringes that occur between the place where the plates touch and the edge of the sheet of paper (thickness = $4.10 \times 10^{-5}$ m). (b) Explain why there is a dark fringe where the plates touch.

**Reasoning** A bright fringe occurs wherever there is constructive interference, as determined by any phase changes due to reflection and the effects of the thickness of the air wedge. We examine the phase changes and the effects of the thickness separately. There is no phase change upon reflection for wave 1, since this light travels from a larger (glass) toward a smaller (air) refractive index. In contrast, there is a half-wavelength phase change for wave 2, since the ordering of the refractive indices is reversed at the lower air/glass boundary where reflection occurs. The net phase change due to reflection for waves 1 and 2, then, is equivalent to a half wavelength. Now we combine any extra distance traveled by ray 2 with this half wavelength and determine the condition for the constructive interference that creates the bright fringes. Constructive interference occurs whenever the *combination* yields an integer number of wavelengths. At nearly perpendicular incidence, the extra travel distance for wave 2 is approximately twice the thickness $t$ of the wedge at any point, so the condition for constructive interference is

$$\underbrace{2t}_{\substack{\text{Extra distance} \\ \text{traveled by} \\ \text{wave 2}}} + \underbrace{\tfrac{1}{2}\lambda_{\text{film}}}_{\substack{\text{Half-wavelength} \\ \text{net phase change} \\ \text{due to reflection}}} = \underbrace{\lambda_{\text{film}}, 2\lambda_{\text{film}}, 3\lambda_{\text{film}}, \ldots}_{\substack{\text{Condition for} \\ \text{constructive interference}}}$$

Subtracting the term $\tfrac{1}{2}\lambda_{\text{film}}$ from the left-hand side of this equation and from each term on the right-hand side yields

$$2t = \underbrace{\tfrac{1}{2}\lambda_{\text{film}}, \tfrac{3}{2}\lambda_{\text{film}}, \tfrac{5}{2}\lambda_{\text{film}}, \ldots}_{(m + \frac{1}{2})\lambda_{\text{film}} \quad m = 0, 1, 2, 3, \ldots}$$

Therefore,

$$t = \frac{(m + \tfrac{1}{2})\lambda_{\text{film}}}{2} \qquad m = 0, 1, 2, 3, \ldots$$

In this expression, note that the "film" is a film of air. Since the refractive index of air is nearly one, $\lambda_{\text{film}}$ is virtually the same as that in a vacuum, so $\lambda_{\text{film}} = 552$ nm.

**Solution**

**(a)** When $t$ equals the thickness of the paper holding the plates apart, the corresponding value of $m$ can be obtained from the equation above:

$$m = \frac{2t}{\lambda_{\text{film}}} - \frac{1}{2} = \frac{2(4.10 \times 10^{-5}\,\text{m})}{552 \times 10^{-9}\,\text{m}} - \frac{1}{2} = 148$$

Since the first bright fringe occurs when $m = 0$, the number of bright fringes is $m + 1 = \boxed{149}$.

**(b)** Where the plates touch, there is a dark fringe because of destructive interference between the light waves represented by rays 1 and 2. Destructive interference occurs because the thickness of the wedge is zero here and the only difference between the rays is the half-wavelength phase change due to reflection from the lower plate.

Another type of air wedge can also be used to determine the degree to which the surface of a lens or mirror is spherical. When an accurate spherical surface is put in contact with an optically flat plate, as in Figure 27.13a, the circular interference fringes shown in part b of the figure can be observed. The circular fringes are called *Newton's rings*. They arise in the same way that the straight fringes arise in Figure 27.12a.

# 27.4 **The Michelson Interferometer**

An interferometer is an apparatus that can be used to measure the wavelength of light by utilizing interference between two light waves. One particularly famous interferometer is that developed by Albert A. Michelson (1852–1931). The Michelson interferometer uses reflection to set up conditions where two light waves interfere. Figure 27.14 presents a schematic drawing of the instrument. Waves emitted by the monochromatic light source strike a *beam splitter*, so called because it splits the beam of light into two parts. The beam splitter is a glass plate, the far side of which is coated with a thin layer of silver that reflects part of the beam upward as wave A in the drawing. The coating is so thin, however, that it also allows the remainder of the beam to pass directly through as wave F. Wave A strikes an adjustable mirror and reflects back on itself. It again crosses the beam

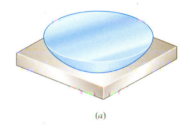

*(a)*

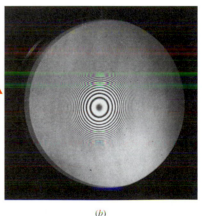

*(b)*

**Figure 27.13** (a) The air wedge between a convex spherical glass surface and an optically flat plate leads to (b) a pattern of circular interference fringes that is known as Newton's rings. (Courtesy Bausch & Lomb)

**The physics of**
the Michelson interferometer.

Adjustable
mirror

$D_F$

Wave A

Thin silver
coating

$D_A$

Light
source

Wave F

Beam
splitter

Compensating
plate

Fixed
mirror

Wave A   Wave F

Viewing
telescope

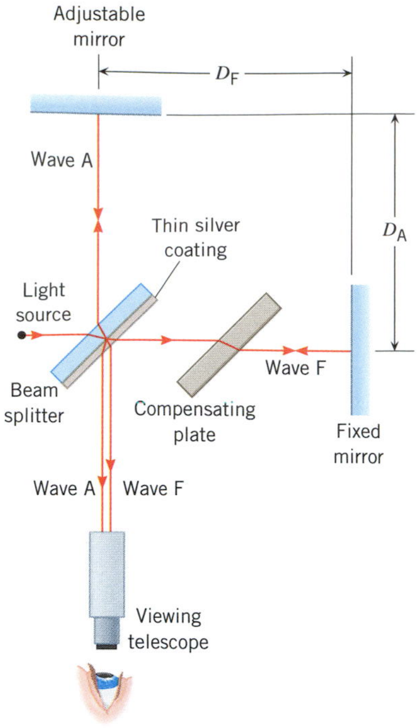

**Figure 27.14** A schematic drawing of a Michelson interferometer.

splitter and then enters the viewing telescope. Wave F strikes a fixed mirror and returns, to be partly reflected into the viewing telescope by the beam splitter. Note that wave A passes through the glass plate of the beam splitter three times in reaching the viewing scope, while wave F passes through it only once. The compensating plate in the path of wave F has the same thickness as the beam splitter plate and ensures that wave F also passes three times through the same thickness of glass on the way to the viewing scope. Thus, an observer who views the superposition of waves A and F through the telescope sees constructive or destructive interference, depending only on the difference in path lengths $D_A$ and $D_F$ traveled by the two waves.

Now suppose the mirrors are perpendicular to each other, the beam splitter makes a 45° angle with each, and the distances $D_A$ and $D_F$ are equal. Waves A and F travel the same distance, and the field of view in the telescope is uniformly bright due to constructive interference. However, if the adjustable mirror is moved away from the telescope by a distance of $\frac{1}{4}\lambda$, wave A travels back and forth by an amount that is twice this value, leading to an extra distance of $\frac{1}{2}\lambda$. Then, the waves are out of phase when they reach the viewing scope, destructive interference occurs, and the viewer sees a dark field. If the adjustable mirror is moved farther, full brightness returns as soon as the waves are in phase and interfere constructively. The in-phase condition occurs when wave A travels a total extra distance of $\lambda$ relative to wave F. Thus, as the mirror is continuously moved, the viewer sees the field of view change from bright to dark, then back to bright, and so on. The amount by which $D_A$ has been changed can be measured and related to the wavelength of the light, since a bright field changes into a dark field and back again each time $D_A$ is changed by a half wavelength. (The back-and-forth change in distance is $\lambda$.) If a sufficiently large number of wavelengths are counted in this manner, the Michelson interferometer can be used to obtain a very accurate value for the wavelength from the measured changes in $D_A$.

## 27.5 Diffraction

In the previous sections of this chapter we have used the principle of linear superposition to investigate some interference phenomena associated with light waves. We will now use the principle of linear superposition to explore three more topics—diffraction, resolving power, and the diffraction grating—all of which involve interference phenomena.

As we have seen in Section 17.3, ***diffraction*** is the bending of waves around obstacles or the edges of an opening. In Figure 27.15 which is similar to Figure 17.10, sound waves are leaving a room through an open doorway. Because the exiting sound waves bend, or diffract, around the edges of the opening, a listener outside the room can hear the sound even when standing around the corner from the doorway.

Diffraction is an interference effect, and the Dutch scientist Christian Huygens (1629–1695) developed a principle that is useful in explaining why diffraction arises. ***Huygens' principle*** describes how a wave front that exists at one instant gives rise to the wave front that exists later on. This principle states that:

> ***Every point on a wave front acts as a source of tiny wavelets that move forward with the same speed as the wave; the wave front at a later instant is the surface that is tangent to the wavelets.***

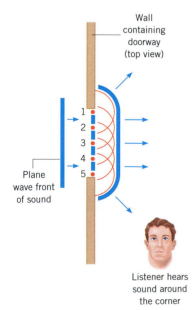

Wall
containing
doorway
(top view)

1
2
3
4
5

Plane
wave front
of sound

Listener hears
sound around
the corner

**Figure 27.15** Sound bends or diffracts around the edges of a doorway, so even a person who is not standing directly in front of the opening can hear the sound. The five red points within the doorway act as sources and emit the five Huygens wavelets shown in red.

We begin by using Huygens' principle to explain the diffraction of sound waves in Figure 27.15. The drawing shows the top view of a plane wave front of sound approaching a doorway and identifies five points on the wave front just as it is leaving the opening. According to Huygens' principle, each of these points acts as a source of wavelets, which are shown as red circular arcs at some moment after they are emitted. The tangent to the wavelets from points 2, 3, and 4 indicates that in front of the doorway the wave front is flat and moving straight ahead. But at the edges, points 1 and 5 are the last points that produce wavelets. Huygens' principle suggests that in conforming to the curved shape of the wavelets near the edges, the new wave front moves into regions that it would not reach otherwise. The sound wave, then, bends or diffracts around the edges of the doorway.

Huygens' principle applies not just to sound waves, but to all kinds of waves. For instance, light has a wave-like nature and, consequently, exhibits diffraction. Therefore, you

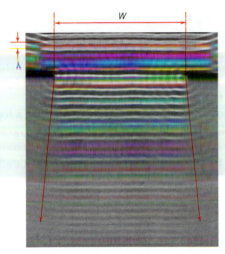

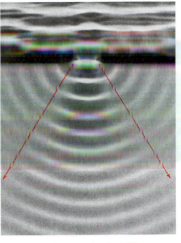

(a) Smaller value for λ/W, less diffraction.

(b) Larger value for λ/W, more diffraction.

**Figure 27.16** These photographs show water waves (horizontal lines) approaching an opening whose width $W$ is greater in (a) than in (b). In addition, the wavelength λ of the waves is smaller in (a) than in (b). Therefore, the ratio λ/W increases from (a) to (b) and so does the extent of the diffraction, as the red arrows indicate. (Courtesy Education Development Center)

may ask, "Since I can hear around the edges of a doorway, why can't I also see around them?" As a matter of fact, light waves do bend around the edges of a doorway. However, the degree of bending is extremely small, so the diffraction of light is not enough to allow you to see around the corner.

As we will learn, the extent to which a wave bends around the edges of an opening is determined by the ratio λ/W, where λ is the wavelength of the wave and $W$ is the width of the opening. The photographs in Figure 27.16 illustrate the effect of this ratio on the diffraction of water waves. The degree to which the waves are diffracted or bent is indicated by the two red arrows in each photograph. In part a, the ratio λ/W is small because the wavelength (as indicated by the distance between the wave fronts) is small relative to the width of the opening. The wave fronts move through the opening with little bending or diffraction into the regions around the edges. In part b, the wavelength is larger and the width of the opening is smaller. As a result, the ratio λ/W is larger, and the degree of bending becomes more pronounced, with the wave fronts penetrating more into the regions around the edges of the opening.

Based on the pictures in Figure 27.16, we might expect that light waves of wavelength λ will bend or diffract appreciably when they pass through an opening whose width $W$ is small enough to make the ratio λ/W sufficiently large. This is indeed the case, as Figure 27.17 illustrates. In this picture, it is assumed that parallel rays (or plane wave fronts) of light fall on a very narrow slit and illuminate a viewing screen that is located far

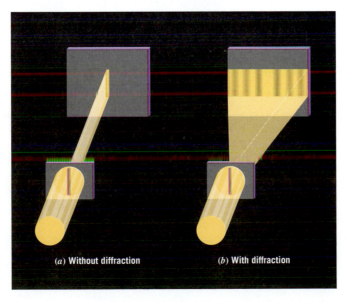

(a) Without diffraction        (b) With diffraction

**Figure 27.17** (a) If light were to pass through a very narrow slit *without* being diffracted, only the region on the screen directly opposite the slit would be illuminated. (b) Diffraction causes the light to bend around the edges of the slit into regions it would not otherwise reach, forming a pattern of alternating bright and dark fringes on the screen. The slit width has been exaggerated for clarity.

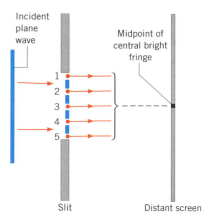

Figure 27.18  A plane wave front is incident on a single slit. This top view of the slit shows five sources of Huygens wavelets. The wavelets travel toward the midpoint of the central bright fringe on the screen, as the red rays indicate. The screen is very far from the slit.

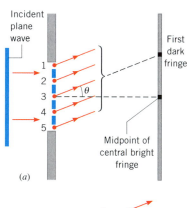

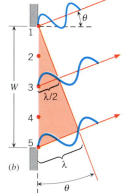

Figure 27.19  These drawings pertain to single-slit diffraction and show how destructive interference leads to the first dark fringe on either side of the central bright fringe. For clarity, only one of the dark fringes is shown. The screen is very far from the slit.

from the slit. Part *a* of the drawing shows what would happen if light were *not* diffracted: it would pass through the slit without bending around the edges and produce an image of the slit on the screen. Part *b* shows what actually happens. The light diffracts around the edges of the slit and brightens regions on the screen that are not directly opposite the slit. The diffraction pattern on the screen consists of a bright central band, accompanied by a series of narrower faint fringes that are parallel to the slit itself.

To help explain how the pattern of diffraction fringes arises, Figure 27.18 shows a top view of a plane wave front approaching the slit and singles out five sources of Huygens wavelets. Consider how the light from these five sources reaches the midpoint on the screen. To simplify things, the screen is assumed to be so far from the slit that the rays from each Huygens source are nearly parallel.* Then, all the wavelets travel virtually the same distance to the midpoint, arriving there in phase. As a result, constructive interference creates a bright central fringe on the screen, directly opposite the slit.

The wavelets emitted by the Huygens sources in the slit can also interfere destructively on the screen, as Figure 27.19 illustrates. Part *a* shows light rays directed from each source toward the first dark fringe. The angle $\theta$ gives the position of this dark fringe relative to the line between the midpoint of the slit and the midpoint of the central bright fringe. Since the screen is very far from the slit, the rays from each Huygens source are nearly parallel and oriented at nearly the same angle $\theta$, as in part *b* of the drawing. The wavelet from source 1 travels the shortest distance to the screen, while that from source 5 travels the farthest. Destructive interference creates the first dark fringe when the extra distance traveled by the wavelet from source 5 is exactly one wavelength, as the colored right triangle in the drawing indicates. Under this condition, the extra distance traveled by the wavelet from source 3 at the center of the slit is exactly one-half of a wavelength.

Therefore, wavelets from sources 1 and 3 in Figure 27.19*b* are exactly out of phase and interfere destructively when they reach the screen. Similarly, a wavelet that originates slightly below source 1 cancels a wavelet that originates the same distance below source 3. Thus, each wavelet from the upper half of the slit cancels a corresponding wavelet from the lower half, and no light reaches the screen. As can be seen from the colored right triangle, the angle $\theta$ locating the first dark fringe is given by sin $\theta = \lambda/W$, where $W$ is the width of the slit.

Figure 27.20 shows the condition that leads to destructive interference at the second dark fringe on either side of the midpoint on the screen. In reaching the screen, the light from source 5 now travels a distance of two wavelengths farther than the light from source 1. Under this condition, the wavelet from source 5 travels one wavelength farther than the wavelet from source 3, and the wavelet from source 3 travels one wavelength farther than the wavelet from source 1. Therefore, each half of the slit can be treated as the entire slit was in the previous paragraph; all the wavelets from the top half interfere destructively with each other, and all the wavelets from the bottom half do likewise. As a result, no light from either half reaches the screen, and another dark fringe occurs. The colored triangle in the drawing shows that this second dark fringe occurs when sin $\theta = 2\lambda/W$. Similar arguments hold for the third- and higher-order dark fringes, with the general result being

**Dark fringes for single-slit diffraction**
$$\sin \theta = m\frac{\lambda}{W} \qquad m = 1, 2, 3, \ldots \qquad (27.4)$$

Between each pair of dark fringes there is a bright fringe due to constructive interference. The brightness of the fringes is related to the light intensity, just as loudness is related to sound intensity. The intensity of the light at any location on the screen is the amount of light energy per second per unit area that strikes the screen there. Figure 27.21 gives a graph of the light intensity along with a photograph of a single-slit diffraction pattern. The central bright fringe, which is approximately twice as wide as the other bright fringes, has by far the greatest intensity.

The width of the central fringe provides some indication of the extent of the diffraction, as Example 6 illustrates.

*When the rays are parallel, the diffraction is called Fraunhofer diffraction in tribute to the German optician Joseph von Fraunhofer (1787–1826). When the rays are not parallel, the diffraction is referred to as Fresnel diffraction, named for the French physicist Augustin Jean Fresnel (1788–1827).

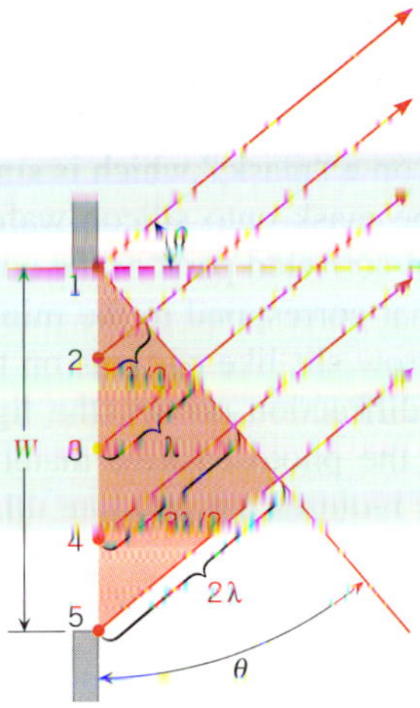

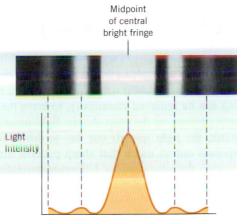

**Figure 27.20** In a single-slit diffraction pattern, multiple dark fringes occur on either side of the central bright fringe. This drawing shows how destructive interference creates the second dark fringe on a very distant screen.

**Figure 27.21** The photograph shows a single-slit diffraction pattern, with a bright and wide central fringe. The higher-order bright fringes are much less intense than the central fringe, as the graph indicates. (From Michel Cagnet, et al., *Atlas of Optical Phenomena*, Springer-Verlag, Berlin.)

## Example 6 Single-Slit Diffraction

Light passes through a slit and shines on a flat screen that is located $L = 0.40$ m away (see Figure 27.22). The width of the slit is $W = 4.0 \times 10^{-6}$ m. The distance between the middle of the central bright fringe and the first dark fringe is $y$. Determine the width $2y$ of the central bright fringe when the wavelength of the light in a vacuum is (a) $\lambda = 690$ nm (red) and (b) $\lambda = 410$ nm (violet).

**Reasoning** The width of the central bright fringe is determined by two factors. One is the angle $\theta$ that locates the first dark fringe on either side of the midpoint. The other is the distance $L$ between the screen and the slit. Larger values for $\theta$ and $L$ lead to a wider central bright fringe. Larger values of $\theta$ mean greater diffraction and occur when the ratio $\lambda/W$ is larger. Thus, we expect the width of the central bright fringe to be greater when $\lambda$ is greater.

**Solution**

(a) The angle $\theta$ in Equation 27.4 locates the first dark fringe when $m = 1$; $\sin \theta = (1)\lambda/W$. Therefore,

$$\theta = \sin^{-1}\left(\frac{\lambda}{W}\right) = \sin^{-1}\left(\frac{690 \times 10^{-9} \text{ m}}{4.0 \times 10^{-6} \text{ m}}\right) = 9.9°$$

According to Figure 27.22, $\tan \theta = y/L$, so

$$y = L \tan \theta = (0.40 \text{ m}) \tan 9.9° = 0.070 \text{ m}$$

The width of the central fringe, then, is $\boxed{2y = 0.14 \text{ m}}$.

(b) Repeating the calculation above with $\lambda = 410$ nm shows that $\theta = 5.9°$ and $\boxed{2y = 0.083 \text{ m}}$. As expected, the width $2y$ is greater in part (a), where the wavelength $\lambda$ is greater.

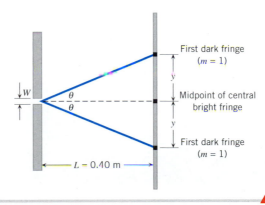

**Figure 27.22** The distance $2y$ is the width of the central bright fringe.

In the production of computer chips it is important to minimize the effects of diffraction. As Figure 23.31 illustrates, such chips are very small and yet contain enormous numbers of electronic components. Such miniaturization is achieved using the techniques of photolithography. The patterns on the chip are created first on a "mask," which is similar to a photographic slide. Light is then directed through the mask onto silicon wafers that have been coated with a photosensitive material. The light-activated parts of the coating can be removed chemically, to leave the ultrathin lines that correspond to the miniature patterns on the chip. As the light passes through the narrow slit-like patterns on the mask, the light spreads out due to diffraction. If excessive diffraction occurs, the light spreads out so much that sharp patterns are not formed on the photosensitive material coating the silicon wafer. Ultraminiaturization of the patterns requires the absolute minimum of diffraction, and currently this is achieved by using ultraviolet light, which has a wavelength shorter than that of visible light. The shorter the wavelength $\lambda$, the smaller the ratio $\lambda/W$, and the less the diffraction, as illustrated in Example 6. Intensive research programs are under way to develop the technique of X-ray lithography. The wavelengths of X-rays are much shorter than those of ultraviolet light and, thus, will reduce diffraction even more, allowing further miniaturization.

Another example of diffraction can be seen when light from a point source falls on an opaque disk, such as a coin (Figure 27.23). The effects of diffraction modify the dark shadow cast by the disk in several ways. First, the light waves diffracted around the circular edge of the disk interfere constructively at the center of the shadow to produce a small bright spot. There are also circular bright fringes in the shadow area. In addition, the boundary between the circular shadow and the lighted screen is not sharply defined but consists of concentric bright and dark fringes. The various fringes are analogous to those produced by a single slit and are due to interference between Huygens wavelets that originate from different points near the edge of the disk.

**The physics of**
producing computer chips using photolithography.

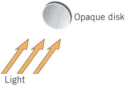

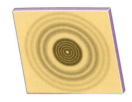

Opaque disk

Light

**Figure 27.23** The diffraction pattern formed by an opaque disk consists of a small bright spot in the center of the dark shadow, circular bright fringes within the shadow, and concentric bright and dark fringes surrounding the shadow.

## 27.6 *Resolving Power*

Figure 27.24 shows three photographs of an automobile's headlights taken at progressively greater distances from the camera. In parts *a* and *b*, the two separate headlights can be seen clearly. In part *c*, however, the car is so far away that the headlights are barely distinguishable and appear almost as a single light. The ***resolving power*** of an optical instrument, such as a camera, is its ability to distinguish between two closely spaced objects. If a camera with a higher resolving power had taken these pictures, the photograph in part *c* would have shown two distinct and separate headlights. Any instrument used for viewing objects that are close together must have a high resolving power. This is true, for example, for a telescope used to view distant stars or for a microscope used to view tiny organisms. We will now see that diffraction occurs when light passes through the circular, or nearly circular, openings that admit light into cameras, telescopes, microscopes, and human eyes. The resulting diffraction pattern places a natural limit on the resolving power of these instruments.

**Figure 27.24** These automobile headlights were photographed at various distances from the camera, closest in part (*a*) and farthest in part (*c*). In part (*c*), the headlights are so far away that they are barely distinguishable. (© Truax/The Image Finders)

(*a*)

(*b*)

(*c*)

Figure 27.25 shows the diffraction pattern created by a small circular opening when the viewing screen is far from the opening. The pattern consists of a central bright circular region, surrounded by alternating bright and dark circular fringes. These fringes are analogous to the rectangular fringes that a single slit produces. The angle $\theta$ in the picture locates the first circular dark fringe relative to the central bright region and is given by

$$\sin\theta = 1.22\frac{\lambda}{D} \qquad (27.5)$$

where $\lambda$ is the wavelength of the light and $D$ is the diameter of the opening. This expression is similar to Equation 27.4 for a slit ($\sin\theta = \lambda/W$, when $m = 1$) and is valid when the distance to the screen is much larger than the diameter $D$.

An optical instrument with the ability to resolve two closely spaced objects can produce images of them that can be identified separately. Think about the images on the film when light from two widely separated point objects passes through the circular aperture of a camera. As Figure 27.26 illustrates, each image is a circular diffraction pattern, but the two patterns do not overlap and are completely resolved. On the other hand, if the objects are sufficiently close together, the intensity patterns created by the diffraction overlap, as Figure 27.27a suggests. In fact, if the overlap is extensive, it may no longer be possible to distinguish the patterns separately. In such a case, the picture from a camera would show a single blurred object instead of two separate objects. In Figure 27.27b the diffraction patterns overlap, but not enough to prevent us from seeing that two objects are present. Ultimately, then, diffraction limits the ability of an optical instrument to produce distinguishable images of objects that are close together.

It is useful to have a criterion for judging whether two closely spaced objects will be resolved by an optical instrument. Figure 27.27a presents the **Rayleigh criterion** for resolution, first proposed by Lord Rayleigh (1842–1919):

> *Two point objects are just resolved when the first dark fringe in the diffraction pattern of one falls directly on the central bright fringe in the diffraction pattern of the other.*

The minimum angle $\theta_{min}$ between the two objects in the drawing is that given by Equation 27.5. If $\theta_{min}$ is small and is expressed in radians, $\sin\theta_{min} \approx \theta_{min}$. Then, Equation 27.5 can be rewritten as

$$\theta_{min} \approx 1.22\frac{\lambda}{D} \qquad (\theta_{min} \text{ in radians}) \qquad (27.6)$$

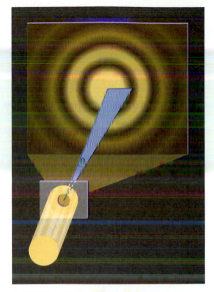

**Figure 27.25** When light passes through a small circular opening, a circular diffraction pattern is formed on a screen. The angle $\theta$ locates the first dark fringe relative to the central bright region. The intensities of the bright fringes and the diameter of the opening have been exaggerated for clarity.

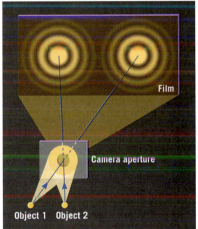

**Figure 27.26** When light from two point objects passes through the circular aperture of a camera, two circular diffraction patterns are formed as images on the film. The images here are completely separated or resolved because the objects are widely separated.

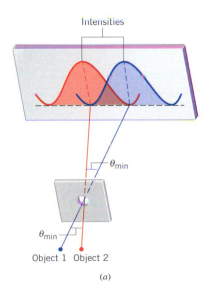

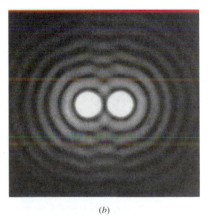

(a)                                    (b)

**Figure 27.27** (a) According to the Rayleigh criterion, two point objects are just resolved when the first dark fringe (zero intensity) of one of the images falls on the central bright fringe (maximum intensity) of the other image. (b) This photograph shows two overlapping but still resolvable diffraction patterns. (From Michel Cagnet et al., *Atlas of Optical Phenomena*, Springer-Verlag, Berlin.)

For a given wavelength $\lambda$ and aperture diameter $D$, this result specifies the smallest angle that two point objects can subtend at the aperture and still be resolved. According to Equation 27.6, optical instruments designed to resolve closely spaced objects (small values of $\theta_{min}$) must utilize the smallest possible wavelength and the largest possible aperture diameter. For example, when short-wavelength ultraviolet light is collected by its large 2.4-m-diameter mirror, the Hubble Space Telescope is capable of resolving two closely spaced stars that have an angular separation of about $\theta_{min} = 1 \times 10^{-7}$ rad. This angle is equivalent to resolving two objects only 1 cm apart when they are $1 \times 10^5$ m (about 62 miles) from the telescope. Example 7 deals with the resolving power of the human eye.

### Example 7 The Human Eye Versus the Eagle's Eye

*The physics of* comparing human eyes and eagle eyes.

(a) A hang glider is flying at an altitude of $H = 120$ m. Green light (wavelength = 555 nm in vacuum) enters the pilot's eye through a pupil that has a diameter of $D = 2.5$ mm. Determine how far apart two point objects must be on the ground if the pilot is to have any hope of distinguishing between them (see Figure 27.28). (b) An eagle's eye has a pupil with a diameter of $D = 6.2$ mm. Repeat part (a) for an eagle flying at the same altitude as the glider.

**Reasoning** According to the Rayleigh criterion, the two objects must be separated by a distance $s$ sufficient to subtend an angle $\theta_{min} \approx 1.22\lambda/D$ at the pupil of the pilot's eye.* Remembering to use radians for the angle, we will relate $\theta_{min}$ to the altitude $H$ and the desired distance $s$.

**Solution**

(a) Using the Rayleigh criterion, we find

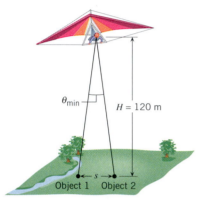

$\theta_{min}$   $H = 120$ m

Object 1   Object 2

$$\theta_{min} \approx 1.22\frac{\lambda}{D} = 1.22\left(\frac{555 \times 10^{-9}\text{ m}}{2.5 \times 10^{-3}\text{ m}}\right) = 2.7 \times 10^{-4}\text{ rad} \qquad (27.6)$$

According to Equation 8.1, $\theta_{min}$ in radians is $\theta_{min} \approx s/H$, so

$$s \approx \theta_{min} H = (2.7 \times 10^{-4}\text{ rad})(120\text{ m}) = \boxed{0.032\text{ m}}$$

**Figure 27.28** The Rayleigh criterion can be used to estimate the smallest distance $s$ that can separate two objects on the ground, if a person on a hang glider is to have any hope of distinguishing one object from the other.

(b) Since the pupil of an eagle's eye is larger than that of a human eye, diffraction creates less of a limitation for the eagle. A calculation like that above, using $D = 6.2$ mm, reveals that the diffraction limit for the eagle is $\boxed{s = 0.013\text{ m}}$.

Many optical instruments have a resolving power that exceeds that of the human eye. The typical camera does, for instance. Conceptual Example 8 compares the abilities of the human eye and a camera to resolve two closely spaced objects.

*Problem solving insight*
The minimum angle $\theta_{min}$ between two objects that are just resolved must be expressed in radians, not degrees, when using Equation 27.6 ($\theta_{min} \approx 1.22\ \lambda/D$).

### Conceptual Example 8 What You See Is Not What You Get

The French postimpressionist artist Georges Seurat developed a technique of painting in which dots of color are placed close together on the canvas. From sufficiently far away the individual dots are not distinguishable, and the images in the picture take on a more normal appearance. Figure 27.29 shows a person in an art museum, looking at one of Seurat's paintings. Suppose that this person stands close to the painting, then backs up until the dots just become indistinguishable to his eyes and takes a picture from this position. When he gets home and has the film developed, however, he can see the individual dots in the enlarged photograph. Why does the camera resolve the individual dots, while his eyes do not? Assume that light enters his eyes through pupils that have diameters of 2.5 mm and enters the camera through an aperture or opening with a diameter of 25 mm.

**Reasoning and Solution** To answer this question, we turn to the Rayleigh criterion for resolving two point objects—namely, $\theta_{min} \approx 1.22\ \lambda/D$. According to the Rayleigh criterion, a

---

*In applying the Rayleigh criterion, we use the given vacuum wavelength because it is nearly identical to the wavelength in air. We use the wavelength in air or vacuum, even though the diffraction occurs within the eye, where the index of refraction is about $n = 1.36$ and the wavelength is $\lambda_{eye} = \lambda_{vacuum}/n$, according to Equation 27.3. The reason is that in entering the eye, the light is refracted according to Snell's law, which also includes an effect due to the index of refraction. If the angle of incidence is small, the effect of $n$ in Snell's law cancels the effect of $n$ in Equation 27.3, to a good degree of approximation.

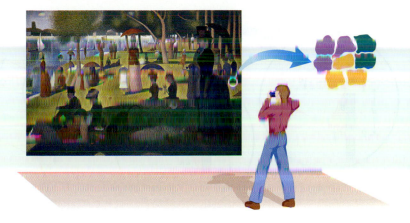

**Figure 27.29** This person is about to take a photograph of a famous painting by Georges Seurat, who developed the technique of using tiny dots of color to construct his images. Conceptual Example 8 discusses what the person sees when the film is developed. (Georges Seurat, *A Sunday on La Grande Jatte*. © 2003. All rights reserved. Reproduced with permission of The Art Institute of Chicago.)

larger value of the diameter $D$ of the opening means that the value for $\theta_{min}$ is smaller, so that an optical instrument can be farther away and still resolve the objects. In other words, a larger value of $D$ means a greater resolving power. For the eye and the camera the diameters are $D_{eye} = 2.5$ mm and $D_{camera} = 25$ mm, so the diameter for the camera is ten times larger than that for the eye. Thus, at the distance at which the eye loses its ability to resolve the individual dots in the painting, the camera can easily resolve the dots. As discussed in the footnote to Example 7, we can ignore the effect on the wavelength of the index of refraction of the material from which the eye is made. We conclude, then, that *the camera succeeds in resolving the dots in the painting while the eye fails because the camera gathers light through a much larger opening.*

**Related Homework:** *Problem 32*

## 27.7 *The Diffraction Grating*

Diffraction patterns of bright and dark fringes occur when monochromatic light passes through a single or double slit. Fringe patterns also result when light passes through more than two slits, and an arrangement consisting of a large number of parallel, closely spaced slits called a *diffraction grating,* has proved very useful. Gratings with as many as 40 000 slits per centimeter can be made, depending on the production method. In one method a diamond-tipped cutting tool is used to inscribe closely spaced parallel lines on a glass plate, the spaces between the lines serving as the slits. In fact, the number of slits per centimeter is often quoted as the number of lines per centimeter.

**The physics of** the diffraction grating.

Figure 27.30 illustrates how light travels to a distant viewing screen from each of five slits in a grating and forms the central bright fringe and the first-order bright fringes on either side. Higher-order bright fringes are also formed but are not shown in the drawing. Each bright fringe is located by an angle $\theta$ relative to the central fringe. These bright fringes are sometimes called the *principal fringes* or *principal maxima*, since they are places where

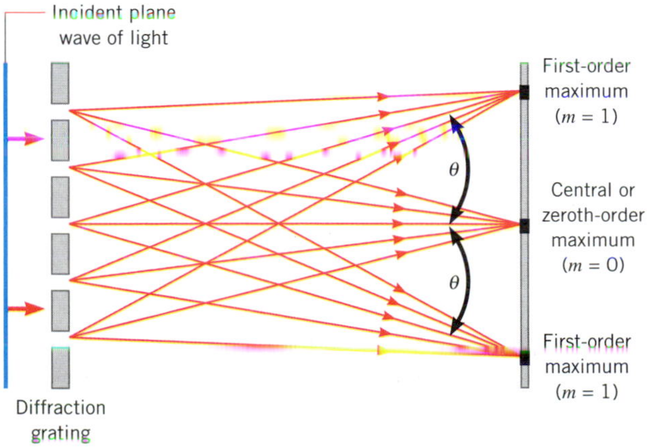

Incident plane wave of light

First-order maximum ($m = 1$)

$\theta$

Central or zeroth-order maximum ($m = 0$)

$\theta$

First-order maximum ($m = 1$)

Diffraction grating

**Figure 27.30** When light passes through a diffraction grating, a central bright fringe ($m = 0$) and higher-order bright fringes ($m = 1, 2, \ldots$) form when the light falls on a distant viewing screen.

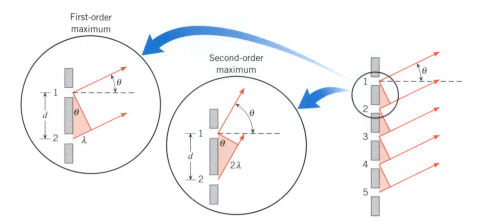

First-order maximum

Second-order maximum

**Figure 27.31** The conditions shown here lead to the first- and second-order intensity maxima in the diffraction pattern produced by the diffraction grating on the right.

the light intensity is a maximum. The term "principal" distinguishes them from other, much less bright fringes that are referred to as secondary fringes or secondary maxima.

Constructive interference creates the principal fringes. To show how, we assume the screen is far from the grating, so that the rays remain nearly parallel while the light travels toward the screen, as in Figure 27.31. In reaching the place on the screen where the first-order maximum is located, light from slit 2 travels a distance of one wavelength farther than light from slit 1. Similarly, light from slit 3 travels one wavelength farther than light from slit 2, and so forth, as emphasized by the four colored right triangles on the right-hand side of the drawing. For the first-order maximum, the enlarged view of one of these right triangles shows that constructive interference occurs when $\sin \theta = \lambda/d$, where $d$ is the separation between the slits. The second-order maximum forms when the extra distance traveled by light from adjacent slits is two wavelengths, so that $\sin \theta = 2\lambda/d$. The general result is

**Principal maxima of a diffraction grating**

$$\sin \theta = m\frac{\lambda}{d} \qquad m = 0, 1, 2, 3, \ldots \qquad (27.7)$$

The separation $d$ between the slits can be calculated from the number of slits per centimeter of grating; for instance, a grating with 2500 slits per centimeter has a slit separation of $d = (1/2500)$ cm $= 4.0 \times 10^{-4}$ cm. Equation 27.7 is identical to Equation 27.1 for the double slit. A grating, however, produces bright fringes that are much *narrower* or *sharper* than those from a double slit, as the intensity patterns in Figure 27.32 reveal. Note also from the drawing that between the principal maxima of a diffraction grating there are secondary maxima with much smaller intensities.

The next example illustrates the ability of a grating to separate the components in a mixture of colors.

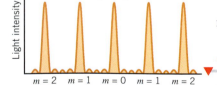

Grating (5 slits)

$m = 2 \quad m = 1 \quad m = 0 \quad m = 1 \quad m = 2$

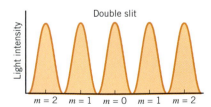

Double slit

$m = 2 \quad m = 1 \quad m = 0 \quad m = 1 \quad m = 2$

**Figure 27.32** The bright fringes produced by a diffraction grating are much narrower than those produced by a double slit. Note the three small secondary bright fringes between the principal bright fringes of the grating. For a large number of slits, these secondary fringes become very small.

### Example 9 Separating Colors with a Diffraction Grating

A mixture of violet light ($\lambda = 410$ nm in vacuum) and red light ($\lambda = 660$ nm in vacuum) falls on a grating that contains $1.0 \times 10^4$ lines/cm. For each wavelength, find the angle $\theta$ that locates the first-order maxima.

**Reasoning** Before Equation 27.7 can be used here, a value for the separation $d$ between the slits is needed: $d = 1/(1.0 \times 10^4$ lines/cm$) = 1.0 \times 10^{-4}$ cm, or $1.0 \times 10^{-6}$ m. For violet light, the angle $\theta_{violet}$ for the first-order maxima ($m = 1$) is given by $\sin \theta_{violet} = m\lambda_{violet}/d$, with an analogous equation applying for the red light.

**Solution** For violet light, the angle locating the first-order maxima is

$$\theta_{violet} = \sin^{-1}\frac{\lambda_{violet}}{d} = \sin^{-1}\left(\frac{410 \times 10^{-9} \text{ m}}{1.0 \times 10^{-6} \text{ m}}\right) = \boxed{24°}$$

For red light, a similar calculation with $\lambda_{red} = 660 \times 10^{-9}$ m shows that $\boxed{\theta_{red} = 41°}$. Because $\theta_{violet}$ and $\theta_{red}$ are different, separate first-order bright fringes are seen for violet and red light on a viewing screen.

If the light in Example 9 had been sunlight, the angles for the first-order maxima would cover all values in the range between 24° and 41°, since sunlight contains all colors or wavelengths between violet and red. Consequently, a rainbow-like dispersion of the colors would be observed to either side of the central fringe on a screen, as can be seen in Figure 27.33. This drawing shows that the spectrum of colors associated with the m = 2 order is completely separate from that of the m = 1 order. For higher orders, however, the spectra from adjacent orders may overlap (see problem 43). The central maximum (m = 0) is white because all the colors overlap there.

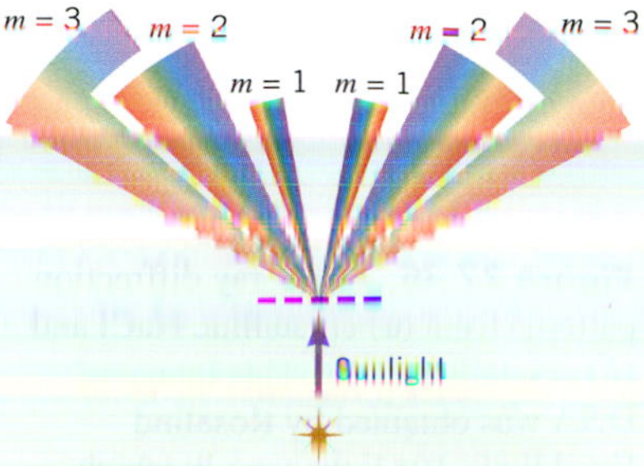

An instrument designed to measure the angles at which the principal maxima of a grating occur is called a grating spectroscope. With a measured value of the angle, calculations such as those in Example 9 can be turned around to provide the corresponding value of the wavelength. As we will point out in Chapter 30, the atoms in a hot gas emit discrete wavelengths, and determining the values of these wavelengths is one important technique used to identify the atoms. Figure 27.34 shows the principle of a grating spectroscope. The slit that admits light from the source (e.g., a hot gas) is located at the focal point of the collimating lens, so the light rays striking the grating are parallel. The telescope is used to detect the bright fringes and, hence, to measure the angle θ.

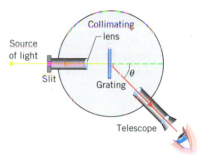

**Figure 27.33** When sunlight falls on a diffraction grating, a rainbow of colors is produced at each principal maximum (m = 1, 2, . . .). The central maximum (m = 0), however, is white but is not shown in the drawing.

*The physics of* **a grating spectroscope.**

**Figure 27.34** A grating spectroscope.

*The physics of* **X-ray diffraction.**

## 27.8 X-Ray Diffraction

Not all diffraction gratings are commercially made. Nature also creates diffraction gratings, although these gratings do not look like an array of closely spaced slits. Instead, nature's gratings are the arrays of regularly spaced atoms that exist in crystalline solids. For example, Figure 27.35 shows the structure of a crystal of ordinary salt (NaCl). Typically, the atoms in a crystalline solid are separated by distances of about $1.0 \times 10^{-10}$ m, so we might expect a crystalline array of atoms to act like a grating with roughly this "slit" spacing for electromagnetic waves of the appropriate wavelength. Assuming that sin θ = 0.5 and that m = 1 in Equation 27.7, then 0.5 = λ/d. A value of $d = 1.0 \times 10^{-10}$ m in this equation gives a wavelength of $\lambda = 0.5 \times 10^{-10}$ m. This wavelength is much shorter than that of visible light and falls in the X-ray region of the electromagnetic spectrum. (See Figure 24.9.)

A diffraction pattern does indeed result when X-rays are directed onto a crystalline material, as Figure 27.36a illustrates for a crystal of NaCl. The pattern consists of a complicated arrangement of spots because a crystal has a complex three-dimensional structure. It is from patterns such as this that the spacing between atoms and the nature of the crystal structure can be determined. X-ray diffraction has also been applied with great success toward understanding the structure of biologically important molecules, such as

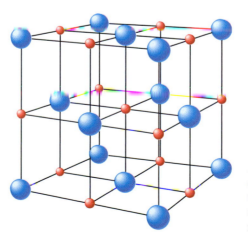

**Figure 27.35** In this drawing of the crystalline structure of sodium chloride, the small red spheres represent positive sodium ions, and the large blue spheres represent negative chloride ions.

**Figure 27.36** The X-ray diffraction patterns from (*a*) crystalline NaCl and (*b*) crystalline DNA. This image of DNA was obtained by Rosalind Franklin in 1953, the year in which Watson and Crick discovered DNA's structure. (*a*: Courtesy Edwin Jones, University of South Carolina; *b*: © Omikron/ Photo Researchers)

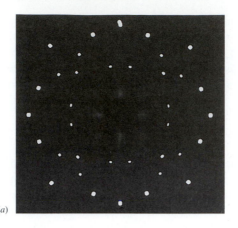

(*a*)

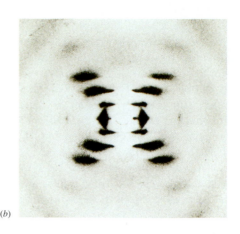

(*b*)

proteins and nucleic acids. One of the most famous results was the discovery in 1953 by Watson and Crick that the structure of the nucleic acid DNA is a double helix. X-ray diffraction patterns such as that in Figure 27.36*b* played the pivotal role in their research.

# Concept Summary

This summary presents an abridged version of the chapter, including the important equations and all available learning aids. For convenient reference, the learning aids (including the text's examples) are placed next to or immediately after the relevant equation or discussion. The following learning aids may be found on-line at **www.wiley.com/college/cutnell**:

| | |
|---|---|
| **Interactive LearningWare** examples are solved according to a five-step interactive format that is designed to help you develop problem-solving skills. | **Concept Simulations** are animated versions of text figures or animations that illustrate important concepts. You can control parameters that affect the display, and we encourage you to experiment. |
| **Interactive Solutions** offer specific models for certain types of problems in the chapter homework. The calculations are carried out interactively. | **Self-Assessment Tests** include both qualitative and quantitative questions. Extensive feedback is provided for both incorrect and correct answers, to help you evaluate your understanding of the material. |

| Topic | Discussion | Learning Aids |
|---|---|---|

### 27.1 The Principle of Linear Superposition

**Principle of linear superposition**

The principle of linear superposition states that when two or more waves are present simultaneously in the same region of space, the resultant disturbance is the sum of the disturbances from the individual waves.

**Constructive interference**

Constructive interference occurs at a point when two waves meet there crest-to-crest and trough-to-trough, thus reinforcing each other. When two waves that start out in phase and have traveled some distance meet at a point, constructive interference occurs whenever the travel distances are the same or differ by any integer number of wavelengths: $\ell_2 - \ell_1 = m\lambda$, where $\ell_1$ and $\ell_2$ are the distances traveled by the waves, and $m = 0, 1, 2, 3, \ldots$.

**Destructive interference**

Destructive interference occurs at a point when two waves meet there crest-to-trough, thus mutually canceling each other. When two waves that start out in phase and have traveled some distance meet at a point, destructive interference occurs whenever the travel distances differ by any odd integer number of half-wavelengths: $\ell_2 - \ell_1 = (m + \frac{1}{2})\lambda$, where $\ell_1$ and $\ell_2$ are the distances traveled by the waves, and $m = 0, 1, 2, 3, \ldots$.

**Coherent sources**

Two sources are coherent if the waves they emit maintain a constant phase relation. In other words, the waves do not shift relative to one another as time passes. If constructive and destructive interference are to be observed, coherent sources are necessary.

### 27.2 Young's Double-Slit Experiment

**Double-slit experiment**

In Young's double-slit experiment, light passes through a pair of closely spaced narrow slits and produces a pattern of alternating bright and dark fringes on a

| Topic | Discussion | Learning Aids |
|-------|-----------|---------------|

viewing screen. The fringes arise because of constructive and destructive interference. The angle $\theta$ that locates the $m$th-order bright fringe is given by

**Bright fringes of a double slit**

$$\sin \theta = \frac{m\lambda}{d} \qquad m = 0, 1, 2, 3, \ldots \qquad (27.1)$$

*Examples 1, 2*
*Interactive LearningWare 27.1*
*Interactive Solution 27.7*

where $\lambda$ is the wavelength of the light, and $d$ is the spacing between the slits. The angle that locates the $m$th dark fringe is given by

**Dark fringes of a double slit**

$$\sin \theta = \frac{(m + \frac{1}{2})\lambda}{d} \qquad m = 0, 1, 2, 3, \ldots \qquad (27.2)$$

### 27.3 Thin-Film Interference

Constructive and destructive interference of light waves can occur with thin films of transparent materials. The interference occurs between light waves that reflect from the top and bottom surfaces of the film.

One important factor in thin-film interference is the thickness of a film relative to the wavelength of the light within the film. The wavelength $\lambda_{\text{film}}$ within a film is

**Wavelength within a film**

$$\lambda_{\text{film}} = \frac{\lambda_{\text{vacuum}}}{n} \qquad (27.3)$$

*Examples 3, 4, 5*
*Interactive LearningWare 27.2*
*Interactive Solution 27.17*

where $\lambda_{\text{vacuum}}$ is the wavelength in a vacuum, and $n$ is the index of refraction of the film.

**Phase change due to reflection**

A second important factor is the phase change that can occur when light reflects at each surface of the film:

1. When light travels through a material with a smaller index of refraction toward a material with a larger index of refraction, reflection at the boundary occurs along with a phase change that is equivalent to one-half a wavelength in the film.

2. When light travels through a material with a larger index of refraction toward a material with a smaller index of refraction, there is no phase change upon reflection at the boundary.

### 27.4 The Michelson Interferometer

**Michelson interferometer**

An interferometer is an instrument that can be used to measure the wavelength of light by employing interference between two light waves. The Michelson interferometer splits the light into two beams, one of which travels to a fixed mirror, reflects from it, and returns. The other beam travels to a movable mirror, reflects from it, and returns. When the two returning beams are combined, interference is observed, the amount of which depends on the travel distances.

 **Use *Self-Assessment Test 27.1* to evaluate your understanding of Sections 27.1–27.4.**

### 27.5 Diffraction

**Diffraction**

Diffraction is the bending of waves around obstacles or around the edges of an opening. Diffraction is an interference effect that can be explained with the aid of Huygens' principle. This principle states that every point on a wave front acts as a source of tiny wavelets that move forward with the same speed as the wave; the wave front at a later instant is the surface that is tangent to the wavelets.

**Huygens' principle**

When light passes through a single narrow slit and falls on a viewing screen, a pattern of bright and dark fringes is formed because of the superposition of Huygens wavelets. The angle $\theta$ that specifies the $m$th dark fringe on either side of the central bright fringe is given by

*Concept Simulation 27.1*

**Dark fringes for single-slit diffraction**

$$\sin \theta = m\frac{\lambda}{W} \qquad m = 1, 2, 3, \ldots \qquad (27.4)$$

*Example 6*
*Interactive LearningWare 27.3*

where $\lambda$ is the wavelength of the light and $W$ is the width of the slit.

| Topic | Discussion | Learning Aids |
|-------|------------|---------------|

### 27.6 Resolving Power

**Resolving power**

The resolving power of an optical instrument is the ability of the instrument to distinguish between two closely spaced objects. Resolving power is limited by the diffraction that occurs when light waves enter an instrument, often through a circular opening.

**Rayleigh criterion**

The Rayleigh criterion specifies that two point objects are just resolved when the first dark fringe in the diffraction pattern of one falls directly on the central bright fringe in the diffraction pattern of the other. According to this specification, the minimum angle (in radians) that two point objects can subtend at a circular aperture of diameter $D$ and still be resolved as separate objects is

**Minimum angle for resolving two point objects**

$$\theta_{min} \approx 1.22 \frac{\lambda}{D} \qquad (\theta_{min} \text{ in radians}) \qquad (27.6)$$

**Examples 7, 8**
**Interactive Solution 27.35**

where $\lambda$ is the wavelength of the light.

### 27.7 The Diffraction Grating

**Diffraction grating**

A diffraction grating consists of a large number of parallel, closely spaced slits. When light passes through a diffraction grating and falls on a viewing screen, the light forms a pattern of bright and dark fringes. The bright fringes are referred to as principal maxima and are found at an angle $\theta$ such that

**Principal maxima of a diffraction grating**

$$\sin \theta = m \frac{\lambda}{d} \qquad m = 0, 1, 2, 3, \ldots \qquad (27.7)$$

**Example 9**

where $\lambda$ is the wavelength of the light and $d$ is the separation between two adjacent slits.

### 27.8 X-Ray Diffraction

A diffraction pattern forms when X-rays are directed onto a crystalline material. The pattern arises because the regularly spaced atoms in a crystal act like a diffraction grating. Because the spacing is extremely small, on the order of $1 \times 10^{-10}$ m, the wavelength of the electromagnetic waves must also be very small—hence, the use of X-rays. The crystal structure of a material can be determined from its X-ray diffraction pattern.

 **Use Self-Assessment Test 27.2 to evaluate your understanding of Sections 27.5–27.8.**

# Problems

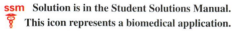

**Section 27.1 The Principle of Linear Superposition,**
**Section 27.2 Young's Double-Slit Experiment**

**1. ssm** A flat observation screen is placed at a distance of 4.5 m from a pair of slits. The separation on the screen between the central bright fringe and the first-order bright fringe is 0.037 m. The light illuminating the slits has a wavelength of 490 nm. Determine the slit separation.

**2.** In a Young's double-slit experiment, the seventh dark fringe is located 0.025 m to the side of the central bright fringe on a flat screen, which is 1.1 m away from the slits. The separation between the slits is $1.4 \times 10^{-4}$ m. What is the wavelength of the light being used?

**3.** A rock concert is being held in an open field. Two loudspeakers are separated by 7.00 m. As an aid in arranging the seating, a test is conducted in which both speakers vibrate in phase and produce an 80.0-Hz bass tone simultaneously. The speed of sound is 343 m/s. A reference line is marked out in front of the speakers, perpendicular to the midpoint of the line between the speakers. Relative to either side of this reference line, what is the smallest angle that locates the places where destructive interference occurs? People seated in these places would have trouble hearing the 80.0-Hz bass tone.

**4.** Two in-phase sources of waves are separated by a distance of 4.00 m. These sources produce identical waves that have a wavelength of 5.00 m. On the line between them, there are two places at which the same type of interference occurs. (a) Is it constructive or destructive interference, and (b) where are the places located?

**5. ssm** In a Young's double-slit experiment, the angle that locates the second-order bright fringe is 2.0°. The slit separation is $3.8 \times 10^{-5}$ m. What is the wavelength of the light?

**\* 6.** Review Conceptual Example 2 before attempting this problem. Two slits are 0.158 mm apart. A mixture of red light (wavelength = 665 nm) and yellow-green light (wavelength = 565 nm) falls on the slits. A flat observation screen is located 2.24 m away. What is the distance on the screen between the third-order red fringe and the third-order yellow-green fringe?

**\* 7.** Refer to **Interactive Solution 27.7** at **www.wiley.com/college/cutnell** for help in solving this problem. In a Young's double-slit experiment the separation $y$ between the second-order bright fringe and the central bright fringe on a flat screen is 0.0180 m, when the light has a wavelength of 425 nm. Assume that the angles that locate the fringes on the screen are small enough so that $\sin \theta \approx \tan \theta$. Find the separation $y$ when the light has a wavelength of 585 nm.

**\*\* 8.** In Young's experiment a mixture of orange light (611 nm) and blue light (471 nm) shines on the double slit. The centers of the first-order bright blue fringes lie at the outer edges of a screen that is located 0.500 m away from the slits. However, the first-order bright orange fringes fall off the screen. By how much and in which direction (toward or away from the slits) should the screen be moved, so that the centers of the first-order bright orange fringes just appear on the screen? It may be assumed that $\theta$ is small, so that $\sin \theta \approx \tan \theta$.

**\*\* 9. ssm www** A sheet of plastic ($n$ = 1.60) covers *one* slit of a double slit (see the drawing). When the double slit is illuminated by monochromatic light ($\lambda_{vacuum}$ = 586 nm), the center of the screen appears dark rather than bright. What is the minimum thickness of the plastic?

*Problem 9*

### Section 27.3 Thin-Film Interference

**10.** Light of wavelength 691 nm (in vacuum) is incident perpendicularly on a soap film ($n$ = 1.33) suspended in air. What are the two smallest nonzero film thicknesses (in nm) for which the reflected light undergoes constructive interference?

**11. ssm** A nonreflective coating of magnesium fluoride ($n$ = 1.38) covers the glass ($n$ = 1.52) of a camera lens. Assuming that the coating prevents reflection of yellow-green light (wavelength in vacuum = 565 nm), determine the minimum nonzero thickness that the coating can have.

**12.** For background on this problem, review Conceptual Example 4. A mixture of yellow light (wavelength = 580 nm in vacuum) and violet light (wavelength = 410 nm in vacuum) falls perpendicularly on a film of gasoline that is floating on a puddle of water. For both wavelengths, the refractive index of gasoline is $n$ = 1.40 and that of water is $n$ = 1.33. What is the minimum nonzero thickness of the film in a spot that looks (a) yellow and (b) violet because of destructive interference?

**13.** A mixture of red light ($\lambda_{vacuum}$ = 661 nm) and green light ($\lambda_{vacuum}$ = 551 nm) shines perpendicularly on a soap film ($n$ = 1.33) that has air on either side. What is the minimum nonzero thickness of the film, so that destructive interference causes it to look red in reflected light?

**14.** Review Conceptual Example 4 before beginning this problem. A soap film with different thicknesses at different places has an unknown refractive index $n$ and air on both sides. In reflected light it looks multicolored. One region looks yellow because destructive interference has removed blue ($\lambda_{vacuum}$ = 469 nm) from the reflected light, while another looks magenta because destructive interference has removed green ($\lambda_{vacuum}$ = 555 nm). In these regions the film has the minimum nonzero thickness $t$ required for the destructive interference to occur. Find the ratio $t_{magenta}/t_{yellow}$.

**\* 15. ssm www** Orange light ($\lambda_{vacuum}$ = 611 nm) shines on a soap film ($n$ = 1.33) that has air on either side of it. The light strikes the film perpendicularly. What is the minimum thickness of the film for which constructive interference causes it to look bright in reflected light?

**\* 16.** Review Conceptual Example 4 before attempting this problem. A film of gasoline ($n$ = 1.40) floats on water ($n$ = 1.33). Yellow light (wavelength = 580 nm in vacuum) shines perpendicularly on this film. (a) Determine the minimum nonzero thickness of the film, such that the film appears bright yellow due to constructive interference. (b) Repeat part (a), assuming that the gasoline film is on glass ($n$ = 1.52) instead of water.

**\* 17.** Consult **Interactive Solution 27.17** at **www.wiley.com/college/cutnell** to review a model for solving this problem. A film of oil lies on wet pavement. The refractive index of the oil exceeds that of the water. The film has the minimum nonzero thickness such that it appears dark due to destructive interference when viewed in red light (wavelength = 640.0 nm in vacuum). Assuming that the visible spectrum extends from 380 to 750 nm, for which visible wavelength(s) (in vacuum) will the film appear bright due to constructive interference?

**\*\* 18.** A piece of curved glass has a radius of curvature of 10.0 m and is used to form Newton's rings, as in Figure 27.13. Not counting the dark spot at the center of the pattern, there are one hundred dark fringes, the last one being at the outer edge of the curved piece of glass. The light being used has a wavelength of 654 nm in vacuum. What is the radius of the outermost dark ring in the pattern?

### Section 27.5 Diffraction

**19. ssm** A diffraction pattern forms when light passes through a single slit. The wavelength of the light is 675 nm. Determine the angle that locates the first dark fringe when the width of the slit is (a) $1.8 \times 10^{-4}$ m and (b) $1.8 \times 10^{-6}$ m.

**20.** A slit whose width is $4.30 \times 10^{-5}$ m is located 1.32 m from a flat screen. Light shines through the slit and falls on the screen. Find the width of the central fringe of the diffraction pattern when the wavelength of the light is 635 nm.

**21.** A single slit has a width of $2.1 \times 10^{-6}$ m and is used to form a diffraction pattern. Find the angle that locates the second dark fringe when the wavelength of the light is (a) 430 nm and (b) 660 nm.

**22.** Light of wavelength 668 nm passes through a slit $6.73 \times 10^{-6}$ m wide and falls on a screen that is 1.85 m away. What is the distance on the screen from the center of the central bright fringe to the third dark fringe on either side?

**23. ssm www** Light shines through a single slit whose width is $5.6 \times 10^{-4}$ m. A diffraction pattern is formed on a flat screen located 4.0 m away. The distance between the middle of the central bright fringe and the first dark fringe is 3.5 mm. What is the wavelength of the light?

**\* 24.** How many dark fringes will be produced on either side of the central maximum if light ($\lambda$ = 651 nm) is incident on a single slit that is $5.47 \times 10^{-6}$ m wide?

**\* 25. ssm** The central bright fringe in a single-slit diffraction pattern has a width that equals the distance between the screen and the slit. Find the ratio $\lambda/W$ of the wavelength of the light to the width of the slit.

**\* 26.** The width of a slit is $2.0 \times 10^{-5}$ m. Light with a wavelength of 480 nm passes through this slit and falls on a screen that is located 0.50 m away. In the diffraction pattern, find the width of the bright fringe that is next to the central bright fringe.

** **27.** In a single-slit diffraction pattern, the central fringe is 450 times as wide as the slit. The screen is 18 000 times farther from the slit than the slit is wide. What is the ratio $\lambda/W$, where $\lambda$ is the wavelength of the light shining through the slit and $W$ is the width of the slit? Assume that the angle that locates a dark fringe on the screen is small, so that $\sin \theta \approx \tan \theta$.

### Section 27.6 Resolving Power

**28.** You are looking down at the earth from inside a jetliner flying at an altitude of 8690 m. The pupil of your eye has a diameter of 2.00 mm. Determine how far apart two cars must be on the ground if you are to have any hope of distinguishing between them in (a) red light (wavelength = 665 nm in vacuum) and (b) violet light (wavelength = 405 nm in vacuum).

**29.** It is claimed that some professional baseball players can see which way the ball is spinning as it travels toward home plate. One way to judge this claim is to estimate the distance at which a batter can first hope to resolve two points on opposite sides of a baseball, which has a diameter of 0.0738 m. (a) Estimate this distance, assuming that the pupil of the eye has a diameter of 2.0 mm and the wavelength of the light is 550 nm in vacuum. (b) Considering that the distance between the pitcher's mound and home plate is 18.4 m, can you rule out the claim based on your answer to part (a)?

**30.** Two stars are $3.7 \times 10^{11}$ m apart and are equally distant from the earth. A telescope has an objective lens with a diameter of 1.02 m and just detects these stars as separate objects. Assume that light of wavelength 550 nm is being observed. Also assume that diffraction effects, rather than atmospheric turbulence, limit the resolving power of the telescope. Find the maximum distance that these stars could be from the earth.

**31.** **ssm** Late one night on a highway, a car speeds by you and fades into the distance. Under these conditions the pupils of your eyes have diameters of about 7.0 mm. The taillights of this car are separated by a distance of 1.2 m and emit red light (wavelength = 660 nm in vacuum). How far away from you is this car when its taillights appear to merge into a single spot of light because of the effects of diffraction?

**32.** Review Conceptual Example 8 as background for this problem. In addition to the data given there, assume that the dots in the painting are separated by 1.5 mm and that the wavelength of the light is $\lambda_{\text{vacuum}} = 550$ nm. Find the distance at which the dots can just be resolved by (a) the eye and (b) the camera.

**33.** Astronomers have discovered a planetary system orbiting the star Upsilon Andromedae, which is at a distance of $4.2 \times 10^{17}$ m from the earth. One planet is believed to be located at a distance of $1.2 \times 10^{11}$ m from the star. Using visible light with a vacuum wavelength of 550 nm, what is the minimum necessary aperture diameter that a telescope must have so that it can resolve the planet and the star?

* **34.** The pupil of an eagle's eye has a diameter of 6.0 mm. Two field mice are separated by 0.010 m. From a distance of 176 m, the eagle sees them as one unresolved object and dives toward them at a speed of 17 m/s. Assume that the eagle's eye detects light that has a wavelength of 550 nm in a vacuum. How much time passes until the eagle sees the mice as separate objects?

* **35.** Refer to **Interactive Solution 27.35** at **www.wiley.com/college/cutnell** to review a method by which this problem can be solved. You are using a microscope to examine a blood sample. Recall from Section 26.12 that the sample should be placed just outside the focal point of the objective lens of the microscope. (a) If the specimen is being illuminated with light of wavelength $\lambda$ and the diameter of the

objective equals its focal length, determine the closest distance between two blood cells that can just be resolved. Express your answer in terms of $\lambda$. (b) Based on your answer to (a), should you use light with a longer wavelength or a shorter wavelength if you wish to resolve two blood cells that are even closer together?

** **36.** Two concentric circles of light emit light whose wavelength is 555 nm. The larger circle has a radius of 4.0 cm, and the smaller circle has a radius of 1.0 cm. When taking a picture of these lighted circles, a camera admits light through an aperture whose diameter is 12.5 mm. What is the maximum distance at which the camera can (a) distinguish one circle from the other and (b) reveal that the inner circle is a circle of light rather than a solid disk of light?

### Section 27.7 The Diffraction Grating

**37.** **ssm** The diffraction gratings discussed in the text are transmission gratings because light *passes through* them. There are also gratings in which the light *reflects from* the grating to form a pattern of fringes. Equation 27.7 also applies to a reflection grating with straight parallel lines when the incident light shines perpendicularly on the grating. The surface of a compact disc (CD) has a multicolored appearance because it acts like a reflection grating and spreads sunlight into its colors. The arms of the spiral track on the CD are separated by $1.1 \times 10^{-6}$ m. Using Equation 27.7, estimate the angle that corresponds to the first-order maximum for a wavelength of (a) 660 nm (red) and (b) 410 nm (violet).

**38.** A diffraction grating produces a first-order bright fringe that is 0.0894 m away from the central bright fringe on a flat screen. The separation between the slits of the grating is $4.17 \times 10^{-6}$ m, and the distance between the grating and the screen is 0.625 m. What is the wavelength of the light shining on the grating?

**39.** For a wavelength of 420 nm, a diffraction grating produces a bright fringe at an angle of 26°. For an unknown wavelength, the same grating produces a bright fringe at an angle of 41°. In both cases the bright fringes are of the same order $m$. What is the unknown wavelength?

**40.** A diffraction grating is 1.50 cm wide and contains 2400 lines. When used with light of a certain wavelength, a third-order maximum is formed at an angle of 18.0°. What is the wavelength (in nm)?

**41.** **ssm** The wavelength of the laser beam used in a compact disc player is 780 nm. Suppose that a diffraction grating produces first-order tracking beams that are 1.2 mm apart at a distance of 3.0 mm from the grating. Estimate the spacing between the slits of the grating.

* **42.** A diffraction grating has 2604 lines per centimeter, and it produces a principal maximum at $\theta = 30.0°$. The grating is used with light that contains all wavelengths between 410 and 660 nm. What is (are) the wavelength(s) of the incident light that could have produced this maximum?

* **43.** **ssm** The same diffraction grating is used with two different wavelengths of light, $\lambda_A$ and $\lambda_B$. The fourth-order principal maximum of light A exactly overlaps the third-order principal maximum of light B. Find the ratio $\lambda_A/\lambda_B$.

* **44.** Three, and only three, bright fringes can be seen on either side of the central maximum when a grating is illuminated with light ($\lambda = 510$ nm). What is the maximum number of lines/cm for the grating?

** **45.** There are 5620 lines per centimeter in a grating that is used with light whose wavelength is 471 nm. A flat observation screen is located at a distance of 0.750 m from the grating. What is the minimum width that the screen must have so the *centers* of all the principal maxima formed on either side of the central maximum fall on the screen?

# Chapter 28 Special Relativity

## 28.1 Events and Inertial Reference Frames

In the theory of special relativity, an *event*, such as the launching of the space shuttle in Figure 28.1, is a physical "happening" that occurs at a certain place and time. In this drawing two observers are watching the lift-off, one standing on the earth and one in an airplane that is flying at a constant velocity relative to the earth. To record the event, each observer uses a *reference frame* that consists of a set of *x, y, z* axes (called a *coordinate system*) and a clock. The coordinate systems are used to establish where the event occurs, and the clocks to specify when. Each observer is at rest relative to his own reference frame. However, the earth-based observer and the airborne observer are moving relative to each other and so are their respective reference frames.

The theory of special relativity deals with a "special" kind of reference frame, called an *inertial reference frame.* As Section 4.2 discusses, an inertial reference frame is one in which Newton's law of inertia is valid. That is, if the net force acting on a body is zero, the body either remains at rest or moves at a constant velocity. In other words, the acceleration of such a body is zero when measured in an inertial reference frame. Rotating and otherwise accelerating reference frames are not inertial reference frames. The earth-based reference frame in Figure 28.1 is not quite an inertial frame because it is subjected to centripetal accelerations as the earth spins on its axis and revolves around the sun. In most situations, however, the effects of these accelerations are small, and we can neglect them. To the extent that the earth-based reference frame is an inertial frame, so is the plane-based reference frame, because the plane moves at a constant velocity relative to the earth. The next section discusses why inertial reference frames are important in relativity.

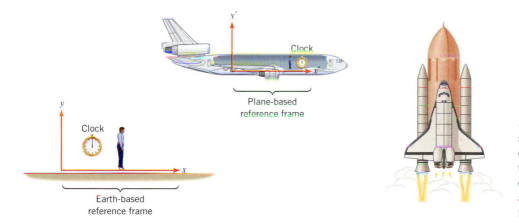

**Figure 28.1** Using an earth-based reference frame, an observer standing on the earth records the location and time of an event (the space shuttle lift-off). Likewise, an observer in the airplane uses a plane-based reference frame to describe the event.

## 28.2 The Postulates of Special Relativity

Einstein built his theory of special relativity on two fundamental assumptions or postulates about the way nature behaves.

■ **THE POSTULATES OF SPECIAL RELATIVITY**

1. *The Relativity Postulate.* The laws of physics are the same in every inertial reference frame.
2. *The Speed of Light Postulate.* The speed of light in a vacuum, measured in any inertial reference frame, always has the same value of *c,* no matter how fast the source of light and the observer are moving relative to each other.

It is not too difficult to accept the relativity postulate. For instance, in Figure 28.1 each observer, using his own inertial reference frame, can make measurements on the motion of the space shuttle. The relativity postulate asserts that both observers find their data to be consistent with Newton's laws of motion. Similarly, both observers find that the behavior of the electronics on the space shuttle is described by the laws of electromagnetism. According to the relativity postulate, *any inertial reference frame is as good as any other for expressing the laws of physics.* As far as inertial reference frames are concerned, nature does not play favorites.

Since the laws of physics are the same in all inertial reference frames, there is no experiment that can distinguish between an inertial frame that is at rest and one that is moving at a constant velocity. When you are seated on the aircraft in Figure 28.1, for instance, it is just as valid to say that you are at rest and the earth is moving as it is to say the converse. It is not possible to single out one particular inertial reference frame as being at "absolute rest." Consequently, it is meaningless to talk about the "absolute velocity" of an object—that is, its velocity measured relative to a reference frame at "absolute rest." Thus, the earth moves relative to the sun, which itself moves relative to the center of our galaxy. And the galaxy moves relative to other galaxies, and so on. According to Einstein, only the relative velocity between objects, not their absolute velocities, can be measured and is physically meaningful.

Whereas the relativity postulate seems plausible, the speed of light postulate defies common sense. For instance, Figure 28.2 illustrates a person standing on the bed of a truck that is moving at a constant speed of 15 m/s relative to the ground. Now, suppose you are standing on the ground and the person on the truck shines a flashlight at you. The person on the truck observes the speed of light to be $c$. What do you measure for the speed of light? You might guess that the speed of light would be $c + 15$ m/s. However, this guess is inconsistent with the speed of light postulate, which states that all observers in inertial reference frames measure the speed of light to be $c$—nothing more, nothing less. Therefore, you must also measure the speed of light to be $c$, the same as that measured by the person on the truck. According to the speed of light postulate, the fact that the flashlight is moving has no influence whatsoever on the speed of the light approaching you. This property of light, although surprising, has been verified many times by experiment.

Since waves, such as water waves and sound waves, require a medium through which to propagate, it was natural for scientists before Einstein to assume that light did too. This hypothetical medium was called the *luminiferous ether* and was assumed to fill all of space. Furthermore, it was believed that light traveled at the speed $c$ only when measured with respect to the ether. According to this view, an observer moving relative to the ether would measure a speed for light that was slower or faster than $c$, depending on whether the observer moved with or against the light, respectively. During the years 1883–1887, however, the American scientists A. A. Michelson and E. W. Morley carried out a series of famous experiments whose results were not consistent with the ether theory. Their results indicated that the speed of light is indeed the same in all inertial reference frames and does not depend on the motion of the observer. These experiments, and others, led eventually to the demise of the ether theory and the acceptance of the theory of special relativity.

The remainder of this chapter reexamines, from the viewpoint of special relativity, a number of fundamental concepts that have been discussed in earlier chapters from the viewpoint of classical physics. These concepts are time, length, momentum, kinetic energy, and the addition of velocities. We will see that each is modified by special relativity in a way that depends on the speed $v$ of a moving object relative to the speed $c$ of light in a vacuum. When the object moves slowly ($v$ is much smaller than $c$ [$v \ll c$]), the modi-

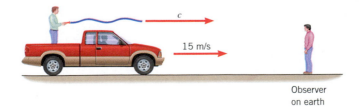

**Figure 28.2** Both the person on the truck and the observer on the earth measure the speed of the light to be $c$, regardless of the speed of the truck.

fication is negligibly small, and the classical version of each concept provides an accurate description of reality. However, when the object moves so rapidly that $v$ is an appreciable fraction of the speed of light, the effects of special relativity must be considered.

It is important to realize that the modifications imposed by special relativity do not imply that the classical concepts of time, length, momentum, kinetic energy, and the addition of velocities, as developed by Newton and others, are wrong. They are just limited to speeds that are very small compared to the speed of light. In contrast, the relativistic view of the concepts applies to all speeds between zero and the speed of light.

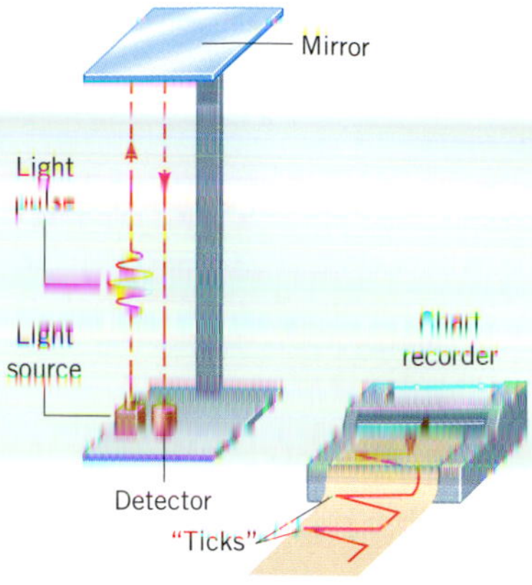

*Figure 28.3* A light clock.

## 28.3 The Relativity of Time: Time Dilation

### TIME DILATION

Common experience indicates that time passes just as quickly for a person standing on the ground as it does for an astronaut in a spacecraft. In contrast, special relativity reveals that the person on the ground measures time passing more slowly for the astronaut than for himself. We can see how this curious effect arises with the help of the clock illustrated in Figure 28.3, which uses a pulse of light to mark time. A short pulse of light is emitted by a light source, reflects from a mirror, and then strikes a detector that is situated next to the source. Each time a pulse reaches the detector, a "tick" registers on the chart recorder, another short pulse of light is emitted, and the cycle repeats. Thus, the time interval between successive "ticks" is marked by a beginning event (the firing of the light source) and an ending event (the pulse striking the detector). The source and detector are so close to each other that the two events can be considered to occur at the same location.

Suppose two identical clocks are built. One is kept on earth, and the other is placed aboard a spacecraft that travels at a constant velocity relative to the earth. The astronaut is at rest with respect to the clock on the spacecraft and, therefore, sees the light pulse move along the up/down path shown in Figure 28.4a. According to the astronaut, the time interval $\Delta t_0$ required for the light to follow this path is the distance $2D$ divided by the speed of light $c$; $\Delta t_0 = 2D/c$. To the astronaut, $\Delta t_0$ is the time interval between the "ticks" of the spacecraft clock—that is, the time interval between the beginning and ending events of the clock. An earth-based observer, however, does *not* measure $\Delta t_0$ as the time interval between these two events. Since the spacecraft is moving, the earth-based observer sees the light pulse follow the diagonal path shown in red in part $b$ of the drawing. This path is longer than the up/down path seen by the astronaut. But light travels at the *same speed* $c$ for both observers, in accord with the speed of light postulate. Therefore, the earth-based observer measures a time interval $\Delta t$ between the two events that is *greater* than the time interval $\Delta t_0$ measured by the astronaut. In other words, the earth-based observer, using his own earth-based clock to measure the performance of the

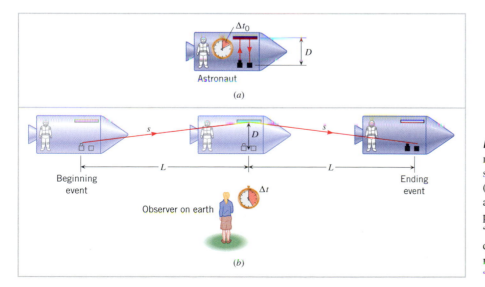

*Figure 28.4* (*a*) The astronaut measures the time interval $\Delta t_0$ between successive "ticks" of his light clock. (*b*) An observer on earth watches the astronaut's clock and sees the light pulse travel a greater distance between "ticks" than it does in part *a*. Consequently, the earth-based observer measures a time interval $\Delta t$ between "ticks" that is greater than $\Delta t_0$.

astronaut's clock, finds that the astronaut's clock runs slowly. This result of special relativity is known as **time dilation.** (To *dilate* means to expand, and the time interval $\Delta t$ is "expanded" relative to $\Delta t_0$.)

The time interval $\Delta t$ that the earth-based observer measures in Figure 28.4*b* can be determined as follows. While the light pulse travels from the source to the detector, the spacecraft moves a distance $2L = v\Delta t$ to the right, where $v$ is the speed of the spacecraft relative to the earth. From the drawing it can be seen that the light pulse travels a total diagonal distance of $2s$ during the time interval $\Delta t$. Applying the Pythagorean theorem, we find that

$$2s = 2\sqrt{D^2 + L^2} = 2\sqrt{D^2 + \left(\frac{v\,\Delta t}{2}\right)^2}$$

But the distance $2s$ is also equal to the speed of light times the time interval $\Delta t$, so $2s = c\,\Delta t$. Therefore,

$$c\,\Delta t = 2\sqrt{D^2 + \left(\frac{v\,\Delta t}{2}\right)^2}$$

Squaring this result and solving for $\Delta t$ gives

$$\Delta t = \frac{2D}{c}\frac{1}{\sqrt{1 - \dfrac{v^2}{c^2}}}$$

But $2D/c = \Delta t_0$, the time interval between successive "ticks" of the spacecraft's clock as measured by the astronaut. With this substitution, the equation for $\Delta t$ can be expressed as

**Time dilation**

$$\Delta t = \frac{\Delta t_0}{\sqrt{1 - \dfrac{v^2}{c^2}}} \tag{28.1}$$

The symbols in this formula are defined as follows:

$\Delta t_0$ = proper time interval, which is the interval between two events as measured by an observer who is at rest with respect to the events and who views them as occurring *at the same place*

$\Delta t$ = dilated time interval, which is the interval measured by an observer who is in motion with respect to the events and who views them as occurring at *different places*

$v$ = relative speed between the two observers

$c$ = speed of light in a vacuum

For a speed $v$ that is less than $c$, the term $\sqrt{1 - v^2/c^2}$ in Equation 28.1 is less than 1, and the dilated time interval $\Delta t$ is greater than $\Delta t_0$. Example 1 shows this time dilation effect.

### Example 1    Time Dilation

The spacecraft in Figure 28.4 is moving past the earth at a constant speed $v$ that is 0.92 times the speed of light. Thus, $v = (0.92)(3.0 \times 10^8 \text{ m/s})$, which is often written as $v = 0.92c$. The astronaut measures the time interval between successive "ticks" of the spacecraft clock to be $\Delta t_0 = 1.0$ s. What is the time interval $\Delta t$ that an earth observer measures between "ticks" of the astronaut's clock?

**Reasoning**  Since the clock on the spacecraft is moving relative to the earth observer, the earth observer measures a greater time interval $\Delta t$ between "ticks" than does the astronaut, who is at rest relative to the clock. The dilated time interval $\Delta t$ can be determined from the time dilation relation, Equation 28.1.

**Solution**  The dilated time interval is

$$\Delta t = \frac{\Delta t_0}{\sqrt{1 - \dfrac{v^2}{c^2}}} = \frac{1.0 \text{ s}}{\sqrt{1 - \left(\dfrac{0.92c}{c}\right)^2}} = \boxed{2.6 \text{ s}}$$

From the point of view of the earth-based observer, the astronaut is using a clock that is running slowly, because the earth-based observer measures a time between "ticks" that is longer (2.6 s) than what the astronaut measures (1.0 s).

Present-day spacecrafts fly nowhere near as fast as the craft in Example 1. Yet circumstances exist in which time dilation can create appreciable errors if not accounted for. The Global Positioning System (GPS), for instance, uses highly accurate and stable atomic clocks on board each of 24 satellites orbiting the earth at speeds of about 4000 m/s. These clocks make it possible to measure the time it takes for electromagnetic waves to travel from a satellite to a ground-based GPS receiver. From the speed of light and the times measured for signals from three or more of the satellites, it is possible to locate the position of the receiver (see Section 5.5). The stability of the clocks must be better than one part in $10^{13}$ to ensure the positional accuracy demanded of the GPS. Using Equation 28.1 and the speed of the GPS satellites, we can calculate the difference between the dilated time interval and the proper time interval as a fraction of the proper time interval and compare the result to the stability of the GPS clocks:

*The physics of*
**the Global Positioning System and special relativity.**

$$\frac{\Delta t - \Delta t_0}{\Delta t_0} = \frac{1}{\sqrt{1 - v^2/c^2}} - 1 = \frac{1}{\sqrt{1 - (4000 \text{ m/s})^2/(3.00 \times 10^8 \text{ m/s})^2}} - 1$$

$$= \frac{1}{1.1 \times 10^{10}}$$

This result is approximately one thousand times greater than the GPS-clock stability of one part in $10^{13}$. Thus, if not taken into account, time dilation would cause an error in the measured position of the earth-based GPS receiver roughly equivalent to that caused by a thousand-fold degradation in the stability of the atomic clocks.

## PROPER TIME INTERVAL

In Figure 28.4 both the astronaut and the person standing on the earth are measuring the time interval between a beginning event (the firing of the light source) and an ending event (the light pulse striking the detector). For the astronaut, who is at rest with respect to the light clock, the two events occur at the same location. (Remember, we are assuming that the light source and detector are so close together that they are considered to be at the same place.) Being at rest with respect to a clock is the usual or "proper" situation, so the time interval $\Delta t_0$ measured by the astronaut is called the ***proper time interval.*** In general, the proper time interval $\Delta t_0$ between two events is the time interval measured by an observer who is at rest relative to the events and sees them at the *same location* in space. On the other hand, the earth-based observer does not see the two events occurring at the same location in space, since the spacecraft is in motion. The time interval $\Delta t$ that the earth-based observer measures is, therefore, not a proper time interval in the sense that we have defined it.

To understand situations involving time dilation, it is essential to distinguish between $\Delta t_0$ and $\Delta t$. It is helpful if one first identifies the two events that define the time interval. These may be something other than the firing of a light source and the light pulse striking a detector. Then determine the reference frame in which the two events occur at the same place. An observer at rest in this reference frame measures the proper time interval $\Delta t_0$.

## SPACE TRAVEL

One of the intriguing aspects of time dilation occurs in conjunction with space travel. Since enormous distances are involved, travel to even the closest star outside our solar system would take a long time. However, as the following example shows, the travel time can be considerably less for the passengers than one might guess.

### Example 2 Space Travel

Alpha Centauri, a nearby star in our galaxy, is 4.3 light-years away. This means that, as measured by a person on earth, it would take light 4.3 years to reach this star. If a rocket leaves for Alpha Centauri and travels at a speed of $v = 0.95c$ relative to the earth, by how much will the passengers have aged, according to their own clock, when they reach their destination? Assume that the earth and Alpha Centauri are stationary with respect to one another.

*The physics of*
**space travel and special relativity.**

**Reasoning** The two events in this problem are the departure from earth and the arrival at Alpha Centauri. At departure, earth is just outside the spaceship. Upon arrival at the destination, Alpha Centauri is just outside. Therefore, relative to the passengers, the two events occur at the same place—namely, just outside the spaceship. Thus, the passengers measure the proper time interval $\Delta t_0$ on their clock, and it is this interval that we must find. For a person left behind on earth, the events occur at *different places*, so such a person measures the dilated time interval $\Delta t$ rather than the proper time interval. To find $\Delta t$ we note that the time to travel a given distance is inversely proportional to the speed. Since it takes 4.3 years to traverse the distance between earth and Alpha Centauri at the speed of light, it would take even longer at the slower speed of $v = 0.95c$. Thus, a person on earth measures the dilated time interval to be $\Delta t = (4.3 \text{ years})/0.95 = 4.5$ years. This value can be used with the time-dilation equation to find the proper time interval $\Delta t_0$.

**Solution** Using the time-dilation equation, we find that the proper time interval by which the passengers judge their own aging is

$$\Delta t_0 = \Delta t \sqrt{1 - \frac{v^2}{c^2}} = (4.5 \text{ years}) \sqrt{1 - \left(\frac{0.95c}{c}\right)^2} = \boxed{1.4 \text{ years}}$$

Thus, the people aboard the rocket will have aged by only 1.4 years when they reach Alpha Centauri, and not the 4.5 years an earthbound observer has calculated.

## VERIFICATION OF TIME DILATION

A striking confirmation of time dilation was achieved in 1971 by an experiment carried out by J. C. Hafele and R. E. Keating.* They transported very precise cesium-beam atomic clocks around the world on commercial jets. Since the speed of a jet plane is considerably less than $c$, the time-dilation effect is extremely small. However, the atomic clocks were accurate to about $\pm 10^{-9}$ s, so the effect could be measured. The clocks were in the air for 45 hours, and their times were compared to reference atomic clocks kept on earth. The experimental results revealed that, within experimental error, the readings on the clocks on the planes were different from those on earth by an amount that agreed with the prediction of relativity.

The behavior of subatomic particles called *muons* provides additional confirmation of time dilation. These particles are created high in the atmosphere, at altitudes of about 10 000 m. When at rest, muons exist only for about $2.2 \times 10^{-6}$ s before disintegrating. With such a short lifetime, these particles could never make it down to the earth's surface, even traveling at nearly the speed of light. However, *a large number of muons do reach the earth*. The only way they can do so is to live longer because of time dilation, as Example 3 illustrates.

### *Example 3* The Lifetime of a Muon

The average lifetime of a muon at rest is $2.2 \times 10^{-6}$ s. A muon created in the upper atmosphere, thousands of meters above sea level, travels toward the earth at a speed of $v = 0.998c$. Find, on the average, (a) how long a muon lives according to an observer on earth, and (b) how far the muon travels before disintegrating.

**Reasoning** The two events of interest are the generation and subsequent disintegration of the muon. When the muon is at rest, these events occur at the same place, so the muon's average (at rest) lifetime of $2.2 \times 10^{-6}$ s is a proper time interval $\Delta t_0$. When the muon moves at a speed $v = 0.998c$ relative to the earth, an observer on the earth measures a dilated lifetime $\Delta t$ that is given by Equation 28.1. The average distance $x$ traveled by a muon, as measured by an earth observer, is equal to the muon's speed times the dilated time interval.

**Solution**

(a) The observer on earth measures a dilated lifetime given by

$$\Delta t = \frac{\Delta t_0}{\sqrt{1 - \frac{v^2}{c^2}}} = \frac{2.2 \times 10^{-6} \text{ s}}{\sqrt{1 - \left(\frac{0.998c}{c}\right)^2}} = \boxed{35 \times 10^{-6} \text{ s}} \tag{28.1}$$

---

* J. C. Hafele and R. E. Keating, "Around-the-World Atomic Clocks: Observed Relativistic Time Gains," *Science,* Vol. 177, July 14, 1972, p. 168.

**(b)** The distance traveled by the muon before it disintegrates is

$$s = v\Delta t = (0.998)(3.00 \times 10^8 \text{ m/s})(35 \times 10^{-6} \text{ s}) = \boxed{1.0 \times 10^4 \text{ m}}$$

Thus, the dilated, or extended, lifetime provides sufficient time for the muon to reach the surface of the earth. If its lifetime were only $2.2 \times 10^{-6}$ s, a muon would travel only 660 m before disintegrating and could never reach the earth.

# 28.4 The Relativity of Length: Length Contraction

Because of time dilation, observers moving at a constant velocity relative to each other measure different time intervals between two events. For instance, Example 2 in the previous section illustrates that a trip from earth to Alpha Centauri at a speed of $v = 0.95c$ takes 4.5 years according to a clock on earth, but only 1.4 years according to a clock in the rocket. These two times differ by the factor $\sqrt{1 - v^2/c^2}$. Since the times for the trip are different, one might ask whether the observers measure different distances between earth and Alpha Centauri. The answer, according to special relativity, is yes. After all, both the earth-based observer and the rocket passenger agree that the relative speed between the rocket and earth is $v = 0.95c$. Since speed is distance divided by time and the time is different for the two observers, it follows that the distances must also be different, if the relative speed is to be the same for both individuals. Thus, the earth observer determines the distance to Alpha Centauri to be $L_0 = v\Delta t = (0.95c)(4.5 \text{ years}) = 4.3$ light-years. On the other hand, a passenger aboard the rocket finds the distance is only $L = v\Delta t_0 = (0.95c)(1.4 \text{ years}) = 1.3$ light-years. The passenger, measuring the shorter time, also measures the shorter distance. This shortening of the distance between two points is one example of a phenomenon known as *length contraction.*

The relation between the distances measured by two observers in relative motion at a constant velocity can be obtained with the aid of Figure 28.5. Part *a* of the drawing shows the situation from the point of view of the earth-based observer. This person measures the time of the trip to be $\Delta t$, the distance to be $L_0$, and the relative speed of the rocket to be $v = L_0/\Delta t$. Part *b* of the drawing presents the point of view of the passenger, for whom the rocket is at rest, and the earth and Alpha Centauri appear to move by at a speed $v$. The passenger determines the distance of the trip to be $L$, the time to be $\Delta t_0$, and the relative speed to be $v = L/\Delta t_0$. Since the relative speed computed by the passenger equals that computed by the earth-based observer, it follows that $v = L/\Delta t_0 = L_0/\Delta t$. Using this result and the time-dilation equation, Equation 28.1, we obtain the following relation between $L$ and $L_0$:

*Length
contraction*
$$L = L_0 \sqrt{1 - \frac{v^2}{c^2}} \qquad (28.2)$$

The length $L_0$ is called the ***proper length;*** it is the length (or distance) between two points *as measured by an observer at rest with respect to them.* Since $v$ is less than $c$, the term $\sqrt{1 - v^2/c^2}$ is less than 1, and $L$ is less than $L_0$. It is important to note that this length contraction occurs only along the direction of the motion. Those dimensions that are perpendicular to the motion are not shortened, as the next example discusses.

**Figure 28.5** (*a*) As measured by an observer on the earth, the distance to Alpha Centauri is $L_0$, and the time required to make the trip is $\Delta t$. (*b*) According to the passenger on the spacecraft, the earth and Alpha Centauri move with speed $v$ relative to the craft. The passenger measures the distance and time of the trip to be $L$ and $\Delta t_0$, respectively, both quantities being less than those in part *a*.

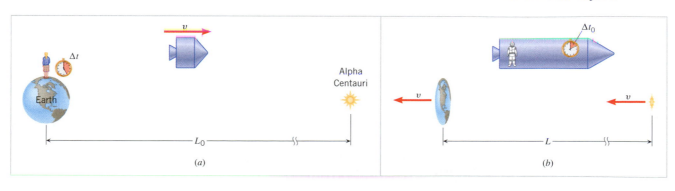

(*a*)

(*b*)

### *Example 4*   The Contraction of a Spacecraft

An astronaut, using a meter stick that is at rest relative to a cylindrical spacecraft, measures the length and diameter of the spacecraft to be 82 and 21 m, respectively. The spacecraft moves with a constant speed of $v = 0.95c$ relative to the earth, as in Figure 28.5. What are the dimensions of the spacecraft, as measured by an observer on earth?

**Reasoning** The length of 82 m is a proper length $L_0$, since it is measured using a meter stick that is at rest relative to the spacecraft. The length $L$ measured by the observer on earth can be determined from the length-contraction formula, Equation 28.2. On the other hand, the diameter of the spacecraft is perpendicular to the motion, so the earth observer does not measure any change in the diameter.

**Solution** The length $L$ of the spacecraft, as measured by the observer on earth, is

$$L = L_0 \sqrt{1 - \frac{v^2}{c^2}} = (82 \text{ m})\sqrt{1 - \left(\frac{0.95c}{c}\right)^2} = \boxed{26 \text{ m}}$$

Both the astronaut and the observer on earth measure the same value for the diameter of the spacecraft: $\boxed{\text{Diameter} = 21 \text{ m}}$. Figure 28.5*a* shows the size of the spacecraft as measured by the earth observer, and part *b* shows the size measured by the astronaut.

When dealing with relativistic effects we need to distinguish carefully between the criteria for the proper time interval and the proper length. The proper time interval $\Delta t_0$ between two events is the time interval measured by an observer who is at rest relative to the events and sees them occurring at the *same place*. All other moving inertial observers will measure a larger value for this time interval. The proper length $L_0$ of an object is the length measured by an observer who is *at rest* with respect to the object. All other moving inertial observers will measure a shorter value for this length. The observer who measures the proper time interval may not be the same one who measures the proper length. For instance, Figure 28.5 shows that the astronaut measures the proper time interval $\Delta t_0$ for the trip between earth and Alpha Centauri, whereas the earth-based observer measures the proper length (or distance) $L_0$ for the trip.

It should be emphasized that the word "proper" in the phrases proper time and proper length does *not* mean that these quantities are the correct or preferred quantities in any absolute sense. If this were so, the observer measuring these quantities would be using a preferred reference frame for making the measurement, a situation that is prohibited by the relativity postulate. According to this postulate, there is no preferred inertial reference frame. When two observers are moving relative to each other at a constant velocity, each measures the other person's clock to run more slowly than his own, and each measures the other person's length, along that person's motion, to be contracted.

## 28.5   *Relativistic Momentum*

Thus far we have discussed how time intervals and distances between two events are measured by observers moving at a constant velocity relative to each other. Special relativity also alters our ideas about momentum and energy.

Recall from Chapter 7 that when two or more objects interact, the principle of conservation of linear momentum applies if the system of objects is isolated. This principle states that the total linear momentum of an isolated system remains constant at all times. (An isolated system is one in which the sum of the external forces acting on the objects is zero.) The conservation of linear momentum is a law of physics and, in accord with the relativity postulate, is valid in all inertial reference frames. That is, when the total linear momentum is conserved in one inertial reference frame, it is conserved in all inertial reference frames.

As an example of momentum conservation, suppose several people are watching two billiard balls collide on a frictionless pool table. One person is standing next to the pool table, and the other is moving past the table with a constant velocity. Since the two balls constitute an isolated system, the relativity postulate requires that both observers must find the total linear momentum of the two-ball system to be the same before, during, and after the collision. For this kind of situation, Section 7.1 defines the classical linear momentum **p** of an object to be the product of its mass $m$ and velocity **v**. As a result, the magnitude of the classical momentum is $p = mv$. As long as the speed of an object is considerably smaller than the speed of light, this definition is adequate. However, when the speed approaches the speed of light, an analysis of the collision shows that the total linear momentum is not conserved in all inertial reference frames if one defines linear momentum simply as the product of mass and velocity. In order to preserve the conservation of linear momentum, it is necessary to modify this definition. The theory of special relativity reveals that the magnitude of the *relativistic momentum* must be defined as in Equation 28.3:

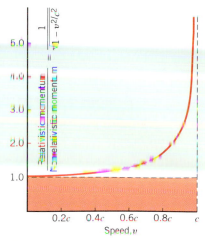

*Magnitude of the relativistic momentum*

$$p = \frac{mv}{\sqrt{1 - \dfrac{v^2}{c^2}}} \qquad (28.3)$$

The total relativistic momentum of an isolated system is conserved in all inertial reference frames.

From Equation 28.3, we can see that the magnitudes of the relativistic and nonrelativistic momenta differ by the same factor of $\sqrt{1 - v^2/c^2}$ that occurs in the time-dilation and length-contraction equations. Since this factor is always less than 1 and occurs in the denominator in Equation 28.3, the relativistic momentum is always larger than the nonrelativistic momentum. To illustrate how the two quantities differ as the speed $v$ increases, Figure 28.6 shows a plot of the ratio of the momentum magnitudes (relativistic/nonrelativistic) as a function of $v$. According to Equation 28.3, this ratio is just $1/\sqrt{1 - v^2/c^2}$. The graph shows that for speeds attained by ordinary objects, such as cars and planes, the relativistic and nonrelativistic momenta are almost equal because their ratio is nearly 1. Thus, at speeds much less than the speed of light, either the nonrelativistic momentum or the relativistic momentum can be used to describe collisions. On the other hand, when the speed of the object becomes comparable to the speed of light, the relativistic momentum becomes significantly greater than the nonrelativistic momentum and must be used. Example 5 deals with the relativistic momentum of an electron traveling close to the speed of light.

**Figure 28.6** This graph shows how the ratio of the magnitude of the relativistic momentum to the magnitude of the nonrelativistic momentum increases as the speed of an object approaches the speed of light.

### Example 5   The Relativistic Momentum of a High-Speed Electron

The particle accelerator at Stanford University (Figure 28.7) is three kilometers long and accelerates electrons to a speed of 0.999 999 999 7$c$, which is very nearly equal to the speed of light. Find the magnitude of the relativistic momentum of an electron that emerges from the accelerator, and compare it with the nonrelativistic value.

**Reasoning and Solution** The magnitude of the electron's relativistic momentum can be obtained from Equation 28.3 if we recall that the mass of an electron is $m = 9.11 \times 10^{-31}$ kg:

$$p = \frac{mv}{\sqrt{1 - \dfrac{v^2}{c^2}}} = \frac{(9.11 \times 10^{-31}\ \text{kg})(0.999\ 999\ 999\ 7c)}{\sqrt{1 - \dfrac{(0.999\ 999\ 999\ 7c)^2}{c^2}}} = \boxed{1 \times 10^{-17}\ \text{kg} \cdot \text{m/s}}$$

This value for the magnitude of the momentum agrees with that measured experimentally. The relativistic momentum is greater than the nonrelativistic momentum by a factor of

$$\frac{1}{\sqrt{1 - \dfrac{v^2}{c^2}}} = \frac{1}{\sqrt{1 - \dfrac{(0.999\ 999\ 999\ 7c)^2}{c^2}}} = \boxed{4 \times 10^4}$$

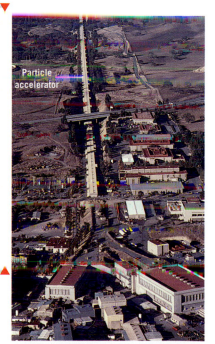

Particle accelerator

## 28.6   The Equivalence of Mass and Energy
### THE TOTAL ENERGY OF AN OBJECT

One of the most astonishing results of special relativity is that mass and energy are equivalent, in the sense that a gain or loss of mass can be regarded equally well as a gain or loss of energy. Consider, for example, an object of mass $m$ traveling at a speed $v$. Einstein

**Figure 28.7** The Stanford three-kilometer linear accelerator accelerates electrons almost to the speed of light. (© Bill Marsh/Photo Researchers)

showed that the *total energy* $E$ of the moving object is related to its mass and speed by the following relation:

**Total energy**
**of an object**

$$E = \frac{mc^2}{\sqrt{1 - \dfrac{v^2}{c^2}}}$$  (28.4)

To gain some understanding of Equation 28.4, consider the special case in which the object is at rest. When $v = 0$ m/s, the total energy is called the *rest energy $E_0$*, and Equation 28.4 reduces to Einstein's now-famous equation:

**Rest energy**
**of an object**

$$E_0 = mc^2$$  (28.5)

The rest energy represents the energy equivalent of the mass of an object at rest. As Example 6 shows, even a small mass is equivalent to an enormous amount of energy.

### Example 6 The Energy Equivalent of a Golf Ball

A 0.046-kg golf ball is lying on the green. (a) Find the rest energy of the golf ball. (b) If this rest energy were used to operate a 75-W light bulb, for how many years could the bulb stay on?

**Reasoning** The rest energy $E_0$ that is equivalent to the mass $m$ of the golf ball is found from the relation $E_0 = mc^2$. The power used by the bulb is 75 W, which means that it consumes 75 J of energy per second. If the entire rest energy of the ball were available for use, the bulb could stay on for a time equal to the rest energy divided by the power.

**Solution**

(a) The rest energy of the ball is

$$E_0 = mc^2 = (0.046 \text{ kg})(3.0 \times 10^8 \text{ m/s})^2 = \boxed{4.1 \times 10^{15} \text{ J}}$$  (28.5)

(b) This rest energy can keep the bulb burning for a time $t$ given by

$$t = \frac{\text{Rest energy}}{\text{Power}} = \frac{4.1 \times 10^{15} \text{ J}}{75 \text{ W}} = 5.5 \times 10^{13} \text{ s}$$  (6.10b)

Since one year contains $3.2 \times 10^7$ s, we find $\boxed{t = 1.7 \times 10^6 \text{ yr}}$, or 1.7 million years!

When an object is accelerated from rest to a speed $v$, the object acquires kinetic energy in addition to its rest energy. The total energy $E$ is the sum of the rest energy $E_0$ and the kinetic energy KE, or $E = E_0 + $ KE. Therefore, the kinetic energy is the difference between the object's total energy and its rest energy. Using Equations 28.4 and 28.5, we can write the kinetic energy as

$$\text{KE} = E - E_0 = mc^2 \left( \frac{1}{\sqrt{1 - \dfrac{v^2}{c^2}}} - 1 \right)$$  (28.6)

This equation is the relativistically correct expression for the kinetic energy of an object of mass $m$ moving at speed $v$.

Equation 28.6 looks nothing like the kinetic energy expression introduced in Chapter 6—namely, KE $= \frac{1}{2}mv^2$. However, for speeds much less than the speed of light ($v \ll c$), the relativistic equation for the kinetic energy reduces to KE $= \frac{1}{2}mv^2$, as can be seen by using the binomial expansion* to represent the square root term in Equation 28.6:

$$\frac{1}{\sqrt{1 - \dfrac{v^2}{c^2}}} = 1 + \frac{1}{2}\left(\frac{v^2}{c^2}\right) + \frac{3}{8}\left(\frac{v^2}{c^2}\right)^2 + \cdots$$

Suppose $v$ is much smaller than $c$—say $v = 0.01c$. The second term in the expansion has the value $\frac{1}{2}(v^2/c^2) = 5.0 \times 10^{-5}$, while the third term has the much smaller value

---

* The binomial expansion states that $(1 - x)^n = 1 - nx + n(n - 1)x^2/2 + \cdots$. In our case, $x = v^2/c^2$ and $n = -1/2$.

$\frac{3}{8}(v^2/c^2)^2 = 3.8 \times 10^{-9}$. The additional terms are smaller still, so if $v \ll c$, we can neglect the third and additional terms in comparison with the first and second terms. Substituting the first two terms into Equation 28.6 gives

$$KE \approx mc^2\left(1 + \frac{1}{2}\frac{v^2}{c^2} - 1\right) = \tfrac{1}{2}mv^2$$

which is the familiar form for the kinetic energy. However, Equation 28.6 gives the correct kinetic energy for all speeds and must be used for speeds near the speed of light, as in Example 7.

### Example 7 A High-Speed Electron

An electron ($m = 9.109 \times 10^{-31}$ kg) is accelerated from rest to a speed of $v = 0.9995c$ in a particle accelerator. Determine the electron's (a) rest energy, (b) total energy, and (c) kinetic energy in millions of electron volts or MeV.

**Reasoning and Solution**

(a) The electron's rest energy is

$$E_0 = mc^2 = (9.109 \times 10^{-31}\ \text{kg})(2.998 \times 10^8\ \text{m/s})^2 = 8.187 \times 10^{-14}\ \text{J} \qquad (28.5)$$

Since 1 eV $= 1.602 \times 10^{-19}$ J, the electron's rest energy is

$$(8.187 \times 10^{-14}\ \text{J})\left(\frac{1\ \text{eV}}{1.602 \times 10^{-19}\ \text{J}}\right) = \boxed{5.11 \times 10^5\ \text{eV} \quad \text{or} \quad 0.511\ \text{MeV}}$$

(b) The total energy of an electron traveling at a speed of $v = 0.9995c$ is

$$E = \frac{mc^2}{\sqrt{1 - \dfrac{v^2}{c^2}}} = \frac{(9.109 \times 10^{-31}\ \text{kg})(2.998 \times 10^8\ \text{m/s})^2}{\sqrt{1 - \left(\dfrac{0.9995c}{c}\right)^2}} \qquad (28.4)$$

$$= \boxed{2.59 \times 10^{-12}\ \text{J} \quad \text{or} \quad 16.2\ \text{MeV}}$$

(c) The kinetic energy is the difference between the total energy and the rest energy:

$$KE = E - E_0 = 2.59 \times 10^{-12}\ \text{J} - 8.2 \times 10^{-14}\ \text{J} \qquad (28.6)$$

$$= \boxed{2.51 \times 10^{-12}\ \text{J} \quad \text{or} \quad 15.7\ \text{MeV}}$$

For comparison, if the kinetic energy of the electron had been calculated from $\frac{1}{2}mv^2$, a value of only 0.26 MeV would have been obtained.

Since mass and energy are equivalent, any change in one is accompanied by a corresponding change in the other. For instance, life on earth is dependent on electromagnetic energy (light) from the sun. Because this energy is leaving the sun (see Figure 28.8), there is a decrease in the sun's mass. Example 8 illustrates how to determine this decrease.

### Example 8 The Sun Is Losing Mass

The sun radiates electromagnetic energy at the rate of $3.92 \times 10^{26}$ W. (a) What is the change in the sun's mass during each second that it is radiating energy? (b) The mass of the sun is $1.99 \times 10^{30}$ kg. What fraction of the sun's mass is lost during a human lifetime of 75 years?

**Reasoning** Since 1 W = 1 J/s, the amount of electromagnetic energy radiated during each second is $3.92 \times 10^{26}$ J. Thus, during each second, the sun's rest energy decreases by this amount. The change $\Delta E_0$ in the sun's rest energy is related to the change $\Delta m$ in its mass by $\Delta E_0 = (\Delta m)c^2$, according to Equation 28.5.

**Solution**

(a) For each second that the sun radiates energy, the change in its mass is

$$\Delta m = \frac{\Delta E_0}{c^2} = \frac{3.92 \times 10^{26}\ \text{J}}{(3.00 \times 10^8\ \text{m/s})^2} = \boxed{4.36 \times 10^9\ \text{kg}}$$

Over 4 billion kilograms of mass are lost by the sun during each second.

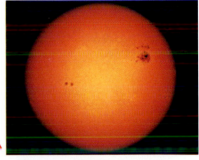

Visible light image.

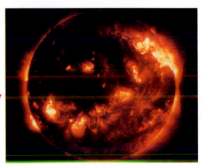

X-ray image.

**Figure 28.8** The sun emits electromagnetic energy over a broad portion of the electromagnetic spectrum. These photographs were obtained using that energy in the indicated regions of the spectrum. (*Top photo:* © Mark Marten/NASA/Photo Researchers; *bottom photo:* Dr. Leon Golub/Photo Researchers)

**(b)** The amount of mass lost by the sun in 75 years is

$$\Delta m = (4.36 \times 10^9 \text{ kg/s}) \left( \frac{3.16 \times 10^7 \text{ s}}{1 \text{ year}} \right) (75 \text{ years}) = 1.0 \times 10^{19} \text{ kg}$$

Although this is an enormous amount of mass, it represents only a tiny fraction of the sun's total mass:

$$\frac{\Delta m}{m_{\text{sun}}} = \frac{1.0 \times 10^{19} \text{ kg}}{1.99 \times 10^{30} \text{ kg}} = \boxed{5.0 \times 10^{-12}}$$

Any change in the energy of a system causes a change in the mass of the system according to $\Delta E_0 = (\Delta m)c^2$. It does not matter whether the change in energy is due to a change in electromagnetic energy, potential energy, thermal energy, or so on. Although any change in energy gives rise to a change in mass, in most instances the change in mass is too small to be detected. For instance, when 4186 J of heat is used to raise the temperature of 1 kg of water by 1 C°, the mass changes by only $\Delta m = \Delta E_0/c^2 = (4186 \text{ J})/(3.00 \times 10^8 \text{ m/s})^2 = 4.7 \times 10^{-14}$ kg. Conceptual Example 9 illustrates further how a change in the energy of an object leads to an equivalent change in its mass.

### Conceptual Example 9  When Is a Massless Spring Not Massless?

Figure 28.9a shows the top view of a spring lying on a horizontal table. The spring is initially unstrained and assumed to be massless. Suppose that the spring is either stretched or compressed by an amount $x$ from its unstrained length, as part b of the drawing shows. Is the mass of the spring still zero, or has it changed? And, if the mass has changed, is the change greater, smaller, or the same when the spring is stretched rather than compressed?

**Reasoning and Solution** Whenever a spring is stretched or compressed, its elastic potential energy changes. In Section 10.3 we showed that the elastic potential energy of an ideal spring is equal to $\frac{1}{2}kx^2$, where $k$ is the spring constant and $x$ is the amount of stretch or compression. Consistent with the theory of special relativity, any change in the total energy of a system, including a change in the elastic potential energy, is equivalent to a change in the mass of the system. Thus, the mass of a strained spring is greater than that of an unstrained (and massless) spring. Furthermore, since the elastic potential energy depends on $x^2$, the increase in mass of the spring is the same whether it is compressed or stretched, provided the magnitude of $x$ is the same in both cases.

The increase is exceedingly small because it is equal to the elastic potential energy divided by $c^2$ and $c^2$ is so large.

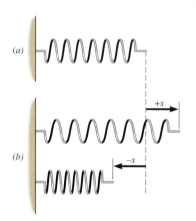

**Figure 28.9** (a) This spring is unstrained and assumed to have no mass. (b) When the spring is either stretched or compressed by an amount $x$, it gains elastic potential energy and, hence, mass.

It is also possible to transform matter itself into other forms of energy. For example, the positron (see Section 31.4) has the same mass as an electron but an opposite electrical charge. If these two particles of matter collide, they are completely annihilated, and a burst of high-energy electromagnetic waves is produced. Thus, matter is transformed into electromagnetic waves, the energy of the electromagnetic waves being equal to the total energies of the two colliding particles. The medical diagnostic technique known as positron emission tomography or PET scanning depends on the electromagnetic energy produced when a positron and an electron are annihilated (see Section 32.6).

The transformation of electromagnetic waves into matter also happens. In one experiment, an extremely high-energy electromagnetic wave, called a gamma ray (see Section 31.4), passes close to the nucleus of an atom. If the gamma ray has sufficient energy, it can create an electron and a positron. The gamma ray disappears, and the two particles of matter appear in its place. Except for picking up some momentum, the nearby nucleus remains unchanged. The process in which the gamma ray is transformed into the two particles is known as *pair production*.

### THE RELATION BETWEEN TOTAL ENERGY AND MOMENTUM

It is possible to derive a useful relation between the total relativistic energy $E$ and the relativistic momentum $p$. We begin by rearranging Equation 28.3 for the momentum, to obtain

$$\frac{m}{\sqrt{1 - v^2/c^2}} = \frac{p}{v}$$

With this substitution, Equation 28.4 for the total energy becomes

$$E = \frac{mc^2}{\sqrt{1 - v^2/c^2}} = \frac{pc^2}{v} \quad \text{or} \quad \frac{v}{c} = \frac{pc}{E}$$

Using this expression to replace $v/c$ in Equation 28.4 gives

$$E = \frac{mc^2}{\sqrt{1 - v^2/c^2}} = \frac{mc^2}{\sqrt{1 - p^2c^2/E^2}} \quad \text{or} \quad E^2 = \frac{m^2c^4}{1 - p^2c^2/E^2}$$

Solving this expression for $E^2$ shows that

$$E^2 = p^2c^2 + m^2c^4 \tag{28.7}$$

## THE SPEED OF LIGHT IS THE ULTIMATE SPEED

One of the important consequences of the theory of special relativity is that objects with mass cannot reach the speed of light in a vacuum. Thus, the speed of light represents the ultimate speed. To see that this speed limitation is a consequence of special relativity, consider Equation 28.6, which gives the kinetic energy of a moving object. As $v$ approaches the speed of light $c$, the $\sqrt{1 - v^2/c^2}$ term in the denominator approaches zero. Hence, the kinetic energy becomes infinitely large. However, the work–energy theorem (Chapter 6) tells us that an infinite amount of work would have to be done to give the object an infinite kinetic energy. Since an infinite amount of work is not available, we are left with the conclusion that objects with mass cannot attain the speed of light $c$.

## 28.7 *The Relativistic Addition of Velocities*

The velocity of an object relative to an observer plays a central role in special relativity, and to determine this velocity, it is sometimes necessary to add two or more velocities together. Figure 28.10 illustrates a truck moving at a constant velocity of $v_{TG} = +15$ m/s relative to an observer standing on the ground, where the plus sign denotes a direction to the right. Suppose someone on the truck throws a baseball toward the observer at a velocity of $v_{BT} = +8.0$ m/s relative to the truck. We might conclude that the observer on the ground sees the ball approaching at a velocity of $v_{BG} = v_{BT} + v_{TG} = 8.0$ m/s + 15 m/s = +23 m/s. These symbols have the following meaning:

$$v_{\boxed{BG}} = \text{velocity of the } \boxed{\text{Baseball}} \text{ relative to the } \boxed{\text{Ground}} = +23 \text{ m/s}$$

$$v_{\boxed{BT}} = \text{velocity of the } \boxed{\text{Baseball}} \text{ relative to the } \boxed{\text{Truck}} = +8.0 \text{ m/s}$$

$$v_{\boxed{TG}} = \text{velocity of the } \boxed{\text{Truck}} \text{ relative to the } \boxed{\text{Ground}} = +15.0 \text{ m/s}$$

Although the result that $v_{BG} = +23$m/s seems reasonable, careful measurements would show that it is not quite right. According to special relativity, the equation $v_{BG} = v_{BT} + v_{TG}$ is not valid for the following reason. If the velocity of the truck had a magnitude sufficiently close to the speed of light, the equation would predict that the observer on the earth could see the baseball moving faster than the speed of light. This is not possible, since no object with a finite mass can move faster than the speed of light.

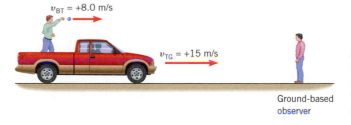

$v_{BT} = +8.0$ m/s

$v_{TG} = +15$ m/s

Ground-based observer

**Figure 28.10** The truck is approaching the ground-based observer at a relative velocity of $v_{TG} = +15$ m/s. The velocity of the baseball relative to the truck is $v_{BT} = +8.0$ m/s.

For the case where the truck and ball are moving along the same straight line, the theory of special relativity reveals that the velocities are related according to

$$v_{BG} = \frac{v_{BT} + v_{TG}}{1 + \dfrac{v_{BT}v_{TG}}{c^2}}$$

The subscripts in this equation have been chosen for the specific situation shown in Figure 28.10. For the general situation, the relative velocities are related by the **velocity-addition formula:**

*Velocity addition*
$$v_{AB} = \frac{v_{AC} + v_{CB}}{1 + \dfrac{v_{AC}v_{CB}}{c^2}} \tag{28.8}$$

where all the velocities are assumed to be constant and the symbols have the following meanings:

$v_{\boxed{AB}}$ = velocity of $\boxed{\text{object A}}$ relative to $\boxed{\text{object B}}$

$v_{\boxed{AC}}$ = velocity of $\boxed{\text{object A}}$ relative to $\boxed{\text{object C}}$

$v_{\boxed{CB}}$ = velocity of $\boxed{\text{object C}}$ relative to $\boxed{\text{object B}}$

For motion along a straight line, the velocities can have either positive or negative values, depending on whether they are directed along the positive or negative direction. Furthermore, switching the order of the subscripts changes the sign of the velocity, so, for example, $v_{BA} = -v_{AB}$.

Equation 28.8 differs from the nonrelativistic formula ($v_{AB} = v_{AC} + v_{CB}$) by the presence of the $v_{AC}v_{CB}/c^2$ term in the denominator. This term arises because of the effects of time dilation and length contraction that occur in special relativity. When $v_{AC}$ and $v_{CB}$ are small compared to $c$, the $v_{AC}v_{CB}/c^2$ term is small compared to 1, so the velocity-addition formula reduces to $v_{AB} \approx v_{AC} + v_{CB}$. However, when either $v_{AC}$ or $v_{CB}$ is comparable to $c$, the results can be quite different, as Example 10 illustrates.

### Example 10 The Relativistic Addition of Velocities

Imagine a hypothetical situation in which the truck in Figure 28.10 is moving relative to the ground with a velocity of $v_{TG} = +0.8c$. A person riding on the truck throws a baseball at a velocity relative to the truck of $v_{BT} = +0.5c$. What is the velocity $v_{BG}$ of the baseball relative to a person standing on the ground?

**Reasoning** The observer on the ground does *not* see the baseball approaching at $v_{BG} = 0.5c + 0.8c = 1.3c$. This cannot be because the speed of the ball would then exceed the speed of light. The velocity-addition formula gives the correct velocity, which has a magnitude less than the speed of light.

**Solution** The ground-based observer sees the ball approaching with a velocity of

$$v_{BG} = \frac{v_{BT} + v_{TG}}{1 + \dfrac{v_{BT}v_{TG}}{c^2}} = \frac{0.5c + 0.8c}{1 + \dfrac{(0.5c)(0.8c)}{c^2}} = \boxed{0.93c} \tag{28.8}$$

Example 10 discusses how the speed of a baseball is viewed by observers in different inertial reference frames. The next example deals with a similar situation, except that the baseball is replaced by the light of a laser beam.

### Conceptual Example 11 The Speed of a Laser Beam

Figure 28.11 shows an intergalactic cruiser approaching a hostile spacecraft. The velocity of the cruiser relative to the spacecraft is $v_{CS} = +0.7c$. Both vehicles are moving at a constant velocity. The cruiser fires a beam of laser light at the enemy. The velocity of the laser beam relative to the cruiser is $v_{LC} = +c$. (a) What is the velocity of the laser beam $v_{LS}$ relative to the renegades aboard the spacecraft? (b) At what velocity do the renegades aboard the spacecraft see the laser beam move away from the cruiser?

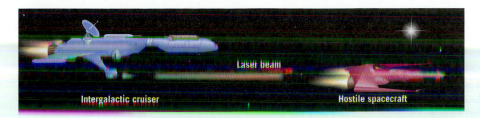

**Figure 28.11** An intergalactic cruiser, closing in on a hostile spacecraft, fires a beam of laser light.

**Reasoning and Solution**

**(a)** Since both vehicles move at a constant velocity, each constitutes an inertial reference frame. According to the speed of light postulate, *all* observers in inertial reference frames measure the speed of light in a vacuum to be $c$. Thus, the renegades aboard the hostile spacecraft see the laser beam travel toward them at the speed of light, even though the beam is emitted from the cruiser, which itself is moving at seven-tenths the speed of light.

**(b)** The renegades aboard the spacecraft see the cruiser approach them at a relative velocity of $v_{CS} = +0.7c$, and they also see the laser beam approach them at a relative velocity of $v_{LS} = +c$. Both these velocities are measured relative to the *same* inertial reference frame — namely, that of the spacecraft. Therefore, the renegades aboard the spacecraft see the laser beam move away from the cruiser at a velocity that is the difference between these two velocities, or $+c - (+0.7c) = +0.3c$. The velocity-addition formula, Equation 28.8, is not applicable here because both velocities are measured relative to the *same* inertial reference frame (the spacecraft's reference frame). The velocity-addition formula can be used only when the velocities are measured relative to different inertial reference frames.

**Related Homework:** *Problem 34*

It is a straightforward matter to show that the velocity-addition formula is consistent with the speed of light postulate. Consider Figure 28.12, which shows a person riding on a truck and holding a flashlight. The velocity of the light, relative to the person on the truck, is $v_{LT} = +c$. The velocity $v_{LG}$ of the light relative to the observer standing on the ground is given by the velocity-addition formula as

$$v_{LG} = \frac{v_{LT} + v_{TG}}{1 + \dfrac{v_{LT}v_{TG}}{c^2}} = \frac{c + v_{TG}}{1 + \dfrac{cv_{TG}}{c^2}} = \frac{(c + v_{TG})c}{(c + v_{TG})} = c$$

Thus, the velocity-addition formula indicates that the observer on the ground and the person on the truck both measure the speed of light to be $c$, independent of the relative velocity $v_{TG}$ between them. This is exactly what the speed of light postulate states.

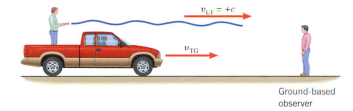

$v_{LT} = +c$

$v_{TG}$

Ground-based observer

**Figure 28.12** The speed of the light emitted by the flashlight is $c$ relative to both the truck and the observer on the ground.

# Concept Summary

This summary presents an abridged version of the chapter, including the important equations and all available learning aids. For convenient reference, the learning aids (including the text's examples) are placed next to or immediately after the relevant equation or discussion. The following learning aids may be found on-line at **www.wiley.com/college/cutnell:**

**Interactive LearningWare** examples are solved according to a five-step interactive format that is designed to help you develop problem-solving skills.

**Concept Simulations** are animated versions of text figures or animations that illustrate important concepts. You can control parameters that affect the display, and we encourage you to experiment.

**Interactive Solutions** offer specific models for certain types of problems in the chapter homework. The calculations are carried out interactively.

**Self-Assessment Tests** include both qualitative and quantitative questions. Extensive feedback is provided for both incorrect and correct answers, to help you evaluate your understanding of the material.

| *Topic* | *Discussion* | *Learning Aids* |
|---|---|---|

### 28.1 Events and Inertial Reference Frames

**Reference frame**

An event is a physical "happening" that occurs at a certain place and time. To record the event an observer uses a reference frame that consists of a coordinate system and a clock. Different observers may use different reference frames.

**Inertial reference frame**

The theory of special relativity deals with inertial reference frames. An inertial reference frame is one in which Newton's law of inertia is valid. Accelerating reference frames are not inertial reference frames.

### 28.2 The Postulates of Special Relativity

**Relativity postulate**

The theory of special relativity is based on two postulates. The relativity postulate states that the laws of physics are the same in every inertial reference frame.

**Speed of light postulate**

The speed of light postulate says that the speed of light in a vacuum, measured in any inertial reference frame, always has the same value of $c$, no matter how fast the source of the light and the observer are moving relative to each other.

### 28.3 The Relativity of Time: Time Dilation

**Proper time interval**

The proper time interval $\Delta t_0$ between two events is the time interval measured by an observer who is at rest relative to the events and views them occurring at the same place. An observer who is in motion with respect to the events and

**Dilated time interval**

who views them as occurring at different places measures a dilated time interval $\Delta t$. The dilated time interval is greater than the proper time interval, according to the time-dilation equation:

**Concept Simulation 28.1**

**Time-dilation equation**

$$\Delta t = \frac{\Delta t_0}{\sqrt{1 - \dfrac{v^2}{c^2}}} \quad (28.1)$$

**Examples 1, 2, 3**
**Interactive Solution 28.5**

In this expression, $v$ is the relative speed between the observer who measures $\Delta t_0$ and the observer who measures $\Delta t$.

### 28.4 The Relativity of Length: Length Contraction

**Proper length**

The proper length $L_0$ between two points is the length measured by an observer who is at rest relative to the points. An observer moving with a relative speed $v$ parallel to the line between the two points does not measure the proper length. Instead, such

**Contracted length**

an observer measures a contracted length $L$ given by the length-contraction formula:

**Example 4**
**Interactive LearningWare 28.1**

**Length-contraction formula**

$$L = L_0 \sqrt{1 - \frac{v^2}{c^2}} \quad (28.2)$$

**Interactive Solution 28.13**

Length contraction occurs only along the direction of the motion. Those dimensions that are perpendicular to the motion are not shortened.

The observer who measures the proper length may not be the observer who measures the proper time interval.

 Use *Self-Assessment Test 28.1* to evaluate your understanding of Sections 28.1–28.4.

### 28.5 Relativistic Momentum

An object of mass $m$, moving with speed $v$, has a relativistic momentum whose magnitude $p$ is given by

**Relativistic momentum**

$$p = \frac{mv}{\sqrt{1 - \dfrac{v^2}{c^2}}} \quad (28.3)$$ **Example 5**

### 28.6 The Equivalence of Mass and Energy

Energy and mass are equivalent. The total energy $E$ of an object of mass $m$, moving at speed $v$, is

**Total energy**

$$E = \frac{mc^2}{\sqrt{1 - \dfrac{v^2}{c^2}}} \quad (28.4)$$

| Topic | Discussion | Learning Aids |
|---|---|---|

**Rest energy**

The rest energy $E_0$ is the total energy of an object at rest ($v = 0$ m/s):

$$E_0 = mc^2 \qquad (28.5) \quad \textbf{Examples 6, 8, 9}$$

The total energy of an object is the sum of its rest energy and its kinetic energy $KE$, so $E = E_0 + KE$. Therefore, the kinetic energy is

**Kinetic energy**

$$KE = E - E_0 = mc^2\left(\frac{1}{\sqrt{1 - \dfrac{v^2}{c^2}}} - 1\right) \qquad (28.6) \quad \textbf{Example 7}$$

**Relation between total energy and momentum**

The relativistic total energy and momentum are related according to

$$E^2 = p^2c^2 + m^2c^4 \qquad (28.7)$$

**The ultimate speed**

Objects with mass cannot attain the speed of light, which is the ultimate speed for such objects.

### 28.7 The Relativistic Addition of Velocities

According to special relativity, the velocity addition formula specifies how the relative velocities of moving objects are related. For objects that move along the same straight line, this formula is

**Velocity-addition formula**

$$v_{AB} = \frac{v_{AC} + v_{CB}}{1 + \dfrac{v_{AC}v_{CB}}{c^2}} \qquad (28.8) \quad \textbf{Examples 10, 11}$$

where $v_{AB}$ is the velocity of object A relative to object B, $v_{AC}$ is the velocity of object A relative to object C, and $v_{CB}$ is the velocity of object C relative to object B. The velocities can have positive or negative values, depending on whether they are directed along the positive or negative direction. Furthermore, switching the order of the subscripts changes the sign of the velocity, so that, for example, $v_{BA} = -v_{AB}$.

 **Use Self-Assessment Test 28.2 to evaluate your understanding of Sections 28.5–28.7.**

# Problems

*Before doing any calculations involving time dilation or length contraction, it is useful to identify which observer measures the proper time interval $\Delta t_0$ or the proper length $L_0$.*

**ssm** Solution is in the Student Solutions Manual.   **www** Solution is available on the World Wide Web at www.wiley.com/college/cutnell
✇ This icon represents a biomedical application.

**Section 28.3 The Relativity of Time: Time Dilation**

**1. ssm** A radar antenna is rotating at an angular speed of 0.25 rad/s, as measured on earth. To an observer moving past the antenna at a speed of 0.80$c$, what is its angular speed?

**2.** A Klingon spacecraft has a speed of 0.75$c$ with respect to the earth. The Klingons measure 37.0 h for the time interval between two events on the earth. What value for the time interval would they measure if their ship had a speed of 0.94$c$ with respect to the earth?

**3. ssm** A law enforcement officer in an intergalactic "police car" turns on a red flashing light and sees it generate a flash every 1.5 s. A person on earth measures that the time between flashes is 2.5 s. How fast is the "police car" moving relative to the earth?

**4.** Suppose that you are traveling on board a spacecraft that is moving with respect to the earth at a speed of 0.975$c$. You are breathing at a rate of 8.0 breaths per minute. As monitored on earth, what is your breathing rate?

*\* **5.** Interactive Solution 28.5 at www.wiley.com/college/cutnell illustrates one way to model this problem. A 6.00-kg object oscillates back and forth at the end of a spring whose spring constant is 76.0 N/m. An observer is traveling at a speed of $1.90 \times 10^8$ m/s relative to the fixed end of the spring. What does this observer measure for the period of oscillation?

*\* **6.** An astronaut travels at a speed of 7800 m/s relative to the earth, a speed that is very small compared to $c$. According to a clock on the earth, the trip lasts 15 days. Determine the *difference* (in seconds) between the time recorded by the earth clock and the astronaut's clock. [*Hint:* When $v \ll c$, the following approximation is valid: $\sqrt{1 - v^2/c^2} \approx 1 - \frac{1}{2}(v^2/c^2)$.]

*\*\* **7.** As observed on earth, a certain type of bacterium is known to double in number every 24.0 hours. Two cultures of these bacteria are prepared, each consisting initially of one bacterium. One culture is left on earth and the other placed on a rocket that

travels at a speed of 0.866c relative to the earth. At a time when the earthbound culture has grown to 256 bacteria, how many bacteria are in the culture on the rocket, according to an earth-based observer?

### Section 28.4 The Relativity of Length: Length Contraction

**8.** A tourist is walking at a speed of 1.3 m/s along a 9.0-km path that follows an old canal. If the speed of light in a vacuum were 3.0 m/s, how long would the path be, according to the tourist?

**9. ssm www** How fast must a meter stick be moving if its length is observed to shrink to one-half of a meter?

**10. Interactive LearningWare 28.1** at **www.wiley.com/college/cutnell** reviews the concepts that play roles in this problem. The distance from earth to the center of our galaxy is about 23 000 ly (1 ly = 1 light-year = $9.47 \times 10^{15}$ m), as measured by an earth-based observer. A spaceship is to make this journey at a speed of 0.9990c. According to a clock on board the spaceship, how long will it take to make the trip? Express your answer in years (1 yr = $3.16 \times 10^7$ s).

**11. ssm** A UFO streaks across the sky at a speed of 0.90c relative to the earth. A person on earth determines the length of the UFO to be 230 m along the direction of its motion. What length does the person measure for the UFO when it lands?

**12.** Suppose you are traveling in space and pass a rectangular landing pad on a planet. Your spacecraft has a speed of 0.85c relative to the planet and moves in a direction parallel to the length of the pad. While moving, you measure the length to be 1800 m and the width to be 1500 m. What are the dimensions of the landing pad according to the engineer who built it?

**13. Interactive Solution 28.13** at **www.wiley/com/college/cutnell** illustrates one approach to solving this problem. A space traveler moving at a speed of 0.70c with respect to the earth makes a trip to a distant star that is stationary relative to the earth. He measures the length of this trip to be 6.5 light-years. What would be the length of this same trip (in light-years) as measured by a traveler moving at a speed of 0.90c with respect to the earth?

* **14.** As the drawing shows, a carpenter on a space station has constructed a 30.0° ramp. A rocket moves past the space station with a relative speed of 0.730c in a direction parallel to side x. What does a person aboard the rocket measure for the angle of the ramp?

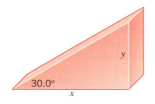

* * **15. ssm www** A rectangle has the dimensions of 3.0 m × 2.0 m when viewed by someone at rest with respect to it. When you move past the rectangle along one of its sides, the rectangle looks like a square. What dimensions do you observe when you move at the same speed along the adjacent side of the rectangle?

* * **16.** A woman and a man are on separate rockets, which are flying parallel to each other and have a relative speed of 0.940c. The woman measures the same value for the length of her own rocket and for the length of the man's rocket. What is the ratio of the value that the man measures for the length of his own rocket to the value he measures for the length of the woman's rocket?

### Section 28.5 Relativistic Momentum

**17.** At what speed is the magnitude of the relativistic momentum of a particle three times the magnitude of the nonrelativistic momentum?

**18.** A jetliner has a mass of $1.2 \times 10^5$ kg and flies at a speed of 140 m/s. (a) Find the magnitude of its momentum. (b) If the speed of light in a vacuum had the hypothetical value of 170 m/s, what would be the magnitude of the jetliner's momentum?

**19. ssm** A rocket of mass $1.40 \times 10^5$ kg has a relativistic momentum the magnitude of which is $3.15 \times 10^{13}$ kg·m/s. How fast is the rocket traveling?

**20.** A woman is 1.6 m tall and has a mass of 55 kg. She moves past an observer with the direction of the motion parallel to her height. The observer measures her relativistic momentum to have a magnitude of $2.0 \times 10^{10}$ kg·m/s. What does the observer measure for her height?

* **21.** Starting from rest, two skaters "push off" against each other on smooth level ice, where friction is negligible. One is a woman and one is a man. The woman moves away with a velocity of +2.5 m/s relative to the ice. The mass of the woman is 54 kg, and the mass of the man is 88 kg. Assuming that the speed of light is 3.0 m/s, so that the relativistic momentum must be used, find the recoil velocity of the man relative to the ice. (*Hint: This problem is similar to Example 6 in Chapter 7.*)

### Section 28.6 The Equivalence of Mass and Energy

**22.** An electron and a positron each have a mass of $9.11 \times 10^{-31}$ kg. They collide and both vanish, with only electromagnetic radiation appearing after the collision. If each particle is moving at a speed of 0.20c relative to the laboratory before the collision, determine the energy of the electromagnetic radiation.

**23. ssm** Determine the ratio of the relativistic kinetic energy to the nonrelativistic kinetic energy ($\frac{1}{2}mv^2$) when a particle has a speed of (a) $1.00 \times 10^{-3}c$ and (b) 0.970c.

**24.** Suppose one gallon of gasoline produces $1.1 \times 10^8$ J of energy, and this energy is sufficient to operate a car for twenty miles. An aspirin tablet has a mass of 325 mg. If the aspirin could be converted completely into thermal energy, how many miles could the car go on a single tablet?

**25.** A nuclear power reactor generates $3.0 \times 10^9$ W of power. In one year, what is the change in the mass of the nuclear fuel due to the energy being taken from the reactor?

**26.** Four kilograms of water are heated from 20.0 °C to 60.0 °C. (a) How much heat is required to produce this change in temperature? [The specific heat capacity of water is 4186 J/(kg·C°).] (b) By how much does the mass of the water increase?

**27. ssm** How much work must be done on an electron to accelerate it from rest to a speed of 0.990c?

* **28.** An object has a total energy of $5.0 \times 10^{15}$ J and a kinetic energy of $2.0 \times 10^{15}$ J. What is the magnitude of the object's relativistic momentum?

* **29.** How close would two stationary electrons have to be positioned so that their total mass is twice what it is when the electrons are very far apart?

### 28.7 The Relativistic Addition of Velocities

**30.** Spaceship Y is between spaceship X and spaceship Z. Spaceship Y is moving toward spaceship Z at a speed of 0.68c. Spaceship Z is moving toward spaceship X at a speed of 0.42c. Assuming that all of the spaceships are moving at constant velocities, so they are inertial reference frames, find the speed of spaceship Y with respect to spaceship X.

**31. ssm** A spacecraft approaching the earth launches an exploration vehicle. After the launch, an observer on earth sees the spacecraft approaching at a speed of 0.50c and the exploration vehicle approaching at a speed of 0.70c. What is the speed of the exploration vehicle relative to the spacecraft?

**32.** Galaxy A is moving away from us with a speed of 0.75c relative to the earth. Galaxy B is moving away from us in the opposite direction with a relative speed of 0.55c. Assume that the earth and the galaxies are moving at constant velocities, so they are inertial reference frames. How fast is galaxy A moving according to an observer in galaxy B?

* **33. ssm www** The crew of a rocket that is moving away from the earth launches an escape pod, which they measure to be 45 m long. The pod is launched toward the earth with a speed of 0.55c relative to the rocket. After the launch, the rocket's speed relative to the earth is 0.75c. What is the length of the escape pod as determined by an observer on earth?

* **34.** Refer to Conceptual Example 11 as an aid in solving this problem. An intergalactic cruiser has two types of guns: a photon cannon that fires a beam of laser light and an ion gun that shoots ions at a velocity of 0.950c relative to the cruiser. The cruiser closes in on an alien spacecraft at a velocity of 0.800c relative to this spacecraft. The captain fires both types of guns. At what velocity do the aliens see (a) the laser light and (b) the ions approach them? At what velocity do the aliens see (c) the laser light and (d) the ions move away from the cruiser?

**35.** Two atomic particles approach each other in a head-on collision. Each particle has a mass of $2.16 \times 10^{-25}$ kg. The speed of each particle is $2.10 \times 10^8$ m/s when measured by an observer standing in the laboratory. (a) What is the speed of one particle as seen by the other particle? (b) Determine the relativistic momentum of one particle, as it would be observed by the other.

# Chapter 29  Particles and Waves

## 29.1  The Wave–Particle Duality

The ability to exhibit interference effects is an essential characteristic of waves. For instance, Section 27.2 discusses Young's famous experiment in which light passes through two closely spaced slits and produces a pattern of bright and dark fringes on a screen (see Figure 27.3). The fringe pattern is a direct indication that interference is occurring between the light waves coming from each slit.

One of the most incredible discoveries of twentieth-century physics is that particles can also behave like waves and exhibit interference effects. For instance, Figure 29.1 shows a version of Young's experiment performed by directing *a beam of electrons* onto a double slit. In this experiment, the screen is like a television screen and glows wherever an electron strikes it. Part *a* of the drawing indicates the pattern that would be seen on the screen if each electron, behaving strictly as a particle, were to pass through one slit or the other and strike the screen. The pattern would consist of an image of each slit. Part *b* shows the pattern actually observed, which consists of bright and dark fringes, reminiscent of the pattern obtained when light waves pass through a double slit. The fringe pattern indicates that the electrons are exhibiting interference effects that are associated with waves.

But how can electrons behave like waves in the experiment shown in Figure 29.1*b*? And what kind of waves are they? The answers to these profound questions will be discussed later in this chapter. For the moment, we intend only to emphasize that the concept of an electron as a tiny discrete particle of matter does not account for the fact that the electron can behave as a wave in some circumstances. In other words, the electron exhibits a dual nature, with both particle-like characteristics and wave-like characteristics.

Here is another interesting question: If a particle can exhibit wave-like properties, can waves exhibit particle-like behavior? As the next three sections reveal, the answer is yes. In fact, experiments that demonstrated the particle-like behavior of waves were performed near the beginning of the twentieth century, before the experiments that demonstrated the wave-like properties of the electrons. Scientists now accept the **wave–particle duality** as an essential part of nature:

> **Waves can exhibit particle-like characteristics, and particles can exhibit wave-like characteristics.**

Section 29.2 begins the remarkable story of the wave–particle duality by discussing the electromagnetic waves that are radiated by a perfect blackbody. It is appropriate to begin with blackbody radiation, because it provided the first link in the chain of experimental evidence leading to our present understanding of the wave–particle duality.

## 29.2  Blackbody Radiation and Planck's Constant

All bodies, no matter how hot or cold, continuously radiate electromagnetic waves. For instance, we see the glow of very hot objects because they emit electromagnetic waves in the visible region of the spectrum. Our sun, which has a surface temperature of about 6000 K, appears yellow, while the cooler star Betelgeuse has a red-orange appearance due to its lower surface temperature of 2900 K. However, at relatively low temperatures, cooler objects emit visible light waves only weakly and, as a result, do not appear to be glowing. Certainly the human body, at only 310 K, does not emit enough visible light to be seen in the dark with the unaided eye. But the body does emit electromagnetic waves in the infrared region of the spectrum, and these can be detected with infrared-sensitive devices.

At a given temperature, the intensities of the electromagnetic waves emitted by an object vary from wavelength to wavelength throughout the visible, infrared, and other re-

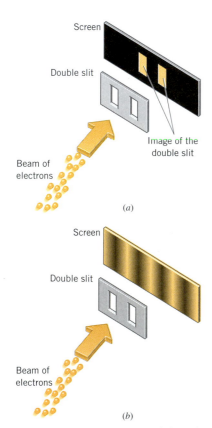

**Figure 29.1** (*a*) If electrons behaved as discrete particles with no wave properties, they would pass through one or the other of the two slits and strike the screen, causing it to glow and produce exact images of the slits. (*b*) In reality, the screen reveals a pattern of bright and dark fringes, similar to the pattern produced when a beam of light is used and interference occurs between the light waves coming from each slit.

gions of the spectrum. Figure 29.2 illustrates how the intensity per unit wavelength depends on wavelength for a perfect blackbody emitter. As Section 13.3 discusses, a perfect blackbody at a constant temperature absorbs and reemits all the electromagnetic radiation that falls on it. The two curves in the drawing show that at a higher temperature the maximum emitted intensity per unit wavelength increases and shifts to shorter wavelengths, toward the visible region of the spectrum. In accounting for the shape of these curves, the German physicist Max Planck (1858–1947) took the first step toward our present understanding of the wave–particle duality.

In 1900 Planck calculated the blackbody radiation curves, using a model that represents a blackbody as a large number of atomic oscillators, each of which emits and absorbs electromagnetic waves. To obtain agreement between the theoretical and experimental curves, Planck assumed that the energy $E$ of an atomic oscillator could have only the discrete values of $E = 0$, $hf$, $2hf$, $3hf$, and so on. In other words, he assumed that

$$E = nhf \qquad n = 0, 1, 2, 3, \ldots \qquad (29.1)$$

where $n$ is either zero or a positive integer, $f$ is the frequency of vibration (in hertz), and $h$ is a constant now called **Planck's constant.*** Planck's constant has a value of

$$h = 6.626\ 068\ 76 \times 10^{-34}\ \text{J} \cdot \text{s}$$

The radical feature of Planck's assumption was that the energy of an atomic oscillator could have only discrete values ($hf$, $2hf$, $3hf$, etc.), with energies in between these values being forbidden. Whenever the energy of a system can have only certain definite values, and nothing in between, the energy is said to be *quantized*. This quantization of the energy was unexpected on the basis of the traditional physics of the time. However, it was soon realized that energy quantization had wide-ranging implications.

Conservation of energy requires that the energy carried off by the radiated electromagnetic waves must equal the energy lost by the atomic oscillators in Planck's model. Suppose, for example, that an oscillator with an energy of $3hf$ emits an electromagnetic wave. According to Equation 29.1, the next smallest allowed value for the energy of the oscillator is $2hf$. In such a case, the energy carried off by the electromagnetic wave would have the value of $hf$, equaling the amount of energy lost by the oscillator. Thus, Planck's model for blackbody radiation sets the stage for the idea that electromagnetic energy occurs as a collection of discrete amounts or packets of energy, the energy of a packet being equal to $hf$. Einstein made the proposal that light consists of such energy packets.

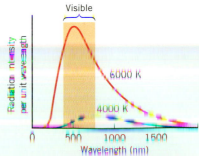

**Figure 29.2** The electromagnetic radiation emitted by a perfect blackbody has an intensity per unit wavelength that varies from wavelength to wavelength, as each curve indicates. At the higher temperature, the intensity per unit wavelength is greater, and the maximum occurs at a shorter wavelength.

## 29.3 *Photons and the Photoelectric Effect*

The total energy $E$ and the linear momentum $\mathbf{p}$ are fundamental concepts in physics. We have seen in Chapters 6 and 7 how they apply to moving particles, such as electrons and protons. The total energy of a (nonrelativistic) particle is the sum of its kinetic energy (KE) and potential energy (PE), or $E = \text{KE} + \text{PE}$. The magnitude $p$ of the particle's momentum is the product of its mass $m$ and speed $v$, or $p = mv$. We will now discuss the fact that electromagnetic waves are composed of particle-like entities called **photons** and that the ideas of energy and momentum also apply to them. However, as we will see, the equations defining photon energy ($E = hf$) and momentum ($p = h/\lambda$) are different from those for a particle.

Experimental evidence that light consists of photons comes from a phenomenon called the **photoelectric effect,** in which electrons are emitted from a metal surface when light shines on it. Figure 29.3 illustrates the effect. The electrons are emitted if the light being used has a sufficiently high frequency. The ejected electrons move toward a positive electrode called the *collector* and cause a current to register on the ammeter. Because the electrons are ejected with the aid of light, they are called **photoelectrons.** As will be discussed shortly, a number of features of the photoelectric effect could not be explained solely with the ideas of classical physics.

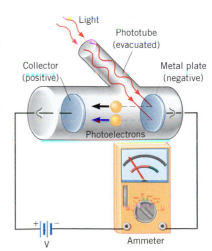

**Figure 29.3** In the photoelectric effect, light with a sufficiently high frequency ejects electrons from a metal surface. These photoelectrons, as they are called, are drawn to the positive collector, thus producing a current.

---

* It is now known that the energy of a harmonic oscillator is $E = (n + \frac{1}{2})hf$, the extra term of $\frac{1}{2}$ being unimportant to the present discussion.

In 1905 Einstein presented an explanation of the photoelectric effect that took advantage of Planck's work concerning blackbody radiation. It was primarily for his theory of the photoelectric effect that he was awarded the Nobel Prize in physics in 1921. In his photoelectric theory, Einstein proposed that light of frequency $f$ could be regarded as a collection of discrete packets of energy (photons), each packet containing an amount of energy $E$ given by

*Energy of a photon* $$E = hf \qquad (29.2)$$

where $h$ is Planck's constant. The light energy given off by a light bulb, for instance, is carried by photons. The brighter the bulb, the greater is the number of photons emitted per second. Example 1 estimates the number of photons emitted per second by a typical light bulb.

### Example 1 Photons from a Light Bulb

In converting electrical energy into light energy, a sixty-watt incandescent light bulb operates at about 2.1% efficiency. Assuming that all the light is green light (vacuum wavelength = 555 nm), determine the number of photons per second given off by the bulb.

**Reasoning** The number of photons emitted per second can be found by dividing the amount of light energy emitted per second by the energy $E$ of one photon. The energy of a single photon is $E = hf$, according to Equation 29.2. The frequency $f$ of the photon is related to its wavelength $\lambda$ by Equation 16.1 as $f = c/\lambda$.

**Solution** At an efficiency of 2.1%, the light energy emitted per second by a sixty-watt bulb is $(0.021)(60.0 \text{ J/s}) = 1.3 \text{ J/s}$. The energy of a single photon is

$$E = hf = \frac{hc}{\lambda} = \frac{(6.63 \times 10^{-34} \text{ J} \cdot \text{s})(3.00 \times 10^8 \text{ m/s})}{555 \times 10^{-9} \text{ m}} = 3.58 \times 10^{-19} \text{ J}$$

Therefore,

$$\text{Number of photons emitted per second} = \frac{1.3 \text{ J/s}}{3.58 \times 10^{-19} \text{ J/photon}} = \boxed{3.6 \times 10^{18} \text{ photons/s}}$$

According to Einstein, when light shines on a metal, a photon can give up its energy to an electron in the metal. If the photon has enough energy to do the work of removing the electron from the metal, the electron can be ejected. The work required depends on how strongly the electron is held. For the *least strongly* held electrons, the necessary work has a minimum value $W_0$ and is called the **work function** of the metal. If a photon has energy in excess of the work needed to remove an electron, the excess appears as kinetic energy of the ejected electron. Thus, the least strongly held electrons are ejected with the maximum kinetic energy $KE_{max}$. Einstein applied the conservation-of-energy principle and proposed the following relation to describe the photoelectric effect:

$$\underbrace{hf}_{\substack{\text{Photon} \\ \text{energy}}} = \underbrace{KE_{max}}_{\substack{\text{Maximum} \\ \text{kinetic energy} \\ \text{of ejected} \\ \text{electron}}} + \underbrace{W_0}_{\substack{\text{Minimum} \\ \text{work needed to} \\ \text{eject electron}}} \qquad (29.3)$$

According to this equation, $KE_{max} = hf - W_0$, which is plotted in Figure 29.4, with $KE_{max}$ along the $y$ axis and $f$ along the $x$ axis. The graph is a straight line that crosses the $x$ axis at $f = f_0$. At this frequency, the electron departs from the metal with no kinetic energy ($KE_{max} = 0 \text{ J}$). According to Equation 29.3, when $KE_{max} = 0 \text{ J}$ the energy $hf_0$ of the incident photon is equal to the work function $W_0$ of the metal: $hf_0 = W_0$.

The photon concept provides an explanation for a number of features of the photoelectric experiment that are difficult to explain without photons. It is observed, for instance, that only light with a frequency above a certain minimum value $f_0$ will eject electrons. If the frequency of the light is below this value, no electrons are ejected, regardless of how intense the light is. The next example determines the minimum frequency value for a silver surface.

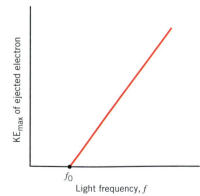

**Figure 29.4** Photons can eject electrons from a metal when the light frequency is above a minimum value $f_0$. For frequencies above this value, ejected electrons have a maximum kinetic energy $KE_{max}$ that is linearly related to the frequency, as the graph shows.

## Example 2  The Photoelectric Effect for a Silver Surface

The work function for a silver surface is $W_0 = 4.73$ eV. Find the minimum frequency that light must have to eject electrons from this surface.

**Reasoning** The minimum frequency $f_0$ is that frequency at which the photon energy equals the work function $W_0$ of the metal, so the electron is ejected with zero kinetic energy. Since $1$ eV $= 1.60 \times 10^{-19}$ J, the work function expressed in joules is $W_0 = (4.73$ eV$)[(1.60 \times 10^{-19}$ J$)/(1$ eV$)] = 7.57 \times 10^{-19}$ J. Using Equation 29.3, we find

$$hf_0 = \underbrace{KE_{max}}_{= 0 \text{ J}} + W_0 \quad \text{or} \quad f_0 = \frac{W_0}{h}$$

**Solution** The minimum frequency $f_0$ is

$$f_0 = \frac{W_0}{h} = \frac{7.57 \times 10^{-19} \text{ J}}{6.63 \times 10^{-34} \text{ J} \cdot \text{s}} = \boxed{1.14 \times 10^{15} \text{ Hz}}$$

Photons with frequencies less than $f_0$ do not have enough energy to eject electrons from a silver surface. Since $\lambda_0 = c/f_0$, the wavelength of this light is $\lambda_0 = 263$ nm, which is in the ultraviolet region of the electromagnetic spectrum.

> **Problem solving insight**
> The work function of a metal is the minimum energy needed to eject an electron from the metal. An electron that has received this minimum energy has no kinetic energy once outside the metal.

Another significant feature of the photoelectric effect is that the maximum kinetic energy of the ejected electrons remains the same when the intensity of the light increases, provided the light frequency remains the same. As the light intensity increases, more photons per second strike the metal, and consequently more electrons per second are ejected. However, since the frequency is the same for each photon, the energy of each photon is also the same. Thus, the ejected electrons always have the same maximum kinetic energy.

Whereas the photon model of light explains the photoelectric effect satisfactorily, the electromagnetic wave model of light does not. Certainly, it is possible to imagine that the electric field of an electromagnetic wave would cause electrons in the metal to oscillate and tear free from the surface when the amplitude of oscillation becomes large enough. However, were this the case, higher-intensity light would eject electrons with a greater maximum kinetic energy, a fact that experiment does not confirm. Moreover, in the electromagnetic wave model, a relatively long time would be required with low-intensity light before the electrons would build up a sufficiently large oscillation amplitude to tear free. Instead, experiment shows that even the weakest light intensity causes electrons to be ejected almost instantaneously, provided the frequency of the light is above the minimum value $f_0$. The failure of the electromagnetic wave model to explain the photoelectric effect does not mean that the wave model should be abandoned. However, we must recognize that the wave model does not account for all the characteristics of light. The photon model also makes an important contribution to our understanding of the way light behaves when it interacts with matter.

Because a photon has energy, the photon can eject an electron from a metal surface when it interacts with the electron. However, a photon is different from a normal particle. A normal particle has a mass and can travel at speeds up to, but not equal to, the speed of light. A photon, on the other hand, travels at the speed of light in a vacuum and does not exist as an object at rest. The energy of a photon is entirely kinetic in nature, because it has no rest energy and no mass. To show that a photon has no mass, we rewrite Equation 28.4 for the total energy $E$ as

$$E\sqrt{1 - \frac{v^2}{c^2}} = mc^2$$

The term $\sqrt{1 - (v^2/c^2)}$ is zero because a photon travels at the speed of light, $v = c$. Since the energy $E$ of the photon is finite, the left side of the equation above is zero. Thus, the right side must also be zero, so $m = 0$ kg and the photon has no mass.

One of the most exciting and useful applications of the photoelectric effect is the charge-coupled device (CCD). An array of these devices is used instead of film in digital cameras (see Figure 29.5) to capture images in the form of many small groups of electrons. CCD arrays are also used in digital camcorders and electronic scanners, and they

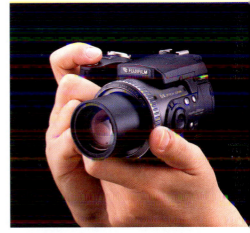

*(a)*

*(b)*

**Figure 29.5** (a) Digital cameras like this one use an array of charge-coupled devices instead of film to capture an image. (© Getty Images News and Sport Services) (b) Images taken by a digital camera can be easily downloaded to a computer and sent to your friends via the Internet. (© Michael Newman/ PhotoEdit)

**The physics of**
**charge-coupled devices**
**and digital cameras.**

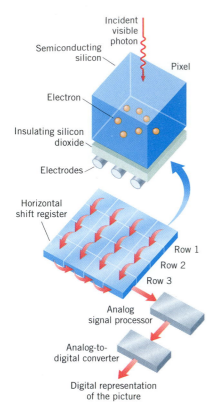

**Figure 29.6** A CCD array can be used to capture photographic images using the photoelectric effect.

**The physics of**
**a safety feature of garage door openers.**

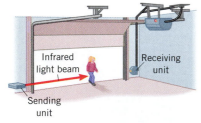

**Figure 29.7** When an obstruction prevents the infrared light beam from reaching the receiving unit, the current in the receiving unit drops. This drop in current is detected by an electronic circuit that stops the downward movement of the door and then causes it to rise.

**The physics of**
**photoevaporation and star formation.**

provide the method of choice with which astronomers capture those spectacular images of the planets and the stars. For use with visible light, a CCD array consists of a sandwich of semiconducting silicon, insulating silicon dioxide, and a number of electrodes, as Figure 29.6 shows. The array is divided into many small sections or pixels, sixteen of which are shown in the drawing. Each pixel captures a small part of a picture. Digital cameras for consumers (rather than professionals) have between one and five million pixels, depending on price. The greater the number of pixels, the better is the resolution of the photograph. The blow-up in Figure 29.6 shows a single pixel. Incident photons of visible light strike the silicon and generate electrons via the photoelectric effect. The range of energies of the visible photons is such that one electron is released when a photon interacts with a silicon atom. The electrons are trapped within a pixel because of a positive voltage applied to the electrodes beneath the insulating layer. Thus, the number of electrons that are released and trapped is proportional to the number of photons striking the pixel. In this fashion, each pixel in the CCD array accumulates an accurate representation of the light intensity at that point on the image. Color information is provided using red, green, or blue filters or a system of prisms to separate the colors. Astronomers use CCD arrays not only in the visible region of the electromagnetic spectrum but in other regions as well.

In addition to trapping the photoelectrons, the electrodes beneath the pixels are used to read out the electron representation of the picture. By changing the positive voltages applied to the electrodes, it is possible to cause all of the electrons trapped in one row of pixels to be transferred to the adjacent row. In this fashion, for instance, row 1 in Figure 29.6 is transferred into row 2, row 2 into row 3, and row 3 into the bottom row, which serves a special purpose. The bottom row functions as a horizontal shift register, from which the contents of each pixel can be shifted to the right, one at a time, and read into an analog signal processor. This processor senses the varying number of electrons in each pixel in the shift register as a kind of wave that has a fluctuating amplitude. After another shift in rows, the information in the next row is read out, and so forth. The output of the analog signal processor is sent to an analog-to-digital converter, which produces a digital representation of the image in terms of the zeros and ones that computers recognize.

Another application of the photoelectric effect depends on the fact that the moving photoelectrons in Figure 29.3 constitute a current—a current that changes as the intensity of the light changes. All automatic garage door openers have a safety feature that prevents the door from closing when it encounters an obstruction (person, vehicle, etc.). As Figure 29.7 illustrates, a sending unit transmits an invisible (infrared) beam across the opening of the door. The beam is detected by a receiving unit that contains a photodiode. A photodiode is a type of *p-n* junction diode (see Section 23.5). When infrared photons strike the photodiode, electrons bound to the atoms absorb the photons and become liberated. These liberated, mobile electrons cause the current in the photodiode to increase. When a person walks through the beam, the light is momentarily blocked from reaching the receiving unit, and the current in the photodiode decreases. The change in current is sensed by electronic circuitry that immediately stops the downward motion of the door and then causes it to rise up.

Figure 29.8*b* shows a central portion of the Eagle Nebula, a giant star-forming region some 7000 light-years from earth. The photo was taken by the Hubble Space Telescope and reveals towering clouds of molecular gas and dust, in which there is dramatic evidence of the energy carried by photons. These clouds extend more than a light-year from base to tip and are the birthplace of stars. A star begins to form within a cloud when the gravitational force pulls together sufficient gas to create a high-density "ball." When the gaseous ball becomes sufficiently dense, thermonuclear fusion (see Section 32.5) occurs at its core, and the star begins to shine. The newly born stars are buried within the cloud and cannot be seen from earth. However, the process of photoevaporation allows astronomers to see many of the high-density regions where stars are being formed. Photoevaporation is the process in which high-energy, ultraviolet (UV) photons from hot stars outside the cloud heat it up, much like microwave photons heat food in a microwave oven. Figure 29.8*b* shows streamers of gas photoevaporating from the cloud as it is illuminated by stars located beyond the photograph's upper edge. As photoevaporation proceeds, globules of gas that are denser than their surroundings are exposed. The globules

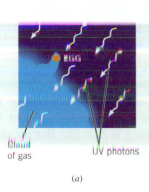

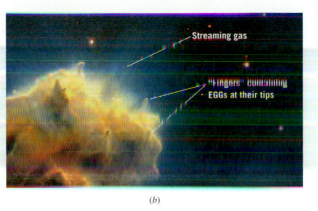

Streaming gas

"Fingers" containing EGGs at their tips

Cloud of gas

UV photons

(a)

(b)

**Figure 29.8** (a) This drawing illustrates the photoevaporation that is occurring in the photograph in part (b). (b) A photo of a portion of the Eagle Nebula taken by the Hubble Space Telescope. Photoevaporation produces finger-like projections on the surface of the gas clouds in the Eagle Nebula. At the fingertips are high density evaporating gaseous globules (EGGs). (Courtesy NASA)

are known as *evaporating gaseous globules* (EGGs), and they are slightly larger than our solar system. The drawing in part *a* of Figure 29.8 shows that the EGGs shade the gas and dust behind them from the UV photons, creating the many finger-like protrusions seen on the surface of the cloud. Astronomers believe that some of these EGGs contain young stars within them. In some cases, so much gas has boiled off that a newborn star can be seen on the surface of an EGG.

# 29.4 The Momentum of a Photon and the Compton Effect

Although Einstein presented his photon model for the photoelectric effect in 1905, it was not until 1923 that the photon concept began to achieve widespread acceptance. It was then that the American physicist Arthur H. Compton (1892–1962) used the photon model to explain his research on the scattering of X-rays by the electrons in graphite. X-rays are high-frequency electromagnetic waves and, like light, they are composed of photons.

Figure 29.9 illustrates what happens when an X-ray photon strikes an electron in a piece of graphite. Like two billiard balls colliding on a pool table, the X-ray photon scatters in one direction after the collision, and the electron recoils in another direction. Compton observed that the scattered photon has a frequency $f'$ that is smaller than the frequency $f$ of the incident photon, indicating that the photon loses energy during the collision. In addition, he found that the difference between the two frequencies depends on the angle $\theta$ at which the scattered photon leaves the collision. The phenomenon in which an X-ray photon is scattered from an electron, with the scattered photon having a smaller frequency than the incident photon, is called the **Compton effect**.

In Section 7.3 the collision between two objects is analyzed using the fact that the total kinetic energy and the total linear momentum of the objects are the same before and after the collision. Similar analysis can be applied to the collision between a photon and an electron. The electron is assumed to be initially at rest and essentially free— that is, not bound to the atoms of the material. According to the principle of conservation of energy,

$$ hf = hf' + \text{KE} \qquad (29.4) $$

| Energy of incident photon | Energy of scattered photon | Kinetic energy of recoil electron |

where the relation $E = hf$ has been used for the photon energies. It follows, then, that $hf' = hf - \text{KE}$, which shows that the energy and corresponding frequency $f'$ of the scattered photon are less than the energy and frequency of the incident photon, just as Compton observed. Since $\lambda' = c/f'$ (Equation 16.1), the wavelength of the scattered X-rays is larger than that of the incident X-rays.

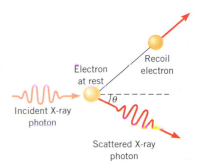

Electron at rest

Recoil electron

Incident X-ray photon

$\theta$

Scattered X-ray photon

**Figure 29.9** In an experiment performed by Arthur H. Compton, an X-ray photon collides with a stationary electron. The scattered photon and the recoil electron depart the collision in different directions.

For an initially stationary electron, conservation of total linear momentum requires that

$$\begin{matrix} \text{Momentum of} \\ \text{incident photon} \end{matrix} = \begin{matrix} \text{Momentum of} \\ \text{scattered photon} \end{matrix} + \begin{matrix} \text{Momentum of} \\ \text{recoil electron} \end{matrix} \quad (29.5)$$

To find an expression for the magnitude $p$ of the photon's momentum, we use Equations 28.3 and 28.4. According to these equations, the momentum of any particle is $p = mv/\sqrt{1 - (v^2/c^2)}$ and its total energy is $E = mc^2/\sqrt{1 - (v^2/c^2)}$. Dividing these two equations, we find that $p/E = v/c^2$. Since a photon travels at the speed of light, $v = c$ and $p/E = 1/c$. Therefore, the momentum of a photon is $p = E/c$. But the energy of a photon is $E = hf$, while the wavelength is $\lambda = c/f$. Therefore, the magnitude of the momentum is

$$p = \frac{hf}{c} = \frac{h}{\lambda} \quad (29.6)$$

Using Equations 29.4, 29.5, and 29.6, Compton showed that the difference between the wavelength $\lambda'$ of the scattered photon and the wavelength $\lambda$ of the incident photon is related to the scattering angle $\theta$ by

$$\lambda' - \lambda = \frac{h}{mc}(1 - \cos\theta) \quad (29.7)$$

In this equation $m$ is the mass of the electron. The quantity $h/(mc)$ is called the **Compton wavelength of the electron** and has the value $h/(mc) = 2.43 \times 10^{-12}$ m. Since $\cos\theta$ varies between $+1$ and $-1$, the shift $\lambda' - \lambda$ in the wavelength can vary between zero and $2h/(mc)$, depending on the value of $\theta$, a fact observed by Compton.

The photoelectric effect and the Compton effect provide compelling evidence that light can exhibit particle-like characteristics attributable to energy packets called photons. But what about the interference phenomena discussed in Chapter 27, such as Young's double-slit experiment and single-slit diffraction, which demonstrate that light behaves as a wave? Does light have two distinct personalities, in which it behaves like a stream of particles in some experiments and like a wave in others? The answer is yes, for physicists now believe that this wave–particle duality is an inherent property of light. Light is a far more interesting (and complex) phenomenon than just a stream of particles or an electromagnetic wave.

In the Compton effect the electron recoils because it gains some of the photon's momentum. In principle, then, the momentum that photons have can be used to make other objects move. Conceptual Example 3 considers a propulsion system for an interstellar spaceship that is based on the momentum of a photon.

### *Conceptual Example 3*
### Solar Sails and Interstellar Spaceships

One propulsion method for interstellar spaceships found in science fiction novels uses a large sail. The intent is that sunlight striking the sail creates a force that pushes the ship away from the sun (Figure 29.10), much as the wind propels a sailboat. Does such a design have any hope of working and, if so, should the surface of the sail facing the sun be shiny like a mirror or black, in order to produce the greatest possible force?

**Reasoning and Solution** There is certainly reason to believe that the design might work. In Conceptual Example 3 in Chapter 7, we found that hailstones striking the roof of a car exert a force on it because they have momentum. Photons also have momentum, so, like the hailstones, they can apply a force to the sail.

As in Chapter 7, we will be guided by the impulse–momentum theorem in assessing the force. This theorem (Equation 7.4) states that when a net force acts on an object, the impulse of the net force is equal to the change in momentum of the object. Greater impulses lead to greater forces for a given time interval. Thus, when a photon collides with the sail, the photon's momentum changes because of the force applied by the sail to the photon. Newton's action–reaction law indicates that the photon simultaneously applies a force of equal magnitude, but opposite direction, to the sail. It is this reaction force that propels the spaceship, and it will be greater when the momentum change experienced by the photon is greater. The surface of

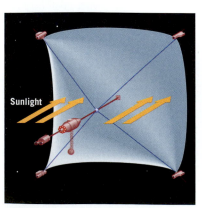

**Figure 29.10** A solar sail provides the propulsion for this interstellar spaceship.

Sunlight

the sail facing the sun, then, should be such that it causes the largest possible momentum change for the impinging photons.

As we found in Section 13.3, radiation reflects from a shiny mirror-like surface and is absorbed by a black surface. Now, consider a photon that strikes the sail perpendicularly. In reflecting from a mirror-like surface, its momentum changes from its value in the forward direction to a value of the same magnitude in the reverse direction. This is a greater change than what occurs when the photon is absorbed by a black surface. Then, the momentum changes only from its value in the forward direction to a value of zero. Consequently, the surface of the sail facing the sun should be shiny in order to produce the greatest possible propulsion force. A shiny surface causes the photons to bounce like the hailstones on the roof of the car and, in so doing, apply a greater force to the sail.

**Related Homework:** *Problem 20*

# 29.5 *The de Broglie Wavelength and the Wave Nature of Matter*

As a graduate student in 1923, Louis de Broglie (1892–1987) made the astounding suggestion that since light waves could exhibit particle-like behavior, particles of matter should exhibit wave-like behavior. De Broglie proposed that all moving matter has a wavelength associated with it, just as a wave does. The notions of energy, momentum, and wavelength are applicable to particles as well as to waves.

De Broglie made the explicit proposal that the wavelength $\lambda$ of a particle is given by the same relation (Equation 29.6) that applies to a photon:

*De Broglie wavelength*

$$\lambda = \frac{h}{p} \tag{29.8}$$

where $h$ is Planck's constant and $p$ is the magnitude of the relativistic momentum of the particle. Today, $\lambda$ is known as the ***de Broglie wavelength*** of the particle.

Confirmation of de Broglie's suggestion came in 1927 from the experiments of the American physicists Clinton J. Davisson (1881–1958) and Lester H. Germer (1896–1971) and, independently, those of the English physicist George P. Thomson (1892–1975). Davisson and Germer directed a beam of electrons onto a crystal of nickel and observed that the electrons exhibited a diffraction behavior, analogous to that seen when X-rays are diffracted by a crystal (see Section 27.8 for a discussion of X-ray diffraction). The wavelength of the electrons revealed by the diffraction pattern matched that predicted by de Broglie's hypothesis, $\lambda = h/p$. More recently, Young's double-slit experiment has been performed with electrons and reveals the effects of wave interference illustrated in Figure 29.1.

Particles other than electrons can also exhibit wave-like properties. For instance, neutrons are sometimes used in diffraction studies of crystal structure. Figure 29.11

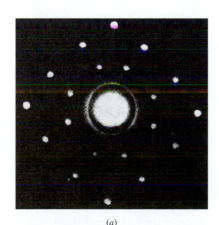
(a)

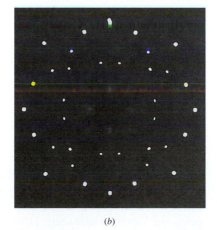
(b)

**Figure 29.11** (*a*) The neutron diffraction pattern (Wollan, Shull and Marney, *Phys. Rev.* 73:527, 1948) and (*b*) the X-ray diffraction pattern for a crystal of sodium chloride (NaCl). (Courtesy Edwin Jones, University of South Carolina)

compares the neutron diffraction pattern and the X-ray diffraction pattern caused by a crystal of rock salt (NaCl).

Although all moving particles have a de Broglie wavelength, the effects of this wavelength are observable only for particles whose masses are very small, on the order of the mass of an electron or a neutron, for instance. Example 4 illustrates why.

### Example 4    The de Broglie Wavelength of an Electron and of a Baseball

Determine the de Broglie wavelength for (a) an electron (mass = $9.1 \times 10^{-31}$ kg) moving at a speed of $6.0 \times 10^{6}$ m/s and (b) a baseball (mass = 0.15 kg) moving at a speed of 13 m/s.

**Reasoning** In each case, the de Broglie wavelength is given by Equation 29.8 as Planck's constant divided by the magnitude of the momentum. Since the speeds are small compared to the speed of light, we can ignore relativistic effects and express the magnitude of the momentum as the product of the mass and the speed.

**Solution**

(a) Since the magnitude $p$ of the momentum is the product of the mass $m$ of the particle and its speed $v$, we have $p = mv$. Using this expression in Equation 29.8 for the de Broglie wavelength, we obtain

$$\lambda = \frac{h}{p} = \frac{h}{mv} = \frac{6.63 \times 10^{-34} \text{ J} \cdot \text{s}}{(9.1 \times 10^{-31} \text{ kg})(6.0 \times 10^{6} \text{ m/s})} = \boxed{1.2 \times 10^{-10} \text{ m}}$$

A de Broglie wavelength of $1.2 \times 10^{-10}$ m is about the size of the interatomic spacing in a solid, such as the nickel crystal used by Davisson and Germer, and, therefore, leads to the observed diffraction effects.

(b) A calculation similar to that in part (a) shows that the de Broglie wavelength of the baseball is $\boxed{\lambda = 3.3 \times 10^{-34} \text{ m}}$. This wavelength is incredibly small, even by comparison with the size of an atom ($10^{-10}$ m) or a nucleus ($10^{-14}$ m). Thus, the ratio $\lambda/W$ of this wavelength to the width $W$ of an ordinary opening, such as a window, is so small that the diffraction of a baseball passing through the window cannot be observed.

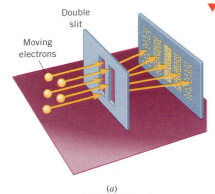

(a)

(b)    After 100 electrons

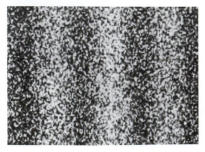

(c)    After 3000 electrons

(d)    After 70 000 electrons

**Figure 29.12** In this electron version of Young's double-slit experiment, the characteristic fringe pattern becomes recognizable only after a sufficient number of electrons have struck the screen. (A. Tonomura, J. Endo, T. Matsuda, and T. Kawasaki, *Am. J. Phys.* 57(2): 117, Feb. 1989)

The de Broglie equation for particle wavelength provides no hint as to what kind of wave is associated with a particle of matter. To gain some insight into the nature of this wave, we turn our attention to Figure 29.12. Part *a* shows the fringe pattern on the screen when electrons are used in a version of Young's double-slit experiment. The bright fringes occur in places where particle waves coming from each slit interfere constructively, while the dark fringes occur in places where the waves interfere destructively.

When an electron passes through the double-slit arrangement and strikes a spot on the screen, the screen glows at that spot, and parts *b*, *c*, and *d* of Figure 29.12 illustrate how the spots accumulate in time. As more and more electrons strike the screen, the spots eventually form the fringe pattern that is evident in part *d*. Bright fringes occur where there is a high probability of electrons striking the screen, and dark fringes occur where there is a low probability. Here lies the key to understanding particle waves. *Particle waves are waves of probability,* waves whose magnitude at a point in space gives an indication of the probability that the particle will be found at that point. At the place where the screen is located, the pattern of probabilities conveyed by the particle waves causes the fringe pattern to emerge. The fact that no fringe pattern is apparent in part *b* of the figure does not mean that there are no probability waves present; it just means that too few electrons have struck the screen for the pattern to be recognizable.

The pattern of probabilities that leads to the fringes in Figure 29.12 is analogous to the pattern of light intensities that is responsible for the fringes in Young's original experiment with light waves (see Figure 27.3). Section 24.4 discusses the fact that the intensity of the light is proportional to either the square of the electric field strength or the square of the magnetic field strength of the wave. In an analogous fashion in the case of particle waves, the probability is proportional to the square of the magnitude $\Psi$ (Greek letter Psi) of the wave. $\Psi$ is referred to as the *wave function* of the particle.

In 1925 the Austrian physicist Erwin Schrödinger (1887–1961) and the German physicist Werner Heisenberg (1901–1976) independently developed theoretical frameworks for determining the wave function. In so doing, they established a new branch of physics called **quantum mechanics**. The word "quantum" refers to the fact that in the world of the atom, where particle waves must be considered, the particle energy is quantized, so only certain energies are allowed. To understand the structure of the atom and the phenomena related to it, quantum mechanics is essential, and the Schrödinger equation for calculating the wave function is now widely used. In the next chapter, we will explore the structure of the atom based on the ideas of quantum mechanics.

## 29.6 The Heisenberg Uncertainty Principle

As the previous section discusses, the bright fringes in Figure 29.12 indicate the places where there is a high probability of an electron striking the screen. And since there are a number of bright fringes, there is more than one place where each electron has some probability of hitting. As a result, it is not possible to specify in advance exactly where on the screen an individual electron will hit. All we can do is speak of the probability that the electron may end up in a number of different places. No longer is it possible to say, as Newton's laws would suggest, that a single electron, fired through the double slit, will travel directly forward in a straight line and strike the screen. This simple model just does not apply when a particle as small as an electron passes through a pair of closely spaced narrow slits. Because the wave nature of particles is important in such circumstances, we lose the ability to predict with 100% certainty the path that a single particle will follow. Instead, only the average behavior of large numbers of particles is predictable, and the behavior of any individual particle is uncertain.

To see more clearly into the nature of the uncertainty, consider electrons passing through a single slit, as in Figure 29.13. After a sufficient number of electrons strike the screen, a diffraction pattern emerges. The electron diffraction pattern consists of alternating bright and dark fringes and is analogous to that for light waves shown in Figure 27.21. Figure 29.13 shows the slit and locates the first dark fringe on either side of the central bright fringe. The central fringe is bright because electrons strike the screen over the entire region between the dark fringes. If the electrons striking the screen outside the central bright fringe are neglected, the extent to which the electrons are diffracted is given by the angle $\theta$ in the drawing. To reach locations within the central bright fringe, some electrons must have acquired momentum in the $y$ direction, despite the fact that they enter the slit traveling along the $x$ direction and have no momentum in the $y$ direction to start with. The figure illustrates that the $y$ component of the momentum may be as large as $\Delta p_y$. The notation $\Delta p_y$ indicates the difference between the maximum value of the $y$ component of the momentum after the electron passes through the slit and its value of zero before the electron passes through the slit. $\Delta p_y$ represents the *uncertainty* in the $y$ component of the momentum in that the $y$ component may have any value from zero to $\Delta p_y$.

It is possible to relate $\Delta p_y$ to the width $W$ of the slit. To do this, we assume that Equation 27.4, which applies to light waves, also applies to particle waves whose de Broglie

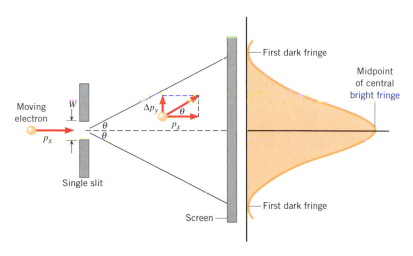

Moving electron

$W$

Single slit

$p_x$

$\theta$

$\Delta p_y$

$p_x$

$\theta$

First dark fringe

Midpoint of central bright fringe

First dark fringe

Screen

**Figure 29.13** When a sufficient number of electrons pass through a single slit and strike the screen, a diffraction pattern of bright and dark fringes emerges. (Only the central bright fringe is shown.) This pattern is due to the wave nature of the electrons and is analogous to that produced by light waves.

wavelength is $\lambda$. This equation, $\sin\theta = \lambda/W$, specifies the angle $\theta$ that locates the first dark fringe. If $\theta$ is small, $\sin\theta \approx \tan\theta$. Moreover, Figure 29.13 indicates that $\tan\theta = \Delta p_y/p_x$, where $p_x$ is the $x$ component of the momentum of the electron. Therefore, $\Delta p_y/p_x \approx \lambda/W$. But $p_x = h/\lambda$ according to de Broglie's equation, so that

$$\frac{\Delta p_y}{p_x} = \frac{\Delta p_y}{h/\lambda} \approx \frac{\lambda}{W}$$

As a result,

$$\Delta p_y \approx \frac{h}{W} \tag{29.9}$$

which indicates that a smaller slit width leads to a larger uncertainty in the $y$ component of the electron's momentum.

It was Heisenberg who first suggested that the uncertainty $\Delta p_y$ in the $y$ component of the momentum is related to the uncertainty in the $y$ position of the electron as the electron passes through the slit. Since the electron can pass through anywhere within the width $W$, the uncertainty in the $y$ position of the electron is $\Delta y = W$. Substituting $\Delta y$ for $W$ in Equation 29.9 shows that $\Delta p_y \approx h/\Delta y$ or $(\Delta p_y)(\Delta y) \approx h$. The result of Heisenberg's more complete analysis is given below in Equation 29.10 and is known as the **Heisenberg uncertainty principle.** Note that the Heisenberg principle is a general principle with wide applicability. It does not just apply to the case of single slit diffraction, which we have used here for the sake of convenience.

■ **THE HEISENBERG UNCERTAINTY PRINCIPLE**

**Momentum and position**

$$(\Delta p_y)(\Delta y) \geq \frac{h}{4\pi} \tag{29.10}$$

$\Delta y$ = uncertainty in a particle's position along the $y$ direction
$\Delta p_y$ = uncertainty in the $y$ component of the linear momentum of the particle

**Energy and time**

$$(\Delta E)(\Delta t) \geq \frac{h}{4\pi} \tag{29.11}$$

$\Delta E$ = uncertainty in the energy of a particle when the particle is in a certain state
$\Delta t$ = time interval during which the particle is in the state

The Heisenberg uncertainty principle places limits on the accuracy with which the momentum and position of a particle can be specified simultaneously. These limits are not just limits due to faulty measuring techniques. They are fundamental limits imposed by nature, and there is no way to circumvent them. Equation 29.10 indicates that $\Delta p_y$ and $\Delta y$ cannot both be arbitrarily small at the same time. If one is small, then the other must be large, so that their product equals or exceeds ($\geq$) Planck's constant divided by $4\pi$. For example, if the position of a particle is known exactly, so that $\Delta y$ is zero, then $\Delta p_y$ is an infinitely large number, and the momentum of the particle is completely uncertain. Conversely, if we assume that $\Delta p_y$ is zero, then $\Delta y$ is an infinitely large number, and the position of the particle is completely uncertain. In other words, the Heisenberg uncertainty principle states that it is impossible to specify precisely both the momentum and position of a particle at the same time.

There is also an uncertainty principle that deals with energy and time, as expressed by Equation 29.11. The product of the uncertainty $\Delta E$ in the energy of a particle and the time interval $\Delta t$ during which the particle remains in a given energy state is greater than or equal to Planck's constant divided by $4\pi$. Therefore, the shorter the lifetime of a particle in a given state, the greater is the uncertainty in the energy of that state.

Example 5 shows that the uncertainty principle has significant consequences for the motion of tiny particles such as electrons but has little effect on the motion of macroscopic objects, even those with as little mass as a Ping-Pong ball.

## Example 5 The Heisenberg Uncertainty Principle

Assume that the position of an object is known so precisely that the uncertainty in the position is only $\Delta y = 1.5 \times 10^{-11}$ m. (a) Determine the minimum uncertainty in the momentum of the object. Find the corresponding minimum uncertainty in the speed of the object, if the object is (b) an electron (mass $= 9.1 \times 10^{-31}$ kg) and (c) a Ping-Pong ball (mass $= 2.2 \times 10^{-3}$ kg).

**Reasoning** The minimum uncertainty $\Delta p_y$ in the $y$ component of the momentum is given by the Heisenberg uncertainty principle as $\Delta p_y = h/(4\pi\Delta y)$, where $\Delta y$ is the uncertainty in the position of the object along the $y$ direction. Both the electron and the Ping-Pong ball have the same uncertainty in their momenta because they have the same uncertainty in their positions. However, these objects have very different masses. As a result, we will find that the uncertainty in the speeds of these objects is very different.

**Solution**

(a) The minimum uncertainty in the $y$ component of the momentum is

$$\Delta p_y = \frac{h}{4\pi\Delta y} = \frac{6.63 \times 10^{-34}\ \text{J}\cdot\text{s}}{4\pi(1.5 \times 10^{-11}\ \text{m})} = \boxed{3.5 \times 10^{-24}\ \text{kg}\cdot\text{m/s}} \qquad (29.10)$$

(b) Since $\Delta p_y = m\Delta v_y$, the minimum uncertainty in the speed of the electron is

$$\Delta v_y = \frac{\Delta p_y}{m} = \frac{3.5 \times 10^{-24}\ \text{kg}\cdot\text{m/s}}{9.1 \times 10^{-31}\ \text{kg}} = \boxed{3.8 \times 10^{6}\ \text{m/s}}$$

Thus, the small uncertainty in the $y$ position of the electron gives rise to a large uncertainty in the speed of the electron.

(c) The uncertainty in the speed of the Ping-Pong ball is

$$\Delta v_y = \frac{\Delta p_y}{m} = \frac{3.5 \times 10^{-24}\ \text{kg}\cdot\text{m/s}}{2.2 \times 10^{-3}\ \text{kg}} = \boxed{1.6 \times 10^{-21}\ \text{m/s}}$$

Because the mass of the Ping-Pong ball is relatively large, the small uncertainty in its $y$ position gives rise to an uncertainty in its speed that is much smaller than that for the electron. Thus, in contrast to the electron, we can know simultaneously where the ball is and how fast it is moving, to a very high degree of certainty.

Example 5 emphasizes how the uncertainty principle imposes different uncertainties on the speeds of an electron (small mass) and a Ping-Pong ball (large mass). For objects like the ball, which have relatively large masses, the uncertainties in position and speed are so small that they have no effect on our ability to determine simultaneously where such objects are and how fast they are moving. The uncertainties calculated in Example 5 depend on more than just the mass, however. They also depend on Planck's constant, which is a very small number. It is interesting to speculate about what life would be like if Planck's constant were much larger than $6.63 \times 10^{-34}$ J·s. Conceptual Example 6 deals with just such speculation.

## Conceptual Example 6 What If Planck's Constant Were Large?

A bullet leaving the barrel of a gun is analogous to an electron passing through the single slit in Figure 29.13. With this analogy in mind, what would hunting be like if Planck's constant had a relatively large value instead of its small value of $6.63 \times 10^{-34}$ J·s?

**Reasoning and Solution** It takes two to hunt, the hunter and the hunted. Consider the hunter first and what it would mean to take aim if Planck's constant were large. The bullet moving down the barrel of the gun has a de Broglie wavelength given by Equation 29.8 as $\lambda = h/p$. This wavelength is large, since Planck's constant $h$ is now large. (We are assuming that the magnitude $p$ of the bullet's momentum is not abnormally large.) Remember from Section 27.5 that the diffraction of light waves through an opening is greater when the wavelength is greater, other things being equal. We expect, therefore, that with $\lambda$ being comparable to the opening in the barrel, the hunter will have to contend with appreciable diffraction when the bullet leaves the barrel. This means that, in spite of being aimed directly at the target, the bullet may come nowhere near it. In Figure 29.13, for comparison, diffraction allows the electron to strike the screen on either side of the maximum intensity point of the central bright fringe. Under such circumstances, the hunter might as well not bother aiming. He'll have just as much success by scattering his shots around randomly, hoping to hit something.

Now consider the target and what it means that the hunter "aimed directly at the target." This must mean that the target's location is known, at least approximately. In other words, the uncertainty $\Delta y$ in the target's position is reasonably small. But the uncertainty principle indicates that the minimum uncertainty in the target's momentum is $\Delta p = h/(4\pi\Delta y)$. Since Planck's constant is now large, $\Delta p$ is also large. The target, therefore, may have a large momentum and be moving rapidly. Even if the bullet happens to reach the place where the hunter aimed, the target may no longer be there. Hunting would be vastly different indeed, if Planck's constant had a very large value.

**Related Homework:** *Problem 34*

# Concept Summary

This summary presents an abridged version of the chapter, including the important equations and all available learning aids. For convenient reference, the learning aids (including the text's examples) are placed next to or immediately after the relevant equation or discussion. The following learning aids may be found on-line at **www.wiley.com/college/cutnell**:

| | |
|---|---|
| **Interactive LearningWare** examples are solved according to a five-step interactive format that is designed to help you develop problem-solving skills. | **Concept Simulations** are animated versions of text figures or animations that illustrate important concepts. You can control parameters that affect the display, and we encourage you to experiment. |
| **Interactive Solutions** offer specific models for certain types of problems in the chapter homework. The calculations are carried out interactively. | **Self-Assessment Tests** include both qualitative and quantitative questions. Extensive feedback is provided for both incorrect and correct answers, to help you evaluate your understanding of the material. |

| Topic | Discussion | Learning Aids |
|---|---|---|
| | **29.1 The Wave–Particle Duality** | |
| | **29.2 Blackbody Radiation and Planck's Constant** | |
| Wave–particle duality | The wave–particle duality refers to the fact that a wave can exhibit particle-like characteristics and a particle can exhibit wave-like characteristics. | |
| Perfect blackbody | At a constant temperature, a perfect blackbody absorbs and reemits all the electromagnetic radiation that falls on it. Max Planck calculated the emitted radiation intensity per unit wavelength as a function of wavelength. In his theory, Planck assumed that a blackbody consists of atomic oscillators that can have only discrete, or quantized, energies. Planck's quantized energies are given by | |
| Energies of atomic oscillators | $$E = nhf \qquad n = 0, 1, 2, 3, \ldots \qquad (29.1)$$ where $h$ is Planck's constant ($6.63 \times 10^{-34}$ J·s) and $f$ is the vibration frequency of an oscillator. | |
| | **29.3 Photons and the Photoelectric Effect** | |
| | All electromagnetic radiation consists of photons, which are packets of energy. The energy of a photon is | **Example 1** |
| Energy of a photon | $$E = hf \qquad (29.2)$$ where $h$ is Planck's constant and $f$ is the frequency of the photon. A photon has no mass and always travels at the speed of light $c$ in a vacuum. | **Interactive LearningWare 29.1** |
| Photoelectric effect<br>Work function | The photoelectric effect is the phenomenon in which light shining on a metal surface causes electrons to be ejected from the surface. The work function $W_0$ of a metal is the minimum work that must be done to eject an electron from the metal. In accordance with the conservation of energy, the electrons ejected from a metal have a maximum kinetic energy $\text{KE}_{\text{max}}$ that is related to the energy $hf$ of the incident photon and the work function of the metal by | |
| Conservation of energy and the photoelectric effect | $$hf = \text{KE}_{\text{max}} + W_0 \qquad (29.3)$$ | **Example 2**<br>**Concept Simulation 29.1** |

 **Use *Self-Assessment Test 29.1* to evaluate your understanding of Sections 29.1–29.3.**

| Topic | Discussion | Learning Aids |
|---|---|---|

### 29.4 The Momentum of a Photon and the Compton Effect

The magnitude $p$ of a photon's momentum is

**Magnitude of a photon's momentum**

$$p = \frac{h}{\lambda}$$ (29.6) Interactive Solution 29.17

where $h$ is Planck's constant and $\lambda$ is the wavelength of the photon.

**Compton effect**

The Compton effect is the scattering of a photon by an electron in a material, the scattered photon having a smaller frequency and, hence, a smaller energy than the incident photon. Part of the photon's energy and momentum are transferred to the recoiling electron. The difference between the wavelength $\lambda'$ of the scattered photon and the wavelength $\lambda$ of the incident photon is related to the scattering angle $\theta$ by

**Wavelength difference in the Compton effect**

$$\lambda' - \lambda = \frac{h}{mc}(1 - \cos \theta)$$ (29.7) Example 3

**Compton wavelength of the electron**

where $m$ is the mass of the electron. The quantity $h/(mc)$ is known as the Compton wavelength of the electron.

### 29.5 The de Broglie Wavelength and the Wave Nature of Matter

The de Broglie wavelength $\lambda$ of a particle is

**De Broglie wavelength**

$$\lambda = \frac{h}{p}$$ (29.8) Example 4
Interactive Solution 29.29

where $h$ is Planck's constant and $p$ is the magnitude of the relativistic momentum of the particle. Because of its wavelength, a particle can exhibit wave-like characteristics. The wave associated with a particle is a wave of probability.

### 29.6 The Heisenberg Uncertainty Principle

The Heisenberg uncertainty principle places limits on our knowledge about the behavior of a particle. The uncertainty principle indicates that

**Uncertainty principle—momentum and position**

$$(\Delta p_y)(\Delta y) \geq \frac{h}{4\pi}$$ (29.10) Examples 5, 6

where $\Delta y$ is the uncertainty in the particle's position along the $y$ direction, and $\Delta p_y$ is the uncertainty in the $y$-component of the linear momentum of the particle.

The uncertainty principle also states that

**Uncertainty principle—energy and time**

$$(\Delta E)(\Delta t) \geq \frac{h}{4\pi}$$ (29.11)

where $\Delta E$ is the uncertainty in the energy of a particle when it is in a certain state, and $\Delta t$ is the time interval during which the particle is in the state.

 **Use Self-Assessment Test 29.2 to evaluate your understanding of Sections 29.4–29.6.**

## Problems

*In working these problems, ignore relativistic effects.*

**ssm** Solution is in the Student Solutions Manual.    **www** Solution is available on the World Wide Web at www.wiley.com/college/cutnell
☤ This icon represents a biomedical application.

### Section 29.3 Photons and the Photoelectric Effect

**1. ssm** Ultraviolet light is responsible for sun tanning. Find the wavelength (in nm) of an ultraviolet photon whose energy is $6.4 \times 10^{-19}$ J.

**2.** The dissociation energy of a molecule is the energy required to break apart the molecule into its separate atoms. The dissociation energy for the cyanogen molecule is $1.22 \times 10^{-18}$ J. Suppose that

this energy is provided by a single photon. Determine the (a) wavelength and (b) frequency of the photon. (c) In what region of the electromagnetic spectrum (see Figure 24.9) does this photon lie?

**3.** An FM radio station broadcasts at a frequency of 98.1 MHz. The power radiated from the antenna is $5.0 \times 10^4$ W. How many photons per second does the antenna emit?

**4.** The maximum wavelength that an electromagnetic wave can have and still eject electrons from a metal surface is 485 nm. What is the work function $W_0$ of this metal? Express your answer in electron volts.

**5. ssm** Ultraviolet light with a frequency of $3.00 \times 10^{15}$ Hz strikes a metal surface and ejects electrons that have a maximum kinetic energy of 6.1 eV. What is the work function (in eV) of the metal?

**6.** Light is shining perpendicularly on the surface of the earth with an intensity of 680 W/m². Assuming all the photons in the light have the same wavelength (in vacuum) of 730 nm, determine the number of photons per second per square meter that reach the earth.

**7.** A magnesium surface has a work function of 3.68 eV. Electromagnetic waves with a wavelength of 215 nm strike the surface and eject electrons. Find the maximum kinetic energy of the ejected electrons. Express your answer in electron volts.

**\*8.** Light is incident on the surface of metallic sodium, whose work function is 2.3 eV. The maximum speed of the photoelectrons emitted by the surface is $1.2 \times 10^6$ m/s. What is the wavelength of the light?

**\*9. ssm** Consult **Interactive LearningWare 29.1** at **www.wiley.com/college/cutnell** for background material relating to this problem. An owl has good night vision because its eyes can detect a light intensity as small as $5.0 \times 10^{-13}$ W/m². What is the minimum number of photons per second that an owl eye can detect if its pupil has a diameter of 8.5 mm and the light has a wavelength of 510 nm?

**\*10.** A proton is located at a distance of 0.420 m from a point charge of $+8.30\ \mu C$. The repulsive electric force moves the proton until it is at a distance of 1.58 m from the charge. Suppose that the electric potential energy lost by the system is carried off by a photon that is emitted during the process. What is its wavelength?

**\*\*11. ssm www** (a) How many photons (wavelength = 620 nm) must be absorbed to melt a 2.0-kg block of ice at 0 °C into water at 0 °C? (b) On the average, how many $H_2O$ molecules does one photon convert from the ice phase to the water phase?

**\*\*12.** A laser emits $1.30 \times 10^{18}$ photons per second in a beam of light that has a diameter of 2.00 mm and a wavelength of 514.5 nm. Determine (a) the average electric field strength and (b) the average magnetic field strength for the electromagnetic wave that constitutes the beam.

**Section 29.4 The Momentum of a Photon and the Compton Effect**

**13.** The microwaves used in a microwave oven have a wavelength of about 0.13 m. What is the momentum of a microwave photon?

**14.** A light source emits a beam of photons, each of which has a momentum of $2.3 \times 10^{-29}$ kg·m/s. (a) What is the frequency of the photons? (b) To what region of the electromagnetic spectrum do the photons belong? Consult Figure 24.9 if necessary.

**15. ssm** Incident X-rays have a wavelength of 0.3120 nm and are scattered by the "free" electrons in graphite. The scattering angle in Figure 29.9 is $\theta = 135.0°$. What is the magnitude of the momentum of (a) the incident photon and (b) the scattered photon? (For accuracy, use $h = 6.626 \times 10^{-34}$ J·s and $c = 2.998 \times 10^8$ m/s.)

**16.** In a Compton scattering experiment, the incident X-rays have a wavelength of 0.2685 nm, and the scattered X-rays have a wavelength of 0.2703 nm. Through what angle $\theta$ in Figure 29.9 are the X-rays scattered?

**17.** Refer to **Interactive Solution 29.17** at **www.wiley.com/college/cutnell** for help in solving this problem. An incident X-ray photon of wavelength 0.2750 nm is scattered from an electron that is initially at

rest. The photon is scattered at an angle of $\theta = 180.0°$ in Figure 29.9 and has a wavelength of 0.2825 nm. Use the conservation of linear momentum to find the momentum gained by the electron.

**\*18.** The X-rays detected at a scattering angle of $\theta = 163°$ in Figure 29.9 have a wavelength of 0.1867 nm. Find (a) the wavelength of an incident photon, (b) the energy of an incident photon, (c) the energy of a scattered photon, and (d) the kinetic energy of the recoil electron. (For accuracy, use $h = 6.626 \times 10^{-34}$ J·s and $c = 2.998 \times 10^8$ m/s.)

**\*19. ssm www** What is the maximum amount by which the wavelength of an incident photon could change when it undergoes Compton scattering from a nitrogen molecule ($N_2$)?

**\*\*20.** Review Conceptual Example 3 before attempting this problem. A beam of visible light has a wavelength of 395 nm and shines perpendicularly on a surface. As a result, there are $3.0 \times 10^{18}$ photons per second striking the surface. By using the impulse–momentum theorem (Section 7.1), obtain the average force that this beam applies to the surface when (a) the surface is a mirror, so the momentum of each photon is reversed after reflection, and (b) the surface is black, so each photon is absorbed and there are no reflected photons.

**Section 29.5 The de Broglie Wavelength and the Wave Nature of Matter**

**21.** A honeybee (mass = $1.3 \times 10^{-4}$ kg) is crawling at a speed of 0.020 m/s. What is the de Broglie wavelength of the bee?

**22.** The interatomic spacing in a crystal of table salt is 0.282 nm. This crystal is being studied in a neutron diffraction experiment, similar to the one that produced the photograph in Figure 29.11a. How fast must a neutron (mass = $1.67 \times 10^{-27}$ kg) be moving to have a de Broglie wavelength of 0.282 nm?

**23. ssm** The de Broglie wavelength of a proton in a particle accelerator is $1.30 \times 10^{-14}$ m. Determine the kinetic energy (in joules) of the proton.

**24.** How fast does a proton have to be moving in order to have the same de Broglie wavelength as an electron that is moving at $4.50 \times 10^6$ m/s?

**25.** Recall from Section 14.3 that the average kinetic energy of an atom in a monatomic ideal gas is given by $\overline{KE} = \frac{3}{2}kT$, where $k = 1.38 \times 10^{-23}$ J/K and $T$ is the Kelvin temperature of the gas. Determine the de Broglie wavelength of a helium atom (mass = $6.65 \times 10^{-27}$ kg) that has the average kinetic energy at room temperature (293 K).

**26.** Neutrons ($m = 1.67 \times 10^{-27}$ kg) are being emitted by two objects. The neutrons are moving with a speed of $2.80 \times 10^3$ m/s and pass through a circular aperture whose diameter is 0.100 mm. This situation is analogous to that shown in Figure 27.27a for light. According to the Rayleigh criterion, what is the minimum angle $\theta_{min}$ (in radians) between the two objects, such that they are just resolved using neutrons?

**\*27. ssm** The width of the central bright fringe in a diffraction pattern on a screen is identical when either electrons or red light (vacuum wavelength = 661 nm) pass through a single slit. The distance between the screen and the slit is the same in each case and is large compared to the slit width. How fast are the electrons moving?

**\*28.** An electron, starting from rest, accelerates through a potential difference of 418 V. What is the final de Broglie wavelength of the electron, assuming that its final speed is much less than the speed of light?

**\*29.** Consult **Interactive Solution 29.29** at **www.wiley.com/college/cutnell** to explore a model for solving this problem. In a television picture tube the electrons are accelerated from rest through a potential difference $V$. Just before an electron strikes the screen, its de Broglie wavelength is $0.900 \times 10^{-11}$ m. What is the potential difference?

**30.** The kinetic energy of a particle is equal to the energy of a photon. The particle moves at 5.0% of the speed of light. Find the ratio of the photon wavelength to the de Broglie wavelength of the particle.

### Section 29.6 The Heisenberg Uncertainty Principle

**31. ssm www** In the lungs there are tiny sacs of air, which are called alveoli. The average diameter of one of these sacs is 0.25 mm. Consider an oxygen molecule (mass $= 5.3 \times 10^{-26}$ kg) trapped within a sac. What is the minimum uncertainty in the velocity of this oxygen molecule?

**32.** An electron is trapped within a sphere whose diameter is $6.0 \times 10^{-15}$ m (about the size of the nucleus of an oxygen atom). What is the minimum uncertainty in the electron's momentum?

**33.** Consider a line that is 2.5 m long. A moving object is somewhere along this line, but its position is not known. (a) Find the minimum uncertainty in the momentum of the object. Find the minimum uncertainty in the object's velocity, assuming that the object is (b) a golf ball (mass $= 0.045$ kg) and (c) an electron.

**34.** Review Conceptual Example 6 as background for this problem. When electrons pass through a single slit, as in Figure 29.13, they form a diffraction pattern. As Section 29.6 discusses, the central bright fringe extends to either side of the midpoint, according to an angle $\theta$ given by $\sin \theta = \lambda/W$, where $\lambda$ is the de Broglie wavelength of the electron and $W$ is the width of the slit. When $\lambda$ is the same size as $W$, $\lambda = 90°$, and the central fringe fills the entire observation screen. In this case, an electron passing through the slit has roughly the same probability of hitting the screen either straight ahead or anywhere off to one side or the other. Now, imagine yourself in a world where Planck's constant is large enough so you exhibit similar effects when you walk through a 0.90-m-wide doorway. Your mass is 82 kg and you walk at a speed of 0.50 m/s. How large would Planck's constant have to be in this hypothetical world?

**35. ssm www** Particles pass through a single slit of width 0.200 mm (see Figure 29.13). The de Broglie wavelength of each particle is 633 nm. After the particles pass through the slit, they spread out over a range of angles. Use the Heisenberg uncertainty principle to determine the minimum range of angles.

**36.** Suppose the minimum uncertainty in the position of a particle is equal to its de Broglie wavelength. If the particle has an average speed of $4.5 \times 10^5$ m/s, what is the minimum uncertainty in its speed?

# Chapter 30 The Nature of the Atom

## 30.1 Rutherford Scattering and the Nuclear Atom

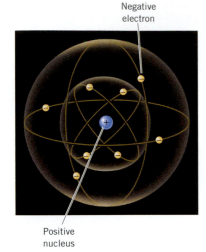

Figure 30.1 In the nuclear atom a small positively charged nucleus is surrounded at relatively large distances by a number of electrons.

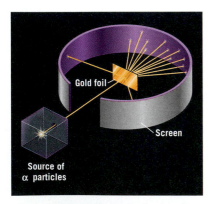

Figure 30.2 A Rutherford scattering experiment in which $\alpha$ particles are scattered by a thin gold foil. The entire apparatus is located within a vacuum chamber (not shown).

An atom contains a small, positively charged nucleus (radius $\approx 10^{-15}$ m), which is surrounded at relatively large distances (radius $\approx 10^{-10}$ m) by a number of electrons, as Figure 30.1 illustrates. In the natural state, an atom is electrically neutral because the nucleus contains a number of protons (each with a charge of $+e$) that equals the number of electrons (each with a charge of $-e$). This model of the atom is universally accepted now and is referred to as the *nuclear atom.*

The nuclear atom is a relatively recent idea. In the early part of the twentieth century a widely accepted model, due to the English physicist Joseph J. Thomson (1856–1940), pictured the atom very differently. In Thomson's view there was no nucleus at the center of an atom. Instead, the positive charge was assumed to be spread throughout the atom, forming a kind of "paste" or "pudding," in which the negative electrons were suspended like "plums."

The "plum-pudding" model was discredited in 1911 when the New Zealand physicist Ernest Rutherford (1871–1937) published experimental results that the model could not explain. As Figure 30.2 indicates, Rutherford and his co-workers directed a beam of alpha particles ($\alpha$ particles) at a thin metal foil made of gold. Alpha particles are positively charged particles (the nuclei of helium atoms, although this was not recognized at the time) emitted by some radioactive materials. If the "plum-pudding" model were correct, the $\alpha$ particles would be expected to pass nearly straight through the foil. After all, there is nothing in this model to deflect the relatively massive $\alpha$ particles, since the electrons have a comparatively small mass and the positive charge is spread out in a diluted "pudding." Using a zinc sulfide screen, which flashed briefly when struck by an $\alpha$ particle, Rutherford and co-workers were able to determine that not all the $\alpha$ particles passed straight through the foil. Instead, some were deflected at large angles, even backward. Rutherford himself said, "It was almost as incredible as if you had fired a fifteen-inch shell at a piece of tissue and it came back and hit you." Rutherford concluded that the positive charge, instead of being distributed thinly and uniformly throughout the atom, was concentrated in a small region called the nucleus.

But how could the electrons in a nuclear atom remain separated from the positively charged nucleus? If the electrons were stationary, they would be pulled inward by the attractive electric force of the nuclear charge. Therefore, the electrons must be moving around the nucleus in some fashion, like planets revolving around the sun. In fact, the nuclear model of the atom is sometimes referred to as the "planetary" model. The dimensions of the atom, however, are such that it contains a larger fraction of empty space than our solar system does, as Conceptual Example 1 discusses.

### Conceptual Example 1   Atoms Are Mostly Empty Space

In the planetary model of the atom, the nucleus (radius $\approx 1 \times 10^{-15}$ m) is analogous to the sun (radius $\approx 7 \times 10^8$ m). Electrons orbit (radius $\approx 1 \times 10^{-10}$ m) the nucleus like the earth orbits (radius $\approx 1.5 \times 10^{11}$ m) the sun. If the dimensions of the solar system had the same proportions as those of the atom, would the earth be closer to or farther away from the sun than it actually is?

**Reasoning and Solution** The radius of an electron orbit is one hundred thousand times larger than the radius of the nucleus: $(1 \times 10^{-10}$ m$)/(1 \times 10^{-15}$ m$) = 10^5$. If the radius of the earth's orbit were one hundred thousand times larger than the radius of the sun, the earth's orbit would have a radius of $10^5 \times (7 \times 10^8$ m$) = 7 \times 10^{13}$ m. This is more than four hundred times greater than the actual orbital radius of $1.5 \times 10^{11}$ m, so **the earth would be much far-**

*ther away from the sun.* In fact, it would be more than ten times farther from the sun than Pluto is, which is our most distant planet and has an orbital radius of about $6 \times 10^{12}$ m. An atom, then, contains a much greater fraction of empty space than does our solar system.

**Related Homework:** *Problem 2*

Although the planetary model of the atom is easy to visualize, it too is fraught with difficulties. For instance, an electron moving on a curved path has a centripetal acceleration, as Section 5.2 discusses. And when an electron is accelerating, it radiates electromagnetic waves, as Section 24.1 discusses. These waves carry away energy. With their energy constantly being depleted, the electrons would spiral inward and eventually collapse into the nucleus. Since matter is stable, such a collapse does not occur. Thus, the planetary model, although providing a more realistic picture of the atom than the "plum-pudding" model, must be telling only part of the story. The full story of atomic structure is fascinating, and the next section describes another aspect of it.

## 30.2  Line Spectra

We have seen in Sections 13.3 and 29.2 that all objects emit electromagnetic waves, and we will see in Section 30.3 how this radiation arises. For a solid object, such as the hot filament of a light bulb, these waves have a continuous range of wavelengths, some of which are in the visible region of the spectrum. The continuous range of wavelengths is characteristic of the entire collection of atoms that make up the solid. In contrast, individual atoms, free of the strong interactions that are present in a solid, emit only certain specific wavelengths rather than a continuous range. These wavelengths are characteristic of the atom and provide important clues about its structure. To study the behavior of individual atoms, low-pressure gases are used in which the atoms are relatively far apart.

A low-pressure gas in a sealed tube can be made to emit electromagnetic waves by applying a sufficiently large potential difference between two electrodes located within the tube. With a grating spectroscope like that in Figure 27.34, the individual wavelengths emitted by the gas can be separated and identified as a series of bright fringes. The series of fringes is called a *line spectrum* because each bright fringe appears as a thin rectangle (a "line") resulting from the large number of parallel, closely spaced slits in the grating of the spectroscope. Figure 30.3 shows the visible parts of the line spectra for neon and mercury. The specific visible wavelengths emitted by neon and mercury give neon signs and mercury vapor street lamps their characteristic colors.

**The physics of** neon signs and mercury vapor street lamps.

**Figure 30.3** The line spectra for neon and mercury, along with the continuous spectrum of the sun. The dark lines in the sun's spectrum are called Fraunhofer lines, three of which are marked by arrows. (Courtesy Bausch & Lomb)

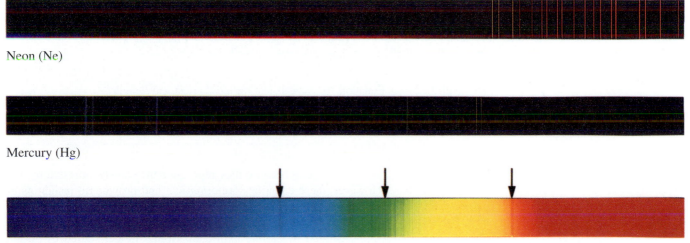

Neon (Ne)

Mercury (Hg)

Solar absorption spectrum (Fraunhofer lines)

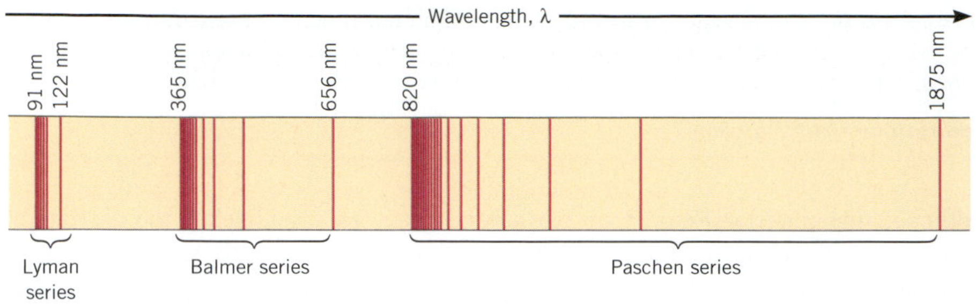

Wavelength, λ

91 nm  122 nm    365 nm    656 nm    820 nm    1875 nm

Lyman series        Balmer series              Paschen series

**Figure 30.4** Line spectrum of atomic hydrogen. Only the Balmer series lies in the visible region of the electromagnetic spectrum.

The simplest line spectrum is that of atomic hydrogen, and much effort has been devoted to understanding the pattern of wavelengths that it contains. Figure 30.4 illustrates in schematic form some of the groups or series of lines in the spectrum of atomic hydrogen. The group of lines in the visible region is known as the *Balmer series,* in recognition of Johann J. Balmer (1825–1898), a Swiss schoolteacher who found an empirical equation that gave the values for the observed wavelengths. This equation is given next, along with similar equations that apply to the *Lyman series* at shorter wavelengths and the *Paschen series* at longer wavelengths, which are also shown in the drawing:

*Lyman series* $\qquad \dfrac{1}{\lambda} = R\left(\dfrac{1}{1^2} - \dfrac{1}{n^2}\right) \qquad n = 2, 3, 4, \ldots$ (30.1)

*Balmer series* $\qquad \dfrac{1}{\lambda} = R\left(\dfrac{1}{2^2} - \dfrac{1}{n^2}\right) \qquad n = 3, 4, 5, \ldots$ (30.2)

*Paschen series* $\qquad \dfrac{1}{\lambda} = R\left(\dfrac{1}{3^2} - \dfrac{1}{n^2}\right) \qquad n = 4, 5, 6, \ldots$ (30.3)

In these equations, the constant term $R$ has the value of $R = 1.097 \times 10^7$ m$^{-1}$ and is called the *Rydberg constant*. An essential feature of each group of lines is that there are a long and a short wavelength limit, with the lines being increasingly crowded toward the short wavelength limit. Figure 30.4 also gives the wavelength limits for each series, and Example 2 determines them for the Balmer series.

## Example 2   The Balmer Series

Find (a) the longest and (b) the shortest wavelengths of the Balmer series.

**Reasoning** Each wavelength in the series corresponds to one value for the integer $n$ in Equation 30.2. Longer wavelengths are associated with smaller values of $n$. The longest wavelength occurs when $n$ has its smallest value of $n = 3$. The shortest wavelength arises when $n$ has a very large value, so that $1/n^2$ is essentially zero.

**Solution**

**(a)** With $n = 3$, Equation 30.2 reveals that for the longest wavelength

$$\frac{1}{\lambda} = R\left(\frac{1}{2^2} - \frac{1}{n^2}\right) = (1.097 \times 10^7 \text{ m}^{-1})\left(\frac{1}{2^2} - \frac{1}{3^2}\right) = 1.524 \times 10^6 \text{ m}^{-1}$$

or $\boxed{\lambda = 656 \text{ nm}}$

**(b)** With $1/n^2 = 0$, Equation 30.2 reveals that for the shortest wavelength

$$\frac{1}{\lambda} = (1.097 \times 10^7 \text{ m}^{-1})\left(\frac{1}{2^2} - 0\right) = 2.743 \times 10^6 \text{ m}^{-1} \quad \text{or} \quad \boxed{\lambda = 365 \text{ nm}}$$

Equations 30.1–30.3 are useful because they reproduce the wavelengths that hydrogen atoms radiate. However, these equations are empirical and provide no insight as to *why* certain wavelengths are radiated and others are not. It was the great Danish physicist, Niels Bohr (1885–1962), who provided the first model of the atom that predicted the discrete wavelengths emitted by atomic hydrogen. Bohr's model started us on the way toward understanding how the structure of the atom restricts the radiated wavelengths to certain values. In 1922 Bohr received the Nobel Prize in physics for his accomplishment.

## 30.3 The Bohr Model of the Hydrogen Atom

In 1913 Bohr presented a model that led to equations such as Balmer's for the wavelengths that the hydrogen atom radiates. Bohr's theory begins with Rutherford's picture of an atom as a nucleus surrounded by electrons moving in circular orbits. In his theory, Bohr made a number of assumptions and combined the new quantum ideas of Planck and Einstein with the traditional description of a particle in uniform circular motion.

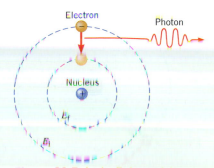

**Figure 30.5** In the Bohr model, a photon is emitted when the electron drops from a larger, higher-energy orbit (energy $= E_i$) to a smaller, lower-energy orbit (energy $= E_f$).

Adopting Planck's idea of quantized energy levels (see Section 29.2), Bohr hypothesized that in a hydrogen atom there can be only certain values of the total energy (electron kinetic energy plus potential energy). These allowed energy levels correspond to different orbits for the electron as it moves around the nucleus, the larger orbits being associated with larger total energies. Figure 30.5 illustrates two of the orbits. In addition, Bohr assumed that an electron in one of these orbits *does not* radiate electromagnetic waves. For this reason, the orbits are called **stationary orbits** or **stationary states**. Bohr recognized that radiationless orbits violated the laws of physics, as they were then known. But the assumption of such orbits was necessary because the traditional laws indicated that an electron radiates electromagnetic waves as it accelerates around a circular path, and the loss of the energy carried by the waves would lead to the collapse of the orbit.

To incorporate Einstein's photon concept (see Section 29.3), Bohr theorized that a photon is emitted only when the electron *changes* orbits from a larger one with a higher energy to a smaller one with a lower energy, as Figure 30.5 indicates. But how do electrons get into the higher-energy orbits in the first place? They get there by picking up energy when atoms collide, which happens more often when a gas is heated, or by acquiring energy when a high voltage is applied to a gas.

When an electron in an initial orbit with a larger energy $E_i$ changes to a final orbit with a smaller energy $E_f$, the emitted photon has an energy of $E_i - E_f$, consistent with the law of conservation of energy. But according to Einstein, the energy of a photon is $hf$, where $f$ is its frequency and $h$ is Planck's constant. As a result, we find that

$$E_i - E_f = hf \tag{30.4}$$

Since the frequency of an electromagnetic wave is related to the wavelength by $f = c/\lambda$, Bohr could use Equation 30.4 to determine the wavelengths radiated by a hydrogen atom. First, however, he had to derive expressions for the energies $E_i$ and $E_f$.

### THE ENERGIES AND RADII OF THE BOHR ORBITS

To derive an expression for the quantized energy levels, Bohr brought together concepts from the older classical physics and the newer modern physics that emerged at the beginning of the twentieth century.

For an electron of mass $m$ and speed $v$ in an orbit of radius $r$, the total energy is the kinetic energy (KE $= \frac{1}{2}mv^2$) of the electron plus the electric potential energy EPE. The potential energy is the product of the charge $(-e)$ on the electron and the electric potential produced by the positive nuclear charge, in accord with Equation 19.3. We assume that the nucleus contains $Z$ protons,* for a total nuclear charge of $+Ze$. The electric potential at a distance $r$ from a point charge of $+Ze$ is given as $+kZe/r$ by Equation 19.6, where $k = 8.988 \times 10^9$ N·m²/C². The electric potential energy is, then, EPE $= (-e)(+kZe/r)$. Consequently, the total energy $E$ of the atom is

$$E = \text{KE} + \text{EPE}$$

$$= \frac{1}{2}mv^2 - \frac{kZe^2}{r} \tag{30.5}$$

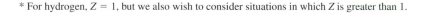

* For hydrogen, $Z = 1$, but we also wish to consider situations in which $Z$ is greater than 1.

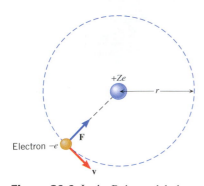

**Figure 30.6** In the Bohr model, the electron is in uniform circular motion around the nucleus. The centripetal force **F** is the electrostatic force of attraction that the positive nuclear charge exerts on the electron.

But a centripetal force of magnitude $mv^2/r$ (Equation 5.3) acts on a particle in uniform circular motion. As Figure 30.6 indicates, the centripetal force is provided by the electrostatic force of attraction **F** that the protons in the nucleus exert on the electron. According to Coulomb's law (Equation 18.1), the magnitude of the electrostatic force is $F = kZe^2/r^2$. Therefore, $mv^2/r = kZe^2/r^2$, or

$$mv^2 = \frac{kZe^2}{r} \tag{30.6}$$

We can use this relation to eliminate the term $mv^2$ from Equation 30.5, with the result that

$$E = \frac{1}{2}\left(\frac{kZe^2}{r}\right) - \frac{kZe^2}{r}$$

$$= -\frac{kZe^2}{2r} \tag{30.7}$$

The total energy of the atom is negative because the negative electric potential energy is larger in magnitude than the positive kinetic energy.

A value for the radius $r$ is needed, if Equation 30.7 is to be useful. To determine $r$, Bohr made an assumption about the orbital angular momentum of the electron. The angular momentum $L$ is given by Equation 9.10 as $L = I\omega$, where $I = mr^2$ is the moment of inertia of the electron moving on its circular path and $\omega = v/r$ is the angular speed of the electron in radians per second. Thus, the angular momentum is $L = (mr^2)(v/r) = mvr$. Bohr conjectured that the angular momentum can assume only certain discrete values; in other words, $L$ is quantized. He postulated that the allowed values are integer multiples of Planck's constant divided by $2\pi$:

$$L_n = mv_n r_n = n\frac{h}{2\pi} \qquad n = 1, 2, 3, \ldots \tag{30.8}$$

Solving this equation for $v_n$ and substituting the result into Equation 30.6 lead to the following expression for the radius $r_n$ of the $n$th Bohr orbit:

$$r_n = \left(\frac{h^2}{4\pi^2 mke^2}\right)\frac{n^2}{Z} \qquad n = 1, 2, 3, \ldots \tag{30.9}$$

With $h = 6.626 \times 10^{-34}$ J·s, $m = 9.109 \times 10^{-31}$ kg, $k = 8.988 \times 10^9$ N·m²/C², and $e = 1.602 \times 10^{-19}$ C, this expression reveals that

**Radii for Bohr orbits (in meters)** 
$$r_n = (5.29 \times 10^{-11} \text{ m})\frac{n^2}{Z} \qquad n = 1, 2, 3, \ldots \tag{30.10}$$

Therefore, in the hydrogen atom ($Z = 1$) the smallest Bohr orbit ($n = 1$) has a radius of $r_1 = 5.29 \times 10^{-11}$ m. This particular value is called the **Bohr radius.** Figure 30.7 shows the first three Bohr orbits for the hydrogen atom.

The expression for the radius of a Bohr orbit can be substituted into Equation 30.7 to show that the corresponding total energy for the $n$th orbit is

$$E_n = -\left(\frac{2\pi^2 mk^2 e^4}{h^2}\right)\frac{Z^2}{n^2} \qquad n = 1, 2, 3, \ldots \tag{30.11}$$

Substituting values for $h$, $m$, $k$, and $e$ into this expression yields

**Bohr energy levels in joules** 
$$E_n = -(2.18 \times 10^{-18} \text{ J})\frac{Z^2}{n^2} \qquad n = 1, 2, 3, \ldots \tag{30.12}$$

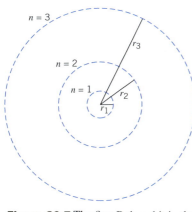

**Figure 30.7** The first Bohr orbit in the hydrogen atom has a radius $r_1 = 5.29 \times 10^{-11}$ m. The second and third Bohr orbits have radii $r_2 = 4r_1$ and $r_3 = 9r_1$, respectively.

Often, atomic energies are expressed in electron volts rather than joules. Since $1.60 \times 10^{-19}$ J = 1 eV, Equation 30.12 can be rewritten as

**Bohr energy levels in electron volts** 
$$E_n = -(13.6 \text{ eV})\frac{Z^2}{n^2} \qquad n = 1, 2, 3, \ldots \tag{30.13}$$

## ENERGY LEVEL DIAGRAMS

It is useful to represent the energy values given by Equation 30.13 on an *energy level diagram*, as in Figure 30.8. In this diagram, which applies to the hydrogen atom ($Z = 1$), the highest energy level corresponds to $n = \infty$ in Equation 30.13 and thus an energy of 0 eV. This is the energy of the atom when the electron is completely removed ($r = \infty$) from the nucleus and is at rest. In contrast, the lowest energy level corresponds to $n = 1$ and has a value of $-13.6$ eV. The lowest energy level is called the *ground state*, to distinguish it from the higher levels, which are called *excited states*. Observe how the energies of the excited states come closer and closer together as $n$ increases.

The electron in a hydrogen atom at room temperature spends most of its time in the ground state. To raise the electron from the ground state ($n = 1$) to the highest possible excited state ($n = \infty$), 13.6 eV of energy must be supplied. Supplying this amount of energy removes the electron from the atom, producing the positive hydrogen ion $H^+$. The energy needed to remove the electron is called the *ionization energy*. Thus, the Bohr model predicts that the ionization energy of atomic hydrogen is 13.6 eV, in excellent agreement with the experimental value. In Example 3 the Bohr model is applied to doubly ionized lithium.

**Figure 30.8** Energy level diagram for the hydrogen atom.

### Example 3  The Ionization Energy of Li$^{2+}$

The Bohr model does not apply when more than one electron orbits the nucleus because it does not account for the electrostatic force that one electron exerts on another. For instance, an electrically neutral lithium atom (Li) contains three electrons in orbit around a nucleus that includes three protons ($Z = 3$), and Bohr's analysis is not applicable. However, the Bohr model can be used for the doubly charged positive ion of lithium (Li$^{2+}$) that results when two electrons are removed from the neutral atom, leaving only one electron to orbit the nucleus. Obtain the ionization energy that is needed to remove the remaining electron from Li$^{2+}$.

**Reasoning** The lithium ion Li$^{2+}$ contains three times the positive nuclear charge that the hydrogen atom contains. Therefore, the orbiting electron is attracted more strongly to the nucleus in Li$^{2+}$ than to the nucleus in the hydrogen atom. As a result, we expect that more energy is required to ionize Li$^{2+}$ than the 13.6 eV required for atomic hydrogen.

**Solution** The Bohr energy levels for Li$^{2+}$ are given by Equation 30.13 with $Z = 3$; $E_n = -(13.6 \text{ eV})(3^2/n^2)$. Therefore, the ground state ($n = 1$) energy is

$$E_1 = -(13.6 \text{ eV})\frac{3^2}{1^2} = -122 \text{ eV}$$

In order to remove the electron from Li$^{2+}$, 122 eV of energy must be supplied: Ionization energy $= 122$ eV. This value for the ionization energy agrees well with the experimental value of 122.4 eV and, as expected, is greater than the 13.6 eV required for atomic hydrogen.

## THE LINE SPECTRA OF THE HYDROGEN ATOM

To predict the wavelengths in the line spectrum of the hydrogen atom, Bohr combined his ideas about atoms (electron orbits are stationary orbits and the angular momentum of an electron is quantized) with Einstein's idea of the photon.

As applied by Bohr, the photon concept is inherent in Equation 30.4, $E_i - E_f = hf$, which states that the frequency $f$ of the photon is proportional to the difference between two energy levels of the hydrogen atom. If we substitute Equation 30.11 for the total energies $E_i$ and $E_f$ into Equation 30.4 and recall from Equation 16.1 that $f = c/\lambda$, we obtain the following result:

$$\frac{1}{\lambda} = \frac{2\pi^2 m k^2 e^4}{h^3 c}(Z^2)\left(\frac{1}{n_f^2} - \frac{1}{n_i^2}\right) \qquad (30.14)$$

$$n_i, n_f = 1, 2, 3, \ldots \quad \text{and} \quad n_i > n_f$$

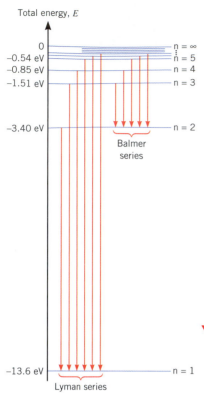

Total energy, $E$

0
$-0.54$ eV    n = ∞   n = 5
$-0.85$ eV    n = 4
$-1.51$ eV    n = 3

$-3.40$ eV    n = 2

Balmer
series

$-13.6$ eV    n = 1

Lyman series

**Figure 30.9** The Lyman and Balmer series of lines in the hydrogen atom spectrum correspond to transitions that the electron makes between higher and lower energy levels, as indicated here.

**Problem solving insight**
In the line spectrum of atomic hydrogen, all lines in a given series (e.g., the Lyman series) are identified by a single value of the quantum number $n_f$ for the *lower energy level into which an electron falls.* Each line in a given series, however, corresponds to a different value of the quantum number $n_i$ for the *higher energy level where an electron originates.*

**The physics of**
absorption lines in the sun's spectrum.

With the known values for $h$, $m$, $k$, $e$, and $c$, it can be seen that $2\pi^2 mk^2 e^4/(h^3 c) = 1.097 \times 10^7$ m$^{-1}$, in agreement with the Rydberg constant $R$ that appears in Equations 30.1–30.3. The agreement between the theoretical and experimental values of the Rydberg constant was a major accomplishment of Bohr's theory.

With $Z = 1$ and $n_f = 1$, Equation 30.14 reproduces Equation 30.1 for the Lyman series. Thus, Bohr's model shows that the Lyman series of lines occurs when electrons make transitions from higher energy levels with $n_i = 2, 3, 4, \ldots$ to the first energy level where $n_f = 1$. Figure 30.9 shows these transitions. Notice that when an electron makes a transition from $n_i = 2$ to $n_f = 1$, the longest wavelength photon in the Lyman series is emitted, since the energy change is the smallest possible. When an electron makes a transition from the highest level where $n_i = \infty$ to the lowest level where $n_f = 1$, the shortest wavelength is emitted, since the energy change is the largest possible. Since the higher energy levels are increasingly close together, the lines in the series become more and more crowded toward the short wavelength limit, as can be seen in Figure 30.4. Figure 30.9 also shows the energy level transitions for the Balmer series, where $n_i = 3, 4, 5, \ldots$, and $n_f = 2$. In the Paschen series (see Figure 30.4) $n_i = 4, 5, 6, \ldots$, and $n_f = 3$. The next example deals further with the line spectrum of the hydrogen atom.

### Example 4   The Brackett Series for Atomic Hydrogen

In the line spectrum of atomic hydrogen there is also a group of lines known as the Brackett series. These lines are produced when electrons, excited to high energy levels, make transitions to the $n = 4$ level. Determine (a) the longest wavelength in this series and (b) the wavelength that corresponds to the transition from $n_i = 6$ to $n_f = 4$. (c) Refer to Figure 24.9 and identify the spectral region in which these lines are found.

**Reasoning** The longest wavelength corresponds to the transition that has the smallest energy change, which is between the $n_i = 5$ and $n_f = 4$ levels in Figure 30.8. The wavelength for this transition, as well as that for the transition from $n_i = 6$ to $n_f = 4$, can be obtained from Equation 30.14.

**Solution**

(a) Using Equation 30.14 with $Z = 1$, $n_i = 5$, and $n_f = 4$, we find that

$$\frac{1}{\lambda} = (1.097 \times 10^7 \text{ m}^{-1})(1^2) \left( \frac{1}{4^2} - \frac{1}{5^2} \right) = 2.468 \times 10^5 \text{ m}^{-1} \quad \text{or} \quad \boxed{\lambda = 4051 \text{ nm}}$$

(b) The calculation here is similar to that in part (a):

$$\frac{1}{\lambda} = (1.097 \times 10^7 \text{ m}^{-1})(1^2) \left( \frac{1}{4^2} - \frac{1}{6^2} \right) = 3.809 \times 10^5 \text{ m}^{-1} \quad \text{or} \quad \boxed{\lambda = 2625 \text{ nm}}$$

(c) According to Figure 24.9, these lines lie in the $\boxed{\text{infrared region}}$ of the spectrum.

The various lines in the hydrogen atom spectrum are produced when electrons change from higher to lower energy levels and photons are emitted. Consequently, the spectral lines are called *emission lines.* Electrons can also make transitions in the reverse direction, from lower to higher levels, in a process known as absorption. In this case, an atom absorbs a photon that has precisely the energy needed to produce the transition. Thus, if photons with a continuous range of wavelengths pass through a gas and then are analyzed with a grating spectroscope, a series of dark *absorption lines* appear in the continuous spectrum. The dark lines indicate the wavelengths that have been removed by the absorption process. Such absorption lines can be seen in Figure 30.3 in the spectrum of the sun, where they are called Fraunhofer lines, after their discoverer. They are due to atoms, located in the outer and cooler layers of the sun, that absorb radiation coming from the interior. The interior portion of the sun emits a continuous spectrum of wavelengths, since it is too hot for individual atoms to retain their structures.

The Bohr model provides a great deal of insight into atomic structure. However, this model is now known to be oversimplified and has been superseded by a more detailed picture provided by quantum mechanics and the Schrödinger equation (see Section 30.5).

# 30.4 De Broglie's Explanation of Bohr's Assumption about Angular Momentum

Of all the assumptions Bohr made in his model of the hydrogen atom, perhaps the most puzzling is the one about the angular momentum of the electron [$L_n = mv_n r_n = nh/(2\pi)$; $n = 1, 2, 3, \ldots$]. Why should the angular momentum have only those values that are integer multiples of Planck's constant divided by $2\pi$? In 1923, ten years after Bohr's work, de Broglie pointed out that his own theory for the wavelength of a moving particle could provide an answer to this question.

In de Broglie's way of thinking, the electron in its circular Bohr orbit must be pictured as a particle wave. And like waves traveling on a string, particle waves can lead to standing waves under resonant conditions. Section 17.5 discusses these conditions for a string. Standing waves form when the total distance traveled by a wave down the string and back is one wavelength, two wavelengths, or any integer number of wavelengths. The total distance around a Bohr orbit of radius $r$ is the circumference of the orbit or $2\pi r$. By the same reasoning, then, the condition for standing particle waves for the electron in a Bohr orbit would be

$$2\pi r = n\lambda \qquad n = 1, 2, 3, \ldots$$

where $n$ is the number of whole wavelengths that fit into the circumference of the circle. But according to Equation 29.8 the de Broglie wavelength of the electron is $\lambda = h/p$, where $p$ is the magnitude of the electron's momentum. If the speed of the electron is much less than the speed of light, the momentum is $p = mv$, and the condition for standing particle waves becomes $2\pi r = nh/(mv)$. A rearrangement of this result gives

$$mvr = n\frac{h}{2\pi} \qquad n = 1, 2, 3, \ldots$$

which is just what Bohr assumed for the angular momentum of the electron. As an example, Figure 30.10 illustrates the standing particle wave on a Bohr orbit for which $2\pi r = 4\lambda$.

De Broglie's explanation of Bohr's assumption about angular momentum emphasizes an important fact—namely, that particle waves play a central role in the structure of the atom. Moreover, the theoretical framework of quantum mechanics provides the basis for determining the wave function $\Psi$ (Greek letter Psi) that represents a particle wave. The next section deals with the picture that quantum mechanics gives for atomic structure, a picture that supersedes the Bohr model. In any case, the Bohr expression for the energy levels (Equation 30.11) can be applied when a single electron orbits the nucleus, whereas the theoretical framework of quantum mechanics can be applied, in principle, to atoms that contain an arbitrary number of electrons.

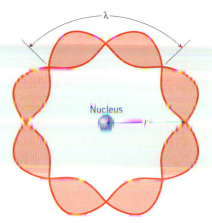

**Figure 30.10** De Broglie suggested standing particle waves as an explanation for Bohr's angular momentum assumption. Here, a standing particle wave is illustrated on a Bohr orbit where four de Broglie wavelengths fit into the circumference of the orbit.

# 30.5 The Quantum Mechanical Picture of the Hydrogen Atom

The picture of the hydrogen atom that quantum mechanics and the Schrödinger equation provide differs in a number of ways from the Bohr model. The Bohr model uses a single integer number $n$ to identify the various electron orbits and the associated energies. Because this number can have only discrete values, rather than a continuous range of values, $n$ is called a **quantum number.** In contrast, quantum mechanics reveals that four different quantum numbers are needed to describe each state of the hydrogen atom. These four are described below.

1. **The principal quantum number $n$.** As in the Bohr model, this number determines the total energy of the atom and can have only integer values: $n = 1, 2, 3, \ldots$. In fact, the Schrödinger equation predicts* that the energy of the hydrogen atom is identical to the energy obtained from the Bohr model: $E_n = -(13.6 \text{ eV}) Z^2/n^2$.

* This prediction requires that small relativistic effects and small interactions within the atom be ignored, and assumes that the hydrogen atom is not located in an external magnetic field.

2. **The orbital quantum number $\ell$.** This number determines the angular momentum of the electron due to its orbital motion. The values that $\ell$ can have depend on the value of $n$, and only the following integers are allowed:

$$\ell = 0, 1, 2, \ldots, (n - 1)$$

For instance, if $n = 1$, the orbital quantum number can have only the value $\ell = 0$, but if $n = 4$, the values $\ell = 0, 1, 2$, and 3 are possible. The magnitude $L$ of the angular momentum of the electron is

$$L = \sqrt{\ell(\ell + 1)} \, \frac{h}{2\pi} \qquad (30.15)$$

3. **The magnetic quantum number $m_\ell$.** The word "magnetic" is used here because an externally applied magnetic field influences the energy of the atom, and this quantum number is used in describing the effect. Since the effect was discovered by the Dutch physicist Pieter Zeeman (1865–1943), it is known as the *Zeeman effect.* When there is no external magnetic field, $m_\ell$ plays no role in determining the energy. In either event, the magnetic quantum number determines the component of the angular momentum along a specific direction, which is called the $z$ direction by convention. The values that $m_\ell$ can have depend on the value of $\ell$, with only the following positive and negative integers being permitted:

$$m_\ell = -\ell, \ldots, -2, -1, 0, +1, +2, \ldots, +\ell$$

For example, if the orbital quantum number is $\ell = 2$, then the magnetic quantum number can have the values $m_\ell = -2, -1, 0, -1$, and $+2$. The component $L_z$ of the angular momentum in the $z$ direction is

$$L_z = m_\ell \frac{h}{2\pi} \qquad (30.16)$$

4. **The spin quantum number $m_s$.** This number is needed because the electron has an intrinsic property called spin angular momentum. Loosely speaking, we can view the electron as spinning while it orbits the nucleus, analogous to the way the earth spins as it orbits the sun. There are two possible values for the spin quantum number of the electron:

$$m_s = +\tfrac{1}{2} \quad \text{or} \quad m_s = -\tfrac{1}{2}$$

Sometimes the phrases "spin up" and "spin down" are used to refer to the directions of the spin angular momentum associated with the values for $m_s$.

Table 30.1 summarizes the four quantum numbers that are needed to describe each state of the hydrogen atom. One set of values for $n$, $\ell$, $m_\ell$, and $m_s$ corresponds to one state. As the principal quantum number $n$ increases, the number of possible combinations of the four quantum numbers rises rapidly, as Example 5 illustrates.

**Table 30.1** *Quantum Numbers for the Hydrogen Atom*

| Name | Symbol | Allowed Values |
|---|---|---|
| Principal quantum number | $n$ | $1, 2, 3, \ldots$ |
| Orbital quantum number | $\ell$ | $0, 1, 2, \ldots, (n - 1)$ |
| Magnetic quantum number | $m_\ell$ | $-\ell, \ldots, -2, -1, 0, +1, +2, \ldots, +\ell$ |
| Spin quantum number | $m_s$ | $-\tfrac{1}{2}, +\tfrac{1}{2}$ |

## Example 5 Quantum Mechanical States of the Hydrogen Atom

▼

Determine the number of possible states for the hydrogen atom when the principal quantum number is (a) $n = 1$ and (b) $n = 2$.

**Reasoning** Each different combination of the four quantum numbers summarized in Table 30.1 corresponds to a different state. We begin with the value for $n$ and find the allowed values for $\ell$. Then, for each $\ell$ value we find the possibilities for $m_\ell$. Finally, $m_s$ may be $+\frac{1}{2}$ or $-\frac{1}{2}$ for each group of values for $n$, $\ell$, and $m_\ell$.

**Solution**

(a) The diagram below shows the possibilities for $\ell$, $m_\ell$, and $m_s$ when $n = 1$:

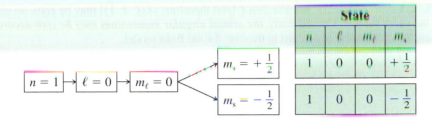

| State | | | |
|---|---|---|---|
| $n$ | $\ell$ | $m_\ell$ | $m_s$ |
| 1 | 0 | 0 | $+\frac{1}{2}$ |
| 1 | 0 | 0 | $-\frac{1}{2}$ |

Thus, there are two different states for the hydrogen atom. In the absence of an external magnetic field, these two states have the same energy, since they have the same value of $n$.

(b) When $n = 2$, there are eight possible combinations for the values of $n$, $\ell$, $m_\ell$, and $m_s$, as the diagram below indicates:

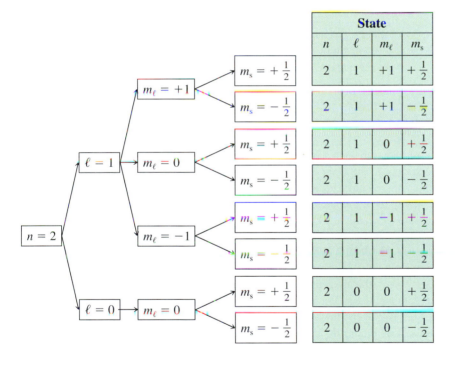

| State | | | |
|---|---|---|---|
| $n$ | $\ell$ | $m_\ell$ | $m_s$ |
| 2 | 1 | +1 | $+\frac{1}{2}$ |
| 2 | 1 | +1 | $-\frac{1}{2}$ |
| 2 | 1 | 0 | $+\frac{1}{2}$ |
| 2 | 1 | 0 | $-\frac{1}{2}$ |
| 2 | 1 | -1 | $+\frac{1}{2}$ |
| 2 | 1 | -1 | $-\frac{1}{2}$ |
| 2 | 0 | 0 | $+\frac{1}{2}$ |
| 2 | 0 | 0 | $-\frac{1}{2}$ |

With the same value of $n = 2$, all eight states have the same energy when there is no external magnetic field.

Quantum mechanics provides a more accurate picture of atomic structure than does the Bohr model. It is important to realize that the two pictures differ substantially, as Conceptual Example 6 illustrates.

## *Conceptual Example 6*  The Bohr Model Versus Quantum Mechanics

Consider two hydrogen atoms. There are no external magnetic fields present, and the electron in each atom has the same energy. According to the Bohr model and to quantum mechanics, is it possible for the electrons in these atoms (a) to have zero orbital angular momentum and (b) to have different orbital angular momenta?

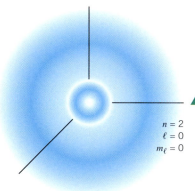

Most probable distance for the electron

$n = 1$
$\ell = 0$
$m_\ell = 0$

**Figure 30.11** The electron probability cloud for the ground state ($n = 1$, $\ell = 0$, $m_\ell = 0$) of the hydrogen atom.

**Reasoning and Solution**

**(a)** In both the Bohr model and quantum mechanics, the energy is proportional to $1/n^2$, according to Equation 30.13, where $n$ is the principal quantum number. Moreover, the value of $n$ may be $n = 1, 2, 3, \ldots$ , and may not be zero. *In the Bohr model, the fact that n may not be zero means that it is not possible for the orbital angular momentum to be zero* because the angular momentum is proportional to $n$, according to Equation 30.8. In the quantum mechanical picture the magnitude of the orbital angular momentum is proportional to $\sqrt{\ell(\ell + 1)}$, as given by Equation 30.15. Here, $\ell$ is the orbital quantum number and may take on the values $\ell = 0, 1, 2, \ldots , (n - 1)$. We note that $\ell$ [and therefore $\sqrt{\ell(\ell + 1)}$] may be zero, no matter what the value for $n$ is. Consequently, *the orbital angular momentum may be zero according to quantum mechanics,* in contrast to the case for the Bohr model.

**(b)** If the electrons have the same energy, they have the same value for the principal quantum number $n$. *In the Bohr model, this means that they cannot have different values for the orbital angular momentum $L_n$,* since $L_n = nh/(2\pi)$, according to Equation 30.8. In quantum mechanics, the energy is also determined by $n$ when external magnetic fields are absent, but the orbital angular momentum is determined by $\ell$. Since $\ell = 0, 1, 2, \ldots , (n - 1)$, different values of $\ell$ are compatible with the same value of $n$. For instance, if $n = 2$ for both electrons, one of them could have $\ell = 0$, while the other could have $\ell = 1$. *According to quantum mechanics, then, the electrons could have different orbital angular momenta, even though they have the same energy.*

The following table summarizes the discussion from parts (a) and (b):

|  | Bohr Model | Quantum Mechanics |
|---|---|---|
| (a) For a given $n$, can the angular momentum ever be zero? | No | Yes |
| (b) For a given $n$, can the angular momentum have different values? | No | Yes |

**Related Homework:** *Problem 26*

---

According to the Bohr model, the $n$th orbit is a circle of radius $r_n$, and every time the position of the electron in this orbit is measured, the electron is found exactly at a distance $r_n$ away from the nucleus. This simplistic picture is now known to be incorrect, and the quantum mechanical picture of the atom has replaced it. Suppose the electron is in a quantum mechanical state for which $n = 1$, and we imagine making a number of measurements of the electron's position with respect to the nucleus. We would find that its position is uncertain, in the sense that there is a probability of finding the electron sometimes very near the nucleus, sometimes very far from the nucleus, and sometimes at intermediate locations. The probability is determined by the wave function $\Psi$, as Section 29.5 discusses. We can make a three-dimensional picture of our findings by marking a dot at each location where the electron is found. More dots occur at places where the probability of finding the electron is higher, and after a sufficient number of measurements, a picture of the quantum mechanical state emerges. Figure 30.11 shows the spatial distribution for an electron in a state for which $n = 1$, $\ell = 0$, and $m_\ell = 0$. This picture is constructed from so many measurements that the individual dots are no longer visible but have merged to form a kind of probability "cloud" whose density changes gradually from place to place. The dense regions indicate places where the probability of finding the electron is higher, and the less dense regions indicate places where the probability is lower. Also indicated in Figure 30.11 is the radius where quantum mechanics predicts the greatest probability per unit radial distance of finding the electron in the $n = 1$ state. This radius matches exactly the radius of $5.29 \times 10^{-11}$ m found for the first Bohr orbit.

For a principal quantum number of $n = 2$, the probability clouds are different than for $n = 1$. In fact, more than one cloud shape is possible because with $n = 2$ the orbital quantum number can be either $\ell = 0$ or $\ell = 1$. Although the value of $\ell$ does not affect the energy of the hydrogen atom, the value does have a significant effect on the shape of the probability clouds. Figure 30.12$a$ shows the cloud for $n = 2$, $\ell = 0$, and $m_\ell = 0$. Part $b$ of

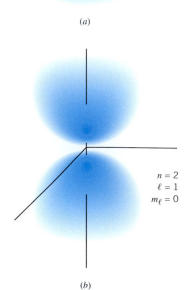

$n = 2$
$\ell = 0$
$m_\ell = 0$

*(a)*

$n = 2$
$\ell = 1$
$m_\ell = 0$

*(b)*

**Figure 30.12** The electron probability clouds for the hydrogen atom when (a) $n = 2$, $\ell = 0$, $m_\ell = 0$ and (b) $n = 2$, $\ell = 1$, $m_\ell = 0$.

the drawing shows that when $n = 2$, $\ell = 1$, and $m_\ell = 0$, the cloud has a two-lobe shape with the nucleus at the center between the lobes. For larger values of $n$, the probability clouds become increasingly complex and are spread out over larger volumes of space.

The probability cloud picture of the electron in a hydrogen atom is very different from the well-defined orbit of the Bohr model. The fundamental reason for this difference is to be found in the Heisenberg uncertainty principle, as Conceptual Example 7 discusses.

## Conceptual Example 7
### The Uncertainty Principle and the Hydrogen Atom

In the Bohr model of the hydrogen atom, the electron in the ground state ($n = 1$) is in an orbit that has a radius of exactly $5.29 \times 10^{-11}$ m. Furthermore, as calculated in Example 3 in Chapter 18, the speed of the electron in this orbit is exactly $2.18 \times 10^6$ m/s. Considering the Heisenberg uncertainty principle, is this a realistic picture of atomic structure?

**Reasoning and Solution** The Bohr model indicates that the electron is located exactly at a radius of $5.29 \times 10^{-11}$ m, so the uncertainty $\Delta y$ in its radial position is zero. We will now show that an uncertainty of $\Delta y = 0$ m is not consistent with the Heisenberg principle, so that the Bohr picture is not realistic. What does the Heisenberg principle [Equation 29.10, $(\Delta p_y)(\Delta y) \geq h/(4\pi)$] say about $\Delta y$? According to the principle, the minimum uncertainty in the radial position of the electron is $\Delta y = h/(4\pi \Delta p_y)$, where $\Delta p_y$ is the uncertainty in the momentum. The magnitude of the electron's momentum is the product of its mass $m$ and speed $v$, so that $\Delta p_y = \Delta(mv) = m \Delta v$. It seems reasonable to assume that the uncertainty $\Delta v$ in the speed should be less than $2.18 \times 10^6$ m/s. This means that the speed is somewhere between zero and twice the value known to apply to the ground state in the Bohr model. A much larger uncertainty would mean that the electron could have so much energy that it would not likely remain in orbit. With $\Delta v = 2.18 \times 10^6$ m/s and $m = 9.11 \times 10^{-31}$ kg for the mass of the electron, we find that the minimum uncertainty in the radial position of the electron is $\Delta y = h/(4\pi m \Delta v) = 2.7 \times 10^{-11}$ m. This uncertainty is quite large, because it is about one-half of the Bohr radius. According to the uncertainty principle, then, the radial position of the electron can be anywhere from approximately one-half to one and a half times the Bohr radius. *The single-radius orbit of the Bohr model does not correctly represent this aspect of reality at the atomic level.* Quantum mechanics, however, does correctly represent it in terms of a probability cloud picture of atomic structure.

# 30.6 The Pauli Exclusion Principle and the Periodic Table of the Elements

Except for hydrogen, all electrically neutral atoms contain more than one electron, the number being given by the atomic number $Z$ of the element. In addition to being attracted by the nucleus, the electrons repel each other. This repulsion contributes to the total energy of a multiple-electron atom. As a result, the one-electron energy expression for hydrogen, $E_n = -(13.6 \text{ eV}) Z^2/n^2$, does not apply to other neutral atoms. However, the simplest approach for dealing with a multiple-electron atom still uses the four quantum numbers $n$, $\ell$, $m_\ell$, and $m_s$.

Detailed quantum mechanical calculations reveal that the energy level of each state of a multiple-electron atom depends on both the principal quantum number $n$ and the orbital quantum number $\ell$. Figure 30.13 illustrates that the energy generally increases as $n$ increases, but there are exceptions, as the drawing indicates. Furthermore, for a given $n$, the energy also increases as $\ell$ increases.

In a multiple-electron atom, all electrons with the same value of $n$ are said to be in the same *shell*. Electrons with $n = 1$ are in a single shell (sometimes called the K shell), electrons with $n = 2$ are in another shell (the L shell), those with $n = 3$ are in a third shell (the M shell), and so on. Those electrons with the same values for both $n$ and $\ell$ are often referred to as being in the same *subshell*. The $n = 1$ shell consists of a single $\ell = 0$ subshell. The

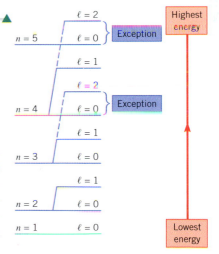

**Figure 30.13** When there is more than one electron in an atom, the total energy of a given state depends on the principal quantum number $n$ and the orbital quantum number $\ell$. Generally, the energy increases with increasing $n$ and, for a fixed $n$, with increasing $\ell$. There are exceptions to the general rule, however, as indicated here. For clarity, levels for $n = 6$ and higher are not shown.

$n = 2$ shell has two subshells, one with $\ell = 0$ and one with $\ell = 1$. Similarly, the $n = 3$ shell has three subshells, one with $\ell = 0$, one with $\ell = 1$, and one with $\ell = 2$.

In the hydrogen atom near room temperature, the electron spends most of its time in the lowest energy level or ground state—namely, in the $n = 1$ shell. Similarly, when an atom contains more than one electron and is near room temperature, the electrons spend most of their time in the lowest energy levels possible. The lowest energy state for an atom is called the *ground state.* However, when a multiple-electron atom is in its ground state, not every electron is in the $n = 1$ shell in general because the electrons obey a principle discovered by the Austrian physicist Wolfgang Pauli (1900–1958).

■ **THE PAULI EXCLUSION PRINCIPLE**

No two electrons in an atom can have the same set of values for the four quantum numbers $n$, $\ell$, $m_\ell$, and $m_s$.

Suppose two electrons in an atom have three quantum numbers that are identical: $n = 3$, $m_\ell = 1$, and $m_s = -\frac{1}{2}$. According to the exclusion principle, it is not possible for each to have $\ell = 2$, for example, since each would then have the same four quantum numbers. Each electron must have a different value for $\ell$ (for instance, $\ell = 1$ and $\ell = 2$) and, consequently, would be in a different subshell. With the aid of the Pauli exclusion principle, we can determine which energy levels are occupied by the electrons in an atom in its ground state, as the next example demonstrates.

### Example 8  Ground States of Atoms

Determine which of the energy levels in Figure 30.13 are occupied by the electrons in the ground state of hydrogen (1 electron), helium (2 electrons), lithium (3 electrons), beryllium (4 electrons), and boron (5 electrons).

**Reasoning** In the ground state of an atom the electrons are in the lowest available energy levels. Consistent with the Pauli exclusion principle, they fill those levels "from the bottom up"—that is, from the lowest to the highest energy.

**Solution** As the colored dot (●) in Figure 30.14 indicates, the electron in the hydrogen atom (H) is in the $n = 1$, $\ell = 0$ subshell, which has the lowest possible energy. A second electron is present in the helium atom (He), and both electrons can have the quantum numbers $n = 1$, $\ell = 0$, and $m_\ell = 0$. However, in accord with the Pauli exclusion principle, each electron must have a different spin quantum number, $m_s = +\frac{1}{2}$ for one electron and $m_s = -\frac{1}{2}$ for the other. Thus, the drawing shows both electrons in the lowest energy level.

The third electron that is present in the lithium atom (Li) would violate the exclusion principle if it were also in the $n = 1$, $\ell = 0$ subshell, no matter what the value for $m_s$ is. Thus, the $n = 1$, $\ell = 0$ subshell is filled when occupied by two electrons. With this level filled, the $n = 2$, $\ell = 0$ subshell becomes the next lowest energy level available and is where the third electron of lithium is found (see Figure 30.14). In the beryllium atom (Be), the fourth electron is in the $n = 2$, $\ell = 0$ subshell, along with the third electron. This is possible, since the third and fourth electrons can have different values for $m_s$.

With the first four electrons in place as just discussed, the fifth electron in the boron atom (B) cannot fit into the $n = 1$, $\ell = 0$ or the $n = 2$, $\ell = 0$ subshell without violating the exclusion principle. Therefore, the fifth electron is found in the $n = 2$, $\ell = 1$ subshell, which is the next available energy level with the lowest energy, as Figure 30.14 indicates. For this electron, $m_\ell$ can be $-1$, 0, or $+1$, and $m_s$ can be $+\frac{1}{2}$ or $-\frac{1}{2}$ in each case. However, in the absence of an external magnetic field, all six of these possibilities correspond to the same energy.

**Figure 30.14** The electrons (●) in the ground state of an atom fill the available energy levels "from the bottom up"—that is, from the lowest to the highest energy, consistent with the Pauli exclusion principle. The ranking of the energy levels in this figure is meant to apply for a given atom only.

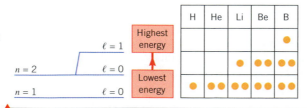

Because of the Pauli exclusion principle, there is a maximum number of electrons that can fit into an energy level or subshell. Example 8 shows that the $n = 1$, $\ell = 0$ subshell can hold at most two electrons. The $n = 2$, $\ell = 1$ subshell, however, can hold six electrons because with $\ell = 1$, there are three possibilities for $m_\ell$ ($-1$, 0, and $+1$), and for each of these, the value of $m_s$ can be $+\frac{1}{2}$ or $-\frac{1}{2}$. In general, $m_\ell$ can have the values 0, $\pm 1$, $\pm 2$, . . . , $\pm\ell$, for $2\ell + 1$ possibilities. Since each of these can be combined with two possibilities for $m_s$, the total number of different combinations for $m_\ell$ and $m_s$ is $2(2\ell + 1)$. This, then, is the maximum number of electrons the $\ell$th subshell can hold, as Figure 30.15 summarizes.

For historical reasons, there is a widely used convention in which each subshell of an atom is referred to by a letter rather than by the value of its orbital quantum number $\ell$. For instance, an $\ell = 0$ subshell is called an s subshell. An $\ell = 1$ subshell and an $\ell = 2$ subshell are known as p and d subshells, respectively. The higher values of $\ell = 3$, 4, and so on, are referred to as f, g, and so on, in alphabetical sequence, as Table 30.2 indicates.

This convention of letters is used in a shorthand notation that is convenient for indicating simultaneously the principal quantum number $n$, the orbital quantum number $\ell$, and the number of electrons in the $n$, $\ell$ subshell. An example of this notation follows.

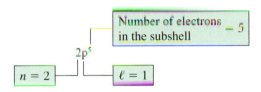

With this notation, the arrangement or configuration of the electrons in an atom can be specified efficiently. For instance, in Example 8, we found that the electron configuration for boron has two electrons in the $n = 1$, $\ell = 0$ subshell, two in the $n = 2$, $\ell = 0$ subshell, and one in the $n = 2$, $\ell = 1$ subshell. In shorthand notation this arrangement is expressed as $1s^2\,2s^2\,2p^1$. Table 30.3 gives the ground-state electron configurations written in this fashion for elements containing up to thirteen electrons. The first five entries are those worked out in Example 8.

Each entry in the periodic table of the elements often includes the ground-state electronic configuration, as Figure 30.16 illustrates for argon. To save space, only the configuration of the outermost electrons and unfilled subshells is specified, using the shorthand notation just discussed. Originally the periodic table was developed by the Russian chemist Dmitri Mendeleev (1834–1907) on the basis that certain groups of elements

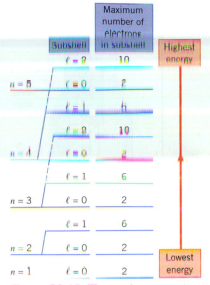

**Figure 30.15** The maximum number of electrons that the $\ell$th subshell can hold is $2(2\ell + 1)$.

**Table 30.2  The Convention of Letters Used to Refer to the Orbital Quantum Number**

| Orbital Quantum Number $\ell$ | Letter |
|---|---|
| 0 | s |
| 1 | p |
| 2 | d |
| 3 | f |
| 4 | g |
| 5 | h |

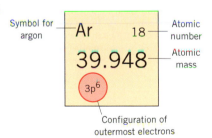

**Figure 30.16** The entries in the periodic table of the elements often include the ground-state configuration of the outermost electrons.

**Table 30.3  Ground-State Electronic Configurations of Atoms**

| Element | Number of Electrons | Configuration of the Electrons |
|---|---|---|
| Hydrogen (H) | 1 | $1s^1$ |
| Helium (He) | 2 | $1s^2$ |
| Lithium (Li) | 3 | $1s^2\,2s^1$ |
| Beryllium (Be) | 4 | $1s^2\,2s^2$ |
| Boron (B) | 5 | $1s^2\,2s^2\,2p^1$ |
| Carbon (C) | 6 | $1s^2\,2s^2\,2p^2$ |
| Nitrogen (N) | 7 | $1s^2\,2s^2\,2p^3$ |
| Oxygen (O) | 8 | $1s^2\,2s^2\,2p^4$ |
| Fluorine (F) | 9 | $1s^2\,2s^2\,2p^5$ |
| Neon (Ne) | 10 | $1s^2\,2s^2\,2p^6$ |
| Sodium (Na) | 11 | $1s^2\,2s^2\,2p^6\,3s^1$ |
| Magnesium (Mg) | 12 | $1s^2\,2s^2\,2p^6\,3s^2$ |
| Aluminum (Al) | 13 | $1s^2\,2s^2\,2p^6\,3s^2\,3p^1$ |

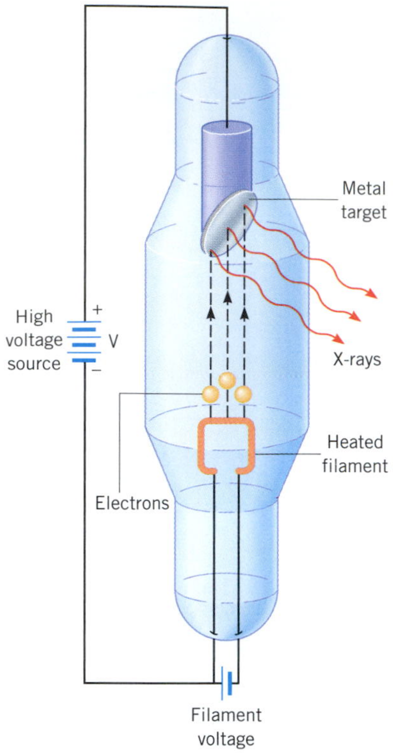

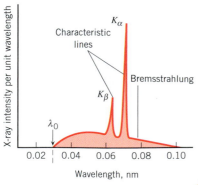

**Figure 30.17** In an X-ray tube, electrons are emitted by a heated filament, accelerate through a large potential difference $V$, and strike a metal target. The X-rays originate when the electrons interact with the target.

**The physics of X-rays.**

**Problem solving insight**
Equation 30.13 for the Bohr energy levels $[E_n = -(13.6 \text{ eV})Z^2/n^2, n = 1]$ can be used in rough calculations of the energy levels involved in the production of $K_\alpha$ X-rays. In this equation, however, the atomic number $Z$ must be reduced by one, to account approximately for the shielding of one K-shell electron by the other K-shell electron.

**Figure 30.18** When a molybdenum target is bombarded with electrons that have been accelerated from rest through a potential difference of 45 000 V, this X-ray spectrum is produced. The vertical axis is not to scale.

exhibit similar chemical properties. There are eight of these groups, plus the transition elements in the middle of the table, including the lanthanide series and the actinide series. The similar chemical properties within a group can be explained on the basis of the configurations of the outer electrons of the elements in the group. Thus, quantum mechanics and the Pauli exclusion principle offer an explanation for the chemical behavior of the atoms. The full periodic table can be found on the inside of the back cover.

## 30.7  X-Rays

X-rays were discovered by Wilhelm K. Roentgen (1845–1923), a Dutch physicist who performed much of his work in Germany. X-rays can be produced when electrons, accelerated through a large potential difference, collide with a metal target made from molybdenum or platinum, for example. The target is contained within an evacuated glass tube, as Figure 30.17 shows. A plot of X-ray intensity per unit wavelength versus the wavelength looks similar to Figure 30.18 and consists of sharp peaks or lines superimposed on a broad continuous spectrum. The sharp peaks are called characteristic lines or *characteristic X-rays* because they are characteristic of the target material. The broad continuous spectrum is referred to as *Bremsstrahlung* (German for "braking radiation") and is emitted when the electrons decelerate or "brake" upon hitting the target.

In Figure 30.18 the characteristic lines are marked $K_\alpha$ and $K_\beta$ because they involve the $n = 1$ or K shell of a metal atom. If an electron with enough energy strikes the target, one of the K-shell electrons can be knocked out. An electron in one of the outer shells can then fall into the K shell, and an X-ray photon is emitted in the process. Example 9 shows that a large potential difference is needed to operate an X-ray tube, so the electrons impinging on the metal target have sufficient energy to generate the characteristic X-rays.

### Example 9  Operating an X-Ray Tube

Strictly speaking, the Bohr model does not apply to multiple-electron atoms, but it can be used to make estimates. Use the Bohr model to estimate the minimum energy that an incoming electron must have to knock a K-shell electron entirely out of an atom in a platinum ($Z = 78$) target in an X-ray tube.

**Reasoning** According to the Bohr model, the energy of a K-shell electron is given by Equation 30.13 with $n = 1$: $E_n = -(13.6 \text{ eV}) Z^2/n^2$. When striking a platinum target, an incoming electron must have at least enough energy to raise the K-shell electron from this low energy level up to the 0-eV level that corresponds to a very large distance from the nucleus. Only then will the incoming electron knock the K-shell electron entirely out of a target atom.

**Solution** The energy of the Bohr $n = 1$ level is

$$E_1 = -(13.6 \text{ eV}) \frac{Z^2}{n^2} = -(13.6 \text{ eV}) \frac{77^2}{1^2} = -8.1 \times 10^4 \text{ eV}$$

In this calculation we have used 77 rather than 78 for the value of $Z$. In so doing, we account approximately for the fact that each of the two K-shell electrons applies a repulsive force to the other. This repulsive force tends to balance the attractive force of one nuclear proton. In effect, one electron shields the other from the force of that proton. Therefore, to raise the K-shell electron up to the 0-eV level, the minimum energy for an incoming electron is $\boxed{8.1 \times 10^4 \text{ eV}}$. One electron volt is the kinetic energy acquired when an electron accelerates from rest through a potential difference of one volt. Thus, a potential difference of 81 000 V must be applied to the X-ray tube.

The $K_\alpha$ line in Figure 30.18 arises when an electron in the $n = 2$ level falls into the vacancy that the impinging electron has created in the $n = 1$ level. Similarly, the $K_\beta$ line arises when an electron in the $n = 3$ level falls to the $n = 1$ level. Example 10 determines an estimate for the $K_\alpha$ wavelength of platinum.

**Example 10   The $K_\alpha$ Characteristic X-Ray for Platinum**

Use the Bohr model to estimate the wavelength of the $K_\alpha$ line in the X-ray spectrum of platinum ($Z = 78$).

**Reasoning** This example is very similar to Example 4, which deals with the emission line spectrum of the hydrogen atom. As in that example, we use Equation 30.14, this time with the initial value of $n$ being $n_i = 2$ and the final value being $n_f = 1$. As in Example 9, a value of 77 rather than 78 is used for $Z$ to account approximately for the shielding effect of the single K-shell electron in canceling out the attraction of one nuclear proton.

**Solution** Using Equation 30.14, we find that

$$\frac{1}{\lambda} = (1.097 \times 10^7 \text{ m}^{-1})(77^2)\left(\frac{1}{1^2} - \frac{1}{2^2}\right) = 4.9 \times 10^{10} \text{ m}^{-1} \quad \text{or}$$

$$\boxed{\lambda = 2.0 \times 10^{-11} \text{ m}}$$

This answer is close to an experimental value of $1.9 \times 10^{-11}$ m.

Another interesting feature of the X-ray spectrum in Figure 30.18 is the sharp cutoff that occurs at a wavelength of $\lambda_0$ on the short-wavelength side of the Bremsstrahlung. This cutoff wavelength is independent of the target material but depends on the energy of the impinging electrons. An impinging electron cannot give up any more than all of its kinetic energy when decelerated by the metal target in an X-ray tube. Thus, at most, an emitted X-ray photon can have an energy equal to the kinetic energy KE of the electron and a frequency given by Equation 29.2 as $f = (\text{KE})/h$, where $h$ is Planck's constant. But the kinetic energy acquired by an electron in accelerating from rest through a potential difference $V$ is $eV$, according to earlier discussions in Section 19.2; $V$ is the potential difference applied across the X-ray tube. Thus, the maximum photon frequency is $f_0 = (eV)/h$. Since $f_0 = c/\lambda_0$, a maximum frequency corresponds to a minimum wavelength, which is the cutoff wavelength $\lambda_0$:

$$\lambda_0 = \frac{hc}{eV} \tag{30.17}$$

Figure 30.18, for instance, assumes a potential difference of 45 000 V, which corresponds to a cutoff wavelength of

$$\lambda_0 = \frac{(6.63 \times 10^{-34} \text{ J} \cdot \text{s})(3.00 \times 10^8 \text{ m/s})}{(1.60 \times 10^{-19} \text{ C})(45 \ 000 \text{ V})} = 2.8 \times 10^{-11} \text{ m}$$

The medical profession began using X-rays for diagnostic purposes almost immediately after their discovery. When a conventional X-ray is obtained, the patient is typically positioned in front of a piece of photographic film, and a single burst of radiation is directed through the patient and onto the film. Since the dense structure of bone absorbs X-rays much more than soft tissue does, a shadow-like picture is recorded on the film. As useful as such pictures are, they have an inherent limitation. The image on the film is a superposition of all the "shadows" that result as the radiation passes through one layer of body material after another. Interpreting which part of a conventional X-ray corresponds to which layer of body material is very difficult.

The technique known as CAT scanning or CT scanning has greatly extended the ability of X-rays to provide images from specific locations within the body. The acronym CAT stands for computerized axial tomography or computer-assisted tomography, and the shorter version CT stands for computerized tomography. In this technique a series of X-ray images are obtained as indicated in Figure 30.19. A number of X-ray beams form a "fanned out" array of radiation and pass simultaneously through the patient. Each of the beams is detected on the other side by a detector, which records the beam intensity. The various intensities are different, depending on the nature of the body material through which the beams have passed. The feature of CAT scanning that leads to dramatic improvements over the conven-

**The physics of** CAT scanning.

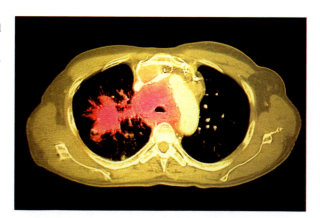

**Figure 30.19** (*a*) In CAT scanning, a "fanned-out" array of X-ray beams is sent through the patient from different orientations. (*b*) A patient being positioned in a CAT scanner. (© Hank Morgan/Photo Researchers)

(*a*)                    (*b*)

tional technique is that the X-ray source can be rotated to different orientations, so that the "fanned-out" array of beams can be sent through the patient from various directions. Figure 30.19*a* singles out two directions for illustration. In reality many different orientations are used, and the intensity of each beam in the array is recorded as a function of orientation. The way in which the intensity of a beam changes from one orientation to another is used as input to a computer. The computer then constructs a highly resolved image of the cross-sectional slice of body material through which the "fan" of radiation has passed. In effect, the CAT scanning technique makes it possible to take an X-ray picture of a cross-sectional "slice" that is perpendicular to the body's long axis. In fact, the word "axial" in the phrase "computerized axial tomography" refers to the body's long axis. Figure 30.20*a* shows a two-dimensional CAT scan of the type discussed here, and part *b* of the figure shows a three-dimensional version.

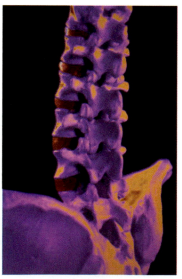

**Figure 30.20** (*a*) This color-enhanced two-dimensional CAT scan of the chest shows a view perpendicular to the spine. The heart is in the center of the image, and the lungs are the dark spaces within the yellow region. The irregular pink shape in the lung on the left side of the image is a cancerous growth. (© Scott Camazine/Photo Researchers) (*b*) This color-enhanced three-dimensional CAT scan shows the pelvis and part of the spine, including the intervertebral discs (red). (© Collection CNRI/Phototake)

## 30.8  *The Laser*

**The physics of the laser.**

The laser is one of the most useful inventions of the twentieth century. Today, there are many types of lasers, and most of them work in a way that depends directly on the quantum mechanical structure of the atom.

When an electron makes a transition from a higher energy state to a lower energy state, a photon is emitted. The emission process can be one of two types, spontaneous or stimulated. In ***spontaneous emission*** (see Figure 30.21*a*), the photon is emitted spontaneously, in a random direction, without external provocation. In ***stimulated emission*** (see Figure 30.21*b*), an incoming photon induces, or stimulates, the electron to change energy levels. To produce stimulated emission, however, the incoming photon must have an energy that exactly matches the difference between the energies of the two levels—namely, $E_i - E_f$. Stimulated emission is similar to a resonant process, in which the incoming photon "jiggles" the electron at just the frequency to which it is particularly sensitive and causes the change between energy levels. This frequency is given by Equation 30.4 as $f = (E_i - E_f)/h$. The operation of lasers depends on stimulated emission.

Stimulated emission has three important features. First, one photon goes in and two photons come out (see Figure 30.21b). In this sense, the process amplifies the number of photons. In fact, this is the origin of the word "laser," which is an acronym for light amplification by the stimulated emission of radiation. Second, the emitted photon travels in the same direction as the incoming photon. Third, the emitted photon is exactly in step with or has the same phase as the incoming photon. In other words, the two electromagnetic waves that these two photons represent are coherent (see Section 17.2) and are locked in step with one another. In contrast, two photons emitted by the filament of an incandescent light bulb are emitted independently. They are not coherent, since one does not stimulate the emission of the other.

Although stimulated emission plays a pivotal role in a laser, other factors are also important. For instance, an external source of energy must be available to excite electrons into higher energy levels. The energy can be provided in a number of ways, including intense flashes of ordinary light and high-voltage discharges. If sufficient energy is delivered to the atoms, more electrons will be excited to a higher energy level than remain in a lower level, a condition known as a *population inversion*. Figure 30.22 compares a normal energy level population with a population inversion. The population inversions used in lasers involve a higher energy state that is *metastable,* in the sense that electrons remain in it for a much longer period of time than they do in an ordinary excited state ($10^{-3}$ s versus $10^{-8}$ s, for example). The requirement of a metastable higher energy state is essential, so that there is more time to enhance the population inversion.

Figure 30.23 shows the widely used helium/neon laser. To sustain the necessary population inversion, a high voltage is discharged across a low-pressure mixture of 15% helium and 85% neon contained in a glass tube. The laser process begins when an atom, via spontaneous emission, emits a photon parallel to the axis of the tube. This photon, via stimulated emission, causes another atom to emit two photons parallel to the tube axis. These two photons, in turn, stimulate two more atoms, yielding four photons. Four yield eight, and so on, in a kind of avalanche. To ensure that more and more photons are created by stimulated emission, both ends of the tube are silvered to form mirrors that

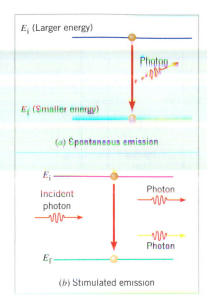

$E_i$ (Larger energy)

Photon

$E_f$ (Smaller energy)

(a) Spontaneous emission

$E_i$

Incident photon

Photon

$E_f$

Photon

(b) Stimulated emission

**Figure 30.21** (*a*) Spontaneous emission of a photon occurs when the electron (●) makes an unprovoked transition from a higher to a lower energy level, the photon departing in a random direction. (*b*) Stimulated emission of a photon occurs when an incoming photon with the correct energy induces an electron to change energy levels, the emitted photon traveling in the same direction as the incoming photon.

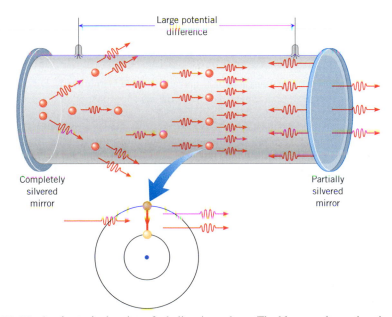

Large potential difference

Completely silvered mirror

Partially silvered mirror

**Figure 30.23** A schematic drawing of a helium/neon laser. The blow-up shows the stimulated emission that occurs when an electron in a neon atom is induced to change from a higher- to a lower-energy level.

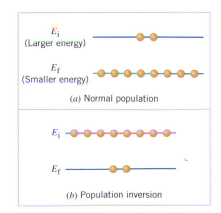

$E_i$ (Larger energy)

$E_f$ (Smaller energy)

(a) Normal population

$E_i$

$E_f$

(b) Population inversion

**Figure 30.22** (*a*) In a normal situation at room temperature, most of the electrons in atoms are found in a lower or ground-state energy level. (*b*) If an external energy source is provided to excite electrons into a higher energy level, a population inversion can be created in which more electrons are in the higher level than in the lower level.

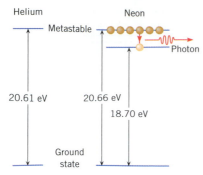

**Figure 30.24** These energy levels are involved in the operation of a helium/neon laser.

reflect the photons back and forth through the helium/neon mixture. One end is only partially silvered, however, so that some of the photons can escape from the tube to form the laser beam. When the stimulated emission involves only a single pair of energy levels, the output beam has a single frequency or wavelength and is said to be monochromatic.

A laser beam is also exceptionally narrow. The width is determined by the size of the opening through which the beam exits, and very little spreading-out occurs, except that due to diffraction around the edges of the opening. A laser beam does not spread much because any photons emitted at an angle with respect to the tube axis are quickly reflected out the sides of the tube by the silvered ends (see Figure 30.23). These ends are carefully arranged to be perpendicular to the tube axis. Since all the power in a laser beam can be confined to a narrow region, the intensity, or power per unit area, can be quite large.

Figure 30.24 shows the pertinent energy levels for a helium/neon laser. By coincidence, helium and neon have nearly identical metastable higher energy states, respectively located 20.61 and 20.66 eV above the ground state. The high-voltage discharge across the gaseous mixture excites electrons in helium atoms to the 20.61-eV state. Then, when an excited helium atom collides inelastically with a neon atom, the 20.61 eV of energy is given to an electron in the neon atom, along with 0.05 eV of kinetic energy from the moving atoms. As a result, the electron in the neon atom is raised to the 20.66-eV state. In this fashion, a population inversion is sustained in the neon, relative to an energy level that is 18.70 eV above the ground state. In producing the laser beam, stimulated emission causes electrons in neon to drop from the 20.66-eV level to the 18.70-eV level. The energy change of 1.96 eV corresponds to a wavelength of 633 nm, which is in the red region of the visible spectrum.

The helium/neon laser is not the only kind of laser. There are many different types, including the ruby laser, the argon-ion laser, the carbon dioxide laser, the gallium arsenide solid-state laser, and chemical dye lasers. Depending on the type and whether the laser operates continuously or in pulses, the available beam power ranges from milliwatts to megawatts. Since lasers provide coherent monochromatic electromagnetic radiation that can be confined to an intense narrow beam, they are useful in a wide variety of situations. Today they are used to reproduce music in compact disc players, to weld parts of automobile frames together, to measure distances accurately in surveying, to transmit telephone conversations and other forms of communication over long distances, and to study molecular structure.

# Concept Summary

This summary presents an abridged version of the chapter, including the important equations and all available learning aids. For convenient reference, the learning aids (including the text's examples) are placed next to or immediately after the relevant equation or discussion. The following learning aids may be found on-line at **www.wiley.com/college/cutnell**:

| | |
|---|---|
| **Interactive LearningWare** examples are solved according to a five-step interactive format that is designed to help you develop problem-solving skills. | **Concept Simulations** are animated versions of text figures or animations that illustrate important concepts. You can control parameters that affect the display, and we encourage you to experiment. |
| **Interactive Solutions** offer specific models for certain types of problems in the chapter homework. The calculations are carried out interactively. | **Self-Assessment Tests** include both qualitative and quantitative questions. Extensive feedback is provided for both incorrect and correct answers, to help you evaluate your understanding of the material. |

| **Topic** | **Discussion** | **Learning Aids** |
|---|---|---|

### 30.1 Rutherford Scattering and the Nuclear Atom

**Nuclear atom**

The idea of a nuclear atom originated in 1911 as a result of experiments by Ernest Rutherford in which α particles were scattered by a thin metal foil. The phrase "nuclear atom" refers to the fact that an atom consists of a small, positively charged nucleus surrounded at relatively large distances by a number of    **Example 1**

| Topic | Discussion | Learning Aids |
|---|---|---|

electrons, whose total negative charge equals the positive nuclear charge when the atom is electrically neutral.

### 30.2 Line Spectra

**Line spectrum**

A line spectrum is a series of discrete electromagnetic wavelengths emitted by the atoms of a low-pressure gas that is subjected to a sufficiently high potential difference. Certain groups of discrete wavelengths are referred to as "series." The following equations can be used to determine the wavelengths in three of the series that are found in the line spectrum of atomic hydrogen:

**Lyman series**

$$\textit{Lyman series} \quad \frac{1}{\lambda} = R\left(\frac{1}{1^2} - \frac{1}{n^2}\right) \quad n = 2, 3, 4, \ldots \quad (30.1)$$

**Balmer series**

$$\textit{Balmer series} \quad \frac{1}{\lambda} = R\left(\frac{1}{2^2} - \frac{1}{n^2}\right) \quad n = 3, 4, 5, \ldots \quad (30.2) \quad \textbf{Example 2}$$

**Paschen series**

$$\textit{Paschen series} \quad \frac{1}{\lambda} = R\left(\frac{1}{3^2} - \frac{1}{n^2}\right) \quad n = 4, 5, 6, \ldots \quad (30.3)$$

The constant term $R$ is called the Rydberg constant and has the value $R = 1.097 \times 10^7 \text{ m}^{-1}$.

### 30.3 The Bohr Model of the Hydrogen Atom

The Bohr model applies to atoms or ions that have only a single electron orbiting a nucleus containing $Z$ protons. This model assumes that the electron exists

**Stationary orbit**

in circular orbits that are called stationary orbits because the electron does not radiate electromagnetic waves while in them. According to this model, a photon is emitted only when an electron changes from an orbit with a higher energy $E_i$ to an orbit with a lower energy $E_f$. The orbital energies and the photon frequency $f$ are related according to

**Photon frequency**

$$E_i - E_f = hf \quad (30.4)$$

where $h$ is Planck's constant. The model also assumes that the orbital angular momentum $L_n$ of the electron can only have the following discrete values:

**Angular momentum**

$$L_n = n\frac{h}{2\pi} \quad n = 1, 2, 3, \ldots \quad (30.8)$$

With these assumptions, it can be shown that the $n$th Bohr orbit has a radius $r_n$ and is associated with a total energy $E_n$ of

**Orbital radius**

$$r_n = (5.29 \times 10^{-11} \text{ m})\frac{n^2}{Z} \quad n = 1, 2, 3, \ldots \quad (30.10)$$

**Total orbital energy**

$$E_n = -(13.6 \text{ eV})\frac{Z^2}{n^2} \quad n = 1, 2, 3, \ldots \quad (30.13)$$

**Ionization energy**

The ionization energy is the energy needed to remove an electron completely from an atom. The Bohr model predicts that the wavelengths comprising the line spectrum emitted by a hydrogen atom can be calculated from

**Example 3**

**Interactive LearningWare 30.1**

**Line spectrum wavelengths**

$$\frac{1}{\lambda} = RZ^2\left(\frac{1}{n_f^2} - \frac{1}{n_i^2}\right) \quad (30.14)$$

$$n_i, n_f = 1, 2, 3, \ldots \quad \text{and} \quad n_i > n_f$$

**Example 4**

**Concept Simulation 30.1**

### 30.4 De Broglie's Explanation of Bohr's Assumption about Angular Momentum

Louis de Broglie proposed that the electron in a circular Bohr orbit should be considered as a particle wave and that standing particle waves around the orbit offer an explanation of the angular momentum assumption in the Bohr model.

### 30.5 The Quantum Mechanical Picture of the Hydrogen Atom

Quantum mechanics describes the hydrogen atom in terms of the following four quantum numbers:

**Principal quantum number**

(1) the principal quantum number $n$, which can have the integer values $n = 1, 2, 3, \ldots$

**Examples 5, 6**

| Topic | Discussion | Learning Aids |
|---|---|---|
| **Orbital quantum number** | (2) the orbital quantum number $\ell$, which can have the integer values $\ell = 0,$ $1, 2, \ldots , (n-1)$ | |
| **Magnetic quantum number** | (3) the magnetic quantum number $m_\ell$, which can have the positive and negative values $m_\ell = -\ell, \ldots , -2, -1, 0, +1, +2, \ldots , +\ell$ | **Interactive Solution 30.51** |
| **Spin quantum number** | (4) the spin quantum number $m_s$, which, for an electron, can be either $m_s = +\frac{1}{2}$ or $m_s = -\frac{1}{2}$ | |
| | According to quantum mechanics, an electron does not reside in a circular orbit but, rather, has some probability of being found at various distances from the nucleus. | **Example 7** |

 Use **Self-Assessment Test 30.1** to evaluate your understanding of Sections 30.1–30.5.

### 30.6 The Pauli Exclusion Principle and the Periodic Table of the Elements

| Topic | Discussion | Learning Aids |
|---|---|---|
| **Pauli exclusion principle** | The Pauli exclusion principle states that no two electrons in an atom can have the same set of values for the four quantum numbers $n$, $\ell$, $m_\ell$, and $m_s$. This principle determines the way in which the electrons in multiple-electron atoms are distributed into shells (defined by the value of $n$) and subshells (defined by the value of $\ell$). | **Example 8** |
| **Convention of letters** | The following notation is used to refer to the orbital quantum numbers: s denotes $\ell = 0$, p denotes $\ell = 1$, d denotes $\ell = 2$, f denotes $\ell = 3$, g denotes $\ell = 4$, h denotes $\ell = 5$, and so on. | |
| | The arrangement of the periodic table of the elements is related to the Pauli exclusion principle. | |

### 30.7 X-Rays

| Topic | Discussion | Learning Aids |
|---|---|---|
| **X-rays** | X-rays are electromagnetic waves emitted when high-energy electrons strike a metal target contained within an evacuated glass tube. The emitted X-ray spectrum of wavelengths consists of sharp "peaks" or "lines", called characteristic X-rays, superimposed on a broad continuous range of wavelengths called Bremsstrahlung. The $K_\alpha$ characteristic X-ray is emitted when an electron in the $n = 2$ level (L shell) drops into a vacancy in the $n = 1$ level (K shell). Similarly, the $K_\beta$ characteristic X-ray is emitted when an electron in the $n = 3$ level (M shell) drops into a vacancy in the $n = 1$ level (K shell). The minimum wavelength, or cutoff wavelength $\lambda_0$, of the Bremsstrahlung is determined by the kinetic energy of the electrons striking the target in the X-ray tube, according to | **Example 9** **Example 10** **Interactive LearningWare 30.2** **Interactive Solution 30.35** |
| **Characteristic X-rays** | | |
| **Bremsstrahlung** | | |
| **Cutoff wavelength** | | |

$$\lambda_0 = \frac{hc}{eV} \qquad (30.17)$$

where $h$ is Planck's constant, $c$ is the speed of light in a vacuum, $e$ is the magnitude of the charge on an electron, and $V$ is the potential difference applied across the X-ray tube.

### 30.8 The Laser

 Use **Self-Assessment Test 30.2** to evaluate your understanding of Sections 30.6–30.8.

# Problems

*In working these problems, ignore relativistic effects.*

 **ssm** Solution is in the Student Solutions Manual.     **www** Solution is available on the World Wide Web at www.wiley.com/college/cutnell
**V** This icon represents a biomedical application.

### Section 30.1 Rutherford Scattering and the Nuclear Atom

**1. ssm** The nucleus of the hydrogen atom has a radius of about $1 \times 10^{-15}$ m. The electron is normally at a distance of about $5.3 \times 10^{-11}$ m from the nucleus. Assuming the hydrogen atom is a sphere with a radius of $5.3 \times 10^{-11}$ m, find (a) the volume of the atom, (b) the volume of the nucleus, and (c) the percentage of the volume of the atom that is occupied by the nucleus.

**2.** Review Conceptual Example 1 and use the information therein as an aid in working this problem. Suppose you're building a scale model of the hydrogen atom, and the nucleus is represented by a ball of radius 3.2 cm (somewhat smaller than a baseball). How many miles away (1 mi = $1.61 \times 10^5$ cm) should the electron be placed?

**3.** The nucleus of a hydrogen atom is a single proton, which has a radius of about $1.0 \times 10^{-15}$ m. The single electron in a hydrogen atom normally orbits the nucleus at a distance of $5.3 \times 10^{-11}$ m. What is the ratio of the density of the hydrogen nucleus to the density of the complete hydrogen atom?

**4.** In a Rutherford scattering experiment a target nucleus has a diameter of $1.4 \times 10^{-14}$ m. The incoming $\alpha$ particle has a mass of $6.64 \times 10^{-27}$ kg. What is the kinetic energy of an $\alpha$ particle that has a de Broglie wavelength equal to the diameter of the target nucleus? Ignore relativistic effects.

**\* 5. ssm** The nucleus of a copper atom contains 29 protons and has a radius of $4.8 \times 10^{-15}$ m. How much work (in electron volts) is done by the electric force as a proton is brought from infinity, where it is at rest, to the "surface" of a copper nucleus?

**\* 6.** There are $Z$ protons in the nucleus of an atom, where $Z$ is the atomic number of the element. An $\alpha$ particle carries a charge of $+2e$. In a scattering experiment, an $\alpha$ particle, heading directly toward a nucleus in a metal foil, will come to a halt when all the particle's kinetic energy is converted to electric potential energy. In such a situation, how close will an $\alpha$ particle with a kinetic energy of $5.0 \times 10^{-13}$ J come to a gold nucleus ($Z = 79$)?

### Section 30.2 Line Spectra
### Section 30.3 The Bohr Model of the Hydrogen Atom

**7. ssm** It is possible to use electromagnetic radiation to ionize atoms. To do so, the atoms must absorb the radiation, the photons of which must have enough energy to remove an electron from an atom. What is the longest radiation wavelength that can be used to ionize the ground-state hydrogen atom?

**8. Concept Simulation 30.1** at www.wiley.com/college/cutnell reviews the concepts on which the solution to this problem depends. The electron in a hydrogen atom is in the first excited state, when the electron acquires an additional 2.86 eV of energy. What is the quantum number $n$ of the state into which the electron moves?

**9.** In the line spectrum of atomic hydrogen there is also a group of lines known as the Pfund series. These lines are produced when electrons, excited to high energy levels, make transitions to the $n = 5$ level. Determine (a) the longest wavelength and (b) the shortest wavelength in this series. (c) Refer to Figure 24.9 and identify the region of the electromagnetic spectrum in which these lines are found.

**10.** A singly ionized helium atom (He$^+$) has only one electron in orbit about the nucleus. What is the radius of the ion when it is in the second excited state?

**11. ssm** For a doubly ionized lithium atom Li$^{2+}$ ($Z = 3$), what is the principal quantum number of the state in which the electron has the same total energy as a ground-state electron has in the hydrogen atom?

**12.** In the hydrogen atom the radius of orbit B is sixteen times greater than the radius of orbit A. The total energy of the electron in orbit A is $-3.40$ eV. What is the total energy of the electron in orbit B?

**13.** Consider the Bohr energy expression (Equation 30.13) as it applies to singly ionized helium He$^+$ ($Z = 2$) and doubly ionized lithium Li$^{2+}$ ($Z = 3$). This expression predicts equal electron energies for these two species for certain values of the quantum number $n$ (the quantum number is different for each species). For quantum numbers less than or equal to 9, what are the lowest three energies (in electron volts) for which the helium energy level is equal to the lithium energy level?

**14.** A hydrogen atom is in the ground state. It absorbs energy and makes a transition to the $n = 3$ excited state. The atom returns to the ground state by emitting two photons. What are their wavelengths?

**\* 15. ssm** A wavelength of 410.2 nm is emitted by the hydrogen atoms in a high-voltage discharge tube. What are the initial and final values of the quantum number $n$ for the energy level transition that produces this wavelength?

**\* 16.** Doubly ionized lithium Li$^{2+}$ ($Z = 3$) and triply ionized beryllium Be$^{3+}$ ($Z = 4$) each emit a line spectrum. For a certain series of lines in the lithium spectrum, the shortest wavelength is 40.5 nm. For the same series of lines in the beryllium spectrum, what is the shortest wavelength?

**\* 17. ssm** The Bohr model can be applied to singly ionized helium He$^+$ ($Z = 2$). Using this model, consider the series of lines that is produced when the electron makes a transition from higher energy levels into the $n_f = 4$ level. Some of the lines in this series lie in the visible region of the spectrum (380–750 nm). What are the values of $n_i$ for the energy levels from which the electron makes the transitions corresponding to these lines?

**\* 18. Interactive LearningWare 30.1** at www.wiley.com/college/cutnell reviews the concepts that play roles in this problem. A hydrogen atom emits a photon that has momentum with a magnitude of $5.452 \times 10^{-27}$ kg·m/s. This photon is emitted because the electron in the atom falls from a higher energy level into the $n = 1$ level. What is the quantum number of the level from which the electron falls? Use a value of $6.626 \times 10^{-34}$ J·s for Planck's constant.

**\*\* 19.** A diffraction grating is used in the first order to separate the wavelengths in the Balmer series of atomic hydrogen. (Section 27.7 discusses diffraction gratings.) The grating and an observation screen (see Figure 27.30) are separated by a distance of 81.0 cm. You may assume that $\theta$ is small, so $\sin\theta \approx \theta$ when radian measure is used for $\theta$. How many lines per centimeter should the grating have, so that the longest and the next-to-the-longest wavelengths in the series will be separated by 3.00 cm on the screen?

**\*\* 20.** (a) Derive an expression for the speed of the electron in the $n$th Bohr orbit, in terms of $Z$, $n$, and the constants $k$, $e$, and $h$. For the hydrogen atom, determine the speed in (b) the $n = 1$ orbit and (c) the $n = 2$ orbit. (d) Generally, when speeds are less than one-tenth the speed of light, the effects of special relativity can be ignored. Are the speeds found in (b) and (c) consistent with ignoring relativistic effects in the Bohr model?

## Section 30.5 The Quantum Mechanical Picture of the Hydrogen Atom

**21. ssm www** The principal quantum number for an electron in an atom is $n = 6$, and the magnetic quantum number is $m_\ell = 2$. What possible values for the orbital quantum number $\ell$ could this electron have?

**22.** A hydrogen atom is in its second excited state. Determine, according to quantum mechanics, (a) the total energy (in eV) of the atom, (b) the magnitude of the maximum angular momentum the electron can have in this state, and (c) the maximum value that the $z$ component $L_z$ of the angular momentum can have.

**23.** The orbital quantum number for the electron in a hydrogen atom is $\ell = 5$. What is the smallest possible value (algebraically) for the total energy of this electron? Give your answer in electron volts.

**24.** It is known that the possible values for the magnetic quantum number $m_\ell$ are $-4$, $-3$, $-2$, $-1$, $0$, $+1$, $+2$, $+3$, and $+4$. Determine the orbital quantum number and the smallest possible value of the principal quantum number.

**\* 25.** The total orbital angular momentum of the electron in a hydrogen atom has a magnitude of $L = 3.66 \times 10^{-34}$ J·s. In the quantum mechanical picture of the atom, what values can the angular momentum component $L_z$ have?

**\* 26.** Review Conceptual Example 6 as background for this problem. For the hydrogen atom, the Bohr model and quantum mechanics both give the same value for the energy of the $n$th state. However, they do not give the same value for the orbital angular momentum $L$. (a) For $n = 1$, determine the values of $L$ [in units of $h/(2\pi)$] predicted by the Bohr model and quantum mechanics. (b) Repeat part (a) for $n = 3$, noting that quantum mechanics permits more than one value of $\ell$ when the electron is in the $n = 3$ state.

**\*\* 27. ssm www** An electron is in the $n = 5$ state. What is the smallest possible value for the angle between the $z$ component of the orbital angular momentum and the orbital angular momentum?

## Section 30.6 The Pauli Exclusion Principle and the Periodic Table of the Elements

**28.** Two of the three electrons in a lithium atom have quantum numbers of $n = 1$, $\ell = 0$, $m_\ell = 0$, $m_s = +\frac{1}{2}$ and $n = 1$, $\ell = 0$, $m_\ell = 0$, $m_s = -\frac{1}{2}$. What quantum numbers can the third electron have if the atom is in (a) its ground state and (b) its first excited state?

**29.** In the style indicated in Table 30.3, write down the ground-state electronic configuration of manganese Mn ($Z = 25$). Refer to Figure 30.15 for the order in which the subshells fill.

**30.** Referring to Figure 30.15 for the order in which the subshells fill and following the style used in Table 30.3, determine the ground-state electronic configuration for cadmium Cd ($Z = 48$).

**31. ssm** When an electron makes a transition between energy levels of an atom, there are no restrictions on the initial and final values of the principal quantum number $n$. According to quantum mechanics, however, there is a rule that restricts the initial and final values of the orbital quantum number $\ell$. This rule is called a *selection rule* and states that $\Delta\ell = \pm 1$. In other words, when an electron makes a transition between energy levels, the value of $\ell$ can only increase or decrease by one. The value of $\ell$ may not remain the same or increase or decrease by more than one. According to this rule, which of the following energy level transitions are allowed: (a) 2s → 1s, (b) 2p → 1s, (c) 4p → 2p, (d) 4s → 2p, and (e) 3d → 3s?

**\* 32.** What is the atom with the smallest atomic number that contains the same number of electrons in its s subshells as it does in its d subshell? Refer to Figure 30.15 for the order in which the subshells fill.

## Section 30.7 X-Rays

**33. ssm** Molybdenum has an atomic number of $Z = 42$. Using the Bohr model, estimate the wavelength of the $K_\alpha$ X-ray.

**34.** The atomic number of lead is $Z = 82$. According to the Bohr model, what is the energy (in joules) of a $K_\alpha$ X-ray photon?

**35. Interactive Solution 30.35** at **www.wiley.com/college/cutnell** provides one model for solving problems such as this one. An X-ray tube is being operated at a potential difference of 52.0 kV. What is the Bremsstrahlung wavelength that corresponds to 35.0% of the kinetic energy with which an electron collides with the metal target in the tube?

**36.** The $K_\beta$ characteristic X-ray line for tungsten has a wavelength of $1.84 \times 10^{-11}$ m. What is the difference in energy between the two energy levels that give rise to this line? Express the answer in (a) joules and (b) electron volts.

**\* 37. ssm** An X-ray tube contains a silver ($Z = 47$) target. The high voltage in this tube is increased from zero. Using the Bohr model, find the value of the voltage at which the $K_\alpha$ X-ray just appears in the X-ray spectrum.

**\* 38.** The highest-energy X-rays produced by an X-ray tube have a wavelength of $1.20 \times 10^{-10}$ m. What is the speed of the electrons in Figure 30.17 just before they strike the metal target?

## Section 30.8 The Laser

**39.** A pulsed laser emits light in a series of short pulses, each having a duration of 25.0 ms. The average power of each pulse is 5.00 mW, and the wavelength of the light is 633 nm. Find (a) the energy of each pulse and (b) the number of photons in each pulse.

**40.** A laser peripheral iridotomy is a procedure for treating an eye condition known as narrow-angle glaucoma, in which pressure buildup in the eye can lead to loss of vision. A neodymium YAG laser (wavelength = 1064 nm) is used in the procedure to punch a tiny hole in the peripheral iris, thereby relieving the pressure buildup. In one application the laser delivers $4.1 \times 10^{-3}$ J of energy to the iris in creating the hole. How many photons does the laser deliver?

**41. ssm www** A laser is used in eye surgery to weld a detached retina back into place. The wavelength of the laser beam is 514 nm, and the power is 1.5 W. During surgery, the laser beam is turned on for 0.050 s. During this time, how many photons are emitted by the laser?

**42.** The dye laser used in the treatment of congenital capillary malformations known as port-wine stains has a wavelength of 585 nm. A carbon dioxide laser produces a wavelength of $1.06 \times 10^{-5}$ m. What is the minimum number of photons that the carbon dioxide laser must produce to deliver at least as much or more energy to a target as does a single photon from the dye laser?

**\* 43.** Fusion is the process by which the sun produces energy. One experimental technique for creating controlled fusion utilizes a solid-state laser that emits a wavelength of 1060 nm and can produce a power of $1.0 \times 10^{14}$ W for a pulse duration of $1.1 \times 10^{-11}$ s. In contrast, the helium/neon laser used at the checkout counter in a bar-code scanner emits a wavelength of 633 nm and produces a power of about $1.0 \times 10^{-3}$ W. How long (in days) would the helium/neon laser have to operate to produce the same number of photons that the solid-state laser produces in $1.1 \times 10^{-11}$ s?

# Chapter 31  Nuclear Physics and Radioactivity

## 31.1  Nuclear Structure

Atoms consist of electrons in orbit about a central nucleus. As we have seen in Chapter 30, the electron orbits are quantum mechanical in nature and have interesting characteristics. Little has been said about the nucleus, however. Since the nucleus is interesting in its own right, we now consider it in greater detail.

The nucleus of an atom consists of neutrons and protons, collectively referred to as **nucleons.** The **neutron,** discovered in 1932 by the English physicist James Chadwick (1891–1974), carries no electrical charge and has a mass slightly larger than that of a proton (see Table 31.1).

The number of protons in the nucleus is different in different elements and is given by the **atomic number Z.** In an electrically neutral atom, the number of nuclear protons equals the number of electrons in orbit around the nucleus. The number of neutrons in the nucleus is $N$. The total number of protons and neutrons is referred to as the **atomic mass number A** because the total nuclear mass is *approximately* equal to $A$ times the mass of a single nucleon:

$$\underbrace{A}_{\substack{\text{Number of protons} \\ \text{and neutrons} \\ \text{(atomic mass number} \\ \text{or nucleon number)}}} = \underbrace{Z}_{\substack{\text{Number of} \\ \text{protons} \\ \text{(atomic number)}}} + \underbrace{N}_{\substack{\text{Number of} \\ \text{neutrons}}} \qquad (31.1)$$

Sometimes $A$ is also called the **nucleon number.** A shorthand notation is often used to specify $Z$ and $A$ along with the chemical symbol for the element. For instance, the nuclei of all naturally occurring aluminum atoms have $A = 27$, and the atomic number for aluminum is $Z = 13$. In shorthand notation, then, the aluminum nucleus is specified as $^{27}_{13}\text{Al}$. The number of neutrons in an aluminum nucleus is $N = A - Z = 14$. In general, for an element whose chemical symbol is X, the symbol for the nucleus is

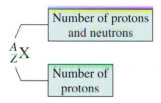

For a proton the symbol is $^1_1\text{H}$, since the proton is the nucleus of a hydrogen atom. A neutron is denoted by $^1_0\text{n}$. In the case of an electron we use $^{\ 0}_{-1}\text{e}$, where $A = 0$ because an electron is not composed of protons or neutrons and $Z = -1$ because the electron has a negative charge.

Nuclei that contain the same number of protons, but a different number of neutrons, are known as **isotopes.** Carbon, for example, occurs in nature in two stable forms. In most

**Table 31.1  Properties of Select Particles**

| Particle | Electric Charge (C) | Mass | |
|---|---|---|---|
| | | Kilograms (kg) | Atomic Mass Units (u) |
| Electron | $-1.60 \times 10^{-19}$ | $9.109\,382 \times 10^{-31}$ | $5.485\,799 \times 10^{-4}$ |
| Proton | $+1.60 \times 10^{-19}$ | $1.672\,622 \times 10^{-27}$ | $1.007\,276$ |
| Neutron | $0$ | $1.674\,927 \times 10^{-27}$ | $1.008\,665$ |
| Hydrogen atom | $0$ | $1.673\,534 \times 10^{-27}$ | $1.007\,825$ |

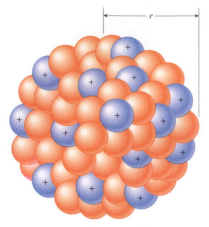

**Figure 31.1** The nucleus is approximately spherical (radius = r) and contains protons (⊕) clustered closely together with neutrons (●).

carbon atoms (98.90%), the nucleus is the $^{12}_{6}\text{C}$ isotope and consists of six protons and six neutrons. A small fraction (1.10%), however, contains nuclei that have six protons and seven neutrons, namely, the $^{13}_{6}\text{C}$ isotope. The percentages given here are the natural abundances of the isotopes. The atomic masses in the periodic table are average atomic masses, taking into account the abundances of the various isotopes.

The protons and neutrons in the nucleus are clustered together to form an approximately spherical region, as Figure 31.1 illustrates. Experiment shows that the radius r of the nucleus depends on the atomic mass number A and is given approximately in meters by

$$r \approx (1.2 \times 10^{-15} \text{ m})A^{1/3} \tag{31.2}$$

The radius of the aluminum nucleus (A = 27), for example, is $r \approx (1.2 \times 10^{-15} \text{ m})27^{1/3} = 3.6 \times 10^{-15}$ m. Equation 31.2 leads to an important conclusion concerning the nuclear density of different atoms, as Conceptual Example 1 discusses.

### Conceptual Example 1  Nuclear Density

▼

It is well known that lead and oxygen contain different atoms and that the density of solid lead is much greater than that of gaseous oxygen. Using Equation 31.2, decide whether the density of the *nucleus* in a lead atom is greater than, approximately equal to, or less than that in an oxygen atom.

**Reasoning and Solution** We know from Equation 11.1 that density is mass divided by volume. The total mass M of a nucleus is approximately equal to the number of nucleons A times the mass m of a single nucleon, since the masses of a proton and a neutron are nearly the same: $M \approx Am$. The values of A are different for lead and oxygen atoms, however. The volume V of the nucleus is approximately spherical and has a radius r, so that $V = \frac{4}{3}\pi r^3$. But $r^3$ is proportional to the number of nucleons A, as we can see by cubing each side of Equation 31.2. Therefore, the volume V is also proportional to A. Thus, when the total mass M is divided by the volume V, the factor of A appears in both the numerator and denominator and is eliminated algebraically from the result, no matter what the value of A is. We find, then, that **the density of the nucleus in a lead atom is approximately the same as it is in an oxygen atom.** In general, Equation 31.2 implies that the nuclear density has nearly the same value in all atoms. The difference in densities between solid lead and gaseous oxygen arises mainly because of the difference in how closely the atoms themselves are packed together in the solid and gaseous phases.

**Related Homework:** *Problems 8, 9*

▲

## 31.2  *The Strong Nuclear Force and the Stability of the Nucleus*

Two positive charges that are as close together as they are in a nucleus repel one another with a very strong electrostatic force. What, then, keeps the nucleus from flying apart? Clearly, some kind of attractive force must hold the nucleus together, since many kinds of naturally occurring atoms contain stable nuclei. The gravitational force of attraction between nucleons is too weak to counteract the repulsive electric force, so a different type of force must hold the nucleus together. This force is the *strong nuclear force* and is one of only three fundamental forces that have been discovered, fundamental in the sense that all forces in nature can be explained in terms of these three. The gravitational force is also one of these forces, as is the electroweak force (see Section 31.5).

Many features of the strong nuclear force are well known. For example, it is almost independent of electric charge. At a given separation distance, nearly the same nuclear force of attraction exists between two protons, between two neutrons, or between a proton and a neutron. The range of action of the strong nuclear force is extremely short, with the force of attraction being very strong when two nucleons are as close as $10^{-15}$ m and essentially zero at larger distances. In contrast, the electric force between two protons decreases to zero only gradually as the separation distance increases to large values and, therefore, has a relatively long range of action.

The limited range of action of the strong nuclear force plays an important role in the stability of the nucleus. For a nucleus to be stable, the electrostatic repulsion between the protons must be balanced by the attraction between the nucleons due to the strong nuclear force. But one proton repels all other protons within the nucleus, since the electrostatic force has such a long range of action. In contrast, a proton or a neutron attracts only its nearest neighbors via the strong nuclear force. As the number $Z$ of protons in the nucleus increases under these conditions, the number $N$ of neutrons has to increase even more, if stability is to be maintained. Figure 31.2 shows a plot of $N$ versus $Z$ for naturally occurring elements that have stable nuclei. For reference, the plot also includes the straight line that represents the condition $N = Z$. With few exceptions, the points representing stable nuclei fall above this reference line, reflecting the fact that the number of neutrons becomes greater than the number of protons as the atomic number $Z$ increases.

As more and more protons occur in a nucleus, there comes a point when a balance of repulsive and attractive forces cannot be achieved by an increased number of neutrons. Eventually, the limited range of action of the strong nuclear force prevents extra neutrons from balancing the long-range electric repulsion of extra protons. The stable nucleus with the largest number of protons ($Z = 83$) is that of bismuth, $^{209}_{83}$Bi, which contains 126 neutrons. All nuclei with more than 83 protons (e.g., uranium, $Z = 92$) are unstable and spontaneously break apart or rearrange their internal structures as time passes. This spontaneous disintegration or rearrangement of internal structure is called *radioactivity,* first discovered in 1896 by the French physicist Henri Becquerel (1852–1908). Section 31.4 discusses radioactivity in greater detail.

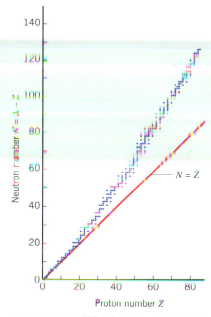

**Figure 31.2** With few exceptions, the naturally occurring stable nuclei have a number $N$ of neutrons that equals or exceeds the number $Z$ of protons. Each dot in this plot represents a stable nucleus.

# 31.3 The Mass Defect of the Nucleus and Nuclear Binding Energy

Because of the strong nuclear force, the nucleons in a stable nucleus are held tightly together. Therefore, energy is required to separate a stable nucleus into its constituent protons and neutrons, as Figure 31.3 illustrates. The more stable the nucleus is, the greater is the amount of energy needed to break it apart. The required energy is called the *binding energy* of the nucleus.

Two ideas that we have studied previously come into play as we discuss the binding energy of a nucleus. These are mass (Section 4.2) and the rest energy of an object (Section 28.6). In Einstein's theory of special relativity, mass and energy are equivalent. A change $\Delta m$ in the mass of a system is equivalent to a change $\Delta E_0$ in the rest energy of the system by an amount $\Delta E_0 = (\Delta m)c^2$, where $c$ is the speed of light in a vacuum. Thus, in Figure 31.3, the binding energy used to disassemble the nucleus appears as extra mass of the separated nucleons. In other words, the sum of the individual masses of the separated protons and neutrons is greater by an amount $\Delta m$ than the mass of the stable nucleus. The difference in mass $\Delta m$ is known as the *mass defect* of the nucleus.

As Example 2 shows, the binding energy of a nucleus can be determined from the mass defect according to Equation 31.3:

$$\text{Binding energy} = (\text{Mass defect})c^2 = (\Delta m)c^2 \qquad (31.3)$$

## Example 2 The Binding Energy of the Helium Nucleus

The most abundant isotope of helium has a $^4_2$He nucleus whose mass is $6.6447 \times 10^{-27}$ kg. For this nucleus, find (a) the mass defect and (b) the binding energy.

**Reasoning** The symbol $^4_2$He indicates that the helium nucleus contains $Z = 2$ protons and $N = 4 - 2 = 2$ neutrons. To obtain the mass defect $\Delta m$, we first determine the sum of the individual masses of the separated protons and neutrons. Then we subtract from this sum the mass of the $^4_2$He nucleus. Finally, we use Equation 31.3 to calculate the binding energy from the value for $\Delta m$.

**Solution**

(a) Using data from Table 31.1, we find that the sum of the individual masses of the nucleons is

$$\underbrace{2(1.6726 \times 10^{-27} \text{ kg})}_{\text{Two protons}} + \underbrace{2(1.6749 \times 10^{-27} \text{ kg})}_{\text{Two neutrons}} = 6.6950 \times 10^{-27} \text{ kg}$$

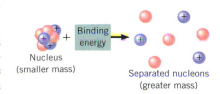

**Figure 31.3** Energy, called the binding energy, must be supplied to break the nucleus apart into its constituent protons and neutrons. Each of the separated nucleons is at rest and out of the range of the forces of the other nucleons.

This value is greater than the mass of the intact $^4_2$He nucleus, and the mass defect is

$$\Delta m = 6.6950 \times 10^{-27} \text{ kg} - 6.6447 \times 10^{-27} \text{ kg} = \boxed{0.0503 \times 10^{-27} \text{ kg}}$$

**(b)** According to Equation 31.3, the binding energy is

$$\frac{\text{Binding}}{\text{energy}} = (\Delta m)c^2 = (0.0503 \times 10^{-27} \text{ kg})(3.00 \times 10^8 \text{ m/s})^2 = 4.53 \times 10^{-12} \text{ J}$$

Usually, binding energies are expressed in energy units of electron volts instead of joules (1 eV = $1.60 \times 10^{-19}$ J):

$$\frac{\text{Binding}}{\text{energy}} = (4.53 \times 10^{-12} \text{ J})\left(\frac{1 \text{ eV}}{1.60 \times 10^{-19} \text{ J}}\right) = 2.83 \times 10^7 \text{ eV} = \boxed{28.3 \text{ MeV}}$$

In this result, one million electron volts is denoted by the unit MeV. The value of 28.3 MeV is more than two million times greater than the energy required to remove an orbital electron from an atom.

---

In calculations such as that in Example 2, it is customary to use the **atomic mass unit** (u) instead of the kilogram. As introduced in Section 14.1, the atomic mass unit is one-twelfth of the mass of a $^{12}_6$C atom of carbon. In terms of this unit, the mass of a $^{12}_6$C atom is exactly 12 u. Table 31.1 also gives the masses of the electron, the proton, and the neutron in atomic mass units. For future calculations, the energy equivalent of one atomic mass unit can be determined by observing that the mass of a proton is $1.6726 \times 10^{-27}$ kg or 1.0073 u, so that

$$1 \text{ u} = (1 \text{ u})\left(\frac{1.6726 \times 10^{-27} \text{ kg}}{1.0073 \text{ u}}\right) = 1.6605 \times 10^{-27} \text{ kg}$$

and

$$\Delta E_0 = (\Delta m)c^2 = (1.6605 \times 10^{-27} \text{ kg})(2.9979 \times 10^8 \text{ m/s})^2 = 1.4924 \times 10^{-10} \text{ J}$$

In electron volts, therefore, one atomic mass unit is equivalent to

$$1 \text{ u} = (1.4924 \times 10^{-10} \text{ J})\left(\frac{1 \text{ eV}}{1.6022 \times 10^{-19} \text{ J}}\right) = 9.315 \times 10^8 \text{ eV} = 931.5 \text{ MeV}$$

Data tables for isotopes give masses in atomic mass units. Typically, however, the given masses are not nuclear masses. They are *atomic masses*—that is, the masses of neutral atoms, including the mass of the orbital electrons. Example 3 deals again with the $^4_2$He nucleus and shows how to take into account the effect of the orbital electrons when using such data to determine binding energies.

### Example 3   The Binding Energy of the Helium Nucleus, Revisited

The atomic mass of $^4_2$He is 4.0026 u, and the atomic mass of $^1_1$H is 1.0078 u. Using atomic mass units instead of kilograms, obtain the binding energy of the $^4_2$He nucleus.

**Reasoning** To determine the binding energy, we calculate the mass defect in atomic mass units and then use the fact that one atomic mass unit is equivalent to 931.5 MeV of energy. The mass of 4.0026 u for $^4_2$He *includes the mass of the two electrons in the neutral helium atom*. To calculate the mass defect, we must subtract 4.0026 u from the sum of the individual masses of the nucleons, including the mass of the electrons. As Figure 31.4 illustrates, the

**Figure 31.4** Data tables usually give the mass of the neutral atom (including the orbital electrons) rather than the mass of the nucleus. When data from such tables are used to determine the mass defect of a nucleus, the mass of the orbital electrons must be taken into account, as this drawing illustrates for the $^4_2$He isotope of helium. See Example 3.

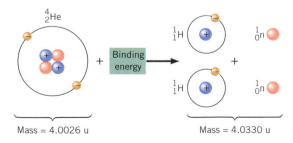

electron mass will be included if the masses of two hydrogen atoms are used in the calculation instead of the masses of two protons. The mass of a $^1_1H$ hydrogen atom is given in Table 31.1 as 1.0078 u, and the mass of a neutron as 1.0087 u.

**Solution** The sum of the individual masses is

$$\underbrace{2(1.0078\ u)}_{\substack{\text{Two hydrogen}\\\text{atoms}}} + \underbrace{2(1.0087\ u)}_{\text{Two neutrons}} = 4.0330\ u$$

The mass defect is $\Delta m = 4.0330\ u - 4.0026\ u = 0.0304\ u$. Since 1 u is equivalent to 931.5 MeV, the binding energy is $\boxed{\text{Binding energy} = 28.3\ \text{MeV}}$, which matches that obtained in Example 2.

To see how the nuclear binding energy varies from nucleus to nucleus, it is necessary to compare the binding energy for each nucleus on a per-nucleon basis. The graph shown in Figure 31.5 shows a plot in which the binding energy divided by the nucleon number $A$ is plotted against the nucleon number itself. In the graph, the peak for the $^4_2He$ isotope of helium indicates that the $^4_2He$ nucleus is particularly stable. The binding energy per nucleon increases rapidly for nuclei with small masses and reaches a maximum of approximately 8.7 MeV/nucleon for a nucleon number of about $A = 60$. For greater nucleon numbers, the binding energy per nucleon decreases gradually. Eventually, the binding energy per nucleon decreases enough so there is insufficient binding energy to hold the nucleus together. Nuclei more massive than the $^{209}_{83}Bi$ nucleus of bismuth are unstable and hence radioactive.

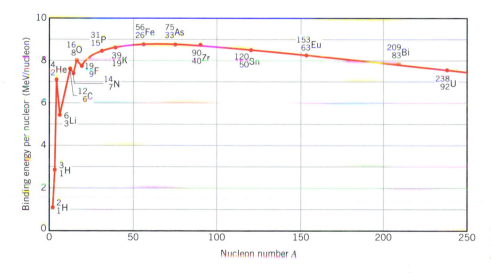

**Figure 31.5** A plot of binding energy per nucleon versus the nucleon number $A$.

## 31.4 Radioactivity

When an unstable or radioactive nucleus disintegrates spontaneously, certain kinds of particles and/or high-energy photons are released. These particles and photons are collectively called "rays." Three kinds of rays are produced by naturally occurring radioactivity: **α rays, β rays,** and **γ rays.** They are named according to the first three letters of the Greek alphabet, alpha ($\alpha$), beta ($\beta$), and gamma ($\gamma$), to indicate the extent of their ability to penetrate matter. $\alpha$ rays are the least penetrating, being blocked by a thin ($\approx 0.01$ mm) sheet of lead, whereas $\beta$ rays penetrate lead to a much greater distance ($\approx 0.1$ mm). $\gamma$ rays are the most penetrating and can pass through an appreciable thickness ($\approx 100$ mm) of lead.

The nuclear disintegration process that produces $\alpha$, $\beta$, and $\gamma$ rays must obey the laws of physics that we have studied in previous chapters. Recall that these laws are called conservation laws because each of them deals with a property (such as mass/energy, electric charge, linear momentum, and angular momentum) that is conserved or does not

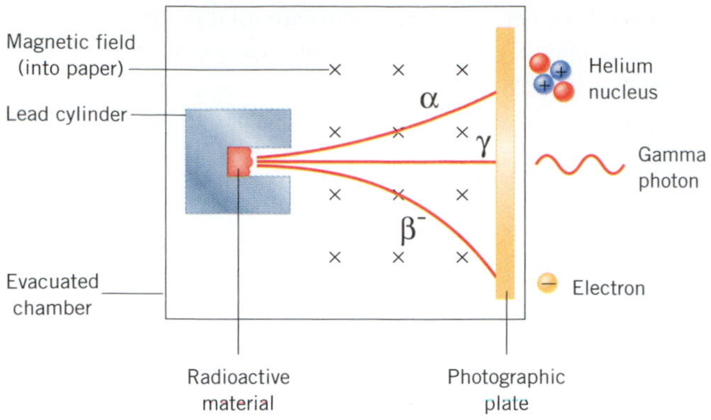

**Figure 31.6** $\alpha$ and $\beta$ rays are deflected by a magnetic field and, therefore, consist of moving charged particles. $\gamma$ rays are not deflected by a magnetic field and, consequently, must be uncharged.

change during a process. To the first four conservation laws we now add a fifth, the conservation of nucleon number. In all radioactive decay processes it has been observed that the number of nucleons (protons plus neutrons) present before the decay is equal to the number of nucleons after the decay. Therefore, the number of nucleons is conserved during a nuclear disintegration. As applied to the disintegration of a nucleus, the conservation laws require that the energy, electric charge, linear momentum, angular momentum, and nucleon number that a nucleus possesses must remain unchanged when it disintegrates into nuclear fragments and accompanying $\alpha$, $\beta$, or $\gamma$ rays.

The three types of radioactivity that occur naturally can be observed in a relatively simple experiment. A piece of radioactive material is placed at the bottom of a narrow hole in a lead cylinder. The cylinder is located within an evacuated chamber, as Figure 31.6 illustrates. A magnetic field is directed perpendicular to the plane of the paper, and a photographic plate is positioned to the right of the hole. Three spots appear on the developed plate, which are associated with the radioactivity of the nuclei in the material. Since moving particles are deflected by a magnetic field only when they are electrically charged, this experiment reveals that two types of radioactivity ($\alpha$ and $\beta$ rays, as it turns out) consist of charged particles, whereas the third type ($\gamma$ rays) does not.

## $\alpha$ DECAY

When a nucleus disintegrates and produces $\alpha$ rays, it is said to undergo **$\alpha$ decay.** Experimental evidence shows that $\alpha$ rays consist of positively charged particles, each one being the $^4_2$He nucleus of helium. Thus, an $\alpha$ particle has a charge of $+2e$ and a nucleon number of $A = 4$. Since the grouping of 2 protons and 2 neutrons in a $^4_2$He nucleus is particularly stable, as we have seen in connection with Figure 31.5, it is not surprising that an $\alpha$ particle can be ejected as a unit from a more massive unstable nucleus.

Figure 31.7 shows the disintegration process for one example of $\alpha$ decay:

$$^{238}_{92}\text{U} \longrightarrow {}^{234}_{90}\text{Th} + {}^4_2\text{He}$$

| Parent nucleus (uranium) | Daughter nucleus (thorium) | $\alpha$ particle (helium nucleus) |

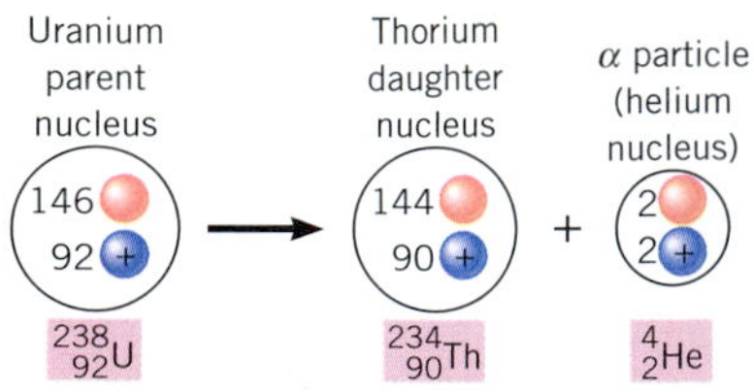

**Figure 31.7** $\alpha$ decay occurs when an unstable parent nucleus emits an $\alpha$ particle and in the process is converted into a different, or daughter, nucleus.

The original nucleus is referred to as the *parent nucleus* (P), and the nucleus remaining after disintegration is called the *daughter nucleus* (D). Upon emission of an $\alpha$ particle, the uranium $^{238}_{92}$U parent is converted into the $^{234}_{90}$Th daughter, which is an isotope of thorium. The parent and daughter nuclei are different, so $\alpha$ decay converts one element into another, a process known as **transmutation.**

Electric charge is conserved during $\alpha$ decay. In Figure 31.7, for instance, 90 of the 92 protons in the uranium nucleus end up in the thorium nucleus, and the remaining 2 protons are carried off by the $\alpha$ particle. The total number of 92, however, is the same before and after disintegration. $\alpha$ decay also conserves the number of nucleons, because the number is the same before (238) and after (234 + 4) disintegration. Consistent with the conservation of electric charge and nucleon number, the general form for $\alpha$ decay is

**α Decay**
$$^{A}_{Z}P \longrightarrow \, ^{A-4}_{Z-2}D \, + \, ^{4}_{2}He$$

| Parent nucleus | Daughter nucleus | α particle (helium nucleus) |

When a nucleus releases an $\alpha$ particle, the nucleus also releases energy. In fact, the energy released by radioactive decay is responsible, in part, for keeping the interior of the earth hot and, in some places, even molten. The following example shows how the conservation of mass/energy can be used to determine the amount of energy released in $\alpha$ decay.

### Example 4   $\alpha$ Decay and the Release of Energy

The atomic mass of uranium $^{238}_{92}U$ is 238.0508 u, that of thorium $^{234}_{90}Th$ is 234.0436 u, and that of an $\alpha$ particle $^{4}_{2}He$ is 4.0026 u. Determine the energy released when $\alpha$ decay converts $^{238}_{92}U$ into $^{234}_{90}Th$.

**Reasoning** Since energy is released during the decay, the combined mass of the $^{234}_{90}Th$ daughter nucleus and the $\alpha$ particle is less than the mass of the $^{238}_{92}U$ parent nucleus. The difference in mass is equivalent to the energy released. We will determine the difference in mass in atomic mass units and then use the fact that 1 u is equivalent to 931.5 MeV.

**Solution** The decay and the masses are shown below:

$$^{238}_{92}U \longrightarrow \, ^{234}_{90}Th \, + \, ^{4}_{2}He$$

$$238.0508 \text{ u} \qquad \underbrace{234.0436 \text{ u} \qquad 4.0026 \text{ u}}_{238.0462 \text{ u}}$$

The decrease in mass is 238.0508 u $-$ 238.0462 u $=$ 0.0046 u. As usual, the masses are atomic masses and include the mass of the orbital electrons. But this causes no error here because the same total number of electrons is included for $^{238}_{92}U$, on the one hand, and for $^{234}_{90}Th$ plus $^{4}_{2}He$, on the other. Since 1 u is equivalent to 931.5 MeV, the released energy is $\boxed{4.3 \text{ MeV}}$.

When $\alpha$ decay occurs as in Example 4, the energy released appears as kinetic energy of the recoiling $^{234}_{90}Th$ nucleus and the $\alpha$ particle, except for a small portion carried away as a $\gamma$ ray. Conceptual Example 5 discusses how the $^{234}_{90}Th$ nucleus and the $\alpha$ particle share in the released energy.

### Conceptual Example 5
### How Energy Is Shared During the $\alpha$ Decay of $^{238}_{92}U$

In Example 4, the energy released by the $\alpha$ decay of $^{238}_{92}U$ is found to be 4.3 MeV. Since this energy is carried away as kinetic energy of the recoiling $^{234}_{90}Th$ nucleus and the $\alpha$ particle, it follows that $KE_{Th} + KE_{\alpha} = 4.3$ MeV. However, $KE_{Th}$ and $KE_{\alpha}$ are not equal. Which particle carries away more kinetic energy, the $^{234}_{90}Th$ nucleus or the $\alpha$ particle?

**Reasoning and Solution** Kinetic energy depends on the mass $m$ and speed $v$ of a particle, since $KE = \frac{1}{2}mv^2$. The $^{234}_{90}Th$ nucleus has a much greater mass than the $\alpha$ particle, and since the kinetic energy is proportional to the mass, it is tempting to conclude that the $^{234}_{90}Th$ nucleus has the greater kinetic energy. This conclusion is not correct, however, since it does not take into account the fact that the $^{234}_{90}Th$ nucleus and the $\alpha$ particle have different speeds after the decay. In fact, we expect the thorium nucleus to recoil with the smaller speed precisely *because* it has the greater mass. The decaying $^{238}_{92}U$ is like a father and his young daughter on ice skates, pushing off against one another. The more massive father recoils with much less speed than the daughter. We can use the principle of conservation of linear momentum to verify our expectation.

As Section 7.2 discusses, the conservation principle states that the total linear momentum of an isolated system remains constant. An isolated system is one for which the vector sum of the external forces acting on the system is zero, and the decaying $^{238}_{92}U$ nucleus fits this description. It is stationary initially, and since momentum is mass times velocity, its initial momentum is zero. In its final form, the system consists of the $^{234}_{90}Th$ nucleus and the $\alpha$

particle and has a final total momentum of $m_{\text{Th}}v_{\text{Th}} + m_{\alpha}v_{\alpha}$. According to momentum conservation, the initial and final values of the total momentum of the system must be the same, so that $m_{\text{Th}}v_{\text{Th}} + m_{\alpha}v_{\alpha} = 0$. Solving this equation for the velocity of the thorium nucleus, we find that $v_{\text{Th}} = -m_{\alpha}v_{\alpha}/m_{\text{Th}}$. Since $m_{\text{Th}}$ is much greater than $m_{\alpha}$, we can see that the speed of the thorium nucleus is less than the speed of the $\alpha$ particle. Moreover, the kinetic energy depends on the square of the speed and only the first power of the mass. As a result of its much greater speed, ***the $\alpha$ particle has the greater kinetic energy.***

**Related Homework:** *Problem 24*

**The physics of**
**radioactivity and smoke detectors.**

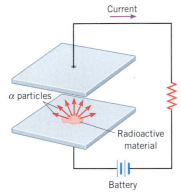

**Figure 31.8** A smoke detector.

One widely used application of $\alpha$ decay is in smoke detectors. Figure 31.8 illustrates how a smoke detector operates. Two small and parallel metal plates are separated by a distance of about one centimeter. A tiny amount of radioactive material at the center of one of the plates emits $\alpha$ particles, which collide with air molecules. During the collisions, the air molecules are ionized to form positive and negative ions. The voltage from a battery causes one plate to be positive and the other negative, so that each plate attracts ions of opposite charge. As a result there is a current in the circuit attached to the plates. The presence of smoke particles between the plates reduces the current, since the ions that collide with a smoke particle are usually neutralized. The drop in current that smoke particles cause is used to trigger an alarm.

## $\beta$ DECAY

The $\beta$ rays in Figure 31.6 are deflected by the magnetic field in a direction opposite to that of the positively charged $\alpha$ rays. Consequently, these $\beta$ rays, which are the most common kind, consist of negatively charged particles or $\beta^-$ particles. Experiment shows that $\beta^-$ particles are electrons. As an illustration of $\beta^-$ decay, consider the thorium $^{234}_{90}$Th nucleus, which decays by emitting a $\beta^-$ particle, as in Figure 31.9:

$$\underset{\substack{\text{Parent}\\\text{nucleus}\\\text{(thorium)}}}{^{234}_{90}\text{Th}} \longrightarrow \underset{\substack{\text{Daughter}\\\text{nucleus}\\\text{(protactinium)}}}{^{234}_{91}\text{Pa}} + \underset{\substack{\beta^- \text{ particle}\\\text{(electron)}}}{^{\phantom{0}0}_{-1}\text{e}}$$

$\beta^-$ decay, like $\alpha$ decay, causes a transmutation of one element into another. In this case, thorium $^{234}_{90}$Th is converted into protactinium $^{234}_{91}$Pa. The law of conservation of charge is obeyed, since the net number of positive charges is the same before (90) and after (91 − 1) the $\beta^-$ emission. The law of conservation of nucleon number is obeyed, since the nucleon number remains at $A = 234$. The general form for $\beta^-$ decay is

**$\beta^-$ Decay**

$$\underset{\substack{\text{Parent}\\\text{nucleus}}}{^{A}_{Z}\text{P}} \longrightarrow \underset{\substack{\text{Daughter}\\\text{nucleus}}}{^{A}_{Z+1}\text{D}} + \underset{\substack{\beta^- \text{ particle}\\\text{(electron)}}}{^{\phantom{0}0}_{-1}\text{e}}$$

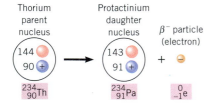

**Figure 31.9** $\beta$ decay occurs when a neutron in an unstable parent nucleus decays into a proton and an electron, the electron being emitted as the $\beta^-$ particle. In the process, the parent nucleus is transformed into the daughter nucleus.

The electron emitted in $\beta^-$ decay does *not* actually exist within the parent nucleus and is *not* one of the orbital electrons. Instead, the electron is created when a neutron decays into a proton and an electron; when this occurs, the proton number of the parent nucleus increases from $Z$ to $Z + 1$ and the nucleon number remains unchanged. The electron is usually fast-moving and escapes from the atom, leaving behind a positively charged atom.

Example 6 illustrates that energy is released during $\beta^-$ decay, just as it is during $\alpha$ decay, and that the conservation of mass/energy applies.

## Example 6 $\beta^-$ Decay and the Release of Energy

The atomic mass of thorium $^{234}_{90}$Th is 234.043 59 u, and the atomic mass of protactinium $^{234}_{91}$Pa is 234.043 30 u. Find the energy released when $\beta^-$ decay changes $^{234}_{90}$Th into $^{234}_{91}$Pa.

**Reasoning and Solution** To find the energy released, we follow the usual procedure of determining how much the mass has decreased because of the decay and then calculating the equivalent energy. The decay and the masses are shown below:

$$\underset{234.043\ 59\ \text{u}}{^{234}_{90}\text{Th}} \longrightarrow \underset{234.043\ 30\ \text{u}}{^{234}_{91}\text{Pa}} + {^{\phantom{0}0}_{-1}\text{e}}$$

When the $^{234}_{90}$Th nucleus of a thorium atom is converted into a $^{234}_{91}$Pa nucleus, the number of orbital electrons remains the same, so the resulting protactinium atom is missing one orbital electron. However, the given mass includes all 91 electrons of a neutral protactinium atom. In effect, then, the value of 234.043 30 u for $^{234}_{91}$Pa already includes the mass of the $\beta^-$ particle. The mass decrease that accompanies the $\beta^-$ decay is 234.043 59 u − 234.043 30 u = 0.000 29 u. The equivalent energy (1 u = 931.5 MeV) is $\boxed{0.27 \text{ MeV}}$. This is the maximum kinetic energy that the emitted electron can have.

**Problem solving insight**
In $\beta^-$ decay, be careful not to include the mass of the electron ($^{0}_{-1}$e) twice. As discussed here for the daughter atom ($^{234}_{91}$Pa), the atomic mass already includes the mass of the emitted electron.

A second kind of $\beta$ decay sometimes occurs.* In this process the particle emitted by the nucleus is a *positron* rather than an electron. A positron, also called a $\beta^+$ particle, has the same mass as an electron but carries a charge of $+e$ instead of $-e$. The disintegration process for $\beta^+$ decay is

**$\beta^+$ Decay**

$$^{A}_{Z}\text{P} \longrightarrow \ _{Z-1}^{A}\text{D} + \ ^{0}_{1}\text{e}$$

| Parent nucleus | Daughter nucleus | $\beta^+$ particle (positron) |

The emitted positron does *not* exist within the nucleus but, rather, is created when a nuclear proton is transformed into a neutron. In the process, the proton number of the parent nucleus decreases from $Z$ to $Z - 1$, and the nucleon number remains the same. As with $\beta^-$ decay, the laws of conservation of charge and nucleon number are obeyed, and there is a transmutation of one element into another.

## $\gamma$ DECAY

The nucleus, like the orbital electrons, exists only in discrete energy states or levels. When a nucleus changes from an excited energy state (denoted by an asterisk *) to a lower energy state, a photon is emitted. The process is similar to the one discussed in Section 30.3 for the photon emission that leads to the hydrogen atom line spectrum. With nuclear energy levels, however, the photon has a much greater energy and is called a $\gamma$ ray. The $\gamma$ decay process is written as follows:

**$\gamma$ Decay**

$$^{A}_{Z}\text{P*} \longrightarrow \ ^{A}_{Z}\text{P} + \gamma$$

| Excited energy state | Lower energy state | $\gamma$ ray |

$\gamma$ decay does *not* cause a transmutation of one element into another. In the next example the wavelength of one particular $\gamma$ ray photon is determined.

### Example 7   The Wavelength of a Photon Emitted During $\gamma$ Decay

What is the wavelength of the 0.186 MeV $\gamma$-ray photon emitted by radium $^{226}_{88}$Ra?

**Reasoning**  The photon energy is the difference between two nuclear energy levels. Equation 30.4 gives the relation between the energy level separation $\Delta E$ and the frequency $f$ of the photon as $\Delta E = hf$. Since $f\lambda = c$, the wavelength of the photon is $\lambda = hc/\Delta E$.

**Problem solving insight**
The energy $\Delta E$ of a $\gamma$-ray photon, like that of photons in other regions of the electromagnetic spectrum (visible, infrared, microwave, etc.), is equal to the product of Planck's constant $h$ and the frequency $f$ of the photon: $\Delta E = hf$.

**Solution**  First we must convert the photon energy into joules:

$$\Delta E = (0.186 \times 10^6 \text{ eV}) \left( \frac{1.60 \times 10^{-19} \text{ J}}{1 \text{ eV}} \right) = 2.98 \times 10^{-14} \text{ J}$$

The wavelength of the photon is

$$\lambda = \frac{hc}{\Delta E} = \frac{(6.63 \times 10^{-34} \text{ J} \cdot \text{s})(3.00 \times 10^8 \text{ m/s})}{2.98 \times 10^{-14} \text{ J}} = \boxed{6.67 \times 10^{-12} \text{ m}}$$

---

* A third kind of $\beta$ decay also occurs in which a nucleus pulls in or captures one of the orbital electrons from outside the nucleus. The process is called **electron capture**, or **K capture**, since the electron normally comes from the innermost or K shell.

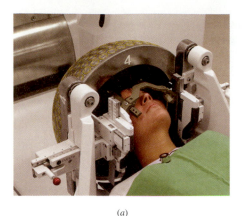
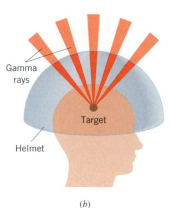

(a)                              (b)

**Figure 31.10** (a) In Gamma Knife radiosurgery, a protective metal helmet containing 201 small holes is placed over the patient's head. (Courtesy Elekta Instruments, Inc.) (b) The holes focus the beams of γ rays to a tiny target within the brain.

**The physics of**
**Gamma Knife radiosurgery.**

**The physics of**
**an exercise thallium heart scan.**

Gamma Knife radiosurgery is becoming a very promising medical procedure for treating certain problems of the brain, including benign and cancerous tumors, as well as blood vessel malformations. The procedure, which involves no knife at all, uses powerful, highly focused beams of γ rays aimed at the tumor or malformation. The γ rays are emitted by a radioactive cobalt-60 source. As Figure 31.10a illustrates, the patient wears a protective metal helmet that is perforated with 201 small holes. Part b of the figure shows that the holes focus the γ rays to a single tiny target within the brain. The target tissue thus receives a very intense dose of radiation and is destroyed, while the surrounding healthy tissue is undamaged. Gamma Knife surgery is a noninvasive, painless, and bloodless procedure that is often performed under local anesthesia. Hospital stays are 70 to 90 percent shorter than with conventional surgery, and patients often return to work within a few days.

An exercise thallium heart scan is a test that uses radioactive thallium to produce images of the heart muscle. When combined with an exercise test, such as walking on a treadmill, the thallium scan helps identify regions of the heart that are not receiving enough blood. The scan is especially useful in diagnosing the presence of blockages in the coronary arteries, the vessels that supply oxygen-rich blood to the heart muscle. During the test, a small amount of thallium is injected into a vein while the patient walks on a treadmill. The thallium attaches itself to the red blood cells and is carried throughout the body. The thallium enters the heart muscle by way of the coronary arteries and collects in the cells of the heart muscle that come into contact with the blood. The thallium isotope used, $^{201}_{81}$Tl, emits γ rays, which a special camera records. Since the thallium reaches those regions of the heart that have an adequate blood supply, lesser amounts show up in areas where the blood flow has been reduced due to partially blocked arteries (see Figure 31.11). A second set of images is taken several hours later, while the patient is resting. These images help differentiate between regions of the heart that temporarily do not receive enough blood (the blood flow returns to normal after the exercise) and regions that are permanently damaged due to, for example, a previous heart attack (the blood flow does not return to normal).

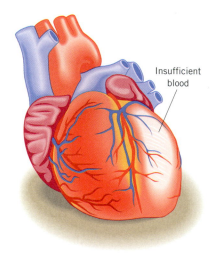

Insufficient blood

**Figure 31.11** An exercise thallium heart scan indicates regions of the heart that receive insufficient blood during exercise.

## 31.5  *The Neutrino*

When a β particle is emitted by a radioactive nucleus, energy is simultaneously released, as Example 6 illustrates. Experimentally, however, it is found that most β particles do not have enough kinetic energy to account for all the energy released. If a β particle carries away only part of the energy, where does the remainder go? The question puzzled physicists until 1930, when Wolfgang Pauli proposed that part of the energy is carried away by another particle that is emitted along with the β particle. This additional particle is called the **neutrino,** and its existence was verified experimentally in 1956. The Greek letter nu (ν) is used to symbolize the neutrino. For instance, the $\beta^-$ decay of thorium $^{234}_{90}$Th (see Section 31.4) is more correctly written as

$$^{234}_{90}\text{Th} \longrightarrow {}^{234}_{91}\text{Pa} + {}^{0}_{-1}\text{e} + \overline{\nu}$$

The bar above the $\nu$ is included because the neutrino emitted in this particular decay process is an antimatter neutrino or antineutrino. A normal neutrino ($\nu$ without the bar) is emitted when $\beta^+$ decay occurs.

The neutrino has zero electric charge and is extremely difficult to detect because it interacts very weakly with matter. For example, the average neutrino can penetrate one light-year of lead (about $9.5 \times 10^{15}$ m) without interacting with it. Thus, even though trillions of neutrinos pass through our bodies every second, they have no effect. Although difficult, it is possible to detect neutrinos. Figure 31.12 shows the Super-Kamiokande neutrino detector in Japan. It is located 915 m underground and consists of a steel cylindrical tank, ten stories tall, whose inner wall is lined with 11 000 photomultiplier tubes. The tank is filled with 12.5 million gallons of ultrapure water. Neutrinos colliding with the water molecules produce light patterns that the photomultiplier tubes detect.

One of the major scientific questions of our time is whether neutrinos have mass. The question is important because neutrinos are so plentiful in the universe. Even a very small mass could account for a significant portion of the mass in the universe and, possibly, have an effect on the formation of galaxies. In 1998 the Super-Kamiokande detector yielded the first strong, but indirect, evidence that neutrinos do indeed have mass. (The mass of the electron neutrino appears to be a tiny fraction of the mass of an electron.) This finding implies that neutrinos travel at less than the speed of light. If the neutrino's mass were zero, like that of a photon, it would travel at the speed of light.

The emission of neutrinos and $\beta$ particles involves a force called *the weak nuclear force* because it is much weaker than the strong nuclear force. It is now known that the weak nuclear force and the electromagnetic force are two different manifestations of a single, more fundamental force, the **electroweak force.** The theory for the electroweak force was developed by Sheldon Glashow (1932– ), Abdus Salam (1926–1996), and Steven Weinberg (1933– ), who shared a Nobel Prize for their achievement in 1979. The electroweak force, the gravitational force, and the strong nuclear force are the three fundamental forces in nature.

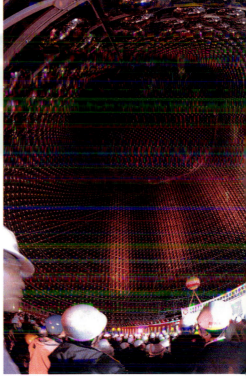

**Figure 31.12** The Super-Kamiokande neutrino detector in Japan consists of a steel cylindrical tank containing 12.5 million gallons of ultrapure water. Its inner wall is lined with 11 000 photomultiplier tubes. (© Kyodo News International)

## 31.6 *Radioactive Decay and Activity*

The question of which radioactive nucleus in a group of nuclei disintegrates at a given instant is decided like the drawing of numbers in a state lottery; individual disintegrations occur randomly. As time passes, the number $N$ of parent nuclei decreases, as Figure 31.13 shows. This graph of $N$ versus time indicates that the decrease occurs in a smooth fashion, with $N$ approaching zero after enough time has passed. To help describe the graph, it is useful to define the **half-life** $T_{1/2}$ of a radioactive isotope as the time required for one-half of the nuclei present to disintegrate. For example, radium $^{226}_{88}$Ra has a half-life of 1600 years, because it takes this amount of time for one-half of a given quantity of this isotope to disintegrate. In another 1600 years, one-half of the remaining radium atoms will have disintegrated, leaving only one-fourth of the original number intact. In Figure 31.13, the number of nuclei present at time $t = 0$ s is $N = N_0$, and the number present at $t = T_{1/2}$ is $N = \frac{1}{2}N_0$. The number present at $t = 2T_{1/2}$ is $N = \frac{1}{4}N_0$, and so on. The value of the half-life depends on the nature of the radioactive nucleus. Values ranging from a fraction of a second to billions of years have been found. See Table 31.2 on p. 664.

Radon $^{222}_{86}$Rn is a naturally occurring radioactive gas produced when radium $^{226}_{88}$Ra undergoes $\alpha$ decay. There is a nationwide concern about radon as a health hazard because radon in the soil is gaseous and can enter the basement of homes through cracks in the foundation. Once inside, the concentration of radon can rise markedly, depending on the type of housing construction and the concentration of radon in the surrounding soil. Radon gas decays into daughter nuclei that are also radioactive. The radioactive nuclei can attach to dust and smoke particles that can be inhaled, and they remain in the lungs to release tissue-damaging radiation. Prolonged exposure to high levels of radon can lead to lung cancer. Since radon gas concentrations can be measured with inexpensive monitoring devices, it is recommended that all homes be tested for radon. Example 8 deals with the half-life of radon $^{222}_{86}$Rn.

**The physics of** radioactive radon gas in houses.

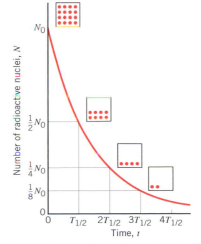

**Figure 31.13** The half-life $T_{1/2}$ of a radioactive decay is the time in which one-half of the radioactive nuclei disintegrate.

**Table 31.2** *Some Half-lives for Radioactive Decay*

| Isotope | | Half-life |
|---|---|---|
| Polonium | $^{214}_{84}\text{Po}$ | $1.64 \times 10^{-4}$ s |
| Krypton | $^{89}_{36}\text{Kr}$ | 3.16 min |
| Radon | $^{222}_{86}\text{Rn}$ | 3.83 da |
| Strontium | $^{90}_{38}\text{Sr}$ | 28.8 yr |
| Radium | $^{226}_{88}\text{Ra}$ | $1.6 \times 10^3$ yr |
| Carbon | $^{14}_{6}\text{C}$ | $5.73 \times 10^3$ yr |
| Uranium | $^{238}_{92}\text{U}$ | $4.47 \times 10^9$ yr |
| Indium | $^{115}_{49}\text{In}$ | $4.41 \times 10^{14}$ yr |

**Example 8** **The Radioactive Decay of Radon Gas**

Suppose $3.0 \times 10^7$ radon atoms are trapped in a basement at the time the basement is sealed against further entry of the gas. The half-life of radon is 3.83 days. How many radon atoms remain after 31 days?

**Reasoning** During each half-life, the number of radon atoms is reduced by a factor of two. Thus, we determine the number of half-lives there are in a period of 31 days and reduce the number of radon atoms by a factor of two for each one.

**Solution** In a period of 31 days there are (31 days)/(3.83 days) = 8.1 half-lives. In 8 half-lives the number of radon atoms is reduced by a factor of $2^8 = 256$. Ignoring the difference between 8 and 8.1 half-lives, we find that the number of atoms remaining is $(3.0 \times 10^7)/256 = \boxed{1.2 \times 10^5}$.

The *activity* of a radioactive sample is the number of disintegrations per second that occur. Each time a disintegration occurs, the number $N$ of radioactive nuclei decreases. As a result, the activity can be obtained by dividing $\Delta N$, the change in the number of nuclei, by $\Delta t$, the time interval during which the change takes place; the average activity over the time interval $\Delta t$ is the magnitude of $\Delta N/\Delta t$, or $|\Delta N/\Delta t|$. Since the decay of any individual nucleus is completely random, the number of disintegrations per second that occurs in a sample is proportional to the number of radioactive nuclei present, so that

$$\frac{\Delta N}{\Delta t} = -\lambda N \tag{31.4}$$

where $\lambda$ is a proportionality constant referred to as the *decay constant.* The minus sign is present in this equation because each disintegration decreases the number $N$ of nuclei originally present.

The SI unit for activity is the *becquerel* (Bq); one becquerel equals one disintegration per second. Activity is also measured in terms of a unit called the *curie* (Ci), in honor of Marie (1867–1934) and Pierre (1859–1906) Curie, the discoverers of radium and polonium. Historically, the curie was chosen as a unit because it is roughly the activity of one gram of pure radium. In terms of becquerels,

$$1 \text{ Ci} = 3.70 \times 10^{10} \text{ Bq}$$

The activity of the radium put into the dial of a watch to make it glow in the dark is about $4 \times 10^4$ Bq, and the activity used in radiation therapy for cancer treatment is approximately a billion times greater, or $4 \times 10^{13}$ Bq.

The mathematical expression for the graph of $N$ versus $t$ shown in Figure 31.13 can be obtained from Equation 31.4 with the aid of calculus. The result for the number $N$ of radioactive nuclei present at time $t$ is

$$N = N_0 e^{-\lambda t} \tag{31.5}$$

assuming that the number at $t = 0$ s is $N_0$. The exponential $e$ has the value $e = 2.718 \ldots$, and many calculators provide the value of $e^x$. We can relate the half-life $T_{1/2}$ of a radioactive nucleus to its decay constant $\lambda$ in the following manner. By substituting $N = \frac{1}{2}N_0$ and $t = T_{1/2}$ into Equation 31.5, we find that $\frac{1}{2} = e^{-\lambda T_{1/2}}$. Taking the natural logarithm of both sides of this equation reveals that $\ln 2 = \lambda T_{1/2}$ or

$$T_{1/2} = \frac{\ln 2}{\lambda} = \frac{0.693}{\lambda} \tag{31.6}$$

The following example illustrates the use of Equations 31.5 and 31.6.

**Example 9** **The Activity of Radon $^{222}_{86}\text{Rn}$**

As in Example 8, suppose there are $3.0 \times 10^7$ radon atoms ($T_{1/2} = 3.83$ days or $3.31 \times 10^5$ s) trapped in a basement. (a) How many radon atoms remain after 31 days? Find the activity (b) just after the basement is sealed against further entry of radon and (c) 31 days later.

**Reasoning** The number $N$ of radon atoms remaining after a time $t$ is given by $N = N_0 e^{-\lambda t}$, where $N_0 = 3.0 \times 10^7$ is the original number of atoms when $t = 0$ s and $\lambda$ is the decay constant. The decay constant is related to the half-life $T_{1/2}$ of the radon atoms by $\lambda = 0.693/T_{1/2}$. The activity can be obtained from Equation 31.4, $\Delta N/\Delta t = -\lambda N$.

**Solution**

**(a)** The decay constant is

$$\lambda = \frac{0.693}{T_{1/2}} = \frac{0.693}{3.83 \text{ days}} = 0.181 \text{ days}^{-1} \qquad (31.6)$$

and the number $N$ of radon atoms remaining after 31 days is

$$N = N_0 e^{-\lambda t} = (3.0 \times 10^7)e^{-(0.181 \text{ days}^{-1})(31 \text{ days})} = \boxed{1.1 \times 10^5} \qquad (31.5)$$

This value is slightly different from that found in Example 8 because there we ignored the difference between 8.0 and 8.1 half-lives.

**(b)** The activity can be obtained from Equation 31.4, provided the decay constant is expressed in reciprocal seconds:

$$\lambda = \frac{0.693}{T_{1/2}} = \frac{0.693}{3.31 \times 10^5 \text{ s}} = 2.09 \times 10^{-6} \text{ s}^{-1} \qquad (31.6)$$

Thus, the number of disintegrations per second is

$$\frac{\Delta N}{\Delta t} = -\lambda N = -(2.09 \times 10^{-6} \text{ s}^{-1})(3.0 \times 10^7) = -63 \text{ disintegrations/s} \qquad (31.4)$$

The activity is the magnitude of $\Delta N/\Delta t$, so initially $\boxed{\text{Activity} = 63 \text{ Bq}}$.

**(c)** From part (a), the number of radioactive nuclei remaining at the end of 31 days is $N = 1.1 \times 10^5$, and reasoning similar to that in part (b) reveals that $\boxed{\text{Activity} = 0.23 \text{ Bq}}$.

▲

# 31.7 *Radioactive Dating*

One important application of radioactivity is the determination of the age of archeological or geological samples. If an object contains radioactive nuclei when it is formed, then the decay of these nuclei marks the passage of time like a clock, half of the nuclei disintegrating during each half-life. If the half-life is known, a measurement of the number of nuclei present today relative to the number present initially can give the age of the sample. According to Equation 31.4, the activity of a sample is proportional to the number of radioactive nuclei, so one way to obtain the age is to compare present activity with initial activity. A more accurate way is to determine the present number of radioactive nuclei with the aid of a mass spectrometer.

The present activity of a sample can be measured, but how is it possible to know what the original activity was, perhaps thousands of years ago? Radioactive dating methods entail certain assumptions that make it possible to estimate the original activity. For instance, the radiocarbon technique utilizes the $^{14}_{6}C$ isotope of carbon, which undergoes $\beta^-$ decay with a half-life of 5730 yr. This isotope is present in the earth's atmosphere at an equilibrium concentration of about one atom for every $8.3 \times 10^{11}$ atoms of normal carbon $^{12}_{6}C$. It is often assumed* that this value has remained constant over the years because $^{14}_{6}C$ is created when cosmic rays interact with the earth's upper atmosphere, a production method that offsets the loss via $\beta^-$ decay. Moreover, nearly all living organisms ingest the equilibrium concentration of $^{14}_{6}C$. However, once an organism dies, metabolism no longer sustains the input of $^{14}_{6}C$, and $\beta^-$ decay causes half of the $^{14}_{6}C$ nuclei to disintegrate every 5730 years. Example 10 illustrates how to determine the $^{14}_{6}C$ activity of one gram of carbon in a living organism.

*The physics of radioactive dating.*

* The assumption that the $^{14}_{6}C$ concentration has always been at its present equilibrium value has been evaluated by comparing $^{14}_{6}C$ ages with ages determined by counting tree rings. More recently, ages determined using the radioactive decay of uranium $^{238}_{92}U$ have been used for comparison. These comparisons indicate that the equilibrium value of the $^{14}_{6}C$ concentration has indeed remained constant for the past 1000 years. However, from there back about 30 000 years, it appears that the $^{14}_{6}C$ concentration in the atmosphere was larger than its present value by up to 40%. As a first approximation we ignore such discrepancies.

### *Example 10* $^{14}_6$C Activity Per Gram of Carbon in a Living Organism

(a) Determine the number of carbon $^{14}_6$C atoms present for every gram of carbon $^{12}_6$C in a living organism. Find (b) the decay constant and (c) the activity of this sample.

**Reasoning** The total number of carbon $^{12}_6$C atoms in one gram of carbon $^{12}_6$C is equal to the corresponding number of moles times Avogadro's number (see Section 14.1). Since there is only one $^{14}_6$C atom for every $8.3 \times 10^{11}$ atoms of $^{12}_6$C, the number of $^{14}_6$C atoms is equal to the total number of $^{12}_6$C atoms divided by $8.3 \times 10^{11}$. The decay constant $\lambda$ for $^{14}_6$C is $\lambda = 0.693/T_{1/2}$, where $T_{1/2}$ is the half-life. The activity is equal to the magnitude of $\Delta N/\Delta t$, which is equal to the decay constant times the number of $^{14}_6$C atoms present, according to Equation 31.4.

**Solution**

(a) One gram of carbon $^{12}_6$C (atomic mass = 12 u) is equivalent to 1.0/12 mol. Since Avogadro's number is $6.02 \times 10^{23}$ atoms/mol and since there is one $^{14}_6$C atom for every $8.3 \times 10^{11}$ atoms of $^{12}_6$C, the number of $^{14}_6$C atoms is

$$\text{Number of } ^{14}_6\text{C atoms for every 1.0 gram of carbon } ^{12}_6\text{C} = \left(\frac{1.0}{12} \text{ mol}\right)\left(6.02 \times 10^{23} \frac{\text{atoms}}{\text{mol}}\right)\left(\frac{1}{8.3 \times 10^{11}}\right)$$

$$= \boxed{6.0 \times 10^{10} \text{ atoms}}$$

(b) Since the half-life of $^{14}_6$C is 5730 yr ($1.81 \times 10^{11}$ s), the decay constant is

$$\lambda = \frac{0.693}{T_{1/2}} = \frac{0.693}{1.81 \times 10^{11} \text{ s}} = \boxed{3.83 \times 10^{-12} \text{ s}^{-1}} \tag{31.6}$$

(c) Equation 31.4 indicates that $\Delta N/\Delta t = -\lambda N$, so the magnitude of $\Delta N/\Delta t$ is $\lambda N$.

$$\text{Activity of } ^{14}_6\text{C for every 1.0 gram of carbon } ^{12}_6\text{C in a living organism} = \lambda N = (3.83 \times 10^{-12} \text{ s}^{-1})(6.0 \times 10^{10} \text{ atoms}) = \boxed{0.23 \text{ Bq}}$$

An organism that lived thousands of years ago presumably had an activity of about 0.23 Bq per gram of carbon. When the organism died, the activity began decreasing. From a sample of the remains, the current activity per gram of carbon can be measured and compared to the value of 0.23 Bq to determine the time that has transpired since death. This procedure is illustrated in Example 11.

### *Example 11* The Iceman

On September 19, 1991, German tourists on a walking trip in the Italian Alps found a Stone Age traveler, later called the Iceman, whose body had become trapped in a glacier. Figure 31.14 shows the well-preserved remains that were dated using the radiocarbon method. Material found with the body had a $^{14}_6$C activity of about 0.121 Bq per gram of carbon. Find the age of the Iceman's remains.

**Reasoning** According to Equation 31.5, the number of nuclei remaining at time $t$ is $N = N_0 e^{-\lambda t}$. Multiplying both sides of this expression by the decay constant $\lambda$ and recognizing that the product of $\lambda$ and $N$ is the activity $A$, we find that $A = A_0 e^{-\lambda t}$, where $A_0 = 0.23$ Bq is the activity at time $t = 0$ s for one gram of carbon. The decay constant $\lambda$ can be determined from the value of 5730 yr for the half-life of $^{14}_6$C, using Equation 31.6. With known values for $A_0$ and $\lambda$, the given activity of $A = 0.121$ Bq per gram of carbon can be used to find $t$, the Iceman's age.

**Solution** For $^{14}_6$C, the decay constant is $\lambda = 0.693/T_{1/2} = 0.693/(5730 \text{ yr}) = 1.21 \times 10^{-4} \text{ yr}^{-1}$. Since $A = 0.121$ Bq and $A_0 = 0.23$ Bq, the age can be determined from

$$A = 0.121 \text{ Bq} = (0.23 \text{ Bq})e^{-(1.21 \times 10^{-4} \text{ yr}^{-1})t}$$

Taking the natural logarithm of both sides of this result gives

$$\ln\left(\frac{0.121 \text{ Bq}}{0.23 \text{ Bq}}\right) = -(1.21 \times 10^{-4} \text{ yr}^{-1})t$$

which gives an age for the sample of $\boxed{t = 5300 \text{ yr}}$.

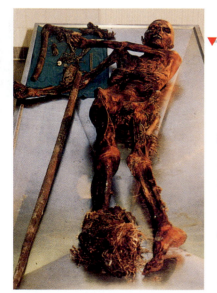

**Figure 31.14** These remains of the Iceman were discovered in the ice of a glacier in the Italian Alps in 1991. Radiocarbon dating reveals his age (see Example 11). (© Corbis Sygma)

Radiocarbon dating is not the only radioactive dating method. For example, other methods utilize uranium $^{238}_{92}$U, potassium $^{40}_{19}$K, and lead $^{210}_{82}$Pb. For such methods to be useful, the half-life of the radioactive species must be neither too short nor too long relative to the age of the sample to be dated, as Conceptual Example 12 discusses.

## Conceptual Example 12  Dating a Bottle of Wine

A bottle of red wine is thought to have been sealed about 5 years ago. The wine contains a number of different kinds of atoms, including carbon, oxygen, and hydrogen. Each of these has a radioactive isotope. The radioactive isotope of carbon is the familiar $^{14}_{6}$C, with a half-life of 5730 yr. The radioactive isotope of oxygen is $^{15}_{8}$O and has a half-life of 122.2 s. The radioactive isotope of hydrogen is $^{3}_{1}$H and is called tritium; its half-life is 12.33 yr. The activity of each of these isotopes is known at the time the bottle was sealed. However, only one of the isotopes is useful for determining the age of the wine accurately. Which is it?

**Reasoning and Solution** In a dating method that measures the activity of a radioactive isotope, the age of the sample is related to the change in the activity during the time period in question. Here the expected age is about 5 years. This period is only a small fraction of the 5730-yr half-life of $^{14}_{6}$C. As a result, relatively few of the $^{14}_{6}$C nuclei would decay during the wine's life, and the measured activity would change little from its initial value. To obtain an accurate age from such a small change would require prohibitively precise measurements. The $^{15}_{8}$O isotope is not very useful, either; the difficulty is its relatively short half-life of 122.2 s. During a 5-year period, so many half-lives of 122.2 s would occur that the activity would decrease to a vanishingly small level. It would not be even possible to measure it. The only remaining option is **the tritium isotope of hydrogen.** The expected age of 5 yr is long enough relative to the half-life of 12.33 yr that a measurable change in activity will occur, but not so long that the activity will have completely vanished for all practical purposes.

**Related Homework:** *Problem 42*

## 31.8  *Radioactive Decay Series*

When an unstable parent nucleus decays, the resulting daughter nucleus is sometimes also unstable. If so, the daughter then decays and produces its own daughter, and so on, until a completely stable nucleus is produced. This sequential decay of one nucleus after another is called a *radioactive decay series*. Examples 4–6 discuss the first two steps of a series that begins with uranium $^{238}_{92}$U:

$$\text{Uranium} \qquad \text{Thorium}$$
$$^{238}_{92}\text{U} \longrightarrow {}^{234}_{90}\text{Th} + {}^{4}_{2}\text{He}$$
$$\longrightarrow {}^{234}_{91}\text{Pa} + {}^{0}_{-1}\text{e}$$
$$\text{Protactinium}$$

Furthermore, Example 8 deals with radon $^{222}_{86}$Rn, which is formed down the line in the $^{238}_{92}$U radioactive decay series. Figure 31.15 on p. 668 shows the entire series. At several points branches occur because more than one kind of decay is possible for an intermediate species. Ultimately, however, the series ends with lead $^{206}_{82}$Pb, which is stable.

The $^{238}_{92}$U series and other such series are the only sources of some of the radioactive elements found in nature. Radium $^{226}_{88}$Ra, for instance, has a half-life of 1600 yr, which is short enough that all the $^{226}_{88}$Ra created when the earth was formed billions of years ago has now disappeared. The $^{238}_{92}$U series provides a continuing supply of $^{226}_{88}$Ra, however.

## 31.9  *Radiation Detectors*

There are a number of devices that can be used to detect the particles and photons ($\gamma$ rays) emitted when a radioactive nucleus decays. Such devices detect the ionization that these particles and photons cause as they pass through matter.

**The physics of radiation detectors.**

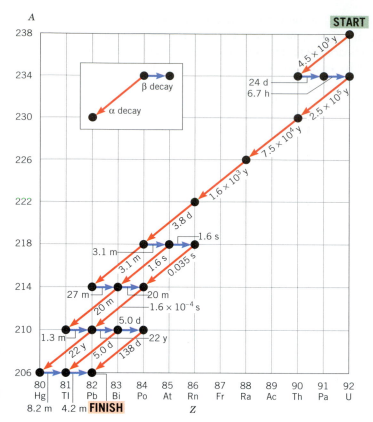

**Figure 31.15** The radioactive decay series that begins with uranium $^{238}_{92}U$ and ends with lead $^{206}_{82}Pb$. Half-lives are given in seconds (s), minutes (m), hours (h), days (d), or years (y). The inset in the upper left identifies the type of decay that each nucleus undergoes.

The most familiar detector is the **Geiger counter**, which Figure 31.16 illustrates. The Geiger counter consists of a gas-filled metal cylinder. The $\alpha$, $\beta$, or $\gamma$ rays enter the cylinder through a thin window at one end. $\gamma$ rays can also penetrate directly through the metal. A wire electrode runs along the center of the tube and is kept at a high positive voltage (1000–3000 V) relative to the outer cylinder. When a high-energy particle

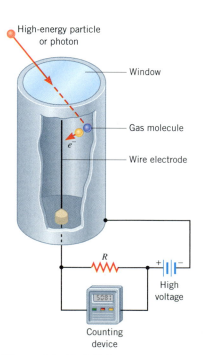

**Figure 31.16** A Geiger counter.

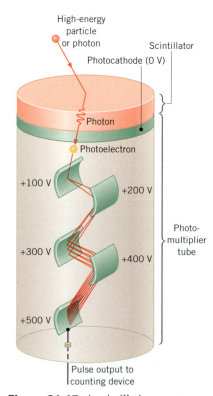

**Figure 31.17** A scintillation counter.

or photon enters the cylinder, it collides with and ionizes a gas molecule. The electron produced from the gas molecule accelerates toward the positive wire, ionizing other molecules in its path. Additional electrons are formed, and an avalanche of electrons rushes toward the wire, leading to a pulse of current through the resistor R. This pulse can be counted or made to produce a "click" in a loudspeaker. The number of counts or clicks is related to the number of disintegrations that produced the particles or photons.

The **scintillation counter** is another important radiation detector. As Figure 31.17 indicates, this device consists of a scintillator mounted on a photomultiplier tube. Often the scintillator is a crystal (e.g., sodium iodide) containing a small amount of impurity (thallium), but plastic, liquid, and gaseous scintillators are also used. In response to ionizing radiation, the scintillator emits a flash of visible light. The photons of the flash then strike the photocathode of the photomultiplier tube. The photocathode is made of a material that emits electrons because of the photoelectric effect. These photoelectrons are then attracted to a special electrode kept at a voltage of about +100 V relative to the photocathode. The electrode is coated with a substance that emits several additional electrons for every electron striking it. The additional electrons are attracted to a second similar electrode (voltage = +200 V) where they generate even more electrons. Commercial photomultiplier tubes contain as many as 15 of these special electrodes, so photoelectrons resulting from the light flash of the scintillator lead to a cascade of electrons and a pulse of current. As in a Geiger tube, the current pulses can be counted.

Ionizing radiation can also be detected with several types of **semiconductor detectors**. Such devices utilize n- and p-type materials (see Section 23.5), and their operation depends on the electrons and holes formed in the materials as a result of the radiation. One of the main advantages of semiconductor detectors is their ability to discriminate between two particles with only slightly different energies.

A number of instruments provide a pictorial representation of the path that high-energy particles follow after they are emitted from unstable nuclei. In a **cloud chamber,** a gas is cooled just to the point where it will condense into droplets, provided nucleating agents are available on which the droplets can form. When a high-energy particle, such as an α particle or a β particle, passes through the gas, the ions it leaves behind serve as nucleating agents, and droplets form along the path of the particle. A **bubble chamber** works in a similar fashion, except it contains a liquid that is just at the point of boiling. Tiny bubbles form along the trail of a high-energy particle passing through the liquid. The paths revealed in a cloud or bubble chamber can be photographed to provide a permanent record of the event. Figure 31.18 shows a photograph of the tracks in a bubble chamber. A **photographic emulsion** also can be used directly to produce a record of the path taken by a particle of ionizing radiation. The ions formed as the particle passes through the emulsion cause silver to be deposited along the track when the emulsion is developed.

**Figure 31.18** A photograph showing particle tracks in a bubble chamber.
(© LBL/Photo Researchers)

# Concept Summary

This summary presents an abridged version of the chapter, including the important equations and all available learning aids. For convenient reference, the learning aids (including the text's examples) are placed next to or immediately after the relevant equation or discussion. The following learning aids may be found on-line at **www.wiley.com/college/cutnell**:

| | |
|---|---|
| **Interactive LearningWare** examples are solved according to a five-step interactive format that is designed to help you develop problem-solving skills. | **Concept Simulations** are animated versions of text figures or animations that illustrate important concepts. You can control parameters that affect the display, and we encourage you to experiment. |
| **Interactive Solutions** offer specific models for certain types of problems in the chapter homework. The calculations are carried out interactively. | **Self-Assessment Tests** include both qualitative and quantitative questions. Extensive feedback is provided for both incorrect and correct answers, to help you evaluate your understanding of the material. |

| Topic | Discussion | Learning Aids |
|---|---|---|

### 31.1 Nuclear Structure

**Nucleons**
**Atomic number**

The nucleus of an atom consists of protons and neutrons, which are collectively referred to as nucleons. A neutron is an elecrically neutral particle whose mass is slightly larger than that of the proton. The atomic number $Z$ is the number of protons in the nucleus. The atomic mass number $A$ (or nucleon number) is the total number of protons and neutrons in the nucleus:

**Atomic mass number (or nucleon number)**

$$A = Z + N \qquad (31.1)$$

where $N$ is the number of neutrons. For an element whose chemical symbol is X, the symbol for the nucleus is $^A_Z$X.

**Isotopes**

Nuclei that contain the same number of protons, but a different number of neutrons, are called isotopes.

The approximate radius (in meters) of a nucleus is given by

**Radius of a nucleus**

$$r \approx (1.2 \times 10^{-15} \text{ m}) A^{1/3} \qquad (31.2) \quad \textbf{Example 1}$$

### 31.2 The Strong Nuclear Force and the Stability of the Nucleus

**Strong nuclear force**

The strong nuclear force is the force of attraction between nucleons (protons and neutrons) and is one of the three fundamental forces of nature. This force balances the electrostatic force of repulsion between protons and holds the nucleus together. The strong nuclear force has a very short range of action and is almost independent of electric charge.

### 31.3 The Mass Defect of the Nucleus and Nuclear Binding Energy

The binding energy of a nucleus is the energy required to separate the nucleus into its constituent protons and neutrons. The binding energy is equal to

**Binding energy**

$$\text{Binding energy} = (\Delta m)c^2 \qquad (31.3) \quad \textbf{Examples 2, 3}$$

**Mass defect**

where $\Delta m$ is the mass defect of the nucleus and $c$ is the speed of light in a vacuum. The mass defect is the amount by which the sum of the individual masses of the protons and neutrons exceeds the mass of the intact nucleus.

**Atomic mass unit**

When specifying nuclear masses, it is customary to use the atomic mass unit (u). One atomic mass unit has a mass of $1.6605 \times 10^{-27}$ kg and is equivalent to an energy of 931.5 MeV.

### 31.4 Radioactivity

**Radioactivity**

Unstable nuclei spontaneously decay by breaking apart or rearranging their internal structure in a process called radioactivity. Naturally occurring radioactivity produces $\alpha$, $\beta$, and $\gamma$ rays. $\alpha$ rays consist of positively charged particles, each particle being the $^4_2$He nucleus of helium. The general form for $\alpha$ decay is

**$\alpha$ Decay**

$$\underbrace{^A_Z\text{P}}_{\substack{\text{Parent}\\\text{nucleus}}} \longrightarrow \underbrace{^{A-4}_{Z-2}\text{D}}_{\substack{\text{Daughter}\\\text{nucleus}}} + \underbrace{^4_2\text{He}}_{\substack{\alpha \text{ particle}\\ \text{(helium nucleus)}}}$$

**Examples 4, 5**
**Interactive Solution 31.25**

The most common kind of $\beta$ ray consists of negatively charged particles, or $\beta^-$ particles, which are electrons. The general form for $\beta^-$ decay is

**$\beta^-$ Decay**

$$\underbrace{^A_Z\text{P}}_{\substack{\text{Parent}\\\text{nucleus}}} \longrightarrow \underbrace{^A_{Z+1}\text{D}}_{\substack{\text{Daughter}\\\text{nucleus}}} + \underbrace{^{\ 0}_{-1}\text{e}}_{\substack{\beta^- \text{ particle}\\ \text{(electron)}}}$$

**Example 6**

**Positron**

$\beta^+$ decay produces another kind of $\beta$ ray, which consists of positively charged particles, or $\beta^+$ particles. A $\beta^+$ particle, also called a positron, has the same mass as an electron, but carries a charge of $+e$ instead of $-e$.

**Transmutation**

If a radioactive parent nucleus disintegrates into a daughter nucleus that has a different atomic number, such as that which occurs in $\alpha$ and $\beta$ decay, one element has been converted into another element, the conversion being referred to as a transmutation.

| Topic | Discussion | Learning Aids |
|-------|-----------|---------------|

γ rays are high-energy photons emitted by a radioactive nucleus. The general form for γ decay is

**γ Decay**

$$_Z^A P^* \longrightarrow {_Z^A P} + \gamma$$

Excited energy state · Lower energy state · γ ray

**Example 7**
**Interactive LearningWare 31.1**

γ decay does not cause a transmutation of one element into another.

### 31.5 The Neutrino

**Neutrino**

The neutrino is an electrically neutral particle that is emitted along with β particles and has a mass that is much, much smaller than the mass of an electron.

 **Use Self-Assessment Test 31.1 to evaluate your understanding of Sections 31.1–31.5.**

### 31.6 Radioactive Decay and Activity

**Half-life**
**Activity**

The half-life of a radioactive isotope is the time required for one-half of the nuclei present to disintegrate or decay. The activity is the number of disintegrations per second that occur. Activity is the magnitude of $\Delta N/\Delta t$, where $\Delta N$ is the change in the number $N$ of radioactive nuclei and $\Delta t$ is the time interval during which the change occurs. In other words, activity is $|\Delta N/\Delta t|$. The SI unit for activity is the becquerel (Bq), one becquerel being one disintegration per second. Activity is also measured in a unit called the curie (Ci); 1 Ci = $3.70 \times 10^{10}$ Bq.

**Example 8**

Radioactive decay obeys the following relation

$$\frac{\Delta N}{\Delta t} = -\lambda N \qquad (31.4)$$

**Interactive Solution 31.37**

**Decay constant**

where $\lambda$ is the decay constant. This equation can be solved by the methods of integral calculus to show that

**Number of nuclei remaining as a function of time**

$$N = N_0 e^{-\lambda t} \qquad (31.5)$$

**Concept Simulation 31.1**

where $N_0$ is the original number of nuclei. The decay constant $\lambda$ is related to the half-life $T_{1/2}$ according to

**Decay constant and half-life**

$$\lambda = \frac{0.693}{T_{1/2}} \qquad (31.6)$$

**Example 9**

### 31.7 Radioactive Dating

If an object contained radioactive nuclei when it was formed, then the decay of these nuclei can be used to determine the age of the object. One way to obtain the age is to relate the present activity $A$ of an object with its initial activity $A_0$:

**Example 10**

**Activity as a function of time**

$$A = A_0 e^{-\lambda t}$$

**Example 11**

where $\lambda$ is the decay constant and $t$ is the age of the object. For radiocarbon dating that uses the $_6^{14}C$ isotope of carbon, the initial activity is often assumed to be $A_0 = 0.23$ Bq.

**Example 12**

### 31.8 Radioactive Decay Series

The sequential decay of one nucleus after another is called a radioactive decay series. A decay series starts with a radioactive nucleus and ends with a completely stable nucleus. Figure 31.15 illustrates one such series that begins with uranium $_{92}^{238}U$ and ends with lead $_{82}^{206}Pb$.

### 31.9 Radiation Detectors

A number of devices are used to detect α and β particles, as well as γ rays. These include the Geiger counter, the scintillation counter, semiconductor detectors, cloud and bubble chambers, and photographic emulsions.

 **Use Self-Assessment Test 31.2 to evaluate your understanding of Sections 31.6–31.9.**

# Problems

*The data given for atomic masses in these problems include the mass of the electrons orbiting the nucleus of the electrically neutral atom.*

**ssm**   Solution is in the Student Solutions Manual.     **www**   Solution is available on the World Wide Web at www.wiley.com/college/cutnell

    This icon represents a biomedical application.

**Section 31.1 Nuclear Structure,**
**Section 31.2 The Strong Nuclear Force and the Stability**
**of the Nucleus**

**1. ssm** In each of the following cases, what element does the symbol X represent and how many neutrons are in the nucleus: (a) $^{195}_{78}$X, (b) $^{32}_{16}$X, (c) $^{63}_{29}$X, (d) $^{11}_{5}$X, and (e) $^{239}_{94}$X? Use the periodic table on the inside of the back cover as needed.

**2.** In electrically neutral atoms, how many: (a) protons are in the uranium $^{238}_{92}$U nucleus, (b) neutrons are in the mercury $^{202}_{80}$Hg nucleus, and (c) electrons are in orbit about the niobium $^{93}_{41}$Nb nucleus?

**3.** By what factor does the nucleon number of a nucleus have to increase in order for the nuclear radius to double?

**4.** What is the radius of a nucleus of titanium $^{48}_{22}$Ti?

**5. ssm www** The largest stable nucleus has a nucleon number of 209, and the smallest has a nucleon number of 1. If each nucleus is assumed to be a sphere, what is the ratio (largest/smallest) of the surface areas of these spheres?

**\*6.** One isotope (X) contains an equal number of protons and neutrons. Another isotope (Y) of the same element has twice the number of neutrons as the first isotope does. Determine the ratio $r_Y/r_X$ of the nuclear radii of the isotopes.

**\*7.** An unknown nucleus contains 70 neutrons and has twice the volume of the nickel $^{60}_{28}$Ni nucleus. Identify the unknown nucleus in the form $^A_Z$X. Use the periodic table on the inside of the back cover as needed.

**\*\*8.** Conceptual Example 1 provides some useful background for this problem. (a) Determine an approximate value for the density (in kg/m³) of the nucleus. (b) If a BB (radius = 2.3 mm) from an air rifle had a density equal to the nuclear density, what mass would the BB have? (c) Assuming the mass of a supertanker is about $1.5 \times 10^8$ kg, how many "supertankers" of mass would this hypothetical BB have?

**\*\*9. ssm www** Refer to Conceptual Example 1 for a discussion of nuclear densities. A neutron star is composed of neutrons and has a density that is approximately the same as that of a nucleus. What is the radius of a neutron star whose mass is 0.40 times the mass of the sun?

**Section 31.3 The Mass Defect of the Nucleus and Nuclear Binding Energy** *(Note: The atomic mass for hydrogen $^1_1$H is 1.007 825 u, including the mass of one electron.)*

**10.** Mercury $^{202}_{80}$Hg has an atomic mass of 201.970 617 u. Obtain the binding energy *per nucleon* (in MeV/nucleon).

**11. ssm** Use the binding energy per nucleon curve in Figure 31.5 to determine the total binding energy of $^{16}_8$O.

**12.** For lead $^{206}_{82}$Pb (atomic mass = 205.974 440 u) obtain (a) the mass defect in atomic mass units, (b) the binding energy (in MeV), and (c) the binding energy per nucleon (in MeV).

**13.** The earth revolves around the sun, and the two represent a bound system that has a binding energy of $2.6 \times 10^{33}$ J. Suppose the earth and sun were completely separated, so that they were infinitely far apart and at rest. What would be the difference between the mass of the separated system and that of the bound system?

**14.** Find the binding energy (in MeV) for lithium $^7_3$Li (atomic mass = 7.016 003 u).

**\*15. ssm** Two isotopes of a certain element have binding energies that differ by 5.03 MeV. The isotope with the larger binding energy contains one more neutron than the other isotope. Find the difference in atomic mass between the two isotopes.

**\*16.** (a) Energy is required to separate a nucleus into its constituent nucleons, as Figure 31.3 indicates; this energy is the *total* binding energy of the nucleus. In a similar way one can speak of the energy that binds a single nucleon to the remainder of the nucleus. For example, separating nitrogen $^{14}_7$N into nitrogen $^{13}_7$N and a neutron takes energy equal to the binding energy of the neutron, as shown below:

$$^{14}_7\text{N} + \text{Energy} \longrightarrow {}^{13}_7\text{N} + {}^1_0\text{n}$$

Find the energy (in MeV) that binds the neutron to the $^{14}_7$N nucleus by considering the mass of $^{13}_7$N (atomic mass = 13.005 738 u) and the mass of $^1_0$n (atomic mass = 1.008 665 u), as compared to the mass of $^{14}_7$N (atomic mass = 14.003 074 u). (b) Similarly, one can speak of the energy that binds a single proton to the $^{14}_7$N nucleus:

$$^{14}_7\text{N} + \text{Energy} \longrightarrow {}^{13}_6\text{C} + {}^1_1\text{H}$$

Following the procedure outlined in part (a), determine the energy (in MeV) that binds the proton (atomic mass = 1.007 825 u) to the $^{14}_7$N nucleus. (c) Which nucleon is more tightly bound, the neutron or the proton?

**Section 31.4 Radioactivity**

**17.** $\alpha$ decay occurs for each of the following nuclei. Write the decay process for each, including the chemical symbols and values for $Z$ and $A$ for the daughter nuclei: (a) $^{212}_{84}$Po and (b) $^{232}_{92}$U.

**18.** Complete the following decay processes by stating what the symbol "X" represents (X = $\alpha$, $\beta^-$, $\beta^+$, or $\gamma$): (a) $^{211}_{82}$Pb $\rightarrow$ $^{211}_{83}$Bi + X, (b) $^{11}_6$C $\rightarrow$ $^{11}_5$B + X, (c) $^{231}_{90}$Th* $\rightarrow$ $^{231}_{90}$Th + X, and (d) $^{210}_{84}$Po $\rightarrow$ $^{206}_{82}$Pb + X.

**19.** For the following nuclei, each undergoing $\beta^-$ decay, write the decay process, identifying each daughter nucleus with its chemical symbol and values for $Z$ and $A$: (a) $^{14}_6$C and (b) $^{212}_{82}$Pb.

**20.** Osmium $^{191}_{76}$Os (atomic mass = 190.960 920 u) is converted into Iridium $^{191}_{77}$Ir (atomic mass = 190.960 584 u) via $\beta^-$ decay. What is the energy (in MeV) released in this process?

**21. ssm** Write the $\beta^-$ decay process for $^{35}_{16}$S, including the chemical symbol and values for $Z$ and $A$.

**22.** Find the energy released when lead $^{211}_{82}$Pb (atomic mass = 210.988 735 u) undergoes $\beta^-$ decay to become bismuth $^{211}_{83}$Bi (atomic mass = 210.987 255 u).

**23. ssm www** Write the $\beta^+$ decay process for each of the following nuclei, being careful to include $Z$ and $A$ and the proper chemical symbol for each daughter nucleus: (a) $^{18}_9$F and (b) $^{15}_8$O.

**\*24.** Review Conceptual Example 5 as background for this problem. The $\alpha$ decay of uranium $^{238}_{92}$U produces thorium $^{234}_{90}$Th (atomic mass = 234.0436 u). In Example 4, the energy released in this decay is determined to be 4.3 MeV. Determine how much of this energy is carried away by the recoiling $^{234}_{90}$Th daughter nucleus and how much by the $\alpha$ particle (atomic mass = 4.002 603 u). Assume that the energy of each particle is kinetic energy, and ignore the small amount of energy carried away by the $\gamma$ ray that is also emitted. In addition, ignore relativistic effects.

* **25.** Refer to **Interactive Solution 31.25** at **www.wiley.com/college/ cutnell** to review a model for solving this type of problem. Polonium $^{210}_{84}$Po (atomic mass = 209.982 848 u) undergoes $\alpha$ decay. Assuming that all the released energy is in the form of kinetic energy of the $\alpha$ particle (atomic mass = 4.002 603 u) and ignoring the recoil of the daughter nucleus (lead $^{206}_{82}$Pb, 205.974 440 u), find the speed of the $\alpha$ particle. Ignore relativistic effects.

* **26.** Determine the symbol $^A_Z$X for the parent nucleus whose $\alpha$ decay produces the same daughter as the $\beta^-$ decay of thallium $^{208}_{81}$Tl.

** **27. ssm** Find the energy (in MeV) released when $\beta^+$ decay converts sodium $^{22}_{11}$Na (atomic mass = 21.994 434 u) into neon $^{22}_{10}$Ne (atomic mass = 21.991 383 u). Notice that the atomic mass for $^{22}_{11}$Na includes the mass of 11 electrons, whereas the atomic mass for $^{22}_{10}$Ne includes the mass of only 10 electrons.

** **28. Interactive LearningWare 31.1** at **www.wiley.com/college/ cutnell** reviews the concepts that are involved in this problem. An isotope of beryllium (atomic mass = 7.017 u) emits a $\gamma$ ray and recoils with a speed of $2.19 \times 10^4$ m/s. Assuming that the beryllium nucleus is stationary to begin with, find the wavelength of the $\gamma$ ray.

### Section 31.6 Radioactive Decay and Activity

**29.** In 9.0 days the number of radioactive nuclei decreases to one-eighth the number present initially. What is the half-life (in days) of the material?

**30.** The isotope $^{224}_{88}$Ra of radium has a decay constant of $2.19 \times 10^{-6}$ s$^{-1}$. What is the half-life (in days) of this isotope?

**31. ssm** How many half-lives are required for the number of radioactive nuclei to decrease to one-millionth of the initial number?

**32.** Iodine $^{131}_{53}$I is used in diagnostic and therapeutic techniques in the treatment of thyroid disorders. This isotope has a half-life of 8.04 days. What percentage of an initial sample of $^{131}_{53}$I remains after 30.0 days?

**33.** A device used in radiation therapy for cancer contains 0.50 g of cobalt $^{60}_{27}$Co (59.933 819 u). The half-life of $^{60}_{27}$Co is 5.27 yr. Determine the activity of the radioactive material.

**34.** Strontium $^{90}_{38}$Sr has a half-life of 28.5 yr. It is chemically similar to calcium, enters the body through the food chain, and collects in the bones. Consequently, $^{90}_{38}$Sr is a particularly serious health hazard. How long (in years) will it take for 99.9900% of the $^{90}_{38}$Sr released in a nuclear reactor accident to disappear?

**35. ssm** If the activity of a radioactive substance is initially 398 disintegrations/min and two days later it is 285 disintegrations/min, what is the activity four days later still, or six days after the start? Give your answer in disintegrations/min.

* **36.** To make the dial of a watch glow in the dark, $1.000 \times 10^{-9}$ kg of radium $^{226}_{88}$Ra is used. The half-life of this isotope is $1.60 \times 10^3$ yr. How many kilograms of radium *disappear* while the watch is in use for fifty years?

* **37.** Refer to **Interactive Solution 31.37** at **www.wiley.com/college/ cutnell** for one approach to solving this problem. To see why one curie of activity was chosen to be $3.7 \times 10^{10}$ Bq, determine the activity (in disintegrations per second) of one gram of radium $^{226}_{88}$Ra ($T_{1/2} = 1.6 \times 10^3$ yr).

* **38.** A sample of ore containing radioactive strontium $^{90}_{38}$Sr has an activity of $6.0 \times 10^5$ Bq. The atomic mass of strontium is 89.908 u and its half life is 28.5 yr. How many grams of strontium are in the sample?

* **39. ssm** Two radioactive nuclei A and B are present in equal numbers to begin with. Three days later, there are three times as many A nuclei as there are B nuclei. The half-life of species B is 1.50 days. Find the half-life of species A.

### Section 31.7 Radioactive Dating

**40.** An archeological specimen containing 9.2 g of carbon has an activity of 1.6 Bq. How old (in years) is the specimen?

**41. ssm** The practical limit to ages that can be determined by radiocarbon dating is about 41 000 yr. In a 41 000 yr-old sample, what percentage of the original $^{14}_6$C atoms remains?

**42.** Review Conceptual Example 12 before starting to solve this problem. The number of unstable nuclei remaining after a time $t = 5.00$ yr is $N$, and the number present initially is $N_0$. Find the ratio $N/N_0$ for (a) $^{14}_6$C (half-life = 5730 yr), (b) $^{15}_8$O (half-life = 122.2 s; use $t = 1.00$ h, since otherwise the answer is out of the range of your calculator), and (c) $^3_1$H (half-life = 12.33 yr). Verify that your answers are consistent with the reasoning in Conceptual Example 12.

**43.** The shroud of Turin is a religious artifact known since the Middle Ages. In 1988 its age was measured using the radiocarbon dating technique, which revealed that the shroud could not have been made before 1200 AD. Of the $^{14}_6$C nuclei that were present in the living matter from which the shroud was made, what percentage remained in 1988?

**44.** The half-life for the $\alpha$ decay of uranium $^{238}_{92}$U is $4.47 \times 10^9$ yr. Determine the age (in years) of a rock specimen that contains sixty percent of its original number of $^{238}_{92}$U atoms.

* **45. ssm** When any radioactive dating method is used, experimental error in the measurement of the sample's activity leads to error in the estimated age. In an application of the radiocarbon dating technique to certain fossils, an activity of 0.10 Bq per gram of carbon is measured to within an accuracy of $\pm$ten percent. Find the age of the fossils and the maximum error (in years) in the value obtained. Assume that there is no error in the 5730-year half-life of $^{14}_6$C nor in the value of 0.23 Bq per gram of carbon in a living organism.

** **46.** (a) A sample is being dated by the radiocarbon technique. If the sample were uncontaminated, its activity would be 0.011 Bq per gram of carbon. Find the true age (in years) of the sample. (b) Suppose the sample is contaminated, so that only 98.0% of its carbon is ancient carbon. The remaining 2.0% is fresh carbon, in the sense that the $^{14}_6$C it contains has not had any time to decay. Assuming that the lab technician is unaware of the contamination, what apparent age (in years) would be determined for the sample?

# Chapter 32 Ionizing Radiation, Nuclear Energy, and Elementary Particles

**The physics of**
the biological effects of ionizing radiation.

## 32.1 Biological Effects of Ionizing Radiation

*Ionizing radiation* consists of photons and/or moving particles that have sufficient energy to knock an electron out of an atom or molecule, thus forming an ion. The photons usually lie in the ultraviolet, X-ray, or γ-ray regions of the electromagnetic spectrum (see Figure 24.9), whereas the moving particles can be the α and β particles emitted during radioactive decay. An energy of roughly 1 to 35 eV is needed to ionize an atom or molecule, and the particles and γ rays emitted during nuclear disintegration often have energies of several million eV. Therefore, a single α particle, β particle, or γ ray can ionize thousands of molecules.

Nuclear radiation is potentially harmful to humans because the ionization it produces can alter significantly the structure of molecules within a living cell. The alterations can lead to the death of the cell and even the organism itself. Despite the potential hazards, ionizing radiation is used in medicine for diagnostic and therapeutic purposes, such as locating bone fractures and treating cancer. The hazards can be minimized only if the fundamentals of radiation exposure, including dose units and the biological effects of radiation, are understood.

*Exposure* is a measure of the ionization produced in air by X-rays or γ rays, and it is defined in the following manner. A beam of X-rays or γ rays is sent through a mass $m$ of dry air at standard temperature and pressure (STP: 0 °C, 1 atm pressure). In passing through the air, the beam produces positive ions whose total charge is $q$. Exposure is defined as the total charge per unit mass of air: exposure = $q/m$. The SI unit for exposure is coulombs per kilogram (C/kg). However, the first radiation unit to be defined was the *roentgen* (R), and it is still used today. With $q$ expressed in coulombs (C) and $m$ in kilograms (kg), the exposure in roentgens is given by

$$\text{Exposure (in roentgens)} = \left(\frac{1}{2.58 \times 10^{-4}}\right)\frac{q}{m} \tag{32.1}$$

Thus, when X-rays or γ rays produce an exposure of one roentgen, $q = 2.58 \times 10^{-4}$ C of positive charge are produced in $m = 1$ kg of dry air:

$$1\,\text{R} = 2.58 \times 10^{-4}\,\text{C/kg} \qquad \text{(dry air, at STP)}$$

Since the concept of exposure is defined in terms of the ionizing abilities of X-rays and γ rays in air, it does not specify the effect of radiation on living tissue. For biological purposes, the *absorbed dose* is a more suitable quantity because it is the energy absorbed from the radiation per unit mass of absorbing material:

$$\text{Absorbed dose} = \frac{\text{Energy absorbed}}{\text{Mass of absorbing material}} \tag{32.2}$$

The SI unit of absorbed dose is the *gray* (Gy), which is the unit of energy divided by the unit of mass: 1 Gy = 1 J/kg. Equation 32.2 is applicable to all types of radiation and absorbing media. Another unit is often used for absorbed dose—namely, the *rad* (rd). The word "rad" is an acronym for **r**adiation **a**bsorbed **d**ose. The rad and the gray are related by

$$1\,\text{rad} = 0.01\,\text{gray}$$

The amount of biological damage produced by ionizing radiation is different for different kinds of radiation. For instance, a 1-rad dose of neutrons is far more likely to produce eye cataracts than a 1-rad dose of X-rays. To compare the damage caused by different types of radiation, *the relative biological effectiveness* (RBE) is used.* The relative biological effectiveness of a particular type of radiation is the ratio of the dose of 200-keV X-rays needed to produce a certain biological effect to the dose of the radiation needed to produce the same biological effect:

$$\text{Relative biological effectiveness (RBE)} = \frac{\text{The dose of 200-keV X-rays that produces a certain biological effect}}{\text{The dose of radiation that produces the same biological effect}} \qquad (32.3)$$

The RBE depends on the nature of the ionizing radiation and its energy, as well as the type of tissue being irradiated. Table 32.1 lists some typical RBE values for different kinds of radiation, assuming that an "average" biological tissue is being irradiated. A value of RBE = 1 for $\gamma$ rays and $\beta^-$ particles indicates that they produce the same biological damage as do 200-keV X-rays. The larger RBE values for protons, $\alpha$ particles, and fast neutrons indicate that they cause substantially more damage. The RBE is often used in conjunction with the absorbed dose to reflect the damage-producing character of the radiation. The product of the absorbed dose in rads (not in grays) and the RBE is the *biologically equivalent dose.*

$$\text{Biologically equivalent dose (in rems)} = \frac{\text{Absorbed dose}}{\text{(in rads)}} \times \text{RBE} \qquad (32.4)$$

The unit for the biologically equivalent dose is the *rem*, short for **r**oentgen **e**quivalent, **m**an. Example 1 illustrates the use of the biologically equivalent dose.

### Example 1 Comparing Absorbed Doses of $\gamma$ Rays and Neutrons

A biological tissue is irradiated with $\gamma$ rays that have an RBE of 0.70. The absorbed dose of $\gamma$ rays is 850 rd. The tissue is then exposed to neutrons whose RBE is 3.5. The biologically equivalent dose of the neutrons is the same as that of the $\gamma$ rays. What is the absorbed dose of neutrons?

**Reasoning** The biologically equivalent doses of the neutrons and the $\gamma$ rays are the same. Therefore, the tissue damage produced in each case is the same. However, the RBE of the neutrons is larger than the RBE of the $\gamma$ rays by a factor of 3.5/0.70 = 5.0. Consequently, we will find that the absorbed dose of the neutrons is only one-fifth as great as that of the $\gamma$ rays.

**Solution** According to Equation 32.4, the biologically equivalent dose is the product of the absorbed dose (in rads) and the RBE; it is the same for the $\gamma$ rays and the neutrons. Therefore, we have

$$\text{Biologically equivalent dose} = (\text{Absorbed dose})_{\gamma\,\text{rays}}\,\text{RBE}_{\gamma\,\text{rays}} = (\text{Absorbed dose})_{\text{neutrons}}\,\text{RBE}_{\text{neutrons}}$$

Solving for the absorbed dose of the neutrons gives

$$(\text{Absorbed dose})_{\text{neutrons}} = (\text{Absorbed dose})_{\gamma\,\text{rays}} \left(\frac{\text{RBE}_{\gamma\,\text{rays}}}{\text{RBE}_{\text{neutrons}}}\right) = (850\text{ rd})\left(\frac{0.70}{3.5}\right) = \boxed{170\text{ rd}}$$

Everyone is continually exposed to background radiation from natural sources, such as cosmic rays (high-energy particles that come from outside the solar system), radioactive materials in the environment, radioactive nuclei (primarily carbon $^{14}_{6}\text{C}$ and potassium $^{40}_{19}\text{K}$) within our own bodies, and radon. Table 32.2 lists the average biologically equivalent doses received from these sources by a person in the United States. According to this table, radon is a major contributor to the natural background radiation. Radon is an odorless radioactive gas and poses a health hazard because, when inhaled, it can damage the lungs and cause cancer. Radon is found in soil and rocks and enters houses through cracks and crevices in the foundation. The amount of radon in the soil varies greatly throughout the country, with some localities having significant amounts and others having virtually none. Accordingly, the dose that any individual receives can vary widely from the average value of 200 mrem/yr given in Table 32.2 (1 mrem = $10^{-3}$ rem). In many houses, the entry of radon can be reduced significantly by sealing the foundation against entry of the gas and providing good ventilation so it does not accumulate.

* The RBE is sometimes called the *quality factor* (QF).

**Table 32.1** *Relative Biological Effectiveness (RBE) for Various Types of Radiation*

| Type of Radiation | RBE |
|---|---|
| 200-keV X-rays | 1 |
| $\gamma$ rays | 1 |
| $\beta^-$ particles (electrons) | 1 |
| Protons | 10 |
| $\alpha$ particles | 10–20 |
| Neutrons | |
| Slow | 2 |
| Fast | 10 |

**Table 32.2** *Average Biologically Equivalent Doses of Radiation Received by a U. S. Resident*[a]

| Source of Radiation | Biologically Equivalent Dose (mrem/yr)[b] |
|---|---|
| *Natural background radiation* | |
| Cosmic rays | 28 |
| Radioactive earth and air | 28 |
| Internal radioactive nuclei | 39 |
| Inhaled radon | $\approx$200 |
| *Man-made radiation* | |
| Consumer products | 10 |
| Medical/dental diagnostics | 39 |
| Nuclear medicine | 14 |
| Rounded total: | 360 |

[a] National Council on Radiation Protection and Measurement, Report No. 93, "Ionizing Radiation Exposure of the Population of the United States," 1987.

[b] 1 mrem = $10^{-3}$ rem.

To the natural background of radiation, a significant amount of man-made radiation has been added, mostly from medical/dental diagnostic X-rays. Table 32.2 indicates an average total dose of 360 mrem/yr from all sources.

The effects of radiation on humans can be grouped into two categories, according to the time span between initial exposure and the appearance of physiological symptoms: (1) short-term or acute effects that appear within a matter of minutes, days, or weeks, and (2) long-term or latent effects that appear years, decades, or even generations later.

*Radiation sickness* is the general term applied to the acute effects of radiation. Depending on the severity of the dose, a person with radiation sickness can exhibit nausea, vomiting, fever, diarrhea, and loss of hair. Ultimately, death can occur. The severity of radiation sickness is related to the dose received, and in the following discussion the biologically equivalent doses quoted are whole-body, single doses. A dose less than 50 rem causes no short-term ill effects. A dose between 50 and 300 rem brings on radiation sickness, the severity increasing with increasing dosage. A whole-body dose in the range of 400–500 rem is classified as an $LD_{50}$ dose, meaning that it is a lethal dose (LD) for about 50% of the people so exposed; death occurs within a few months. Whole-body doses greater than 600 rem result in death for almost all individuals.

Long-term or latent effects of radiation may appear as a result of high-level brief exposure or low-level exposure over a long period of time. Some long-term effects are hair loss, eye cataracts, and various kinds of cancer. In addition, genetic defects caused by mutated genes may be passed on from one generation to the next.

Because of the hazards of radiation, the federal government has established dose limits. The permissible dose for an individual is defined as the dose, accumulated over a long period of time or resulting from a single exposure, that carries negligible probability of a severe health hazard. Federal standards (1991) state that an individual in the general population should not receive more than 500 mrem of man-made radiation each year, *exclusive* of medical sources. A person exposed to radiation in the workplace (e.g., a radiation therapist) should not receive more than 5 rem per year from work-related sources.

## 32.2  *Induced Nuclear Reactions*

Section 31.4 discusses how a radioactive parent nucleus disintegrates spontaneously into a daughter nucleus. It is also possible to bring about, or induce, the disintegration of a stable nucleus by striking it with another nucleus, an atomic or subatomic particle, or a $\gamma$-ray photon. A *nuclear reaction* is said to occur whenever the incident nucleus, particle, or photon causes a change to occur in a target nucleus.

In 1919, Ernest Rutherford observed that when an $\alpha$ particle strikes a nitrogen nucleus, an oxygen nucleus and a proton are produced. This nuclear reaction is written as

$$\underbrace{{}^{4}_{2}\text{He}}_{\substack{\text{Incident} \\ \alpha \text{ particle}}} + \underbrace{{}^{14}_{7}\text{N}}_{\substack{\text{Nitrogen} \\ \text{(target)}}} \longrightarrow \underbrace{{}^{17}_{8}\text{O}}_{\text{Oxygen}} + \underbrace{{}^{1}_{1}\text{H}}_{\text{Proton, } p}$$

Because the incident $\alpha$ particle induces the transmutation of nitrogen into oxygen, this reaction is an example of an *induced nuclear transmutation.*

Nuclear reactions are often written in a shorthand form. For example, the reaction above is designated by ${}^{14}_{7}\text{N}\ (\alpha, p)\ {}^{17}_{8}\text{O}$. The first and last symbols represent the initial and final nuclei, respectively. The symbols within the parentheses denote the incident $\alpha$ particle (on the left) and the small emitted particle or proton $p$ (on the right). Some other induced nuclear transmutations are listed below, together with the equivalent shorthand notations:

| Nuclear Reaction | Notation |
|---|---|
| ${}^{1}_{0}\text{n} + {}^{10}_{5}\text{B} \rightarrow {}^{7}_{3}\text{Li} + {}^{4}_{2}\text{He}$ | ${}^{10}_{5}\text{B}\ (n, \alpha)\ {}^{7}_{3}\text{Li}$ |
| $\gamma + {}^{25}_{12}\text{Mg} \rightarrow {}^{24}_{11}\text{Na} + {}^{1}_{1}\text{H}$ | ${}^{25}_{12}\text{Mg}\ (\gamma, p)\ {}^{24}_{11}\text{Na}$ |
| ${}^{1}_{1}\text{H} + {}^{13}_{6}\text{C} \rightarrow {}^{14}_{7}\text{N} + \gamma$ | ${}^{13}_{6}\text{C}\ (p, \gamma)\ {}^{14}_{7}\text{N}$ |

Induced nuclear reactions, like the radioactive decay process discussed in Section 31.4, obey the conservation laws of physics. We have discussed these laws in previous chapters and now note the important role that the conservation laws play in all nuclear processes. In particular, both the total electric charge of the nucleons and the total number of nucleons are conserved during an induced nuclear reaction. The fact that these quantities are conserved makes it possible to identify the nucleus produced in a reaction, as the next example illustrates.

### Example 2 An Induced Nuclear Transmutation

An $\alpha$ particle strikes an aluminum $^{27}_{13}\text{Al}$ nucleus. As a result, an unknown nucleus $^A_Z\text{X}$ and a neutron $^1_0\text{n}$ are produced:

$$^4_2\text{He} + {}^{27}_{13}\text{Al} \longrightarrow {}^A_Z\text{X} + {}^1_0\text{n}$$

Identify the nucleus produced, including its atomic number $Z$ (the number of protons) and its atomic mass number $A$ (the number of nucleons).

**Reasoning** The total electric charge of the nucleons is conserved, so that we can set the total number of protons before the reaction equal to the total number after the reaction. The total number of nucleons is also conserved, so that we can set the total number before the reaction equal to the total number after the reaction. These two conserved quantities will allow us to identify the nucleus $^A_Z\text{X}$.

**Solution** The conservation of total electric charge and total number of nucleons leads to the equations listed below:

| Conserved Quantity | Before Reaction | | After Reaction |
|---|---|---|---|
| Total electric charge (number of protons) | $2 + 13$ | $=$ | $Z + 0$ |
| Total number of nucleons | $4 + 27$ | $=$ | $A + 1$ |

Solving these equations for $Z$ and $A$ gives $Z = 15$ and $A = 30$. Since $Z = 15$ identifies the element as phosphorus (P), the nucleus produced is $\boxed{^{30}_{15}\text{P}}$.

Induced nuclear transmutations can be used to produce isotopes that are not found naturally. In 1934, Enrico Fermi suggested a method for producing elements with a higher atomic number than uranium ($Z = 92$). These elements—neptunium ($Z = 93$), plutonium ($Z = 94$), americium ($Z = 95$), and so on—are known as *transuranium elements,* and none occurs naturally. They are created in a nuclear reaction between a suitably chosen lighter element and a small incident particle, usually a neutron or an $\alpha$ particle. For example, Figure 32.1 shows a reaction that produces plutonium from uranium. A neutron is captured by a uranium $^{238}_{92}\text{U}$ nucleus, producing $^{239}_{92}\text{U}$ and a $\gamma$ ray. The $^{239}_{92}\text{U}$ nucleus is radioactive and decays with a half-life of 23.5 min into neptunium $^{239}_{93}\text{Np}$. Neptunium is also radioactive and disintegrates with a half-life of 2.4 days into plutonium $^{239}_{94}\text{Pu}$. Plutonium is the final product and has a half-life of 24 100 yr.

The neutrons that participate in nuclear reactions can have kinetic energies that cover a wide range. In particular, those that have a kinetic energy of about 0.04 eV or less are called **thermal neutrons.** The name derives from the fact that such a relatively small kinetic energy is comparable to the average translational kinetic energy of a molecule at room temperature.

## 32.3 Nuclear Fission

In 1939 four German scientists, Otto Hahn, Lise Meitner, Fritz Strassmann, and Otto Frisch, made an important discovery that ushered in the atomic age. They found that a uranium nucleus, after absorbing a neutron, splits into two fragments, each with a smaller mass than the original nucleus. The splitting of a massive nucleus into two less massive fragments is known as *nuclear fission.*

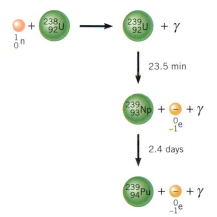

**Figure 32.1** An induced nuclear reaction is shown in which $^{238}_{92}\text{U}$ is transmuted into the transuranium element plutonium $^{239}_{94}\text{Pu}$.

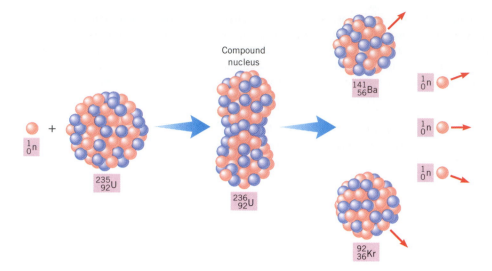

**Figure 32.2** A slowly moving neutron causes the uranium nucleus $^{235}_{92}$U to fission into barium $^{141}_{56}$Ba, krypton $^{92}_{36}$Kr, and three neutrons.

Figure 32.2 shows a fission reaction in which a uranium $^{235}_{92}$U nucleus is split into barium $^{141}_{56}$Ba and krypton $^{92}_{36}$Kr nuclei. The reaction begins when $^{235}_{92}$U absorbs a slowly moving neutron, creating a "compound nucleus," $^{236}_{92}$U. The compound nucleus disintegrates quickly into $^{141}_{56}$Ba, $^{92}_{36}$Kr, and three neutrons according to the following reaction:

$$\underbrace{^{1}_{0}n + ^{235}_{92}U}_{} \longrightarrow \underbrace{^{236}_{92}U}_{\substack{\text{Compound}\\\text{nucleus}\\\text{(unstable)}}} \longrightarrow \underbrace{^{141}_{56}Ba}_{\text{Barium}} + \underbrace{^{92}_{36}Kr}_{\text{Krypton}} + \underbrace{3^{1}_{0}n}_{\text{3 neutrons}}$$

This reaction is only one of the many possible reactions that can occur when uranium fissions. For example, another reaction is

$$^{1}_{0}n + ^{235}_{92}U \longrightarrow \underbrace{^{236}_{92}U}_{\substack{\text{Compound}\\\text{nucleus}\\\text{(unstable)}}} \longrightarrow \underbrace{^{140}_{54}Xe}_{\text{Xenon}} + \underbrace{^{94}_{38}Sr}_{\text{Strontium}} + \underbrace{2^{1}_{0}n}_{\text{2 neutrons}}$$

Some reactions produce as many as 5 neutrons; however, the average number produced per fission is 2.5.

When a neutron collides with and is absorbed by a uranium nucleus, the uranium nucleus begins to vibrate and becomes distorted. The vibration continues until the distortion becomes so severe that the attractive strong nuclear force can no longer balance the electrostatic repulsion between the nuclear protons. At this point, the nucleus bursts apart into fragments, which carry off energy, primarily in the form of kinetic energy. The energy carried off by the fragments is enormous and was stored in the original nucleus primarily in the form of electric potential energy. An average of roughly 200 MeV of energy is released per fission. This energy is approximately $10^{8}$ times greater than the energy released per molecule in an ordinary chemical reaction, such as the combustion of gasoline or coal. Example 3 demonstrates how to estimate the energy released during the fission of a nucleus.

### Example 3   The Energy Released During Nuclear Fission

Estimate the amount of energy released when a massive nucleus ($A = 240$) fissions.

**Reasoning** Figure 31.5 shows that the binding energy of a nucleus with $A = 240$ is about 7.6 MeV per nucleon. We assume that this nucleus fissions into two fragments, each with $A \approx 120$. According to Figure 31.5, the binding energy of the fragments increases to about 8.5 MeV per nucleon. Consequently, when a massive nucleus fissions, there is a release of about 8.5 MeV − 7.6 MeV = 0.9 MeV of energy per nucleon.

**Solution** Since there are 240 nucleons involved in the fission process, the total energy released per fission is approximately (0.9 MeV/nucleon)(240 nucleons) $\approx$ $\boxed{200 \text{ MeV}}$.

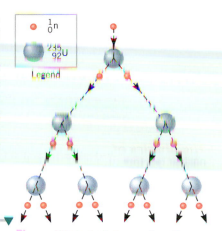

Virtually all naturally occurring uranium is composed of two isotopes. These isotopes and their natural abundances are $^{238}_{92}U$ (99.275%) and $^{235}_{92}U$ (0.720%). Although $^{238}_{92}U$ is by far the most abundant isotope, the probability that it will capture a neutron and fission is very small. For this reason, $^{238}_{92}U$ is not the isotope of choice for generating nuclear energy. In contrast, the isotope $^{235}_{92}U$ readily captures a neutron and fissions, *provided the neutron is a thermal neutron* (kinetic energy $\approx 0.04$ eV or less). The probability of a thermal neutron causing $^{235}_{92}U$ to fission is about five hundred times greater than a neutron whose energy is relatively high—say, 1 MeV. Thermal neutrons can also be used to fission other nuclei, such as plutonium $^{239}_{94}Pu$. Conceptual Example 4 deals with one of the reasons why thermal neutrons are useful for inducing nuclear fission.

## Conceptual Example 4
### Thermal Neutrons Versus Thermal Protons or Alpha Particles

Why is it possible for a thermal neutron (i.e., one with a relatively small kinetic energy) to penetrate a nucleus, whereas a proton or an $\alpha$ particle would need a much larger amount of energy to penetrate the same nucleus?

**Reasoning and Solution** To penetrate a nucleus, a particle such as a neutron, a proton, or an $\alpha$ particle must have enough kinetic energy to do the work of overcoming any repulsive force that it encounters along the way. A repulsive force can arise because protons in the nucleus are electrically charged. *Since a neutron is electrically neutral, however, it encounters no electrostatic force of repulsion as it approaches the nuclear protons and, hence, needs relatively little energy to reach the nucleus.* In contrast, a proton and an $\alpha$ particle both carry positive charge, and each encounters an electrostatic force of repulsion as it approaches the nuclear protons. It is true that each also experiences the attractive strong nuclear force from the nuclear protons and neutrons. But the strong nuclear force has an extremely short range of action and, therefore, comes into play only when an impinging particle reaches the nucleus. In comparison, the electrostatic force of repulsion has a long range of action and is encountered by an incoming charged particle throughout the entire journey to the target nucleus. Consequently, *an impinging proton or $\alpha$ particle needs a relatively large amount of kinetic energy to do the work of overcoming the repulsive electrostatic force.*

**Related Homework:** *Problem 40*

The fact that the uranium fission reaction releases 2.5 neutrons, on the average, makes it possible for a self-sustaining series of fissions to occur. As Figure 32.3 illustrates, each neutron released can initiate another fission event, resulting in the emission of still more neutrons, followed by more fissions, and so on. A *chain reaction* is a series of nuclear fissions whereby some of the neutrons produced by each fission cause additional fissions. During an uncontrolled chain reaction, it would not be unusual for the number of fissions to increase a thousandfold within a few millionths of a second. With an average energy of about 200 MeV being released per fission, an uncontrolled chain reaction can generate an incredible amount of energy in a very short time, as happens in an atomic bomb (which is actually a *nuclear* bomb).

By limiting the number of neutrons in the environment of the fissile nuclei, it is possible to establish a condition whereby each fission event contributes, on average, only *one neutron* that fissions another nucleus (see Figure 32.4). In this manner, the chain reaction and the rate of energy production are *controlled*. The controlled fission chain reaction is the principle behind nuclear reactors used in the commercial generation of electric power.

## 32.4 *Nuclear Reactors*

A nuclear reactor is a type of furnace in which energy is generated by a controlled fission chain reaction. The first nuclear reactor was built by Enrico Fermi in 1942, on the floor of a squash court under the west stands of Stagg Field at the University of Chicago. Today, there are a number of kinds and sizes of reactors, and many have the same three basic components: fuel elements, a neutron moderator, and control rods. Figure 32.5 illustrates these components.

**Figure 32.3** A chain reaction. For clarity, it is assumed that each fission generates two neutrons (2.5 neutrons are actually liberated on the average). The fission fragments are not shown.

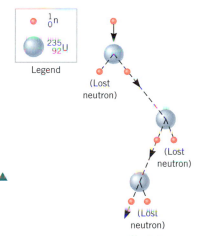

**Figure 32.4** In a controlled chain reaction, only one neutron, on average, from each fission event causes another nucleus to fission. As a result, energy is released at a steady or controlled rate.

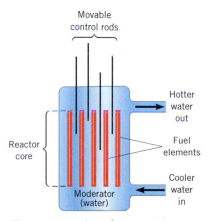

**Figure 32.5** A nuclear reactor consists of fuel elements, control rods, and a moderator (in this case, water).

The *fuel elements* contain the fissile fuel and, for example, may be thin rods about 1 cm in diameter. In a large power reactor there may be thousands of fuel elements placed close together, the entire region of fuel elements being known as the *reactor core.*

Uranium $^{235}_{92}$U is a common reactor fuel. Since the natural abundance of this isotope is only about 0.7%, there are special uranium-enrichment plants to increase the percentage. Most commercial reactors use uranium in which the amount of $^{235}_{92}$U has been enriched to about 3%.

Whereas neutrons with energies of about 0.04 eV (or less) readily fission $^{235}_{92}$U, the neutrons released during the fission process have significantly greater energies of several MeV or so. Consequently, a nuclear reactor must contain some type of material that will decrease or moderate the speed of such energetic neutrons so they can readily fission additional $^{235}_{92}$U nuclei. The material that slows down the neutrons is called a *moderator.* One commonly used moderator is water. When an energetic neutron leaves a fuel element, the neutron enters the surrounding water and collides with water molecules. With each collision, the neutron loses an appreciable fraction of its energy and slows down. Once slowed down to thermal energy by the moderator, a process that takes less than $10^{-3}$ s, the neutron is capable of initiating a fission event upon reentering a fuel element.

If the output power from a reactor is to remain constant, only one neutron from each fission event must trigger a new fission, as Figure 32.4 suggests. When each fission leads to one additional fission—no more or no less—the reactor is said to be *critical.* A reactor normally operates in a critical condition, because then it produces a steady output of energy. The reactor is *subcritical* when, on average, the neutrons from each fission trigger *less than one* subsequent fission. In a subcritical reactor, the chain reaction is not self-sustaining and eventually dies out. When the neutrons from each fission trigger *more than one* additional fission, the reactor is *supercritical.* During a supercritical condition, the energy released by a reactor increases. If left unchecked, the increasing energy can lead to a partial or total meltdown of the reactor core, with the possible release of radioactive material into the environment.

Clearly, a control mechanism is needed to keep the reactor in its normal or critical state. This control is accomplished by a number of *control rods* that can be moved into and out of the reactor core (see Figure 32.5). The control rods contain an element, such as boron or cadmium, that readily absorbs neutrons without fissioning. If the reactor becomes supercritical, the control rods are automatically moved farther into the core to absorb the excess neutrons causing the condition. In response, the reactor returns to its critical state. Conversely, if the reactor becomes subcritical, the control rods are partially withdrawn from the core. Fewer neutrons are absorbed, more neutrons are available for fission, and the reactor again returns to its critical state.

Figure 32.6 illustrates a pressurized water reactor. In such a reactor, the heat generated within the fuel rods is carried away by water that surrounds the rods. To remove as much heat as possible, the water temperature is allowed to rise to a high value (about

**The physics of** nuclear reactors.

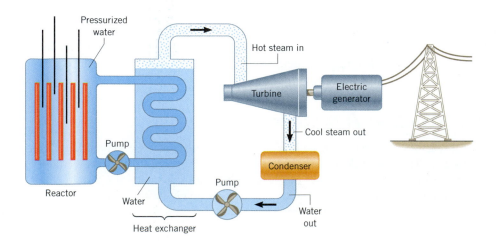

**Figure 32.6** Diagram of a nuclear power plant that uses a pressurized water reactor.

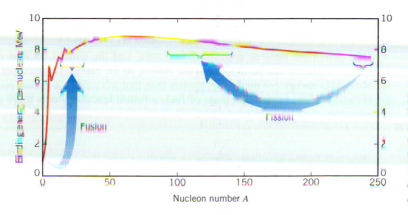

**Figure 32.7** When fission occurs, a massive nucleus divides into two fragments whose binding energy per nucleon is greater than that of the original nucleus. When fusion occurs, two low-mass nuclei combine to form a more massive nucleus whose binding energy per nucleon is greater than that of the original nuclei.

300 °C). To prevent boiling, which occurs at 100 °C at 1 atmosphere of pressure, the water is pressurized in excess of 150 atmospheres. The hot water is pumped through a heat exchanger, where heat is transferred to water flowing in a second, closed system. The heat transferred to the second system produces steam that drives a turbine. The turbine is coupled to an electric generator, whose output electric power is delivered to consumers via high-voltage transmission lines. After exiting the turbine, the steam is condensed back into water that is returned to the heat exchanger.

## 32.5 Nuclear Fusion

In Example 3 in Section 32.3, the binding-energy-per-nucleon curve is used to estimate the amount of energy released in the fission process. As summarized in Figure 32.7, the massive nuclei at the right end of the curve have a binding energy of about 7.6 MeV per nucleon. The less massive fission fragments are near the center of the curve and have a binding energy of approximately 8.5 MeV per nucleon. The energy released per nucleon by fission is the difference between these two values, or about 0.9 MeV per nucleon.

A glance at the far left end of the diagram in Figure 32.7 suggests another means of generating energy. Two nuclei with very low mass and relatively small binding energies per nucleon could be combined or "fused" into a single, more massive nucleus that has a greater binding energy per nucleon. This process is called *nuclear fusion.* A substantial amount of energy can be released during a fusion reaction, as Example 5 shows.

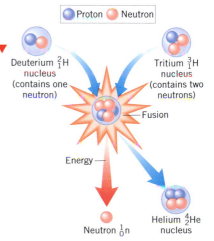

### Example 5 The Energy Released During Nuclear Fusion

Two isotopes of hydrogen, $^2_1\text{H}$ (deuterium, D) and $^3_1\text{H}$ (tritium, T), fuse to form $^4_2\text{He}$ and a neutron according to the following reaction:

$$^2_1\text{H} + ^3_1\text{H} \longrightarrow ^4_2\text{He} + ^1_0\text{n}$$

Determine the energy released by this fusion reaction, which is illustrated in Figure 32.8.

**Reasoning** Energy is released, so the total mass of the final nuclei is less than the total mass of the initial nuclei. To determine the energy released, we find the amount (in atomic mass units u) by which the total mass has decreased. Then, we use the fact that 1 u is equivalent to 931.5 MeV of energy, as determined in Section 31.3. This approach is the same as that used in Section 31.4 for radioactive decay.

**Solution** The masses of the initial and final nuclei in this reaction, as well as that of the neutron, are

**Figure 32.8** Deuterium and tritium are fused together to form a helium nucleus ($^4_2\text{He}$). The result is the release of an enormous amount of energy, mainly carried by a single high-energy neutron ($^1_0\text{n}$).

| Initial Masses | | Final Masses | |
|---|---|---|---|
| $^2_1\text{H}$ | 2.0141 u | $^4_2\text{He}$ | 4.0026 u |
| $^3_1\text{H}$ | 3.0161 u | $^1_0\text{n}$ | 1.0087 u |
| Total: | 5.0302 u | Total: | 5.0113 u |

The decrease in mass, or the mass defect, is $\Delta m = 5.0302$ u $-$ 5.0113 u $= 0.0189$ u. Since 1 u is equivalent to 931.5 MeV, the energy released is $\boxed{17.6 \text{ MeV}}$.

The deuterium nucleus contains 2 nucleons, and the tritium nucleus contains 3. Thus, there are 5 nucleons that participate in the fusion, so the energy released per nucleon is about 3.5 MeV. This energy per nucleon is greater than that released in a fission process ($\approx 0.9$ MeV per nucleon). Thus, for a given mass of fuel, a fusion reaction yields more energy than a fission reaction.

---

Because fusion reactions release so much energy, there is considerable interest in fusion reactors, although to date no commercial units have been constructed. The difficulties in building a fusion reactor arise mainly because the two low-mass nuclei must be brought sufficiently near each other so that the short-range strong nuclear force can pull them together, leading to fusion. But each nucleus has a positive charge and repels the other electrically. For the nuclei to get sufficiently close in the presence of the repulsive electric force, they must have large kinetic energies, and hence large temperatures, to start with. For example, a temperature of around a hundred million °C is needed to start the deuterium–tritium reaction discussed in Example 5.

Reactions that require such extremely high temperatures are called ***thermonuclear reactions.*** The most important thermonuclear reactions occur in stars, such as our own sun. The energy radiated by the sun comes from deep within its core, where the temperature is high enough to initiate the fusion process. One group of reactions thought to occur in the sun is the *proton–proton* cycle, which is a series of reactions whereby six protons form a helium nucleus, two positrons, two $\gamma$ rays, two protons, and two neutrinos. The energy released by the proton–proton cycle is about 25 MeV (see problem 33).

Man-made fusion reactions have been carried out in a fusion-type nuclear bomb—commonly called a hydrogen bomb. In a hydrogen bomb, the fusion reaction is ignited by a fission bomb using uranium or plutonium. The temperature produced by the fission bomb is sufficiently high to initiate a thermonuclear reaction where, for example, hydrogen isotopes are fused into helium, releasing even more energy. For fusion to be useful as a commercial energy source, the energy must be released in a steady, controlled manner—unlike that in a bomb. To date, scientists have not succeeded in constructing a fusion device that produces more energy on a continual basis than is expended in operating the device. A fusion device uses a high temperature to start a reaction, and under such a condition, all the atoms are completely ionized to form a *plasma* (a gas composed of charged particles, like $^2_1\text{H}^+$ and $e^-$). The problem is to confine the hot plasma for a long enough time so that collisions among the ions can lead to fusion.

**The physics of** nuclear fusion using magnetic confinement.

One ingenious method of confining the plasma is called *magnetic confinement* because it uses a magnetic field to contain and compress the charges in the plasma. Charges moving in the magnetic field are subject to magnetic forces. As the forces increase, the associated pressure builds, and the temperature rises. The gas becomes a superheated plasma, ultimately fusing when the pressure and temperature are high enough.

**The physics of** nuclear fusion using inertial confinement.

Another type of confinement scheme, known as *inertial confinement,* is also being developed. Tiny, solid pellets of fuel are dropped into a container. As each pellet reaches the center of the container, a number of high-intensity laser beams strike the pellet simultaneously. The heating causes the exterior of the pellet to vaporize almost instantaneously. However, the inertia of the vaporized atoms keeps them from expanding outward as fast as the vapor is being formed. As a result, high pressures, high densities, and high temperatures are achieved at the center of the pellet, thus causing fusion. A variant of inertial confinement fusion, called "Z pinch," is under development at Sandia National Laboratories in New Mexico. This device would also implode tiny fuel pellets, but without using lasers. Instead, scientists are using a cylindrical array of fine tungsten wires that are connected to a gigantic capacitor. When the capacitor is discharged, a huge current is sent through the wires. The heated wires vaporize almost instantly, generating a hot gas of ions, or plasma. The plasma is driven inward upon itself by the huge magnetic field produced by the current. The compressed plasma becomes super hot and generates a gigantic X-ray pulse. This pulse, it is hoped, would implode the solid fuel pellets to temperatures and pressures at which fusion would occur.

When compared to fission, fusion has some attractive features as an energy source. As we have seen in Example 5, fusion yields more energy than fission, for a given mass of fuel. Moreover, one type of fuel, $^2_1\text{H}$ (deuterium), is found in the waters of the oceans and is plentiful, cheap, and relatively easy to separate from the common $^1_1\text{H}$ isotope of hydrogen. Fissile materials like naturally occurring uranium $^{235}_{92}\text{U}$ are much less available, and supplies could be depleted within a century or two. However, the commercial use of fusion to provide cheap energy remains in the future.

## 32.6 Elementary Particles

### SETTING THE STAGE

By 1932 the electron, the proton, and the neutron had been discovered and were thought to be nature's three *elementary particles,* in the sense that they were the basic building blocks from which all matter is constructed. Experimental evidence obtained since then, however, shows that several hundred additional particles exist, and scientists no longer believe that the proton and the neutron are elementary particles.

Most of these new particles have masses greater than the electron's mass, and many are more massive than protons or neutrons. Virtually all the new particles are unstable and decay in times between about $10^{-6}$ and $10^{-23}$ s.

Often, new particles are produced by accelerating protons or electrons to high energies and letting them collide with a target nucleus. For example, Figure 32.9 shows a collision between an energetic proton and a stationary proton. If the incoming proton has sufficient energy, the collision produces an entirely new particle, the *neutral pion* ($\pi^0$). The $\pi^0$ particle lives for only about $0.8 \times 10^{-16}$ s before it decays into two $\gamma$-ray photons. Since the pion did not exist before the collision, it was created from part of the incident proton's energy. Because a new particle such as the neutral pion is often created from energy, it is customary to report the mass of the particle in terms of its equivalent *rest energy* (see Equation 28.5). Often, energy units of MeV are used. For instance, detailed analyses of experiments reveal that the mass of the $\pi^0$ particle is equivalent to a rest energy of 135.0 MeV. For comparison, the more massive proton has a rest energy of 938.3 MeV. Analyses of experiments also provide the electric charge and other properties of particles created in high-energy collisions. In the limited space available here, it is not possible to describe all the new particles that have been found. However, we will highlight some of the more significant discoveries.

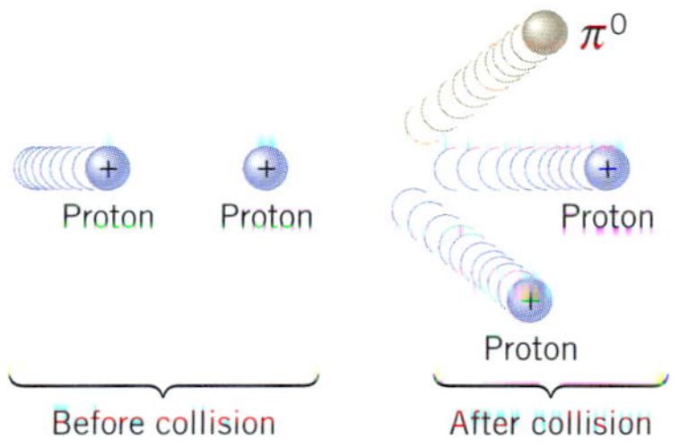

**Figure 32.9** When an energetic proton collides with a stationary proton, a neutral pion ($\pi^0$) is produced. Part of the energy of the incident proton goes into creating the pion.

### NEUTRINOS

In 1930, Wolfgang Pauli suggested that a particle called the *neutrino* (now known as the electron neutrino) should accompany the $\beta$ decay of a radioactive nucleus. As Section 31.5 discusses, the neutrino has no electric charge, has a very small mass (a tiny fraction of the mass of an electron), and travels at speeds approaching (but less than) the speed of light. Neutrinos were finally discovered in 1956. Today, neutrinos are created in abundance in nuclear reactors and particle accelerators and are thought to be plentiful in the universe.

### POSITRONS AND ANTIPARTICLES

The year 1932 saw the discovery of the *positron* (a contraction for "positive electron"). The positron has the same mass as the electron but carries an opposite charge of $+e$. A collision between a positron and an electron is likely to annihilate both particles, converting them into electromagnetic energy in the form of $\gamma$ rays. For this reason, positrons never coexist with ordinary matter for any appreciable length of time. The mutual annihilation of a positron and an electron lies at the heart of an important medical diagnostic technique, as Conceptual Example 6 discusses.

## Conceptual Example 6  Positron Emission Tomography

Positron emission tomography, or PET scanning, as it is known, utilizes positrons in the following way. Certain radioactive isotopes decay by positron emission—for example, oxygen $^{15}_{8}$O. Such isotopes are injected into the body, where they collect at specific sites. The positron ($^{0}_{1}$e) emitted during the decay of the isotope encounters an electron ($^{0}_{-1}$e) in the body tissue almost at once. The resulting mutual annihilation produces two γ-ray photons ($^{0}_{1}$e + $^{0}_{-1}$e → γ + γ), which are detected by devices mounted on a ring around the patient, as Figure 32.10a shows. The two photons strike oppositely positioned detectors and, in doing so, reveal the line on which the annihilation occurred. Such information leads to a computer-generated image that can be useful in diagnosing abnormalities at the site where the radioactive isotope collected (see Figure 32.11). How does the principle of conservation of linear momentum explain the fact that the photons strike detectors located opposite one another?

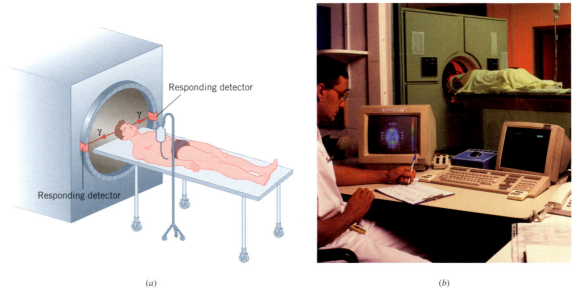

(a)                                                    (b)

**Figure 32.10** (a) In positron emission tomography, or PET scanning, a radioactive isotope is injected into the body. The isotope that is used decays by emitting a positron, which annihilates an electron in the body tissue, producing two γ-ray photons. These photons strike detectors mounted on opposite sides of a ring that surrounds the patient. (b) PET scanning is being used to monitor metabolic activity in the brain. (© CC Studio/Photo Researchers)

**Reasoning and Solution** The momentum conservation principle states that the total linear momentum of an isolated system remains constant (see Section 7.2). Therefore, the total momentum of the γ-ray photons will be equal to the total momentum of the positron and electron before annihilation, provided that the system is isolated. An isolated system is one on which no net external force acts. The positron and the electron do exert electrostatic forces on one another, however, since they carry electric charges. But these are internal, not external, forces and cannot change the total linear momentum of the two-particle system. The total photon momentum, then, must equal the total momentum of the positron and the electron, which is nearly zero, to the extent that the two particles have much less momentum than the photons do. With a total linear momentum of zero, the momentum vector of one photon must point opposite to the momentum vector of the other photon. Thus, the two photons depart from the annihilation site traveling in opposite directions and strike oppositely located detectors.

**Related Homework:** *Problem 37*

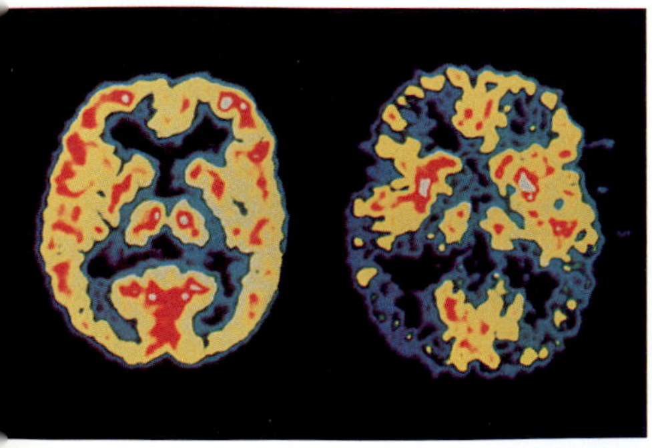

Healthy
brain                    Diseased
brain

**Figure 32.11** Positron emission tomography (PET) provides an important medical diagnostic technique. The PET scan of the diseased brain shown here reveals Alzheimer's disease. The colors are computer-generated. (© National Institutes of Health/Photo Researchers)

The positron is an example of an antiparticle, and after its discovery, scientists came to realize that for every type of particle there is a corresponding type of antiparticle. The antiparticle is a form of matter that has the same mass as the particle but carries an opposite electric charge (e.g., the electron–positron pair) or has a magnetic moment that is oriented in an opposite direction relative to the spin (e.g., the neutrino–antineutrino pair). A few electrically neutral particles, like the photon and the neutral pion ($\pi^0$), are their own antiparticles.

# MUONS AND PIONS

In 1937, the American physicists S. H. Neddermeyer and C. D. Anderson discovered a new charged particle whose mass was about 207 times greater than the mass of the electron. The particle is designated by the Greek letter $\mu$ (mu) and is known as a *muon*. There are two muons that have the same mass but opposite charge; the particle $\mu^-$ and its antiparticle $\mu^+$. The $\mu^-$ muon has the same charge as the electron, whereas the $\mu^+$ muon has the same charge as the positron. Both muons are unstable, with a lifetime of $2.2 \times 10^{-6}$ s. The $\mu^-$ muon decays into an electron ($\beta^-$), a muon neutrino ($\nu_\mu$), and an electron antineutrino ($\bar{\nu}_e$), according to the following reaction:

$$\mu^- \longrightarrow \beta^- + \nu_\mu + \bar{\nu}_e$$

The $\mu^+$ muon decays into a positron ($\beta^+$), a muon antineutrino ($\bar{\nu}_\mu$), and an electron neutrino ($\nu_e$):

$$\mu^+ \longrightarrow \beta^+ + \bar{\nu}_\mu + \nu_e$$

Muons interact with protons and neutrons via the weak nuclear force (see Section 31.5).

The Japanese physicist Hideki Yukawa (1907–1981) predicted in 1935 that *pions* exist, but they were not discovered until 1947. Pions come in three varieties: one that is positively charged, the negatively charged antiparticle with the same mass, and the neutral pion, mentioned earlier, which is its own antiparticle. The symbols for these pions are, respectively, $\pi^+$, $\pi^-$, and $\pi^0$. The charged pions are unstable and have a lifetime of $2.6 \times 10^{-8}$ s. The decay of a charged pion almost always produces a muon:

$$\pi^- \longrightarrow \mu^- + \bar{\nu}_\mu$$
$$\pi^+ \longrightarrow \mu^+ + \nu_\mu$$

As mentioned earlier, the neutral pion $\pi^0$ is also unstable and decays into two $\gamma$-ray photons, the lifetime being $0.8 \times 10^{-16}$ s. The pions are of great interest because, unlike the muons, the pions interact with protons and neutrons via the strong nuclear force.

# CLASSIFICATION OF PARTICLES

It is useful to group the known particles into three families—the photons, the leptons, and the hadrons—as Table 32.3 on p. 686 summarizes. This grouping is made according to the nature of the force by which a particle interacts with other particles. The *photon family,* for instance, has only one member, the photon. The photon interacts only with charged particles, and the interaction is only via the *electromagnetic force.* No other particle behaves in this manner.

The *lepton family* consists of particles that interact by means of the *weak nuclear force.* Leptons can also exert gravitational and (if the leptons are charged) electromagnetic forces on other particles. The four better-known leptons are the electron, the muon, the electron neutrino $\nu_e$, and the muon neutrino $\nu_\mu$. Table 32.3 lists these particles together with their antiparticles. Recently, two other leptons have been discovered, the tau particle ($\tau$) and its neutrino ($\nu_\tau$), bringing the number of particles in the lepton family to six.

The *hadron family* contains the particles that interact by means of the *strong nuclear force and the weak nuclear force.* Hadrons can also interact by gravitational and electromagnetic forces, but at short distances ($\leq 10^{-15}$ m) the strong nuclear force dominates. Among the hadrons are the proton, the neutron, and the pions. As Table 32.3 indicates, most hadrons are short-lived. The hadrons are subdivided into two groups, the *mesons* and the *baryons,* for a reason that will be discussed in connection with the idea of quarks.

# QUARKS

As more and more hadrons were discovered, it became clear that they were not all elementary particles. The suggestion was made that the hadrons are made up of smaller, more elementary particles called *quarks.* In 1963, a quark theory was advanced independently by M. Gell–Mann (1929–  ) and G. Zweig (1937–  ). The theory proposed that there are three quarks and three corresponding antiquarks, and that hadrons are constructed from combinations of these. Thus, the quarks are elevated to the status of ele-

**Table 32.3    Some Particles and Their Properties**

| Family | Particle | Particle Symbol | Antiparticle Symbol | Rest Energy (MeV) | Lifetime (s) |
|---|---|---|---|---|---|
| **Photon** | Photon | $\gamma$ | Self[a] | 0 | Stable |
| **Lepton** | Electron | $e^-$ or $\beta^-$ | $e^+$ or $\beta^+$ | 0.511 | Stable |
| | Muon | $\mu^-$ | $\mu^+$ | 105.7 | $2.2 \times 10^{-6}$ |
| | Tau | $\tau^-$ | $\tau^+$ | 1777 | $2.9 \times 10^{-13}$ |
| | Electron neutrino | $\nu_e$ | $\bar{\nu}_e$ | $\approx 0$ | Stable |
| | Muon neutrino | $\nu_\mu$ | $\bar{\nu}_\mu$ | $\approx 0$ | Stable |
| | Tau neutrino | $\nu_\tau$ | $\bar{\nu}_\tau$ | $\approx 0$ | Stable |
| **Hadron** | | | | | |
| *Mesons* | | | | | |
| | Pion | $\pi^+$ | $\pi^-$ | 139.6 | $2.6 \times 10^{-8}$ |
| | | $\pi^0$ | Self[a] | 135.0 | $8.4 \times 10^{-17}$ |
| | Kaon | $K^+$ | $K^-$ | 493.7 | $1.2 \times 10^{-8}$ |
| | | $K^0_S$ | $\bar{K}^0_S$ | 497.7 | $8.9 \times 10^{-11}$ |
| | | $K^0_L$ | $\bar{K}^0_L$ | 497.7 | $5.2 \times 10^{-8}$ |
| | Eta | $\eta^0$ | Self[a] | 547.3 | $< 10^{-18}$ |
| *Baryons* | | | | | |
| | Proton | $p$ | $\bar{p}$ | 938.3 | Stable |
| | Neutron | $n$ | $\bar{n}$ | 939.6 | 886 |
| | Lambda | $\Lambda^0$ | $\bar{\Lambda}^0$ | 1116 | $2.6 \times 10^{-10}$ |
| | Sigma | $\Sigma^+$ | $\bar{\Sigma}^-$ | 1189 | $8.0 \times 10^{-11}$ |
| | | $\Sigma^0$ | $\bar{\Sigma}^0$ | 1193 | $7.4 \times 10^{-20}$ |
| | | $\Sigma^-$ | $\bar{\Sigma}^+$ | 1197 | $1.5 \times 10^{-10}$ |
| | Omega | $\Omega^-$ | $\Omega^+$ | 1672 | $8.2 \times 10^{-11}$ |

[a] The particle is its own antiparticle.

mentary particles for the hadron family. The particles in the photon and lepton families are considered to be elementary, and as such they are not composed of quarks.

The three quarks were named *up* ($u$), *down* ($d$), and *strange* ($s$), and were assumed to have, respectively, fractional charges of $+\frac{2}{3}e$, $-\frac{1}{3}e$, and $-\frac{1}{3}e$. In other words, a quark possesses a charge magnitude smaller than that of an electron, which has a charge of $-e$. Table 32.4 lists the symbols and electric charges of these quarks and the corresponding antiquarks. Experimentally, quarks should be recognizable by their fractional charges, but in spite of an extensive search for them, free quarks have never been found.

According to the original quark theory, the mesons are different from the baryons, because each meson consists of only two quarks—a quark and an antiquark—whereas a baryon contains three quarks. For instance, the $\pi^-$ pion (a meson) is composed of a $d$ quark and a $\bar{u}$ antiquark, $\pi^- = d + \bar{u}$, as Figure 32.12 shows. These two quarks combine to give the $\pi^-$ pion a net charge of $-e$. Similarly, the $\pi^+$ pion is a combination of the $\bar{d}$

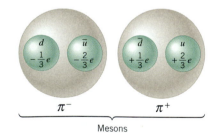

**Figure 32.12** According to the original quark model of hadrons, all mesons consist of a quark and an antiquark, whereas baryons contain three quarks.

**Table 32.4    Quarks and Antiquarks**

| | Quarks | | Antiquarks | |
|---|---|---|---|---|
| Name | Symbol | Charge | Symbol | Charge |
| Up | $u$ | $+\frac{2}{3}e$ | $\bar{u}$ | $-\frac{2}{3}e$ |
| Down | $d$ | $-\frac{1}{3}e$ | $\bar{d}$ | $+\frac{1}{3}e$ |
| Strange | $s$ | $-\frac{1}{3}e$ | $\bar{s}$ | $+\frac{1}{3}e$ |
| Charmed | $c$ | $+\frac{2}{3}e$ | $\bar{c}$ | $-\frac{2}{3}e$ |
| Top | $t$ | $+\frac{2}{3}e$ | $\bar{t}$ | $-\frac{2}{3}e$ |
| Bottom | $b$ | $-\frac{1}{3}e$ | $\bar{b}$ | $+\frac{1}{3}e$ |

and $u$ quarks, $\pi = \bar{d} + u$. In contrast, protons and neutrons, being baryons, consist of three quarks. A proton contains the combination $d + u + u$, and a neutron contains the combination $d + d + u$ (see Figure 32.12). These groups of three quarks give the correct charges for the proton and neutron.

The original quark model was extremely successful in predicting not only the correct charges for the hadrons, but other properties as well. However, in 1974 a new particle, the $J/\psi$ meson, was discovered. This meson has a rest energy of 3097 MeV, much larger than the rest energies of other known mesons. The existence of the $J/\psi$ meson could be explained only if a new quark–antiquark pair existed; this new quark was called *charmed* (*c*). With the discovery of more and more particles, it has been necessary to postulate a fifth and a sixth quark; their names are *top* (*t*) and *bottom* (*b*), although some scientists prefer to call these quarks *truth* and *beauty*. Today, there is firm evidence for all six quarks, each with its corresponding antiquark. All of the hundreds of the known hadrons can be accounted for in terms of these six quarks and their antiquarks.

In addition to electric charge, quarks also have other properties. For example, each quark possesses a characteristic called *color,* for which there are three possibilities: blue, green, or red. The corresponding possibilities for the antiquarks are antiblue, antigreen, and antired. The use of the term "color" and the specific choices of blue, green, and red are arbitrary, for the visible colors of the electromagnetic spectrum have nothing to do with quark properties. The quark property of color, however, is important, because it brings the quark model into agreement with the Pauli exclusion principle and enables the model to account for experimental observations that are otherwise difficult to explain.

## THE STANDARD MODEL

The various elementary particles that have been discovered can interact via one or more of the following four forces: the gravitational force, the strong nuclear force, the weak nuclear force, and the electromagnetic force. In particle physics, the phrase *"the standard model"* refers to the currently accepted explanation for the strong nuclear force, the weak nuclear force, and the electromagnetic force. In this model, the strong nuclear force between quarks is described in terms of the concept of color, the theory being referred to as quantum chromodynamics. According to the standard model, the weak nuclear force and the electromagnetic force are separate manifestations of a single even more fundamental force, referred to as the electroweak force, as we have seen in Section 31.5.

In the standard model, our understanding of the building blocks of matter follows the hierarchical pattern illustrated in Figure 32.13. Molecules, such as water ($H_2O$) and glucose ($C_6H_{12}O_6$), are composed of atoms. Each atom consists of a nucleus that is surrounded by a cloud of electrons. The nucleus, in turn, is made up of protons and neutrons, which are composed of quarks.

**Figure 32.13** The current view of how matter is composed of basic units, starting with a molecule and ending with a quark. The approximate sizes of each unit are also listed.

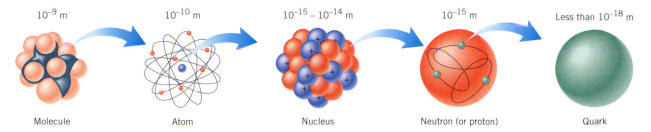

| $10^{-9}$ m | $10^{-10}$ m | $10^{-15} – 10^{-14}$ m | $10^{-15}$ m | Less than $10^{-18}$ m |
| Molecule | Atom | Nucleus | Neutron (or proton) | Quark |

## 32.7 *Cosmology*

*Cosmology* is the study of the structure and evolution of the universe. In this study, both the very large and the very small aspects of the universe are important. Astronomers, for example, study stars located at enormous distances from the earth, up to billions of light-years away. In contrast, particle physicists focus their efforts on the very small elementary particles ($10^{-18}$ m or smaller) that comprise matter. The synergy between the work of astronomers and that of particle physicists has led to significant advances in our understanding of the universe. Central to that understanding is the belief that the universe is expanding, and we begin by discussing the evidence that justifies this belief.

## THE EXPANDING UNIVERSE AND THE BIG BANG

*The physics of*
*an expanding universe.*

The idea that the universe is expanding originated with the astronomer Edwin P. Hubble (1889–1953). He found that light reaching the earth from distant galaxies is Doppler-shifted toward greater wavelengths—that is, toward the red end of the visible spectrum. As Section 24.5 discusses, this type of Doppler shift results when the observer and the source of the light are moving away from each other. The speed at which a galaxy is receding from the earth can be determined from the measured Doppler shift in wavelength. Hubble found that a galaxy located at a distance $d$ from the earth recedes from the earth at a speed $v$ given by

*Hubble's law* $$v = Hd \qquad (32.5)$$

where $H$ is a constant known as the Hubble parameter. In other words, the recession speed is proportional to the distance $d$, so that more distant galaxies are moving away from the earth at greater speeds. Equation 32.5 is referred to as Hubble's law.

Hubble's picture of an expanding universe does not mean that the earth is at the center of the expansion. In fact, there is no literal center. Imagine a loaf of raisin bread expanding as it bakes. Each raisin moves away from every other raisin, without any single one acting as a center for the expansion. Galaxies in the universe behave in a similar fashion. Observers in other galaxies would see distant galaxies moving away, just as we do.

*The physics of*
*"dark energy."*

Not only is the universe expanding, it is doing so at an accelerated rate, according to recent astronomical measurements of the brightness of supernovas, or exploding stars. To account for the accelerated rate, astronomers have postulated that a kind of "dark energy" pervades the universe. The normal gravitational force between galaxies slows the rate at which they are moving away from each other. The dark energy, however, gives rise to a force that counteracts gravity and pushes galaxies apart. As yet, little is known about the properties of dark energy.

Experimental measurements by astronomers indicate that an approximate value for the Hubble parameter is

$$H = 0.022 \, \frac{\text{m}}{\text{s} \cdot \text{light-year}}$$

The value for the Hubble parameter is believed to be accurate within 10%. Scientists are very interested in obtaining an accurate value for $H$ because it can be related to an age for the universe, as the next example illustrates.

### Example 7 An Age for the Universe

▼

Determine an estimate of the age of the universe using Hubble's law.

**Reasoning** Consider a galaxy currently located at a distance $d$ from the earth. According to Hubble's law, this galaxy is moving away from us at a speed of $v = Hd$. At an earlier time, therefore, this galaxy must have been closer. We can imagine, in fact, that in the remote past the separation distance was relatively small and that the universe originated at such a time. To estimate the age of the universe, we calculate the time it has taken the galaxy to recede to its present position. For this purpose, time is simply distance divided by speed, or $t = d/v$.

**Solution** Using Hubble's law and the fact that a distance of 1 light-year is $9.46 \times 10^{15}$ m, we estimate the age of the universe to be

$$t = \frac{d}{v} = \frac{d}{Hd} = \frac{1}{H}$$

$$t = \frac{1}{0.022 \, \dfrac{\text{m}}{\text{s} \cdot \text{light-year}}} = \frac{1}{\left(0.022 \, \dfrac{\text{m}}{\text{s} \cdot \text{light-year}}\right)\left(\dfrac{1 \text{ light-year}}{9.46 \times 10^{15} \text{ m}}\right)}$$

$$= 4.3 \times 10^{17} \text{ s} \quad \text{or} \quad \boxed{1.4 \times 10^{10} \text{ yr}}$$

▲

The idea presented in Example 7, that our galaxy and other galaxies in the universe were very close together at some earlier instant in time, lies at the heart of the **Big Bang theory**. This theory postulates that the universe had a definite beginning in a cataclysmic event, sometimes called the primeval fireball. Dramatic evidence supporting the theory was discovered in 1965 by Arno A. Penzias (1933– ) and Robert W. Wilson (1936– ). Using a radio telescope, they discovered that the earth is being bathed in weak electromagnetic waves in the microwave region of the spectrum (wavelength = 7.35 cm, see Figure 24.9). They observed that the intensity of these waves is the same, no matter where in the sky they pointed their telescope, and concluded that the waves originated outside of our galaxy. This microwave background radiation, as it is called, represents radiation left over from the Big Bang and is a kind of cosmic afterglow. Subsequent measurements have confirmed the research of Penzias and Wilson and shown that the microwave radiation is consistent with a perfect blackbody (see Sections 13.3 and 29.2) radiating at a temperature of 2.7 K, in agreement with theoretical analysis of the Big Bang. In 1978, Penzias and Wilson received a Nobel Prize for their discovery.

## THE STANDARD MODEL FOR THE EVOLUTION OF THE UNIVERSE

Based on the recent experimental and theoretical research in particle physics, scientists have proposed an evolutionary sequence of events following the Big Bang. This sequence is known as the *standard cosmological model* and is illustrated in Figure 32.14.

Immediately following the Big Bang, the temperature of the universe was incredibly high, about $10^{32}$ K. During this initial period, the three fundamental forces (the gravitational force, the strong nuclear force, and the electroweak force) all behaved as a single unified force. Very quickly, in about $10^{-43}$ s, the gravitational force took on a separate identity all its own, as Figure 32.14 indicates. Meanwhile, the strong nuclear force and the electroweak force continued to act as a single force, which is sometimes referred to as the GUT force. GUT stands for the Grand Unified Theory that presumably would explain such a force. Slightly later, at about $10^{-35}$ s after the Big Bang, the GUT force separated into the strong nuclear force and the electroweak force (see Figure 32.14), the universe expanding and cooling somewhat to a temperature of roughly $10^{28}$ K. From this point on, the strong nuclear force behaved as we know it today, while the electroweak force maintained its identity. In this scenario, note that the weak nuclear force and the electromagnetic force have not yet manifested themselves as separate entities. The disappearance of the electroweak force and the appearance of the weak nuclear force and the electromagnetic force eventually occurred at approximately

**Figure 32.14** According to the standard cosmological model, the universe has evolved as illustrated here. In this model, the universe is presumed to have originated with a cataclysmic event known as the Big Bang. The times shown are those following this event.

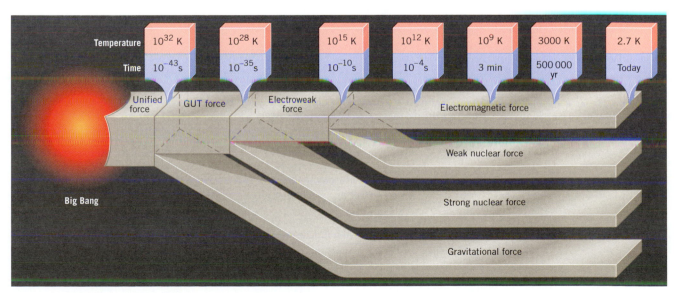

$10^{-10}$ s after the Big Bang, when the temperature of the expanding universe had cooled to about $10^{15}$ K.

From the Big Bang up until the strong nuclear force separated from the GUT force at a time of $10^{-35}$ s, all particles of matter were similar, and there was no distinction between quarks and leptons. After this time, quarks and leptons became distinguishable. Eventually the quarks and antiquarks formed hadrons, such as protons and neutrons and their antiparticles. By a time of $10^{-4}$ s after the Big Bang, however, the temperature had cooled to approximately $10^{12}$ K, and the hadrons had mostly disappeared. Protons and neutrons survived only as a very small fraction of the total number of particles, the majority of which were leptons such as electrons, positrons, and neutrinos. Like most of the hadrons before them, most of the electrons and positrons eventually disappeared. However, they did leave behind a relatively small number of electrons to join the small number of protons and neutrons at a time of about 3 min following the Big Bang. At this time the temperature of the expanding universe had decreased to about $10^9$ K, and small nuclei such as that of helium began forming. Later, when the universe was about 500 000 years old and the temperature had dropped to near 3000 K, hydrogen and helium atoms began forming. As the temperature decreased further, stars and galaxies formed, and today we find a temperature of 2.7 K characterizing the cosmic background radiation of the universe.

# Concept Summary

This summary presents an abridged version of the chapter, including the important equations and all available learning aids. For convenient reference, the learning aids (including the text's examples) are placed next to or immediately after the relevant equation or discussion. The following learning aids may be found on-line at **www.wiley.com/college/cutnell**:

| | |
|---|---|
| **Interactive LearningWare** examples are solved according to a five-step interactive format that is designed to help you develop problem-solving skills. | **Concept Simulations** are animated versions of text figures or animations that illustrate important concepts. You can control parameters that affect the display, and we encourage you to experiment. |
| **Interactive Solutions** offer specific models for certain types of problems in the chapter homework. The calculations are carried out interactively. | **Self-Assessment Tests** include both qualitative and quantitative questions. Extensive feedback is provided for both incorrect and correct answers, to help you evaluate your understanding of the material. |

| *Topic* | *Discussion* | *Learning Aids* |
|---|---|---|

### 32.1  *Biological Effects of Ionizing Radiation*

Ionizing radiation

Ionizing radiation consists of photons and/or moving particles that have enough energy to ionize an atom or molecule. Exposure is a measure of the ionization produced in air by X-rays or $\gamma$ rays. When a beam of X-rays or $\gamma$ rays is sent through a mass $m$ of dry air (0 °C, 1 atm pressure) and produces positive ions whose total charge is $q$, the exposure in coulombs per kilogram (C/kg) is

Exposure (in C/kg)

$$\text{Exposure (in coulombs per kilograms)} = \frac{q}{m}$$

With $q$ in coulombs (C) and $m$ in kilograms (kg), the exposure in roentgens is

Exposure (in roentgens)

$$\text{Exposure (in roentgens)} = \left(\frac{1}{2.58 \times 10^{-4}}\right)\frac{q}{m} \qquad (32.1)$$

The absorbed dose is the amount of energy absorbed from the radiation per unit mass of absorbing material:

Absorbed dose

$$\text{Absorbed dose} = \frac{\text{Energy absorbed}}{\text{Mass of absorbing material}} \qquad (32.2)$$

The SI unit of absorbed dose is the gray (Gy); 1 Gy = 1 J/kg. However, the rad (rd) is another unit that is often used: 1 rd = 0.01 Gy.

The amount of biological damage produced by ionizing radiation is different for different types of radiation. The relative biological effectiveness (RBE) is the absorbed dose of 200-keV X-rays required to produce a certain biological

| Topic | Discussion | Learning Aids |
|-------|-----------|---------------|
| | effect divided by the dose of a given type of radiation that produces the same biological effect. | |
| **Relative biological effectiveness** | $$\text{RBE} = \frac{\text{The dose of 200-keV X-rays that produces a certain biological effect}}{\text{The dose of radiation that produces the same biological effect}} \quad (32.3)$$ | |
| | The biologically equivalent dose (in rems) is the product of the absorbed dose (in rads) and the RBE: | |
| **Biologically equivalent dose** | $$\begin{array}{c}\text{Biologically equivalent dose} \\ \text{(in rems)}\end{array} = \begin{array}{c}\text{Absorbed dose} \\ \text{(in rads)}\end{array} \times \text{RBE} \quad (32.4)$$ | **Example 1** |

### 32.2 Induced Nuclear Reactions

| Topic | Discussion | Learning Aids |
|-------|-----------|---------------|
| **Induced nuclear reaction** **Induced nuclear transmutation** | An induced nuclear reaction occurs whenever a target nucleus is struck by an incident nucleus, an atomic or subatomic particle, or a $\gamma$-ray photon and undergoes a change as a result. An induced nuclear transmutation is a reaction in which the target nucleus is changed into a nucleus of a new element. | |
| **Nuclear reactions obey conservation laws** | All nuclear reactions (induced or spontaneous) obey the conservation laws of physics as they relate to mass/energy, electric charge, linear momentum, angular momentum, and nucleon number. | **Example 2** |
| **Shorthand notation for nuclear reactions** | Nuclear reactions are often written in a shorthand form, such as $^{14}_{7}\text{N}(\alpha, p)\,^{17}_{8}\text{O}$. The first and last symbols $^{14}_{7}\text{N}$ and $^{17}_{8}\text{O}$ denote, respectively, the initial and final nuclei. The symbols within the parentheses denote the incident $\alpha$ particle (on the left) and the small emitted particle or proton $p$ (on the right). | |
| **Thermal neutron** | A thermal neutron is one that has a kinetic energy of about 0.04 eV. | |

### 32.3 Nuclear Fission

| Topic | Discussion | Learning Aids |
|-------|-----------|---------------|
| **Nuclear fission** **Chain reaction** **Controlled chain reaction** | Nuclear fission occurs when a massive nucleus splits into two less massive fragments. Fission can be induced by the absorption of a thermal neutron. When a massive nucleus fissions, energy is released because the binding energy per nucleon is greater for the fragments than for the original nucleus. Neutrons are also released during nuclear fission. These neutrons can, in turn, induce other nuclei to fission and lead to a process known as a chain reaction. A chain reaction is said to be controlled if each fission event contributes, on average, only one neutron that fissions another nucleus. | **Example 3** **Interactive LearningWare 32.1** **Example 4** **Concept Simulation 32.1** |

### 32.4 Nuclear Reactors

| Topic | Discussion | Learning Aids |
|-------|-----------|---------------|
| **Nuclear reactor** **Fuel elements** **Neutron moderator** **Control rods** **Critical state** **Subcritical state** **Supercritical state** | A nuclear reactor is a device that generates energy by a controlled chain reaction. Many reactors in use today have the same three basic components: fuel elements, a neutron moderator, and control rods. The fuel elements contain the fissile fuel, and the entire region of fuel elements is known as the reactor core. The neutron moderator is a material (water, for example) that slows down the neutrons released in a fission event to thermal energies so they can initiate additional fission events. Control rods contain material that readily absorbs neutrons without fissioning. They are used to keep the reactor in its normal, or critical, state, in which each fission event leads to one additional fission, no more, no less. The reactor is subcritical when, on average, the neutrons from each fission trigger less than one subsequent fission. The reactor is supercritical when, on average, the neutrons from each fission trigger more than one additional fission. | |

### 32.5 Nuclear Fusion

| Topic | Discussion | Learning Aids |
|-------|-----------|---------------|
| **Nuclear fusion** **Magnetic confinement** **Inertial confinement** | In a fusion process, two nuclei with smaller masses combine to form a single nucleus with a larger mass. Energy is released by fusion when the binding energy per nucleon is greater for the larger nucleus than for the smaller nuclei. Fusion reactions are said to be thermonuclear because they require extremely high temperatures to proceed. Current studies of nuclear fusion utilize either magnetic confinement or inertial confinement to contain the fusing nuclei at the high temperatures that are necessary. | **Example 5** **Interactive Solution 32.29** |

| Topic | Discussion | Learning Aids |
|---|---|---|

### 32.6 Elementary Particles

Subatomic particles are divided into three families: the photon family (which consists only of the photon), the lepton family (which includes the electron), and the hadron family (which includes the proton and the neutron).

~~~mentary particles are the basic building blocks of matter. All members of the ~~~oton and lepton families are elementary particles.

The quark theory proposes that the hadrons are not elementary particles but are composed of elementary particles called quarks. Currently, the hundreds of hadrons can be accounted for in terms of six quarks (up, down, strange, charmed, top, and bottom) and their antiquarks.

**Example 6**

**The standard model**

The standard model consists of two parts: (1) the currently accepted explanation for the strong nuclear force in terms of the quark concept of "color" and (2) the theory of the electroweak interaction.

### 32.7 Cosmology

**Cosmology**

Cosmology is the study of the structure and evolution of the universe. Our universe is expanding. The speed $v$ at which a distant galaxy recedes from the earth is given by Hubble's law:

**Hubble's law**

$$v = Hd \qquad (32.5)$$

**Example 7**

where $H = 0.022$ m/(s·light-year) is called the Hubble parameter and $d$ is the distance of the galaxy from the earth.

**Hubble parameter**

**Big Bang theory**

The Big Bang theory postulates that the universe had a definite beginning in a cataclysmic event, sometimes called the primeval fireball. The radiation left over from this event is in the microwave region of the electromagnetic spectrum, and it is consistent with a perfect blackbody radiating at a temperature of 2.7 K, in agreement with theoretical analysis of the Big Bang.

**Standard cosmological model**

The standard cosmological model for the evolution of the universe is summarized in Figure 32.14.

*Use* **Self-Assessment Test 32.1** *to evaluate your understanding of Sections 32.1–32.7.*

# Problems

**ssm** Solution is in the Student Solutions Manual.    **www** Solution is available on the World Wide Web at www.wiley.com/college/cutnell

This icon represents a biomedical application.

### Section 32.1 Biological Effects of Ionizing Radiation

**1. ssm www** A 75-kg person is exposed to 45 mrem of $\alpha$ particles (RBE = 12). How much energy has this person absorbed?

**2.** Neutrons (RBE = 2.0) and $\alpha$ particles have the same biologically equivalent dose. However, the absorbed dose of the neutrons is six times the absorbed dose of the $\alpha$ particles. What is the RBE for the $\alpha$ particles?

**3.** A film badge worn by a radiologist indicates that she has received an absorbed dose of $2.5 \times 10^{-5}$ Gy. The mass of the radiologist is 65 kg. How much energy has she absorbed?

**4.** During an X-ray examination, a person is exposed to radiation at a rate of $3.1 \times 10^{-5}$ grays per second. The exposure time is 0.10 s, and the mass of the exposed tissue is 1.2 kg. Determine the energy absorbed.

**5. ssm** A beam of $\gamma$ rays passes through $4.0 \times 10^{-3}$ kg of dry air and generates $1.7 \times 10^{12}$ ions, each with a charge of $+e$. What is the exposure (in roentgens)?

**6.** A beam of particles is directed at a 0.015-kg tumor. There are $1.6 \times 10^{10}$ particles per second reaching the tumor, and the energy of each particle is 4.0 MeV. The RBE for the radiation is 14. Find the biologically equivalent dose given to the tumor in 25 s.

**\*7.** A 2.0-kg tumor is being irradiated by a radioactive source. The tumor receives an absorbed dose of 12 Gy in a time of 850 s. Each disintegration of the radioactive source produces a particle that enters the tumor and delivers an energy of 0.40 MeV. What is the activity $\Delta N/\Delta t$ (see Section 31.6) of the radioactive source?

**\*8.** A beam of nuclei is used for cancer therapy. Each nucleus has an energy of 130 MeV, and the relative biological effectiveness (RBE) of this type of radiation is 16. The beam is directed onto a 0.17-kg tumor, which receives a biologically equivalent dose of 180 rem. How many nuclei are in the beam?

**\*9. ssm** What absorbed dose (in rad) of $\gamma$ rays is required to change a block of ice at 0.0 °C into steam at 100.0 °C?

## Section 32.2 Induced Nuclear Reactions

**10.** Identify the unknown species $^A_Z X$ in the nuclear reaction $^{22}_{11}Na$ $(d, \alpha)$ $^A_Z X$, where $d$ stands for the deuterium isotope $^2_1H$ of hydrogen.

**11. ssm** A nitrogen $^{14}_7N$ nucleus absorbs a deuterium $^2_1H$ nucleus during a nuclear reaction. What is the name, atomic number, and nucleon number of the compound nucleus?

**12.** Write the reactions below in the shorthand form discussed in the text.

(a) $^1_0n + ^{14}_7N \longrightarrow ^{14}_6C + ^1_1H$

(b) $^1_0n + ^{238}_{92}U \longrightarrow ^{239}_{92}U + \gamma$

(c) $^1_0n + ^{24}_{12}Mg \longrightarrow ^{23}_{11}Na + ^2_1H$

**13.** Complete the following nuclear reactions, assuming that the unknown quantity signified by the question mark is a single entity:

(a) $^{43}_{20}Ca$ $(\alpha, ?)^{46}_{21}Sc$     (d) ? $(\alpha, p)$ $^{17}_8O$

(b) $^9_4Be$ $(?, n)$ $^{12}_6C$     (e) $^{55}_{25}Mn$ $(n, \gamma)$ ?

(c) $^9_4Be$ $(p, \alpha)$ ?

**14.** A neutron causes $^{232}_{90}Th$ to change according to the reaction

$$^1_0n + ^{232}_{90}Th \longrightarrow ^A_Z X + \gamma$$

(a) Identify the unknown nucleus $^A_Z X$, giving its atomic mass number $A$, its atomic number $Z$, and the symbol X for the element. (b) The $^A_Z X$ nucleus subsequently undergoes $\beta^-$ decay, and its daughter does too. Identify the final nucleus, giving its atomic mass number, atomic number, and name.

\* **15. ssm** Consider the induced nuclear reaction $^2_1H + ^{14}_7N \rightarrow$ $^{12}_6C + ^4_2He$. The atomic masses are $^2_1H$ (2.014 102 u), $^{14}_7N$ (14.003 074 u), $^{12}_6C$ (12.000 000 u), and $^4_2He$ (4.002 603 u). Determine the energy (in MeV) released when the $^{12}_6C$ and $^4_2He$ nuclei are formed in this manner.

\* **16.** During a nuclear reaction, an unknown particle is absorbed by a copper $^{63}_{29}Cu$ nucleus, and the reaction products are $^{62}_{29}Cu$, a neutron, and a proton. What is the name, atomic number, and nucleon number of the *compound nucleus?*

## Section 32.3 Nuclear Fission, Section 32.4 Nuclear Reactors

**17.** $^{235}_{92}U$ absorbs a thermal neutron and fissions into rubidium $^{93}_{37}Rb$ and cesium $^{141}_{55}Cs$. What nucleons are produced by the fission, and how many are there?

**18.** How many neutrons are produced when $^{235}_{92}U$ fissions in the following way: $^1_0n + ^{235}_{92}U \rightarrow ^{132}_{50}Sn + ^{101}_{42}Mo + $ neutrons?

**19. ssm** When a $^{235}_{92}U$ (235.043 924 u) nucleus fissions, about 200 MeV of energy is released. What is the ratio of this energy to the rest energy of the uranium nucleus?

**20. Interactive LearningWare 32.1** at **www.wiley.com/college/cutnell** reviews the concepts that lie at the heart of this problem. What energy (in MeV) is liberated by the following fission reaction?

$$\underbrace{^1_0n}_{1.009 \text{ u}} + \underbrace{^{235}_{92}U}_{235.044 \text{ u}} \longrightarrow \underbrace{^{141}_{56}Ba}_{140.914 \text{ u}} + \underbrace{^{92}_{36}Kr}_{91.926 \text{ u}} + \underbrace{3^1_0n}_{3(1.009 \text{ u})}$$

**21.** Neutrons released by a fission reaction must be slowed by collisions with the moderator nuclei before the neutrons can cause further fissions. Suppose a 1.5-MeV neutron leaves each collision with 65% of its incident energy. How many collisions are required to reduce the neutron's energy to at least 0.040 eV, which is the energy of a thermal neutron?

**22.** The energy released by each fission within the core of a nuclear reactor is $2.0 \times 10^2$ MeV. The number of fissions occurring each

second is $2.0 \times 10^{19}$. Determine the power (in watts) that the reactor generates.

\* **23. ssm** When 1.0 kg of coal is burned, about $3.0 \times 10^7$ J of energy is released. If the energy released per $^{235}_{92}U$ fission is $2.0 \times 10^2$ MeV, how many kilograms of coal must be burned to produce the same energy as 1.0 kg of $^{235}_{92}U$?

\* **24.** Suppose the $^{239}_{94}Pu$ nucleus fissions into two fragments whose mass ratio is 0.32 : 0.68. With the aid of Figure 32.7, estimate the energy (in MeV) released during this fission.

\* **25.** (a) If each fission of a $^{235}_{92}U$ nucleus releases about $2.0 \times 10^2$ MeV of energy, determine the energy (in joules) released by the complete fissioning of 1.0 gram of $^{235}_{92}U$. (b) How many grams of $^{235}_{92}U$ are consumed in one year, in order to supply the energy needs of a household that uses 30.0 kWh of energy per day, on the average?

\*\* **26.** A 20.0 kiloton atomic bomb releases as much energy as 20.0 kilotons of TNT (1.0 kiloton of TNT releases about $5.0 \times 10^{12}$ J of energy). Recall that about $2.0 \times 10^2$ MeV of energy is released when each $^{235}_{92}U$ nucleus fissions. (a) How many $^{235}_{92}U$ nuclei are fissioned to produce the bomb's energy? (b) How many grams of uranium are fissioned? (c) What is the equivalent mass (in grams) of the bomb's energy?

\*\* **27. ssm www** A nuclear power plant is 25% efficient, meaning that 25% of the power it generates goes into producing electricity. The remaining 75% is wasted as heat. The plant generates $8.0 \times 10^8$ watts of electric power. If each fission releases $2.0 \times 10^2$ MeV of energy, how many kilograms of $^{235}_{92}U$ are fissioned per year?

## Section 32.5 Nuclear Fusion

**28.** Consider the fusion of $^1_1H$ (mass = 1.0078 u) with $^{12}_6C$ (mass = 12.0000 u) to form $^{13}_7N$ (mass = 13.0057 u). Determine the energy (in MeV) released in this reaction, part of which is carried away by a $\gamma$ ray.

**29. Interactive Solution 32.29** at **www.wiley.com/college/cutnell** illustrates one way to approach problems of this type. The fusion of two deuterium nuclei ($^2_1H$, mass = 2.0141 u) can yield a helium nucleus ($^3_2He$, mass = 3.0160 u) and a neutron ($^1_0n$, mass = 1.0087 u). What is the energy (in MeV) released in this reaction?

**30.** Tritium ($^3_1H$) is a rare isotope of hydrogen that can be produced by the fusion reaction

$$\underbrace{^1_Z X}_{1.0087 \text{ u}} + \underbrace{^A_1 Y}_{2.0141 \text{ u}} \longrightarrow \underbrace{^3_1 H}_{3.0161 \text{ u}} + \gamma$$

(a) Determine the atomic mass number $A$, the atomic number $Z$, and the names X and Y of the unknown particles. (b) Using the masses given in the reaction, determine how much energy (in MeV) is released by this reaction.

\* **31. ssm** Imagine your car is powered by a fusion engine in which the following reaction occurs: $3^2_1H \rightarrow ^4_2He + ^1_1H + ^1_0n$. The masses are $^2_1H$ (2.0141 u), $^4_2He$ (4.0026 u), $^1_1H$ (1.0078 u), and $^1_0n$ (1.0087 u). The engine uses $6.1 \times 10^{-6}$ kg of deuterium $^2_1H$ fuel. If one gallon of gasoline produces $2.1 \times 10^9$ J of energy, how many gallons of gasoline would have to be burned to equal the energy released by all the deuterium fuel?

\* **32.** Deuterium ($^2_1H$) is an attractive fuel for fusion reactions because it is abundant in the waters of the oceans. In the oceans, about 0.015% of the hydrogen atoms in the water ($H_2O$) are deuterium atoms. (a) How many deuterium atoms are there in one kilogram of water? (b) If each deuterium nucleus produces about 7.2 MeV in a fusion reaction, how many kilograms of ocean water would be needed to supply the energy needs of the United States for one year, estimated to be $9.3 \times 10^{19}$ J?

** **33.** The proton–proton cycle thought to occur in the sun consists of the following sequence of reactions:

$$(1) \; {}^1_1\text{H} + {}^1_1\text{H} \longrightarrow {}^2_1\text{H} + {}^0_1\text{e} + \nu$$

$$(2) \; {}^1_1\text{H} + {}^2_1\text{H} \longrightarrow {}^3_2\text{He} + \gamma$$

$$(3) \; {}^3_2\text{He} + {}^3_2\text{He} \longrightarrow {}^4_2\text{He} + {}^1_1\text{H} + {}^1_1\text{H}$$

In these reactions ${}^0_1\text{e}$ is a positron (mass = 0.000 549 u), $\nu$ is a neutrino (mass $\approx$ 0 u), and $\gamma$ is a gamma ray photon (mass = 0 u). Note that reaction (3) uses two ${}^3_2\text{He}$ nuclei, which are formed by *two* reactions of type (1) and *two* reactions of type (2). Verify that the proton–proton cycle generates about 25 MeV of energy. The atomic masses are ${}^1_1\text{H}$ (1.007 825 u), ${}^2_1\text{H}$ (2.014 102 u), ${}^3_2\text{He}$ (3.016 030 u), and ${}^4_2\text{He}$ (4.002 603 u). Be sure to account for the fact that there are two electrons in two hydrogen atoms, whereas there is one electron in a single deuterium (${}^2_1\text{H}$) atom. The mass of one electron is 0.000 549 u.

## Section 32.6 Elementary Particles

**34.** The main decay mode for the negative pion is $\pi^- \rightarrow \mu^- + \bar{\nu}_\mu$. Find the energy (in MeV) released in this decay. Consult Table 32.3 as needed.

**35.** **ssm** The lambda particle $\Lambda^0$ has an electric charge of zero. It is a baryon and, hence, is composed of three quarks. They are all different. One of these quarks is the up quark $u$, and there are no anti-quarks present. Make a list of the three possibilities for the quarks contained in $\Lambda^0$. (Other information is needed to decide which one of these possibilities is the $\Lambda^0$ particle.)

**36.** A high-energy proton collides with a stationary proton, and the reaction $p + p \rightarrow n + p + \pi^+$ occurs. The rest energy of the $\pi^+$ pion is 139.6 MeV. Ignore momentum conservation and find the minimum energy (in MeV) the incident proton must have.

**37.** Review Conceptual Example 6 as background for this problem. An electron and its antiparticle annihilate each other, producing two $\gamma$-ray photons. The kinetic energies of the particles are negligible. For each photon, determine (a) its energy (in MeV), (b) its wavelength, and (c) the magnitude of its momentum.

**38.** The $K^-$ particle has a charge of $-e$ and contains one quark and one antiquark. (a) Which quarks can the particle *not* contain? (b) Which antiquarks can the particle *not* contain?

* **39.** **ssm** **www** Suppose a neutrino is created and has an energy of 35 MeV. (a) Assuming the neutrino, like the photon, has no mass and travels at the speed of light, find the momentum of the neutrino. (b) Determine the de Broglie wavelength of the neutrino.

* **40.** Review Conceptual Example 4 as background for this problem. An energetic proton is fired at a stationary proton. For the reaction to produce new particles, the two protons must approach each other to within a distance of about $8.0 \times 10^{-15}$ m. The moving proton must have a sufficient speed to overcome the repulsive Coulomb force. What must be the minimum initial kinetic energy (in MeV) of the proton?

# Appendix A
# Powers of Ten and Scientific Notation

In science, very large and very small decimal numbers are conveniently expressed in terms of powers of ten, some of which are listed below:

$$10^3 = 10 \times 10 \times 10 = 1000 \qquad 10^{-3} = \frac{1}{10 \times 10 \times 10}$$
$$= 0.001$$

$$10^2 = 10 \times 10 = 100 \qquad 10^{-2} = \frac{1}{10 \times 10} = 0.01$$

$$10^1 = 10 \qquad 10^{-1} = \frac{1}{10} = 0.1$$

$$10^0 = 1$$

Using powers of ten, we can write the radius of the earth in the following way, for example:

$$\text{Earth radius} = 6\,380\,000 \text{ m} = 6.38 \times 10^6 \text{ m}$$

The factor of ten raised to the sixth power is ten multiplied by itself six times, or one million, so the earth's radius is 6.38 million meters. Alternatively, the factor of ten raised to the sixth power indicates that the decimal point in the term 6.38 is to be moved six places *to the right* to obtain the radius as a number without powers of ten.

For numbers less than one, negative powers of ten are used. For instance, the Bohr radius of the hydrogen atom is

$$\text{Bohr radius} = 0.000\,000\,000\,0529 \text{ m} = 5.29 \times 10^{-11} \text{ m}$$

The factor of ten raised to the minus eleventh power indicates that the decimal point in the term 5.29 is to be moved eleven places *to the left* to obtain the radius as a number without powers of ten. Numbers expressed with the aid of powers of ten are said to be in *scientific notation.*

Calculations that involve the multiplication and division of powers of ten are carried out as in the following examples:

$$(2.0 \times 10^6)(3.5 \times 10^3) = (2.0 \times 3.5) \times 10^{6+3} = 7.0 \times 10^9$$

$$\frac{9.0 \times 10^7}{2.0 \times 10^4} = \left(\frac{9.0}{2.0}\right) \times 10^7 \times 10^{-4}$$

$$= \left(\frac{9.0}{2.0}\right) \times 10^{7-4} = 4.5 \times 10^3$$

The general rules for such calculations are

$$\frac{1}{10^n} = 10^{-n} \tag{A-1}$$

$$10^n \times 10^m = 10^{n+m} \qquad \text{(Exponents added)} \tag{A-2}$$

$$\frac{10^n}{10^m} = 10^{n-m} \qquad \text{(Exponents subtracted)} \tag{A-3}$$

where $n$ and $m$ are any positive or negative number.

Scientific notation is convenient because of the ease with which it can be used in calculations. Moreover, scientific notation provides a convenient way to express the significant figures in a number, as Appendix B discusses.

# Appendix B
# Significant Figures

The number of *significant figures* in a number is the number of digits whose values are known with certainty. For instance, a person's height is measured to be 1.78 m, with the measurement error being in the third decimal place. All three digits are known with certainty, so that the number contains three significant figures. If a zero is given as the last digit to the right of the decimal point, the zero is presumed to be significant. Thus, the number 1.780 m contains four significant figures. As another example, consider a distance of 1500 m. This number contains only two significant figures, the one and the five. The zeros immediately to the left of the unexpressed decimal point are not counted as significant figures. However, zeros located between significant figures are significant, so a distance of 1502 m contains four significant figures.

Scientific notation is particularly convenient from the point of view of significant figures. Suppose it is known that a certain distance is fifteen hundred meters, to four significant figures. Writing the number as 1500 m presents a problem because it implies that only two significant figures are known. In contrast, the scientific notation of $1.500 \times 10^3$ m has the advantage of indicating that the distance is known to four significant figures.

When two or more numbers are used in a calculation, the number of significant figures in the answer is limited by the number of significant figures in the original data. For instance, a rectangular garden with sides of 9.8 m and 17.1 m has an area of (9.8 m)(17.1 m). A calculator gives 167.58 m² for this product. However, one of the original lengths is known only to two significant figures, so the final answer is limited to only two significant figures and should be rounded off to 170 m². In general, **when numbers are multiplied or divided, the number of significant figures in the final answer equals the smallest number of significant figures in any of the original factors.**

The number of significant figures in the answer to an addition or a subtraction is also limited by the original data. Consider the total distance along a biker's trail that consists of three segments with the distances shown as follows:

$$
\begin{array}{rr}
& 2.5 \ \text{km} \\
& 11 \quad\ \text{km} \\
& 5.26 \ \text{km} \\
\hline
\text{Total} & 18.76 \ \text{km}
\end{array}
$$

The distance of 11 km contains no significant figures to the right of the decimal point. Therefore, neither does the sum of the three distances,

and the total distance should not be reported as 18.76 km. Instead, the answer is rounded off to 19 km. In general, *when numbers are added or subtracted, the last significant figure in the answer occurs in the last column (counting from left to right) containing a number that results from a combination of digits that are all significant.* In the answer of 18.76 km, the eight is the sum of $2 + 1 + 5$, each digit being significant. However, the seven is the sum of $5 + 0 + 2$, and the zero is not significant, since it comes from the 11-km distance, which contains no significant figures to the right of the decimal point.

# Appendix C
# Algebra

## C.1  Proportions and Equations

Physics deals with physical variables and the relations between them. Typically, variables are represented by the letters of the English and Greek alphabets. Sometimes, the relation between variables is expressed as a proportion or inverse proportion. Other times, however, it is more convenient or necessary to express the relation by means of an equation, which is governed by the rules of algebra.

If two variables are **directly proportional** and one of them doubles, then the other variable also doubles. Similarly, if one variable is reduced to one-half its original value, then the other is also reduced to one-half its original value. In general, if $x$ is directly proportional to $y$, then increasing or decreasing one variable by a given factor causes the other variable to change in the same way by the same factor. This kind of relation is expressed as $x \propto y$, where the symbol $\propto$ means "is proportional to."

Since the proportional variables $x$ and $y$ always increase and decrease by the same factor, the ratio of $x$ to $y$ must have a constant value, or $x/y = k$, where $k$ is a constant, independent of the values for $x$ and $y$. Consequently, a proportionality such as $x \propto y$ can also be expressed in the form of an equation: $x = ky$. The constant $k$ is referred to as a **proportionality constant.**

If two variables are **inversely proportional** and one of them increases by a given factor, then the other decreases by the same factor. An inverse proportion is written as $x \propto 1/y$. This kind of proportionality is equivalent to the following equation: $xy = k$, where $k$ is a proportionality constant, independent of $x$ and $y$.

## C.2  Solving Equations

Some of the variables in an equation typically have known values, and some do not. It is often necessary to solve the equation so that a variable whose value is unknown is expressed in terms of the known quantities. **In the process of solving an equation, it is permissible to manipulate the equation in any way, as long as a change made on one side of the equals sign is also made on the other side.** For example, consider the equation $v = v_0 + at$. Suppose values for $v$, $v_0$, and $a$ are available, and the value of $t$ is required. To solve the equation for $t$, we begin by subtracting $v_0$ from *both* sides:

$$
\begin{array}{rcl}
v & = & v_0 + at \\
-v_0 & = & -v_0 \\
\hline
v - v_0 & = & at
\end{array}
$$

Next, we divide both sides of $v - v_0 = at$ by the quantity $a$:

$$
\frac{v - v_0}{a} = \frac{at}{a} = (1)t
$$

On the right side, the $a$ in the numerator divided by the $a$ in the denominator equals one, so that

$$
t = \frac{v - v_0}{a}
$$

It is always possible to check the correctness of the algebraic manipulations performed in solving an equation by substituting the answer back into the original equation. In the previous example, we substitute the answer for $t$ into $v = v_0 + at$:

$$
v = v_0 + a\left(\frac{v - v_0}{a}\right) = v_0 + (v - v_0) = v
$$

The result $v = v$ implies that our algebraic manipulations were done correctly.

Algebraic manipulations other than addition, subtraction, multiplication, and division may play a role in solving an equation. The same basic rule applies, however: Whatever is done to the left side of an equation must also be done to the right side. As another example, suppose it is necessary to express $v_0$ in terms of $v$, $a$, and $x$, where $v^2 = v_0^2 + 2ax$. By subtracting $2ax$ from both sides, we isolate $v_0^2$ on the right:

$$
\begin{array}{rcl}
v^2 & = & v_0^2 + 2ax \\
-2ax & = & -2ax \\
\hline
v^2 - 2ax & = & v_0^2
\end{array}
$$

To solve for $v_0$, we take the positive and negative square root of *both* sides of $v^2 - 2ax = v_0^2$:

$$
v_0 = \pm\sqrt{v^2 - 2ax}
$$

## C.3  Simultaneous Equations

When more than one variable in a single equation is unknown, additional equations are needed if solutions are to be found for all of the unknown quantities. Thus, the equation $3x + 2y = 7$ cannot be solved by itself to give unique values for both $x$ and $y$. However, if $x$ and $y$ also (i.e., simultaneously) obey the equation $x - 3y = 6$, then both unknowns can be found.

There are a number of methods by which such simultaneous equations can be solved. One method is to solve one equation for $x$ in terms of $y$ and substitute the result into the other equation to obtain an expression containing only the single unknown variable $y$. The equation $x - 3y = 6$, for instance, can be solved for $x$ by adding $3y$ to each side, with the result that $x = 6 + 3y$. The substitution of this expression for $x$ into the equation $3x + 2y = 7$ is shown below:

$$3x + 2y = 7$$
$$3(6 + 3y) + 2y = 7$$
$$18 + 9y + 2y = 7$$

We find, then, that $18 + 11y = 7$, a result that can be solved for $y$:

$$
\begin{array}{rr}
18 + 11y = & 7 \\
-18 & -18 \\
\hline
11y = & -11
\end{array}
$$

Dividing both sides of this result by 11 shows that $y = -1$. The value of $y = -1$ can be substituted in either of the original equations to obtain a value for $x$:

$$
\begin{array}{rcl}
x - 3y & = & 6 \\
x - 3(-1) & = & 6 \\
x + 3 & = & 6 \\
-3 & & -3 \\
\hline
x & = & 3
\end{array}
$$

## C.4  The Quadratic Formula

Equations occur in physics that include the square of a variable. Such equations are said to be *quadratic* in that variable, and often can be put into the following form:

$$ax^2 + bx + c = 0 \qquad (C\text{-}1)$$

where $a$, $b$, and $c$ are constants independent of $x$. This equation can be solved to give the **quadratic formula,** which is

$$x = \frac{-b \pm \sqrt{b^2 - 4ac}}{2a} \qquad (C\text{-}2)$$

The $\pm$ in the quadratic formula indicates that there are two solutions. For instance, if $2x^2 - 5x + 3 = 0$, then $a = 2$, $b = -5$, and $c = 3$. The quadratic formula gives the two solutions as follows:

**Solution 1:**
**Plus sign**
$$x = \frac{-b + \sqrt{b^2 - 4ac}}{2a}$$
$$= \frac{-(-5) + \sqrt{(-5)^2 - 4(2)(3)}}{2(2)}$$
$$= \frac{+5 + \sqrt{1}}{4} = \frac{3}{2}$$

**Solution 2:**
**Minus sign**
$$x = \frac{-b - \sqrt{b^2 - 4ac}}{2a}$$
$$= \frac{-(-5) - \sqrt{(-5)^2 - 4(2)(3)}}{2(2)}$$
$$= \frac{+5 - \sqrt{1}}{4} = 1$$

# Appendix D
# Exponents and Logarithms

Appendix A discusses powers of ten, such as $10^3$, which means ten multiplied by itself three times, or $10 \times 10 \times 10$. The three is referred to as an **exponent.** The use of exponents extends beyond powers of ten. In general, the term $y^n$ means the factor $y$ is multiplied by itself $n$ times. For example, $y^2$, or $y$ squared, is familiar and means $y \times y$. Similarly, $y^5$ means $y \times y \times y \times y \times y$.

The rules that govern algebraic manipulations of exponents are the same as those given in Appendix A (see Equations A-1, A-2, and A-3) for powers of ten:

$$\frac{1}{y^n} = y^{-n} \qquad (D\text{-}1)$$

$$y^n y^m = y^{n+m} \qquad \text{(Exponents added)} \qquad (D\text{-}2)$$

$$\frac{y^n}{y^m} = y^{n-m} \qquad \text{(Exponents subtracted)} \qquad (D\text{-}3)$$

To the three rules above we add two more that are useful. One of these is

$$y^n z^n = (yz)^n \qquad (D\text{-}4)$$

The following example helps to clarify the reasoning behind this rule:

$$3^2 5^2 = (3 \times 3)(5 \times 5) = (3 \times 5)(3 \times 5) = (3 \times 5)^2$$

The other additional rule is

$$(y^n)^m = y^{nm} \qquad \text{(Exponents multiplied)} \qquad (D\text{-}5)$$

To see why this rule applies, consider the following example:

$$(5^2)^3 = (5^2)(5^2)(5^2) = 5^{2+2+2} = 5^{2\times3}$$

Roots, such as a square root or a cube root, can be represented with fractional exponents. For instance,

$$\sqrt{y} = y^{1/2} \quad \text{and} \quad \sqrt[3]{y} = y^{1/3}$$

In general, the $n$th root of $y$ is given by

$$\sqrt[n]{y} = y^{1/n} \qquad (D\text{-}6)$$

The rationale for Equation D-6 can be explained using the fact that $(y^n)^m = y^{nm}$. For instance, the fifth root of $y$ is the number that, when multiplied by itself five times, gives back $y$. As shown below, the term $y^{1/5}$ satisfies this definition:

$$(y^{1/5})(y^{1/5})(y^{1/5})(y^{1/5})(y^{1/5}) = (y^{1/5})^5 = y^{(1/5)\times5} = y$$

Logarithms are closely related to exponents. To see the connection between the two, note that it is possible to express any number $y$ as another number $B$ raised to the exponent $x$. In other words,

$$y = B^x \qquad (D\text{-}7)$$

The exponent $x$ is called the **logarithm** of the number $y$. The number $B$ is called the **base number.** One of two choices for the base number is usually used. If $B = 10$, the logarithm is known as the *common logarithm,* for which the notation "log" applies:

**Common logarithm**          $y = 10^x$   or   $x = \log y$          (D-8)

If $B = e = 2.718 \ldots$, the logarithm is referred to as the *natural logarithm,* and the notation "ln" is used:

**Natural logarithm**          $y = e^z$   or   $z = \ln y$          (D-9)

The two kinds of logarithms are related by

$$\ln y = 2.3026 \log y \qquad \text{(D-10)}$$

Both kinds of logarithms are often given on calculators.

The logarithm of the product or quotient of two numbers $A$ and $C$ can be obtained from the logarithms of the individual numbers according to the rules below. These rules are illustrated here for natural logarithms, but they are the same for any kind of logarithm.

$$\ln (AC) = \ln A + \ln C \qquad \text{(D-11)}$$

$$\ln \left( \frac{A}{C} \right) = \ln A - \ln C \qquad \text{(D-12)}$$

Thus, the logarithm of the product of two numbers is the sum of the individual logarithms, and the logarithm of the quotient of two numbers is the difference between the individual logarithms. Another useful rule concerns the logarithm of a number $A$ raised to an exponent $n$:

$$\ln A^n = n \ln A \qquad \text{(D-13)}$$

Rules D-11, D-12, and D-13 can be derived from the definition of the logarithm and the rules governing exponents.

# Appendix E
# Geometry and Trigonometry

## E.1  *Geometry*
### ANGLES

Two angles are equal if
1. They are vertical angles (see Figure E1).
2. Their sides are parallel (see Figure E2).

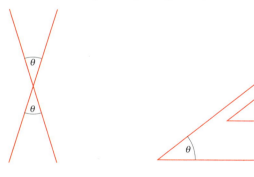

**Figure E1**                    **Figure E2**

3. Their sides are mutually perpendicular (see Figure E3).

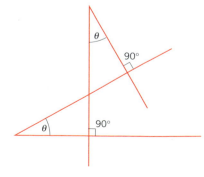

**Figure E3**

### TRIANGLES

1. The **sum of the angles** of any triangle is 180° (see Figure E4).

$\alpha + \beta + \gamma = 180°$

**Figure E4**

2. A **right triangle** has one angle that is 90°.
3. An **isosceles triangle** has two sides that are equal.
4. An **equilateral triangle** has three sides that are equal. Each angle of an equilateral triangle is 60°.
5. Two triangles are **similar** if two of their angles are equal (see Figure E5). The corresponding sides of similar triangles are proportional to each other:

$$\frac{a_1}{a_2} = \frac{b_1}{b_2} = \frac{c_1}{c_2}$$

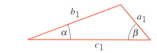

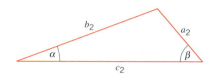

**Figure E5**

6. Two similar triangles are **congruent** if they can be placed on top of one another to make an exact fit.

# CIRCUMFERENCES, AREAS, AND VOLUMES OF SOME COMMON SHAPES

**1.** Triangle of base $b$ and altitude $h$ (see Figure E6):

$$\text{Area} = \tfrac{1}{2}bh$$

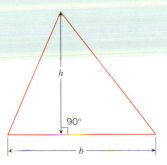

*Figure E6*

**2.** Circle of radius $r$:

$$\text{Circumference} = 2\pi r$$
$$\text{Area} = \pi r^2$$

**3.** Sphere of radius $r$:

$$\text{Surface area} = 4\pi r^2$$
$$\text{Volume} = \tfrac{4}{3}\pi r^3$$

**4.** Right circular cylinder of radius $r$ and height $h$ (see Figure E7):

$$\text{Surface area} = 2\pi r^2 + 2\pi rh$$
$$\text{Volume} = \pi r^2 h$$

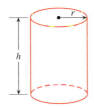

*Figure E7*

# E.2 *Trigonometry*

## BASIC TRIGONOMETRIC FUNCTIONS

**1.** For a right triangle, the sine, cosine, and tangent of an angle $\theta$ are defined as follows (see Figure E8):

$$\sin\theta = \frac{\text{Side opposite }\theta}{\text{Hypotenuse}} = \frac{h_o}{h}$$

$$\cos\theta = \frac{\text{Side adjacent to }\theta}{\text{Hypotenuse}} = \frac{h_a}{h}$$

$$\tan\theta = \frac{\text{Side opposite }\theta}{\text{Side adjacent to }\theta} = \frac{h_o}{h_a}$$

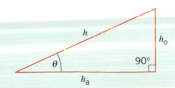

*Figure E8*

**2.** The secant ($\sec\theta$), cosecant ($\csc\theta$), and cotangent ($\cot\theta$) of an angle $\theta$ are defined as follows:

$$\sec\theta = \frac{1}{\cos\theta} \qquad \csc\theta = \frac{1}{\sin\theta} \qquad \cot\theta = \frac{1}{\tan\theta}$$

# TRIANGLES AND TRIGONOMETRY

**1.** The **Pythagorean theorem** states that the square of the hypotenuse of a right triangle is equal to the sum of the squares of the other two sides (see Figure E8):

$$h^2 = h_o{}^2 + h_a{}^2$$

**2.** The **law of cosines** and the **law of sines** apply to any triangle, not just a right triangle, and they relate the angles and the lengths of the sides (see Figure E9):

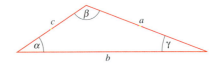

*Figure E9*

*Law of cosines* $\qquad c^2 = a^2 + b^2 - 2ab\cos\gamma$

*Law of sines* $\qquad \dfrac{a}{\sin\alpha} = \dfrac{b}{\sin\beta} = \dfrac{c}{\sin\gamma}$

# OTHER TRIGONOMETRIC IDENTITIES

**1.** $\sin(-\theta) = -\sin\theta$
**2.** $\cos(-\theta) = \cos\theta$
**3.** $\tan(-\theta) = -\tan\theta$
**4.** $(\sin\theta)/(\cos\theta) = \tan\theta$
**5.** $\sin^2\theta + \cos^2\theta = 1$
**6.** $\sin(\alpha \pm \beta) = \sin\alpha\cos\beta \pm \cos\alpha\sin\beta$

$$\text{If } \alpha = 90°,\ \sin(90° \pm \beta) = \cos\beta$$
$$\text{If } \alpha = \beta,\ \sin 2\beta = 2\sin\beta\cos\beta$$

**7.** $\cos(\alpha \pm \beta) = \cos\alpha\cos\beta \mp \sin\alpha\sin\beta$

$$\text{If } \alpha = 90°,\ \cos(90° \pm \beta) = \mp\sin\beta$$
$$\text{If } \alpha = \beta,\ \cos 2\beta = \cos^2\beta - \sin^2\beta = 1 - 2\sin^2\beta$$

# Answers to Odd-Numbered Problems

## Chapter 1
1. (a) $5 \times 10^{-3}$ g
   (b) 5 mg
   (c) $5 \times 10^3$ μg
3. (a) 5700 s
   (b) 86 400 s
5. 1.37 lb
7. (a) correct
   (b) not correct
   (c) not correct
   (d) correct
   (e) correct
9. $3.13 \times 10^8$ m$^3$
11. 5.5 km
13. 80.1 km, 25.9° south of west
15. 3.73 m
17. 35.3°
19. 99° opposite the 190-cm side, $3.0 \times 10^1$ degrees opposite the 95-cm side, 51° opposite the 150-cm side
21. 200 N due east or 600 N due west
23. 0.90 km, 56° north of west
25. (a) 5.31 km, south
   (b) 5.31 km, north
27. (a) 5600 newtons
   (b) along the dashed line
29. (a) 8.6 units
   (b) 34.9° north of west
   (c) 8.6 units
   (d) 34.9° south of west
31. (a) 15.8 m/s
   (b) 6.37 m/s
33. (a) 147 km
   (b) 47.9 km
35. (a) $5.70 \times 10^2$ newtons
   (b) 33.6° south of west
37. (a) 25.0°
   (b) 34.8 newtons
39. (a) 322 newtons
   (b) 209 newtons
   (c) 279 newtons
41. 7.1 m, 9.9° north of east
43. 3.00 m, 42.8° above the $-x$ axis
45. (a) 2.7 km
   (b) $6.0 \times 10^1$ degrees north of east
47. (a) 10.4 units
   (b) 12.0 units
49. (a) 178 units
   (b) 164 units

## Chapter 2
1. 0.80 s
3. (a) 12.4 km
   (b) 8.8 km, due east
5. (a) 464 m/s
   (b) 1040 mi/h
7. 52 m
9. $7.2 \times 10^3$ m
11. 2.1 s
13. (a) 4.0 s
   (b) 4.0 s
15. 3.44 m/s, due west
17. +30.0 m/s
19. 4.5 m
21. (a) 11.4 s
   (b) 44.7 m/s$^2$
23. (a) $1.7 \times 10^2$ cm/s$^2$
   (b) 0.15 s
25. $-3.1$ m/s$^2$
27. 0.74 m/s
29. 39.2 m
31. 0.87 m/s$^2$, in the same direction as the velocity
33. 1.2
35. (a) 13 m/s
   (b) 0.93 m/s$^2$
37. 91.5 m/s
39. $-1.5$ m/s$^2$
41. 1.1 s
43. 6.12 s
45. 1.7 s
47. 8.96 m
49. 0.767 m/s
51. $2.0 \times 10^1$ m
53. 10.6 m
55. 6.0 m below the top of the cliff
57. The answer is in graphical form.
59. 1.9 m/s$^2$ (segment A), 0 m/s$^2$ (segment B), 3.3 m/s$^2$ (segment C)
61. $-8.3$ km/h$^2$
63. (a) 6.6 s
   (b) 5.3 m/s

## Chapter 3
1. $8.8 \times 10^2$ m
3. 8600 m
5. 242 m/s
7. 5.4 m/s
9. 27.0°
11. (a) $2.99 \times 10^4$ m/s
   (b) $2.69 \times 10^4$ m/s
13. 2.40 m
15. (a) 1.78 s
   (b) 20.8 m/s
17. 14.1 m/s
19. (a) $6.0 \times 10^1$ m
   (b) 290 m
21. 30.0 m
23. (a) 85 m
   (b) 610 m
25. (a) 1.1 s
   (b) 1.3 s

27. 24 buses
29. 1.7 s
31. 48 m
33. 11 m/s
35. 14.7 m/s
37. 14.9 m
39. 21.9 m/s, 40.0°
41. 0.141° and 89.860°
43. $D = 850$ m, $H = 31$ m
45. $v_{0B} = 8.79$ m/s, $\theta_B = 81.5°$

## Chapter 4
1. 93 N
3. $3.5 \times 10^4$ N
5. 130 N
7. 37 N
9. (a) 3.6 N
   (b) 0.40 N
11. 1.83 m/s$^2$, left
13. 30.9 m/s$^2$, 27.2° above the $+x$ axis
15. 0.78 m, 21° south of east
17. 18.4 N, 68° north of east
19. $1.8 \times 10^{-7}$ N
21. (a) $W = 1.13 \times 10^3$ N, $m = 115$ kg
   (b) $W = 0$ N, $m = 115$ kg
23. 0.223 m/s$^2$
25. $1.76 \times 10^{24}$ kg
27. (a) 3.75 m/s$^2$
   (b) $2.4 \times 10^2$ N
29. 4.7 kg
31. 0.0050
33. 0.414 $L$
35. $7.3 \times 10^2$ N
37. The block will move. $a = 3.72$ m/s$^2$
39. 0.444
41. (a) 390 N
   (b) 7.7 m/s, direction is toward second base
43. (a) 0.980 m/s$^2$, direction is opposite to the direction of motion
   (b) 29.5 m
45. 68°
47. (a) $7.40 \times 10^5$ N
   (b) $1.67 \times 10^9$ N
49. 9.70 N for each force
51. 929 N
53. $1.00 \times 10^2$ N, 53.1° south of east
55. 62 N
57. 16.3 N
59. 286 N
61. 406 N
63. 18.0 m/s$^2$, 56.3° above the $+x$ axis
65. 1730 N, due west
67. (a) 2.99 m/s$^2$
   (b) 129 N
69. 6.6 m/s

**71. (a)** 1.3 N
　　**(b)** 6.5 N
**73.** 29 400 N
**75.** 0.14 m/s$^2$
**77. (a)** 4.25 m/s$^2$
　　**(b)** 1080 N
**79.** 0.265 m
**81. (a)** $\Delta T_A = 0$ N, $\Delta T_B = -4.7$ N,
　　　$\Delta T_C = 0$ N
　　**(b)** $\Delta T_A = 0$ N, $\Delta T_B = 0$ N,
　　　$\Delta T_C = +4.7$ N
**83.** 1.2 s
**85. (a)** 13.7 N
　　**(b)** 1.37 m/s$^2$

## Chapter 5

**1.** 160 s
**3.** 6.9 m/s$^2$
**5.** 332 m
**7. (a)** $5.0 \times 10^1$ m/s$^2$
　　**(b)** zero
　　**(c)** $2.0 \times 10^1$ m/s$^2$
**9.** 2.2
**11.** 0.68 m/s
**13. (a)** 0.189 N
　　**(b)** 4.00
**15.** 12 m/s
**17.** 3500 N
**19. (a)** 3510 N
　　**(b)** 14.9 m/s
**21.** $2.0 \times 10^1$ m/s
**23.** 184 m
**25.** $2.12 \times 10^6$ N
**27.** $1.33 \times 10^4$ m/s
**29.** $4.20 \times 10^4$ m/s
**31.** 1/27
**33.** $2.45 \times 10^4$ N
**35. (a)** 912 m
　　**(b)** 228 m
　　**(c)** 2.50 m/s$^2$

## Chapter 6

**1.** $-2.6 \times 10^6$ J
**3. (a)** 2980 J
　　**(b)** 3290 J
**5.** 42.8°
**7. (a)** 54.9 N
　　**(b)** 1060 J
　　**(c)** $-1060$ J
　　**(d)** 0 J
**9.** 45 N
**11.** $2.07 \times 10^3$ N
**13. (a)** $3.1 \times 10^3$ J
　　**(b)** $2.2 \times 10^2$ J
**15. (a)** 38 J
　　**(b)** $3.8 \times 10^3$ N
**17.** $6.4 \times 10^5$ J
**19.** 18%
**21.** $1.4 \times 10^{11}$ J
**23.** 10.9 m/s
**25. (a)** $-3.0 \times 10^4$ J
　　**(b)** The resistive force is not a
　　　conservative force.

**27.** $2.39 \times 10^5$ J
**29.** $5.24 \times 10^5$ J
**31.** $2.3 \times 10^4$ J
**33. (a)** 28.3 m/s
　　**(b)** 28.3 m/s
　　**(c)** 28.3 m/s
**35.** 4.13 m
**37.** 4.8 m/s
**39.** 3.29 m/s
**41.** 6.33 m
**43.** 40.8 kg
**45.** $-4.51 \times 10^4$ J
**47.** 16.5 m
**49.** $-1.21 \times 10^6$ J
**51. (a)** 2.8 J
　　**(b)** 35 N
**53.** 4.17 m/s
**55. (a)** $3.3 \times 10^4$ W
　　**(b)** $5.1 \times 10^4$ W
**57.** $3.6 \times 10^6$ J
**59.** $3.0 \times 10^3$ W
**61.** $6.7 \times 10^2$ N
**63. (a)** 93 J
　　**(b)** 0 J
　　**(c)** 2.3 m/s
**65. (a)** Bow 1 requires more work.
　　**(b)** 25 J
**67. (a)** $1.50 \times 10^2$ J
　　**(b)** 7.07 m/s

## Chapter 7

**1.** $-8.7$ kg·m/s
**3. (a)** $+1.7$ kg·m/s
　　**(b)** $+570$ N
**5.** 11 N·s, in the same direction as the
　　average force
**7.** $+69$ N
**9.** 3.7 N·s
**11.** 4.28 N·s, upward
**13.** 960 N
**15.** $-1.5 \times 10^{-4}$ m/s
**17. (a)** Bonzo, since he has the recoil
　　　velocity with the smaller
　　　magnitude
　　**(b)** 1.7
**19. (a)** 77.9 m/s
　　**(b)** 45.0 m/s
**21.** $m_1 = 1.00$ kg, $m_2 = 1.00$ kg
**23.** $+547$ m/s
**25.** 84 kg
**27.** 3.00 m
**29.** $+9.09$ m/s
**31. (a)** $-0.400$ m/s (5.00-kg ball),
　　　$+1.60$ m/s (7.50-kg ball)
　　**(b)** $+0.800$ m/s
**33. (a)** $+8.9$ m/s
　　**(b)** $-3.6 \times 10^4$ N·s
　　**(c)** 5.9 m
**35. (a)** 73.0°
　　**(b)** 4.28 m/s
**37. (a)** $+3.17$ m/s
　　**(b)** 0.0171

**39. (a)** 5.56 m/s
　　**(b)** $-2.83$ m/s (1.50-kg ball),
　　　$+2.73$ m/s (4.60-kg ball)
　　**(c)** 0.409 m (1.50-kg ball),
　　　0.380 m (4.60-kg ball)
**41.** $4.67 \times 10^6$ m
**43.** $6.46 \times 10^{-11}$ m

## Chapter 8

**1.** 13 rad/s
**3.** 63.7 grad
**5.** $6.4 \times 10^{-3}$ rad/s$^2$
**7.** 492 rad/s
**9.** 825 m
**11.** 336 m/s
**13.** 6.05 m
**15.** 25 rev
**17.** 157.3 rad/s
**19. (a)** $1.2 \times 10^4$ rad
　　**(b)** $1.1 \times 10^2$ s
**21.** 28 rad/s
**23.** 12.5 s
**25.** $1.95 \times 10^4$ rad
**27.** 7.37 s
**29.** 157 m/s
**31.** 22 rev/s
**33. (a)** $4.66 \times 10^2$ m/s
　　**(b)** 70.6°
**35. (a)** 1.25 m/s
　　**(b)** 7.98 rev/s
**37.** 14.8 rad/s
**39. (a)** 9.00 m/s$^2$
　　**(b)** radially inward
**41. (a)** 2.5 m/s$^2$
　　**(b)** 3.1 m/s$^2$
**43.** $1/\sqrt{3}$
**45.** 1.00 rad
**47.** 8.71 rad/s$^2$
**49.** 693 rad
**51.** 28.0 rad/s
**53.** 11.8 rad
**55.** 20.6 rad

## Chapter 9

**1.** 4.2 N·m
**3.** 843 N
**5.** 1.3
**7. (a)** $\tau = FL$
　　**(b)** $\tau = FL$
　　**(c)** $\tau = FL$
**9.** 0.667 m
**11.** 196 N (force on each hand),
　　96 N (force on each foot)
**13.** 0.591
**15. (a)** 27 N, to the left
　　**(b)** 27 N, to the right
　　**(c)** 27 N, to the right
　　**(d)** 143 N, downward and to the left
　　　(79° below the horizontal)
**17.** 1200 N, to the left
**19. (a)** $1.60 \times 10^5$ N
　　**(b)** $4.20 \times 10^5$ N
**21.** 37.6°

**23. (a)** $1.21 \times 10^3$ N
  **(b)** $1.01 \times 10^3$ N, downward
**25.** 51.4 N
**27.** 1.7 m
**29.** 1.25 kg·m$^2$
**31. (a)** $-11$ N·m
  **(b)** $-9.2$ rad/s$^2$
**33. (a)** 0.131 kg·m$^2$
  **(b)** $3.6 \times 10^{-4}$ kg·m$^2$
  **(c)** 0.149 kg·m$^2$
**35. (a)** 5.94 rad/s$^2$
  **(b)** 44.0 N
**37.** 0.78 N
**39. (a)** 2.67 kg·m$^2$
  **(b)** 1.16 m
**41.** 2.12 s
**43. (a)** $v_{T1} = 12.0$ m/s, $v_{T2} = 9.00$ m/s,
  $v_{T3} = 18.0$ m/s
  **(b)** $1.08 \times 10^3$ J
  **(c)** 60.0 kg·m$^2$
  **(d)** $1.08 \times 10^3$ J
**45.** $6.1 \times 10^5$ rev/min
**47.** 2/7
**49.** 3/4
**51.** 2.0 m
**53.** 4.4 kg·m$^2$
**55.** 0.26 rad/s
**57.** 8% increase
**59.** 0.17 m

## Chapter 10

**1.** 237 N
**3. (a)** 7.44 N
  **(b)** 7.44 N
**5.** 0.012 m
**7.** $2.29 \times 10^{-3}$ m
**9.** 0.79
**11.** 0.240 m
**13. (a)** $1.00 \times 10^3$ N/m
  **(b)** 0.340
**15.** $3.5 \times 10^4$ N/m
**17. (a)** 0.080 m
  **(b)** 1.6 rad/s
  **(c)** 2.0 N/m
  **(d)** 0 m/s
  **(e)** 0.20 m/s$^2$
**19.** 696 N/m
**21.** 4.3 kg
**23.** 0.806
**25.**

| h (meters) | KE | PE (gravity) | PE (elastic) | E |
|---|---|---|---|---|
| 0 | 0 J | 0 J | 8.76 J | 8.76 J |
| 0.200 | 1.00 J | 3.92 J | 3.84 J | 8.76 J |
| 0.400 | 0 J | 7.84 J | 0.92 J | 8.76 J |

**27. (a)** 58.8 N/m
  **(b)** 11.4 rad/s
**29.** $7.18 \times 10^{-2}$ m
**31.** 14 m/s
**33. (a)** $9.0 \times 10^{-2}$ m
  **(b)** 2.1 m/s
**35.** 1.25 m/s (11.2-kg block),
  0.645 m/s (21.7-kg block)

**37.** $2.37 \times 10^3$ N/m
**39.** 0.99 m
**41.** 6.0 m/s$^2$
**43.** 0.816
**45.** 7R/5
**47.** $3.7 \times 10^{-5}$ m
**49.** $1.6 \times 10^5$ N
**51.** 260 m
**53.** $1.4 \times 10^{-6}$
**55.** $-2.8 \times 10^{-4}$
**57.** $6.6 \times 10^4$ N
**59. (a)** $6.3 \times 10^{-2}$ m
  **(b)** $7.3 \times 10^{-2}$ m
**61.** $4.6 \times 10^{-4}$
**63.** $1.0 \times 10^{-3}$ m
**65.** $-4.4 \times 10^{-5}$

## Chapter 11

**1.** 317 m$^2$
**3.** 3400 N
**5.** $6.6 \times 10^6$ kg
**7.** 1.9 gal
**9.** 63%
**11.** $1.1 \times 10^3$ N
**13.** $4.33 \times 10^6$ Pa
**15.** 24 blocks
**17.** 0.750 m
**19.** $1.2 \times 10^5$ Pa
**21. (a)** $3.5 \times 10^6$ N
  **(b)** $1.2 \times 10^6$ N
**23. (a)** $2.45 \times 10^5$ Pa
  **(b)** $1.73 \times 10^5$ Pa
**25.** $7.0 \times 10^5$ Pa
**27.** $1.19 \times 10^5$ Pa
**29.** 0.74 m
**31.** $3.8 \times 10^5$ N
**33. (a)** 93.0 N
  **(b)** 94.9 N
**35.** $8.50 \times 10^5$ N·m
**37.** $5.7 \times 10^{-2}$ m
**39.** 4.89 m
**41.** $2.7 \times 10^{-4}$ m$^3$
**43.** $2.04 \times 10^{-3}$ m$^3$
**45.** 0.20 m
**47.** $7.6 \times 10^{-2}$ m
**49.** $5.28 \times 10^{-2}$ m and $6.20 \times 10^{-2}$ m
**51.** 1.91 m/s
**53. (a)** 0.18 m
  **(b)** 0.14 m
**55. (a)** $1.6 \times 10^{-4}$ m$^3$/s
  **(b)** $2.0 \times 10^1$ m/s
**57. (a)** 150 Pa
  **(b)** The pressure inside the roof is greater
  than that outside the roof, so there is a
  net outward force.
**59.** $1.92 \times 10^5$ N
**61. (a)** $2.48 \times 10^5$ Pa
  **(b)** $1.01 \times 10^5$ Pa
  **(c)** 0.342 m$^3$/s
**63.** 33 m/s
**65. (a)** 14 m/s
  **(b)** 0.98 m$^3$/s

**67.** 9600 N
**69.** 7.78 m/s

## Chapter 12

**1.** 0.2 C°
**3. (a)** 10.0 °C and 40.0 °C
  **(b)** 283.2 K and 313.2 K
**5.** $-459.67$ °F
**7.** $-164$ °C
**9.** 0.084 m
**11.** 1500 m
**13.** 110 C°
**15.** $-2.82 \times 10^{-4}$ m
**17.** 49 °C
**19.** 2.0027 s
**21. (a)** tension
  **(b)** $9.6 \times 10^7$ N/m$^2$
**23.** 0.6
**25.** $3.1 \times 10^{-3}$ m$^3$
**27.** $2.5 \times 10^{-7}$ m$^3$
**29.** 18
**31.** 0.33 gal
**33.** 9.0 mm
**35.** 45 atm
**37.** 6.9
**39.** 36.2 °C
**41.** $230
**43.** 21.03 °C
**45.** 940 °C
**47.** 21 barrels
**49.** 0.016 C°
**51.** $3.9 \times 10^5$ J
**53.** 9.3 kg
**55.** $9.49 \times 10^{-3}$ kg
**57.** $1.85 \times 10^5$ J
**59. (a)** $3.0 \times 10^{20}$ J
  **(b)** 3.2 years
**61.** $1.9 \times 10^4$ J/kg
**63.** $2.6 \times 10^{-3}$ kg
**65.** $3.50 \times 10^2$ m/s

## Chapter 13

**1.** $8.0 \times 10^2$ J/s
**3.** 14 h
**5.** $2.0 \times 10^{-3}$ m
**7.** 17
**9. (a)** 21 °C
  **(b)** 18 °C
**11. (a)** 130 °C
  **(b)** 830 J
  **(c)** 237 °C
**13.** 287 °C
**15. (a)** 2.0
  **(b)** 0.61
**17.** 14.5 da
**19.** 0.3
**21.** 320 K
**23.** 0.70
**25.** 12

## Chapter 14

1. Aluminum
3. (a) 294.307 u
   (b) $4.887 \times 10^{-25}$ kg
5. $1.07 \times 10^{-22}$ kg
7. $2.6 \times 10^{-10}$ m
9. $1.0 \times 10^3$ kg
11. $2.3 \times 10^{-2}$ mol
13. (a) 201 mol
    (b) $1.21 \times 10^5$ Pa
15. $2.5 \times 10^{21}$
17. 882 K
19. 925 K
21. $5.9 \times 10^4$ g
23. 0.205
25. $6.19 \times 10^5$ Pa
27. 308 K
29. (a) 46.3 m²/s²
    (b) 40.1 m²/s²
31. $1.2 \times 10^4$ m/s
33. $3.9 \times 10^5$ J
35. 2820 m
37. $4.0 \times 10^1$ Pa

## Chapter 15

1. (a) −261 J
   (b) Work is done on the system.
3. (a) −87 J
   (b) +87 J
5. 32 miles
7. 13 000 J
9. (a) 3100 J
   (b) negative
11. 0.24 m
13. $3.0 \times 10^5$ Pa
15. The answer is a proof.
17. $4.99 \times 10^{-6}$
19. (a) 0 J
    (b) $-6.1 \times 10^3$ J
    (c) 310 K
21. 0.66
23. 1.81
25. (a) $-8.00 \times 10^4$ J
    (b) Heat flows out of the gas.
27. 19.3
29. (a) 327 K
    (b) 0.132 m³
31. 2400 J
33. (a) $1.1 \times 10^4$ J
    (b) $1.8 \times 10^4$ J
35. 5/2
37. (a) 60.0%
    (b) 40.0%
39. $2.38 \times 10^4$ J
41. 0.631
43. (a) 8600 J
    (b) 3100 J
45. $e = e_1 + e_2 - e_1 e_2$
47. 0.21
49. (a) 1260 K
    (b) $1.74 \times 10^4$ J
51. lowering the temperature of the cold reservoir

53. (a) 0.360
    (b) $1.3 \times 10^{13}$ J
55. 5.7 C°
57. $3.24 \times 10^4$ J
59. $5.86 \times 10^5$ J
61. 9.03
63. 1.4 K
65. (a) $2.0 \times 10^1$
    (b) $1.5 \times 10^4$ J
67. $1.26 \times 10^3$ K
69. 11.6 J/K
71. (a) $+3.68 \times 10^3$ J/K
    (b) $+1.82 \times 10^4$ J/K
    (c) The vaporization process creates more disorder.
73. (a) $+8.0 \times 10^2$ J/K
    (b) The entropy of the universe should increase.

## Chapter 16

1. 0.083 Hz
3. 0.49 m
5. 0.25 m
7. 78 cm
9. $5.0 \times 10^1$ s
11. (a) 1.1 m/s
    (b) 6.55 m
13. 64 N
15. 600 m/s
17. 7.7 m/s²
19. $m_1 = 28.7$ kg, $m_2 = 14.3$ kg
21. $3.26 \times 10^{-3}$ s
23. $y = (0.37$ m$) \sin (2.6 \pi t + 0.22 \pi x)$, where x and y are in meters and t is in seconds.
25. (a) $A = 0.45$ m, $f = 4.0$ Hz, $\lambda = 2.0$ m, $v = 8.0$ m/s
    (b) −x direction
27. 2.5 N
29. 1730 m/s
31. (a) $7.87 \times 10^{-3}$ s
    (b) 4.12
33. 2.06
35. (a) metal wave first, water wave second, air wave third
    (b) Second sound arrives 0.059 s later, and third sound arrives 0.339 s later.
37. 690 rad/s
39. tungsten
41. 650 m
43. $8.0 \times 10^2$ m/s
45. 239 m/s
47. 0.404 m
49. 1.98%
51. $2.4 \times 10^{-5}$ W/m²
53. 6.5 W
55. $7.6 \times 10^3$ W/m²
57. 2.6 s
59. 0.316 W/m²
61. 1.3
63. 56.2 W
65. 2.6
67. 0.84 s

69. 2.39 dB
71. 17 m/s
73. 860 Hz
75. 1350 Hz
77. 22 m/s
79. 209 m

## Chapter 17

1. (a) 2 cm
   (b) 1 cm
3. The answer is a series of drawings.
5. 107 Hz
7. 3.89 m
9. 3.90 m, 1.55 m, 6.25 m
11. (a) 44°
    (b) 0.10 m
13. $1.5 \times 10^4$ Hz
15. 3.7°
17. 8 Hz
19. (a) 50 kHz
    (b) 90 kHz
21. 8 Hz
23. $1.10 \times 10^2$ Hz
25. 0.46 m
27. 171 N
29. (a) 180 m/s
    (b) 1.2 m
    (c) 150 Hz
31. 0.077 m
33. 20.8° and 53.1°
35. (a) $f_2 = 800$ Hz, $f_3 = 1200$ Hz, $f_4 = 1600$ Hz
    (b) $f_2 = 800$ Hz, $f_3 = 1200$ Hz, $f_4 = 1600$ Hz
    (c) $f_3 = 1200$ Hz, $f_5 = 2000$ Hz, $f_7 = 2800$ Hz
37. 0.50 m
39. 602 Hz
41. 6.1 m
43. $1.68 \times 10^5$ Pa

## Chapter 18

1. $1.5 \times 10^{13}$
3. $1.6 \times 10^{13}$
5. (a) $+1.5 q$
   (b) $+4 q$
   (c) $+4 q$
7. 120 N
9. 0.14 N
11. (a) both positive or both negative
    (b) $1.7 \times 10^{-16}$ C
13. (a) $4.56 \times 10^{-8}$ C
    (b) $3.25 \times 10^{-6}$ kg
15. $3.8 \times 10^{12}$
17. $+2.0 \mu$C
19. $-3.3 \times 10^{-6}$ C
21. $3.5 \times 10^{-5}$ C
23. (a) 15.4°
    (b) 0.813 N
25. 1.8 N due east
27. $6.5 \times 10^3$ N/C downward
29. (a) $-6.2 \times 10^7$ N/C
    (b) $+2.9 \times 10^8$ N/C

**31.** 1.3 m
**33. (a)** positive
   **(b)** $2.53 \times 10^7$ protons
**35.** $3.9 \times 10^6$ N/C, in the $+y$ direction
**37.** 35 N/C
**39.** $|q_1| = 0.716\ q$, $|q_2| = 0.0895\ q$
**41.** 0.577
**43.** $3.25 \times 10^{-8}$ C
**45.** 61°
**47. (a)** 350 N·m²/C
   **(b)** 460 N·m²/C
**49. (a)** $2.3 \times 10^5$ N·m²/C
   **(b)** $2.3 \times 10^5$ N·m²/C
   **(c)** $2.3 \times 10^5$ N·m²/C
**51. (a)** The flux through the face in the
   $x, z$ plane at $y = 0$ m is $-6.0 \times$
   $10^1$ N·m²/C. The flux through the
   face parallel to the $x, z$ plane at $y =$
   0.20 m is $+6.0 \times 10^1$ N·m²/C.
   The flux through each of the
   remaining four faces is zero.
   **(b)** 0 N·m²/C
**53.** The answer is a proof.

## Chapter 19

**1.** $1.1 \times 10^{-20}$ J
**3. (a)** $+2.00 \times 10^{-14}$ J
   **(b)** $2.00 \times 10^{-14}$ J
**5.** $9.4 \times 10^7$ m/s
**7.** 67 hp
**9.** 339 V
**11.** $+3.6 \times 10^{-9}$ C
**13.** $-4.35 \times 10^{-18}$ J
**15.** $-4.7 \times 10^{-2}$ J
**17.** $-q/\sqrt{2}$
**19.** $-0.746$ J
**21.** $1.53 \times 10^{-14}$ m
**23.** 0.0342 m
**25. (a)** $-2q/3$
   **(b)** $-2q$
**27.** 18 000 V
**29.** $3.5 \times 10^4$ V
**31.** $+9.0 \times 10^3$ V
**33. (a)** 179 V
   **(b)** 143 V
   **(c)** 155 V
**35.** 0.213 J
**37.** $1.1 \times 10^3$ V
**39.** 5.3
**41. (a)** 33 J
   **(b)** 8500 W
**43.** 52 V
**45.** $1.2 \times 10^{-8}$ J
**47.** The answer is a proof.

## Chapter 20

**1. (a)** 2.6 C
   **(b)** 310 J
**3.** 0.21 A

**5.** 82 Ω
**7.** $6.2 \times 10^4$ J
**9. (a)** $4.7 \times 10^{13}$
   **(b)** 17 C°
**11.** 1.64
**13.** $9.9 \times 10^{-3}$ m
**15.** $-34.6$ °C
**17.** 189 Ω
**19.** $L_{\text{tungsten}}/L_{\text{carbon}} = 70$
**21.** $6.0 \times 10^2$ W
**23.** 8.9 h
**25. (a)** 1300 W
   **(b)** 480 W
   **(c)** 1.4
**27.** 250 °C
**29.** 33 °C
**31.** 1.77 A
**33. (a)** $9.0 \times 10^2$ W
   **(b)** $1.8 \times 10^3$ W
**35.** $92
**37.** 2.0 h
**39.** 32 Ω
**41. (a)** 10.0 V
   **(b)** 5.00 V
**43. (a)** 145 Ω
   **(b)** 74 V
**45. (a)** 15.5 V
   **(b)** 14.2 W
**47. (a)** 35 Ω
   **(b)** $5.0 \times 10^1$ Ω
**49.** 190 Ω
**51.** 5.3 Ω
**53. (a)** 4.57 A
   **(b)** 1450 W
**55. (a)** 3.6 Ω
   **(b)** 33 A, the breaker will open
**57.** $3.58 \times 10^{-8}$ m²
**59.** $1.0 \times 10^2$ Ω
**61.** 4.6 Ω
**63.** 25 Ω
**65.** 6.00 Ω, 0.545 Ω, 3.67 Ω, 2.75 Ω,
   2.20 Ω, 1.50 Ω, 1.33 Ω, 0.833 Ω
**67.** 0.054 Ω
**69.** 30
**71.** 24.0 V
**73. (a)** 0.38 A
   **(b)** $2.0 \times 10^1$ V
   **(c)** $B$
**75.** 33 A
**77.** 0.75 V; left end
**79.** 0.94 V; $D$
**81.** $3.43 \times 10^3$ Ω
**83.** 0.450 Ω
**85. (a)** 30.0 V
   **(b)** 28.1 V
**87.** 2.0 μF
**89.** 1.7 μF
**91.** The answer is a proof.
**93.** $C_0$
**95.** $1.2 \times 10^{-2}$ s
**97.** $4.1 \times 10^{-7}$ F
**99.** 0.29 s

## Chapter 21

**1.** 75.1° and 105°
**3.** $3.7 \times 10^{-12}$ N
**5.** $4.1 \times 10^{-3}$ m/s
**7.** 19.7°
**9. (a)** due south
   **(b)** $2.55 \times 10^{14}$ m/s²
**11. (a)** The answer is a drawing.
   **(b)** $2.6 \times 10^{-2}$ T
   **(c)** $2.7 \times 10^{16}$ m/s²
**13.** $1.5 \times 10^{-8}$ s
**15. (a)** $7.2 \times 10^6$ m/s
   **(b)** $3.5 \times 10^{-13}$ N
**17.** $1.63 \times 10^{-2}$ m
**19. (a)** $\theta = 0°$
   **(b)** 0.29 m
**21.** 0.16 T
**23.** $8.7 \times 10^{-3}$ s
**25.** $9.6 \times 10^4$ m/s
**27.** 8.1 N
**29.** 0.96 N (top and bottom sides),
   0 N (left and right sides)
**31.** 57.6°
**33. (a)** left to right
   **(b)** $1.1 \times 10^{-2}$ m
**35.** 14 A
**37. (a)** 24 A·m²
   **(b)** 4.8 N·m
**39.** 0.062 m
**41. (a)** 170 N·m
   **(b)** increase
**43.** 8.3 N
**45.** $1.2 \times 10^{-5}$ A·m²
**47.** $8.0 \times 10^{-5}$ T
**49.** $1.9 \times 10^{-4}$ N·m
**51.** 190 A
**53.** $H = 2.1\ R$
**55.** 8.6 A, The current in the outer coil has an
   opposite direction relative to that in the
   inner coil.
**57.** $1.04 \times 10^{-2}$ T
**59.** $I_3$ is directed out of the paper;
   $I_3/I = 2.$
**61. (a)** $1.1 \times 10^{-5}$ T
   **(b)** $4.4 \times 10^{-6}$ T
**63.** The answer is a proof.

## Chapter 22

**1.** 150 m/s
**3.** 0.065 V
**5.** (rod A) emf = 0 V; (rod B) emf = 1.6 V
   and end 2 is positive; (rod C) emf = 0 V
**7. (a)** 3.3 m/s
   **(b)** 4.6 N
**9. (a)** 0.23 kg
   **(b)** $-1.8$ J
   **(c)** 1.8 J
**11.** 70.5°
**13. (a)** $1.2 \times 10^{-4}$ Wb
   **(b)** $3.2 \times 10^{-4}$ Wb
   **(c)** $2.0 \times 10^{-4}$ Wb

15. (two triangular ends) 0 Wb; (bottom surface) 0 Wb, (1.2 m $\times$ 0.30 m surface) 0.090 Wb; (1.2 m $\times$ 0.50 m surface) 0.090 Wb
17. 1.5 m$^2$/s
19. 8.6 $\times$ 10$^{-5}$ T
21. 4.8 s
23. 6.6 $\times$ 10$^{-2}$ J
25. (a) 3.6 $\times$ 10$^{-3}$ V
    (b) 2.0 $\times$ 10$^{-3}$ m$^2$/s; area must be shrunk
27. 2100 rad/s
29. (a) left to right
    (b) right to left
31. (a) clockwise
    (b) clockwise
33. There is no induced current.
35. 12 V
37. 0.30 T
39. (a) 2.4 Hz
    (b) 15 rad/s
    (c) 0.62 T
41. 102 V
43. (a) 0.80 A
    (b) 8.00 A
    (c) 4.40 A
45. 1.5 $\times$ 10$^9$ J
47. 1.4 V
49. 220
51. 2.80 $\times$ 10$^{-4}$ H
53. $M = \mu_0 \pi N_1 N_2 R_2^2/(2R_1)$
55. 1.0 $\times$ 10$^1$ W
57. step-down transformer, 1/12
59. (a) 7.0 $\times$ 10$^5$ W
    (b) 7.0 $\times$ 10$^1$ W
61. The answer is a proof.

## Chapter 23

1. 126 Hz
3. 2.7 $\times$ 10$^{-6}$ F
5. 5.00 $\times$ 10$^{-2}$ s
7. 3/2
9. 8.0 $\times$ 10$^1$ Hz
11. 176 mH
13. 0.17 V
15. 83.9 V
17. 38 V
19. (a) 0.925 A
    (b) 31.8°
21. 270 Hz
23. (a) 29.0 V
    (b) −0.263 A
25. 0.651 W
27. (a) 352 Hz
    (b) 15.5 A
29. 3.1 kHz
31. 2.8 kHz
33. (a) 2.94 $\times$ 10$^{-3}$ H
    (b) 4.84 Ω
    (c) 0.163
35. (a) 4/√3
    (b) 1/√3

## Chapter 24

1. 0.015 m
3. 1.5 $\times$ 10$^{-4}$ H
5. The answers are in graphical form.
7. 1.4 $\times$ 10$^{17}$ Hz
9. 3.7 $\times$ 10$^4$
11. 1.25 m
13. 1.5 $\times$ 10$^{10}$ Hz
15. 1.3 $\times$ 10$^6$ m
17. 540 rev/s
19. 8.75 $\times$ 10$^5$
21. 3.8 $\times$ 10$^2$ W/m$^2$
23. 0.07 N/C
25. (a) 183 N/C
    (b) 6.10 $\times$ 10$^{-7}$ T
27. 5600 W
29. 3.93 $\times$ 10$^{26}$ W
31. (a) receding
    (b) 3.1 $\times$ 10$^6$ m/s
33. (a) 6.175 $\times$ 10$^{14}$ Hz
    (b) 6.159 $\times$ 10$^{14}$ Hz
35. 71.6°
37. 14 W/m$^2$
39. 20

## Chapter 25

1. 55°
3. 14°
5. 10°
7. 11.3°, 45.0°, 71.6°, 135°, 162°
9. 33.7°
11. (a) The image distance is 3.0 $\times$ 10$^1$ cm behind the mirror.
    (b) The image height is 5.0 cm.
13. (a) The image distance is 16.7 cm behind the mirror.
    (b) The image height is 6.67 cm.
15. 10.9 cm
17. +74 cm
19. (a) 290 cm
    (b) −8.9 cm
    (c) upside down
21. +22 cm
23. (a) convex
    (b) 24.0 cm
25. −3
27. +42.0 cm
29. (a) The answer is a proof.
    (b) The answer is a proof.

## Chapter 26

1. 2.00 $\times$ 10$^8$ m/s
3. 1.198
5. 2.0 $\times$ 10$^{-11}$ s
7. 1.82
9. (a) 43°
    (b) 31°
11. 3.0 m
13. 1.92 $\times$ 10$^8$ m/s
15. 1.65
17. 1.19 mm

19. The answer is a derivation.
21. 3.23 cm
23. 1.54
25. 1.51
27. (a) B
    (b) A
29. 42.67°
31. 25.0°
33. 34.39°
35. 1.52
37. The answer is a proof.
39. 0.35°
41. (red ray) 44.6°; (violet ray) 45.9°
43. (red ray) 52.7°; (violet ray) 56.2°
45. $d_i = 18$ cm
47. 2.8
49. (a) $d_i = -75$ cm, $m = 2.5$
    (b) $d_i = -75.0$ cm, $m = 2.50$
51. (a) −0.00625 m
    (b) −0.0271 m
53. (a) converging
    (b) 2$f$
    (c) 2$f$
55. 48 cm
57. +35 cm and +90.5 cm
59. 0.13 m to the right of the second lens
61. (a) 4.00 cm to the left of the diverging lens
    (b) −0.167
    (c) virtual
    (d) inverted
    (e) smaller
63. 11.8 cm
65. (a) 18.1 cm
    (b) real
    (c) inverted
67. (a) 35.2 cm
    (b) 32.9 cm
69. −220 cm
71. −9.2 m
73. (a) 11.8 cm
    (b) 47.8 cm
75. (a) −4.5 m
    (b) 0.50 m
77. 3.7
79. 0.13 m
81. 6.3 cm
83. 0.64 rad
85. 0.81 cm
87. 0.435 cm
89. 0.261 cm
91. 1.1 m
93. (a) 1.3-diopter lens
    (b) 0.86 m
    (c) −8.5
95. −24

## Chapter 27

1. 6.0 $\times$ 10$^{-5}$ m
3. 17.8°
5. 660 nm

**7.** 0.0248 m
**9.** 487 nm
**11.** 102 nm
**13.** 207 nm
**15.** 115 nm
**17.** 427 nm
**19. (a)** 0.21°
   **(b)** 22°
**21. (a)** 24°
   **(b)** 39°
**23.** 490 nm
**25.** 0.447
**27.** 0.013
**29. (a)** 220 m
   **(b)** No.
**31.** $1.0 \times 10^4$ m
**33.** 2.3 m
**35. (a)** $1.22\lambda$
   **(b)** shorter
**37. (a)** 37°
   **(b)** 22°
**39.** 630 nm
**41.** $4.0 \times 10^{-6}$ m
**43.** 3/4
**45.** 1.95 m

## Chapter 28

**1.** 0.15 rad/s
**3.** $2.4 \times 10^8$ m/s
**5.** 2.28 s
**7.** 16
**9.** $2.60 \times 10^8$ m/s
**11.** 530 m
**13.** 4.0 light-years
**15.** 3.0 m × 1.3 m
**17.** $2.83 \times 10^8$ m/s
**19.** $1.80 \times 10^8$ m/s
**21.** $-2.0$ m/s
**23. (a)** 1.0
   **(b)** 6.6
**25.** 1.1 kg
**27.** $5.0 \times 10^{-13}$ J
**29.** $1.40 \times 10^{-15}$ m
**31.** $0.31c$
**33.** 42 m
**35. (a)** $2.82 \times 10^8$ m/s
   **(b)** $1.8 \times 10^{-16}$ kg · m/s

## Chapter 29

**1.** 310 nm
**3.** $7.7 \times 10^{29}$ photons/s
**5.** 6.3 eV
**7.** 2.10 eV
**9.** 73 photons/s
**11. (a)** $2.1 \times 10^{24}$ photons
   **(b)** 32 molecules/photon
**13.** $5.1 \times 10^{-33}$ kg · m/s

**15. (a)** $2.124 \times 10^{-24}$ kg · m/s
   **(b)** $2.096 \times 10^{-24}$ kg · m/s
**17.** $4.755 \times 10^{-24}$ kg · m/s
**19.** $9.50 \times 10^{-17}$ m
**21.** $2.6 \times 10^{-28}$ m
**23.** $7.77 \times 10^{-13}$ J
**25.** $7.38 \times 10^{-11}$ m
**27.** $1.10 \times 10^3$ m/s
**29.** $1.86 \times 10^4$ V
**31.** $4.0 \times 10^{-6}$ m/s
**33. (a)** $2.1 \times 10^{-35}$ kg · m/s
   **(b)** $4.7 \times 10^{-34}$ m/s
   **(c)** $2.3 \times 10^{-5}$ m/s
**35.** $-0.0144° \le \theta \le +0.0144°$

## Chapter 30

**1. (a)** $6.2 \times 10^{-31}$ m³
   **(b)** $4 \times 10^{-45}$ m³
   **(c)** $7 \times 10^{-13}$%
**3.** $1.5 \times 10^{14}$
**5.** $-8.7 \times 10^6$ eV
**7.** 91.2 nm
**9. (a)** 7458 nm
   **(b)** 2279 nm
   **(c)** infrared
**11.** 3
**13.** $-13.6$ eV, $-3.40$ eV, $-1.51$ eV
**15.** $n_i = 6$ and $n_f = 2$
**17.** $6 \le n_i \le 19$
**19.** 2180 lines/cm
**21.** 2, 3, 4, and 5
**23.** $-0.378$ eV
**25.** $\pm 3.16 \times 10^{-34}$ J · s,
   $\pm 2.11 \times 10^{-34}$ J · s,
   $\pm 1.05 \times 10^{-34}$ J · s, 0 J · s
**27.** 26.6°
**29.** $1s^2\,2s^2\,2p^6\,3s^2\,3p^6\,4s^2\,3d^5$
**31. (a)** not permitted
   **(b)** permitted
   **(c)** not permitted
   **(d)** permitted
   **(e)** not permitted
**33.** $7.230 \times 10^{-11}$ m
**35.** $6.83 \times 10^{-11}$ m
**37.** 21 600 V
**39. (a)** $1.25 \times 10^{-4}$ J
   **(b)** $3.98 \times 10^{14}$ J
**41.** $1.9 \times 10^{17}$
**43.** 21 days

## Chapter 31

**1. (a)** X = Pt, 117 neutrons
   **(b)** X = S, 16 neutrons
   **(c)** X = Cu, 34 neutrons
   **(d)** X = B, 6 neutrons
   **(e)** X = Pu, 145 neutrons

**3.** 8
**5.** 35.2
**7.** $^{120}_{50}$Sn
**9.** $9.4 \times 10^3$ m
**11.** 128 MeV
**13.** $2.9 \times 10^{16}$ kg
**15.** 1.003 27 u
**17. (a)** $^{212}_{84}\text{Po} \rightarrow {}^{208}_{82}\text{Pb} + {}^4_2\text{He}$
   **(b)** $^{232}_{92}\text{U} \rightarrow {}^{228}_{90}\text{Th} + {}^4_2\text{He}$
**19. (a)** $^{14}_{6}\text{C} \rightarrow {}^{14}_{7}\text{N} + {}^{0}_{-1}\text{e}$
   **(b)** $^{212}_{82}\text{Pb} \rightarrow {}^{212}_{83}\text{Bi} + {}^{0}_{-1}\text{e}$
**21.** $^{35}_{16}\text{S} \rightarrow {}^{35}_{17}\text{Cl} + {}^{0}_{-1}\text{e}$
**23. (a)** $^{18}_{9}\text{F} \rightarrow {}^{18}_{8}\text{O} + {}^{0}_{+1}\text{e}$
   **(b)** $^{15}_{8}\text{O} \rightarrow {}^{15}_{7}\text{N} + {}^{0}_{+1}\text{e}$
**25.** $1.61 \times 10^7$ m/s
**27.** 1.82 MeV
**29.** 3.0 days
**31.** 19.9
**33.** $2.1 \times 10^{13}$ Bq
**35.** 146 disintegrations/min
**37.** $3.7 \times 10^{10}$ Bq
**39.** 7.23 days
**41.** 0.70%
**43.** 90.9%
**45.** 6900 yr, maximum error is 900 yr

## Chapter 32

**1.** $2.8 \times 10^{-3}$ J
**3.** $1.6 \times 10^{-3}$ J
**5.** 0.26 roentgens
**7.** $4.4 \times 10^{11}$ s$^{-1}$
**9.** $3.01 \times 10^8$ rd
**11.** oxygen $^{16}_{8}$O
**13. (a)** proton, $^1_1$H
   **(b)** alpha particle, $^4_2$He
   **(c)** lithium, $^6_3$Li
   **(d)** nitrogen, $^{14}_{7}$N
   **(e)** manganese, $^{56}_{25}$Mn
**15.** 13.6 MeV
**17.** 2 neutrons
**19.** $9.0 \times 10^{-4}$
**21.** 41
**23.** $2.7 \times 10^6$ kg
**25. (a)** $8.2 \times 10^{10}$ J
   **(b)** 0.48 g
**27.** 1200 kg
**29.** 3.3 MeV
**31.** 1.0 gal
**33.** released energy = 24.7 MeV
**35.** possibility #1 = $u, d, s$
   possibility #2 = $u, d, b$
   possibility #3 = $u, s, b$
**37. (a)** 0.513 MeV
   **(b)** $2.43 \times 10^{-12}$ m
   **(c)** $2.73 \times 10^{-22}$ kg · m/s
**39. (a)** $1.9 \times 10^{-20}$ kg · m/s
   **(b)** $3.5 \times 10^{-14}$ m

# Index

## SI Units

| Quantity | Name of Unit | Symbol | Expression in Terms of Other SI Units | Quantity | Name of Unit | Symbol | Expression in Terms of Other SI Units |
|---|---|---|---|---|---|---|---|
| Length | meter | m | Base unit | Pressure, stress | pascal | Pa | $N/m^2$ |
| Mass | kilogram | kg | Base unit | Viscosity | — | — | $Pa \cdot s$ |
| Time | second | s | Base unit | Electric charge | coulomb | C | $A \cdot s$ |
| Electric current | ampere | A | Base unit | Electric field | — | — | N/C |
| Temperature | kelvin | K | Base unit | Electric potential | volt | V | J/C |
| Amount of substance | mole | mol | Base unit | Resistance | ohm | $\Omega$ | V/A |
| Velocity | — | — | m/s | Capacitance | farad | F | C/V |
| Acceleration | — | — | $m/s^2$ | Inductance | henry | H | $V \cdot s/A$ |
| Force | newton | N | $kg \cdot m/s^2$ | Magnetic field | tesla | T | $N \cdot s/(C \cdot m)$ |
| Work, energy | joule | J | $N \cdot m$ | Magnetic flux | weber | Wb | $T \cdot m^2$ |
| Power | watt | W | J/s | Specific heat capacity | — | — | $J/(kg \cdot K)$ or $J/(kg \cdot C°)$ |
| Impulse, momentum | — | — | $kg \cdot m/s$ | Thermal conductivity | — | — | $J/(s \cdot m \cdot K)$ or $J/(s \cdot m \cdot C°)$ |
| Plane angle | radian | rad | m/m | Entropy | — | — | J/K |
| Angular velocity | — | — | rad/s | Radioactive activity | becquerel | Bq | $s^{-1}$ |
| Angular acceleration | — | — | $rad/s^2$ | Absorbed dose | gray | Gy | J/kg |
| Torque | — | — | $N \cdot m$ | Exposure | — | — | C/kg |
| Frequency | hertz | Hz | $s^{-1}$ | | | | |
| Density | — | — | $kg/m^3$ | | | | |

## The Greek Alphabet

| | | | | | | | | | | |
|---|---|---|---|---|---|---|---|---|---|---|
| Alpha | A | $\alpha$ | Iota | I | $\iota$ | Rho | P | $\rho$ |
| Beta | B | $\beta$ | Kappa | K | $\kappa$ | Sigma | $\Sigma$ | $\sigma$ |
| Gamma | $\Gamma$ | $\gamma$ | Lambda | $\Lambda$ | $\lambda$ | Tau | T | $\tau$ |
| Delta | $\Delta$ | $\delta$ | Mu | M | $\mu$ | Upsilon | Y | $\upsilon$ |
| Epsilon | E | $\epsilon$ | Nu | N | $\nu$ | Phi | $\Phi$ | $\phi$ |
| Zeta | Z | $\zeta$ | Xi | $\Xi$ | $\xi$ | Chi | X | $\chi$ |
| Eta | H | $\eta$ | Omicron | O | $o$ | Psi | $\Psi$ | $\psi$ |
| Theta | $\Theta$ | $\theta$ | Pi | $\Pi$ | $\pi$ | Omega | $\Omega$ | $\omega$ |

$R = PV/T$

1.98 cal

$8.48 \times 10^5$ kgm/T

8.31441 Pa m³ k⁻¹ mol⁻¹

8.31441 J k⁻¹ mol⁻¹